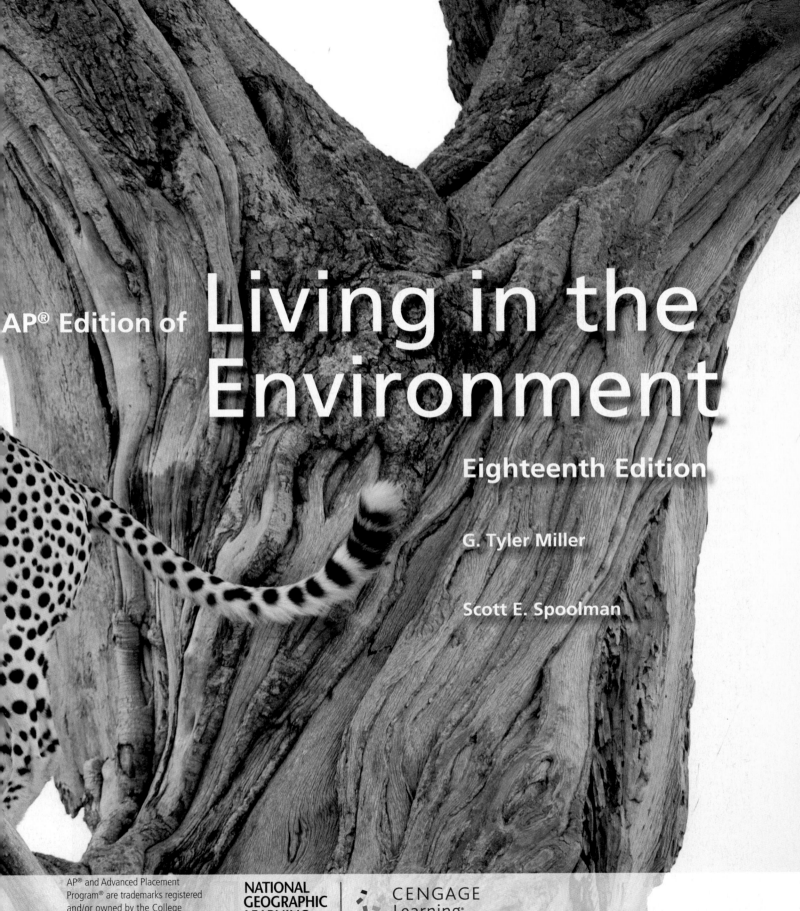

AP® Edition of

Living in the Environment

Eighteenth Edition

G. Tyler Miller

Scott E. Spoolman

NATIONAL GEOGRAPHIC LEARNING | **CENGAGE** Learning·

Australia • Brazil • Mexico • Singapore • United Kingdom • United States

National Geographic Learning | Cengage Learning

Living in the Environment, **AP® Edition, Eighteenth Edition**
G. Tyler Miller, Scott E. Spoolman

Senior Product Team Manager: Yolanda Cossio

Content Developer: Jake Warde

Content Coordinator: Shannon Holt

Product Assistant: Kellie Petruzzelli

Media Developer: Alexandria Brady

Executive Brand Manager: Nicole Hamm

Senior Market Development Manager: Tom Ziolkowski

Content Project Manager: Harold P. Humphrey

Senior Art Director: Pamela Galbreath

Manufacturing Planner: Karen Hunt

Senior Rights Acquisitions Specialist: Dean Dauphinais

Production Service: Dan Fitzgerald, Graphic World Inc.

Photo Researcher: Christina Ciaramella, PreMedia Global

Text Researcher: Melissa Tomaselli, PreMedia Global

Copy Editor: Graphic World Inc.

Illustrator: Patrick Lane, ScEYEnce Studios

Text and Cover Designer: Jeanne Calabrese

Cover and Title Page Image: © Frans Lanting/National Geographic Creative

Compositor: Graphic World Inc.

Associate Content Developer, Advanced and Elective Products Program: Ashley Bargende

Senior Media Developer, Advanced and Elective Products Program: Philip Lanza

Library of Congress Control Number: 2013945510

High School Edition:

ISBN-13: 978-1-285-19728-9

ISBN-10: 1-285-19728-3

Cengage Learning
200 First Stamford Place, 4th Floor
Stamford, CT 06902
USA

Cengage Learning is a leading provider of customized learning solutions with office locations around the globe, including Singapore, the United Kingdom, Australia, Mexico, Brazil, and Japan. Locate your local office at **www.cengage.com/global**.

Cengage Learning products are represented in Canada by Nelson Education, Ltd.

To learn more about Cengage Learning Solutions, visit **www.cengage.com**.

To find online supplements and other instructional support, please visit **www.cengagebrain.com**.

AP® and Advanced Placement Program® are trademarks registered and/or owned by the College Board, which was not involved in the production of, and does not endorse, this product.

FSC
www.fsc.org
MIX
Paper from responsible sources
FSC® C084269

Printed in the United States of America
6 7 8 9 20 19 18 17

BRIEF CONTENTS

Aleksander Bolbot/Shutterstock.com

v

CONTENTS

AP® and Advanced Placement Program® are trademarks registered and/or owned by the College Board, which was not involved in the production of, and does not endorse, this product.

Aleksander Bolbot/Shutterstock.com

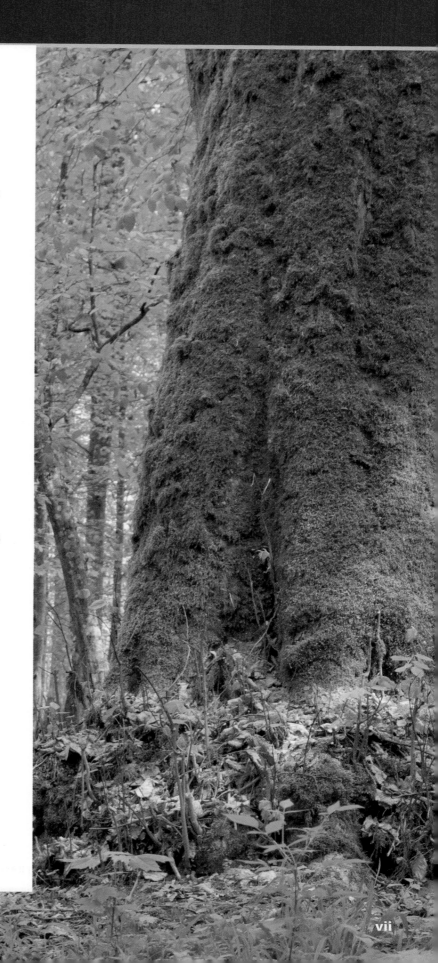

For Teachers

We wrote this book to help teachers achieve three important goals: *first,* to explain to their students the basics of environmental science; *second,* to help their students in using this scientific foundation to understand the environmental problems that we face and to evaluate possible solutions to them; and *third,* to inspire their students to make a difference in how we treat the earth on which our lives and economies depend, and thus in how we treat ourselves and our descendants.

We view environmental problems and possible solutions to them through the lens of *sustainability*—the integrating theme of this book. We believe that most people can live comfortable and fulfilling lives and that societies will be more prosperous and peaceful when sustainability becomes one of the chief measures by which personal choices and public policies are made. We introduce this book with a vision of such a more sustainable future in the Core Case Study of Chapter 1. Our belief in such a future is foundational to this textbook, and we consistently challenge students to work toward attaining it.

For this reason, we are happy to announce our new partnership with *The National Geographic Society,* which shares our goals, as reflected in its statement of purpose: *Inspiring people to care about the planet.* One result of this new collaboration is the addition of many stunning and informative photographs, numerous maps, and several new stories of National Geographic Explorers—people who are making a positive difference in the world. With these new tools, we continue to tell of the good news from various fields of environmental science, hoping to inspire young people to commit themselves to making our world a more sustainable place to live for their own and future generations.

What's New in This Edition?

- *Our new partnership with National Geographic* has given us access to hundreds of amazing photographs, numerous maps, and inspiring stories of *National Geographic Explorers*—people who are leading the way in environmental science, education, or entrepreneurial enterprises.

- A *stunning new design* with a National Geographic look that enhances visual learning.

- *Campus Sustainability boxes:* short descriptions about what selected U.S. colleges and universities are doing to make their institutions more sustainable. These stories are complemented by a new Core Case Study in Chapter 24 that summarizes several other such efforts.

- *Three social science principles of sustainability.* These complement the three scientific principles of sustainability

that we have long used to explain how life on Earth has sustained itself for billions of years, and they act as guidelines for making a possible transition to more sustainable economies and societies.

- *New Core Case Studies* for 18 of the book's 25 chapters that serve as an integrating theme throughout each chapter. They bring important real-world stories to the forefront for use in applying those chapters' concepts and principles.

- *Two new end-of-chapter exercises: Doing Environmental Science* and *Global Environment Watch* research projects give students challenging new ways to apply the material.

Sustainability Is the Integrating Theme of This Book

Sustainability, a watchword of the 21st century for those concerned about the environment, is the overarching theme of this textbook. You can see the sustainability emphasis by looking at the Brief Contents (p. v).

Six principles of sustainability play a major role in carrying out this book's sustainability theme. These principles are introduced in Chapter 1. They are depicted in Figure 1-2 (p. 6), in Figure 1-5 (p. 9), and on the back cover of the student edition and are used throughout the book, with each reference marked in the margin by (see pp. 62 and 218).

We use the following five major subthemes to integrate material throughout this book (see diagram on back cover of the student edition).

- *Natural Capital.* Sustainability depends on the natural resources and ecosystem services that support all life and economies. See Figures 1-3, p. 7, and 10-4, p. 220.

- *Natural Capital Degradation.* We describe how human activities can degrade natural capital. See Figures 1-7, p. 11, and 7-17, p. 160.

- *Solutions.* We present existing and proposed solutions to environmental problems in a balanced manner and challenge students to use critical thinking to evaluate them. See Figures 10-16, p. 227, and 18-26, p. 496.

- *Trade-offs.* The search for solutions involves trade-offs, because any solution requires weighing advantages against disadvantages. Our Trade-offs diagrams located in several chapters present the benefits and drawbacks of various environmental technologies and solutions to environmental problems. See Figures 12-19, p. 293, and 15-11, p. 383.

- *Individuals Matter.* Throughout the book, Individuals Matter boxes and some of the Case Studies describe what various scientists and concerned citizens (including several National Geographic Explorers) have done to help us work toward sustainability (see pp. 82, 240, and 303). Also, a number of What Can You Do? diagrams describe how readers can deal with the problems we face (see Figures 9-12, p. 202, and 13-28, p. 341). Eight especially important things individuals can do are summarized in Figure 25-14 (p. 696).

Other Key Features of This Textbook

- **Up-to-Date Coverage.** Our textbooks have been widely praised for keeping users up to date in the rapidly changing field of environmental science. We have used thousands of articles and reports published in 2010–2013 to update the information and concepts in this book. Major new or updated topics include planetary boundaries that indicate ecological tipping points (Science Focus 3.3, p. 72); hydraulic fracturing (fracking) in oil and natural gas production and its harmful effects (pp. 379–380 and 383–385); and the rising threat of ocean acidification (Science Focus 11.2, p. 252), along with dozens of other important topics.

- **Concept-Centered Approach.** To help students focus on the main ideas, we built each major chapter section around a key question and one or two key concepts, which state the section's most important take-away messages. In each chapter, all key questions are listed at the front of the chapter, and each chapter section begins with its key question and concepts (see pp. 29 and 31). Also, the concept applications are highlighted and referenced throughout each chapter.

- **Science-Based Coverage.** Chapters 2–8 cover scientific principles important to the course and discuss how scientists work (see Brief Contents, p. v). Important environmental science topics are explored in depth in Science Focus boxes distributed among the chapters throughout the book (see pp. 94 and 203) and integrated throughout the book in various Case Studies (see pp. 238 and 256) and in numerous figures.

- **Global Perspective.** This book also provides a global perspective, first on the ecological level, revealing how all the world's life is connected and sustained within the biosphere, and second, through the use of information and images from around the world. This includes more than 80 maps in the basic text and in Supplement 6. Half of these maps are new and more than half of the new maps are from National Geographic. At the end of each chapter is a Global Environment Watch exercise that applies this global perspective (see p. 245).

- **Core Case Studies.** Each chapter opens with a Core Case Study (see pp. 190 and 278), which is applied throughout the chapter. These applications are indicated by the notation (**Core Case Study**) wherever they occur (see pp. 202, 281, and 301). Each chapter ends with a *Tying It All Together* box (see pp. 213 and 312), which connects the Core Case Study and other material in the chapter to some or all of the principles of sustainability.

- **Case Studies.** In addition to the 25 Core Case Studies, more than 70 additional Case Studies (see pp. 92, 200, and 331) appear throughout the book (and are listed in the Detailed Contents, pp. vi–xv). Each of these provides an in-depth look at specific environmental problems and their possible solutions. We also have included very brief descriptions of efforts on several college campuses to study or apply principles of sustainability in our new *Campus Sustainability* stories that appear in several of the book's chapters (see pp. 210 and 270).

- **Critical Thinking.** The Preface for Students (p. xxiii) describes critical thinking skills, and specific critical thinking exercises are used throughout the book in several ways:
  - As more than 100 *Thinking About* exercises that ask students to analyze material immediately after it is presented (see pp. 35 and 264).
  - In all *Science Focus* boxes.
  - In dozens of *Connections* boxes that stimulate critical thinking by exploring the often surprising connections related to environmental problems (see pp. 18 and 195).
  - In the captions of many of the book's figures (see Figures 3-15, p. 63, and 9-8, p. 198).
  - In end-of-chapter questions (see pp. 214 and 314).

- **Visual Learning.** With a new design heavily influenced by material from National Geographic and more than 400 photographs—two-thirds of them new and 20% of them from the archives of National Geographic—this is the most visually appealing environmental science textbook available (see Figures 3-21, p. 71; 7-16, p. 159; and 10-18, p. 229). Also new to this edition is the inclusion of more than 200 additional small photos as insets in various diagrams. Add in the more than 130 diagrams—34 of them new or improved in this edition—each designed to present complex ideas in understandable ways relating to the real world (see Figures 3-3, p. 54; 3-17, p. 66; and 4-2, p. 79), and you have one of the most visually informative textbooks available.

- **Flexibility.** To meet the diverse needs of hundreds of widely varying environmental science courses, we have designed a highly flexible book that allows teachers to vary the order of chapters and sections within chapters without exposing students to terms and concepts that could confuse them. We recommend that teachers start with Chapter 1, which defines basic terms and gives an overview of sustainability, population, pollution, resources, and economic development issues that are discussed throughout the book. This provides a springboard for teachers to use other chapters in almost any order. One often-used strategy is to follow Chapter 1 with Chapters 2–8, which introduce basic science and ecological concepts. Teachers can then use the remaining chapters in any order desired. Some teachers follow Chapter 1 with any or all of Chapters 23, 24, and 25 on environmental economics, politics, and worldviews, respectively, before proceeding to the chapters on basic science and ecological concepts. We provide a second level of flexibility in seven Supplements (see p. xv in the Detailed Contents and p. S1), which teachers can assign as desired to meet the needs of their specific courses. Examples include environmental history of the United States (Supplement 3), basic chemistry (Supplement 4), weather basics (Supplement 5), maps (Supplement 6, see Figure 5, p. S30, and Figure 6, p. S32), and basic environmental data and data analysis (Supplement 7, see Figure 7, p. S67, and Figure 10, p. S68).

- **In-Text Study Aids.** Each chapter begins with a list of *Key Questions* showing how the chapter is organized (see p. 401). When a new *key term* is introduced and defined, it is printed in boldface type, and all such terms are summarized in the glossary at the end of the book. More than 100 *Thinking About* exercises reinforce learning by asking students to think critically about the implications of various environmental issues and solutions immediately after they are discussed in the text (see p. 409). The captions of many figures contain similar questions that get students to think about the figure content (see Figure 16-16, p. 416). In their reading, students also encounter *Connections* boxes, which briefly describe connections between human activities and environmental consequences, environmental and social issues, and environmental issues and solutions (see p. 423). Finally, the text of each chapter wraps up with three *Big Ideas* (see p. 435), which summarize and reinforce three of the major take-away messages from each chapter, and a *Tying It All Together* section that relates the Core Case Study and other chapter content to the principles of sustainability (see p. 436). Again, this reinforces the main messages of the chapter along with the themes of sustainability to give students a stronger understanding of how it all ties together.

Each chapter ends with a *Chapter Review* section containing a detailed set of review questions that include all the chapter's key terms in bold type; *Critical Thinking* questions that encourage students to think about and apply what they have learned to their lives; *Doing Environmental Science*—an exercise that will help students to experience the work of various environmental scientists; a *Global Environment Watch* exercise taking students to Cengage's GREENR site, where they can use this tool for interesting research related to chapter content; and a *Data Analysis* or *Ecological Footprint Analysis* problem built around ecological footprint data or some other environmental data set. (See pp. 436–439.) And at the end of the book, we have included a comprehensive glossary that includes definitions of all key terms as well as many other terms that are important to environmental science.

Supplements for Teachers

- **PowerLecture.** This CD, available to adopters, allows you to create custom lecture outlines using all of the figures and photos from the text, videos, and animations. PowerPoint's editing tools allow use of slides from other lectures, modification or removal of figure labels and leaders, insertion of your own slides, saving slides as JPEG images, and preparation of lectures for use on the Web. Content is also available online.

- **Teacher's Resource Guide for the Advanced Placement Program®.** Available to adopters. The Teacher's Resource Guide has been thoughtfully revised and updated to make creating your lesson plans even easier. Some of the features include updated answers to the end-of-chapter questions, suggested answers for the review questions and the quantitative exercises, and a revised video reference list with Web resources.

- **Test Bank.** Available to adopters. The Test Bank contains thousands of questions and answers in a variety of formats, including multiple choice, true/false, fill-in-the-blank, critical thinking questions, and essay questions.

- **Transparencies.** Online Transparency Correlation Guide. This guide correlates the transparency set created for *Living in the Environment 17e*, *Environmental Science 13e*, *Sustaining the Earth 10e*, and *Essentials of Ecology 6e* to the new editions of these texts: *Living in the Environment 18e*, *Environmental Science 14e*, *Sustaining the Earth 11e*, and *Essentials of Ecology 7e*.

- **AP® Aplia.** Aplia™ is a Cengage Learning online homework system dedicated to improving learning by increasing student effort and engagement. Aplia makes it easy for teachers to assign frequent online homework assignments. Aplia provides students with prompt and detailed feedback to help them learn as they work through the questions, and features interactive tutorials to fully engage them in learning course concepts. Automatic grading and powerful assessment tools give teachers real-time reports of student progress, participation, and performance, and

Aplia's easy-to-use course management features let teachers flexibly administer course announcements and materials online. With Aplia, students will show up to class fully engaged and prepared, and teachers will have more time to do what they do best. . . teach. Please contact your sales representative for available discounts.

- **Global Environment Watch.** Updated several times a day, the Global Environment Watch is a focused portal into GREENR—the Global Reference on the Environment, Energy, and Natural Resources—an ideal one-stop site for classroom discussion and research projects. This resource center keeps courses up to date with the most current news on the environment. Users get access to information from trusted academic journals, news outlets, and magazines, as well as statistics, an interactive world map, videos, primary sources, case studies, podcasts, and much more.

- **Virtual Field Trips in Environmental Issues.** This supplement brings the field to you, with dynamic panoramas, videos, photographs, maps, and quizzes covering important topics within environmental science. A case study approach covers the issues of keystone species, the role of climate change in extinctions, invasive species, the evolution of a species in relation to its environment, and an ecosystem approach to sustaining biodiversity. Students are engaged, interacting with real issues to help them think critically about the world around them.

Help Us Improve This Book or Its Supplements

Let us know how you think this book can be improved. If you find any errors, bias, or confusing explanations, please e-mail us about them at:

- mtg89@hotmail.com
- spoolman@tds.net

Most errors can be corrected in subsequent printings of this edition, as well as in future editions.

Acknowledgments

We wish to thank the many students and teachers who have responded so favorably to the 17 previous editions of *Living in the Environment*, AP Edition the 14 editions of *Environmental Science*, the 10 editions of *Sustaining the Earth*, and the 6 editions of *Essentials of Ecology*, and who have corrected errors and offered many helpful suggestions for improvement. We are also deeply indebted to the more than 300 reviewers, who pointed out errors and suggested many important improvements in the various editions of these three books.

It takes a village to produce a textbook, and the members of the talented production team, listed on the copy-

right page, have made vital contributions. Our special thanks go to development editor Jake Warde, production editors Hal Humphrey and Dan Fitzgerald, designer Pam Galbreath, copy editor Chris DeVito, compositor Craig Beffa, photo researcher Christina Ciaramella, artist Patrick Lane, media developer Alexandria Brady, assistant editor Alexis Glubka, product assistant Kellie Petruzzelli, and Cengage Learning's hardworking sales staff. Finally, we are very fortunate to have the guidance, inspiration, and unfailing support of Life Sciences Senior Product Team Manager Yolanda Cossio and her dedicated team of highly talented people who have made this and other book projects such a pleasure to work on.

G. Tyler Miller

Scott E. Spoolman

Guest Essayists

Guest essays by the following authors are available online: **M. Kat Anderson**, ethnoecologist with the National Plant Center of the USDA's Natural Resource Conservation Center; **Lester R. Brown**, president, Earth Policy Institute; **Alberto Ruz Buenfil**, environmental activist, writer, and performer; **Robert D. Bullard**, professor of sociology and director of the Environmental Justice Resource Center at Clark Atlanta University; **Michael Cain**, ecologist and adjunct professor at Bowdoin College; **Herman E. Daly**, senior research scholar at the School of Public Affairs, University of Maryland; **Lois Marie Gibbs**, director, Center for Health, Environment, and Justice; **Garrett Hardin**, professor emeritus (now deceased) of human ecology, University of California, Santa Barbara; **John Harte**, professor of energy and resources, University of California, Berkeley; **Paul G. Hawken**, environmental author and business leader; **Jane Heinze-Fry**, environmental educator; **Paul F. Kamitsuja**, infectious disease expert and physician; **Amory B. Lovins**, energy policy consultant and director of research, Rocky Mountain Institute; **Bobbi S. Low**, professor of resource ecology, University of Michigan; **John J. Magnuson**, Director Emeritus of the Center for Limnology, University of Wisconsin, Madison; **Lester W. Milbrath**, director of the research program in environment and society, State University of New York, Buffalo; **Peter Montague**, director, Environmental Research Foundation; **Norman Myers**, tropical ecologist and consultant in environment and development; **David W. Orr**, professor of environmental studies, Oberlin College; **Noel Perrin**, adjunct professor of environmental studies, Dartmouth College; **David Pimentel**, professor of insect ecology and agricultural sciences, Cornell University; **John Pichtel**, Ball State University; **Andrew C. Revkin**, environmental author and environmental reporter for the New York Times; **Vandana Shiva**, physicist, educator, environmental consultant; **Nancy Wicks**, ecopioneer and director of Round Mountain Organics; and **Donald Worster**, environmental historian and professor of American history, University of Kansas.

Dr. Dean Goodwin and his colleagues Berry Cobb, Deborah Stevens, Jeannette Adkins, Jim Lehner, Judy Treharne, Lonnie Miller, and Tom Mowbray provided excellent contributions to the Data Analysis and Ecological Footprint Analysis exercises. Mary Jo Burchart of Oakland Community College wrote the in-text Global Environment Watch exercises.

Cumulative List of Reviewers

Barbara J. Abraham, Hampton College; Donald D. Adams, State University of New York at Plattsburgh; Larry G. Allen, California State University, Northridge; Susan Allen-Gil, Ithaca College; James R. Anderson, U.S. Geological Survey; Mark W. Anderson, University of Maine; Kenneth B. Armitage, University of Kansas; Samuel Arthur, Bowling Green State University; Gary J. Atchison, Iowa State University; Thomas W. H. Backman, Lewis-Clark State College; Marvin W. Baker, Jr., University of Oklahoma; Virgil R. Baker, Arizona State University; Stephen W. Banks, Louisiana State University in Shreveport; Ian G. Barbour, Carleton College; Albert J. Beck, California State University, Chico; Eugene C. Beckham, Northwood University; Diane B. Beechinor, Northeast Lakeview College; W. Behan, Northern Arizona University; David Belt, Johnson County Community College; Keith L. Bildstein, Winthrop College; Andrea Bixler, Clarke College; Jeff Bland, University of Puget Sound; Roger G. Bland, Central Michigan University; Grady Blount II, Texas A&M University, Corpus Christi; Lisa K. Bonneau, University of Missouri–Kansas City; Georg Borgstrom, Michigan State University; Arthur C. Borror, University of New Hampshire; John H. Bounds, Sam Houston State University; Leon F. Bouvier, Population Reference Bureau; Daniel J. Bovin, Université Laval; Jan Boyle, University of Great Falls; James A. Brenneman, University of Evansville; Michael F. Brewer, Resources for the Future, Inc.; Mark M. Brinson, East Carolina University; Dale Brown, University of Hartford; Patrick E. Brunelle, Contra Costa College; Terrence J. Burgess, Saddleback College North; David Byman, Pennsylvania State University, Worthington–Scranton; Michael L. Cain, Bowdoin College; Lynton K. Caldwell, Indiana University; Faith Thompson Campbell, Natural Resources Defense Council, Inc.; John S. Campbell, Northwest College; Ray Canterbery, Florida State University; Ted J. Case, University of San Diego; Ann Causey, Auburn University; Richard A. Cellarius, Evergreen State University; William U. Chandler, Worldwatch Institute; F. Christman, University of North Carolina, Chapel Hill; Lu Anne Clark, Lansing Community College; Preston Cloud, University of California, Santa Barbara; Bernard C. Cohen, University of Pittsburgh; Richard A. Cooley, University of California, Santa Cruz; Dennis J. Corrigan; George Cox, San Diego State University; John D. Cunningham, Keene State College; Herman E. Daly, University of Maryland; Raymond F. Dasmann, University of California, Santa Cruz; Kingsley Davis, Hoover Institution; Edward E. DeMartini, University of California, Santa Barbara; James Demastes, University of Northern Iowa; Charles E. DePoe, Northeast Louisiana University; Thomas R. Detwyler, University of Wisconsin; Bruce DeVantier, Southern Illinois University at Carbondale; Peter H. Diage, University of California, Riverside; Stephanie Dockstader, Monroe Community College; Lon D. Drake, University of Iowa; Michael Draney, University of Wisconsin–Green Bay; David DuBose, Shasta College; Dietrich Earnhart, University of Kansas; Robert East, Washington & Jefferson College; T. Edmonson, University of Washington; Thomas Eisner, Cornell University; Michael Esler, Southern Illinois University; David E. Fairbrothers, Rutgers University; Paul P. Feeny, Cornell University; Richard S. Feldman, Marist College; Vicki Fella-Pleier, La Salle University; Nancy Field, Bellevue Community College; Allan Fitzsimmons, University of Kentucky; Andrew J. Friedland, Dartmouth College; Kenneth O. Fulgham, Humboldt State University; Lowell L. Getz, University of Illinois at Urbana–Champaign; Frederick F. Gilbert, Washington State University; Jay Glassman, Los Angeles Valley College; Harold Goetz, North Dakota State University; Srikanth Gogineni, Axia College of University of Phoenix; Jeffery J. Gordon, Bowling Green State University; Eville Gorham, University of Minnesota; Michael Gough, Resources for the Future; Ernest M. Gould, Jr., Harvard University; Peter Green, Golden West College; Katharine B. Gregg, West Virginia Wesleyan College; Paul K. Grogger, University of Colorado at Colorado Springs; L. Guernsey, Indiana State University; Ralph Guzman, University of California, Santa Cruz; Raymond Hames, University of Nebraska, Lincoln; Robert Hamilton IV, Kent State University, Stark Campus; Raymond E. Hampton, Central Michigan University; Ted L. Hanes, California State University, Fullerton; William S. Hardenbergh, Southern Illinois University at Carbondale; John P. Harley, Eastern Kentucky University; Neil A. Harriman, University of Wisconsin, Oshkosh; Grant A. Harris, Washington State University; Harry S. Hass, San Jose City College; Arthur N. Haupt, Population Reference Bureau; Denis A. Hayes, environmental consultant; Stephen Heard, University of Iowa; Gene Heinze-Fry, Department of Utilities, Commonwealth of Massachusetts; Jane Heinze-Fry, environmental educator; John G. Hewston, Humboldt State University; David L. Hicks, Whitworth College; Kenneth M. Hinkel, University of Cincinnati; Eric Hirst, Oak Ridge National Laboratory; Doug Hix, University of Hartford; S. Holling, University of British Columbia; Sue Holt, Cabrillo College; Donald Holtgrieve, California State University, Hayward; Michelle Homan, Gannon University; Michael H. Horn, California State University, Fullerton; Mark A. Hornberger, Bloomsburg University; Marilyn Houck, Pennsylvania State University; Richard D. Houk, Winthrop College; Robert J. Huggett, College of William and Mary; Donald Huisingh, North Carolina State University; Catherine Hurlbut, Florida Community College at Jacksonville; Marlene K. Hutt, IBM; David R. Inglis, University of Massachusetts; Robert Janiskee, University of South Carolina; Hugo H. John, University of Connecticut; Brian A. Johnson, University of Pennsylvania, Bloomsburg; David I. Johnson, Michigan State University; Mark Jonasson, Crafton Hills College; Zoghlul Kabir, Rutgers, New Brunswick; Agnes Kadar, Nassau Community College; Thomas L. Keefe, Eastern Kentucky University; David Kelley, University of St. Thomas; William E. Kelso, Louisiana State University; Nathan Keyfitz, Harvard University; David Kidd, University of New Mexico; Pamela S. Kimbrough; Jesse Klingebiel, Kent School; Edward J. Kormondy, University of Hawaii–Hilo/West Oahu College; John V. Krutilla, Resources for the Future, Inc.; Judith Kunofsky, Sierra Club; E. Kurtz; Theodore Kury, State University of New York at Buffalo; Troy A. Ladine, East Texas Baptist University; Steve Ladochy, University of Winnipeg; Anna J. Lang, Weber State University; Mark B. Lapping, Kansas State University; Michael L. Larsen, Campbell University; Linda Lee, University of Connecticut; Tom Leege, Idaho Department of Fish and Game; Maureen Leupold, Genesee Community College; William S. Lindsay, Monterey Peninsula College; E. S. Lindstrom, Pennsylvania State University; M. Lippiman, New York University Medical Center; Valerie A. Liston, University of Minnesota; Dennis Livingston, Rensselaer Polytechnic Institute; James P. Lodge, air pollution consultant; Raymond C. Loehr, University of Texas at Austin; Ruth Logan, Santa Monica City College; Robert D. Loring, DePauw University; Paul F. Love, Angelo State University; Thomas Lovering, University of California, Santa Barbara; Amory B. Lovins, Rocky Mountain Institute; Hunter Lovins, Rocky Mountain Institute; Gene A. Lucas, Drake University; Claudia Luke, University of California, Berkeley; David Lynn; Timothy F. Lyon, Ball State University; Stephen Malcolm, Western Michigan University; Melvin G. Marcus, Arizona State University; Gordon E. Matzke, Oregon State University; Parker Mauldin, Rockefeller Foundation; Marie McClune, The Agnes Irwin School (Rosemont, Pennsylvania); Theodore R. McDowell, California State University; Vincent E. McKelvey, U.S. Geological Survey; Robert T. McMaster, Smith College; John G. Merriam, Bowling Green State University; A. Steven Messenger, Northern Illinois University; John Meyers, Middlesex Community College; Raymond W. Miller, Utah State University; Arthur B. Millman, University of Massachusetts, Boston; Sheila Miracle, Southeast Kentucky Community & Technical College; Fred Montague, University of Utah; Rolf Monteen, California Polytechnic State University; Debbie Moore, Troy University Dothan Campus; Michael K. Moore, Mercer University; Ralph Morris, Brock University, St. Catherine's, Ontario, Canada; Angela Morrow, Auburn University; William W. Murdoch, University of California, Santa Barbara; Norman Myers, environmental consultant; Brian C. Myres, Cypress Col-

lege; A. Neale, Illinois State University; Duane Nellis, Kansas State University; Jan Newhouse, University of Hawaii, Manoa; Jim Norwine, Texas A&M University, Kingsville; John E. Oliver, Indiana State University; Mark Olsen, University of Notre Dame; Carol Page, copy editor; Bill Paletski, Penn State University; Eric Pallant, Allegheny College; Charles F. Park, Stanford University; Richard J. Pedersen, U.S. Department of Agriculture, Forest Service; David Pelliam, Bureau of Land Management, U.S. Department of the Interior; Murray Paton Pendarvis, Southeastern Louisiana University; Dave Perault, Lynchburg College; Rodney Peterson, Colorado State University; Julie Phillips, De Anza College; John Pichtel, Ball State University; William S. Pierce, Case Western Reserve University; David Pimentel, Cornell University; Peter Pizor, Northwest Community College; Mark D. Plunkett, Bellevue Community College; Grace L. Powell, University of Akron; James H. Price, Oklahoma College; Marian E. Reeve, Merritt College; Carl H. Reidel, University of Vermont; Charles C. Reith, Tulane University; Erin C. Rempala, San Diego City College; Roger Revelle, California State University, San Diego; L. Reynolds, University of Central Arkansas; Ronald R. Rhein, Kutztown University of Pennsylvania; Charles Rhyne, Jackson State University; Robert A. Richardson, University of Wisconsin; Benjamin F. Richason III, St. Cloud State University; Jennifer Rivers, Northeastern University; Ronald Robberecht, University of Idaho; William Van B. Robertson, School of Medicine, Stanford University; C. Lee Rockett, Bowling Green State University; Terry D. Roelofs, Humboldt State University; Daniel Ropek, Columbia George Community College; Christopher Rose, California Polytechnic State University; Richard G. Rose, West Valley College; Stephen T. Ross, University of Southern Mississippi; Robert E. Roth, Ohio State University; Dorna Sakurai, Santa Monica College; Arthur N. Samel, Bowling Green State University; Shamili Sandiford, College of DuPage; Floyd Sanford, Coe College; David Satterthwaite, I.E.E.D., London; Stephen W. Sawyer, University of Maryland; Arnold Schecter, State University of New York; Frank Schiavo, San Jose State University; William H. Schlesinger, Ecological Society of America; Stephen H. Schneider, National Center for Atmospheric Research; Clarence A. Schoenfeld, University of Wisconsin, Madison; Madeline Schreiber, Virginia Polytechnic Institute; Henry A. Schroeder, Dartmouth Medical School; Lauren A. Schroeder, Youngstown State University; Norman B. Schwartz, University of Delaware; George Sessions, Sierra College; David J. Severn, Clement Associates; Don Sheets, Gardner-Webb University; Paul Shepard, Pitzer College and Claremont Graduate School; Michael P. Shields, Southern Illinois University at Carbondale; Kenneth Shiovitz; F. Siewert, Ball State University; E. K. Silbergold, Environmental Defense Fund; Joseph L. Simon, University of South Florida; William E. Sloey, University of Wisconsin, Oshkosh; Robert L. Smith, West Virginia University; Val

Smith, University of Kansas; Howard M. Smolkin, U.S. Environmental Protection Agency; Patricia M. Sparks, Glassboro State College; John E. Stanley, University of Virginia; Mel Stanley, California State Polytechnic University, Pomona; Richard Stevens, Monroe Community College; Norman R. Stewart, University of Wisconsin, Milwaukee; Frank E. Studnicka, University of Wisconsin, Platteville; Chris Tarp, Contra Costa College; Roger E. Thibault, Bowling Green State University; Nathan E. Thomas, University of South Dakota; William L. Thomas, California State University, Hayward; Shari Turney, copy editor; John D. Usis, Youngstown State University; Tinco E. A. van Hylckama, Texas Tech University; Robert R. Van Kirk, Humboldt State University; Donald E. Van Meter, Ball State University; Rick Van Schoik, San Diego State University; Gary Varner, Texas A&M University; John D. Vitek, Oklahoma State University; Harry A. Wagner, Victoria College; Lee B. Waian, Saddleback College; Warren C. Walker, Stephen F. Austin State University; Thomas D. Warner, South Dakota State University; Kenneth E. F. Watt, University of California, Davis; Alvin M. Weinberg, Institute of Energy Analysis, Oak Ridge Associated Universities; Brian Weiss; Margery Weitkamp, James Monroe High School (Granada Hills, California); Anthony Weston, State University of New York at Stony Brook; Raymond White, San Francisco City College; Douglas Wickum, University of Wisconsin, Stout; Charles G. Wilber, Colorado State University; Nancy Lee Wilkinson, San Francisco State University; John C. Williams, College of San Mateo; Ray Williams, Rio Hondo College; Roberta Williams, University of Nevada, Las Vegas; Samuel J. Williamson, New York University; Dwina Willis, Freed-Hardeman University; Ted L. Willrich, Oregon State University; James Winsor, Pennsylvania State University; Fred Witzig, University of Minnesota at Duluth; Martha Wolfe, Elizabethtown Community and Technical College; George M. Woodwell, Woods Hole Research Center; Todd Yetter, University of the Cumberlands; Robert Yoerg, Belmont Hills Hospital; Hideo Yonenaka, San Francisco State University; Brenda Young, Daemen College; Anita Závodská, Barry University; Malcolm J. Zwolinski, University of Arizona.

For Students

Students who can begin early in their lives to think of things as connected, even if they revise their views every year, have begun the life of learning.

Mark Van Doren

Why Is It Important to Study Environmental Science?

Welcome to **environmental science**—an *interdisciplinary* study of how the earth works, how we interact with the earth, and how we can deal with the environmental problems we face. Because environmental issues affect every part of your life, the concepts, information, and issues discussed in this book and the course you are taking will be useful to you now and throughout your life.

Understandably, we are biased, but *we strongly believe that environmental science is the single most important course that you could take.* What could be more important than learning about the earth's life-support system, how our choices and activities affect it, and how we can reduce our growing environmental impact? Evidence indicates strongly that we will have to learn to live more sustainably by reducing our degradation of the planet's life-support system. We hope this book and the learning opportunities available to you online will inspire you to become involved in this change in the way we view and treat the earth, which sustains us, our economies, and all other living things.

You Can Improve Your Study and Learning Skills

Maximizing your ability to learn involves trying to *improve your study and learning skills.* Here are some suggestions for doing so:

- ■ *Develop a passion for learning.*
- ■ *Get organized.*
- ■ *Make daily to-do lists.* Put items in order of importance, focus on the most important tasks, and assign a time to work on these items. Shift your schedule as needed to accomplish the most important items.
- ■ *Set up a study routine in a distraction-free environment.* Study in a quiet, well-lit space. Take breaks every hour or so. During each break, take several deep breaths and move around; this will help you to stay more alert and focused.
- ■ *Avoid procrastination.* Do not fall behind on your reading and other assignments. Set aside a particular time for studying each day and make it a part of your daily routine.
- ■ *Make hills out of mountains.* It is psychologically difficult to read an entire book, read a chapter in a book, write a paper, or cram to study for a test.

Instead, break these large tasks (mountains) down into a series of small tasks (hills). Each day, read a few pages of a book or chapter, write a few paragraphs of a paper, and review what you have studied and learned.

- ■ *Ask and answer questions as you read.* For example, "What is the main point of a particular subsection or paragraph?" "How does it relate to the key question and key concepts addressed in each major chapter section?"
- ■ *Focus on key terms.* Use the glossary in your textbook to look up the meaning of terms or words you do not understand. This book shows all key terms in **bold** type and lesser, but still important, terms in *italicized* type. The MindTap online edition of this text provides direct links to definitions for all bold-type terms. The *Chapter Review* questions at the end of each chapter also include the chapter's key terms in bold. Flash cards for testing your mastery of key terms for each chapter are available on the website for this book, or you can make your own.
- ■ *Interact with what you read.* You could highlight key sentences and paragraphs and make notes in the margins. You might also mark important pages that you want to return to. The MindTap edition supports extensive note-taking features.
- ■ *Review to reinforce learning.* Before each class session, review the material you learned in the previous session and read the assigned material.
- ■ *Become a good note taker.* Learn to write down the main points and key information from any lecture using your own shorthand system. Review, fill in, and organize your notes as soon as possible after each class.
- ■ *Check what you have learned.* At the end of each chapter, you will find review questions that cover all of the key material in each chapter section. We suggest that you try to answer each of these questions after studying each chapter section.
- ■ *Write out answers to questions to focus and reinforce learning.* Write down your answers to the critical thinking questions found in the *Thinking About* boxes throughout the chapters, in many figure captions, and at the end of each chapter. These questions are designed to inspire you to think critically about key ideas and connect them to other ideas and to your own life. Also, write down your answers to all chapter-ending review questions. Additional quizzes can be found online as well. Save your answers for review and test preparation.
- ■ *Use the buddy system.* Study with a friend or become a member of a study group to compare notes, review material, and prepare for tests. Explaining

something to someone else is a great way to focus your thoughts and reinforce your learning. Attend any review sessions offered by teachers or teaching assistants.

- **Learn your teacher's test style.** Does your teacher emphasize multiple-choice, fill-in-the-blank, true-or-false, factual, or essay questions? How much of the test will come from the textbook and how much from lecture material? Adapt your learning and studying methods to this style.

- **Become a good test taker.** Avoid cramming. Eat well and get plenty of sleep before a test. Arrive on time or early. Calm yourself and increase your oxygen intake by taking several deep breaths. (Do this also about every 10–15 minutes while taking the test.) Look over the test and answer the questions you know well first. Then work on the harder ones. Use the process of elimination to narrow down the choices for multiple-choice questions. For essay questions, organize your thoughts before you start writing. If you have no idea what a question means, make an educated guess. You might earn some partial credit and avoid getting a zero. Another strategy for getting some credit is to show your knowledge and reasoning by writing something like this: "If this question means so and so, then my answer is _____."

- **Take time to enjoy life.** Every day, take time to laugh and enjoy nature, beauty, and friendship.

You Can Improve Your Critical Thinking Skills

Critical thinking involves developing skills to analyze information and ideas, judge their validity, and make decisions. Critical thinking helps you to distinguish between facts and opinions, evaluate evidence and arguments, and take and defend informed positions on issues. It also helps you to integrate information and see relationships and to apply your knowledge to dealing with new and different problems, as well as to your own lifestyle choices. Here are some basic skills for learning how to think more critically.

- **Question everything and everybody.** Be skeptical, as any good scientist is. Do not believe everything you hear and read, including the content of this textbook, without evaluating the information you receive. Seek other sources and opinions.

- **Identify and evaluate your personal biases and beliefs.** Each of us has biases and beliefs taught to us by our parents, teachers, friends, role models, and our own experience. What are your basic beliefs, values, and biases? Where did they come from? What assumptions are they based on? How sure are you that your beliefs, values, and assumptions are right and why?

According to the American psychologist and philosopher William James, "A great many people think they are thinking when they are merely rearranging their prejudices."

- **Be open-minded and flexible.** Be open to considering different points of view. Suspend judgment until you gather more evidence, and be willing to change your mind. Recognize that there may be a number of useful and acceptable solutions to a problem and that very few issues are either black or white. Try to take the viewpoints of those you disagree with. Understand that there are trade-offs involved in dealing with any environmental issue, as you will learn in reading this book.

- **Be humble about what you know.** Some people are so confident in what they know that they stop thinking and questioning. To paraphrase American writer Mark Twain, "It's what we know is true, but just ain't so, that hurts us."

- **Find out how the information related to an issue was obtained.** Are the statements you heard or read based on firsthand knowledge and research or on hearsay? Are unnamed sources used? Is the information based on reproducible and widely accepted scientific studies or on preliminary scientific results that may be valid but need further testing? Is the information based on a few isolated stories or experiences or on carefully controlled studies that have been reviewed by experts in the field involved? Is it based on unsubstantiated and dubious scientific information or beliefs?

- **Question the evidence and conclusions presented.** What are the conclusions or claims based on the information you're considering? What evidence is presented to support them? Does the evidence support them? Is there a need to gather more evidence to test the conclusions? Are there other, more reasonable conclusions?

- **Try to uncover differences in basic beliefs and assumptions.** On the surface, most arguments or disagreements involve differences of opinion about the validity or meaning of certain facts or conclusions. Scratch a little deeper and you will find that many disagreements are based on different (and often hidden) basic assumptions concerning how we look at and interpret the world around us. Uncovering these basic differences can allow the parties involved to understand one another's viewpoints and to agree to disagree about their basic assumptions, beliefs, or principles.

- **Try to identify and assess any motives on the part of those presenting evidence and drawing conclusions.** What is their expertise in this area? Do they have any unstated assumptions, beliefs, biases, or values? Do they have a personal agenda? Can they benefit financially or politically from acceptance of their evidence and conclusions? Would investigators with different basic assumptions or beliefs take the same data and come to different conclusions?

- **Expect and tolerate uncertainty.** Recognize that scientists cannot establish absolute proof or certainty about anything. However, the reliable results of science have a high degree of certainty.

- **Check the arguments you hear and read for logical fallacies and debating tricks.** Here are six of many examples of such debating tricks: *First,* attack the presenter of an argument rather than the argument itself. *Second,* appeal to emotion rather than facts and logic. *Third,* claim that if one piece of evidence or one conclusion is false, then all other related pieces of evidence and conclusions are false. *Fourth,* say that a conclusion is false because it has not been scientifically proven (scientists never prove anything absolutely, but they can often establish high degrees of certainty). *Fifth,* inject irrelevant or misleading information to divert attention from important points. *Sixth,* present only either/or alternatives when there may be a number of options.

- **Do not believe everything you read on the Internet.** The Internet is a wonderful and easily accessible source of information that includes alternative explanations and opinions on almost any subject or issue—much of it not available in the mainstream media and scholarly articles. Blogs of all sorts have become a major source of information, even more important than standard news media for some people. However, because the Internet is so open, anyone can post anything they want to some blogs and other websites with no editorial control or review by experts. As a result, evaluating information on the Internet is one of the best ways to put into practice the principles of critical thinking discussed here. Use and enjoy the Internet, but think critically and proceed with caution.

- **Develop principles or rules for evaluating evidence.** Develop a written list of principles to serve as guidelines for evaluating evidence and claims. Continually evaluate and modify this list on the basis of your experience.

- **Become a seeker of wisdom, not a vessel of information.** Many people believe that the main goal of their education is to learn as much as they can by gathering more and more information. We believe that the primary goal is to learn how to sift through mountains of facts and ideas to find the few *nuggets of wisdom* that are especially useful for understanding the world and for making decisions. This book is full of facts and numbers, but they are useful only to the extent that they lead to an understanding of key ideas, scientific laws, theories, concepts, and connections. The major

goals of the study of environmental science are to find out how nature works and sustains itself *(environmental wisdom)* and to use *principles of environmental wisdom* to help make human societies and economies more sustainable, more just, and more beneficial and enjoyable for all. As writer Sandra Carey observed, "Never mistake knowledge for wisdom. One helps you make a living; the other helps you make a life."

To help you practice critical thinking, we have supplied questions throughout this book, found within each chapter in brief boxes labeled *Thinking About,* in the captions of many figures, and at the end of each chapter. There are no right or wrong answers to many of these questions. A good way to improve your critical thinking skills is to compare your answers with those of your classmates and to discuss how you arrived at your answers.

Use the Learning Tools We Offer in This Book

We have included a number of tools throughout this textbook that are intended to help you improve your learning skills and apply them. First, consider the *Key Questions* list at the beginning of each chapter section. You can use these to preview a chapter and to review the material after you've read it.

Next, note that we use three different special notations throughout the text. Each chapter opens with a **Core Case Study**, and each time we tie material within the chapter back to this core case, we note it in bold, colored type as we did in this sentence. You will also see two icons appearing regularly in the text margins. When you see the *sustainability* icon, you will know that you have just read something that relates directly to the overarching theme of this text, summarized by our six **principles of sustainability**, which are introduced in Figures 1-2, p. 6, and 1-5, p. 9, and which appear on the back cover of the student edition. The *Good News* icon appears near each of many examples of successes that people have had in dealing with the environmental challenges we face.

We also include several brief *Connections* boxes to show you some of the often surprising connections between environmental problems or processes and some of the products and services we use every day or some of the activities we partake in. These, along with the *Thinking About* boxes scattered throughout the text (both designated by the *Consider This. . .* heading), are intended to get you to think carefully about activities and choices we take for granted and how they might be affecting the environment.

At the end of each chapter, we list what we consider to be the *three big ideas* that you should take away from the chapter. Following that list in each chapter is a *Tying It All Together* box. This feature quickly reviews the Core Case Study and how chapter material relates to it, and it explains how the principles of sustainability can be applied to deal with challenges discussed in the **Core Case Study** and throughout the chapter.

We have also included a *Chapter Review* section at the end of each chapter with questions listed for each chapter section. These questions cover all of the key material and key terms in each chapter. A variety of other exercises and projects follow this review section at the end of each chapter.

Finally, at the back of the book, we have included a comprehensive glossary. It includes definitions of all the book's key terms, as well as definitions of many other important terms.

Know Your Own Learning Style

People have different ways of learning and it can be helpful to know your own learning style. *Visual learners* learn best from reading and viewing illustrations and diagrams. *Auditory learners* learn best by listening and discussing. They might benefit from reading aloud while studying and using a tape recorder in lectures for study and review. *Logical learners* learn best by using concepts and logic to uncover and understand a subject rather than relying mostly on memory.

This book and its supporting website material contain plenty of tools for all types of learners. Visual learners can benefit from using flash cards (available online) to memorize key terms and ideas. This is a highly visual book with many carefully selected photographs and diagrams designed to illustrate important ideas, concepts, and processes. For logical learners, the book is organized by key concepts that are revisited throughout any chapter and related carefully to other concepts, major principles, and case studies and other examples. We urge you to become aware of your own learning style and make the most of these various tools.

This Book Presents a Positive, Realistic Environmental Vision of the Future

Our goal is to present a positive vision of our environmental future based on realistic optimism. To do so, we strive not only to present the facts about environmental issues, but also to give a balanced presentation of different viewpoints. We consider the advantages and disadvantages of various technologies and proposed solutions to environmental problems. We argue that environmental solutions usually require *trade-offs* among opposing parties, and that the best solutions are *win-win* solutions. Such solutions are achieved when people with different viewpoints work together to come up with a solution that both sides can live with. And we present the good news as well as the bad news about efforts to deal with environmental problems.

One cannot study a subject as important and complex as environmental science without forming conclusions, opinions, and beliefs. However, we argue that any such results should be based on use of critical thinking to evaluate conflicting positions and to understand the trade-offs involved in most environmental solutions. To that end, we emphasize critical thinking throughout this textbook, and we encourage you to develop a practice of applying critical thinking to everything you read and hear, both in school and throughout your life.

Help Us Improve This Book

Researching and writing a book that covers and connects the numerous major concepts from the wide variety of environmental science disciplines is a challenging and exciting task. Almost every day, we learn about some new connection in nature. However, in a book this complex, there are bound to be some errors—some typographical mistakes that slip through and some statements that you might question, based on your knowledge and research. We invite you to contact us to correct any errors you find, point out any bias you see, and suggest ways to improve this book. Please e-mail your suggestions to Tyler Miller at mtg89@hotmail.com or Scott Spoolman at spoolman@tds.net.

Now start your journey into this fascinating and important study of how the earth's life-support system works and how we can leave our planet in a condition at least as good as what we now enjoy. Have fun.

Supplements for Students

You have a large variety of electronic and other supplemental materials available to you to help you take your learning experience beyond this textbook:

- ***Fast Track to a 5 [9781285750941] Preparing for the AP® Environmental Science Examination.*** This test-prep manual, written by David Hong of Diamond Bar High School, Diamond Bar, California, and Karen Lionberger of Chattahoochee High School, Alpharetta, Georgia, includes an introduction for students on taking the AP® Exam, a diagnostic test to help students assess their level of preparedness, review chapters following the AP® topic list, with multiple-choice and free-response questions for each chapter, and two complete practice tests. *Fast Track to a 5* may be purchased either with the text or separately.

- ***AP® Aplia.*** Is a Cengage Learning online homework system dedicated to improving learning by increasing student effort and engagement. Aplia provides you with prompt and detailed feedback to help you learn as you work through the questions, and features interactive tutorials to fully engage you in learning course concepts.

- ***The Interactive eBook*** for *Living in the Environment,* AP Edition. Retains the full-color look and feel of the printed text and includes highlighting and note-taking.

- ***Global Environment Watch.*** Updated several times a day, the Global Environment Watch is a focused portal into GREENR—the Global Reference on the Environment, Energy, and Natural Resources—an ideal one-stop site for classroom discussion and research projects. This resource center keeps courses up-to-date with the most current news on the environment. Users get access to information from trusted academic journals, news outlets, and magazines, as well as statistics, an interactive world map, videos, primary sources, case studies, podcasts, and much more.

- **New! *Virtual Field Trips in Environmental Issues.*** *Virtual Field Trips in Environmental Issues* brings the field to you, with dynamic panoramas, videos, photographs, maps, and quizzes covering important topics within environmental science. A case study approach covers the issues of *keystone species, climate change's role in extinctions, invasive species, the evolution of a species due to its environment,* and *an ecosystem approach to sustaining biodiversity.* Students are engaged, interacting with real issues to help them think critically about the world around them.

Other student learning tools include the following:

- *Essential Study Skills for Science Students* by Daniel D. Chiras. This book includes chapters on developing good study habits, sharpening memory, getting the most out of lectures, labs, and reading assignments, improving test-taking abilities, and becoming a critical thinker. Available for students on teacher's request.

- *Lab Manual.* Edited by Edward Wells, this lab manual includes both hands-on and data analysis labs to help your students develop a range of skills.

- *What Can You Do?* This guide presents students with a variety of ways that they can affect the environment, and shows them how to track the effect their actions have on their carbon footprint. Available for students on teacher's request.

Topic	Correlation to Living in the Environment
I. EARTH SYSTEMS AND RESOURCES	
A. Earth Science Concepts	
Geologic time scale	pages 8, 83, 86
plate tectonics	pages 86, 351–353
earthquakes	pages 86, 364–369
volcanism	pages 86, 364, S55
seasons	page 145
solar intensity and latitude	page 145
B. The Atmosphere	
Composition	pages 53, 475–476
structure	pages 53, 475–476
weather and climate	pages 145–146, 475, 507–510, S19–S23
atmospheric circulation and the Coriolis Effect	pages 145–148
atmosphere-ocean interactions	pages 145–148, S22
ENSO	pages S20, S21
C. Global Water Resources and Use	
Freshwater/saltwater	pages 169, 316, 319–321, 326, 329–330, 332–333, 335, 336–340, 341–343
ocean circulation	pages 145–146
agricultural, industrial, and domestic use	pages 310–321
surface and groundwater issues	pages 178–182, 184, 319–321
global problems	pages 316, 319–321, 326, 329–330, 332–333, 335, 336–340, 341–343, 542–544, 545–547, 548–553, 554–557, 558–561, 562–564
conservation	pages 336–340, 341–343
D. Soil and Soil Dynamics	
Rock cycle	pages 357–360
formation	pages 290, 351–352
composition	pages 290, 351–352
physical and chemical properties	page 290
main soil types	page 290
erosion and other soil problems	pages 288–290, 304–305
soil conservation	pages 304–310
II. THE LIVING WORLD	
A. Ecosystem Structure	
Biological populations and communities	pages 58, 109–110, 111–114
ecological niches	pages 91–94, 103
interactions among species	pages 103–104
keystone species	pages 76, 94–96, 100
species diversity and edge effects	pages 90, 91
major terrestrial and aquatic biomes	pages 148–160, 169, 170–177, 178–182
B. Energy Flow	
Photosynthesis and cellular respiration	pages 54–57, 62–63
food webs and trophic levels	pages 59–61
ecological pyramids	pages 59–61

Topic	Correlation to *Living in the Environment*

G. Renewable Energy

Solar energy	pages 414–419, 632
solar electricity	pages 417–419, S60, S61
hydrogen fuel cells	pages 430–432
biomass	pages 424–426
wind energy	pages 421–423, 436, S62
small-scale hydroelectric	page 420
ocean waves and tidal energy	pages 420–421
geothermal	pages 427–430, S63
environmental advantages/disadvantages	pages 413–419, 420–421, 421–423, 424–426, 427–429, 430–431

VI. POLLUTION

A. Pollution Types

Air pollution:

Sources—primary and secondary	pages 476–485
major air pollutants	pages 476–485
measurement units	pages 476–485
smog	pages 148, 481–484
acid deposition—causes and effects	pages 485–488
heat islands and temperature inversions	pages 483–484
indoor air pollution	pages 488–491
remediation and reduction strategies	pages 493–496
Clean Air Act and other relevant laws	pages 493–494, S8

Noise pollution:

Sources	pages 612–613
effects	page 613
control measures	page 613

Water pollution:

Types	pages 544, 545–548, 548–553, 554–558, 558–562, 562–570
sources, causes, and effects	pages 545–548, 548–553, 554–558, 558–562, 562–570
cultural eutrophication	pages 550–552
groundwater pollution	pages 554–557
maintaining water quality; water purification	pages 556–557
sewage treatment/septic systems	pages 565–566
Clean Water Act and other relevant laws	pages 557–558, 668

Solid waste:

Types	pages 576, 577–579
disposal	pages 580–582, 582–586, 587–588, 588–590, 590–596
reduction	pages 590–596, 596–600

B. Impacts on the Environment and Human Health

Hazards to human health:

Environmental risk analysis	pages 442, 443–453
acute and chronic effects	pages 443–453, 453–457
dose-response relationships	pages 458–459
air pollutants	pages 479–480, 481–484, 485–488
smoking and other risks	pages 466–467

Taking on the challenge and rigor of an Advanced Placement course can be an enriching experience in your academic career. Whether you are taking an AP course at your school or you are working on an AP course independently, the stage is set for an immeasurable intellectual experience. As the school year progresses, you will gain a deeper understanding of the scientific processes, concepts, and methodologies required to understand the mechanisms that drive our natural world. Environmental science is an interdisciplinary study by which you will acquire the ability to identify and analyze both natural and anthropogenic environmental problems. By the end of this course you will also be able to assess the relative risks associated with these environmental problems as well as suggest more sustainable solutions and alternatives to the current conventional methods that cause environmental damage. This course will provide an exhilarating and awakening experience for you through the study and understanding of today's global environmental problems and their solutions.

Sometime during the spring, as May approaches, the AP exam will start to feel more like a tangible obstacle you will soon face. At times, the amount of material you must master to be successful on the Advanced Placement exam may seem somewhat overwhelming or intimidating. If you are dreading the AP exam, just know that you are not the only one. However, there are some simple exam tips and review steps you can follow that will get you on the fast track to making that 5.

First and foremost, it is imperative to maintain the motivation and a positive attitude to guide your preparation. If you manage your time effectively, you will meet this one major challenge—mastery of a large amount of material in a short period of time. You should develop a review schedule to help direct your study time so that you spread out your review over the course of at least several weeks to a month prior to the exam rather than relying on just a few short days of cramming. Sticking to your review schedule and allowing enough practice time prior to the exam will help you master the course material and will also give you an invaluable amount of confidence on the day of the exam.

Before the Examination

First and foremost you need to make sure you are registered to take the AP exam. Most schools have already taken care of this aspect of the exam for you, but you

should check with your AP teacher or coordinator to make sure your paperwork has been submitted to the College Board. If you have a documented disability and will need testing accommodations on the exam, you should ensure that it has been reported by your AP coordinator to the College Board. If you are studying AP independently, call AP Services at the College Board for the name of the local AP coordinator, who will help you through the registration process. You should also check your calendar for conflicts and speak to your school's AP coordinator about making arrangements to take an alternate AP exam on a later date. The College Board will allow students with more than one AP exam at the same time on the same date to take an alternate exam on a later date. Other acceptable circumstances that warrant an alternate exam include conflicts with IB or state-mandated exams and religious holidays. Conflicts with athletic or academic contests, and even family vacations are also acceptable; however, there is an additional charge for taking an alternate exam for those reasons.

Just like with any test, the night before the exam you should get plenty of sleep so that you have enough energy to stay focused during the entire three hours of testing. This is not the time to stay up studying all night. Instead, focus on briefly reviewing broad concepts such as common solutions to major environmental problems or frequently used vocabulary terms. The night before the test is a good time to get together everything you will need the next morning for the exam.

What will you need on the morning of the exam? First you need to eat a good breakfast so that you have plenty of energy for the test—being tired can actually cause you to give up too easily on a question and cost you points on the exam. For the exam you will need a sharpened No. 2 pencil to answer the multiple-choice section. Also, you will need a blue or black ink pen for your essays on the free-response portion of the exam. It is a good idea to take two or three pencils and pens so you have plenty of backup choices if one fails to work properly. It is also a good idea to take a watch with you in case the testing room does not have a clock. It will be important to be aware of the time on both your multiple-choice section and the free response to allow you to utilize your time effectively. Make sure you take with you anything that has been requested by your AP coordinator—such as your Social Security number, photo ID, or registration ticket. Remember that calculators are NOT allowed on the AP Environmental Science exam. Cell phones are also not permitted in the exam rooms.

Finally, it is important that you are completely comfortable while taking the exam. Wear clothes and shoes that are comfortable. Some students also like to bring a jacket or sweatshirt in case the testing room is too cold and becomes uncomfortable to test in. It is also a good idea to bring a snack to eat during the break. You will have a short break between the multiple-choice portion, which is 90 minutes long, and the free-response portion, which is also 90 minutes long. Three hours of testing is a long time and you don't want to be hungry going into the essay section of your test.

The AP Environmental Science examination consists of two sections: Section I has one hundred multiple-choice questions; Section II has four free-response questions. You will have ninety minutes for each section. There will be a short break between the two sections of the exam. You must write an essay for each of the four questions. You should monitor the time during both sections of the exam and move nimbly through both sections so you do not get bogged down on one multiple-choice question or on one essay.

Strategies for the Multiple-Choice Section

Here are some important rules of thumb to help you work your way through the multiple-choice questions:

- **Terminology.** Each question has a stem that may be in the form of a question or statement, and five options that are the possible answers to the question. Exactly one of the five options is correct.

- **There is one correct answer; however. . .** Sometimes, the "correct" answer is "the best" among the options given. In that case, more than one answer may actually be correct, but one is the best option.

- **Manage the time.** With 90 minutes to answer 100 questions, you have less than one minute to answer each question. Check your pace; try to answer 25 questions every 20 minutes. This will leave 10 minutes at the end of the exam to look back over your answers.

- **Cover the options.** Try reading the stem with the options covered. Answer in your mind, and then select the option that is closest to your answer.

- **Carefully and completely read the entire question.** The item writers may have intentionally included attractive options for students who are rushing through the exam.

- **Write in your multiple-choice booklet.** As you eliminate options, cross them out by drawing a line through them. On questions with a negative stem or roman numerals (see question types below) write a T or an F next to the options as you work through the question. If you skip an extremely difficult or long question, circle it and return to it after you have answered all of the easy questions. If it helps you to do so, underline key words in a question that help you to focus on what is important in the question. Use the test booklet as scratch paper for any questions that require calculations. Calculator use is not permitted on the AP Environmental Science exam.

- **Fill in your answer document carefully.** If you must erase, do so completely. Bring a white plastic eraser to use on your answer document. Your exam will be scored by a machine and any stray marks may be read as answers. When you skip questions, don't forget to skip the corresponding lines on your answer sheet, or you will waste time erasing and moving answers later.

- **Answer every question.** Points are no longer deducted for incorrect answers on AP exams, so you should answer every question.

- **Guessing advice.** If you must guess, make a random selection from the options that you did not eliminate. Avoid picking an option that you do not know is correct, if it only "sounds right" to you. Picking options that sound right could lead you into a trap set by the item writer for a student, who like you, knows a little bit, but not quite enough about the question to select the one best answer. If you must make a guess, do so by making an unbiased random selection from the options that remain after you have eliminated those you are certain are wrong.

Types of Multiple-Choice Questions

There are a few different types of multiple-choice questions on the AP Environmental Science Exam.

Options-First Questions

The multiple-choice section of the AP Environmental Science Exam usually begins with about 20 options-first questions. These questions require you to use the same set of options for a series of questions. You may use the same option as the answer for more than one of the questions, and some may not be used at all. A good way to think about this type of question is as a series of questions that all happen to have the same set of options. If you think about them in that way, you will never wonder if an option can be correct more than once, or if you will use all of the options. For example:

(A) Mercury
(B) Carbon monoxide
(C) Lead
(D) Ozone
(E) Sulfur dioxide

1. A secondary air pollutant that is formed in photo-chemical smog
2. A contaminant of coal that accumulates in the tissues of some species of fish
3. Reacts in air to form acids which can fall to the ground as acid rain

Answers:

1–D. Ozone is the only secondary pollutant among the five options.
2–A. Mercury is a contaminant in coal that vaporizes when the coal is burned. After falling back to earth, the mercury is washed into aquatic ecosystems where it may be assimilated in the tissues of fish.
3–E. Sulfur dioxide reacts with other substances in the air to form acids that can fall to the ground as acid rain.

Traditional Questions

Nearly all of the multiple-choice questions will be traditional multiple-choice questions. These questions will be straightforward questions with five options and one correct answer. These questions simply require you to read the question and select the most correct answer. For example:

1. Which of the following is an asphyxiant produced during incomplete combustion reactions?

   (A) Mercury
   (B) Carbon monoxide
   (C) Lead
   (D) Ozone
   (E) Sulfur dioxide

   Answer: B. Carbon monoxide is produced during incomplete combustion and it is an asphyxiant.

Negative Stem Questions

You should expect to encounter few of this type of question, but be prepared for them. They may be the hardest questions on the exam. If the answer is not immediately obvious to you, you may want to circle this type of question and return to it later. In negative-stem questions, all of the options are correct except for one. These questions usually begin with a statement like, "All of the following are correct EXCEPT. . ." or "Which of the following is NOT. . ." In either case, there will be four options that are in the correct context, and one that is not—that is the correct answer. One way to approach these questions is as a series of true/false questions. There is only one false answer, and that is the correct answer. It will help you to keep track of which options are true and false if you mark them with a T or an F as you read. For example:

1. Which of the following is NOT one of the criteria air pollutants monitored by the U.S. EPA?

   (A) Mercury
   (B) Carbon monoxide
   (C) Lead
   (D) Ozone
   (E) Sulfur dioxide

   Answer: A. Mercury is not one of the criteria air pollutants monitored by the U.S. EPA. The criteria air pollutants are carbon monoxide, lead, ozone, sulfur dioxide, nitrogen oxides, and particulate matter. If you marked the options with a T or F, answer A should be the only option with an F next to it.

Roman Numeral Questions

You should also expect to encounter relatively few Roman numeral questions. These questions are not as difficult as the negative stem questions because you are looking for correct answers. In Roman numeral questions, there are usually three or four answers labeled with Roman numerals. One or more of the answers could be correct. You are then provided with a series of options from which you select the correct Roman numeral or set of Roman numerals. The option with the Roman numerals followed by the correct answers is correct. For example:

1. Which of the following are toxic heavy metals?

   (I) Mercury
   (II) Carbon monoxide
   (III) Lead

   (A) I only
   (B) II only
   (C) III only
   (D) I and II only
   (E) I and III only

   Answer: E. Mercury and lead are toxic heavy metals, and carbon monoxide is not a toxic heavy metal, which makes Roman numerals I and III the only two correct answers.

Least and Most Likely Questions

This type of question may have two or more options that are correct, but only one that is the least or most likely. When answering these questions, don't answer until you have read through all of the options, and if more than one is correct, rank the likeliness of each correct option by placing a number next to the option.

1. If released or formed in the atmosphere, which of the following will least likely result in long-term environmental damage to an ecosystem?

   (A) Mercury
   (B) Carbon monoxide
   (C) Lead
   (D) Ozone
   (E) Sulfur dioxide

   Answer: B. Carbon monoxide is not persistent, nor does it bioaccumulate. Unlike ozone or sulfur dioxide, carbon monoxide does not produce long-term ecological damage.

Data Interpretation and Analysis Questions

In this type of question, there is a data set, graph, or chart for you to review and interpret. These questions may require that you do some calculations to determine the answer. Often there is more than one question that makes use of the data set, graph, or chart. You should do all scratch work in the test booklet. Calculators are not permitted on the AP Environmental Science exam.

U.S. Airborne Lead Air Quality 1980–2005	
Year	Lead Concentration ($\mu g/m^3$)
1980	1.25
1985	0.67
1990	0.29
1995	0.11
2000	0.09
2005	0.09

1. The percent change in the airborne lead concentration between 1980 and 2000 was closest to which of the following?

   (A) 10%
   (B) 17%
   (C) 81%
   (D) 87%
   (E) 93%

   Answer: E. To determine the percent change: take the difference between the two values, divide that difference by the original value, and multiply by 100. In this case $1.25 - 0.09 = 1.16$ divided by 1.25 equals 0.928 multiplied by 100 equals 92.8%, which is closest to 93%. You should be prepared to do arithmetic like this on your AP Environmental Science exam. Remember, no calculators, so practice your math!

2. Which of the following best explains the trend in the data of U.S. airborne lead concentrations?

   (A) The inclusion of lead in the EPA's list of Criteria Air Pollutants
   (B) The phaseout of the use of lead-based paints
   (C) The widespread use of catalytic converters
   (D) The elimination of lead additives in gasoline
   (E) The use of wet scrubbers on coal-fired power plants

   Answer: D. The plummet in airborne lead concentrations during the 1980s and 1990s is due to the phaseout of the use of tetraethyl lead as an additive in gasoline. Note that more than one of the options are reasonable reasons why lead emissions were reduced; this question illustrates the second bullet in the strategies above: "**There is one** correct answer; however. . . Sometimes, the 'correct' answer is 'the best' among the options given."

Free-Response Questions

The free-response portion of the AP Environmental Science exam is 90 minutes long. During this time you must answer four questions. This allows you approximately 22 minutes for each essay. It is very important that you watch your time on each essay and not to get wrapped up in just one. Remember that these essays account for 40% of your exam score and therefore each essay is worth 10% of your overall grade.

While individual free-response questions may have several sub-parts, keep in mind that the total of all parts is worth 10 points. Typically, each sub-part of a free-response question will be worth 2 to 4 points, depending on complexity and number of parts of the question. It is possible for the points available to total more than ten, but the maximum score for any free-response question is ten. Different parts of a free-response question often are not of equal difficulty, but usually one or two parts will be simpler than others. Answer the easiest part of the question first and do not omit an entire question just because some parts seem difficult. Partial credit is always available on such questions, but only if an answer is given and knowledge about the question being asked is demonstrated.

Each question in the free-response section will be broken down into separate parts such as: (a), (b), (c), etc. Even individual parts to a question can be further broken down, for example:

   (a) Calculate the following by using statistics given in the table.
      i. How many gallons of water did the family use in one month?
      ii. What is the annual cost of electricity for operating their dishwasher?

Types of Free-Response Questions
Data-Set Questions

You will have at least one question that is a data-set question, often referred to as the math question, and it will center around a graph, diagram, table/chart, or statistics given in the introductory paragraph. Approximately 50% of your points on this type of question will be earned by doing arithmetical calculations. Even though this is considered the math question, these questions typically have one to two parts you can earn points on that do not deal directly with any calculations or graphing but rather with the environmental issue or concept the question refers to. So, if you don't feel you can do the math part of the question, **don't give up**! Read the remaining sections of the free-response and answer the portions that are conceptual and not mathematical. Remember to always **show your work** (only in the pink answer booklet) for any calculation and give the appropriate units with each number written.

Reminder: Calculators are not permitted on the AP Environmental Science exam so you will need to be comfortable using basic algebra such as multiplying and dividing exponents and scientific notation.

Document-Based Question (DBQ)

You will have one question that is prompted with some type of document that will pertain to the questions being asked. It is important to remember these prompts are primarily used to introduce the questions. Although you may need to use some numbers or information provided in the document to help answer a question, do not expect to pull all of your answers directly from the passage given. A common mistake that students make is to simply repeat information given in the document as their answer. As mentioned, the purpose of the document is to introduce the question or questions. Students are expected to demonstrate comprehension by adding information learned from readings or class.

Synthesis and Evaluation

There are two synthesis/evaluation questions on the free-response section of the AP Environmental Science exam. It may be easiest to understand what a synthesis/evaluation question is by understanding what it is not. That is, if you determine that a question is not a math question, and it is not a DBQ, then it is a synthesis/evaluation question. A synthesis/evaluation question will require you to write an essay in which you demonstrate your knowledge about one of the numerous topics of study in the AP Environmental Science course. Your ability to earn points in these essays will be determined by your knowledge of the subject matter. In rare cases, you may be required to perform a simple calculation in a synthesis/evaluation question; however, the calculation will likely be worth only one or two points. These questions will usually have a small paragraph to introduce a topic and then 4–5 questions or prompts pertaining to that topic. As with all free-response questions you may be required to identify, describe, explain, discuss, etc. Be sure to identify the verbs used to ask the question (identify, describe, discuss, etc.) and follow the free-response tips listed below when writing your essay. An example of what these types of questions may look like:

1. The harvesting of wood for fuelwood, and charcoal made from fuelwood, is increasing at an alarming rate. This is causing the deforestation of many areas, particularly in developing countries. Many of these coun-

tries are now cutting down trees at a rate 10–20% faster than they are being replanted.

(a) Identify TWO uses of fuelwood in developing countries.
(b) Describe how deforestation impacts the global carbon cycle.
(c) Identify and describe one environmental problem and one economic benefit which result from clear-cutting forests.
(d) Discuss TWO sustainable forestry practices that could be implemented in these countries.
(e) Write an argument in support of the practice of sustainably harvesting fuelwood.

Vocabulary Used in Free-Response Questions

You can earn a maximum of 10 points on each of the four free-response questions. You should pay close attention to the word choice used in each part of the free response. Keying in on what type of verb is being used in the question can help you maximize both your time and points on your essay section.

Identify or List

This term is asking you to give a specific object, advantage, disadvantage, cause, solution, etc. It is important to always write in prose on every question. Even if it may seem like you can answer this type of question in one word—don't. You must write the answer in a complete sentence to get any credit.

For example, the question above states, "Identify two uses of fuelwood in developing countries." The answer can be as short as, "Two uses of fuelwood in developing countries are for home heating and cooking." However, you would earn no points for simply writing "heating and cooking."

Describe or Explain

These terms are asking you to further illustrate your answer by using details beyond just identifying an object or solution. Often you will see these terms used with the exact number of solutions or objects you are to describe. A good description will usually take more than one sentence.

For example, the question may state "Describe TWO benefits of using fuelwood." A good response to this prompt is, "Two benefits of the use of fuelwood are the ability for people to heat their homes and to cook their food without the need of fossil fuels. Unlike our conventional fuel sources, if managed sustainably, fuelwood can be a renewable resource."

Or

For example, the question above states, "Describe how deforestation impacts the global carbon cycle." A good response to this prompt is, "Deforestation removes trees that function as carbon reservoirs by absorbing carbon dioxide during photosynthesis. This will increase carbon dioxide levels in the atmosphere."

Note: It is also possible to have a combination of two verb choices in one question. They may ask you to either "Identify and describe. . . ." or "Identify and explain. . . ." You should simply follow the rules of both verbs as given in examples above.

Discuss

Here you need to really elaborate on your answer to the question. Often, you will receive 2–3 points for this type of question; therefore, it is important to write additional detail and go beyond a simple description or explanation in your answer. A good discussion will usually take several sentences or an entire paragraph.

For example, the question asks, "Discuss a benefit to using fuelwood over current conventional fossil fuels." A good response to this prompt is, "A benefit of using fuelwood over fossil fuels is the reduction in sulfur dioxide emissions associated with coal burning practices. By decreasing sulfur dioxide emissions we also reduce acid deposition problems that cause tissue damage to trees in terrestrial ecosystems and cause fish kills due to lowered pH in aquatic systems." Remember, your discussion questions require you to not only identify and define concepts but to further elaborate on the issue at hand.

Or

For example, the question above states, "Discuss TWO sustainable forestry practices that could be implemented in these countries." A good response to this prompt is, "One practice is the use of tree plantations, which can grow fast-growing tree species and be managed so that the rate of tree harvesting is equal to the rate of replenishment. This will reduce the need to cut native forests. A second practice is selective cutting, which will remove some, but not all of the trees from forests. Trees can then be replanted in the areas from which they were removed and the forest allowed to recover before being cut again. This will minimize the area of tree cover being lost in native forests."

Write an Argument

An argument is a series of statements all in support of a stated position on an issue. This is the only time that students may be rewarded for writing a lengthy list. An argument is the most *extensive* and detailed response that you could be called upon to write and it will likely be at least one paragraph in length.

For example, the question above states, "Write an argument in support of the practice of sustainably har-

vesting fuelwood." A good response to this prompt is, "Harvesting fuelwood sustainably avoids using practices like clear-cutting forests. Clear-cutting causes soil erosion, and runoff into waterways causing sedimentation, which decreases primary productivity, increases fish kills due to suffocation, and results in poor water quality for people who depend on the waterways for their domestic water use. Furthermore, when the rate of tree harvesting exceeds the rate of replanting, tracts of forest that function as carbon dioxide reservoirs are no longer available, which will increase carbon dioxide levels in the atmosphere and further global climate change."

Free-Response Tips

- **Don't go green on the APES free-response section.** For your free-response portion of the exam, you will be given a green booklet with the questions in it and a pink answer booklet with the questions and the space in which you will write your answers. The green insert stays at your local school and will never be seen by an AP exam grader. So, make sure all of your calculations, scratch work, and answers are written in your pink answer booklet—especially for any mathematical calculations.

- **Collect your thoughts.** Read each question carefully, thinking about the verb types being used. Take a few minutes to organize your answer in your mind before you begin.

- **Write in prose.** Always answer in complete sentences. Make sure you are writing clearly and large enough for your reader to read your answer easily. Outline forms or bulleted lists are not acceptable and will not be graded.

- **Be careful about creativity.** Shy away from using examples or solutions that are specific to your local region or are uniquely your own ideas. Remember, answers on the rubric for the free-response section must be able to be applied to every single student across the country and around the world that takes this exam. Therefore, you should only use commonly known examples or solutions. Avoid fabricating information as well. It is a waste of time and will not earn any points.

- **Be time-conscious.** You only have approximately 22 minutes per essay. If you work on the essays you feel most comfortable with first, it can increase the amount of time you have for the ones you don't know as well. Do not get bogged down working on "one hard one."

- **Strike out to save time.** If you make a mistake, don't waste valuable time and lose your momentum on a question by attempting to completely obliterate or erase your work; cross or strike out your original answer with a single line or an "X" and keep moving forward.

- **Read the verbs in the question carefully.** Note any verbs such as *identify, describe, explain,* and *discuss.* Make sure once you are done answering that you ACTUALLY answered **all** the parts to the question.

- **Read the question carefully.** You may be prompted to describe an environmental problem in one part of a question, an economic problem in another, and an environmental benefit in a third. Underline or circle the key terms "environmental" and "economic," and be certain to provide an appropriate response. Note: a good rule of thumb is to read "environmental" as "ecological" and to write about how the topic being addressed in the question affects the abundance, diversity, or distribution of life.

- **Label your answers** so there is no doubt where your answer is. All questions have very specific parts such as a, b, c, or i, ii, iii. This will also help you ensure that you have finished answering the question and can move on or come back as needed.

- **Do not restate the question** in your answer—it is a waste of time. You also do not need to have any introduction or conclusion parts to your essay. Remember—you cannot win over your reader. He or she is simply looking for correct statements that demonstrate knowledge of the concept.

- **Follow through with your thoughts or examples** by defining or explaining any key terms you use. Many students fail to get points because they didn't finish their thoughts. An easy way to accomplish this is to define any key terms you use in the essay and provide examples to illustrate what you mean.

- **Do not write long lists as your answer**! If the question says "Identify TWO" then identify only two items. The graders are instructed to grade only the first two items even if a student writes a longer list. Thus, if two items are requested, and a student lists six items of which the first two are incorrect, no points will be given even if the last four of the six answers were correct.

- **Do not use clichés** for answers (for example, "there is no away in pollution" or "not in my backyard"). Instead you should scientifically explain your answer to get the points.

- **Be concise**—do not tell the reader everything you know on the topic. Make sure you are only answering what they ask for so you do not waste valuable time that could be used on other essays.

- **Answer the question that you were asked…**not just any question to which you happen to know the answer. Sometimes, in an effort to fill space, students will launch into a dissertation that is not relevant to the question that was asked. Don't waste valuable time writing about all the things you know on another topic that is not related to the question.

- **A picture is worth 1,000 words only if they ask for it.** Draw a picture or diagram only if it is specifically asked for in the question.

- **Neatness counts.** Messy or illegible writing is very difficult to grade. You do not want to make it hard on your AP reader to award you points on your essay. It is **always best** if they can easily find and read your answers for each essay question. If they cannot decipher your handwriting, then awarding points is more difficult.

- **Math-based free-response questions.** When asked to do calculations **show your work**. You normally need to show your work and **label units** to receive credit on math questions. Also, if you have no idea how to do the calculations portion of the free response, **keep reading**! A math-based free-response question almost always contains a part of the question you could answer without the calculations being done. Make sure all of your work and answers are in your **pink** answer booklet!

- **Practice your math!** Every AP Environmental Science student should be comfortable working with metric prefixes, decimals, percentages, fractions, algebra, exponents, and scientific notations.

G. TYLER MILLER

G. Tyler Miller has written 62 textbooks for introductory courses in environmental science, basic ecology, energy, and environmental chemistry. Since 1975, Miller's books have been the most widely used textbooks for environmental science in the United States and throughout the world. They have been used by almost 3 million students and have been translated into eight languages.

Miller has a professional background in chemistry, physics, and ecology. He has PhD from the University of Virginia and has received two honorary doctoral degrees for his contributions to environmental education. He taught college for 20 years, developed one of the nation's first environmental studies programs, and developed an innovative interdisciplinary undergraduate science program before deciding to write environmental science textbooks full time in 1975. Currently, he is the president of Earth Education and Research, devoted to improving environmental education.

He describes his hopes for the future as follows:

If I had to pick a time to be alive, it would be the next 75 years. Why? First, there is overwhelming scientific evidence that we are in the process of seriously degrading our own life-support system. In other words, we are living unsustainably. Second, within your lifetime we have the opportunity to learn how to live more sustainably by working with the rest of nature, as described in this book.

I am fortunate to have three smart, talented, and wonderful sons—Greg, David, and Bill. I am especially privileged to have Kathleen as my wife, best friend, and research associate. It is inspiring to have a brilliant, beautiful (inside and out), and strong woman who cares deeply about nature as a lifemate. She is my hero. I dedicate this book to her and to the earth.

SCOTT E. SPOOLMAN

Scott Spoolman is a writer and textbook editor with more than 30 years of experience in educational publishing. He has worked with Tyler Miller since 2003 as a contributing editor on earlier editions of *Living in the Environment*, *Environmental Science*, and *Sustaining the Earth*. With Norman Myers, he also coauthored *Environmental Issues and Solutions: A Modular Approach*.

Spoolman holds a master's degree in science journalism from the University of Minnesota. He has authored numerous articles in the fields of science, environmental engineering, politics, and business. He worked as an acquisitions editor on a series of college forestry textbooks. He has also worked as a consulting editor in the development of over 70 college and high school textbooks in fields of the natural and social sciences.

In his free time, he enjoys exploring the forests and waters of his native Wisconsin along with his family—his wife, environmental educator Gail Martinelli, and his children, Will and Katie.

Spoolman has the following to say about his collaboration with Tyler Miller.

I am honored to be working with Tyler Miller as a coauthor to continue the Miller tradition of thorough, clear, and engaging writing about the vast and complex field of environmental science. I share Tyler Miller's passion for ensuring that these textbooks and their multimedia supplements will be valuable tools for students and teachers. To that end, we strive to introduce this interdisciplinary field in ways that will be informative and sobering, but also tantalizing and motivational.

If the flip side of any problem is indeed an opportunity, then this truly is one of the most exciting times in history for students to start an environmental career. Environmental problems are numerous, serious, and daunting, but their possible solutions generate exciting new career opportunities. We place high priorities on inspiring students with these possibilities, challenging them to maintain a scientific focus, pointing them toward rewarding and fulfilling careers, and in doing so, working to help sustain life on the earth.

My Environmental Journey — *G. Tyler Miller*

My environmental journey began in 1966 when I heard a lecture on population and pollution problems by Dean Cowie, a biophysicist with the U.S. Geological Survey. It changed my life. I told him that if even half of what he said was valid, I would feel ethically obligated to spend the rest of my career teaching and writing to help students learn about the basics of environmental science. After spending six months studying the environmental literature, I concluded that he had greatly underestimated the seriousness of these problems.

I developed an undergraduate environmental studies program and in 1971 published my first introductory environmental science book, an interdisciplinary study of the connections between energy laws (thermodynamics), chemistry, and ecology. In 1975, I published the first edition of *Living in the Environment*. Since then, I have completed multiple editions of this textbook, and of three others derived from it, along with other books.

Beginning in 1985, I spent ten years in the deep woods living in an adapted school bus that I used as an environmental science laboratory and writing environmental science textbooks. I evaluated the use of passive solar energy design to heat the structure; buried earth tubes to bring in air cooled by the earth (geothermal cooling) at a cost of about $1 per summer; set up active and passive systems to provide hot water; installed an energy-efficient instant hot water heater powered by LPG; installed energy-efficient windows and appliances and a composting (waterless)

toilet; employed biological pest control; composted food wastes; used natural planting (no grass or lawnmowers); gardened organically; and experimented with a host of other potential solutions to major environmental problems that we face.

I also used this time to learn and think about how nature works by studying the plants and animals around me. My experience from living in nature is reflected in much of the material in this book. It also helped me to develop the six simple principles of sustainability that serve as the integrating theme for this textbook and to apply these principles to living my life more sustainably.

I came out of the woods in 1995 to learn about how to live more sustainably in an urban setting where most people live. Since then, I have lived in two urban villages, one in a small town and one within a large metropolitan area.

Since 1970, my goal has been to use a car as little as possible. Since I work at home, I have a "low-pollute commute" from my bedroom to a chair and a laptop computer. I usually take one airplane trip a year to visit my sister and my publisher.

As you will learn in this book, life involves a series of environmental trade-offs. Like most people, I still have a large environmental impact, but I continue to struggle to reduce it. I hope you will join me in striving to live more sustainably and sharing what you learn with others. It is not always easy, but it sure is fun.

Cengage Learning's Commitment to Sustainable Practices

We the authors of this textbook and Cengage Learning, the publisher, are committed to making the publishing process as sustainable as possible. This involves four basic strategies:

- *Using sustainably produced paper.* The book publishing industry is committed to increasing the use of recycled fibers, and Cengage Learning is always looking for ways to increase this content. Cengage Learning works with paper suppliers to maximize the use of paper that contains only wood fibers that are certified as sustainably produced, from the growing and cutting of trees all the way through paper production.

- *Reducing resources used per book.* The publisher has an ongoing program to reduce the amount of wood pulp, virgin fibers, and other materials that go into each

sheet of paper used. New, specially designed printing presses also reduce the amount of scrap paper produced per book.

- *Recycling.* Printers recycle the scrap paper that is produced as part of the printing process. Cengage Learning also recycles waste cardboard from shipping cartons, along with other materials used in the publishing process.

- *Process improvements.* In years past, publishing has involved using a great deal of paper and ink for the writing and editing of manuscripts, copyediting, reviewing page proofs, and creating illustrations. Almost all of these materials are now saved through use of electronic files. Very little paper and ink were used in the preparation of this textbook.

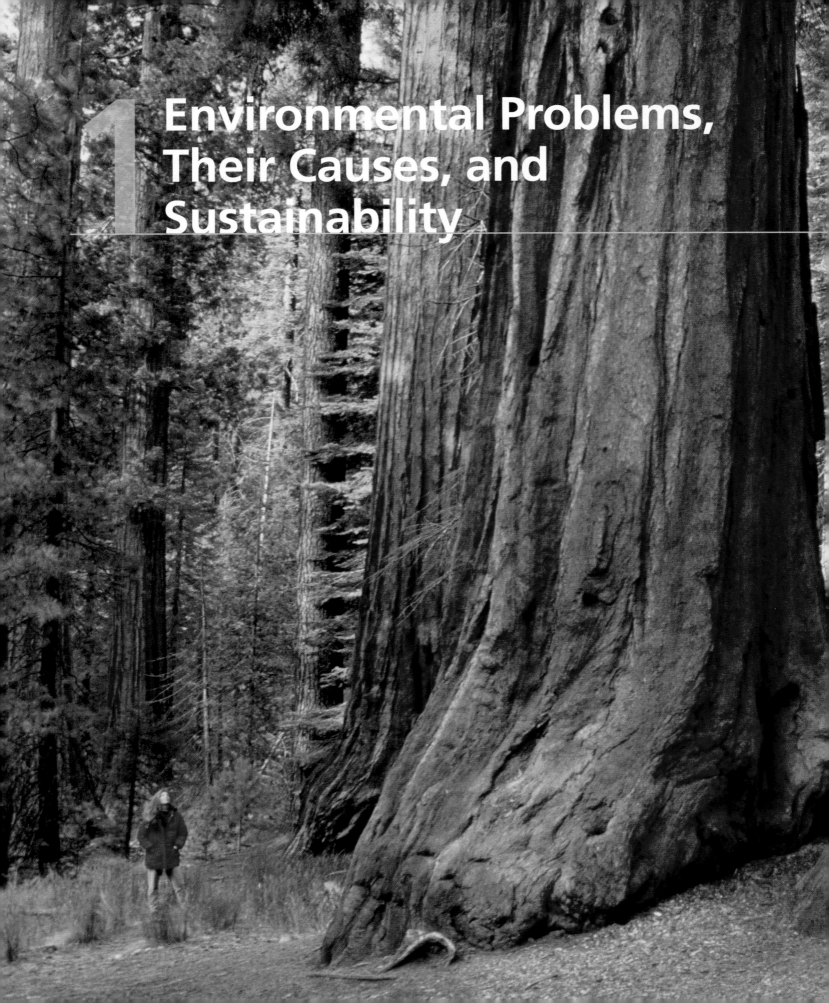

1 Environmental Problems, Their Causes, and Sustainability

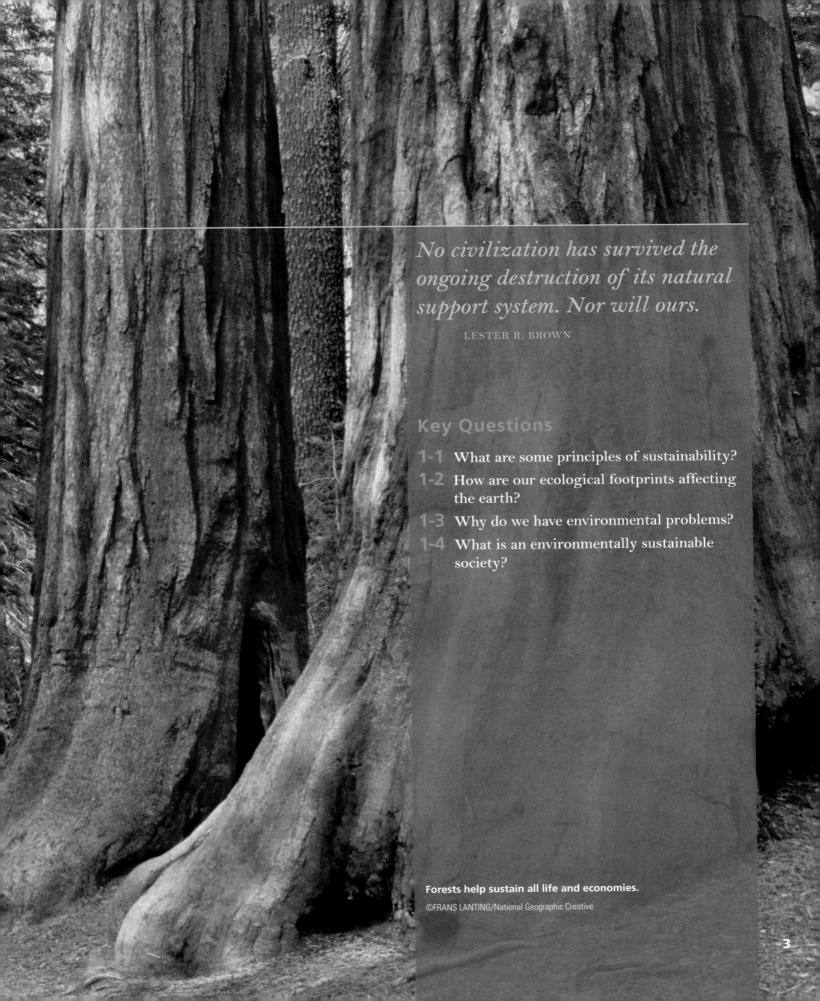

No civilization has survived the ongoing destruction of its natural support system. Nor will ours.

LESTER R. BROWN

Key Questions

1-1 What are some principles of sustainability?

1-2 How are our ecological footprints affecting the earth?

1-3 Why do we have environmental problems?

1-4 What is an environmentally sustainable society?

Forests help sustain all life and economies.

©FRANS LANTING/National Geographic Creative

Mostovyi Sergii Igorevich/Shutterstock.com

Figure 1-1 These parents—like Emily and Michael in our fictional vision of a possible world in 2065—are teaching their children about some of the world's environmental problems (left) and helping them to enjoy the wonders of nature (right).

©JOEL SARTORE/National Geographic Creative

Michael Rodriguez and Emily Briggs graduated from college in 2018. Michael earned a master's degree in environmental education, became a high-school teacher, and loved teaching environmental science. Emily established a thriving practice as an environmental lawyer.

Michael and Emily met while doing volunteer work for an environmental organization, got married, and had a child. They taught her about some of the world's environmental problems (Figure 1-1, left) and about the joys of nature that they had experienced as children (Figure 1-1, right). This led their daughter to become involved in working toward a more sustainable world.

When Michael and Emily were growing up, there had been increasing signs of stress on the land, air, water, and wildlife from the harmful environmental impacts of a growing population consuming more resources. However, by 2018, a small but growing number of people had begun shifting to more environmentally sustainable lifestyles.

In 2065, Emily and Michael celebrated the birth of their grandchild. He was born into a world where the loss of species and the degradation of land had slowed to a trickle. The atmosphere, oceans, lakes, and rivers were cleaner, 80% of the solid wastes were reused or recycled, and energy waste had been cut in half.

By 2050, significant atmospheric warming and the resulting climate change had occurred as many climate scientists had projected in the 1990s. However, the threat of further climate change and air and water pollution had begun to decrease because of greatly reduced energy waste and the gradual shift in human use of energy resources from oil and coal to cleaner energy from the sun, wind, flowing water, and other renewable resources.

By 2065, farmers producing most of the world's food had shifted to more sustainable farming practices that helped to conserve water and protect and renew

much of the planet's vital topsoil. In addition, the human population had peaked at about 8 billion in 2050 and then had begun a slow decline, lessening human pressure on the earth's life-support systems. In 2065, Emily and Michael felt a great sense of pride, knowing that they and their child and countless others had helped to bring about these improvements so that current and future generations could live more sustainably on this marvelous planet that is our only home.

Sustainability is the capacity of the earth's natural systems and human cultural systems to survive, flourish, and adapt to changing environmental conditions into the very long-term future. It is the overarching theme of this textbook, as well as a focal point for understanding the environmental problems we face and for exploring possible solutions. Our goal is to present to you a realistic and hopeful vision of what could be.

1-1 What Are Some Principles of Sustainability?

CONCEPT 1-1A
Nature has been sustained for billions of years by relying on solar energy, biodiversity, and chemical cycling.

CONCEPT 1-1B
Our lives and economies depend on energy from the sun and on natural resources and ecosystem services (*natural capital*) provided by the earth.

CONCEPT 1-1C
We could shift toward living more sustainably by applying full-cost pricing, searching for win-win solutions, and committing to preserving the earth's life-support system for future generations.

Environmental Science Is a Study of Connections in Nature

The **environment** is everything around us. It includes the living and the nonliving things (air, water, and energy) with which we interact in a complex web of relationships that connect us to one another and to the world we live in. Despite our many scientific and technological advances, we are utterly dependent on the earth for clean air and water, food, shelter, energy, fertile soil, and everything else in the planet's *life-support system.*

This textbook is an introduction to **environmental science**, an *interdisciplinary* study of how humans interact with the living and nonliving parts of their environment. It integrates information and ideas from the *natural sciences* such as biology, chemistry, and geology; the *social sciences* such as geography, economics, and political science; and the *humanities* such as ethics. The three goals of environmental science are **(1)** to learn how life on the earth has survived and thrived, **(2)** to understand how we interact with the environment, and **(3)** to find ways to deal with environmental problems and live more sustainably.

A key component of environmental science is **ecology**, the biological science that studies how living things interact with one another and with their environment. These living things are called **organisms**. Each organism belongs to a **species**, a group of organisms that has a unique set of characteristics that distinguish it from other groups of organisms.

A major focus of ecology is the study of ecosystems. An **ecosystem** is a set of organisms within a defined area or volume that interact with one another and with their environment of nonliving matter and energy. For example, a forest ecosystem consists of plants (especially trees; see chapter-opening photo), animals, and various other organisms that decompose organic materials, all interacting with one another, with solar energy, and with the chemicals in the forest's air, water, and soil.

We should not confuse environmental science and ecology with **environmentalism**, a social movement dedicated to trying to protect the earth's life-support systems for all forms of life. Environmentalism is practiced more in the political and ethical arenas than in the realm of science.

Three Scientific Principles of Sustainability

How has an incredible variety of life on the earth been sustained for at least 3.5 billion years in the face of catastrophic changes in environmental conditions? Such changes included gigantic meteorites impacting the earth, ice ages lasting for hundreds of millions of years, and long warming periods during which melting ice raised sea levels by hundreds of feet.

The latest version of our species has been around for only about 200,000 years—less than the blink of an eye relative to the 3.5 billion years that life has existed on the earth. Yet, there is mounting scientific evidence that, as we have expanded into and dominated almost all of the earth's ecosystems during that short time, we have seriously degraded these natural systems that support our species and all other life forms, as well as our economies. Thus, the newest challenge for the human species is to learn how to live more sustainably (**Core Case Study**), and some scientists argue that we have no time to waste in doing so.

Many scientists contend that the earth is the only real example of a sustainable system. Our science-based research leads us to believe that three major natural factors have played the key roles in the long-term sustainability of life on this planet, as summarized below and in Figure 1-2 (**Concept 1-1A**). We use these three **scientific principles of sustainability**, or *lessons from nature,* throughout the book to suggest how we might move toward a more sustainable future.

- **Dependence on solar energy:** The sun warms the planet and provides energy that plants use to produce **nutrients**, or the chemicals necessary for their own life processes along with those of most other animals, including humans. The sun also powers indirect forms of **solar energy** such as wind and flowing water, which we use to produce electricity.
- **Biodiversity** (short for *biological diversity*): **Biodiversity** is the variety of genes, organisms, species, and ecosystems in which organisms exist and interact. The interactions among species, especially the feeding relationships, provide vital ecosystem services and keep any population from growing too large. Biodiversity also provides countless ways for life to adapt to changing environmental conditions, even catastrophic changes that wipe out large numbers of species.
- **Chemical cycling: Chemical cycling**, or **nutrient cycling**, is the circulation of chemicals necessary for life from the environment (mostly from soil and

Figure 1-2 Three scientific **principles of sustainability** based on how nature has sustained a huge variety of life on the earth for 3.5 billion years, despite drastic changes in environmental conditions (Concept 1-1A).

Solar Energy

Chemical Cycling

Biodiversity

© Cengage Learning

water) through organisms and back to the environment. Because the earth receives no new supplies of these chemicals, organisms must recycle them continuously in order to survive. This means that there is little waste in nature, other than in the human world, because the wastes of any organism become nutrients or raw materials for other organisms.

Sustainability Has Certain Key Components

Sustainability, the central integrating theme of this book, has several critical components that we use as subthemes. One such component is **natural capital**—the natural resources and natural services that keep us and other species alive and support human economies (Figure 1-3).

Natural resources are materials and energy in nature that are essential or useful to humans. They are often classified as *inexhaustible resources* (such as energy from the sun and wind), *renewable resources* (such as air, water,

topsoil, plants, and animals), or *nonrenewable* or *depletable resources* (such as copper, oil, and coal). **Natural services**, or **ecosystem services**, are processes provided by healthy ecosystems. Examples include purification of air and water, renewal of topsoil, and pollination, which support life and human economies at no monetary cost to us. For example, forests help to purify air and water, regulate climate, reduce soil erosion, and provide countless species with a place to live.

One vital natural service is chemical, or nutrient, cycling—one of the three scientific **principles of sustainability**. An important component of nutrient cycling is *topsoil*—a vital natural resource that provides us and most other land-dwelling species with food. Without nutrient cycling in topsoil, life as we know it could not exist on the earth's land.

Natural capital is also supported by energy from the sun—another of the scientific **principles of sustainability** (Figure 1-2). Thus, our lives and economies depend on

Natural Capital

Natural Capital = Natural Resources + Ecosystem Services

Solar energy

Air
Air purification
Climate control
UV protection (ozone layer)

Renewable energy (sun, wind, water flows)

Life (biodiversity)
Population control
Pest control

Water
Water purification
Waste treatment

Nonrenewable minerals (iron, sand)

Soil
Soil renewal

Land
Food production
Nutrient recycling

Natural gas
Oil

Nonrenewable energy (fossil fuels)

Coal seam

Natural resources

Ecosystem services

© Cengage Learning

Figure 1-3 Natural capital consists of natural resources (blue) and natural or ecosystem services (orange) that support and sustain the earth's life and human economies (**Concept 1-1A**).

energy from the sun, and on natural resources and natural services *(natural capital)* provided by the earth (**Concept 1-1B**).

A second component of sustainability—and another subtheme of this text—is to recognize that many human activities can *degrade natural capital* by using normally renewable resources such as trees and topsoil faster than nature can restore them and by overloading the earth's normally renewable air and water systems with pollution and wastes. For example, in some parts of the world, we are replacing diverse and naturally sustainable forests (Figure 1-4) with crop plantations that can be sustained only with large inputs of water, fertilizer, and pesticides. We are also adding harmful chemicals and wastes to some rivers, lakes, and oceans faster than these bodies of water can cleanse themselves through natural processes.

This leads us to a third component of sustainability: *solutions.* While environmental scientists search for scientific solutions to problems such as the unsustainable degra-

dation of forests and other forms of natural capital, social scientists are looking for economic and political solutions. For example, a scientific solution to the problems of depletion of forests is to stop burning or cutting down biologically diverse, mature forests (Figure 1-4). A scientific solution to the problem of pollution of rivers is to prevent the excessive dumping of harmful chemicals and wastes into streams and to allow them to recover naturally. However, to implement such solutions, governments might have to enact and enforce environmental laws and regulations.

The search for solutions often involves conflicts. For example, when a scientist argues for protecting a natural forest on government-owned land to help preserve its important diversity of plants and animals, the timber company that had planned to harvest the trees in that forest might protest. Dealing with such conflicts often involves making *trade-offs,* or compromises—another component of sustainability. For example, the timber company might be persuaded to plant and harvest trees in an area that it had

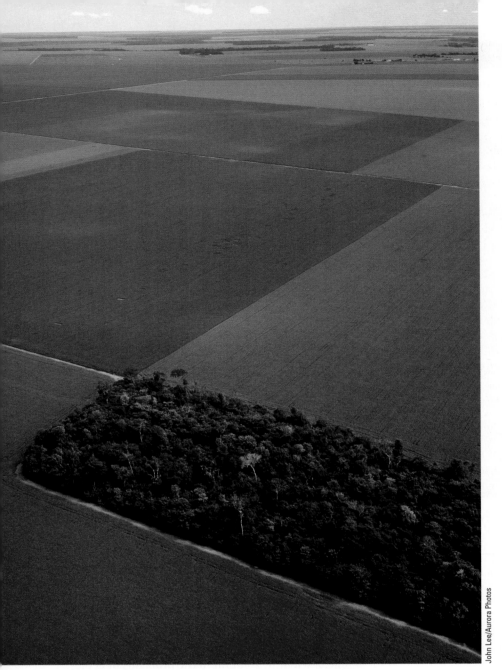

Other Principles of Sustainability Come from the Social Sciences

Our search for solutions and trade-offs to environmental problems has led us to propose three **social science principles of sustainability** (Figure 1-5), derived from studies of economics, political science, and ethics. We believe that these, along with our three *scientific principles of sustainability* (Figure 1-2), can serve as general guidelines for living more sustainably.

The social science principles of sustainability are

- **Full-cost pricing** (from economics): Many economists urge us to find ways to include the harmful environmental and health costs of producing and using goods and services in their market prices—a practice called **full-cost pricing**. This would give consumers better information about the environmental impacts of their lifestyles, and it would allow them to make more informed choices about the goods and services they use.

- **Win-win solutions** (from political science): We can learn to work together in dealing with environmental problems by focusing on solutions that will benefit the largest possible number of people, as well as the environment. This means shifting from an *I win, you lose* approach to a *we both win* approach (*win-win* solutions), and to an *I win, you win, and the earth wins* approach (*win-win-win* solutions).

- **A responsibility to future generations** (from ethics): We should accept our responsibility to leave the planet's life-support systems in at least as good a shape as what we now enjoy, for future generations.

Figure 1-4 Small remaining area of once diverse Amazon rain forest surrounded by vast simplified soybean fields in the Brazilian state of Motto Grosso.

John Lee/Aurora Photos

already cleared or degraded, instead of clearing the trees in an undisturbed diverse natural forest. In return, the government might give the company a *subsidy,* or financial support, to meet some of the costs for planting the trees.

In making a shift toward sustainability, the daily actions of each and every individual are important. In other words, *individuals matter*—another subtheme of this book. History shows that almost all of the significant changes in human systems have come from the bottom up, through the collective actions of individuals and from individuals inventing more sustainable ways of doing things. Thus, *sustainability begins with actions at personal and local levels.*

We will explore these principles further in this chapter and apply them throughout this textbook. For quick reference, you can find all six principles of sustainability on the back cover of this book.

Some Resources Are Renewable and Some Are Not

From a human standpoint, a **resource** is anything that we can obtain from the environment to meet our needs and wants. Some resources, such as solar energy, wind, surface water, and edible wild plants, are directly available for use. Other resources, such as petroleum, minerals, underground water, and cultivated plants, become useful to us only with some effort and technological ingenuity. For example, petroleum was merely a mysterious, oily fluid until we learned how to convert it into gasoline, heating oil, and other products.

Solar energy is called an **inexhaustible resource** because its continuous supply is expected to last for at least 6 billion years, until the sun dies. A **renewable resource** is one that can be replenished by natural processes within hours to centuries, as long as we do not use it up faster than natural processes can renew it. Examples include forests, grasslands, fishes, fertile topsoil, clean air, and freshwater.

The highest rate at which we can use a renewable resource indefinitely without reducing its available supply is called its **sustainable yield**. During most of the 10,000 years since we invented agriculture, civilization has lived on the sustainable yield of the earth's natural systems. But in recent decades we have been living unsustainably by degrading and depleting the earth's natural capital at an accelerating rate to fuel a growing population and ever-increasing resource consumption. As a result, in parts of the world, we are overharvesting forests, overgrazing grasslands, overfishing oceans, overdrawing underground water deposits (aquifers), and overloading the air and water with harmful wastes and pollutants.

Nonrenewable or **exhaustible resources** exist in a fixed quantity, or *stock,* in the earth's crust. On a time scale of millions to billions of years, geologic processes

©PETE MCBRIDE/National Geographic Creative

Figure 1-6 It would take more than a million years for natural processes to replace the coal that was removed within a couple of decades from this strip mine in Wyoming (USA).

Left: ©Minerva Studio/Shutterstock.com. Center: mikeledray/Shutterstock.com. Right: ©Yuri Arcurs/Shutterstock.com.

Figure 1-5 Three social science **principles of sustainability** can help us make a transition to a more environmentally and economically sustainable future.

ECONOMICS
Full-cost pricing

POLITICS
Win-win results

ETHICS
Responsibility to future generations

can renew such resources. However, on the much shorter human time scale of hundreds to thousands of years, we can deplete these resources much faster than nature can form them. Such exhaustible stocks include *energy resources* such as oil and coal (Figure 1-6), *metallic mineral resources* such as copper and aluminum, and *nonmetallic mineral resources* such as salt and sand.

As we deplete such resources, human ingenuity can often find substitutes. However, sometimes there is no acceptable or affordable substitute for a resource.

From an environmental and sustainability viewpoint, the priorities for more sustainable use of nonrenewable resources should be, in order: **R**efuse (don't use), **R**educe (use less), **R**euse, and **R**ecycle. Each of these steps helps to extend supplies and to reduce the environmental impacts of using these resources. According to a number of environmental scientists, we already know how to reuse or recycle at least 80% of the metal, glass, and other

GOOD NEWS

nonrenewable resources that we use. However, we cannot recycle or reuse nonrenewable energy resources such as oil, natural gas, and coal. Once burned, the concentrated energy they contain is dispersed as heat into the environment and is no longer available to us.

Countries Differ in Their Resource Use and Environmental Impact

The United Nations (UN) classifies the world's countries as economically more developed or less developed, based primarily on their average income per person. In using these classifications in this textbook, we do not mean to imply that either type of country is superior to the other. We are simply distinguishing between countries in terms

of their economic activity because this measure is highly useful for studying the resource use and environmental impacts of nations and regions.

More-developed countries—industrialized nations with high average income—have 17% of the world's population and include the United States, Canada, Japan, Australia, and most European countries. All other nations, in which 83% of the world's people live, are classified as **less-developed countries**, most of them in Africa, Asia, and Latin America. Some are *middle-income, moderately developed countries* such as China, India, Brazil, Thailand, and Mexico. Others are *low-income, least-developed countries,* including Congo, Haiti, Nigeria, and Nicaragua. (Figure 3, p. S28, in Supplement 6 is a map of high-, upper-middle-, lower-middle-, and low-income countries.)

1-2 How Are Our Ecological Footprints Affecting the Earth?

CONCEPT 1-2

As our ecological footprints grow, we are depleting and degrading more of the earth's natural capital.

We Are Living Unsustainably

The earth is a beautiful, largely blue and white island of water, land, and life that moves around the sun at about 107,000 kilometers per hour (67,000 miles per hour). The life on this planet has been sustained for billions of years by solar energy, chemical cycling, and biodiversity (the scientific **principles of sustainability**).

However, a large and growing body of scientific evidence indicates that the human species, a relative newcomer on the planet, is violating these sustainability principles. In fact, we are living unsustainably by wasting, depleting, and degrading the earth's natural capital (Figure 1-3) at an accelerating rate—a process known as **environmental degradation**, summarized in Figure 1-7. Scientists also refer to this as **natural capital degradation**.

In many parts of the world, renewable forests are shrinking (Figure 1-4), deserts are expanding, topsoil is eroding, and suburbs are replacing croplands. In addition, the lower atmosphere is warming, floating ice and glaciers are melting at unexpected rates, sea levels are rising, ocean acidity is increasing, and floods, droughts, severe weather, and forest fires are more frequent in some areas. In a number of regions, rivers are running dry, harvests of many species of fish are dropping sharply, and coral reefs are dying. Species are becoming extinct at least 100 times faster than in prehuman times, and extinction rates are projected to increase by at least another 100-fold during this century.

In 2005, the UN released its *Millennium Ecosystem Assessment,* a 4-year study by 1,360 experts from 95 countries. According to this study, human activities have degraded or overused about 60% of the earth's natural

or ecosystem services (Figure 1-3, orange boxes), mostly since 1950. The report's summary statement warned that "human activity is putting such a strain on the natural functions of Earth that the ability of the planet's ecosystems to sustain future generations can no longer be taken for granted." The report also concluded that we have scientific, economic, and political solutions to these problems that we could implement within a few decades in order to make the transition to a more sustainable future within your lifetime (**Core Case Study**), as you will learn in reading this book.

GOOD NEWS

Pollution Comes from a Number of Sources

A major environmental problem is **pollution**, which is contamination of the environment by a chemical or other agent such as noise or heat to a level that is harmful to the health, survival, or activities of humans or other organisms. Polluting substances, or *pollutants,* can enter the environment naturally, such as from volcanic eruptions, or through human activities, such as the burning of coal and gasoline, and the dumping of chemicals into rivers, lakes, and oceans. At a high enough concentration in the air, in water, or in our bodies, almost any chemical can cause harm and be classified as a pollutant.

The pollutants we produce come from two types of sources. **Point sources** are single, identifiable sources. Examples are the smokestack of a coal-burning power or industrial plant (Figure 1-8), the drainpipe of a factory, and the exhaust pipe of an automobile. **Nonpoint sources** are dispersed and often difficult to identify. Examples are pesticides blown from the land into the air and the runoff of fertilizers, pesticides, and trash from the land into streams and lakes (Figure 1-9). It is much easier and cheaper to identify and control or prevent pollution from point sources than from widely dispersed nonpoint sources.

Natural Capital Degradation

Degradation of Normally Renewable Natural Resources

Climate change

Shrinking forests

Degraded wildlife habitat

Air pollution

Species extinction

Soil erosion

Water pollution

Aquifer depletion

Declining ocean fisheries

© Cengage Learning

Figure 1-7 Natural capital degradation: Examples of the degradation of normally renewable natural resources and natural services (Figure 1-3) in parts of the world, mostly as a result of growing populations and rising rates of resource use per person.

We have tried to deal with pollution in two very different ways. One approach is **pollution cleanup**, which involves cleaning up or diluting pollutants after we have produced them. For example, we can dilute and reduce the local effects of air pollutants from smokestacks (Figure 1-8) by releasing them high into the atmosphere where winds can carry them to downwind areas. However, while tall smokestacks can reduce local air pollution they can increase air pollution in downwind areas.

The other approach is **pollution prevention**, efforts focused on greatly reducing or eliminating the production of pollutants. For example, we can enact pollution-control laws that ban, or set low levels for, the emission of various pollutants into the atmosphere or into bodies of water. Many environmental scientists say that pollution prevention is a key to a more sustainable future because it works better and in the long run is cheaper than cleanup.

Dudarev Mikhail/Shutterstock.com

Figure 1-8 *Point-source air pollution* from smokestacks in a coal-burning industrial plant.

Igor Jandric/Shutterstock.com

Figure 1-9 The trash in this river came from a large area of land and is an example of *nonpoint-source water pollution*.

Figure 1-10 *Patterns of natural resource consumption:* The photo on the left shows a makeshift home for a poor family in a Mumbai, India, slum with few possessions. The photo on the right shows a typical suburban home for a U.S. family in Maryland that is filled with their possessions.

The Tragedy of the Commons: Degrading Commonly Shared Renewable Resources

Some renewable resources, known as *open-access renewable resources,* can be used by almost anyone. Examples are the atmosphere, the open ocean, and its fishes. Other examples of less open, but often *shared resources,* are grasslands and forests. Many of these commonly held resources have been environmentally degraded. In 1968, biologist Garrett Hardin (1915–2003) called such degradation the *tragedy of the commons.*

Degradation of a shared or open-access resource occurs because each user of the resource reasons, "If I do not use this resource, someone else will. The little bit that I use or pollute is not enough to matter, and anyway, it's a renewable resource." When the number of users is small, this logic works. Eventually, however, the cumulative effect of many people trying to exploit a shared resource can degrade it and eventually exhaust or ruin it, and then no one can benefit from it. That is the tragedy.

There are two major ways to deal with this difficult problem. One is to use a shared or open-access renewable resource at a rate well below its estimated sustainable yield by using less of the resource, regulating access to the resource, or doing both. For example, governments can establish laws and regulations limiting the annual harvests of various types of ocean fishes or regulating the amount of pollutants that we add to the atmosphere or the oceans.

The other way is to convert shared renewable resources to private ownership. The reasoning is that if you own something, you are more likely to protect your investment. However, history shows that shared renewable resources such as forests and soils can be quickly degraded,

once they are privately owned. Also, this approach is not practical for open-access resources such as the atmosphere, the oceans, and our global life-support system, which cannot be divided up and sold as private property.

Ecological Footprints: Our Environmental Impacts

Many poor people in less-developed countries struggle to survive. Their individual use of resources and the resulting environmental impact is low and is devoted mostly to meeting their basic needs and living in makeshift housing (Figure 1-10, left). However, a large number of such people in a given area can have a large total environmental impact. By contrast, many individuals in more-developed nations enjoy **affluence**, or wealth, which allows them to consume large amounts of resources far beyond their basic needs and live in large homes filled with possessions (Figure 1-10, right). Their environmental impact is more a product of their rates of resource use per person.

When people use renewable resources, it can result in natural capital degradation (Figure 1-7), pollution, and wastes. We can think of this harmful environmental impact as an **ecological footprint**—the amount of land and water needed to supply a person or an area with renewable resources such as food and water, and that are needed to absorb and recycle the wastes and pollution produced by such resource use. (The developers of this tool for measuring environmental impacts chose to focus on renewable resources, although the use of nonrenewable resources also contributes to our ecological footprints.)

The **per capita ecological footprint** is the average ecological footprint of an individual in a given country or

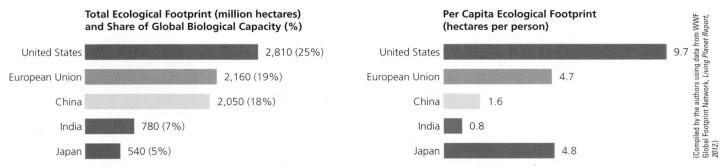

Total Ecological Footprint (million hectares) and Share of Global Biological Capacity (%)

United States	2,810 (25%)
European Union	2,160 (19%)
China	2,050 (18%)
India	780 (7%)
Japan	540 (5%)

Per Capita Ecological Footprint (hectares per person)

United States	9.7
European Union	4.7
China	1.6
India	0.8
Japan	4.8

(Compiled by the authors using data from WWF Global Footprint Network, *Living Planet Report*, 2012.)

Figure 1-11 **Natural capital use and degradation:** These graphs show the total and per capita ecological footprints of selected countries.

area. We can use various simple online tools to estimate our ecological footprints. Figure 1-11 compares the total and per capita ecological footprints for selected countries. Figure 1-12 shows the human ecological impact in different parts of the world. In Supplement 6, see a map of the ecological footprint for North America (Figure 8, p. S33).

If the total ecological footprint for a city, a country, or the world is larger than its *biological capacity* to replenish its renewable resources and absorb the resulting wastes and pollution, it is said to have an *ecological deficit*. In other words, its people are living unsustainably by depleting their natural capital instead of living off the renewable

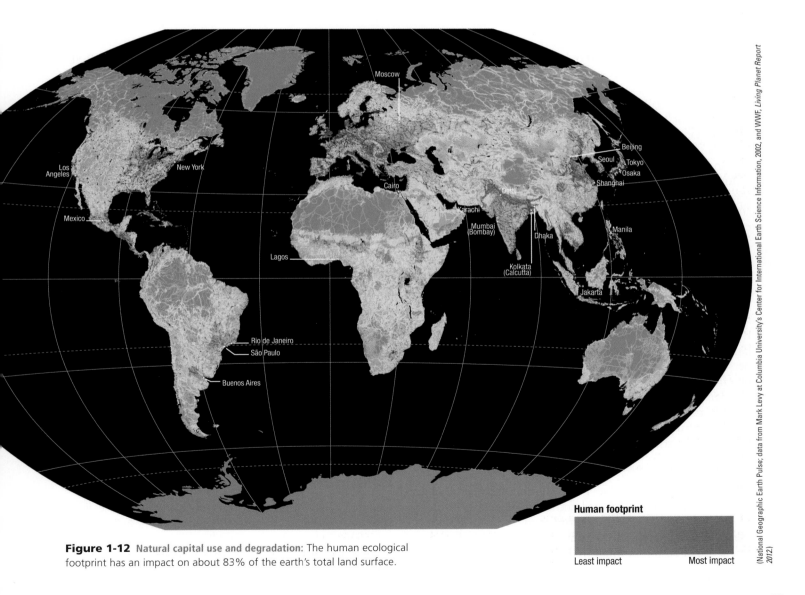

Figure 1-12 **Natural capital use and degradation:** The human ecological footprint has an impact on about 83% of the earth's total land surface.

Human footprint

Least impact — Most impact

(National Geographic Earth Pulse, data from Mark Levy at Columbia University's Center for International Earth Science Information, 2002, and WWF, *Living Planet Report*, 2012.)

(Compiled by the authors using data from WWF, *Living Planet Report 2012*.)

Figure 1-13 **Natural capital degradation:** Estimated number of planet Earths needed to indefinitely support our ecological footprints today (left), in 2030 (middle), and today if everyone were to have the same ecological footprint as the average American now has. ***Question:*** If we are living beyond the earth's renewable biological capacity, why do you think the human population and per capita resource consumption are still growing rapidly?

supply of resources provided by such capital. Globally we are running up a huge ecological deficit (Figure 1-13). (In Supplement 6, see Figure 4, p. S29 for a map of countries that are either ecological debtors or ecological creditors.)

Ecological footprint data and models, though imperfect, provide useful rough estimates of our individual, national, and global environmental impacts. We can use existing and emerging technologies and economic tools to reduce the size of our ecological footprints over the next few decades to the point where we need only one earth to support the human population (**Core Case Study**). We discuss ways to make this shift throughout this book.

IPAT Is Another Environmental Impact Model

In the early 1970s, scientists Paul Ehrlich and John Holdren developed a simple model showing how population size (P), affluence, or resource consumption per person (A), and the beneficial and harmful environmental effects of technologies (T) help to determine the environmental impact (I) of human activities. This provides another rough estimate of how much humanity is degrading the natural capital it depends on. We can summarize this model by the simple equation:

$$\text{Impact (I)} = \text{Population (P)} \times \text{Affluence (A)} \times \text{Technology (T)}$$

Figure 1-14 shows the relative importance of these three factors in less-developed and more-developed countries. While the ecological footprint model emphasizes the use of renewable resources, this model includes the per capita use of both renewable and nonrenewable resources.

Some forms of technology such as polluting factories, gas-guzzling motor vehicles, and coal-burning power plants increase environmental impact by raising the harmful T factor (Figure 1-14, red arrows). However, other technologies reduce environmental impact by decreasing the T factor (Figure 1-14, green arrows). Examples are pollution control and prevention technologies, fuel-efficient cars, and wind turbines and solar cells that generate electricity with a low environmental impact.

In most less-developed countries, the key factor in total environmental impact (Figure 1-14, top) is usually population size (P) as a growing number of poor people struggle to stay alive. For example, while each individual in a rural village needs only a small amount of wood for heating and cooking, the total population of the village can eventually clear most or all of the trees in the nearby area for firewood. In more-developed countries, the key factor in overall environmental impact is affluence (A), which can result in high rates of per capita resource use, pollution, and resource depletion (Figure 1-14, bottom). Affluence is also a key factor in less-developed countries whose economies and populations are growing (see the following Case Study).

CASE STUDY
China's Growing Number of Affluent Consumers

About 1.4 billion affluent consumers put immense pressure on the earth's renewable and nonrenewable natural capital. Most of these middle-class consumers live in more-developed countries, but an estimated 300 million of

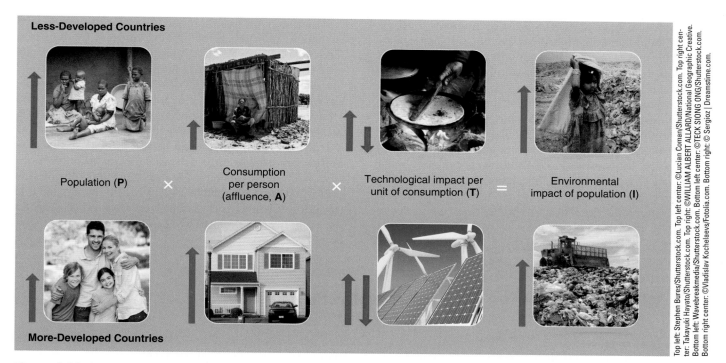

Less-Developed Countries

Population (**P**) × Consumption per person (affluence, **A**) × Technological impact per unit of consumption (**T**) = Environmental impact of population (**I**)

More-Developed Countries

Figure 1-14 This simple model demonstrates how three factors—population size, affluence (resource use per person), and technology—help to determine the environmental impacts of populations in less-developed countries (top) and more-developed countries (bottom). Red arrows show generalized harmful impacts and green arrows show generalized beneficial impacts.

© Cengage Learning

them—a number almost the size of the U.S. population—live in China, and this number may double by 2020.

China has the world's largest population and second-largest economy. It is the world's leading consumer of wheat, rice, meat, coal, fertilizer, steel, cement, and oil. China also leads the world in the production of goods such as televisions, cell phones, and refrigerators. It has also produced more wind turbines than any other country and will soon become the world's largest producer of solar cells and fuel-efficient cars. Between 2010 and 2025, China expects to build 10 cities the size of New York City.

Now, after 20 years of industrialization, China contains two-thirds of the world's most polluted cities. Some of its major rivers are choked with waste and pollution, and some areas of its coastline are basically devoid of fishes and other ocean life. A massive cloud of air pollution, largely generated in China, affects China and other Asian countries, areas of the Pacific Ocean, and even parts of the West Coast of North America.

Suppose that China's economy continues to grow at an overall rapid rate and its population size reaches 1.5 billion by around 2025, as projected by UN population experts. Environmental policy expert Lester R. Brown estimates that if such projections are accurate, China will need two-thirds of the world's current grain harvest, twice the amount of paper now consumed in the world, and more than all the oil currently produced in the world.

According to Brown:

The western economic model—the fossil fuel–based, automobile-centered, throwaway economy—is not going to work for China. . . or for the other 3 billion people in developing countries who are also dreaming the "American dream."

For more details on China's growing ecological footprint, see the online Guest Essay by Norman Myers on this topic.

Cultural Changes Can Grow or Shrink Our Ecological Footprints

Until about 10,000 to 12,000 years ago, we were mostly *hunter–gatherers* who obtained food by hunting wild animals or scavenging their remains, and gathering wild plants. Early hunter–gatherers lived in small groups, consumed few resources, had few possessions, and moved as needed to find enough food for their survival.

Since then, three major cultural changes have occurred. *First* was the *agricultural revolution,* which began 10,000–12,000 years ago when humans learned how to grow and breed plants and animals for food, clothing, and other purposes. *Second* was the *industrial–medical revolution,* beginning about 275 years ago when people invented machines for the large-scale production of goods in factories. This involved learning how to get energy from fossil fuels (such as coal and oil) and how to grow large quantities of food in an efficient manner. It also included medical advances that have allowed a growing number of people to live longer and healthier lives. Finally, the *information–globalization revolution* began about 50 years ago, when we developed new technologies for gaining rapid access to all kinds of information and resources on a global scale.

Each of these three cultural changes gave us more energy and new technologies with which to alter and control more of the planet's resources to meet our basic needs and increasing wants. They also allowed expansion of the human population, mostly because of larger food supplies and longer life spans. In addition, they each resulted in greater resource use, pollution, and environmental degradation as they allowed us to dominate the planet and expand our ecological footprints (Figure 1-13).

On the other hand, some technological leaps have enabled us to begin shrinking our ecological footprints by reducing our use of energy and matter resources and our production of wastes and pollution. For example, use of the energy-efficient compact fluorescent and LED lightbulbs and energy-efficient cars and buildings is on the rise. Many environmental scientists and other analysts see such developments as evidence of an emerging fourth major cultural change in the form of a **sustainability revolution**, in which we could learn to live more sustainably with smaller ecological footprints, during this century (**Core Case Study**). A key factor in making this transition involves asking ourselves: How much stuff do we really need in order to live healthful, comfortable, and happy lives?

1-3 Why Do We Have Environmental Problems?

CONCEPT 1-3A
Major causes of environmental problems are population growth, unsustainable resource use, poverty, avoidance of full-cost pricing, and increasing isolation from nature.

CONCEPT 1-3B
Our environmental worldviews play a key role in determining whether we live unsustainably or more sustainably.

Experts Have Identified Several Causes of Environmental Problems

According to a number of environmental and social scientists, the major causes of the environmental problems we face are **(1)** population growth, **(2)** wasteful and unsustainable resource use, **(3)** poverty, **(4)** failure to include the harmful environmental costs of goods and services

in their market prices, and **(5)** increasing isolation from nature (Figure 1-15) (Concept 1-3A).

We discuss each of these causes in detail in later chapters of this book. Let us begin with a brief overview of them.

The Human Population Is Growing at a Rapid Rate

Exponential growth occurs when a quantity such as the human population increases at a fixed percentage per unit of time, such as 1% or 2% per year. Exponential growth starts off slowly, but after only a few doublings, it grows to enormous numbers because each doubling is twice the total of all earlier growth.

Here is an example of the immense power of exponential growth. Fold a piece of paper in half to double its thickness. If you could continue doubling the thickness of

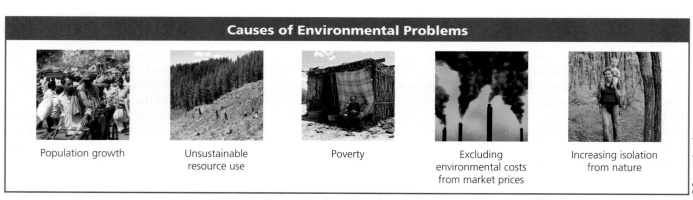

Causes of Environmental Problems

Population growth | Unsustainable resource use | Poverty | Excluding environmental costs from market prices | Increasing isolation from nature

© Cengage Learning

Figure 1-15 Environmental and social scientists have identified five basic causes of the environmental problems we face (Concept 1-3A). *Question:* For each of these causes, what are two environmental problems that result?

Left: Jeremy Richards/Shutterstock.com. Left center: steve estvanik/Shutterstock.com. Center: ©Lucian Coman/Shutterstock.com. Right center: El Greco/Shutterstock.com. Right: ©Maxim Tupikov/Shutterstock.com.

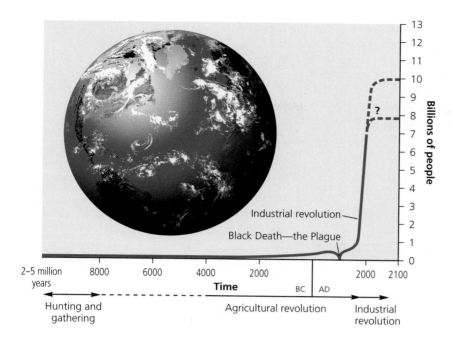

Figure 1-16 *Exponential growth:* The J-shaped curve represents past exponential world population growth, with projections to 2100 showing possible population stabilization as the J-shaped curve of growth changes to an S-shaped curve. (This figure is not to scale.)

(Compiled by the authors using data from the World Bank, United Nations, and Population Reference Bureau.)

the paper 50 times, it would be thick enough to almost reach the sun—150 million kilometers (93 million miles) away! Hard to believe, isn't it?

The human population has been growing exponentially (Figure 1-16). There are now about 7.1 billion people on the earth with about 84 million more people added each year. There could be 9.6 billion of us by 2050. This projected addition of 2.5 billion more people within your lifetime is more than 8 times the current U.S. population and 2 times that of China, the world's most populous nation.

CONSIDER THIS. . .

CONNECTIONS Exponential Growth and Doubling Time: The Rule of 70

The doubling time of the human population or any factor can be calculated by using the rule of 70: doubling time (years) = 70/annual growth rate (%). The world's population is growing at about 1.2% per year. At this rate how long will it take to double its size? If you have an investment that grows by 10% a year, how long will it take to double your money?

No one knows how many people the earth can support indefinitely, and at what level of average resource consumption per person, without seriously degrading the ability of the earth to support us, our economies, and the millions of other forms of life. However, our expanding total and per capita ecological footprints (Figures 1-12 and 1-13) are disturbing warning signs.

We could slow population growth with the goal of having it level off at around 8 billion by 2050, as suggested in the **Core Case Study**, instead of growing to 9.6 billion or more. Some ways to do this include reducing poverty through economic development, promoting family planning, and elevating the status of women, as discussed in Chapter 6.

Affluence Has Harmful and Beneficial Environmental Effects

The lifestyles of many consumers in more-developed countries and in less-developed countries such as China and India are built on growing affluence, which results in high levels of resource consumption and the resulting environmental degradation, wastes, and pollution.

These harmful environmental effects of affluence can be dramatic. In its 2012 *Living Planet Report*, the World Wildlife Fund (WWF) estimated that the United States is responsible for almost half of the global ecological footprint. The average American consumes about 30 times as much as the average Indian and 100 times as much as the average person in the world's poorest countries. According to some ecological footprint estimates, it takes about 27 large tractor-trailer loads of resources per year to support one typical American, and in 2012, there were 315 million Americans.

The problem is that providing each of these tractor-trailer loads represents environmental degradation, such as air pollution and water pollution from factories and motor vehicles and land degradation from the mining of raw materials used to make the products we consume. This is why higher levels of consumption expand a person's ecological footprint. Another downside to wealth is that it allows affluent consumers to obtain their resources from almost anywhere in the world without seeing the harmful environmental and health impacts of their high-consumption lifestyles.

On the other hand, affluence can allow for better education, which can lead people to become more concerned about environmental quality. It also provides money for developing technologies to reduce pollution, environ-

Figure 1-17 Poor settlers in Peru have cleared and burned this small plot of tropical rain forest in the Amazon and planted it with maize seedlings.

mental degradation, and resource waste. As a result, in the United States and most other affluent countries, the air is clearer, drinking water is purer, and most rivers and lakes are cleaner than they were in the 1970s. In addition, the food supply is more abundant and safer, the incidence of life-threatening infectious diseases has been greatly reduced, life spans are longer, and some endangered species are being rescued from extinction hastened by human activities. **GOOD NEWS**

These improvements in environmental quality were achieved because of greatly increased scientific research and technological advances financed by affluence. Education also spurred many citizens to insist that businesses and governments work toward improving environmental quality.

Poverty Has Harmful Environmental and Health Effects

Poverty is a condition in which people are unable to fulfill their basic needs for adequate food, water, shelter, health care, and education. According to the World Bank, about 900 million people—almost three times the U.S. population—live in *extreme poverty* (Figure 1-10, left), struggling to live on the equivalent of less than $1.25 a day. This is less than what many people spend for a bottle of water or a cup of coffee. About one of every three, or 2.6 billion, of the world's people struggles to live on less than $2.25 a day. Could you do this?

Poverty causes a number of harmful environmental and health effects. The daily lives of the world's poorest people are focused on getting enough food, water, and cooking and heating fuel to survive. Desperate for short-term survival, these individuals collectively degrade forests (Figure 1-17), topsoil, grasslands, fisheries, and wildlife. They do not have the luxury of worrying about long-term environmental quality or sustainability. Even though the poor in less-developed countries use very few resources per person, the large population size in some countries leads to a high overall environmental impact (Figure 1-14, top).

However, poverty does not necessarily lead to environmental degradation. Some of the world's poor people have learned how to take care of their environment as a part of their long-term survival strategy. For example, many small-scale farmers in African countries plant and nurture trees and work to conserve the soils that they depend on.

CONSIDER THIS. . .

CONNECTIONS Poverty and Population Growth

To many poor people, having more children is a matter of survival. Their children help them gather fuel (mostly wood and animal dung), haul drinking water, and tend crops and livestock. The children also help to care for their parents in their old age (their 40s or 50s in the poorest countries) because they do not have social security, health care, and retirement funds. This is largely why populations in some of the poorest less-developed countries continue to grow at high rates.

While poverty can increase some types of environmental degradation, the reverse is also true. Pollution and environmental degradation have a severe impact on the poor and can worsen their poverty. Consequently, many of the world's poor people die prematurely from several preventable health problems. One such problem is *malnutrition* caused by a lack of protein and other nutrients needed for good health (Figure 1-18). The resulting weakened condition can increase an individual's chances of death from starvation and normally nonfatal ailments such as diarrhea and measles.

A second health problem is limited access to adequate sanitation facilities and clean drinking water. More than one-third of the world's people have no real bathroom facilities and are forced to use backyards, alleys, ditches, and streams. As a result, about one of every eight of the world's people get water for drinking, washing, and cooking from sources polluted by human and animal feces. A third health problem is severe respiratory disease that people get from breathing the smoke of open fires or poorly vented stoves used for heating and cooking inside their dwellings. This indoor air pollution kills about 2.4 million people a year, according to the World Health Organization.

In 2010, the World Health Organization estimated that one or more of these factors, mostly related to poverty, cause premature death for about 7 million children under the age of 5 each year. Some hopeful news is that this number of annual deaths is down from about 10 million in 1990. Even so, every day an average of at least 19,000 young children die prematurely from these causes. This is equivalent to *95 fully loaded 200-passenger airliners crashing every day with no survivors*. The news media rarely cover this ongoing human tragedy.

Rowan Gillson/Superstock

Figure 1-18 One of every three children younger than age 5 in less-developed countries, including this starving child in Bangladesh, suffers from severe malnutrition caused by a lack of calories and protein.

CONSIDER THIS. . .

THINKING ABOUT The Poor, the Affluent, and Rapidly Increasing Population Growth

Some see the rapid population growth of the poor in less-developed countries as the primary cause of our environmental problems (Figure 1-14, top). Others say that the much higher resource use per person in more-developed countries is the more important factor (Figure 1-14, bottom). Which factor do you think is more important? Why?

Prices of Goods and Services Do Not Include Harmful Environmental Costs

Another basic cause of environmental problems has to do with how goods and services are priced in the marketplace. Companies using resources to provide goods for consumers generally are not required to pay for most of the harmful environmental and health costs of supplying such goods. For example, timber companies pay the cost of clear-cutting forests but do not pay for the resulting environmental degradation and loss of wildlife habitat. The primary goal of a company is to maximize profits for its owners or stockholders. Indeed, it would be economic suicide for a company to add these costs to its prices unless government regulations were to create a level economic playing field by using taxes or regulations to require all businesses to pay for the environmental and health costs of producing their products.

Because the prices of goods and services do not include most of their harmful environmental and human health costs, consumers have no effective way to evaluate the harmful effects, on their own health and on the earth's life-support systems, of producing and using these goods and services. For example, producing and using gasoline result in air pollution and other problems that damage the environment and people's health. Scientists and economists have estimated that the real cost of gasoline to U.S. consumers would be about $3.18 per liter ($12 per gallon) if the estimated short- and long-term harmful environmental and health costs were included in its pump price.

Partly because of this inadequate pricing system, many people do not understand the importance and truly high value of the natural resources and natural services that make up the earth's natural capital (Figure 1-3). This lack of information is a major reason for why we are degrading these key components of our life-support system (Figure 1-7).

Another problem arises when governments (taxpayers) give companies *subsidies* such as tax breaks and pay-

Figure 1-19 These ecotourists atop endangered Asian elephants in India's Kaziranga National Park are learning about threatened barasingha deer and numerous other species.

ments to assist them with using resources to run their businesses. This helps to create jobs and stimulate economies, but environmentally harmful subsidies encourage the depletion and degradation of natural capital. (See the online Guest Essay on this topic by Norman Myers.)

We could live more sustainably by finding ways to include in market prices the harmful environmental and health costs of the goods and services that we use. Implementing such full-cost pricing would be a way to apply one of the three social science **principles of sustainability** (Figure 1-5). Two ways to do this over the next two decades would be to shift from environmentally harmful government subsidies to environmentally beneficial subsidies, and to tax pollution and waste heavily while reducing taxes on income and wealth. We discuss such *subsidy shifts* and *tax shifts* in Chapter 23.

We Are Increasingly Isolated from Nature

Today, three out of four people in the more-developed countries and one of every two people in the world live in urban areas, and this shift from rural to urban living is continuing at a rapid pace. Our artificial urban environments and our increasing use of cell phones, computers, and other electronic devices isolate us from the natural world that provides the food, water, and most of the raw materials used to produce the consumer goods that we depend on.

Thus, it is not surprising that many people do not know the full story of where their food, water, and other goods come from. Similarly, many people are unaware of the amounts of wastes and pollutants they produce, of where they go, and of how these wastes and pollutants affect the environment. Many do not understand that life on earth has been sustained largely by the recycling of wastes—one of the scientific **principles of sustainability** (Figure 1-2).

According to some analysts, many of us are increasingly suffering from *nature deficit disorder*. They suggest that by not having enough contacts with the natural world (chapter-opening photo and Figure 1-19) a person can be more likely to suffer from stress, have health problems, show unwarranted irritability or aggression, and be less adaptable to changes in life. These analysts also argue that this disorder helps to explain why we are rapidly degrading our life-support system. They ask: How will we shrink our ecological footprints and live more sustainably if we do not experience and understand our utter dependence on the earth's natural systems and the natural capital they provide for us?

People Have Different Views about Environmental Problems and Their Solutions

Another challenge we face is that people differ over the seriousness of the world's environmental problems and what we should do to help solve them. Differing opinions about environmental problems and solutions arise mostly out of differing environmental worldviews. Your **environmental worldview** is your set of assumptions and values reflecting how you think the world works and what you think your role in the world should be. **Environmental ethics**, which is the study of our various beliefs about what is right and wrong with how we treat the environment, provides important ways to examine our worldviews. Here are some important *ethical questions* relating to the environment:

- Why should we care about the environment?
- Are we the most important beings on the planet or are we just one of the earth's millions of different forms of life?
- Do we have an obligation to see that our activities do not cause the extinction of other species? Should we try to protect all species or only some? How do we decide which to protect?
- Do we have an ethical obligation to pass on to future generations the extraordinary natural world in a condition that is at least as good as what we inherited?
- Should every person be entitled to equal protection from environmental hazards regardless of race, gender, age, national origin, income, social class, or any other factor? This is the central ethical and political

issue for what is known as the *environmental justice* movement. (See the online Guest Essay on this topic by Robert D. Bullard.)

- How do we promote sustainability?

CONSIDER THIS...

THINKING ABOUT Our Responsibilities

How would you answer each of the questions above? Compare your answers with those of your classmates. Record your answers and, at the end of this course, return to these questions to see if your answers have changed.

People with widely differing environmental worldviews can take the same data, be logically consistent with it, and arrive at quite different answers to such questions because they start with different assumptions and moral, ethical, or religious beliefs. Environmental worldviews are discussed in detail in Chapter 25, but here is a brief introduction.

The **planetary management worldview** holds that we are separate from and in charge of nature, that nature exists mainly to meet our needs and increasing wants, and that we can use our ingenuity and technology to manage the earth's life-support systems, mostly for our benefit, into the distant future.

The **stewardship worldview** holds that we can and should manage the earth for our benefit, but that we have an ethical responsibility to be caring and responsible managers, or *stewards*, of the earth. It says we should encourage environmentally beneficial forms of economic growth and development and discourage environmentally harmful forms.

The **environmental wisdom worldview** holds that we are part of, and dependent on, nature and that the earth's life-support system exists for all species, not just for us. According to this view, our success depends on learning how the earth sustains itself (Figure 1-2 or back cover of this book) and integrating such *environmental wisdom* into the ways we think and act.

1-4 What Is an Environmentally Sustainable Society?

CONCEPT 1-4
Living sustainably means living off the earth's natural income without depleting or degrading the natural capital that supplies it.

The More Environmentally Sustainable Societies Protect Natural Capital and Live Off Its Income

According to most environmental scientists, our ultimate goal should be to achieve an **environmentally sustainable society**—one that meets the current and future

basic resource needs of its people in a just and equitable manner without compromising the ability of future generations to meet their basic needs (**Core Case Study**).

Imagine that you win $1 million in a lottery. Suppose you invest this money (your capital) and earn 10% interest per year. If you live on just the interest, or the income made by your capital, you will have a sustainable annual income of $100,000 that you can spend each year indefinitely without depleting your capital. However, if you spend $200,000 per year, while still allowing interest to accumulate, all of your money will be gone early in the seventh year. Even if you spend only $110,000 per year and allow the interest to accumulate, you will be bankrupt within 25 years.

Tuy Sereivathana: Elephant Protector

Courtesy of Tom Dusenbery

As a young child in rural Cambodia, Tuy Sereivathana developed a respect for nature, became fascinated with elephants, and decided that he wanted to be an elephant protector.

Since 1970, Cambodia's rain forest cover has dropped from over 70% of the country's land area to 3% primarily because of population growth, rapid development, illegal logging, and warfare. This severe loss of forests forced elephants to search for food and water on farmlands, and it set up a conflict between elephants and poor farmers who killed the elephants to protect their food supply. Elephants have also been killed illegally for their valuable ivory tusks.

Since 1995, Sereivathana, with a master's degree in forestry, has been on a mission to accomplish two goals. One is to raise the population of Cambodia's endangered Asian elephants from less than 400 in 2010 to 1,000 by 2030. The other is to show poor farmers that protecting elephants and other forms of wildlife can help them escape poverty.

Sereivathana directs the Cambodian Elephant Conservation Group, devoted to reducing poaching and helping farmers work together to use low-cost and innovative ways to protect their crops without having to kill elephants. Affectionately known as Uncle Elephant, he helped farmers set up nighttime lookouts for elephants. He taught villagers to scare raiding elephants away by using foghorns and fireworks and using solar-powered electric fences to mildly shock them. He also encouraged farmers to stop growing watermelons and bananas, which elephants love, and to grow crops such as eggplant and chile peppers that elephants shun. Measures such as these have lowered crop losses, and farmers have benefited economically.

Since 2005, mostly because of Sereivathana's efforts, no elephants have been killed in Cambodia as a result of conflicts with humans. His community-based model for elephant and wildlife conservation is being used in neighboring communities and in other countries such as Indonesia and Vietnam. His model is an outstanding example of applying the win-win principle of sustainability (Figure 1-5).

In 2010, Sereivathana was one of the six recipients of the Goldman Environmental Prize (often dubbed the "Nobel Prize for the environment"). In 2011 he was named a National Geographic Emerging Explorer.

Background photo: Ekkachai/Shutterstock.com

The lesson here is an old one: *Protect your capital and live on the income it provides.* Deplete or waste your capital and you will move from a sustainable to an unsustainable lifestyle.

The same lesson applies to our use of the earth's natural capital (Figure 1-3)—the global trust fund that nature has provided for us, for future generations, and for the earth's other species. *Living sustainably* means living on **natural income**, the renewable resources such as plants, animals, soil, clean air, and clean water, provided by the earth's natural capital. It also means not depleting or degrading the earth's natural capital, which supplies this income, and providing the human population with adequate and equitable access to this natural income for the foreseeable future (**Concept 1-4**).

🔍 **CONSIDER THIS. . .**

CAMPUS SUSTAINABILITY* Arizona State University's School of Sustainability

Wrigley Hall houses Arizona State University's School of Sustainability, the first comprehensive degree-granting program in sustainability in the United States for both undergrads and graduate students. Established in 2007, it is an interdisciplinary program focused on finding practical solutions to environmental problems while considering important economic and social factors. The courses emphasize learning by doing, researching with faculty members, interacting with the business world and with K–12 students, committing to community service, and leadership training.

© 2013 Arizona Board of Regents on behalf of Arizona State University. Used with permission.

A More Sustainable Future Is Possible

Moving toward environmental sustainability will require an overall attitude that combines environmental wisdom with compassion for all forms of life, including humans. Some people seem to be born with this sort of attitude (see Individuals Matter 1.1). Others learn it as they experience nature in various ways.

Making a shift toward a more sustainable future will involve some tough challenges. However, here are two

---

*In several chapters of this book, through this Campus Sustainability feature, we are highlighting the efforts of certain colleges and universities in various areas of sustainability.

pieces of good news: *First,* research by social scientists suggests that it takes only 5–10% of the population of a community, a country, or the world to bring about major social change. *Second,* such research also shows that significant social change can occur in a much shorter time than most people think. Anthropologist Margaret Mead summarized our potential for social change: "Never doubt that a small group of thoughtful, committed citizens can change the world. Indeed, it is the only thing that ever has."

GOOD NEWS

Evidence from the physical sciences and the social sciences indicates that we have perhaps 50 years and no more than 100 years to make a new cultural shift from unsustainable living to more sustainable living, if we start now. One of the goals of this book is to provide a realistic vision of a more environmentally sustainable future (**Core Case Study**) based on energizing, realistic hope rather than on immobilizing fear, gloom, and doom.

Three strategies for reducing our ecological footprints, helping to sustain the earth's natural capital, and making a transition to more sustainable lifestyles and economies are summarized in the *three big ideas* of this chapter:

Big Ideas

- A more sustainable future will require that we rely more on energy from the sun and other renewable energy sources, protect biodiversity through the preservation of natural capital, and avoid disrupting the earth's vitally important chemical cycles.

- A major goal for becoming more sustainable is full-cost pricing—the inclusion of harmful environmental and health costs in the market prices of goods and services.

- We will benefit ourselves and future generations if we commit ourselves to finding win-win-win solutions to our problems and to leaving the planet's life-support system in at least as good a shape as what we now enjoy.

Pecold/Shutterstock.com

We face an array of serious environmental problems. This book is about understanding these problems and finding *solutions* to them. A key to most solutions is to apply the three scientific **principles of sustainability** (Figure 1-2) and the three social science **principles of sustainability** (Figure 1-5) to the design of our economic and social systems, as well as to our individual lifestyles. We can use such strategies to try to slow the rapidly increasing losses of biodiversity, to sharply reduce production of wastes and pollution, to switch to more sustainable sources of energy, and to promote more sustainable forms of agriculture and other uses of land and water. We can also use these principles to sharply reduce poverty and slow human population growth.

Suppose that we begin to make environmental choices during this century that promote sustainability, as the fictional characters Emily and Michael and people like them did in the **Core Case Study** that opens this chapter. Then, chances are that we will help to create an extraordinary and more sustainable future for ourselves, for future generations, and for most other forms of life on our planetary home. If we get it wrong, we face ecological disruption that could set humanity back for centuries and wipe out as many as half of the world's species, as well as much of the human population.

You have the good fortune to be a member of the 21st century's *transition generation* that will decide which path humanity takes. This means confronting the urgent challenges presented by the major environmental problems discussed in this book. It is an incredibly exciting and challenging time to be alive as we struggle to develop a more sustainable relationship with this planet that is our only home.

GOOD NEWS

Chapter Review

Core Case Study

1. Summarize the authors' vision of a more sustainable world, which could be attainable by 2065.

Section 1-1

2. What are the three key concepts for this section? Define **sustainability**. Define **environment**. Distinguish among **environmental science**, **ecology**, and **environmentalism**. Distinguish between an **organism** and a **species**. What is an **ecosystem**? What are three **scientific principles of sustainability** derived from how the natural world works? What is **solar energy** and why is it important to life on the earth? What is **biodiversity** and why is it important to life on the earth? Define **nutrients**. Define **chemical** or **nutrient cycling** and explain why it is important to life on the earth.

3. Define **natural capital**. Define **natural resources** and **natural services,** or **ecosystem services,** and give two examples of each. Give three examples of how we are degrading natural capital. Explain how finding solutions to environmental problems involves making trade-offs. Explain why individuals matter in dealing with the environmental problems we face. What are three **social science principles of sustainability**? What is **full-cost pricing** and why is it important?

4. What is a **resource**? Distinguish between an **inexhaustible resource** and a **renewable resource** and give an example of each. What is the **sustainable yield** of a renewable resource? Define and give two examples of a **nonrenewable** or **exhaustible resource**. Explain why the suggested priorities for more sustainable use of nonrenewable resources are, in order: **R**efuse, **R**educe, **R**euse, and **R**ecycle. What percentage of the metals and other nonrenewable materials that we use could be reused or recycled? Distinguish between **more-developed countries** and **less-developed countries** and give an example of a high-income, middle-income, and low-income country.

Section 1-2

5. What is the key concept for this section? Define and give three examples of **environmental degradation (natural capital degradation)**. About what percentage of the earth's natural or ecosystem services has been degraded by human activities? Define **pollution**. Distinguish between **point sources** and **nonpoint sources** of pollution and give an example of each. Distinguish between **pollution cleanup** and **pollution prevention** and give an example of each. What is the *tragedy of the commons*? What are two ways to deal with this effect? Explain why they don't work for some systems.

6. What is **affluence**? What is an **ecological footprint**? What is a **per capita ecological footprint**? Compare the total and per capita ecological footprints of the United States and China. Use the ecological footprint concept to explain how we are living unsustainably in terms of the estimated number of planet Earths that we need to sustain ourselves now and in the future.

7. What is the IPAT model for estimating our environmental impact? Explain how we can use this model to estimate the impacts of the human populations in less-developed and more-developed countries. Describe the environmental impacts of China's new affluent consumers. Describe three major cultural changes that have occurred since humans were hunter–gatherers and how they have increased our overall environmental impact. What would a **sustainability revolution** involve?

Section 1-3

8. What are the two key concepts for this section? Identify five basic causes of the environmental problems that we face. What is **exponential growth**? What is the rule of 70? What is the current size of the human population? How many people are added each year? How many people might be here by 2050? How do Americans, Indians, and the average people in the poorest countries compare in terms of average consumption per person? What are two types of environmental damage resulting from growing affluence?

How can affluence help us to solve environmental problems? What is **poverty** and what are three of its harmful environmental and health effects? About how many of the world's people struggle to live on the equivalent of $1.25 a day? How many try to live on $2.25 a day? Explain the connection between poverty and population growth. List three major health problems suffered by many of the world's poor.

9. Explain how excluding the harmful environmental and health costs of production from the prices of goods and services affects the environmental problems we face. What is the connection between government subsidies, resource use, and environmental degradation? What are two ways to include the harmful environmental and health costs of the goods and services that we use in their market prices? Explain how lack of knowledge of the nature and importance of natural capital and our increasing isolation from nature can intensify the environmental problems that we face. What is an **environmental worldview**? What are **environmental ethics**? Distinguish among the **planetary management**, **stewardship**, and **environmental wisdom worldviews**.

Section 1-4

10. What is the key concept for this section? What is an **environmentally sustainable society**? What is **natural income** and what does it mean to live off of natural income? Describe Tuy Sereivathana's efforts to prevent elephants from becoming extinct in Cambodia and to reduce the country's poverty. What are two pieces of good news about making the transition to a more sustainable society? Based on the three scientific **principles of sustainability** and the three social science **principles of sustainability**, what are three important ways to make a transition to sustainability as summarized in this chapter's *three big ideas*? Explain how we can use the six principles of sustainability to move us closer to the vision of a more sustainable world described in the **Core Case Study** that opens this chapter.

Note: Key terms are in bold type. Knowing the meanings of these terms will help you in the course you are taking.

Critical Thinking

1. Do you think you are living unsustainably? Explain. If so, what are the three most environmentally unsustainable components of your lifestyle? List two ways in which you could apply each of the three scientific **principles of sustainability** (Figure 1-2) and each of the three social science **principles of sustainability** (Figure 1-5) to making your lifestyle more environmentally sustainable.

2. Do you believe that a vision such as the one described in the **Core Case Study** that opens this chapter is possible? Why or why not? What, if anything, do you believe will be different from that vision of the future? Explain. If your vision of what it will be like in 2065 is sharply different from that in the **Core Case Study**, write a description of your vision. Compare your answers to this question with those of your classmates.

3. For each of the following actions, state one or more of the three scientific **principles of sustainability** that are involved: **(a)** recycling aluminum cans; **(b)** using a rake instead of a leaf blower; **(c)** walking or bicycling to class instead of driving; **(d)** taking your own reusable bags to the grocery store to carry your purchases home; **(e)** volunteering to help restore a prairie; and **(f)** lobbying elected officials to require that at least 20% of your country's electricity be produced with renewable wind and solar power by 2020.

4. Explain why you agree or disagree with the following propositions:
   a. Stabilizing population is not desirable because, without more consumers, economic growth would stop.
   b. The world will never run out of resources because we can use technology to find substitutes and to help us reduce resource waste.

5. What do you think when you read that the average American consumes 30 times more resources than the average citizen of India? Are you skeptical, indifferent, sad, helpless, guilty, concerned, or outraged by this fact? Do you think that these differences in consumption have led to problems? If so, describe them and propose some possible solutions.

6. When you read that at least 19,000 children age 5 and younger die each day (13 per minute) from preventable malnutrition and infectious disease, how does it make you feel? Can you think of something that you and others could do to address this problem? What might that be?

7. Explain why you agree or disagree with each of the following statements: **(a)** humans are superior to other forms of life; **(b)** humans are in charge of the earth; **(c)** the value of other forms of life depends only on whether or not they are useful to humans; **(d)** based on records of past extinctions and the history of life on the earth over the last 3.5 billion years, biologists hypothesize that all forms of life eventually become extinct, and we should not worry about whether our activities hasten their extinction; **(e)** all forms of life have an inherent right to exist; **(f)** all economic growth is good; **(g)** nature has an almost unlimited storehouse of resources for human use; **(h)** technology can solve our environmental problems; **(i)** I do not believe I have any obligation to future generations; and **(j)** I do not believe I have any obligation to other forms of life.

8. What are the basic beliefs within your environmental worldview? Record your answer. Then, at the end of this course, return to your answer to see if your environmental worldview has changed. Are the beliefs included in your environmental worldview consistent with the answers you gave to Question 7 above? Are your actions that affect the environment consistent with your environmental worldview? Explain.

Doing Environmental Science

Estimate your own ecological footprint by using one of the many estimator tools available on the Internet. Is your ecological footprint larger or smaller than you thought it would be, according to this estimate? Why do you think this is so? List three ways in which you could reduce your ecological footprint. Try one of them for a week, and write a report on this change.

Global Environment Watch Exercise

Using the world maps in Figure 1, p. S24 and Figure 3, p. S28 in Supplement 6, choose one more-developed country and one less-developed country to compare their ecological footprints (found under Quick Facts on the country portal). Click on the ecological footprint number to view a graph of both the ecological footprint and biocapacity of each country. Using those graphs, determine which country appears to be living more sustainably. What would be some reasons for the differences in their levels of sustainability?

Ecological Footprint Analysis

If the *ecological footprint per person* of a country or the world is larger than its *biological capacity per person* to replenish its renewable resources and absorb the resulting waste products and pollution, the country or the world is said to have an *ecological deficit*. If the reverse is true, the country or the world has an *ecological credit* or *reserve*. Use the data below to calculate the ecological deficit or credit for the countries listed and for the world. (For a map of ecological creditors and debtors see Figure 4, p. S29, in Supplement 6.)

Place	Per Capita Ecological Footprint (hectares per person)	Per Capita Biological Capacity (hectares per person)	Ecological Credit (+) or Debit (−) (hectares per person)
World	2.2	1.8	−0.4
United States	9.8	4.7	
China	1.6	0.8	
India	0.8	0.4	
Russia	4.4	0.9	
Japan	4.4	0.7	
Brazil	2.1	9.9	
Germany	4.5	1.7	
United Kingdom	5.6	1.6	
Mexico	2.6	1.7	
Canada	7.6	14.5	

Compiled by the authors using data from WWF *Living Planet Report 2006*.

1. Which two countries have the largest ecological deficits? Why do you think they have such large deficits?

2. Which two countries have an ecological credit? Why do you think each of these countries has an ecological credit?

3. Rank the countries in order from the largest to the smallest per capita ecological footprint.

1. Pollution from which of the following is an example of point source pollution?
   a. Fertilizers from agricultural fields
   b. Runoff from Midwestern farms
   c. Oil spills from leaky pipelines
   d. Pesticides from farmlands
   e. Smoke from an industrial plant smokestack

2. Countries such as China increasingly have populations that are attaining a middle-class status. This has led to all of the following EXCEPT
   a. increased harmful effects of affluenza.
   b. more people purchasing cell phones, cars, and appliances.
   c. a decrease in environmental impact as residents become increasingly aware of the effects of pollution.
   d. two-thirds of the world's most polluted cities being in China.
   e. less arable land available for farming.

Use the graphs below to answer questions 3–6.

3. India's per capita ecological footprint is approximately
   a. 1/10 of China's ecological footprint.
   b. 1/5 of China's ecological footprint.
   c. 1/2 of China's ecological footprint.
   d. 2X China's ecological footprint.
   e. equal to China's ecological footprint.

4. In the year 2000, what was the earth's ecological footprint?
   a. 1.2 earths
   b. 0.5 earth
   c. 2.0 earths
   d. 1.5 earths
   e. 1 earth

5. All of the following ways would help reduce the projected ecological footprint EXCEPT
   a. using shared resources at rates well below their estimated sustainable yields.
   b. converting open-access resources to private ownership.
   c. reusing resources over and over.
   d. recycling nonrenewable metallic resources.
   e. continuing reliance on the current fossil fuel–based, automobile-centered economy.

6. From the bar graph you can infer that Africa's ecological footprint is no greater than
   a. 5%.
   b. 8%.
   c. 12%.
   d. 26%.
   e. 20%.

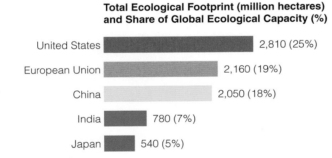

Total Ecological Footprint (million hectares) and Share of Global Ecological Capacity (%)

United States — 2,810 (25%)
European Union — 2,160 (19%)
China — 2,050 (18%)
India — 780 (7%)
Japan — 540 (5%)

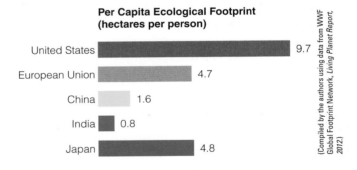

Per Capita Ecological Footprint (hectares per person)

United States — 9.7
European Union — 4.7
China — 1.6
India — 0.8
Japan — 4.8

(Compiled by the authors using data from WWF Global Footprint Network, *Living Planet Report, 2012*.)

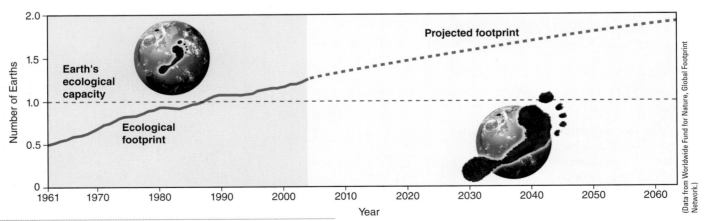

Number of Earths

Earth's ecological capacity

Ecological footprint

Projected footprint

Year

(Data from Worldwide Fund for Nature, Global Footprint Network.)

7. Examples of the three principles of sustainability include:

   1. Solar energy, which enables plants to photosynthesize and helps humans produce electricity
   2. Organisms and their ecosystems, the natural services they provide, and the ability of organisms to adapt to changes in their environment
   3. Cultural changes such as the agricultural revolution and the industrial revolution, which in many ways enhanced human living conditions but had negative effects on the environment
   4. The continuous cycling of chemicals, without which life would not exist

   a. 1 and 2
   b. 1, 2, and 3
   c. 1, 2, and 4
   d. 1, 3, and 4
   e. All of the above

Use the graph at right to answer questions 8–10.

8. In what year did the human population reach 6 billion?
   a. 8000 BC
   b. 4000 BC
   c. 2000 BC
   d. 1000 AD
   e. 2000 AD

9. The graph illustrates which type of growth?
   a. Exponential
   b. Linear
   c. Logistic
   d. Parabolic
   e. Inverse

10. Both dotted lines assume that population will
   a. increase at a linear rate.
   b. decrease.
   c. increase exponentially.
   d. stabilize.
   e. collapse.

Refer to the figure at right to answer question 11.

11. The figure illustrates
   I. the conservation of energy.
   II. a cycle.
   III. the conservation of matter.
   a. I only
   b. I and II
   c. II only
   d. I and III
   e. I, II, and III

12. Solutions to environmental problems are not always easy. This is because
   a. there is only one correct answer to environmental problems.
   b. it means that people will suffer.
   c. once a solution is suggested it can never be changed.
   d. laws must always be written to solve the environmental problem.
   e. any solution has both advantages and disadvantages.

13. Which of the below is an example of the tragedy of the commons?
   a. Treaties allowing for the limited taking of whales for research purposes
   b. Offshore clam beds in public areas depleted below replenishment levels by overfishing
   c. Individuals catching their legal limit of fish on a daily basis
   d. Shrimp boats accidentally catching turtles in their nets
   e. A large storm smothering a large oyster bed with a new sand bar

2 Science, Matter, Energy, and Systems

Science is built up of facts, as a house is built of stones; but an accumulation of facts is no more a science than a heap of stones is a house.

HENRI POINCARÉ

Key Questions

2-1 What do scientists do?

2-2 What is matter and what happens when it undergoes change?

2-3 What is energy and what happens when it undergoes change?

2-4 What are systems and how do they respond to change?

Scientist studying biodiversity in a tropical rain forest in Indonesia by using a light to trap moths.

Tim Laman/National Geographic Creative

Suppose a logging company plans to cut down all of the trees on a hillside behind your house. You are very concerned and want to know about the possible harmful environmental effects of this action.

One way to learn about such effects is to conduct a *controlled experiment,* just as environmental scientists do. They begin by identifying key *variables,* such as water loss and soil nutrient content, that might change after the trees are cut down. Then, they set up two groups. One is the *experimental group,* in which a chosen variable is changed in a known way. The other is the *control group,* in which the chosen variable is not changed. They then compare the results from the two groups.

In 1963, botanist F. Herbert Bormann, forest ecologist Gene Likens, and their colleagues began carrying out such a controlled experiment. Their goal was to compare the loss of water and soil nutrients from an area of uncut forest (the *control*

site) with one that had been stripped of its trees (the *experimental site*).

They built V-shaped concrete dams across the creeks at the bottoms of several forested valleys in the Hubbard Brook Experimental Forest in New Hampshire (Figure 2-1). The dams were designed so that all surface water leaving each forested valley had to flow across a dam, where scientists could measure its volume and dissolved nutrient content.

First, the researchers measured the amounts of water and dissolved soil nutrients flowing from an undisturbed forested area in one of the valleys (the control site, Figure 2-1, left). These measurements showed that an undisturbed mature forest is very efficient at storing water and retaining chemical nutrients in its soils.

Next, they set up an experimental forest area in a nearby valley (Figure 2-1, right). One winter, they cut down all the trees and shrubs in that valley, left them where they fell, and sprayed the area with herbicides

to prevent the regrowth of vegetation. Then, for 3 years, they compared outflow of water and nutrients in this experimental site with those in the control site.

The scientists found that, with no plants to help absorb and retain water, the amount of water flowing out of the deforested valley increased by 30–40%. As this excess water ran rapidly over the ground, it eroded soil and carried dissolved nutrients out of the topsoil in the deforested site. Overall, the loss of key soil nutrients from the experimental forest was 6–8 times that in the nearby uncut control forest.

In this chapter, you will learn more about how scientists study nature and about the matter and energy that make up the world within and around us. You will also learn about three *scientific laws,* or rules of nature, that govern the changes that matter and energy undergo. And you will learn the important difference between a scientific hypothesis and a scientific theory.

© Cengage Learning

Figure 2-1 This controlled field experiment measured the loss of water and soil nutrients from a forest due to deforestation. The forested valley (left) was the control site; the cutover valley (right) was the experimental site.

2-1 What Do Scientists Do?

CONCEPT 2-1
Scientists collect data and develop hypotheses, theories, models, and laws about how nature works.

Science Is a Search for Order in Nature

Science is an attempt to discover how nature works and to use that knowledge to describe what is likely to happen in nature. It is based on the assumption that events in the physical world follow orderly cause-and-effect patterns that can be understood through careful observation, measurements, experimentation, and modeling. Figure 2-2 summarizes the scientific process.

There is nothing mysterious about this scientific process. You use it all the time in making decisions. As the famous physicist Albert Einstein put it, "The whole of science is nothing more than a refinement of everyday thinking."

Scientists Use Observations, Experiments, and Models to Answer Questions about How Nature Works

Here is a more formal outline of the steps scientists often take in trying to understand the natural world, although they do not always follow the steps in the order listed. The outline is based on the scientific experiment carried out by Bormann and Likens (**Core Case Study**), which illustrates the nature of the scientific process shown in Figure 2-2.

- *Identify a problem.* Bormann and Likens identified the loss of water and soil nutrients from cutover forests as a problem worth studying.
- *Find out what is known about the problem.* They searched the scientific literature to find out what scientists knew about both the retention and the loss of water and soil nutrients in forests.
- *Ask a question to investigate.* The scientists asked, "How does clearing forested land affect its ability to store water and retain soil nutrients?"
- *Perform an experiment and collect and analyze data to answer the question.* To collect **data**—information needed to answer their questions—scientists often perform experiments and make observations and measurements, as Bormann and Likens did (Figure 2-1). (Sometimes, they simply observe and measure natural phenomena to collect data without doing an experiment.)
- *Propose a hypothesis to explain the data.* Scientists suggest a **scientific hypothesis**—a possible and testable answer to a scientific question or explanation of what scientists observe in nature. Bormann and Likens

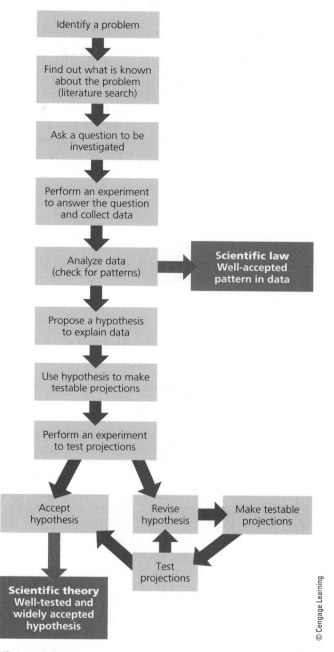

Figure 2-2 This diagram illustrates the general process that scientists use for discovering and testing ideas about how the natural world works.

came up with the following hypothesis to explain their data: When a forest is cleared of its vegetation and exposed to rain and melting snow, it retains less water and loses large quantities of soil nutrients.

- *Use the hypothesis to make projections that can be tested.* Scientists make projections about what should happen if their hypothesis is correct and then run experiments to test the projections. Bormann and Likens projected that if their hypothesis was valid for nitrogen, then a cleared forest should also lose other soil nutrients such as phosphorus over a similar time period and under similar weather conditions.

- *Test the projections with further experiments or observations.* To test their projection, Bormann and Likens repeated their controlled experiment and measured the phosphorus content of the soil. Another way to test projections is to use a **model**, an approximate representation or simulation of a system.
- *Accept or revise the hypothesis.* After Bormann and Likens confirmed that the soil in a cleared forest also loses phosphorus, they measured losses of other soil nutrients, which further supported their hypothesis. The research of other scientists also supported the hypothesis. A well-tested and widely accepted scientific hypothesis or a group of related hypotheses is called a **scientific theory**. The research by Bormann and Likens and other scientists led to a widely accepted scientific theory that trees and other plants hold soil in place and help it to retain water and nutrients needed by the plants for their growth.

Scientists Are Curious and Skeptical, and They Demand Evidence

Four important features of the scientific process are *curiosity, skepticism, reproducibility,* and *peer review.* Good scientists are extremely curious about how nature works, and they are keen observers of what is happening in nature (see Individuals Matter 2.1). Scientists tend to be highly skeptical of new data and hypotheses. They say, "Show me your evidence and explain the reasoning behind the scientific ideas or hypotheses that you propose to explain your data." Any evidence that scientists gather should also be reproducible. In other words, other scientists should be able to get the same results if they run the same experiments.

Science is a community effort, and an important part of the scientific process is **peer review**. It involves scientists openly publishing details of the methods they used, the results of their experiments, and the reasoning behind their hypotheses for other scientists working in the same field (their peers) to evaluate.

For example, Bormann and Likens (**Core Case Study**) submitted the results of their forest experiments to a respected scientific journal. Before publishing this report, the journal's editors asked other soil and forest experts to review it. Other scientists have repeated the measurements of soil content in undisturbed and cleared forests of the same type and also in different types of forests, and their results have also been subjected to peer review. In addition, computer models of forest systems have been used to evaluate this problem, with the results also subjected to peer review.

Scientific knowledge advances in this self-correcting way, with scientists continually questioning the measurements and data produced by their peers. They also collect new data and sometimes come up with new and better hypotheses (Science Focus 2.1).

Critical Thinking and Creativity Are Important in Science

Scientists use logical reasoning and critical thinking skills (p. xxiv) to learn about the natural world. Thinking critically involves four important steps:

1. Be skeptical about everything you read or hear.
2. Look at the evidence and evaluate it and any related information, along with inputs and opinions from a variety of reliable sources.
3. Be open to many viewpoints and evaluate each one before coming to a conclusion.
4. Identify and evaluate your personal assumptions, biases, and beliefs. As the American psychologist and philosopher William James observed, "A great many people think they are thinking when they are merely rearranging their prejudices."

Logic and critical thinking are very important tools in science, but imagination, creativity, and intuition are just as vital. According to physicist Albert Einstein, "There is no completely logical way to a new scientific idea."

Scientific Theories and Laws Are the Most Important and Certain Results of Science

The real goal of scientists is to develop theories and laws, based on facts and data that explain how the natural world works, as illustrated in the quotation that opens this chapter. *We should never take a scientific theory lightly.* It has been tested widely, is supported by extensive evidence, and is accepted as being a useful explanation of some phenomenon by most scientists in a particular field or related fields of study.

Because of this rigorous testing process, scientific theories are rarely overturned unless new evidence discredits them or scientists come up with better explanations. So when you hear someone say, "Oh, that's just a theory," you will know that he or she does not have a clear understanding of what a scientific theory is. In sports terms, developing a widely accepted scientific theory is roughly equivalent to winning a gold medal in the Olympics.

Another important and reliable outcome of science is a **scientific law**, or **law of nature**—a well-tested and widely accepted description of what we find happening repeatedly and in the same way in nature. An example is the *law of gravity.* After making many thousands of observations and measurements of objects falling from different heights, scientists developed the following scientific law: all objects fall to the earth's surface at predictable speeds.

We can break a society's law, for example, by driving faster than the speed limit. But *we cannot break a*

JENS SCHLUETER/AFP/Getty Images

Jane Goodall: Chimpanzee Researcher and Protector

Jane Goodall is a primatologist and environmental educator with a PhD from Cambridge University. She is also a National Geographic Explorer-in-Residence Emeritus. At age 26, she began a 50-year career of studying chimpanzee social and family life in the Gombe Stream Game Reserve in the African country of Tanzania. She is shown above with one of the chimpanzees she studied. By carefully observing the chimps while living with them, she discovered that chimps have more complex social interactions than had previously been known. For example, she found that they can use tools, eat meat as well as vegetation, and carry on extended fights with one another.

One of her major scientific discoveries was that chimpanzees have tool-making skills. She observed some chimpanzees modifying twigs or blades of grass and then poking them into termite mounds. When the termites latched on to these primitive tools, the chimpanzees pulled them out and ate the termites. Goodall has also observed that chimps can learn simple sign language, do simple arithmetic, play computer games, develop relationships, and worry about and protect one another.

In 1977, she established the Jane Goodall Institute, a nonprofit organization that works to preserve great ape populations and their habitats. Her research encouraged the Tanzanian government to convert the game preserve into the Gombe Stream National Park, which her institute now supports.

Goodall has been a strong advocate for animal rights and has led campaigns against sport hunting, keeping animals in zoos, and using them for medical research. In 1991 she started Roots and Shoots, an environmental education program for youth that is active in more than 100 countries. She has received many awards and prizes for her scientific contributions and conservation efforts. She has also written 23 books for adults and children and produced 14 films about the lives and importance of chimpanzees.

In 2011, the movie *Jane's Journey* was made about her life's work. Now in her late 70s, Goodall still spends nearly 300 days a year traveling and educating people throughout the world about chimpanzees and the need to protect the environment. She says, "I can't slow down. . . . If we're not raising new generations to be better stewards of the environment, what's the point?"

Background photo: namatae/Shutterstock.com

SCIENCE FOCUS 2.1

SOME REVISIONS IN A POPULAR SCIENTIFIC HYPOTHESIS

For years, the story of Easter Island has been used in textbooks as an example of how humans can seriously degrade their own life-support system and as a warning about what we are doing to our life-support system.

What happened on this small island in the South Pacific is a story about environmental degradation and the demise of an ancient civilization of Polynesians living there. Years ago, scientists studied the island and its remains, including more than 300 huge statues (Figure 2-A). They hypothesized that over time, the Polynesians began living unsustainably as their population grew, and they used the island's forest and soil resources faster than they could be renewed. They further hypothesized that when the forests were depleted, there was no firewood for cooking or keeping warm and no wood for building large canoes in order to leave the island. They also hypothesized that, with the forest cover gone, soils eroded, crop yields plummeted, famine struck, the population dwindled, and the civilization collapsed.

In 2006, anthropologist Terry L. Hunt evaluated the accuracy of past measurements and other evidence and carried out new research to reevaluate the hypothesis about what happened on Easter Island. He used his data to formulate an alternative hypothesis to try to explain the human tragedy on Easter Island, and he came to some new conclusions. *First,* the Polynesians arrived on the island about 800 years

ago, not 2,900 years ago, as had been thought. *Second,* their population size probably never exceeded 3,000, contrary to the earlier estimate of up to 15,000.

Third, the Polynesians did use the island's trees and other vegetation in an unsustainable manner, and visitors reported that by 1722, most of the island's trees were gone. However, one question not answered by this earlier hypothesis was, why did the trees never grow back? Recent evidence and Hunt's new hypothesis suggest that rats (which either came along with the original settlers as stowaways or were brought along as a source of protein for the long voyage) played a key role in the island's permanent deforestation. Over the years, the rats multiplied rapidly into the millions and devoured the seeds that would have regenerated the forests. According to this new hypothesis, the rats played a key role in the fall of the civilization on Easter Island.

This story is an excellent example of how science works. The gathering of new scientific data and the reevaluation of older data led to a revised hypothesis that challenged earlier thinking about the decline of civilization on Easter Island. As a result, the tragedy may not be as clear an example of human-caused ecological collapse as was once thought.

Note that the original Easter Island story was a scientific hypothesis, not a widely tested and accepted scientific

Figure 2-A These and several thousand other statues were created by an ancient civilization of Polynesians on Easter Island. Some of them are as tall as a five-story building and weigh as much as 89 metric tons (98 tons).

modestlife/Shutterstock.com

theory. And Hunt's research presents another scientific hypothesis based on new data. Further research may convert Hunt's hypothesis to the status of a scientific theory, or more research may lead to other insights into what happened on this island. This is how science works.

Critical Thinking

Does the new doubt about the original Easter Island hypothesis mean that we should not be concerned about using resources unsustainably on the island in space that we call Earth? Explain.

scientific law, unless we discover new evidence that leads to changes in the law.

The Results of Science Can Be Tentative, Reliable, or Unreliable

Sometimes, preliminary scientific results that capture news headlines have not been widely tested and accepted by peer review. They are not yet considered reliable, and can be thought of as **tentative science** or **frontier science**. Some of these results and hypotheses will be vali-

dated and classified as reliable and some will be discredited and classified as unreliable. At the frontier stage, it is normal for scientists to disagree about the meaning and accuracy of data and the validity of hypotheses and results. This is how scientific knowledge advances.

By contrast, **reliable science** consists of data, hypotheses, models, theories, and laws that are widely accepted by all or most of the scientists who are considered experts in the field under study. The results of reliable science are based on the self-correcting process of testing, open peer review, and debate. New evidence and better hypotheses may discredit or alter widely accepted scientific theories, although this is rare. But until that happens, those theories are considered to be the results of reliable science.

Scientific hypotheses and results that are presented as reliable without having undergone the rigors of widespread peer review, or that have been discarded as a result of peer review, are considered to be **unreliable science**. Here are some critical thinking questions you can use to uncover unreliable science:

- Was the experiment well designed? Did it involve a control group? (**Core Case Study**)
- Does the proposed hypothesis explain the data?
- Are there no other, more reasonable explanations of the data?
- Are the investigators unbiased in their interpretations of the results? Did their funding come from unbiased sources?
- Have the data and conclusions been subjected to peer review?
- Are the conclusions of the research widely accepted by other experts in this field?

If "yes" is the answer to each of these questions, then you can classify the results as reliable science. Otherwise, the results may represent tentative science that needs further testing and evaluation, or you can classify them as unreliable science.

Science Has Some Limitations

Environmental science and science in general have four important limitations. *First,* scientists cannot prove or disprove anything absolutely, because there is always some degree of uncertainty in scientific measurements, observations, and models. Instead, scientists try to establish that a particular scientific theory or law has a very high *probability* or *certainty* (at least 90%) of being useful for understanding some aspect of the natural world.

Many scientists do not use the word *proof* because it implies "absolute proof" to people who don't understand how science works. For example, most scientists will rarely say something like, "Cigarettes cause lung cancer." Rather, they might say, "Overwhelming evidence from thousands of studies indicates that people who smoke regularly for many years have a greatly increased chance of developing lung cancer."

🔍 **CONSIDER THIS. . .**

THINKING ABOUT Scientific Proof

Does the fact that science can never prove anything absolutely mean that its results are not valid or useful? Explain.

A *second* limitation of science is that scientists are human and thus are not totally free of bias about their own results and hypotheses. However, the high standard of evidence required through peer review helps to uncover or greatly reduce personal bias and expose occasional cheating by scientists who falsify their results.

A *third* limitation—especially important to environmental science—is that many systems in the natural world involve a huge number of variables with complex interactions. This makes it difficult, too costly, and too time consuming to test one variable at a time in controlled experiments such as the one described in the **Core Case Study** that opens this chapter. To try to deal with this problem, scientists develop *mathematical models* that can take into account the interactions of many variables. Running such models on high-speed computers can sometimes overcome the limitations of testing each variable individually, saving both time and money. In addition, scientists can use computer models to simulate global experiments on phenomena such as climate change that cannot be done in a controlled physical experiment.

A *fourth* limitation of science involves the use of statistical tools. For example, there is no way to measure accurately how many metric tons of soil are eroded annually worldwide. Instead, scientists use statistical sampling and other mathematical methods to estimate such numbers. However, such results should not be dismissed as "only estimates" because they can indicate important trends.

Despite these limitations, science is the most useful way that we have of learning about how nature works and projecting how it might behave in the future. But we still know too little about how the earth works, about its present state of environmental health, and about the current and future environmental impacts of our activities.

2-2 What Is Matter and What Happens When It Undergoes Change?

CONCEPT 2-2A

Matter consists of elements and compounds, which in turn are made up of atoms, ions, or molecules.

CONCEPT 2-2B

Whenever matter undergoes a physical or chemical change, no atoms are created or destroyed (the law of conservation of matter).

Matter Consists of Elements and Compounds

To begin our study of environmental science, we look at matter—the stuff that makes up life and its environment. **Matter** is anything that has mass and takes up space. It can exist in three *physical states*—solid, liquid, and gas— and two *chemical forms*—elements and compounds.

Figure 2-3 Mercury (left) and gold (right) are chemical elements. Each has a unique set of properties and cannot be broken down into simpler substances.

Table 2-1 Chemical Elements Used in This Book			
Element	**Symbol**	**Element**	**Symbol**
arsenic	As	lead	Pb
bromine	Br	lithium	Li
calcium	Ca	mercury	Hg
carbon	C	nitrogen	N
copper	Cu	phosphorus	P
chlorine	Cl	sodium	Na
fluorine	F	sulfur	S
gold	Au	uranium	U

© Cengage Learning

An **element** is a fundamental type of matter that has a unique set of properties and cannot be broken down into simpler substances by chemical means. For example, the elements gold (Figure 2-3, left) and mercury (Figure 2-3, right) cannot be broken down chemically into any other substance. Chemists arrange the known elements on the basis of their chemical behavior in what is called the Periodic Table of Elements (see Figure 1, p. S12, Supplement 4).

Some matter is composed of one element, such as mercury or gold (Figure 2-3). And some elements such as hydrogen (H), nitrogen (N), oxygen (O), and chlorine (Cl) are found in nature as combinations of two of their atoms represented by the chemical formulas H_2, N_2, O_2, and Cl_2. However, most matter consists of **compounds**, combinations of two or more different elements held together in fixed proportions. For example, water is a compound made of the elements hydrogen and oxygen that have chemically combined with one another. (See Supplement 4, p. S12, for an expanded discussion of basic chemistry.)

To simplify things, chemists represent each element by a one- or two-letter symbol. Table 2-1 lists the elements and their symbols that you need to know to understand the material in this book.

Atoms, Molecules, and Ions Are the Building Blocks of Matter

The most basic building block of matter is an **atom**, the smallest unit of matter into which an element can be divided and still have its distinctive chemical properties. The idea that all elements are made up of atoms is called the **atomic theory** and it is the most widely accepted scientific theory in chemistry.

Atoms are incredibly small. For example, more than 3 million hydrogen atoms could sit side by side on the period at the end of this sentence. If you could view atoms with a supermicroscope, you would find that each different type of atom contains a certain number of three types of *subatomic particles:* **neutrons (n)** with no electrical charge; **protons (p)**, each with a positive electrical charge (+); and **electrons (e)**, each with a negative electrical charge (−).

Each atom has an extremely small center called the **nucleus**, which contains one or more protons and, in most cases, one or more neutrons. Outside of the nucleus we find one or more electrons in rapid motion (Figure 2-4). We cannot determine the exact locations of the electrons. Instead, scientists can estimate the *probability* that they will be found at various locations outside the nucleus in certain spatial patterns that are called *electron probability clouds*. This is somewhat like saying that there are a number of bees flying around inside a cloud. We do not know their exact locations, but the cloud represents an area in which there is a high probability of finding them.

Each atom in its basic form has equal numbers of positively charged protons and negatively charged electrons. Because these electrical charges cancel one another, *an atom in its basic form has no net electrical charge.*

Each element has a unique **atomic number** equal to the number of protons in the nucleus of its atom. Carbon (C), with 6 protons in its nucleus, has an atomic number of 6, whereas uranium (U), a much larger atom, has 92 protons in its nucleus and thus an atomic number of 92.

Because electrons have so little mass compared to protons and neutrons, *most of an atom's mass is concentrated in its nucleus.* The mass of an atom is described by its **mass number**, the total number of neutrons and protons in its

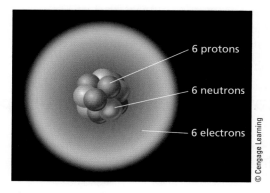

Figure 2-4 This is a greatly simplified model of a carbon-12 atom. It consists of a nucleus containing six protons, each with a positive electrical charge, and six neutrons with no electrical charge. Six negatively charged electrons are found outside its nucleus.

Table 2-2 Chemical Ions Used in This Book

Positive Ion	Symbol	Components
hydrogen ion	H^+	One hydrogen atom, one positive charge
sodium ion	Na^+	One sodium atom, one positive charge
calcium ion	Ca^{2+}	One calcium atom, two positive charges
aluminum ion	Al^{3+}	One aluminum atom, three positive charges
ammonium ion	NH_4^+	One nitrogen atom, four hydrogen atoms, one positive charge

Negative Ion	Symbol	Components
chloride ion	Cl^-	One chlorine atom, one negative charge
hydroxide ion	OH^-	One oxygen atom, one hydrogen atom, one negative charge
nitrate ion	NO_3^-	One nitrogen atom, three oxygen atoms, one negative charge
carbonate ion	CO_3^{2-}	One carbon atom, three oxygen atoms, two negative charges
sulfate ion	SO_4^{2-}	One sulfur atom, four oxygen atoms, two negative charges
phosphate ion	PO_4^{3-}	One phosphorus atom, four oxygen atoms, three negative charges

© Cengage Learning

nucleus. For example, a carbon atom with 6 protons and 6 neutrons in its nucleus has a mass number of 12, and a uranium atom with 92 protons and 143 neutrons in its nucleus has a mass number of 235 (92 + 143 = 235).

Each atom of a particular element has the same number of protons in its nucleus. But the nuclei of atoms of a particular element can vary in the number of neutrons they contain and, therefore, in their mass numbers. The forms of an element having the same atomic number but different mass numbers are called **isotopes** of that element. Scientists identify isotopes by attaching their mass numbers to the name or symbol of the element. For example, the three most common isotopes of carbon are carbon-12 (with six protons and six neutrons), carbon-13 (with six protons and seven neutrons), and carbon-14 (with six protons and eight neutrons). Carbon-12 makes up about 98.9% of all naturally occurring carbon.

A second building block of matter is a **molecule**, a combination of two or more atoms of the same or different elements held together by forces called *chemical bonds*. Molecules are the basic building blocks of many compounds (see Figure 3, p. S14, in Supplement 4 for examples). An example of a molecule is that of water, or H_2O, which consists of two atoms of hydrogen and one atom of oxygen held together by chemical bonds. Another example is methane, or CH_4 (the major component of natural gas), which consists of four atoms of hydrogen and one atom of carbon.

A third building block of some types of matter is an **ion**—an atom or a group of atoms with one or more net positive or negative electrical charges. Like atoms, ions are made up of protons, neutrons, and electrons. (See p. S13 in Supplement 4 for details on how ions form.) Chemists use a superscript after the symbol of an ion to indicate how many positive or negative electrical charges it has, as shown in Table 2-2.

The nitrate ion (NO_3^-) is a nutrient essential for plant growth. Figure 2-5 shows measurements of the loss of nitrate ions from the deforested area (Figure 2-1, right) in the controlled experiment run by Bormann and Likens (**Core Case Study**). Numerous chemical analyses of the water flowing through the dam at the cleared forest site showed an average 60-fold rise in the concentration of NO_3^- compared to water running off the forested site. After a few years, however, vegetation began growing back in the cleared valley and nitrate levels in its runoff returned to normal levels.

Ions are also important for measuring a substance's **acidity** in a water solution, a chemical characteristic that helps determine how a substance dissolved in water will interact with and affect its environment. The acidity of a water solution is based on the comparative amounts of hydrogen ions (H^+) and hydroxide ions (OH^-) contained in a particular volume of the solution. Scientists use **pH** as

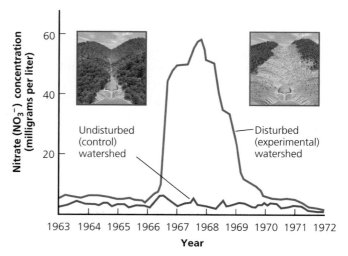

Figure 2-5 This graph shows the loss of nitrate ions (NO_3^-) from a deforested watershed in the Hubbard Brook Experimental Forest (**Core Case Study**).

(Based on data from F. H. Bormann and Gene Likens.)

Na⁺   Cl⁻

Figure 2-6 A solid crystal of an ionic compound such as sodium chloride (NaCl) consists of a three-dimensional array of oppositely charged ions held together by the strong forces of attraction between oppositely charged ions.

© Cengage Learning

Table 2-3 Compounds Used in This Book			
Compound	Formula	Compound	Formula
sodium chloride	NaCl	methane	CH_4
sodium hydroxide	NaOH	glucose	$C_6H_{12}O_6$
carbon monoxide	CO	water	H_2O
carbon dioxide	CO_2	hydrogen sulfide	H_2S
nitric oxide	NO	sulfur dioxide	SO_2
nitrogen dioxide	NO_2	sulfuric acid	H_2SO_4
nitrous oxide	N_2O	ammonia	NH_3
nitric acid	HNO_3	calcium carbonate	$CaCO_3$

© Cengage Learning

a measure of acidity. Pure water (not tap water or rainwater) has an equal number of H^+ and OH^- ions. It is called a *neutral solution* and has a pH of 7. An *acidic solution* has more hydrogen ions than hydroxide ions and has a pH less than 7. A *basic solution* has more hydroxide ions than hydrogen ions and has a pH greater than 7. (See Figure 4, p. S15, in Supplement 4 for more details.)

Chemists use a **chemical formula** to show the number of each type of atom or ion in a compound. This shorthand contains the symbol for each element present (Table 2-1) and uses subscripts to show the number of atoms or ions of each element in the compound's basic structural unit. Examples of compounds and their formulas encountered in this book are sodium chloride (NaCl) and water (H_2O, read as "H-two-O"). Sodium chloride is an ionic compound with a three-dimensional array of sodium ions (Na^+) and chloride ions (Cl^-) (Figure 2-6). These and other compounds important to our study of environmental science are listed in Table 2-3.

You might want to mark these pages containing Tables 2-1, 2-2, and 2-3, because they show the key elements, ions, and compounds used in this book. Think of them as lists of some main chemical characters in the story of matter that makes up the natural world.

Organic Compounds Are the Chemicals of Life

Plastics (Figure 2-7), table sugar, vitamins, aspirin, penicillin, and most of the chemicals in your body are called **organic compounds**, because they contain at least two carbon atoms combined with atoms of one or more other elements. All other compounds are called **inorganic compounds**. One exception, methane (CH_4), has only one carbon atom but is considered an organic compound.

The millions of known organic (carbon-based) compounds include the following:

- *Hydrocarbons:* compounds of carbon and hydrogen atoms. One example is methane (CH_4), the main component of natural gas and the simplest organic compound. Another is octane (C_8H_{18}), a major component of gasoline.
- *Chlorinated hydrocarbons:* compounds of carbon, hydrogen, and chlorine atoms. An example is the insecticide DDT ($C_{14}H_9Cl_5$).
- *Simple carbohydrates (simple sugars):* certain types of compounds of carbon, hydrogen, and oxygen atoms. An example is glucose ($C_6H_{12}O_6$), which most plants and animals break down in their cells to obtain energy. (For more details, see Figure 5, p. S16, in Supplement 4.)

Larger and more complex organic compounds, essential to life, are composed of *macromolecules.* Some of these molecules are called *polymers,* formed when a number of simple organic molecules *(monomers)* are linked together by chemical bonds—somewhat like rail cars linked in a freight train. The three major types of organic polymers are

- *complex carbohydrates* such as cellulose and starch, which consist of two or more monomers of simple sugars such as glucose (see Figure 5, p. S16, in Supplement 4), important sources of energy in the food we eat;
- *proteins* formed by monomers called *amino acids* (see Figure 6, p. S16, in Supplement 4), important for building certain tissues in our bodies; and
- *nucleic acids* (DNA and RNA) formed by monomers called *nucleotides* (see Figures 7 and 8, pp. S16 and S17, in Supplement 4), and key chemicals in the reproductive processes of many organisms.

Lipids, which include fats and waxes, are not made of monomers but are a fourth type of macromolecule essential for life (see Figure 9, p. S17, in Supplement 4).

Figure 2-7 Plastics, which are found in an amazing variety of widely used and useful products, consist of organic compounds.

Matter Comes to Life through Cells, Genes, and Chromosomes

All organisms are composed of one or more **cells**—the fundamental structural and functional units of life. They are minute compartments covered with a thin membrane, and within them, the processes of life occur. The idea that all living things are composed of cells is called the *cell theory* and it is the most widely accepted scientific theory in biology.

Above, we mentioned nucleotides in DNA (see Figures 7 and 8, pp. S16 and S17, in Supplement 4). Within some DNA molecules are certain sequences of nucleotides called **genes**. Each of these distinct pieces of DNA contains instructions, or codes, called *genetic information*, for making specific proteins. Each of these coded units of genetic information leads to a specific **trait**, or characteristic, passed on from parents to offspring during reproduction in an animal or plant.

In turn, thousands of genes make up a single **chromosome**, a double helix DNA molecule (see Figure 8, p. S17, in Supplement 4) wrapped around some proteins. Genetic information coded in your chromosomal DNA is what makes you different from an oak leaf, an alligator, or a mosquito, and from your parents. The relationships of genetic material to cells are depicted in Figure 2-8.

Matter Undergoes Physical, Chemical, and Nuclear Changes

When a sample of matter undergoes a **physical change**, there is no change in its *chemical composition*. A piece of aluminum foil cut into small pieces is still aluminum foil. When solid water (ice) melts and when liquid water boils, the resulting liquid water and water vapor are still made up of H_2O molecules.

When a **chemical change**, or **chemical reaction**, takes place, there is a change in the chemical composition of the substances involved. Chemists use a *chemical equation* (and a process called *balancing the equation*, see p. S17 in Supplement 4) to show how chemicals are rear-

Figure 2-8 This diagram shows the relationships among cells, nuclei, chromosomes, DNA, and genes.

Photo: ©Flashon Studio/Shutterstock.com

A human body contains trillions of cells, each with an identical set of genes.

Each human cell (except for red blood cells) contains a nucleus.

Each cell nucleus has an identical set of chromosomes, which are found in pairs.

A specific pair of chromosomes contains one chromosome from each parent.

Each chromosome contains a long DNA molecule in the form of a coiled double helix.

Genes are segments of DNA on chromosomes that contain instructions to make proteins—the building blocks of life.

© Cengage Learning

ranged in a chemical reaction. For example, coal is made up almost entirely of the element carbon (C). When coal is burned completely in a power plant, the solid carbon in the coal combines with oxygen gas (O_2) from the atmosphere to form the gaseous compound carbon dioxide (CO_2). Chemists use the following shorthand chemical equation to represent this chemical reaction:

Reactant(s) ⟶ **Product(s)**

Carbon + Oxygen ⟶ Carbon dioxide + **Energy**

C + O_2 ⟶ CO_2 + **Energy**

C + O O ⟶ O C O + Energy

Black solid   *Colorless gas*   *Colorless gas*

© Cengage Learning

Radioactive decay

Nuclear fission

Nuclear fusion

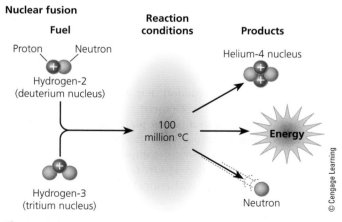

Figure 2-9 There are three types of nuclear changes: natural radioactive decay (top), nuclear fission (middle), and nuclear fusion (bottom).

In addition to physical and chemical changes, matter can undergo three types of **nuclear change**, or change in the nuclei of its atoms (Figure 2-9). **Radioactive decay** occurs when the nuclei of unstable isotopes spontaneously emit fast-moving chunks of matter (alpha particles or beta particles), high-energy radiation (gamma rays), or both at a fixed rate (Figure 2-9, top). **Nuclear fission** occurs when the nuclei of certain isotopes with large mass numbers (such as uranium-235) are split apart into lighter nuclei when struck by a neutron and release energy. Each fission releases neutrons, which can cause more nuclei to fission. This cascade of fissions can result in a chain reaction that releases an enormous amount of energy in a short time (Figure 2-9, middle). **Nuclear fusion** occurs when two nuclei of lighter atoms, such as hydrogen, are forced together at extremely high temperatures until they fuse to form a heavier nucleus and release a tremendous amount of energy (Figure 2-9, bottom).

We Cannot Create or Destroy Atoms: The Law of Conservation of Matter

We can change elements and compounds from one physical or chemical form to another, but we cannot create or destroy any of the atoms involved in any physical or chemical change. All we can do is rearrange the atoms, ions, or molecules into different spatial patterns (physical changes) or chemical combinations (chemical changes). These facts, based on many thousands of measurements, describe a scientific law known as the **law of conservation of matter**: Whenever matter undergoes a physical or chemical change, no atoms are created or destroyed (Concept 2-2B).

CONSIDER THIS...

CONNECTIONS Waste and the Law of Conservation of Matter

The law of conservation of matter means we can never really throw anything away because the atoms in any form of matter cannot be destroyed as it undergoes physical or chemical changes. Stuff that we put out in the trash may be buried in a sanitary landfill, but we have not really thrown it away because the atoms in this waste material will always be around in one form or another. We can burn trash, but we then end up with ash that must be put somewhere, and with gases emitted by the burning that can pollute the air. We can reuse or recycle some materials and chemicals, but the law of conservation of matter means we will always face the problem of what to do with some quantity of the wastes and pollutants we produce because their atoms cannot be destroyed.

© Cengage Learning

2-3 What Is Energy and What Happens When It Undergoes Change?

CONCEPT 2-3A
Whenever energy is converted from one form to another in a physical or chemical change, no energy is created or destroyed (first law of thermodynamics).

CONCEPT 2-3B
Whenever energy is converted from one form to another in a physical or chemical change, we end up with lower-quality or less-usable energy than we started with (second law of thermodynamics).

Figure 2-10 Kinetic energy, created by the gaseous molecules in a mass of moving air, turns the blades of this wind turbine. The turbine then converts this kinetic energy to electrical energy, which is another form of kinetic energy.

Energy Comes in Many Forms

Suppose you find this book on the floor and you pick it up and put it on your desktop. To do this you have to use a certain amount of muscular force or work to move the book from one place to another. In scientific terms, work is done when any object is moved a certain distance (work = force × distance). Also, whenever you touch a hot object such as a stove, heat flows from the stove to your finger. Both of these examples involve **energy**: the capacity to do work or to transfer heat.

There are two major types of energy: *moving energy* (called kinetic energy) and *stored energy* (called potential energy). Matter in motion has **kinetic energy**, or energy associated with motion. Examples are flowing water, a car speeding down the highway, electricity (electrons flowing through a wire or other conducting material), and wind (a mass of moving air that we can use to produce electricity, as shown in Figure 2-10).

Another form of kinetic energy is **heat**, or **thermal energy**, the total kinetic energy of all moving atoms, ions, or molecules in an object, a body of water, or the atmosphere. If the atoms, ions, or molecules in a sample of matter move faster, it will become warmer. When two objects at different temperatures come in contact with one another, heat flows from the warmer object to the cooler object. You learned this the first time you touched a hot stove.

In another form of kinetic energy, called **electromagnetic radiation**, energy travels in the form of a *wave* as a result of changes in electrical and magnetic fields. There are many different forms of electromagnetic radiation (Figure 2-11), each having a different *wavelength* (the distance between successive peaks or troughs in the wave) and *energy content*. Forms of electromagnetic radiation with short wavelengths, such as gamma rays, X-rays, and ultraviolet (UV) radiation, have more energy than do forms with longer wavelengths, such as visible light and

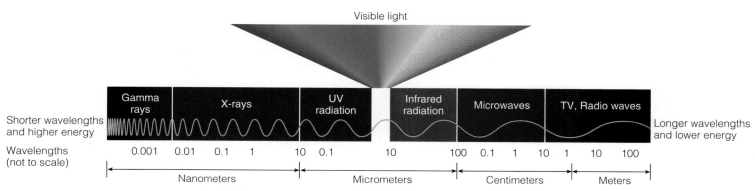

Animated Figure 2-11 The electromagnetic spectrum consists of a range of electromagnetic waves, which differ in wavelength (the distance between successive peaks or troughs) and energy content.

© Cengage Learning

Figure 2-12 The water stored in this reservoir behind a dam in the U.S. state of Tennessee has potential energy, which becomes kinetic energy when the water flows through channels built into the dam where it spins a turbine and produces electricity—another form of kinetic energy.

Bryan Busovicki/Shutterstock.com

infrared (IR) radiation. Visible light makes up most of the spectrum of electromagnetic radiation emitted by the sun.

The other major type of energy is **potential energy**, which is stored and potentially available for use. Examples of this type of energy include a rock held in your hand, the water in a reservoir behind a dam, and the chemical energy stored in the carbon atoms of coal or in molecules of food that you eat.

We can change potential energy to kinetic energy. If you hold this book in your hand, it has potential energy. However, if you drop it on your foot, the book's potential energy changes to kinetic energy. When a car engine burns gasoline, the potential energy stored in the chemical bonds of the gasoline molecules changes into kinetic energy that propels the car, and into heat that flows into the environment. When water in a reservoir flows through channels in a dam (Figure 2-12), its potential energy becomes kinetic energy that we can use to spin turbines in the dam to produce electricity—another form of kinetic energy.

Renewable and Nonrenewable Energy

Scientists divide energy resources into two major categories: renewable energy and nonrenewable energy. **Renewable energy** is energy gained from resources that are replenished by natural processes in a relatively short time. Examples are solar energy, available somewhere on the earth all of the time, firewood from trees, wind, moving water, and heat that comes from the earth's interior (geothermal energy).

Nonrenewable energy is energy from resources that can be depleted and are not replenished by natural processes within a human time scale. Examples are energy produced by the burning of oil, coal, and natural gas, and nuclear energy released when the nuclei of atoms of uranium fuel are split apart.

About 99% of the energy that we use to survive—the energy that keeps us warm and supports the plants that we and other organisms eat—comes from the sun at no cost to us. This is in keeping with the solar energy **principle of sustainability** (see Figure 1-2, p. 6 or back cover). Without this essentially inexhaustible solar energy, the earth would be frozen and life as we know it would not exist.

This direct input of solar energy produces several other indirect forms of renewable solar energy. Examples are *wind* (Figure 2-10), *hydropower* (falling and flowing water, Figure 2-12), and *biomass* (solar energy converted to chemical energy and stored in the tissues of trees and other plants) that can be burned to provide heat.

Commercial energy—energy that is sold in the marketplace—makes up the remaining 1% of the energy we use to supplement the earth's direct input of solar energy. About 87% of the commercial energy used in the world and 87% of that used in the United States come from burning nonrenewable **fossil fuels**, or oil, coal, and natural gas (Figure 2-13). They are called fossil fuels because they were formed over millions of years as layers of the decaying remains of ancient plants and animals were exposed to intense heat and pressure within the earth's crust.

Some Types of Energy Are More Useful Than Others

Energy quality is a measure of the capacity of a type of energy to do useful work. **High-quality energy** is concentrated energy that has a high capacity to do useful work. Examples are very high-temperature heat, concentrated sunlight, high-speed wind, and the energy released when we burn gasoline or coal.

Figure 2-13 Fossil fuels: Nonrenewable oil, coal, and natural gas (left, center, and right, respectively) supply most of the commercial energy that we use to supplement renewable energy from the sun.

Left: Andrea Danti/Shutterstock.com. Middle: Tom Mc Nemar/Shutterstock.com. Right: Olga Utlyakova/Shutterstock.com.

By contrast, **low-quality energy** is energy that is so dispersed that it has little capacity to do useful work. For example, the enormous number of moving molecules in the atmosphere or in an ocean together have such low-quality energy and such a low temperature that we cannot use them to move things or to heat things to high temperatures.

Energy Changes Are Governed by Two Scientific Laws

After observing and measuring energy being changed from one form to another in millions of physical and chemical changes, scientists have summarized their results in the **first law of thermodynamics**, also known as the **law of conservation of energy**. According to this scientific law, whenever energy is converted from one form to another in a physical or chemical change, no energy is created or destroyed (Concept 2-3A).

This scientific law tells us that no matter how hard we try or how clever we are, we cannot get more energy out of a physical or chemical change than we put in. This is one of nature's basic rules that we cannot violate.

Because the first law of thermodynamics states that energy cannot be created or destroyed, but only converted from one form to another, you may be tempted to think we will never have to worry about running out of energy. Yet if you fill a car's tank with gasoline and drive around or run your computer battery down, something has been lost. What is it? The answer is *energy quality,* the amount of energy available for performing useful work.

Thousands of experiments have shown that whenever energy is converted from one form to another in a physical or chemical change, we end up with lower-quality or less useable energy than we started with (Concept 2-3B). This is a statement of the **second law of thermodynamics**. The resulting low-quality energy usually takes the form of heat that flows into the environment. In the environment, the random motion of air or water molecules further disperses this heat, decreasing its temperature to the point where its energy quality is too low to do much useful work.

In other words, *when energy is changed from one form to another, it always goes from a more useful form to a less useful form.* No one has ever found a violation of this fundamental scientific law.

⊙ CONSIDER THIS. . .

CONNECTIONS Can We Recycle or Reuse Energy?

We can recycle and reuse various forms of matter such as paper and aluminum. However, because of the second law of thermodynamics, we can never recycle or reuse high-quality energy to perform useful work. Once the concentrated, high-quality energy in a serving of food, a full tank of gasoline, or a chunk of uranium nuclear fuel is released, it is degraded to low-quality heat and dispersed into the environment.

Scientists estimate that about 84% of the energy used in the United States is either unavoidably wasted because of the second law of thermodynamics (41%) or unnecessarily wasted (43%). Thus, thermodynamics teaches us an important lesson: the cheapest and quickest way to get more energy is to stop wasting almost half the energy we use. One way to do this is to improve our *energy efficiency,* which means getting more work out of the energy we use.

For example, only 5% of the electrical energy used by most incandescent lightbulbs (Figure 2-14, left) produces light, while the other 95% ends up as low-quality waste heat in the environment. There are much more efficient alternatives (Figure 2-14, center and right). We could also use more energy-efficient motor vehicle engines, power plants, and appliances, as we discuss in Chapter 16.

Figure 2-14 Three generations of increasingly energy-efficient lightbulbs: an incandescent bulb (left), a fluorescent bulb (center), and an LED bulb (right).

2-4 What Are Systems and How Do They Respond to Change?

CONCEPT 2-4

Systems have inputs, flows, and outputs of matter and energy, and feedback can affect their behavior.

Systems Respond to Change through Feedback Loops

A **system** is a set of components that function and interact in some regular way. The human body, a river, an economy, and the earth are all systems.

Most systems have the following key components: **inputs** of matter and energy from the environment, **flows** or **throughputs** of matter and energy within the system, and **outputs** of matter and energy to the environment (Figure 2-15) (Concept 2-4).

A system can become unsustainable if the throughputs of matter and energy resources exceed the abilities of the system's environment to provide the required resource inputs and to absorb or dilute the system's outputs of matter and energy. One of the most powerful tools used by environmental scientists to study how the components of systems interact is computer modeling (Science Focus 2.2).

When people ask you for *feedback,* they are usually seeking your response to something they said or did. They

might feed your response back into their mental processes to help them decide whether and how to change what they are saying or doing.

Similarly, most systems are affected in one way or another by **feedback**, any process that increases (positive feedback) or decreases (negative feedback) a change to a system (Concept 2-4). Such a process, called a **feedback loop**, occurs when an output of matter, energy, or information is fed back into the system as an input and leads to changes in that system. Note that, unlike the human brain, most systems do not consciously decide how to respond to feedback. Nevertheless, feedback can affect the behavior of systems.

A **positive feedback loop** causes a system to change further in the same direction (Figure 2-16). In the Hubbard Brook experiments, for example (**Core Case Study**), researchers found that when vegetation was removed from a stream valley, flowing water from precipitation caused erosion and losses of nutrients, which caused more vegetation to die. With even less vegetation to hold soil in place, flowing water caused even more erosion and nutrient loss, which caused even more plants to die.

Such accelerating positive feedback loops are of great concern in several areas of environmental science. One of the most alarming is the melting of polar ice, which has occurred as the temperature of the atmosphere has risen during the past few decades. As that ice melts, there is less of it to reflect sunlight, and more water that is exposed to sunlight. Because water is darker than ice, it absorbs more solar energy, making the polar areas warmer and causing the ice to melt faster, thus exposing more water. The melting of polar ice is therefore accelerating, causing a number of serious problems that we explore further in Chapter 19. If a system gets locked into an accelerating positive feedback loop, it can reach a breaking point that can destroy the system or suddenly change its behavior.

A **negative**, or **corrective, feedback loop** causes a system to change in the opposite direction from which it

Figure 2-15 A greatly simplified model of a system.

© Cengage Learning

Inputs (from environment) · Throughputs · Outputs (to environment)
Energy resources · Matter resources · Information · System processes · Work or products · Waste and pollution · Heat

Figure 2-16 A *positive feedback loop.* Decreasing vegetation in a valley causes increasing erosion and nutrient losses that in turn cause more vegetation to die, resulting in more erosion and nutrient losses. **Question:** Can you think of another positive feedback loop in nature?

© Cengage Learning

Decreasing vegetation...

...which causes more vegetation to die.

...leads to erosion and nutrient loss...

is moving. A simple example is a thermostat, a device that controls how often and how long a heating or cooling system runs (Figure 2-17). When the furnace in a house turns on and begins heating the house, we can set the thermostat to turn the furnace off when the temperature in the house reaches the set number. The house then stops getting warmer and starts to cool.

🔍 **CONSIDER THIS. . .**

THINKING ABOUT The Hubbard Brook Experiments and Feedback Loops

How might experimenters have employed a negative feedback loop to stop, or correct, the positive feedback loop that resulted in increasing erosion and nutrient losses in the Hubbard Brook experimental forest (**Core Case Study**)?

An important example of a negative feedback loop is the recycling and reuse of some resources such as aluminum. For example, an aluminum can is an output of a mining and manufacturing system. When we recycle the can, that output becomes an input. This reduces the amount of aluminum ore that we must mine and process to make aluminum cans. It also reduces the harmful environmental impacts of the mining and processing of aluminum ore. Such a negative feedback loop therefore can help reduce the harmful environmental impacts of human activities by decreasing the use of matter and energy resources and the amount of pollution and solid waste produced by the use of such resources. It is an example of applying the chemical cycling **principle of sustainability** (see Figure 1-2, p. 6 or back cover).

It Can Take a Long Time for a System to Respond to Feedback

A complex system will often show a **time delay**, or a lack of response during a period of time between the input of a feedback stimulus and the system's response to it. For example, scientists could plant trees in a degraded area such as the Hubbard Brook experimental forest to slow erosion and nutrient losses (**Core Case Study**). But it would take years for the trees and other vegetation to grow in order to accomplish this purpose.

Time delays can allow an environmental problem to build slowly until it reaches a *threshold level,* or **tipping point**—the point at which a fundamental shift in the

behavior of a system occurs. Reaching a tipping point is somewhat like stretching a rubber band. We can get away with stretching it to several times its original length. But at some point, we reach an irreversible tipping point where the rubber band breaks.

Prolonged delays dampen the negative feedback mechanisms that might slow, prevent, or halt environmental problems. In the Hubbard Brook example (**Core Case Study**), if soil erosion and nutrient losses had reached a certain point where the land could no longer support vegetation, then a tipping point would have been reached and it would have been too late to plant trees in order to try to restore the system. In Chapter 3, we discuss several major environmental tipping points that we may already have exceeded or will likely exceed in the near future.

System Effects Can Be Amplified through Synergy

A **synergistic interaction**, or **synergy**, occurs when two or more processes interact so that the combined effect is greater than the sum of their separate effects. For example, scientific studies reveal such an interaction between smoking and inhaling asbestos particles. Nonsmokers who

SCIENCE FOCUS 2.2

THE USEFULNESS OF MODELS

Scientists use *models,* or simulations, to learn how systems work. Mathematical models are especially useful when there are many interacting variables, when the time frame of events being modeled is long, and when controlled experiments are impossible or too expensive to conduct. One of our most powerful and useful technologies is mathematical modeling with high-speed supercomputers.

Making a mathematical model usually requires that the modelers go through three steps many times. *First,* they identify the major components of the system and how they interact, and develop mathematical equations that summarize this information. In succeeding runs, these equations are steadily refined. *Second,* modelers use a high-speed computer to describe the likely behavior of the system based on

the equations. *Third,* they compare the system's projected behavior with known information about its actual behavior. They keep doing this until the model mimics the past and current behavior of the system.

After building and testing a mathematical model, scientists can use it to project what is *likely* to happen under a variety of conditions. In effect, they use mathematical models to answer *if–then* questions: "*If* we do such and such, *then* what is likely to happen now and in the future?" This process can give us a variety of projections or scenarios of possible outcomes based on different assumptions. Mathematical models (like all other models) are no better than the assumptions on which they are built and the data we feed into them.

Scientists applied this process of model-building to the data collected by

researchers Bormann and Likens in their Hubbard Brook experiments (**Core Case Study**). These scientists created mathematical models based on the Hubbard Brook data to describe a forest and to project what might happen to soil nutrients and other variables if the forest were disturbed or cut down.

Other areas of environmental science in which computer modeling is becoming increasingly important include studies of the complex systems that govern climate change, deforestation, biodiversity loss, and the oceans.

Critical Thinking

What are two limitations of computer models? Does the existence of limitations mean that we should not rely on such models? Explain. What are the alternatives?

House warms

Temperature reaches desired setting and furnace goes off

Furnace on

House cools

Furnace off

Temperature drops below desired setting and furnace goes on

Figure 2-17 A *negative feedback loop.* When a house being heated by a furnace gets to a certain temperature, its thermostat is set to turn off the furnace, and the house begins to cool instead of continuing to get warmer. When the house temperature drops below the set point, this information is fed back to turn the furnace on until the desired temperature is reached again.

© Cengage Learning

are exposed to asbestos particles for long periods of time increase their risk of getting lung cancer fivefold. But people who smoke and are exposed to asbestos have 50 times the risk that nonsmokers have of getting lung cancer.

On the other hand, synergy can be helpful. You may find that you are able to study longer or run farther if you do these activities with a studying or running partner. Your physical and mental systems can do a certain amount of work on their own. But the synergistic effect of you and your partner working together can make your individual systems capable of accomplishing more in the same amount of time. When individuals work together to find and implement win-win solutions to environmental problems, they are applying one of the social science **principles of sustainability** (see Figure 1-5, p. 9 or back cover).

Big Ideas

- You cannot really throw anything away. According to the *law of conservation of matter*, no atoms are created or destroyed whenever matter undergoes a physical or chemical change. Thus, we cannot do away with matter; we can only change it from one physical state or chemical form to another.

- You cannot get something for nothing. According to the *first law of thermodynamics*, or the *law of conservation of energy*, whenever energy is converted from one form to another in a physical or chemical change, no energy is created or destroyed. This means that in causing such changes, we cannot get more energy out than we put in.

- You cannot break even. According to the *second law of thermodynamics*, whenever energy is converted from one form to another in a physical or chemical change, we always end up with lower-quality or less usable energy than we started with.

TYING IT ALL TOGETHER — The Hubbard Brook Forest Experiment and Sustainability

steve estvanik/Shutterstock.com

The controlled experiment discussed in the **Core Case Study** that opened this chapter revealed that clearing a mature forest degrades some of its natural capital (see Figure 1-3, p. 7, and photo at left). Specifically, the loss of trees and vegetation altered the ability of the forest to retain and recycle water and other critical plant nutrients—a crucial ecological function based on one of the three scientific **principles of sustainability** (see Figure 1-2, p. 6 or back cover). In other words, the uncleared forest (Figure 2-1, left) was a more sustainable system than a similar area of cleared forest (Figure 2-1, right).

This clearing of vegetation also violated the other two scientific **principles of sustainability**. For example, the cleared forest lost most of its plants that had produced food for the forest's animals by using solar energy and that supplied nutrients to the soil when they died. As a result, many of the forest's key nutrients were lost instead of being recycled. And the loss of plants and the resulting loss of animals reduced the life-sustaining biodiversity of the cleared forest.

Humans clear forests to harvest timber, grow crops, build settlements, and expand cities. The key question is, how far can we go in expanding our ecological footprints (see Figure 1-13, p. 14) without threatening the quality of life for our own species and for the other species that help to keep us alive and support our economies? To live more sustainably, we need to find and maintain a balance between preserving undisturbed natural systems and the natural resources and ecosystem services they provide and modifying other natural systems for our use.

Chapter Review

Core Case Study

1. Describe the controlled scientific experiment carried out in the Hubbard Brook Experimental Forest.

Section 2-1

2. What is the key concept for this section? What is **science**? List the steps involved in a scientific process. What is **data**? What is a **model**? Distinguish among a **scientific hypothesis**, a **scientific theory**, and a **scientific law (law of nature)**. Summarize Jane Goodall's scientific and educational achievements. What is **peer review** and why is it important?

3. Explain why scientific theories are not to be taken lightly and why people often use the term *theory* incorrectly. Explain why scientific theories and laws are the most important and most certain results of science.

4. Distinguish among **tentative science (frontier science)**, **reliable science**, and **unreliable science**. What are four limitations of science in general and environmental science in particular?

Section 2-2

5. What are the two key concepts for this section? What is **matter**? Distinguish between an **element** and a **compound** and give an example of each. Define **atoms**, **molecules**, and **ions** and give an example of each. What is the **atomic theory**? Distinguish among **protons (p)**, **neutrons (n)**, and **electrons (e)**. What is the **nucleus** of an atom? Distinguish between the **atomic number** and the **mass number** of an element. What is an **isotope**? What is **acidity**? What is **pH**?

6. What is a **chemical formula**? Distinguish between **organic compounds** and **inorganic compounds** and give an example of each. Distinguish among complex carbohydrates, proteins, nucleic acids, and lipids. What is a **cell**? Define **gene**, **trait**, and **chromosome**.

7. Define and distinguish between a **physical change** and a **chemical change (chemical reaction)** and give an example of each. What is a **nuclear change**? Define and explain the differences among natural **radioactive decay**, **nuclear fission**, and **nuclear fusion**. What is the **law of conservation of matter** and why is it important?

Section 2-3

8. What are the two key concepts for this section? What is **energy**? Distinguish between **kinetic energy** and **potential energy** and give an example of each. What is **heat (thermal energy)**? Define and give two examples of **electromagnetic radiation**. Define and distinguish between **renewable energy** and **nonrenewable energy**. What are **fossil fuels** and how are they formed? Why are they nonrenewable? What is **energy quality**? Distinguish between **high-quality energy** and **low-quality energy** and give an example of each. What is the **first law of thermodynamics (law of conservation of energy)** and why is it important? What is the **second law of thermodynamics** and why is it important? Explain why the second law means that we can never recycle or reuse high-quality energy.

9. Define and give an example of a **system**. Distinguish among the **inputs, flows (throughputs)**, and **outputs** of a system. Why are scientific models useful? What is **feedback**? What is a **feedback loop**? Distinguish between a **positive feedback loop** and a **negative (corrective) feedback loop** in a system, and give an example of each. Define **time delay** and **synergistic interaction (synergy)**, give an example of each, and explain how they can affect systems. What is a **tipping point**?

10. What are this chapter's *three big ideas*? Explain how the Hubbard Brook Experimental Forest controlled experiments illustrated the three scientific **principles of sustainability**.

Note: Key terms are in bold type.

Critical Thinking

1. What ecological lesson can we learn from the controlled experiment on the clearing of forests described in the **Core Case Study** that opened this chapter?

2. Suppose you observe that all of the fish in a pond have disappeared. Describe how you might use the scientific process described in the **Core Case Study** and in Figure 2-2 to determine the cause of this fish kill.

3. Respond to the following statements:
   a. Scientists have not absolutely proven that anyone has ever died from smoking cigarettes.
   b. The *natural greenhouse effect theory*—that certain gases such as water vapor and carbon dioxide help to warm the lower atmosphere—is not a reliable idea because it is just a scientific theory.

4. A tree grows and increases its mass. Explain why this is not a violation of the law of conservation of matter.

5. If there is no "away" where organisms can get rid of their wastes, due to the law of conservation of matter, why is the world not filled with waste matter?

6. Suppose someone wants you to invest money in an automobile engine, claiming that it will produce more energy than is found in the fuel used to run it. What would be your response? Explain.

7. Use the second law of thermodynamics to explain why we can use oil only once as a fuel, or in other words, why we cannot recycle its high-quality energy.

8. Imagine that for one day (a) you have the power to revoke the law of conservation of matter, and (b) you have the power to violate the first law of thermodynamics. For each of these scenarios, list three ways in which you would use your new power. Explain your choices.

Doing Environmental Science

Find (a) a newspaper or magazine article or a report on the Web that attempts to discredit a scientific hypothesis because it has not been proven, or (b) a report of a new scientific hypothesis that has the potential be controversial. Analyze the piece by doing the following: (1) determine its source (author or organization); (2) detect an alternative hypothesis, if any, that is offered by the author; (3) determine the primary objective of the author (for example, to debunk the original hypothesis, to state an alternative hypothesis, or to raise new questions, and so on); (4) summarize the evidence given by the author(s) for his or her position; and (5) compare the authors' evidence with the evidence for the original hypothesis. Write a report summarizing your analysis and compare it with those of your classmates.

Global Environment Watch Exercise

Search *Easter Island* and under the "News" section, click on "View All." Read current articles on what happened on Easter Island and explain in your own words whether or not the Easter Islanders were living sustainably. How could they have changed their ways in order to live more sustainably in their environment?

Data Analysis

Consider the graph on the right that shows loss of calcium from the experimental cutover site of the Hubbard Brook Experimental Forest (**Core Case Study**) compared with that of the control site. Note that this figure is very similar to Figure 2-5, which compares loss of nitrates from the two sites. After studying this graph, answer the questions below.

1. In what year did the calcium loss from the experimental site begin a sharp increase? In what year did it peak? In what year did it again level off?

2. In what year were the calcium losses from the two sites closest together? In the span of time between 1963 and 1972, did they ever get that close again?

3. Does this graph support the hypothesis that cutting the trees from a forested area causes the area to lose nutrients more quickly than leaving the trees in place? Explain.

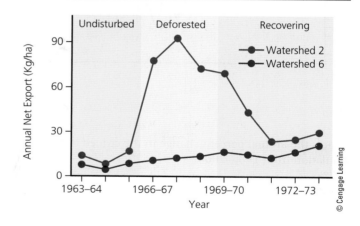

Use the graphs of the Chesapeake Bay, Maryland, striped bass and blue crab population data below to answer questions 1 and 2.

Striped Bass

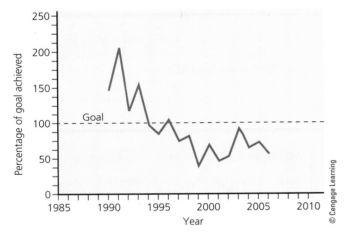

Blue Crab

1. What is the annual percent change in the striped bass population from 1985–2000?
   a. 7%
   b. 15%
   c. 100%
   d. 110%
   e. 1500%

2. What might be one possible reason for the greater fluctuations in the blue crab population?
   a. Blue crabs are more highly influenced by rainfall patterns than striped bass.
   b. Blue crabs have a much shorter life span (2 years) than striped bass and naturally have more variability in their population numbers.
   c. Global warming patterns have caused an increase in the Chesapeake Bay temperatures causing increased crab spawning rates.
   d. Increased sediment flow has disrupted striped bass migratory patterns causing a decline in their numbers.
   e. Eutrophic dead zones in the Chesapeake Bay causing low dissolved oxygen levels have caused a decline in crab numbers.

3. Which of the below is an example of a high-quality energy source?
   a. Atmospheric oxygen gas that humans need to survive
   b. Recycled components of discarded computers and other electronic devices
   c. Gasoline used to power automobiles
   d. Wood from a fallen tree used to keep a campfire burning on a cold night
   e. Iron molecules retrieved from the world's oceans

4. Which of the below is NOT an example of a macromolecule essential to life?
   a. Glucose (monosaccharide) formed by photosynthesis
   b. Keratin (protein) used in structural support
   c. Triglyceride used in cells to store energy
   d. DNA used by cells to carry the genetic code
   e. Starch used by most plants to store energy

5. The energy you use to walk from point A to point B is a result of several energy transformations from one form to another beginning with energy from the sun. The net result of these transformations is
   a. a loss of energy.
   b. fewer electrons.
   c. an increase in heat energy.
   d. an increase in useable energy.
   e. a decrease in entropy.

© Cengage Learning

6. The warming of the oceans causing less carbon dioxide to be soluble in them, and at the same time leaving more carbon dioxide in the atmosphere causing an increased warming of the atmosphere, is an example of a
   a. model of a system showing a paradigm shift.
   b. negative feedback system.
   c. positive feedback system.
   d. scientific principle of sustainability.
   e. natural law.

7. A scientist made the following statement in 2008:

   The striped bass Young of the Year (YOY) index for 2008, an annual measurement of the number of juvenile striped bass taken in the Maryland portion of the Chesapeake Bay, is one of the lowest recorded since data was recorded beginning in 1990. The 2008 YOY index for striped bass was 3.2 while the long-term average is 11.7.

   (Kennebec Journal Morning Sentinel; "Chesapeake striped bass population down for '08" 10/29/2008; http://morningsentinel. mainetoday.com/sports/stories/227216717.html)

   This statement is an example of
   a. possible human bias research introduced into scientific studies.
   b. scientists often trying to prove their results.
   c. a limitation of science in that actual populations are difficult to measure.
   d. a scientific statement that attempts to answer ethical questions of overfishing.
   e. an example of how a theory can become a law over time.

8. Which of the below organic molecules is **incorrectly** paired with its function or role?
   a. Methane–component of natural gas
   b. DDT–chlorinated hydrocarbon (insecticide)
   c. Starch–complex carbohydrate for energy storage
   d. Chromosomes–energy storage
   e. Atrazine–herbicide that blocks photosynthesis

9. The installation of ice booms on the Niagara River may well have changed the erosion pattern along the river. The irreversible loss of an island from over 100 acres in size to less than 3 acres is an example of
   a. a negative feedback loop.
   b. the principle of sustainability.
   c. a paradigm shift.
   d. the irreversible nature of human change.
   e. a tipping point.

Questions 10–14 refer to the description of an experiment below.

Ecologists designed an experiment to determine whether nitrates or phosphates are more limiting to algae growth. A lake in Canada (Lake 226) was divided into two equal-sized sections by a vinyl curtain. Each sub-basin of the lake was then fertilized, one with nitrates and carbon and the other with phosphates, nitrates, and carbon. The amount of phytoplankton in the lake was then measured in each side, with the side of the lake with phosphates added showing a greater rate of phytoplankton growth.

10. Frequently, studies in nature are difficult to set up and may lack some aspects of more traditional laboratory based experiments. What element of this experiment could be considered to be lacking?
    a. An independent variable   d. Repeated trials
    b. Constants   e. A dependent variable
    c. A control

11. Which of the below is the independent variable in this experiment?
    a. The location of the lakes
    b. The two lakes separated by a vinyl curtain
    c. The amount of phytoplankton growth
    d. The depth of the lakes
    e. The phosphate added to one side of the lake

12. Identify the constants in this experiment.
    a. The depth of the lakes
    b. The nitrates and carbon added to both lakes
    c. The amount of phosphate added
    d. The amount of phytoplankton growth
    e. The species of fish in both lakes compared to other lakes

13. Which of the below would be a valid hypothesis for this experiment?
    a. The amount of phytoplankton growth will vary with the depth of the lake.
    b. If more phosphate is added to one side of the lake, then there will be more phytoplankton growth on that side.
    c. If one side of the lake receives more sunlight, then that side will have more phytoplankton growth.
    d. Phytoplankton grows faster with more nutrients added.
    e. If scientists change the amount of nutrients, then the growth of phytoplankton will change.

14. What concept below best describes the process being studied in this experiment?
    a. The greenhouse effect
    b. Cultural eutrophication
    c. Denitrification in response to fertilizers
    d. Red tide events
    e. Acid rain deposition

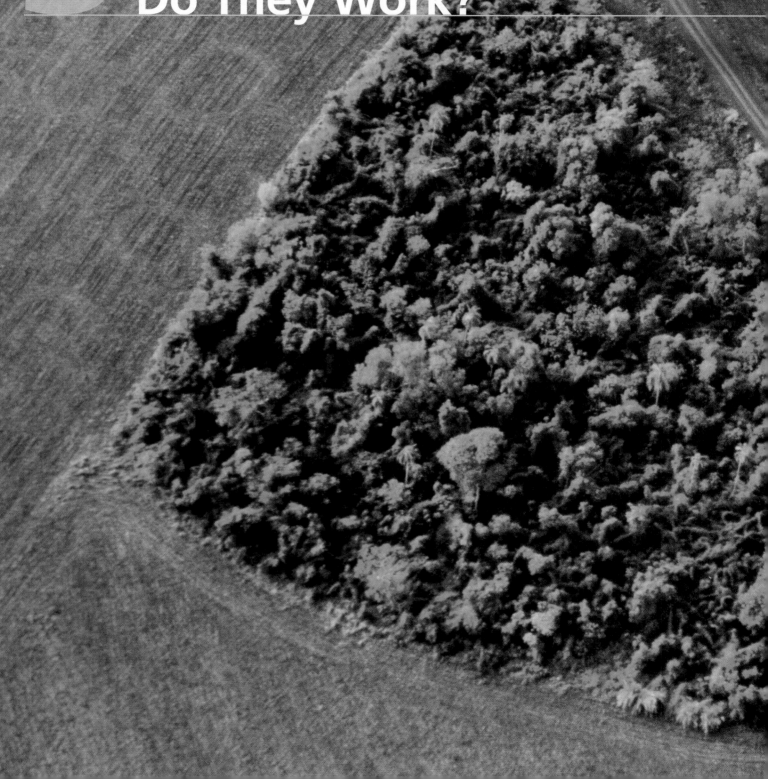

3 Ecosystems: What Are They and How Do They Work?

To halt the decline of an ecosystem, it is necessary to think like an ecosystem.

DOUGLAS WHEELER

Key Questions

3-1 How does the earth's life-support system work?

3-2 What are the major components of an ecosystem?

3-3 What happens to energy in an ecosystem?

3-4 What happens to matter in an ecosystem?

3-5 How do scientists study ecosystems?

Tropical rain forest remnant surrounded by farmland near Iguacu National Park, Brazil.

Frans Lanting/National Geographic Creative

Tropical rain forests are found near the earth's equator and contain an incredible variety of life. These lush forests are warm year-round and have high humidity and heavy rainfall almost daily. Although they cover only about 2% of the earth's land surface, studies indicate that they contain up to half of the world's known terrestrial plant and animal species. For these reasons, they make an excellent natural laboratory for the study of *ecosystems*—communities of organisms interacting with one another and with the physical environment of matter and energy in which they live.

So far, at least half of these forests have been destroyed or disturbed by humans cutting down trees, growing crops, grazing cattle, and building settlements (see chapter-opening photo and Figure 3-1), and the degradation of these centers of biodiversity is increasing. Ecologists warn that without strong protective measures, most of these forests will probably be gone or severely degraded by the end of this century.

So why should we care that tropical rain forests are disappearing? Scientists give three reasons. *First,* clearing these forests will reduce the earth's vital biodiversity by destroying or degrading the habitats of many of the unique plant and animal species that live in them, which could lead to their early extinction.

Second, the destruction of these forests is helping to accelerate atmospheric warming, and thus projected climate change (as discussed in more detail in Chapter 19). The reason is that eliminating large areas of trees faster than they can grow back decreases the forests' ability to remove human-generated emissions of the gas carbon dioxide (CO_2) from the atmosphere, which contributes to atmospheric warming and projected climate change.

Third, large-scale rain forest loss can change regional weather patterns in ways that can prevent the return of diverse tropical rain forests in cleared or severely degraded areas. When this irreversible *ecological tipping point* is reached, tropical rain forests in such areas become much less diverse tropical grasslands.

Ecologists study an ecosystem to learn how its variety of organisms interact with their living (*biotic*) environment of other organisms and with their nonliving (*abiotic*) environment of soil, water, other forms of matter, and energy, mostly from the sun. In effect, ecologists study *connections in nature.* For example, they have found that tropical rain forests and other ecosystems recycle nutrients and provide humans and other organisms with essential natural services and natural resources, in keeping with the chemical cycling **principle of sustainability** (see Figure 1-2, p. 6 or back cover). In this chapter, we look more closely at how tropical rain forests, and ecosystems in general, work. We also examine how human activities such as the clear-cutting of forests can disrupt the cycling of nutrients within ecosystems and the flow of energy through them.

17 Jun 1975

6 May 2003

Photos: United Nations Environment Programme

Figure 3-1 **Natural capital degradation:** Satellite image of the loss of tropical rain forest, cleared for farming, cattle grazing, and settlements, near the Bolivian city of Santa Cruz between June 1975 (left) and May 2003 (right).

3-1 How Does the Earth's Life-Support System Work?

CONCEPT 3-1A
The four major components of the earth's life-support system are the atmosphere (air), the hydrosphere (water), the geosphere (rock, soil, and sediment), and the biosphere (living things).

CONCEPT 3-1B
Life is sustained by the flow of energy from the sun through the biosphere, the cycling of nutrients within the biosphere, and gravity.

Earth's Life-Support System Has Four Major Components

The earth's life-support system consists of four main spherical systems (Figure 3-2) that interact with one another—the atmosphere (air), the hydrosphere (water), the geosphere (rock, soil, and sediment), and the biosphere (living things) (**Concept 3-1A**).

The **atmosphere** is a thin spherical envelope of gases surrounding the earth's surface. Its inner layer, the **troposphere**, extends only about 17 kilometers (11 miles) above sea level at the tropics and about 7 kilometers (4 miles) above the earth's north and south poles. It contains the air we breathe, consisting mostly of nitrogen (78% of the total volume) and oxygen (21%). Most of the remaining 1% of the air consists of water vapor, carbon dioxide, and methane. The next layer, reaching from 17 to 50 kilometers (11–31 miles) above the earth's surface, is called the **stratosphere**. Its lower portion holds enough ozone (O_3) gas to filter out about 95% of the sun's harmful *ultraviolet (UV) radiation*. This global sunscreen allows life to exist on the surface of the planet.

The **hydrosphere** is made up of all of the water on or near the earth's surface. It is found as *water vapor* in the atmosphere, as *liquid water* on the surface and underground, and as *ice*—polar ice, icebergs, glaciers, and ice in frozen soil-layers called *permafrost*. The oceans, which cover about 71% of the globe, contain about 97% of the earth's water.

The **geosphere** consists of the earth's intensely hot *core*, a thick *mantle* composed mostly of rock, and a thin outer *crust*. Most of the geosphere is located in the earth's interior. Its upper portion contains nonrenewable *fossil fuels*—coal, oil, and natural gas—and minerals that we use, as well as renewable soil chemicals (nutrients) that organisms need in order to live, grow, and reproduce.

The **biosphere** consists of the parts of the atmosphere, hydrosphere, and geosphere where life is found. If the earth were an apple, the biosphere would be no thicker than the apple's skin. One important goal of environmental science is to understand the key interactions that occur within this thin layer of air, water, soil, and organisms.

Three Factors Sustain the Earth's Life

Life on the earth depends on three interconnected factors (**Concept 3-1B**):

1. The *one-way flow of high-quality energy* from the sun, through living things in their feeding interactions, into the environment as low-quality energy (mostly heat dispersed into air or water at a low temperature), and eventually to outer space as heat (Figure 3-3). This is in keeping with the solar energy **principle of sustainability** (see Figure 1-2, p. 6 or back cover). No round-trips are allowed because high-quality energy cannot be recycled, according to the second law of thermodynamics (see Chapter 2, p. 43). As this solar energy interacts with carbon dioxide (CO_2) and other gases in the troposphere, it warms the troposphere—a process known as the **greenhouse effect**. Without this natural process, the earth would be too cold to support the forms of life we find here today.

2. The *cycling of nutrients* (the atoms, ions, and molecules needed for survival by living organisms) through parts of the biosphere. Because the earth does not get

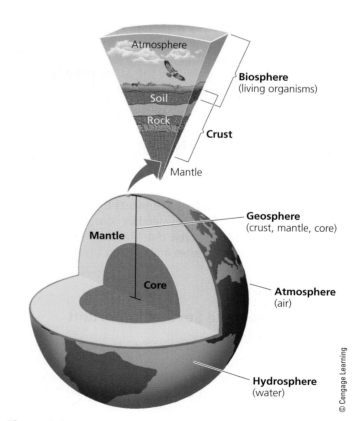

Figure 3-2 Natural capital: The earth consists of a land sphere (*geosphere*), an air sphere (*atmosphere*), a water sphere (*hydrosphere*), and a life sphere (*biosphere*) (**Concept 3-1A**).

© Cengage Learning

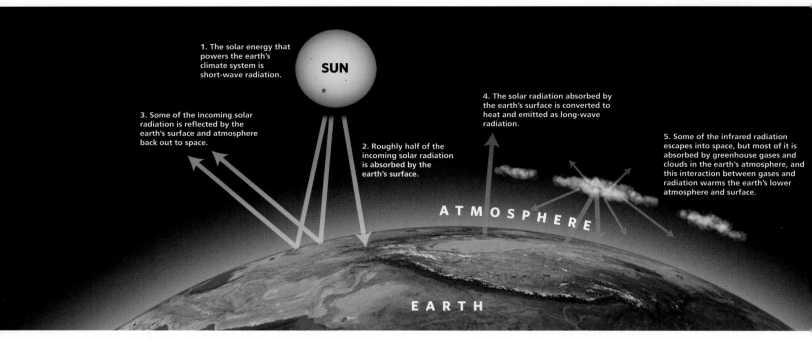

Animated Figure 3-3 **Greenhouse Earth.** High-quality solar energy flows from the sun to the earth. It is degraded to lower-quality energy (mostly heat) as it interacts with the earth's air, water, soil, and life-forms, and eventually returns to space. Certain gases in the earth's atmosphere retain enough of the sun's incoming energy as heat to warm the planet in what is known as the *greenhouse effect.*

© Cengage Learning

The following labels appear within the figure:

1. The solar energy that powers the earth's climate system is short-wave radiation.

2. Roughly half of the incoming solar radiation is absorbed by the earth's surface.

3. Some of the incoming solar radiation is reflected by the earth's surface and atmosphere back out to space.

4. The solar radiation absorbed by the earth's surface is converted to heat and emitted as long-wave radiation.

5. Some of the infrared radiation escapes into space, but most of it is absorbed by greenhouse gases and clouds in the earth's atmosphere, and this interaction between gases and radiation warms the earth's lower atmosphere and surface.

SUN

ATMOSPHERE

EARTH

---

significant inputs of matter from space, its essentially fixed supply of nutrients must be continually recycled to support life in keeping with the chemical cycling **principle of sustainability** (see Figure 1-2, p. 6 or back cover). The law of con-servation of matter (see Chapter 2, p. 40) governs this nutrient cycling process.

3. *Gravity,* which allows the planet to hold onto its atmo-sphere and helps to enable the movement and cycling of chemicals through air, water, soil, and organisms.

3-2 What Are the Major Components of an Ecosystem?

CONCEPT 3-2
Some organisms produce the nutrients they need, others get the nutrients they need by consuming other organisms, and some recycle nutrients back to producers by decomposing the wastes and remains of other organisms.

Ecosystems Have Several Important Components

Ecology is the science that focuses on how organisms interact with one another and with their nonliving envi-ronment of matter and energy. Scientists classify matter into levels of organization ranging from atoms to galax-ies. Ecologists study interactions within and among five of these levels—**organisms**, **populations**, **communities**, **ecosystems**, and the **biosphere**, which are illustrated and defined in Figure 3-4.

The biosphere and its ecosystems are made up of liv-ing *(biotic)* and nonliving *(abiotic)* components. Examples of nonliving components are water, air, nutrients, rocks, heat, and solar energy. Living components include plants, animals, microbes, and all other organisms. Figure 3-5 is a greatly simplified diagram of some of the living and non-living components of a terrestrial ecosystem.

Ecologists assign every type of organism in an ecosystem to a *feeding level,* or **trophic level**, depending on its source of food or nutrients. We can broadly classify the living organisms that transfer energy and nutrients from one tro-phic level to another within an ecosystem as *producers* and *consumers.*

Producers, sometimes called **autotrophs** (self-feeders), make the nutrients they need from compounds

Biosphere	Parts of the earth's air, water, and soil where life is found
Ecosystem	A community of different species interacting with one another and with their nonliving environment of matter and energy
Community	Populations of different species living in a particular place, and potentially interacting with each other
Population	A group of individuals of the same species living in a particular place
Organism	An individual living being
Cell	The fundamental structural and functional unit of life
Molecule	Chemical combination of two or more atoms of the same or different elements
Atom	Smallest unit of a chemical element that exhibits its chemical properties

© Cengage Learning

Animated Figure 3-4 This diagram illustrates some levels of the organization of matter in nature. Ecology focuses on the top five of these levels.

and energy obtained from their environment (Concept 3-2). In a process called **photosynthesis**, plants typically capture about 1% of the solar energy that falls on their leaves and use it in combination with carbon dioxide and water to form organic molecules, including energy-rich carbohydrates (such as glucose, $C_6H_{12}O_6$), which store the chemical energy plants need. Although

hundreds of chemical changes take place during photosynthesis, we can summarize the overall reaction as follows:

carbon dioxide + water + **solar energy** → glucose + oxygen

$$6\,CO_2 + 6\,H_2O + \textbf{solar energy} \rightarrow C_6H_{12}O_6 + 6\,O_2$$

(See Supplement 4, p. S17, for information on how to balance chemical equations such as this one.)

On land, most producers are trees and other green plants (Figure 3-6, left). In freshwater and ocean ecosystems, algae and aquatic plants growing near shorelines are the major producers (Figure 3-6, right). In open water, the dominant producers are *phytoplankton*—mostly microscopic organisms that float or drift in the water. The map in Figure 5 on p. S30 of Supplement 6 shows the distribution of green plants, or *plant biomass*, over the earth's land and water.

Most of the earth's producers are photosynthetic organisms that get their energy indirectly from the sun and convert it to chemical energy stored in their cells. However, some producer bacteria live in the dark in extremely hot water around fissures on the ocean floor. Their source of energy is heat from the earth's interior, or *geothermal energy,* and they represent an exception to the solar energy **principle of sustainability**.

Besides producers, all other organisms in an ecosystem are **consumers**, or **heterotrophs** ("other-feeders"), that cannot produce the nutrients they need through photosynthesis or other processes (Concept 3-2). They get their nutrients by feeding on other organisms (producers or other consumers) or their remains. In other words, all consumers (including humans) depend on producers for their nutrients.

There are several types of consumers. **Primary consumers**, or **herbivores** (plant eaters), are animals that eat mostly green plants. Examples are caterpillars, giraffes, and *zooplankton,* or tiny sea animals that feed on phytoplankton. **Carnivores** (meat eaters) are animals that feed on the flesh of other animals. Some carnivores such as spiders, lions (Figure 3-7), and most small fishes are **secondary consumers** that feed on the flesh of herbivores. Other carnivores such as tigers, hawks, and killer whales (orcas) are **tertiary** (or higher-level) **consumers** that feed on the flesh of other carnivores. Some of these relationships are shown in Figure 3-5. **Omnivores** such as pigs, rats, and humans eat plants and other animals.

Animated Figure 3-5 Key living and nonliving components of an ecosystem in a field.

© Cengage Learning

Labels in figure: Precipitation, Oxygen (O$_2$), Carbon dioxide (CO$_2$), Producer, Secondary consumer (fox), Primary consumer (rabbit), Producers, Water, Decomposers, Soluble mineral nutrients

🔍 **CONSIDER THIS...**

THINKING ABOUT What You Eat

When you ate your most recent meal, were you an herbivore, a carnivore, or an omnivore?

Decomposers are consumers that, in the process of obtaining their own nutrients, release nutrients from the wastes or remains of plants and animals and then return those nutrients to the soil, water, and air for reuse by producers (Concept 3-2). Most decomposers are bacteria and fungi (Figure 3-8, left). Other consumers, called **detritus feeders**, or **detritivores**, feed on the wastes or dead bodies of other organisms; these wastes are called *detritus* (dee-TRI-tus), which means debris. Examples are earthworms, some insects, and vultures (Figure 3-8, right).

Hordes of detritivores and decomposers can transform a fallen tree trunk into wood particles and, finally, into simple inorganic molecules that plants can absorb as nutrients (Figure 3-9). Thus, in natural ecosystems the wastes and dead bodies of organisms serve as resources for other organisms, as the nutrients that make life possible are continually recycled, in keeping with the chemical cycling **principle of sustainability** (see Figure 1-2, p. 6 or back cover). As a result, *there is very little waste of nutrients in nature*. However, only a small percentage of the huge amounts of waste materials that human societies produce are recycled. This explains why, unlike other species, the human species is a major violator of the chemical cycling principle.

Decomposers and detritivores, many of which are microscopic organisms (Science Focus 3.1), are the key to nutrient cycling. Without them, the planet would be overwhelmed with plant litter, animal wastes, dead animal bodies, and garbage.

Figure 3-6 Some producers such as trees and other plants (left) live on land. Others such as green algae (right) live in water.

Aleksander Bolbot/Shutterstock.com

Lance Rider/Shutterstock.com

Figure 3-7 This lioness (a carnivore) is feeding on a freshly killed zebra (an herbivore) in Kenya, Africa.

Organisms Get Their Energy in Different Ways

Producers, consumers, and decomposers use the chemical energy stored in glucose and other organic compounds to fuel their life processes. In most cells, this energy is released by **aerobic respiration**, which uses oxygen to convert glucose (or other organic nutrient molecules) back into carbon dioxide and water. The net effect of the hundreds of steps in this complex process is represented by the following chemical reaction:

glucose + oxygen → carbon dioxide + water + **energy**

$$C_6H_{12}O_6 + 6 O_2 \rightarrow 6 CO_2 + 6 H_2O + \textbf{energy}$$

Notice that, although the detailed steps differ, the net chemical change for aerobic respiration is the opposite of that for photosynthesis (p. 55).

Some decomposers get the energy they need by breaking down glucose (or other organic compounds) in the *absence* of oxygen. This form of cellular respiration is called **anaerobic respiration**, or **fermentation**. Instead of carbon dioxide and water, the end products of this process are compounds such as methane gas (CH_4, the main component of natural gas), ethyl alcohol (C_2H_6O), acetic acid ($C_2H_4O_2$, the key component of vinegar), and hydrogen sulfide (H_2S, a highly poisonous gas that smells like rotten eggs). Note that all organisms get their energy from aerobic or anaerobic respiration but only plants carry out photosynthesis.

To summarize, ecosystems and the biosphere are sustained through a combination of *one-way energy flow* from the sun through these systems and the *nutrient cycling* of key materials within them (**Concept 3-1B**)—in keeping with two of the scientific **principles of sustainability** (Figure 3-10).

Figure 3-8 These sulfur tuft fungi feeding on a tree stump (left) are decomposers. The vultures and Marabou storks (above), eating the carcass of an animal that was killed by another animal, are detritivores.

MANY OF THE WORLD'S MOST IMPORTANT ORGANISMS ARE INVISIBLE TO US

They are everywhere. Trillions can be found inside your body, on your skin, in a handful of soil, and in a cup of ocean water. They are *microbes,* or *microorganisms,* catchall terms for many thousands of species of bacteria, protozoa, fungi, and floating phytoplankton. Though most of them are too small to be seen with the naked eye, they are the biological rulers of the earth.

Microbes do not get the respect they deserve. Most of us view them primarily as threats to our health in the form of infectious bacteria or fungi that cause athlete's foot and other skin diseases, and protozoa that cause diseases such as malaria. But these harmful microbes are in the minority.

In 2012 scientist Philip Tarr and other researchers identified more than 10,000 species of bacteria, fungi, and other microbes that live in or on our bodies. Many of them provide us with vital services. Bacteria in our intestinal tracts help break down the food we eat, and microbes

in our noses help prevent harmful bacteria from reaching our lungs. According to recent research, the greater the diversity of the bacterial zoo in our stomachs, the better our health is likely to be. We are alive largely because of these multitudes of microbes toiling away completely out of sight.

Bacteria and other microbes help to purify the water we drink by breaking down any plant and animal wastes in the water. Bacteria and fungi (such as yeast) also help to produce foods such as bread, cheese, yogurt, soy sauce, beer, and wine. Bacteria and fungi in the soil decompose organic wastes into nutrients that can be taken up by plants that are then eaten by humans and many other plant eaters. Without these tiny creatures, we would go hungry and be up to our necks in waste matter.

Microbes, particularly phytoplankton in the ocean, also provide much of the plan-

et's oxygen and help to regulate the earth's temperature by removing some of the carbon dioxide produced when we burn coal, natural gas, and gasoline. Scientists are working on using microbes to develop new medicines and fuels. Genetic engineers are inserting genetic material into existing microorganisms to convert them to microbes that can be used to clean up polluted water and soils.

Some microorganisms assist us in controlling plant diseases and populations of insect species that attack our food crops. By relying more on these microbes for pest control, we could reduce the use of potentially harmful chemical pesticides. In other words, microbes are a vital part of the earth's natural capital.

Critical Thinking

What are two advantages that microbes have over humans for thriving in the world?

Figure 3-9 Various detritivores and decomposers (mostly fungi and bacteria) can "feed on" or digest parts of a log and eventually convert its complex organic chemicals into simpler inorganic nutrients that can be taken up by producers.

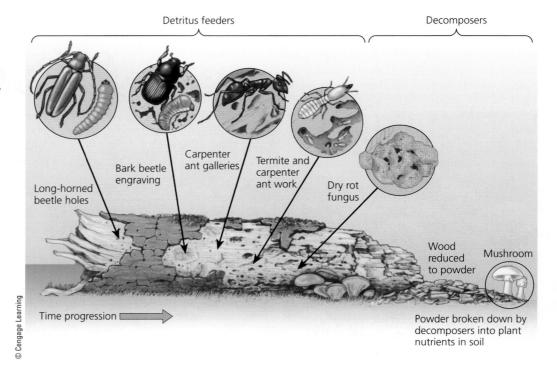

Detritus feeders

Decomposers

Long-horned beetle holes

Bark beetle engraving

Carpenter ant galleries

Termite and carpenter ant work

Dry rot fungus

Wood reduced to powder

Mushroom

Powder broken down by decomposers into plant nutrients in soil

Time progression

© Cengage Learning

Animated Figure 3-10 Natural capital: The main components of an ecosystem are energy, chemicals, and organisms. Nutrient cycling and the flow of energy—first from the sun, then through organisms, and finally into the environment as low-quality heat—link these components.

© Cengage Learning

🔍 **CONSIDER THIS. . .**

THINKING ABOUT Chemical Cycling and the Law of Conservation of Matter

Explain the relationship between chemical cycling in ecosystems and the law of conservation of matter (see Chapter 2, p. 40).

3-3 What Happens to Energy in an Ecosystem?

CONCEPT 3-3

As energy flows through ecosystems in food chains and webs, the amount of chemical energy available to organisms at each successive feeding level decreases.

Energy Flows through Ecosystems in Food Chains and Food Webs

The chemical energy stored as nutrients in the bodies and wastes of organisms flows through ecosystems from one trophic (feeding) level to another. For example, a plant uses solar energy to store chemical energy in a leaf. A caterpillar eats the leaf, a robin eats the caterpillar, and a hawk eats the robin. Decomposers and detritus feeders consume the wastes and remains of all members of this and other food chains and return their nutrients to the soil for reuse by producers.

A sequence of organisms, each of which serves as a source of nutrients or energy for the next, is called a **food chain**. It determines how chemical energy and nutrients move along the same pathways from one organism to another through the trophic levels in an ecosystem—primarily through photosynthesis, feeding, and decomposition—as shown in Figure 3-11. Every use and transfer of energy by organisms involves a loss of some degraded high-quality energy to the environment as heat, in accordance with the second law of thermodynamics.

In natural ecosystems, most consumers feed on more than one type of organism, and most organisms are eaten

or decomposed by more than one type of consumer. Because of this, organisms in most ecosystems form a complex network of interconnected food chains called a **food web** (Figure 3-12). We can assign trophic levels in food webs just as we can in food chains. Food chains and webs show how producers, consumers, and decomposers are connected to one another as energy flows through trophic levels in an ecosystem.

Usable Energy Decreases with Each Link in a Food Chain or Web

Each trophic level in a food chain or web contains a certain amount of **biomass**, the dry weight of all organic matter contained in its organisms. In a food chain or web, chemical energy stored in biomass is transferred from one trophic level to another.

As a result of the second law of thermodynamics, energy transfer through food chains and food webs is not very efficient because, with each energy transfer from one trophic level to another, some usable chemical energy is degraded and lost to the environment as low-quality heat. In other words, as energy flows through ecosystems in food chains and webs, there is a decrease in the amount of high-quality chemical energy available to organisms at each succeeding feeding level (Concept 3-3).

The percentage of usable chemical energy transferred as biomass from one trophic level to the next varies, depend-

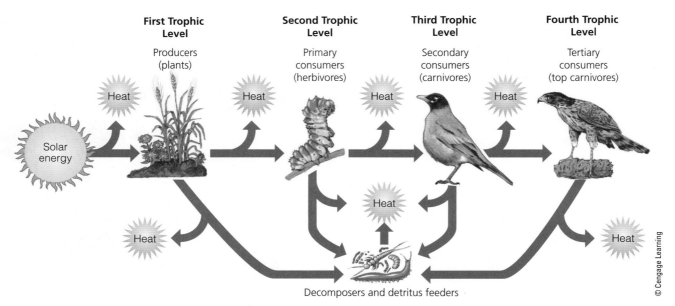

Animated Figure 3-11 In a food chain, chemical energy in nutrients flows through various trophic levels. **Question:** Think about what you ate for breakfast. At what level or levels on a food chain were you eating?

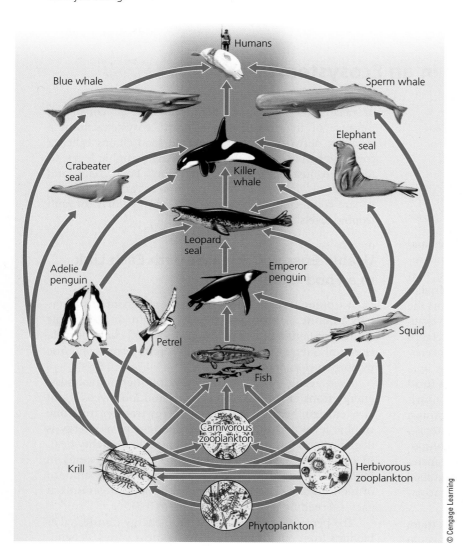

ing on what types of species and ecosystems are involved. The more trophic levels there are in a food chain or web, the greater is the cumulative loss of usable chemical energy as it flows through the trophic levels. The **pyramid of energy flow** in Figure 3-13 illustrates this energy loss for a simple food chain, assuming a 90% energy loss with each transfer.

CONSIDER THIS. . .

THINKING ABOUT Energy Flow and the Second Law of Thermodynamics

Explain the relationship between the second law of thermodynamics (see Chapter 2, p. 43) and the flow of energy through a food chain or web.

The large loss in chemical energy between successive trophic levels explains why food chains and webs rarely have more than four or five trophic levels. In most cases, too little chemical energy is left after four or five transfers to support organisms feeding at these high trophic levels. Thus, there are far fewer tigers in tropical rain forests (**Core Case Study**) and other areas than there are insects.

Animated Figure 3-12 This is a greatly simplified *food web* found in the southern hemisphere. The shaded middle area shows a simple food chain that is part of these complex interacting feeding relationships. Many more participants in the web, including an array of decomposer and detritus feeder organisms, are not shown here. **Question:** Can you imagine a food web of which you are a part? Try drawing a simple diagram of it.

Usable energy available
at each trophic level
(in kilocalories)

Tertiary consumers (human) 10

Secondary consumers (perch) 100

Primary consumers (zooplankton) 1,000

Producers (phytoplankton) 10,000

Heat

Heat

Heat

Heat

Decomposers

Heat

© Cengage Learning

Animated Figure 3-13 This model is a generalized *pyramid of energy flow* that shows the decrease in usable chemical energy available at each succeeding trophic level in a food chain or web. The model assumes that with each transfer from one trophic level to another, there is a 90% loss of usable energy to the environment in the form of low-quality heat. Calories and joules are used to measure energy. 1 kilocalorie = 1,000 calories = 4,184 joules. **Question:** Why is a vegetarian diet more energy efficient than a meat-based diet?

🔍 **CONSIDER THIS...**

CONNECTIONS Energy Flow and Feeding People

Energy flow pyramids explain why the earth could support more people if they all ate at lower trophic levels by consuming grains, vegetables, and fruits directly rather than passing such crops through another trophic level and eating herbivores such as cattle. About two-thirds of the world's people survive primarily by eating wheat, rice, and corn at the first trophic level because most of them cannot afford to eat much meat.

Some Ecosystems Produce Plant Matter Faster Than Others Do

The amount, or mass, of living organic material (biomass) that a particular ecosystem can support is determined by how much solar energy its producers can capture and store as chemical energy and by how rapidly they can do so. **Gross primary productivity (GPP)** is the *rate* at which an ecosystem's producers (usually plants) convert solar energy into chemical energy in the form of biomass found in their tissues. It is usually measured in terms of energy production per unit area over a given time span, such as kilocalories per square meter per year (kcal/m²/yr). The map in Figure 7 on p. S33 of Supplement 6 shows the gross primary productivity across North America.

To stay alive, grow, and reproduce, producers must use some of the chemical energy stored in the biomass they make for their own respiration. **Net primary productivity (NPP)** is the *rate* at which producers use photosynthesis to produce and store chemical energy *minus* the *rate* at which they use some of this stored chemical energy through aerobic respiration. NPP measures how fast producers can make the chemical energy that is stored in their tissues and that is potentially available to other organisms (consumers) in an ecosystem.

Gross primary productivity is similar to the *rate* at which you make money, or the number of dollars you earn per year. Net primary productivity is similar to the amount of money earned per year that you can spend after subtracting your work expenses such as the costs of transportation, clothes, food, and supplies.

Ecosystems and aquatic life zones differ in their NPP, as illustrated in Figure 3-14. Despite its low NPP, the open ocean produces more of the earth's biomass per year than any other ecosystem or life zone. This occurs because the enormous volume of the global ocean, which covers 71% of the earth's surface, contains huge numbers of producers, including phytoplankton.

Tropical rain forests have a very high NPP because they have a large number and variety of producer trees and other plants. When such forests are cleared (see chapter-opening photo and **Core Case Study**) or burned to make way for crops or for grazing cattle, there is a sharp drop in the NPP and a loss of many of the diverse array of plant and animal species.

As we have seen, producers are the source of all nutrients in an ecosystem that are available for the producers themselves and for the consumers and decomposers that feed on them. Only the biomass represented by NPP is available as nutrients for consumers, and they use only a portion of it. Thus, *the planet's NPP ultimately limits the number of consumers (including humans) that can survive on the earth.* This is an important lesson from nature.

🔍 **CONSIDER THIS...**

CONNECTIONS Humans and Earth's Net Primary Productivity

Peter Vitousek, Stuart Rojstaczer, and other ecologists estimate that humans now use, waste, or destroy about 38% of the earth's total potential NPP. This is a remarkably high value, considering that the human population makes up less than 1% of the total biomass of all of the earth's consumers that depend on producers for their nutrients.

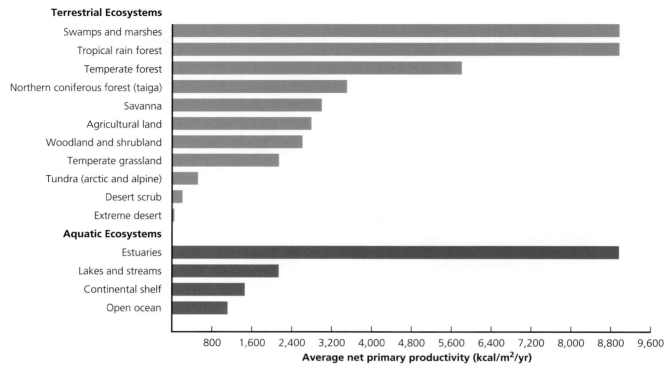

Terrestrial Ecosystems

Average net primary productivity (kcal/m²/yr)

Figure 3-14 The estimated annual average *net primary productivity* in major life zones and ecosystems is expressed in this graph as kilocalories of energy produced per square meter per year (kcal/m²/yr). **Question:** What are the three most productive and the three least productive systems?

(Compiled by the authors using data from R. H. Whittaker, *Communities and Ecosystems*, 2nd ed., New York: Macmillan, 1975.)

3-4 What Happens to Matter in an Ecosystem?

CONCEPT 3-4
Matter, in the form of nutrients, cycles within and among ecosystems and the biosphere, and human activities are altering these chemical cycles.

Nutrients Cycle within and among Ecosystems

The elements and compounds that make up nutrients move continually through air, water, soil, rock, and living organisms within ecosystems, as well as in the biosphere in cycles called **nutrient cycles**, or *biogeochemical cycles* (literally, life-earth-chemical cycles). This is in keeping with the chemical cycling **principle of sustainability** (see Figure 1-2, p. 6 or back cover). These cycles, which are driven directly or indirectly by incoming solar energy and by the earth's gravity, include the hydrologic (water), carbon, nitrogen, phosphorus, and sulfur cycles. They are important components of the earth's natural capital (see Figure 1-3, p. 7), and human activities are altering them (Concept 3-4).

As nutrients move through their biogeochemical cycles, they may accumulate in certain portions of

the cycles and remain there for varying periods of time. These temporary storage sites such as the atmosphere, the oceans and other bodies of water, and underground deposits are called *reservoirs*.

🔍 CONSIDER THIS. . .

CONNECTIONS Nutrient Cycles and Life
Nutrient cycles connect past, present, and future forms of life. Some of the carbon atoms in your skin may once have been part of an oak leaf, a dinosaur's skin, or a layer of limestone rock. Your grandmother, rock star Bono, or a hunter–gatherer who lived 25,000 years ago may have inhaled some of the nitrogen molecules you just inhaled.

The Water Cycle

Water (H_2O) is an amazing substance (Science Focus 3.2) that is necessary for life on the earth. The **hydrologic cycle**, or **water cycle**, collects, purifies, and distributes the earth's fixed supply of water, as shown in Figure 3-15.

The water cycle is powered by energy from the sun and involves three major processes—evaporation, precipitation, and transpiration. Incoming solar energy causes *evaporation*, or the conversion of water from liquid to vapor from the earth's oceans, lakes, rivers, and soil. This

Condensation

Ice and snow

Condensation

Transpiration from plants

Precipitation to land

Evaporation of surface water

Evaporation from ocean

Runoff

Lakes and reservoirs

Precipitation to ocean

Runoff

Increased runoff on land covered with crops, buildings, and pavement

Infiltration and percolation into aquifer

Increased runoff from cutting forests and filling wetlands

Runoff

Groundwater in aquifers

Overpumping of aquifers

Water pollution

Runoff

Ocean

☐ Natural process
☐ Natural reservoir
■ Human impacts
▶ Natural pathway
▶ Pathway affected by human activities

© Cengage Learning

Animated Figure 3-15 **Natural capital:** This is a simplified model of the *water cycle*, or *hydro-logic cycle*, in which water circulates in various physical forms within the biosphere. The red arrows and boxes identify major effects of human activities on this cycle. **Question:** What are three ways in which your lifestyle directly or indirectly affects the hydrologic cycle?

water vapor rises into the atmosphere, where it condenses into droplets, and gravity then draws the water back to the earth's surface as *precipitation* (rain, snow, sleet, and dew). Over land, about 90% of the water that reaches the atmosphere evaporates from the surfaces of plants, through a process called *transpiration,* and from the soil.

Water that returns to the earth's surface as precipitation takes various paths. Most precipitation falling on terrestrial ecosystems becomes **surface runoff.** This water flows into streams, which eventually carry water to lakes and oceans, from which it can evaporate to repeat the cycle. Some surface water also seeps into the upper layers

of soils where it is used by plants, and some evaporates from the soils back into the atmosphere.

Some precipitation is converted to ice that is stored in *glaciers* (Figure 3-16), usually for long periods of time. Some precipitation sinks through soil and permeable rock formations to underground layers of rock, sand, and gravel called **aquifers,** where it is stored as **groundwater.**

A small amount of the earth's water ends up in the living components of ecosystems. As producers, plants absorb some of this water through their roots, most of which evaporates from plant leaves back into the atmosphere during transpiration. Some of the water that plants absorb combines

WATER'S UNIQUE PROPERTIES

Water (H_2O) is a remarkable substance with a unique combination of properties:

- *Forces of attraction,* called *hydrogen bonds* (Figure 3-A), *hold water molecules together*—the major factor determining water's distinctive properties.
- *Water exists as a liquid over a wide temperature range because of the hydrogen bonds between its molecules.* If water did not have a high boiling point, the oceans would have evaporated long ago.
- *Liquid water changes temperature slowly because it can store a large amount of heat without a large change in its own temperature.* This high heat storage capacity helps protect living organisms from temperature changes, moderates the earth's climate, and makes water an excellent coolant for car engines and power plants.
- *It takes a large amount of energy to evaporate water because of its hydrogen bonds.* Water absorbs large amounts of heat as it changes into water vapor and releases this heat as the vapor condenses back to liquid water. This helps to distribute heat throughout the world and to determine regional and local climates. It also makes evaporation a cooling process—explaining why you feel cooler when perspiration evaporates from your skin.
- *Liquid water can dissolve a variety of compounds* (see Figure 2, p. S13, in Supplement 4). It carries dissolved nutrients into the tissues of living organisms,

flushes waste products out of those tissues, serves as an all-purpose cleanser, and helps to remove and dilute the water-soluble wastes of civilization. This property also means that water-soluble wastes can easily pollute water.

- *Water filters out wavelengths of the sun's ultraviolet radiation that would harm some aquatic organisms.* However, down to a certain depth, it is transparent to sunlight that is necessary for photosynthesis.
- *The hydrogen bonds between water molecules also allow liquid water to adhere to a solid surface.* This enables narrow columns of water to rise through a plant from its roots to its leaves (a process called *capillary action*).
- *Unlike most liquids, water expands when it freezes.* This means that ice floats on water because it has a lower density (mass per unit of volume) than liquid water has. Otherwise, lakes and streams in cold climates would freeze solid, losing most of their aquatic life. Because water expands on freezing, it can break pipes, crack a car's engine

block (if it doesn't contain antifreeze), break up pavement, and fracture rocks.

Slightly negative charge

Hydrogen bonds

Slightly positive charge

Figure 3-A *Hydrogen bond:* Because of the slightly unequal sharing of electrons in an H_2O molecule, its oxygen atoms hold a slightly negative charge and its hydrogen atoms hold a slightly positive charge. Since opposite electrical charges attract one another, the hydrogen atoms of one water molecule are attracted to the oxygen atoms in other water molecules. These fairly weak forces of attraction *between* molecules (represented by the dashed lines) are called *hydrogen bonds.*

© Cengage Learning

Critical Thinking

Pick two of the special properties listed above and, for each property, explain how life on the earth would be different if it did not exist.

with carbon dioxide during photosynthesis to produce high-energy organic compounds such as carbohydrates. Eventually these compounds are broken down in plant cells, which release the water back into the environment. Consumers get their water from their food and by drinking it.

Because water dissolves many nutrient compounds, it is a major medium for transporting nutrients within and between ecosystems. Water is also the primary sculptor of the earth's landscape as it flows over and wears down

rock over millions of years with results such as the Grand Canyon in the southwestern United States.

Throughout the hydrologic cycle, many natural processes purify water. Evaporation and subsequent precipitation act as a natural distillation process that removes impurities dissolved in water. Water flowing aboveground through streams and lakes, and underground in aquifers, is naturally filtered and partially purified by chemical and biological processes—mostly by the actions of decomposer

Figure 3-16 This glacier in Patagonia, Argentina, stores water for long periods of time as part of the hydrologic cycle.

bacteria—as long as these natural processes are not overloaded. Thus, *the hydrologic cycle can be viewed as a cycle of natural renewal of water quality.*

Only about 0.024% of the earth's vast water supply is available to humans and other species as liquid freshwater in accessible groundwater deposits and in lakes, rivers, and streams (see Figure 25, p. S50, in Supplement 6). The rest is too salty for us to use, is stored as ice, or is too deep underground to extract at affordable prices using current technology.

Humans alter the water cycle in three major ways (see the red arrows and boxes in Figure 3-15). *First,* we withdraw large quantities of freshwater from rivers, lakes, and aquifers sometimes faster than nature can replace it. As a result, some aquifers are being depleted and some rivers no longer flow to the ocean. *Second,* we clear vegetation from land for agriculture, mining, road building, and other activities, and cover much of the land with buildings, concrete, and asphalt. This increases runoff and reduces infiltration that would normally recharge groundwater supplies. *Third,* we drain and fill wetlands for farming and urban development. Left undisturbed, wetlands provide the natural service of flood control, acting like sponges to absorb and hold overflows of water from drenching rains or rapidly melting snow.

CONSIDER THIS...

CONNECTIONS Clearing a Rain Forest Can Affect Local Weather and Climate

Clearing vegetation can alter weather patterns by reducing transpiration, especially in dense tropical rain forests (**Core Case Study**). Because so many plants in such a forest transpire water into the atmosphere, vegetation is the primary source of local rainfall. Cutting down large areas of forest raises ground temperatures (because it reduces shade) and can reduce local rainfall so much that the forest cannot grow back. If this occurs over a large area for three or more decades, the climate of the affected area can change, and much less diverse tropical grasslands can replace biologically diverse forests.

The Carbon Cycle

Carbon is the basic building block of the carbohydrates, fats, proteins, DNA, and other organic compounds necessary for life. Various compounds of carbon circulate through the biosphere, the atmosphere, and parts of the hydrosphere, in the **carbon cycle** shown in Figure 3-17.

The carbon cycle is based on carbon dioxide (CO_2) gas, which makes up about 0.039% of the volume of the earth's atmosphere and is also dissolved in water. Carbon dioxide (along with water vapor in the water cycle) is a key component of the atmosphere's thermostat. If the carbon cycle removes too much CO_2 from the atmosphere, the atmosphere will cool, and if it generates too much CO_2, the atmosphere will get warmer. Thus, even slight changes in this cycle caused by natural or human factors can affect the earth's climate and ultimately help to determine the types of life that can exist in various places.

Terrestrial producers remove CO_2 from the atmosphere and aquatic producers remove it from the water. These producers then use the CO_2 and photosynthesis to produce complex carbohydrates such as glucose ($C_6H_{12}O_6$).

The cells in oxygen-consuming producers, consumers, and decomposers then carry out aerobic respiration. This process breaks down glucose and other complex organic compounds to produce CO_2 in the atmosphere and water for reuse by producers. Because of this linkage between *photosynthesis* in producers and *aerobic respiration* in producers, consumers, and decomposers, these processes circulate carbon in the biosphere. Oxygen and hydrogen—the other elements in carbohydrates—cycle almost in step with carbon.

Some carbon atoms take a long time to recycle. Decomposers release the carbon stored in the bodies of dead organisms on land back into the air as CO_2, which can remain in the atmosphere for 100 years or more. And in water, decomposers release carbon that can be stored as insoluble carbonates in bottom sediment for very long periods of time. Indeed, marine sediments are the earth's largest store of carbon. Over millions of years, buried deposits of dead plant matter and bacteria were compressed between layers of sediment, where high pressure and heat converted them to carbon-containing *fossil fuels* such as coal, oil, and natural gas (see Figure 2-13, p. 43, and Figure 3-17).

This long-stored carbon was not released to the atmosphere as CO_2 until fossil fuels were extracted and burned. Small portions of these deposits have also been exposed to air by long-term geological processes that can take place over millions of years. However, in only a few hundred years, we have extracted and burned huge quantities of fossil fuels that took millions of years to form. This is why, on a human time scale, fossil fuels are nonrenewable resources.

We are altering the carbon cycle (see the red arrows in Figure 3-17), mostly by adding large amounts of carbon dioxide to the atmosphere (see Figure 14, p. S70, Supplement 7) when we burn carbon-containing fossil fuels (especially coal to produce electricity). We also alter the cycle by clearing carbon-absorbing vegetation from forests,

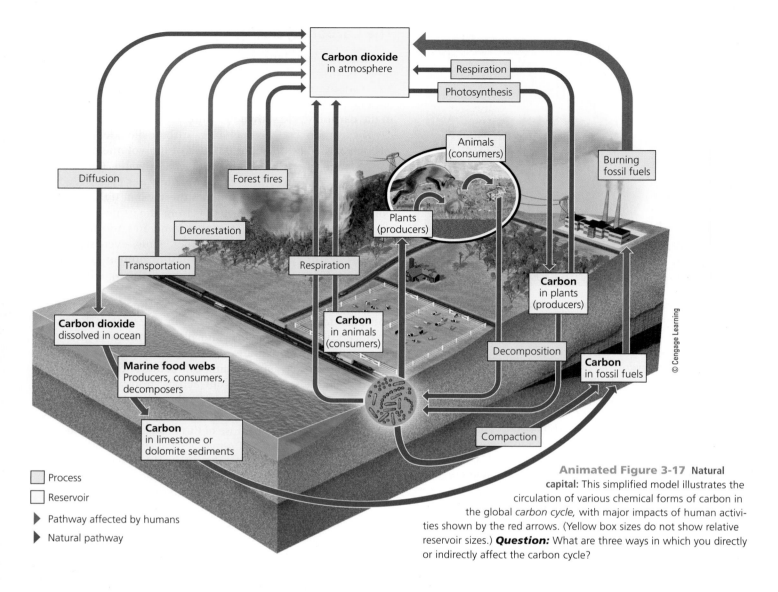

Diffusion

Carbon dioxide
in atmosphere

Respiration

Photosynthesis

Forest fires

Animals
(consumers)

Burning
fossil fuels

Deforestation

Plants
(producers)

Transportation

Respiration

Carbon
in plants
(producers)

Carbon dioxide
dissolved in ocean

Carbon
in animals
(consumers)

Marine food webs
Producers, consumers,
decomposers

Decomposition

Carbon
in fossil fuels

© Cengage Learning

Carbon
in limestone or
dolomite sediments

Compaction

Process

Reservoir

Pathway affected by humans

Natural pathway

Animated Figure 3-17 **Natural capital:** This simplified model illustrates the circulation of various chemical forms of carbon in the global *carbon cycle*, with major impacts of human activities shown by the red arrows. (Yellow box sizes do not show relative reservoir sizes.) **Question:** What are three ways in which you directly or indirectly affect the carbon cycle?

especially tropical forests, faster than it can grow back (**Core Case Study**). Human activities are altering both the rate of energy flow and the cycling of nutrients within the carbon cycle. In other words, humanity has a large and growing *carbon footprint* that makes up a significant part of our overall ecological footprints (see Figure 1-13, p. 14).

Computer models of the earth's climate systems indicate that increased concentrations of atmospheric CO_2 and other greenhouse gases including methane (CH_4) are very likely to warm the atmosphere by enhancing the planet's natural greenhouse effect (which is why they are called greenhouse gases), and thus to change the earth's climate during this century, as we discuss in Chapter 19.

The Nitrogen Cycle: Bacteria in Action

The major reservoir for nitrogen is the atmosphere. Chemically unreactive nitrogen gas (N_2) makes up 78% of the volume of the atmosphere. Nitrogen is a crucial component of proteins, many vitamins, and nucleic acids such as DNA (see Figure 8, p. S17, in Supplement 4). However, N_2 cannot be absorbed and used directly as a nutrient by multicellular plants or animals.

Two natural processes convert, or *fix,* N_2 into compounds that plants and animals can use as nutrients. One such process is electrical discharges, or lightning, taking place in the atmosphere. The other takes place in aquatic systems, in soil, and in the roots of some plants, where specialized bacteria, called *nitrogen-fixing bacteria,* complete this conversion as part of the **nitrogen cycle**, which is depicted in Figure 3-18.

The nitrogen cycle consists of several major steps. In *nitrogen fixation,* specialized bacteria in soil, as well as blue-green algae (cyanobacteria) in aquatic environments, combine gaseous N_2 with hydrogen to make ammonia (NH_3). The bacteria use some of the ammonia they produce as a nutrient and excrete the rest into the soil or water. Some of the ammonia is converted to ammonium ions (NH_4^+) that plants can use as a nutrient.

Ammonia that is not taken up by plants may undergo *nitrification.* In this process, specialized soil bacteria such as

Legend:
- [ ] Process
- [ ] Reservoir
- ▶ Pathway affected by humans
- ▶ Natural pathway

Nitrogen in atmosphere

Denitrification by bacteria

Nitrogen fixation by bacteria

Electrical storms

Nitrogen oxides from burning fuel and using inorganic fertilizers

Nitrogen in animals (consumers)

Commercial nitrogen fertilizer

Nitrogen in plants (producers)

Nitrates from fertilizer runoff and decomposition

Decomposition

Uptake by plants

Nitrate in soil

Nitrogen loss to deep ocean sediments

Nitrogen in ocean sediments

Nitrification by bacteria

Ammonia in soil

Animated Figure 3-18 **Natural capital:** This is a simplified model of the circulation of various chemical forms of nitrogen in the global *nitrogen cycle*, with major human impacts shown by the red arrows. (Yellow box sizes do not show relative reservoir sizes.) **Question:** What are two ways in which the carbon cycle and the nitrogen cycle are linked?

the Rhizobium bacteria convert most of the NH_3 and NH_4^+ in the soil to *nitrate ions* (NO_3^-), which are easily taken up by the roots of plants. The plants then use these forms of nitrogen to produce various amino acids, proteins, nucleic acids, and vitamins. Animals that eat plants eventually consume these nitrogen-containing compounds, as do detritus feeders and decomposers.

Plants and animals return nitrogen-rich organic compounds to the environment as both wastes and cast-off particles of tissues such as leaves, skin, or hair, and through their bodies when they die and are decomposed or eaten by detritus feeders. Vast armies of specialized decomposer bacteria convert this detritus into simpler nitrogen-containing inorganic compounds such as ammonia (NH_3) and water-soluble salts containing ammonium ions (NH_4^+).

In *denitrification,* specialized bacteria in waterlogged soil and in the bottom sediments of lakes, oceans, swamps, and bogs convert NH_3 and NH_4^+ back into nitrate ions, and then into nitrogen gas (N_2), which is released to the atmosphere to begin the nitrogen cycle again.

We intervene in the nitrogen cycle in several ways (see the red arrows in Figure 3-18) that affect what goes on in the atmosphere and in aquatic systems. We

add large amounts of nitrogen oxides to the atmosphere when we burn gasoline and other fuels and when we use commercial nitrate fertilizers. For example, we add nitric oxide (NO) into the atmosphere when N_2 and O_2 combine as we burn any fuel at high temperatures, such as in car, truck, and jet engines. In the atmosphere, this gas can be converted to nitrogen dioxide gas (NO_2) and nitric acid vapor (HNO_3), which can return to the earth's surface as damaging *acid deposition,* commonly called *acid rain.*

We also add nitrous oxide (N_2O) to the atmosphere through the action of anaerobic bacteria on commercial nitrogen-containing fertilizer or organic animal manure applied to the soil. This greenhouse gas can warm the atmosphere and take part in reactions that deplete stratospheric ozone, which keeps most of the sun's harmful ultraviolet radiation from reaching the earth's surface (as we discuss in Chapter 18).

We are also removing large amounts of nitrogen (N_2) from the atmosphere faster than the cycle can replace it. This N_2 is removed primarily for use in industrial processes that convert it to ammonia (NH_3) and ammonium ions (NH_4^+) used in fertilizers.

Process

Reservoir

▶ Pathway affected by humans

▶ Natural pathway

Phosphates in sewage

Phosphates in mining waste

Runoff

Runoff

Phosphates in fertilizer

Runoff

Plate tectonics

Sea birds

Erosion

Phosphate in rock (fossil bones, guano)

Ocean food webs

Animals (consumers)

Phosphate dissolved in water

Phosphate in shallow ocean sediments

Phosphate in deep ocean sediments

© Cengage Learning

Plants (producers)

Bacteria

Figure 3-19 **Natural capital:** This is a simplified model of the circulation of various chemical forms of phosphorus (mostly phosphates) in the *phosphorus cycle*, with major human impacts shown by the red arrows. (Yellow box sizes do not show relative reservoir sizes.) **Questions:** What are two ways in which the phosphorus cycle and the nitrogen cycle are linked? What are two ways in which the phosphorus cycle and the carbon cycle are linked?

We upset the nitrogen cycle in aquatic ecosystems by adding excess nitrates (NO_3^-) to bodies of water through agricultural runoff of fertilizers and animal manure and through discharges from municipal sewage treatment systems. This can cause excessive growth of algae that can disrupt aquatic systems.

According to the 2005 Millennium Ecosystem Assessment, since 1950, human activities have more than doubled the annual release of nitrogen from the land into the rest of the environment. This is mostly from the greatly increased use of inorganic fertilizers to grow crops. The amount released is projected to double again by 2050 (see Figure 16, p. S70, in Supplement 7).

This excessive input of nitrogen into the air and water contributes to pollution and other problems to be discussed in later chapters. Nitrogen overload is a serious and growing local, regional, and global environmental problem that has attracted little attention. Princeton University physicist Robert Socolow calls for countries around the world to work out some type of nitrogen management agreement to help prevent this problem from reaching crisis levels.

🔍 **CONSIDER THIS. . .**

THINKING ABOUT **The Nitrogen Cycle and Tropical Deforestation**

What effects might the clearing and degrading of tropical rain forests (**Core Case Study**) have on the nitrogen cycle in these forest ecosystems and on any nearby aquatic systems? (See Figure 2-1, p. 30, and Figure 2-5, p. 37.)

The Phosphorus Cycle

Compounds of phosphorus (P) circulate through water, the earth's crust, and living organisms in the **phosphorus cycle**, depicted in Figure 3-19. Most of these compounds contain *phosphate* ions (PO_4^{3-}), which serve as an important nutrient. In contrast to the cycles of water, carbon, and nitrogen, the phosphorus cycle does not include the atmosphere. The major reservoir for phosphorus is phosphate salts containing PO_4^{3-} in terrestrial rock formations and ocean-bottom sediments. The phosphorus cycle is slow compared to the water, carbon, and nitrogen cycles.

As water runs over exposed rocks, it slowly erodes away inorganic compounds that contain phosphate ions.

Labels in figure:
- Sulfur dioxide in atmosphere
- Smelting
- Burning coal
- Refining fossil fuels
- Sulfuric acid and Sulfate deposited as acid rain
- Dimethyl sulfide a bacteria by-product
- Sulfur in animals (consumers)
- Mining and extraction
- Sulfur in plants (producers)
- Uptake by plants
- Sulfur in ocean sediments
- Decay
- Decay
- Decay
- Sulfur in soil, rock, and fossil fuels
- © Cengage Learning

Legend:
- Process
- Reservoir
- Pathway affected by humans
- Natural pathway

Animated Figure 3-20 Natural capital: This is a simplified model of the circulation of various chemical forms of sulfur in the *sulfur cycle,* with major impacts of human activities shown by the red arrows. (Yellow box sizes do not show relative reservoir sizes.) **Question:** What are two ways in which the sulfur cycle is linked to each of the phosphorus, nitrogen, and carbon cycles?

The running water carries these phosphate ions into the soil where they can be absorbed by the roots of plants and by other producers. Phosphate compounds are also transferred by food webs from producers to consumers, eventually including detritus feeders and decomposers. In both producers and consumers, phosphates are a component of biologically important molecules such as nucleic acids (see Figure 7, p. S16, in Supplement 4) and energy transfer molecules such as ADP and ATP (see Figure 11, p. S18, in Supplement 4). Phosphate is also a major component of vertebrate bones and teeth.

Phosphate can be lost from the cycle for long periods of time when it is washed from the land into streams and rivers and is carried to the ocean. There it can be deposited as marine sediment and remain trapped for millions of years. Over time, geological processes can uplift and expose these seafloor deposits, from which phosphate can be eroded to start the cycle again.

Because most soils contain little phosphate, the lack of it often limits plant growth on land unless phosphorus (as phosphate salts mined from the earth) is applied to the soil as a fertilizer. Lack of phosphorus also limits the growth of producer populations in many freshwater streams and lakes because phosphate salts are only slightly soluble in water and thus do not release many phosphate ions that producers need as nutrients.

Human activities are affecting the phosphorus cycle (as shown by the red arrows in Figure 3-19). One such activity is the removal of large amounts of phosphate from the earth to make fertilizer. Also, by clearing tropical forests (**Core Case Study**), we reduce phosphate levels in tropical soils. Topsoil that is eroded from fertilized crop fields, lawns, and golf courses carries large quantities of phosphate ions into streams, lakes, and oceans. There they stimulate the growth of producers such as algae and various aquatic plants. Phosphate-rich runoff from the land often produces huge populations of algae, which then upset chemical cycling and other processes in bodies of water.

The Sulfur Cycle

Sulfur circulates through the biosphere in the **sulfur cycle,** shown in Figure 3-20. Much of the earth's sulfur is stored underground in rocks and minerals and in the form of sulfate (SO_4^{2-}) salts buried deep under ocean sediments.

Sulfur also enters the atmosphere from several natural sources. Hydrogen sulfide (H_2S)—a colorless, highly poisonous gas with a rotten-egg smell—is released from active volcanoes and from organic matter broken down by anaerobic decomposers in flooded swamps, bogs, and tidal flats. Sulfur dioxide (SO_2), a colorless and suffocating gas, also comes from volcanoes.

Particles of sulfate (SO_4^{2-}) salts, such as ammonium sulfate, enter the atmosphere from sea spray, dust storms, and forest fires. Plant roots absorb sulfate ions and incorporate the sulfur as an essential component of many proteins.

Certain marine algae produce large amounts of volatile dimethyl sulfide, or DMS (CH_3SCH_3). Tiny droplets of DMS serve as nuclei for the condensation of water into droplets found in clouds. In this way, changes in DMS emissions can affect cloud cover and climate.

In the atmosphere, DMS is converted to sulfur dioxide, some of which in turn is converted to sulfur trioxide gas (SO_3) and to tiny droplets of sulfuric acid (H_2SO_4). DMS also reacts with other atmospheric chemicals such as ammonia to produce tiny particles of sulfate salts. These droplets and particles fall to the earth as components of *acid deposition,* which along with other air pollutants can harm trees and aquatic life.

In the oxygen-deficient environments of flooded soils, freshwater wetlands, and tidal flats, specialized bacteria convert sulfate ions to sulfide ions (S^{2-}). The sulfide ions can then react with metal ions to form insoluble metallic sulfides, which are deposited as rock or metal ores (often extracted by mining and converted to various metals), and the cycle continues.

Human activities have affected the sulfur cycle primarily by releasing large amounts of sulfur dioxide (SO_2) into the atmosphere (as shown by the red arrows in Figure 3-20). We release sulfur to the atmosphere in three ways. *First,* we burn sulfur-containing coal and oil to produce electric power. *Second,* we refine sulfur-containing oil (petroleum) to make gasoline, heating oil, and other useful products. *Third,* we extract metals such as copper, lead, and zinc from sulfur-containing compounds in rocks that are mined for these metals. In the atmosphere, SO_2 is converted to droplets of sulfuric acid (H_2SO_4) and particles of sulfate (SO_4^{2-}) salts, which return to the earth as acid deposition, which in turn can damage ecosystems.

3-5 How Do Scientists Study Ecosystems?

CONCEPT 3-5

Scientists use both field research and laboratory research, as well as mathematical and other models, to learn about ecosystems.

Some Scientists Study Nature Directly

Scientists use both field and laboratory research and mathematical and other models to learn about ecosystems (Concept 3-5). *Field research,* sometimes called "muddy-boots biology," involves going into forests and other natural settings to observe and measure the structure of ecosystems and what happens in them (see Chapter 2 opening photo). Most of what we know about ecosystems has come from such research.

Scientists have been particularly creative in studying tropical forests (**Core Case Study**). In a few cases, ecologists have erected tall construction cranes that provide them access to the canopies of tropical forests. This, along with rope walkways between treetops (Figure 3-21), has helped them to identify and observe the rich diversity of species living or feeding in these treetop habitats.

Sometimes ecologists carry out controlled experiments by isolating and changing a variable in part of an area and comparing the results with nearby unchanged areas. A good example of this is reported in the Core Case Study of Chapter 2 (p. 30).

Scientists also use aircraft and satellites equipped with sophisticated cameras and other *remote sensing* devices to scan and collect data on the earth's surface. Then they use *geographic information system (GIS)* software to capture, store, analyze, and display such information. Such software can store geographic and ecological spatial data electronically as numbers or as images. For example, a GIS can convert digital satellite images into global, regional, and local maps showing variations in vegetation, gross primary productivity, air pollution emissions, and many other variables.

Scientists can also attach tiny radio transmitters to animals and use global positioning systems (GPS) to learn about the animals by tracking where they go and how far they go. This technology is very important for studying endangered species. **Green Careers:** GIS analyst; remote sensing analyst

Some Scientists Study Ecosystems in the Laboratory

Since the 1960s, ecologists have increasingly supplemented field research by using *laboratory research*—setting up, observing, and making measurements of model ecosystems and populations under laboratory conditions (Figure 3-22). They have created such simplified systems in containers such as culture tubes, bottles, aquariums, and greenhouses, and in indoor and outdoor chambers where they can control temperature, light, CO_2, humidity, and other variables.

Such systems make it easier for scientists to carry out controlled experiments. In addition, laboratory experiments often are quicker and less costly than similar exper-

Figure 3-21 Scientists learn about tropical forests by using a system of ropes and pulleys to move among the treetops.

Bill Hatcher/National Geographic Creative

iments in the field. But there is a catch: scientists must consider how well their scientific observations and measurements in a simplified, controlled system under laboratory conditions reflect what actually takes place under more complex and dynamic conditions found in nature. Thus, the results of laboratory research must be coupled with and supported by field research.

⊙ CONSIDER THIS. . .

THINKING ABOUT Greenhouse Experiments and Tropical Rain Forests

How would you design an experiment that includes an experimental group and a control group and uses a greenhouse to determine how clearing a patch of tropical rain forest vegetation might affect the temperature above the cleared patch?

Some Scientists Use Models to Simulate Ecosystems

Since the late 1960s, ecologists have developed mathematical and other types of models that simulate ecosystems. Computer simulations can help scientists understand large and very complex systems, such as lakes, oceans, forests, and the earth's climate system, that cannot be adequately studied and modeled in field and laboratory research. Scientists are learning a lot about how the earth works by feeding data into increasingly sophisticated models of the earth's systems and running them

on high-speed supercomputers. **Green Careers:** ecosystem modeler

Of course, simulations and projections made with ecosystem models are no better than the data and assumptions used to develop the models. Ecologists must determine the relationships among key variables that they will use to develop and test ecosystem models. They must also do careful field and laboratory research to get *baseline data*, or beginning measurements, of the variables they are studying in the world's major terrestrial and aquatic ecosystems.

JOEL SARTORE/National Geographic Creative

Figure 3-22 These biologists are studying frogs in a laboratory in Ecuador.

TESTING PLANETARY BOUNDARIES: FROM HOLOCENE TO ANTHROPOCENE

For most of the past 10,000 years, we have been living in an era called the **Holocene**—a period of relatively stable climate and other environmental conditions following a glacial period. This general stability has allowed the human population to grow and to develop agriculture and to flourish within societies that have taken over a large and growing share of the earth's resources.

According to Will Steffen, Paul J. Crutzen, Johan Rockström, and other scientists, since the industrial revolution began around 1750, we have entered an era called the **Anthropocene**. In this new era, humans have become major agents of change in the functioning of the earth's life-support system as their ecological footprints have spread over the earth (Figure 1-12, p. 13, and Figure 1-13, p. 14).

In 2009, a group of 28 internationally renowned scientists, led by Johan Rockström of the Stockholm Resilience Centre, designated nine major *planetary boundaries,* or ecological tipping points (see p. 45). In 2012, ecologist Steven Running suggested adding a tenth planetary boundary based on the growing use of the global net primary productivity by humans. These scientists have hypothesized that if we exceed these boundaries, we could trigger abrupt and long-lasting or irreversible environmental changes within ten major systems that play key roles in earth's ability to sustain life (Figure 3-B). Exceeding some of these boundaries will have local and regional effects. However, exceeding any of the climate change, ocean acidification, or stratospheric ozone boundaries would likely have long-lasting global effects.

So far, the limit values of these planetary boundaries are uncertain, and this presents a major problem. We need to do much more research to fill in the missing data on these planetary boundaries and on how our exceeding them could affect the health of humans, other species, and the earth's life-support systems, as well as human economies and political systems. This research could help us to develop strategies for staying within planetary boundaries and for dealing with the effects on systems where we have exceeded such boundaries.

In 2009, the U.S. National Aeronautics and Space Administration and Cisco, a large computer networking company, teamed up to develop a *planetary skin*—a sort of global nervous system that will integrate land, air, and space-based sensors throughout the world. This pilot project will provide scientists, the public, government agencies, and private businesses with valuable baseline data on the state of the earth, which we can use to help us make decisions on how to reduce our environmental impacts.

GOOD NEWS

We Need to Learn More about the Health of the World's Ecosystems

We need baseline data on the condition of the world's ecosystems to see how they are changing, to develop effective strategies for preventing or slowing their degradation, and to avoid ecological tipping points beyond which these systems will be disrupted or destroyed. By analogy, your doctor needs baseline data on your blood pressure, weight, and the functioning of your organs and other systems, which can be obtained through basic tests. If your health declines in some way, the doctor can run new tests and compare the results with the baseline data to identify any changes and to determine any necessary treatments.

According to the 2005 Millennium Ecosystem Assessment, scientists have less than half of the basic ecological data they need in order to evaluate the status of ecosystems in the United States. Even fewer data are available for most other parts of the world. Ecologists have called for a massive program to develop baseline data for the world's ecosystems in order to try to identify and avoid certain *planetary boundaries,* or ecological tipping points related to the earth's major natural systems (Science Focus 3.3).

Big Ideas

- Life is sustained by the flow of energy from the sun through the biosphere, the cycling of nutrients within the biosphere, and gravity.

- Some organisms produce the nutrients they need, others survive by consuming other organisms, and still others live on the wastes and remains of organisms while recycling nutrients that are used again by producer organisms.

- Human activities are altering the flow of energy through food chains and webs and the cycling of nutrients within ecosystems and the biosphere.

Figure 3-B Planetary boundaries for ten major components of the earth's life-support system. A team of scientists estimated that human activities have exceeded the boundary limits for three systems (shown in red) and are close to the limits for five other systems (shown in orange). There is not enough information to evaluate the other two systems (shown in white).

(Compiled by the authors using data from Johan Rockström, Paul Crutzen, and James Hansen, et al., 2009, "Planetary Boundaries: Exploring the Safe Operating Space for Humanity," *Ecology and Society,* vol. 14, no. 2, p. 32.)

Photo: Sailorr/Shutterstock.com

Phosphorus Cycle
(Amount of phosphorus entering global oceans per year)

Ocean Acidification
(Carbonate concentration in global oceans)

Nitrogen Cycle
(Amount of N_2 removed from atmosphere per year)

Freshwater Use
(Global consumption of water per year)

Carbon Cycle Climate Change
(Atmospheric concentration of CO_2)

Land Use
(Percentage of global land surfaces converted to farming)

Biodiversity Loss
(Number of species extinctions per year)

Ozone Depletion
(Stratospheric ozone concentration)

Chemical Pollution
(Levels of persistent organic pollutants, toxic heavy metals, plastics, endocrine disruptors, and radioactive wastes in the environment)

Atmospheric Aerosols
(Regional atmospheric concentration of particulates)

TYING IT ALL TOGETHER   Tropical Rain Forests and Sustainability

This chapter began with a discussion of the importance of the world's incredibly diverse tropical rain forests (**Core Case Study**). These ecosystems showcase the functioning of the three scientific **principles of sustainability**, which apply as well to the world's other ecosystems.

First, producers within rain forests rely on *solar energy* to produce a vast amount of biomass through photosynthesis. *Second,* species living in the forests take part in, and depend on, the *cycling of nutrients* and the flow of energy within the forests and throughout the biosphere. *Third,* tropical rain forests contain a huge and vital part of the earth's *biodiversity,* and interactions among species living in these forests help to sustain these complex ecosystems.

We also discussed recent research on the dangers of our exceeding key planetary boundaries through activities that could leave many natural systems, including tropical rain forests, unstable and unsustainable. In particular, exceeding the climate change, ocean acidity, and stratospheric ozone boundaries could have long-lasting global effects. In many of the chapters to follow, we examine these threats further, and we consider ways in which we can apply the six **principles of sustainability** in seeking to stay within the key planetary boundaries and to live more sustainably.

Chapter Review

Core Case Study

1. What are three harmful effects of the clearing and degradation of tropical rain forests (**Core Case Study**)?

Section 3-1

2. What are the two key concepts for this section? Define and distinguish among the **atmosphere**, **troposphere**, **stratosphere**, **hydrosphere**, **geosphere**, and **biosphere**. What three interconnected factors sustain life on the earth? Describe the flow of energy to and from the earth. What is the **greenhouse effect** and why is it important?

Section 3-2

3. What is the key concept for this section? Define **ecology**. Define **organism**, **population**, and **community** and give an example of each. Define and distinguish between an **ecosystem** and the **biosphere**.

4. Distinguish between the living and nonliving components in ecosystems and give two examples of each. What are two other terms used to describe these ecosystem components?

5. What is a **trophic level**? Distinguish among **producers (autotrophs)**, **consumers (heterotrophs)**, **decomposers**, and **detritus feeders (detritivores)**, and give an example of each in an ecosystem. Describe the processes of **photosynthesis**. Distinguish among **primary consumers (herbivores)**, **carnivores**, **secondary consumers**, **tertiary consumers**, and **omnivores**, and give an example of each. Explain why there is little waste in nature and why humans are a major exception to this natural process.

6. Explain the importance of microbes. Distinguish between **aerobic respiration** and **anaerobic respiration (fermentation)**. What two processes sustain ecosystems and the biosphere and how are they linked?

Section 3-3

7. What is the key concept for this section? Define and distinguish between a **food chain** and a **food web**.

Explain what happens to energy as it flows through food chains and webs. What is **biomass**? What is the **pyramid of energy flow**? Why are there more insects than tigers in the world? Why don't most of the world's people eat meat?

8. Distinguish between **gross primary productivity (GPP)** and **net primary productivity (NPP)**, and explain their importance.

Section 3-4

9. What is the key concept for this section? What happens to matter in an ecosystem? What is a **nutrient cycle**? Explain how nutrient cycles connect past, present, and future life. Describe the **hydrologic cycle**, or **water cycle**. Summarize the unique properties of water. What three major processes are involved in the water cycle? What is **surface runoff**? Define **groundwater**. What is an **aquifer**? What percentage of the earth's water supply is available to humans and other species as liquid freshwater? Explain how human activities are affecting the water cycle in three ways. Explain how clearing a rain forest can affect local weather and climate (**Core Case Study**). Describe the **carbon**, **nitrogen**, **phosphorus**, and **sulfur cycles**, and explain how human activities are affecting each cycle.

Section 3-5

10. What is the key concept for this section? Describe three ways in which scientists study ecosystems. Explain why we need much more basic data about the structure and condition of the world's ecosystems. Distinguish between the **Holocene** and **Anthropocene** eras. List ten planetary boundaries designated by scientists. Which three of these boundaries have already been exceeded, according to the scientists? What are this chapter's *three big ideas*? How are the three scientific **principles of sustainability** showcased in tropical rain forests?

Note: Key terms are in bold type.

Critical Thinking

1. How would you explain the importance of tropical rain forests (**Core Case Study**) to people who think that such forests have no connection to their lives?

2. Explain **(a)** why the flow of energy through the biosphere depends on the cycling of nutrients, and

(b) why the cycling of nutrients depends on gravity (Concept 3-1B).

3. Explain why microbes are so important. List two beneficial effects and two harmful effects of microbes on your health. Write a brief description of what you

think would happen to you if microbes were eliminated from the earth.

4. Make a list of the foods you ate for lunch or dinner today. Trace each type of food back to a particular producer species. Describe the sequence of feeding levels that led to your feeding.

5. Use the second law of thermodynamics (see Chapter 2, p. 43) to explain why many poor people in less-developed countries live on a mostly vegetarian diet.

6. What changes might take place in the hydrologic cycle if the earth's climate becomes **(a)** hotter or **(b)** cooler? In each case, what are two ways in which these changes might affect your life?

7. What would happen to an ecosystem if **(a)** all of its decomposers and detritus feeders were eliminated, **(b)** all of its producers were eliminated, and **(c)** all of its insects were eliminated? Could a balanced ecosystem exist with only producers and decomposers and no consumers such as humans and other animals? Explain.

8. For each of the earth's ten major planetary boundaries (Figure 3-B), describe how our exceeding that boundary might affect **(a)** you, **(b)** any child you might have, and **(c)** any grandchild you might have.

Doing Environmental Science

Visit a nearby terrestrial ecosystem or aquatic life zone and try to identify major producers, primary and secondary consumers, detritus feeders, and decomposers. Take notes and describe at least one example of each of these types of organisms. Make a simple sketch showing how these organisms might be related to each other or to other organisms in a food chain or food web.

Global Environment Watch Exercise

Search for *Nitrogen Cycle* and look for information on how humans are affecting the nitrogen cycle. Specifically, look for impacts on the atmosphere and on human health from emissions of nitrogen oxides, and look for the harmful ecological effects of the runoff of nitrate fertilizers into rivers and lakes. Make a list of these impacts and use this information to review your daily activities. Find three things that you do regularly that contribute to these impacts.

Data Analysis

Recall that net primary productivity (NPP) is the *rate* at which producers can make the chemical energy that is stored in their tissues and that is potentially available to other organisms (consumers) in an ecosystem. In Figure 3-14, it is expressed as units of energy (kilocalories, or *kcal*) produced in a given area (square meters, or m^2) over a period of time (a year). Look again at Figure 3-14 and consider the differences in NPP among various ecosystems. Then answer the following questions:

1. What is the approximate NPP of a tropical rain forest in kcal/m²/yr? Which terrestrial ecosystem produces about one-third of that rate? Which aquatic ecosystem has about the same NPP as a tropical rain forest?

2. Early in the 20th century, large areas of temperate forestland in the United States were cleared to make way for agricultural land. For each unit of this forest area that was cleared and replaced by farmland, about how much NPP was lost?

3. Why do you think deserts and grasslands have dramatically lower NPP than swamps and marshes?

4. About how many times more NPP do estuaries produce, compared to lakes and streams? Why do you think this is so?

1. A tipping point in the disappearance of the tropical rain forests would be
   a. the change in regional weather patterns after clearing the forests that prevents their return.
   b. the clearing of rain forests for agricultural lands that adds carbon dioxide to the atmosphere.
   c. the loss of trees causing less carbon dioxide to be absorbed from the atmosphere.
   d. the loss of habitat for endangered species as the rain forests are cleared.
   e. the clearing of land for a new road to be built through the rain forest.

2. The best example of a population is
   a. a group of cats known as *Felis concolor* living and interbreeding in one area at the same time.
   b. a community of different species interacting with one another and their nonliving environment.
   c. groups of different species living in a specific place and potentially interacting with each other.
   d. the movement of nitrogen from a terrestrial system into the atmosphere.
   e. the rabbits who are preyed upon by bobcats.

3. Which is NOT an important ecological role of insects?
   a. Pollinating almond and fruit trees in California so the farmers themselves do not have to pollinate the flowers
   b. Eating insect pests that may irritate or harm other insects
   c. Providing a nutritious food source for other insects and for animals such as bats and frogs
   d. Fertilizing the soil with nutrients from their waste material and when they decompose
   e. Transmitting tropical disease to humans who visit locations between the Tropic of Cancer and the Tropic of Capricorn

Questions 4–7 refer to the compounds listed below.
   a. Nitrogen gas (N_2)
   b. Oxygen gas (O_2)
   c. Water (H_2O)
   d. Phosphate (PO_4^{3-})
   e. Methane (CH_4)

4. A greenhouse gas produced by raising cattle.

5. Can be stored in marine sediments for long periods of time.

6. A greenhouse gas emitted into the atmosphere through transpiration or evaporation.

7. A limiting factor for most plant growth in lakes and terrestrial systems.

8. Organisms live within a range of tolerance that determines where they live. Which of the below is NOT a range of tolerance?
   a. Trout populations live in areas with a high dissolved oxygen content and a narrow range of temperature.
   b. Coral polyps need a symbiotic algae living within them.
   c. Desert cacti can thrive in areas with low levels of rainfall.
   d. Blue crabs in an estuary can tolerate wide swings of salinity and temperature.
   e. Royal ferns require moist areas and are only found in wetlands.

Question 9 refers to the data in the table shown below.

Ecosystem Type	NPP (g/m²/yr)	Energy Provided to 1° Consumer (Cal./gram)
Tropical rain forest	2,200	4

9. Applying the 10% rule, if a shrew living in a tropical rain forest has a daily requirement of 500 Cal., how many square meters of rain forest are needed to support one shrew that feeds only on snails which are primary consumers?
   a. 0.6 m²
   b. 1.1 m²
   c. 207.4 m²
   d. 401.5 m²
   e. 2073.9 m²

10. The end product of the wood in a log being decomposed by fungi is
    a. nitrogen gas.
    b. carbon dioxide gas and water.
    c. heat energy.
    d. coal.
    e. carbon monoxide gas.

11. Which of the below is NOT a human intervention in the nitrogen cycle?
    a. Large amounts of nitric oxide (NO) are released into the atmosphere by smokestacks which can cause acid rain.
    b. Nitrous oxide (N_2O) is added to the atmosphere through the action of bacteria on livestock waste.
    c. Large amounts of nitrogen are released into the atmosphere as forests are cleared.
    d. Large amounts of nitrates are added to the Chesapeake Bay causing a eutrophic dead zone.
    e. Specialized bacteria convert ammonia into nitrate and then into nitrogen gas reducing nitrogen available to plants.

12. The phosphorus cycle includes all of the following EXCEPT it
    a. is found in rock as fossil bones and guano.
    b. is a fast-moving atmospheric cycle.
    c. is a limiting factor for many plants.
    d. can be deposited as marine sediment and not be released for millions of years.
    e. is returned to the soil as fertilizer and can cause algae blooms.

4 Biodiversity and Evolution

Few problems are less recognized, but more important than, the accelerating disappearance of the earth's biological resources. In pushing other species to extinction, humanity is busy sawing off the limb on which it is perched.

PAUL EHRLICH

Key Questions

4-1 What is biodiversity and why is it important?

4-2 How does the earth's life change over time?

4-3 How do geological processes and climate change affect evolution?

4-4 How do speciation, extinction, and human activities affect biodiversity?

4-5 What is species diversity and why is it important?

4-6 What roles do species play in ecosystems?

Monarch butterfly on a flower.

Ariel Bravy/Shutterstock.com

Amphibians—frogs, toads, and salamanders—were among the earliest vertebrates (animals with backbones) to emerge from the earth's waters and live on the land. Historically, they have been able to adjust to and survive environmental changes more effectively than many other species. However, the amphibian world is changing rapidly.

An amphibian lives part of its life in water and part on land. Now, many of the 6,700 or more amphibian species are having difficulty adapting to rapid changes that have taken place in their water and land habitats during the past few decades. Such changes have resulted primarily from human activities such as use of pesticides and other chemicals that become water pollutants.

Since 1980, populations of hundreds of amphibian species throughout the world have declined or vanished (Figure 4-1). According to the International Union for Conservation of Nature (IUCN), about 33% of all known amphibian species are threatened with extinction, and more than 40% of all known amphibian species are declining.

No single cause has been identified to explain the declines of many amphibian species. However, scientists have identified a number of factors that affect amphibians at various points in their life cycles. For example, frog eggs have no shells to protect frog embryos from water pollutants, and adult frogs are often exposed to insecticides contained in the many insects they eat. We explore these and other factors later in this chapter.

Why should we care if some amphibian species become extinct? Scientists give three reasons. *First,* amphibians are sensitive biological *indicators* of changes in environmental conditions such as habitat loss, air and water pollution, ultraviolet (UV) radiation, and climate change. The growing threats to their survival indicate that environmental conditions are deteriorating in many parts of the world.

Second, adult amphibians play important ecological roles in biological communities. For example, amphibians eat more insects (including mosquitoes) than do birds. In some habitats, the extinction of certain amphibian species could lead to extinction of certain species of other amphibians, aquatic insects, reptiles, birds, fish, and mammals that feed on amphibians or their larvae.

Third, amphibians are a genetic storehouse of pharmaceutical products waiting to be discovered. For example, compounds in secretions from the skin of certain amphibians have been isolated and used as painkillers and antibiotics, and in treatments for burns and heart disease.

Many scientists believe that the threats to amphibians present a warning about a number of environmental threats to biodiversity. In this chapter, we will learn about biodiversity, about how it has arisen and why it is important, and about how it is threatened. We will also look at some possible solutions to these problems.

Figure 4-1 These are specimens of some of the nearly 200 amphibian species that have gone extinct since 1970.

Joel Sartore/National Geographic Creative

4-1 What Is Biodiversity and Why Is It Important?

CONCEPT 4-1

The biodiversity found in genes, species, ecosystems, and ecosystem processes is vital to sustaining life on earth.

Biodiversity Is a Crucial Part of the Earth's Natural Capital

Biological diversity, or **biodiversity**, is the variety of the earth's *species,* or varying life-forms, the genes they contain, the ecosystems in which they live, and the ecosystem processes such as energy flow and nutrient cycling that sustain all life (Figure 4-2). Acting together, these four components of the earth's biodiversity provide us with the ecosystem services (Figure 1-3, orange items, p. 7) that sustain us and our economies.

Recall that a **species** is a group of organisms with a set of characteristics that distinguish it from other groups of organisms, and in sexually reproducing organisms, individuals must be able to mate and produce fertile offspring in order to be grouped within a species.

We do not know how many species there are on the earth. Estimates range from 8 million to 100 million. In 2011, a team of biologists led by Camilo Mora and Boris Worm estimated that there are about 7–10 million species. Other scientific estimates put the number much higher.

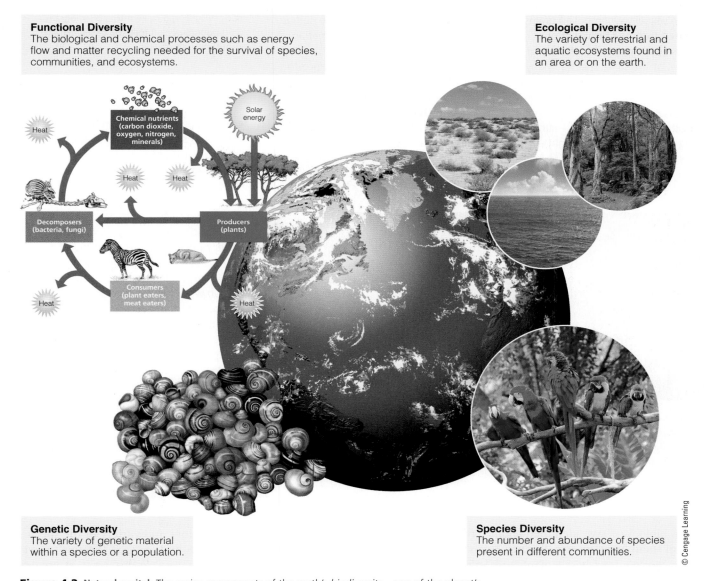

Functional Diversity
The biological and chemical processes such as energy flow and matter recycling needed for the survival of species, communities, and ecosystems.

Ecological Diversity
The variety of terrestrial and aquatic ecosystems found in an area or on the earth.

Genetic Diversity
The variety of genetic material within a species or a population.

Species Diversity
The number and abundance of species present in different communities.

© Cengage Learning

Figure 4-2 **Natural capital:** The major components of the earth's *biodiversity*—one of the planet's most important renewable resources and a key component of its natural capital (see Figure 1-3, p. 7).
Question: Why do you think we should protect the earth's biodiversity from our actions?

INSECTS PLAY A VITAL ROLE IN OUR WORLD

We classify many insect species as *pests* because they compete with us for food, spread human diseases such as malaria, bite or sting us, and invade our lawns, gardens, and houses. Some people fear insects and many think the only good bug is a dead bug. They fail to recognize the vital roles insects play in helping to sustain life on earth.

For example, *pollination* is a vital ecosystem service that allows flowering plants to reproduce sexually when pollen grains are transferred from the flower of one plant to a receptive part of the flower of another plant of the same species. Many of the earth's plant species depend on insects to pollinate their flowers (chapter-opening photo and Figure 4-A, left).

Insects that eat other insects—such as the praying mantis (Figure 4-A, right)—help to control the populations of at least

half the species of insects we call pests. This free pest control service is an important part of the earth's natural capital. Some insects also play a key role in loosening and renewing the topsoil that supports plant life on land. Others such as the dung beetle recycle animal wastes (dung) often by rolling it into balls (Figure 4-B) and burying it for use as food.

Insects have been around for at least 400 million years—about 2,000 times longer than the latest version of the human species. Some species reproduce at an astounding rate and can rapidly develop new genetic traits such as resistance to pesticides. They also have an exceptional ability to evolve into new species when faced with changing environmental conditions, and many species are now being challenged to do so because of environmental changes brought about by the rap-

Figure 4-B A dung beetle rolls a ball of dung. Dung beetles can roll up to 10 times their body weight—equivalent to a 68-kilogram (150-pound) person rolling a 680-kilogram (1,500-pound) boulder.

idly expanding human population and its growing resource use per person.

The scientific study of insects is called *entomology*—a field that includes a vast number of subspecialties and applications. Entomologists are expanding their research in areas related to environmental threats to insect populations. Environmental changes, many of them caused by human activities, are making such threats clearer every year. For example, entomologist Diana Cox-Foster of Pennsylvania State University (USA) is studying the decline of honeybees, which are extremely important pollinators. Her lab work focuses on *colony collapse disorder*, the name given to the disappearances of many bee colonies in recent years. This disorder is threatening to disrupt whole ecosystems that depend on bees for pollination, as well as much of the human food supply. We discuss this serious environmental problem more fully in Chapter 9.

Critical Thinking

Identify three insect species not discussed above that benefit your life.

Figure 4-A *Importance of insects:* Bees (left) and numerous other insects pollinate flowering plants that serve as food for many plant eaters, including humans. This praying mantis, which is eating a moth (right), and many other insect species help to control the populations of most of the insect species we classify as pests.

So far, biologists have identified about 2 million species—most of them being insects (Science Focus 4.1). Up to half of the world's land-based plant and animal species live in tropical rain forests. Scientists believe that most of the unidentified species live in the planet's rain forests and in the largely unexplored oceans.

Species diversity, the number and variety of the species present in any biological community, is the most obvious component of biodiversity. We discuss it in greater detail in Section 4-5. Another important component is *genetic diversity,* the variety of genes found in a population or in a species (Figure 4-3). The earth's many spe-

Figure 4-3 *Genetic diversity* among individuals in this population of a species of Caribbean snail is reflected in the variations in shell color and banding patterns. Genetic diversity can also include other variations such as slight differences in chemical makeup, sensitivity to various chemicals, and behavior.

cies contain a vast variety of genes, which enable life on the earth to survive and adapt to dramatic environmental changes.

Ecosystem diversity—the earth's variety of deserts, grasslands, forests, mountains, oceans, lakes, rivers, and wetlands—is another major component of biodiversity. Each of these ecosystems is a storehouse of genetic and species diversity. Biologists have classified the terrestrial (land) ecosystems into **biomes**—large regions such as forests, deserts, and grasslands with distinct climates and certain species (especially vegetation) adapted to them. Figure 4-4 shows different major biomes along the 39th parallel spanning the United States. We discuss biomes in more detail in Chapter 7.

Yet another important component of biodiversity is *functional diversity*—the variety of processes such as energy flow and matter cycling that occur within ecosystems (see Figure 3-10, p. 59) as species interact with one another in food chains and webs.

The earth's biodiversity is a vital part of the natural capital (see Figure 1-3, p. 7) that helps to keep us alive and supports our economies. With the help of technology, we use biodiversity to provide us with food, wood, fibers, energy from wood and biofuels, and medicines. Biodiversity also plays critical roles in providing us with the ecosystem services that preserve the quality of the air and water, maintain the fertility of topsoil, decompose and recycle wastes, and control populations of species that we call pests. The four components of biodiversity also increase the stability of ecosystems and increase the resistance of ecosystems to harmful invasive species. We owe much of what we know about biodiversity to a fairly small number of researchers, such as Edward O. Wilson (Individuals Matter 4.1).

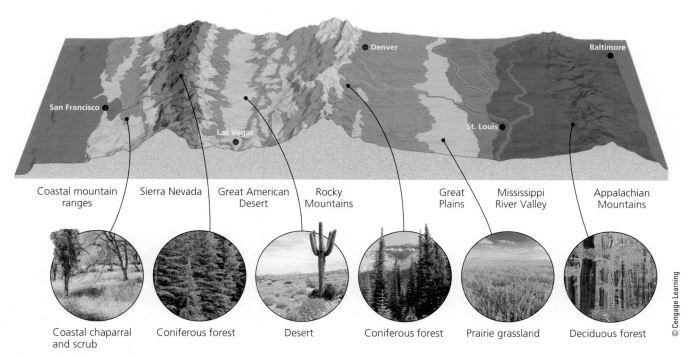

Coastal mountain ranges   Sierra Nevada   Great American Desert   Rocky Mountains   Great Plains   Mississippi River Valley   Appalachian Mountains

Coastal chaparral and scrub   Coniferous forest   Desert   Coniferous forest   Prairie grassland   Deciduous forest

Figure 4-4 The major biomes found along the 39th parallel across the United States show a variety of ecosystems. The differences in tree and other plant species reflect changes in climate, mainly differences in average annual precipitation and temperature.

Edward O. Wilson: A Champion of Biodiversity

Jim Harrison

As a boy growing up in the southeastern United States, Edward O. Wilson became interested in insects at the age of nine. He has said, "Every kid has a bug period. I never grew out of mine."

Before entering college, Wilson had decided he would specialize in the study of ants. After he grew fascinated with that tiny organism, and throughout his long career, he steadily widened his focus to include the entire biosphere. He now spends much of his time studying, writing, and speaking about biodiversity and working on Harvard University's *Encyclopedia of Life,* an online database for the planet's known and named species.

During his long career, Wilson has taken on many challenges. He and other researchers working together discovered how ants communicate using chemicals called pheromones. He also studied the complex social behavior of ants and has compiled much of his work into the epic volume, *The Ants,* published in 1990. His ant research has been applied to the study and understanding of other social organisms, including humans. He has also proposed the hypothesis that humans have a natural affinity for wildlife and wild places—a concept he calls *biophilia* (or love of life).

One of Wilson's landmark works is *The Diversity of Life,* published in 1992, in which he put together the principles and practical issues of biodiversity more completely than anyone had to that point. Wilson is now deeply involved in writing and lecturing about the need for global conservation efforts and is promoting the goal of completing a global survey of biodiversity. He has won more than 100 national and international awards and has written 25 books, two of which won the Pulitzer Prize for General Nonfiction. About the importance of biodiversity, he writes:

"Until we get serious about exploring biological diversity. . . science and humanity at large will be flying blind inside the biosphere. . . . How can we save Earth's life forms from extinction if we don't even know what most of them are?"

Background photo: Christian Musat/Shutterstock.com

4-2 How Does the Earth's Life Change over Time?

CONCEPT 4-2A

The scientific theory of evolution explains how life on the earth changes over time due to changes in the genes of populations.

CONCEPT 4-2B

Populations evolve when genes mutate and give some individuals genetic traits that enhance their abilities to survive and to produce offspring with these traits (natural selection).

Biological Evolution by Natural Selection Explains How Life Changes over Time

Most of what we know about the long history of life on the earth comes from **fossils**: mineralized or petrified replicas of skeletons, bones, teeth, shells, leaves, and seeds, or impressions of such items found in rocks (Figure 4-5). Scientists also drill core samples from glacial ice at the earth's poles and on mountaintops, and examine the signs of ancient life found at different layers in these cores.

The entire body of evidence gathered using these methods, which is called the *fossil record,* is uneven and incomplete. Some forms of life left no fossils, and some fossils have decomposed. The fossils found so far represent probably only 1% of all species that have ever lived. Trying to reconstruct the development of life with so little evidence is the work of *paleontology*—a challenging scientific detective game. **Green Careers:** paleontologist

How did we end up with such an amazing array of species? The scientific answer involves **biological evolution** (or simply **evolution**): the process whereby the earth's life changes over time through changes in the genes of populations of organisms in succeeding generations (Concept 4-2A). According to the **theory of evolution**, all species evolved from earlier, ancestral species. In other words, life comes from life.

The idea that organisms change over time and are descended from a single common ancestor has been around in one form or another since the early Greek philosophers. But no one had developed a convincing explanation of how this could happen until 1858 when naturalists Charles Darwin (1809–1882) and Alfred Russel Wallace (1823–1913) independently proposed the concept of *natural selection* as a mechanism for biological evolution. Darwin meticulously gathered evidence for this idea and published it in 1859 in his book, *On the Origin of Species by Means of Natural Selection.*

Darwin and Wallace observed that individual organisms must struggle constantly to survive by getting enough food, water, and other resources, to avoid being eaten, and to reproduce. They also observed that individuals in a population with a specific advantage over other individuals in that population were more likely to survive and produce offspring that had the same specific advantage. The advantage was due to a characteristic, or *trait*, possessed by these individuals but not by others of their kind.

Based on these observations, Darwin and Wallace described a process called **natural selection**, in which individuals with certain traits are more likely to survive and reproduce under a particular set of environmental conditions than are those without the traits (Concept 4-2B). The scientists concluded that these survival traits would become more prevalent in future populations of the species as individuals with those traits became more numerous and passed their traits on to their offspring.

A huge body of evidence supports this idea. As a result, *biological evolution through natural selection* has become an important scientific theory that generally explains how life has changed over the past 3.5 billion years and why life is so diverse today. However, there are still many unanswered questions that generate scientific debate about the details of evolution by natural selection.

Ira Block/National Geographic Creative

Figure 4-5 This fossil shows the mineralized remains of an early ancestor of the present-day horse. It roamed the earth more than 35 million years ago. Note that you can also see fish skeletons on this fossil.

Mutations and Changes in the Genetic Makeup of Populations Lead to Biological Evolution by Natural Selection

The process of biological evolution by natural selection involves changes in a population's genetic makeup through successive generations. Note that *populations—not individuals—evolve by becoming genetically different.*

The first step in this process is the development of *genetic variability,* or variety in the genetic makeup of individuals in a population. This occurs through **mutations**: changes in the DNA molecules of a gene in any cell that can be inherited by offspring. Most mutations result from random changes that occur in coded genetic instructions when DNA molecules (see Figure 9, p. S17, in Supplement 4) are copied each time a cell divides and whenever an organism reproduces. Some mutations also occur from exposure to external agents such as radioactivity and natural and human-made chemicals (called *mutagens*).

Mutations can occur in any cell, but only those that take place in genes of reproductive cells are passed on to offspring. Sometimes, such a mutation can result in a new genetic trait, called a *heritable trait,* which can be passed from one generation to the next. In this way, populations develop differences among individuals, including genetic variability.

The next step in biological evolution is *natural selection,* in which environmental conditions favor some individuals over others. The favored individuals possess heritable traits that give them some advantage over other individuals in a given population. Such a trait is called an **adaptation**, or **adaptive trait**—any heritable trait that improves the ability of an individual organism to survive and to reproduce at a higher rate than other individuals in a population are able to do under prevailing environmental conditions. For example, in the face of snow and cold, a few gray wolves in a population that have thicker fur might live longer and thus produce more offspring than do those without thicker fur. As those longer-lived wolves mate, genes for thicker fur spread throughout the population and individuals with those genes increase in number

and pass this helpful trait on to more offspring. Thus, the scientific concept of natural selection explains how populations adapt to changes in environmental conditions.

Another important example of natural selection at work is the evolution of *genetic resistance*—the ability of one or more organisms in a population to tolerate a chemical designed to kill it. Such resistance develops fairly quickly in populations of organisms that produce large numbers of offspring, such as many species of bacteria and insects.

For example, certain bacteria (Figure 4-6a) have developed genetic resistance to widely used antibacterial drugs, or *antibiotics,* which have become a force of natural selection. Often, when such drugs are used (Figure 4-6b), a few bacteria that are genetically resistant to them survive and rapidly produce more offspring than the bacteria that were killed by the drug could have produced (Figure 4-6c). Thus, the antibiotic eventually loses its effectiveness as genetically resistant bacteria keep reproducing while those that are susceptible to the drug die off (Figure 4-6d). (We discuss this form of genetic resistance further in Chapter 17.)

One way to summarize the process of biological evolution by natural selection is: *Genes mutate, individuals are selected, and populations evolve such that they are better adapted to survive and reproduce under existing environmental conditions* (**Concept 4-2B**).

A remarkable example of species evolution by natural selection is *Homo sapiens sapiens.* We have evolved certain traits that have allowed us to dominate most of the earth (Case Study that follows).

CASE STUDY

How Did Humans Become Such a Powerful Species?

Like many other species, humans have survived and thrived because we have certain traits that allow us to adapt to and modify parts of the environment to increase our chances of surviving and reproducing.

Evolutionary biologists attribute our success to three adaptations: *strong opposable thumbs* that allowed us to grip

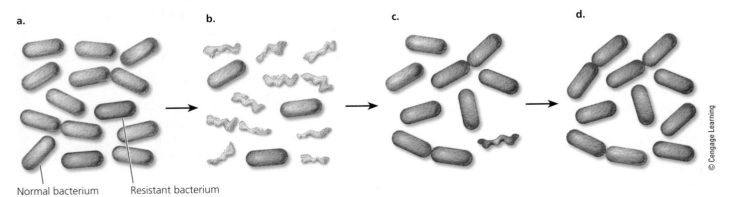

a.   b.   c.   d.

Normal bacterium   Resistant bacterium

© Cengage Learning

Figure 4-6 *Evolution by natural selection.* A population of bacteria **(a)** is exposed to an antibiotic, which **(b)** kills most individuals, but none of those possessing a trait that makes them resistant to the drug (shown in red). The resistant bacteria multiply **(c)** and eventually, **(d)** replace all or most of the nonresistant bacteria.

and use tools better than the few other animals that have thumbs could do; an *ability to walk upright,* which gave us agility and freed up our hands for many uses; and a *complex brain,* which allowed us to develop many skills, including the ability to use speech and to read and write to transmit complex ideas (Figure 4-7).

These adaptations have helped us to develop tools, weapons, protective devices, and technologies that extend our limited senses of sight, hearing, and smell. Thus, in an eye-blink of the 3.5-billion-year history of life on earth, we have developed powerful technologies and taken over much of the earth's net primary productivity for our own use. At the same time, we have degraded much of the planet's life-support system as our ecological footprints have grown (see Figure 1-13, p. 14).

However, adaptations that make a species successful during one period of time may not be enough to ensure the species' survival when environmental conditions change. This is no less true for humans, and some environmental conditions are now changing rapidly, largely due to our own actions. (We focus on several such changes in later chapters.)

One of our adaptations—our powerful brain—may enable us to live more sustainably by understanding and copying the ways in which nature has sustained itself for billions of years, despite major changes in environmental conditions.

Adaptation through Natural Selection Has Limits

In the not-too-distant future, will adaptations to new environmental conditions through natural selection allow our skin to become more resistant to the harmful effects of UV radiation, our lungs to cope with air pollutants, and our livers to better detoxify pollutants in our bodies?

According to scientists in this field, the answer is *no* because of two limitations on adaptation through natural selection. *First,* a change in environmental conditions can lead to such an adaptation only for genetic traits already present in a population's gene pool or for traits resulting from mutations, which occur randomly.

Second, even if a beneficial heritable trait is present in a population, the population's ability to adapt may be limited by its reproductive capacity. Populations of genetically diverse species that reproduce quickly—such as weeds,

Figure 4-8 One type of carnivorous plant is the Venus flytrap. A fly or other small insect, entering a hinged opening on the plant's leaf, touches trigger hairs that cause the opening to snap shut in less than a second and trap its prey. The plant then uses enzymes to digest its prey over a period of 1–2 weeks.

Mary Lane/Shutterstock.com

mosquitoes, rats, bacteria, and cockroaches—often adapt to a change in environmental conditions in a short time (days to years). By contrast, species that cannot produce large numbers of offspring rapidly—such as elephants, tigers, sharks, and humans—take a much longer time (typically thousands or even millions of years) to adapt through natural selection.

Three Common Myths about Evolution through Natural Selection

Evolution experts have identified three common misconceptions about biological evolution through natural selection. One is that "survival of the fittest" means "survival of the strongest." To biologists, *fitness* is a measure of reproductive success, not strength. Thus, the fittest individuals are those that leave the most descendants.

Another misconception is that organisms develop certain traits because they need them. For example, certain plants, called *carnivorous plants* (Figure 4-8), feed on insects not because they once needed to in order to survive. Rather, some ancestors of these plants had characteristics that enabled them to trap insects and to draw nutrients from them. This trait gave them an advantage over other plants in an environment where there were lots of insects available, and this enabled them to produce more offspring that had such characteristics. Thus, over time, their populations grew and continued to evolve in this way.

A third misconception is that evolution by natural selection involves some grand plan of nature in which species become more perfectly adapted. From a scientific standpoint, no plan or goal for genetic perfection has been identified in the evolutionary process.

© Cengage Learning

Figure 4-7 *Homo sapiens sapiens* has had three advantages over other mammals that helped the species to survive certain pressures of natural selection and become a dominant species on planet Earth.

4-3 How Do Geological Processes and Climate Change Affect Evolution?

CONCEPT 4-3
Tectonic plate movements, volcanic eruptions, earthquakes, and climate change have shifted wildlife habitats, wiped out large numbers of species, and created opportunities for the evolution of new species.

Geological Processes Affect Natural Selection

The earth's surface has changed dramatically over its long history. Scientists have discovered that huge flows of molten rock within the earth's interior have broken its surface into a series of gigantic solid plates, called *tectonic plates*. For hundreds of millions of years, these plates have drifted slowly on the planet's mantle (Figure 4-9).

Rock and fossil evidence indicates that 200–250 million years ago, all of the earth's present-day continents were connected in a super continent called Pangaea (Figure 4-9, left). About 135 million years ago, Pangaea began splitting apart as the earth's tectonic plates moved, eventually resulting in the present-day locations of the continents (Figure 4-9, right).

The fact that tectonic plates drift has had two important effects on the evolution and distribution of life on the earth. *First,* the locations of continents and oceanic basins have greatly influenced the earth's climate and thus have helped to determine where plants and animals can live. *Second,* the movement of continents has allowed species to move, adapt to new environments, and form new species through natural selection. When continents join together, populations can disperse to new areas and adapt to new environmental conditions. When continents separate and when islands are formed, populations must evolve under isolated conditions or become extinct.

Adjoining tectonic plates that are grinding along slowly next to one another sometimes shift quickly. Such sudden movement of tectonic plates can cause *earthquakes,* which can also affect biological evolution by causing fissures in the earth's crust that can separate and isolate populations of species. Over long periods of time, this can lead to the formation of new species as each isolated population changes genetically in response to new environmental conditions. *Volcanic eruptions* that occur along the boundaries of tectonic plates can also affect biological evolution by destroying habitats and reducing, isolating, or wiping out populations of species (**Concept 4-3**).

Climate Change and Catastrophes Affect Natural Selection

Throughout its history, the earth's climate has changed drastically. At times, it has cooled and covered much of the earth with glacial ice. At other times it has warmed, melted that ice, and drastically raised sea levels, which in turn increased the total area covered by the oceans and reduced the earth's total land area. Such alternating periods of cooling and heating have led to the advance and retreat of ice sheets at high latitudes over much of the northern hemisphere, most recently about 18,000 years ago (Figure 4-10).

These long-term climate changes have had a major effect on biological evolution by determining where different types of plants and animals can survive and thrive, and by changing the locations of different types of ecosystems such as deserts, grasslands, and forests (**Concept 4-3**). Some species became extinct because the climate changed too rapidly for them to adapt and survive, and new species evolved to take over their ecological roles.

Another force affecting natural selection has been catastrophic events such as collisions between the earth and large asteroids. There may have been many of these collisions during the 3.5 billion years of life on earth.

Figure 4-9 Over millions of years, the earth's continents have moved very slowly on several gigantic tectonic plates. ***Question:*** How might an area of land splitting apart cause the extinction of a species?

18,000 years before present

Northern Hemisphere
Ice coverage

Modern day
(August)

Legend
Continental ice
Sea ice
Land above sea level

(Compiled by the authors using data from the National Oceanic and Atmospheric Administration.)

Figure 4-10 These maps of North America show the large-scale changes in glacial ice coverage during the past 18,000 years. **Question:** What are two characteristics of an animal and two characteristics of a plant that natural selection would have favored as these ice sheets (left) advanced?

Such impacts have caused widespread destruction of ecosystems and wiped out large numbers of species. On the other hand, they have also caused shifts in the locations of ecosystems and created opportunities for the evolution of new species.

On a long-term basis, the three scientific **principles of sustainability** (see Figure 1-2, p. 6 or back cover), especially the biodiversity principle (Figure 4-2), have enabled life on the earth to adapt to drastic changes in environmental conditions.

4-4 How Do Speciation, Extinction, and Human Activities Affect Biodiversity?

CONCEPT 4-4A
As environmental conditions change, the balance between the formation of new species and the extinction of existing species determines the earth's biodiversity.

CONCEPT 4-4B
Human activities are decreasing biodiversity by causing the extinction of many species and by destroying or degrading habitats needed for the development of new species.

How Do New Species Evolve?

Under certain circumstances, natural selection can lead to an entirely new species. In this process, called **speciation**, one species splits into two or more different species. For sexually reproducing organisms, a new species forms when one population of a species has evolved to the point where its members can no longer breed and produce fertile offspring with members of another population that did not change or that evolved differently.

The most common way in which speciation occurs, especially among sexually reproducing species, is when a barrier or distant migration separates two or more populations of a species and prevents the flow of genes between them. This happens in two phases: first geographic isolation, and then reproductive isolation.

Geographic isolation occurs when different groups of the same population of a species become physically isolated from one another for a long period of time. For example, part of a population may migrate in search of food and then begin living as a separate population in another area with different environmental conditions. Populations can also be separated by a physical barrier (such as a mountain range, stream, or road), a volcanic

Figure 4-11 These poison dart frogs vary in coloration, partly because they were exposed to different environmental conditions.

eruption, tectonic plate movements, or winds or flowing water that carry a few individuals to a distant area. These separated populations can develop quite different characteristics. For example, populations of poison dart frogs (**Core Case Study**) living on different islands or in different parts of a region can have dramatic differences in coloration, as shown in Figure 4-11.

In **reproductive isolation**, mutation and change by natural selection operate independently in the gene pools of geographically isolated populations. If this process continues long enough, members of the geographically and reproductively isolated populations of sexually reproducing species may become so different in genetic makeup

that they cannot produce live, fertile offspring if they are rejoined and attempt to interbreed. As a result, one species has become two, and speciation has occurred (Figure 4-12).

Humans are playing an increasing role in the process of speciation. We have learned to shuffle genes from one species to another through **artificial selection** and, more recently, through **genetic engineering** (Science Focus 4.2).

All Species Eventually Become Extinct

Another process affecting the number and types of species on the earth is **extinction**, the process in which an entire species ceases to exist (also referred to as *biological extinction;* when a species becomes extinct over a large region, but not globally, it is called *local extinction*). When environmental conditions change dramatically or rapidly, a population of a species faces three possible futures: *adapt* to the new conditions through natural selection, *migrate* (if possible) to another area with more favorable conditions, or *become extinct*.

Species that are found in only one area, called **endemic species**, are especially vulnerable to extinction. They exist on islands and in other unique areas, especially in tropical rain forests where most species have highly specialized roles. For these reasons, they are unlikely to be able to migrate or adapt in the face of rapidly changing environmental conditions. Many of these endangered species are amphibians (**Core Case Study**). One example is the golden toad (Figure 4-13), which apparently became extinct in 1989 even though it lived in the well-protected Monteverde Cloud Forest Reserve in the mountains of Costa Rica.

Fossils and other scientific evidence indicate that all species eventually become extinct, but drastic changes in environmental conditions can eliminate large groups of species relatively rapidly. Throughout most of the earth's long history, species have disappeared at a low rate, called the **background extinction rate**. Based on the fossil record and analysis of ice cores, biologists estimate that

Figure 4-12 *Geographic isolation* can lead to reproductive isolation, divergence of gene pools, and speciation.

© Cengage Learning

Early fox population → Spreads northward and southward and separates

Northern population

Southern population

Arctic Fox Adapted to cold through heavier fur, short ears, short legs, and short nose. White fur matches snow for camouflage.

Different environmental conditions lead to different selective pressures and evolution into two different species.

Gray Fox Adapted to heat through lightweight fur and long ears, legs, and nose, which give off more heat.

SCIENCE FOCUS 4.2

CHANGING THE GENETIC TRAITS OF POPULATIONS

We have used artificial selection to change the genetic characteristics of populations with similar genes. In this process, we select one or more desirable genetic traits in the population of a plant or animal such as a type of wheat, fruit (Figure 4-C), or dog. Then we use *selective breeding*, or *crossbreeding*, to generate populations of the species containing large numbers of individuals with the desired traits.

Note that artificial selection involves crossbreeding between genetic varieties of the same species or between species that are genetically close to one another, and thus it is not a form of speciation. Most of the grains, fruits, and vegetables we eat are produced by artificial selection. It has also given us food crops with higher yields, cows that give more milk, trees that grow faster, and many different types of dogs and cats. But traditional crossbreeding is a slow process and it can be used only on species that are close to one another genetically.

Now scientists are using genetic engineering to speed up our ability to manipulate genes. In this process, scientists alter an organism's genetic material by adding, deleting, or changing segments of its DNA to produce desirable traits or to eliminate undesirable ones. It enables scientists to transfer genes between different species that would not interbreed in nature. For example, we can put genes from a cold-water fish species into a tomato plant to give it properties that enable it to resist cold weather.

Figure 4-D These mice are both 6 months old. The one on the left is normal, while the mouse on the right has had a human growth hormone gene inserted into its cells. Mice with this gene grow 2–3 times faster than, and twice as large as, mice without it. **Question:** How do you think the creation of such species might change the process of evolution by natural selection?

Scientists have used genetic engineering to develop modified crop plants, new drugs, pest-resistant plants, and animals that grow rapidly (Figure 4-D). They have also created genetically engineered bacteria to extract minerals such as copper from their underground ores and to clean up spills of oil and other toxic pollutants.

Critical Thinking

If genetic engineering were more widely applied to plants and animals in the future, how might it affect the evolutionary process? What benefits and harms might result?

Pear   Apple

Desired trait (color)

Crossbreeding

Offspring

Best result

Crossbreeding

New offspring

Desired result

Figure 4-C Artificial selection involves the crossbreeding of species that are close to one another genetically. In this example, similar fruits are being crossbred.

© Cengage Learning

---

the average annual background extinction rate has been about 0.0001% of all species per year, which amounts to 1 species lost for every million species on the earth per year. At this rate, if there were 10 million species on the earth, about 10 of them, on average, would go extinct every year.

In contrast to the background extinction rate, a **mass extinction** is a significant rise in extinction rates above the background level. In such a catastrophic, widespread, and often global event, large groups of species (25–95% of all species) are wiped out. Fossil and geological evidence indicate that there have been at least three and probably five mass extinctions (at intervals of 20–60 million years) dur-

ing the past 500 million years. Scientists point out that there have been numerous other times when extinction rates have gone well above the background extinction rate.

A mass extinction provides an opportunity for the evolution of new species that can fill unoccupied ecological roles or newly created ones. As a result, evidence indicates that each occurrence of mass extinction has been followed by an increase in species diversity over several million years as new species have arisen to occupy new habitats or to exploit newly available resources.

As environmental conditions change, the balance between formation of new species (speciation) and

Figure 4-13 This male golden toad lived in Costa Rica's high-altitude Monteverde Cloud Forest Reserve. The species became extinct in 1989 apparently because its habitat dried up.

extinction of existing species determines the earth's biodiversity (Concept 4-4A). The existence of millions of species today means that speciation, on average, has kept ahead of extinction. However, many scientists argue that higher extinction rates and other evidence indicate that we are experiencing the beginning of a new mass extinction.

There is also considerable evidence that much of the current rise in the extinction rate and the resulting loss of biodiversity are primarily due to human activities (Concept 4-4B), as our ecological footprints spread over the planet (see Figure 1-12, p. 13, and Figure 1-13, p. 14). Research indicates that the largest cause of the rising rate of species extinctions is the loss, fragmentation, and degradation of habitats. These losses occur as we cultivate more land to grow crops and clear more forestland for farming, ranching, and settlement (see Figure 3-1, p. 52). We examine this issue further in Chapters 9 and 10.

4-5 What Is Species Diversity and Why Is It Important?

CONCEPT 4-5
Species diversity is a major component of biodiversity and tends to increase the sustainability of some ecosystems.

Species Diversity Includes the Variety and Abundance of Species in a Particular Place

An important characteristic of a community and the ecosystem to which it belongs is its **species diversity**, or the number and variety of species it contains. One important component of species diversity is *species richness,* the number of different species in a given area. For example, an area of natural temperate rain forest (Figure 4-14, left) has a much higher species richness than a tree farm (Figure 4-14, right), which is usually planted with one species and hosts just a few other species that migrate in.

For any given community, another component of species diversity is *species evenness,* a measure of the relative abundance, or the comparative numbers of individuals of each species present. The more even the numbers of individuals in each species in a community, the higher the species evenness in that community. For example, a tree farm (Figure 4-14, right) with a large number of individuals of one species of trees and low numbers of individuals from the other plant species has low species evenness. However, most temperate rain forests (Figure 4-14, left) have high species evenness, because they contain similar numbers of individuals from each of many different species.

The species diversity of communities varies with their geographical location. For most terrestrial plants and animals, species diversity (primarily species richness) is highest in the tropics and declines as we move from the equa-

tor toward the poles. The most species-rich environments are tropical rain forests, large tropical lakes, coral reefs, and the ocean-bottom zone.

Species-Rich Ecosystems Tend to Be Productive and Sustainable

Ecologists have been conducting research to answer two important questions: *First,* is plant productivity higher in species-rich ecosystems? *Second,* does species richness enhance the *stability,* or *sustainability,* of an ecosystem?

Research suggests that the answers to both questions may be *yes,* and from that research, two hypotheses have emerged. According to the first, the more diverse an ecosystem is, the more productive it will be. That is, with a greater variety of producer species, an ecosystem will produce more plant biomass, which in turn will support a greater variety of consumer species.

The second hypothesis is that greater species richness and productivity will make an ecosystem more stable or sustainable. According to this hypothesis, a complex ecosystem, with many different species (high species richness) and the resulting variety of feeding paths, has more ways to respond to most environmental disturbances and stresses because it does not have all its eggs in one basket. According to biologist Edward O. Wilson, "There's a common sense element to this: the more species you have, the more likely you're going to have an insurance policy for the whole ecosystem."

Ecologist David Tilman and his colleagues at the University of Minnesota did research that supported both hypotheses. They found that communities with high plant species richness produced a certain amount of biomass more consistently than did communities with fewer spe-

Figure 4-14 This area of natural temperate rain forest in Washington's Olympic National Park (left) has a much higher number of species (higher species richness) and higher species evenness than this tree farm planted in Oregon (right) has.

Left: Natalia Bratslavsky/Shutterstock.com; Right: Knowlesgal. . . | Dreamstime.com

cies. The species-rich communities were also less affected by drought and more resistant to invasions by insect species. Later laboratory studies involved setting up artificial ecosystems in growth chambers where researchers could control and manipulate key variables such as temperature, light, and atmospheric gas concentrations. These studies have supported Tilman's findings.

Ecologists hypothesize that in a species-rich ecosystem, each species can exploit a different portion of the resources available. For example, some plants will bloom early and others will bloom late. Some have shallow roots to absorb water and nutrients in topsoil, and others use longer roots to tap into deeper soils. A number of studies support this hypothesis, although some do not.

There is debate among scientists about how much species richness is needed to help sustain various ecosystems. Some research suggests that the average annual net primary productivity of an ecosystem reaches a peak with 10–40 producer species. Many ecosystems contain more than 40 producer species, but do not necessarily produce more biomass or reach a higher level of stability. While there is more research to do in this area, most ecologists now accept as a useful hypothesis the idea that species richness appears to increase the productivity and stability, or sustainability, of an ecosystem (Concept 4-5).

4-6 What Roles Do Species Play in Ecosystems?

CONCEPT 4-6A
Each species plays a specific ecological role called its *niche*.

CONCEPT 4-6B
Any given species may play one or more of four important roles—native, nonnative, indicator, or keystone—in a particular ecosystem.

Each Species Plays a Role in Its Ecosystem

An important principle of ecology is that *each species has a specific role to play in the ecosystems where it is found* (Concept 4-6A). Scientists describe the role that a species plays in its ecosystem as its **ecological niche**, or simply **niche**. It is a species' way of life in a community and includes everything that affects its survival and reproduction, such as how much water and sunlight it needs, how much space it requires, what it feeds on, what feeds on it, and the temperatures and other conditions it can tolerate. A species' niche should not be confused with its **habitat**, which is the place where it lives. Its niche is its pattern of living.

Scientists use the niches of species to classify them mostly as *generalists* or *specialists*. **Generalist species** have broad niches (Figure 4-15, right curve). They can live in many different places, eat a variety of foods, and often tolerate a wide range of environmental conditions. Flies, cockroaches, mice, rats, white-tailed deer, and humans are generalist species.

In contrast, **specialist species** occupy narrow niches (Figure 4-15, left curve). They may be able to live in only

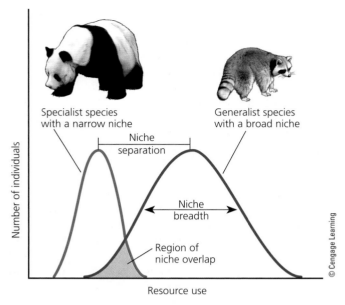

Figure 4-15 Specialist species such as the giant panda have a narrow niche (left curve) and generalist species such as the raccoon have a broad niche (right curve).

one type of habitat, use just one or only a few types of food, or tolerate a narrow range of climatic and other environmental conditions. For example, some shorebirds occupy specialized niches, feeding on crustaceans, insects, and other organisms found on sandy beaches and their adjoining coastal wetlands (Figure 4-16).

Because of their narrow niches, specialists are more prone to extinction when environmental conditions change. For example, China's *giant panda* (Case Study that follows) is highly endangered because of a combination of

habitat loss, low birth rate, and its specialized diet consisting mostly of bamboo.

Is it better to be a generalist or a specialist? It depends. When environmental conditions are fairly constant, as in a tropical rain forest, specialists have an advantage because they have fewer competitors. But under rapidly changing environmental conditions, the more adaptable generalist usually is better off than the specialist.

CONSIDER THIS. . .

THINKING ABOUT Amphibians' Niches

Do you think that most amphibian species (**Core Case Study**) occupy specialist or generalist niches? Explain.

CASE STUDY

The Giant Panda—A Highly Endangered Specialist

The *giant panda* (Figure 4-17) is among the most threatened of all species, rated as *endangered* by the IUCN. According to the World Wildlife Fund (WWF) there are 1,600 to 3,000 giant pandas left in the wild, most of them in China.

Pandas evolved to live in the forests of China where bamboo, their main food source, grows thickly among evergreen trees. A typical panda spends 12 hours a day eating bamboo stalks and leaves.

The need for bamboo makes the panda a specialist species. For this reason, the main threat to pandas is destruction and fragmentation of their forest habitat where the bamboo grows. The pandas are now limited to six remaining mountain forests of central and southwestern China. Some scientists warn that projected climate change due to atmospheric warming will limit the range of bamboo forests, making the problem even worse.

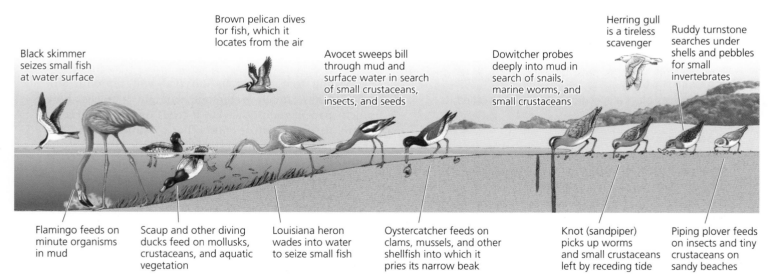

Black skimmer seizes small fish at water surface

Brown pelican dives for fish, which it locates from the air

Avocet sweeps bill through mud and surface water in search of small crustaceans, insects, and seeds

Dowitcher probes deeply into mud in search of snails, marine worms, and small crustaceans

Herring gull is a tireless scavenger

Ruddy turnstone searches under shells and pebbles for small invertebrates

Flamingo feeds on minute organisms in mud

Scaup and other diving ducks feed on mollusks, crustaceans, and aquatic vegetation

Louisiana heron wades into water to seize small fish

Oystercatcher feeds on clams, mussels, and other shellfish into which it pries its narrow beak

Knot (sandpiper) picks up worms and small crustaceans left by receding tide

Piping plover feeds on insects and tiny crustaceans on sandy beaches

Figure 4-16 Various bird species in a coastal wetland occupy specialized feeding niches. This specialization reduces competition and allows for sharing of limited resources.

© Cengage Learning

Figure 4-17
This giant panda is eating a stalk from a bamboo tree. Bamboo stalks and leaves make up about 95% of the diet of this specialist species.

Mike Flippo/Shutterstock.com

The giant panda is also endangered because of its low reproductive rate. Pandas have a very short breeding season, and females give birth to only one or two cubs every 2–3 years. If twins are born, usually only one survives, because the mother usually cannot produce enough milk for two cubs. Another threat is poaching—the illegal killing of giant pandas for their fur.

The government of China, along with international conservation groups, has been trying to address these threats to pandas. By 2011, the government had set aside more than 50 panda reserves in about half of the remaining panda habitat area. The government estimated that about 60% of the remaining wild panda population was thus protected. In addition, according to the WWF, there were more than 300 pandas in captivity on mainland China and in 30 other countries.

GOOD NEWS

Groups like the IUCN and WWF call for protecting larger areas of forest and for connecting fragmented areas with forested corridors through which the pandas could move between seasons to gain more access to bamboo. They also call for more patrolling to control poaching and illegal logging in protected areas.

Species Can Play Four Major Roles within Ecosystems

Niches can be classified further in terms of specific roles that certain species play within ecosystems. Ecologists describe *native, nonnative, indicator,* and *keystone* roles. Any given species may play one or more of these four roles in a particular ecosystem (**Concept 4-6B**).

Native species are those species that normally live and thrive in a particular ecosystem. Other species that migrate into, or are deliberately or accidentally introduced into, an ecosystem are called **nonnative species**, also referred to as *invasive, alien,* and *exotic species.*

People often think of nonnative species as threatening. In fact, most introduced and domesticated plant species such as food crops and flowers and animals such as chickens, cattle, and fish from around the world are beneficial to us. However, some nonnative species can compete with and reduce a community's native species, causing unintended and unexpected consequences. In 1957, for example, Brazil imported wild African honeybees (Figure 4-18) to help increase honey production. Instead of helping, the bees displaced some native honeybee populations, which had the effect of reducing the honey supply.

Since then, these nonnative honeybee species—popularly known as "killer bees"—have moved northward into Central America and parts of the southwestern and southeastern United States. The wild African bees are not the fearsome killers portrayed in some horror movies. However, they are aggressive and unpredictable and have killed thousands of domesticated animals and an estimated 1,000 people in the western hemisphere, many of whom were allergic to bee stings.

Nonnative species can spread rapidly if they find a new location with favorable conditions. In their new niches, these species often do not face the predators and diseases they face in their native niches, or they may be able to out-compete some native species in their new locations. We will examine this environmental threat in greater detail in Chapter 9.

Indicator Species Serve as Biological Smoke Alarms

Species that provide early warnings of damage to a community or an ecosystem are called **indicator species**. For example, in the **Core Case Study** that opened this chapter, we learned that some amphibians are classified as indicator species. The decline of an amphibian population can indicate the presence of parasites, disease-causing

John Lindsay-Smith/Shutterstock.com

Figure 4-18 Wild African honeybees, popularly known as "killer bees," were imported into Brazil and their populations have since spread widely.

SCIENTISTS ARE SEARCHING FOR THE CAUSES OF AMPHIBIAN DECLINES

Herpetologists, the scientists who study frogs and other amphibians (Figure 4-E), have identified a number of factors—both natural and human-caused—that affect these species at various points in their life cycles. One of the natural causes is *parasites* such as flatworms, which feed on the amphibian eggs laid in water and apparently have caused an increase in the number of births of amphibians with limbs missing or with extra limbs.

Some herpetologists hypothesize that *viral and fungal diseases,* especially the chytrid fungus that attacks the skin of frogs, are reducing the frogs' ability to take in water through their skin. This leads to death from dehydration. Such diseases can spread fairly easily, because adults of many amphibian species congregate in large numbers to breed.

As with other threatened species, one of the major threats to amphibians is *habitat loss and fragmentation.* This is almost completely a human-caused problem resulting from the clearing of forests and the draining and filling of freshwater wetlands for farming and urban development.

Another threat is *prolonged drought,* which can dry up breeding pools that frogs and other amphibians depend on for reproduction and in early life. There is evidence that human activities are accelerating climate change by adding to atmospheric warming (as we discuss in Chapter 19) and this is helping to prolong natural

droughts in many areas of the world. A 2005 study found an apparent correlation between climate change caused by atmospheric warming and the extinction of about two-thirds of the 110 known species of harlequin frogs in tropical forests of Central and South America.

Another human-influenced problem is *higher levels of UV radiation,* which can harm embryos of amphibians in shallow ponds, as well as adults basking in the sun for warmth. Historically, such radiation has been screened by ozone in the stratosphere, but during the past few decades, ozone-depleting chemicals released into the atmosphere from human sources have destroyed some of the protective ozone. International action has been taken to reduce the threat of stratospheric ozone depletion, but it will take about 50 years for ozone levels to recover to those of 1960.

Pollution and *overhunting* are two other human-caused threats to amphibians. Frogs and other species are increasingly exposed to pesticides in ponds and in the bodies of insects that they consume. This can make them more vulnerable to bacterial, viral, and fungal diseases and to some parasites. Also, frogs are hunted for their leg meat, especially in Asia and France, where they are overhunted in many areas.

Yet another threat to amphibians being studied by scientists is the invasion of their habitats by *nonnative predators and competitors,* such as certain fish spe-

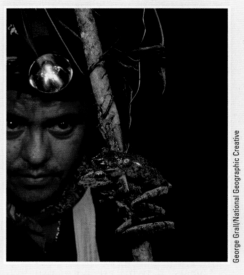

Figure 4-E This researcher is watching frogs mating in a tropical forest in Brazil.

George Grall/National Geographic Creative

cies. Some of this immigration is natural, but humans accidentally or deliberately transport many species to amphibian habitats, another problem that we discuss further in Chapter 9.

Most herpetologists believe that a combination of these factors, which vary from place to place, probably is responsible for most of the decline and disappearances among amphibian species.

Critical Thinking

Of the factors listed above, which three do you think could be most effectively controlled by human efforts?

---

microbes, or pollution in the local environment (Science Focus 4.3), as well as habitat destruction and fragmentation. It can also be a sign of the effects of climate change (see Figure 4-13).

Birds are excellent biological indicators because they are found almost everywhere and are affected quickly by environmental changes such as the loss or fragmentation of their habitats and the introduction of chemical pesticides. Some butterflies are also indicator species because their association with various plant species makes them vulnerable to habitat loss and fragmentation. The popula-

tions of many bird and butterfly species are declining—a problem we explore more fully in Chapter 9.

Keystone Species Play Critical Roles in Their Ecosystems

A keystone is the wedge-shaped stone placed at the top of a stone archway. Remove this stone and the arch collapses. In some communities and ecosystems, ecologists hypothesize that certain species play a similar role. **Keystone spe-**

Figure 4-19 *Keystone species:* The American alligator plays an important ecological role in its marsh and swamp habitats in the southeastern United States.

Martha Marks/Shutterstock.com

cies are species whose roles have a large effect on the types and abundance of other species in an ecosystem.

Keystone species often exist in relatively limited numbers in their ecosystems, but the effects that they have there are often much larger than their numbers would suggest. And because of their often-smaller numbers, some keystone species are more vulnerable to extinction than other species are.

Keystone species can play several critical roles in helping to sustain ecosystems. One such role is the *pollination* of flowering plant species by butterflies (see chapter-opening photo), honeybees (Figure 4-A), hummingbirds, bats, and other species. In addition, *top predator* keystone species feed on and help to regulate the populations of other species. Examples are the wolf, leopard, lion, some shark species, and the American alligator (see Case Study that follows).

The loss of a keystone species can lead to population crashes and extinctions of other species in a community that depends on them for certain ecosystem services. This is why it so important for scientists to identify and protect keystone species.

CASE STUDY

The American Alligator—A Keystone Species That Almost Went Extinct

The American alligator (Figure 4-19) is a keystone species because it plays a number of important roles in the ecosystems where it is found in the southeastern United States. This species has outlived the dinosaurs and survived many challenges to its existence.

In the 1930s, hunters began killing large numbers of these animals for their exotic meat and their soft belly skin, used to make expensive shoes, belts, and pocketbooks. Other people hunted alligators for sport or out of dislike for the large reptile. By the 1960s, hunters and poachers had wiped out 90% of the alligators in the state of Louisiana, and the alligator population in the Florida Everglades was also near extinction.

Those who did not care much for the alligator were probably not aware of its important ecological role—its *niche*—in subtropical wetland ecosystems. Alligators dig deep depressions, or gator holes. These depressions hold freshwater during dry spells, serve as refuges for aquatic life, and supply freshwater and food for fishes, insects, snakes, turtles, birds, and other animals.

The large nesting mounds that alligators build provide nesting and feeding sites for some herons and egrets, and red-bellied turtles lay their eggs in old gator nests. In addition, alligators eat large numbers of gar, a predatory fish, which helps to maintain populations of game fish such as bass and bream that the gar eat.

As alligators create gator holes and nesting mounds, they help to keep shore and open water areas free of invading vegetation. Without this free ecosystem service, freshwater ponds and coastal wetlands where alligators live would be filled in with shrubs and trees, and dozens of species would disappear from these ecosystems.

In 1967, the U.S. government placed the American alligator on the endangered species list. By 1977, because it was protected, its populations had made a strong enough comeback to be removed from the endangered species list.

Today, there are well over a million alligators in Florida. In fact, the population has recovered to the point where the state allows property owners to kill alligators that stray onto their land. To conservation biologists, the comeback of the American alligator is an important success story in wildlife conservation.

CONSIDER THIS...

THINKING ABOUT The American Alligator and Biodiversity

What are two ways in which the American alligator supports one or more of the four components of biodiversity (Figure 4-2) within its environment?

Big Ideas

- Populations evolve when genes mutate and give some individuals genetic traits that enhance their abilities to survive and to produce offspring with these traits (natural selection).

- Human activities are degrading the earth's vital biodiversity by hastening the extinction of species and by disrupting habitats needed for the development of new species.

- Each species plays a specific ecological role (its ecological niche) in the ecosystem where it is found.

TYING IT ALL TOGETHER   Amphibians and Sustainability

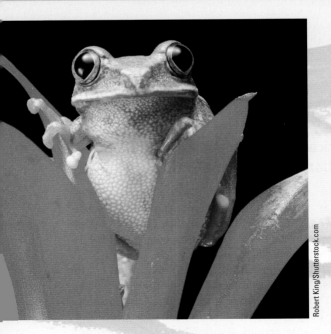

Robert King/Shutterstock.com

The **Core Case Study** on amphibians at the beginning of this chapter shows that the importance of a species does not always match the public's perception of it. While many people worry about the fates of the giant panda and polar bears, most are not aware that many species of amphibians are declining dramatically or that they play key roles in their ecosystems, such as controlling insect populations. Their extinction would soon be followed by the disappearances of many species of birds, fish, and mammals that feed on them.

In this chapter, we studied the importance of biodiversity—the numbers and varieties of species found in different parts of the world, along with genetic, ecosystem, and functional diversity. We also studied the process whereby all species came to be, according to the scientific theory of biological evolution through natural selection. Taken together, biodiversity and evolution represent two vital and irreplaceable forms of natural capital.

Finally, we examined the variety of roles played by species in ecosystems. For example, we saw that some species, including many amphibians, are indicator species that clue us in to the presence of threats to biodiversity, to ecosystems, and to the biosphere. Others such as the American alligator are keystone species that play vital roles in sustaining the ecosystems where they live.

Ecosystems and the variety of species they contain are functioning examples of the three scientific **principles of sustainability** (see Figure 1-2, p. 6 or back cover) in action. They depend on solar energy and on the cycling of nutrients. Ecosystems also help to sustain biodiversity.

Chapter Review

Core Case Study

1. Describe the threats to many of the world's amphibian species (**Core Case Study**) and explain why we should avoid hastening the extinction of any amphibian species through our activities.

Section 4-1

2. What is the key concept for this section? Define **biodiversity (biological diversity)** and list and describe its four major components. What is the importance of biodiversity? Define **species**. Summarize the importance of insects. Define and give three examples of **biomes**. Summarize the scientific contributions of Edward O. Wilson.

Section 4-2

3. What are the two key concepts for this section? What is a **fossil** and why are fossils important for understanding the history of life? What is **biological evolution** (or **evolution**)? State the **theory of evolution**. What is **natural selection**? What is a **mutation** and what role do mutations play in evolution by natural selection? What is an **adaptation (adaptive trait)**? How did humans become such a powerful species? What are two limitations on evolution by natural selection? What are three myths about evolution through natural selection?

Section 4-3

4. What is the key concept for this section? Describe how geologic processes can affect natural selection. How can climate change and catastrophes such as asteroid impacts affect natural selection?

Section 4-4

5. What are the two key concepts for this section? Define **speciation**. Distinguish between **geographic isolation** and **reproductive isolation**, and explain how they can lead to the formation of a new species. Define and distinguish between **artificial selection** and **genetic engineering** and give an example of each.

6. What is **extinction**? What is an **endemic species** and why can such a species be vulnerable to extinction? Define and distinguish between **background extinction rate** and **mass extinction**.

Section 4-5

7. What is the key concept for this section? Define **species diversity** and distinguish between species richness and species evenness. Explain why species-rich ecosystems tend to be productive and sustainable.

Section 4-6

8. What are the two key concepts for this section? Define and distinguish between an **ecological niche** (or **niche**) and a **habitat**. Distinguish between **generalist species** and **specialist species** and give an example of each. Why has the fact that the giant panda is a specialist species led to its classification as an endangered species?

9. Define and distinguish among **native**, **nonnative**, **indicator**, and **keystone species** and give an example of each. What major ecological role do many amphibian species play (**Core Case Study**)? List six factors that contribute to the threats of extinction for frogs and other amphibians. Describe the role of the American alligator as a keystone species.

10. What are this chapter's *three big ideas*? How are ecosystems and the variety of species they contain related to the three scientific **principles of sustainability**?

Note: Key terms are in bold type.

Critical Thinking

1. How might we and other species be affected if all amphibians (**Core Case Study**) were to go extinct?

2. What role does each of the following processes play in helping to implement the three scientific **principles of sustainability**: **(a)** natural selection, **(b)** speciation, and **(c)** extinction?

3. How would you respond to someone who tells you that:
   a. he or she does not believe in biological evolution because it is "just a theory"?
   b. we should not worry about air pollution because natural selection will enable humans to develop lungs that can detoxify pollutants?

4. Is the human species a keystone species? Explain. If humans were to become extinct, what are three species that might also become extinct and three species whose populations would probably grow?

5. How would you respond to someone who says that because extinction is a natural process, we should not worry about the loss of biodiversity when species become extinct largely as a result of our activities?

6. List three ways in which you could apply Concept 4-4B to making your lifestyle more environmentally sustainable.

7. Congratulations! You are in charge of the future evolution of life on the earth. What are the three things that you would consider to be the most important to do?

8. If you were forced to choose between saving the giant panda from extinction and saving amphibians, which would you choose? Explain.

Doing Environmental Science

Study an ecosystem of your choice, such as a meadow, a patch of forest, a garden, or an area of wetland. (If you cannot do this physically, do so virtually by reading about an ecosystem online or in a library.) Determine and list five major plant species and five major animal species in your ecosystem. Write hypotheses about **(a)** which of these species, if any, are indicator species and **(b)** which of them, if any, are keystone species. Explain how you arrived at your hypotheses. Then design an experiment to test each of your hypotheses, assuming you would have unlimited means to carry them out.

Global Environment Watch Exercise

Search for *Amphibians* to find out more about the current state of these species with regard to threats to their existence (**Core Case Study**). What actions are being taken by various nations and organizations to protect amphibians? Write a short summary report on your research.

Data Analysis

The following table is a sample of a very large body of data reported by R. A. Alford and S. J. Richards in their book *Extinction in our Times–Global Amphibian Decline*. It compares various areas of the world in terms of the number of amphibian species found and the number of amphibian species that were endemic, or unique to each area. Scientists like to know these percentages because endemic species tend to be more vulnerable to extinction than do non-endemic species. Study the table below and then answer the questions that follow it.

Area	Number of Species	Number of Endemic Species	Percentage Endemic
Pacific/Cascades/Sierra Nevada Mountains—North America	52	43	
Southern Appalachian Mountains—USA	101	37	
Southern Coastal Plain—USA	68	27	
Southern Sierra Madre—Mexico	118	74	
Highlands of Western Central America	126	70	
Highlands of Costa Rica and Western Panama	133	68	
Tropical Southern Andes Mountains—Bolivia and Peru	132	101	
Upper Amazon Basin—Southern Peru	102	22	

1. Fill in the fourth column to calculate the percentage of amphibian species that are endemic to each area.

2. What two areas have the highest numbers of endemic species? Name the two areas with the highest percentages of endemic species.

3. What two areas have the lowest numbers of endemic species? What two areas have the lowest percentages of endemic species?

4. What two areas have the highest percentages of non-endemic species?

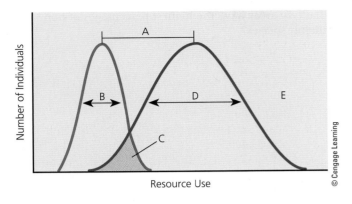

Questions 1–3 refer to the diagram above.

1. Which part of the diagram depicts the resource use that would be in competition if a new species moves into an area and is in direct competition with another species?
   a. A
   b. B
   c. C
   d. D
   e. E

2. The niche of the black bear which is an omnivore would be best represented as
   a. A.
   b. B.
   c. C.
   d. D.
   e. E.

3. The area under the curve depicted by letter B would be best described as
   a. the productivity of a producer such as grass.
   b. the productivity of a tropical rain forest.
   c. the niche of a specialist like the panda bear that eats only bamboo.
   d. the competition between a producer and a consumer.
   e. how far the niche of two species of animals are separated from each other.

4. Which of the below is the best summary of evolution?
   a. The need for organisms to evolve a more perfect form.
   b. The strongest organisms will survive to reproduce.
   c. The change in the genetic makeup of successive generations of a species.
   d. The grand plan of nature determines how organisms will evolve.
   e. The organisms have desires for particular traits and so evolve them.

5. Keystone species such as alligators in the Southeast U.S. have an important role and effect on their environment through
   a. having large numbers to regulate the producers.
   b. playing a critical role in sustaining their ecosystem.
   c. shaping the community to enhance it for other species.
   d. controlling bacterial levels preventing the infection of other species.
   e. eliminating all potential predators in the ecosystem.

Questions 6–9 refer to the terms below.

   a. indicator species
   b. keystone species
   c. foundation species
   d. native species
   e. specialist species

6. As pollution levels in streams rise, many aquatic insects such as the mayfly quickly disappear. For this reason, many aquatic insects are studied intensively.

7. The beaver transforms its environment from streams to ponds and swamps, allowing a diverse collection of organisms to thrive that would not normally be able to survive.

8. Kelp (a brown algae) forms large beds, creating a habitat for many fishes and shellfishes.

9. The *Chitymomma* is an Agave that regionally helps or is used to define the Chihuahuan Desert of Northern Mexico and the Southwestern United States.

10. Protists are best described as
    a. prokaryotes such as cyanobacteria, consisting of single cells with no nucleus.
    b. single-celled bacteria such as thermophiles that live in extreme conditions such as oxygen-free sediments in swamps, water with very high salinity, or heat such as hydrothermal vents.
    c. eukaryotes, mostly single-celled organisms with a nucleus, such as diatoms and protozoans, some of which cause diseases in humans.
    d. usually multicelled; many are decomposers such as molds and mildews.
    e. multicelled organisms such as mosses and ferns, which can be annuals or perennials.

11. Atrazine is an herbicide that blocks photosynthesis and is frequently cited as a cause of mutations in frogs and other amphibians. Atrazine is often applied at rates of 2.9×10^7 ppb on agricultural fields, yet the EPA limit for atrazine in drinking water is 3 ppb. If a farmer applies atrazine on a field at the rate of 4 pounds per acre, and sprays an average farm of 1,000 acres, how many gallons of water could be contaminated at a rate of 3 ppb? (1 gallon of water weighs 8.34 pounds)
    a. 4.2×10^{12} gallons (Lake Seneca, the deepest Finger Lake in NY)
    b. 2.5×10^{12} gallons
    c. 3.70×10^{11} gallons
    d. 1.60×10^{11} gallons
    e. 2.1×10^{10} gallons (Lake Otisco, one of the smaller Finger Lakes in NY)

12. Which of the below are mechanisms of speciation?
    I. Reproductive isolation
    II. Mutation
    III. Natural selection
    a. I only
    b. II only
    c. I and II only
    d. II and III only
    e. I, II, and III

13. The best example of an endemic species is
    a. the golden toad, extinct since 1989, which lived in the Monteverde Cloud Forest Reserve.
    b. amphibians that live in ponds in agricultural areas.
    c. reindeer that live north of the Arctic Circle.
    d. humans that live in the Andes Mountains of South America.
    e. the gray wolf, which is locally extinct in the American northeast.

5 Biodiversity, Species Interactions, and Population Control

In looking at nature, never forget that every single organic being around us may be said to be striving to increase its numbers.

CHARLES DARWIN, 1859

Key Questions

5-1 How do species interact?

5-2 How do communities and ecosystems respond to changing environmental conditions?

5-3 What limits the growth of populations?

A clownfish gains protection by living among sea anemones and helps protect the anemones from some of their predators.

cbpix/Shutterstock.com

Southern sea otters (Figure 5-1, left) live in giant kelp forests (Figure 5-1, right) in shallow waters along parts of the Pacific coast of North America. Most of the remaining members of this endangered species are found off the western coast of the United States between the cities of Santa Cruz and Santa Barbara, California.

Southern sea otters are fast and agile swimmers that dive to the ocean bottom looking for shellfish and other prey, including sea urchins. On the surface they swim on their backs and feed on their prey, using their bellies as a table (Figure 5-1, left). Each day, a sea otter consumes 20–35% of its weight in clams, mussels, crabs, sea urchins, abalone, and about 40 other species of bottom-dwelling organisms.

It is estimated that between 13,000 and 20,000 southern sea otters once lived in California's coastal waters. By the early 1900s, they were hunted almost to extinction in this region by fur traders who killed them for their thick, luxurious fur.

Commercial fisherman also killed otters because they viewed them as competitors in the hunt for valuable abalone and other shellfish.

Since that time, this population has generally grown from a low of about 50 in 1938 to an estimated 2,800 in 2012. Their partial recovery got a boost in 1977 when the U.S. Fish and Wildlife Service declared the species endangered in most of its range, with a total population of only 1,850 individuals. Despite such progress, the population has a long way to go to justify removing it from the endangered species list.

Why should we care about the southern sea otters of California? One reason is *ethical:* many people believe it is wrong to allow human activities to cause the extinction of a species. Another reason is that people love to look at these appealing and highly intelligent animals as they play in the water. As a result, they help to generate millions of dollars a year

in tourism revenues. A third reason—and a key reason in our study of environmental science—is that biologists classify them as a *keystone species* (p. 94). Scientists hypothesize that in the absence of southern sea otters, sea urchins and other kelp-eating species would probably destroy the Pacific coast kelp forests and much of the rich biodiversity they support.

Biodiversity is an important part of the earth's natural capital and is the focus of one of the three scientific **principles of sustainability** (see Figure 1-2, p. 6 or back cover). In this chapter, we will look at two factors that affect biodiversity: how species interact and help control one another's population sizes and how communities, ecosystems, and populations of species respond to changes in environmental conditions.

Figure 5-1 An endangered southern sea otter in Monterey Bay, California (USA) uses a stone to crack the shells of clams that it feeds on (left). It lives in a giant bed of seaweed called kelp (right).

Left: © Xfkirsten | Dreamstime.com; Right: Paul Whitted/Shutterstock.com

5-1 How Do Species Interact?

CONCEPT 5-1
Five types of interactions among species—interspecific competition, predation, parasitism, mutualism, and commensalism—affect the resource use and population sizes of species.

Most Species Compete with One Another for Certain Resources

Ecologists have identified five basic types of interactions among species as they share limited resources such as food, shelter, and space. These types of interactions are called *interspecific competition, predation, parasitism, mutualism,* and *commensalism,* and they all have significant effects on the population sizes of the species in an ecosystem and on how they use resources (**Concept 5-1**).

The most common interaction among species is **interspecific competition**, which occurs when members of two or more species interact to use the same limited resources such as food, water, light, and space. While fighting for resources does occur, most interspecific competition involves the ability of one species to become more efficient than another species in obtaining the resources it needs.

Recall that each species plays a role in its ecosystem called its *ecological niche* (p. 91 and Figure 4-15, p. 92). When two species compete with one another for the same resources, their niches *overlap.* The greater this overlap, the more intense is their competition for key resources.

If one species can take over the largest share of one or more key resources, each of the other competing species must move to another area (if possible), adapt by shifting its feeding habits or behavior through natural selection to reduce or alter its niche, suffer a sharp population decline, or become extinct in that area.

Humans compete with many other species for space, food, and other resources. As our ecological footprints grow and spread (see Figure 1-13, p. 14), we are taking over or degrading the habitats of many other species and depriving them of resources they need in order to survive.

Some Species Evolve Ways to Share Resources

Over a time scale long enough for natural selection to occur, populations of some species develop adaptations that allow them to reduce or avoid competition with other species for resources. One way this happens is through **resource partitioning**. It occurs when species competing for similar scarce resources evolve specialized traits that allow them to share resources by using parts of them, using them at different times, or using them in different ways.

Figure 5-2 shows resource partitioning by some insect-eating bird species. In this case, their adaptations allow them to reduce competition by feeding in different portions of certain spruce trees and by feeding on different insect species.

Blackburnian Warbler   Black-throated Green Warbler   Cape May Warbler   Bay-breasted Warbler   Yellow-rumped Warbler

Figure 5-2 *Sharing the wealth:* Resource partitioning among five species of insect-eating warblers in the spruce forests of the U.S. state of Maine. Each species spends at least half its feeding time in its associated yellow-highlighted areas of these spruce trees.

(Based on R. H. MacArthur, "Population Ecology of Some Warblers in Northeastern Coniferous Forests," *Ecology* 36 (1958): 533–536, 1958.)

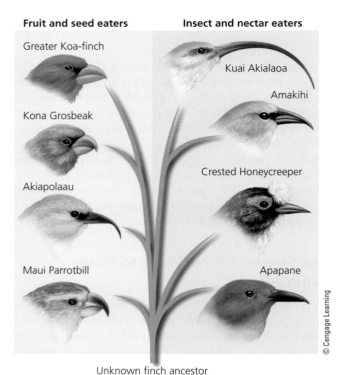

Fruit and seed eaters

Greater Koa-finch

Kona Grosbeak

Akiapolaau

Maui Parrotbill

Insect and nectar eaters

Kuai Akialaoa

Amakihi

Crested Honeycreeper

Apapane

Unknown finch ancestor

© Cengage Learning

Figure 5-3 *Specialist species of honeycreepers:* Through natural selection, different species of honeycreepers have shared resources by evolving specialized beaks to take advantage of certain types of food such as insects, seeds, fruits, and nectar from certain flowers. **Question:** Look at each bird's beak and take a guess at what sort of food that bird might eat.

Another example of resource partitioning through natural selection involves birds called honeycreepers that live in the U.S. state of Hawaii (Figure 5-3). And Figure 4-16 (p. 92) shows how the evolution of specialized feeding niches has reduced competition for resources among bird species in a coastal wetland.

Consumer Species Feed on Other Species

In **predation**, a member of one species (the **predator**) feeds directly on all or part of a living organism (the **prey**) as part of a food web. Together, the two different species, such as a brown bear (the preda-

tor, or hunter) and a salmon (the prey, or hunted), form a **predator–prey relationship** (Figure 5-4). Such relationships are also shown in Figure 3-7 (p. 57).

🔍 **CONSIDER THIS. . .**

CONNECTIONS Grizzly Bears and Moths

During the summer months, the grizzly bears of the Greater Yellowstone ecosystem in the western United States eat huge amounts of army cutworm moths, which huddle in masses high on remote mountain slopes. In this predator–prey interaction, one grizzly bear can dig out and lap up as many as 40,000 cutworm moths in a day. Consisting of 50–70% fat, the moths offer a nutrient that the bear can store in its fatty tissues and draw on during its winter hibernation.

In a giant kelp forest ecosystem, sea urchins prey on kelp, a form of seaweed (Science Focus 5.1). However, as a keystone species (see Chapter 4, p. 94), southern sea otters (**Core Case Study**) prey on the sea urchins and thus help keep them from destroying the kelp forests.

Predators have a variety of methods that help them to capture prey. *Herbivores* can simply walk, swim, or fly to the plants they feed on. *Carnivores* feeding on mobile prey have two main options: *pursuit* and *ambush*. Some, such as the cheetah, catch prey by running fast. Others, such as the American bald eagle, can fly and have keen eyesight. Still others cooperate in capturing their prey. For example, female African lions often hunt together to prey on zebras (Figure 3-7, p. 57), wildebeest, antelopes, and other fast-running large animals of the open savanna grasslands.

Other predators use *camouflage* to hide in plain sight and ambush their prey. For example, praying mantises (see Figure 4-A, right, p. 80) sit on flowers or plants of a color similar to their own and ambush visiting insects. White ermines (a type of weasel), snowy owls, and arctic foxes (Figure 5-5) hunt their prey in snow-covered areas.

Steve Hilebrand/U.S. Fish and Wildlife Service

Figure 5-4 *Predator–prey relationship:* This brown bear (the predator) in the U.S. state of Alaska has captured and will feed on this salmon (the prey).

SCIENCE FOCUS 5.1

THREATS TO KELP FORESTS

A kelp forest is composed of large concentrations of a seaweed called *giant kelp*. Anchored to the ocean floor, its long blades grow toward the sunlit surface waters (Figure 5-1, right). Under good conditions, the blades can grow 0.6 meter (2 feet) in a day and the plant can grow as high as a ten-story building. The blades are very flexible and can survive all but the most violent storms and waves.

Kelp forests are one of the most biologically diverse ecosystems found in marine waters, supporting large numbers of marine plants and animals. These forests help reduce shore erosion by blunting the force of incoming waves and trapping some of the outgoing sand.

Kelp plants are prey to sea urchins (Figure 5-A). Large populations of these predators can rapidly devastate a kelp forest because they eat the bases of young kelp plants. Scientific studies by biologists including James Estes of the University of California at Santa Cruz indicate that the southern sea otter is a keystone species (**Core Case Study**) that helps to sustain kelp forests by controlling populations of sea urchins. An adult southern sea otter (Figure 5-1, left) can eat as many as 1,500 sea urchins a day. That is equivalent to about 200 quarter-pound hamburgers a day for a 70-kilogram (150-pound) person. Also, a 2012 study by Estes and ecologist Chris Wilmers indicated that kelp forests absorb 12 times more CO_2 when otters are present than when they are not.

A second threat to kelp forests is polluted water running off the land and into the coastal waters where kelp forests grow. The pollutants in this runoff include pesticides and herbicides, which can kill kelp plants and other kelp forest species and upset the food webs in these aquatic forests. Another runoff pollutant is fertil-izer. Its plant nutrients (mostly nitrates) can cause excessive growth of algae and other plants, which block some of the sunlight needed to support the growth of giant kelp.

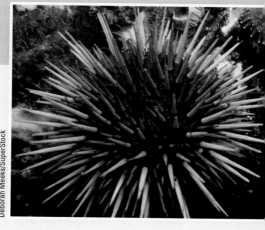

Figure 5-A This purple sea urchin inhabits the coastal waters of the U.S. state of California and feeds on kelp.

Deborah Meeks/SuperStock

Critical Thinking

List three ways to protect giant kelp forests and southern sea otters.

Paul Nicklen/National Geographic Creative

Figure 5-5 A white arctic fox hunts its prey by blending into its snowy background to try to avoid being detected.

People camouflage themselves to hunt wild game and use camouflaged traps to capture wild animals. Some predators use *chemical warfare* to attack their prey. For example, some spiders and poisonous snakes use venom to paralyze their prey and to deter their predators.

Prey species have evolved many ways to avoid predators, including abilities to run, swim, or fly fast, and some have highly developed senses of sight, sound, or smell that alert them to the presence of predators. Other avoidance adaptations include protective shells (as on armadillos and turtles), thick bark (on giant sequoias), spines (on porcupines), and thorns (on cacti and rose bushes). Many lizards have brightly colored tails that break off when they are attacked, often giving them enough time to escape.

Other prey species use the camouflage of certain shapes or colors. Some insect species have shapes that look like twigs (Figure 5-6a), or bird droppings on leaves. A leaf insect can be almost invisible against its background (Figure 5-6b), as can an arctic hare in its white winter fur.

Chemical warfare is another common strategy. Some prey species discourage predators with chemicals that are *poisonous* (oleander plants), *irritating* (stinging nettles and bombardier beetles, Figure 5-6c), *foul smelling* (skunks and stinkbugs), or *bad tasting* (buttercups and monarch butterflies, Figure 5-6d). When attacked, some species of squid and octopus emit clouds of black ink, allowing them to escape by confusing their predators.

Many bad-tasting, bad-smelling, toxic, or stinging prey species have evolved *warning coloration,* brightly colored advertising that helps experienced predators to recognize and avoid them. They flash a warning: "Eating me is risky." Examples are the brilliantly colored, foul-tasting monarch butterflies (Figure 5-6d) and poisonous frogs (Figure 5-6e and Figure 4-11, p. 88). When a bird such as a blue jay eats a monarch butterfly, it usually vomits and learns to avoid them.

🔍 CONSIDER THIS...

CONNECTIONS Coloration and Dangerous Species

Biologist Edward O. Wilson gives us two rules, based on coloration, for evaluating possible danger from any unknown animal species we encounter in nature. *First,* if it is small and strikingly beautiful, it is probably poisonous. *Second,* if it is strikingly beautiful and easy to catch, it is probably deadly.

Some butterfly species gain protection by looking and acting like other, more dangerous species, a protective device known as *mimicry.* For example, the nonpoisonous viceroy butterfly (Figure 5-6f) mimics the monarch butterfly. Other prey species use *behavioral strategies* to avoid predation. Some attempt to scare off predators by puffing up (blowfish), spreading their wings (peacocks), or mimicking a predator (Figure 5-6h). Some moths have wings that look like the eyes of much larger animals (Figure 5-6g). Other prey species gain some protection by living in large groups such as schools of fish and herds of antelope.

At the individual level, members of the predator species benefit from their predation and members of the

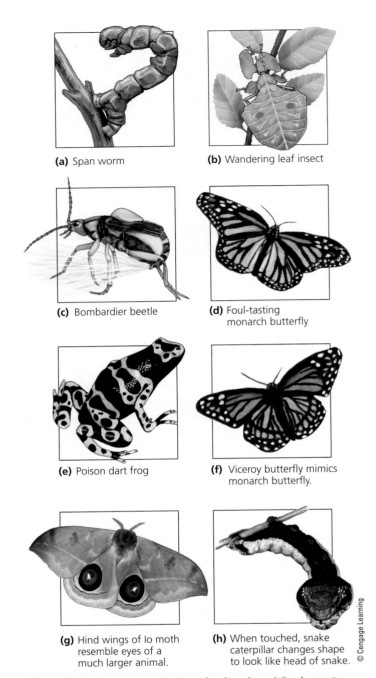

(a) Span worm **(b)** Wandering leaf insect

(c) Bombardier beetle **(d)** Foul-tasting monarch butterfly

(e) Poison dart frog **(f)** Viceroy butterfly mimics monarch butterfly.

(g) Hind wings of Io moth resemble eyes of a much larger animal. **(h)** When touched, snake caterpillar changes shape to look like head of snake.

© Cengage Learning

Figure 5-6 These prey species have developed specialized ways to avoid their predators: (a, b) *camouflage,* (c, d, e) *chemical warfare,* (d, e, f) *warning coloration,* (f) *mimicry,* (g) *deceptive looks,* and (h) *deceptive behavior.*

prey species are harmed. At the population level, predation plays a role in evolution by natural selection. Animal predators, for example, tend to kill the sick, weak, aged, and least fit members of a prey population because they are the easiest to catch. Individuals with better defenses against predation thus tend to survive longer and leave more offspring with adaptations that can help them avoid predation.

Figure 5-7 *Coevolution:* This bat is using ultrasound to hunt a moth. As the bats evolve traits to increase their chance of getting a meal, the moths evolve traits to help them avoid being eaten.

Some people view certain animal predators with contempt. When a hawk tries to capture and feed on a rabbit, some root for the rabbit. Yet the hawk, like all predators, is merely trying to get enough food for itself and its young. In doing so, it plays an important ecological role in controlling rabbit populations.

Interactions between Predator and Prey Species Can Drive Each Other's Evolution

Predator and prey populations can exert intense natural selection pressures on one another. Over time, as a prey species develops traits that make it more difficult to catch, its predators face selection pressures that favor traits increasing their ability to catch their prey. Then the prey species must get better at eluding the more effective predators.

When populations of two different species interact in such a way over a long period of time, changes in the gene pool of one species can lead to changes in the gene pool of the other. Such changes can help both competing species to become more competitive or to avoid or reduce competition. Biologists call this natural selection process **coevolution**.

For example, bats prey on certain species of moths (Figure 5-7), and they hunt at night using *echolocation* to navigate and to locate their prey. They emit pulses of extremely high-frequency, high-intensity sound that bounce off objects, and they capture the returning echoes that tell them where their prey is located. As a countermeasure, certain moth species have evolved ears that are especially sensitive to the sound frequencies that bats use to find them. When they hear the bat frequencies, they try to escape by

Figure 5-8 *Parasitism:* This blood-sucking, parasitic sea lamprey has attached itself to an adult lake trout from the Great Lakes (USA, Canada).

dropping to the ground or flying evasively. Some bat species have evolved ways to counter this defense by changing the frequency of their sound pulses. In turn, some moths have evolved their own high-frequency clicks to jam the bats' echolocation systems. Some bat species have then adapted by turning off their echolocation systems and using the moths' clicks to locate their prey.

Some Species Feed off Other Species by Living on or inside Them

Parasitism occurs when one species (the *parasite*) feeds on another organism (the *host*), usually by living on or inside the host. In this relationship, the parasite benefits and the host is often harmed but not immediately killed.

A parasite usually is much smaller than its host and rarely kills it. However, most parasites remain closely associated with their hosts, draw nourishment from them, and may gradually weaken them over time.

Some parasites such as tapeworms live inside their hosts. Other parasites such as mistletoe plants and blood-sucking sea lampreys (Figure 5-8) attach themselves to the outsides of their hosts. Some parasites, including fleas and ticks, move from one host to another while others, such as tapeworms, spend their adult lives within a single host.

From the host's point of view, parasites are harmful. But from the population perspective, parasites can promote biodiversity by helping to keep the populations of their hosts in check.

In Some Interactions, Both Species Benefit

In **mutualism**, two species behave in ways that benefit both by providing each with food, shelter, or some other resource. One example is pollination of flowering plants by species such as honeybees, hummingbirds, and butterflies (see Chapter 4 opening photo, p. 76) that feed on the nectar of flowers.

Figure 5-9 shows an example of a mutualistic relationship that combines *nutrition* and *protection*. It involves

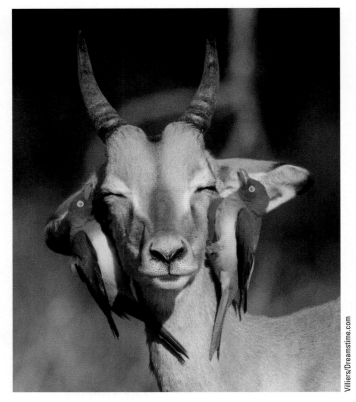

Figure 5-9 *Mutualism:* Oxpeckers feed on parasitic ticks that infest animals such as this impala and warn of approaching predators.

Figure 5-10 In an example of *commensalism,* this pitcher plant is attached to a branch of a tree without penetrating or harming the tree. This carnivorous plant feeds on insects that become trapped inside it.

birds that ride on the backs of large animals such as African buffalo, elephants, rhinoceroses, and impalas (Figure 5-9). The birds remove and eat parasites and pests (such as ticks and flies) from the animals' bodies and often make noises warning the larger animals when predators are approaching.

Another example of mutualism involves clownfish, which usually live within sea anemones (see chapter-opening photo), whose tentacles sting and paralyze most fish that touch them. The clownfish, which are not harmed by the tentacles, gain protection from predators and feed on the waste matter left from the anemones' meals. The sea anemones benefit because the clownfish protect them from some of their predators and parasites.

In *gut inhabitant mutualism,* armies of bacteria in the digestive systems of animals help to break down (digest) the animals' food. In turn, the bacteria receive a sheltered habitat and food from their hosts. Hundreds of millions of bacteria in your gut secrete enzymes that help you digest the food you eat.

It is tempting to think of mutualism as an example of cooperation between species. In reality, the species in a mutualistic interaction benefit one another unintentionally and are in it for themselves.

In Some Interactions, One Species Benefits and the Other Is Not Harmed

Commensalism is an interaction that benefits one species but has little, if any, beneficial or harmful effect on the other. One example involves plants called *epiphytes* (air plants), which attach themselves to the trunks or branches of trees (Figure 5-10) in tropical and subtropical forests. Epiphytes benefit by having a solid base on which to grow. They also live in an elevated spot that gives them better access to sunlight, water from the humid air and rain, and nutrients falling from the tree's upper leaves and limbs. Their presence apparently does not harm the tree. Similarly, birds benefit by nesting in trees, generally without harming them.

5-2 How Do Communities and Ecosystems Respond to Changing Environmental Conditions?

CONCEPT 5-2
The structure and species composition of communities and ecosystems change in response to changing environmental conditions through a process called *ecological succession*.

Communities and Ecosystems Change over Time: Ecological Succession

The types and numbers of species in biological communities and ecosystems change in response to changing environmental conditions such as a fires, volcanic eruptions, climate change, and the clearing of forests to plant crops. The normally gradual change in species composition in a given area is called **ecological succession** (Concept 5-3).

Ecologists recognize two main types of ecological succession, depending on the conditions present at the beginning of the process. **Primary ecological succession** involves the gradual establishment of communities of different species in lifeless areas where there is no soil in a terrestrial ecosystem or no bottom sediment in an aquatic ecosystem. Examples include bare rock exposed by a retreating glacier (Figure 5-11), newly cooled lava, an abandoned highway or parking lot, and a newly created shallow pond or reservoir. Primary succession usually takes hundreds to thousands of years because of the need to build up fertile soil or aquatic sediments to provide the nutrients needed to establish a plant community.

The other, more common type of ecological succession is called **secondary ecological succession**, in which a series of communities or ecosystems with different species develop in places containing soil or bottom sediment. This type of succession begins in an area where an ecosystem has been disturbed, removed, or destroyed, but some soil or bottom sediment remains. Candidates for secondary succession include abandoned farmland (Figure 5-12), burned or cut forests, heavily polluted streams, and flooded land. Because some soil or sediment is present, new vegetation can begin to grow, usually within a few weeks. It begins with the germination of seeds already in the soil and seeds imported by wind or in the droppings of birds and other animals.

Ecological succession is an important ecosystem service that tends to increase the biodiversity of communities and ecosystems by increasing species richness and interactions among species. Such interactions in turn enhance sustainability by promoting population control and by increasing the complexity of food webs, which enhances energy flow and nutrient cycling. As part of the earth's natural capital, both primary and secondary ecological succession are examples of *natural ecological restoration*.

Ecologists have identified three factors that affect how and at what rate succession occurs. One is *facilitation*, in which one set of species makes an area suitable for spe-

© Cengage Learning

Figure 5-11 *Primary ecological succession:* Over almost a 1,000 years, these plant communities developed, starting on bare rock exposed by a retreating glacier on Isle Royal, Michigan (USA) in western Lake Superior. The details of this process vary from one site to another.

Exposed rocks

Lichens and mosses

Small herbs and shrubs

Heath mat

Jack pine, black spruce, and aspen

Balsam fir, paper birch, and white spruce forest community

Time

Animated Figure 5-12 *Natural ecological restoration of disturbed land:* This diagram shows the undisturbed secondary ecological succession of plant communities on an abandoned farm field in the U.S. state of North Carolina. It took 150–200 years after the farmland was abandoned for the area to become covered with a mature oak and hickory forest.

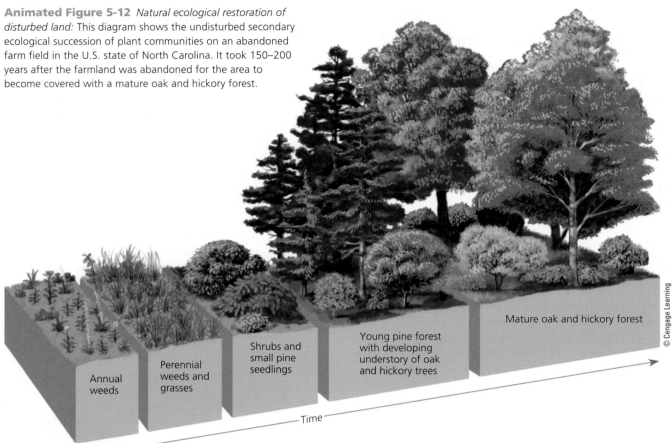

Annual weeds

Perennial weeds and grasses

Shrubs and small pine seedlings

Young pine forest with developing understory of oak and hickory trees

Mature oak and hickory forest

Time

© Cengage Learning

cies with different niche requirements, and often less suitable for itself. For example, as lichens and mosses gradually build up soil on a rock in primary succession, herbs and grasses can move in and crowd out the lichens and mosses.

A second factor is *inhibition*, in which some species hinder the establishment and growth of other species. For example, the needles dropping off some pine trees make the soil beneath the trees too acidic for most other plants to grow there. A third factor is *tolerance*, in which plants in the late stages of succession succeed because they are not in direct competition with other plants for key resources. Shade-tolerant plants, for example, can live in shady forests because they do not need as much sunlight as the trees above them do.

Ecological Succession Does Not Follow a Predictable Path

According to the traditional view, ecological succession proceeds in an orderly sequence along an expected path until a certain stable type of *climax community* occupies an area. On land, such a community is dominated by a few long-lived plant species, often within a mature forest (Figures 5-11 and 5-12), and is in balance with its environment. This

equilibrium model of succession is what ecologists once meant when they talked about the *balance of nature*.

Over the last several decades, many ecologists have changed their views about balance and equilibrium in nature. There is a general tendency for succession to lead to more complex, diverse, and presumably sustainable ecosystems. However, a close look at almost any terrestrial community or ecosystem reveals that it consists of an ever-changing mosaic of patches of vegetation in different stages of succession. The current scientific view is that we cannot predict a given course of succession or view it as inevitable progress toward an ideally adapted climax plant community or ecosystem. Rather, ecological succession reflects the ongoing struggle by different species for enough light, water, nutrients, food, space, and other key resources. Most ecologists now recognize that mature, late-successional ecosystems are in a state of continual disturbance and change.

Living Systems Are Sustained through Constant Change

All living systems, from a cell to the biosphere, are constantly changing in response to changing environmental conditions. However, living systems contain complex processes that interact to provide some degree of stability, or sustainability, over each system's expected life span. This *stability*, or capacity to withstand external stress and disturbance, is maintained only by constant change in response to changing environmental conditions. For example, in a mature tropical rain forest, some trees die and others take their places. However, unless the forest is cut, burned, or otherwise destroyed, you would still recognize it as a tropical rain forest 50 or 100 years from now.

It is useful to distinguish between two aspects of stability or sustainability in living systems. One is **inertia**, or **persistence**: the ability of a living system such as a grassland or a forest to survive moderate disturbances. A second factor is **resilience**: the ability of a living terrestrial system to be restored through secondary ecological succession after a more severe disturbance.

Evidence suggests that some ecosystems have one of these properties but not the other. For example, tropical rain forests have high species richness and high inertia and thus are resistant to significant change or damage. But once a large tract of tropical rain forest is cleared or severely damaged, the resilience of the resulting degraded forest ecosystem may be so low that it reaches an ecological tipping point after which it may not be restored by secondary ecological succession. One reason for this is that most of the nutrients in a typical rain forest are stored in its vegetation, not in the topsoil, as in most other terrestrial ecosystems. Once the nutrient-rich vegetation is gone, daily rains can remove most of the remaining soil nutrients and thus prevent the return of a tropical rain forest on a large cleared area.

By contrast, grasslands are much less diverse than most forests, and consequently they have low inertia and can burn easily. However, because most of their plant matter is stored in underground roots, these ecosystems have high resilience and can recover quickly after a fire, as their root systems produce new grasses. Grassland can be destroyed only if its roots are plowed up and something else is planted in its place, or if it is severely overgrazed by livestock or other herbivores.

5-3 What Limits the Growth of Populations?

CONCEPT 5-3
No population can grow indefinitely because of limitations on resources and because of competition among species for those resources.

Most Populations Live in Clumps

A **population** is a group of interbreeding individuals of the same species (Figure 5-13). Figure 5-14 shows three ways in which the members of a population are typically distributed or dispersed in their habitat. Most populations live together in *clumps* (Figure 5-14a) such as packs of wolves, schools of fish (Figure 5-13), and flocks of birds. Southern sea otters (**Core Case Study**) (Figure 5-1, left), for example, are usually found in groups known as rafts or pods ranging in size from a few to several hundred animals.

Why clumps? Several reasons: *First,* the resources a species needs vary greatly in availability from place to place, so the species tends to cluster where the resources are available. *Second,* individuals moving in groups have a better chance of encountering patches or clumps of resources, such as water and vegetation, than they would searching for the resources on their own. *Third,* living in groups provides some protection from predators. *Fourth,* living in packs gives some predator species a better chance of getting a meal.

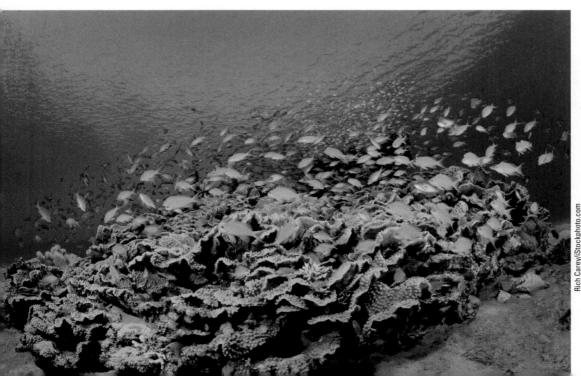

Figure 5-13 A population, or school, of Anthias fish on coral in Australia's Great Barrier Reef.

Rich Carey/iStockphoto.com

a. Clumped (elephants)

b. Uniform (creosote bush)

c. Random (dandelions)

Figure 5-14 Three general habitat *dispersion patterns* for a population's individuals.

Left: EcoPrint/Shutterstock.com. Center: kenkistler/Shutterstock.com. Right: Nataly Lukhanina/Shutterstock.com.

Populations Can Grow, Shrink, or Remain Stable

Four variables—*births, deaths, immigration,* and *emigration*—govern changes in population size. A population increases through birth and immigration (arrival of individuals from outside the population) and decreases through death and emigration (departure of individuals from the population):

Population change = (Births + Immigration) − (Deaths + Emigration)

A population's **age structure**—its distribution of individuals among various age groups—can have a strong effect on how rapidly it grows or declines. Age groups are usually described in terms of organisms not mature enough to reproduce (the *prereproductive stage*), those capable of reproduction (the *reproductive stage*), and those too old to reproduce (the *postreproductive stage*).

The size of a population will likely increase if it is made up mostly of individuals in their reproductive stage, or soon to enter this stage. In contrast, a population dominated by individuals in their postreproductive stage will tend to decrease over time.

Some Factors Can Limit Population Size

Different species and their populations thrive under different physical and chemical conditions. Some need bright sunlight; others flourish in shade. Some need a hot environment; others prefer a cool or cold one. Some do best under wet conditions; others thrive in dry conditions.

Each population in an ecosystem has a **range of tolerance** to variations in its physical and chemical environment, as shown in Figure 5-15. Individuals within a population may also have slightly different tolerance ranges for temperature or other physical or chemical factors because of small differences in their genetic makeup, health, and age. For example, a trout population may do best within a narrow band of temperatures (*optimum level* or *range*), but a few individuals can survive above and below that band. Of course, if the water becomes much too hot or too cold, none of the trout can survive.

A number of physical or chemical factors can help to determine the number of organisms in a population. Sometimes one or more factors, known as **limiting factors**, are more important than other factors in regulating population growth. This ecological principle is called the **limiting factor principle**: *Too much or too little of any physical or chemical factor can limit or prevent the growth of a population, even if all other factors are at or near the optimal range of tolerance.*

On land, precipitation often is the limiting factor. For example, low precipitation levels in desert ecosystems limit desert plant growth. Soil nutrients also can act as a limiting factor on land. Suppose a farmer plants corn in phosphorus-poor soil. Even if water, nitro-

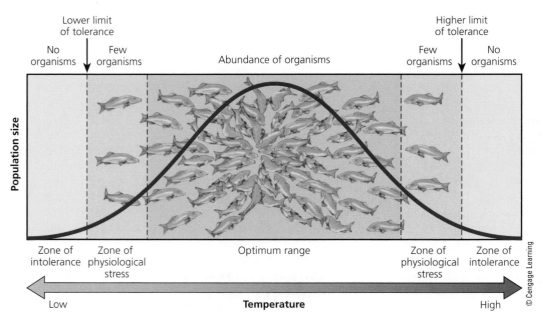

Figure 5-15 Range of tolerance for a population of trout to changes in water temperature.

© Cengage Learning

gen, potassium, and other nutrients are at optimal levels, the corn will stop growing when it uses up the available phosphorus.

Too much of a physical or chemical factor can also be limiting. For example, too much water or fertilizer can kill plants. Temperature can be a limiting factor as well. Both high and low temperatures can limit the survival and population sizes of various terrestrial species, especially plants and many insect species.

Important limiting physical factors for populations in *aquatic life zones,* or water-filled areas that support life, include temperature (Figure 5-15), sunlight, nutrient availability, acidity, and the low levels of oxygen gas in the water *(dissolved oxygen content).* Another limiting factor in aquatic life zones is *salinity*—the amounts of various inorganic minerals or salts dissolved in a given volume of water.

Another factor that can limit the sizes of some populations is **population density**, the number of individuals in a population found within a defined area or volume. Some limiting factors become more important as a population's density increases. For example, in a dense population, parasites and diseases can spread more easily and have the effect of controlling population growth. On the other hand, a higher population density can help sexually reproducing individuals to find mates more easily in order to produce offspring and increase the size of a population.

Different Species Have Different Reproductive Patterns

Different reproductive patterns help to ensure the long-term survival of species. Some species have many, usually small, offspring and give them little or no parental care or protection. Examples include algae, bacteria, and most insects. These species reproduce at an early age and overcome typically massive losses of offspring by producing so many offspring that a few will likely survive to reproduce many more offspring and keep this reproductive pattern going.

At the other extreme are species that tend to reproduce later in life and have a small number of offspring with longer life spans. Typically, the offspring of mammals with this reproductive strategy develop inside their mothers (where they are safe), and are born fairly large. After birth, they mature slowly and are cared for and protected by one or both parents, and in some cases by living in herds or groups, until they reach reproductive age and begin the cycle again.

Most large mammals (such as elephants, whales, and humans) follow this reproductive pattern. Many of these species—especially those with long times between generations and with low reproductive rates, such as elephants, rhinoceroses, and sharks—are vulnerable to extinction. Most organisms have reproductive patterns between these two extremes.

No Population Can Grow Indefinitely: J-Curves and S-Curves

Some species have an incredible ability to increase their numbers. Members of such populations typically reproduce at an early age, have many offspring each time they reproduce, and reproduce many times, with short intervals between successive generations. For example, with no controls on its population growth, a species of bacteria that can reproduce every 20 minutes would generate enough offspring to form a 0.3-meter-deep (1-foot-deep) layer over the surface of the entire earth in only 36 hours—a dramatic illustration of the potential of exponential growth.

Fortunately, this will not happen. Research reveals that regardless of their reproductive strategy, no population of a species can grow indefinitely because of limitations on resources and competition with populations of other species for those resources (Concept 5-2). In the real world, a rapidly growing population of any species eventually reaches some size limit imposed by one or more limiting factors such as the availability of light, water, temperature, space, or nutrients, or by exposure to predators or infectious diseases.

There are always limits to population growth in nature. For example, one reason California's southern sea otters (**Core Case Study**) face extinction is that they cannot reproduce rapidly (Science Focus 5.2).

Environmental resistance is the combination of all factors that act to limit the growth of a population. It largely determines an area's **carrying capacity**: the maximum population of a given species that a particular habitat can sustain indefinitely. The growth rate of a population decreases as its size nears the carrying capacity of its environment because resources such as food, water, and space begin to dwindle.

A population with few, if any, limitations on its resource supplies can grow exponentially at a fixed rate or percentage per year. Exponential growth starts slowly but then accelerates as the population increases, because as its base size increases, so does the number added to the population each year. Plotting the number of individuals against time yields a J-shaped growth curve. Figure 5-16 shows such a curve for sheep on an island south of Australia in the early 19th century.

Exponential growth (left third of the curve in Figure 5-16) occurs when a population has essentially unlimited resources to support its growth. Such exponential growth is eventually converted to *logistic growth* (center of the curve in Figure 5-16), in which the growth rate decreases as the population becomes larger and faces environmental resistance. Over time, the population size typically stabilizes at or near the *carrying capacity* of its environment, which results in a sigmoid (S-shaped) population growth curve. Depending on resource availability, the size of a population often fluctuates around its carrying capacity (right third of the curve in Figure 5-16). Also, unlike in

WHY DO CALIFORNIA'S SOUTHERN SEA OTTERS FACE AN UNCERTAIN FUTURE?

The southern sea otter (**Core Case Study**) cannot rapidly increase its numbers for several reasons. Female southern sea otters reach sexual maturity between 2 and 5 years of age, can reproduce until age 15, and typically each produce only one pup a year.

The population size of southern sea otters has fluctuated in response to changes in environmental conditions. One such change has been a rise in populations of the orcas (killer whales) that feed on them. Scientists hypothesize that orcas began feeding more on southern sea otters when populations of their normal prey, sea lions and seals, began declining. In 2010 and 2011, the number of southern sea otters killed or injured by sharks increased for reasons that scientists are trying to understand.

Another factor may be parasites known to breed in the intestines of cats. Scientists hypothesize that some southern sea otters may be dying because coastal area cat owners flush feces-laden cat litter down their toilets or dump it in storm drains that empty into coastal waters. The feces contain parasites that then infect the otters.

Otters are also being threatened by blooms of toxic algae that are fed by urea, a key ingredient in fertilizer that washes into coastal waters. Other pollutants released by human activities are PCBs and other fat-soluble toxic chemicals that can accumulate in the tissues of the shellfish on which otters feed, and this proves fatal to otters. The facts that southern sea otters feed at high trophic levels and live close to the shore make them vulnerable to these and other pollutants in coastal waters. Scientists note that as an indicator species, the southern sea otter is revealing the degraded condition of its coastal water habitat.

Some southern sea otters die when they encounter oil spilled from ships. The entire California southern sea otter population could be wiped out by a large oil spill from a single tanker off the state's central coast or by the rupture of an offshore oil well, should drilling for oil be allowed off this coast.

The factors listed here, mostly resulting from human activities, plus a fairly low reproductive rate, have hindered the ability of the endangered southern sea otter to rebuild its population (Figure 5-B). According to the U.S. Geological Survey, the California southern sea otter population would have to reach at least 3,090 for 3 years in a row before it could be considered for removal from the endangered species list.

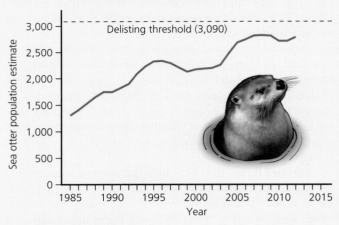

Figure 5-B Changes in the population size of southern sea otters off the coast of the U.S. state of California, 1983–2012.

(Compiled by the authors using data from U.S. Geological Survey.)

Critical Thinking

How would you design a controlled experiment to test the hypothesis that cat litter flushed down toilets might be killing southern sea otters?

---

the simple model in Figure 5-16, the carrying capacity can vary over time due to changing conditions.

Changes in habitat or other environmental conditions can reduce the populations of some species while increasing the populations of others (see the following Case Study).

CASE STUDY

Exploding White-Tailed Deer Populations in the United States

By 1900, habitat destruction and uncontrolled hunting had reduced the white-tailed deer (Figure 5-17) population in the United States to about 500,000 animals. In the 1920s and 1930s, laws were passed to protect the remaining deer. Hunting was restricted and predators, including wolves and mountain lions that preyed on the deer, were nearly eliminated.

These protections worked, and for some suburbanites and farmers, perhaps too well. Today there are over 25 million white-tailed deer in the United States. During the last 50 years, large numbers of Americans have moved into the wooded habitat of deer where suburban areas have expanded. By gardening and landscaping around their homes, people have provided deer with flowers, shrubs, garden crops, and other plants that they like to eat.

Deer prefer to live in the edge areas of forests and woodlots for security, and they go to nearby fields,

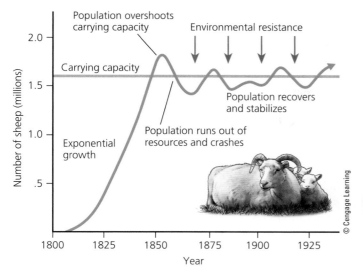

Animated Figure 5-16 Growth of a sheep population on the island of Tasmania between 1800 and 1925. After sheep were introduced in 1800, their population grew exponentially thanks to an ample food supply and few predators. By 1855, they had overshot the land's carrying capacity. Their numbers then stabilized and fluctuated around a carrying capacity of about 1.6 million sheep.

Figure 5-17 White-tailed deer populations in the United States have been growing.

orchards, lawns, and gardens for food. Thus, a suburban neighborhood can be an all-you-can-eat paradise for white-tailed deer. Their populations in such areas have soared. In some woodlands, they are consuming native ground-cover vegetation and this has allowed nonnative weed species to take over. The deer also help to spread Lyme disease (carried by deer ticks) to humans.

In addition, each year about 1.5 million deer–vehicle collisions injure at least 10,000 Americans and kill at least 200—the highest of all death tolls resulting from encounters with wild animals. By comparison, between 2000 and 2010, deer–vehicle collisions killed about 2,000 Americans, compared to 28 killed by bears, and 10 killed by sharks.

There are no easy answers to the deer population problem in the suburbs. Changes in hunting regulations that allow for the killing of more female deer have cut down the overall deer population. However, this has had a limited effect on deer populations in suburban areas, because it is too dangerous to allow widespread hunting with guns in such populated communities. Some areas have hired experienced and licensed archers who use bows and arrows to help reduce deer numbers. To protect nearby residents the archers hunt from elevated tree stands and only shoot their arrows downward.

Some communities spray the scent of deer predators or of rotting deer meat in edge areas to scare off deer. Others use electronic equipment that emits high-frequency sounds, which humans cannot hear, for the same purpose. Some home owners surround their gar-

dens and yards with high, black plastic mesh fencing that is invisible from a distance.

Deer can be trapped and moved from one area to another, but this is expensive and must be repeated whenever they move back into an area. Also, there are questions concerning where to move the deer and how to pay for such programs.

We can also put deer on birth control. Darts loaded with contraceptives can be shot into female deer to hold down their birth rates. However, this is expensive and must be repeated every year. One possibility is an experimental, single-shot contraceptive vaccine that lasts for several years. Another approach is to trap dominant males and use chemical injections to sterilize them. Both of these approaches will require years of testing.

Meanwhile, suburbanites can expect deer to chow down on their shrubs, flowers, and garden plants unless they can protect their properties with high, deer-proof fences, repellents, or other methods. Deer have to eat every day just as we do. Suburban dwellers could consider not planting trees, shrubs, and flowers that attract deer around their homes.

CONSIDER THIS. . .

THINKING ABOUT White-Tailed Deer

Some people blame the white-tailed deer for invading farms and suburban yards and gardens to eat food that humans have made easily available to them. Others say humans are mostly to blame because they have invaded deer territory, eliminated most of the predators that kept deer populations under control, and provided the deer with plenty to eat in their lawns, gardens, and crop fields. Which view do you hold? Why? Do you see a solution to this problem?

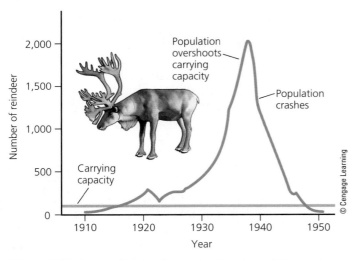

Figure 5-18 Exponential growth, overshoot, and population crash of reindeer introduced onto the small Bering Sea island of St. Paul in 1910.

When a Population Exceeds Its Carrying Capacity It Can Crash

Some populations do not make a smooth transition from exponential growth to logistic growth. Instead, they use up their resource supplies and temporarily *overshoot*, or exceed, the carrying capacity of their environment. In such cases, the population suffers a sharp decline, called *dieback,* or **population crash**, unless part of the population can switch to new resources or move to an area that has more resources. Such a crash occurred when reindeer were introduced onto a small island in the Bering Sea (Figure 5-18).

The carrying capacity of any given area is not fixed. In some areas, it can rise or decline seasonally and from year to year because of variations in weather, such as a drought that causes decreases in available vegetation. Other factors include the presence or absence of predators and an abundance or scarcity of competitors.

Humans Are Not Exempt from Nature's Population Controls

Humans are not exempt from population crashes. Ireland experienced such a crash after a fungus destroyed its potato crop in 1845. About 1 million people died from hunger or diseases related to malnutrition, and 3 million people migrated to other countries, especially the United States.

During the 14th century, the *bubonic plague* spread through densely populated European cities and killed at least 25 million people. The bacterium causing this disease normally lives in rodents. It was transferred to humans by fleas that fed on infected rodents and then bit humans. The disease spread like wildfire through crowded cities, where sanitary conditions were poor and rats were abundant. Today, several antibiotics, not available until recently, can be used to treat bubonic plague.

Currently, the world is experiencing a global epidemic of AIDS, caused by infection with the human immunodeficiency virus (HIV). Between 1981 and 2011, AIDS killed more than 30 million people and continues to claim another 1.7 million lives each year—an average of 3 deaths per minute.

So far, technological, social, and other cultural changes have expanded the earth's carrying capacity for the human species. We have increased food production and used large amounts of energy and matter resources to occupy formerly uninhabitable areas, to expand agriculture, and to control the populations of other species that compete with us for resources. Some say we can keep expanding our ecological footprint in this way indefinitely, mostly because of our technological ingenuity. Others say that sooner or later, we will reach the limits that nature eventually imposes on any population that exceeds or degrades its resource base. We discuss these issues in Chapter 6.

Big Ideas

- Certain interactions among species affect their use of resources and their population sizes.
- Changes in environmental conditions cause communities and ecosystems to gradually alter their species composition and population sizes (ecological succession).
- There are always limits to population growth in nature.

fred goldstein/shutterstock.com

Before the arrival of European settlers on the western coast of North America, the southern sea otter population was part of a complex ecosystem made up of kelp, bottom-dwelling creatures, whales, and other species depending on one another for survival. Giant kelp forests served as food and shelter for sea urchins. Sea otters ate the sea urchins and other kelp eaters. Some species of whales and sharks ate the otters. And detritus from all these species helped to maintain the giant kelp forests. Each of these interacting populations was kept in check by—and helped to sustain—all the others.

When European settlers arrived and began hunting the otters for their pelts, they probably didn't know much about the intricate web of life beneath the ocean surface. But with the effects of overhunting, people realized they had done more than simply take sea otters. They had torn the web, disrupted an entire ecosystem, and triggered a loss of valuable natural resources and ecosystem services, including biodiversity.

Populations of most plants and animals depend, directly or indirectly, on solar energy, and all populations play roles in the cycling of nutrients in the ecosystems where they live. In addition, the biodiversity found in the variety of species in different terrestrial and aquatic ecosystems provides alternative paths for energy flow and nutrient cycling, better opportunities for natural selection as environmental conditions change, and natural population control mechanisms. When we disrupt these paths, we violate all three scientific **principles of sustainability**. In this chapter, we focused on how the biodiversity principle promotes sustainability, provides a variety of species to restore damaged ecosystems through ecological succession, and limits the sizes of populations.

Chapter Review

Core Case Study

1. Explain how southern sea otters act as a keystone species in their environment (**Core Case Study**). Explain why we should care about protecting this species from extinction that could result primarily from human activities.

Section 5-1

2. What is the key concept for this section? Define **interspecific competition**. Define and give two examples of **resource partitioning** and explain how it can increase species diversity. Define **predation** and distinguish between a **predator** species and a **prey** species and give an example of each. What is a **predator–prey relationship**?

3. Explain why we should help to preserve kelp forests. Describe three ways in which predators can increase their chances of feeding on their prey and three ways in which prey species can avoid their predators. Define and give an example of **coevolution**.

4. Define **parasitism**, **mutualism**, and **commensalism** and give an example of each. Explain how each of these species interactions, along with predation, can affect the population sizes of species in ecosystems.

Section 5-2

5. What is the key concept for this section? What is **ecological succession**? Distinguish between **primary ecological succession** and **secondary ecological succession** and give an example of each. List three factors that can affect how ecological succession occurs. Explain why succession does not follow a predictable path.

6. Explain how living systems achieve some degree of sustainability by undergoing constant change in response to changing environmental conditions. In terms of the stability of ecosystems, distinguish between **inertia (persistence)** and **resilience** and give an example of each.

7. What is the key concept for this section? Define **population**. Why do most populations live in clumps? List four variables that govern changes in population size. Write an equation showing how these variables interact. What is a population's **age structure** and what are the three major age groups called? Define **range of tolerance**. Define **limiting factor** and give an example. State the **limiting factor principle**. Define **population density** and explain how some limiting factors can become more important as a population's density increases. Describe two different reproductive strategies that can enhance the long-term survival of a species.

8. Explain why no population can grow indefinitely. Distinguish between the **environmental resistance** and the **carrying capacity** of an environment, and use these concepts to explain why there are always limits to population growth in nature. Why is the recovery of southern sea otters a slow one, and what factors are threatening this recovery? Describe the exploding white-tailed deer population problem in the United States and discuss options for dealing with it.

9. Define and give an example of a **population crash**. Explain why humans are not exempt from nature's population controls.

10. What are this chapter's *three big ideas*? Explain how the interactions of plant and animal species in any ecosystem are related to the scientific **principles of sustainability**.

Note: Key terms are in bold type.

Critical Thinking

1. What difference would it make if the southern sea otter (**Core Case Study**) became extinct primarily because of human activities? What are three things we could do to help prevent the extinction of this species?

2. Use the second law of thermodynamics (see Chapter 2, p. 43) to help explain why predators are generally less abundant than their prey. In your explanation, make use of the pyramid of energy flow (see Figure 3-13, p. 61).

3. How would you reply to someone who argues that we should not worry about the effects that human activities have on natural systems because ecological succession will heal the wounds of such activities and restore the balance of nature?

4. How would you reply to someone who contends that efforts to preserve natural systems are not worthwhile because nature is largely unpredictable?

5. Explain why most species with a high capacity for population growth (such as bacteria, flies, and cockroaches) tend to have small individuals, while those with a low capacity for population growth (such as humans, elephants, and whales) tend to have large individuals.

6. Which reproductive strategy do most species of insect pests and harmful bacteria use? Why does this make it difficult for us to control their populations?

7. List three factors that have limited human population growth in the past and that we have overcome. Explain how we overcame each of these factors. List two factors that may limit human population growth in the future. Do you think that we are close to reaching those limits? Explain.

8. If the human species were to suffer a population crash, what are three species that might move in to occupy part of our ecological niche? Explain why this might happen.

Doing Environmental Science

Visit a nearby land area, such as a partially cleared or burned forest, a grassland, or an abandoned crop field, and record signs of secondary ecological succession. Take notes on your observations and formulate a hypothesis about what sort of disturbance led to this succession. Include your thoughts about whether this disturbance was natural or caused by humans. Study the area carefully to see whether you can find patches that are at different stages of succession and record your thoughts about what sorts of disturbances have caused these differences. You might want to research the topic of ecological succession in such an area.

Global Environment Watch Exercise

Search for *kelp forests,* and use the results to find sources of information about how a warmer ocean, as a result of climate change, might affect California's coastal kelp beds on which the southern sea otters depend (**Core Case Study**). Write a report on what you found. Try to include informa-tion on current effects of warmer water on the kelp beds as well as projections about future effects. Also, summarize any information you might find on possible ways to prevent harm to these kelp beds.

Data Analysis

The graph below shows changes in the size of an emperor penguin population in terms of numbers of breeding pairs on the island of Terre Adelie in the Antarctic. Use the graph to answer the questions below.

(Compiled by the authors using data from Woods Hole Oceanographic Institute and *Nature,* May 10, 2001.)

1. Assuming that the penguin population fluctuates around the carrying capacity, what was the approximate carrying capacity of the island for the penguin population from 1960 to 1975? What was the approximate carrying capacity of the island for the penguin population from 1980 to 2010?

2. What was the percentage decline in the penguin population from 1975 to 2010?

Questions 1–5 refer to the species interactions below.

   a. Interspecific competition
   b. Predation
   c. Parasitism
   d. Mutualism
   e. Commensalism

1. Three species of grasses compete for sunlight in a grassy lawn.

2. A species of ant living in the thorns of the Acacia tree protect the tree from herbivores and feed off of lipid-rich food-bodies on the tree.

3. Barnacles adhering to the skin of a whale. The barnacle benefits by being transported to new sources of food and the presence of barnacles does not appear to hamper or enhance the survival of the animals carrying them.

4. Chipmunks, squirrels, and other animals eat the pine nuts that pine trees produce.

5. A species of pseudoscorpion disperses by concealing itself under the wings of large beetles. The pseudoscorpions gain the advantage of being dispersed over wide areas while being protected from predators. The beetle is unaffected by the presence of the hitchhikers.

Questions 6 and 7 refer to the graph below of an otter population off the coast of California.

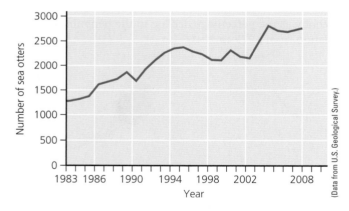

(Data from U.S. Geological Survey.)

6. What generalized trend can be made from 1983 to 2007 from this graph?
   a. The otter is declining more than growing.
   b. The otter is growing at an exponential rate.
   c. The otter has reached its carrying capacity.
   d. The otter is showing a slow rate of population increase.
   e. The otter is an endangered species.

7. What is the percent change in the otter population from 1995 to 2000?
   a. −300
   b. −87
   c. −13.0
   d. −0.13
   e. 0.0

8. Which of the following is a good example of coevolution?
   a. An interaction between species that benefits each, such as oxpeckers eating ticks and flies found on the backs of elephants
   b. The sea lamprey attaching itself to fish for nourishment, such as is occurring in the Great Lakes
   c. A bromeliad attaching itself to tropical trees to gather sunlight, water, and nutrients
   d. Populations of the common garter snake have evolved a resistance to the toxins of their prey, newts of the genus *Taricha;* as the newts continue to evolve more potent toxins, the garter snakes evolve an increased amount of resistance to the poison over time as well
   e. Several species of birds, all feeding on the same species of tree, evolve traits that enable them to eat at different levels of the tree or by eating different kinds of prey species

9. Which of the following has NOT contributed to the explosion of the white-tailed deer population?
   a. Elimination of many of their natural predators
   b. Hunting laws restricting their hunting
   c. Suburbanization eliminating many forests and creating many edge habitats
   d. Trapping of deer and moving them out of areas where they are not wanted
   e. Many people planting their properties with edible flowers and shrubs

Question 10 refers to the diagram below.

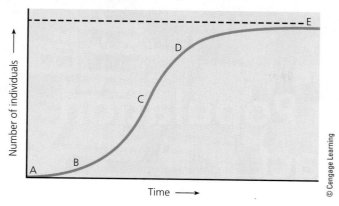

© Cengage Learning

10. Which point on the curve above reflects the selection most like that of an elephant, a K-selected species that has reached its carrying capacity?

11. Which of the below is undergoing secondary succession?
    a. A boulder field left from the retreat of a glacier
    b. Recently cooled lava from a volcano
    c. An agricultural field recently cut for hay
    d. An area newly covered with sand dunes
    e. A newly created small pond

12. Which of the below is an example of density-dependent population control?
    a. Humans clearing a forested area
    b. A forest fire burning a woodland tract
    c. Red-tailed hawks feeding on rabbits
    d. An avalanche clearing a hillside of all trees and shrubs
    e. Pollution from a city killing invertebrates in a local stream

13. When does a tipping point occur?
    a. A population can become weak or die off if the optimal amount of physical or chemical factors no longer exists in their habitat.
    b. There is a limit to the number of organisms that can be sustained in a particular habitat.
    c. Each year, more and more of the sea ice melts in the Arctic. In less than 5 years there will most likely be a summer when there is no sea ice in the Arctic and this will have large ramifications for the ecology of the area that is dependent on sea ice.
    d. The animal population rapidly changes, as when 26 reindeer were introduced to the island of St. Paul in 1910. By 1935, the population grew to more than 2,000 reindeer. Because of a lack of resources, the population dropped to eight reindeer by 1950.
    e. Vegetation grows in response to a change in the environment, such as when a forest is cut down and species gradually return to the area, which may then become fully forested.

6

The Human Population
and Its Impact

The problems to be faced are vast and complex, but come down to this: 7 billion people are breeding exponentially. The process of fulfilling their wants and needs is stripping earth of its biotic capacity to support life; a climactic burst of consumption by a single species is overwhelming the skies, earth, waters, and fauna.

PAUL HAWKEN

Key Questions

6-1 How do environmental scientists think about human population growth?

6-2 What factors influence the size of the human population?

6-3 How does a population's age structure affect its growth or decline?

6-4 How can we slow human population growth?

Jaipur, India.

Eastland Photo/Alamy

121

It took about 200,000 years—from the time that the latest version of our species *Homo sapiens sapiens* evolved, to the 1920s—for our population to reach an estimated 2 billion. It took less than 50 years to add the second 2 billion people (by about 1974), and 25 years to add the third 2 billion (by 1999). Twelve years later, in late 2011, we topped 7 billion (Figure 6-1).

So why does it matter that there are now 7.1 billion people on the earth—almost 3 times as many as there were in 1950? Some say it doesn't matter and they contend that we can develop new technologies that could easily support 16 billion people. Many scientists disagree and argue that the current exponential growth of the human population is unsustainable. They point out that as our population grows and as we use more of the earth's natural resources, our ecological footprints also expand (see Figure 1-13, p. 14) and degrade the natural capital that keeps us alive and supports our lifestyles and economies.

In 2009, a team of internationally renowned scientists estimated that we have exceeded the global boundary limits for three of the major components of the earth's life-support system and are close to the boundary limits for five other components (Figure 3-B, p. 73). Some scientists say that because of our strong effect on natural systems, we now live in a new *Anthropocene era* in which our collective actions are moving key parts of our life-support system into danger zones.

Three major factors account for the rapid rise of the human population. *First,* the emergence of early and modern agriculture about 10,000 years ago allowed us to feed more people with each passing century. *Second,* we have developed technologies that have helped us to expand into almost all of the planet's climate zones and habitats (see Figure 1-12, p. 13). *Third,* our death rates have dropped sharply because of improved sanitation and health care and the development of antibiotics and vaccines to help control infectious diseases.

So what is a sustainable level for the human population? Population experts have made low, medium, and high projections for how big the human population will be by the end of this century (Figure 6-1). No one knows whether any of these population sizes are sustainable or for how long.

In this chapter, we examine population growth trends, the environmental impacts of the growing population, proposals for dealing with human population growth, and some stories about countries where population growth has been stabilized.

Figure 6-1 The human population has grown exponentially—showing slow growth throughout most of history and shooting up at a rapid rate within the last 200 years. This graph also shows projections to 2100 that range from 8 billion to 16 billion. (This figure is not to scale.)

Photo: ©NASA

6-1 How Do Environmental Scientists Think about Human Population Growth?

CONCEPT 6-1
The continuing rapid growth of the human population and its impacts on natural capital raise questions about how long the human population can keep growing.

Human Population Growth Shows Certain Trends

Estimates of what the global human population is likely to be in 2050 range from 7.8 billion to 10.8 billion people, with a medium projection of 9.6 billion people. For the year 2100, the projected population size ranges from 8 billion to 16 billion (Figure 6-1). These varying estimates depend on a variety of factors that we consider in more detail later in this chapter. However, *demographers*, or population experts, recognize three important growth trends.

First of all, in recent decades, the rate of population growth has slowed (Figure 6-2), but the world's population is still growing (Figure 6-1) at a rate of about 1.2%. This may not seem like much but in 2012 this growth added about 84 million people to the population—an average of more than 230,000 people each day, or almost 3 more people every second.

Second, demographers recognize that, geographically, human population growth is unevenly distributed and this pattern is expected to continue (Figure 6-3). About 2% of the 84 million new arrivals on the planet in 2012 were added to the world's more-developed coun-

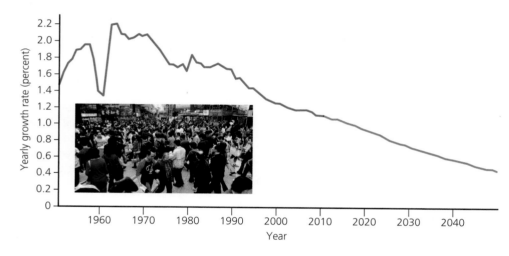

Figure 6-2 The annual growth *rate* of world population has generally dropped since the 1960s, but the population has continued to grow (Figure 6-1).

(Compiled by the authors using data from United Nations Population Division, U.S. Census Bureau, and Population Reference Bureau.)

Photo: JUSTIN GUARIGLIA/National Geographic Creative

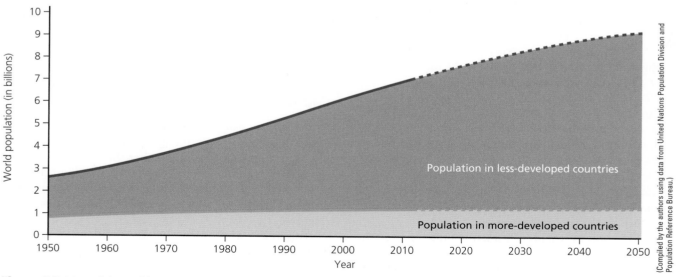

Figure 6-3 Most of the world's population growth between 1950 and 2012 took place in the world's less-developed countries. This gap has been projected to increase between 2012 and 2050.

HOW LONG CAN THE HUMAN POPULATION KEEP GROWING?

To survive and provide resources for growing numbers of people, humans have modified, cultivated, built on, and degraded a large and increasing portion of the earth's natural systems (Figure 6-A). As a result, our ecological footprints have spread across much of the globe (see Figure 1-12, p. 13) and are projected to expand significantly further by 2030. Some scientists estimate that by then, we would need the equivalent of two planet Earths to sustain the projected population of 9.6 billion people (see Figure 1-13, p. 14).

Scientific studies of the populations of other species tell us that *no population can continue growing indefinitely* (p. 113). How long can we continue to avoid the reality of the earth's carrying capacity for our species by sidestepping many of the factors that sooner or later limit the growth of any population?

The debate over this important question has been going on since 1798 when Thomas Malthus, a British economist, hypothesized that the human population tends to grow exponentially, while food supplies tend to increase more slowly at a linear rate. So far, Malthus has been proven wrong. Food production has grown at an exponential rate instead of at a linear rate because of genetic and other technological advances in industrialized food production.

Environmental scientists are reexamining arguments such as those of Malthus about possible limits to the growth of human populations and economies. One view is that we have exceeded some of those limits, with too many people collectively degrading the earth's life-support system (see Figure 3-B, p. 73). To some scientists and other analysts, the key problem is *overpopulation* because of the sheer number of people in less-developed countries (see Figure 1-14, top, p. 15), which have 82% of the world's population. To others, the key factor is *overconsumption* in affluent, more-developed countries because of their high rates of resource use per person (see Figure 1-14, bottom).

Another view of population growth is that, so far, technological advances have allowed us to overcome the environmental limits that all populations of other species face and that this has had the effect of increasing the earth's carrying capacity for our species. They point out that average life expectancy in most of the world has been steadily rising despite dire warnings from some environmental scientists that we are seriously degrading our life-support system.

Some of these analysts argue that because of our technological ingenuity, there are few, if any, limits to human population growth and resource use per person. They believe that we can continue ever-increasing economic growth and avoid serious damage to our life-support systems by making technological advances in areas such as food production and medicine, and by finding substitutes for resources that we are depleting. As a result, they see no need to slow the world's population growth.

Proponents of slowing and eventually stopping population growth point out that

tries, where population is growing exponentially at about 0.1% a year. The other 98% were added to the world's middle- and low-income, less-developed countries, where the population is growing exponentially, 14 times faster at about 1.4% a year, on average.

At least 95% of the 2.6 billion people projected to be added to the world's population between 2012 and 2050 will be born into the least-developed countries, which are not equipped to deal with the pressures of such rapid growth. The world's three most populous countries are, in order, China, India, and the United States. In 2012, more than one of every three persons on the earth lived in China (with 19% of the world's population) or India (with 18%).

A third important trend in human population growth is the movement of people from rural areas to cities. More than half of the world's people now live in *urban areas,* or cities (see chapter-opening photo) and their surrounding suburbs. The great majority of these urban dwellers live in less-developed countries where resources for dealing with rapidly growing populations are limited. (We cover this and other *urbanization trends* in Chapter 22.)

Human Population Growth Impacts Natural Capital

As the human population grows, so does the global total human ecological footprint (see Figure 1-13, p. 14), and the bigger this footprint, the higher the overall impact on the earth's natural capital. The 2005 Millennium Ecosystem Assessment concluded that human activities have degraded about 60% of the earth's ecosystem services (Figure 6-4).

This raises a question that many scientists are now studying: How many people can the earth support indefinitely (Concept 6-1)? Some analysts contend that we need to define the planet's **cultural carrying capacity**: the maximum number of people who could live in reason-

Figure 6-A Natural capital degradation: Fertile topsoil is an irreplaceable form of natural capital necessary for supplying the world's people with food. According to a joint survey by the UN Environment Programme (UNEP) and the World Resources Institute, topsoil is eroding faster than it forms on about 38% of the world's cropland, including this farmland in Iowa.

Lynn Betts/USDA Natural Resources Conservation Service

we are not providing the basic necessities for about 1.4 billion people—one of every five on the planet—who struggle to survive on the equivalent of about $1.25 per day. This raises a serious question: How will we meet the basic needs of the additional 2.6 billion people projected to be alive in 2050?

Proponents of slowing growth also warn of two potentially serious consequences that we could face if we do not sharply lower birth rates. First, death rates might increase because of declining health and environmental conditions and increasing social disruption in some areas, as is already happening in parts of Africa. A worst-case scenario for such a trend is a crash of the human population to a more sustainable level, perhaps as low as 2 billion. Second, resource use and degradation of normally renewable resources may intensify as more consumers increase their already large ecological footprints in more-developed countries and in rapidly developing countries such as China, India, and Brazil.

So far, advances in food production and health care have staved off widespread population declines. But there is extensive and growing evidence that we are steadily depleting and degrading much of the earth's irreplaceable natural capital (see Figure 1-3, p. 7) that supports us. We can get away with this for awhile, because the earth's life-support system is very complex and resilient, and because of built-in time delays between disturbances to the system and responses to them from within the system. But like unseen termites eating away the foundation of a house, at some point, such disturbances could reach a *tipping* or *breaking point* (see Figure 3-B, p. 73) beyond which there could be damaging and long-lasting change.

No one knows how close we are to environmental limits that, many scientists say, eventually will control the size of the human population. However, these scientists argue that human population growth is a vital scientific, political, economic, and ethical issue that we must confront.

Critical Thinking

How close do you think we are to the environmental limits of human population growth? Very close, moderately close, or far away? Explain.

able freedom and comfort indefinitely, without decreasing the ability of the earth to sustain future generations. (See the online Guest Essay by Garrett Hardin on this topic.) This issue has long been a topic of scientific and political debate (Science Focus 6.1).

Animated Figure 6-4 We humans have altered the natural systems that sustain our lives and economies in at least eight major ways to meet the increasing needs and wants of our growing population (Concept 6-1). **Questions:** In your daily living, do you think you contribute directly or indirectly to any of these harmful environmental impacts? Which ones? Explain.

Top: ©Dirk Ercken/Shutterstock.com. Center: ©Fulcanelli/Shutterstock.com. Bottom: © Werner Stoffberg/Shutterstock.com.

Natural Capital Degradation

Altering Nature to Meet Our Needs

Reducing biodiversity

Increasing use of net primary productivity

Increasing genetic resistance in pest species and disease-causing bacteria

Eliminating many natural predators

Introducing harmful species into natural communities

Using some renewable resources faster than they can be replenished

Disrupting natural chemical cycling and energy flow

Relying mostly on polluting and climate-changing fossil fuels

© Cengage Learning

6-2 What Factors Influence the Size of the Human Population?

CONCEPT 6-2A
Population size increases through births and immigration, and decreases through deaths and emigration.

CONCEPT 6-2B
The average number of children born to the women in a population (total fertility rate) is the key factor that determines population size.

The Human Population Can Grow, Decline, or Remain Fairly Stable

The basics of global population change are quite simple. If there are more births than deaths during a given period of time, the earth's population increases, and when the opposite is true, it decreases. When the number of births equals the number of deaths during a particular time period, population size does not change.

Instead of using the total numbers of births and deaths per year, demographers use the **crude birth rate** (the number of live births per 1,000 people in a population in a given year) and the **crude death rate** (the number of deaths per 1,000 people in a population in a given year).

Human populations grow or decline in particular countries, cities, or other areas through the interplay of three factors: *births (fertility), deaths (mortality),* and *migration.* We can calculate the **population change** of an area by subtracting the number of people leaving a population (through death and emigration) from the number entering it (through birth and immigration) during a specified period of time (usually 1 year) (**Concept 6-2A**):

$$\text{Population change} = (\text{Births} + \text{Immigration}) - (\text{Deaths} + \text{Emigration})$$

When births plus immigration exceed deaths plus emigration, a population grows; when the reverse is true, a population declines. (see Figure 18, p. S44, in Supplement 6 for a map showing various percentage rates of population growth in the world's countries in 2012.)

Women Are Having Fewer Babies but the World's Population Is Still Growing

Another measurement used in population studies is the **fertility rate**, a measure of how many children are born in a population over a set period of time. Here, we consider two types of fertility rates, the first being the **replacement-level fertility rate**: the average number of children that couples in a population must bear to replace themselves. It is slightly higher than two children per couple (2.1 in more-developed countries and as high as 2.5 in some less-developed countries), mostly because some children die before reaching their reproductive

years. Any fertility rate above the replacement level will cause a population to grow.

If we were to reach the replacement-level fertility rate for the world, would it bring an immediate halt to population growth? No, because the number of *future* parents alive has grown dramatically. If all of today's couples were to have an average of 2.1 children, they would not be contributing to population growth. But if all of today's girl children were to grow up to have an average of 2.1 children as well, the world's population would continue to grow for 50 years or more (assuming death rates do not rise) because there are so many girls under age 15 who will be moving into their reproductive years.

The second type of fertility rate, and the key factor affecting human population growth and size, is the **total fertility rate (TFR)**: the average number of children born to the women in a population during their reproductive years (**Concept 6-2B**). Between 1955 and 2012, the average TFR for more-developed countries dropped from 2.8 to 1.6, and the average TFR for less-developed countries dropped from 6.2 to 2.6. The latter number is projected to continue dropping. If the global TFR, which was 2.4 in 2012, were to drop to 2.1, the world's population would continue to grow for about 50 years and then would level off and gradually decline.

Between 1955 and 2012, the global TFR dropped from 5 to 2.4. Those who support slowing the world's population growth view this as good news. However, to eventually halt population growth, the global TFR will have to drop to 2.1. (See Figure 19, p. S45, in Supplement 6 for a map showing how TFRs vary globally.)

CASE STUDY

The U.S. Population—Third-Largest and Growing

The population of the United States grew from 76 million in 1900 to 314 million by 2012, despite oscillations in the country's TFR and population growth rate (Figure 6-5). It took the country 139 years to reach a population of 100 million people, 52 years to add another 100 million (by 1967), and only 39 years to add the third 100 million (by 2006). During the period of high birth rates between 1946 and 1964, known as the *baby boom,* 79 million people were added to the U.S. population. At the peak of the baby boom in 1957, the average TFR was 3.7 children per woman. In 2012, and in most years since 1972, it has been at or below 2.1 children per woman (1.9 in 2012), compared to a global TFR of 2.4.

The drop in the TFR has slowed the rate of population growth in the United States. But the country's population is still growing faster than those of most other more-developed countries and it is not close to leveling off. According to the U.S. Census Bureau, about 2.3 mil-

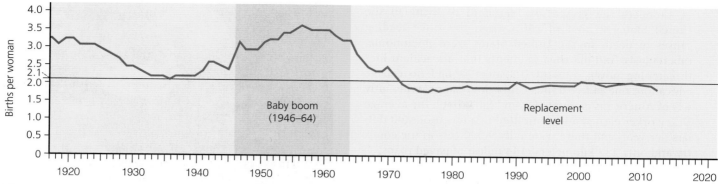

Figure 6-5 The graph shows the total fertility rates for the United States between 1917 and 2012. **Question:** The U.S. fertility rate has declined and remained at or below replacement levels since 1972. So why is the population of the United States still increasing?

(Compiled by the authors using data from Population Reference Bureau and U.S. Census Bureau.)

lion people were added to the U.S. population in 2012. About 1.6 million (70% of the total) were added because there were that many more births than deaths, and about 0.7 million (30% of the total) were immigrants.

Immigration has become a political issue in the United States. The country was founded by immigrants and since 1820, it has admitted almost twice as many immigrants and refugees as all other countries combined. The number of legal immigrants (including refugees) has varied during different periods because of changes in immigration laws and rates of economic growth (Figure 6-6).

In addition to the fourfold increase in population growth since 1900, some amazing changes in lifestyles took place in the United States during the 20th century (Figure 6-7), which led to Americans living longer. Along with this came dramatic increases in per capita resource use and a much larger total and per capita ecological footprint.

Figure 6-6 Legal immigration to the United States, 1820–2006 (the last year for which data are available). The large increase in immigration since 1989 resulted mostly from the Immigration Reform and Control Act of 1986, which granted legal status to certain illegal immigrants who could show they had been living in the country before January 1, 1982.

Photo: Andresr/Shutterstock.com

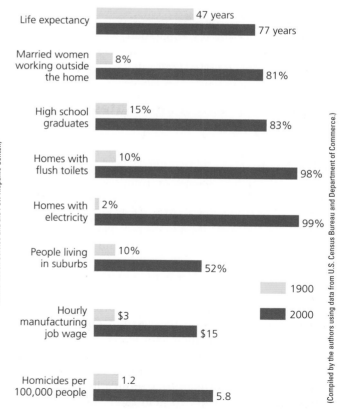

Figure 6-7 Some major changes took place in the United States between 1900 and 2000. **Question:** Which two of these changes do you think had the biggest impacts on the U.S. ecological footprint?

Compare the present-day standard of living in the United States to that of 1907 when the three leading causes of death in the United States were pneumonia, tuberculosis, and diarrhea (ailments that are seldom life-threatening now); 90% of U.S. doctors had no college education; one of five adults could not read or write; only 6% of Americans graduated from high school; the average U.S. worker earned a few hundred to a few thousand dollars per year; and there were only 9,000 cars in the country and only 232 kilometers (144 miles) of paved roads.

The U.S. Census Bureau projects that the U.S. population is likely to increase from 314 million in 2012 to 400 million by 2050—an addition of 86 million more Americans over four decades. Because of a high per-person rate of resource use and the resulting waste and pollution, each addition to the U.S. population has an enormous environmental impact (see the map in Figure 8, p. S33, in Supplement 6). Recall that the environmental impact of a population is obtained by multiplying the effects of three factors (see Figure 1-14, p. 15, bottom): population size, affluence (and resulting high rates of resource use per person), and technology. The United States has by far the world's largest total and per capita ecological footprint (see Figure 1-11, p. 13), mostly because of the size of its population multiplied by its very high rate of resource use per person. This explains why some analysts consider the United States to be the world's most overpopulated country.

 CONSIDER THIS. . .

THINKING ABOUT The U.S. Population

Considering this information, do you believe that the United States is the world's most overpopulated country? Explain.

Several Factors Affect Birth Rates and Fertility Rate

Many factors affect a country's average birth rate and TFR. One is the *importance of children as a part of the labor force,* especially in less-developed countries. This is a major part of why it makes sense for many poor couples in those countries to have a large number of children. They need help with hauling daily drinking water (Figure 6-8), gathering wood for heating and cooking, and tending crops and livestock.

Another economic factor is the *cost of raising and educating children.* Birth and fertility rates tend to be lower in more-developed countries, where raising children is much more costly because they do not enter the labor force until they are in their late teens or twenties. In the United States, it costs more than $235,000 to raise a middle-class child from birth to age 18. By contrast, many children in poor countries receive little education and instead have to work to help their families survive (Figure 6-9).

The *availability of, or lack of, private and public pension systems* can influence the decisions of some couples on how

Figure 6-8 These young girls in India are carrying water.

many children to have, especially the poor in less-developed countries. Pensions reduce a couple's need to have several children to help support them in old age.

Also, there are more *infant deaths* in poorer countries, and over time, this has affected cultural norms related to family size in these countries. The more children a couple has, the more likely it is that at least a few will survive and grow to adulthood.

Urbanization plays a role in birth rate trends. People living in urban areas usually have better access to family planning services and tend to have fewer children than do those living in the rural areas of poorer countries.

Another important factor is the *educational and employment opportunities available for women.* Total fertility rates tend to be low when women have access to education and paid employment outside the home. In less-developed countries, a woman with no education typically has two more children than does a woman with a high school education. In most societies, better-educated women tend to marry later and have fewer children.

Average age at marriage (or, more precisely, the average age at which a woman has her first child) also plays a role. Women normally have fewer children when their average age at marriage is 25 or older.

Birth rates and TFRs are also affected by the *availability of legal abortions.* According to the World Health Organization and the Guttmacher Institute, each year, about 210 million women become pregnant and at least 42 million of them get abortions—about 22 million of them legal and the other 20 million illegal (and often unsafe). The *availability of reliable birth control methods* also allows women to control the number and spacing of the children they have.

Religious beliefs, traditions, and cultural norms also play a role. In some countries, these factors favor large families as many people strongly oppose abortion and some forms of birth control.

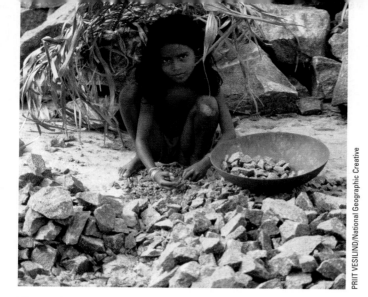

Figure 6-9 This young girl is breaking granite into gravel in the Kerala State of India.

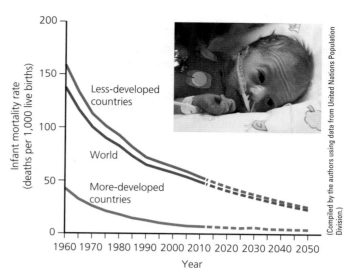

Figure 6-10 Infant mortality rates for the world's more-developed countries and less-developed countries, 1950–2012, with projections to 2050 based on medium population projections.

Photo: Gert Vrey/Shutterstock.com

CONSIDER THIS. . .

CONNECTIONS Preference for Male Children and Social Problems in China

In China, where families are strongly encouraged to have only one child, there is a strong preference for male children, partly because daughters are likely to marry and leave their parents. Some pregnant Chinese women use ultrasound to determine the gender of their fetuses and get an abortion if the child is female. As a result, baby boys are sometimes kidnapped and sold to families that want boys, and some young girls have been kidnapped and sold as brides for single men. The Chinese government estimates that by 2030, about 30 million Chinese men will not be able to find wives and if the economy slows, they might have poor job prospects. Sociologists say this is a recipe for social upheaval.

Several Factors Affect Death Rates

The rapid growth of the world's population over the past 100 years is not primarily the result of a rise in the birth rate. Instead, it has been caused largely by declining death rates, especially in less-developed countries. More people in some of these countries are living longer and fewer infants are dying because of increased food supplies, improvements in food distribution, better nutrition, medical advances such as immunizations and antibiotics, improved sanitation, and safer water supplies.

A useful indicator of the overall health of people in a country or region is **life expectancy**, which for any given year is the average number of years a person born in that year can be expected to live. Between 1955 and 2012, the average global life expectancy increased from 48 years to 70 years. In 2012, Japan had the world's longest life expectancy of 83 years. In the world's poorest countries, life expectancy is 55 years or less. **GOOD NEWS**

Between 1900 and 2012, the average U.S. life expectancy increased from 47 years to 79 years. However, the United States ranks 32nd among nations in life expectancy, down from 5th in 1950. Research indicates that a key factor in this ranking is poor health, even though the United States leads the world in health-care costs per person.

Another important indicator of overall health in a population is its **infant mortality rate**, the number of babies out of every 1,000 born who die before their first birthday. It is viewed as one of the best measures of a society's quality of life because it reflects a country's general level of nutrition and health care. A high infant mortality rate usually indicates insufficient food *(undernutrition)*, poor nutrition *(malnutrition),* and a high incidence of infectious disease. Infant mortality also affects the TFR. In areas with low infant mortality rates, women tend to have fewer children because fewer of their children die at an early age.

Infant mortality rates in most countries have declined dramatically since 1965 (Figure 6-10) but vary widely among different countries (see Figure 20, p. S45, Supplement 6). Even so, every year, more than 4 million infants (most of them in less-developed countries) die of *preventable* causes during their first year of life. This average of nearly 11,000 mostly unnecessary infant deaths per day is equivalent to 55 jet airliners, each loaded with 200 infants younger than age 1, crashing *every day* with no survivors—a tragedy rarely reported in the news.

In 1900 the U.S. infant mortality rate was 165. In 2012 it was 6.0. This sharp decline was a major factor in the marked increase in U.S. average life expectancy during this period. However, in 2012 the United States ranked 44th among all nations in terms of infant mortality rates for two main reasons: the generally inadequate health care for poor women during pregnancy, as well as for their babies after birth, and drug addiction among many pregnant women.

PROJECTING POPULATION CHANGE

Estimates of the human population size in 2050 range from 7.8 billion to 10.8 billion people—a difference of 3 billion. Why this wide range of estimates? The answer is that there are countless factors that demographers have to consider when making projections for any country or region of the world.

First, they have to determine the reliability of current population estimates. Many of the more-developed countries such as the United States have fairly good estimates. But many countries have poor knowledge of their population size. Some countries even deliberately inflate or deflate the numbers for economic or political purposes.

Second, demographers make assumptions about trends in fertility—that is, whether women in a country will, on average, have fewer babies or more babies in the future. The demographers might assume that fertility is declining by a certain percentage per year. If the estimate of the rate of declining fertility is off by a few percentage points, the resulting percentage increase in population can be magnified over a number of years, and be quite different from the projected population size increase.

For example, UN demographers assumed that Kenya's fertility rate would decline, and based on that, in 2002, they projected that Kenya's total population would be 44 million by 2050. In reality, the fertility rate rose from 4.7 to 4.8 children per woman. The population momentum was therefore more powerful than expected, and in 2010, the UN revised its projection for Kenya's population in 2050 to 70.8 million, which was 60% higher than its earlier projection.

Third, population projections are made by a variety of organizations employing demographers. UN projections tend to be cited most often. But the U.S. Census Bureau also makes world population projections, as does the International Institute for Applied Systems Analysis (IIASA) and the U.S. Population Reference Bureau. The World Bank has also made projections in the past and maintains a database. All of these organizations use differing sets of data and differing methods, and their projections vary (Figure 6-B).

Critical Thinking

If you were in charge of the world and making decisions about resource use based on population projections, which of the projections in Figure 6-B would you rely on? Explain.

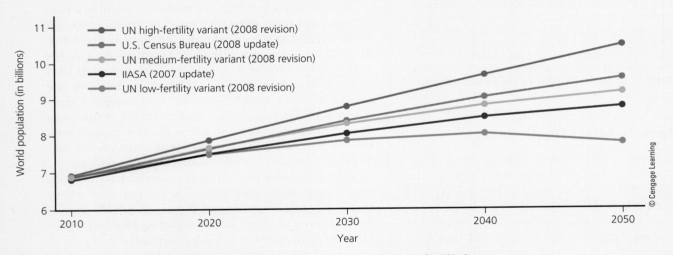

Figure 6-B World population projections to 2050 from three different organizations: the UN, the U.S. Census Bureau, and IIASA. Note that the uppermost, middle, and lowermost curves of these five projections are all from the UN, each assuming a different level of fertility.

Migration Affects an Area's Population Size

The third factor in population change is **migration**: the movement of people into *(immigration)* and out of *(emigration)* specific geographic areas. Most people migrating from one area or country to another seek jobs and economic improvement. But many are driven to migrate by religious persecution, ethnic conflicts, political oppression, or war.

According to a United Nations (UN) study and another study by environmental scientist Norman Myers, in 2008 there were at least 40 million *environmental refugees*—people who had to leave their homes because of water or food shortages, soil erosion, or some other form of environmental degradation or depletion. It is difficult to find more recent, reliable estimates of the number of such refugees, but in 2012, the UN projected that by 2020, there will be 50 million environmental refugees.

Estimates of any population's future numbers can vary considerably. These varying estimates depend mostly on TFR projections. However, demographers also have to make assumptions about death rates, migration, and a number of other variables. If their assumptions are wrong, their population forecasts can be way off the mark (Science Focus 6.2).

6-3 How Does a Population's Age Structure Affect Its Growth or Decline?

CONCEPT 6-3
The numbers of males and females in young, middle, and older age groups determine how fast a population grows or declines.

A Population's Age Structure Helps Us to Make Projections

An important factor determining whether the population of a country increases or decreases is its **age structure**: the numbers or percentages of males and females in young, middle, and older age groups in that population (**Concept 6-3**).

Population experts construct a population *age-structure diagram* by plotting the percentages or numbers of males and females in the total population in each of three age categories: *prereproductive* (ages 0–14), consisting of individuals normally too young to have children; *reproductive* (ages 15–44), consisting of those normally able to have children; and *postreproductive* (ages 45 and older), with individuals normally too old to have children. Figure 6-11 presents generalized age-structure diagrams for countries with rapid, slow, zero, and negative population growth rates.

A country with a large percentage of its people younger than age 15 (represented by a wide base in Figure 6-11, far left) will experience rapid population growth unless death rates rise sharply. Because of this *demographic momentum,* the number of births in such a country will rise for several decades even if women have an average of only one or two children each, due to the large number of girls entering their prime reproductive years.

In 2012, about 26% of the world's population—29% in the less-developed countries and 16% in more-developed countries—was under age 15. By 2025, the world's current 1.8 billion people under age 15—roughly one of every four persons on the planet—will move into their prime reproductive years. The dramatic differences in population age structure between less-developed and more-developed countries (Figure 6-12) show why most future human population growth will take place in less-developed countries (Figure 6-3).

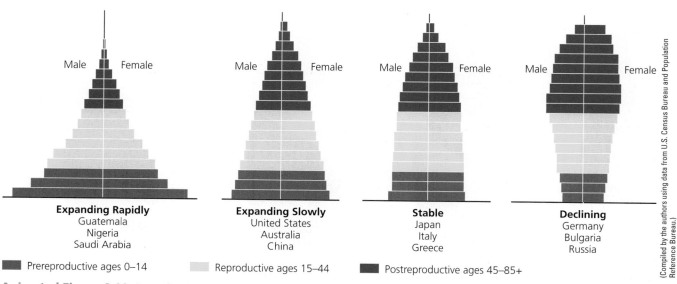

Animated Figure 6-11 Generalized population age-structure diagrams for countries with rapid (1.5–3%), slow (0.3–1.4%), zero (0–0.2%), and negative (declining) population growth rates. ***Question:*** Which of these diagrams best represents the country where you live?

(Compiled by the authors using data from U.S. Census Bureau and Population Reference Bureau.)

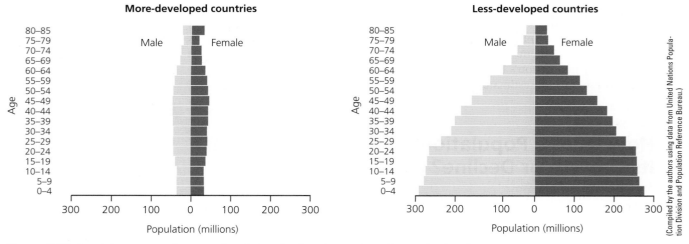

Figure 6-12 Population structure by age and sex in less-developed countries and more-developed countries for 2011. **Question:** If all girls under 15 were to have only one child during their lifetimes, how do you think these structures would change over time?

The global population of seniors—people who are 65 and older—is projected to triple by 2050, when one of every six people will be a senior. (See the Case Study that follows.) This graying of the world's population is due largely to declining birth rates and medical advances that have extended life spans. In 2012, the three nations with the largest percentage of their population age 65 or older were, in order, Japan, Germany, and Italy. In such countries, the number of working adults is shrinking in proportion to the number of seniors, which in turn is slowing the growth of tax revenues in these countries. Some analysts worry about how such societies will support their growing populations of seniors.

(Compiled by the authors using data from United Nations Population Division and Population Reference Bureau.)

The American Baby Boom

Changes in the distribution of a country's age groups have long-lasting economic and social impacts. For example, consider the American baby boom, which added 79 million people to the U.S. population between 1946 and 1964. Over time, this group looks like a bulge moving up through the country's age structure, as shown in Figure 6-13.

For decades, members of the baby-boom generation have strongly influenced the U.S. economy because they make up about 36% of all adult Americans. Baby boomers created the youth market in their teens and twen-

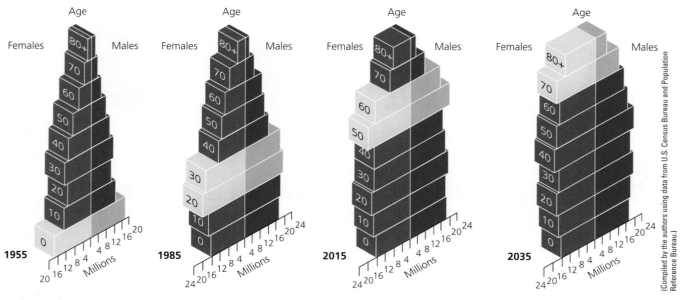

Animated Figure 6-13 Age-structure charts tracking the baby-boom generation in the United States, 1955, 1985, 2015 (projected), and 2035 (projected).

(Compiled by the authors using data from U.S. Census Bureau and Population Reference Bureau.)

ties and are now creating the late middle age and senior markets. In addition to having this economic impact, the baby-boom generation plays an increasingly important role in deciding who gets elected to public office and what laws are passed or weakened.

Since 2011, when the first baby boomers began turning 65, the number of Americans older than age 65 has grown at the rate of about 10,000 a day and will do so through 2030. This process has been called the *graying of America*. As the number of working adults declines in proportion to the number of seniors, there may be political pressure from baby boomers to increase tax revenues to help support the growing senior population. This could lead to economic and political conflicts between younger and older Americans.

🔍 CONSIDER THIS...

CONNECTIONS Baby Boomers, the U.S. Work Force, and Immigration

According to the U.S. Census Bureau, after 2020, much higher immigration levels will be needed to supply enough workers as baby boomers retire. According to a recent study by the UN Population Division, if the United States wants to maintain its current ratio of workers to retirees, it will need to absorb an average of 10.8 million immigrants every year—more than 10 times the current immigration level—through 2050.

Populations Made Up Mostly of Older People Can Decline Rapidly

As the percentage of people age 65 or older increases, more countries will begin experiencing population declines. If population decline is gradual, its harmful effects usually can be managed. However, some countries are experiencing fairly rapid declines and feeling such effects more severely.

Japan has the world's highest percentage of elderly people (above age 65) and the world's lowest percentage of young people (below age 15). In 2012, Japan's population was 128 million. By 2050, its population is projected to be 95.5 million, a 25% drop. As its population declines, there will be fewer adults working and paying taxes to support an increasing elderly population. Because Japan discourages immigration, it may face a bleak economic future. As a result, some have called for the country to rely more on robots to do its manufacturing jobs and on selling robots in the global economy to help support its aging population.

In China, the growth in numbers of children has slowed because of its one-child policy. As a result, the average age of China's population has been increasing over the past two decades at one of the fastest rates ever recorded. While China's population is not yet declining, the UN estimates that by 2025, China is likely to have too few young workers to support its rapidly aging population. This graying of the Chinese population could lead to a declining work force, higher wages for workers, limited funds for supporting continued economic development, and fewer children and grandchildren to care for the growing number of elderly people. These concerns and other factors may slow economic growth and have led to some relaxation of China's one-child population control policy.

Figure 6-14 lists some of the problems associated with rapid population decline. Countries currently faced with rapidly declining populations include Japan, Russia, Germany, Bulgaria, Hungary, Ukraine, Serbia, Greece, Portugal, and Italy.

Populations Can Decline Due to a Rising Death Rate: The AIDS Tragedy

A large number of deaths from AIDS can disrupt a country's social and economic structure by removing significant numbers of young adults from its population. According to the World Health Organization, between 1981 and 2012, AIDS killed more than 30 million people (617,000 in the United States).

Unlike hunger and malnutrition, which kill mostly infants and children, AIDS kills primarily young adults and leaves many children orphaned, some of whom are also infected with HIV, the virus that can lead to AIDS. Worldwide, AIDS is the leading cause of death for people of ages 15–49.

This pandemic has had a devastating effect in some countries, and has changed their population age structures

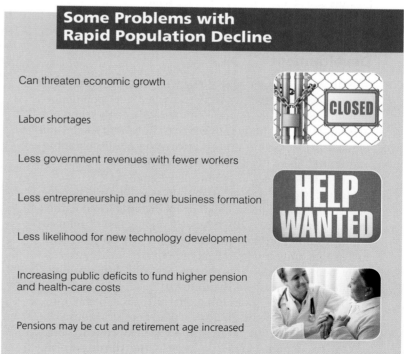

Some Problems with Rapid Population Decline

- Can threaten economic growth
- Labor shortages
- Less government revenues with fewer workers
- Less entrepreneurship and new business formation
- Less likelihood for new technology development
- Increasing public deficits to fund higher pension and health-care costs
- Pensions may be cut and retirement age increased

Figure 6-14 Rapid population decline can cause several problems. **Question:** Which two of these problems do you think are the most important?

Top: ©Slavoljub Pantelic/Shutterstock.com. Center: lofoto/Shutterstock.com. Bottom: Yuri Arcurs/Shutterstock.com.

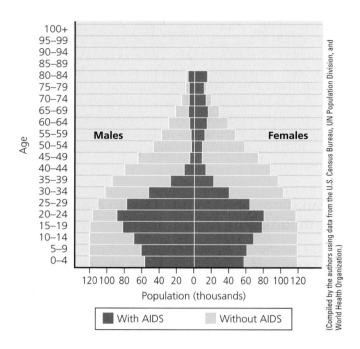

(Compiled by the authors using data from the U.S. Census Bureau, UN Population Division, and World Health Organization.)

Figure 6-15 In Botswana, more than 25% of people ages 15–49 were infected with HIV in 2011. This figure shows two projected age structures for Botswana's population in 2020—one including the possible effects of the AIDS epidemic (red bars), and the other not including those effects (yellow bars). **Question:** How might this affect Botswana's economic development?

(Figure 6-15). This has had a number of harmful effects. One is a sharp drop in average life expectancy, especially in several southern African countries where 15–26% of all people between ages 15 and 49 are infected with HIV. Another is the loss of productive young-adult workers and trained personnel such as scientists, farmers, engineers, and teachers, as well as government, business, and health-care workers. The essential services they could provide are therefore lacking, and there are fewer taxpayers and fewer workers available to support the very young and the elderly. Many experts have called for the creation of a massive international program to help countries ravaged by AIDS.

6-4 How Can We Slow Human Population Growth?

CONCEPT 6-4
We can slow human population growth by reducing poverty, elevating the status of women, and encouraging family planning.

The First Step Is to Promote Economic Development

Scientific studies and experience have shown that the three most effective ways to slow or stop population growth are to reduce poverty, primarily through economic development and universal primary education; to elevate the status of women; and to encourage family planning and reproductive health care (**Concept 6-4**). Let's begin by looking at the role of economic development.

In the world's most desperately poor countries, couples tend to have more children for reasons listed earlier in this chapter (p. 128). Thus, in many less-developed countries, total fertility and population growth rates tend to be high, and large numbers of poor people are increasingly being crowded into unsanitary and difficult living conditions in slums and shantytowns.

Demographers, examining the birth and death rates of western European countries that became industrialized during the 19th century, have developed a hypothesis of population change known as the **demographic transition**: As countries become industrialized and economically developed, their populations tend to grow more slowly. According to the hypothesis, this transition takes place in four stages, as shown in Figure 6-16.

Some analysts believe that most of the world's less-developed countries will make a demographic transition over the next few decades, mostly because newer technologies will help them to develop economically and to raise their per capita incomes. Other analysts fear that rapid population growth, extreme poverty, and increasing environmental degradation and resource depletion in some low-income, less-developed countries could leave these countries stuck in stage 2 of the demographic transition.

Another factor that could hinder the demographic transition in some less-developed countries is that since 1985, economic assistance from more-developed countries has generally dropped. The resulting shortage of funds, coupled with poor economies in some countries, could leave large numbers of people trapped in poverty, which could in turn keep population growth rates high in such countries. Many experts argue that more-developed countries should help less-developed nations to make the demographic transition by aiding them in their economic development. They contend that such an effort could help to stabilize the global population, thereby making for a more sustainable world.

Empowering Women Can Help to Slow Population Growth

A number of studies show that women tend to have fewer children if they are educated, have the ability to control their own fertility, earn an income of their own, and live in societies that do not suppress their rights. Although women make up roughly half of the world's population,

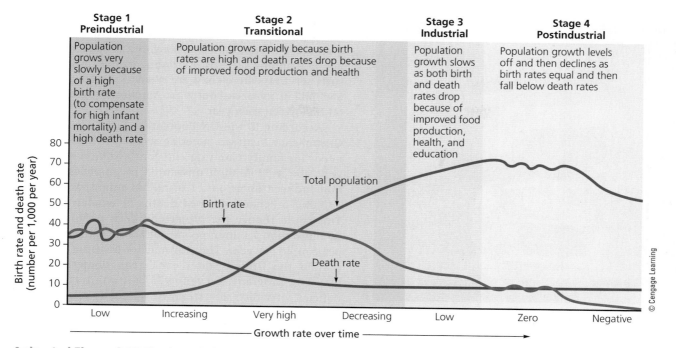

Stage 1
Preindustrial

Stage 2
Transitional

Stage 3
Industrial

Stage 4
Postindustrial

Population grows very slowly because of a high birth rate (to compensate for high infant mortality) and a high death rate

Population grows rapidly because birth rates are high and death rates drop because of improved food production and health

Population growth slows as both birth and death rates drop because of improved food production, health, and education

Population growth levels off and then declines as birth rates equal and then fall below death rates

Total population

Birth rate

Death rate

Birth rate and death rate (number per 1,000 per year)

80
70
60
50
40
30
20
10
0

Low   Increasing   Very high   Decreasing   Low   Zero   Negative

Growth rate over time

© Cengage Learning

Animated Figure 6-16 The *demographic transition,* which a country can experience as it becomes industrialized and more economically developed, can take place in four stages. **Question:** At what stage is the country where you live?

in most societies they have fewer rights and educational and economic opportunities than men have.

Women do almost all of the world's domestic work and child care for little or no pay and provide more unpaid health care (within their families) than do all of the world's organized health-care services combined. In rural areas of Africa, Latin America, and Asia, women do 60–80% of the work associated with growing food, gathering, and hauling wood (Figure 6-17) and animal dung for use as fuel, and hauling water. As one Brazilian woman observed, "For poor women, the only holiday is when you are asleep."

While women account for 66% of all hours worked, they receive only 10% of the world's income and own just 2% of the world's land. They also make up 70% of the world's poor and 66% of its 800 million illiterate adults. Because sons are more valued than daughters in many societies, girls are often kept at home to work instead of being sent to school. Globally, the number of school-age girls who do not attend elementary school is more than 900 million—almost 3 times the entire U.S. population. Poor women who cannot read often have an average of five to seven children, compared to two or fewer children in societies where almost all women can read.

However, an increasing number of women in less-developed countries are taking charge of their lives and reproductive behavior. As it expands, such bottom-up change driven by individual women will play an important role in stabilizing populations, reducing poverty and environmental degradation, and allowing more access to basic human rights.

Family Planning Can Provide Several Benefits

Family planning provides educational and clinical services that help couples choose how many children to have and when to have them. Such programs vary from culture to culture, but most of them provide information on birth spacing, birth control, and health care for pregnant women and infants (Figure 6-18).

Figure 6-17 This woman in Nepal is bringing home firewood. Typically, she spends 2 hours a day, 2 or 3 times a week, on this task.

Iv Nikolny/Shutterstock.com

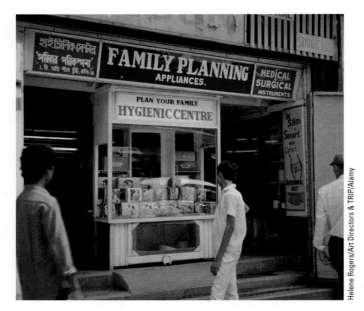

Figure 6-18 This family planning center is located in Kolkata (Calcutta), India.

Family planning has been a major factor in reducing the number of births throughout most of the world. It has also reduced the number of abortions performed each year and has decreased the numbers of mothers and fetuses dying during pregnancy, according to studies by the UN Population Division and other population agencies.

Such studies indicate that family planning is responsible for a drop of at least 55% in total fertility rates (TFRs) in less-developed countries, from 6.0 in 1960 to 2.6 in 2012. For example, family planning, coupled with economic development, played a major role in the sharp drop in the TFR in Bangladesh from around 6.8 to 2.3 by 2012. Between 1971 and 2012, Thailand also used family planning to cut its annual population growth rate from 3.2% to 0.5%, and to reduce its TFR from 6.4 to 1.6. According to the UN, had there not been the sharp drop in TFRs since the 1970s, with all else being equal, the world's population today would be about 8.5 billion instead of 7 billion.

Family planning also has financial benefits. Studies have shown that each dollar spent on family planning in countries such as Thailand, Egypt, and Bangladesh saves $10–$16 in health, education, and social-service costs by preventing unwanted births.

Despite these successes, certain problems have hindered progress in some countries. There are two major problems. *First*, according to the UN Population Fund, about 42% of all pregnancies in less-developed countries are unplanned and about 26% end with abortion. So ensuring access to voluntary contraception would play a key role in stabilizing the populations and reducing the number of abortions in such countries.

Second, an estimated 215 million couples in less-developed countries want to limit their number of children and determine their spacing, but they lack access to family planning services. According to the UN, providing such services where they are needed would cost about $6.7 billion a year—about what Americans together spend on Halloween each year.

According to the UN Population Fund and the Alan Guttmacher Institute, meeting women's current unmet needs for family planning and contraception could prevent about 53 million unwanted pregnancies, 24 million induced abortions, 1.6 million infant deaths, and 142,000 pregnancy-related deaths of women per year. This could reduce the projected global population size by more than 1 billion people, at an average cost of $20 per couple per year. The Guttmacher Institute estimates that in the United States each year, domestic family planning programs prevent 973,000 unintended pregnancies, of which an estimated 406,000 would end in abortion.

Some analysts call for expanding family planning programs to educate men about the importance of having fewer children and taking more responsibility for raising them. Proponents also call for greatly increased research in order to develop more effective birth control methods for men.

The experiences of countries such as Japan, Thailand, Bangladesh, South Korea, Taiwan, and China show that a country can achieve or come close to replacement-level fertility within a decade or two. Thus, the real story of the past 50 years has been the sharp reduction in the *rate* of population growth (Figure 6-2) resulting from a combination of the reduction of poverty through economic development, empowerment of women, and the promotion of family planning. However, the global population is still growing fast enough to add possibly several billion more people during this century (Figure 6-1).

CASE STUDY

Slowing Population Growth in India

For six decades, India has tried to control its population growth with only modest success. The world's first national family planning program began in India in 1952, when its population was nearly 400 million. In 2012, after 60 years of population control efforts, India had 1.26 billion people—the world's second largest population. Much of this increase occurred because the country's life expectancy rose from 38 years of age in 1952 to 65 in 2012, mostly as a result of declining death rates.

In 1952, India added 5 million people to its population. In 2012, it added 19 million—more than any other country. Also, 31% of India's population is under age 15, which sets the country up for further rapid population growth. The United Nations projects that by 2030, India will be the world's most populous country, and by 2050 it will have a population of 1.69 billion.

Figure 6-19 These people are shopping at Kolkata's (Calcutta's) South City Mall, the largest shopping mall in India.

India has the world's fourth largest economy and a thriving and rapidly growing middle class of more than 100 million people—a number nearly equal to a third of the U.S. population. This growing class of consumers will enlarge India's ecological footprint, as more Indians use more resources with every passing year (Figure 6-19).

However, the country faces a number of serious poverty, malnutrition, and environmental problems that could worsen as its population continues to grow rapidly. About one-fourth of all people in India's cities live in slums, and prosperity and progress have not touched many of the nearly 650,000 rural villages where more than two-thirds of India's population lives. Nearly half of the country's labor force is unemployed or underemployed and 42% of its population lives in extreme poverty (Figure 6-20).

For decades, the Indian government has provided family planning services throughout the country and has strongly promoted a smaller average family size. Even so, Indian women have an average of 2.5 children.

Two factors help to account for larger families in India. *First,* most poor couples believe they need several children to work and care for them in old age. *Second,* the strong cultural preference in India for male children means that some couples keep having children until they produce one or more boys. The result: even though 90% of Indian couples have access to at least one modern birth control method, only 47% actually use one (compared to 85% in China).

India also faces critical resource and environmental problems. With 18% of the world's people, India has just 2.3% of the world's land resources and 2% of its forests. About half the country's cropland is degraded as a result of soil erosion and overgrazing. In addition, more than two-thirds of its water is seriously polluted, sanitation services often are inadequate, and many of its major cities suffer from serious air pollution.

India is undergoing rapid economic growth, which is expected to accelerate. This not only will help many people in India, but it will also put more pressure on the country's and the earth's natural capital. On the other hand, economic growth may help India to slow its population growth by accelerating its demographic transition.

CASE STUDY

Slowing Population Growth in China: A Success Story

China is the world's most populous country, with 1.35 billion people (Figure 6-21), followed by India with 1.26 billion and the United States with 314 million. In 2011, the U.S. Census Bureau projected that if current trends continue, China's population will increase to about 1.4 billion by 2026 and then will begin a slow decline to as low as 750 million by the end of this century.

In the 1960s, China's large population was growing so rapidly that there was a serious threat of mass starvation. To avoid this, government officials decided to take measures that eventually led to the establishment of the

Figure 6-20 Homeless people in Kolkata, India in 2011.

Figure 6-21
Thousands of people crowd a street in China, where almost one-fifth of all the people on the planet live.

world's most extensive, intrusive, and strict family planning and birth control program.

China's goal has been to sharply reduce population growth by promoting one-child families. The government provides contraceptives, sterilizations, and abortions for married couples. In addition, married couples pledging to have no more than one child receive a number of benefits, including better housing, more food, free health care, salary bonuses, and preferential job opportunities for their child. Couples who break their pledge lose such benefits.

Since this government-controlled program began, China has made impressive efforts to feed its people and bring its population growth under control. Between 1972 and 2012, the country cut its birth rate in half and reduced the average number of children born to its women from 5.7 to 1.5, compared to 1.9 children per woman in the United States.

Since 1980, China has undergone rapid industrialization and economic growth. According to the Earth Policy Institute, between 1990 and 2010, this reduced the number of people living in extreme poverty by almost 500 million. It also helped at least 300 million Chinese—a number almost equal to the entire U.S. population—to become middle-class consumers. Over time, China's rapidly growing middle class will consume more resources per person, expanding China's ecological footprint (see Figure 1-11, p. 13) within its own borders and in other parts of the world that provide it with resources. This will put a strain on China's and the earth's natural capital unless China steers a course toward more environmentally sustainable economic development.

Some have criticized China for having such a strict population control policy. However, government officials say that the alternative was mass starvation. They estimate that China's one-child policy has reduced its population size by as many as 400 million people.

Big Ideas

- The human population is growing rapidly and may soon bump up against environmental limits.

- Even if population growth is not a problem, the increasing use of resources per person is expanding the overall human ecological footprint and putting a strain on the earth's resources.

- We can slow human population growth by reducing poverty, elevating the status of women, and encouraging family planning.

This chapter began with a discussion of the fact that the world's human population has now passed 7 billion (**Core Case Study**). We noted that this is a result of exponential population growth and that many environmental scientists believe such growth to be unsustainable for the long run. We briefly considered some of the environmental problems brought on by exponential human population growth. We looked at factors that influence the growth of populations, as well as at how some countries have made progress in controlling population growth.

In the first six chapters of this book, you have learned how ecosystems and species have been sustained throughout the earth's history, in keeping with the three scientific **principles of sustainability**, by nature's reliance on solar energy, nutrient cycling, and biodiversity (see Figure 1-2, p. 6 or back cover). These three principles can guide us in dealing with the problems brought on by population growth and decline. That is, by employing solar and other renewable-energy technologies more widely, we can cut pollution and emissions of climate-changing gases that are increasing as the population grows. By reusing and recycling more materials, we could cut our waste and reduce our ecological footprints. And in focusing on preserving biodiversity, we could help to sustain the life-support system on which we and all other species depend.

Chapter Review

Core Case Study

1. Summarize the story of how human population growth has surpassed 7 billion and explain why this is significant to many environmental scientists (**Core Case Study**). List three factors that account for the rapid increase in the world's human population over the past 200 years.

Section 6-1

2. What is the key concept for this section? What is the range of estimates for the size of the human population in 2050? Summarize the three major population growth trends recognized by demographers. About how many people are added to the world's population each year? What are the world's three most populous countries? List eight major ways in which we have altered natural systems to meet our needs. Define **cultural carrying capacity**. Summarize the debate over whether and how long the human population can keep growing.

Section 6-2

3. What are the two key concepts for this section? Define and distinguish between **crude birth rate** and **crude death rate**. List three variables that affect the growth and decline of human populations. Explain how a given area's **population change** is calculated. Define **fertility rate**, and distinguish between the **replacement-level fertility rate** and the **total fertility rate (TFR)**. How has the global TFR changed since 1955?

4. Summarize the story of population growth in the United States and explain why it is high compared to population growth in most other more-developed countries. About how much of the annual U.S. population growth is due to immigration? List six changes in lifestyles that have taken place in the United States during the 20th century, leading to a rise in per capita resource use. What is the end effect of such changes in terms of the U.S. ecological footprint?

5. List nine factors that affect birth rates and fertility rates. Explain why there are more boys than girls in some countries. Define **life expectancy** and **infant mortality rate** and explain how they affect the population size of a country. Why does the United States have a lower life expectancy and higher infant mortality rate than a number of other more-developed countries?

6. What is **migration**? What are environmental refugees and how quickly are their numbers growing? Describe three major factors that demographers have to consider in making population projections.

Section 6-3

7. What is the key concept for this section? What is the **age structure** of a population? Explain how age structure affects population growth and economic growth. Describe the American baby boom and some of its economic and social effects. What are some problems related to rapid population decline due to an aging population? How has the AIDS epidemic affected the age structure of some countries, especially in Africa?

Section 6-4

8. What is the key concept for this section? What is the **demographic transition** and what are its four stages? What factors could hinder some less-developed countries from making this transition?

9. Explain how the reduction of poverty and empowerment of women can help countries to slow their population growth. What is **family planning** and how can it help to stabilize populations? Describe India's efforts to control its population growth. Describe China's population control program and compare it with that of India.

10. What are this chapter's *three big ideas*? Summarize the story of human population growth and explain how the three scientific **principles of sustainability** (see Figure 1-2, p. 6 or back cover) can guide us in dealing with the problems that stem from population growth and decline.

Note: Key terms are in bold type.

Critical Thinking

1. Do you think that the global population of 7.1 billion (**Core Case Study**) is too large? Explain. If your answer was *yes*, what do you think should be done to slow human population growth? If your answer was *no*, do you believe that there is a population size that would be too big? Explain.

2. If you could say hello to a new person every second without taking a break and working around the clock, how many people could you greet in 1 day? How many in a year? How long would it take you to greet the 84 million people who were added to the world's population this year? How many years would it take you to greet all 7.1 billion people on the planet?

3. Which of the three major environmental worldviews summarized on p. 21 do you believe underlie the two major positions on whether the world is overpopulated (Science Focus 6.1)?

4. Should everyone have the right to have as many children as they want? Explain. Is your belief on this issue consistent with your environmental worldview?

5. Is it rational for a poor couple in a less-developed country such as India to have four or five children? Explain.

6. Identify a major local, national, or global environmental problem, and describe the role that population growth plays in this problem.

7. Some people believe the most important environmental goal is to sharply reduce the rate of population growth in less-developed countries, where at least 92% of the world's population growth is expected to take place between now and 2050. Others argue that the most serious environmental problems stem from high levels of resource consumption per person in more-developed countries, which have much larger ecological footprints per person than do less-developed countries. What is your view on this issue? Explain.

8. Experts have identified population growth as one of the major causes of the environmental problems we face. The population of the United States is growing faster than that of any other more-developed country. However, this fact is rarely discussed, and the U.S. government has no official policy for slowing U.S. population growth. Why do think this is so? Do you think there should be such a policy? If so, explain your thinking and list three steps you would take as a leader to slow U.S. population growth. If not, explain your thinking.

Doing Environmental Science

Prepare an age-structure diagram for your community. You will need to estimate how many people belong in each age category (see p. 131). To do this, interview a randomly drawn sample of the population to find out their ages and then divide your sample into age groups. (Be sure to interview equal numbers of males and females.) Then find out the total population of your community and apply the percentages for each age group from your sample to the whole population in order to make your estimates. Create your diagram and then use it to project future population trends. Write a report in which you discuss some economic, social, and environmental effects that might result from these trends.

Global Environment Watch Exercise

Find three different projections for the size of the global population in 2050 (**Core Case Study**). Explain how the projections were made. To do this, try to find out the assumptions behind each of the projections with regard to total fertility rates, crude death rates, infant mortality rates, life expectancies, and other factors. Based on your reading, choose the projection that you believe to be the closest to reality, and explain why you chose this projection.

Data Analysis

The chart below shows selected population data for two different countries, A and B. Study the chart and answer the questions that follow.

	Country A	Country B
Population (millions)	144	82
Crude birth rate (number of live births per 1,000 people per year)	43	8
Crude death rate (number of deaths per 1,000 people per year)	18	10
Infant mortality rate (number of babies per 1,000 born who die in first year of life)	100	3.8
Total fertility rate (average number of children born to women during their child-bearing years)	5.9	1.3
% of population under 15 years old	45	14
% of population older than 65 years	3.0	19
Average life expectancy at birth	47	79
% urban	44	75

© Cengage Learning

1. Calculate the rates of natural increase (due to births and deaths, not counting immigration) for the populations of country A and country B. Based on these calculations and the data in the table, for each of the countries, suggest whether it is a more-developed country or a less-developed country and explain the reasons for your answers.

2. Describe where each of the two countries may be in the stages of demographic transition (Figure 6-16). Discuss factors that could hinder either country from progressing to later stages in the demographic transition.

3. Explain how the percentages of people under 15 years of age in each country could affect its per capita and total ecological footprints.

1. A medium population growth projection calls for a population of 9.3 billion people in 2050 up from the current population of 6.7 billion. What percent increase is this?
   a. 15
   b. 39
   c. 72
   d. 100
   e. 139

2. In 1798 Thomas Malthus hypothesized the collapse of the human population. What factor proved Malthus wrong and led to the continued increases in the human population?
   a. Use of fossil fuels
   b. Increased genetic resistance of pest species
   c. Reliance on renewable resources
   d. Food production increased at an exponential rate
   e. Modern health care doubling the birth rate worldwide

3. The infant mortality rate is NOT
   a. used to calculate the population growth rate.
   b. lower in the United States than in any other country.
   c. in developing countries most commonly a result of diarrheal diseases caused by polluted water.
   d. higher in countries where women are empowered by education and equality.
   e. an indication of the availability of nutritious food and health in a country.

4. Why is replacement-level fertility slightly higher than two children per couple?
   a. Some people do not marry.
   b. Some children die before reaching reproductive years.
   c. Use of contraceptives is increasing in developed countries.
   d. Death rates are high in developing countries.
   e. Death rates exceed birth rates in stable countries.

5. Taiwan, Cuba, and most of Europe have lower infant mortality rates than the U.S., in part because of
   a. drug addiction among pregnant women.
   b. high death rate among teenagers.
   c. increased education among teenage women.
   d. more teenage pregnancies.
   e. increased use of contraception and fewer births.

6. One of the most important factors in helping to stabilize a developing country's population and reducing environmental degradation is to
   a. decrease the rate at which people may emigrate.
   b. reduce the infant mortality rate by making available more medicine for both infants and pregnant women.
   c. empower and educate women.
   d. teach people the importance of recycling materials used in their everyday lives.
   e. ensure that people have enough food to eat by establishing local plots of land for rent, thereby enabling local communities to grow their own food.

7. Which of the below is an example of a positive use of nature's natural capital?
   a. Elimination of many natural predators of humans
   b. Increasing use of the earth's net primary productivity to feed humans
   c. Increased reliance on renewable energy resources
   d. Introduction of species into new communities to control pests
   e. Modification of the earth's geochemical cycles for energy uses

8. Which of the below is the correct order of demographic transition states that a country experiences when industrializing?
   a. Iron age → agrarian → preindustrial → industrial
   b. Nomadic → preindustrial → industrial → postindustrial
   c. Nomadic → agrarian → preindustrial → industrial
   d. Preindustrial → transitional → industrial → postindustrial
   e. Agrarian → preindustrial → transitional → industrial

9. Which of the below has been utilized in both India and China to attempt to reduce population growth rates?
   a. Limiting immigration
   b. Increasing emigration
   c. Attempting to prevent the transition to an industrial state
   d. Importing new technologies into the country
   e. Family planning and education

7

Climate and Biodiversity

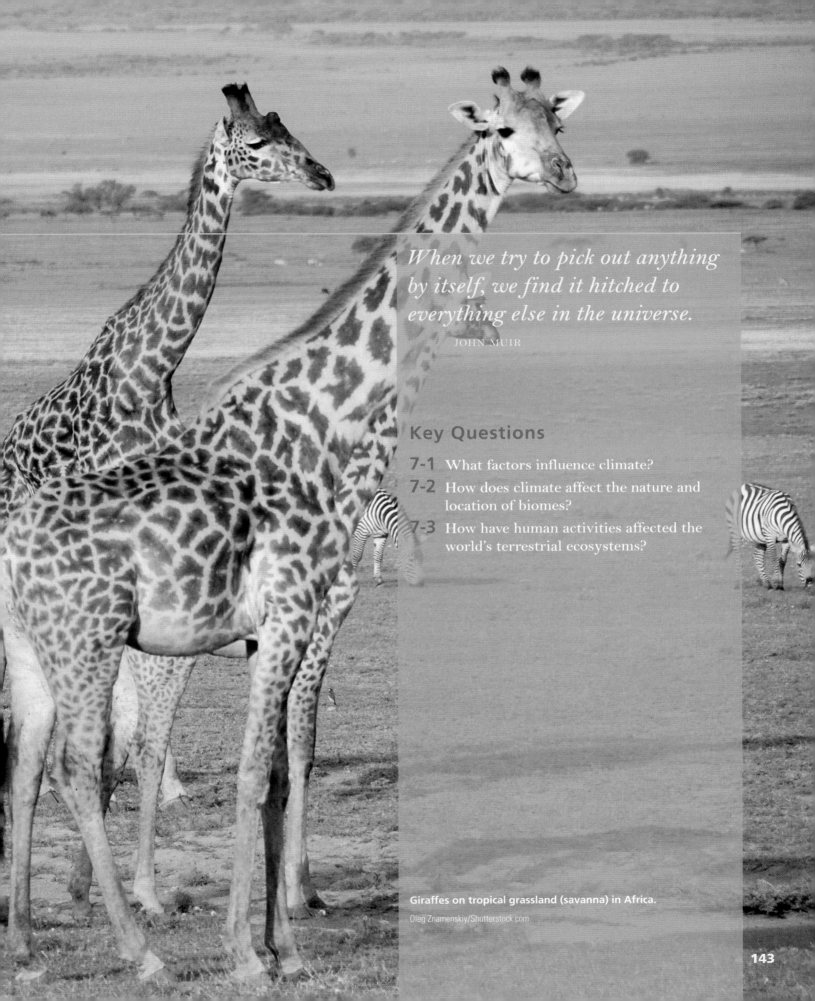

When we try to pick out anything by itself, we find it hitched to everything else in the universe.

JOHN MUIR

Key Questions

7-1 What factors influence climate?

7-2 How does climate affect the nature and location of biomes?

7-3 How have human activities affected the world's terrestrial ecosystems?

Giraffes on tropical grassland (savanna) in Africa.

Oleg Znamenskiy/Shutterstock.com

The earth hosts a great diversity of species and *habitats,* or places where these species can live. Some species live in *terrestrial,* or land, habitats such as deserts, grasslands (see chapter-opening photo), and forests—the three major types of terrestrial ecosystems, also called *biomes.* Other species live in water-covered habitats in *aquatic life zones,* such as rivers, lakes, and oceans. In this chapter, we look at key terrestrial habitats, and in the next chapter, we look at major aquatic habitats.

Why do forests grow on some areas of the earth's land while deserts form in other areas? The answers lie largely in differences in *climate,* the average atmospheric conditions in a region over a period of time ranging from at least three decades to thousands of years. Differences in climate result mostly from long-term differences in average annual precipitation and temperature. These differences lead to three major types of climate—*tropical* (areas near the equator, receiving the most intense sunlight), *polar*

(areas near the earth's poles, receiving the least intense sunlight), and *temperate* (areas between the tropic and polar regions).

Throughout these regions, we find different types of ecosystems, vegetation, and animals adapted to the various climate conditions. For example, scattered throughout the temperate areas of the globe, we find *temperate deciduous forests* (Figure 7-1). Such forests typically see warm summers, cold winters, and abundant precipitation—rain in summer and snow in winter months. They are dominated by a few species of *broadleaf deciduous trees* such as oak, hickory, maple, aspen, and birch. Animal species living in these forests include predators such as wolves, foxes, and wildcats. They feed on herbivores such as white-tailed deer (see Figure 5-17, p. 115), squirrels, rabbits, opossums, and mice. Warblers, robins, and other bird species live in these forests during the spring and summer, mating and raising their young.

These species are adapted to the conditions of temperate forests. For example, most of the trees' leaves, after developing their vibrant colors in the fall (Figure 7-1, left), drop off the trees. This allows the trees to survive the cold winters by becoming dormant (Figure 7-1, right). Each spring, they sprout new leaves and spend their summers growing and producing until the cold weather returns. Some forest mammals, including bears, spend their summers storing fat and then hibernating during winter, sleeping the coldest months away. Many bird species migrate to warmer climates in the late fall to avoid the cold and snow.

Temperate deciduous forests once covered the eastern half of the United States and western Europe. But as these areas became urbanized, most of the original forests were cleared. Today, on a worldwide basis, this terrestrial ecosystem has been disturbed more than any other terrestrial system by human activities.

In this chapter, we examine the key role that climate plays in the location and formation of forests and all other major terrestrial ecosystems that make up an important part of the earth's terrestrial biodiversity.

Figure 7-1 A temperate deciduous forest in fall (left) and in winter (right).

Left: © Marc von Hacht/Shutterstock.com. Right: © Waltraud Ingerl/Istockphoto.com.

7-1 What Factors Influence Climate?

CONCEPT 7-1

Key factors that influence an area's climate are incoming solar energy, the earth's rotation, global patterns of air and water movement, gases in the atmosphere, and the earth's surface features.

The Earth Has Many Different Climates

It is important to understand the difference between weather and climate. **Weather** is a set of physical conditions of the lower atmosphere, including temperature, precipitation, humidity, wind speed, cloud cover, and other factors, in a given area over a period of hours or days. (Supplement 5, p. S19, introduces you to the basics of weather.)

Weather differs from **climate**, which is the general pattern of atmospheric conditions in a given area over periods ranging from at least three decades to thousands of years. In other words, climate is the sum of weather conditions in a given area, averaged over a long period of time.

We know a lot about the daily weather where we live, but we might not know as much about the climate where we live or of how it has changed over the past 30 to 100 years. To get such an understanding, you would first have to find and plot data on the average temperature and average precipitation in your area from year to year over at least the last three decades. Then you would have to note whether these numbers have generally gone up, stayed the same, or gone down.

Based on this type of analysis, scientists have described the various regions of the earth according to their climates. Figure 7-2 shows these major climate zones along with the major **ocean currents**—mass movements of surface water driven by winds blowing over the oceans. These currents help to determine regional climates and are a key component of the earth's natural capital (see Figure 1-3, p. 7).

Climate varies among the earth's different regions primarily because, over long periods of time, patterns of global air circulation and ocean currents distribute heat and precipitation unevenly between the tropics and other

Polar (ice) · Subarctic (snow) · Cool temperate · Highland · Warm ocean current · River

Warm temperate · Dry · Tropical · Major upwelling zones · Cold ocean current

© Cengage Learning

Animated Figure 7-2 **Natural capital:** This generalized map of the earth's current climate zones also shows the major ocean currents and upwelling areas (where currents bring nutrients from the ocean bottom to the surface). **Question:** Based on this map, what is the general type of climate where you live?

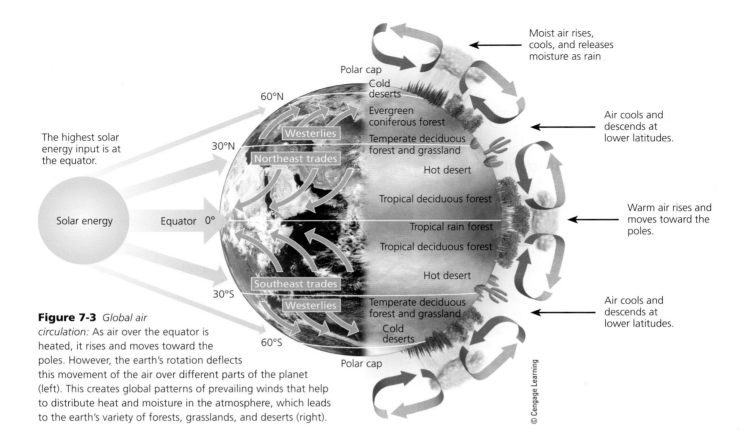

The highest solar energy input is at the equator.

Solar energy

Equator 0°

60°N

30°N

30°S

60°S

Westerlies

Northeast trades

Southeast trades

Westerlies

Polar cap
Cold deserts
Evergreen coniferous forest
Temperate deciduous forest and grassland
Hot desert
Tropical deciduous forest
Tropical rain forest
Tropical deciduous forest
Hot desert
Temperate deciduous forest and grassland
Cold deserts
Polar cap

Moist air rises, cools, and releases moisture as rain

Air cools and descends at lower latitudes.

Warm air rises and moves toward the poles.

Air cools and descends at lower latitudes.

© Cengage Learning

Figure 7-3 *Global air circulation:* As air over the equator is heated, it rises and moves toward the poles. However, the earth's rotation deflects this movement of the air over different parts of the planet (left). This creates global patterns of prevailing winds that help to distribute heat and moisture in the atmosphere, which leads to the earth's variety of forests, grasslands, and deserts (right).

parts of the world (Figure 7-3). Three major factors affect the circulation of air in the lower atmosphere:

1. *Uneven heating of the earth's surface by the sun.* Air is heated much more at the equator, where the sun's rays strike directly, than at the poles, where sunlight strikes at an angle and spreads out over a much greater area (Figure 7-3, left). These differences in the input of solar energy to the atmosphere help explain why tropical regions near the equator are hot, why polar regions are cold, and why temperate regions in between generally have both warm and cool temperatures (Figure 7-2). The intense input of solar radiation in tropical regions leads to greatly increased evaporation of moisture from forests, grasslands, and bodies of water. As a result, tropical regions normally receive more precipitation than do other areas of the earth.

2. *Rotation of the earth on its axis.* As the earth rotates around its axis, the equator spins faster than the regions to its north and south. As a result, heated air masses, rising above the equator and moving north and south to cooler areas, are deflected in different ways over different parts of the planet's surface (Figure 7-3, left). The atmosphere over these different areas is divided into huge regions called *cells*, distinguished by the direction of air movement. The differing directions of air movement are called *prevailing winds*—major surface winds that blow almost continuously and help

to distribute heat and moisture over the earth's surface and to drive ocean currents.

3. *Properties of air, water, and land.* Heat from the sun evaporates ocean water and transfers heat from the oceans to the atmosphere, especially near the hot equator. This evaporation of water creates giant cyclical convection cells that circulate air, heat, and moisture both vertically and from place to place in the atmosphere, as shown in Figure 7-3, right, and in Figure 7-4.

Driven by prevailing winds and the earth's rotation, the earth's major ocean currents (Figure 7-2) help to redistribute heat from the sun, thereby influencing climate and vegetation, especially near coastal areas. This heat and differences in water *density* (mass per unit volume) create warm and cold ocean currents. Prevailing winds and irregularly shaped continents interrupt these currents and cause them to flow in roughly circular patterns between the continents, clockwise in the northern hemisphere and counterclockwise in the southern hemisphere.

Water also moves vertically in the oceans as denser water sinks while less dense water rises. This creates a connected loop of deep and shallow ocean currents (which are separate from those shown in Figure 7-2). This loop acts somewhat like a giant conveyor belt that moves heat from the surface to the deep sea and trans-

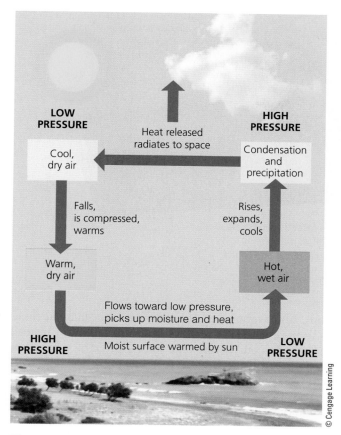

Figure 7-4 Energy is transferred by *convection* in the atmosphere—the process by which warm, wet air rises, then cools and releases heat and moisture as precipitation (right side and top, center). Then the cooler, denser, and drier air sinks, warms up, and absorbs moisture as it flows across the earth's surface (left side and bottom) to begin the cycle again.

fers warm and cold water between the tropics and the poles (Figure 7-5).

The ocean and the atmosphere are strongly linked in two ways: ocean currents are affected by winds in the atmosphere, and heat from the ocean affects atmospheric circulation. One example of the interactions between the ocean and the atmosphere is the *El Niño–Southern Oscillation*, or *ENSO* (see Figure 4, p. S21, in Supplement 5). This large-scale weather phenomenon occurs every few years when prevailing winds in the tropical Pacific Ocean weaken and change direction. The resulting above-average warming of Pacific waters alters the weather over at least two-thirds of the earth for 1 or 2 years (see Figure 5, p. S21, in Supplement 5).

The earth's air circulation patterns, prevailing winds, and configuration of continents and oceans are all factors in the formation of six giant convection cells (like the one shown in Figure 7-4), three of them south of the equator and three north of the equator. These cells lead to an irregular distribution of climates and of the resulting deserts, grasslands, and forests, as shown in Figure 7-3, right (**Concept 7-1**).

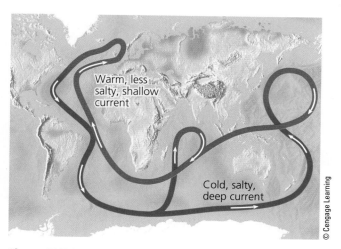

Figure 7-5 A connected loop of deep and shallow ocean currents transports warm and cool water to various parts of the earth.

Greenhouse Gases Warm the Lower Atmosphere

As energy flows from the sun to the earth, some of it is reflected by the earth's surface back into the atmosphere (see Figure 3-3, p. 54). Molecules of certain gases in the atmosphere, including water vapor (H_2O), carbon dioxide (CO_2), methane (CH_4), and nitrous oxide (N_2O), absorb some of this solar energy and release a portion of it as infrared radiation (heat) that warms the lower atmosphere. Thus, these gases, called **greenhouse gases**, play a role in determining the lower atmosphere's average temperatures and thus the earth's climates.

The earth's surface also absorbs much of the solar energy that strikes it and transforms it into longer-wavelength infrared radiation, which then rises into the lower atmosphere. Some of this heat escapes into space, but some is absorbed by molecules of greenhouse gases and emitted into the lower atmosphere as even longer-wavelength infrared radiation (see Figure 2-11, p. 41). Some of this released energy radiates into space, and some adds to the warming of the lower atmosphere and the earth's surface. Together, these processes result in a natural warming of the troposphere, called the **greenhouse effect** (see Figure 3-3, p. 54). Without this natural warming effect, the earth would be a very cold and mostly lifeless planet.

Human activities such as the burning of fossil fuels, clearing of forests, and growing of crops release carbon dioxide, methane, and nitrous oxide into the atmosphere. A considerable body of scientific evidence, combined with climate model projections, indicates that continued inputs of these greenhouse gases into the atmosphere from human activities are likely to enhance the earth's natural greenhouse effect and change the earth's climate during this century. If this occurs, it will likely alter temperature and precipitation patterns, raise average sea levels, and shift areas where we can grow crops and where some

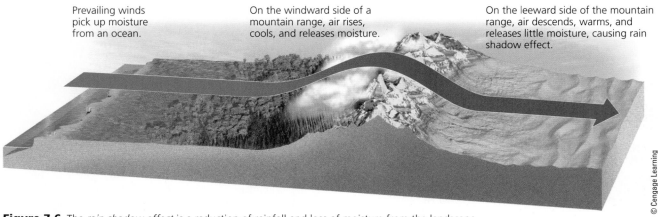

Prevailing winds pick up moisture from an ocean.

On the windward side of a mountain range, air rises, cools, and releases moisture.

On the leeward side of the mountain range, air descends, warms, and releases little moisture, causing rain shadow effect.

© Cengage Learning

Figure 7-6 The *rain shadow effect* is a reduction of rainfall and loss of moisture from the landscape on the leeward side of a mountain. Warm, moist air in onshore winds loses most of its moisture as rain and snow that fall on the windward slopes of a mountain range. This leads to semiarid and arid conditions on the leeward side of the mountain range and on the land beyond.

types of plants and animals (including humans) can live, as discussed more fully in Chapter 19.

The Earth's Surface Features Affect Local Climates

Various topographic features of the earth's surface can create local and regional climatic conditions that differ from the general climate in some regions. For example, mountains interrupt the flow of prevailing surface winds and the movement of storms. When moist air blowing inland from an ocean reaches a mountain range, it is forced upward. As it rises, it cools and expands, and then loses most of its moisture as rain and snow that fall on the windward slope of the mountain.

As the drier air mass passes over the mountaintops, it flows down the leeward slopes (facing away from the wind), and warms up. This increases its ability to hold moisture, but the air releases little moisture and instead tends to dry out plants and soil below. This process is called the **rain shadow effect** (Figure 7-6), and over many decades, it results in *semiarid* or *arid* conditions on the leeward side of a high mountain range. Sometimes this effect leads to the formation of deserts such as Death Valley, a part of the Mojave Desert found in parts of the U.S. states of California, Nevada, Utah, and Arizona.

Cities also create distinct microclimates. Bricks, concrete, asphalt, and other building materials absorb and hold heat, and buildings block wind flow. Motor vehicles and the heating and cooling systems of buildings release large quantities of heat and pollutants. As a result, cities on average tend to have more haze and smog, higher temperatures that make them *heat islands,* and lower wind speeds than the surrounding countryside.

7-2 How Does Climate Affect the Nature and Location of Biomes?

CONCEPT 7-2
Differences in long-term average annual precipitation and temperature lead to the formation of tropical, temperate, and cold deserts, grasslands, and forests, and largely determine their locations.

Climate Helps to Determine Where Terrestrial Organisms Can Live

Differences in climate (Figure 7-2) help to explain why one area of the earth's land surface is a desert, another a grassland, and another a forest (see Figure 2, p. S26, Supplement 6). Furthermore, different combinations of varying average annual precipitation and temperatures,

along with global air circulation and ocean currents, lead to the formation of tropical (hot), temperate (moderate), and polar (cold) deserts, grasslands, and forests, as summarized in Figure 7-7 (**Concept 7-2**).

Climate and vegetation vary according to *latitude* and also according to *elevation,* or height above sea level. If you climb a tall mountain, from its base to its summit, you can observe changes in plant life similar to those you would encounter in traveling from the equator to the earth's northern polar region (Figure 7-8).

Figure 7-9 shows how scientists have divided the world into several major **biomes**—large terrestrial regions, each characterized by a certain type of climate and a certain combination of dominant plant life. The variety of terrestrial biomes and aquatic systems is one of

Figure 7-7 **Natural capital:** Average precipitation and average temperature, acting together as limiting factors over a long time, help to determine the type of desert, grassland, or forest in any particular area, and thus the types of plants, animals, and decomposers found in that area (assuming it has not been disturbed by human activities).

Figure 7-8 Biomes and climate both change with elevation (left), as well as with latitude (right).

the four components of the earth's biodiversity (see Figure 4-2, p. 79)—a vital part of the earth's natural capital.

By comparing Figure 7-9 with Figure 7-2, you can see how the world's major biomes vary with climate. Figure 4-4 (p. 81) shows how major biomes along the 39th parallel in the United States are related to different climates. See the maps in Supplement 6 (pp. S32–S39) that show the major biomes in North America (Figure 6), South America (Figure 10), Europe (Figure 11), Asia (Figure 12), Africa (Figure 13), and Oceania and Australia (Figure 14).

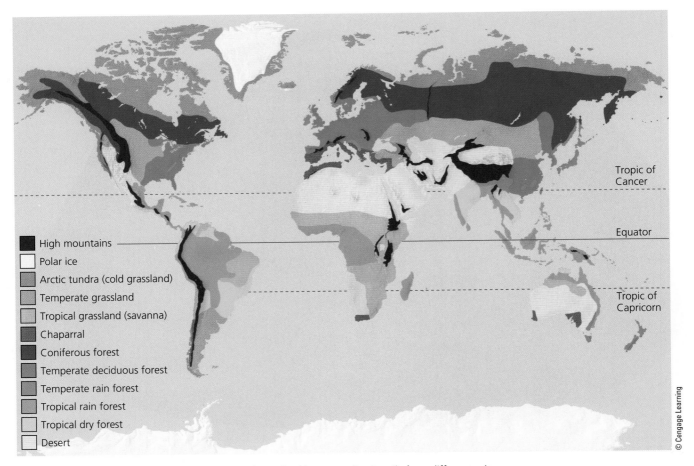

High mountains
Polar ice
Arctic tundra (cold grassland)
Temperate grassland
Tropical grassland (savanna)
Chaparral
Coniferous forest
Temperate deciduous forest
Temperate rain forest
Tropical rain forest
Tropical dry forest
Desert

Tropic of Cancer

Equator

Tropic of Capricorn

© Cengage Learning

Animated Figure 7-9 Natural capital: The earth's major *biomes* result primarily from differences in climate.

On maps such as the one in Figure 7-9, biomes are shown with sharp boundaries, and each biome is covered with one general type of vegetation. In reality, biomes are not uniform. They consist of a *mosaic of patches,* each with somewhat different biological communities but with similarities typical of the biome. These patches occur primarily because of the irregular distribution of the resources needed by plants and animals and because human activities have removed or altered the natural vegetation in many areas.

CONSIDER THIS...

THINKING ABOUT Biomes, Climate, and Human Activities

Use Figure 7-2 to determine the general type of climate where you live and Figure 7-9 to determine the general type of biome that should exist where you live. Then use Figure 1-12, p. 13 and Figure 8 in Supplement 6, p. S33 to determine how human ecological footprints have affected the biome where you live.

There Are Three Major Types of Deserts

In a *desert,* annual precipitation is low and often scattered unevenly throughout the year. During the day, the baking sun warms the ground and evaporates water from plant leaves and from the soil. But at night, most of the

heat stored in the ground radiates quickly into the atmosphere. Desert soils have little vegetation and moisture to help store the heat and the skies above deserts are usually clear. This explains why in a desert you might roast during the day but shiver at night.

A combination of low rainfall and varying average temperatures creates a variety of desert types—tropical, temperate, and cold (Figures 7-7 and 7-10 and **Concept 7-2**). *Tropical deserts* (Figure 7-10, top photo) such as the Sahara and the Namib of Africa are hot and dry most of the year (Figure 7-10, top graph). They have few plants and a hard, windblown surface strewn with rocks and some sand. They are the deserts we often see in the movies.

In *temperate deserts* (Figure 7-10, center photo) such as the Sonoran Desert in southeastern California, southwestern Arizona, and northwestern Mexico, daytime temperatures are high in summer and low in winter and there is more precipitation than in tropical deserts (Figure 7-10, center graph). The sparse vegetation primarily consists of widely dispersed, drought-resistant shrubs and cacti or other succulents adapted to the lack of water and temperature variations.

In *cold deserts* such as the Gobi Desert in Mongolia, vegetation is sparse (Figure 7-10, bottom photo).

Tropical desert

Temperate desert

Cold desert

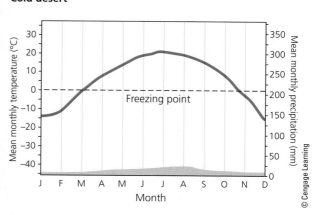

Figure 7-10 These climate graphs track the typical variations in annual temperature (red) and precipitation (blue) in tropical, temperate, and cold deserts. Top photo: a *tropical desert* in Morocco. Center photo: a *temperate desert* in southeastern California, with saguaro cactus, a prominent species in this ecosystem. Bottom photo: a *cold desert,* Mongolia's Gobi Desert, where Bactrian camels live.
Question: Which month of the year has the highest temperature and which month has the lowest rainfall for each of the three types of deserts?

STAYING ALIVE IN THE DESERT

Adaptations for survival in the desert have two themes: *beat the heat* and *every drop of water counts.*

Desert plants have evolved a number of strategies based on such adaptations. During long hot and dry spells, plants such as mesquite and creosote drop their leaves to survive in a dormant state. *Succulent* (fleshy) *plants* such as the saguaro ("sah-WAH-ro") cactus (Figure 7-A and Figure 7-10, middle photo) have three adaptations: they have no leaves, which can lose water to the atmosphere through *transpiration;* they store water and synthesize food in their expandable, fleshy tissue; and they reduce water loss by opening their pores only at night to take up carbon dioxide (CO_2). The spines of these and many other desert plants guard them from being eaten by herbivores seeking the precious water they hold.

Some desert plants use deep roots to tap into groundwater. Others such as prickly pear and saguaro cacti use widely spread shallow roots to collect water after brief showers and store it in their spongy tissues.

Some desert plants conserve water by having wax-coated leaves that reduce water loss. Others such as annual wild-flowers and grasses store much of their biomass in seeds that remain inactive, sometimes for years, until they receive enough water to germinate. Shortly after a rain, these seeds germinate, grow, and carpet some deserts with dazzling arrays of colorful flowers (Figure 7-A) that last for up to a few weeks.

Most desert animals are small. Some beat the heat by hiding in cool burrows or rocky crevices by day and coming out at night or in the early morning. Others become dormant during periods of extreme heat or drought. Some larger animals such as camels (Figure 7-10, bottom photo) can drink massive quantities of water when it is available and store it in their fat for use as needed. Also, the camel's thick fur actually helps it to keep cool because the air spaces in the fur insulate the camel's skin against the outside heat. And camels do not sweat, which reduces their water loss through evaporation. Kangaroo rats never drink water. They get the water they need by breaking down fats in seeds that they consume.

Insects and reptiles such as rattlesnakes have thick outer coverings to minimize water loss through evaporation, and their wastes are dry feces and a dried concen-

Figure 7-A After a brief rain, these wildflowers bloomed in this temperate desert in Picacho Peak State Park in the U.S. state of Arizona.

trate of urine. Many spiders and insects get their water from dew or from the food they eat.

Critical Thinking

What are three steps you would take to survive in the open desert if you had to?

Winters are cold, summers are warm or hot, and precipitation is low (Figure 7-10, bottom graph). In all types of deserts, plants and animals have evolved adaptations that help them to stay cool and to get enough water to survive (Science Focus 7.1).

Desert ecosystems are fragile because they experience slow plant growth, low species diversity, slow nutrient cycling (due to low bacterial activity in the soils), and very little water. Their soils take from decades to centuries to recover from disturbances such as the growing problem of off-road vehicle traffic, which can also destroy the habitats for a variety of animals that live underground. The lack of vegetation, especially in tropical and polar deserts, also makes them vulnerable to heavy erosion from sandstorms.

There Are Three Major Types of Grasslands

Grasslands occur primarily in the interiors of continents in areas that are too moist for deserts to form and too dry for forests to grow (Figure 7-9). Grasslands persist because of a combination of seasonal drought, grazing by large herbivores, and occasional fires—all of which keep shrubs and trees from growing in large numbers.

The three main types of grasslands—tropical, temperate, and cold (arctic tundra)—result from combinations of low average precipitation and varying average temperatures (Concept 7-2) (Figures 7-7 and 7-11).

One type of tropical grassland, called a *savanna* (Figure 7-11, top photo), contains widely scattered clumps of trees. This biome usually has warm temperatures year-

Tropical grassland (savanna)

Temperate grassland (prairie)

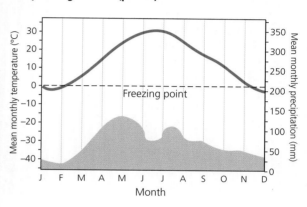

Cold grassland (arctic tundra)

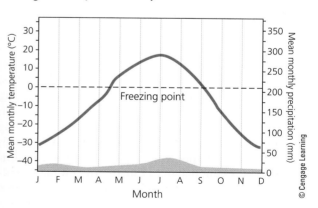

Figure 7-11 These climate graphs track the typical variations in annual temperature (red) and precipitation (blue) in tropical, temperate, and cold (arctic tundra) grasslands. Top photo: *savanna (tropical grassland)* in Kenya, Africa, with zebras grazing. Center photo: *prairie (temperate grassland)* in the U.S. state of Illinois, with wildflowers in bloom. Bottom photo: *arctic tundra (cold grassland)* in Alaska's Arctic National Wildlife Refuge in summer. **Question:** Which month of the year has the highest temperature and which month has the lowest rainfall for each of the three types of grasslands?

Figure 7-12 **Natural capital degradation:** This intensively cultivated cropland is an example of the replacement of biologically diverse temperate grasslands (such as in the center photo of Figure 7-11) with a monoculture crop.

round and alternating dry and wet seasons (Figure 7-11, top graph).

Tropical savannas in East Africa are home to *grazing* (primarily grass-eating) and *browsing* (twig- and leaf-nibbling) hoofed animals, including wildebeests, gazelles, zebras (Figure 7-11, top photo), giraffes (chapter-opening photo), and antelopes, as well as their predators such as lions, hyenas, and humans. Herds of these grazing and browsing animals migrate to find water and food in response to seasonal and year-to-year variations in rainfall (Figure 7-11, blue region in top graph) and food availability. Savanna plants, like those in deserts, are adapted to survive drought and extreme heat; many have deep roots that can tap into groundwater.

🔍 CONSIDER THIS. . .

CONNECTIONS Grassland Niches and Feeding Habits

As an example of differing niches, some large herbivores have evolved specialized eating habits that minimize competition among species for the vegetation found on the savanna. For example, giraffes eat leaves and shoots from the tops of trees, elephants eat leaves and branches farther down, wildebeests prefer short grasses, and zebras graze on longer grasses and stems.

In a *temperate grassland*, winters can be bitterly cold, summers are hot and dry, and annual precipitation is fairly sparse and falls unevenly throughout the year (Figure 7-11, center graph). Because the aboveground parts of most of the grasses die and decompose each year, organic matter accumulates to produce deep, fertile topsoil. This topsoil is held in place by a thick network of the grasses' intertwined roots (unless the topsoil is plowed up, which exposes it to high winds found in these biomes). This biome's grasses are adapted to periodic droughts and to fires that burn the plant parts above the ground but do not harm the roots, from which new grass can grow.

In the mid-western and western areas of the United States, we find two types of temperate grasslands depending primarily on average rainfall: *short-grass prairies* (Figure 7-11, center photo) and the *tallgrass prairies* (which get more rain). In all prairies, winds blow almost continuously and evaporation is rapid, often leading to fires in the summer and fall. This combination of winds and fires helps to maintain such grasslands by hindering tree growth. Many of the world's natural temperate grasslands have been converted to farmland, because their fertile soils are useful for growing crops (Figure 7-12) and grazing cattle.

Cold grasslands, or *arctic tundra* (Russian for "marshy plain"), lie south of the arctic polar ice cap (Figure 7-9). During most of the year, these treeless plains are bitterly cold (Figure 7-11, bottom graph), swept by frigid winds, and covered with ice and snow. Winters are long with few hours of daylight, and the scant precipitation falls primarily as snow.

Under the snow, this biome is carpeted with a thick, spongy mat of low-growing plants, primarily grasses, mosses, lichens, and dwarf shrubs. Trees and tall plants cannot survive in the cold and windy tundra because they would lose too much of their heat. Most of the annual growth of the tundra's plants occurs during the 7- to 8-week summer, when there is daylight almost around the clock.

One outcome of the extreme cold is the formation of **permafrost**, underground soil in which captured water stays frozen for more than two consecutive years. During the brief summer, the permafrost layer keeps melted snow and ice from draining into the ground. As a consequence, many shallow lakes, marshes, bogs, ponds, and other seasonal wetlands form when snow and frozen surface soil melt on the waterlogged tundra. Hordes of mosquitoes, black flies, and other insects thrive in these shallow surface pools. They serve as food for large colonies of migratory birds (especially waterfowl) that migrate from the south to nest and breed in the bogs and ponds.

Animals in this biome survive the intense winter cold through adaptations such as thick coats of fur (arctic wolf, arctic fox, and musk oxen) or feathers (snowy owl) or living underground (arctic lemming). In the summer, caribou (often called reindeer) and other types of deer migrate to the tundra to graze on its vegetation.

Tundra is a fragile biome. Most tundra soils formed about 17,000 years ago when glaciers began retreating after the last Ice Age (see Figure 4-10, p. 87). These soils usually are nutrient poor. Because of the short growing season, tundra soil and vegetation recover very slowly from damage or disturbance. Human activities in the arctic tundra—primarily on and around oil drilling sites, pipelines, mines, and military bases—leave scars that persist for centuries.

Another type of tundra, called *alpine tundra,* occurs above the limit of tree growth but below the permanent snow line on high mountains (Figure 7-8, left). The vegetation is similar to that found in arctic tundra, but it receives more sunlight than arctic vegetation gets. During the brief summer, alpine tundra can be covered with an array of beautiful wildflowers.

There Are Three Major Types of Forests

Forests are lands that are dominated by trees. The three main types of forests—*tropical, temperate* (**Core Case Study**), and *cold* (northern coniferous, or boreal)—result from combinations of varying precipitation levels and varying average temperatures (Concept 7-2) (Figures 7-7 and 7-13).

Tropical rain forests (Figure 7-13, top photo) are found near the equator (Figure 7-9), where hot, moisture-laden air rises and dumps its moisture (Figure 7-3). These lush forests have year-round, uniformly warm temperatures, high humidity, and almost daily heavy rainfall (Figure 7-13, top graph). This fairly constant warm, wet climate is ideal for a wide variety of plants and animals.

Tropical rain forests are dominated by *broadleaf evergreen plants,* which keep most of their leaves year-round. The tops of the trees form a dense *canopy* (Figure 7-13, top photo) that blocks most light from reaching the forest floor. For this reason, there is little vegetation on the forest floor. Many of the plants that do live at the ground level have enormous leaves to capture what little sunlight filters down to them.

Some trees are draped with vines (called *lianas*) that reach for the treetops to gain access to sunlight. In the canopy, the vines grow from one tree to another, providing walkways for many species living there. When a large tree is cut down, its network of lianas can pull down other trees.

Tropical rain forests have a very high net primary productivity (see Figure 3-14, p. 62). They are teeming with life and possess incredible biological diversity. Although tropical rain forests cover only about 2% of the earth's land surface, ecologists estimate that they contain at least 50% of the known terrestrial plant and animal species. For example, a single tree in these forests may support several thousand different insect species (see Chapter 2 opening photo, p. 28). Plants from tropical rain forests are a source of chemicals, many of which have been used as blueprints for making most of the world's prescription drugs.

Rain forest species occupy a variety of specialized niches in distinct layers, which help to enable these forests' great biodiversity (high species richness). Vegetation layers are structured, for the most part, according to the plants' needs for sunlight, as shown in Figure 7-14. Much of the animal life, particularly insects, bats, and birds, lives in the sunny canopy layer, with its abundant shelter and supplies of leaves, flowers, and fruits. To study life in the treetops, ecologists climb trees and build platforms and boardwalks in the upper canopy.

Dropped leaves, fallen trees, and dead animals decompose quickly in tropical rain forests because of the warm, moist conditions and the hordes of decomposers. About 90% of the nutrients released by this rapid decomposition are quickly taken up and stored by trees, vines, and other plants. Nutrients that are not taken up are soon leached from the thin topsoil by the almost daily rainfall. As a result, very little plant litter builds up on the ground. The resulting lack of fertile soil helps to explain why rain forests are not good places to clear and grow crops or graze cattle on a sustainable basis.

So far, at least half of these forests have been destroyed or disturbed by human activities such as farming and cattle production, and the pace of this destruction and degradation is increasing (see Chapter 3 Core Case Study, p. 52). Ecologists warn that without strong protective

Tropical rain forest

Temperate deciduous forest

Northern coniferous forest (boreal forest, taiga)

Figure 7-13 These climate graphs track the typical variations in annual temperature (red) and precipitation (blue) in tropical, temperate, and cold (northern coniferous, or boreal) forests. Top photo: the closed canopy of a *tropical rain forest* in Costa Rica. Middle photo: a *temperate deciduous forest* in the autumn near Hamburg, Germany (see also Figure 7-1, left). Bottom photo: a *northern coniferous forest* in Canada's Jasper National Park. **Question:** Which month of the year has the highest temperature and which month has the lowest rainfall for each of the three types of forests?

Figure 7-14 Specialized plant and animal niches are *stratified*, or arranged roughly in layers, in a tropical rain forest. Filling such specialized niches enables species to avoid or minimize competition for resources and results in the coexistence of a great variety of species.

measures, most of these forests, along with their rich biodiversity and other very valuable ecosystem services, could be gone by the end of this century.

The second major type of forest, the *temperate deciduous forest*, is the subject of this chapter's **Core Case Study** (see middle photo of Figure 7-13). Because they have cooler temperatures and fewer decomposers than tropical forests have, these forests also have a slower rate of decomposition. As a result, they accumulate a thick layer of slowly decaying leaf litter, which becomes a storehouse of nutrients. Figure 7-15 shows two of the animal species found in this biome.

On a global basis, temperate forests have been degraded by various human activities, especially logging and urban expansion, more than any other terrestrial biome. However, within 100–200 years, forests of this type that have been cleared can return through secondary ecological succession (see Figure 5-12, p. 110).

Cold or *northern coniferous forests* (Figure 7-13, bottom photo) are also called *boreal forests* or *taigas* ("TIE-guhs").

These forests are found just south of the arctic tundra in northern regions across North America, Asia, and Europe (Figure 7-9) and above certain altitudes in the Sierra Nevada and Rocky Mountain ranges of the United States. In this subarctic climate, winters are long and extremely cold; in the northernmost taigas, winter sunlight is available only 6–8 hours per day. Summers are short, with cool to warm temperatures (Figure 7-13, bottom graph), and the sun shines as long as 19 hours a day during mid-summer.

Most boreal forests are dominated by a few species of *coniferous* (cone-bearing) *evergreen trees* such as spruce, fir, cedar, hemlock, and pine that keep most of their leaves (or needles) year-round. Most of these species have small, needle-shaped, wax-coated leaves that can withstand the intense cold and drought of winter, when snow blankets the ground. Plant diversity is low because few species can survive the winters when soil moisture is frozen.

Beneath the stands of trees in these forests is a deep layer of partially decomposed conifer needles. Decomposi-

Figure 7-15 The eastern gray squirrel (left) and the broad-winged hawk (right), which preys on the gray squirrel, are found in North America's temperate deciduous forest ecosystems (**Core Case Study**).

Left: GEORGE GRALL/National Geographic Creative; Right: Pr2is/Dreamstime.com

tion is slow because of low temperatures, the waxy coating on the needles, and high soil acidity. The decomposing conifer needles make the thin, nutrient-poor topsoil acidic, which prevents most other plants (except certain shrubs) from growing on the forest floor.

This biome contains a variety of wildlife. Year-round residents include bears, wolves, moose, lynx, and many burrowing rodent species. Caribou spend the winter in the taiga and the summer in the arctic tundra (Figure 7-11, bottom). During the brief summer, warblers and other insect-eating birds feed on hordes of flies, mosquitoes, and caterpillars.

Coastal coniferous forests or *temperate rain forests* are found in scattered coastal temperate areas with ample rainfall or moisture from dense ocean fogs. Thick stands of these forests with large conifers such as Sitka spruce, Douglas fir, and redwoods once dominated undisturbed areas of these biomes along the coast of North America, from Canada to northern California in the United States.

Mountains Play Important Ecological Roles

Some of the world's most spectacular environments are high on *mountains* (Figure 7-16), steep or high-elevation lands that cover about one-fourth of the earth's land surface (Figure 7-9). Mountains are places where dramatic changes in altitude, slope, climate, soil, and vegetation take place over very short distances (Figure 7-8, left).

About 1.2 billion people (17% of the world's population) live in mountain ranges or in their foothills, and 4 billion people (56% of the world's population) depend on mountain systems for all or some of their water. Because of the steep slopes, mountain soils are easily eroded when the vegetation holding them in place is

removed by natural disturbances such as landslides and avalanches, or by human activities such as timber cutting and agriculture. Many mountains are *islands of biodiversity* surrounded by a sea of lower-elevation landscapes transformed by human activities.

Mountains play important ecological roles. They contain the majority of the world's forests, which are habitats for much of the planet's terrestrial biodiversity. They often are habitats for *endemic species,* those that are found nowhere else on earth. They also serve as sanctuaries for animals that are capable of migrating to higher altitudes and surviving in such environments. Every year, more of these animals are driven from lowland areas to mountain habitats by human activities and by a warming climate.

🔍 CONSIDER THIS. . .

CONNECTIONS Mountains and Climate

Mountains help to regulate the earth's climate. Many mountaintops are covered with glacial ice and snow that reflect some solar radiation back into space, which helps to cool the earth. However, many of the world's mountain glaciers are melting, primarily because the atmosphere has gotten warmer over the last three decades. While glaciers reflect solar energy, the darker rocks exposed by melting glaciers absorb that energy. This helps to warm the atmosphere above them, which melts more ice and warms the atmosphere more—in an escalating cycle of change.

Finally, mountains play a critical role in the hydrologic cycle (see Figure 3-15, p. 63) by serving as major storehouses of water. During winter, precipitation is stored as ice and snow. In the warmer weather of spring and summer, much of this snow and ice melts, releasing water to streams for use by wildlife and by humans for drinking and for irrigating crops. As the atmosphere has warmed over the last three decades, some mountaintop snow packs and glaciers are melting earlier in the spring each year.

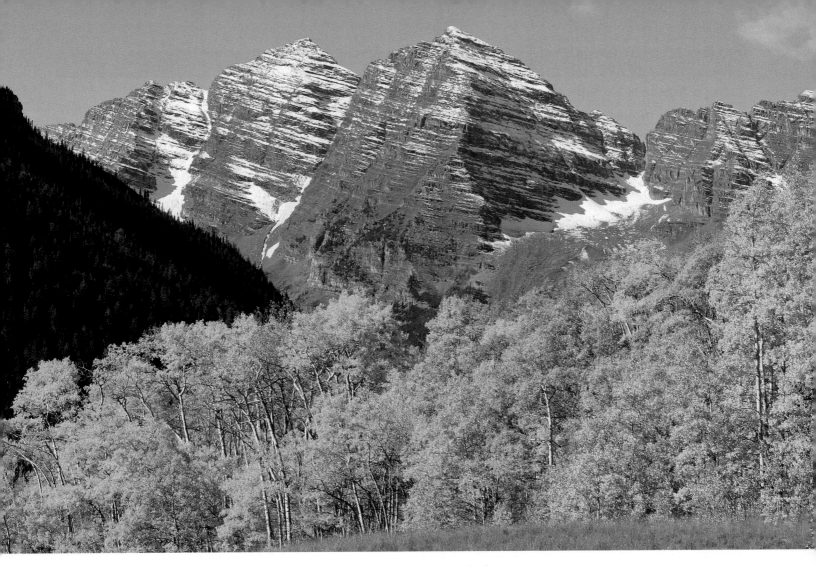

Figure 7-16 Mountains, such as this one in Aspen, Colorado, play important ecological roles.

CHARLES KOGOD/National Geographic Creative

This is lowering food production in certain areas, because much of the water needed throughout the summer to irrigate crops gets released too quickly and too early.

Scientific measurements and climate models indicate that a large number of the world's mountaintop glaciers may disappear during this century if the atmosphere keeps getting warmer as projected. This could force many people to move from their homelands in search of new water supplies and places to grow their crops. Despite the ecological, economic, and cultural importance of mountain ecosystems, protecting them has not been a high priority for governments or for many environmental organizations.

7-3 How Have Human Activities Affected the World's Terrestrial Ecosystems?

CONCEPT 7-3
Human activities are disrupting ecosystem and economic services provided by many of the earth's deserts, grasslands, forests, and mountains.

Humans Have Disturbed Much of the Earth's Land

According to the 2005 Millennium Ecosystem Assessment and later updates of such research, about 60% of the world's major terrestrial ecosystems are being degraded or used unsustainably, as the human ecologi-

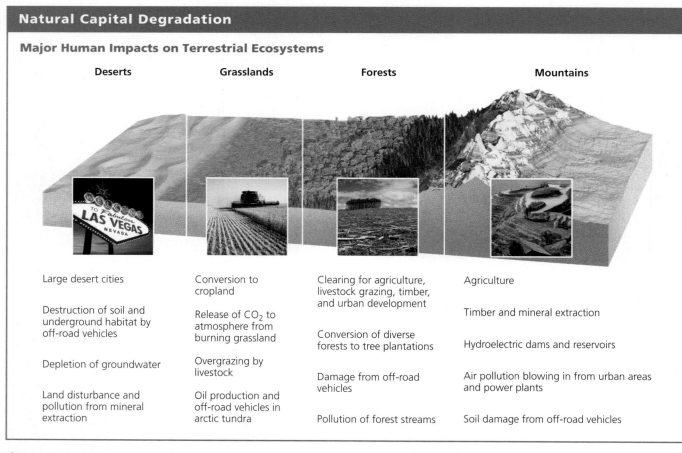

Natural Capital Degradation

Major Human Impacts on Terrestrial Ecosystems

Deserts	Grasslands	Forests	Mountains
Large desert cities	Conversion to cropland	Clearing for agriculture, livestock grazing, timber, and urban development	Agriculture
Destruction of soil and underground habitat by off-road vehicles	Release of CO_2 to atmosphere from burning grassland	Conversion of diverse forests to tree plantations	Timber and mineral extraction
Depletion of groundwater	Overgrazing by livestock	Damage from off-road vehicles	Hydroelectric dams and reservoirs
Land disturbance and pollution from mineral extraction	Oil production and off-road vehicles in arctic tundra	Pollution of forest streams	Air pollution blowing in from urban areas and power plants
			Soil damage from off-road vehicles

© Cengage Learning

Figure 7-17 Human activities have had major impacts on the world's deserts, grasslands, forests, and mountains (Concept 7-3), as summarized here. **Question:** For each of these biomes, which two of the impacts listed do you think are the most harmful?

Left: ©somchaij/Shutterstock.com. Left center: Orientaly/Shutterstock.com. Right center: © Eppic | Dreamstime.com. Right: ©Vasik Olga/Shutterstock.com.

Figure 7-18 **Natural capital degradation:** Projected status of biodiversity in 2018, resulting from ongoing threats of habitat loss and fragmentation, invasions of nonnative species, pollution, climate change, and unsustainable exploitation.

National Geographic; Conservation International

cal footprint gets bigger and spreads across the globe (see Figure 1-12, p. 13, and Figure 1-13, p. 14). Figure 7-17 summarizes some of the human impacts on the world's deserts, grasslands, forests, and mountains (Concept 7-3). Figure 7-18 shows the projected status of biodiversity in 2018 due to long-term threats from human activities.

How long can we keep eating away at these terrestrial forms of natural capital without threatening our economies and the long-term survival of our own and many other species? No one knows. But there are increasing signs that we need to come to grips with this vital issue.

Michael Nichols/National Geographic Creative

J. Michael Fay: Defender of Threatened Wild Places and National Geographic Explorer-in-Residence

Conservationist J. Michael Fay is a biologist with the Wild-life Conservation Society. For much of his life, he has been driven by a passion to explore and preserve some of the earth's vital and threatened wild places. To do this, he has gathered important ecological data about several major ecosystems by flying over and photographing them and by walking through these areas.

In 1996, Fay flew over the forests of the African countries of Gabon and the Congo and studied a vast corridor of forest spanning the two countries. In 1997, he hiked more than 3,200 kilometers (2,000 miles) through this corridor and talked with people to learn more about what he had seen from the air. The data that he collected were used by the government of Gabon to create a system of 13 protected national parks. In 2004, Fay spent 8 months flying over the entire continent of Africa, documenting its ecosystems and the impacts of human activities on them. Many of the 116,000 photographs he took can be seen on Google.

In 2008, Fay turned his attention to California's coastal redwood forests. He spent a year hiking 2,900 kilometers (1,900 miles) through this one-of-kind ecosystem, collecting ecological data with the goal of helping to protect it. This was one more example of how Fay, a modern-day explorer, uses his scientific knowledge and his passion for conservation to help us focus on the urgent need to preserve the earth's remaining wild places.

Background photo: Sharon Eisenzopf/Shutterstock.com

a. 10,000 years ago **b.** 5,000 years ago **c.** 200 years ago

(Compiled by the authors using data from P.A. Delcourt, 1981, and D.M. Gates, 1993.)

■ Warm temperate evergreen forest	■ Mixed conifer and hardwood forest
■ Cool temperate deciduous forest	■ Mixed spruce and pine forest

■ Boreal forest   ■ Tundra
■ Prairie

Figure 7-19 As temperatures rose during the last 18,000 years, huge quantities of ice melted and the global climate changed. This led to shifts in the sizes and locations of several major biomes in eastern North America. **Question:** Based on these maps, how might you expect the sizes and locations of these biomes to change if the earth's average atmospheric temperature rises by about 2–5 C° (4–9 F°) during this century?

Many environmental scientists call for a global effort to better understand the nature and state of the world's major ecosystems and to use such scientific data to protect the world's remaining wild areas from development (see Individuals Matter 7.1). In addition, they call for us to restore many of the land areas that have been degraded, especially in areas that are rich in biodiversity. However, such efforts are highly controversial because of the timber, mineral, fossil fuel, and other resources found on or under many of the earth's remaining wild land areas. These issues are discussed in Chapter 10.

Sizes and Locations of Biomes Can Change

The locations and sizes of the world's biomes shown in Figure 7-9 are not fixed and they change as the earth's climate changes. For example, the climate has changed drastically since the end of the last ice age, about 18,000 years ago (see Figure 4-10, p. 87). During the warmer interglacial period of the last 10,000 years, the earth's average atmospheric temperature rose by about 5 C° (9 F°). This changed the locations and sizes of biomes in several parts of the world, including North America (Figure 7-19).

Considerable scientific evidence indicates that human activities are likely to raise the average atmospheric temperature by 2–5 C° (4–9 F°) during this century, mostly as a result of human activities. This will likely change the sizes and locations of many biomes and alter the ecological map of the earth's land areas. These changes could also wipe out many species and degrade important ecosystem services. If these scientific projections are correct, such changes will take place within 100 years or so, instead of within thousands of years as they have in the past. This gives us and other species very little time to deal with such projected changes.

Big Ideas

- Differences in climate, based mostly on long-term differences in average temperature and precipitation, largely determine the types and locations of the earth's deserts, grasslands, and forests.

- The earth's terrestrial ecosystems provide important ecosystem and economic services.

- Human activities are degrading and disrupting many of the ecosystem and economic services provided by the earth's terrestrial ecosystems.

A Temperate Deciduous Forest and Sustainability

mihalec/Shutterstock.com

In this chapter we discussed the influence of climate on terrestrial biodiversity in the formation of biomes—deserts, grasslands, and forests. In the **Core Case Study**, we focused on the *temperate deciduous forest* biome, noting how the plants and animals that have evolved within such forests are adapted to climatic conditions there. Thus we saw that *climate*—the weather conditions in a given area averaged over periods of time ranging from three decades to thousands of years—plays a key role in determining the nature of terrestrial ecosystems, as well as the life-forms that live in those systems.

These relationships are in keeping with the three scientific **principles of sustainability** (see Figure 1-2, p. 6 or back cover). The earth's dynamic climate system helps to distribute heat from solar energy and to recycle the earth's nutrients. This in turn helps to generate and support the biodiversity found in the earth's various biomes.

Scientists have made some progress in understanding the ecology of the world's terrestrial systems, as well as how the vital ecosystem and economic services they provide are being degraded and disrupted. One of the major lessons from their research is: *in nature, everything is connected.* According to these scientists, we urgently need more research on the components and workings of the world's biomes, on how they are interconnected, and on which connections are in the greatest danger of being disrupted by human activities. With such information, we will have a clearer picture of how our activities affect the natural capital that supports the earth's life and of what we can do to help sustain that natural capital on which we and all other species depend.

Chapter Review

Core Case Study

1. Describe a temperate deciduous forest (**Core Case Study**) and explain why it serves as an example of how differences in climate lead to the formation of different types of ecosystems.

Section 7-1

2. What is the key concept for this section? Distinguish between **weather** and **climate**. Define **ocean currents**. Describe three major factors that determine how air circulates in the lower atmosphere. Explain how varying combinations of precipitation and temperature, along with global air circulation and ocean currents, lead to the formation of various types of forests, grasslands, and deserts.

3. Define and give four examples of a **greenhouse gas**. What is the **greenhouse effect** and why is it important to the earth's life and climate?

4. What is the **rain shadow effect** and how can it lead to the formation of deserts? Why do cities tend to have more haze and smog, higher temperatures, and lower wind speeds than the surrounding countryside?

Section 7-2

5. What is the key concept for this section? Describe how climate and vegetation vary with latitude and elevation. What is a **biome**? Explain why there are three major types of each of the major biomes (deserts, grasslands, and forests). Explain why biomes are not uniform. Describe how climate and vegetation vary with latitude and elevation.

6. Describe how the three major types of deserts differ in their climate and vegetation. Why are desert ecosystems fragile? How do desert plants and animals survive?

7. Explain how the three major types of grasslands differ in their climate and vegetation. What is a savanna? Why have many of the world's temperate grasslands disappeared? What is **permafrost**?

8. Explain how the three major types of forests differ in their climate and vegetation. Why is biodiversity so high in tropical rain forests? Why do most soils in tropical rain forests hold few plant nutrients? Why do temperate deciduous forests typically have a thick layer of decaying litter? How do most species of coniferous evergreen trees survive the cold winters in boreal forests? What are coastal coniferous or temperate rain forests? What important ecological roles do mountains play?

Section 7-3

9. What is the key concept for this section? About what percentage of the world's major terrestrial ecosystems are being degraded or used unsustainably? Summarize the ways in which human activities have affected the world's deserts, grasslands, forests, and mountains. How is a warming climate likely to change the earth's biomes?

10. What are this chapter's *three big ideas*? Summarize the connections between climate and terrestrial ecosystems, and explain how these connections are in keeping with the three scientific **principles of sustainability** (see Figure 1-2, p. 6 or back cover).

Note: Key terms are in bold type.

Critical Thinking

1. Why do you think temperate deciduous forests (**Core Case Study**) are among the biomes most extensively disturbed by human activities?

2. Describe the roles of temperature and precipitation in determining what parts of the earth's land are covered with **(a)** desert, **(b)** arctic tundra, **(c)** temperate grassland, **(d)** tropical rain forest, and **(e)** temperate deciduous forest (**Core Case Study**).

3. Why do deserts and arctic tundra support a much smaller number and variety of animals than do tropical forests? Why do most animals in a tropical rain forest live in its trees?

4. How might the distribution of the world's forests, grasslands, and deserts shown in Figure 7-9 differ if the prevailing winds shown in Figure 7-3 did not exist?

5. Which biomes are best suited for **(a)** raising crops and **(b)** grazing livestock? Use the three scientific **principles of sustainability** to come up with three guidelines for growing crops and grazing livestock more sustainably in these biomes.

6. What type of biome do you live in? (If you live in a developed area, what type of biome was the area before it was developed?) List three ways in which your lifestyle could be contributing to the degradation of this biome. What changes could you make in order to reduce your contribution, if any?

7. You are a defense attorney arguing in court for sparing a tropical rain forest from being cut down. Give your three best arguments for the defense of this ecosystem. Do the same for the case of temperate deciduous forest (**Core Case Study**).

8. Congratulations! You are in charge of the world. What are the three most important features of your plan for helping to sustain the earth's terrestrial biodiversity and ecosystem services?

Doing Environmental Science

Using Google Earth, find an undeveloped ecosystem somewhere on the planet that you can zoom in on to observe some features. For example, find a forest or grassland area that is undeveloped. Take careful notes on what you see. Then find such an example of a biome that has been developed, study it, and write a detailed comparison of your undeveloped and developed biome samples. If possible, check in on these systems once during your course term and again at the end of the term and describe any changes to the systems that you can observe. Suggest a hypothesis to explain any major changes.

Global Environment Watch Exercise

Search for *tropical rain forests* and use the topic portal to find information on **(a)** trends in the global rate of destruction of these forest; **(b)** what areas of the world are seeing rising rates of destruction and what areas are seeing falling rates; and **(c)** what is being done to protect them in various areas. Write a report on your findings.

Data Analysis

In this chapter, you learned how long-term variations in average temperatures and average precipitation play a major role in determining the types of deserts, forests, and grasslands found in different parts of the world. Below are typical annual climate graphs for a tropical grassland (savanna) in Africa and a temperate grassland in the Midwestern United States.

Tropical grassland (savanna)

Temperate grassland (prairie)

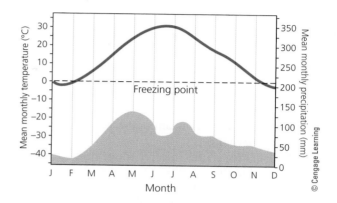

1. In what month (or months) does the most precipitation fall in each of these areas?

2. What are the driest months in each of these areas?

3. What is the coldest month in the tropical grassland?

4. What is the warmest month in the temperate grassland?

1. Long-term differences in which two variables are the primary determinants of climate?
   a. Temperature and cloud cover
   b. Precipitation and temperature
   c. Precipitation and soil type
   d. Temperature and soil type
   e. Soil type and cloud cover

2. Which of the below is an example of a change in weather?
   a. A large area changes from a grassland to a desert.
   b. While climbing a mountain, one moves through several different biomes.
   c. A thunderstorm forms and drops 1.5 cm of rain.
   d. Increased greenhouse gases in the atmosphere cause atmospheric warming.
   e. Desertification of an area causes the area to see increased annual temperatures.

Questions 3–6 refer to the diagram of surface currents in the oceans.

© Cengage Learning

3. A current responsible for moving heat away from the equator.

4. The reversal of this current is associated with ENSO events.

5. Cape Cod juts out from the Eastern coast of North America. This current keeps the waters off Cape Cod very cool.

6. This current causes offshore upwellings of nutrient rich water that is responsible for plankton rich waters sustaining a large anchovy fishery.

7. As one climbs a mountain, the elevation change causes a change in biomes. Which of the below is a change caused by elevation?
   a. Biomes become more diverse because of the decrease in temperature.
   b. Precipitation increases as air rises and water vapor condenses, resulting in vegetation that thrives with a large amount of moisture.
   c. Temperature is highest on mountaintops because they are closer to the sun and therefore receive more solar radiation.
   d. The land becomes arid, resulting in the growth of different types of vegetation.
   e. The change in mountainous climate is very similar to the change in climate as latitude decreases.

8. Which of the below is a true statement about biomes?
   a. Biomes have not been affected by human presence because the human population is not dense enough to impact them.
   b. Although biomes have distinctive vegetation growing in them, basically the same kinds of animals live in all of them.
   c. All desert biomes are hot and dry.
   d. Biomes have a range of temperatures and precipitation that result in specific kinds of plants that have adapted to those conditions.
   e. Mountainous areas all over the world contain the same biomes.

Questions 9–11 refer to the climate graphs below. The red line represents temperature and the blue represents precipitation.

A

B

C

D

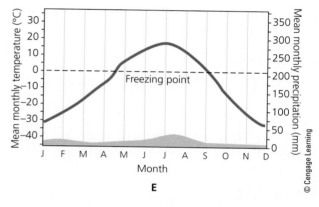

E

9. This biome is characterized by subfreezing temperatures much of the year, low precipitation rates, and the formation of permafrost.

10. This biome found in equatorial regions has large amounts of precipitation but has low levels of nutrients in the soil.

11. This biome often has snow, yet the average monthly temperature is always greater than the freezing point. Trees in this biome lose their leaves every fall, producing colorful displays.

12. Which of the below is NOT an adaptation found in plants living in the desert?
   a. Deep roots to tap into groundwater
   b. Waxy-coated leaves preventing water loss
   c. Storing much of their biomass as seeds and remaining inactive most of the year
   d. Breaking down stored fats to produce needed water
   e. Opening stomata (pores) only at night to absorb needed carbon dioxide

8 Aquatic Biodiversity

*The sea, once it casts its spell,
holds one in its net of wonders
forever.*

JACQUES-YVES COUSTEAU

Key Questions

8-1 What is the general nature of aquatic
systems?

8-2 Why are marine aquatic systems
important?

8-3 How have human activities affected marine
ecosystems?

8-4 Why are freshwater ecosystems important?

8-5 How have human activities affected
freshwater ecosystems?

Coral reef off the coast of Fiji in the South Pacific ocean

Coral reefs form in clear, warm coastal waters in tropical areas. These stunningly beautiful natural wonders (see chapter-opening photo) are among the world's oldest, most diverse, and most productive ecosystems.

Coral reefs are formed by massive colonies of tiny animals called *polyps* (close relatives of jellyfish). They slowly build reefs by secreting a protective crust of limestone (calcium carbonate) around their soft bodies. When the polyps die, their empty crusts remain behind as a platform for more reef growth. The resulting elaborate network of crevices, ledges, and holes serves as calcium carbonate "condominiums" for a variety of marine animals.

Coral reefs are the result of a mutually beneficial relationship between polyps and tiny single-celled algae called *zooxanthellae* ("zoh-ZAN-thel-ee") that live in the tissues of the polyps. The algae provide the polyps with food and oxygen through photosynthesis and help produce the coral's calcium carbonate skeleton. Algae also give the reefs their stunning coloration. The polyps, in turn, provide the algae with a well-protected home and some of their nutrients.

Although shallow- and deep-water coral reefs occupy only about 0.2% of the ocean floor, they provide important ecosystem and economic services. They act as natural barriers that help to protect 15% of the world's coastlines from erosion caused by battering waves and storms. These reefs provide habitats for one-quarter of all marine organisms, and they produce about one-tenth of the global fish catch.

Studies by the Global Coral Reef Monitoring Network and other scientist groups estimate that since the 1950s, some 45–53% of the world's shallow coral reefs have been destroyed or degraded by pollution and other stressors, and another 25–33% could be lost within 20–40 years. Also, deep coral reefs that are thousands

Rainer von Brandis/iStockphoto.com

Figure 8-1 This bleached coral has lost most of its algae because of changes in the environment such as warming of the waters and deposition of sediments.

of years old are being destroyed by large numbers of trawler fishing boats that drag huge weighted nets across the ocean bottom. One result of such stresses is *coral bleaching* (Figure 8-1). Pollution or water that is too warm can cause the algae, on which corals depend for food, to die off. Without food, the coral polyps die, leaving behind a white skeleton of calcium carbonate.

Coral reefs are vulnerable to damage because they grow slowly and are disrupted easily. Runoff of soil and other materials from the land can cloud the water and block the sunlight that the algae in shallow reefs need for photosynthesis.

Also, the water in which shallow reefs live must have a temperature of 18–30°C (64–86°F) and cannot be too acidic. This explains why the two major long-term threats to coral reefs are projected *climate change,* which could raise the water temperature above this limit in most reef areas, and *ocean acidification,* which could make it harder for polyps to build reefs and could even dissolve some of their calcium carbonate formations.

In this chapter, we explore ocean and freshwater ecosystems and the threats they face. We also consider some measures we can all take to help sustain these vital systems.

8-1 What Is the General Nature of Aquatic Systems?

CONCEPT 8-1A
Saltwater and freshwater aquatic life zones cover almost three-fourths of the earth's surface, with oceans dominating the planet.

CONCEPT 8-1B
The key factors determining biodiversity in aquatic systems are temperature, dissolved oxygen content, availability of food, and access to light and nutrients necessary for photosynthesis.

Most of the Earth Is Covered with Water

When viewed from outer space, the earth appears to be almost completely covered with water (Figure 8-2). Saltwater covers about 71% of the earth's surface, and freshwater occupies roughly another 2.2%.

Although the *global ocean* is a single and continuous body of water, geographers divide it into four large areas—the Atlantic, Pacific, Arctic, and Indian Oceans—separated by the continents. The largest ocean is the Pacific, which contains more than half of the earth's water and covers one-third of the earth's surface. Together, the oceans hold almost 98% of the earth's water. Each of us is connected to, and utterly dependent on, the earth's global ocean through the water cycle (see Figure 3-15, p. 63).

The aquatic equivalents of biomes are called **aquatic life zones**—saltwater and freshwater portions of the biosphere that can support life. The distribution of many aquatic organisms is determined largely by the water's *salinity*—the amounts of various salts such as sodium chloride (NaCl) dissolved in a given volume of water. As a result, aquatic life zones are classified into two major types: **saltwater** or **marine life zones** (oceans and their bays, estuaries, coastal wetlands, shorelines, coral reefs, and mangrove forests) and **freshwater life zones** (lakes, rivers, streams, and inland wetlands). Some systems such as estuaries are a mix of saltwater and freshwater, but scientists classify them as marine systems for purposes of discussion.

Figure 8-2 *The ocean planet:* The salty oceans cover 90% of the planet's ocean hemisphere (left) and nearly half of its land–ocean hemisphere (right) (**Concept 8-1A**).

Ocean hemisphere   Land–ocean hemisphere

© Cengage Learning

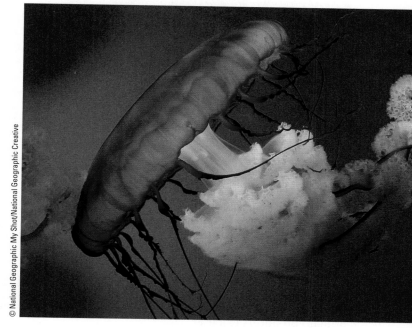

© National Geographic My Shot/National Geographic Creative

Figure 8-3 Jellyfish are drifting zooplankton that use their long tentacles with stinging cells to stun or kill their prey. Jellyfish populations are expanding globally and taking over many marine ecosystems because excessive nutrient runoff from the land has spurred the growth of algae on which small jellyfish feed, and because many of their predators have been removed by overfishing.

Aquatic Species Drift, Swim, Crawl, and Cling

Saltwater and freshwater life zones contain several major types of organisms. One type consists of **plankton**, which can be divided into three groups. The first group consists of drifting organisms called *phytoplankton* ("FY-toe-plankton"), which includes many types of algae. These tiny aquatic plants and even smaller *ultraplankton*—the second group of plankton—are the producers that make up the base of most aquatic food chains and webs (see Figure 3-12, p. 60). Through photosynthesis, they produce about half of the earth's oxygen, on which we depend for survival.

The third group is made up of drifting animals called *zooplankton* ("ZOH-uh-plank-ton"), which feed on phytoplankton and on other zooplankton (see Figure 3-12, p. 60). The members of this group range in size from single-celled protozoa to large invertebrates such as jellyfish (Figure 8-3).

A second major type of aquatic organism is **nekton**, strongly swimming consumers such as fish, turtles, and whales. The third type, **benthos**, consists of bottom-dwellers such as oysters and sea stars (Figure 8-4), which anchor themselves to ocean-bottom structures; clams and worms, which burrow into the sand or mud; and lobsters

Figure 8-4 Bottom-dwelling starfish, four types of which appear here, use their tube feet to attach themselves to surfaces on the ocean bottom and move slowly in search of food.

and crabs, which walk about on the sea floor. A fourth major type is **decomposers** (mostly bacteria), which break down organic compounds in the dead bodies and wastes of aquatic organisms into nutrients that aquatic primary producers can use.

Key factors determining the types and numbers of organisms found in different areas of the ocean are *temper-*

ature, dissolved oxygen content, availability of food, and *availability of light and nutrients required for photosynthesis,* such as carbon (as dissolved CO_2 gas), nitrogen (as NO_3^-), and phosphorus (mostly as PO_4^{3-}) (**Concept 8-1B**).

In deep aquatic systems, photosynthesis is largely confined to the upper layer—the *euphotic* or *photic zone,* through which sunlight can penetrate. The depth of the euphotic zone in oceans and deep lakes is reduced when the water is clouded by excessive growth of algae—called *algal blooms*—that results from nutrient overloads. This cloudiness, called **turbidity**, can occur naturally, such as from algal growth. It can also be caused by soil and other sediments being carried by rain and melting snow from cleared land into adjoining bodies of water. This is one of the problems plaguing shallow coral reefs (**Core Case Study**), as excessive turbidity due to silt runoff prevents photosynthesis and causes the corals to die.

In shallow systems such as small open streams, lake edges, and ocean shorelines, ample supplies of nutrients for primary producers are usually available, which tends to make these areas high in biodiversity. By contrast, in most areas of the open ocean, nitrates, phosphates, iron, and other nutrients are often in short supply, and this limits net primary productivity (NPP) (see Figure 3-14, p. 62).

8-2 Why Are Marine Aquatic Systems Important?

CONCEPT 8-2
Saltwater ecosystems provide major ecosystem and economic services and are irreplaceable reservoirs of biodiversity.

Oceans Provide Vital Ecosystem and Economic Services

Oceans provide enormously valuable ecosystem and economic services (Figure 8-5). One estimate of the combined value of these goods and services from all marine coastal ecosystems is over $12 trillion per year.

As land dwellers, we have a distorted and limited view of the aquatic wilderness that covers most of the earth's surface. We know more about the surface of the moon than we know about the oceans. According to aquatic scientists, the scientific investigation of poorly understood marine and freshwater aquatic systems could yield immense ecological and economic benefits.

Marine aquatic systems are enormous reservoirs of biodiversity. They include many different ecosystems, which host a great variety of species, genes, and biological and chemical processes, thus helping to sustain the four major components of the earth's biodiversity (see Figure 4-2, p. 79). Marine life is found in three major *life zones:* the coastal zone, the open sea, and the ocean bottom (Figure 8-6).

Natural Capital

Marine Ecosystems

Ecosystem Services	Economic Services
Oxygen supplied through photosynthesis	Food
Water purification	Energy from waves and tides
Climate moderation	Pharmaceuticals
CO_2 absorption	Harbors and transportation routes
Nutrient cycling	Recreation and tourism
Reduced storm damage (mangroves, barrier islands, coastal wetlands)	Employment
Biodiversity: species and habitats	Minerals

Figure 8-5 Marine systems provide a number of important ecosystem and economic services (**Concept 8-2**). *Questions:* Which two ecosystem services and which two economic services do you think are the most important? Why?

Top: Willyam Bradberry/Shutterstock.com. Bottom: James A. Harris/Shutterstock.com.

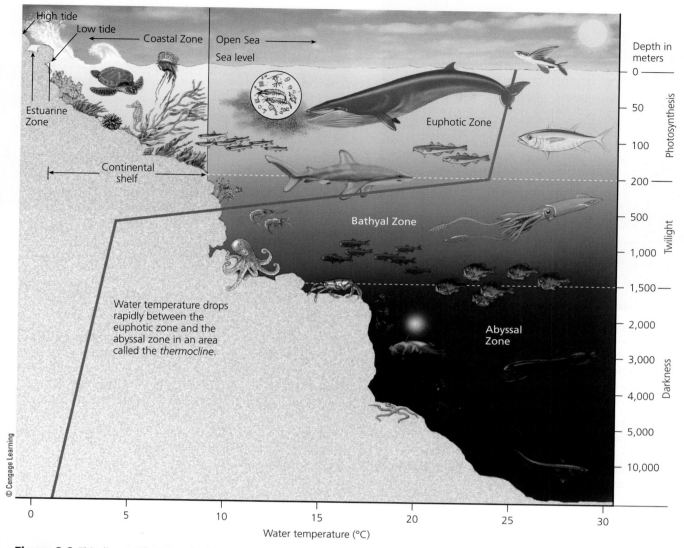

Figure 8-6 This diagram illustrates the major life zones and vertical zones (not drawn to scale) in an ocean. Actual depths of zones may vary. Available light determines the euphotic, bathyal, and abyssal zones. Temperature zones also vary with depth, shown here by the red line. ***Question:*** How is an ocean like a rain forest? (*Hint:* see Figure 7-14, p. 157.)

The **coastal zone** is the warm, nutrient-rich, shallow water that extends from the high-tide mark on land to the gently sloping, shallow edge of the *continental shelf* (the submerged part of the continents). It makes up less than 10% of the world's ocean area, but it contains 90% of all marine species and is the site of most large commercial marine fisheries. This zone's aquatic systems include estuaries, coastal marshes, mangrove forests, and coral reefs.

Estuaries and Coastal Wetlands Are Highly Productive

An **estuary** is where a river meets the sea (Figure 8-7). It is a partially enclosed body of water where seawater mixes with the river's freshwater, as well as nutrients and pollutants in runoff from the land.

Estuaries are associated with **coastal wetlands**— coastal land areas covered with water all or part of the year. These wetlands include *coastal marshes* (Figure 8-8) and *mangrove forests* (Figure 8-9). They are some of the earth's most productive ecosystems (see Figure 3-14, p. 62) because of high nutrient inputs from rivers and from adjoining land, rapid circulation of nutrients by tidal flows, and ample sunlight penetrating the shallow waters. For example, mangrove forests around the world host 69 different species of trees that can grow in salty water. They provide habitat, food, and nursery sites for a variety of fishes and other aquatic species.

Sea-grass beds are another component of coastal marine biodiversity (Figure 8-10). They consist of at least 60 species of plants that grow underwater in estuaries and shallow waters along most continental coastlines.

Figure 8-7 This satellite photo shows a view of an *estuary* taken from space. A sediment plume (turbidity caused by runoff) forms at the mouth of Madagascar's Betsiboka River as it flows through the estuary and into the Mozambique Channel.

Figure 8-8 This coastal marsh is located on Cape Cod, Massachusetts (USA).

These highly productive and physically complex systems support a variety of marine species. Like other coastal systems, they owe their high NPP (see Figure 3-14, p. 62) to ample supplies of sunlight and plant nutrients that flow from land and are distributed by wind and ocean currents.

These coastal aquatic systems provide important ecosystem and economic services. They help to maintain water quality in tropical coastal zones by filtering toxic pollutants, excess plant nutrients, and sediments and by absorbing other pollutants. They provide food, habitats, and nursery sites for a variety of aquatic and terrestrial species. They also reduce storm damage and coastal erosion by absorbing waves and storing excess water produced by storms and tsunamis.

Rocky and Sandy Shores Host Different Types of Organisms

The gravitational pull of the moon and sun causes *tides* to rise and fall about every 6 hours in most coastal areas. The area of shoreline between low and high tides is called the **intertidal zone**. Organisms living in this zone must be able to avoid being swept away or crushed by waves, and must deal with being immersed during high tides and left high and dry (and much hotter) at low tides. They must also survive changing levels of salinity when heavy rains dilute saltwater. To deal with such stresses, most intertidal organisms hold on to something, dig in, or hide in protective shells.

Figure 8-9 This is a mangrove forest on the coast of Queensland, Australia.

Figure 8-10 Sea-grass beds, such as this one near the coast of San Clemente Island, California, support a variety of marine species.

© James Forte/National Geographic Creative

On some coasts, steep *rocky shores* are pounded by waves. The numerous pools and other habitats in these intertidal zones contain a great variety of species that occupy different niches in response to daily and seasonal changes in environmental conditions such as temperature, water flows, and salinity (Figure 8-11, top).

Other coasts have gently sloping *barrier beaches,* or *sandy shores,* that support other types of marine organisms (Figure 8-11, bottom), most of which keep hidden from view and survive by burrowing, digging, and tunneling in the sand. These sandy beaches and their adjoining coastal wetlands are also home to a variety of shorebirds that have evolved in specialized niches to feed on crustaceans, insects, and other organisms (see Figure 4-16, p. 92). Many of these same species also live on *barrier islands*—low, narrow, sandy islands that form offshore, parallel to nearby coastlines.

Undisturbed barrier beaches generally have one or more rows of natural sand dunes in which the sand is held in place by the roots of plants, usually grasses. These dunes are the first line of defense against the ravages of the sea. These areas are attractive to real estate developers, and such real estate is scarce and valuable. However, coastal developers frequently remove the protective dunes or build behind the first set of dunes, covering them with buildings and roads. Large storms can then flood and even sweep away seaside construction and severely erode the sandy beaches.

Coral Reefs Are Amazing Centers of Biodiversity

As we noted in the **Core Case Study**, coral reefs are among the world's oldest and most diverse and productive ecosystems (see chapter-opening photo). These amazing centers of aquatic biodiversity are the marine equivalents of tropical rain forests, with complex interactions among their diverse populations of species.

Worldwide, coral reefs are being damaged and destroyed at an alarming rate by a variety of human activities. A growing threat, as mentioned in this chapter's **Core Case Study**, is **ocean acidification**—the increasing levels of acid in the world's oceans. This occurs because the oceans absorb about a third of the CO_2 emitted into the atmosphere by human activities, especially the burning of carbon-containing fossil fuels. The CO_2 reacts with ocean water to form a weak acid and decreases the levels of carbonate ions (CO_3^{2-}) needed to form coral and the shells and skeletons of organisms such as crabs, oysters, and some phytoplankton.

The Open Sea and the Ocean Floor Host a Variety of Species

The sharp increase in water depth at the edge of the continental shelf separates the coastal zone from the vast volume of the ocean called the **open sea**. Primarily on the basis of the penetration of sunlight, this deep blue sea is divided into three *vertical zones* (Figure 8-6). Temperatures also change with depth (Figure 8-6, red line) and we can use them to define zones that help to determine species diversity in these zones.

The *euphotic zone* is the brightly lit upper zone, where drifting phytoplankton carry out about 40% of the world's photosynthetic activity. Nutrient levels are low and levels of dissolved oxygen are high in the euphotic zone. One exception to this is those areas called *upwelling zones,* where ocean currents driven by coastal winds or by differences in water temperature at different depths bring water up from the deepest waters (see Figure 7-2, p. 145). Upwellings carry nutrients from the ocean bottom to the surface for use by producers, and thus these zones contain high levels of nutrients. Large, fast-swimming predatory fishes such as swordfish, sharks, and bluefin tuna populate the euphotic zone. They feed on secondary and higher-level consumers, which are supported directly or indirectly by producers.

Figure 8-11 *Living between the tides:* Some organisms with specialized niches are found in various zones on rocky shore beaches (top) and barrier or sandy beaches (bottom). Organisms are not drawn to scale.

SCIENCE FOCUS 8.1

WE ARE STILL LEARNING ABOUT THE OCEAN'S BIODIVERSITY

Scientists have long assumed that open-ocean waters contained few microbial life forms. But a 2007 report challenged that assumption and drastically increased our knowledge of the ocean's genetic diversity.

A team of scientists led by J. Craig Venter (Figure 8-A) took 2 years to conduct a census (an estimated count based on sampling) of ocean microbes. They sailed around the world, stopping every 320 kilometers (200 miles) to pump seawater through extremely fine filters, from which they gathered data on bacteria, viruses, and other microbes. It was the most thorough of such censuses ever conducted.

Using a supercomputer, they counted genetic coding for 6 million new proteins—double the number that had previously been known. They also reported that they were discovering new genes and proteins at the same rate at the end of their voyage as they had at the start of it, meaning there is still much more of this biodiversity to discover.

This means that the ocean contains a much higher diversity of microbial life than had previously been thought. Ocean-water microbes play an important role in the absorption of carbon by the ocean, as well as in the ocean food web. Since the 2007 report, Venter has led more expeditions to continue the sampling in other areas, including the waters of the Baltic, Mediterranean, and Black Seas.

Critical Thinking

Why is it that the *rate* of discovery of new genes and proteins was important to Venter and his colleagues? Explain.

© Mark Wilson/Getty Images

Figure 8-A J. Craig Venter

The *bathyal zone* is the dimly lit middle zone, which receives little sunlight and therefore does not contain photosynthesizing producers. Zooplankton and smaller fishes, many of which migrate to feed on the surface at night, populate this zone.

The lowest zone, called the *abyssal zone,* is dark and very cold. There is no sunlight to support photosynthesis, and this zone has little dissolved oxygen. Nevertheless, the deep ocean floor is teeming with life, because it contains enough nutrients to support a large number of species. Most organisms of the deep waters and ocean floor get their food from showers of dead and decaying organisms—called *marine snow*—drifting down from upper, lighted levels of the ocean.

Some abyssal-zone organisms, including many types of worms, are *deposit feeders,* which take mud into their guts and extract nutrients from it. Others such as oysters, clams, and sponges are *filter feeders,* which pass water through or over their bodies and extract nutrients from it.

NPP is quite low in the open sea, except in upwelling areas. However, because the open sea covers so much of the earth's surface, it makes the largest contribution to the earth's overall NPP. Also, scientists are learning that the open sea contains more biodiversity than it was thought to hold until a few years ago (see Science Focus 8.1, above).

8-3 How Have Human Activities Affected Marine Ecosystems?

CONCEPT 8-3

Human activities threaten aquatic biodiversity and disrupt ecosystem and economic services provided by saltwater systems.

Human Activities Are Disrupting and Degrading Marine Ecosystems

Human activities are disrupting and degrading some ecosystem and economic services provided by marine aquatic systems, especially coastal marshes, shorelines, mangrove forests, and coral reefs (Concept 8-3).

For example, according to a 2008 United Nations (UN) Food and Agriculture Organization report, at least one-fifth of the world's mangrove forests were eliminated between 1980 and 2005, mostly to make way for human coastal developments. Also, since 1980, about 29% of the world's sea-grass beds have been lost to pollution and other disturbances.

In 2008, the U.S. National Center for Ecological Analysis and Synthesis (NCEAS) used computer models to analyze and provide the first-ever comprehensive map of the effects of 17 different types of human activities on the world's oceans. In this 4-year study, an international team of scientists found that human activities have heavily affected 41% of the world's ocean area. No area of the oceans has been left completely untouched, according to the report.

In their desire to live near a coast, many people are unwittingly taking part in this degradation of natural capital. In 2012, about 45% of the world's population and more than half of the U.S. population lived along or near coasts and these percentages are rising rapidly.

Major threats to marine systems from human activities include:

- Coastal development, which destroys and pollutes coastal habitats
- Runoff of nonpoint sources of pollutants such as silt, fertilizers, pesticides, and livestock wastes
- Point-source pollution such as sewage from cruise ships and spills from oil tankers
- Pollution and degradation of coastal wetlands and estuaries (see Case Study at right)
- Overfishing, which depletes populations of commercial fish species
- Use of fishing trawlers that drag weighted nets across the ocean bottom, degrading and destroying its habitats
- Invasive species, introduced by humans, which can deplete populations of native aquatic species and cause economic damage
- Ocean warming
- Ocean acidification

A major threat that is very alarming to some marine scientists is that of projected climate change, enhanced by human activities, which is warming and acidifying the oceans and slowly melting land-based glaciers in Greenland and other parts of the world. This could cause a rise in sea levels during this century that would destroy shallow coral reefs and flood coastal marshes and many coastal cities.

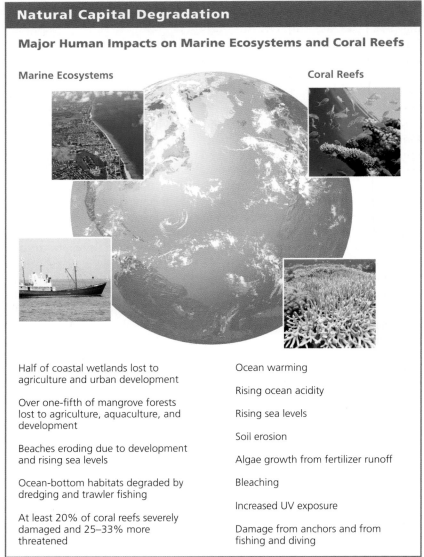

Natural Capital Degradation

Major Human Impacts on Marine Ecosystems and Coral Reefs

Marine Ecosystems — Coral Reefs

Half of coastal wetlands lost to agriculture and urban development

Over one-fifth of mangrove forests lost to agriculture, aquaculture, and development

Beaches eroding due to development and rising sea levels

Ocean-bottom habitats degraded by dredging and trawler fishing

At least 20% of coral reefs severely damaged and 25–33% more threatened

Ocean warming

Rising ocean acidity

Rising sea levels

Soil erosion

Algae growth from fertilizer runoff

Bleaching

Increased UV exposure

Damage from anchors and from fishing and diving

© Cengage Learning

Figure 8-12 Human activities are having major harmful impacts on all marine ecosystems (left) and particularly on coral reefs (right) (Concept 8-3). **Questions:** Which two of the threats to marine ecosystems do you think are the most serious? Why? Which two of the threats to coral reefs do you think are the most serious? Why?

Top left: Jorg Hackemann/Shutterstock.com. Top right: Rich Carey/Shutterstock.com. Bottom left: Piotr Marcinski/Shutterstock.com. Bottom right: Rostislav Ageev/Shutterstock.com.

A second threat, which some scientists view as more serious than the threat of climate change, is ocean acidification, known as the "other CO_2 problem." The level of acids in the world's oceans has increased by about 30% since the beginning of the Industrial Revolution and is projected to continue increasing during this century by 100–150%.

This trend is a threat to coral reefs (**Core Case Study**) and to phytoplankton and many shellfish that form their shells from calcium carbonate. It could kill off much of the oceans' phytoplankton that form the base of the entire marine food web. A 2010 study led by Boris Worm concluded that since 1950, the average global concentration of phytoplankton in the upper ocean had declined by 40% and that this decline was steadily continuing at about 1%

per year. Some scientists consider this decline a major threat to the entire global marine ecosystem. According to a 2007 study by Ove Hoegh-Guldberg and 16 other scientists, unless we take action soon to significantly reduce CO_2 emissions, the oceans may be too acidic and too warm for most of the world's coral reefs to survive this century, and the important ecosystem and economic services they provide will be lost.

Figure 8-12 shows some of the effects of these human impacts on marine systems in general (left) and on coral reefs in particular (right). We examine some of these impacts more closely in Chapters 11 and 19.

CONSIDER THIS...

THINKING ABOUT Coral Reef Destruction

How might the loss of most of the world's remaining tropical coral reefs (**Core Case Study**) affect your life and the lives of any children or grandchildren you might have? What are two things you could do to help reduce this loss?

CASE STUDY

The Chesapeake Bay—An Estuary in Trouble

Since 1960, the Chesapeake Bay (Figure 8-13)—the largest estuary in the United States—has been in serious trouble from water pollution, mostly because of human activities. One problem is population growth. Between 1940 and 2012, the number of people living in the Chesapeake Bay area grew from 3.7 million to more than 17 million, and could reach 18 million by 2020, according to estimates by the Chesapeake Bay Program.

The estuary receives wastes from point and nonpoint sources scattered throughout its huge drainage basin, which lies in parts of six states and the District of Columbia (Figure 8-13). The shallow bay has become a huge pollution sink because only 1% of the waste entering it is flushed into the Atlantic Ocean. Phosphate and nitrate levels have been high in many parts of the bay, causing algal blooms that deplete the oxygen dissolved in the waters, making them unsuitable for most forms of aquatic life. Commercial harvests of the bay's once-abundant oysters and crabs, as well as several important fish species, fell sharply after the 1960s because of a combination of pollution, overfishing, and disease.

Point sources, primarily sewage treatment plants and industrial plants, account for 60% by weight of the phosphates. Nonpoint sources—mostly runoff of fertilizer, animal wastes from urban, suburban, and agricultural land, and deposition of pollutants from the atmosphere—account for 60% by weight of the nitrates. In addition, runoff of sediment, mostly from soil erosion, harms the submerged grasses on which crabs and young fish depend. Runoff also increases when trees near the bay are cut down for development.

A century ago, oysters were so abundant that they filtered and cleaned the Chesapeake's entire volume

Figure 8-13 The Chesapeake Bay is severely degraded as a result of water pollution and atmospheric deposition of air pollutants.

of water every 3 days. This important form of natural capital provided by these *keystone species* (see Chapter 4, pp. 94–96) helped remove or reduce excess nutrients and algal blooms. Now the oyster population has been reduced to the point where this filtration process takes a year and the keystone role of this species has been severely weakened.

In 1983, the United States implemented the Chesapeake Bay Program. In this ambitious attempt at *integrated coastal management*, citizens' groups, communities, state legislatures, and the federal government worked together to reduce pollution inputs into the bay. One strategy of the program was to establish land-use regulations to reduce agricultural and urban runoff in the bay's drainage area. Other strategies included banning phosphate detergents, upgrading sewage treatment plants, and monitoring industrial discharges more closely. Some adjoining wetlands have been restored and large areas of the bay were replanted with sea grasses to help filter out excessive nutrients and other pollutants.

In 2008, after 25 years of effort costing almost $6 billion, the Chesapeake Bay Program had failed to meet its goals. This was because of increased population and development, a drop in state and federal funding, and a lack of cooperation and enforcement among local, state, and federal officials. That year, the Chesapeake Bay Foundation reported that the bay's water quality was "very poor" and only 21% of the established goals had been met.

However, in 2011, a team of scientists led by Rebecca R. Murphy reported encouraging news. After analyzing 60 years of water-quality data, the scientists found that the sizes of oxygen-depleted zones that had been growing every summer in the bay had leveled off in the 1980s and had been declining since then. This meant that the integrated efforts to reduce inputs of fertilizers, animal wastes, and other pollutants were having a positive effect on the bay.

The other good news is that crab populations in the Chesapeake Bay have rebounded due to a set of measures put in place in 2008 by the states of Maryland and Virginia. While the bay's blue crab population was on the verge of collapse in 2003, it is now growing again to sustainable levels. As one researcher noted, efforts such as these must continue at the same or higher levels if the Chesapeake Bay Program is to meet its goals.

🔍 **CONSIDER THIS. . .**

THINKING ABOUT **The Chesapeake Bay**

What are three ways in which Chesapeake Bay area residents could apply the three scientific **principles of sustainability** (see Figure 1-2, p. 6 or back cover) to try to improve the environmental quality of the bay?

8-4 Why Are Freshwater Ecosystems Important?

CONCEPT 8-4
Freshwater ecosystems provide major ecosystem and economic services, and they are irreplaceable reservoirs of biodiversity.

Water Stands in Some Freshwater Systems and Flows in Others

Precipitation that does not sink into the ground or evaporate becomes **surface water**—freshwater that flows or is stored in bodies of water on the earth's surface. *Freshwater life zones* include *standing* (lentic) bodies of freshwater such as lakes, ponds, and inland wetlands, and *flowing* (lotic) systems such as streams and rivers. Surface water that flows into such bodies of water is called **runoff**. A **watershed**, or **drainage basin**, is the land area that delivers runoff, sediment, and dissolved substances to a stream, lake, or wetland.

Although freshwater systems cover less than 2.2% of the earth's surface, they provide a number of important ecosystem and economic services (Figure 8-14).

Lakes are large natural bodies of standing water formed when precipitation, runoff, streams, rivers, and groundwater seepage fill depressions in the earth's surface. Causes of such depressions include glaciation (as in Lake Louise in Alberta, Canada), displacement of the earth's crust (Lake Nyasa in East Africa), and volcanic activity (Crater Lake, Figure 8-15). Drainage areas on surrounding land supply lakes with water from rainfall, melting snow, and streams.

Freshwater lakes vary tremendously in size, depth, and nutrient content. Deep lakes normally consist of four distinct zones that are defined by their depth and distance from shore (Figure 8-16). The top layer, called the *littoral* ("LIT-tore-el") *zone,* is near the shore and consists of the shallow sunlit waters to the depth at which rooted plants stop growing. It has a high level of biodiversity because of ample sunlight and inputs of nutrients from the surrounding land. Species living in the littoral zone include many rooted plants; animals such as turtles, frogs, and crayfish; and fish such as bass, perch, and carp.

Natural Capital

Freshwater Systems

Ecosystem Services	Economic Services
Climate moderation	Food
Nutrient cycling	Drinking water
Waste treatment	
Flood control	Irrigation water
Groundwater recharge	Hydroelectricity
Habitats for many species	Transportation corridors
Genetic resources and biodiversity	Recreation
Scientific information	Employment

© Cengage Learning

Figure 8-14 Freshwater systems provide many important ecosystem and economic services (Concept 8-4). **Questions:** Which two ecosystem services and which two economic services do you think are the most important? Why?

Top: © Galyna Andrushko/Shutterstock.com. Bottom: © Kletr/Shutterstock.com.

The next layer is the *limnetic* ("lim-NET-ic") *zone,* the open, sunlit surface layer away from the shore that extends to the depth penetrated by sunlight. This is the main photosynthetic zone of the lake, the layer that produces the food and oxygen that support most of the lake's consumers. Its most abundant organisms are microscopic phytoplankton and zooplankton. Some large species of fish spend most of their time in this zone, with occasional visits to the littoral zone to feed and reproduce.

Figure 8-15 Crater Lake, in the U.S. state of Oregon, formed in the crater of an extinct volcano.

Next comes the *profundal* ("pro-FUN-dahl") *zone,* a layer of deep, open water where it is too dark for photosynthesis. Without sunlight and plants, oxygen levels are often low here. Fishes adapted to the lake's cooler and darker water are found in this zone.

The bottom layer of the lake is called the *benthic* ("BEN-thic") *zone,* inhabited mostly by decomposers, detritus feeders, and some species of fish (benthos). The benthic zone is nourished mainly by dead matter that falls from the littoral and limnetic zones and by sediment washing into the lake.

Some Lakes Have More Nutrients Than Others

Ecologists classify lakes according to their nutrient content and primary productivity. Lakes that have a small supply of plant nutrients are called **oligotrophic lakes**. This type of lake (Figure 8-15) is often deep and has steep banks. Glaciers and mountain streams supply water to many such lakes, bringing little in the way of sed-

Animated Figure 8-16 This diagram illustrates the distinct zones of life in a fairly deep temperate-zone lake. *Question:* How are deep lakes like tropical rain forests? (*Hint:* See Figure 7-14, p. 157.)

iment or microscopic life to cloud the water. These lakes usually have crystal-clear water and small populations of phytoplankton and fish species (such as smallmouth bass and trout). Because of their low levels of nutrients, these lakes have a low NPP.

Over time, sediments, organic material, and inorganic nutrients wash into most oligotrophic lakes, and plants grow and decompose to form bottom sediments. A lake with a large supply of nutrients is called a **eutrophic lake** (Figure 8-17). Such lakes typically are shallow and have murky brown or green water with high turbidity. Because of their high levels of nutrients, these lakes have a high NPP.

Human inputs of nutrients through the atmosphere and from urban and agricultural areas within a lake's watershed can accelerate the eutrophication of the lake. This process, called **cultural eutrophication**, often puts excessive nutrients into lakes. Many lakes fall somewhere between the two extremes of nutrient enrichment. They are called **mesotrophic lakes**.

Freshwater Streams and Rivers Carry Large Volumes of Water

In drainage basins, water accumulates in small streams that join to form rivers. Some rivers join with other rivers and, altogether, the planet's streams and rivers carry huge amounts of water from highlands to lakes and oceans (Figure 8-18).

Painted turtle

Blue-winged teal

Green frog

Muskrat

Pond snail

Plankton

Littoral zone

Diving beetle

Limnetic zone

Profundal zone

Northern pike

Benthic zone

Yellow perch

Bloodworms

© Cengage Learning

Figure 8-17 This eutrophic lake, found in western New York State, has received large flows of plant nutrients. As a result, its surface is covered with mats of algae.

In many areas, streams begin in mountainous or hilly areas, which collect and release water falling to the earth's surface as rain or as snow that melts during warm seasons. The downward flow of surface water and groundwater from mountain highlands to the sea typically takes place in three aquatic life zones characterized by different environmental conditions: the *source zone,* the *transition zone,* and the *floodplain zone* (Figure 8-18). Rivers and streams can differ somewhat from this generalized model.

In the narrow *source zone* (Figure 8-18, left), headwater streams are usually shallow, cold, clear, and swiftly flowing (Figure 8-18, photo inset). As this turbulent water flows and tumbles downward over obstacles such as rocks, waterfalls, and rapids, it dissolves large amounts of oxygen from the air. Most of these streams are not very productive because of a lack of nutrients and primary producers. Their nutrients come primarily from organic matter, mostly leaves, branches, and the bodies of living and dead insects that fall into the stream from nearby land.

The source zone is populated by cold-water fish species (such as trout in some areas), which need lots of dissolved oxygen. Many fishes and other animals in fast-flowing headwater streams have compact and flattened bodies that allow them to live under stones. Others have streamlined and muscular bodies that allow them to swim in the rapid, strong currents. Most of the plants in this zone are algae and mosses attached to rocks and other surfaces under water.

In the *transition zone* (Figure 8-18, center), headwater streams merge to form wider, deeper, and warmer streams that flow down gentler slopes with fewer obstacles. They can be more turbid (containing suspended sediment) and slower flowing than headwater streams, and they tend to have less dissolved oxygen. The warmer water and other conditions in this zone support more producers, as well as cool-water and warm-water fish species (such as black bass) with slightly lower oxygen requirements.

Figure 8-18 There are three zones in the downhill flow of water: the *source zone,* which contains *headwater* streams found in highlands and mountains (see photo inset); the *transition zone,* which contains wider, lower-elevation streams; and the *floodplain zone,* which contains rivers that empty into larger rivers or into the ocean.

As streams flow downhill, they shape the land through which they pass. Over millions of years, the friction of moving water has leveled mountains and cut deep canyons, and rocks and soil removed by the water have been deposited as sediment in low-lying areas. In these *floodplain zones* (Figure 8-18, right), streams join into wider and deeper rivers that flow across broad, flat valleys. Water in this zone usually has higher temperatures and less dissolved oxygen than water in the two higher zones. The slow-moving rivers sometimes support fairly large populations of producers such as algae and cyanobacteria, as well as rooted aquatic plants along the shores.

Because of increased erosion and runoff over a larger area, water in the floodplain zone often is muddy and contains high concentrations of suspended particulate matter (silt). The main channels of these slow-moving, wide, and murky rivers support distinctive varieties of fishes (such as carp and catfish), whereas their backwaters support species similar to those found in lakes. At its mouth, a river may divide into many channels as it flows through its *delta*—an area at the mouth of a river that was built up by deposited sediment and contains coastal wetlands and estuaries (Figure 8-7).

Coastal deltas and wetlands, as well as inland wetlands and floodplains, are important parts of the earth's natural capital (see Figure 1-3, p. 7). They absorb and slow the velocity of floodwaters from coastal storms, hurricanes, and tsunamis and provide habitats for a diversity of marine life (see the Case Study that follows).

CONSIDER THIS...

CONNECTIONS Stream Water Quality and Watershed Land

Streams receive most of their nutrients from bordering land ecosystems. Such nutrients come from falling leaves, animal feces, insects, and other forms of biomass washed into streams during heavy rainstorms or by melting snow. Chemicals and other substances flowing off the land can also pollute streams. Thus, the levels and types of nutrients and pollutants in a stream depend on what is happening in the stream's watershed.

CASE STUDY

River Deltas and Coastal Wetlands—Vital Components of Natural Capital Now in Jeopardy

Coastal deltas, mangrove forests, and coastal wetlands provide considerable natural protection against flood and wave damage from coastal storms, hurricanes, typhoons, and tsunamis.

When we remove or degrade these natural speed bumps and sponges, any damage from a natural disaster such as a hurricane or typhoon is intensified. As a result, flooding in places like New Orleans, Louisiana (USA), other parts of the U.S. Gulf Coast, and Venice, Italy, are largely self-inflicted unnatural disasters. For example, Louisiana, which contains about 40% of all coastal wetlands in the lower 48 states, has lost more than a fifth of

such wetlands since 1950 to oil and gas wells and other forms of coastal development.

Humans have built dams, levees, and hydroelectric power plants on many of the world's rivers to control water flows and to generate electricity. This helps to reduce flooding along rivers, but it also reduces flood protection provided by the coastal deltas and wetlands. Because river sediments are deposited in the reservoirs behind dams, the river deltas do not get their normal inputs of sediment to build them back up, and they naturally sink into the sea.

As a result, 24 of the world's 33 major river deltas are sinking rather than rising and their protective coastal wetlands are flooding, according to a 2009 study by geologist James Syvitski and his colleagues. The study found that 85% of the world's sinking deltas have experienced severe flooding in recent years, and that global delta flooding is likely to increase by 50% by the end of this century. This is because of dams and other human-made structures that reduce the flow of silt and also the projected rise in sea levels resulting from climate change. It poses a serious threat to the roughly 500 million people in the world who live on river deltas.

For example, the Mississippi River once delivered huge amounts of sediments to its delta each year. But the multiple dams, levees, and canals built in this river system funnel much of this sediment load through the wetlands and out into the Gulf of Mexico. Instead of building up delta lands, this causes them to *subside*, or to sink. Other human activities that are leading to such subsidence include the extraction of groundwater, oil, and natural gas. As many of the river delta's freshwater wetlands have been lost to this subsidence, saltwater from the Gulf has intruded and killed many plants that depended on the river water, further degrading this coastal aquatic system.

This subsidence helps to explain why the city of New Orleans, Louisiana (Figure 8-19), has long been 3 meters (10 feet) below sea level. Dams and levees were built to help protect the city from flooding. However, in 2005, the powerful winds and waves from Hurricane Katrina overwhelmed these defenses. They are being rebuilt, but subsidence will probably put New Orleans 6 meters (20 feet) below sea level at some point in the future. Add to this the reduced protection from coastal and inland wetlands and barrier islands, and you have a recipe for a major and much more damaging unnatural disaster if the area is struck by another major hurricane.

To make matters worse, global sea levels have risen almost 0.3 meters (1 foot) since 1900 and are projected to rise another 0.3–0.9 meter (1–3 feet) by the end of this century. This is because the projected warming of the atmosphere will also warm the ocean, causing its waters to expand, and melt glaciers and other land-based ice, which will add to the ocean's volume. Such a rising sea level would put many of the world's coastal areas, including New Orleans and most of Louisiana's present-day coast, under water (Figure 8-20).

Figure 8-19 Much of the U.S. city of New Orleans, Louisiana, was flooded by the storm surge that accompanied Hurricane Katrina, which made landfall just east of the city on August 29, 2005.

©NOAA

The good news is that we now understand how our building of dams and other structures on rivers can affect their deltas and associated coastal wetlands. We know how the engineering of rivers can degrade or eliminate the ecosystem services provided by these freshwater systems and the marine coastal systems they feed. Environmental scientists argue that we must use such knowledge to sustain, rather than degrade or destroy, these

GOOD NEWS

components of natural capital and their vital ecosystem and economic services.

Freshwater Inland Wetlands Are Vital Sponges

Inland wetlands are lands located away from coastal areas that are covered with freshwater all or part of the time—excluding lakes, reservoirs, and streams. They include *marshes* (Figure 8-21, left), *swamps* (Figure 8-21, right), and *prairie potholes* (which are depressions carved out by ancient glaciers). Other examples of inland wetlands are *floodplains*, which receive excess water during heavy rains and floods.

Some wetlands are covered with water year-round. Others, called *seasonal wetlands*, remain under water or are soggy for only a short time each year. The latter include prairie potholes, floodplain wetlands, and arctic tundra (see Figure 7-11, bottom, p. 153). Some can stay dry for years before water covers them again. In such cases, scientists must use the composition of the soil or the presence of certain plants (such as cattails, bulrushes, or red maples) to determine that a particular area is a wetland. Wetland plants are highly productive because of an abundance of nutrients available to them. Many wetlands are important habitats for game fishes, muskrats, otters, beavers, migratory waterfowl, and other bird species.

Inland wetlands provide a number of free ecosystem and economic services, which include:

- filtering and degrading toxic wastes and pollutants;
- reducing flooding and erosion by absorbing storm water and releasing it slowly, and by absorbing overflows from streams and lakes;
- helping to sustain stream flows during dry periods;
- helping to recharge groundwater aquifers;
- helping to maintain biodiversity by providing habitats for a variety of species;
- supplying valuable products such as fishes and shellfish, blueberries, cranberries, wild rice, and timber; and
- providing recreation for birdwatchers, nature photographers, boaters, anglers, and waterfowl hunters.

Figure 8-20 The areas in red represent projected coastal flooding that would result from a 1-meter (3-foot) rise in sea level due to projected climate change by the end of this century.

Used by permission from Jonathan Overpeck and Jeremy Weiss, University of Arizona

CONSIDER THIS...

THINKING ABOUT Inland Wetlands

Which two ecosystem services and which two economic services provided by inland wetlands do you believe are the most important? Why? List two ways in which our daily activities directly or indirectly degrade inland wetlands.

individuals matter 8.1

Alexandra Cousteau: Environmental Advocate, Filmmaker, and National Geographic Emerging Explorer

© Courtesy of Alexandra Cousteau

Alexandra Cousteau is proud of her heritage as granddaughter of Captain Jacques-Yves Cousteau and daughter of Philippe Cousteau. Her father and grandfather were legendary underwater explorers who brought the mysteries and wonders of the oceans into living rooms around the world with their films and books. She is also determined to build her own legacy as an advocate focused on water-related environmental issues.

The focus of Cousteau's life is to advocate the importance of conservation and sustainable management of water in order to preserve a healthy planet. Her global initiatives seek to inspire and empower individuals to protect not only the ocean and its inhabitants, but also the human communities that rely on freshwater resources. She seeks to make water one of the defining issues of this century, arguing: "We inhabit a water planet, and unless we protect, manage, and restore that resource, the future will be a very different place from the one we imagine today."

To that end, she is carrying on her family's proud tradition of storytelling. She says, "We evolved as a storytelling species, but the environmental community hasn't fully leveraged this approach. . . . That's why my grandfather was so successful. He was a master storyteller."

However, Cousteau is exploring a territory not even imagined by her grandfather—that of social networking and other emerging modes of communication. She is studying such trends along with video games and fantasy sports networks to see what makes them so popular and successful, and she believes that environmental advocates can use such new media tools to inform people about how their actions affect our water resources and the environments we inhabit. As an example, she cites the smart phone apps that help people to make sustainable choices in seafood when ordering from a restaurant menu or buying seafood at a grocery store.

One of Cousteau's major projects is Blue Legacy, an organization dedicated to "telling the story of our Water Planet" and shaping society's dialogue to include water as one of the defining issues of the 21st century by leading the conversation on the importance of "watershed-first thinking" and inspiring mainstream audiences to fully participate in the conservation and restoration of the water in their local communities. It is a source for articles and short films about a variety of issues related to water issues. Her website also features her many expeditions around the world, exploring and filming the intersection between water and our society. Led by Cousteau, these are explorations of ecosystems such as the Great Lakes and the Chesapeake Bay (background photo)—systems that are greatly stressed by various human activities.

Background photo: Lh8971/Dreamstime.com

a.

b.

Figure 8-21 This great white egret lives in an inland marsh in the Florida Everglades (left). This cypress swamp (right) is located in South Carolina.

8-5 How Have Human Activities Affected Freshwater Ecosystems?

CONCEPT 8-5
Human activities threaten biodiversity and disrupt ecosystem and economic services provided by freshwater lakes, rivers, and wetlands.

Human Activities Are Disrupting and Degrading Freshwater Systems

Human activities are disrupting and degrading many of the ecosystem and economic services provided by freshwater rivers, lakes, and wetlands (Concept 8-5) in four major ways. *First,* dams and canals restrict the flows of about 40% of the world's 237 largest rivers. This alters or destroys terrestrial and aquatic wildlife habitats along these rivers and in their coastal deltas and estuaries. By reducing the flow of sediments to river deltas, these structures also lead to degraded coastal wetlands and greater damage from coastal storms (Case Study, p. 181). *Second,* flood-control levees and dikes built along rivers disconnect the rivers from their floodplains, destroy aquatic habitats, and alter or degrade the functions of adjoining wetlands.

A *third* major human impact on freshwater systems comes from cities and farms, which add pollutants and excess plant nutrients to streams, rivers, and lakes. For example, runoff of nutrients into a lake (cultural eutrophication, Figure 8-17) causes explosions in the populations of algae and cyanobacteria, which deplete the lake's dissolved oxygen. When these organisms die and sink to the lake bottom, decomposers go to work and further deplete the oxygen in deeper waters. Fishes and other species may then die off, which can mean a major loss in biodiversity.

Fourth, many inland wetlands have been drained or filled to grow crops or have been covered with concrete, asphalt, and buildings. More than half of the inland wetlands estimated to have existed in the continental United States during the 1600s no longer exist. About 80% of lost wetlands were drained to grow crops. The rest were lost to mining, logging, oil and gas extraction, highway construction, and urban development. The heavily farmed U.S. state of Iowa has lost about 99% of its original inland wetlands.

This loss of natural capital has been an important factor in greater and more frequent flood damage in parts of the United States. Many other countries have suffered similar losses. For example, 80% of all inland wetlands in Germany and France have been destroyed.

When we look further into human impacts on aquatic systems in Chapter 11, we will also explore possible solutions to environmental problems that result from these impacts, as well as ways to help sustain aquatic biodiversity. This is an area that will offer great opportunities to young scientists and other professionals in the years to come. (See Individuals Matter 8.1.)

Big Ideas

- Saltwater and freshwater aquatic life zones cover almost three-fourths of the earth's surface, and oceans dominate the planet.

- The earth's aquatic systems provide important ecosystem and economic services.

- Certain human activities threaten biodiversity and disrupt ecosystem and economic services provided by aquatic systems.

Vlad61/Shutterstock.com

This chapter's **Core Case Study** pointed out the ecological and economic importance of the world's incredibly diverse coral reefs. They are living examples of the three scientific **principles of sustainability** (see Figure 1-2, p. 6 or back cover) in action. They thrive on solar energy, participate in the cycling of carbon and other chemicals, and sustain a great deal of aquatic biodiversity.

In this chapter, we have also seen that coral reefs and other aquatic systems are being severely stressed by a variety of human activities. Research shows that when such harmful human activities are reduced, some coral reefs and other endangered aquatic systems can recover fairly quickly.

As with terrestrial systems, scientists have made a start in understanding the ecology of the world's aquatic systems and how humans are degrading and disrupting the vital ecosystem and economic services they provide, but we still know far too little about the vital parts of the earth's life-support system. In studying these systems, scientists have again found that *in nature, everything is connected*. They argue that we urgently need more research on the components and workings of the world's aquatic life zones, on how they are interconnected, and on which systems are in the greatest danger of being disrupted by human activities.

We can take the lessons on how life in aquatic ecosystems has sustained itself for billions of years and use these scientific **principles of sustainability** to help sustain our own systems, as well as the ecosystems on which we depend. By relying more on solar energy and less on fossil fuels, we could drastically cut pollution of aquatic systems and CO_2 emissions that are causing ocean warming and ocean acidification. By reusing and recycling more of the materials and chemicals we use, we could further reduce pollution and disruption of the chemical cycling within aquatic systems. And by learning more about aquatic biodiversity and its importance, we could go a long way toward preserving it and sustaining its valuable ecosystem services.

Chapter Review

Core Case Study

1. What are **coral reefs** and why should we care about them? What is coral bleaching? What are the major threats to coral reefs?

Section 8-1

2. What are the two key concepts for this section? What percentage of the earth's surface is covered with water? What is an **aquatic life zone**? Distinguish between a **saltwater (marine) life zone** and a **freshwater life zone**, and give two examples of each. Define **plankton** and describe three types of plankton. Distinguish among **nekton**, **benthos**, and **decomposers** and give an example of each. List five factors that determine the types and numbers of organisms found in the three layers of aquatic life zones. What is **turbidity** and how does it occur? Describe one of its harmful impacts.

Section 8-2

3. What is the key concept for this section? What major ecosystem and economic services are provided by marine systems? What are the three major life zones in an ocean? Define and distinguish between the **coastal zone** and the **open sea**. Distinguish between an **estuary** and a **coastal wetland** and explain why each has high net primary productivity. Explain the ecological and economic importance of coastal marshes, mangrove forests, and sea-grass beds.

4. What is the **intertidal zone**? Distinguish between rocky and sandy shores and describe some of the organisms often found on each type of shoreline. Explain the importance of coral reefs. What is **ocean acidification** and why is it a threat to coral reefs? Describe the three major zones in the open sea. Why does the open sea have a low net primary productivity? What have scientists recently learned about the ocean's biodiversity?

Section 8-3

5. What is the key concept for this section? List five human activities that pose major threats to marine systems and eight human activities that threaten coral reefs. Explain why the Chesapeake Bay is an estuary in trouble. What is being done about some of its problems?

Section 8-4

6. What is the key concept for this section? Define **surface water**, **runoff**, and **watershed (drainage basin)**. What major ecosystem and economic services do freshwater systems provide? What is a **lake**? What four zones are found in deep lakes? Distinguish among **oligotrophic**, **eutrophic**, and **mesotrophic lakes**. What is **cultural eutrophication**?

7. Describe the three zones that a stream passes through as it flows from highlands to lower elevations. Explain how the building of dams and other structures on rivers can affect the river deltas and associated coastal wetlands. How do these effects in turn threaten human coastal communities?

8. Give three examples of **inland wetlands** and describe the ecological and economic importance of such wetlands.

Section 8-5

9. What is the key concept for this section? What are four ways in which human activities are disrupting and degrading freshwater systems? Describe losses of inland wetlands in the United States in terms of the area of wetlands lost and the resulting loss of ecosystem and economic services.

10. What are this chapter's three big ideas? How do coral reefs showcase the three scientific **principles of sustainability**? How can we use these three principles to help sustain the earth's vital aquatic ecosystems?

Note: Key terms are in **bold** type.

Critical Thinking

1. What are three steps that governments and private interests could take to protect the world's remaining coral reefs (**Core Case Study**)?

2. Can you think of any ways in which you might be contributing to the degradation of a nearby or distant aquatic ecosystem? Describe the system and how your actions might be affecting it. What are three things you could do to reduce your impact?

3. You are a defense attorney arguing in court for protecting a coral reef (**Core Case Study**) from harmful human activities. Give your three most important arguments for the defense of this ecosystem.

4. How would you respond to someone who argues that we should use the deep portions of the world's oceans to deposit our radioactive and other hazardous wastes because the deep oceans are vast and are located far away from human habitats? Give reasons for your response.

5. Why is increasing ocean acidification considered a very serious problem? If acid levels in the ocean rise sharply during your lifetime, how might this affect you? Can you think of ways in which you might be contributing to this problem? What could you do to reduce your impact?

6. Suppose a developer builds a housing complex overlooking a coastal marsh (Figure 8-8) and the result is pollution and degradation of the marsh. Describe the effects of such a development on the wildlife in the marsh, assuming at least one species is eliminated as a result.

7. Suppose you have a friend who owns property that includes a freshwater wetland and the friend tells you she is planning to fill the wetland to make more room for her lawn and garden. What would you say to this friend?

8. Congratulations! You are in charge of the world. What are the three most important features of your plan to help sustain the earth's aquatic biodiversity?

Doing Environmental Science

Using Google Earth, find a relatively undisturbed aquatic ecosystem somewhere on the planet that you can zoom in on to observe some features. For example, find a coastal wetland or a river system that seems to be undisturbed by human activities on adjoining land, such as farming or urban development. Write a description of what you see. Then find an example of a comparable system that has been disturbed by human activities and write a detailed comparison of these two areas. If possible, check in on these systems once during your course term and again at the end of the term and describe any changes to the system that you can observe. Suggest a hypothesis to explain major changes.

Global Environment Watch Exercise

Search for *Coral reefs* and use the topic portal to find information on **(a)** trends in the global rate of coral destruction; **(b)** what areas of the world are seeing rising rates and what areas are seeing falling rates; and **(c)** what is being done to protect them in various areas. Write a report on your findings.

Data Analysis

Some 45–53% of the world's shallow coral reefs have been destroyed or severely damaged (**Core Case Study**). A number of factors have played a role in this serious loss of aquatic biodiversity, including ocean warming, sediment from coastal soil erosion, excessive algal growth from fertilizer runoff, coral bleaching, rising sea levels, ocean acidification, overfishing, and damage from hurricanes.

In 2005, scientists Nadia Bood, Melanie McField, and Rich Aronson conducted research to evaluate the recovery of coral reefs in Belize from the combined effects of mass bleaching and Hurricane Mitch in 1998. Some of these reefs are in protected waters where no fishing is allowed. The researchers speculated that reefs in waters where no fishing is allowed should recover faster than reefs in waters where fishing is allowed. The graph to the left shows some of the data they collected from three highly protected (unfished) sites and three unprotected (fished) sites to evaluate their hypothesis. Study this graph and answer the questions below.

1. By about what percentage did the mean coral cover drop in the protected (unfished) reefs between 1997 and 1999?

2. By about what percentage did the mean coral cover drop in the protected (unfished) reefs between 1997 and 2005?

3. By about what percentage did the coral cover drop in the unprotected (fished) reefs between 1997 and 1999?

4. By about what percentage did the coral cover change in the unprotected (fished) reefs between 1997 and 2005?

5. Do these data support the hypothesis that coral reef recovery should occur faster in areas where fishing is prohibited? Explain.

This graph tracks the effects of restricting fishing on the recovery of unfished and fished coral reefs damaged by the combined effects of mass bleaching and Hurricane Mitch in 1998.

(Compiled by the authors with data from Melanie McField, et al., *Status of Caribbean Coral Reefs after Bleaching and Hurricanes in 2005*, NOAA, 2008. Report available at www.coris.noaa.gov/activities/caribbean_rpt/.)

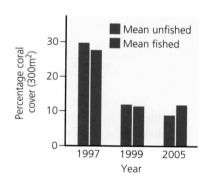

1. Which element below will be most helpful for reducing erosion from wave action to the coastline of an island in the ocean?
   a. A coral reef, because it is in the shallow water near the shoreline
   b. A sandy beach, because sand is easily replaced after a storm
   c. An offshore pelagic zone, because the large organisms that live there help lessen the size of the waves
   d. A sand dune, because it protects the oceanic coastline from damaging waves and wind
   e. A salt marsh, because this is where saltwater and freshwater mix

2. Zooplankton are animal larvae that
   I. are primary consumers that feed on phytoplankton.
   II. are secondary consumers that feed on other zooplankton.
   III. are photosynthetic bacteria responsible for most of the ocean surface's primary productivity.
   a. I only
   b. II only
   c. III only
   d. I and II only
   e. II and III only

3. Which of the below factors is NOT a primary determining factor for whether an organism is found on the surface or in the middle of the water column in the ocean?
   a. Temperature
   b. Dissolved oxygen
   c. Availability of light
   d. Salinity
   e. Nutrient availability

4. Which area below will have the lowest net primary productivity/km²?
   a. Estuaries, because most organisms cannot survive in water of varying salinity and depth
   b. Lake edges, because people often populate lake shorelines
   c. Ocean shorelines, because the waves sweep organisms out to the open sea
   d. Small streams, because they are not large enough to provide the necessary nutrients
   e. Open ocean, because of the lack of many nutrients such as nitrates and iron

5. The primary force responsible for tidal action is the
   a. gravitational pull of the moon, because although it is smaller than earth, it is very close to earth.
   b. rotation of the earth on its axis, because the rate of rotation is so fast.
   c. tilt of the earth on its axis, because ocean water will rush to the tilted side.
   d. gravitational pull of the sun, because although it is very far away, it is massive in size.
   e. uneven heating, because the heated water becomes less dense and is therefore more easily moved by the wind.

6. Which of the below is the correct path that a grain of sand would take as it is eroded from a mountain, carried downstream, and deposited on the ocean floor?
   a. Littoral zone → aphotic zone → estuary → benthos
   b. Estuary → littoral zone → pelagic zone → abyssal
   c. Pelagic zone → estuary → benthos → photic zone
   d. Benthos → estuary → profundal → limnetic
   e. Littoral zone → estuary → abyssal → limnetic

7. Where are photosynthetic organisms least likely to be found?
   a. In estuaries, where freshwater and saltwater mix and where there are many nutrients
   b. In the euphotic zone, with low nutrient levels and high oxygen levels
   c. In the abyssal zone, which is very cold and with little dissolved oxygen, where many organisms live and feed upon dead and decaying organic particles
   d. In the intertidal zone, where organisms must adapt to alternating high and low tides, as well as changing salinity levels
   e. In coral reefs, because the coral uses all the light and nutrients to survive

8. Which area of the oceans is most used by people for recreation?
   a. Abyssal plain, because scuba divers enjoy seeing the light-producing organisms and the shipwrecks
   b. Littoral zone, because it is near the shoreline and is easily accessible for recreational activities
   c. Upwelling zones, because the upward movement makes water sports more competitive
   d. Primary dune, because the slopes are enjoyed by skateboarders and they will not be hurt if they fall
   e. Profundal zone, because it is deep and dark and pleasant for water skiers

Questions 9–12 refer to the human impacts on eco-systems listed below.

   a. Coral reef damage
   b. Overfishing
   c. Coastal development
   d. Cultural eutrophication from release of excess nutrients
   e. Population growth and burning of fossil fuels

9. Causes large areas of little to no oxygen in the Chesapeake Bay and the Gulf of Mexico.

10. Causes large areas of beach loss or erosion or migration of beaches to new locations.

11. Caused by boats improperly anchoring.

12. Causes ocean warming and pH reduction, leading to coral bleaching.

9 Sustaining Biodiversity: Saving Species and Ecosystem Services

The last word in ignorance is the person who says of an animal or plant: "What good is it?" . . . If the land mechanism as a whole is good, then every part of it is good, whether we understand it or not. Harmony with land is like harmony with a friend; you cannot cherish his right hand and chop off his left.

ALDO LEOPOLD

Key Questions

9-1 What role do humans play in the loss of species and ecosystem services?

9-2 Why should we care about sustaining species and the ecosystem services they provide?

9-3 How do humans accelerate species extinction and degradation of ecosystem services?

9-4 How can we sustain wild species and their ecosystem services?

Endangered wild Siberian tiger.

In meadows, forests, farm fields, and gardens around the world, industrious honeybees (Figure 9-1) flit from one flowering plant to another. They spend their days collecting nectar and pollen that they take back to their hives where the typical bee lives with 30,000 to 100,000 other bees. They feed their young the protein-rich pollen collected from the flowers, and the adults live on the honey made from the collected nectar and stored in the hive's wax honeycombs.

Honeybees play a key role in providing us with one of nature's most important ecosystem services: *pollination,* the transfer of pollen within and among flowering plants that enables them to produce seeds and fruit. Bees pollinate many flower species and some of our most important food crops, including many vegetables, fruits, and tree nuts such as almonds. Globally, about one-third of the human food supply comes from insect-pollinated plants, and European honeybees are responsible for 80% of that pollination.

Many U.S. growers rent European honeybees from commercial beekeepers who together truck about 2.7 million hives to farms across the country to pollinate different crops at bloom times. These growers have largely replaced the earth's free pollination service provided by a diversity of bees and other pollinators with reliance mostly on a single bee species.

Some producers believe we need this industrialized pollination system in order to grow enough food. Others see such heavy dependence on a single bee species as a potentially dangerous violation of the earth's biodiversity **principle of sustainability**. They argue that this dependence will put food supplies at risk if the European honeybees decline.

In fact, bee experts assembled by the U.S. National Academy of Sciences have reported a 30% drop in populations of European honeybee in the United States since the 1980s. Since 2006, this situation has worsened in the United States and in parts of Europe as massive numbers of European honeybees have been disappearing from their colonies during the winter and not returning as expected in the spring—a phenomenon called **colony collapse disorder (CCD)**. In each year since 2006, about 30–50% of the European honeybee colonies in the United States have suffered from CCD. Researchers are searching for the causes of this problem and for ways to deal with it.

Scientists project that during this century, human activities, especially those that contribute to habitat loss and climate change, are likely to play a key role in the extinction of one-fourth to one-half of the world's identified plant and animal species (Figure 9-2). Many scientists view this threat to the earth's vital ecosystem services as one of the most serious and long-lasting environmental and economic problems we face. In this chapter, we discuss the causes of this problem and possible ways to deal with it.

Figure 9-1 European honeybee drawing nectar from a flower.

© DARLYNE A. MURAWSKI/National Geographic Creative

9-1 What Role Do Humans Play in the Loss of Species and Ecosystem Services?

CONCEPT 9-1

Species are becoming extinct 100 to 1,000 times faster than they were before modern humans arrived on earth, and by the end of this century, the extinction rate is projected to be 10,000 times higher than that background rate.

Extinctions Are Natural but Sometimes They Increase Sharply

When a species can no longer be found anywhere on the earth, it has suffered **biological extinction**. The disappearance of any species, and especially those that play keystone roles (see Chapter 4, pp. 94–96), is irreversible. The loss of a keystone species or a major reduction in its populations can lead to population declines or extinctions of species with strong connections to such species and to a breakdown in ecosystem services that depend on those connections.

Ecologists refer to such a series of changes as a *trophic cascade*. For example, a 2011 report by ecologist James A. Estes and his colleagues pointed out that sharp declines in the populations of top predators such as sea otters (see Chapter 5 Core Case Study, p. 102), lions, tigers, wolves, and some shark species can affect populations of the species on which they feed. Such declines can also result in degradation of habitats and ecosystem services such as chemical cycling and energy flows (see Chapter 4 Case Study, p. 95).

The extinction of many species in a relatively short period of geologic time is called a **mass extinction**. Geologic, fossil, and other records indicate that the earth has experienced five mass extinctions, when 50–95% of the world's species appear to have become extinct. After each mass extinction, the earth's overall biodiversity eventually returned to equal or higher levels, but each recovery required millions of years. Such long-term recovery is an example of the biodiversity **principle of sustainability** in action.

The causes of past mass extinctions are poorly understood but probably involved global changes in environmental conditions. Examples are sustained and significant global warming or cooling, large changes in sea level, and catastrophes such as multiple large-scale volcanic eruptions. One hypothesis is that the last mass extinction, which took place about 65 million years ago, occurred after a large asteroid hit the planet and spewed huge amounts of dust and debris into the atmosphere. This could have reduced the input of solar energy and cooled the planet long enough to wipe out the dinosaurs and many other forms of life.

Some Human Activities Hasten Extinctions and Threaten Ecosystem Services

Extinction is a natural process. Scientists who study it base much of their work on an estimated **background extinction rate**—the rate that existed before modern humans evolved some 200,000 years ago—which scientists believe was about 1 species per year for every 1 million wild species living on the earth.

However, scientific evidence indicates that extinction rates have risen in some areas as human populations have spread over most of the globe, destroying and degrading habitats, consuming huge quantities of resources, and creating large and growing ecological footprints (see Figure 1-13, p. 14). In some areas, these trends are threatening the ecosystem services that sustain our lives and economies. In the words of biodiversity expert Edward O. Wilson (see Individuals Matter 4.1, p. 82), "The natural world is everywhere disappearing before our eyes—cut to pieces, mowed down, plowed under, gobbled up, replaced by human artifacts."

Evidence is piling up that human activities are causing losses in global, regional, and local biodiversity at an increasing rate. Each year, the World Wildlife Fund publishes its annual Living Planet Index (LPI) based on its monitoring of populations of over 2,500 vertebrate species. Between 1970 and 2008, the global LPI fell by 28% with a 31% drop in temperate areas and a 61% drop in tropical areas.

Scientists from around the world who conducted the 2005 Millennium Ecosystem Assessment estimated that the current annual rate of species extinction is at least 100 to 1,000 times the estimated background extinction rate (Science Focus 9.1). We have identified about 2 million species so far, but scientists estimate that there are many millions more to identify. For simplicity, let's assume there are 10 million species on earth. Then, at the background extinction rate of 1 species per million per year, about 10 species would disappear naturally each year. However, at today's estimated rate of 100 to 1,000 times the background rate, we are losing between 1,000 and 10,000 species per year, or between 2 and 27 species every day, on average.

Biodiversity researchers project that during this century, the extinction rate is likely to rise to at least 10,000 times the background rate—mostly because of habitat loss and degradation, climate change, and other environmentally harmful effects of human activities (Concept 9-1). At this rate, if there are 10 million species on the earth, then about 100,000 species would be expected to disappear each year—an average of about 274 species per day or about 11 every hour. By the end of this century, most of the big carnivorous cats, including tigers (see chapter-opening photo), cheetahs, and lions, will probably exist only in zoos and

SCIENCE FOCUS 9.1

ESTIMATING EXTINCTION RATES

Figure 9-A Painting of a pair of North American passenger pigeons, which once were one of the world's most abundant bird species. They became extinct in the wild in 1913 mostly because of habitat loss and overhunting.

©LOUIS AGASSI FUERTES/National Geographic Creative

Scientists who try to catalog extinctions, estimate past extinction rates, and project future extinction rates face three problems. *First,* because the natural extinction of a species typically takes a very long time, it is difficult to document. *Second,* we have identified only about 2 million of the world's estimated 7–10 million and perhaps as many as 100 million species. *Third,* scientists know little about the ecological roles of most of the species that have been identified, or about how vulnerable they are to extinction and how their extinction might affect other species and the ecosystem services in the ecosystems where they are found.

One approach to estimating future extinction rates is to study records documenting past rates at which easily observable mammals and birds (Figure 9-A) have become extinct. Most of these extinctions have occurred since humans began to dominate the planet about 10,000 years ago, when we began the shift from hunting and gathering food in the wild to growing our food. This information can be compared with fossil records of extinctions that occurred before that time.

Another approach is to observe how reductions in habitat area affect extinction rates. The *species–area relationship,* studied by Edward O. Wilson (see Individuals Matter 4.1, p. 82) and Robert MacArthur, suggests that, on average, a 90% loss of land habitat in a given area can cause the extinction of about 50% of the species living in that area. Scientists use this model to estimate the number of current and future extinctions in patches or "islands" of shrinking wild habitat that are surrounded by degraded habitats or by rapidly growing human developments.

Scientists also use mathematical models to estimate the risk of a particular species becoming endangered or extinct within a certain period of time. These *population viability analysis* (PVA) models include factors such as trends in population size, past and projected changes in habitat availability, interactions with other species, and genetic factors.

Researchers know that their estimates of extinction rates are based on incomplete data and sampling, and on imperfect models. Thus, they are continually striving to get more and better data and to improve the models they use in order to estimate extinction rates and to project the effects of such extinctions on vital ecosystem services such as pollination (**Core Case Study**).

At the same time, they point to considerable and growing evidence that human activities have accelerated the rate of species extinction and that this rate is increasing. According to these biologists, arguing over the numbers and waiting to get better data and models should not delay our acting now to help prevent extinctions, along with the resulting threats to key ecosystem services, that result mostly from human activities.

Critical Thinking

Does the fact that extinction rates can only be estimated make them unreliable? Why or why not? (*Hint:* See Chapter 2, pp. 34–35.)

small wildlife sanctuaries. And most elephants and rhinoceroses will likely disappear from the wild, along with gorillas, chimpanzees, and orangutans. According to conservation biologist Thomas Lovejoy, we are experiencing the beginning of a tsunami of species extinction and degradation of ecosystem services that is likely to spread over much of the world during this century.

So why is this a big deal? According to biodiversity researchers Edward O. Wilson and Stuart Pimm, at this extinction rate, at least 25% and as many as 50% of the world's current 2 million identified animal and plant species could vanish from the wild by the end of this century, along with many of the millions of unidentified species. This would amount to a sixth mass extinction caused primarily by human activities with much of it taking place within just one century.

These experts note that with the loss of such a huge portion of the planet's biodiversity, we would also likely lose whole ecosystems that depend on the vanishing species, along with the vital ecosystem services they provide, including air and water purification, natural pest control, and pollination (**Core Case Study**). According to the 2005 Millennium Ecosystem Assessment, 15 of 24 major ecosystem services are in decline. If such estimates are only half correct, we can see why many biologists warn that such a massive loss of biodiversity and ecosystem services within the span of a single human lifetime is one of the most important and long-lasting environmental and economic problems we face.

CONSIDER THIS. . .

THINKING ABOUT Extinction

How might your lifestyle change if human activities contribute to the extinction of up to half of the world's identified species during this century? How might this affect the lives of any children or grandchildren you might have? List two aspects of your lifestyle that contribute to this threat to the earth's natural capital.

In fact, Wilson, Pimm, and other extinction experts consider a projected extinction rate of 10,000 times the background extinction rate to be low, for several reasons. *First,* both the rate of extinction and the resulting threats to ecosystem services are likely to increase sharply during the next 50–100 years because of the harmful environmental impacts of the rapidly growing human population and its growing use of resources per person (see Figure 1-14, p. 15).

Second, the current and projected extinction rates in the world's *biodiversity hotspots*—areas that are highly endangered centers of biodiversity—are much higher than the global average. Biodiversity expert Norman Myers and several other researchers urge us to focus our efforts on lowering the rates of extinction in such areas as soon as possible. They see such emergency action as the best and quickest way to prevent much of the earth's biodiversity from being lost during this century. Other scientists urge us to identify and protect areas where vital ecosystem services such as pollination by honeybees (**Core Case Study**) and topsoil formation are being threatened.

Third, we are eliminating, degrading, fragmenting, and simplifying many biologically diverse environments—including tropical forests, coral reefs, wetlands, and estuaries—that serve as potential sites for the emergence of new species. Thus, in addition to greatly increasing the rate of extinction, we may be limiting the long-term recovery of biodiversity by eliminating these places where new species can evolve. In other words, we are also creating a *speciation crisis*. (See the online Guest Essay by Norman Myers on this topic.) Based on what scientists have learned about the recovery of biodiversity after past mass extinctions, it will take 5 million to 10 million years for the earth's processes to replace the projected number of species that are likely to go extinct during this century primarily because of human activities.

In addition, Philip Levin, Donald Levin, and other biologists argue that, while our activities are likely to reduce the speciation rates for some species, they might increase the speciation rates for other rapidly reproducing species such as weeds and rodents, as well as cockroaches and many other species of insects. Rapidly expanding populations of such species could crowd and compete with various other species, further accelerating their extinction and also threatening key ecosystem services.

Endangered and Threatened Species Are Ecological Smoke Alarms

Biologists classify species that are heading toward biological extinction as either *endangered* or *threatened*. An **endangered species** has so few individual survivors that the species could soon become extinct. A **threatened species** (also known as a *vulnerable species*) still has enough remaining individuals to survive in the short term, but because of declining numbers, it is likely to become endangered in the near future.

Figure 9-2 shows four of the 20,219 species listed in 2012 by the International Union for Conservation of Nature (IUCN) as critically endangered, endangered, or

a. Mexican gray wolf: About 42 in the forests of Arizona and New Mexico

b. California condor: 226 in the southwestern United States (up from 9 in 1986)

c. Whooping crane: 437 in North America

d. Sumatran tiger: No more than 500 on the Indonesian island of Sumatra

Figure 9-2 *Endangered natural capital:* These four critically endangered species are threatened with extinction, largely because of human activities. The number below each photo indicates the estimated total number of individuals of that species remaining in the wild, as of 2012.

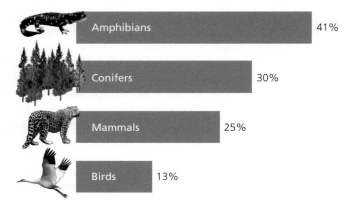

Amphibians	41%
Conifers	30%
Mammals	25%
Birds	13%

Figure 9-3 Endangered natural capital: Comparison of the percentages of various types of known species that are threatened with extinction hastened by human activities as of 2012 (**Concept 9-1**).
Question: Why do you think so many of the world's amphibians are threatened with extinction? (See Chapter 4 Core Case Study, p. 78.)

(Compiled by the authors using data from 2012 IUCN Red List of Threatened Species.)

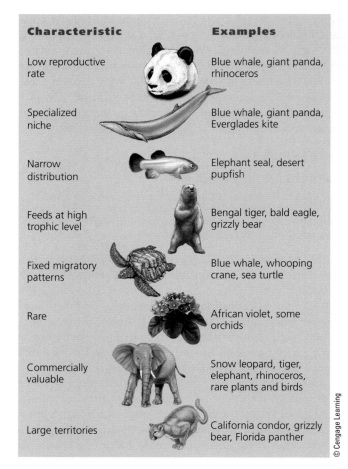

Characteristic	Examples
Low reproductive rate	Blue whale, giant panda, rhinoceros
Specialized niche	Blue whale, giant panda, Everglades kite
Narrow distribution	Elephant seal, desert pupfish
Feeds at high trophic level	Bengal tiger, bald eagle, grizzly bear
Fixed migratory patterns	Blue whale, whooping crane, sea turtle
Rare	African violet, some orchids
Commercially valuable	Snow leopard, tiger, elephant, rhinoceros, rare plants and birds
Large territories	California condor, grizzly bear, Florida panther

Figure 9-4 Certain characteristics can put a species in greater danger of becoming extinct.

© Cengage Learning

threatened. This is an 92% increase over the number listed in 1996. The real number of species in trouble is very likely much higher. This annual assessment by the IUCN shows that some types of species are threatened with extinction hastened by human activities more than others are (Figure 9-3). In 2012 there were 1,143 U.S. species listed as endangered or threatened under the U.S. Endangered Species Act. Again, the real number is very likely much higher.

Some species have characteristics that increase their chances of becoming extinct (Figure 9-4). As biodiversity expert Edward O. Wilson puts it, "The first animal species to go are the big, the slow, the tasty, and those with valuable parts such as tusks and skins."

Species can also become *regionally extinct* in the areas where they are normally found, as well as *functionally extinct* when their populations crash to the point where they can no longer play their functional roles in an ecosystem. The latter is also called *extinction of ecological interactions*, because when a species' numbers drop to a certain point, its interactions with other species are lost or greatly diminished. Important ecosystem services that depend on these interactions might also then be lost or diminished, and this is often difficult to detect until it is too late.

For example, the American alligator is a keystone species in its marsh and swamp habitats of the southeastern United States. (See Case Study, Chapter 4, p. 95.) When its numbers dwindled in the 1960s, certain ecosystem services, such as the building of gator nests, were not performed, and bird species that depended on these nesting sites also declined. After the alligator was placed on the U.S. endangered species list, it made a strong comeback. The ecosystem services it performed were once again available, and its ecosystems recovered.

9-2 Why Should We Care about Sustaining Species and the Ecosystem Services They Provide?

CONCEPT 9-2
We should avoid speeding up the extinction of wild species because of the ecosystem and economic services they provide, because it can take millions of years for nature to recover from large-scale extinctions, and because many people believe that species have a right to exist regardless of their usefulness to us.

Species Are a Vital Part of the Earth's Natural Capital

According to the World Wildlife Fund, only 50,000–60,000 orangutans (Figure 9-5) remain in the wild, most of them in the tropical forests of Indonesia. These highly intelligent animals are disappearing at a rate of more than 1,000–2,000

Figure 9-5 **Natural capital degradation:** These endangered orangutans depend on a rapidly disappearing tropical forest habitat. **Question:** What difference will it make if human activities hasten the extinction of the orangutan?

per year because of illegal smuggling and the clearing of their tropical forest habitat to make way for plantations of oil palms that supply palm oil used in cosmetics, cooking, and the production of biodiesel fuel. An illegally smuggled, live orangutan sells for a street price of up to $10,000. Without urgent protective action, the endangered orangutan may disappear in the wild within the next two decades.

Does it matter that orangutans might soon become extinct mostly because of human activities? Does it matter if some unknown plant or insect in a tropical forest meets the same fate? If all species eventually become extinct, why should we worry about the rate of extinction? New species eventually evolve through speciation to take the places of those species lost through mass extinctions, so why should we care if we greatly speed up the extinction rate over the next 50–100 years?

According to biologists, there are three major reasons why we should work to prevent our activities from causing or hastening the extinction of other species. *First,* the world's species provide vital *ecosystem services* (see Figure 1-3, p. 7) that help to keep us alive and support our economies (Concept 9-2). For example, we depend on honeybees (**Core Case Study**) and other insects for pollination of many food crops and on certain bird species for natural pest control. When species disappear, we can lose such services, and at some point, such losses can threaten our own health and our economies in the affected areas.

Species also depend on other species within their food webs and, thus, by eliminating any species or sharply reducing its populations, especially a species that plays a keystone role (see Chapter 4, pp. 94–95), we can speed up the extinction of other species. This is another way in which, by hastening extinction, we can upset an ecosystem and degrade its important ecosystem services.

🔍 CONSIDER THIS. . .

CONNECTIONS Species and Ecosystem Services

Plant and animal species provide us with services in ways that you might not expect. For example, those that live in streams help to purify the flowing water. Trees and other forest plants produce oxygen, without which we could not survive. And earthworms aerate topsoil that we use for growing our food. These and other amazing ecosystem services are free and available around the clock for us as long as we don't degrade them through pollution or overuse.

Many species also contribute to *economic services* on which we depend (Concept 9-2). For example, various plant species provide economic value as food crops, and from trees, we produce fuelwood, lumber, and paper. *Bioprospectors* search tropical forests and other ecosystems to find plants and animals that scientists can use to make medicinal drugs (Figure 9-6). According to a United Nations University report, 62% of all cancer drugs were derived from the discoveries of bioprospectors. Despite their economic and medicinal potential, less than 0.5% of the world's known plant species have been examined for their medicinal properties.

Rauvolfia
Rauvolfia sepentina,
Southeast Asia
Anxiety, high
blood pressure

Pacific yew
Taxus brevifolia,
Pacific Northwest
Ovarian cancer

Foxglove
Digitalis purpurea,
Europe
Digitalis for heart failure

Cinchona
Cinchona ledogeriana,
South America
Quinine for malaria treatment

Rosy periwinkle
Cathranthus roseus,
Madagascar
Hodgkin's disease,
lymphocytic leukemia

Neem tree
Azadirachta indica,
India
Treatment of many
diseases, insecticide,
spermicide

© Cengage Learning

Figure 9-6 Natural capital: These plant species are examples of *nature's pharmacy*. Once the active ingredients in the plants have been identified, scientists can usually produce them synthetically. The active ingredients in nine of the ten leading prescription drugs originally came from wild organisms.

In 2011, scientists estimated that as many as 10,000 *phytochemicals*—certain chemicals such as antioxidants that occur naturally in plants—have the potential to slow aging, reduce pain, and help us to reduce our weight, prevent various cancers, and control diseases such as diabetes. Many of these potentially beneficial chemicals are available as over-the-counter supplements that are far less expensive than prescription drugs. However, more research is needed to measure their benefits.

By preserving species and their habitats, we can also gain economic benefits from wildlife tourism, or *ecotourism*, which generates more than $1 million per minute in tourist expenditures, worldwide. Conservation biologist Michael Soulé estimates that a male lion living to age 7 generates about $515,000 from tourism in Kenya, but only about $1,000 if it is killed for its skin. Ecotourism is thriving because people enjoy wildlife (Figure 9-7). **GREEN CAREER:** Ecotourism guide

GOOD NEWS

A *second* reason for preventing extinctions caused or hastened by human activities is that analysis of past mass extinctions indicates it will take 5 million to 10 million years for natural speciation to rebuild the biodiversity that is likely to be lost during this century. As a result, any grandchildren we might have and many thousands of future generations are unlikely to be able to depend on the life-sustaining biodiversity and vital ecosystem services that currently support our well-being and our economies.

Third, many people believe that wild species have a right to exist, regardless of their usefulness to us (**Concept 9-2**). According to the stewardship view, we have a responsibility to protect the earth's species—our only known living companions in the universe—from becoming extinct as a result of human activities, and to prevent

ROY TOFT/National Geographic Creative

Figure 9-7 Many species of wildlife such as this endangered hyacinth macaw in Mato Grosso, Brazil, are sources of beauty and pleasure. It is endangered because of habitat loss and illegal capture in the wild by pet traders.

the degradation of the world's ecosystem services and thus the human life-support system.

This ethical viewpoint raises a number of challenging questions. Since we cannot save all species from the harmful consequences of our actions, we have to make choices about which ones to protect. Should we protect more animal species than plant species and, if so, which ones should we protect? Some people support protecting familiar and appealing species such as elephants, whales, tigers, and orangutans (Figure 9-5), but care much less about protecting plants or insects that serve as the base of the food supply for many other species (**Core Case Study**). Others might think little about getting rid of species that most people fear or hate, such as mosquitoes, cockroaches, disease-causing bacteria, snakes, sharks, and bats.

Some scientists argue for halting activities that hasten extinctions because no one has ever seen or studied most of the life-sustaining biodiversity that is being lost. To biologist Edward O. Wilson, carelessly eliminating species is like burning millions of books that we have never read.

9-3 How Do Humans Accelerate Species Extinction and Degradation of Ecosystem Services?

CONCEPT 9-3
The greatest threats to species and ecosystem services are (in order) loss or degradation of habitat, harmful invasive species, human population growth, pollution, climate change, and overexploitation.

Loss of Habitat Is the Single Greatest Threat to Species: Remember HIPPCO

Biodiversity researchers summarize the most important direct causes of extinction and threats to ecosystem services using the acronym **HIPPCO**: **H**abitat destruction, degradation, and fragmentation; **I**nvasive (nonnative) species; **P**opulation growth and increasing use of resources; **P**ollution; **C**limate change; and **O**verexploitation (Concept 9-3).

According to biodiversity researchers, the greatest threat to wild species is habitat loss (Figure 9-8), degradation, and fragmentation. Specifically, deforestation in tropical areas (see Chapter 3 opening photo and Figure 3-1, p. 52) is the greatest threat to species and to the ecosystem services they provide, followed by the destruction and degradation of coastal wetlands and coral reefs (see Chapter 8 opening photo, pp. 166–167), the plowing of grasslands (see Figure 7-12, p. 154), and the pollution of streams, lakes, and oceans.

Island species—many of them found nowhere else on earth—are especially vulnerable to extinction when their habitats are destroyed, degraded, or fragmented, because they have nowhere else to go. This is why the collection of islands that make up the U.S. state of Hawaii is America's "extinction capital"—with 63% of its species at risk.

Habitat fragmentation occurs when a large, intact area of habitat such as a forest or natural grassland is divided, typically by roads, logging operations, crop fields, and urban development, into smaller, isolated patches or *habitat islands*. This process can decrease tree cover in forests and block animal migration routes. It can also divide populations of a species into smaller, increasingly isolated groups that are more vulnerable to predators, competitor species, disease, and catastrophic events such as storms and fires. In addition, habitat fragmentation creates barriers that limit the abilities of some species to disperse and colonize new areas, to locate adequate food supplies, and to find mates.

Most national parks and other nature reserves are habitat islands, many of them surrounded by potentially damaging logging and mining operations, coal-burning power plants, industrial activities, and human settlements. Freshwater lakes are also habitat islands that are especially vulnerable to the introduction of harmful invasive species and pollution from human activities. Can you think of other examples of habitat islands?

We Have Moved Disruptive Species into Some Ecosystems

After habitat loss and degradation, the next biggest threat to animal and plant extinctions and the ecosystem services they provide is the deliberate or accidental introduction of harmful species into ecosystems (Concept 9-3).

Many introductions of nonnative species have been beneficial to us. According to a study by ecologist David Pimentel, nonnative species such as corn, wheat, rice, and other food crops, as well as some species of cattle, poultry, and other livestock, provide more than 98% of the U.S. food supply. Similarly, nonnative tree species are grown in about 85% of the world's tree plantations. Some deliberately introduced species have helped to control pests. And highly beneficial European honeybees (**Core Case Study**) were brought to North America in the 1600s by English settlers who harvested the bees' honey and used the wax from their hives to make candles.

The problem is that, in their new habitats, some introduced species do not face the natural predators, competi-

GOOD NEWS

Indian Tiger

☐ Range 100 years ago
■ Range today

Black Rhino

☐ Range in 1700
■ Range today

African Elephant

☐ Probable range 1600
■ Range today

Asian or Indian Elephant

☐ Former range
■ Range today

(Compiled by the authors using data from International Union for Conservation of Nature and World Wildlife Fund.)

Animated Figure 9-8 Natural capital degradation: These maps reveal the reductions in the ranges of four wildlife species, mostly as the result of severe habitat loss and fragmentation and illegal hunting for some of their valuable body parts. **Question:** Would you support expanding these ranges even though this would reduce the land available for human habitation and farming? Explain.

tors, parasites, viruses, bacteria, or fungi that had helped to control their numbers in their original habitats. Such nonnative species can thus crowd out populations of many native species, disrupt ecosystem services, cause human health problems, and lead to economic losses. When this happens the nonnative species are viewed as harmful *invasive species*. Invasive species rarely cause the global extinction of other species, but they can cause population declines and local and regional extinctions of some native species.

🔍 **CONSIDER THIS. . .**

CONNECTIONS Giant Snails and Meningitis

In 1988, the giant East African land snail was imported to Brazil from East Africa as a cheap substitute for conventional escargot (snails), used as a source of food. It is the world's largest land snail, growing to the size of a human fist and weighing 1 kilogram (2.2 pounds) or more, and it can feed on at least 500 different types of plants. Mating adults lay about 1,200 eggs a year. When export prices for escargot fell, breeders dumped the imported snails into forests and other natural systems. Since then, they have spread widely, devouring many native plants and food crops such as lettuce. They also can carry rat lungworm, a parasite that burrows into the human brain and causes potentially lethal meningitis. Authorities eventually banned imports of the snail, but so far, this invasive species has been unstoppable.

Figure 9-9 shows some of the 7,100 or more invasive species that, after being deliberately or accidentally introduced into the United States, have caused ecological and economic harm. According to the U.S. Fish and Wildlife Service, about 40% of the species listed as endangered in the United States and 95% of those in the U.S. state of Hawaii are on the list because of threats from invasive species.

In 2009, Achim Steiner, head of the UN Environment Program (UNEP), and environmental scientist David Pimentel estimated that, globally, invader species cause $1.4 trillion a year in economic and ecological damages—an average of $2.7 million a minute—and the damages are rising rapidly. An example of such damage can be found in the story of the deliberately introduced kudzu vine (see the following Case Study).

CASE STUDY
The Kudzu Vine and Kudzu Bugs

An example of a deliberately introduced plant species is the *kudzu* ("CUD-zoo") *vine,* which in the 1930s was imported from Japan and planted in the southeastern

United States to help control soil erosion. It worked too well. Kudzu does control erosion, but it grows so rapidly that it engulfs hillsides, gardens, trees, stream banks, cars (Figure 9-10), and anything else in its path. Dig it up or burn it, and it still keeps spreading. It is very difficult to kill, even with the use of grazing goats and herbicides, which can also damage other plants and contaminate water supplies. Scientists have found a common fungus that can kill kudzu within a few hours, but they need to investigate any harmful side effects it may have.

This plant—sometimes called "the vine that ate the South"—has spread throughout much of the southeastern United States. It could spread to the north if the climate gets warmer as scientists project.

Kudzu is considered a menace in the United States, but Asians use a powdered kudzu starch in beverages, gourmet confections, and herbal remedies for a range of diseases. Almost every part of the kudzu plant is edible. Its leaves are delicious when deep-fried and contain high levels of vitamins A and C. Also, ingesting small amounts of kudzu powder can lessen one's desire for alcohol, and thus it could be used to reduce alcoholism and binge

Deliberately Introduced Species

Purple loosestrife

African honeybee ("Killer bee")

Kudzu

Nutria

European wild boar (Feral pig)

Accidentally Introduced Species

Sea lamprey (attached to lake trout)

Argentina fire ant

Burmese python

Formosan termite

Zebra mussel

© Cengage Learning

Figure 9-9 These are some of the estimated 7,100 harmful invasive species that have been deliberately or accidentally introduced into the United States.

Figure 9-10 Kudzu has grown over this car in the U.S. state of Georgia.

drinking. And although kudzu can engulf and kill trees, it might eventually help to save some of them. Researchers at the Georgia Institute of Technology have found that kudzu could be used in place of trees as a source of fiber for making paper. GOOD NEWS

The brown, pea-sized Kudzu bug is another invasive species that was imported from Japan. It breeds in and feeds on patches of kudzu, and it can help to reduce the spread of the vine. However, it spreads even more rapidly than the kudzu vine, and it also feeds on soybeans and thus could pose a major threat to soy crops. When it is disturbed, the bug emits a chemical that can irritate human skin and stain it yellow. It smells like industrial-strength drain cleaner, which is why it is also called a stinkbug.

The very adaptable Kudzu bug is a strong flier and an excellent hitchhiker, catching rides on vehicles that help it to spread to new areas. Since its arrival in 2009 in Georgia, it has spread rapidly throughout much of North Carolina and South Carolina, is moving north and west, and probably will soon be found in areas where soybeans are grown. Some pesticides can kill this bug, but might end up boosting their numbers by promoting genetic resistance to the pesticides (see Figure 4-6, p. 84). Researchers hope to change this bug through genetic engineering in such a way that it will stop eating soybeans. They are also evaluating the use of a wasp whose larvae attack kudzu bug embryos. However, scientists see no immediate way to eradicate this rapidly expanding invader species.

Some Accidentally Introduced Species Can Disrupt Ecosystems

Many unwanted nonnative invaders arrive from other continents as stowaways on aircraft, in the ballast water of tankers and cargo ships, and as hitchhikers on imported products such as wooden packing crates. Cars and trucks can also spread the seeds of nonnative plant species

embedded in their tire treads. Many tourists return home with living plants that can multiply and become invasive. Some of these plants might also contain insects that can invade new areas, multiply rapidly, and threaten crops.

In the 1930s, the extremely aggressive Argentina fire ant (Figure 9-9) was accidentally introduced into the United States in Mobile, Alabama. The ants may have arrived on shiploads of lumber or coffee imported from South America. They can float on water and have no natural predators in the southern United States where they have spread rapidly. Now, the insect has stowed away on exported goods in shipping containers and has invaded other countries, including China, Taiwan, Malaysia, and Australia.

When these ants invade an area, they can wipe out as much as 90% of native ant populations. Spreading mounds containing fire ant colonies are found in many fields and yards in the southeastern United States. Walk on one of these mounds, and as many as 100,000 ants may swarm out of their nest to simultaneously attack you with painful, burning stings. They have killed deer fawns, ground-nesting birds, baby sea turtles, newborn calves, pets, and at least 80 people who were allergic to their venom—some of them infants and elderly people who could not escape the ants.

Widespread pesticide spraying in the 1950s and 1960s temporarily reduced fire ant populations. But this chemical warfare actually hastened the advance of the rapidly multiplying fire ants by reducing populations of many native ant species. Even worse, it promoted development of genetic resistance to pesticides in the fire ants through natural selection (see Figure 4-6, p. 84).

In 2009, pest management scientist Scott Ludwig reported some success in using tiny parasitic flies to reduce fire ant populations. The flies dive-bomb the fire ants and inject eggs inside them, from which their larvae hatch and, within 6 months, eat away the brains of the ants. Then the parasitic fly emerges looking for more fire ants to attack and kill. The researchers say that the flies do not attack native ant species. But more research is needed to see how well this approach will work. GOOD NEWS

In recent years, several U.S. southern states have been invaded by biting, flea-sized *hairy crazy ants,* which are very resistant to most pesticides. They could help to wipe out fire ants but they are harder to control than fire ants. They also invade beehives and thus could be playing a role in honeybee declines in the United States (**Core Case Study**).

CASE STUDY

Burmese Pythons Are Eating Their Way through the Florida Everglades

Burmese pythons, along with African pythons and several species of boa constrictors, have been accidentally introduced in Everglades National Park in the U.S. state of Florida. About a million of these snakes, imported from

Figure 9-11 University of Florida researchers hold a 4.6-meter-long (15-foot-long), 74-kilogram (162-pound) Burmese python captured in Everglades National Park shortly after it had eaten a 1.8-meter-long (6-foot-long) American alligator.

and South African pythons, and yellow anacondas, as well as to move them across state lines. In 2013, more than 1,300 people took part in a 1-month contest to find and remove Burmese pythons in Florida's wetlands.

Wildlife officials estimate that there are many thousands of pythons and constrictors living in the Everglades, and their numbers are increasing rapidly. Research indicates that predation by these snakes is altering the complex food web and ecosystem services of the Everglades. This is an excellent example of what can happen if an invasive top predator ends up in an ecosystem where it has no natural enemies.

Prevention Is the Best Way to Reduce Threats from Invasive Species

Once a harmful nonnative species becomes established in an ecosystem, its removal is almost impossible—somewhat like trying to collect smoke after it has come out of a chimney. Americans are paying more than $160 billion a year to eradicate or control an increasing number of invading species, with not much success. Clearly, the best way to limit the harmful impacts of nonnative species is to prevent them from being introduced into ecosystems.

Scientists suggest several ways to do this, including:

- Funding a massive research program to identify the major characteristics of successful invaders, the types of ecosystems that are vulnerable to invaders, and the natural predators, parasites, bacteria, and viruses that could be used to control populations of established invaders.
- Greatly increasing ground surveys and satellite observations to track invasive plant and animal species, and developing better models for predicting how they will spread and what harmful effects they might have.
- Identifying major harmful invader species and establishing international treaties banning their transfer from one country to another, as is now done for endangered species, while stepping up inspection of imported goods to enforce such bans.
- Requiring cargo ships to discharge their ballast water and to replace it with saltwater at sea before entering ports, or to sterilize such water or to pump nitrogen into the water to displace dissolved oxygen and kill most invader organisms.
- Educating the public about the effects of releasing exotic plants and pets into the environment near where they live.

Africa and Asia, have been sold as pets. After learning that these reptiles do not make good pets, some owners have dumped them into the wetlands of the Everglades.

The Burmese python (Figure 9-11) can live 20–25 years, growing as long as 5 meters (16 feet). It can weigh as much 77 kilograms (170 pounds) and be as big around as a telephone pole. Pythons are hard to find and kill and they reproduce rapidly. They have huge appetites and feed at night, eating a variety of birds and mammals and occasionally other reptiles, including the American alligator. They seize their prey with their sharp teeth, wrap themselves around the prey, and squeeze them to death before feeding on them. They have also been known to eat pet cats and dogs, small farm animals, and geese.

According to a 2012 National Academy of Sciences peer-reviewed study, since 1993, the rapidly growing population of the Burmese pythons in the Everglades has greatly depleted populations of marsh and eastern cottontail rabbits, red and gray foxes, raccoons, Virginia opossums, and white-tailed deer. The pythons also eat a variety of bird species (some of them endangered) and American alligators—a keystone species and top predator in the Everglades ecosystem (see Chapter 4, Case Study, pp. 95–96).

Researchers say that the Burmese python population in Florida's wetlands cannot be controlled. Some fear that the species could spread to other swampy wetlands in the southern half of the United States by finding natural routes to, or by being released to other wetlands by pet owners. In 2012, the U.S. Department of the Interior made it illegal to import Burmese pythons, North

What Can You Do?

Controlling Invasive Species

- Do not buy wild plants and animals or remove them from natural areas.

- Do not release wild pets in natural areas.

- Do not dump aquarium contents or unused fishing bait into waterways or storm drains.

- When camping, use only local firewood.

- Brush or clean pet dogs, hiking boots, mountain bikes, canoes, boats, motors, fishing tackle, and other gear before entering or leaving wild areas.

Figure 9-12 **Individuals matter:** Here are some ways to prevent or slow the spread of harmful invasive species. **Questions:** Which two of these actions do you think are the most important to take? Why? Which of these actions do you plan to take?

Figure 9-12 shows some of the things you can do to help prevent or slow the spread of harmful invasive species.

Population Growth, High Rates of Resource Use, Pollution, and Climate Change Can Cause Species Extinctions

Past and projected *human population growth* and rising rates of *resource use per person* have greatly expanded the human ecological footprint (see Figure 1-13, p. 14). This has eliminated, degraded, and fragmented vast areas of wildlife habitat (Figure 9-8). Acting together, these two growth factors have caused the extinction of many species (Concept 9-3).

Pollution also threatens some species with extinction (Concept 9-3), as has been shown by the unintended effects of certain pesticides. According to the U.S. Fish and Wildlife Service, each year, pesticides kill about one-fifth of the European honeybee colonies that pollinate almost a third of U.S. food crops (**Core Case Study** and Science Focus 9.2). According to the U.S. Fish and Wildlife Service, pesticides also kill more than 67 million birds and 6–14 million fish each year, and they threaten about 20% of the country's endangered and threatened species.

During the 1950s and 1960s, populations of fish-eating birds such as ospreys, brown pelicans, and bald eagles plummeted. A chemical derived from the pesticide DDT was magnified as it moved up through their food web through processes called *bioaccumulation* and *biomagnification* (Figure 9-13). The chemical made these top predator birds' eggshells so fragile that they could not reproduce successfully. Also hard hit in those years, in other ecosystems, were such predatory birds as the prairie falcon, sparrow hawk, and peregrine falcon, which help to control populations of rabbits, ground squirrels, and other crop eaters. Since the U.S. ban on DDT in 1972, most of these bird species have made a comeback.

According to a study by Conservation International, projected *climate change* could help to drive a quarter to half of all land animals and plants to extinction by the end of this century. Scientific studies indicate that the polar bear (see Case Study that follows) is threatened because of higher temperatures and melting sea ice in its polar habitat. Similarly, other species, especially some marine species such as coral polyps, are threatened by a warming environment that is making their habitats unfit for their survival.

However, a 2013 research article by ecologists Roland Jánsson and Christer Nilsson indicated that as global warming expands temperate climate zones, 43 of the 61 species they studied are likely to expand their habitats. On the other hand, populations of some cold-weather species such as the lemming and Artic fox are likely to decrease but are unlikely to become extinct.

Figure 9-13 *Bioaccumulation and biomagnification:* DDT is a fat-soluble chemical that can accumulate in the fatty tissues of animals. In a food chain or web, the accumulated DDT is biologically magnified in the bodies of animals at each higher trophic level, as it was in the case of a food chain in the U.S. state of New York, illustrated here. (Dots in this figure represent DDT.) **Question:** How does this story demonstrate the value of pollution prevention?

DDT in fish-eating birds (ospreys) 25 ppm

DDT in large fish (needlefish) 2 ppm

DDT in small fish (minnows) 0.5 ppm

DDT in zooplankton 0.04 ppm

DDT in water 0.000003 ppm, or 3 ppt

HONEYBEE LOSSES: A SEARCH FOR CAUSES

Over the past 50 years the European honeybee population in the United States has been cut in half. Since 2006, at least one-third of the U.S. population of this species has disappeared because of colony collapse disorder (**Core Case Study**). This problem also occurs in parts of Europe, China, and India.

Scientific research has found a number of possible reasons for this decline, including:

- The *varroa mite* (Figure 9-B), a parasitic insect, can weaken and kill honeybee adults by feeding on their blood and larvae.
- *Harmful interactions between viruses and fungi* found in European honeybees and in almost all colonies suffering from colony collapse disorder. For example, the *Israeli acute paralysis virus,* which can be spread to hives by the varroa mite, and the *nosema fungus* can interact to kill bees by weakening their immune systems.
- *Pesticides.* As honeybees forage for nectar they can come into contact with harmful pesticides that they can carry back to the hives. Some research indicates that *neonictinoids*—among the world's most widely use pesticides—can disrupt the nervous systems of bees and can decrease their ability to find their way back to their hives. Neonictinoids have been found in the pollen and nectar of corn plants serviced by bees and in high-fructose corn syrup that many beekeepers feed their bees after harvesting their honey. Pesticides in the air can also end up in the honeycomb wax and stored pollen in hives. A U.S. researcher at a USDA lab in North Carolina found

more than 170 different pesticides in samples of bees, honeycomb wax, and stored pollen. Exposure to such a cocktail of pesticides can weaken bees' immune systems and make them vulnerable to deadly parasites, viruses, and fungi.

- *Stress and poor nutrition* from being transported long distances around the United States as insect workers for the industrial pollination business (Figure 9-C). Overworking and overstressing honeybees by moving them around the country can weaken their immune systems and make them more vulnerable to death from parasites, viruses, fungi, and pesticides. In natural ecosystems, honeybees gather nectar and pollen from a variety of flowering plants, but industrial worker honeybees feed mostly on pollen or nectar from one

U.S. Department of Agriculture

Figure 9-B The parasitic varroa mite, shown here on a honeybee host, can weaken and kill honeybees.

crop or a small number of crops that may lack the nutrients they need.

The growing consensus among bee researchers is that one or more combinations of the factors listed here are probably responsible for colony collapse disorder.

Critical Thinking

Can you think of some ways in which commercial beekeepers could lessen one or more of the threats described here? Explain.

Figure 9-C European honeybee hive boxes in an acacia orchard. Each year, commercial beekeepers rent and deliver several million hives by truck to farmers throughout the United States.

© Cristi111 | Dreamstime.com

Polar Bears and Climate Change

The world's 20,000–25,000 polar bears are found in 19 subpopulations scattered across the frozen Arctic Circle. About 60% of them are in Canada, and the rest live in arctic areas of Greenland, Norway, Russia, and the U.S. state of Alaska.

Throughout the winter, polar bears hunt for ringed seals on floating sea ice (Figure 9-14) that expands each winter and contracts as the temperature rises during summer. By eating the seals, the bears build up their body fat. In the summer and fall, they live off this fat until hunting resumes when the ice expands again during winter.

Figure 9-14 On floating ice in Svalbard, Norway, a polar bear feeds on its ringed seal prey. **Question:** Do you think it matters that the polar bear may become extinct during this century primarily because of human activities? Explain.

©age footstock/SuperStock

Scientific measurements reveal that the earth's atmosphere has been getting warmer since 1975 and that this warming is occurring twice as fast in the Arctic as in the rest of the world. Thus, arctic ice is melting faster and the average annual area of floating sea ice in the Arctic during the summer is decreasing. The floating winter ice is also breaking up earlier each year, shortening the polar bears' hunting season. In addition, much of the remaining ice is getting too thin to support the weight of an adult polar bear.

These changes mean that polar bears must swim longer distances to find prey and have less time to feed and store fat. As a result, they must fast longer, which weakens them. As females become weaker, their ability to reproduce and keep their young cubs alive declines. In 2008, the U.S. Fish and Wildlife Service placed the Alaskan polar bear on its list of threatened species.

According to the International Union for Conservation of Nature, the world's total polar bear population is likely to decline by 30–35% by 2050. By the end of this century, polar bears might be found only in zoos. However, the evidence for such population projections varies with the different polar bear sub-populations, their locations, and what they eat. Of the 19 sub-populations, in 2011, some were shrinking and some were steady or growing.

The Illegal Killing, Capturing, and Selling of Wild Species Threaten Biodiversity

Some protected species are illegally killed (poached) for their valuable parts or are sold live to collectors. Globally, this illegal trade in wildlife brings in an average of at least $600,000 an hour. Organized crime has moved into illegal wildlife smuggling because of the huge profits involved. Few of the smugglers are caught or punished.

To poachers, a highly endangered, live mountain gorilla (of which there are about 880 left in the wild) is worth $150,000, and the pelt of an endangered giant panda (1,600 to 3,000 left in the wild in China) can bring $100,000. A poached rhinoceros horn (Figure 9-15) can be worth as much as $99,000 per kilogram ($45,000 per pound) on the black market. Their horns are used to make dagger handles in Yemen and Oman. Powdered rhino horn has long been used in traditional medicines for a variety of ailments and as an alleged male aphrodisiac in many Asian countries including China, India, and Vietnam. There is no verifiable evidence for these benefits and the horn is largely made of the same material in fingernails and animal hooves.

Mostly because of poaching, three of the five rhino species—the black rhino (4,880 left in the wild, see Figure 9-8), the Sumatran (2,700 left), and the Javan rhino (about 30 left)—are critically endangered. They have survived for about 40 million years, but poachers very likely could wipe them out within the next 50–80 years. However, during the last century, the number of southern white rhinos in South Africa rose from 50 to 17,400 because people worked to protect their habitat and reduce poaching. GOOD NEWS

Rhinos can be raised on farms and if their horns are carefully cut to no more than 8 centimeters (3 inches) from the base, new horns will grow back in about 2 years. Such sustainable harvesting of rhino horns could help to reduce poaching. Also, park rangers in some rhino areas are using motion-sensitive cameras to photograph and identify poachers.

In 1900, there were an estimated 100,000 tigers in the wilds of Asia in a rapidly shrinking range (Figure 9-8, top left). Today there are about 3,200, half of them in India. Three of the nine subspecies of tigers have become extinct and the other six, including the Siberian (see chapter-opening photo) and Sumatran tigers (Figure 9-2d), are endangered.

This is mostly due to a 90% loss of habitat, caused by rapid human population growth, and poaching, much of which is motivated by poverty. The Indian, or Bengal, tiger is at risk because a coat made from its fur can sell for as much as $100,000 in Tokyo, and tiger skins sell for up to $35,000. The bones and penis of a single tiger can fetch as much as $70,000 in China, the world's biggest market for such illegal items. According to the World Wildlife Fund, without emergency action to curtail poaching and preserve tiger habitat, few if any tigers will be left in the wild by 2022.

Around the globe, the legal and illegal trade in wild species for use as pets is a huge and very profitable business. Many owners of wild pets do not know that, for every live animal captured and sold in the pet market, many others are killed or die in transit. According to the IUCN, more than 60 bird species, mostly parrots

Figure 9-15 A poacher in South Africa killed this critically endangered northern white rhinoceros for its two horns. This species is now extinct in the wild. With a rhino horn worth up to $300,000 on the Asian black market, thieves have been stealing the horns from museums, antique stores, and even private collections. **Question:** What would you say if you could talk to the poacher who killed this animal for its horns?

(Figure 9-7), are endangered or threatened because of the wild bird trade (see the Case Study that follows).

CONSIDER THIS. . .

CONNECTIONS Exotic Pets and Human Diseases

Most people are unaware that some imported exotic animals carry diseases such as hantavirus, Ebola virus, Asian bird flu, herpes B virus (carried by most adult macaques), and salmonella (from pets such as hamsters, turtles, and iguanas). These diseases can spread quite easily from pets to their owners and then to other people.

Other wild species whose populations are depleted because of the pet trade include many amphibians (see Chapter 4 Core Case Study, p. 78), various reptiles, and tropical fishes taken mostly from the coral reefs of Indonesia and the Philippines. Divers catch tropical fish by using plastic squeeze bottles of poisonous cyanide to stun them. For each fish caught alive, many more die. In addition, the cyanide solution kills the coral animals (polyps) that create the reef.

Some exotic plants, especially orchids and cacti (see Figure 7-10, center, p. 151), are endangered because they are gathered, often illegally, and sold to collectors to decorate houses, offices, and landscapes. A mature crested saguaro cactus can earn a cactus rustler as much as $15,000, and an orchid collector might pay $5,000 for a single rare orchid.

According to the U.S. Fish and Wildlife Service, collectors of exotic birds might pay $10,000 for an endangered hyacinth macaw parrot (Figure 9-7) smuggled out of Brazil. However, during its lifetime, a single hyacinth macaw left in the wild could account for as much as $165,000 in ecotourist revenues.

CONSIDER THIS. . .

THINKING ABOUT Collecting Wild Species

Some people believe it is unethical to collect wild animals and plants for display and personal pleasure and that such plants and animals should be left to live in the wild. Explain why you agree or disagree with this view.

A Rising Demand for Bushmeat Threatens Some African Species

For centuries, indigenous people in much of West and Central Africa have sustainably hunted wildlife for *bushmeat* as a source of food. But in the last two decades, bushmeat hunting in some areas has skyrocketed as hunters have tried to provide food for rapidly growing populations or to make a living by supplying restaurants in major cities with exotic meats from gorillas (Figure 9-16) and other species. Logging roads in once-inaccessible forests have made such hunting much easier. As a result, some forests in areas such as Africa's Congo basin are being stripped of many of their monkeys, apes, antelope (the most commonly hunted bushmeat animal), elephants, hippos, and other wild animals.

Bushmeat hunting has driven at least one species—Miss Waldron's red colobus monkey—to complete extinction. It is also a factor in reducing some populations of orangutans (Figure 9-5). Another problem is that the butchering and eating of some forms of bushmeat have helped to spread fatal diseases such as HIV/AIDS and the Ebola virus from animals to humans.

The U.S. Agency for International Development (USAID) is trying to reduce unsustainable hunting for bushmeat in some areas of Africa by introducing alternative sources of food, including farmed fish. They are also showing villagers how to breed large rodents such as cane rats as a source of food. GOOD NEWS

CASE STUDY

A Disturbing Message from the Birds

Approximately 70% of the world's more than 10,000 known bird species are declining in numbers, and much of this decline is related to human activities, summarized by HIPPCO. According to the IUCN 2012 Red List of Endangered Species, roughly one of every eight (13%) of all bird species is threatened with extinction mostly by habitat loss, degradation, and fragmentation (the H in HIPPCO). About three-fourths of the threatened bird species live in forests, many of which are being cleared at a rapid rate, especially in many tropical areas of Asia, Latin America, and Africa.

According to a 2011 study, *State of the Birds*, almost one-third of the more than 800 bird species in the United States are endangered (Figure 9-17), threatened, or in decline, mostly because of habitat loss and degradation, invasive species, and climate change. About one-third of all endangered and threatened bird species in the United States live in Hawaii.

Sharp declines in bird populations have occurred among songbird species that migrate long distances, such as tanagers, orioles, thrushes, vireos, and warblers. These birds nest deep in North American woods in the summer and spend their winters in Central or South America or on the Caribbean Islands. The primary causes of these

Photoshot Holdings Ltd/Alamy

Figure 9-16 *Bushmeat* such as this severed head of an endangered lowland gorilla in the Congo is consumed as a source of protein by local people in parts of West and Central Africa and is sold in national and international marketplaces and served in some restaurants where wealthy patrons regard gorilla meat as a source of status and power. ***Question:*** How, if at all, is this different from killing a cow for food?

population declines appear to be habitat loss and fragmentation of the birds' breeding habitats. In North America, road construction and housing developments result in the breaking up or clearing of woodlands. In Central and South America, tropical forest habitats, mangroves, and wetland forests are suffering the same fate. In addition, the populations of 40% of the world's water birds are in decline because of the global loss of wetlands.

After habitat loss, the intentional or accidental introduction of nonnative species such as bird-eating rats is the second greatest danger, affecting about 28% of the world's threatened birds. Other such invasive species (the I in HIPPCO) include snakes (such as the brown tree snake), and mongooses, which kill hundreds of millions of birds each year.

Population growth, the first P in HIPPCO, also threatens some bird species, as more people spread out over the landscape and increase their use of timber, food, and other resources, destroying or disturbing some bird habitats. The second P in HIPPCO is for pollution, another major threat to birds. Countless birds are exposed to oil spills, pesticides, herbicides, and toxic lead from shotgun pellets that fall into wetlands and from lead sinkers left by anglers.

🔍 CONSIDER THIS...

CONNECTIONS Vultures, Wild Dogs, and Rabies

The IUCN placed four species of carcass-eating vultures found in India on the endangered list after their populations fell by more than 95% during the early 1990s. Scientists discovered that they were dying from kidney failure when they fed on the carcasses of cows that had been given an anti-inflammatory drug to increase milk production. As the vultures died off, huge numbers of cow carcasses, normally a source of food for the vultures, were consumed by wild dogs and rats whose populations the vultures had helped to control by reducing their food supply. As wild dog populations exploded, the number of dogs with rabies also increased. About 48,000 people bitten by the rabid dogs died of rabies. The drug that caused this problem was banned, and by 2012, the vultures were making a slow comeback.

Overexploitation (the O in HIPPCO) of birds and other species is also a major threat to bird populations. Fifty-two of the world's 388 parrot species (Figure 9-7) are threatened partly because so many parrots are captured for the pet trade (often illegally) for sale usually to buyers in Europe and the United States.

Industrialized fishing fleets also pose a threat to birds. At least 23 species of seabirds, including albatrosses, face extinction. Many of these diving birds drown after becoming hooked on baited lines or trapped in huge nets that are set out by fishing boat crews.

Biodiversity scientists view this decline of bird species with alarm. One reason is that birds are excellent *indicator species* because they live in every climate and biome, respond quickly to environmental changes in their habitats, and are relatively easy to track and count. What their decline is indicating to these scientists is widespread environmental degradation.

A second reason is that birds perform critically important economic and ecosystem services in ecosystems throughout

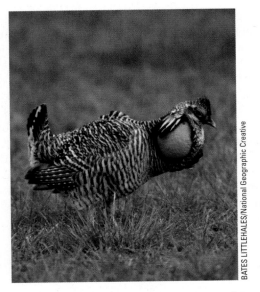

BATES LITTLEHALES/National Geographic Creative

Figure 9-17 **Endangered natural capital:** This endangered Attwater's prairie chicken lives in a wildlife refuge in the U.S. state of Texas.

Çağan Hakkı Şekercioğlu: Protector of Birds and National Geographic Emerging Explorer

Çağan Şekercioğlu, assistant professor at the University of Utah Department of Biology, is a bird expert, conservation ecologist, tropical biologist, and accomplished wildlife photographer. He has seen over 60% of the planet's known bird species in 70 countries, developed a global database on bird ecology, and become an expert on the causes and consequences of bird extinctions around the world. One example of an endangered species he has studied is the great green macaw of South and Central America (shown here in the background photo).

In 2010, this scientist with an undergraduate degree from Harvard and a PhD in biology from Stanford University was listed in the top 1% of world's scientists as measured by the number of times his research was cited in ecology and environmental science reports between 2000 and 2010. In 2011, he was selected as a National Geographic Emerging Explorer and named as Turkey's Scientist of the Year.

In 2007 Şekercioğlu founded KuzeyDoğa, an award-winning ecological research and community-based conservation organization to help conserve and protect the wildlife of northeastern Turkey. One of its goals is to promote ecotourism, especially bird watching, as a way to protect threatened bird species and their habitats and improve economic conditions in the participating Turkish communities. He also developed Turkey's first protected wildlife corridor, which would stretch across the eastern half of the country.

Based on his extensive research Şekercioğlu estimates that the percentage of the world's known bird species that are endangered could nearly double from 13% in 2012 to 25% by the end of this century. According to this champion for the world's biodiversity, "My ultimate goal is to prevent extinctions and consequent collapses of critical ecosystem processes while making sure that human communities benefit from conservation as much as the wildlife they help conserve. . . . I don't see conservation as people versus nature, I see it as a collaboration."

Background photo: ©Courtesy of Dr. Sekercioglu

©Courtesy of Dr. Sekercioglu

the world. For example, many birds play specialized roles in pollination and seed dispersal, especially in tropical areas, so extinctions of these bird species could lead to extinctions of plants that depend on the birds for pollination. Then, some specialized animals that feed mostly on these plants might also become extinct. Such a cascade of extinctions, in turn, could affect our own food supplies and well-being. Biodiversity scientists (Individuals Matter 9.1) urge us to listen more carefully to what birds are telling us about the state of the environment, for their sake, as well as for ours.

9-4 How Can We Sustain Wild Species and Their Ecosystem Services?

CONCEPT 9-4
We can reduce species extinction and sustain ecosystem services by establishing and enforcing national environmental laws and international treaties, creating protected wildlife sanctuaries, and taking precautionary measures to prevent such harm.

International Treaties and National Laws Can Help to Protect Species

Several international treaties and conventions help to protect endangered and threatened wild species (**Concept 9-4**). One of the most far reaching is the 1975 *Convention on International Trade in Endangered Species (CITES)*. This treaty, signed by 178 countries, bans the hunting, capturing, and selling of threatened or endangered species. It lists 926 species that are in danger of extinction and cannot be commercially traded as live specimens or for their parts or products. It also restricts international trade of roughly 5,000 species of animals and 29,000 species of plants that are at risk of becoming threatened.

CITES has helped to reduce the international trade of many threatened animals, including elephants, crocodiles, cheetahs, and chimpanzees. But the effects of this treaty are limited because enforcement varies from country to country (Figure 9-18), and convicted violators often pay only small fines. Also, member countries can exempt themselves from protecting any listed species, and much of the highly profitable illegal trade in wildlife and wildlife products goes on in countries that have not signed the treaty.

Another important treaty is the *Convention on Biological Diversity (CBD)*, ratified or accepted by 193 countries (but as of 2012, not by the United States). It legally commits participating governments to reducing the global rate of biodiversity loss and to equitably sharing the benefits from use of the world's genetic resources. This includes efforts to prevent or control the spread of ecologically harmful invasive species.

This convention is a landmark in international law because it focuses on ecosystems rather than on individual species, and it links biodiversity protection to issues such as the traditional rights of indigenous peoples. However, because some key countries, including the United States, have not ratified it, implementation has been slow. Also, the law contains no severe penalties or other enforcement mechanisms.

National laws are also important for sustaining wild species and ecosystem services. Several countries have strong laws (see the following Case Study), but scientists call for more, stronger, and more widespread national legal efforts.

CASE STUDY
The U.S. Endangered Species Act

The *Endangered Species Act of 1973* (ESA; amended in 1982, 1985, and 1988) was designed to identify and protect endangered species in the United States and abroad (**Concept 9-4**). This act is probably the most far-reaching environmental law ever adopted by any nation, which has made it controversial.

Under the ESA, the National Marine Fisheries Service (NMFS) is responsible for identifying and listing endangered and threatened ocean species, while the U.S. Fish and Wildlife Service (USFWS) is to identify and list all other endangered and threatened species. Any decision by either agency to list or delist a species must be based on biological factors alone, without consideration of economic or political factors. However, the two agencies can use economic factors in deciding whether and how to protect endangered habitat and in developing recovery plans for listed species. The ESA also forbids federal agencies (except the Defense Department) to carry out, fund, or authorize projects that would jeopardize any endangered or threatened species or destroy or modify its critical habitat.

Figure 9-18 These poachers were caught while trying to sell a tiger skin in Madhya Pradesh State, India.

For offenses committed on private lands, fines as high as $100,000 and 1 year in prison can be imposed to ensure protection of the habitats of endangered species, although this provision has rarely been used. This part of the act has been controversial because at least 90% of the listed species live totally or partially on private land. Since 1982, however, the ESA has been amended to give private landowners various economic incentives to help save endangered species living on their lands. The ESA also makes it illegal for Americans to sell or buy any product made from an endangered or threatened species or to hunt, kill, collect, or injure such species in the United States.

Between 1973 and 2012, the number of U.S. species on the official endangered and threatened species lists increased from 92 to 1,476. According to a study by the Nature Conservancy, about 33% of the country's species are at risk of extinction, and 15% of all species are at high risk—far more than the current number listed. In 2012, 77% of the protected species were covered by active recovery plans. Successful recovery plans include those for the American alligator (see Chapter 4, Case Study, pp. 95–96), the gray wolf, the peregrine falcon, the bald eagle, and the brown pelican.

The ESA also requires that all commercial shipments of wildlife and wildlife products enter or leave the country through one of 17 designated airports and ocean ports. The 120 full-time USFWS inspectors can inspect only a small fraction of the more than 200 million wild animals brought legally into the United States annually. Each year, tens of millions of wild animals are also brought in illegally, but few illegal shipments of endangered or threatened animals or plants are confiscated. Even when they are caught, many violators are not prosecuted, and convicted violators often pay only a small fine.

Since 1995, there have been numerous efforts to weaken the ESA and to reduce its already meager annual budget. Opponents of the act contend that it puts the rights and welfare of endangered plants and animals above those of people. Some critics would do away with this act. They call it an expensive failure because only 26 species have recovered enough to be removed from the endangered species list (Figure 9-19), and because 10 became extinct while on the list. Most biologists insist that it has not been a failure, for three main reasons.

First, species are listed only when they face serious danger of extinction. ESA supporters argue that this is similar to a hospital emergency room set up to take only the most desperate cases, often with little hope for recovery. Such a facility could not be expected to save all of its patients.

Second, according to federal data, the conditions of more than half of the listed species are stable or improving, 90% are recovering at rates specified by their recovery plans, and 99% of the protected species are still surviving. A hospital emergency room taking only the most desperate cases and then stabilizing or improving the con-

ditions of more than half of those patients while keeping 99% of them alive and 90% recovering would be considered to be highly effective.

Third, the 2012 budget for protecting endangered species amounted to an average expenditure of about 86 cents per U.S. citizen. To its supporters, it is amazing that the federal agencies, on such a small budget, have managed to stabilize or improve the conditions of more than half of the listed species.

ESA supporters agree that the act can be improved and that federal regulators have sometimes been too heavy handed in enforcing it. They cite a study by the U.S. National Academy of Sciences that recommended three major changes in the way the law is being implemented, in order to make it more scientifically sound and effective:

1. Greatly increase the meager funding for implementing the act.
2. Put more emphasis on developing recovery plans more quickly, as scientists have found that species with recovery plans have a better chance of getting off the endangered list.
3. When a species is first listed, establish the core of its habitat as critical for its survival and give that area the maximum protection.

Most biologists and wildlife conservationists believe that the United States also needs a new law that emphasizes protecting and sustaining biological diversity and ecosystem services rather than focusing mostly on saving individual species. (We examine this idea further in Chapter 10.)

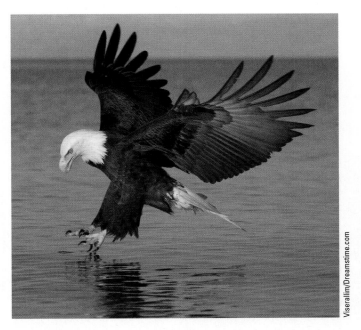

Figure 9-19 The American bald eagle has been removed from the U.S. endangered species list. Here, an eagle is about to catch a fish in its powerful talons.

CAMPUS SUSTAINABILITY

The Endangered Species Lab at Washington State University

How do we save endangered species and prevent extinction processes? That is one of the questions that drives researchers at Washington State University's Endangered Species Lab. Students there conduct field studies in both natural and managed landscapes to explore the roles played by evolution and species behavior in extinction processes. They also study how landscape factors affect the process of reintroduction of threatened and endangered species, seeking to apply such knowledge to efforts to restore native habitats. The researchers have focused on a number of species, including Columbian sharp-tailed grouse, burrowing owls (see photo), northern leopard frogs, a variety of bat species, and the Columbia Basin pygmy rabbit.

We Can Establish Wildlife Refuges and Other Protected Areas

In 1903, President Theodore Roosevelt established the first U.S. federal wildlife refuge at Pelican Island, Florida (Figure 9-20), to help protect birds such as the brown pelican (Figure 9-20, inset photo) from extinction. It took more than a century but this protection worked. In 2009, the brown pelican was removed from the U.S. Endangered Species list. By 2012, there were more than 560 refuges in the National Wildlife Refuge System. Each year, more than 47 million Americans visit these refuges to hunt, fish, hike, and watch birds and other wildlife.

GOOD NEWS

More than three-fourths of the refuges serve as wetland sanctuaries that are vital for protecting migratory waterfowl. More than one-fourth of all U.S. endangered and threatened species have habitats in the refuge system, and some refuges have been set aside specifically for certain endangered species (Concept 9-4). Such areas have helped Florida's key deer, the brown pelican, and the trumpeter swan to recover.

There is also bad news about refuges. According to a General Accounting Office study, activities considered harmful to wildlife, such as mining, oil drilling, and use of off-road vehicles, occur in nearly 60% of the nation's wildlife refuges. Also, a 2008 study prepared for Congress found that, for years, the country's wildlife refuges have received so little funding that a third of them have no staff, and boardwalks, public buildings, and other structures in some refuges are in disrepair.

Biodiversity scientists are urging the U.S. government to set aside more refuges for endangered plants and to significantly increase the long-underfunded budget for the refuge system. They are also calling on the Congress and state legislatures to allow abandoned military lands that contain significant wildlife habitat to be designated as wildlife refuges.

Seed Banks, Botanical Gardens, and Wildlife Farms Can Help to Protect Species

Seed banks preserve genetic information and endangered plant species by storing their seeds in refrigerated, low-humidity environments. More than 1,000 seed banks around the world collectively hold about 3 million samples.

Some species cannot be preserved in seed banks. The banks are of varying quality and expensive to operate, and some can be destroyed by fires and other mishaps. However, the Svalbard Global Seed Vault, a new underground facility on a remote island in the Arctic, will eventually contain 100 million of the world's seeds. It is not vulnerable to power losses, fires, storms, or war.

The world's 1,600 *botanical gardens* and *arboreta* contain living plants that represent almost one-third of the world's known plant species. But they contain only about 3% of the world's rare and threatened plant species and have too little space and funding to preserve most of those species.

We can take pressure off some endangered or threatened species by raising individuals of these species on *farms* for commercial sale. In Florida, for example, alligators are raised on farms for their meat and hides. Butterfly farms established to raise and protect endangered species flourish in Papua New Guinea, where many butterfly species are threatened by development activities. These farms are also used to educate visitors about the need to protect butterfly species.

Zoos and Aquariums Can Protect Some Species

Zoos, aquariums, game parks, and animal research centers are being used to preserve some individuals of critically endangered animal species, with the long-term goal of reintroducing the species into protected wild habitats.

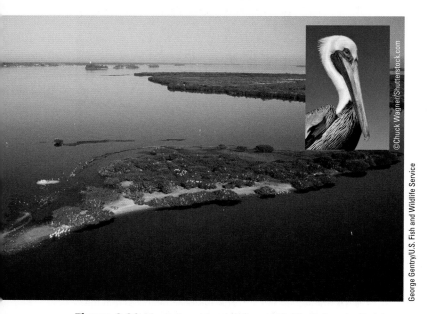

Figure 9-20 The Pelican Island National Wildlife Refuge in Florida was America's first National Wildlife Refuge.

Two techniques for preserving endangered terrestrial species are egg pulling and captive breeding. *Egg pulling* involves collecting wild eggs laid by critically endangered bird species and then hatching them in zoos or research centers. In *captive breeding,* some or all of the wild individuals of a critically endangered species are collected for breeding in captivity, with the aim of reintroducing the offspring into the wild. Captive breeding has been used to save the peregrine falcon and the California condor (Figure 9-2b).

Other techniques for increasing the populations of captive species include artificial insemination, embryo transfer (surgical implantation of eggs of one species into a surrogate mother of another species), use of incubators, and cross fostering (in which the young of a rare species are raised by parents of a similar species). Scientists also use computer databases that hold information on family lineages of endangered zoo animals along with DNA analysis to match individuals for mating—a computer dating service for zoo animals.

The ultimate goal of captive breeding programs is to build up populations to a level where they can be reintroduced into the wild. Successes include the black-footed ferret, the golden lion tamarin (a highly endangered monkey species), the Arabian oryx, and the California condor (Figure 9-2b). However, most reintroductions fail because of lack of suitable habitat, inability of individuals bred in captivity to survive in the wild, and renewed overhunting or poaching.

One problem for captive breeding programs is that a captive population of an endangered species must typically number 100–500 individuals in order for it to avoid extinction through accident, disease, or loss of genetic diversity through inbreeding. Recent genetic research indicates that 10,000 or more individuals are needed for an endangered species to maintain its capacity for biological evolution. Zoos and research centers do not have the funding or space to house such large populations.

Public aquariums (Figure 9-21) that exhibit unusual and attractive species of fish and marine animals such as seals and dolphins help to educate the public about the need to protect such species. However, mostly because of limited funds, public aquariums have not served as effective gene banks for endangered marine species, especially marine mammals that need large volumes of water.

While zoos, aquariums, and botanical gardens perform valuable services, they cannot by themselves solve the growing problem of species extinction. Figure 9-22 lists some things you can do to help deal with this problem.

GOOD NEWS

The Precautionary Principle

Biodiversity scientists call for us to take precautionary action to avoid hastening species extinction and disrupting essential ecosystem services. This approach is based on the **precautionary principle**: When substantial preliminary evidence indicates that an activity can harm human

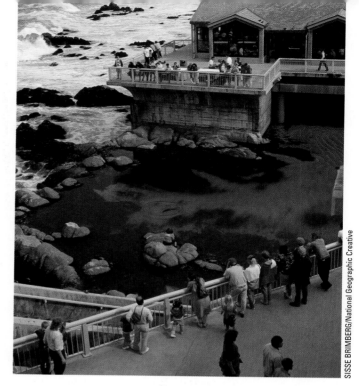

Figure 9-21 The Monterey Bay Aquarium in Monterey, California (USA), contains this tidewater pool, which is used to train rescued sea otter pups to survive in the wild.

SISSE BRIMBERG/National Geographic Creative

health or the environment, we should take precautionary measures to prevent or reduce such harm even if some of the cause-and-effect relationships have not been fully established scientifically. It is based on the commonsense idea behind many adages, including "Better safe than sorry" and "Look before you leap."

Scientists use the precautionary principle to argue for both the preservation of species and protections of entire ecosystems and their ecosystem services, which is the focus of the next chapter. The precautionary principle is also used as a strategy for dealing with other challenges such as preventing exposure to harmful chemicals in the air we breathe, the water we drink, and the food we eat.

Using limited financial and human resources to protect biodiversity based on the precautionary principle involves dealing with three important questions:

1. Should we focus mostly on protecting species or on protecting ecosystems and the ecosystem services they provide for a variety of species, and how do we allocate limited resources between these two priorities?

2. How do we decide which species should get the most attention in our efforts to protect as many species as possible? For example, should we focus on protecting the most threatened species or on protecting keystone species? Protecting species that are appealing to humans, such as tigers (see Figure 9-2d) and orangutans (Figure 9-5), can increase public awareness of the need for wildlife conservation. But some argue that we should instead protect more ecologically important endangered species (see the following Case

What Can You Do?

Protecting Species

- Do not buy furs, ivory products, or other items made from endangered or threatened animal species.

- Do not buy wood or wood products from tropical or old-growth forests.

- Do not buy pet animals or plants taken from the wild.

- Tell friends and relatives what you're doing about this problem.

Figure 9-22 **Individuals matter:** You can help to prevent the extinction of species. **Questions:** Which two of these actions do you believe are the most important? Why?

Figure 9-23 This woman is hand-pollinating pear tree blossoms on Honshu Island, Japan, where overuse of pesticides decimated populations of natural pollinators such as honeybees. This is a time-consuming and expensive alternative to natural pollination.

Study), especially in severely threatened biodiversity hotspots.

3. How do we determine which habitat areas are the most critical to protect?

A 2012 study by an international team of conservation scientists, led by environmental economist Donald McCarthy, estimated that we could sharply reduce the threat of species extinction and protect key conservation sites by spending about $80 billion a year. This investment in sustaining biodiversity is less than 20% of the amount spent globally each year on soft drinks.

CASE STUDY

Protecting Honeybees and Other Pollinators

Failure to protect honeybees could lead to a loss in some areas of the ecosystem service of pollination that they provide. For example, in some parts of Japan and China, overuse of pesticides and other factors have caused farmers to have to pollinate some crops by hand (Figure 9-23)—a time-consuming and expensive undertaking.

Researchers, farmers, and beekeepers are searching for ways to protect European honeybees and other pollinators and to sustain their populations (**Core Case Study**). They are taking several steps, some of which involve applying the precautionary principle.

Some beekeepers are breeding bees that are not as vulnerable to harmful parasitic mites and fungi. They are also improving nutrition for bees that are transported from place to place, for example, by not feeding their bees sugar syrups or pollen substitutes. Some farmers are raising their own honeybee colonies to avoid having to bring in stressed, unhealthy honeybees for pollination.

We can all work to preserve healthy populations of honeybees that help to supply us with at least a third of our food. Ways to do this include buying food from local, USDA-certified organic farmers who do not use pesticides and buying local, organic honey to support organic beekeepers. Some people are setting up their own beehives and producing their own organic honey. Others, by growing their own food organically, are supporting native, wild honeybee populations that benefit from such crops. Many people are replacing their lawns with a variety of native plants that make good sources of nectar for bees, including foxglove, red clover, and bee balm.

Big Ideas

- We are hastening the extinction of wild species and degrading the ecosystem services they provide by destroying and degrading their habitats, introducing harmful invasive species, and increasing human population growth, pollution, climate change, and overexploitation.

- We should avoid causing the extinction of wild species because of the ecosystem and economic services they provide and because their existence should not depend primarily on their usefulness to us.

- We can work to prevent the extinction of species and to protect overall biodiversity and ecosystem services by using laws and treaties, protecting wildlife sanctuaries, and making greater use of the precautionary principle.

Malwina Szweda/Shutterstock.com

In this chapter, we have looked at the human activities that are hastening the extinction of many species and at how we might curtail those activities. We learned that there is considerable evidence that as many as half of the world's wild species could go extinct during this century, largely as a result of human activities that threaten many species and some of the vital ecosystem services they provide. For example, populations of honeybees, vital for pollinating crops that supply much of our food, have been declining for a variety of reasons (**Core Case Study**), many of them related to human activities. One of the key reasons for such problems is that most people are simply unaware of the highly valuable ecosystem and economic services provided by the earth's species.

In keeping with two of the three scientific **principles of sustainability** (see Figure 1-2, p. 6 or back cover), acting to prevent the extinction of species as a result of human activities helps to preserve not only the earth's biodiversity, but also the vital ecosystem services that sustain us, including chemical cycling. Thus, it is not only for these species that we ought to act, but also for the long-term health and well-being of our own species in keeping with the win-win and future generations' **principles of sustainability** (see Figure 1-5, p. 9 or back cover).

Chapter Review

Core Case Study

1. What economic and ecological roles do honeybees play? What is *pollination*? How are human activities contributing to the decline of many populations of European honeybees, and why should we care? What is **colony collapse disorder (CCD)**?

Section 9-1

2. What is the key concept for this section? What is **biological extinction**? Explain what a trophic cascade is and give an example. What is a **mass extinction**? What is the **background extinction rate**, and how do estimated current and projected extinction rates compare with it? What percentage of the earth's land and what percentage of the earth's oceans have been disturbed by human activities? Explain how scientists estimate extinction rates and describe three problems they face. Give three reasons why many extinction experts believe that projected extinction rates are probably on the low side. What percentage of the world's species are likely to go extinct, largely as a result of human activities, during this century? Distinguish between **endangered species** and **threatened species** and give an example of each. List four characteristics that make some species especially vulnerable to extinction.

Section 9-2

3. What is the key concept for this section? What are three reasons for trying to avoid hastening the extinc-

tion of wild species? Describe two economic and two ecological benefits of species diversity. Explain how saving other species and the ecosystem services they provide can help us to save our own species and our cultures and economies.

Section 9-3

4. What is the key concept for this section? What is **HIPPCO**? What is the greatest threat to wild species? What is **habitat fragmentation**? Describe the major effects of habitat loss and fragmentation. Why are island species especially vulnerable to extinction? What are habitat islands?

5. Give two examples of the benefits that have been gained by the introduction of nonnative species. Give two examples of the harmful effects of nonnative species that have been introduced deliberately. Describe the harmful and beneficial effects of introducing the kudzu vine. Give two examples of harmful results of accidental introductions of nonnative species. List four ways to limit the harmful impacts of nonnative species. Explain why prevention is the best way to reduce threats from invasive species and list five ways to implement this strategy. Summarize the roles of population growth, overconsumption, pollution, and climate change in the extinction of wild species. Explain how pesticides such as DDT can accumulate and be biomagnified in food chains and webs.

6. Summarize the roles of population growth, over-consumption, pollution, and climate change in the extinction of wild species. List possible causes of the decline of European honeybee populations in the United States. Describe how human activities threaten polar bears in the Arctic. Why does poaching occur? Give three examples of species that are threatened by this illegal activity. Why are wild tigers likely to disappear within a few decades? What is the connection between infectious diseases in humans and the pet trade? Describe the threat to some forms of wildlife from the increased hunting for bush meat.

7. List the major threats to the world's bird populations and give two reasons for protecting bird species from extinction. Describe environmental explorer Çağan Şekercioğlu's contributions to our understanding of the ecological importance of birds and threats to their extinction.

Section 9-4

8. What is the key concept for this section? Name two international treaties that are used to help protect species. What is the U.S. Endangered Species Act? How successful has it been, and why is it controversial?

9. Summarize the roles and limitations of wildlife refuges, gene banks, botanical gardens, wildlife farms, zoos, and aquariums in protecting some species. Describe the role of captive breeding in efforts to prevent species extinction and give an example of success in returning a nearly extinct species to the wild. What is the **precautionary principle** and how can we use it to help protect wild species and overall biodiversity? What are three important questions related to the use of this principle? List five ways to lessen the environmental threats to European honeybees and other pollinators.

10. What are this chapter's *three big ideas*? How do two of the three scientific **principles of sustainability** apply to protecting honeybees and other wild species from extinction along with protecting the ecosystem services provided by species? What two social science **principles of sustainability** are involved in protecting species from extinction due to human activities?

Note: Key terms are in bold type.

Critical Thinking

1. What are three aspects of your lifestyle that might directly or indirectly contribute to declines in European honeybee populations and the endangerment of other pollinator species (**Core Case Study**)?

2. Give your response to the following statement: "Eventually, all species become extinct. So it does not really matter that the world's remaining tiger species or a tropical forest plant is endangered mostly because of human activities." Be honest about your reaction, and give arguments to support your position.

3. Do you accept the ethical position that each species has the inherent right to survive without human interference, regardless of whether it serves any useful purpose for humans? Explain. Would you extend this right to the *Anopheles* mosquito, which transmits malaria, and to infectious bacteria? Explain.

4. Wildlife ecologist and environmental philosopher Aldo Leopold wrote this with respect to preventing the extinction of wild species: "To keep every cog and wheel is the first precaution of intelligent tinkering." Explain how this statement relates to the material in this chapter.

5. What would you do if fire ants invaded your yard and house? Explain your reasoning behind your course of action. How might your actions affect other species or the ecosystem you are dealing with?

6. How do you think your lifestyle might contribute directly or indirectly to the extinction of some bird species? What are two things that you think should be done to reduce the rate of extinction of bird species?

7. Which of the following statements best describes your feelings toward wildlife?
   a. As long as it stays in its space, wildlife is okay.
   b. As long as I do not need its space, wildlife is okay.
   c. I have the right to use wildlife habitat to meet my own needs.
   d. When you have seen one redwood tree, elephant, or some other form of wildlife, you have seen them all, so lock up a few of each species in a zoo or wildlife park and do not worry about protecting the rest.
   e. Wildlife should be protected in its current ranges.

8. Environmental groups in a heavily forested state want to restrict logging in some areas to save the habitat of an endangered squirrel. Timber company officials argue that the well-being of one type of squirrel is not as important as the well-being of the many families who would be affected if the restriction were to cause the company to lay off hundreds of workers. If you had the power to decide this issue, what would you do and why? Describe any trade-offs included in your solution.

Doing Environmental Science

Identify examples of habitat destruction or degradation in the area in which you live or go to school. Try to determine and record any harmful effects that these activities have had on the populations of one wild plant and one animal species. (Name each of these species and describe how they have been affected.) Do some research on the Internet and/or in a school library on management plans, and then develop a management plan for restoring the habitats and species you have studied. Try to determine whether trade-offs are necessary with regard to the human activities you have observed, and account for these trade-offs in your management plan. Compare your plan with those of your classmates.

Global Environment Watch Exercise

Search for *Extinction,* and scroll to statistics on the portal's page. Click on "Known Causes of Animal Extinction since 1600." You will find four general categories of causes. Thinking about history from 1600 through today, how do you think humans have changed their impact on species in each of these categories? Has the impact increased or decreased over this time period? Give specific examples of changes in this timeframe to support your answers.

Data Analysis

Examine the following data released by the World Resources Institute and answer the questions that follow the table.

Country	Total Land Area in Square Kilometers (Square Miles)	Protected Area as Percent of Total Land Area (2003)	Total Number of Known Breeding Bird Species (1992–2002)	Number of Threatened Breeding Bird Species (2002)	Threatened Breeding Bird Species as Percent of Total Number of Known Breeding Bird Species
Afghanistan	647,668 (250,000)	0.3	181	11	
Cambodia	181,088 (69,900)	23.7	183	19	
China	9,599,445 (3,705,386)	7.8	218	74	
Costa Rica	51,114 (19,730)	23.4	279	13	
Haiti	27,756 (10,714)	0.3	62	14	
India	3,288,570 (1,269,388)	5.2	458	72	
Rwanda	26,344 (10,169)	7.7	200	9	
United States	9,633,915 (3,718,691)	15.8	508	55	

(Compiled by the authors using data from World Resources Institute, *Earth Trends, Biodiversity and Protected Areas, Country Profiles.*)

1. Complete the table by filling in the last column. For example, to calculate this value for Costa Rica, divide the number of threatened breeding bird species by the total number of known breeding bird species and multiply the answer by 100 to get the percentage.

2. Arrange the countries from largest to smallest according to total land area. Does there appear to be any correlation between the size of country and the percentage of threatened breeding bird species? Explain your reasoning.

1. Introduced species often become pests because they
   a. produce only a few young after a long period of gestation.
   b. inhabit only niches that have recently become available after the local extinction of another species.
   c. can live in many niches, easily adapt to many variables, and eat a wide variety of food.
   d. have a low genetic variability, enabling them to live in a wide variety of habitats.
   e. increase biodiversity because they attract more species to the areas in which they settle.

2. The biggest threat to biodiversity is
   a. asteroid strikes, such as what happened 65 million years ago resulting in the extinction of the dinosaurs.
   b. human population growth, because as population increases so does the number of plants and animals they eat, resulting in a decrease of those organisms.
   c. habitat destruction, such as cutting down portions of tropical rain forest, which destroys the niches occupied by tropical organisms.
   d. the proliferation of zoos and botanical gardens, causing people to care less about preserving species in the wild because they are being cared for in those sanctuaries.
   e. an increase in climate change, because it will result in many species moving northward or to higher elevations as they seek out their necessary habitat.

3. The sale of a large forest to a developer causing the building of roads, logging, and urban development would most likely lead to
   a. an increase in the number of indicator species because the loss of species during development would cause remaining species to become indicators.
   b. increased biodiversity because niches are created along roads where trees are cut down and in newly urban areas.
   c. a net increase in available biomass because more sunlight is available for autotrophs.
   d. a loss of biodiversity due to habitat fragmentation, because small populations of species are more vulnerable to predation, competition, and disease.
   e. a decrease in invasive species because they would also have a difficult time surviving in fragmented land that has lost many of its resources.

4. Which of the following would prevent invasive species from being introduced and becoming established?
   I. Inspect foreign goods before they enter the country.
   II. Allow ships to discharge their ballast water as they enter the port.
   III. Establish international treaties banning the transfer of species across borders.
   a. I only
   b. II only
   c. I and II only
   d. II and III only
   e. I and III only

5. Which of the following international treaties bans the hunting, capturing, and selling of threatened or endangered species in 172 countries?
   a. Endangered Species Act (ESA)
   b. The National Marine Fisheries Act
   c. The Convention on International Trade in Endangered Species (CITES)
   d. National Environmental Policy Act (NEPA)
   e. The Marine Mammal Protection Act

6. Biological extinction occurs when
   a. a species becomes extinct in a specific area, such as the gray wolf in the American northeast.
   b. a species is found nowhere on earth, such as the passenger pigeon.
   c. additional species are preyed upon, such as when bats adapt their echolocation ability to hear the moths whose ears had adapted to hear the sound frequencies of bats.
   d. there is a loss of indicator species, such as when the population of butterflies quickly declines in an area where they previously had been plentiful.
   e. an event such as a volcanic eruption results in some members of a species becoming separated from other members, and ultimately each group becomes genetically different and can no longer interbreed and produce offspring who can interbreed.

7. Most extinction experts believe that the extent of biodiversity loss will increase over the next 50 to 100 years because of
   a. the projected growth of the world's human population and resource use per person.
   b. the increase due to poaching and the demand for endangered species.
   c. the reduced input of solar energy cooling the planet and altering the earth's life.
   d. natural disease processes.
   e. less fragmentation of biodiverse regions.

8. The giant panda is prone to ecological and biological extinction because of all of the following reasons EXCEPT
   a. its low reproductive rate, and it is not usual for pandas to successfully breed in captivity so they can be reintroduced to their habitat.
   b. it lives in specific areas in Asia in a narrow distribution.
   c. it is a specialist, living in a specific habitat and eating specific food.
   d. there are limited numbers of giant pandas living in the wild.
   e. it is a top predator because no other large animal in the area eats only bamboo.

9. The best example of biomagnification is
   a. nutrients from a nearby farm are washed into a pond, resulting in an increase of algae and a loss of oxygen, and eventually a dead zone.
   b. after wolves were domesticated, people chose to breed individuals who had characteristics they desired, resulting in a wide variety of dog breeds with different abilities and physical appearances.
   c. individual elephants had traits that enabled then to better survive in their habitat and they became more likely to produce offspring with that same trait, which enabled the offspring to survive and reproduce.
   d. DDT was a pesticide that was used universally during the 1950s and 1960s. This chemical entered the food chain and worked its way up, eventually making the eggs of predatory birds so fragile they could not successfully reproduce.
   e. a road is built, fragmenting a forest and exposing the animals to introduced species, local climate changes, and disease.

Questions 10–12 refer to the following.
   a. Habitat destruction, degradation, and fragmentation
   b. Invasive (nonnative species)
   c. Population and resource use growth
   d. Greenhouse effect
   e. Overexploitation

10. Sale of exotic pets and decorative plants

11. Logging, mining, building of roads, and urban development

12. No natural predators, generalist, and high reproductive rate

10

Sustaining Terrestrial Biodiversity: Saving Ecosystems and Ecosystem Services

There is no solution, I assure you, to save Earth's biodiversity other than preservation of natural environments in reserves large enough to maintain wild populations sustainably.

EDWARD O. WILSON

Key Questions

10-1 What are the major threats to forest ecosystems?

10-2 How should we manage and sustain forests?

10-3 How should we manage and sustain grasslands?

10-4 How should we manage and sustain parks and nature reserves?

10-5 What is the ecosystem approach to sustaining biodiversity and ecosystem services?

Coastal rain forest in Costa Rica's Corcovado National Park.

FRANS LANTING/National Geographic Creative

217

Tropical forests once completely covered Central America's Costa Rica, which is smaller in area than the U.S. state of West Virginia and about one-tenth the size of France. Between 1963 and 1983, politically powerful ranching families cleared much of the country's forests to graze cattle.

Despite such widespread forest loss, tiny Costa Rica is a superpower of biodiversity, with an estimated 500,000 plant and animal species. A single park in Costa Rica is home to more bird species (Figure 10-1, left) than are found in all of North America. This oasis of biodiversity is also home to an amazing variety of other exotic wildlife, including monkeys (Figure 10-1, right), jaguars, lizards, snakes, spiders, and frogs.

This biodiversity results mostly from two factors. One is the country's tropical geographic location, lying between two oceans and having both coastal (see chapter-opening photo) and mountainous regions that provide a variety of microclimates and habitats for wildlife. The other factor is the government's strong conservation efforts.

In the mid-1970s, Costa Rica established a system of nature reserves and national parks that, by 2012, included more than 25% of its land—6% of it reserved for indigenous peoples. Costa Rica now devotes a larger proportion of its land to biodiversity conservation than does any other country, in keeping with the biodiversity **principle of sustainability** (see Figure 1-2, p. 6 or back cover).

To reduce *deforestation,* or the widespread removal of forests, the government has eliminated subsidies for converting forestland to rangeland. Instead, it pays landowners to maintain or restore tree cover. The strategy has worked: Costa Rica has gone from having one of the world's highest deforestation rates to having one of the lowest.

Since 1980, *biodiversity* (see Figure 4-2, p. 79) has emerged as a key concept of biology. Ecologists warn that human population growth, economic development, and poverty are exerting increasing pressure on the earth's ecosystems and on the ecosystem services they provide that help to sustain biodiversity. In 2010, a report by two United Nations environmental bodies warned that unless radical and creative action is taken now to conserve the earth's biodiversity, many local and regional ecosystems that support human lives and livelihoods are at risk of collapsing.

This chapter is devoted to helping us understand the threats to the earth's forests, grasslands, and other storehouses of terrestrial biodiversity, and to seeking ways to help sustain these vital ecosystems. To many scientists, one of our most important challenges is to sustain the world's vital biodiversity and the ecosystem services it provides.

GOOD NEWS

SUSTAINABILITY

Figure 10-1 Costa Rica is one of the world's most biologically rich places. Two of its half-million species are the scarlet macaw parrot (left) and the white-faced capuchin monkey (right).

Left: Vladimir Melnik/Shutterstock.com. Right: © Vilainecre... | Dreamstime.com.

10-1 What Are the Major Threats to Forest Ecosystems?

CONCEPT 10-1A
Forest ecosystems provide ecosystem services far greater in value than the value of raw materials obtained from forests.

CONCEPT 10-1B
Unsustainable cutting and burning of forests, along with diseases and insects, all made worse by projected climate change, are the chief threats to forest ecosystems.

Forests Vary in Their Age, Makeup, and Origins

Natural and planted forests occupy about 31% of the earth's land surface (excluding Greenland and Antarctica). Figure 7-9 (p. 150) shows the distribution of the world's northern coniferous, temperate, and tropical forests.

Forest managers and ecologists classify natural forests into two major types based on their age and structure: old-growth and second-growth forests. An **old-growth forest**, or **primary forest**, is an uncut or regenerated forest that has not been seriously disturbed by human activities or natural disasters for 200 years or more (Figure 10-2 and Chapter 1 opening photo). Old-growth forests are reservoirs of biodiversity because they provide ecological niches for a multitude of wildlife species (see Figure 7-14, p. 157). According to the United Nations Environment Programme (UNEP), they make up about 36% of the world's forests.

A **second-growth forest** is a stand of trees resulting from secondary ecological succession (see Figure 5-12, p. 110). These forests develop after the trees in an area have been removed by human activities, such as clear-cutting for timber or conversion to cropland, or by natural forces such as fire, hurricanes, or volcanic eruptions.

A **tree plantation**, also called a **tree farm** or **commercial forest** (Figure 10-3), is a managed forest containing only one or two species of trees that are all of the same age. They are usually harvested by clear-cutting as soon as they become commercially valuable. The land is then replanted and clear-cut again in a regular cycle. When managed carefully, such plantations can produce wood at a rapid rate and thus increase their owners' profits. Some analysts project that eventually, tree plantations could supply most of the wood used for industrial purposes such as papermaking. This would help to protect the world's remaining old-growth and second-growth forests, as long as they are not cleared to make room for tree plantations.

The downside of tree plantations is that, with only one or two tree species, they are much less biologically diverse

Figure 10-3 The rotation cycle of cutting and regrowth of monoculture tree plantations is short, usually 20–30 years. In tropical countries, where trees can grow more rapidly year-round, the rotation cycle can be 6–10 years. Like most tree plantations, the pine tree plantation in this photo was grown on land that had been cleared of an old-growth or second-growth forest.

Figure 10-2 This protected old-growth rain forest at a high altitude in Monteverde, Costa Rica (**Core Case Study**), is home for a rich diversity of plant and animal species.

Figure 10-4 Forests provide many important ecosystem and economic services (**Concept 10-1A**). ***Question:*** Which two ecosystem services and which two economic services do you think are the most important?

Photo: Val Thoermer/Shutterstock.com

and less sustainable than old-growth and second-growth forests because they violate nature's biodiversity **principle of sustainability**. Tree plantations do not provide the wildlife habitats and ecosystems services such as water storage and purification that diverse natural forests do. Also, repeated cycles of cutting and replanting can eventually deplete the topsoil of nutrients and hinder the regrowth of any type of forest on such land.

Forests Provide Important Economic and Ecosystem Services

We should care about sustaining forests because they provide highly valuable economic and ecosystem services (Figure 10-4 and **Concept 10-1A**). For example, through photosynthesis, forests remove CO_2 from the

atmosphere and store it in organic compounds (biomass). By performing this ecosystem service as a part of the global carbon cycle (see Figure 3-17, p. 66), forests help to stabilize average atmospheric temperatures and the earth's climate. Forests also provide habitats for about two-thirds of the earth's terrestrial species. In addition, according to the UN Food and Agriculture Organization (FAO) and the UNEP, forests are home to more than 300 million people, and about 1 billion people living in extreme poverty depend on forests for their survival.

Along with highly valuable ecosystem services, forests provide us with raw materials, especially wood. More than half of the wood removed from the earth's forests is used as *biofuel* for cooking and heating. The remainder of the harvest, called *industrial wood,* is used primarily to make lumber and paper.

Forests also provide us with important health benefits. For example, traditional medicines, used by 80% of the world's people, are derived mostly from plant species that are native to forests, and chemicals found in tropical forest plants are used as blueprints for making most of the world's prescription drugs (see Figure 9-6, p. 196). Scientists and economists have various ways of estimating the economic value of major ecosystem services provided by the world's forests and other ecosystems (Science Focus 10.1).

There Are Several Ways to Harvest Trees

Because of the immense economic value of forests, the harvesting of wood is one of the world's major industries. The first step in harvesting trees is to build roads for access and timber removal. Even carefully designed logging roads have a number of harmful effects (Figure 10-5)—namely, increased topsoil erosion and sediment runoff into waterways, habitat fragmentation, and loss of biodiversity. Logging roads also expose forests to invasion by nonnative pests, diseases, and wildlife species. And they open once-inaccessible forests to miners, ranchers, farmers, hunters, and off-road vehicles.

Once loggers reach a forest area, they use a variety of methods to harvest the trees (Figure 10-6). With *selective cutting*, intermediate-aged or mature trees in a forest are cut singly or in small groups (Figure 10-6a). Loggers

Figure 10-5 Natural capital degradation: Building roads into previously inaccessible forests is the first step in harvesting timber, but it also paves the way to fragmentation, destruction, and degradation of forest ecosystems.

Old growth / New highway

Cleared plots for grazing / Cleared plots for agriculture / Highway

© Cengage Learning

PUTTING A PRICE TAG ON NATURE'S ECOSYSTEM SERVICES

Currently, forests and other ecosystems are valued mostly for their economic services (Figure 10-4, right). Ecologists and ecological economists call for us to also calculate the monetary value of the ecosystem services provided by forests (Figure 10-4, left), as a way to implement the full-cost pricing **principle of sustainability** (see Figure 1-5, p. 9 or back cover).

In 1997, a team of ecologists, economists, and geographers, led by ecological economist Robert Costanza of the University of Vermont, estimated the monetary worth of the earth's ecosystem services, which can be thought of as *ecological income,* somewhat like interest income earned from a savings account. We can think of the earth's stock of natural resources that provide the ecosystem services as the savings account from which the ecological income flows.

Costanza's team estimated the monetary value of 17 ecosystem services provided by all ecosystems (including, for example, pollination and regulation of atmospheric temperatures) to be at least $33.2 trillion per year—equal to about 39% of the value of all of the goods and services produced in 2012 throughout the world. The researchers also estimated that the amount of money we would need to put into a savings account in order to earn that amount of interest income would be at least $500 trillion—an average of about $70,400 for each person on earth in 2012.

According to Costanza's study, the world's forest ecosystems alone provide us with ecosystem services worth at least $4.7 trillion per year—hundreds of times more than their economic value in terms of lumber, paper, and other wood products (**Concept 10-1A**). Some of these comparative value estimates are shown in Figure 10-A. The researchers pointed out that their estimates were very conservative.

Costanza's team had examined more than 100 studies and a variety of methods used to estimate the values of ecosystems. For example, one method that had been used was to estimate the costs of replacing ecosystems' services such as water purification with technological versions of the same services.

In 2002, Costanza and other researchers reported on a similar analysis comparing the value of preserving natural ecosystems with the values that could be obtained by using such systems to create farmland, to harvest timber, and to create aquaculture ponds, among other uses. The researchers estimated that preserving ecosystems in a global network of nature reserves occupying 15% of the earth's land surface and 30% of the ocean would provide $4.4 trillion to $5.2 trillion worth of ecosystem services—about 100 times the economic value of converting those systems to human uses.

If the estimates from these studies are reasonable, we can draw three important conclusions: **(1)** the earth's ecosystem services are essential for all humans and their economies; **(2)** the economic value of these services is huge; and **(3)** ecosystem services are an ongoing source of ecological income, as long as they are used sustainably.

Critical Thinking

Some analysts believe that we should not try to put economic values on the world's irreplaceable ecosystem services because their value is infinite. Do you agree with this view? Explain. What is the alternative?

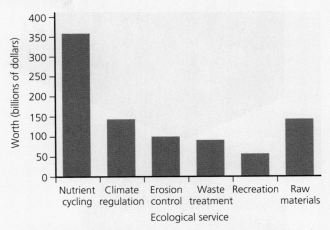

Figure 10-A Estimates of the annual global economic values of some ecosystem services provided by forests, compared to the value of the raw materials they produce (in billions of dollars).

(Compiled by the authors using data from Robert Costanza.)

often remove all the trees from an area in what is called a *clear-cut* (Figure 10-6b and Figure 10-7). Clear-cutting is the most efficient and often the least costly way to harvest trees, but it can do considerable harm to an ecosystem. Figure 10-8 summarizes some advantages and disadvantages of clear-cutting.

A variation of clear-cutting that allows a more sustainable timber yield without widespread destruction is *strip cutting* (Figure 10-6c). It involves clear-cutting a strip of trees along the contour of the land within a corridor narrow enough to allow natural forest regeneration within a few years. After regeneration, loggers cut another strip next to the first, and so on.

One of the major threats to forests is the unsustainable harvesting of trees (Concept 10-B). We examine this growing problem later in this chapter.

a. Selective cutting

Clear stream

b. Clear-cutting

Muddy stream

c. Strip cutting

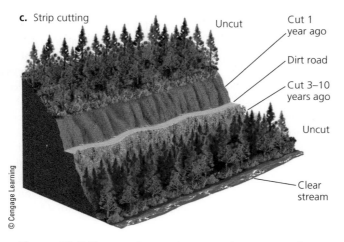

Uncut

Cut 1 year ago

Dirt road

Cut 3–10 years ago

Uncut

Clear stream

Figure 10-6 There are three major ways to harvest trees. **Question:** If you were cutting trees in a forest you owned, which method would you choose and why?

Fire, Insects, and Climate Change Can Threaten Forest Ecosystems

Two types of fires can affect forest ecosystems. *Surface fires* (Figure 10-9, left) usually burn only undergrowth and leaf litter on the forest floor. They may kill seedlings and small trees, but they spare most mature trees and allow most wild animals to escape.

Figure 10-7 Clear-cut forest.

Trade-Offs

Clear-Cutting Forests

Advantages	Disadvantages
Higher timber yields	Reduces biodiversity
Maximum profits in shortest time	Destroys and fragments wildlife habitats
Can reforest with fast-growing trees	Can lead to water pollution, flooding, and erosion, especially on steep slopes

Figure 10-8 There are advantages and disadvantages to clear-cutting a forest. **Questions:** Which single advantage and which single disadvantage do you think are the most important? Why?

Photo: © Kirill Livshitskiy/Shutterstock.com

Occasional surface fires have a number of ecological benefits. They:

- burn away flammable ground material such as dry brush and help to prevent more destructive fires;
- free valuable mineral nutrients tied up in slowly decomposing litter and undergrowth;
- release seeds from the cones of tree species such as lodgepole pines;
- stimulate the germination of certain tree seeds such as those of the giant sequoia and jack pine; and
- help to control destructive insects and tree diseases.

Figure 10-9 Surface fires (left) usually burn only undergrowth and leaf litter on a forest floor. They can help to prevent more destructive crown fires (right) by removing flammable ground material.

Left: David J. Moorhead/The University of Georgia. Right: age fotostock/SuperStock.

Another type of fire, called a *crown fire* (Figure 10-9, right), is an extremely hot fire that leaps from treetop to treetop, burning whole trees. Crown fires usually occur in forests that have not experienced surface fires for several decades, a situation that allows dead wood, leaves, and other flammable ground litter to accumulate. These rapidly burning fires can destroy most vegetation, kill wildlife, increase topsoil erosion, and burn or damage human structures in their paths. As part of a natural cycle, forest fires are not a major threat to forest ecosystems, except in the parts of the world where people intentionally burn forests to clear the land. Another threat to forests is the accidental or deliberate introduction of disease-causing organisms and destructive insects.

On top of these threats, projected climate change could harm many forests. Rising temperatures and increased drought influenced by a warmer atmosphere will likely make many forest areas more suitable for insect pests, which would then multiply and kill more trees. The resulting combination of drier forests and more dead trees could also increase the number and intensity of forest fires (**Concept 10-1B**).

quite high and each year amounts to the clearing of forest areas totaling roughly the area of Costa Rica.

These forest losses are concentrated in less-developed countries, especially those in the tropical areas of Latin America, Indonesia, and Africa. However, scientists are also concerned about the increased clearing of the northern boreal forests of Alaska, Canada, Scandinavia, and Russia, which together make up about one-fourth of the world's forested area.

According to the WRI, if current deforestation rates continue, about 40% of the world's remaining intact forests will have been logged or converted to other uses within two decades if not sooner. Clearing large areas of forests, especially old-growth forests, has important short-

Almost Half of the World's Forests Have Been Cut Down

Deforestation is the temporary or permanent removal of large expanses of forest for agriculture, settlements, or other uses. Surveys by the World Resources Institute (WRI) indicate that during the past 8,000 years, human activities have reduced the earth's virgin or frontier forest cover by about 47%, with most of this loss occurring in the last 60 years (Figure 10-10).

Worldwide, deforestation leads to a net loss of about 52,000 square kilometers (20,000 square miles) of forest per year, according to surveys by the FAO and the WRI. Some good news is that this is down from an annual loss of about 83,000 square kilometers (32,000 square miles) in the 1990s. However, the deforestation rate is still

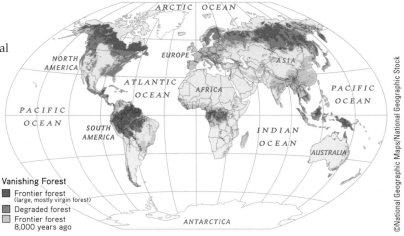

Vanishing Forest
- Frontier forest (large, mostly virgin forest)
- Degraded forest
- Frontier forest 8,000 years ago

Figure 10-10 Global deforestation over the past 8,000 years.

©National Geographic Maps/National Geographic Stock

Natural Capital Degradation

Deforestation

- Water pollution and soil degradation from erosion

- Acceleration of flooding

- Local extinction of specialist species

- Habitat loss for native and migrating species

- Release of CO_2 and loss of CO_2 absorption

© Cengage Learning

Figure 10-11 Deforestation has some harmful environmental effects that can reduce biodiversity and degrade the ecosystem services provided by forests (Figure 10-4, left).

term economic benefits (Figure 10-4, right column), but it also has a number of harmful environmental effects (Figure 10-11), including severe erosion and loss of topsoil (Figure 10-12) that was once renewed largely by forest ecosystems at no cost to us.

In 2011, the FAO reported that the net total forest cover in several countries, including the United States (see the Case Study that follows), changed very little or even increased between 2000 and 2010. Some of the increases resulted from natural reforestation by secondary ecological succession on cleared forest areas and abandoned croplands (see Figure 5-12, p. 110). Other increases in forest cover were due to the spread of commercial tree plantations (Figure 10-3, right) and to a global program, sponsored by the UNEP, to plant billions of trees throughout much of the world—many of them in tree plantations. China now leads the world in new forest cover, mostly due to its plantations of fast-growing trees.

CASE STUDY

Many Cleared Forests in the United States Have Grown Back

Forests cover about 30% of the U.S. land area, providing habitats for more than 80% of the country's wildlife species and containing about two-thirds of the nation's surface water. Today, forests in the United States (including tree plantations) cover more area than they did in 1920. The primary reason is that many of the old-growth forests that were cleared or partially cleared between 1620 and 1920 have grown back naturally through secondary ecological succession (Figure 10-13).

There are now fairly diverse second-growth (and in some cases third-growth) forests in every region of the United States except in much of the West. In 1995, environmental writer Bill McKibben cited forest regrowth in the United States—especially in the East—as "the great environmental success story of the United States, and in some ways, the whole world." Protected forests make up about 40% of the country's total forest area, mostly in the

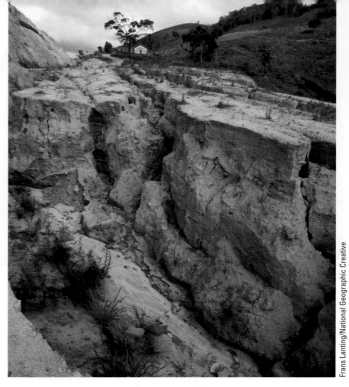

Frans Lanting/National Geographic Creative

Figure 10-12 Severe soil erosion occurred in this area of Brazil after a large area of the tropical Atlantic forest was removed.

National Forest System, which consists of 155 national forests managed by the U.S. Forest Service (USFS) (Figure 10-14).

On the other hand, since the mid-1960s, a large area of the nation's remaining old-growth and fairly diverse second-growth forests has been cut down and replaced with biologically simplified tree plantations. According to biodiversity researchers, this reduces overall forest biodiversity and it can disrupt important ecosystem services.

Tropical Forests Are Disappearing Rapidly

Tropical forests (see Figure 7-13, top, p. 156) cover about 6% of the earth's land area—roughly the area of the continental United States. Climatic and biological data suggest that mature tropical forests once covered at least twice as much area as they do today. Most of this loss of half of the world's tropical forests has taken place since 1950 (see Chapter 3, Core Case Study).

Satellite scans and ground-level surveys indicate that large areas of tropical rain forests and tropical dry forests are being cut rapidly in parts of Africa, Southeast Asia, and South America (see Figure 10-15 and Figure 3-1, p. 52)—especially in Brazil's vast Amazon Basin, which has more than 40% of the world's remaining tropical forests. Such clearing of trees, which absorb carbon dioxide as they grow, helps to hasten climate change because tropical forests absorb and store about one-third of the terrestrial carbon on the planet as part of the carbon cycle (see Figure 3-17, p. 66).

The drier climate increases the risk of more and bigger natural forest fires, which add more climate-changing

a. 1620

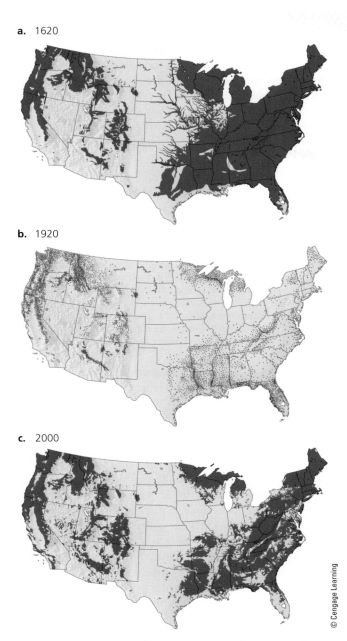

b. 1920

c. 2000

© Cengage Learning

Figure 10-13 In 1620, **(a)** when European settlers were moving to North America, forests covered more than half of the current land area of the continental United States. By 1920, **(b)** most of these forests had been decimated. Since then, a combination of secondary ecological succession and the expansion of commercial forests has resulted in greatly expanded forest cover. In 2000, **(c)** secondary and commercial forests covered about a third of U.S. land in the lower 48 states.

carbon dioxide to the atmosphere and further reduce the number of trees that remove carbon dioxide from the atmosphere. It also dehydrates the topsoil by exposing it to sunlight. The dry topsoil can then be blown away, making it difficult for new plants to become established. This can lead to an irreversible *ecological tipping point* (p. 45), beyond which a forest cannot grow back in the area and is then often replaced by tropical grassland, or savanna. Scientists project that if current burning and deforesta-

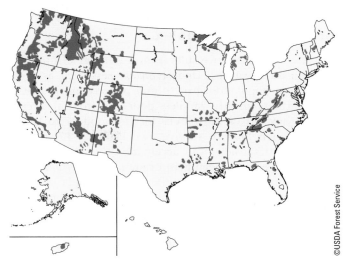

©USDA Forest Service

Figure 10-14 U.S. National Forests. *Question:* Why do think most national forests are in the western half of the United states?

Spacephotos/Age Fotostock

Figure 10-15 Natural capital degradation: Conversion of tropical rain forest into agricultural settlements near Santa Cruz la Sierra, Bolivia, between 1970 and 2000.

tion rates continue, 20–30% of the Amazon Basin will be turned into savanna in the next 50 years, and most of it could become savanna by 2080.

Studies indicate that at least half of the world's known species of terrestrial plants, animals, and insects live in tropical forests. Because of their specialized niches (see Figure 7-14, p. 157) many of these species are highly vulnerable to extinction when their forest habitats are destroyed or degraded. The FAO warns that, at the current rate of global tropical deforestation, as much as 50% of the world's remaining old-growth tropical forests will be gone or severely degraded by the end of this century (**Concept 10-1B**).

Causes of Tropical Deforestation Are Varied and Complex

Tropical deforestation results from a number of underlying and direct causes. Underlying causes, such as pressures from population growth and poverty, push subsistence farmers and the landless poor into tropical forests, where they cut or burn trees for firewood or to try to grow enough food to survive (see Figure 1-17, p. 18). Government subsidies can accelerate other direct causes such as large-scale logging and ranching by reducing the costs of timber harvesting, cattle grazing, and farming on vast plantations of crops such as soybeans and oil palms.

The major direct causes of deforestation vary in different tropical areas. Tropical forests in the Amazon and other South American countries are cleared (see Figure 1-4, p. 8) or burned primarily for cattle grazing and large soybean plantations. In Indonesia, Malaysia, and other areas of Southeast Asia, tropical forests are being replaced with major plantations of oil palm, which produces an oil used in cooking, cosmetics, and biodiesel fuel for motor vehicles (especially in Europe). In Africa, the primary direct cause of deforestation is people clearing plots for small-scale farming and harvesting wood for fuel.

Global trade has also furthered tropical forest degradation. In 2011, tropical forest researcher William Laurance reported that China was by far the biggest consumer of tropical timber. It is the destination for more than half of the world's timber shipments and much of the paper pulp shipped internationally. This includes large volumes of illegally harvested timber, worth about $15 billion a year, according to a 2011 report by Interpol and the World Bank. As a result, huge amounts of wood and paper products are produced in China and shipped all over the world through a process that includes the unsustainable cutting of tropical forests.

The degradation of a tropical forest usually begins when a road is cut deep into the forest interior for logging and settlement (Figure 10-5). Loggers then use selective cutting (Figure 10-6a) to remove the largest and best trees. When these big trees fall, many other trees often fall with them because of their shallow roots and the network of vines connecting the trees in the forest's canopy.

Burning is widely used to clear forest areas for agriculture, settlement, and other purposes. Healthy rain forests do not burn naturally, but roads, settlements, and other developments fragment them. The resulting patches of forest dry out and readily ignite. According to a 2005 study by forest scientists, widespread fires in the Amazon basin are changing weather patterns by raising temperatures and reducing rainfall. The resulting droughts dry out the forests and make them more likely to burn—an example of a runaway positive feedback loop (see Figure 2-16, p. 45).

🔍 CONSIDER THIS. . .

CONNECTIONS Burning Tropical Forests and Climate Change

The burning of tropical forests releases CO_2 into the atmosphere. Rising concentrations of this gas are warming the atmosphere and causing changes in the global climate. Scientists estimate that tropical forest fires account for at least 17% of all human-created greenhouse gas emissions, and that each year, they emit twice as much CO_2 as all of the world's cars and trucks emit. The large-scale burning of the Amazon rain forest accounts for 75% of Brazil's greenhouse gas emissions, making Brazil the world's fourth largest emitter of such gases, according to the National Inventory of Greenhouse Gases. And with these forests gone, even if savanna or second-growth forests replace them, far less CO_2 will be absorbed for photosynthesis, resulting in even more atmospheric warming.

Foreign corporations operating under government concession contracts do much of this logging. After they remove the best timber, they often sell the land to ranchers who burn the remaining timber to clear the land for cattle grazing. Within a few years, their cattle typically overgraze the land and the ranchers move their operations to another forest area. Then they sell the degraded land to farmers, who plow it up for large plantations of crops such as soybeans, or to settlers for small-scale farming. After a few years of crop growing and erosion from rain, the nutrient-poor topsoil is depleted of nutrients. Then the farmers and settlers move on to newly cleared land to repeat this environmentally destructive process. Scientists project that if current burning and deforestation rates continue, much of the Amazon forest will be converted to savanna by this failure to follow the biodiversity **principle of sustainability**. 🌿

🔍 CONSIDER THIS. . .

THINKING ABOUT Tropical Forests

Why should you care if most of the world's remaining tropical forests are burned or cleared or converted to savanna within your lifetime? What are three ways in which this might affect your life or the lives of any children and grandchildren that you might have?

10-2 How Should We Manage and Sustain Forests?

CONCEPT 10-2

We can sustain forests by emphasizing the economic value of their ecosystem services, removing government subsidies that hasten their destruction, protecting old-growth forests, harvesting trees no faster than they are replenished, and planting trees.

We Can Manage Forests More Sustainably

Biodiversity researchers and a growing number of foresters have called for more sustainable forest management. Figure 10-16 lists ways to achieve this goal (Concept 10-2). Certification of sustainably grown timber and of sustainably produced forest products can help con- GOOD NEWS

Solutions

More Sustainable Forestry

- Include ecosystem services of forests in estimates of their economic value

- Identify and protect highly diverse forest areas

- Stop logging in old-growth forests

- Stop clear-cutting on steep slopes

- Reduce road-building in forests and rely more on selective and strip cutting

- Leave most standing dead trees and larger fallen trees for wildlife habitat and nutrient cycling

- Put tree plantations only on deforested and degraded land

- Certify timber grown by sustainable methods

Figure 10-16 There are a number of ways to grow and harvest trees more sustainably (**Concept 10-2**). **Questions:** Which three of these methods of more sustainable forestry do you think are the best methods? Why?

sumers to play their part in reaching this goal (Science Focus 10.2).

Loggers could use more sustainable practices in tropical forests. For example, they can use sustainable selective cutting (Figure 10-6a) and strip cutting (Figure 10-6c) to harvest tropical trees for lumber instead of clear-cutting the forests (Figure 10-6b). They could also be more careful when cutting individual trees, taking care to cut canopy vines (lianas) before felling a tree and using the least obstructed paths to remove the logs. These practices would sharply reduce damage to neighboring trees.

Many economists are urging governments to begin making a shift to more sustainable forest management strategies by phasing out government subsidies and tax breaks that encourage forest degradation and deforestation and replacing them with forest-sustaining economic rewards. This would be an application of the full-cost pricing **principle of sustainability** (see Figure 1-5, p. 9 or back cover) because it would have the effect of raising prices on unsustainably produced timber and wood products. Costa Rica (**Core Case Study**) is taking a lead in using this approach. Governments can also encourage tree planting programs to help restore degraded forests. **GREEN CAREER:** Sustainable forestry

We Can Improve the Management of Forest Fires

In the United States, the Smokey Bear educational campaign undertaken by the Forest Service and the National Advertising Council has likely prevented many forest fires, saved many lives, and prevented billions of dollars in losses of trees, wildlife, and human structures. At the same time, this educational program has convinced much of the public that all forest fires are bad and should be prevented or put out. Ecologists warn that trying to prevent all forest fires can make matters worse by increasing the likelihood of destructive crown fires (Figure 10-9, right) due to the accumulation of highly flammable underbrush and smaller trees in some forests.

Ecologists and forest fire experts have proposed several strategies for reducing fire-related harm to forests and to people who use or live in the forests. One approach is to set small, contained surface fires to remove flammable small trees and underbrush in the highest-risk forest areas. Such *prescribed burns* require careful planning and monitoring to keep them from getting out of control. However, in 2012, research by ecologists William Baker and Mark Williams, which included studies of descriptions of past fires and tree ring data, indicated that natural low-level fires were not as widespread as other studies had indicated and that they rarely prevented more severe fires in much of the western United States. These researchers contend that expensive prescribed burns and thinning should not be used to prevent large fires in most western forests except to protect human structures.

A second strategy is to allow some fires on public lands to burn, thereby removing flammable underbrush and smaller trees, as long as the fires do not threaten human structures and life.

A third approach is to protect houses and other buildings in fire-prone areas by thinning trees and other vegetation in a zone of about 60 meters (200 feet) around them, and eliminating the use of highly flammable construction materials such as wood shingles.

A fourth approach is to thin forest areas that are vulnerable to fire by clearing away small fire-prone trees and underbrush under careful environmental controls. Many forest fire scientists warn that such thinning operations should not remove economically valuable medium-size and large trees for two reasons. *First,* these are usually the most fire-resistant trees. *Second,* their removal encourages dense growth of more flammable young trees and underbrush and leaves behind highly flammable *slash,* the debris left behind by a logging operation. Many of the worst fires in U.S. history burned through cleared forest areas that contained slash. A 2006 study by Forest Service researchers found that thinning forests without using prescribed burning to remove the slash can greatly increase rather than decrease the risk of heavy fire damage. The previously cited research in 2012 by ecologists Baker and Williams also questioned the value of thinning in most western forests.

We Can Reduce the Demand for Harvested Trees

According to the Worldwatch Institute and to forestry analysts, *up to 60% of the wood consumed in the United States is wasted unnecessarily.* This results from inefficient use of construction materials, excessive packaging, overuse of

SCIENCE FOCUS 10.2

CERTIFYING SUSTAINABLY GROWN TIMBER AND PRODUCTS SUCH AS THE PAPER USED IN THIS BOOK

Collins Pine is a forest products company that owns and manages a large area of productive timberland in the northeastern part of the U.S. state of California. Since 1940, the company has used selective cutting to help maintain the ecological and economic sustainability of its timberland.

Since 1993, Scientific Certification Systems (SCS) of Oakland, California, has evaluated the company's timber production. SCS, which is part of the nonprofit Forest Stewardship Council® (FSC®), was formed to develop environmentally sound and responsible forestry practices for use in certifying timber and timber products.

Each year, SCS evaluates Collins Pine's landholdings and has consistently found that their cutting of trees has not exceeded

long-term forest regeneration; roads and harvesting systems have not caused unreasonable ecological damage; topsoil has not been damaged; and downed wood (boles) and standing dead trees (snags) are left to provide wildlife habitat. As a result, SCS judges the company to be a good employer and a good steward of its land and water resources.

The FSC reported that, by 2012, about 5% of the world's forest area in 80 countries had been certified according to FSC standards. The FSC also certifies 5,400 manufacturers and distributors of wood products. The paper used in this book was produced with the use of responsibly managed timber, as certified by the FSC symbol, and contains recycled paper fibers in the body and cover stocks. Figure 10-B

Figure 10-B This Forest Stewardship Council® (FSC®) symbol certifies that the paper used in this textbook was produced from environmentally responsible sources with the use of 10% recycled fibers in the cover and body stocks. It also appears on the back cover of this book.

shows the FSC certification used for this textbook.

Critical Thinking

Should governments provide subsidies or tax breaks for sustainably grown timber to encourage this practice? Explain.

U.S. Department of Agriculture

Figure 10-17 Solutions: The pressure to cut trees to make paper could be greatly reduced by planting and harvesting a fast-growing plant known as kenaf.

junk mail, inadequate paper recycling, and failure to reuse or find substitutes for wooden shipping containers.

One reason for cutting trees is to provide pulp for making paper, but paper can be made by using fiber from sources other than trees. China uses rice straw and other agricultural residues to make much of its paper. Most of the small amount of tree-free paper produced in the United States is made from the fibers of a rapidly growing woody annual plant called *kenaf* (pronounced "kuh-NAHF," Figure 10-17). Kenaf and other nontree fibers such as hemp yield more paper pulp per area of land than tree farms do and require less use of pesticides and herbicides.

According to the USDA, kenaf is "the best option for tree-free papermaking in the United States" and could replace wood-based paper within 20–30 years. However, while timber companies successfully lobby for government subsidies to grow and harvest trees GOOD NEWS

Figure 10-18 Natural capital degradation: Haiti's deforested brown landscape (left) contrasts sharply with the heavily forested green landscape of its neighboring country, the Dominican Republic.

to make paper, there are no major lobbying efforts or subsidies for producing paper from kenaf and other alternatives to trees.

Another way to reduce the demand for tree cutting is to sharply reduce the use of throwaway paper products made from trees. We can instead choose reusable plates, cups, and cloth napkins and handkerchiefs. Recycled paper products—including soft toilet paper—are becoming more available every year.

Humans have always used trees for fuel, but now the demand for wood as fuel is becoming unsustainable in many areas (see the following Case Study).

CASE STUDY
Deforestation and the Fuelwood Crisis

About 50% of the wood harvested globally each year, and 75% of that harvested in less-developed countries, is burned directly for fuel or converted to charcoal fuel. More than 2 billion people in less-developed countries use fuelwood (see Figure 6-17, p. 135) and charcoal made from wood for heating and cooking.

However, most of these countries are suffering from fuelwood shortages because people are cutting trees for fuelwood and forest products 10–20 times faster than new trees are being planted. As the demand for fuelwood in urban areas of some less-developed countries exceeds the sustainable yield of nearby forests, expanding rings of deforested land encircle such cities. The FAO

warns that, by 2050, the demand for fuelwood could easily be 50% greater than the amount that can be sustainably supplied.

For example, Haiti, a country with 10.3 million people, was once a tropical paradise, 60% of it covered with forests. Now it is an ecological disaster. Largely because its trees were cut for fuelwood and to make charcoal, less than 2% of its land is now covered with trees (Figure 10-18). With the trees gone, soils have eroded away in many areas, making it much more difficult to grow crops.

The U.S. Agency for International Development funded the planting of 60 million trees over more than two decades in Haiti. However, the local people cut most of them down for firewood and to make charcoal before they could grow into mature trees. This unsustainable use of natural capital and failure to follow the **principles of sustainability** have led to a downward spiral of environmental degradation, poverty, disease, social injustice, crime, and violence in Haiti.

One way to reduce the severity of the fuelwood crisis in less-developed countries is to establish small plantations of fast-growing fuelwood trees and shrubs around farms and in community woodlots. Another approach to this problem is to encourage more efficient use of wood by providing villagers with more fuel-efficient wood stoves. Villagers could stop using wood if they could get access to solar ovens and electric hotplates powered by solar- or wind-generated electricity. Another option is stoves that burn renewable biomass, such as sun-dried roots of various gourds and squash plants, or methane produced from crop and animal wastes—technologies that could be provided inexpensively. All of these are also low-pollution options that would greatly reduce the number of deaths caused by indoor air pollution from open fires and poorly designed stoves.

In addition, scientists are looking for ways to produce charcoal for heating and cooking without cutting down trees. For example, Professor Amy Smith of the Massachusetts Institute of Technology is developing a way to make charcoal from the fibers in a waste product called bagasse, which is left over from sugar cane processing in Haiti, Brazil, and many other countries.

Countries such as South Korea, China, Nepal, and Senegal have used such methods to reduce fuelwood shortages, while sustaining biodiversity through reforestation and efforts to reduce topsoil erosion. Indeed, the mountainous country of South Korea is a global model for its successful reforestation following severe deforestation during the war between North and South Korea, which ended in 1953. Today, forests cover almost two-thirds of the country, and tree plantations near villages supply fuelwood on a sustainable basis.

There Are Several Ways to Reduce Tropical Deforestation

Analysts have suggested various ways to protect tropical forests and use them more sustainably (Figure 10-19).

At the international level, *debt-for-nature swaps* can make it financially attractive for countries to protect their tropical forests. In such swaps, participating countries act as custodians of protected forest reserves in return for foreign aid or debt relief. In a similar strategy, called *conservation concessions,* governments or private conservation organizations pay nations for agreeing to preserve their natural resources.

National governments can also take important steps to reduce deforestation. Between 2005 and 2011, Brazil cut its deforestation rate by 80% by cracking down on illegal logging and setting aside a conservation reserve in the Amazon Basin that is roughly equal to the size of France. In 2011, the UNEP reported that 75% of the protected areas created around the world between 2003 and 2011 are located in Brazil. Governments can also end subsidies that fund the construction of logging roads and instead subsidize more sustainable forestry (Figure 10-16) and the planting of trees in *reforestation programs.*

Consumers can reduce the demand for unsustainable and illegal logging in tropical forests by buying only wood and wood products that have been FSC-certified (see Science Focus 10.2). For building projects, using recy-

Solutions

Sustaining Tropical Forests

Prevention		Restoration
Protect the most diverse and endangered areas		Encourage regrowth through secondary succession
Educate settlers about sustainable agriculture and forestry		
Subsidize only sustainable forest use		Rehabilitate degraded areas
Protect forests through debt-for-nature swaps and conservation concessions		
Certify sustainably grown timber	Sustainably grown timber	Concentrate farming and ranching in already-cleared areas
Reduce poverty and slow population growth		

© Cengage Learning

Figure 10-19 These are some effective ways to protect tropical forests and to use them more sustainably (Concept 10-2). **Questions:** Which three of these solutions do you think are the best ones? Why?

Top: STILLFX/Shutterstock.com. Center: Manfred Mielke/USDA Forest Service Bugwood.org.

United Nations Environment Programme

Figure 10-20 In 1977, Wangari Maathai (1940–2011) founded the internationally acclaimed Green Belt Movement.

Figure 10-21 These women joined Africa's Green Belt Movement and are planting trees.

©Green Belt Movement

cled waste lumber is an option for some. People are also choosing wood substitutes such as recycled plastic building materials and bamboo.

Around the world, and especially in tropical forest areas, tree planting has become a high priority. The late Wangari Maathai (Figure 10-20), a Nobel Peace Prize winner, promoted and inspired tree planting in her native country of Kenya and throughout the world in what became the Green Belt Movement (Figure 10-21). Her efforts inspired the UNEP to implement a global effort to plant at least 1 billion trees a year beginning in 2006. By 2011, the year Maathai died, about 12.6 billion trees had been planted in 193 countries.

GOOD NEWS

10-3 How Should We Manage and Sustain Grasslands?

CONCEPT 10-3

We can sustain the productivity of grasslands by controlling the numbers and distribution of grazing livestock and by restoring degraded grasslands.

Some Rangelands Are Overgrazed

Grasslands provide many important ecosystem services, including soil formation, erosion control, chemical cycling, storage of atmospheric carbon dioxide in biomass, and maintenance of biodiversity.

After forests, grasslands are the ecosystems most widely used and altered by human activities. **Rangelands** are unfenced grasslands in temperate and tropical climates that supply *forage,* or vegetation for grazing (grass-eating) and browsing (shrub-eating) animals. Cattle, sheep, and goats graze on about 42% of the world's grassland. The 2005 UN Millennium Ecosystem Assessment—a 4-year study by 1,360 experts from 95 countries—estimated that this could increase to 70% by 2050. Livestock also graze in **pastures**, which are managed grasslands or fenced meadows often planted with domesticated grasses or other forage crops such as alfalfa and clover.

Blades of rangeland grass grow from the base, not at the tip as broadleaf plants do. Thus, as long as only the upper half of

the blade is eaten and its lower half remains, rangeland grass is a renewable resource that can be grazed again and again. Moderate levels of grazing are healthy for grasslands, because removal of mature vegetation stimulates rapid regrowth and encourages greater plant diversity.

Overgrazing occurs when too many animals graze for too long, damaging the grasses and their roots, and exceeding the carrying capacity of a rangeland area (Figure 10-22, left). Overgrazing reduces grass cover, exposes the topsoil to erosion by water and wind, and compacts the soil, which diminishes its capacity to hold water. Overgrazing also encourages the invasion of once-productive rangeland by species such as sagebrush, mesquite, cactus, and cheatgrass, which cattle will not eat.

USDA, Natural Resources Conservation Service

Figure 10-22 **Natural capital degradation:** To the left of the fence is overgrazed rangeland. The land to the right of the fence is lightly grazed.

Limited data from FAO surveys in various countries indicate that overgrazing by livestock has caused a loss in productivity in as much as 20% of the world's rangeland.

We Can Manage Rangelands More Sustainably

The most widely used way to manage rangelands more sustainably is to control the number of grazing animals and the duration of their grazing in a given area so the carrying capacity of the area is not exceeded (Concept 10-3). One method for doing this is called *rotational grazing*, in which cattle are confined by portable fencing to one area for a short time (often only 1–2 days) and then moved to a new location.

Cattle tend to aggregate around natural water sources, especially along streams or rivers lined by thin strips of lush vegetation known as *riparian zones*, and around ponds created to provide water for livestock. Overgrazing by cattle can destroy the vegetation in such areas (Figure 10-23, left). Ranchers can protect overgrazed land from further grazing by moving their livestock around and by fencing off these damaged areas, which eventually leads to their natural restoration by ecological succession (Figure 10-23, right). Ranchers can also move cattle around by providing supplemental feed at selected sites and by strategically locating watering ponds and tanks and salt blocks.

A more expensive and less widely used method of rangeland management is to suppress the growth of unwanted invader plants by the use of herbicides, mechanical removal, or controlled burning. A cheaper way to discourage unwanted vegetation in some areas is through controlled, short-term trampling by large numbers of livestock such as sheep, goats, and cattle that destroy the invasive plants' root systems. The least expensive way to deal with degradation of rangelands is to prevent it, by using methods described above and in the following Case Study.

CASE STUDY

Grazing and Urban Development in the American West—Cows or Condos?

The landscape is changing in U.S. ranch country. Since 1980, millions of people have moved to parts of the southwestern United States, and a growing number of ranchers are selling their land to developers. Housing developments, condos, and small "ranchettes" are creeping out from the edges of many southwestern cities and towns. Most people moving to the southwestern states value the landscape for its scenery and recreational opportunities, but uncontrolled urban development can degrade these very qualities.

For decades, some environmental scientists and environmentalists have sought with limited success to reduce overgrazing on these lands and, in particular, to reduce or eliminate livestock grazing permits on public lands. Now, because of the population surge in the Southwest and the resulting development of land, ranchers, ecologists, and environmentalists are joining together to try to preserve a number of cattle ranches as the best hope for sustaining the key remaining grasslands and the habitats they provide for native species.

One preservation strategy involves land trust groups, which pay ranchers for *conservation easements*—deed restrictions that bar future owners from developing the land. These groups are also pressuring local governments to zone the land in order to prevent large-scale development in ecologically fragile rangeland areas.

The main goal of groups that are trying to preserve western lands is to identify areas that are sustainable for grazing, areas that are best for sustainable urban development, and areas that should be neither grazed nor developed. This is in keeping with the win-win **principle of sustainability** (see Figure 1-5, p. 9 or back cover).

Figure 10-23 Natural capital restoration: In the mid-1980s, cattle had degraded the vegetation and soil on this stream bank along the San Pedro River in the U.S. state of Arizona (left). Within 10 years, the area was restored through secondary ecological succession (right) after grazing and off-road vehicle use were banned (Concept 10-3).

Left: U.S. Bureau of Land Management. Right: U.S. Bureau of Land Management.

10-4 How Should We Manage and Sustain Parks and Nature Reserves?

CONCEPT 10-4

Sustaining biodiversity will require more effective protection of existing parks and nature reserves, as well as the protection of much more of the earth's remaining undisturbed land area.

National Parks Face Many Environmental Threats

According to the International Union for the Conservation of Nature (IUCN), there are now more than 6,600 major national parks (see chapter-opening photo) located in more than 120 countries. However, most of these parks are too small to sustain many large animal species. And many parks suffer from invasions by harmful nonnative species that compete with and reduce the populations of native species. Some national parks are so popular that large numbers of visitors are degrading the natural features that make them attractive (see the Case Study that follows).

Parks in less-developed countries have the greatest biodiversity of all the world's parks, but only about 1% of these parklands are protected. Local people in many of these countries enter the parks illegally in search of wood, game animals, and other natural products that they need for their daily survival. Loggers and miners operate illegally in many of these parks, as do wildlife poachers who kill animals to obtain and sell items such as rhino horns (see Figure 9-15, p. 205), elephant tusks, and furs. Park services in most of the less-developed countries have too little money and too few personnel to fight these invasions, either by force or through education.

CASE STUDY

Stresses on U.S. Public Parks

The U.S. National Park System, established in 1912, includes 59 major national parks, sometimes called the country's crown jewels (Figure 10-24), along with 339 monuments and historic sites. States, counties, and cities also operate public parks.

Popularity is one of the biggest problems for many parks. Between 1960 and 2011, the number of recreational visitors to U.S. national parks more than tripled, reaching about 279 million. In some U.S. parks and other public lands, noisy and polluting dirt bikes, dune buggies, jet skis, snowmobiles, and other off-road vehicles destroy or damage fragile vegetation, disturb wildlife, and degrade the aesthetic experience for many visitors. Some visitors expect parks to have grocery stores, laundries, bars, and other such conveniences.

A number of parks also suffer damage from the migration or deliberate introduction of nonnative species. Euro-pean wild boars, imported into the state of North Carolina in 1912 for hunting, threaten vegetation in parts of the very popular Great Smoky Mountains National Park. Nonnative mountain goats in Washington State's Olympic National Park trample and destroy the root systems of native vegetation and accelerate soil erosion.

At the same time, native species—some of them threatened or endangered—are killed in, or illegally removed from, almost half of all U.S. national parks. This is what happened to the gray wolf in Yellowstone National Park until it was successfully reintroduced there after a 50-year absence (Science Focus 10.3).

Many U.S. national parks have become threatened islands of biodiversity surrounded by seas of commercial development. Nearby human activities that threaten wildlife and recreational values in many national parks include mining, logging, livestock grazing, coal-fired power plants, water diversion, and urban development. According to the National Park Service, air pollution, mostly from coal-fired power plants and dense vehicle traffic, degrades scenic views in many U.S. national parks more than 90% of the time.

The National Park Service estimated that in 2011, the national parks had at least an $8 billion backlog for long overdue maintenance and repairs to trails, buildings, and other park facilities. Some analysts say more of these funds could come from private concessionaires who provide campgrounds, restaurants, hotels, and other services for park visitors. They pay the government franchise fees averaging only about 6–7% of their gross receipts, and many large concessionaires with long-term contracts pay as little as 0.75%. Analysts say these percentages could reasonably be increased to around 20%.

Nature Reserves Occupy Only a Small Part of the Earth's Land

Most ecologists and conservation biologists believe the best way to preserve biodiversity is to create a worldwide network of protected areas. A map of the earth's remaining wildlands can be found in Supplement 6, Figure 16, p. S41, and Figure 17, pp. S42–S43, contains a map of the world's protected areas.

As of 2012, less than 13% of the earth's land area (not including Antarctica) was protected either strictly or partially in about 150,000 wildlife refuges, nature reserves, parks, and wilderness areas. This 13% figure is misleading because no more than 6% of the earth's land is strictly protected from potentially harmful human activities. In other words, *we have reserved 94% of the earth's land for human use.*

Conservation biologists call for full protection of at least 20% of the earth's land area in a global system of biodiversity reserves that would include multiple exam-

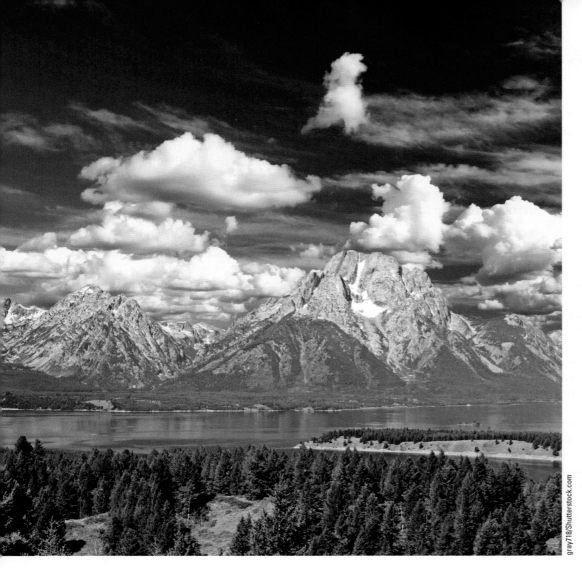

Figure 10-24 Grand Teton National Park in the U.S. state of Wyoming.

tainably without harming the inner core (see the Case Study that follows). Instead of shutting people out of the protected areas and likely creating enemies, this approach enlists local people as partners in protecting a reserve from unsustainable uses such as illegal logging and poaching. It is an application of the biodiversity and win-win **principles of sustainability**.

Another important concept in nature reserve design is the *habitat corridor*—a strip of protected land connecting two reserves that allows animals to migrate from one area to another as needed. By 2012, the United Nations had used these design concepts to create a global network of 621 *biosphere reserves* in 117 countries. However, most biosphere reserves fall short of these design ideals and receive too little funding for their protection and management.

ples of all the earth's biomes (**Concept 10-4**). In 2012, the IUCN estimated that making this investment in protecting a vital part of our life-support system would cost roughly $23 billion a year—more than 4 times the current expenditure.

Most developers and resource extractors oppose protecting even 13% of the earth's remaining undisturbed ecosystems. They contend that these areas might contain valuable resources that would add to current economic growth. Ecologists and conservation biologists disagree. They view protected areas as islands of biodiversity and ecosystem services that help to sustain all life and economies indefinitely and that serve as centers of future evolution. In other words, they serve as an "ecological insurance policy" for us and other species. (See the online Guest Essay on this topic by Norman Myers.)

In establishing nature reserves, the size of the reserve is important, as is the design. Whenever possible, conservation biologists call for using the *buffer zone concept* to design and manage nature reserves. This means strictly protecting an inner core of a reserve, usually by establishing two buffer zones in which local people can extract resources sus-

CASE STUDY

Identifying and Protecting Biodiversity in Costa Rica

For several decades, Costa Rica (**Core Case Study**) has been using government and private research agencies to identify the plants and animals that make it one of the world's most biologically diverse countries. The government has consolidated the country's parks and reserves into several large conservation areas, or *megareserves,* distributed throughout the country (Figure 10-25). Each reserve contains a protected inner core surrounded by two buffer zones that local and indigenous people can use for sustainable logging, crop farming, cattle grazing, hunting, fishing, and ecotourism. They were designed with the goal of sustaining about 80% of the country's rich biodiversity.

In addition to its ecological benefits, Costa Rica's biodiversity conservation strategy has paid off financially. Today, the country's largest source of income is its $1-billion-a-year tourism industry, almost two-thirds of which involves ecotourism (Figure 10-26).

SCIENCE FOCUS 10.3

REINTRODUCING THE GRAY WOLF TO YELLOWSTONE NATIONAL PARK

Around 1800, at least 350,000 gray wolves (Figure 10-C) roamed over about three-quarters of America's lower 48 states, especially in the West. They survived mostly by preying on abundant bison, elk, caribou, and deer. However, between 1850 and 1900, most of them were shot, trapped, or poisoned by ranchers, hunters, and government employees.

Ecologists recognize the important role that this keystone predator species once played in parts of the West, especially in the six northern Rocky Mountain states of Montana, Idaho, Wyoming (where Yellowstone National Park is located), Utah, Oregon, and Washington. The wolves culled herds of bison, elk, moose, and mule deer, and kept down coyote populations. By leaving some of their kills partially uneaten, they provided meat for scavengers such as ravens, bald eagles, ermines, grizzly bears, and foxes.

When wolves declined, herds of plant-browsing elk, moose, and mule deer expanded and devastated vegetation such as willow and aspen trees growing near streams and rivers. This led to increased soil erosion and to declining populations of other wildlife species such as beaver, which eat willow and aspen. This in turn affected species that depended on wetlands created by the beavers.

When Congress passed the U.S. Endangered Species Act in 1973, only a few hundred gray wolves remained outside of Alaska, primarily in Minnesota and Michigan. In 1974, the gray wolf was listed as an endangered species in the lower 48 states.

In 1987, the U.S. Fish and Wildlife Service (USFWS) proposed reintroducing gray wolves into the Yellowstone National Park ecosystem in Wyoming to help restore and sustain biodiversity and to prevent further environmental degradation of the ecosystem. The proposal brought angry protests, some from area ranchers who feared the wolves would leave the park and attack their cattle and sheep. Other objections came from hunters who feared the wolves would kill too many big-game animals, and from mining and logging companies that feared the government would halt their operations on wolf-populated federal lands.

In 1995 and 1996, federal wildlife officials caught gray wolves in Canada and relocated 31 of them in Yellowstone National Park. Scientists estimate that the long-term carrying capacity of the park is 110 to 150 gray wolves. By 2012, the park had about 97 gray wolves.

For more than a decade, wildlife ecologist Robert Crabtree and a number of other scientists have been studying the effects of reintroducing the gray wolf into Yellowstone National Park. They have put radio collars on most of the wolves to gather data and track their movements. They have also studied changes in vegetation and in the populations of various plant and animal species.

This research has suggested that the return of this keystone predator species has sent ecological ripples through the park's ecosystem. It has contributed to a decline in populations of elk, the wolves' primary food source, which had grown too large for the carrying capacity of much of the park. The leftovers of elk killed by wolves have been an important food source for scavengers such as bald eagles and ravens. Also, wary elk are gathering less near streams and rivers, which has helped to spur the regrowth of trees in these areas. This, in turn, has helped to stabilize and shade stream banks, lowering the water temperature and making better habitat for trout. Beavers seeking willow and aspen for food and for dam building materials have returned, and the dams they build establish additional wetlands and create more favorable habitat for aspens.

The wolves have also cut in half the population of coyotes—the top predators in the absence of wolves. This has reduced

Figure 10-C After becoming almost extinct in much of the western United States, the *gray wolf* was listed and protected as an endangered species in 1974.

Tom Kitchin/Tom Stack & Associates

coyote attacks on cattle from surrounding ranches and has led to larger populations of small animals such as ground squirrels, mice, and gophers, which are hunted by coyotes, eagles, and hawks. Overall, this experiment in ecosystem restoration has helped to reestablish and sustain some of the biodiversity that the Yellowstone ecosystem once had.

Between 1974 and 2012, the wolf population in the six Northern Rocky Mountain states and the Great Lakes grew from around 100 to around 6,000 individuals. In 2013, the USFWS proposed removing the gray wolf from the U.S. Endangered Species list in the lower 48 states. The agency considers the return of the gray wolf "one of the world's great conservation successes." If this rule is implemented, protection of this species would be left up to state wildlife agencies. Conservation groups plan to fight such a ruling in court, arguing that it could lead to greatly increased killing of the gray wolf.

Critical Thinking

Do you favor removing federal protection of the gray wolf as an endangered species? Explain.

Figure 10-26 These tourists are exploring a forest canopy via a walkway in Costa Rica's Monteverde Cloud Forest Reserve.

Figure 10-25 Solutions: Costa Rica has created several *mega-reserves*. Green areas are protected natural parklands and yellow areas are the surrounding buffer zones.

CONSIDER THIS. . .

CAMPUS SUSTAINABILITY
Protecting Green Space at Emory University

Emory University is located in Atlanta, Georgia (USA), where, as in many urban areas, green space is being lost to development every year. At Emory, however, more than half of the campus area is designated as green space, off-limits to development. In 2003, the university put in place a "no net loss of forest canopy" policy. It requires that for every tree removed to make way for a building or other development, another tree must be planted. For these and other efforts, Emory has received many notices in publications such as *Sierra Magazine* as being in the forefront of campus sustainability.

Protecting Wilderness Is an Important Way to Preserve Biodiversity

One way to protect undeveloped lands from human exploitation is to set them aside as **wilderness**—land officially designated as an area where natural communities have not been seriously disturbed by humans and where harmful human activities are limited by law (**Concept 10-4**). Theodore Roosevelt, the first U.S. president to set aside protected areas, summarized what we should do with wilderness: "Leave it as it is. You cannot improve it."

Some critics oppose protecting large areas for their scenic and recreational value for a relatively small number of people. They believe this keeps some areas of the planet from being economically useful to large numbers of people who need it now. Most conservation biologists disagree. To them, the most important reasons for pro-

tecting wilderness and other areas from exploitation and degradation involve the long-term needs of all species—to *preserve biodiversity* as a vital part of the earth's natural capital and to *protect wilderness areas as centers for evolution* in response to mostly unpredictable changes in environmental conditions. In other words, protecting wilderness areas is equivalent to investing in a long-term biodiversity insurance policy for the earth's life and ecosystems.

In the United States, conservationists have been trying to save wild areas from development since 1900. Overall, they have fought a losing battle. Not until 1964 did Congress pass the Wilderness Act. It allowed the government to protect undeveloped tracts of public land from development as part of the National Wilderness Preservation System (Figure 10-27).

The area of protected wilderness in the United States grew by nearly 12-fold between 1964 and 2012. Even so, only about 5% of all U.S. land is protected as wilderness—more than 54% of it in Alaska. Only about 2.7% of the land area of the lower 48 states is protected, most of it in the West.

One problem is that only a small number of the 709 wilderness areas in the lower 48 states are large enough to sustain all of the species they contain. Also, the system includes only 81 of the country's 233 distinct ecosystems. Most wilderness areas in the lower 48 states are also threatened habitat islands in a sea of development.

According to a 2008 study by conservation biologists Oliver Pergans and Patricia Zaradic, wilderness protection is being eroded because since the 1980s, fewer people are taking part in outdoor activities such as hiking, camping, fishing, and hunting. Instead, people are more often opting for indoor activities involving video games, the Internet, social networking, and viewing rented movies. Some scientists and other analysts warn that, as activities in the natural world become less popular, there could be less citizen pressure for protecting wilderness and for setting aside nature reserves.

Figure 10-27 The John Muir Wilderness in the U.S. state of California was among the first areas to be designated as wilderness.

10-5 What Is the Ecosystem Approach to Sustaining Biodiversity and Ecosystem Services?

CONCEPT 10-5
We can help to sustain terrestrial biodiversity by identifying and protecting severely threatened areas (biodiversity hotspots), sustaining ecosystem services, restoring damaged ecosystems (using restoration ecology), and sharing with other species much of the land we dominate (using reconciliation ecology).

The Ecosystems Approach: A Five-Point Strategy

Most biologists and wildlife conservationists believe that the best way to keep from hastening the extinction of wild species through human activities is to protect threatened habitats and ecosystem services. This *ecosystems approach* would generally employ the following five-point plan:

1. Map the world's terrestrial ecosystems and create an inventory of the species contained in each of them and the ecosystem services these ecosystems provide.
2. Identify terrestrial ecosystems that are resilient and that can recover if not overwhelmed by harmful human activities, along with ecosystems that are fragile and that need protection.
3. Locate and protect the most endangered terrestrial ecosystems and species, with emphasis on protecting plant biodiversity and ecosystem services.
4. Seek to restore as many degraded ecosystems as possible.
5. Make development *biodiversity-friendly* by providing significant financial incentives (such as tax breaks and write-offs) and technical help to private landowners who agree to help protect endangered ecosystems.

These steps have been applied in varying combinations in several locations around the world, as we discuss further throughout the rest of this chapter.

Protecting Global Biodiversity Hotspots Is an Urgent Priority

To protect as much of the earth's remaining biodiversity as possible, some biodiversity scientists urge the adoption of an *emergency action* strategy to identify and quickly protect **biodiversity hotspots**—areas especially rich in plant species that are found nowhere else and are in great danger of extinction (Concept 10-5). These areas have suffered serious ecological disruption, mostly because of rapid human population growth and the resulting pressure on natural resources and ecosystem services.

Animated Figure 10-28 Endangered natural capital: Biologists have identified these 34 biodiversity hotspots. Compare this map with the global map of the human ecological footprint, shown in Figure 1-12, p. 13. **Question:** Why do you think so many hotspots are located near coastal areas?

(Compiled by the authors using data from Center for Applied Biodiversity Science at Conservation International.)

Environmental scientist Norman Myers first proposed this idea in 1988 (see his online Guest Essay on this topic). Myers and his colleagues at Conservation International relied primarily on diversity of plant species as a way to identify biodiversity hotspots because data on plant diversity were the most readily available. They also viewed plant diversity as an indicator of animal diversity.

Figure 10-28 shows 34 terrestrial biodiversity hotspots identified by biologists. (For a map of hotspots in the United States, see Figure 9, p. S34, in Supplement 6.) Identified biodiversity hotspots cover only a little more than 2% of the earth's land surface, but they contain an estimated 50% of the world's flowering plant species and 42% of all terrestrial vertebrates (mammals, birds, reptiles, and amphibians), according to Conservation International. Yet, on average, only about 5% of the hotspots is truly protected with government funding and law enforcement.

Biodiversity hotspots are also home to a large majority of the world's endangered or critically endangered species, as well as to 1.2 billion people—one-fifth of the world's population. Says Norman Myers, "I can think of no other biodiversity initiative that could achieve so much at a comparatively small cost, as the hotspots strategy." In 2012, the IUCN began publishing its Red List of Ecosystems that are vulnerable, endangered, or critically endangered as a companion to its Red List of Threatened Species.

CASE STUDY

Madagascar: An Endangered Center of Biodiversity

Madagascar, the world's fourth largest island, lies in the Indian Ocean off the east coast of Africa. Most of its numerous species have evolved in near isolation from mainland Africa and all other land areas for at least 40 million years. As a result, roughly 90% of the more than 200,000 plant and animal species (Figure 10-29) found in this Texas-size biodiversity hotspot are found nowhere else on the earth.

Many of Madagascar's plant and animal species are among the world's most endangered, primarily because of habitat loss. People have cut down or burned more than 90% of Madagascar's original forests to get firewood and lumber and to make way for small farms, large terraced rice plantations, or cattle grazing. Only about 17% of the island's original vegetation remains, and as a result, Madagascar is one of the world's most eroded countries. Huge quantities of its precious topsoil have run off its hills, flowing as sediment in its rivers and emptying into its coastal waters (Figure 8-7, p. 172). All of this explains why Madagascar is one of the world's most threatened biodiversity hotspots.

Since 1984, the government, conservation organizations, and scientists (see Individuals Matter 10.1) worldwide have mounted efforts to slow the country's rapid loss of biodiversity. However, such efforts are hampered by Madagascar's rapid population growth. Between 1994 and 2012, its population grew from 12 million to 22 million and is projected to more than double, growing to about 54 million by 2050. The country is also very poor with 90% of its population struggling to survive on the equivalent of less than $2.25 per day. This puts pressure on its dwindling forest resources.

Despite the efforts to preserve Madagascar's biodiversity, less than 3% of its land area is officially protected. To reduce the rapid losses of biodiversity, the country will need to slow its population growth drastically and teach many of its people how to make a living from reforestation, ecotourism, and more sustainable uses of its forests, wildlife, and soil resources.

Protecting Ecosystem Services Is Also an Urgent Priority

Another way to help sustain the earth's biodiversity is to identify and protect areas where vital ecosystem services (see the orange boxed labels in Figure 1-3, p. 7) are being impaired enough to reduce biodiversity and harm local residents.

This approach has received more attention since the release of the 2005 UN Millennium Ecosystem Assessment. It identified key ecosystem services that provide numerous ecological and economic benefits, including those provided by forests (Figure 10-4). The study pointed out that humans are degrading or overusing about 60% of the ecosystem services provided by various ecosystems around the world, and it outlined ways to help sustain these vital services.

The ecosystem services approach recognizes that most of the world's ecosystems are already dominated or influenced by human activities and that such pres-sures are increasing as the human population, urbanization, resource use, and the human ecological footprint all expand (see Figures 1-12, p. 13, and 1-13, p. 14). Proponents of this approach recognize that setting aside and protecting reserves and wilderness areas, especially highly endangered biodiversity hotspots (Figure 10-28) and ecosystems, are vital. However, they contend that such efforts by themselves will not significantly slow the steady erosion of the earth's biodiversity and ecosystem services, nor will they help to reduce the poverty that plays a major role in ecosystem degradation.

Proponents of this strategy would also identify highly stressed *liferaft ecosystems.* These would be areas where poverty levels are high and where a large part of the economy depends on various ecosystem services that are being degraded severely enough to threaten the well-being of people and other forms of life. In such areas, residents, public officials, and conservation scientists would work together to develop strategies to help protect human communities along with natural biodiversity and the ecosystem services that support all life and economies. Thus, instead of pitting nature against people, this approach applies the win-win **principle of sustainability** (see Figure 1-5, p. 9 or back cover).

We Can Rehabilitate and Partially Restore Ecosystems That We Have Damaged

Almost every natural place on the earth has been affected or degraded to some degree by human activities. We can at least partially reverse much of this harm through **ecological restoration**: the process of repairing damage caused by humans to the biodiversity and ecosystem services pro-

Luke Dollar: A National Geographic Emerging Explorer Working to Save Biodiversity in Madagascar

Courtesy of Luke Dollar

Luke Dollar, with a PhD in ecology from Duke University, is Associate Professor of Biology at Pfeiffer University. He is also a research associate with the Duke University Primate Center and founder of the Carnivore Conservation and Research Trust.

For 18 years, this boots-on-the-ground ecologist has been conducting research in Madagascar to help preserve its threatened biodiversity. He is an expert on the fosa (also spelled "fossa") shown here in his arms and in the background photo. This top predator and keystone species is found nowhere else in the world. It is threatened by the loss of about 97% of its habitat, mostly due to deforestation.

Each year, Dollar conducts Earthwatch Institute expeditions, during which students and citizens participate in his research and learn about Madagascar's amazing but endangered biodiversity. He also works to help locals find alternatives to traditional slash-and-burn agriculture in order to help reduce the pressure on Madagascar's remaining forests and wildlife. Dollar says, "I wake up every morning knowing I am one of the luckiest guys on Earth because I am doing exactly what I want to do and it's going to make a difference."

Background photo: Hotshotsworldwide/Dreamstime.com

vided by ecosystems. Examples include replanting forests (Science Focus 10.4), reintroducing native species (Science Focus 10.3), removing invasive species, freeing river flows by removing dams, and restoring grasslands, coral reefs, wetlands, and stream banks (Figure 10-23, right).

Ecological restoration takes time, which is not a problem for the earth, but it could be for our species. During the past 12,000 years and especially during the past 100 years, we have dramatically altered the land and life that took 3.5 billion years to develop on much of the planet. Through

secondary succession (Figure 5-12, p. 110) many disturbed areas of land can return to forest or to other natural ecosystems, but this can take hundreds of years. Some scientists are concerned that much of the damage we have done will not be mended by nature in time for us to be able to depend on the damaged ecosystems for maintaining our life-support system. When we degrade natural capital (see Figure 1-3, p. 7), including the earth's biodiversity and ecosystem services that are crucial parts of our life-support system, in the long run, we threaten ourselves.

ECOLOGICAL RESTORATION OF A TROPICAL DRY FOREST IN COSTA RICA

Costa Rica (**Core Case Study**) is the site of one of the world's largest ecological restoration projects. In the lowlands of its Guanacaste National Park, a small, tropical dry forest was burned, degraded, and fragmented for large-scale conversion of the area to cattle ranches and farms. Now it is being restored and reconnected to a rain forest on nearby mountain slopes. The goal is to eliminate damaging nonnative grasses and reestablish a tropical dry-forest ecosystem during the next 100–300 years.

Daniel Janzen (Figure 10-D), professor of conservation biology at the University of Pennsylvania and a leader in the field of restoration ecology, helped to galvanize international support for this restoration project. He used his own MacArthur Foundation grant money to purchase this Costa Rican land for designation as a national park. He also raised more than $10 million for restoring the park.

Janzen recognizes that ecological restoration and protection of the park will fail unless the people in the surrounding area believe they will benefit from such efforts. His vision is to see that the nearly 40,000 people who live near the park play an essential role in the restoration of the degraded forest, a concept he calls *biocultural restoration*.

By actively participating in the project, local residents reap educational, economic, and environmental benefits. Local farmers are paid to remove nonnative species, to sow large areas with tree seeds, and to plant tree seedlings started in Janzen's lab. Local grade school, high school, and university students and citizens' groups study the park's ecology during field trips. The park's location near the Pan American Highway makes it an ideal area for ecotourism, which stimulates the local economy. GOOD NEWS

The project also serves as a training ground in tropical forest restoration for scientists from all over the world. Research scientists working on the project give guest classroom lectures and lead field trips.

In a few decades, today's Costa Rican children will be running the park and the local political system. If they understand

Figure 10-D Daniel Janzen, Professor of Conservation Biology at the University of Pennsylvania.

the ecological importance of their local environment, they will be more likely to protect and sustain its biological resources. Janzen believes that education, awareness, and involvement—not guards and fences—are the best ways to restore degraded ecosystems and to protect largely intact ecosystems from unsustainable use. This is an application of the biodiversity and win-win **principles of sustainability**.

Critical Thinking

Would such an ecological restoration project be possible in the area where you live? Explain.

By studying how natural ecosystems recover, scientists are learning how to speed up ecological succession processes using a variety of approaches, including the following four:

1. *Restoration:* returning a degraded habitat or ecosystem to a condition as similar as possible to its natural state. The problem is that in some cases, we do not know what the original natural state of an area was, and restoring it to its estimated natural state often is no longer feasible because environmental conditions have changed.
2. *Rehabilitation:* turning a degraded ecosystem into a functional or useful ecosystem without trying to restore it to its original condition. Examples include removing pollutants from abandoned mining or industrial sites and replanting trees to reduce soil erosion in clear-cut forests.
3. *Replacement:* replacing a degraded ecosystem with another type of ecosystem. For example, a degraded forest could be replaced by a productive pasture or tree plantation.
4. *Creating artificial ecosystems:* for example, artificial wetlands have been created in some areas to help reduce flooding and to treat sewage.

Researchers have suggested a science-based, four-step strategy for carrying out most forms of ecological restoration and rehabilitation.

- *First,* identify the causes of the degradation (such as pollution, farming, overgrazing, mining, or invasive species).
- *Second,* stop the abuse by eliminating or sharply reducing these factors. This would include removing toxic soil pollutants, improving depleted soil by adding nutrients and new topsoil, preventing fires, and controlling or eliminating disruptive nonnative species.
- *Third,* if necessary, reintroduce key species to help restore natural ecological processes, as was done with gray wolves in the Yellowstone ecosystem (Science Focus 10.3).

- *Fourth*, protect the area from further degradation and allow secondary ecological succession to occur (Figure 10-23, right).

Some analysts worry that ecological restoration could encourage continuing environmental destruction and degradation by suggesting that any ecological harm we do can be undone. Restoration ecologists disagree. They point out that so far, we have been able to protect only about 5% of the earth's land from the effects of human activities, so ecological restoration is badly needed for many of the world's ecosystems.

We Can Share Areas We Dominate with Other Species

Ecologist Michael L. Rosenzweig suggests that we develop a form of conservation biology called **reconciliation ecology**. This science focuses on inventing, establishing, and maintaining new habitats to conserve species diversity in places where people live, work, or play. In other words, the focus is on learning how to share with other species some of the spaces we dominate.

For example, people can learn how protecting local wildlife and ecosystems can provide economic resources for their communities by encouraging sustainable forms of ecotourism. In the Central American country of Belize, for instance, conservation biologist Robert Horwich has helped to establish a local sanctuary for the black howler monkey. He convinced local farmers to set aside strips of forest to serve as habitats and corridors through which these monkeys can travel. The reserve, run by a local women's cooperative, has attracted ecotourists and biologists. The community has built a black howler museum, and local residents receive income by housing and guiding ecotourists and biological researchers.

However, without proper controls popular ecotourism sites can be degraded when they are overrun with visitors. In addition, developers have erected hotels and other tourist facilities that have added to the degradation, often without adding many jobs for local residents.

In many areas of the world, people are learning how to protect vital insect pollinators, such as native butterflies and bees (see Chapter 9, Core Case Study, p. 190), which are vulnerable to insecticides and habitat loss. Neighborhoods and municipal governments are doing this by agreeing to reduce or eliminate the use of pesticides on their lawns, fields, golf courses, and parks. Neighbors also work together to plant gardens of flowering plants as a source of food for bees and other pollinating insect species. Some neighborhoods and farmers have built devices made of wood and plastic straws that serve as beehives.

People have also worked together to protect bluebirds within human-dominated habitats where most of the bluebirds' nesting trees have been cut down and bluebird populations have declined. Special boxes were designed to accommodate nesting bluebirds, and the North American Bluebird Society has encouraged Canadians and Americans to use these boxes on their properties and to keep house cats away from nesting bluebirds. Now bluebird numbers are rising again.

There are many other examples of individuals and groups working together on projects to restore grasslands, wetlands, streams, and other degraded areas. They involve applying the biodiversity and win-win **principles of sustainability** (see Figure 1-2, p. 6, and Figure 1-5, p. 9, or back cover).

Figure 10-30 lists some ways in which you can help to sustain the earth's terrestrial biodiversity.

What Can You Do?

Sustaining Terrestrial Biodiversity

- Plant trees and take care of them
- Recycle paper and buy recycled paper products
- Buy sustainably produced wood and wood products and wood substitutes such as recycled plastic furniture and decking
- Help restore a degraded forest or grassland
- Landscape your yard with a diversity of native plants

© Cengage Learning

Figure 10-30 Individuals matter: These are some ways in which you can help to sustain terrestrial biodiversity. **Questions:** Which two of these actions do you think are the most important? Why? Which of these things do you already do?

Big Ideas

- The economic values of the important ecosystem services provided by the world's ecosystems are far greater than the value of raw materials obtained from those systems.

- We can manage forests, grasslands, and nature reserves more effectively by protecting more land and by preventing overuse and degradation of these areas and the renewable resources they contain.

- We can sustain terrestrial biodiversity and ecosystem services by protecting biodiversity hotspots and ecosystem services, restoring damaged ecosystems, and sharing with other species much of the land we dominate.

Eduardo Rivero/Shutterstock.com

In this chapter, we looked at how humans are destroying or degrading terrestrial biodiversity in a variety of ecosystems. We also saw how we can reduce this destruction and degradation by using the earth's resources more sustainably. The **Core Case Study** introduced us to what Costa Rica is doing to protect and restore its precious biodiversity—primarily by protecting a larger percentage of its land from unsustainable development (see chapter-opening photo) than any other country. This strategy has also paid off financially by promoting the country's various parks and reserves as centers for ecotourism.

We also discussed the importance of preserving what remains of richly diverse and highly endangered ecosystems (biodiversity hotspots) and of sustaining the earth's ecosystem services. We examined the key strategy of restoring or rehabilitating some of the ecosystems we have degraded (restoration ecology). In addition, we explored ways in which people can share with other species some of the land they occupy in order to help sustain biodiversity (reconciliation ecology).

Preserving biodiversity involves applying the three scientific **principles of sustainability** (see Figure 1-2, p. 6 or back cover, top). First, it means respecting biodiversity and understanding the value of sustaining it. Then, in helping to sustain biodiversity by planting trees, for example, we also help to restore and preserve the flows of energy from the sun through food webs and the cycling of nutrients within ecosystems. Also, if we rely less on fossil fuels and more on direct solar energy and its indirect forms, such as wind and flowing water, we will generate less pollution and interfere less with chemical cycling and other forms of natural capital that sustain biodiversity and our own lives and societies.

We can also apply the three social science **principles of sustainability** (see Figure 1-5, p. 9 or back cover, bottom) to help preserve biodiversity. First, placing economic value on nature's ecosystem services tends to make the preservation of those services more important. Second, we know that people can successfully work together to find win-win solutions to problems of environmental degradation. And third, all of these efforts are driven by our ethical responsibility to sustain biodiversity for current and future generations.

Chapter Review

Core Case Study

1. Summarize the story of Costa Rica's efforts to preserve its rich biodiversity.

Section 10-1

2. What are the two key concepts for this section? Distinguish among **old-growth (primary) forests**, **second-growth forests**, and **tree plantations** (**tree farms** or **commercial forests**). What major ecological and economic benefits do forests provide? Describe the efforts of scientists and economists to put a price tag on the major ecosystem services provided by forests and other ecosystems.

3. Explain how building roads into previously inaccessible forests can harm the forests. Distinguish among selective cutting, clear-cutting, and strip cutting in the harvesting of trees. What are the major advantages and disadvantages of clear-cutting forests? What are two types of forest fires? What are some possible ecological benefits of occasional surface fires? How can the introduction of insects and disease-causing organisms harm forests? Explain some possible effects of climate change on forests.

4. What is **deforestation** and what parts of the world are experiencing the greatest forest losses? List some major harmful environmental effects of deforestation. Describe the encouraging news about reforestation in the United States. Explain how increased reliance on tree plantations can reduce overall forest biodiversity and degrade forest topsoil. How serious is tropical deforestation? Describe some major causes of tropical deforestation. Explain how widespread tropical deforestation can irreversibly convert a tropical forest to a tropical grassland.

Section 10-2

5. What is the key concept for this section? Describe four ways to manage forests more sustainably. What is certified sustainably grown timber? What are four ways to reduce the harm caused by forest fires to forests and to people? What is a prescribed fire? What are three ways to reduce the need to harvest trees? Describe the global fuelwood crisis. What are five ways to protect tropical forests and use them more sustainably? Describe the efforts by Wangari Maathai and the Green Belt Movement to restore forests.

Section 10-3

6. What is the key concept for this section? Distinguish between **rangelands** and **pastures**. What is **overgrazing** and what are its harmful environmental effects? What are three ways to reduce overgrazing and use rangelands more sustainably? Explain why there is a growing conflict between grazing and urban development in the American southwest.

Section 10-4

7. What is the key concept for this section? What are the major environmental threats to national parks in the world and in the United States? Why are many U.S. national parks considered to be threatened islands of biodiversity? Describe some of the ecological effects of reintroducing the gray wolf to Yellowstone National Park. What percentage of the world's land has been set aside and protected as nature reserves, and what percentage do conservation biologists believe should be protected?

8. How should nature reserves be designed and managed? What is the buffer zone concept? What is a biosphere reserve? Describe what Costa Rica has done to identify and protect much of its biodiversity. What is **wilderness** and why is it important? Summarize the controversy over protecting wilderness in the United States.

Section 10-5

9. What is the key concept for this section? Summarize the five-point strategy recommended by biologists for protecting ecosystems. What is a **biodiversity hotspot** and why is it important to protect such areas? About how much of the earth's land surface is occupied by hotspots and what percentages of the world's flowering plants and terrestrial vertebrates live in these areas? Explain the importance of protecting ecosystem services and list three ways to do this.

10. Define **ecological restoration**. What are four approaches to restoration? Summarize the science-based, four-point strategy for carrying out ecological restoration and rehabilitation. Describe the ecological restoration of the Guanacaste National Park forest in Costa Rica. Define and give three examples of **reconciliation ecology**. What are this chapter's *three big ideas*? Explain the relationship between preserving biodiversity as it is done in Costa Rica and the scientific and social science **principles of sustainability**.

Note: Key terms are in bold type.

Critical Thinking

1. Why do you think Costa Rica has set aside a larger percentage of its land for biodiversity conservation than the United States has? Should the United States reserve more of its land for this purpose? Explain.

2. If we fail to protect a much larger percentage of the world's remaining old-growth forests and tropical rain forests, what are three harmful effects that this failure is likely to have on any children and grandchildren you might have?

3. In the early 1990s, Miguel Sanchez, a subsistence farmer in Costa Rica, was offered $600,000 by a hotel developer for a piece of land that he and his family had been using sustainably for many years. The land, which contained an old-growth rain forest and a black sand beach, was surrounded by an area under rapid development. Sanchez refused the offer. Explain how Sanchez's decision was an application of one of the social science **principles of sustainability** (see Figure 1-5, p. 9 or back

cover). What would you have done if you were in Sanchez's position? Explain your thinking.

4. There is controversy over whether Yellowstone National Park should be accessible by snowmobile during winter. Conservationists and backpackers who use cross-country skis and snowshoes for winter excursions in the park are opposed to this. They contend that snowmobiles are noisy and that they pollute the air, destroy vegetation, and disrupt some of the park's wildlife. Proponents say that snowmobiles should be allowed so that snowmobilers can enjoy the park during winter when cars are mostly banned. They point out that new snowmobiles are made to cut pollution and noise. As of 2013, this issue was yet to be resolved on a long-term basis. What is your view on this issue? Explain.

5. In 2009, environmental analyst Lester R. Brown estimated that reforesting the earth and restoring its degraded rangelands would cost about $15 billion a year. Suppose the United States, the world's most afflu-

ent country, agreed to put up half of this money, which would amount to about $23 per American citizen per year. Would you support doing this? Explain. If your answer was yes, what part or parts of the federal budget would you decrease to come up with these funds?

6. Should more-developed countries provide at least half of the money needed to help preserve remaining tropical forests in less-developed countries? Explain. Do you think that the long-term economic and ecological benefits of doing this would outweigh the short-term economic costs? Explain.

7. Are you in favor of establishing more wilderness areas in the United States, especially in the lower 48 states (or in the country where you live)? Explain. What might be some drawbacks of doing this?

8. You are a defense attorney arguing in court for preserving an old-growth forest that developers want to clear for cropland and other uses. Give your three strongest arguments for preserving this ecosystem. How would you counter the argument that preserving the forest would harm the local economy by causing a loss of jobs in the timber industry?

Doing Environmental Science

Pick an area near where you live or go to school that hosts plants and animals. It could be a yard, an abandoned lot, a park, a forest, or some part of your campus. Visit this area at least 3 times and make a survey of the plants and animals that you find there, including any trees, shrubs, groundcover plants, insects, reptiles, amphibians, birds, and mammals. Also, take a small sample of the topsoil and find out what organisms are living there. (Be careful to get permission from whomever owns or manages the land before doing any digging.) Using guidebooks and other resources to help identify different species, record your findings and categorize them into the general types of organisms listed above. Then think of, and record, five ecosystem services that you think some or all of these organisms provide.

Global Environment Watch Exercise

Go to the *Forests and Deforestation* portal and next to the Statistics heading click "View All." On this page, click on "Share of Tropical Deforestation, 2000–2005." Choose one of these countries and research the deforestation in this country further (tip: use the World Map feature). Write a report on your findings and include possible solutions for this problem. Solutions may include those legislated by governments, as well as those being tried by private individuals or companies.

Ecological Footprint Analysis

Use the table below to answer the questions that follow.

Country	Area of Tropical Rain Forest (square kilometers)	Area of Deforestation per Year (square kilometers)	Annual Rate of Tropical Forest Loss
A	1,800,000	50,000	
B	55,000	3,000	
C	22,000	6,000	
D	530,000	12,000	
E	80,000	700	

© Cengage Learning

1. What is the annual rate of tropical rain forest loss, as a percentage of total forest area, in each of the five countries? Answer by filling in the blank column in the table.

2. What is the annual rate of tropical deforestation collectively in all of the countries represented in the table?

3. According to the table, and assuming the rates of deforestation remain constant, which country's tropical rain forest will be completely destroyed first?

4. Assuming the rate of deforestation in country C remains constant, how many years will it take for all of its tropical rain forests to be destroyed?

5. Assuming that a hectare (1.0 hectare = 0.01 square kilometer) of tropical rain forest absorbs 0.85 metric tons (1 metric ton = 2,200 pounds) of carbon dioxide per year, what would be the total annual growth in the carbon footprint (carbon emitted but not absorbed by vegetation because of deforestation) in metric tons of carbon dioxide per year for each of the five countries in the table?

1. The largest area of old-growth forest in the world is located in
   a. the United States.
   b. Russia.
   c. Costa Rica.
   d. Mexico.
   e. China.

2. The best example of selective cutting is
   a. loggers cut small groups of intermediate or mature trees, resulting in less erosion and loss of nutrients on the hillside.
   b. loggers cut all the trees in an area, resulting in a loss of nutrients in the soil and in sediment gathering in the stream downslope.
   c. loggers cut a thin strip of trees along a hillside, the forest naturally regenerates, and loggers then cut a thin strip of trees next to the regenerated area.
   d. in a managed forest containing one or two species of trees, one section is clear-cut and replanted. It will be clear-cut again when the trees are large enough.
   e. the loggers' goal is to maintain forest flora and fauna in a natural state. Therefore, they sometimes do not use fossil fuels to remove cut trees from the forest, but instead use horses to transport the trees.

3. All of the below would result in an area that was recently clear-cut EXCEPT
   a. more water flows on sloped, denuded land, resulting in erosion, which increases the likelihood of mudslides.
   b. a loss of soil nutrients, because of the increase in wind and water erosion.
   c. an increase in erosion, because water flows faster on denuded slopes.
   d. an increase in biodiversity, because land with fewer trees and less sunlight is more inviting to a variety of species.
   e. an increase in water pollution as sediment flows into streams.

4. The best example of a crown fire is
   a. the type of fire that burns seedlings, small trees, and leaf litter, but not mature trees and most animals.
   b. controlled fires that are deliberately set to reduce the occurrence of natural fires.
   c. fire that allows dead wood, leaves, and other flammable ground litter to accumulate.
   d. fire that is deliberately set, frequently in the tropical rain forest, after the trees have been cut down.
   e. fires that occur along newly exposed edges of forest after fragmentation has occurred.

5. The huge demand for fuel wood for cooking in developing countries is leading to a downward spiral of environmental degradation. All of the following have resulted from the overuse of this resource EXCEPT
   a. poverty, because soil erosion in denuded areas results in loss of nutrients, so local farmers cannot grow enough crops.
   b. increased poverty that leads to an increase in disease because people are not getting enough nutritious food to be healthy.
   c. preservation, because as local people see the results of deforestation, they demand that remaining forests be protected.
   d. crime, because poor people will sometimes steal food for their children.
   e. violence, because people without enough food for their families will sometimes fight other people for it.

6. One effect of our overgrazing of the natural rangelands is
   a. increased biodiversity because of the increase in plants cattle will not eat, such as mesquite, sagebrush, and cactus, as well as the insects and mammals they attract.
   b. decreased solar power because soil is compacted and can no longer support solar power cells.
   c. inversion of the human population pyramid as less vegetation is available and ranchers must sell their cattle because of the lack of rangeland available to them.
   d. soil erosion, because cattle damage the grass roots and the grass can no longer regrow.
   e. increased soil porosity, because cattle are not present to compact the land as they walk and stand on it.

7. A farmer in a developing nation will try to put the maximum amount of crops on his land. This farmer would most likely use which method of tree cutting?
   a. The farmer would cut a strip of trees and when they grow back, cut another strip of trees, preventing mass erosion to protect his crops.
   b. The farmer would cut most of the trees and leave those that produce seeds so new trees can grow back.
   c. The farmer would remove all the mature trees and leave the younger ones to provide shade for crops that need shade to survive.
   d. The farmer would cut all the trees in the area regardless of the nutrient loss and erosion that would occur.
   e. The farmer would remove only small groups of older trees so more nutrients would be available for his crops.

8. A habitat corridor has all of the following characteristics EXCEPT
   a. animals that need larger ranges can move between protected areas.
   b. a species whose habitat has become damaged can move to another area.
   c. a lack of predators because prey species can easily escape through the corridors.
   d. animals can migrate from one area to another to find food.
   e. individuals can mate with others in different areas and increase their species' genetic diversity.

9. Biodiversity hotspots are all of the following EXCEPT
   a. areas that contain plant species that are in danger of extinction.
   b. they are protected by an international treaty that was signed by more than 100 countries in 2012.
   c. the areas total just greater than 2% of Earth's land surface.
   d. the hotspots include most of the west coast of South America.
   e. they are areas especially rich in plant species that are found nowhere else.

Questions 10–12 refer to the following ecological restoration processes.
   a. Restoration
   b. Rehabilitation
   c. Replacement
   d. Creating artificial ecosystems
   e. Mitigation

10. Covering a landfill with topsoil and planting with vegetation

11. Converting a Midwestern farm back to original prairie grassland

12. Creating a man-made wetland to help treat sewage

13. Which of the following would be an example of restoration of disturbed lands?
   a. Writing a law that makes it illegal to build on the disturbed lands.
   b. Building a wall to prevent erosion during severe storms such as hurricanes.
   c. Building biological corridors between human habitations.
   d. Converting a forest into a national park.
   e. Replanting a destroyed area back to its natural state.

11 Sustaining Aquatic Biodiversity and Ecosystem Services

The coastal zone may be the single most important portion of our planet. The loss of its biodiversity may have repercussions far beyond our worst fears.

G. CARLETON RAY

Key Questions

11-1 What are the major threats to aquatic biodiversity and ecosystem services?

11-2 How can we protect and sustain marine biodiversity?

11-3 How should we manage and sustain marine fisheries?

11-4 How should we protect and sustain wetlands?

11-5 How should we protect and sustain freshwater lakes, rivers, and fisheries?

11-6 What should be our priorities for sustaining aquatic biodiversity?

Endangered green sea turtle swimming over a coral reef.

George Grall/National Geographic Creative

Sea turtles have been roaming the oceans for more than 100 million years—about 500 times longer than our species has been around. Today all seven species of sea turtles (Figure 11-1) are in danger of becoming extinct, mostly because of human activities taking place during the last 100 years.

Sea turtles spend most of their lives traveling throughout the world's oceans, but adult females almost always return to the beaches where they were born to lay their eggs. They come ashore at night and use their back flippers to dig nests on sand beaches and coastal dunes. Each female lays a clutch of around 100–110 eggs, then buries them and returns to the ocean. After the baby turtles hatch, they dig their way out of the nest, often at night, to scamper toward the water. During this dangerous trip, birds and other predators eat many of them. Only about one of every thousand hatchlings survives to adulthood.

The leatherback sea turtle (Figure 11-1) is one of the most endangered of the seven sea turtle species. Named for its leathery shell, it is the largest of all sea turtles. Adult leatherbacks are 1.8–2 meters (3–6 feet) long and weigh 250–700 kilograms (550–1,500 pounds). They feed mostly on jellyfish (Figure 8-3, p. 169) and help control jellyfish populations. They also swim great distances, migrating across the Atlantic and Pacific Oceans, as scientists have learned by using satellites to track their movements.

While leatherbacks survived the impact of the giant asteroid that probably wiped out the dinosaurs more than 60 million years ago, they and the other sea turtle species may not survive the growing human impact on their environment. A method of fishing called *trawler fishing* has destroyed many of the coral gardens that have served as their feeding grounds. The turtles are hunted for leather, and their eggs are taken for food. They often drown after becoming entangled in fishing nets and lines, as well as in lobster and crab traps.

Pollution of ocean water is another threat. Leatherbacks and some other sea turtles can mistake discarded plastic bags for jellyfish and choke to death trying to eat them. Beachgoers and motor vehicles sometimes crush their nests. Artificial lights can disorient newly hatched baby turtles, which try to find their way to the ocean by moving toward moonlight reflected from the ocean's surface. Going in the wrong direction increases their chances of ending up as food for predators. Add to this the threat of rising sea levels from projected climate change during this century. This will flood many nesting habitats and change ocean currents, which could disrupt the turtles' migration routes.

In this chapter, we will examine the effects of human activities on sea turtles, as well as on many other ocean and freshwater species. We will also explore ways to prevent or lessen these effects in order to help sustain aquatic life—a major component of the biodiversity that sustains our lives and economies.

Loggerhead
119 centimeters

Olive ridley
76 centimeters

Flatback
99 centimeters

Hawksbill
89 centimeters

Leatherback
188 centimeters

Green turtle
124 centimeters

Kemp's ridley
76 centimeters

© Cengage Learning

Figure 11-1 There are seven species of sea turtles, all of them endangered. All but the flatback are found in U.S. waters.

11-1 What Are the Major Threats to Aquatic Biodiversity and Ecosystem Services?

CONCEPT 11-1

Aquatic species and the ecosystem and economic services they provide are threatened by habitat loss, invasive species, pollution, climate change, and overexploitation, all made worse by the growth of the human population and resource use.

We Have Much to Learn about Aquatic Biodiversity

Although we live on a water planet, we have explored only about 5% of the earth's interconnected oceans (see Figure 8-2, p. 169) and know relatively little about marine biodiversity and its many functions. Marine biologist Chris Bowler has estimated that only 1% of the life-forms in the sea have been properly identified and studied. We also have limited knowledge about freshwater biodiversity.

Scientists have observed three general patterns related to marine biodiversity. *First,* the greatest marine biodiversity occurs around coral reefs, in estuaries, and on the deep-ocean floor. *Second,* biodiversity is greater near the coasts than in the open sea because of the larger variety of producers and habitats in coastal areas. *Third,* biodiversity is generally greater in the bottom region of the ocean than in the surface region because of the larger variety of habitats and food sources on the ocean bottom.

The deepest part of ocean, where sunlight does not penetrate (Figure 8-6, p. 171), is the planet's least explored environment. But this is changing. More than 2,400 scientists from 80 countries are working on a 10-year project to catalog the species in this region of the ocean. So far, they have used remotely operated deep-sea vehicles to identify more than 17,000 species living in this ocean zone and they are adding a few thousand new species every year.

Marine ecosystems provide important ecosystem and economic services (see Figure 8-5, p. 170). For example, a 2010 study, *The Economics of Ecosystems and Biodiversity,* done by an international team of economists and scientists, estimated that an area of coral reef roughly equal to the size of a city block provides economic and ecosystem services worth more than $1.2 million a year. Thus, scientific investigation of marine aquatic systems is an exciting research frontier that could lead to immense ecological and economic benefits. Freshwater systems, which occupy only 1% of the earth's surface, also provide important ecosystem and economic services (see Figure 8-14, p. 178).

Human Activities Are Destroying and Degrading Aquatic Habitat

Human domination of nature has now reached the world's vast oceans with an accelerating impact. As with terrestrial biodiversity, the greatest threats to aquatic biodiversity and ecosystem services (Concept 11-1) can be remembered with the aid of the acronym HIPPCO, with H standing for *habitat loss and degradation.* Some 90% of the fish living in the ocean spawn on coral reefs (see Chapter 8 opening photo, p. 166), in coastal wetlands and marshes (see Figure 8-8, p. 172), in mangrove forests (see Figure 8-9, p. 172), or in rivers that empty into the sea.

All of these ecosystems are under intense pressure from human activities (see Figure 8-12, p. 176). For example, United Nations Environment Programme (UNEP) scientists reported in 2010 that a fifth of the world's mangroves have been lost since 1980. They continue to be destroyed for firewood, coastal construction, and shrimp farming. In addition, a 2009 study revealed that 58% of the world's coastal sea-grass beds (see Figure 8-10, p. 173) have been degraded or destroyed, mostly by dredging and coastal development.

Sea-bottom habitats are faring no better, being threatened by dredging operations and trawler fishing boats. Like giant submerged bulldozers, trawlers drag huge nets weighted down with chains and steel plates over the ocean floor to harvest a few species of bottom fish and shellfish (Figure 11-2). Each year, thousands of trawlers scrape and disturb an area of ocean floor many times larger than the annual global total area of forests that are clear-cut. These ocean-floor communities could take decades or centuries to recover. According to marine scientist Elliot Norse, "Bottom trawling is probably the largest human-caused disturbance to the biosphere."

Coral reefs serve as habitat for many hundreds of marine species (see Chapter 8, Core Case Study, p. 168). Coastal development, pollution, ocean warming, and **ocean acidification**—the rising levels of acid in ocean waters due to their absorption of carbon dioxide from the atmosphere—threaten them. The Coral Reef Alliance reported in 2010 that more than a quarter of the world's shallow coral reefs had been destroyed or severely damaged. A 2011 study by the World Resources Institute estimated that currently 75% of the world's shallow reefs are threatened by climate change, overfishing, pollution, and ocean acidification, and that by 2050, some 90% will be threatened. In 2013, coral reef ecologist Daniel Barshis and his colleagues found genes in some coral species that might help them adapt to higher water temperatures projected to result from climate change.

Photos: © Peter J. Auster/National Undersea Research Center

Figure 11-2 Natural capital degradation: These photos show an area of ocean bottom before (left) and after (right) a trawler net scraped it like a gigantic bulldozer. **Question:** What land activities are comparable to this?

Habitat disruption is also a problem in freshwater aquatic zones. The main causes are the damming of rivers and excessive withdrawal of river water for irrigation and urban water supplies. These activities destroy aquatic habitats, degrade water flows, and disrupt freshwater biodiversity.

Invasive Species Are Degrading Aquatic Biodiversity

Another problem that threatens aquatic biodiversity is the deliberate or accidental introduction of hundreds of harmful invasive species (see Figure 9-9, p. 199)—the I in HIPPCO—into coastal waters, wetlands, and lakes throughout the world (**Concept 11-1**). These *bioinvaders* are displacing or causing the extinction of native species and disrupting ecosystem services and human economies. According to the U.S. Fish and Wildlife Service, bioinvaders are blamed for about two-thirds of all fish extinctions in the United States since 1900 and have caused huge economic losses.

Many of these invaders arrive in the ballast water that is stored in tanks in large cargo ships to keep them stable. The ships take in ballast water from one harbor, along with whatever microorganisms and tiny fish species it contains, and dump it into another—an environmentally and economically harmful effect of globalized trade. Even when ballast water is flushed from an oceangoing ship's tank before it enters a harbor—a measure now required in many ports—the ship can still bring invaders that are stuck to its hull.

One invader that worries scientists and fishers on the east coast of North America is a species of *lionfish* native to the western Pacific Ocean (Figure 11-3). Scientists believe it escaped from outdoor aquariums in Miami, Florida, that were damaged by Hurricane Andrew in 1992. Lionfish populations have exploded at the highest rate of any species ever recorded by scientists in this part of the world.

One scientist described the lionfish as "an almost perfectly designed invasive species." It reaches sexual maturity rapidly, has large numbers of offspring, and is protected by venomous spines. It competes with popular reef fish species such as grouper and snapper, taking their food and eating their young. One ray of hope for controlling this population is the fact that the lionfish tastes good. Scientists are hoping to see a growing market for lionfish as seafood.

⊙ CONSIDER THIS. . .

CONNECTIONS Lionfish and Coral Reef Destruction

Researchers have found that lionfish eat at least 50 species of prey fish, including parrotfish, that normally consume enough algae around coral reefs to keep the algae from overgrowing and killing the corals. Scientists warn that, where lionfish are now the dominant species such as in the Bahamas, unchecked algae could overwhelm and destroy some reefs.

In addition to threatening native species, invasive species can disrupt and degrade whole ecosystems and their ecosystem services. This is the focus of study for a growing number of researchers (see Science Focus 11.1).

Population Growth and Pollution Can Reduce Aquatic Biodiversity

In 2010, according to UNEP, about 80% of the world's people were living along or near seacoasts, mostly in large coastal cities. This coastal population growth—the first P in HIPPCO—has added to the already intense pressure on the world's coastal zones, resulting in more pollution, habitat destruction, and other problems (**Concept 11-1**).

The UNEP estimates that about 80% of all ocean pollution—the second P in HIPPCO—comes from land-based coastal activities. Growing inputs of nitrogen, mostly from nitrate fertilizers, into marine and freshwater systems is causing algal blooms (see Figure 8-17, p. 180), lower levels of dissolved oxygen, fish die-offs, and degradation of ecosystem services in many parts of the world.

SCIENCE FOCUS 11.1

HOW INVASIVE CARP HAVE MUDDIED SOME WATERS

Lake Wingra lies within the city of Madison, Wisconsin, surrounded mostly by a forest preserve. The lake contains a number of invasive plant and fish species, including purple loosestrife (see Figure 9-9, p. 199) and common carp. The carp were introduced in the late 1800s and since then have made up as much as half of the fish biomass in the lake. They devour algae called *chara,* which would normally cover the lake bottom and stabilize its sediments. Consequently, fish movements and winds stir these sediments, which accounts for much of the water's excessive *turbidity,* or cloudiness.

Knowing this, Dr. Richard Lathrup, a limnologist (lake scientist) who worked with Wisconsin's Department of Natural Resources, hypothesized that removing the carp would help to restore the natural ecosystem of Lake Wingra. Lathrop speculated that if the carp were removed, the bottom sediments would settle and become stabilized, allowing the water to clear. Clearer water would in turn allow native plants to

receive more sunlight and become reestablished on the lake bottom, replacing purple loosestrife and other invasive plants that now dominate its shallow shoreline waters.

Lathrop and his colleagues installed a thick, heavy vinyl curtain around a 1-hectare (2.5-acre), square-shaped perimeter that extended out from the shore. This barrier hung from buoys on the surface to the bottom of the lake, isolating the volume of water within it. The researchers then removed all of the carp from this study area and began observing results. Within 1 month, the waters within the barrier were noticeably clearer, and within a year, the difference in clarity was dramatic (Figure 11-A) and native plants once again grew in the shallow shoreline waters.

Lathrop notes that removing and keeping carp out of Lake Wingra would be a daunting task, perhaps impossible, but his controlled scientific experiment clearly shows the effects that an invasive species can have on an aquatic ecosystem. And it reminds us that preventing the introduc-

©Mike Kakuska

Figure 11-A *Lake Wingra* in Madison, Wisconsin (USA) became clouded with sediment partly because of the introduction of invasive species such as the common carp. Removal of carp in the experimental area shown here resulted in a dramatic improvement in the clarity of the water.

tion of invasive species in the first place is the best and least expensive way to avoid such effects.

Critical Thinking

What are two other results of this controlled experiment that you might expect? (*Hint:* Think food webs.)

Toxic pollutants from industrial and urban areas can kill some forms of aquatic life by poisoning them. Ocean pollution from plastic items dumped from ships and garbage barges, and left as litter on beaches, kills up to 1 million seabirds and 100,000 mammals (Figure 11-4) and sea turtles each year (**Core Case Study**).

Climate Change Is a Growing Threat

Projected climate change—the C in HIPPCO—threatens aquatic biodiversity (Concept 11-1) and ecosystem services, partly by contributing to rising sea levels. During the past 120 years, average sea levels have risen by about 20 centimeters (8 inches). Computer models developed by climate scientists estimate they will rise another 18–60 centimeters (0.6–2 feet) and perhaps as high as

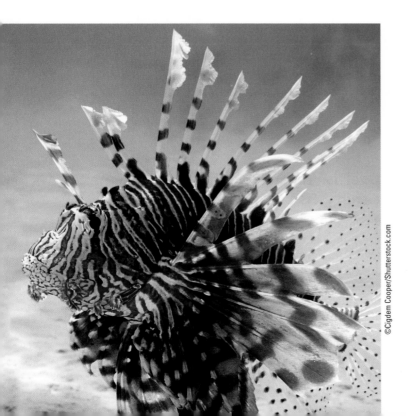

©Cigdem Cooper/Shutterstock.com

Figure 11-3 The *common lionfish* has invaded the eastern coastal waters of North America, where it has few, if any, predators.

OCEAN ACIDIFICATION: THE OTHER CO_2 PROBLEM

Our greatly increased burning of fossil fuels, especially during the last 60 years, has added carbon dioxide (CO_2) to the atmosphere faster than it can be removed by the carbon cycle (see Figure 3-17, p. 66). Research indicates that this has played a key role in the observed increase in the atmosphere's average temperature, especially since 1975. Most climate scientists recognize this temperature increase as a serious environmental problem that will very likely change the earth's climate, especially during the latter half of this century.

However, there is another problem related to CO_2 emissions—that of *ocean acidification,* a term that can be somewhat misleading. Ocean water is *basic,* not *acidic.* This means that it has more hydroxide ions (OH^-) than hydrogen ions (H^+) (see p. S14 and Figure 4, p. S15, in Supplement 4). When atmospheric CO_2 combines with ocean water, it forms carbonic acid (H_2CO_3), a weak acid also found in carbon-

ated drinks. The resulting higher levels of hydrogen ions (H^+) in the water makes the water less basic. Thus, *increasing acidity* of the oceans could also be described as *decreasing basicity.*

The problem is that, as ocean water becomes less basic, the level of carbonate ions (CO_3^{2-}) in the water drops because they react with hydrogen ions (H^+) to form bicarbonate ions (HCO_3^-). Carbonate ions are vital to many aquatic species, including phytoplankton, corals, sea snails, crabs, and oysters, that use them to produce calcium carbonate ($CaCO_3$), the main component of their shells and bones. As carbonate ion levels drop, these species are hindered in forming their shells and skeletons, as revealed in a 2012 study by Lloyd Peck and other ocean scientists. As a result, shell-building species and coral reefs will grow more slowly. And when the hydrogen ion concentration of seawater gets high enough, their calcium carbonate components will begin to dis-

solve. Colder waters with a greater capacity for removing CO_2 from the atmosphere will be affected first (Figure 11-B). Some scientists refer to the resulting thinner and weaker shells and bones of sea creatures as "osteoporosis of the sea."

This problem is already affecting some areas of our economy. On the coast of the U.S. state of Oregon, for example, oyster larvae have died off in large numbers in recent years, making it hard for some oyster growers to stay in business. Preliminary research indicates that these larvae deaths are occurring because carbonate ion levels are too low for oyster larvae to build shells during the first days of their lives.

Such harmful effects are now under intensive study. Previous studies have suggested that higher levels of hydrogen ions (H^+) would be diluted as they circulated evenly throughout the ocean. But to make matters worse, researchers have discovered that higher levels of H^+ and lower levels

---

1.5 meters (5 feet) by 2100 from a combination of thermal expansion as ocean water warms and the partial melting of land-based ice in glaciers and ice sheets.

Such a rise in sea level would destroy more shallow coral reefs, swamp some low-lying islands, drown many highly productive coastal wetlands, and put many coastal areas such as a large part of the U.S. Gulf and East Coasts underwater (Figure 8-20, p. 182). In addition, some Pacific island nations could lose more than half of their protective coastal mangrove forests by 2100, according to a 2006 study by UNEP. Coral reefs and a number of ocean organisms are also threatened by ocean acidification, a process that is closely related to climate change (Science Focus 11.2).

⊙ CONSIDER THIS. . .

CONNECTIONS Protecting Mangroves and Dealing with Climate Change

Protecting mangrove forests and restoring them in areas where they have been destroyed are important ways to reduce the impacts of rising sea levels and storm surges, because mangroves forests can slow storm-driven waves. These ecosystem services will become more important if tropical storms become more intense as a result of projected climate change. Protecting and restoring these natural coastal barriers is much cheaper and more effective than building concrete sea walls or moving threatened coastal towns and cities inland.

Warmer and more acidic ocean water is also stressing phytoplankton, the foundation of the marine food web (Figure 3-12, p. 60). These tiny life-forms produce half of the earth's oxygen and absorb a great deal of the carbon dioxide that we are adding to the atmosphere through the burning of fossil fuels and other activities. A team of scientists led by Boris Worm reported in 2010 that global phytoplankton populations have declined 40% since the 1950s, probably because warmer, more acidic ocean waters make it harder for these plankton to absorb their critical nutrients such as iron.

Overfishing and Overharvesting: Gone Fishing, Fish Gone

The human demand for seafood has been met historically through fishing. A **fishery** is a concentration of a particular wild aquatic species suitable for commercial harvesting in a given ocean area or inland body of water. However, scientists warn that the human demand for seafood has now become unsustainable. Just to keep consuming seafood at our current rate, we will need 2.5 times the area of the earth's oceans, according to the *Fishprint of Nations 2006,* a

National Geographic Maps/National Geographic Creative

of CO_3^{2-} are concentrated in the shallow waters where most coral reefs and shellfish are found (Figure 11-B). A major part of this problem is that the basicity of ocean waters is decreasing so rapidly that most organisms likely will not have time to adapt to them.

Scientists estimate that the oceans have soaked up roughly a third of the CO_2 emitted by human activities, and this has played a key role in the decreasing basicity of the world's oceans. Since we began burning fossil fuels in large quantities during the Industrial Revolution, there has been a 30% rise in the average acidity (actually a 30% decrease in average basicity) of surface ocean water, according to a 2010 UNEP summary of research on this problem. Scientists project that, by 2100, we will see a 150% drop in the average basicity of surface ocean water because of increasing acid levels.

These scientists call for a crash research program to monitor the decreases in basicity and carbonate ion levels of ocean waters around the world. They also want to learn more about which species and ocean habitats are the most vulnerable and which species will be better able to adapt to these changing environmental conditions.

According to most marine scientists, the only way to slow these changes is through a quick and sharp reduction in the use of fossil fuels around the world, which would lessen the massive inputs of CO_2 into the air and from there into the ocean. We can also slow the rise of acid levels in ocean waters by protecting and restoring mangrove forests, sea-grass beds, and coastal wetlands, because these aquatic systems take up and store some of the atmospheric CO_2 that is at the heart of this problem.

Figure 11-B Calcium carbonate levels in ocean waters, calculated from historical data (left), and projected for 2100 (right). Colors shifting from blue to red indicate where waters are becoming less basic. In the late 1800s, when CO_2 began to pile up rapidly in the atmosphere, tropical corals weren't yet affected by ocean acidification. But today carbonate levels have dropped substantially near the Poles, and by 2100, they may be too low even in the tropics for coral reefs to survive.

(Sources: Andrew G. Dickson, Scripps Institution of Oceanography, U.C. San Diego, and Sarah Cooley, Woods Hole Oceanographic Institution. Used by permission from National Geographic.)

Critical Thinking

How might a massive loss of life in the oceans affect life on land? How might it affect your life?

study based on the concept of the human ecological footprint (see **Concept 1-2**, p. 10, and Figure 1-12, p. 13).

Overfishing—the O in HIPPCO—is not new. Archaeological evidence indicates that for thousands of years, humans living in some coastal areas have overharvested fishes, shellfish, seals, turtles, whales, and other marine mammals (**Concept 11-1**). But today, fish are hunted throughout the world's oceans by a global fleet of about 4.4 million fishing boats using advanced technology to locate and harvest fish and shellfish in massive numbers. Research indicates that modern industrial fishing has been a key factor in the depletion of up to 80% of the populations of some wild fish species in only 10–15 years.

Industrial fishing fleets use global satellite positioning equipment, sonar fish-finding devices, huge nets and long fishing lines, spotter planes, and refrigerated factory ships that can process and freeze their enormous catches. These highly efficient fleets have helped to supply the growing demand for seafood (Figure 11-5), but critics say that they are vacuuming the seas, decreasing marine biodiversity, and degrading important marine ecosystem services.

For example, trawlers catch fishes and shellfish—especially cod, flounder, shrimp, and scallops—that live

©Doris Alcorn/U.S. National Maritime Fisheries

Figure 11-4 This Hawaiian monk seal was slowly starving to death before a discarded piece of plastic was removed from its snout.

on or near the ocean floor. They have destroyed vast areas of ocean-bottom habitat (Figure 11-2). Also, these nets often capture endangered sea turtles (Figure 11-6), causing them to drown (**Core Case Study**).

Another fishing method, *purse-seine fishing,* is used to catch surface-dwelling species such as tuna, mackerel,

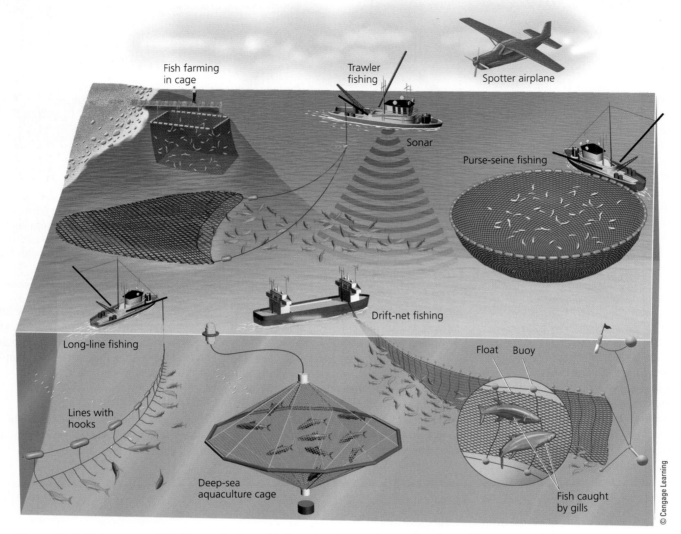

Figure 11-5 Major commercial fishing methods used to harvest various marine species, along with some methods used to raise fish through aquaculture.

anchovies, and herring (Figure 11-7), which tend to feed in schools near the surface or in shallow areas. After a spotter plane locates a school, the fishing vessel encloses it with a large purse-seine net. Some of these nets have killed large numbers of dolphins that swim on the surface above schools of tuna.

Some fishing vessels also use *long-lining*, which involves putting out lines up to 100 kilometers (60 miles) long, hung with thousands of baited hooks to catch swordfish, tuna, sharks, and ocean-bottom species such as halibut and cod. Long lines also hook and kill large numbers of sea turtles (**Core Case Study**), dolphins, and seabirds.

With *drift-net fishing*, fish are caught by drifting nets that can hang as deep as 15 meters (50 feet) below the surface and extend to 64 kilometers (40 miles) long. These nets trap and kill large quantities of unwanted fish, called *bycatch*, along with marine mammals, seabirds, and sea turtles (**Core Case Study**). Nearly one-third of the world's annual fish catch by weight consists of bycatch species that are mostly

thrown overboard dead or dying. This adds to the depletion of these species and it puts stress on some of the species that feed on them. Also, according to a 2009 UNEP report, abandoned and lost nets known as *ghost nets* float beneath the surface in many areas, trapping and drowning aquatic animals for years, before they finally sink or are recovered.

In the Fishprint of Nations study, scientists used the term **fishprint**, defined as the area of ocean needed to sustain the fish consumption of an average person, a nation, or the world. They estimated that all nations together are taking 57% more wild fish than these species' populations can sustain in the long run and that 87% of oceanic commercial fisheries are being fished at or beyond their sustainable yields.

In most cases, overfishing leads to *commercial extinction,* which occurs when it is no longer profitable to continue harvesting the affected species. Overfishing can result in only a temporary depletion of fish stocks, as long as depleted areas and fisheries are allowed to recover. But

Figure 11-6 This green sea turtle died after being caught in a fishing net.

as industrialized fishing fleets take more and more of the world's available fish and shellfish, recovery times for severely depleted populations are increasing and can be two decades or more. Some depleted fisheries may not recover as jellyfish and other invasive species move in and take over their food webs (see the following Case Study).

For example, in the late 1950s, fishing fleets began overfishing the 500-year-old Atlantic cod fishery off the coast of Newfoundland, Canada. They used bottom trawlers to capture larger shares of the stock, reflected in the sharp rise in the graph in Figure 11-7. This extreme overexploitation of the fishery led to a steady decline in the fish catch throughout the 1970s. After a slight recovery in the 1980s, the fishery collapsed and in 1992, it was shut down, putting at least 35,000 fishers and fish processors out of work in more than 500 coastal communities. It was reopened on a limited basis in 1998 but then closed indefinitely in 2003. The Newfoundland fishing industry is now recovering by harvesting snow crab and northern shrimp.

One result of the increasingly efficient global hunt for fish is that larger individuals of commercially valuable wild species—including cod, marlin, swordfish, and tuna—are becoming scarce. Between 1950 and 2006, according to a study led by marine ecologist Boris Worm, 90% or more of these and other large, predatory, open-ocean fishes had disappeared. This resulted in a massive change to ocean food webs and the ecosystem services they provide, taking place in only a few decades.

Figure 11-7 Natural capital degradation: This graph illustrates the collapse of Newfoundland's Atlantic cod fishery.

For example, the *Atlantic,* or *Northern bluefin tuna,* is among the largest of ocean fishes. It is highly prized for making sushi and sashimi, and the demand for those products has resulted in the severe depletion of this species. Japan consumes about 80% of the annual catch of Atlantic blue-

Figure 11-8 *Bluefin tuna ranching:* These bluefin tunas have been corralled in an underwater pen and will be fattened for slaughter to make sushi.

1.5 million whales of all species between 1925 and 1975. This overharvesting drove 8 of the 11 major whale species to commercial extinction before commercial whaling was banned internationally in 1986.

After the ban, the estimated number of whales killed by commercial whale hunts worldwide dropped from 42,480 in 1970 to about 2,800 in 2011. Most of the whales killed now are taken by Norway, Japan (allegedly for scientific purposes), the Faroe Islands, and Iceland, despite the ban. Japan, Norway, and Iceland have worked to overthrow the international moratorium on commercial whaling and to reverse the separate international ban on the buying and selling of whale products.

fins. In 2001, one giant Atlantic bluefin sold for $175,000 in a Tokyo fish market. Such high prices have helped to decrease the populations of this species by 75–80% since the 1970s. As a result, the International Union for Conservation of Nature (IUCN) classifies the Atlantic bluefin tuna as critically endangered. So far, measures to regulate the catch for this species have largely failed.

A practice that some marine scientists strongly criticize is that of *tuna ranching,* in which schools of half-grown Atlantic bluefin and other tuna are herded by the thousands into underwater pens and towed to areas where they are held and fattened for slaughter (Figure 11-8). Richard Ellis, a leading U.S. marine conservationist, described it this way: "If you had to design a way to guarantee the decimation of a breeding population, this would be it: catch the fish before they are old enough to breed and keep them penned up until they are killed."

A number of marine scientists have called for a ban on the fishing and ranching of Atlantic bluefins until the species can recover. However, so far, strong opposition from Japan and from the fishing industry has prevented such a ban.

As commercially valuable species have become scarce, fishers have turned to other species such as sharks, which are now being overfished (see the second following Case Study). Also, as large species are overfished, the fishing industry is shifting to smaller marine species such as herring, sardines, and krill. About 90% of this catch is converted to fishmeal and fish oil, most of which is fed to farmed fish. One researcher referred to this as "stealing the ocean's food supply," because these smaller species make up much of the diet of larger predator fish, seabirds, and toothed whales.

Some marine mammals have also been threatened with extinction by overharvesting. The most prominent examples are whales, including the highly endangered blue whale. Whale harvesting in international waters has followed the classic pattern of a tragedy of the commons (see Chapter 1, p. 12), with whalers killing an estimated

CASE STUDY

The Great Jellyfish Invasion

Jellyfish, which existed before the dinosaurs, are not fish. A jellyfish (see Figure 8-3, p. 169) has no brain, head, heart, eyes, or bones and is basically a cluster of nerves that can detect warmth, food, odors, and vibration. Its generally bell-shaped body is filled with a jelly-like substance. It uses its tentacles dangling from its body to sting or stun prey, to draw food into its mouth, and to defend itself against predators such as other jellyfish, tuna, sharks, swordfish, and sea turtles. Most jellyfish feed on zooplankton, fish eggs, small fish, and other jellyfish.

Jellyfish sizes vary widely. Some are as small as a mosquito. The largest is the Arctic lion's mane jellyfish (Figure 11-9) with a diameter of up to 2.5 meters (8 feet) and tentacles as long as 37 meters (120 feet). It can weigh as much as 150 kilograms (330 pounds). The sting of a jellyfish can cause an itch like that of a bug bite or a more serious burning sensation lasting for several days. However, a sting from the Portuguese man-of-war, the Australian box jelly, or the tiny Irukandji jellyfish can kill a human within minutes.

Jellyfish are often found in large swarms or *blooms* of thousands, even millions of individuals. In recent years, the numbers of these blooms observed by scientists and fishers have been rising and more frequent jellyfish stings have had a harmful economic effect on a number of popular tourist beach areas. Each year, the stings of lethal jellyfish kill dozens of people—far more than the annual average number of people killed by sharks. In addition, jellyfish blooms often cause beach closings, disrupt commercial fishing operations by clogging nets, and close down coal-burning and nuclear power plants by clogging their cooling water intakes. Fish in some ocean aquacul-

Figure 11-9 The lion's mane is the largest of the roughly 1,500 known jellyfish species. It can be as wide as a full-size school bus and about 3 times as long.

ture pens (Figure 11-5) have been wiped out by swarms of jellyfish attracted to the pens by the fish feed and fish waste found there.

Several factors play a role in the rapid rise of jellyfish populations. One is the overfishing of species that eat small jellyfish, including endangered sea turtles (**Core Case Study**). The jellyfish, in turn, feed on fish eggs and larvae, making it harder for the depleted fish stocks to recover. Another factor is inputs of excessive plant nutrients from fertilizer runoff and sewage due to coastal urban development. This spurs the growth of phytoplankton, on which jellyfish feed, which helps them to increase their numbers. This results in oxygen-depleted zones, which jellyfish can tolerate better than most other species.

Warmer waters are also associated with jellyfish blooms, so continuing climate change may result in jellyfish becoming the dominant species in many marine ecosystems around the world. A 2010 UNEP report linked rising jellyfish populations to increasing ocean acidification, which harms some marine species but not jellyfish.

According to Chinese oceanographer Wei Hao and other marine scientists, the startling growth of jellyfish populations threatens to upset marine food webs and eco-

system services and turn some of the most productive seas into jellyfish empires. Once jellyfish take over a marine ecosystem they might dominate it indefinitely.

CASE STUDY

Why Should We Protect Sharks?

Sharks have roamed the world's oceans for more than 400 million years. As *keystone species*, some shark species play crucial roles in helping to keep their ecosystems functioning. Some shark species that feed at or near the tops of food webs remove injured and sick animals from the ocean. Without this ecosystem service provided by some shark species, the oceans would be teeming with dead and dying fish and marine mammals.

More than 400 known species of sharks inhabit the world's oceans. They vary widely in size and behavior, from the goldfish-sized dwarf dog shark to the whale shark (Figure 11-10, left), which can grow to the length of a city bus and weigh as much as two full-grown African elephants.

Many people, influenced by movies, popular novels, and widespread media coverage of shark attacks, think of sharks as people-eating monsters. In reality, the three largest species—the whale shark (Figure 11-10, left),

Figure 11-10 The threatened whale shark (left), which feeds on plankton, is the largest fish in the ocean and is quite friendly to humans. The scalloped hammerhead shark (right) is endangered.

Left: Colin Parker/National Geographic My Shot/National Geographic Creative. Right: Westend61/SuperStock.

Figure 11-11 This endangered loggerhead turtle is escaping a fishing net equipped with a turtle excluder device (TED).

©NOAA

basking shark, and megamouth shark—are gentle giants. These plant-eating sharks swim through the water with their mouths open, filtering out and swallowing huge quantities of phytoplankton.

Media coverage of shark attacks greatly exaggerates the danger from sharks. Every year, members of a few species, including the great white, bull, tiger, oceanic whitetip, and hammerhead sharks, injure 60–75 people and typically kill 6–10 people worldwide. Some of these sharks feed on sea lions and other marine mammals and sometimes mistake swimmers and surfers for their usual prey.

For every shark that injures or kills a person, people kill about 1.2 million sharks. As many as 73 million sharks are caught each year for their valuable fins and then thrown back alive into the water, fins removed, to bleed to death or drown because they can no longer swim. Sharks are also killed for their livers, meat, hides, and jaws, and because we fear them. Each year an estimated 50 million sharks die when fishing lines and nets trap them.

Harvested shark fins are worth as much as $1.2 billion a year. They are widely used in Asia as an ingredient in expensive soup (up to $100 a bowl) and as a pharmaceutical cure-all. According to the wildlife conservation group WildAid, there is no reliable evidence that the fins provide flavor or have any nutritional or medicinal value. The group also warns that consuming shark fins and shark meat can be harmful to human health because they contain very high levels of mercury and other toxins. In 2012, neurologist Deborah Mash found neurotoxins in the fins of sharks of seven different species with links to Parkinson's, Alzheimer's, and Lou Gehrig's diseases.

According to a 2009 IUCN study, 32% of the world's open-ocean shark species are threatened with extinction, including the scalloped hammerhead shark (Figure 11-10, right). Sharks are especially vulnerable to population declines because they grow slowly, mature late, and have only a few offspring per generation. Today, they are among the earth's most vulnerable and least protected animals. And because some sharks are keystone species, their ecosystems and the ecosystem services they provide are also threatened.

It is encouraging that in 2012, French Polynesia created the world's largest shark sanctuary, although protecting sharks in such a large area will be difficult. Also, in 2013, the Convention on International Trade in Endangered Species (CITES) placed trade limits on catches of oceanic whitetip, porbeagle, and three species of hammerhead sharks.

CONSIDER THIS. . .

CONNECTIONS Shark Declines and Scallop Fishing

In 2007, scientists reported that the decline in certain shark populations along the U.S. Atlantic coast led to an explosion in populations of rays and skates, which sharks normally feed on. As a result, the rays and skates were feasting on bay scallops, which resulted in a sharp decline in their population and in the area's bay scallop fishing business.

Extinction of Aquatic Species Is a Growing Threat

Beyond the *commercial extinction* of a number of fisheries, many fish species are also threatened with *biological extinction,* mostly from overfishing, water pollution, wetlands destruction, and excessive removal of water from rivers and lakes. According to the 2011 IUCN Red List of Threatened Species, of all species evaluated, 34% of the world's marine species and 71% of the world's freshwater fish species face extinction within the next 6 to 7 decades. A 2012 analysis by researchers at BirdLife International found that seabirds are more endangered than other types of birds. Some 27% of seabird species are threatened, compared to 12% of all bird species combined.

In 2011, an international panel of 27 marine scientists assessed how human impacts are affecting the world's oceans. Their startling conclusion was that because of a combination of habitat loss, overfishing, pollution, and ocean acidification, marine life is poised to enter a new period of mass extinction, which would eventually affect the world's terrestrial and marine ecosystems and thus the world's economies.

Among the most threatened of all marine species are sea turtles (**Core Case Study**). This is due to a number of factors, including people taking their eggs and loss or degradation of their beach nesting habitat. Female turtles are often blocked from reaching their nesting sites by seawalls, sandbags, and other devices designed to slow beach erosion in developed coastal areas. Also, many adult sea turtles die when commercial and recreational fishers accidentally catch them in nets (Figure 11-6) or when they become entangled in ocean debris such as plastic bags and lost or abandoned fishing nets (ghost nets).

In U.S. waters, the National Marine Fisheries Service requires the use of turtle excluder devices (TEDs) on commercial fishing nets that allow captured sea turtles to escape (Figure 11-11). Since 1990, fishing regulations have reduced accidental sea turtle deaths in U.S. waters by 90%. In 2012, marine biologist Helen **GOOD NEWS** Bailer and her colleagues began using satellites to track highly endangered leatherback turtles and the routes of trawlers that trap and kill many of these turtles. Such data could help regulatory agencies in setting times and places where they might limit trawler fishing to protect the leatherbacks.

11-2 How Can We Protect and Sustain Marine Biodiversity?

CONCEPT 11-2

We can help to sustain marine biodiversity by using laws and economic incentives to protect species, setting aside marine reserves to protect ecosystems and ecosystem services, and using community-based integrated coastal management.

Laws and Treaties and Economic Incentives Can Help to Sustain Aquatic Biodiversity

Protecting marine biodiversity is difficult for several reasons. *First,* the human ecological footprint (see Figure 1-11, p. 13) and the fishprint are expanding so rapidly that it is difficult to monitor their impacts. *Second,* much of the damage to the oceans and other bodies of water is not visible to most people. *Third,* many people incorrectly view the seas as an inexhaustible resource that can absorb an almost infinite amount of waste and pollution and still produce all the seafood we want. *Fourth,* most of the world's ocean area lies outside the legal jurisdiction of any country. Thus, much of it is an open-access resource, subject to overexploitation—a classic case of the tragedy of the commons (see Chapter 1, p. 12).

Nevertheless, there are several ways to protect and sustain marine biodiversity, one of which is the regulatory approach (Concept 11-2). For example, in the United States and a number of other countries, laws have helped to protect sea turtle nesting sites (**Core Case Study**) from egg poachers and destruction by vehicles (Figure 11-12). **GOOD NEWS**

National and international laws and treaties to help protect marine species include the 1975 Convention on International Trade in Endangered Species (CITES), the 1979 Global Treaty on Migratory Species, the U.S. Marine Mammal Protection Act of 1972, the U.S. Endangered Species Act of 1973 (ESA; see Chapter 9, pp. 208–209), the U.S. Whale Conservation and Protection Act of 1976, and the 1995 International Convention on Biological Diversity. The ESA and several international agreements have been used to identify and protect endangered and threatened marine species, including whales, seals, sea lions, and sea turtles. The problem is that with some international agreements, it is hard to get all nations to comply, which can weaken the effectiveness of such agreements. Even when agreements and regulations are enforced, the resulting fines and punishments for violators are often inadequate.

Another way to protect endangered and threatened aquatic species is to use economic incentives (Concept 11-2). For example, according to a World Wildlife Fund study, sea turtles are worth more to coastal communities alive than dead (**Core Case Study**). The report estimates that sea turtle tourism typically brings in almost 3 times more money than the sale of turtle products such as meat, leather, and eggs brings in. In Brazil, a sea turtle program hires ex-poachers to protect rather than exploit the turtle populations. Educating citizens about this issue could inspire communities to protect more turtles.

Marine Sanctuaries Protect Ecosystems and Species

By international law, a country's offshore fishing zone extends to 370 kilometers (200 nautical miles) from its shores. Foreign fishing vessels can take certain quotas of fish within such zones, called *exclusive economic zones,* but only with a government's permission. Ocean areas beyond

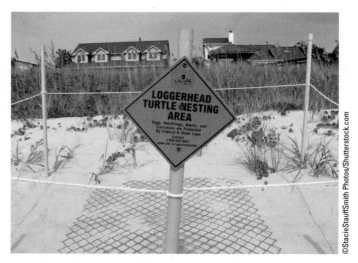

Figure 11-12 Loggerhead sea turtle nests are protected by law in some areas, including Myrtle Beach, South Carolina.

the legal jurisdiction of any country are known as the *high seas*, and laws and treaties pertaining to them are difficult to monitor and enforce.

The United Nations Law of the Sea treaty, which went into effect in 1984, has been signed by 164 countries (but not by the United States). Under this treaty, the world's coastal nations have jurisdiction over 36% of the ocean surface and 90% of the world's fish stocks. Instead of using this treaty to protect their fishing grounds, many governments have promoted overfishing by subsidizing fishing fleets and failing to establish and enforce stricter regulation of fish catches in their coastal waters.

However, some countries are attempting to protect marine biodiversity and to sustain fisheries by establishing marine sanctuaries in their coastal waters. Since 1986, the IUCN has helped such nations to establish a global system of *marine protected areas* (MPAs)—areas of ocean partially protected from human activities. In 2010, the IUCN reported that there were 5,880 MPAs worldwide (355 in U.S. waters), covering about 1.6% of the world's ocean surface.

The number of MPAs is growing, but most of them are only partially protected. Nearly all allow dredging, trawler fishing, and other ecologically harmful resource–extraction activities. And many of them are too small to be effective in protecting larger species. However, since 2007 the U.S. state of California has been establishing the nation's most extensive network of MPAs in which fishing will be banned or strictly limited. In 2011, Costa Rica (see GOOD NEWS Core Case Study, Chapter 10, p. 218) expanded one of its MPAs to help protect a number of marine species, including the critically endangered leatherback sea turtle (**Core Case Study**) and the endangered scalloped hammerhead shark (Figure 11-10, right). And in 2012, Australia announced that it would create the world's largest MPA consisting of the Great Barrier Reef Marine Park and the neighboring Coral Sea Reserve.

Establishing a Global Network of Marine Reserves: An Ecosystem Approach to Marine Sustainability

Many scientists and policy makers call for the widespread use of an *ecosystem approach* focused on protecting and sustaining entire marine ecosystems and their ecosystem services for current and future generations rather than relying mostly on protecting individual species. The goal of this ecological approach is to establish a global network of fully protected *marine reserves*, areas that are declared off-limits to destructive human activities in order to enable their ecosystems to flourish and recover.

This global network would include large reserves on the high seas, especially near extremely productive nutrient upwelling areas (see Figure 7-2, p. 145), and a mixture of smaller reserves in coastal zones that are adjacent to well-managed, sustainable commercial fishing areas. Such protected "underwater wilderness" areas would be

closed to activities such as commercial fishing, dredging, and mining, as well as to waste disposal. Most reserves in the proposed global network would permit less-harmful activities such as recreational boating, shipping, and in some cases, small-scale, nondestructive fishing. However, most reserves would also contain core zones where no human activity would be allowed.

Marine reserves work and they work quickly. Scientific studies show that within fully protected marine reserves, within 2–4 years after strict protection begins, commercially valuable fish populations can double, average fish size can grow by almost a third, fish reproduction can triple, and species diversity can increase by almost one-fourth. Furthermore, these improvements can last for decades (**Concept 11-2**). Research also shows that reserves benefit nearby fisheries, because fish move into and out of the reserves, and currents carry fish larvae produced inside reserves to adjacent fishing grounds, thus GOOD NEWS bolstering the populations there.

In 2012, the Australian government announced that it would establish the world's largest network of fully protected marine reserves around its coasts. Two other giant marine reserves have been created by the United States around the Northwestern Hawaiian Islands and by the United Kingdom around the Chagos Islands in the Indian Ocean. About 3% of U.S. territorial waters have been set aside in 223 marine reserves.

The IUCN and other scientific groups have identified *marine hotspots*—threatened areas in need of full protection because of their importance to marine biodiversity and ecosystem services (Figure 11-13). Despite the importance of such protection, only 0.8% of the world's oceans are fully protected—closed to fishing and other potentially harmful human activities—compared with about 5% of the world's land. In other words, 99.2% of the world's oceans are not effectively protected from harmful human activities. Furthermore, many marine reserves are too small to protect most of the species within them and do not provide adequate protection against illegal fishing, garbage dumping, or pollution that flows from the land into coastal waters.

The trend toward creating huge megareserves is encouraging. However, researchers are struggling to design ways to monitor changes over vast areas and to determine the effectiveness of such reserves in sustaining and rebuilding marine populations. In addition, resource managers are trying to find affordable ways to enforce fishing bans in such vast ocean areas.

Many marine scientists, including National Geographic's Explorer-in-Residence Sylvia Earle (Individuals Matter 11.1), call for protecting at least 30% of the world's oceans as fully protected marine reserves. They also urge that protected corridors be established to connect the global network of marine reserves, especially those in coastal waters. This would also help species to move to different habitats in the process of adapting to the effects of ocean warming, acidification, and many forms of ocean pollution.

Figure 11-13 Natural capital: Ten *marine hotspots*—threatened areas, including coral reefs, and threatened fisheries that provide vital ecosystem services and that marine scientists say should be fully protected from harmful human activities.

(Compiled by the authors using data from International Union for the Conservation of Nature.)

A team of U.K. scientists led by Andrew Balmford studied 83 well-managed marine reserves and concluded that it would cost $12 billion to $14 billion a year to manage reserves covering 30% of the world's oceans. This would roughly equal to the annual subsidies that promote overfishing, which governments currently provide to the global fishing industry.

🔍 **CONSIDER THIS. . .**

THINKING ABOUT Marine Reserves

Do you support setting aside at least 30% of the world's oceans as fully protected marine reserves? Explain. How would this affect your life?

Restoration Helps to Protect Marine Biodiversity but Prevention Is the Key

A dramatic example of marine system restoration is Japan's attempt to restore its largest coral reef—90% of which has died—by seeding it with new corals. Divers drill holes into the dead reefs and insert ceramic discs holding sprigs of fledgling coral. Survival rates of the young corals were only at about 33% in 2009 but were rising, according to the Japanese government. Scientists see this experiment as possibly leading to more global efforts to save reefs by transplanting corals. Figure 11-14 also shows how protection has helped to restore coral reefs near Kanton Island, an atoll located in the South Pacific roughly halfway between Fiji and Hawaii.

GOOD NEWS

While many scientists applaud such efforts to restore aquatic systems, they also note that these projects could fail if the problems that caused the degradation of the systems are not addressed (see Chapter 8, Core Case Study, p. 168). That is why they also argue that we must shift to taking a *prevention approach* toward aquatic ecosystem degradation, which is far less expensive and risky than restoration efforts. And they say we must make this shift soon.

For example, a study by IUCN and scientists from the Nature Conservancy concluded that the world's shallow coral reefs and mangrove forests could survive currently projected climate change if we relieve other threats such as overfishing and pollution. However, while some shallow coral species may be able to adapt to warmer temperatures, they may not have enough time to do this unless we act now to slow down the projected rapid rate of climate change, especially during the latter half of this century. In addition, ocean acidification can slow the growth of coral reefs.

To deal with problems of pollution and overfishing, communities must closely monitor and regulate fishing and coastal land development and greatly reduce pollution from land-based activities. Coastal residents must also think carefully about the chemicals they put on their lawns and the kinds of waste they generate because many of these chemicals will end up in coastal waters.

More important, each of us can make careful choices in purchasing only sustainably harvested or sustainably farmed seafood. People could also reduce their carbon footprints in order to slow ocean acidification along with other

Sylvia Earle—Ocean Protector and National Geographic Explorer-in-Residence

James A. Sugar/National Geographic Creative

Sylvia Earle is one of the world's most respected oceanographers and is a National Geographic Society Explorer-in-Residence. She has taken a leading role in helping us to understand the world's oceans and to protect them. *Time* magazine named her the first Hero for the Planet and the U.S. Library of Congress calls her "a living legend."

Earle has led more than 100 ocean research expeditions and has spent more than 7,000 hours underwater, either diving or descending in research submarines to study ocean life. She has focused her research on the ecology and conservation of marine ecosystems, with an emphasis on developing deep-sea exploration technology.

She is the author of more than 175 publications and has been a participant in numerous radio and television productions. During her long career, Earle has also been the Chief Scientist of the U.S. National Oceanic and Atmospheric Administration (NOAA) and she has founded three companies devoted to developing submarines and other devices for deep-sea exploration and research. She has received more than 100 major international and national honors, including a place in the National Women's Hall of Fame.

These days, Earle is leading a campaign called *Mission Blue* to finance research and to ignite public support for a global network of marine protected areas, which she dubs "hope spots." Her goal is to help save and restore the oceans, which she calls "the blue heart of the planet." She says, "There is still time, but not a lot, to turn things around."

Background photo: ©Igor Kovalchuk/Shutterstock.com

harmful effects of projected climate change on marine biodiversity, as discussed in more detail in Chapter 19.

One strategy emerging in some coastal communities is *integrated coastal management*—a community-based effort to develop and use coastal resources more sustainably (**Concept 11-2**). The overall aim of such programs is for fishers, business owners, developers, scientists, citizens, and politicians to identify shared problems and goals in their use of marine resources. The idea is to develop workable, cost-

effective, and adaptable solutions that will help to preserve biodiversity, ecosystem services, and environmental quality, while also meeting economic and social goals.

This requires all participants to seek reasonable short-term trade-offs that can lead to long-term ecological and economic benefits—an example of applying the win-win **principle of sustainability** (see Figure 1-5, p. 9 or back cover). For example, fishers might have to give up harvesting various fish species in certain

Figure 11-14 Severe bleaching of coral near Kanton Island in the South Pacific (left), and an example of the recovery of coral in a protected area near the island (right).

Brian J. Skerry/National Geographic Creative

Brian J. Skerry/National Geographic Creative

areas until stocks recover enough to restore biodiversity in those areas. This might help to provide fishers with a more sustainable future for their businesses.

Australia manages its huge Great Barrier Reef Marine Park in this way, and more than 100 integrated coastal management programs are being developed throughout the world. Another example of such management in the United States is the Chesapeake Bay Program (see Chapter 8, Case Study, p. 168).

GOOD NEWS

11-3 How Should We Manage and Sustain Marine Fisheries?

CONCEPT 11-3
Sustaining marine fisheries will require improved monitoring of fish and shellfish populations, cooperative fisheries management among communities and nations, reduction of fishing subsidies, and careful consumer choices in buying seafood.

Estimating and Monitoring Fishery Populations Is the First Step

The first step in protecting and sustaining the world's marine fisheries is to make the best possible estimates of their fish and shellfish populations (Concept 11-3). The traditional approach has used a *maximum sustained yield (MSY)* model to project the maximum number of individuals that can be harvested annually from fish or shellfish stocks without causing a population drop. However, the MSY concept has not worked very well because of the difficulty in estimating the populations and growth rates of fish and shellfish stocks. Also, harvesting a particular species at its estimated maximum sustainable level can affect the populations of other target and nontarget marine species.

In recent years, some fishery biologists and managers have begun using the *optimum sustained yield (OSY)* concept. They have attempted to take into account interactions among species and to provide more room for error. Similarly, another approach is *multispecies management* of a number of interacting species, which takes into account their competitive and predator–prey interactions. An even more ambitious approach is to develop complex computer models for managing multispecies fisheries in *large marine systems.* However, it is a political challenge to get groups of nations to cooperate in planning and managing such large systems.

There are uncertainties built into using any of these approaches because the biology of marine species and their interactions is not completely understood, and because of limited data on changing ocean conditions. As a result, many fishery and environmental scientists are increasingly interested in using the *precautionary principle* for managing fisheries and large marine systems. This means sharply reducing fish harvests and closing some overfished areas until they recover and until we have more information about what levels of fishing they can sustain.

Some Communities Cooperate to Regulate Fish Harvests

An obvious step to take in protecting marine biodiversity—and therefore fisheries—is to regulate fishing. Traditionally, many coastal fishing communities have developed allotment and enforcement systems for controlling fish catches in which each fisher gets a share of the total allowable catch. Such *catch-share systems* have sustained fisheries and jobs in many communities for hundreds and sometimes thousands of years.

An example of a successful catch-share system is Norway's Lofoten fishery, one of the world's largest cod fisheries. For 100 years, it has been self-regulated, with no participation by the Norwegian government (Figure 11-15). The key is to guarantee each participant a share of

GOOD NEWS

Figure 11-15 For hundreds of years, fishing villages such as this one on one of the Lofoten Islands in northern Norway have used self-regulated cooperative *catch-share systems* to sustain their fisheries.

the total allowable catch. Another example is the Alaskan halibut fishery, which grew dramatically more profitable after a catch-share system was put in place in 1995.

However, the influx of large state-of-the-art fishing boats and international fishing fleets has weakened the ability of many coastal communities to regulate and sustain local fisheries. Community management systems have often been replaced by *comanagement,* in which coastal communities and the government work together to manage fisheries. Currently, more than 210 of the world's fisheries are comanaged.

With this approach, a central government typically sets quotas for various species and divides the quotas among communities. The government may also limit fishing seasons and regulate the types of fishing gear that can be used to harvest a particular species. Each community then allocates and enforces its quota among its members based on its own rules. Often, communities focus on managing inshore fisheries, and the central government manages the offshore fisheries. When it works, community-based comanagement illustrates that overfishing and the tragedy of the commons (see Chapter 1, p. 12) are not inevitable. **GOOD NEWS**

Government Subsidies Can Encourage Overfishing

Governments around the world give a total of more than $30 billion per year in subsidies to fishers to help them keep their businesses running, according to a 2006 study by fishery experts U. R. Sumaila and Daniel Pauly. Some marine scientists estimate that, each year, $10 billion to $14 billion of these subsidies is spent to encourage overfishing and expansion of the fishing industry. The result is too many boats chasing too few fish. Some argue that such subsidies are not a wise investment because they promote overfishing of targeted fish stocks, which causes

economic losses of about $50 billion a year, according to a 2008 World Bank report.

CONSIDER THIS. . .

THINKING ABOUT Fishing Subsidies

Do you think that government fishing subsidies that promote unsustainable fishing should be eliminated or drastically reduced? Explain. Would your answer be different if your livelihood depended on commercial fishing?

Consumer Choices Can Help Sustain Fisheries and Aquatic Biodiversity

An important component of sustaining aquatic biodiversity and ecosystem services is bottom-up pressure from consumers demanding *sustainable seafood,* which will encourage more responsible fishing practices (**Concept 11-3**). One way to enable this is by labeling fresh and frozen seafood to inform consumers about how and where the fish and shellfish were caught. Another important component is the certification of sustainably caught seafood. The London-based Marine Stewardship Council (MSC) was created in 1999 to support sustainable fishing and to certify sustainably produced seafood. Only certified **GOOD NEWS**

Solutions

Managing Fisheries

Fishery Regulations

Set low catch limits

Improve monitoring and enforcement

Economic Approaches

Reduce or eliminate fishing subsidies

Certify sustainable fisheries

Protection

Establish no-fishing areas

Establish more marine protected areas

Consumer Information

Label sustainably harvested fish

Publicize overfished and threatened species

Bycatch

Use nets that allow escape of smaller fish

Use net escape devices for seabirds and sea turtles

Aquaculture

Restrict locations of fish farms

Improve pollution control

Nonnative Invasions

Kill or filter organisms from ship ballast water

Clean aquatic recreation gear

Figure 11-16 There are a number of ways to manage fisheries more sustainably and protect marine biodiversity. **Questions:** Which four of these solutions do you think are the best ones? Why?

fisheries are allowed to use the MSC's "Fish Forever" eco-label, which certifies that the fish were caught by fishers who used environmentally sound and socially responsible practices. Another approach is to certify and label products of sustainable *aquaculture,* or fish farming operations.

An important way for seafood consumers to help sustain aquatic biodiversity is to choose plant-eating species of fish, such as tilapia, raised through aquaculture. Carnivorous fish raised through aquaculture are fed fishmeal made from wild-caught fish. Some ocean fish species used to make fishmeal are being overfished at higher rates every year. Thus, by choosing herbivorous species, consumers can lessen the demand for this overfishing.

Figure 11-16 summarizes actions that individuals, organizations, and governments can take to help sustain global fisheries, marine biodiversity, and ecosystem services. History shows that most attempts to improve environmental quality and promote environmental sustainability require bottom-up political and economic pressure by concerned citizens. Individuals matter.

11-4 How Should We Protect and Sustain Wetlands?

CONCEPT 11-4
We can maintain the ecosystem and economic services of wetlands by protecting remaining wetlands and restoring degraded wetlands.

Coastal and Inland Wetlands Are Disappearing around the World

Coastal wetlands and marshes (see Figure 8-8, p. 172) and inland wetlands are important reservoirs of aquatic biodiversity that provide vital economic and ecosystem services. Despite their ecological value, the United States has lost more than half of its coastal and inland wetlands since 1900, and other countries have lost even more. New Zealand, for example, has lost 92% of its original coastal wetlands, and Italy has lost 95%.

For centuries, people have drained, filled in, or covered over swamps, marshes, and other wetlands to create rice fields or other cropland, to accommodate expanding cities and suburbs, and to build roads. Wetlands have also been destroyed in order to extract minerals, oil, and natural gas, and to eliminate breeding grounds for insects that cause diseases such as malaria.

To make matters worse, coastal wetlands in many parts of the world will probably be under water before the end of this century because of rising sea levels. If the atmosphere warms as projected, ocean waters will also warm and expand, and some of the world's land-based glaciers and ice sheets will melt, adding water to the ocean. This could seriously degrade the aquatic biodiversity and the ecosystem services provided by coastal wetlands, as well as the commercially important fish and shellfish species and millions of migratory waterfowl and other birds that depend on these wetlands.

We Can Preserve and Restore Wetlands

Scientists, land managers, landowners, and environmental groups are involved in intensive efforts to preserve existing wetlands and restore degraded ones (**Concept 11-4**), and laws have been passed to protect wetlands.

In the United States, zoning laws have been used to steer development away from wetlands. The U.S. government requires a federal permit to fill in wetlands occupying more than 1.2 hectares (3.0 acres) or to deposit dredged material in wetlands. According to the U.S. Fish and Wildlife Service, this law has helped to cut the average annual wetland loss by 80% since 1969. GOOD NEWS However, there are continuing attempts by land developers to weaken such wetlands protection. Only about 6% of the country's remaining inland wetlands are federally protected, and state and local wetland protection is inconsistent and generally weak.

The stated goal of current U.S. federal policy is *zero net loss* in the functioning and value of coastal and inland wetlands. A policy known as *mitigation banking* allows destruction of existing wetlands as long as an equal or greater area of the same type of wetland is created or restored. However, a study by the National Academy of Sciences found that at least half of the attempts to create new wetlands failed to replace lost ones, and most of the created wetlands did not provide the ecosystem services of natural wetlands. The study also found that wetland creation projects often fail to meet the standards set for them and are not adequately monitored.

Creating and restoring wetlands has become a profitable business. Private investment bankers make money by buying wetland areas and restoring or upgrading them by working with the U.S. Army Corps of Engineers and the U.S. Environmental Protection Agency. This creates wetlands banks or credits that the bankers can then sell to developers.

It is difficult to restore or create wetlands (see the following Case Study). Thus, most U.S. wetland banking systems require replacing each hectare of destroyed wetland with two or more hectares of restored or created wetland (Figure 11-17) as a built-in ecological insurance policy.

Ecologists urge that mitigation banking should be used only as a last resort. They also call for making sure that new replacement wetlands are created and evaluated *before* existing wetlands are destroyed. This example of applying the precautionary principle is often the reverse of what is actually done.

Figure 11-17 This human-created wetland is located near Orlando, Florida (USA).

CONSIDER THIS. . .

THINKING ABOUT Wetlands Mitigation

Is it good policy to require that a new wetland be created and evaluated before anyone is allowed to destroy the wetland it is supposed to replace? Explain.

CASE STUDY
Can We Restore the Florida Everglades?

South Florida's Everglades was once a 100-kilometer-wide (62-mile-wide), knee-deep sheet of water flowing slowly south from Lake Okeechobee to Florida Bay (Figure 11-18, red dashed lines). As this shallow body of water—known as the "River of Grass"—trickled south, it created a vast network of wetlands (Figure 11-18, photo) with a variety of wildlife habitats.

To help preserve the wilderness in the lower end of the Everglades system, in 1947, the U.S. government established Everglades National Park, which contains a portion of the remaining Everglades. But this protection effort did not work—as conservationists had predicted—because of a massive water distribution and land development project to the north. Between 1962 and 1971, the U.S. Army Corps of Engineers transformed the wandering 166-kilometer-long (103-mile-long) Kissimmee River into a mostly straight 84-kilometer (52-mile) canal flowing into Lake Okeechobee (Figure 11-18). The canal provided flood control by speeding the flow of water, but it drained large wetlands north of Lake Okeechobee, which farmers then converted to grazing land.

This and other projects have provided south Florida's rapidly growing population with a reliable water supply and flood protection. But as a result, much of the original Everglades has been drained, paved over, polluted by agricultural runoff, and invaded by a number of plant and animal species. The Everglades is now less than half its original size and much of it has dried out, leaving large areas vulnerable to summer wildfires. About 90% of the wading birds in Everglades National Park have vanished and populations of many remaining wading bird species have dropped sharply (Figure 11-19). Populations of other vertebrates, from deer to turtles, are down 75–95%.

By the 1970s, state and federal officials recognized that this huge plumbing project was reducing populations of native plants and wildlife—a major source of tourism revenues for Florida. It was also cutting the water supply for the more than 5 million residents of south Florida. In

Figure 11-18 Florida's Everglades is the site of the world's largest ecological restoration project—an attempt to undo and redo an engineering project that has been destroying this vast wetland and threatening water supplies for south Florida's rapidly growing population.

Photo: ©slowfish/Shutterstock.com

1990, Florida's state government and the U.S. government agreed on a plan for the world's largest ecological restoration project, known as the Comprehensive Everglades Restoration Plan. The U.S. Army Corps of Engineers is supposed to carry out this joint federal and state plan to partially restore the Everglades.

The project has several ambitious goals, including restoration of the curving flow of more than half of the Kissimmee River; removal of 400 kilometers (248 miles) of canals and levees that block natural water flows south of Lake Okeechobee; conversion of large areas of farmland to marshes; the creation of 18 large reservoirs and underground water storage areas to store water for the lower Everglades and for south Florida's population; and building a canal–reservoir system for catching the water now flowing out to sea and pumping it back into the Everglades.

Will this huge ecological restoration project work? It depends not only on the abilities of scientists and engineers but also on prolonged political and economic support from citizens, the state's powerful sugarcane and agricultural industries, and elected state and federal officials. Already, some restrictions on phosphorus discharges from sugarcane plantations have been relaxed, which could worsen pollution problems. The project has had cost overruns and funding shortages and is considerably behind schedule.

🔍 CONSIDER THIS. . .

THINKING ABOUT Everglades Restoration

Do you support carrying out the proposed plan for partially restoring the Florida Everglades, including having the federal government provide half of the funding? Explain.

Figure 11-19 Since 1978, the population of the roseate spoonbill has dropped by more than 50%.

11-5 How Should We Protect and Sustain Freshwater Lakes, Rivers, and Fisheries?

CONCEPT 11-5

Freshwater ecosystems are strongly affected by human activities on adjacent lands, and protection of these ecosystems must include protection of their watersheds.

Freshwater Ecosystems Are in Jeopardy

The ecosystem and economic services provided by many of the world's freshwater lakes, rivers, and fisheries (see Figure 8-14, p. 178) are severely threatened by human activities (**Concept 11-5**). Currently, four of every ten freshwater fish species in North America are considered to be vulnerable, threatened, or endangered, according to a joint study by U.S., Canadian, and Mexican scientists.

As 40% of the world's rivers have been dammed or otherwise engineered, and as many of the world's freshwater wetlands have been destroyed, aquatic species have been crowded out of at least half of the world's freshwater habitat areas. Invasive species, pollution, and climate change threaten the ecosystems of many lakes, rivers, and wetlands. Many freshwater fish stocks are overharvested. And increasing human population pressure and projected climate change during this century will elevate these threats.

Sustaining and restoring the biodiversity and ecosystem services provided by freshwater lakes and rivers is a complex and challenging task, as shown by the following Case Study.

CASE STUDY
Can the Great Lakes Survive Repeated Invasions by Alien Species?

Invasions by nonnative species are a major threat to the biodiversity and ecological functioning of many lakes, as illustrated by what has happened to the five Great Lakes, located on the border between the United States and Canada.

Collectively, the Great Lakes are the world's largest body of freshwater. Since the 1920s, these lakes have been invaded by at least 180 nonnative species and the number keeps rising. Many of the alien invaders arrive on the hulls of, or in bilge-water discharges of, oceangoing ships that have been entering the Great Lakes through the St. Lawrence Seaway since 1959.

One of the biggest threats, the *sea lamprey,* reached the westernmost Great Lakes as early as 1920. This parasite attaches itself to almost any kind of fish and kills the victim by sucking out its blood (see Figure 5-8, p. 107). Over the years, it has depleted populations of many important sport fish species such as lake trout. The United States and Canada keep the lamprey population down by applying a chemical that kills lamprey larvae where they spawn in streams that feed into the lakes—at a cost of about $15 million a year.

In 1986, larvae of the *zebra mussel* (see Figure 9-9, p. 199) arrived in ballast water discharged from a European ship near Detroit, Michigan. This thumbnail-sized mollusk reproduces rapidly and has displaced other mussel species and thus depleted the food supply for some other Great Lakes aquatic species. The mussels have also clogged irrigation pipes, shut down water intake pipes for power plants and city water supplies, fouled beaches, and jammed ships' rudders. They have grown in thick masses on many boat hulls, piers, and other exposed aquatic surfaces (Figure 11-20). This mussel has also spread to freshwater communities in parts of southern Canada and 18 eastern U.S. states. Damages and attempts to control this mussel cost the two countries about $1 billion a year—an average of $114,000 per hour.

Sometimes, nature aids us in controlling an invasive alien species. For example, populations of zebra mussels are declining in some parts of the Great Lakes because a

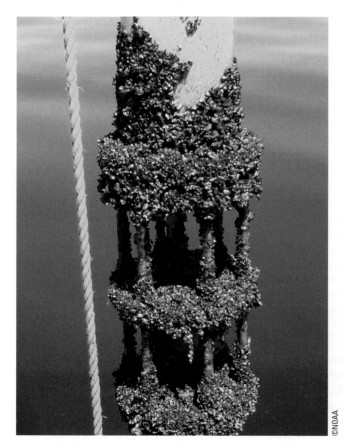

Figure 11-20 These *zebra mussels* are attached to a water current meter in Lake Michigan.

native sponge growing on their shells is preventing them from opening up their shells to breathe. However, it is not clear whether the sponges will effectively control the invasive mussels in the long run and what harmful ecological effects the sponges might have.

In 1989, a larger and potentially more destructive species, the *quagga mussel,* invaded the Great Lakes, probably discharged in the ballast water of a Russian freighter. It can survive at greater depths and tolerate more extreme temperatures than the zebra mussel can. In 2009, scientists reported that quagga mussels had rapidly replaced many other bottom-dwellers in Lake Michigan. This has reduced the food supply for many fish and other species, thus leading to a major disruption of the lake's food web. There is concern that quagga mussels may spread by river transport and eventually colonize eastern U.S. ecosystems such as the Chesapeake Bay (see Chapter 8, Case Study, p. 177) and waterways in parts of Florida. In 2007, it was found to have crossed the United States, probably hitching a ride on a boat or trailer being hauled cross-country. It now resides in the Colorado River and its reservoir system.

The *Asian carp* (Figure 11-21) is the most recent threat to the Great Lakes system. In the 1970s, catfish farmers in the southern United States imported two species of Asian carp to help remove suspended matter and algae from their aquaculture farm ponds. Heavy flooding during the 1990s caused many of these ponds to overflow, which resulted in the release of some of the carp into the Mississippi River. After working their way up the Mississippi River system, these invaders are now close to entering Lake Michigan, if they have not done so already.

These highly prolific fish can quickly grow as long as 1.2 meters (4 feet) and weigh up to 50 kilograms (110 pounds), and they can eat as much as 20% of their own body weight in plankton every day. They can therefore rapidly disrupt lake food webs. They also have the ability to jump clear of the water, and several boaters have been injured after being hit by jumping carp.

These fish have no natural predators in the rivers they have invaded or in the Great Lakes. Joel Brammeler, president of the Alliance for the Great Lakes, warned that "if Asian carp get into Lake Michigan, there is no stopping them."

Managing River Basins Is Complex and Controversial

Rivers and streams provide important ecosystem and economic services (Figure 11-22), but overfishing, pollution, dams, and water withdrawal for irrigation are disrupting these services. Currently, the services are given little or no monetary value when the costs and benefits of dam and reservoir projects are assessed. According to environmental economists, attaching even crudely estimated monetary values to these ecosystem services—an application

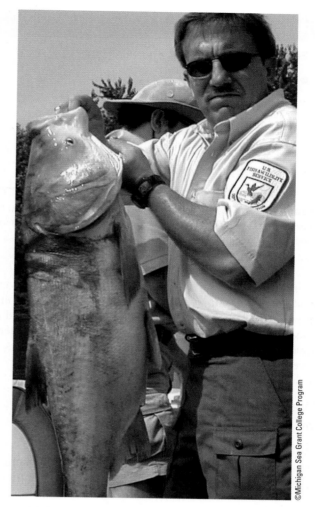

©Michigan Sea Grant College Program

Figure 11-21 Asian carp may be the next major invasive species to threaten the Great Lakes.

of the full-cost pricing **principle of sustainability** (see Figure 1-5, p. 9 or back cover)—would help to sustain them.

An example of such disruption—one that especially illustrates biodiversity loss—is what happened in the Columbia River, which runs through parts of southwestern Canada and the northwestern United States. It has 119 dams, 19 of which are major generators of inexpensive hydroelectric power. It also supplies water for major urban areas and large irrigation projects.

The Columbia River dam system has benefited many people, but it has sharply reduced populations of wild salmon. These migratory fish hatch in the upper reaches of the streams and rivers that form the headwaters of the Columbia River, migrate to the ocean where they spend most of their adult lives, and then swim upstream to return to the place where they were hatched to spawn and die. Dams interrupt their life cycle by interfering with the migration of young fish downstream, and blocking the return of mature fish attempting to swim upstream to their spawning grounds.

© Cengage Learning

Figure 11-23 The upper St. Croix River, flowing on the border between Minnesota and Wisconsin, was one of the first eight rivers to be designated a national wild and scenic river. It hosts a rich diversity of plant and animal species that are protected from riverside development and other harmful human activities.

Natural Capital

Ecosystem Services of Rivers

- Deliver nutrients to sea to help sustain coastal fisheries

- Deposit silt that maintains deltas

- Purify water

- Renew and renourish wetlands

- Provide habitats for wildlife

Figure 11-22 Rivers provide some important ecosystem services. *Questions:* Which two of these services do you believe are the most important? Why?

Since the dams were built, the Columbia River's wild Pacific salmon population has dropped by 94% and nine of the Pacific Northwest salmon species are listed as endangered or threatened. Since 1980, the U.S. federal government has spent more than $3 billion in efforts to save the salmon, but none have been effective.

We Can Protect Freshwater Ecosystems by Protecting Watersheds

Sustaining freshwater aquatic systems begins with understanding that land and water are connected. For example, lakes and streams receive many of their nutrients from the ecosystems of bordering land. Such nutrient inputs come from falling leaves, animal feces, and pollutants generated by people, all of which are washed into bodies of water by rainstorms and melting snow. Therefore, to protect a stream or lake from excessive inputs of nutrients and pollutants, we must protect its watershed (Concept 11-5).

As with marine systems, freshwater ecosystems can be protected through laws, economic incentives, and restoration efforts. For example, restoring and sustaining the ecosystem and economic services of rivers will probably require taking down some dams and restoring river flows. In addition, some scientists and politicians have argued for protecting all remaining free-flowing rivers by prohibiting the construction of dams.

With that in mind, in 1968, the U.S. Congress passed the National Wild and Scenic Rivers Act to establish protection of rivers with outstanding wildlife, geological, scenic, recreational, historical, or cultural values. The law classified *wild rivers* as those that are relatively inaccessible (except by trail), and *scenic rivers* as rivers of great scenic value that are free of dams, mostly undeveloped, and accessible in only a few places by roads. These rivers are now protected from widening, straightening, dredging, filling, and damming (Figure 11-23).

In 2009, the U.S. Congress passed a law increasing the total length of wild and scenic rivers by half. But the Wild and Scenic Rivers System keeps only 3% of U.S. rivers free-flowing and protects less than 1% of the country's total river length.

CONSIDER THIS. . .

CAMPUS SUSTAINABILITY
Maine's College of the Atlantic

The College of the Atlantic (COA) in Bar Harbor, Maine, near Acadia National Park is well known for its student-led projects aimed at helping people to live more sustainably. COA students are working with people from the local community to preserve the Union River

Courtesy College of the Atlantic

watershed near their campus. Like any river ecosystem, the Union River system depends on the environmental quality of its watershed—the land from which water drains to form the river. In working on this project, students are helping to map the watershed and prevent or control pollution within this aquatic system, which in turn helps to preserve the area's aquatic biodiversity.

Freshwater Fisheries Need Better Protection

Sustainable management of freshwater fisheries involves supporting populations of commercial and sport fish species, preventing such species from being overfished, and reducing or eliminating populations of harmful invasive species. The traditional way of managing freshwater fish species is to regulate the time and length of fishing seasons and the number and size of fish that can be taken.

Other techniques include building reservoirs and ponds and stocking them with fish, and protecting and creating fish spawning sites. In addition, some fishery managers try to protect fish habitats from sediment buildup and other forms of pollution. They also work to prevent or reduce large human inputs of plant nutrients that spur the excessive growth of aquatic plants.

Some fishery managers seek to control predators, parasites, and diseases by improving habitats, breeding genetically resistant fish varieties, and using antibiotics and dis-

National Park Service

infectants. Hatcheries can be used to restock ponds, lakes, and streams with prized species such as trout, and entire river basins can be managed to protect valued species such as salmon. However, all of these practices should be based on continuing studies of their effects on aquatic biodiversity and ecosystem services. **GREEN CAREERS:** Limnology and fishery management

11-6 What Should Be Our Priorities for Sustaining Aquatic Biodiversity?

CONCEPT 11-6

Sustaining the world's aquatic biodiversity requires mapping it, protecting aquatic hotspots, creating large and fully protected marine reserves, protecting freshwater ecosystems, and restoring degraded coastal and inland wetlands.

We Can Use an Ecosystem Approach to Sustain Aquatic Biodiversity and Ecosystem Services

Edward O. Wilson and other biodiversity experts have proposed the following priorities for an ecosystem approach to sustaining aquatic biodiversity and ecosystem services (Concept 11-6):

- Complete the mapping of the world's aquatic biodiversity, identifying and locating as many plant and animal species as possible in order to make conservation efforts more precise and cost-effective.
- Identify and preserve the world's aquatic biodiversity hotspots and areas where deteriorating ecosystem services threaten people and many other forms of life.
- Create large and fully protected marine reserves to allow damaged marine ecosystems to recover and to allow fish stocks to be replenished.
- Protect and restore the world's lakes and river systems, which are among the most threatened ecosystems of all.
- Initiate ecological restoration projects worldwide in systems such as coral reefs and inland and coastal wetlands to help preserve the biodiversity and important ecosystem services in those systems.
- Find ways to raise the incomes of people who live on or near protected lands and waters so that they can become partners in the protection and sustainable use of ecosystems.

There is growing evidence that the current harmful effects of human activities on aquatic biodiversity and ecosystem services could be reversed over the next two decades. Doing this will require implementing an ecosystem approach to sustaining both terrestrial and aquatic ecosystems. According to Edward O. Wilson (see Individuals Matter 4.1, p. 82), such a conservation strategy would cost about $30 billion per year—an amount that could be provided by a tax of one penny per cup of coffee consumed in the world each year. **GOOD NEWS**

This strategy for protecting the earth's vital biodiversity and ecosystem services will not be implemented without bottom-up political pressure on elected officials from individual citizens and groups. It will also require cooperation among scientists, engineers, and key people in government and the private sector.

A key part of this strategy will be for individuals to "vote with their wallets" by trying to buy only products and services that do not have harmful impacts on terrestrial and aquatic biodiversity. For example, we can eat lower on the food chain by choosing plant-eating species such as tilapia, carp, and catfish instead of carnivorous species such as salmon and sea bass, and especially, instead of farm-raised species that are typically fed fish-meal made of wild-caught fish. Guides to sustainably produced seafood are available on websites such as that of the Monterrey Bay Aquarium.

Big Ideas

- The world's aquatic systems provide important economic and ecosystem services, and scientific investigation of these poorly understood ecosystems could lead to immense ecological and economic benefits.

- Aquatic ecosystems and fisheries are being severely degraded by human activities that lead to aquatic habitat disruption and loss of biodiversity.

- We can sustain aquatic biodiversity by establishing protected sanctuaries, managing coastal development, reducing water pollution, and preventing overfishing.

Rich Carey/Shutterstock.com

This chapter began with a look at the threatened and endangered species of sea turtles and how human activities are threatening their populations in several ways (**Core Case Study**), including habitat destruction and degradation of vital marine systems such as coral reefs (photo at left). Throughout the chapter, we also examined how these threats along with introductions of invasive species, population pressures, climate change, and overexploitation are harming sea turtles and many other marine and freshwater aquatic species. We looked at how many of the world's fisheries are being depleted.

We also explored possible solutions to these problems. Importantly, we know that when areas of the oceans are left undisturbed, marine ecosystems tend to recover their natural functions, and fish populations rebound fairly quickly. Similarly, it seems that the best approach to sustaining freshwater biodiversity is to use an ecosystem approach in order to use the world's wetlands, lakes, and rivers more sustainably.

We can achieve greater success in sustaining aquatic biodiversity by applying the three scientific **principles of sustainability** (see Figure 1-2, p. 6 or back cover). This means reducing inputs of sediments and excess nutrients, which cloud water, lessen the input of solar energy, and upset aquatic food webs and the natural cycling of nutrients in aquatic systems. It also means valuing aquatic biodiversity and putting a high priority on preserving the biodiversity and ecosystem services of aquatic systems. Applying the social science **principles of sustainability** (see Figure 1-5, p. 9 or back cover) can help us to achieve these goals.

Chapter Review

Core Case Study

1. Describe the status of sea turtles and explain how human activities are threatening their existence (**Core Case Study**).

Section 11-1

2. What is the key concept for this section? How much do we know about the habitats and species that make up the earth's aquatic biodiversity? What are three general patterns of marine biodiversity? Describe the threat to marine biodiversity from bottom trawling. How have coral reefs been threatened? What is **ocean acidification**? What are two causes of disruption of freshwater habitats?

3. Give two examples of threats to marine aquatic systems from invasive species and two examples of the same for freshwater systems. What are two harmful effects on aquatic systems resulting from the growth of the human population in coastal areas? Give two examples of how pollution is affecting aquatic systems. What are three ways in which projected climate change could threaten aquatic biodiversity? Explain how ocean acidification occurs and why it is a serious problem.

4. Define **fishery**. What are three major harmful effects of overfishing? Describe the effects of trawler fishing, purse-seine fishing, long-lining, and drift-net fishing. What is *bycatch*? What is a **fishprint**? Summarize the story of the collapse of the Atlantic cod fishery. Explain why the Atlantic bluefin tuna is seriously endangered. Explain how marine mammals are threatened by overfishing and give an example. Summarize the story of jellyfish invasions. Summarize the arguments for protecting sharks. About what percentage of marine species and what percentage of freshwater species are in danger of extinction?

Section 11-2

5. What is the key concept for this section? How have laws and treaties been used to help sustain aquatic species? What is the main problem that interferes with enforcing international agreements? How can economic incentives help to sustain aquatic biodiversity? Give an example of how this can happen.

6. Explain how marine protected areas and marine reserves can be used to help sustain aquatic biodiversity and ecosystem services. What are *marine hotspots*? What percentage of the world's oceans is strictly protected from harmful human activities in marine reserves? Summarize the contributions of Sylvia Earle to the protection of aquatic biodiversity. Give two examples of how marine systems can be restored. Describe the roles of fishing communities and individual consumers in regulating fishing and coastal development. What is *integrated coastal management*?

Section 11-3

7. What is the key concept for this section? Describe three ways of estimating the sizes of fish populations and list their limitations. How can the precautionary principle be applied in managing fisheries and large marine systems? What are catch-share and co-management systems and how can they help to sustain fisheries? How can government subsidies encourage overfishing? Explain how consumer choices can help to sustain fisheries and aquatic biodiversity and ecosystem services. List five ways to manage global fisheries more sustainably.

Section 11-4

8. What is the key concept for this section? What percentage of the U.S. coastal and inland wetlands has been destroyed since 1900? What are the major threats to wetlands and their ecosystem services? How does the United States attempt to reduce wetland losses? Summarize the story of efforts to restore the Florida Everglades.

Section 11-5

9. What is the key concept for this section? Describe the major threats to the world's rivers and other freshwater systems. Explain how invasions by nonnative species are threatening the Great Lakes. What are some ways to help sustain river systems? What are three ways to protect freshwater habitats and fisheries?

Section 11-6

10. What is the key concept for this section? List six priorities for applying the ecosystem approach to sustaining aquatic biodiversity. What are the three big ideas of this chapter? How can we apply the three scientific **principles of sustainability** in efforts to protect sea turtles from extinction (**Core Case Study**) and in helping to sustain aquatic biodiversity and ecosystem services?

Note: Key terms are in bold type.

Critical Thinking

1. Why could sea turtles (**Core Case Study**) be considered indicator species? What does the plight of the world's sea turtles indicate about the ways in which we treat marine ecosystems?

2. Write a short essay describing how each of the six factors summarized by HIPPCO has affected sea turtles (**Core Case Study**) or could affect them further. Look for and describe connections among these factors. For example, how does one factor enhance the effects of one or more other factors? Suggest ways in which each of the factors could be reduced.

3. What do you think are the three greatest threats to aquatic biodiversity and ecosystem services? For each of them, explain your thinking. Overall, why are aquatic species more vulnerable to extinction hastened by human activities than terrestrial species are? Why is it more difficult to identify and protect endangered marine species than to identify and protect endangered terrestrial species?

4. Why should you be concerned about jellyfish populations taking over large areas of the ocean? Why can jellyfish be considered an indicator species, and what does the explosion of jellyfish populations indicate about marine ecosystems? What are three things you would do to try to slow the changes in the world's aquatic ecosystems that have likely caused the rise of jellyfish populations?

5. Why do you think no-fishing marine reserves recover their biodiversity faster and more effectively than do protected areas where fishing is allowed but restricted? Explain.

6. How might continued overfishing of marine species affect your lifestyle? How could it affect the lives of any children or grandchildren you might have? What are three things you could do to help prevent overfishing?

7. Should fishers who harvest fish from a country's publicly owned waters be required to pay the government fees for the fish they catch? Explain. If your livelihood depended on commercial fishing, would you be for or against such fees?

8. Some scientists consider ocean acidification to be one of the most serious environmental and economic threats that the world faces. How do you contribute to ocean acidification in your daily life? What are three things you could do to help reduce the threat of ocean acidification?

Doing Environmental Science

Pick a coastal area, river, stream, lake, or wetland near where you live and research and write a brief account of its history. Then survey and take notes on its present condition. Has its condition improved or deteriorated during the last 10 years? What governmental or private efforts are being used to protect this aquatic system? Write a report summarizing your findings. Based on your report along with your ecological knowledge of this system, write up some recommendations to policy makers for protecting it. Try presenting your recommendations to one or more local policy makers.

Global Environment Watch Exercise

Search for *Coral Reefs* and use the topic portal to do the following: **(a)** identify two cases of coral reef degradation—one where the reef continues to be degraded and one where a degraded reef is being restored; **(b)** for each case, list the main causes for this degradation; **(c)** for each case, describe any efforts to protect or restore the reef; **(d)** for each case, give your best estimation of whether the reef will survive and explain your reasoning.

Ecological Footprint Analysis

A fishprint provides a measure of a country's fish harvest in terms of area. The unit of area used in fishprint analysis is the global hectare (gha), a unit weighted to reflect the relative ecological productivity of the area fished. When compared with the fishing area's *sustainable biocapacity* (its ability to provide a stable supply of fish year after year, expressed in terms of yield per area), its fishprint indicates whether the country's annual fishing harvest is sustainable.

The fishprint and biocapacity are calculated using the following formulas:

Fishprint in (gha) = metric tons of fish harvested per year/productivity in metric tons per hectare × weighting factor

Biocapacity in (gha) = sustained yield of fish in metric tons per year/ productivity in metric tons per hectare × weighting factor

The following graph shows the earth's total fishprint and biocapacity. Study it and answer the questions that follow.

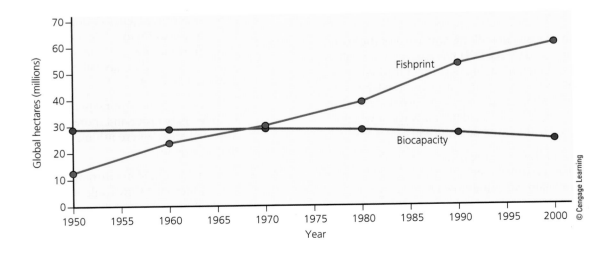

1. Based on the graph,
   a. What is the current status of the global fisheries with respect to sustainability?
   b. In what year did the global fishprint begin to exceed the biological capacity of the world's oceans?
   c. By how much did the global fishprint exceed the biological capacity of the world's oceans in 2000?

2. Assume a country harvests 18 million metric tons of fish annually from an ocean area with an average productivity of 1.3 metric tons per hectare and a weighting factor of 2.68. What is the annual fishprint of that country?

3. If biologists determine that this country's sustained yield of fish is 17 million metric tons per year,
   a. What is the country's sustainable biological capacity?
   b. Is the county's annual fishing harvest sustainable?
   c. To what extent, as a percentage, is the country undershooting or overshooting its biological capacity?

Questions 1–4 refer to the fish harvesting methods below.

   a. Trawler fishing
   b. Purse-seine fishing
   c. Long line fishing
   d. Drift-net fishing
   e. Sonar

1. Enclosing a school of fish with a large net. Used to capture yellow fin tuna.

2. Can be 80 miles long with thousands of baited hooks. This often results in bycatch of many ocean fish species.

3. Dragging huge nets across the bottom of the ocean which are weighted down to harvest bottom fish and shellfish.

4. Often called "ghost-fishing" because these huge nets are left to float in the ocean by themselves for days resulting in a large bycatch.

Use the graph of the collapse of the cod fishery in the northwest Atlantic to answer questions 5–7.

(Data from Millennium Ecosystem Assessment.)

5. Identify the 10-year period when the greatest rate of decline in the fish harvest took place.
   a. 1940–1950
   b. 1950–1960
   c. 1960–1970
   d. 1970–1980
   e. 1990–2000

6. For the 10-year period showing the greatest rate of decline, calculate the rate of decline in the fish harvest in tons per year.
   a. 700,000
   b. 70,000
   c. 7,000
   d. 600,000
   e. 60,000

7. The graph illustrates a concept known as
   a. the tragedy of the commons.
   b. invasive species.
   c. bycatch.
   d. riparian destruction.
   e. biological extinction.

8. All of the following are ecological services provided by coastal wetlands EXCEPT
   a. they are rich in nutrients and are therefore productive places for aquatic biodiversity.
   b. the large amount of vegetation acts as a natural filter for contaminants.
   c. coastal wetlands act as a buffer and help protect the land during hurricanes and tropical storms.
   d. the ground can absorb a great deal of water, reducing local damage from floods.
   e. because they are moist and contain many nutrients, they are important cropland.

9. The U.S. federal policy that allows for the destruction of existing wetlands as long as an equal area of the same type of wetland is created is known as
   a. CERCLA.
   b. the Montreal Protocol.
   c. mitigation banking.
   d. FIRPA.
   e. the Law of the Sea.

10. One reason for the decline in recent years of global fish catch is
  a. bottom trawling scrapes the ocean floor, destroying habitats where many fishes live.
  b. invasive species, especially those introduced in ballast water, prey upon and destroy the habitat of native fishes.
  c. climate change and the resulting rise in sea level are damaging and drowning coral reefs, which are an important habitat for many species of fish.
  d. approximately 4 million fishing boats use modern technology to hunt vast numbers of fishes in the open ocean.
  e. fishes that generate their own light in a process called bioluminescence are more likely to be seen and therefore caught by fishing boats.

Use the graph below to answer the following questions.

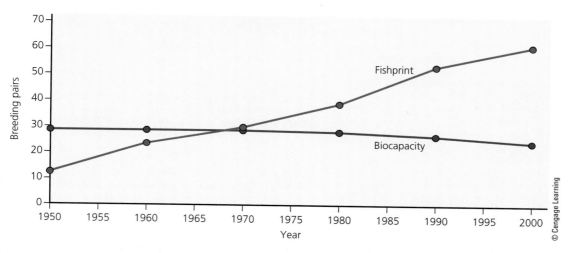

11. What is the relationship between fishprint and biocapacity?
  a. Inversely proportional
  b. Directly proportional
  c. Intermittent

12. In what year was the fishprint 33% above the biocapacity?
  a. 1950
  b. 1960
  c. 1970
  d. 1980
  e. 1990

13. Calculate the approximate percent change of fishprint from 1960 to 1980.
  a. 40%
  b. 50%
  c. 60%
  d. 70%
  e. 80%

12 Food Production and the Environment

There are two spiritual dangers in not owning a farm. One is the danger of supposing that breakfast comes from the grocery, and the other that heat comes from the furnace.

ALDO LEOPOLD

Key Questions

12-1 What is food security and why is it difficult to attain?

12-2 How is food produced?

12-3 What environmental problems arise from industrialized food production?

12-4 How can we protect crops from pests more sustainably?

12-5 How can we improve food security?

12-6 How can we produce food more sustainably?

The Rodale Research Center in Kutztown, Pennsylvania, studies many aspects of agriculture.

JIM RICHARDSON/National Geographic Creative

A *food desert* is an urban area where people have little or no easy access to nutritious food without traveling long distances. In the United States, an estimated 23.5 million people, including 6.5 million children, live in such areas, according to the U.S. Centers for Disease Control and Prevention. People living in these urban neighborhoods tend to rely on convenience stores and fast food restaurants that mainly offer high-calorie, highly processed foods that can lead to higher risks of obesity, diabetes, and heart disease.

Will Allen (Figure 12-1), one of six children of a sharecropper, grew up on a farm in Maryland, but left the farming life for college and a professional basketball career, followed by a successful corporate marketing career. In 1993, Allen had decided to return to his roots, and he bought the last working farm within the city limits of Milwaukee, Wisconsin. It was to become a food oasis in a food desert.

On this small urban plot, Allen developed Growing Power, Inc., an ecologically based farm and a showcase for forms of agriculture that apply all three scientific **principles of sustainability** (see Figure 1-2, p. 6 or back cover). It is powered partly by solar electric and solar hot water systems and makes use of several greenhouses to capture solar energy for growing food throughout the year. The farm produces an amazing diversity of crops—150 varieties of vegetables along with organic herbs and sunflowers. It also produces chickens, turkeys, goats, fish such as tilapia and perch, and honeybees. And the farm's nutrients are recycled in creative ways. For example, wastes from the farmed fish are used as nutrients to raise some of the crops.

The farm's products are sold locally at Growing Power farm stands throughout the region and to various restaurants. Allen also worked with the city to establish the Farm-to-City Market Basket program through which people can sign up for weekly deliveries of organic produce at modest prices.

In addition, Growing Power runs an education program for school children who visit the farm to learn about where their food comes from. Allen also runs a training program for about 1,000 people every year who want to learn organic farming methods. In 2011, the farm partnered with the city of Milwaukee to create 150 new green jobs for unemployed, low-income workers, building greenhouses and growing food organically. Growing Power has expanded, locating another urban farm in a neighborhood of Chicago, Illinois, and setting up satellite training sites in five other states.

For his creative and energetic efforts, Allen has won several prestigious awards, including a MacArthur Fellowship. However, he is most proud of the fact that his urban farm helps to feed more than 10,000 people every year and puts people to work raising good food.

In this chapter we look at different ways to produce food, the environmental effects of food production, and how to produce food more sustainably.

Figure 12-1 In 1996, Will Allen founded Growing Power—an urban farm in Milwaukee, Wisconsin.

©Courtesy of Growing Power, Inc.

12-1 What Is Food Security and Why Is It Difficult to Attain?

CONCEPT 12-1A
Many people in less-developed countries have health problems from not getting enough food, while many people in more-developed countries suffer health problems from eating too much.

CONCEPT 12-1B
The greatest obstacles to providing enough food for everyone are poverty, war, bad weather, climate change, and the harmful environmental effects of industrialized food production.

Poverty Is the Root Cause of Food Insecurity

Food security is the condition under which all or most of the people in a population have daily access to enough nutritious food to live active and healthy lives. Today, we produce more than enough food to meet the basic nutritional needs of every person on the earth. But even with this food surplus, one of every six people in less-developed countries is not getting enough to eat. These people face **food insecurity**—living with chronic hunger

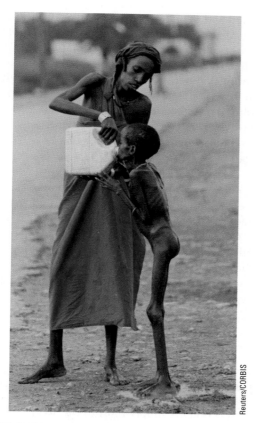

Figure 12-2 This woman is feeding her child, who was starving as a result of famine caused by a civil war that has been going on in Somalia, Africa, most of the time since 1991.

and poor nutrition, which threatens their ability to lead healthy and productive lives (Concept 12-1A).

Most agricultural experts agree that *the root cause of food insecurity is poverty,* which prevents poor people from growing or buying enough food to meet their needs (Figure 12-2 and Figure 1-18, p. 19). This is not surprising given that about half of the world's people are trying to survive on the equivalent of $2.25 a day and one out of six people struggle to get by on $1.25 a day. Other obstacles to food security are war, bad weather (such as prolonged drought, flooding, and heat waves), and climate change (Concept 12-1B), along with corruption and political upheaval.

Each day, there are about 230,000 more people at the dinner table and by 2050, there will likely be at least 2.5 billion more people to feed (see Figure 6-1, p. 122). This projected population increase will be more than twice China's current population and more than 8 times the current U.S. population. Most of these people will live in huge cities in less-developed countries (see Figure 6-3, p. 123) where nearby cropland is scarce and often suffers from eroding soils and irrigation wells going dry. A critical question is, how can we feed the projected 9.6 billion people in 2050 without causing serious harm to the environment? We explore possible answers to this question throughout this chapter.

Many People Suffer from Chronic Hunger and Malnutrition

To maintain good health and resist disease, individuals need fairly large amounts of *macronutrients* (such as carbohydrates, proteins, and fats; see Figures 5, 6, and 9, pp. S16–S17, in Supplement 4), and smaller amounts of *micronutrients*—vitamins, such as A, B, C, and E, and minerals, such as iron, iodine, and calcium.

People who cannot grow or buy enough food to meet their basic energy needs suffer from **chronic undernutrition**, or **hunger** (Concept 12-1A). Most of the world's hungry people live in low-income, less-developed countries where they typically can afford only a low-protein, high-carbohydrate, vegetarian diet consisting mainly of grains such as wheat, rice, or corn. In more-developed countries, people living in food deserts (**Core Case Study**) have a similar problem, except that their diet is heavy on cheap food loaded with fats, sugar, and salt. In both cases, people often suffer from **chronic malnutrition**, a condition in which they do not get enough protein and other key nutrients. This can weaken them, making them more vulnerable to disease, and hindering the normal physical and mental development of children.

According to the United Nations (UN) Food and Agriculture Organization (FAO), in 2010 there were about 1 billion chronically undernourished and malnourished people—one of every seven people on the planet. (See

Figure 23, p. S49, in Supplement 6 for a map of the countries with the most undernourished people.) The FAO estimates that each year, nearly 6 million children younger than age 5 die from chronic hunger and malnutrition and from increased susceptibility to normally nonfatal infectious diseases (such as measles and diarrhea) because of their weakened condition. That is an average of 11 children per minute dying from these causes.

Perhaps the worst form of food shortage is **famine**, which occurs when there is a severe shortage of food in an area and which can result in mass starvation, many deaths, economic chaos, and social disruption. Famines are usually caused by crop failures from drought, flooding, war (Figure 12-2), and other catastrophic events (Concept 12-1B).

Many People Do Not Get Enough Vitamins and Minerals

Many people, most of them in less-developed countries, suffer from a deficiency of one or more vitamins and minerals, usually *vitamin A, iron,* and *iodine* (Concept 12-1A). According to the World Health Organization (WHO), at least 250,000 children younger than age 6, most of them in less-developed countries, go blind every year from a lack of vitamin A. Within a year, more than half of them die.

Having too little *iron* (Fe)—a component of the hemoglobin that transports oxygen in the blood—causes *anemia*, which causes fatigue, makes infection more likely, and increases a woman's chances of dying from hemorrhage in childbirth. According to the WHO, one of every five people in the world—mostly women and children in less-developed countries—suffers from iron deficiency.

The chemical element *iodine* (I) is essential for proper functioning of the thyroid gland, which produces hormones that control the body's rate of metabolism. Chronic lack of iodine can cause stunted growth, mental retardation, and goiter—a severely swollen thyroid gland that can lead to deafness (Figure 12-3). According to the UN, some 600 million people (almost twice the current U.S. population) suffer from goiter, most of them in less-developed countries, especially in South and Southeast Asia. And 26 million children suffer irreversible brain damage every year from lack of iodine. According to the FAO and the WHO, eliminating this serious health problem by adding traces of iodine to salt would cost the equivalent of only 2–3 cents per year for every person in the world.

Figure 12-3 This woman has a massive goiter from a lack of iodine in her diet.

Many People Have Health Problems from Eating Too Much

Overnutrition occurs when food energy intake exceeds energy use and causes excess body fat. Too many calories, too little exercise, or both can cause overnutrition. People who are underfed and underweight and those who are overfed and overweight share similar health problems: *lower life expectancy, greater susceptibility to disease and illness,* and *lower productivity and life quality* (Concept 12-1A).

We live in a world where about 1 billion people face health problems because they do not get enough nutritious food to eat and another 1.6 billion people have health problems because they eat too much, according to the World Health Organization (WHO). The WHO estimates that obesity contributes to the deaths of 2.8 million people per year.

The United States suffers from an overweight and obesity epidemic, mostly because, on average, people consume too much sugar, fat, and salt and lead physically inactive lifestyles. According to the U.S. Centers for Disease Control and Prevention (CDC), 69% of American adults over age 20 are overweight or obese. Studies show that this plays an important role in four of the top ten causes of death in the United States—heart disease, stroke, type 2 diabetes, and some forms of cancer.

12-2 How Is Food Produced?

CONCEPT 12-2
We have used high-input industrialized agriculture and lower-input traditional agriculture to greatly increase food supplies.

Food Production Has Increased Dramatically

Today, three systems supply most of our food, using about 40% of the world's land. *Croplands* produce mostly grains and provide about 77% of the world's food using 11% of its land area. *Rangelands, pastures,* and *feedlots* produce meat and meat products and supply about 16% of the world's food using about 29% of the world's land area. And *fisheries* and *aquaculture* (fish farming) supply fish and shellfish, which make up about 7% of the world's food supply.

These three systems depend on a small number of plant and animal species. Of the estimated 50,000 plant species that people can eat, only 14 of them supply about 90% of the world's food calories. About two of every three people in the world survive primarily by eating one or more of three grain crops—*rice, wheat,* and *corn.* Only a few species of mammals and fish provide most of the world's meat and seafood.

Such food specialization puts us in a vulnerable position, should any of the small number of crop strains, livestock breeds, and fish and shellfish species we depend on become depleted as a result of factors such as disease, environmental degradation, and climate change. This violates the biodiversity **principle of sustainability**, which calls for depending on a variety of food sources as an ecological insurance policy for dealing with changes in environmental conditions that have occurred throughout human history.

Despite such genetic vulnerability, since 1960 there has been a staggering increase in global food production from all three of the major food production systems (Concept 12-2). This was made possible by technological advances such as greater use of tractors and other farm machinery and high-tech fishing equipment (see Figure 11-5, p. 254). Another major advance has been the development of **irrigation**, a mix of methods by which water is supplied to crops by artificial means. Other technological developments include the increasing use of manufactured chemical fertilizers and pesticides to provide plant nutrients and reduce crop losses to pests.

Industrialized Crop Production Relies on High-Input Monocultures

Crops are raised by means of two types of agriculture: industrialized agriculture and subsistence agriculture. **Industrialized agriculture**, or **high-input agriculture**, uses heavy equipment (Figure 12-4) along with large amounts of financial capital, fossil fuels, water, commercial inorganic fertilizers, and pesticides to produce single crops, or *monocultures*. The major goal of industrialized agriculture is to steadily increase each crop's *yield*—the amount of food produced per unit of land. Industrialized agriculture is practiced on 25% of all cropland, mostly in more-developed countries, and produces about 80% of the world's food (Concept 12-2).

Plantation agriculture is a form of industrialized agriculture used primarily in tropical less-developed countries. It involves growing *cash crops* such as bananas, coffee, vegetables, soybeans (mostly to feed livestock; see Figure 1-4, p. 8), sugarcane (to produce sugar and ethanol fuel), and palm oil (used to produce cooking oil and

Figure 12-4 This farmer, harvesting a wheat crop in the Midwestern United States, relies on expensive heavy equipment and uses large amounts of seed, manufactured inorganic fertilizer and pesticides, and fossil fuels to produce the crop.

Brenda Carson/Shutterstock.com

biodiesel fuel; Figure 12-5) on large monoculture plantations, mostly for export to more-developed countries.

A newer form of industrialized agriculture uses large arrays of greenhouses to raise crops indoors. In water-short, arid areas such as parts of Spain, farmers can save water by using greenhouses (Figure 12-6), because they can deliver water more efficiently to greenhouse crops than they can to outdoor crops. In addition, because the water can be purified and recycled, water use and water pollution are reduced sharply in such systems, compared to conventional crop irrigation systems. At Growing Power, Will Allen (**Core Case Study**) provides jobs by hiring local people to build simple greenhouses out of arc-shaped metal hoops covered with plastic *(hoop houses)*—an inexpensive application of the solar energy **principle of sustainability**.

Traditional Agriculture Often Relies on Low-Input Polyculture

Some 2.7 billion people (39% of the world's people) in less-developed countries practice *traditional agriculture*. It provides about one-fifth of the world's food crops on about three-fourths of its cultivated land.

There are two main types of traditional agriculture. **Traditional subsistence agriculture** supplements energy from the sun with the labor of humans and draft animals to produce enough crops for a farm family's survival, with little left over to sell or store as a reserve for hard times. In **traditional intensive agriculture**, farmers increase their inputs of human and draft-animal labor, animal manure for fertilizer, and water to obtain higher crop yields. If the weather cooperates, farmers can produce enough food to feed their families and have some left over to sell for income.

Figure 12-5 These large plantations of oil palms on the island of Borneo in Malaysia were planted in an area once covered with tropical rain forest, as seen on the left side of this photo.

age fotostock/SuperStock.com

Some traditional farmers focus on cultivating a single crop, but many grow several crops on the same plot simultaneously, a practice known as **polyculture**. Such crop diversity—an example of implementing the biodiversity **principle of sustainability** (see Figure 1-2, p. 6 or back cover)—reduces the chance of losing most or all of the year's food supply to pests, bad weather, and other misfortunes.

One type of polyculture is known as *slash-and-burn agriculture* (see Figure 1-17, p. 18). This type of subsistence agriculture involves burning and clearing small plots in tropical forests, growing a variety of crops for a few years until the soil is depleted of nutrients, and then shifting to other plots to begin the process again. Over many centuries, users of this method have learned that each abandoned patch normally has to be left fallow (unplanted) for 10–30 years before the soil becomes fertile enough to grow crops again.

In parts of South America and Africa, some traditional farmers grow as many as 20 different crops together on small cleared plots in tropical forests. The crops rely on sunshine and natural fertilizers such as animal manure for their growth, and they mature at different times. They provide food throughout the year and keep the topsoil covered to reduce erosion from wind and water. Polyculture also lessens the need for fertilizer and water, because root systems at different depths in the soil capture nutrients and moisture efficiently. Synthetic insecticides and herbicides are rarely needed because multiple habitats are created for natural predators of crop-eating insects, and weeds have trouble competing with the multitude and density of crop plants.

©bright/Shutterstock.com

Figure 12-6 Spain uses the world's largest array of greenhouses to grow most of its produce.

Research shows that, on average, such low-input polyculture produces higher yields than does high-input monoculture. For example, ecologists Peter Reich and David Tilman found that carefully controlled polyculture plots with 16 different species of plants consistently outproduced plots with 9, 4, or only 1 type of plant species. Therefore, some analysts argue for greatly increased use of polyculture to produce food more sustainably. The Growing Power farm (**Core Case Study**) practices polyculture by growing a variety of crops in inexpensive greenhouses—an application of the solar energy and biodiversity **principles of sustainability**.

Organic Agriculture Is on the Rise

A fast-growing sector of the U.S. and world economies is **organic agriculture**, in which crops are grown with the use of ecologically sound and sustainable methods and without the use of synthetic pesticides, synthetic inorganic fertilizers, and genetically engineered plants or animals. To be classified as *organically grown,* animals must be raised on 100% organic feed without the use of antibiotics or growth hormones. Growing Power (**Core Case Study**) has become a well-known model for such food production. Figure 12-7 compares organic agriculture with industrialized agriculture.

In the United States, by law, a label of *100 percent organic* (or *USDA Certified Organic*) means that a product is produced only by organic methods and contains all organic ingredients. Products labeled *organic* must contain at least 95% organic ingredients and those labeled *made with organic ingredients* must contain at least 70% organic ingredients. The word *natural* is used on food labels primarily as an advertising ploy and carries no requirement for organic content.

More than two decades of research indicates that organic farming has a number of environmental advantages over conventional industrialized farming. A major advantage is that 100% certified organic food reduces consumer exposure to pesticide residues and to bacteria-resistant antibiotics that can be found in lower-grade organic and nonorganic foods. There is controversy over whether organic food is more nutritious than nonorganic food, although the hope for better nutrition is not the only reason people choose such foods.

On the other hand, conventional agriculture usually can produce higher yields from monoculture crops on smaller areas of land than organic agriculture can. Another drawback is that most organically grown food costs 10–100% more than conventionally produced food (depending on types of food), primarily because organic farming is more labor intensive. However, Will Allen and other Growing Power farmers (**Core Case Study**) are learning how to use sustainable farming methods to get higher yields of a variety of organic crops at affordable prices.

A Closer Look at Industrialized Crop Production

Farmers have two ways to produce more food: farming more land or getting higher yields from existing cropland. Since 1950, about 88% of the increase in global food production has come from using high-input industrialized agriculture to increase crop yields in a process called the **green revolution**.

Figure 12-7 There are major differences between industrialized agriculture and organic agriculture, as summarized here.

Left top: B Brown/Shutterstock.com. Left center: ©Zorandim/Shutterstock.com. Left bottom: ©Art Konovalov/Shutterstock.com. Right top: Noam Armonn/2010. Used under license from Shutterstock.com. Right center: ©Varina and Jay Patel/Shutterstock.com. Right bottom: Adisa/Shutterstock.com.

Total World Grain Production

World Grain Production per Capita

Figure 12-8 These graphs show the growth in worldwide grain production (left) of wheat, corn, and rice, and per capita grain production (right) between 1961 and 2011. **Question:** Why do you think grain production per capita has grown less consistently than total grain production?

(Compiled by the authors using data from U.S. Department of Agriculture, Worldwatch Institute, UN Food and Agriculture Organization, and Earth Policy Institute.)

A green revolution involves three steps. *First,* develop and plant monocultures of selectively bred or genetically engineered high-yield varieties of key crops such as rice, wheat, and corn. *Second,* produce high yields by using large inputs of water, synthetic inorganic fertilizers, and pesticides. *Third,* increase the number of crops grown per year on a plot of land through *multiple cropping*. Between 1950 and 1970, in what was called the *first green revolution,* this high-input approach dramatically raised crop yields in most of the world's more-developed countries, especially the United States (see the Case Study that follows).

A *second green revolution* has been taking place since 1967. Fast-growing varieties of rice and wheat, specially bred for tropical and subtropical climates, have been introduced into middle-income, less-developed countries such as India, China, and Brazil. Producing more food on less land in such countries has helped to protect some biodiversity by preserving large areas of forests, grasslands, wetlands, and easily eroded mountain terrain that might otherwise be used for farming.

Largely because of the two green revolutions, world grain production more than tripled between 1961 and 2011 (Figure 12-8, left). Per capita food production increased by 31% between 1961 and 1985, but since then it has generally declined slightly (Figure 12-8, right).

People directly consume about 46% of the world's grain production. About 34% of it is used to feed livestock and thus is indirectly consumed by people who eat meat and meat products. The remaining 20% (mostly corn) is used to make biofuels for motor vehicles, along with starches and plastics. In 2011, about 40% of the U.S. corn crop was used to produce biofuels (mostly energy-inefficient ethanol). The use of corn to fuel cars instead of to feed livestock and people raised food prices in countries that were importing corn from the United States. This led to lower food security, more hunger, slower economic growth, and political instability, including food riots in several countries.

CASE STUDY

Industrialized Food Production in the United States—The First Green Revolution

In the United States, industrialized farming has evolved into *agribusiness,* as a small number of giant multinational corporations increasingly control the growing, processing, distribution, and sale of food in U.S. and global markets. As a result, the average U.S. farmer now feeds 129 people compared to 19 people in the 1940s. Figure 24, p. S49, in Supplement 6 is a map showing the areas of U.S. land that are farmed. In total annual sales, agriculture is bigger than the country's automotive, steel, and housing industries combined. The entire agricultural system (from farm to grocery store, home, and restaurant) employs more people than any other industry.

Since 1950, U.S. industrialized agriculture has more than doubled the yields of key crops such as wheat, corn, and soybeans without the need for cultivating more land. Such yield increases have kept large areas of U.S. forests, grasslands, and wetlands from being converted to farmland.

People in less-developed countries typically spend up to 40% of their income on food. The world's 1.4 billion poorest people, struggling to live on the equivalent of less than $2.25 a day, typically spend about 70% of their meager income on food. By contrast, because of the efficiency of U.S. agriculture, Americans spend the lowest percentage of income on food in the world—an average of 9% of their disposable income (5% at home and 4% eating out).

However, because of a number of *hidden costs* related to their food consumption, most American consumers are not aware that their actual food costs are much higher than the market prices they pay. Such hidden costs include taxes to pay for farm *subsidies,* or government payments intended to help farmers (mostly producers of corn, wheat, soybeans, and rice) stay in business and increase their yields. Hidden costs also include the costs of pollution and environmental degradation caused by industrialized agriculture (discussed later in this chapter) and higher health insurance bills related to the harmful environmental and health effects of industrialized agriculture. As food expert Michael Pollan puts it, "There is no such thing as cheap food. The real cost of food is paid somewhere. And if it isn't paid at the cash register, it's charged to the environment or to the public purse in the form of subsidies. And it's charged to your health."

Crossbreeding and Genetic Engineering Produce New Varieties of Crops and Livestock

For centuries, farmers and scientists have used *crossbreeding* through *artificial selection* to develop genetically improved varieties of crops and livestock animals (see Science Focus 4.2, p. 89). Such selective breeding in this first *gene revolution* has yielded amazing results. For example, ancient ears of corn were about the size of your little finger, and wild tomatoes were once the size of grapes, but most of the large varieties used now were selectively bred.

Traditional crossbreeding is a slow process, typically taking 15 years or more to produce a commercially valuable new crop variety, and it can combine traits only from species that are genetically similar. Typically, resulting varieties remain useful for only 5–10 years before pests and diseases reduce their yields. Important advances are still being made with this method.

Today, scientists are creating a *second gene revolution* by using *genetic engineering* to develop genetically improved strains of crops and livestock animals. Engineers use a process called *gene splicing* to alter an organism's genetic material through adding, deleting, or changing segments of its DNA (see Figure 8, p. S17, in Supplement 4). The goal of this process is to produce desirable traits or to eliminate undesirable ones by enabling scientists to transfer genes between different species that would not normally interbreed in nature. The resulting organisms are called *genetically modified organisms (GMOs).*

Compared to traditional crossbreeding, developing a new crop variety through gene splicing takes about half as long, usually costs less, and allows for the insertion of genes from almost any other organism into crop or animal cells. Half of the world's genetically modified (GM) crops are planted in the United States and other countries, with large GM crop areas in Argentina and Brazil. However, the European Union severely restricts the use of GM seeds and the importation of products containing GM food.

In the United States, 94% of all soybean, 93% of all rapeseed (for canola oil), and 88% of all corn crops are genetically engineered. According to the U.S. Department of Agriculture (USDA), at least 80% of the food products on U.S. supermarket shelves contain some form of geneti

cally engineered food or ingredients, and that proportion is increasing. In 2012 the Environmental Working Group estimated that each year, the average American adult consumes 88 kilograms (193 pounds) of genetically modified foods. In the United States, GM ingredient information is not included on food labels because law does not require it even though polls indicate that 90% of the public wants to have such information. This is largely because seed companies have successfully lobbied against such a requirement. However, USDA certified 100% organic food, which is labeled as such, makes no use of genetically modified seeds or ingredients.

Bioengineers plan to develop new GM varieties of crops that are resistant to heat, cold, drought, insect pests, parasites, viral diseases, herbicides, and salty or acidic soil. They also hope to develop crop plants that can grow faster and survive with little or no irrigation and with less use of fertilizer and pesticides. Many scientists believe that such innovations hold great promise for helping to improve global food security. Others warn that genetic engineering is not free of drawbacks, which we examine later in this chapter.

Meat Consumption Has Grown Steadily

Meat and animal products such as eggs and milk are sources of high-quality protein and represent the world's second major food-producing system. Between 1950 and 2010, world meat production—mostly beef, pork, and poultry—increased more than sixfold (Figure 12-9) and average consumption per person of meat and meat products more than doubled, according to the FAO. Global meat consumption is likely to more than double again by 2050 as affluence rises and more middle-income people begin consuming more meat and animal products in rapidly developing countries such as China, where meat consumption increased almost ninefold between 1978 and 2012. Currently, China consumes a quarter of all the meat produced in the world.

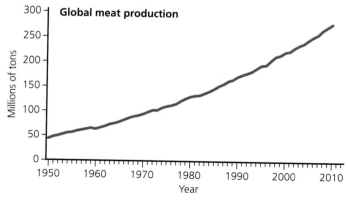

Figure 12-9 Global meat production grew more than sixfold between 1950 and 2010.

(Compiled by the authors using data from USDA, FAO, and Earth Policy Institute.)

Figure 12-10 *Industrialized beef production:* On this cattle feedlot in west Texas, thousands of cattle are fattened on grain for a few months before being slaughtered.

India, Egypt, and the United Arab Emirates are buying or leasing cropland in a number of other countries.

Fish and Shellfish Production Have Risen Dramatically

The world's third major food-producing system consists of fisheries and aquaculture. A **fishery** is a concentration of particular aquatic species suitable for commercial harvesting in a given ocean area or inland body of water. Industrial fishing fleets harvest most of the world's marine catch of wild fish by using a variety of methods (see Figure 11-5, p. 254). Fish and shellfish are also produced through **aquaculture** or **fish farming**—the practice of raising fish in freshwater ponds, lakes, reservoirs, and rice paddies, and in underwater cages in coastal lagoons and estuaries or offshore in deeper ocean waters.

Global seafood production has increased dramatically since 1950 (Figure 12-12) (Concept 12-2). In 2011, about 58% of the world's fish and shellfish were caught mostly by industrial fishing fleets and 42% were produced through aquaculture.

About half of the world's meat comes from livestock grazing on grass in unfenced rangelands and enclosed pastures. The other half is produced through an industrialized factory farm system. It involves raising large numbers of animals bred to gain weight quickly, mostly in crowded *feedlots* (Figure 12-10) or in very crowded pens and cages in huge buildings called *concentrated animal feeding operations (CAFOs)*. Most veal calves, pigs, chickens (Figure 12-11), and turkeys that are raised in more-developed countries spend their lives in CAFOs.

A number of environmental and animal health problems are associated with factory farming. Animals are fed grain, soybeans, fishmeal, or fish oil, and some of this feed is doctored with growth hormones and antibiotics, added to accelerate livestock growth. Because of the crowding, animal wastes and runoff from feedlots and CAFOs create serious impacts on the air and water, which we examine later in this chapter.

As a country's income grows, more of its people tend to eat more meat, much of it produced by feeding large amounts of grain to livestock. The resulting increased demand for grain, often accompanied by a loss of cropland to urban development, can lead to greater reliance on grain imports. Within a few decades, a rapidly developing country can go from producing all the grain it needs to importing most of its grain.

China and India are likely to follow this trend as they become more industrialized and urbanized. For example, China has roughly 20% of the world's population but only 7% of its arable land. Also, since 1998, China's total arable land area has been shrinking because of rapid urbanization. If China were to import just one-fifth of the grain it uses, its imports would equal the amount of grain the United States typically exports each year—roughly half the world's entire grain exports. This is one reason why countries such as China, South Korea, Saudi Arabia,

Figure 12-11 This concentrated chicken feeding operation is located in Iowa (USA). Such operations can house up to 100,000 chickens.

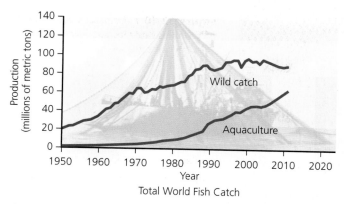

Figure 12-12 World seafood production, including both wild catch (marine and inland) and aquaculture, grew between 1950 and 2011, with the wild catch generally leveling off since 1996 and aquaculture production rising sharply since 1990. **Question:** Why do you think the wild catch has leveled off?

(Compiled by the authors using data from UN Food and Agriculture Organization, Worldwatch Institute, and Earth Policy Institute.)

According to a 2012 report on world fisheries by the FAO, about 57% of the world's ocean fisheries are being harvested at full capacity, and another 30% are over-fished. Some fishery scientists warn that failure to reduce overfishing and ocean acidification and to slow projected climate change could severely threaten food security for roughly 3 billion people who now depend on fish for at least 20% of their animal protein.

Aquaculture, the fastest-growing type of food production, is sometimes called the *blue revolution*. Between 1980 and 2011, the amount of fish produced through aquaculture in 181 countries increased almost 12-fold (Figure 12-12). Aquaculture now provides jobs for 100 million people. Asia accounted for 88% (with China alone accounting for 60%) of the world's aquaculture production in 2011.

Most of the world's aquaculture involves raising species that feed on algae or other plants—mainly carp in China and India, catfish in the United States, tilapia in several countries, and shellfish in a number of coastal countries. However, the farming of meat-eating species

such as shrimp and salmon is growing rapidly, especially in more-developed countries. Such species are often fed fishmeal and fish oil produced from other fish and their wastes. In the next major section, we examine the harmful environmental impacts of aquaculture that could limit its growth.

Industrialized Food Production Requires Huge Inputs of Energy

The industrialization of food production has been made possible by the abundance of nonrenewable oil and natural gas used to run farm machinery and fishing vessels, to pump irrigation water for crops, and to produce synthetic pesticides (mostly from petrochemicals produced when oil is refined) and synthetic inorganic fertilizers. Fossil fuels are also used to process food and transport it long distances within and between countries. For example, in the United States, food travels an average of 2,400 kilometers (1,300 miles) from farm to plate. As a result, agriculture uses about 20% of all the energy used in the United States.

When we consider the energy used to grow, store, process, package, transport, refrigerate, and cook all plant and animal food, it takes about 10 units of fossil fuel energy to put 1 unit of food energy on the table in the United States. Also, according to a study led by ecological economist Peter Tyedmers, the world's fishing fleets use about 12.5 units of energy to put 1 unit of food energy from seafood on the table. In other words, today's systems for producing, processing, transporting, and preparing food are highly dependent on fossil fuels, and together, they result in a large *net energy loss*.

On the other hand, the amount of energy per calorie used to produce crops in the United States has declined by about 50% since the 1970s. One factor in this decline is that the amount of energy used to produce synthetic nitrogen fertilizer has dropped sharply. Another is the rising use of conservation tillage, which sharply reduces the energy use and harmful environmental effects of plowing, as is explained further in the next section of this chapter.

12-3 What Environmental Problems Arise from Industrialized Food Production?

CONCEPT 12-3

Future food production may be limited by soil erosion and degradation, desertification, irrigation water shortages, air and water pollution, climate change, and loss of biodiversity.

Producing Food Has Major Environmental Impacts

Industrialized agriculture allowed farmers to use less land to produce more food, and this has helped to protect biodiversity in many areas by reducing the destruction of forests and grasslands for farming. However, many analysts point out that industrialized agriculture has greater overall harmful environmental impacts than any other human

Natural Capital Degradation

Food Production

Biodiversity Loss	Soil	Water	Air Pollution	Human Health
Conversion of grasslands, forests, and wetlands to crops or rangeland	Erosion	Aquifer depletion	Emissions of greenhouse gases CO_2 from fossil fuel use, N_2O from inorganic fertilizer use, and methane (CH_4) from cattle	Nitrates in drinking water (blue baby)
Fish kills from pesticide runoff	Loss of fertility	Increased runoff, sediment pollution, and flooding from cleared land		Pesticide residues in water, food, and air
Killing of wild predators to protect livestock	Salinization	Pollution from pesticides	Other air pollutants from fossil fuel use and pesticide sprays	Livestock wastes in drinking and swimming water
Loss of agrobiodiversity replaced by monoculture strains	Waterlogging	Algal blooms and fish kills caused by runoff of fertilizers and farm wastes		Bacterial contamination of meat
	Desertification			

Figure 12-13 Food production has a number of harmful environmental effects (Concept 12-3). **Question:** Which item in each of these categories do you believe is the most harmful?

Left: ©Orientaly/Shutterstock.com. Left center: pacopi/Shutterstock.com. Center: W.A. Bannazewski/Visuals Unlimited. Right center: B Brown/Shutterstock.com. Right: Rowan Gillson/Superstock.

activity and these environmental effects may limit future food production (Concept 12-3).

According to a 2010 study by 27 experts assembled by the United Nations Environment Programme (UNEP), agriculture accounts for about 70% of the freshwater removed from aquifers and surface waters, worldwide. It also uses about 38% of the world's ice-free land, emits about 25% of the world's greenhouse gas emissions, and produces about 60% of all water pollution. Figure 12-13 summarizes the harmful effects of industrialized agriculture.

Topsoil Erosion Is a Serious Problem in Parts of the World

Soil is literally the foundation of life on land (Science Focus 12.1). Specifically, the fertile top layer of many soils, called **topsoil**, is a vital component of natural capital, because it stores the water and nutrients needed by plants.

One major problem related to agriculture is **soil erosion**—the movement of soil components, especially

Figure 12-14 Natural capital degradation: Flowing water from rainfall is the leading cause of topsoil erosion as seen on this farm in the U.S. state of Tennessee (left). Severe water erosion can become gully erosion, which has damaged this cropland in western Iowa (right).

Left: Tim McCabe/USDA Natural Resources Conservation Service. Right: ©USDA Natural Resources Conservation Service.

Figure 12-15 Wind is an important cause of topsoil erosion in dry areas that are not covered by vegetation such as this bare crop field in the U.S. state of Iowa.

©Lynn Betts/USDA Natural Resources Conservation Service

surface litter and topsoil, from one place to another by the actions of wind and water. Some topsoil erosion is natural, but much of it is caused by human activities.

Flowing water, the largest cause of erosion, carries away particles of topsoil that have been loosened by rainfall (Figure 12-14, left). Severe erosion of this type leads to the formation of gullies (Figure 12-14, right). Wind also loosens and blows topsoil particles away, especially in areas with a dry climate and relatively flat and exposed land (Figure 12-15).

In undisturbed, vegetated ecosystems, the roots of plants help to anchor the topsoil and to prevent some erosion. But when we remove soil-holding grasses, trees, and other vegetation through activities such as farming (see Figure 7-12, p. 154), deforestation (see Figure 10-12, p. 224), and overgrazing (see Figure 10-22, p. 231), the topsoil can be subject to erosion. A joint survey by the UNEP and the World Resources Institute indicated that topsoil is eroding faster than it forms on about 38% of the world's cropland (Figure 12-16). (See the online Guest Essay on soil erosion by David Pimentel.)

Erosion of topsoil has two major harmful effects. One is *loss of soil fertility* through depletion of plant nutrients in topsoil. The other is *water pollution* in surface waters where eroded topsoil ends up as sediment, which can kill fish and shellfish and clog irrigation ditches, boat channels, reservoirs, and lakes. This pollution problem gets worse when the eroded sediments contain pesticide residues that can be ingested by aquatic organisms and biomagnified within food webs (see Figure 9-13, p. 202).

Drought and Human Activities Are Degrading Drylands

Drylands in regions with arid and semiarid climates occupy about 40% of the world's land area and are home to some 2 billion people. A major threat to food security in some of the world's water-short drylands is **desertification**—the process in which the productive potential of topsoil falls by 10% or more because of a combination of prolonged drought and human activities that expose topsoil to erosion.

Desertification can be *moderate* (with a 10–25% drop in productivity), *severe* (with a drop of 25–50%), or *very severe* (with a drop of more than 50%, usually resulting in huge gullies and sand dunes; see Figure 12-17). Only in extreme cases does desertification lead to what we call desert. But severe desertification can expand existing desert areas or create new deserts.

Over thousands of years, the earth's deserts have expanded and contracted, primarily because of climate change. However, human use of the land, especially for agricultural purposes, has accelerated desertification in some parts of the world mostly because of deforestation, excessive plowing, and overgrazing. Some 40% of the U.S. land area is dry land in the western half of the country—especially in parts of Arizona, New Mexico, and

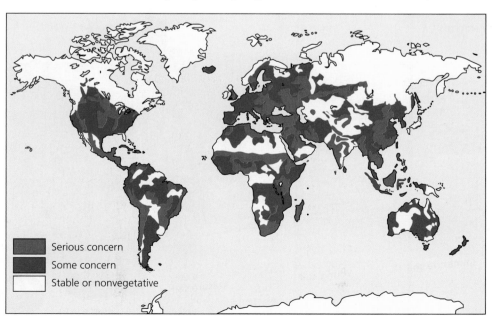

Figure 12-16 Natural capital degradation: Topsoil erosion is a serious problem in some parts of the world. **Question:** Can you see any geographical pattern associated with this problem?

(Compiled by the authors using data from the U.N. Environment Programme and the World Resources Institute.)

Serious concern

Some concern

Stable or nonvegetative

SOIL IS THE FOUNDATION OF LIFE ON LAND

Soil is a complex mixture of eroded rock, mineral nutrients, decaying organic matter, water, air, and billions of living organisms, most of them microscopic decomposers. Soil formation begins when bedrock is slowly broken down into fragments and particles by physical, chemical, and biological processes, called *weathering*. Figure 12-A shows profiles of different-aged soils.

Soil, on which all terrestrial life depends, is a key component of the earth's natural capital. It supplies most of the nutrients needed for plant growth (see Figure 3-5, p. 56). It also purifies and stores water, while organisms living in the soil remove carbon dioxide from the atmosphere and store it as organic carbon compounds, thereby helping to control the earth's climate as part of the carbon cycle (see Figure 3-17, p. 66).

Most soils that have developed over long periods of time, called *mature soils*, contain horizontal layers, or *horizons* (Figure 12-A), each with a distinct texture and composition that vary with different types of soils. Most mature soils have at least three of the four possible horizons. Think of them as the top floors in the geological

building of life under your feet.

The roots of most plants and the majority of a soil's organic matter are concentrated in the soil's two upper layers, the *O horizon* of leaf litter and the *A horizon* of topsoil. In most mature soils, these two layers teem with bacteria, fungi, earthworms, and small insects, all interacting in complex food webs. Bacteria and other decomposer microorganisms, found by the billions in every handful of topsoil, break down some of the soil's complex organic compounds into a porous mixture of the partially decomposed bodies of dead plants and animals, called *humus,* and inorganic materials such as clay, silt, and sand. Soil moisture carrying these dissolved nutrients is drawn up by the roots of plants and transported through stems and into leaves as a key part of the earth's chemical cycling **principle of sustainability** (see Figure 1-2, p. 6 or back cover).

The *B horizon (subsoil)* and the *C horizon (parent material)* contain most of a soil's inorganic matter, mostly broken-down rock consisting of varying mixtures

of sand, silt, clay, and gravel. Much of it is transported by water from the A horizon (Figure 12-A). The C horizon lies on a base of parent material, which is often *bedrock*.

The spaces, or *pores*, between the solid organic and inorganic particles in the upper and lower soil layers contain varying amounts of air (mostly nitrogen and oxygen gas) and water. Plant roots use the oxygen for cellular respiration. As long as the O and A horizons are anchored by vegetation, the soil layers as a whole act as a sponge, storing water and nutrients, and releasing them in a nourishing trickle.

Although topsoil is a renewable resource, it is renewed very slowly, which means it can be depleted. Just 1 centimeter (0.4 inch) of topsoil can take hundreds of years to form, but it can be washed or blown away in a matter of weeks or months when we plow grassland or clear a forest and leave its topsoil unprotected.

Critical Thinking

How does soil contribute to each of the four components of biodiversity described in Figure 4-2, p. 79?

Moss and lichen
Organic debris
Rock fragments
Grasses and small shrubs
Oak tree
Honey fungus
Fern
Wood sorrel
Millipede
Earthworm
O horizon Leaf litter
A horizon Topsoil
Mole
Bacteria
B horizon Subsoil
Fungus
C horizon Parent material
Mite
Nematode
Bedrock
Immature soil
Young soil
Mature soil
Root system
Red earth mite
Beetle larva

© Cengage Learning

Animated Figure 12-A This diagram shows a generalized soil profile and illustrates how soil is formed. **Questions:** What role do you think the tree in this figure plays in soil formation? How might the soil formation process change if the tree were removed?

Figure 12-17 Volunteers are planting vegetation in an effort to slow desertification in this area of Somalia, Africa.

Michael S. Yamashita/National Geographic Creative

Texas. It is subject to desertification, mainly from a combination of drought and overgrazing by livestock.

In the 1930s, much of the topsoil in several dry and windy Midwestern states was lost because of a combination of poor cultivation practices and prolonged drought. The resulting severe wind erosion of topsoil led to crop failures and to the formation of a barren *dust bowl*, with thousands of environmental refugees migrating to other parts of the country.

In 2012, researchers at the Earth Policy Institute warned that excessive overgrazing by livestock (mostly sheep and goats) is creating two new dust bowls. One is in Africa south of the Sahara Desert, and the other straddles northern China and southern Mongolia. According to this report, about 90% of China's grasslands are degraded and suffering from desertification. Since 1950, more than 24,000 Chinese villages have been abandoned because of spreading sands, and these lands are in danger of becoming desert.

Another way to degrade topsoil is to deplete it of key plant nutrients, either by repeatedly using it to grow crops or by allowing livestock to trample and overgraze the land. To make matters worse, according to a 2007 study by the Intergovernmental Panel on Climate Change (IPCC), projected climate change is likely to greatly increase severe and prolonged drought, expand desertification, and threaten food supplies in various areas of the world.

Excessive Irrigation Has Serious Consequences

A major reason for the success in boosting productivity on farms is the use of irrigation, which accounts for about 70% of the water that humanity uses. Currently, the one-fifth of the world's cropland that is irrigated produces about 45% of the world's food.

However, irrigation has a downside. Most irrigation water is a dilute solution of various salts, such as sodium chloride (NaCl), that are picked up as the water flows over or through soil and rocks. Irrigation water that is not absorbed into the topsoil evaporates, leaving behind a thin crust of dissolved mineral salts in the topsoil. Repeated applications of irrigation water in dry climates lead to the gradual accumulation of salts in the upper soil layers—a soil degradation process called **soil salinization**. It stunts crop growth, lowers crop yields, and can eventually kill plants and ruin the land.

The UN estimates that severe soil salinization has reduced yields on at least 10% of the world's irrigated cropland, and that by 2020, 30% of the world's arable land will be salty. The most severe salinization occurs in China, India, Egypt, Pakistan, Mexico, Australia, and Iraq. Salinization affects almost one-fourth of irrigated cropland in the United States, especially in western states (Figure 12-18).

Another problem with irrigation is **waterlogging**, in which water accumulates underground and gradually raises the water table, especially when farmers apply large amounts of irrigation water in an effort to leach salts deeper into the soil. Waterlogging lowers the productivity of crop plants and kills them after prolonged exposure, because it deprives plants of the oxygen they need to survive. At least 10% of the world's irrigated land suffers from this worsening problem.

Perhaps the biggest problem resulting from excessive irrigation in agriculture is that it has contributed to depletion of groundwater and surface water supplies in many areas of the world. We discuss this in Chapter 13.

Agriculture Contributes to Air Pollution and Climate Change

Agricultural activities, including the clearing and burning of forests to raise crops or livestock, create a great deal of air pollution. They also account for more than a quarter of all human-generated emissions of carbon dioxide (CO_2), which is helping to warm the atmosphere in a process of

Figure 12-18 Natural capital degradation: White alkaline salts have displaced crops that once grew on this heavily irrigated land in the U.S. state of Colorado.

climate change that is projected to play an important role in making some areas unsuitable for growing crops during this century.

According to the 2006 FAO study, *Livestock's Long Shadow*, industrialized livestock production generates about 18% of the world's greenhouse gases—more than all of the world's cars, trucks, buses, and planes emit. In particular, cattle and dairy cows release the greenhouse gas methane (CH_4)—with about 25 times the warming potential of CO_2 per molecule—mostly through belching. Along with the methane generated by liquid animal manure stored in feedlot waste lagoons, this accounts for 18% of the global annual emissions of methane. And nitrous oxide (N_2O), with about 300 times the warming capacity of CO_2 per molecule, is released in huge quantities by synthetic inorganic fertilizers, as well as by livestock manure.

Food and Biofuel Production Systems Have Caused Major Losses of Biodiversity

Natural biodiversity and some ecosystem services are threatened when tropical and other forests are cleared (see Figure 10-11, p. 224) and when grasslands are plowed up and replaced with croplands used to produce food and biofuels (Concept 12-3).

For example, one of the fastest-growing threats to the world's biodiversity is the cutting or burning of large areas of tropical forest in Brazil's Amazon Basin and the clearing of areas of its *cerrado,* a huge tropical grassland region south of the Amazon Basin. This land is being burned or cleared for cattle ranches, large plantations of soybeans grown for cattle feed (see Figure 1-4, p. 8), and sugarcane used for making ethanol fuel for cars.

In Indonesia, tropical forests are burned to make way for plantations of oil palm trees (Figure 12-5) increasingly used to produce biodiesel fuel for cars. Such forests are

also being cleared for food production in Africa and many other areas of Asia.

A related problem is the increasing loss of **agrobiodiversity**—the genetic variety of animal and plant species used on farms to produce food. Scientists estimate that since 1900, we have lost 75% of the genetic diversity of agricultural crops that existed then. For example, India once planted 30,000 varieties of rice. Now more than 75% of its rice production comes from only ten varieties and soon, almost all of its production might come from just one or two varieties. In the United States, about 97% of the food plant varieties available to farmers in the 1940s no longer exist, except perhaps in small amounts in seed banks and in the backyards of a few gardeners.

Traditionally, farmers have saved seeds from year to year to save money and to have the ability to grow food in times of famine. Families in India and most other less-developed countries still do this, but in the United States, this tradition is disappearing as more farmers plant seeds for genetically engineered crops. Companies that sell these seeds have patents on them, forbid users to save them, and have sued a number of farmers caught saving such seeds.

In losing agrobiodiversity, we are rapidly shrinking the world's genetic "library," which is critical for increasing food yields. This failure to preserve agrobiodiversity is a serious violation of the biodiversity **principle of sustainability** (see Figure 1-2, p. 6 or back cover).

Individual plants and seeds from endangered varieties of crops and wild plant species important to the world's food supply are stored in about 1,400 refrigerated seed banks, as well as in agricultural research centers and botanical gardens scattered around the world. However, power failures, fires, storms, war, and unintentional disposal of seeds can cause irreversible losses of these stored plants and seeds. More secure seed banks are being built (as we discuss later in this chapter).

However, the seeds of many plants cannot be stored successfully in seed banks. And because stored seeds do not remain alive indefinitely, they must be planted and germinated periodically, and new seeds must be collected for storage. Unless this is done, seed banks become *seed morgues.*

There Is Controversy over Genetically Engineered Foods

While genetic engineering could help to improve food security for some, controversy has arisen over the use of this technology. Its producers and investors see GM food production as a potentially sustainable way to solve world hunger problems. However, some critics consider it potentially dangerous "Frankenfood" that would allow a small number of seed companies to patent genetically modified crops and control most of the world's food production.

Figure 12-19 summarizes the major projected benefits and drawbacks of this new technology.

Some critics recognize the potential benefits of GM crops (Figure 12-19, left) but they point out most of the GM crops developed so far have provided very few of these benefits. In 2011, an international team of scientists and analysts published the *Global Citizens' Report on the State of GMOs*. It calls into serious question industry claims that genetic engineering will increase crop yields, lessen the need for pesticides, and yield drought-tolerant crops. The report also summarized findings indicating that GM crops with built-in toxins, such as Bt toxins, widely used to fend off insects in corn production, could threaten human health by triggering an inflammatory response leading to diseases such as diabetes and heart disease. In addition, herbicide-resistant genetically engineered crops have led to increased herbicide use and to herbicide-resistant superweeds, some of which can grow more than 2 meters (6 feet) tall.

Critics also point out that if GM organisms released into the environment cause some harmful genetic or ecological effects, as some scientists expect, those organisms cannot be recalled. For example, genes in plant pollen from genetically engineered crops can spread among nonengineered species. The new strains can then form hybrids with wild crop varieties, which could reduce the natural genetic biodiversity of the wild strains. This could in turn reduce the gene pool needed to crossbreed new crop varieties and to develop new genetically engineered varieties—another violation of the biodiversity **principle of sustainability** (see Figure 1-2, p. 6 or back cover).

The Ecological Society of America and various critics of genetically engineered crops call for more controlled field experiments and long-term testing to better understand the ecological and health risks, and stricter regulation of this rapidly growing technology.

🔍 **CONSIDER THIS. . .**

CONNECTIONS GM Crops and Organic Food Prices
The spread of GM crop genes by wind carrying pollen from field to field threatens the production of certified organic crops, which must be grown without such genes to be classified as organic. Because organic farmers have to perform expensive tests to detect GMOs or take costly planting measures to prevent the spread of GMOs to their fields from nearby crop fields, they have to raise the prices of their produce.

There Are Limits to Expansion of the Green Revolutions

So far, several factors have limited the success of the green revolutions and may limit them in the future (**Concept 12-3**). Without huge inputs of water and synthetic inorganic fertilizers and pesticides, most green revolution and genetically engineered crop varieties produce yields that are no higher (and are sometimes lower) than those from traditional strains. These high inputs also cost too much for most subsistence farmers in less-developed countries.

Trade-Offs

Genetically Modified Crops and Foods

Projected Advantages	Projected Disadvantages
May need less fertilizer, pesticides, and water	Have unpredictable genetic and ecological effects
Can be resistant to insects, disease, frost, and drought	May put toxins in food
Can grow faster	Can promote pesticide-resistant insects, herbicide-resistant weeds, and plant diseases
May tolerate higher levels of herbicides	Could disrupt seed market and reduce biodiversity

© Cengage Learning

Figure 12-19 Use of genetically modified crops and foods has advantages and disadvantages. **Questions:** Which two advantages and which two disadvantages do you think are the most important? Why?

Top: ©Lenar Musin/Shutterstock.com. Bottom: ©oksix/Shutterstock.com.

Scientists point out that where such inputs do increase yields, there comes a point where yields stop growing because of the inability of crop plants to take up nutrients from additional fertilizer and irrigation water. This helps to explain the slowdown in the rate of growth in global grain yields from an average increase of 2.1% a year between 1950 and 1990 to 1.3% annually between 1990 and 2011.

Can we expand the green revolutions by irrigating more cropland? Since 1978, the amount of irrigated land per person has been declining, and it is projected to fall much more between 2012 and 2050. One reason for this is population growth, which is projected to add 2.6 billion more people between 2012 and 2050. Other factors are wasteful use of irrigation water, soil salinization, and the fact that most of the world's farmers do not have enough money to irrigate their crops. In addition, according to many scientific studies, projected climate change during this century is likely to melt some of the mountain glaciers that provide irrigation and drinking water for many millions of people in China, India, and South America.

Can we increase the food supply by cultivating more land? We have already cleared or converted about 38% of the world's ice-free land surface for use as croplands and pastures. By clearing tropical forests and irrigating arid land, we could more than double the area of the world's cropland. The problem is that such massive clearing of forests would greatly speed up and magnify climate change and biodiversity losses. Also, much of this land has poor soil fertility, steep slopes, or both, and cultivating such land would be expensive and probably not ecologically sustainable.

In addition, during this century, fertile croplands in coastal areas, including many of the major rice-growing floodplains and river deltas in Asia, are likely to be flooded by rising sea levels resulting from projected climate change. Food production could also drop sharply in some major food-producing areas because of longer and more intense droughts and heat waves, also resulting from projected climate change.

Crop yields could be increased with the use of conventional or genetically engineered crops that are more tolerant of drought, which are in the early stages of being tested. Commercial fertilizers have played a role in green revolutions, but their use in most of the more-developed countries has reached a level of diminishing returns in terms of increased crop yields. However, there are parts of the world, especially in Africa, where additional fertilizer could boost crop production.

Industrialized Meat Production Has Harmful Environmental Consequences

Proponents of industrialized meat production point out that it has increased meat supplies, reduced overgrazing, kept food prices down, and yielded higher profits. But environmental scientists point out that feedlots (Figure 12-10) and concentrated animal feeding operations (Figure 12-11) use large amounts of water to grow feed for livestock and to wash away their wastes. In his book *The Food Revolution*, John Robbins estimates that "you'd save more water by not eating a pound of California beef than you would by not showering for a year."

Analysts also point out that meat produced by industrialized agriculture is artificially cheap because most of its harmful environmental and health costs are not included in the market prices of meat and meat products, a violation of the full-cost pricing **principle of sustainability** (see Figure 1-5, p. 9 or back cover). Figure 12-20 summarizes the advantages and disadvantages of industrialized meat production.

When livestock are grazed on open land instead of raised in feedlots, environmental impacts can still be high, especially when forests are cut down or burned to make way for cattle ranches, as is occurring in Brazil's Amazon forests. In 2008, the FAO reported that overgrazing along with soil compaction and erosion by livestock had degraded about 20% of the world's grasslands and pastures. The same report estimated that rangeland grazing and industrialized livestock production caused about 55% of all topsoil erosion and sediment pollution, and fully one-third of the water pollution resulting from the runoff of nitrogen and phosphorus from excessive inputs of synthetic fertilizers.

Trade-Offs

Animal Feedlots

Advantages	Disadvantages
Increased meat production	Animals unnaturally confined and crowded
Higher profits	Large inputs of grain, fishmeal, water, and fossil fuels
Less land use	Greenhouse gas (CO_2 and CH_4) emissions
Reduced overgrazing	Concentration of animal wastes that can pollute water
Reduced soil erosion	Use of antibiotics can increase genetic resistance to microbes in humans
Protection of biodiversity	

© Cengage Learning

Figure 12-20 Use of animal feedlots and confined animal feeding operations has advantages and disadvantages. **Questions:** Which single advantage and which single disadvantage do you think are the most important? Why?

Top: Mikhail Malyshev/Shutterstock.com. Bottom: ©Maria Dryfhout/Shutterstock.com.

CONSIDER THIS...

CONNECTIONS Meat Production and Ocean Dead Zones

Huge amounts of synthetic inorganic fertilizers are used in the Midwestern United States to produce corn for animal feed and to produce ethanol fuel for cars. Much of this fertilizer runs off cropland and eventually goes into the Mississippi River. The added nitrate and phosphate nutrients overfertilize coastal waters in the Gulf of Mexico, where the river flows into the ocean. Each year, this creates a "dead zone" often larger than the U.S. state of Massachusetts. This oxygen-depleted zone threatens one-fifth of the nation's seafood yield. In other words, growing corn in the Midwest, largely to feed cattle and fuel cars, degrades aquatic biodiversity and seafood production in the Gulf of Mexico.

Industrialized meat production makes use of large amounts of fossil fuel energy (mostly from oil), which helps to make it one of the chief sources of air and water pollution and greenhouse gas emissions. A 2011 assessment by the Environmental Working Group of the "cradle-to-grave" carbon footprint of food items found that producing lamb, beef, pork, farmed salmon, or cheese through industrialized agriculture generates 10–20 times more greenhouse gases per unit of weight than does producing common vegetables and grains. According to agricultural science writer Michael Pollan, if all Americans picked 1 day per week to have no meat, the reduction in greenhouse gas emissions would be equivalent to taking 30 to 40 million cars off the road for a year.

Another growing problem is the use of antibiotics in industrialized livestock production facilities. A 2009 study by the Union of Concerned Scientists (UCS) found that about 80% of all antibiotics sold in the United States (and

50% of those in the world) are added to animal feed. This is done to try to prevent the spread of diseases in crowded feedlots and CAFOs and to promote the growth of the animals before they are slaughtered. The UCS study, as well as several others, concluded that this is an important factor in the rise of genetic resistance (see Figure 4-6, p. 84) among many disease-causing microbes. Such resistance can reduce the effectiveness of some antibiotics used to treat infectious diseases in humans, and it can promote the development of new and aggressive disease organisms.

Finally, according to the USDA, animal waste produced by the American meat industry amounts to about 130 times the amount of waste produced by the country's human population. Globally, only about half of all manure is returned to the land as nutrient-rich fertilizer—a violation of the chemical cycling **principle of sustainability**. Much of the other half ends up polluting the air, water, and soil, producing foul odors, and emitting large quantities of climate-changing greenhouse gases into the atmosphere.

Aquaculture Can Harm Aquatic Ecosystems

Figure 12-21 lists the major benefits and drawbacks of aquaculture, which in 2011 accounted for about 42% of global seafood production. Some analysts warn that the harmful environmental effects of aquaculture could limit its future production potential (Concept 12-3).

One major environmental problem associated with aquaculture is that about a third of the wild fish caught from the oceans are used to make the fishmeal and fish oil that are to fed to farmed fish. This is contributing to the depletion of many populations of wild fish that are crucial to the marine food web—a serious threat to marine biodiversity.

Also, this is a very inefficient process. According to marine scientist John Volpe, it takes about 3 kilograms (6.6 pounds) of wild fish to produce 1 kilogram (2.2 pounds) of farmed salmon, and this ratio increases to 5 to 1 for farmed cod and 20 to 1 for farmed tuna.

Another problem is that some fishmeal and fish oil fed to farm-raised fish are contaminated with long-lived toxins such as PCBs and dioxins that are picked up from the ocean floor. Aquaculture producers contend that the concentrations of these chemicals are not high enough to threaten human health.

Fish farms, especially those that raise carnivorous fish such as salmon and tuna, produce large amounts of wastes. Along with pesticides and antibiotics used on fish farms, these wastes can pollute aquatic ecosystems and fisheries.

Aquaculture can also end up promoting the spread of invasive plant species. An Asian kelp called *wakame* or *undaria* is a popular food product raised on some farms. But this invasive seaweed is disrupting coastal aquatic

Trade-Offs

Aquaculture

Advantages	Disadvantages
High efficiency	Large inputs of land, grain, and fishmeal
High yield	Large waste output
Reduces over-harvesting of fisheries	Loss of mangrove forests and estuaries
Jobs and profits	Dense populations vulnerable to disease

© Cengage Learning

Figure 12-21 Use of aquaculture has advantages and disadvantages. **Questions:** Which single advantage and which single disadvantage do you think are the most important? Why?

Top: ©Vladislav Gajic/Shutterstock.com. Bottom: ©Nordling/Shutterstock.com.

systems in several parts of the world. Yet another problem is that farmed fish can escape their pens and mix with wild fish, possibly disrupting the gene pools of wild populations.

Major seed companies are now pushing to use patented genetically modified soybeans as the primary feed for farm-raised fish and shellfish. This could increase water pollution because fish that are fed soy produce more waste than other fish. It would also give a small number of seed companies control over much of the world's seafood production. And it could encourage more deforestation and loss of biodiversity that result when soy plantations replace tropical forests (see Figure 1-4, p. 8).

There is also controversy over the use of a type of farmed salmon that has been genetically engineered, through the combination of growth genes from a Chinook salmon and a sea eel, to grow quickly to the size of wild salmon. The genetically altered salmon requires about 25% less feed per unit of body weight, which could lower its cost and reduce pressure on fish stock used as feed. There is concern that this genetically engineered salmon could escape from fish farms and interbreed with wild salmon. Proponents say this is unlikely because the genetically engineered salmon is sterile.

In Section 12-5 of this chapter, we consider some possible solutions to the serious environmental problems that result from food production, as well as some ways to produce food more sustainably. But first, let's consider a special set of environmental problems and solutions related to protecting food supply systems from pests.

12-4 How Can We Protect Crops from Pests More Sustainably?

CONCEPT 12-4
We can sharply cut pesticide use without decreasing crop yields by using a mix of cultivation techniques, biological pest controls, and small amounts of selected chemical pesticides as a last resort (integrated pest management).

Nature Controls the Populations of Most Pests

A **pest** is any species that interferes with human welfare by competing with us for food, invading our homes, lawns, or gardens, destroying building materials, spreading disease, invading ecosystems, or simply being a nuisance. Worldwide, only about 100 species of plants (weeds), animals (mostly insects), fungi, and microbes cause most of the damage to the crops we grow.

In natural ecosystems and in many polyculture crop fields, *natural enemies* (predators, parasites, and disease organisms) control the populations of most potential pest species. This free ecosystem service is an important part of the earth's natural capital. For example, the world's 30,000 known species of spiders kill far more crop-eating insects every year than humans kill by using chemicals. Most spiders, including the wolf spider (Figure 12-22), do not harm humans.

Figure 12-22 **Natural capital:** This ferocious-looking wolf spider with a grasshopper in its mouth is one of many important insect predators that can be killed by some pesticides.

When we clear forests and grasslands, plant monoculture crops, and douse fields with chemicals that kill pests, we upset many of these natural population checks and balances that help to implement the biodiversity **principle of sustainability**. Then we must devise and pay for ways to protect our monoculture crops, tree plantations, lawns, and golf courses from insects and other pests that nature has helped to control at no charge.

We Use Pesticides to Help Control Pest Populations

We have developed a variety of synthetic **pesticides**—chemicals used to kill or control populations of organisms that we consider undesirable. Common types of pesticides include *insecticides* (insect killers), *herbicides* (weed killers), *fungicides* (fungus killers), and *rodenticides* (rat and mouse killers).

We did not invent the use of chemicals to repel or kill other species. For nearly 225 million years, plants have been producing chemicals called *biopesticides* to ward off, deceive, or poison the insects and herbivores that feed on them. This battle produces a never-ending, ever-changing coevolutionary process: insects and herbivores overcome various plant defenses through natural selection and new plant defenses are favored by natural selection.

In the 1600s, farmers used nicotine sulfate, extracted from tobacco leaves, as an insecticide. Eventually, other *first-generation pesticides*—mainly natural chemicals taken from plants—were developed. Farmers were copying nature's solutions—developed, tested, and modified through natural selection over millions of years—to apply to their pest problems.

A major pest control revolution began in 1939, when entomologist Paul Müller discovered DDT (dichlorodiphenyltrichloroethane)—the first of the so-called *second-generation pesticides* produced in the laboratory. It soon became the world's most-used pesticide, and Müller received the Nobel Prize in Physiology or Medicine in 1948 for his discovery. Since then, chemists have created hundreds of other pesticides by making slight modifications in the molecules of various classes of chemicals.

Some synthetic pesticides, called *broad-spectrum agents,* are toxic to beneficial species as well as to pests. Examples are chlorinated hydrocarbon compounds such as DDT and organophosphate compounds such as malathion and parathion. Others, called *selective,* or *narrow-spectrum, agents,* are effective against a narrowly defined group of organisms. Examples are algicides for algae and fungicides for fungi.

Pesticides vary in their *persistence,* the length of time they remain deadly in the environment. Some, such as DDT and related compounds, remain in the environment for years and can be biologically magnified in food chains and webs (see Figure 9-13, p. 202). Others, such as organophosphates, are active for days or weeks and are not biologically magnified but can be highly toxic to humans.

Some second-generation pesticides have turned out to be highly hazardous for birds and other forms of wildlife. In 1962, biologist Rachel Carson published her famous book *Silent Spring*, sounding a warning that eventually led to strict controls on the use of DDT and several other widely used pesticides.

Since 1950, pesticide use has grown more than 50-fold and most of today's pesticides are 10–100 times more toxic than those used in the 1950s. Since 1970, chemists have continued to develop natural repellents and other biopesticides, again copying nature.

About one-fourth of the pesticides used in the United States are aimed at ridding houses, gardens, lawns, parks, playing fields, swimming pools, and golf courses of insects and other species that we view as pests. According to the U.S. Environmental Protection Agency (EPA), the amount of synthetic pesticides used on the average U.S. homeowner's lawn is 10 times the amount (per unit of land area) typically used on U.S. croplands.

Synthetic Pesticides Provide Several Benefits

Use of synthetic pesticides has its advantages and disadvantages. Proponents contend that the benefits of pesticides (Figure 12-23, left) outweigh their harmful effects (Figure 12-23, right). They point to the following benefits:

- *They have saved human lives.* Since 1945, DDT and other insecticides probably have prevented the premature deaths of at least 7 million people (some say as many as 500 million) from insect-transmitted diseases such as malaria (carried by the *Anopheles* mosquito), bubonic plague (carried by rat fleas), and typhus (carried by body lice and fleas).
- *They have been known to increase food supplies* by reducing food losses from pests, for some crops in some areas.
- *They can help farmers to increase their profits.* Officials of pesticide companies estimate that, under certain conditions, a dollar spent on pesticides can lead to an increase in crop yields worth as much as $4.
- *They work fast.* Pesticides control most pests quickly, have a long shelf life, and are easily shipped and applied.
- *When used properly, the health risks of some pesticides are very low, relative to their benefits,* according to some scientific studies.
- *Newer pesticides are safer to use and more effective than many older ones.* Greater use is being made of chemicals derived originally from plants (biopesticides), which are safer to use and less damaging to the environment than are many older pesticides. Genetic engineering is also being used to develop pest-resistant crop strains

Trade-Offs

Conventional Chemical Pesticides

Advantages	Disadvantages
Expand food supplies	Promote genetic resistance
Raise profits	Can kill pests' natural enemies and harm wildlife and people
Work fast	Can pollute air, water, and land
Are safe if used properly	Are expensive for farmers

© Cengage Learning

Figure 12-23 Use of synthetic pesticides has advantages as well as disadvantages. **Questions:** Which single advantage and which single disadvantage do you think are the most important? Why?

B Brown/Shutterstock.com

and genetically altered crops that produce natural biopesticides.

Synthetic Pesticides Have Several Drawbacks

Opponents of widespread use of synthetic pesticides contend that the harmful effects of these chemicals (Figure 12-23, right) outweigh their benefits (Figure 12-23, left). They cite several problems.

- *They accelerate the development of genetic resistance to pesticides in pest organisms.* Insects breed rapidly, and within 5–10 years (much sooner in tropical areas), they can develop immunity to widely used pesticides through natural selection and then come back stronger than before. Since 1945, about 1,000 species of insects and rodents (mostly rats) and 550 types of weeds and plant diseases have developed genetic resistance to one or more pesticides. Since 1996, the widespread use of glyphosate herbicide has led to at least 15 species of "superweeds" that are genetically resistant to it.
- *They can put farmers on a financial treadmill.* Because of genetic resistance, farmers can find themselves having to pay more and more for a chemical pest control program that can become less and less effective.
- *Some insecticides kill natural predators and parasites that help to control the pest populations.* About 100 of the 300 most destructive insect pests in the United States were minor pests until widespread use of insecticides wiped out many of their natural predators, including spiders (Figure 12-22). (See the Case Study that follows.)
- *Pesticides are usually applied inefficiently and often pollute the environment.* According to the USDA, about 98–99.9% of the insecticides and more than 95% of

the herbicides applied by aerial spraying or ground spraying do not reach the target pests. They end up in the air, surface water, groundwater, bottom sediments, food, and nontarget organisms, including humans, livestock, and wildlife.

- *Some pesticides harm wildlife.* According to the USDA and the U.S. Fish and Wildlife Service, each year, some of the pesticides applied to cropland poison honeybee colonies on which we depend for pollination of many food crops (see Chapter 9, Core Case Study, p. 190). According to a study by the Center for Biological Diversity, pesticides menace one of every three endangered and threatened species in the United States.
- *Some pesticides threaten human health.* The WHO and UNEP have estimated that pesticides annually poison at least 3 million agricultural workers in less-developed countries and at least 300,000 workers in the United States. They also cause 20,000–40,000 deaths per year, worldwide. According to studies by the National Academy of Sciences, pesticide residues in food cause an estimated 4,000–20,000 cases of cancer per year in the United States.

The pesticide industry disputes these claims, arguing that if used as intended, pesticides do not remain in the environment at levels high enough to cause serious environmental or health problems. Figure 12-24 lists some ways to reduce your exposure to synthetic pesticides.

CASE STUDY
Ecological Surprises: The Law of Unintended Consequences

Malaria once infected nine of every ten people in North Borneo, now known as the eastern Malaysian state of Sabah. In 1955, the WHO sprayed the island with dieldrin (a DDT relative) to kill malaria-carrying mosquitoes. The program was so successful that the dreaded disease was nearly eliminated.

Then unexpected things began to happen. The dieldrin also killed other insects, including flies and cockroaches living in houses, which made the islanders happy. Next, small insect-eating lizards living in the houses died after gorging themselves on dieldrin-contaminated insects. Then cats began dying after feeding on the lizards. In the absence of cats, rats flourished and overran the villages. When the residents became threatened by sylvatic plague carried by rat fleas, the WHO parachuted healthy cats onto the island to help control the rats. Operation Cat Drop worked.

But then the villagers' roofs began to fall in. The dieldrin had killed wasps and other insects that fed on a type of caterpillar that had either avoided or was not affected by the insecticide. With most of its predators eliminated, the caterpillar population exploded, munching its way through its favorite food: the leaves used in thatch roofs.

Ultimately, this episode ended well. Both malaria and the unexpected effects of the spraying program were brought

What Can You Do?

Reducing Exposure to Pesticides

- Grow some of your food using organic methods
- Buy certified organic food
- Wash and scrub all fresh fruits and vegetables
- Eat less meat, no meat, or certified organically produced meat
- Before cooking, trim the fat from meat

© Cengage Learning

Figure 12-24 Individuals matter: You can reduce your exposure to pesticides. **Questions:** Which three of these steps do you think are the most important ones to take? Why?

under control. Nevertheless, this chain of unintended and unforeseen events reminds us that whenever we intervene in nature and affect organisms that interact with one another, we need to ask, "Now what will happen?"

Pesticide Use Has Not Consistently Reduced U.S. Crop Losses to Pests

Despite some claims to the contrary, largely because of genetic resistance and the loss of many natural predators, synthetic pesticides have not always succeeded in reducing U.S. crop losses.

When David Pimentel, an expert on insect ecology, evaluated data from more than 300 agricultural scientists and economists, he reached three major conclusions. *First,* between 1942 and 1997, estimated crop losses from insects almost doubled from 7% to 13%, despite a tenfold increase in the use of synthetic insecticides. *Second,* according to the International Food Policy Research Institute, the estimated environmental, health, and social costs of pesticide use in the United States are $5–$10 in damages for every dollar spent on pesticides. *Third,* experience indicates that alternative pest management practices could cut the use of synthetic pesticides by half on 40 major U.S. crops without reducing crop yields (**Concept 12-4**).

The pesticide industry disputes these findings. However, numerous studies and experience support them. Sweden has cut its pesticide use in half with almost no decrease in crop yields. The soup company Campbell's® uses no pesticides on the tomatoes it grows in Mexico, and yields have not dropped.

Laws and Treaties Can Help to Protect Us from the Harmful Effects of Pesticides

More than 25,000 different pesticide products are used in the United States. Three federal agencies, the EPA, the USDA, and the Food and Drug Administration (FDA), regulate the use of these pesticides under the Federal Insecticide, Fungicide, and Rodenticide Act (FIFRA), first passed in 1947 and amended in 1972.

Under FIFRA, the EPA was supposed to assess the health risks of the active ingredients in synthetic pesticide products already in use. However, since 1972, less than 10% of the active ingredients in pesticide products have been tested for chronic health effects. And serious evaluation of the health effects of the 1,200 inactive ingredients is only partially done. The EPA says that the U.S. Congress has not provided them with enough funds to carry out this complex and lengthy evaluation process.

In 1996, Congress passed the Food Quality Protection Act, mostly because of growing scientific evidence and citizen pressure concerning the effects of small amounts of pesticides on children. This act requires the EPA to reduce the allowed levels of pesticide residues in food by a factor of 10 when there is inadequate information on the potentially harmful effects on children.

Between 1972 and 2012, the EPA used FIFRA to ban or severely restrict the use of 64 active pesticide ingredients, including DDT and most other chlorinated hydrocarbon insecticides. However, according to studies by the National Academy of Sciences, federal laws regulating pesticide use are inadequate and poorly enforced by the three agencies. One study found that as much as 98% of the potential risk of developing cancer from pesticide residues on food grown in the United States would be eliminated if EPA standards were as strict for pesticides developed before 1972 as they are for newer pesticides.

CONSIDER THIS. . .

CONNECTIONS Pesticides and Organic Foods: The Dirty Dozen

According to a 2012 report by the Environmental Working Group (EWG), you could reduce your pesticide intake by up to 90% by eating only 100% USDA Certified Organic versions of 12 types of fruits and vegetables that tend to have the highest pesticide residues. These foods, which the EWG calls the "dirty dozen," are apples, celery, cherry tomatoes, cucumbers, grapes, hot peppers, nectarines (imported), peaches, potatoes, spinach, strawberries, and sweet bell peppers. Pesticide proponents say the residue concentrations in those foods are too low to cause harm. But some scientists urge consumers to follow the precautionary principle and buy only 100% USDA Certified Organic versions of these foods.

Although laws within countries protect citizens to some extent, banned or unregistered pesticides may be manufactured in one country and exported to other countries. For example, U.S. pesticide companies make and export to other countries pesticides that have been banned or severely restricted—or never even evaluated—in the United States. Other countries also export banned or unapproved pesticides.

However, in what environmental scientists call a *circle of poison*, or the *boomerang effect*, residues of some banned or unapproved chemicals used in synthetic pesticides exported to other countries can return to the exporting countries on imported food. Winds can also carry persistent pesticides from one country to another.

Environmental and health scientists have urged the U.S. Congress—without success—to ban such exports. Supporters of the exports argue that such sales increase economic growth and provide jobs and that if the United States did not export these pesticides, other countries would.

In 1998, more than 50 countries developed an international treaty that requires exporting countries to have informed consent from importing counties for exports of 22 synthetic pesticides and 5 industrial chemicals. In 2000, more than 100 countries developed an international agreement to ban or phase out the use of 12 especially hazardous persistent organic pollutants (POPs)—9 of them persistent hydrocarbon pesticides such as DDT and other chemically similar pesticides. By 2011, the initial list of 12 chemicals had been expanded to 21. In 2004 the POPS treaty went into effect and by 2012 had been signed or ratified by 172 countries, but not by the United States.

CONSIDER THIS. . .

THINKING ABOUT Exporting Pesticides

Should companies be allowed to export synthetic pesticides that have been banned, severely restricted, or not approved for use in their home countries? Explain.

There Are Alternatives to Synthetic Pesticides

Many scientists argue that we should greatly increase the use of biological, ecological, and other alternative methods for controlling pests and diseases that affect crops and human health (Concept 12-4). Here are some of these alternatives:

- *Fool the pest.* A variety of *cultivation practices* can be used to fake out pests. Examples include rotating the types of crops planted in a field each year and adjusting planting times so that major insect pests either starve or get eaten by their natural predators.
- *Provide homes for pest enemies.* Farmers can increase the use of polyculture, which uses plant diversity to reduce losses to pests by providing habitats for the pests' predators.
- *Implant genetic resistance.* Use genetic engineering to speed up the development of pest- and disease-resistant crop strains (Figure 12-25). But controversy persists over whether the projected advantages of using GM plants outweigh their projected disadvantages (Figure 12-19).
- *Bring in natural enemies.* Use *biological control* by importing natural predators (Figures 12-22 and 12-26), parasites, and disease-causing bacteria and viruses to help regulate pest populations. This approach is nontoxic to other species and is usually less costly than applying pesticides. However, some biological control agents are difficult to mass produce and are often slower acting

Monsanto

Figure 12-25 Solutions: *Genetic engineering* can be used to reduce pest damage. Both of these tomato plants were exposed to destructive caterpillars. The normal plant's leaves are almost gone (left), whereas the genetically altered plant shows little damage (right). **Questions:** Would you have any concerns about eating the genetically engineered tomato? Why or why not?

and more difficult to apply than synthetic pesticides are. Sometimes the agents can multiply and become pests themselves.

- *Use insect perfumes.* Trace amounts of *sex attractants* (called *pheromones*) can be used to lure pests into traps or to attract their natural predators into crop fields. Each of these chemicals attracts only one species. They have little chance of causing genetic resistance and are not harmful to nontarget species. However, they are costly and time-consuming to produce.

- *Bring in the hormones.* Hormones are chemicals produced by animals to control their developmental processes at different stages of life. Scientists have learned how to identify and use hormones that disrupt an insect's normal life cycle, thereby preventing it from reaching maturity and reproducing. Use of insect hormones has some of the same advantages and disadvantages as use of sex attractants has. Also, they take weeks to kill an insect, are often ineffective with large infestations of insects, and sometimes break down before they can act.

- *Reduce the use of synthetic herbicides to control weeds.* Organic farmers control weeds by methods such as crop rotation, mechanical cultivation, hand weeding, and the use of cover crops and mulches.

Integrated Pest Management Is a Component of More Sustainable Agriculture

Many pest control experts and farmers believe the best way to control crop pests is through **integrated pest management (IPM)**, a carefully designed program in which each crop and its pests are evaluated as parts of an ecosystem, and farmers use a combination of cultivation, biological, and chemical tools and techniques, applied in a coordinated process (**Concept 12-4**).

The overall aim of IPM is to reduce crop damage to an economically tolerable level. Each year, crops are rotated, or moved from field to field, in an effort to disrupt pest infestations, and fields are monitored carefully. When an economically damaging level of pests is reached, farmers first use biological methods (natural predators, parasites, and disease organisms) and cultivation controls (such as altering planting time and using large machines to vacuum up harmful bugs). They apply small amounts of synthetic insecticides—preferably biopesticides—only when insect or weed populations reach a threshold where the potential cost of pest damage to crops outweighs the cost of applying the pesticide.

IPM has a good track record. In Sweden and Denmark, farmers have used it to cut their synthetic pesticide use by more than half. In Cuba, where organic farming is used almost exclusively, farmers make extensive use of IPM. In Brazil, IPM has reduced pesticide use on soybeans by as much as 90%.

Scott Bauer/USDA Agricultural Research Service

Figure 12-26 **Natural capital:** In this example of biological pest control, a wasp is parasitizing a gypsy moth caterpillar.

According to the U.S. National Academy of Sciences, these and other experiences show that a well-designed IPM program can reduce synthetic pesticide use and pest control costs by 50–65%, without reducing crop yields and food quality. IPM can also reduce inputs of fertilizer and irrigation water, and slow the development of genetic resistance, because pests are attacked less often and with lower doses of pesticides. IPM is an important form of *pollution prevention* that reduces risks to wildlife and human health and applies the biodiversity **principle of sustainability** (see Figure 1-2, p. 6 or back cover).

Despite its promise, IPM—like any other form of pest control—has some drawbacks. It requires expert knowledge about each pest situation and takes more time than does using conventional pesticides. Methods developed for a crop in one area might not apply to areas with even slightly different growing conditions. Initial costs may be higher, although long-term costs typically are lower than those of using conventional pesticides. Widespread use of IPM has been hindered in the United States and a number of other countries by government subsidies that pay for synthetic chemical pesticides, as well as by opposition from pesticide manufacturers, and a shortage of IPM experts. **GREEN CAREER:** integrated pest management

A growing number of scientists are urging the USDA to use a three-point strategy to promote IPM in the United States. *First,* add a 2% sales tax on synthetic pesticides and use the revenue to fund IPM research and education. *Second,* set up a federally supported IPM demonstration project on at least one farm in every county in the United States. *Third,* train USDA field personnel and county farm agents in IPM so they can help farmers use this alternative. Because these measures would reduce its profits, the pesticide industry has vigorously, and successfully, opposed them.

Several UN agencies and the World Bank have joined together to establish an IPM facility. Its goal is to promote the use of IPM by disseminating information and establishing networks among researchers, farmers, and agricultural extension agents involved in IPM.

12-5 How Can We Improve Food Security?

CONCEPT 12-5
We can improve food security by reducing poverty and chronic malnutrition, relying more on locally grown food, and cutting food waste.

Use Government Policies to Improve Food Production and Security

Agriculture is a financially risky business. Whether farmers have a good or bad year depends on factors over which they have little control: weather, crop prices, crop pests and diseases, interest rates on loans, and global markets.

Governments use two main approaches to influence food production. First, they can *control food prices* by putting a legally mandated upper limit on them in order to keep them artificially low. This makes consumers happy but makes it harder for farmers to make a living.

Second, they can *provide subsidies* by giving farmers price supports, tax breaks, and other financial support to keep them in business and to encourage them to increase food production. However, if government subsidies are too generous and the weather is good, farmers and livestock producers may produce more food than can be sold.

Some analysts call for ending such subsidies. They point to New Zealand, which ended farm subsidies in 1984. After the shock wore off, innovation took over and production of some foods such as milk quadrupled. Brazil has also ended most of its farm subsidies. Some analysts call for replacing traditional subsidies for farmers with subsidies that promote more environmentally sustainable farming practices.

For example, Growing Power's Will Allen (**Core Case Study**) proposed the creation of a public–private institution to be called the Centers for Urban Agriculture. It would be a national training center with a large urban farm for research and development of sustainable farming practices. One goal would be to develop a functioning community food delivery system that could serve a large city and provide jobs, job training, and a nutritious food supply for those most in need of it.

Similarly, government subsidies to fishing fleets can promote overfishing and the reduction of aquatic biodiversity. For example, several governments give the highly destructive bottom-trawling industry (see Figure 11-2, right, p. 250) a total of about $150 million per year in subsidies, which is the main reason fishers who use this practice can stay in business. Many analysts call for replacing those harmful subsidies with ones that promote more sustainable fishing and aquaculture.

Other Government and Private Programs Are Increasing Food Security

Government and private programs that reduce poverty by helping the poor to help themselves can improve food security (**Concept 12-1B**). For example, some programs provide small loans at low interest rates to poor people to help them start businesses or buy land to grow their own food.

Some analysts urge governments to establish special programs focused on saving children from the harmful health effects of poverty. Studies by the United Nations

Children's Fund (UNICEF) indicate that one-half to two-thirds of nutrition-related childhood deaths could be prevented at an average annual cost of $5–$10 per child. This would involve simple measures such as immunizing more children against childhood diseases, preventing dehydration from diarrhea by giving infants a mixture of sugar and salt in their water, and preventing blindness by giving children an inexpensive vitamin A capsule twice a year.

Some farmers and plant breeders are working on preserving a diverse gene pool as another way to improve food security. **GOOD NEWS** For example, an organization called the Global Crop Diversity Trust is seeking to prevent the disappearance of 100,000 varieties of food crops. The trust is working with about 50 seed banks around the world to cultivate and store seeds from endangered varieties of many food plant species. The world's most secure seed bank is the Doomsday Seed Bank, located underground on a frozen Norwegian arctic island. It is being stocked with duplicates of many of the world's seed collections.

There are also many private, mostly nonprofit, organizations that are working to help individuals, communities, and nations to improve their food security. For example, Will Allen (**Core Case Study**) argues that instead of trying to transfer complex technologies such as genetic engineering to less-developed countries, we should be helping them to develop simple, sustainable, local food production and distribution systems that will give them more control over their food security. Another person who is working toward this goal on a more regional basis is National Geographic Explorer Jennifer Burney (Individuals Matter 12.1).

We Can Grow and Buy More Food Locally and Cut Food Waste

According to some experts, one way to increase food security is to grow more of our food locally or regionally, ideally with certified organic farming practices. A growing number of consumers are becoming "locavores" and buying more of their food from local and regional producers in farmers' markets, which provide access to fresher seasonal foods. In 2012, there were more than 5,000 farmers' markets in the United States, triple the number of such markets in 1994, according to the USDA.

In addition, many people are participating in *community-supported agriculture (CSA)* programs in which they buy shares of a local farmer's crop and receive a box of fruits or vegetables each week during the summer and fall. Growing Power (**Core Case Study**) runs such a program for inner-city residents. For many of these people, the organically grown food they get from the urban farm greatly improves their diets and increases their chances of living longer and healthier lives.

By buying locally, people support local economies and farm families. They also help to reduce fossil fuel energy costs for food producers, as well as the greenhouse gas emissions resulting from refrigeration and transportation of food products over long distances.

An increase in the demand for locally grown food could result in more small, diversified farms that produce organic, minimally processed food from plants and animals. Such eco-farming could be one of this century's challenging new careers for many young people. **GREEN CAREER:** small-scale sustainable agriculture

Sustainable agriculture entrepreneurs and ordinary citizens who live in urban areas could grow more of their own food, as the Growing Power farm has shown (**Core Case Study**). According to the USDA, around 15% of the world's food is grown in urban areas, and this percentage could easily be doubled. People are planting gardens and raising chickens in many urban and suburban backyards, growing dwarf fruit trees in large containers of soil, and raising vegetables on rooftops, balconies, and patios. People are also building raised gardening beds in urban parking lots—a growing practice known as *asphalt gardening*.

In a 2011 report, community systems expert Zaid Hassan noted that in Cuba, the government ran a program to turn the capital city of Havana's many vacant lots into urban farms or gardens. As a result, about 41% of the city's area is now used for organic agriculture that generates more than half of Cuba's vegetables.

Many urban schools, colleges, and universities are benefiting greatly from having gardens on school grounds. Not only do the students have a ready source of fresh produce, but they also learn about where their food comes from and how to grow their own food more sustainably. **GOOD NEWS**

🔍 CONSIDER THIS. . .

CAMPUS SUSTAINABILITY
Yale University

Yale University is committed to growing food locally and more sustainably. Students use a plot that takes up less than half the area of a typical city block to produce more than 300 varieties of vegetables, fruits, and flowers that are used on campus and sold locally. According to the school's dining director, up to 49% of the food served on campus is locally grown, and much of it is raised organically.

©Yale Sustainable Food Project/Yale University

In the future, much of our food might be grown in cities within high-rise buildings. Figure 12-27 is an architectural drawing of such a *vertical farm*. This building with crops growing on every floor would put into practice the three scientific **principles of sustainability** (see Figure 1-2, p. 6). Its sloped glass front would bring in sunlight, and excess heat collected in this way could be stored in tanks underneath the building for use as needed. It could also have solar panels for generating electricity on its rooftop or on an overhang shown in the bottom of Figure 12-27. The building could also capture and recycle rainwater for irrigating its wide diversity of crops.

individuals matter 12.1

Jennifer Burney: Environmental Scientist and National Geographic Emerging Explorer

Environmental scientist and National Geographic Emerging Explorer Jennifer Burney has been studying the harmful environmental effects of various methods of food production. She notes that smallholder farmers represent the majority of the world's poorest people and need to boost their productivity for better standards of living and health. She is trying to help them do that without replicating some of the agricultural ills such as wasteful irrigation and fertilizer runoff that are the legacy of large-scale industrial farming in the developed world.

However, in struggling to survive, poor farmers do not have the luxury of worrying about their long-term environmental impacts. That is why Burney has focused on helping people to grow, distribute, and cook their food more sustainably. By analyzing the numerous factors that affect food production, she has been able to isolate problems, define them practically, and find some possible solutions.

For example, in arid sub-Saharan Africa, farmers must depend on rainfall for raising crops on small plots because only 20% of the rainfall flows into streams and aquifers while the rest evaporates. Overpumping can quickly deplete the groundwater. These factors, worsened by drought, make it hard for farmers to feed their families.

In analyzing this problem, Burney found two connected technologies—solar irrigation systems and drip irrigation (background photo)—that could serve as a solution. Drip irrigation systems sip water and drip it directly onto plant roots instead of pumping and dumping it. Solar-powered pumps work without the need for batteries or fuel. On sunny days, when crops need water more, the solar panels speed the pumping; on cloudy days when there is less evaporation, the pumping slows down. Thus, only the amount of water that is needed is pumped on most days.

This has allowed farmers to grow fruits and vegetables on a larger scale and improve their incomes and food security. Burney's work has been greatly appreciated by local farmers. She reports: "Everyone wants to get involved. It's almost become a tourist attraction."

Growing Power/Rendering by The Kubala Washatko Architects, Inc.

Figure 12-27 A vertical farm building that could be used to grow a diversity of crops.

Finally, people can sharply cut food waste as an important component of improving food security (Concept 12-5). A 2011 UN study found that about one-third of all food produced globally is lost during production or thrown away. Environmental scientist Vaclav Smil and other researchers have estimated that Americans throw away 30–50% of their food supply. This wasted food is worth at least $43 billion a year, almost twice as much as the estimated $24 billion needed to eliminate undernutrition and malnutrition in the world. In some less-developed countries, food waste can be high due to spoilage from heat and pests and lack of refrigeration and food storage systems.

12-6 How Can We Produce Food More Sustainably?

CONCEPT 12-6
We can produce food more sustainably by using resources more efficiently, sharply decreasing the harmful environmental effects of industrialized food production, and eliminating government subsidies that promote such harmful impacts.

Many Farmers Are Reducing Soil Erosion

Land used for food production must have fertile topsoil (Figure 12-A), which takes hundreds of years to form. Thus, sharply reducing topsoil erosion is the single most important component of more sustainable agriculture.

Soil conservation involves using a variety of methods to reduce topsoil erosion and restore soil fertility, mostly by keeping the land covered with vegetation. Farmers have used a number of methods to reduce topsoil erosion. For example, *terracing* involves converting steeply sloped land into a series of broad, nearly level terraces that run across the land's contours (Figure 12-28a). Each terrace retains water for crops and reduces topsoil erosion by controlling runoff.

On less steeply sloped land, *contour planting* (Figure 12-28b) can be used to reduce topsoil erosion. It involves plowing and planting crops in rows across the slope of the land rather than up and down. Each row acts as a small dam to help hold topsoil by slowing runoff. Similarly, *strip-cropping* (Figure 12-28b) helps to reduce erosion and to restore soil fertility with alternating strips of a row crop (such as corn or cotton) and another crop that completely covers

a. Lim Yong Hian/Shutterstock.com

b. Ron Nichols/USDA Natural Resources Conservation Service

Figure 12-28 Soil conservation methods include **(a)** terracing; **(b)** contour planting and strip cropping; **(c)** alley cropping; and **(d)** windbreaks between crop fields.

c. Manfred Mielke/USDA Forest Service Bugwood.org

d. Fedorov Oleksiy/Shutterstock.com

the soil, called a *cover crop* (such as alfalfa, clover, oats, or rye). The cover crop traps topsoil that erodes from the row crop and catches and reduces water runoff.

Alley cropping, or *agroforestry* (Figure 12-28c), is another way to slow the erosion of topsoil and to maintain soil fertility. One or more crops, usually legumes or other crops that add nitrogen to the soil, are planted together in alleys between orchard trees or fruit-bearing shrubs, which provide shade. This reduces water loss by evaporation and helps retain and slowly release soil moisture.

Farmers can also establish *windbreaks,* or *shelterbelts,* of trees around crop fields to reduce wind erosion (Figure 12-28d). The trees retain soil moisture, supply wood for fuel, and provide habitats for birds and insects that help with pest control and pollination.

Another way to greatly reduce topsoil erosion is to eliminate or minimize the plowing and tilling of top-

soil and leave crop residues on the ground. This is called *conservation-tillage farming,* accomplished with the use of special tillers and planting machines that inject seeds and fertilizer directly through crop residues into minimally disturbed topsoil. Weeds are controlled with herbicides. This type of farming increases crop yields and greatly reduces soil erosion and water pollution from sediment and fertilizer runoff. It also helps farmers survive prolonged drought by helping to keep more moisture in the soil.

In 2011, farmers used conservation tillage GOOD NEWS on about 63% of U.S. cropland (up from 17% in 1982), helped by the increased use of herbicides. How-

HYDROPONICS: GROWING CROPS WITHOUT SOIL

Plants need sunlight, carbon dioxide (from the air), and mineral nutrients such as nitrogen and phosphorus. Traditionally, farmers have obtained these nutrients from soil. **Hydroponics** involves growing plants by exposing their roots to a nutrient-rich water solution instead of soil, usually inside of a greenhouse (Figure 12-B).

Indoor hydroponic farming yields a number of benefits. Crops can be grown year-round under controlled conditions almost anywhere, regardless of weather conditions. In dense urban areas, crops can be grown on rooftops, underground with artificial lighting (as is now done in Tokyo, Japan), and on floating barges, thus requiring much less land. Because the nutrient and water solution can be recycled, farmers can cut their use of fertilizers and water. Thus, they also reduce water pollution from the runoff of fertilizers into streams or other waterways. Also, in a well-controlled greenhouse environment, there is little or no need for pesticides.

With these advantages, some scientists say we could use hydroponics to produce a larger amount of the world's food without causing most of the serious harmful environmental effects of industrialized food production. However, there are three major reasons why this is not happening. *First,* it takes a lot of money to establish such systems, although they typically are less expensive than conventional systems in the long run. *Second,* many growers fear that hydroponics requires substantial technical knowledge, while in reality it is very similar to traditional gardening and crop production. *Third,* it could threaten the profits of large and politically powerful companies that produce farming-related products such as pesticides, manufactured inorganic fertilizers, and farm equipment.

Despite these obstacles, large hydroponic facilities are found in a number of countries, including New Zealand, Germany, the Netherlands, and the United States. Hydroponics is not likely to replace conventional industrialized agriculture, but with further research and development, it could play an increasing role in helping us to make the transition to more sustainable agriculture over the next several decades.

Figure 12-B These salad greens are being grown hydroponically (without soil) in a greenhouse. The plant roots are immersed in a trough and exposed to nutrients dissolved in running water that can be reused.

Critical Thinking

What are some possible drawbacks to the use of hydroponics? Might any of these drawbacks take away from the sustainability of hydroponics? Explain.

---

ever, one drawback is that the greater use of herbicides is promoting the growth of herbicide-resistant weeds that force farmers to use larger doses of weed killers or, in some cases, to return to plowing. The USDA estimates that using conservation tillage on 80% of U.S. cropland would reduce topsoil erosion by at least 50%. Conservation tillage is used on about 10% of the world's cropland, although it is widely used in some countries, including the United States, Brazil, Argentina, Canada, and Australia.

Another way to conserve topsoil is to grow plants without using soil. Some producers are raising crops in greenhouses, using a system called *hydroponics* (Science Focus 12.2). At the Growing Power farm (**Core Case Study**), Will Allen has developed such a system for raising salad greens and fish together. Wastewater from the fish runs flows into hydroponic troughs where it nourishes the plants. The plant roots filter the water, which is then returned to the fish runs. This closed-loop, chemical-free *aquaponic system* conserves soil, water, and energy while supporting more than 100,000 tilapia and perch, which are sold in local markets along with the salad greens.

Still another way to conserve the earth's topsoil is to retire the estimated one-tenth of the world's marginal cropland that is highly erodible and accounts for the majority of all topsoil erosion. The goal would be to identify *erosion hotspots,* withdraw these areas from cultivation, and plant them with grasses or trees, at least until their topsoil has been renewed. Some countries, such as the United States, have paid farmers to set aside considerable areas of cropland for conservation purposes.

We Can Restore Soil Fertility

The best way to maintain soil fertility is through topsoil conservation. The next best option is to restore some of the lost plant nutrients that have been washed, blown, or leached out of topsoil, or that have been removed by repeated crop harvesting. To do this, farmers can use organic fertilizer derived from plant and animal materials or **synthetic inorganic fertilizer** manufactured of inorganic compounds that contain *nitrogen, phosphorus,* and *potassium* along with trace amounts of other plant nutrients.

There are several types of *organic fertilizers*. One is **animal manure**: the dung and urine of cattle, horses, poultry, and other farm animals. It improves topsoil structure, adds organic nitrogen, and stimulates the growth of beneficial soil bacteria and fungi. Another type, called **green manure**, consists of freshly cut or growing green vegetation that is plowed into the topsoil to increase the organic matter and humus available to the next crop. A third type is **compost**, produced when microorganisms in topsoil break down organic matter such as leaves, crop residues, food wastes, paper, and wood in the presence of oxygen.

The Growing Power farm (**Core Case Study**) depends greatly on its large piles of compost. Will Allen invites local grocers and restaurant owners to send their food wastes to add to the pile. To make this compost, Allen uses millions of red wiggler worms, which reproduce rapidly and eat their own weight in food wastes every day, converting it to plant nutrients. Also, the process of composting generates a considerable amount of heat, which is used to help warm the farm's greenhouses during cold months.

One way to degrade soils is to plant crops such as corn and cotton on the same land several years in a row, a practice that can deplete nutrients—especially nitrogen—in the topsoil. *Crop rotation* is one way to reduce such losses. A farmer plants an area with a nutrient-depleting crop one year, and the next year, plants the same area with legumes, whose root nodules add nitrogen to the soil. This method helps to restore topsoil nutrients while reducing erosion by keeping the topsoil covered with vegetation. A 2012 study done by Iowa State University found that rotating soy and corn crops on a 3- or 4-year cycle produced better yields than did planting the same crop year after year. This method also reduced the need for nitrogen fertilizer and herbicides by up to 88%, cut toxins in groundwater 200-fold, and did not reduce profits.

Many farmers, especially those in more-developed countries, rely on synthetic inorganic fertilizers. The use of these products has grown more than ninefold since 1950, and it now accounts for about 25% of the world's crop yield. While these fertilizers can replace depleted inorganic nutrients, they do not replace organic matter. To completely restore nutrients to topsoil, both inorganic and organic fertilizers must be used. Many scientists are encouraging farmers, especially those in less-developed countries, to make greater use of green manure as a more sustainable way to restore soil fertility.

Another way to restore the fertility of degraded or contaminated soils, especially in polluted urban settings, is to use biological methods. For example, Growing Power farmers (**Core Case Study**) have used red wiggler worms to break down the toxins present in contaminated urban soils and to improve soil fertility for raising food crops.

We Can Reduce Soil Salinization and Desertification

We know how to prevent and deal with soil salinization, as summarized in Figure 12-29. The problem is that most of these solutions are costly in one way or another.

Reducing desertification is not easy. We cannot control the timing and location of prolonged droughts caused by changes in weather patterns. But we can reduce population growth, overgrazing, deforestation, and destructive forms of planting, irrigation, and mining in dryland areas, which have left much land vulnerable to topsoil erosion and thus desertification. We can also work to decrease the human contribution to projected climate change, which is expected to create severe and prolonged droughts in larger areas of the world during this century.

It is possible to restore land suffering from desertification by planting trees and other plants (Figure 12-17) that anchor topsoil and hold water. We can also grow trees and crops together (alley cropping, Figure 12-28c), and establish windbreaks around farm fields (Figure 12-28d).

Some Producers Practice More Sustainable Aquaculture

Scientists and producers are working on ways to make aquaculture more sustainable and to reduce its harmful environmental effects. One such approach is open-ocean aquaculture, which involves raising large carnivorous fish in underwater pens—some as large as a high school gymnasium—located up to 300 kilometers (190 miles) offshore (see Figure 11-5, p. 254), where rapid currents can sweep away fish wastes and dilute them. Some farmed

Figure 12-29 There are ways to prevent soil salinization and ways to clean it up (**Concept 12-6**). **Questions:** Which two of these solutions do you think are the best ones? Why?

USDA Natural Resources Conservation Service

Figure 12-30 The Scandinavian Silver Eel Farm in Helsingborg, Sweden, recirculates its water and converts fish wastes to fertilizers.

Solutions

More Sustainable Aquaculture

■ Protect mangrove forests and estuaries

■ Improve management of wastes

■ Reduce escape of aquaculture species into the wild

■ Set up self-sustaining polyaquaculture systems that combine aquatic plants, fish, and shellfish

■ Certify sustainable forms of aquaculture

Figure 12-31 We can make aquaculture more sustainable and reduce its harmful effects. **Questions:** Which two of these solutions do you think are the best ones? Why?

fish from such operations can escape and breed with wild fish. However, the environmental impact of raising fish far offshore is smaller than that of raising fish near shore and much smaller than that of industrialized commercial fishing.

Other fish farmers are reducing coastal damage from aquaculture by raising shrimp and fish species in inland facilities using zero-discharge freshwater ponds and tanks. In such *recirculating aquaculture systems*, the water used to raise the fish is continually recycled. For example, like the Growing Power aquaponic system (**Core Case Study**), the Scandinavian Silver Eel Farm in Helsingborg, Sweden (Figure 12-30), captures its fish wastes and converts them to fertilizer. This reduces the discharge of polluting wastes and the need for antibiotics and other chemicals used to combat disease. It also eliminates the problem of farmed fish escaping into natural aquatic systems. **GREEN CAREER:** sustainable aquaculture

In the long run, making aquaculture more sustainable will require some fundamental changes for producers and consumers. One such change would be for more consumers to choose fish species that eat algae and other vegetation rather than other fish. Raising carnivorous fishes such as salmon, trout, tuna, grouper, and cod contributes to overfishing and population crashes within species used to feed these carnivores, and will eventually be unsustainable. Raising plant-eating fishes such as carp, tilapia, and catfish avoids this problem. However, it becomes less sustainable when aquaculture producers try to increase yields by feeding fishmeal to such plant-eating species, as many of them are.

Another change would be for fish farmers to emphasize *polyaquaculture*, which has been part of aquaculture for centuries, especially in Southeast Asia. Polyaquaculture operations raise fish or shrimp along with algae, seaweeds, and shellfish in coastal lagoons, ponds, and tanks. The wastes of the fish or shrimp feed the other species, and in the best of these operations, there are just enough wastes from the first group to feed the second group. Polyaquaculture applies the recycling and biodiversity **principles of sustainability**.

Figure 12-31 lists some ways to make aquaculture more sustainable and to reduce its harmful environmental effects.

We Can Produce Meat and Dairy Products More Efficiently

Meat production has a huge environmental impact and meat consumption is the largest factor in the growing ecological footprints of individuals in affluent nations.

Meat production is also highly inefficient. Currently, about 38% of the world's grain harvest and 37% of the global fish catch are used to produce animal protein. If everyone in the world today ate the average amount of meat and fish consumed by Americans, the global annual grain harvest would feed only about 2.5 billion people—a little over a third of the current world population.

A more sustainable form of meat production and consumption would involve shifting from less grain-efficient forms of animal protein, such as beef, pork, and carnivorous fish produced by aquaculture, to more grain-efficient forms, such as poultry and plant-eating farmed fish (Figure 12-32). Such a shift is under way. Since 1996, poultry has taken the lead over beef in the marketplace, and

Figure 12-32 Kilograms of grain required for each kilogram of body weight added for each type of animal.

(Compiled by the authors using data from U.S. Department of Agriculture.)

within a decade or so, herbivorous fish farming may exceed beef production.

Many people regularly have one or two meatless days per week. Others are going further and eliminating most or all meat from their diets, replacing it with a balanced vegetarian diet of fruits, vegetables, and protein-rich foods such as peas, beans, and lentils. Research indicates that people who live on a Mediterranean-style diet that includes mainly poultry, seafood, cheese, raw fruits, vegetables, and olive oil tend to be healthier and live longer than those who eat a lot of red meat.

CONSIDER THIS...

THINKING ABOUT Meat Consumption

Would you be willing to live lower on the food chain by eating much less meat, or even no meat at all? Explain.

We Can Make a Shift to More Sustainable Food Production

Modern industrialized food production yields large amounts of food at prices that are low as long as they do not include the harmful environmental costs of food pro-

duction. But to a growing number of analysts, it is unsustainable, because it violates the three scientific **principles of sustainability** (see Figure 1-2, p. 6 or back cover). It relies heavily on the use of fossil fuels and thus adds greenhouse gases to the atmosphere and contributes to climate change. It also reduces biodiversity and agrobiodiversity and interferes with the cycling of plant nutrients. These facts are hidden from consumers because most of the harmful environmental costs of food production (Figure 12-13) are not included in the market prices of food—a violation of the full-cost pricing **principle of sustainability** (see Figure 1-5, p. 9 or back cover).

Figure 12-33 lists the major components of more sustainable food production. Compared to high-input farming, low-input agriculture produces similar yields with less energy input per unit of yield and lower greenhouse gas emissions. It improves topsoil fertility and reduces topsoil erosion, can often be more profitable for farmers, and can help poor families to feed themselves (Concept 12-6).

One component of more sustainable food production is 100% USDA Certified Organic agriculture (Figure 12-8). Many experts support a shift to organic farming mostly because it sharply reduces the harmful environmental effects of industrialized farming. Figure 12-34 summarizes the major conclusions of a 22-year research study at the

Solutions

More Sustainable Food Production

More	Less
High-yield polyculture	Soil erosion
Organic fertilizers	Soil salinization
Biological pest control	Water pollution
Integrated pest management	Aquifer depletion
Efficient irrigation	Overgrazing
Perennial crops	Overfishing
Crop rotation	Loss of biodiversity and agrobiodiversity
Water-efficient crops	Fossil fuel use
Soil conservation	Greenhouse gas emissions
Subsidies for sustainable farming	Subsidies for unsustainable farming

© Cengage Learning

Figure 12-33 More sustainable, low-input food production has a number of major components. (Concept 12-6). **Questions:** For each list in this diagram (left and right), which two items do you think are the most important? Why?

Top: ©Marko5/Shutterstock.com. Center: © Anhong | Dreamstime.com. Bottom: ©USDA Natural Resources Conservation Service.

Solutions

Organic Farming

- Improves soil fertility
- Reduces soil erosion
- Retains more water in soil during drought years
- Uses about 30% less energy per unit of yield
- Lowers CO_2 emissions
- Reduces water pollution by recycling livestock wastes
- Eliminates pollution from pesticides
- Increases biodiversity above and below ground
- Benefits wildlife such as birds and bats

(Compiled by the authors using data from Paul Mader, David Dubois, and David Pimentel.)

Figure 12-34 More than two decades of research have revealed some of the major advantages of organic farming over conventional industrialized farming.

Top: ©Chamille White/Shutterstock.com. Center: ©Marbury/Shutterstock.com. Bottom: ©Robert Kneschke/Shutterstock.com.

THE LAND INSTITUTE AND PERENNIAL POLYCULTURE

Some scientists call for greater reliance on conventional and organic polycultures of perennial crops as a component of more sustainable agriculture. Such crops can live for many years without having to be replanted and are better adapted to regional soil and climate conditions than most annual crops.

More than three decades ago, plant geneticist Wes Jackson co-founded the Land Institute in the U.S. state of Kansas. One of the institute's goals has been to grow a diverse mixture of edible perennial plants to supplement traditional annual monoculture crops and to help reduce the latter's harmful environmental effects. Examples in this polyculture mix include perennial grasses, plants that add nitrogen to the soil, sunflowers, grain crops, and plants that provide natural insecticides. Some of these plants could also be used as a source of renewable biofuel for motor vehicles.

Researchers are busy trying to improve yields of different varieties of these perennial crops, which are not all as high as yields of annual crops. However, in the U.S. state of Washington, researchers have bred perennial wheat varieties that have a 70% higher yield than today's commercially grown annual wheat varieties.

The Land Institute's approach, called *natural systems agriculture*, copies nature by growing a diversity of perennial crops using organic methods. It has a number of environmental benefits. Because there is no need to till the soil and replant perennials each year, this approach produces much less topsoil erosion and water pollution. It also reduces the need for irrigation because the deep roots of such perennials retain more water than do the shorter roots of annuals (Figure 12-C). There is little or no need for chemical fertilizers and pesticides, and thus little or no pollution from these sources. Perennial polycultures also remove and store more carbon from the atmosphere, and growing them requires less energy than does growing crops in conventional monocultures.

This approach has worked well in some areas of the world. For example, in Malawi, Africa, farmers have greatly raised their crop yields by planting rows of perennial pigeon peas between corn rows. The pea plants have doubled the carbon and nitrogen content in the soils, while increasing soil water retention. These legume plants also provide a welcome source of protein for the farm families.

Some scientists note that adapting perennials to the grand scale of industrialized

Figure 12-C The roots of an annual wheat crop plant (left) are much shorter than those of big bluestem (right), a tallgrass prairie perennial plant.

Photo Courtesy of The Land Institute

crop production will be challenging and time-consuming. However, Wes Jackson calls for governments to promote this and other forms of more sustainable agriculture in order to reduce topsoil erosion, sustain nitrogen nutrients in topsoil, cut the wasteful use of irrigation water, reduce dependence on fossil fuels, and reduce dead zones in coastal areas (Connections, p. 294). He reminds us that "if our agriculture is not sustainable, then our food supply is not sustainable."

Critical Thinking

Why do you think large seed companies generally oppose this form of more sustainable agriculture?

Rodale Institute in Kutztown, Pennsylvania (USA) (see chapter-opening photo), comparing organic and conventional farming.

According to these studies, yields of organic crops in more-developed countries can be as much as 20% lower than yields of conventionally raised crops. This is largely because most agriculture in these countries is done on an industrial scale. However, in less-developed countries, it is done on a smaller scale. A 2008 study by the UNEP surveyed 114 small-scale farms in 24 African countries and found that yields more than doubled where organic farming practices had been used. In some East African operations, the yield increased by 128% with these changes.

Where yields from organic farming are lower than conventional yields, farmers often make up for this by not having to use expensive synthetic pesticides, herbicides, and fertilizers, and they usually receive higher prices for their crops. As a result, the net economic return per unit of land from organic crop production is often equal to or higher than that from conventional crop production.

The Rodale Institute studies also showed that organic farming methods use 28–32% less energy per unit of crop yield and that organically raised corn and soybean crops consistently store more carbon in the soil than conventional crops. These results suggest that organic agriculture can play an important role in slowing projected climate change by reducing fossil fuel use and greenhouse gas emissions and by pulling more CO_2 from the atmosphere.

One drawback of organic farming, compared with industrialized agriculture, is that it requires more human

Figure 12-35 This farm in Bavaria, Germany, gets its electricity from solar cells mounted on its buildings' roofs.

labor to use methods such as integrated pest management, crop rotation, low-till cultivation, and multicropping. However, it could be an important provider of jobs and income, especially for some farmers in less-developed countries. Organic farming is also a source of income for a growing number of younger farmers in more-developed countries who want to apply ecological principles to agriculture and help reduce its harmful environmental impacts.

Another important component of more sustainable agriculture could be to rely less on conventional monoculture and more on polyculture, in which a diversity of organic crops are grown on the same plot. Of particular interest to some scientists is the idea of using polyculture to grow *perennial crops*—crops that grow back year after year on their own (Science Focus 12.3).

Well-designed organic polyculture, like organic monoculture, helps to conserve and replenish topsoil, requires less water, cuts water losses, and reduces the need for fertilizers and pesticides. It also reduces the air and water pollution associated with conventional indus- GOOD NEWS trialized agriculture.

Most proponents of more sustainable agriculture call for using more environmentally sustainable forms of both high-yield polyculture and high-yield monoculture, with increasing emphasis on employing organic farming methods (Concept 12-6). Large-scale industrialized agriculture currently works against the earth by replacing natural biodiversity with engineered monocultures. It changes the earth to suit the crop, but with modifications, it could instead diversify crops to suit the earth's natural processes, in keeping with all three scientific **principles of sustainability**.

Another key to developing more sustainable agriculture is to shift from using fossil fuels to relying more on renewable energy for food production—an important application of the solar energy **principle of sustainability** that has been well demonstrated by the Growing Power farm (**Core Case Study**). To produce the electricity and fuels needed for food production, farmers can make greater use of renewable solar energy (Figure 12-35), wind, flowing water, and biofuels produced from farm wastes in tanks called *biogas digesters*.

Analysts suggest five major strategies to help farmers and consumers to make the transition to more sustainable agriculture over the next 50 years (Concept 12-6). *First,* greatly increase research on more sustainable organic farming and perennial polyculture, and on improving human nutrition. *Second,* establish education and training programs in more sustainable agriculture for students, farmers, and government agricultural officials. *Third,* set up an international fund to give farmers in poor countries access to various types of more sustainable agriculture. *Fourth,* replace government subsidies for environmentally harmful forms of industrialized agriculture with subsidies that encourage more sustainable agriculture. And *fifth,* mount a massive program to educate consumers about the true environmental and health costs of the food they buy.

Proponents of more sustainable food production systems say that over the next five decades, a combination of education and economic policies that reward more sustainable agriculture could lead to such a shift. GOOD NEWS They seek to inform people, especially young consumers, about where their food really comes from, how it is produced, and what the environmentally harmful effects of industrialized food production are.

Figure 12-36 lists ways in which you can promote more sustainable food production.

Figure 12-36 Individuals matter: There are a number of ways to promote more sustainable food production (Concept 12-6). **Questions:** Which three of these actions do you think are the most important? Why?

Big Ideas

- About 1 billion people have health problems because they do not get enough to eat and 1.6 billion people face health problems from eating too much.

- Modern industrialized agriculture has a greater harmful impact on the environment than any other human activity.

- More sustainable forms of food production will greatly reduce the harmful environmental impacts of industrialized food production systems while likely increasing food security.

TYING IT ALL TOGETHER   Growing Power and Sustainability

Baloncici/Shutterstock.com

This chapter began with a look at how Growing Power (**Core Case Study**), an ecologically-based urban farm, is providing a diversity of good food to people living in a food desert. Its founder Will Allen, in demonstrating how food can be grown more sustainably at affordable prices, is showing how to make the transition to more sustainable food production while applying the three scientific **principles of sustainability** (see Figure 1-2, p. 6 or back cover). Modern industrialized agriculture, aquaculture, and other forms of industrialized food production violate all of these principles.

Making the transition to more sustainable food production means relying more on solar and other forms of renewable energy and less on fossil fuels. It also means sustaining chemical cycling through topsoil conservation and by returning crop residues and animal wastes to the soil. It involves helping to sustain natural, agricultural, and aquatic biodiversity by relying on a greater variety of crop and animal strains and seafood, raised by certified organic methods and sold locally, as in farmers' markets (see photo at left). Controlling pest populations through broader use of conventional and perennial polyculture and integrated pest management will also help to sustain biodiversity.

Such efforts will be enhanced if we can control the growth of the human population and our wasteful use of food and other resources. Governments could help these efforts by replacing environmentally harmful agricultural and fishing subsidies and tax breaks with more environmentally beneficial ones. Finally, the transition to more sustainable food production would be accelerated if we could find ways to include the harmful environmental and health costs of food production in the market prices of food. For this reason, and because more sustainable food production would benefit the earth and help to sustain natural capital, it would be in keeping with all three social science **principles of sustainability** (see Figure 1-5, p. 9 or back cover).

Chapter Review

Core Case Study

1. Summarize the benefits that the Growing Power farm (**Core Case Study**) has brought to its community. How does the farm showcase the three scientific **principles of sustainability**?

Section 12-1

2. What are the two key concepts for this section? Define **food security** and **food insecurity**. What is the root cause of food insecurity? Distinguish between **chronic undernutrition (hunger)** and **chronic malnutrition** and describe their harmful effects. What is a **famine**? Describe the effects of diet deficiencies in vitamin A, iron, and iodine. What is **overnutrition** and what are its harmful effects?

Section 12-2

3. What is the key concept for this section? What three systems supply most of the world's food? Define **irrigation**. Define and distinguish among **industrialized agriculture (high-input agriculture)**, **plantation agriculture**, **traditional subsistence agriculture**, and **traditional intensive agriculture**. Define **polyculture** and summarize its benefits. Define **organic agriculture** and compare its main components with those of conventional industrialized agriculture. What is a **green revolution**? Summarize the story of industrialized food production in the United States.

4. Distinguish between crossbreeding through artificial selection and genetic engineering. Describe the second green revolution based on genetic engineering. Describe the growth of industrialized meat production. What are feedlots and CAFOs? What is a **fishery**? What is **aquaculture (fish farming)**? Explain why industrialized food production requires large inputs of energy. Why does it result in a net energy loss?

Section 12-3

5. What is the key concept for this section? List two major benefits of high-yield modern agriculture. Define **soil** and describe its formation and the major layers in mature soils. What is **topsoil** and why is it one of our most important resources? What is **soil erosion** and what are its two major harmful environmental effects? What is **desertification** and what are its harmful environmental effects? Define **soil salinization** and **waterlogging** and describe their harmful environmental effects. What is the biggest problem resulting from excessive use of water for irrigation in agriculture?

6. Summarize industrialized agriculture's contribution to projected climate change. Explain how industrialized food production systems have caused losses in biodiversity. What is **agrobiodiversity** and how is it being affected by industrialized food production? List the advantages and disadvantages of using genetic engineering in food production. What factors can limit green revolutions? Compare the benefits and harmful effects of industrialized meat production. Explain the connection between feeding livestock and the formation of ocean dead zones. Compare the benefits and harmful effects of aquaculture.

Section 12-4

7. What is the key concept for this section? What is a **pest**? Define and give two examples of a **pesticide**. Summarize the advantages and disadvantages of using synthetic pesticides. Describe the use of laws and treaties to help protect U.S. citizens from the harmful effects of pesticides. List seven alternatives to conventional pesticides. Define **integrated pest management (IPM)** and list its advantages.

Section 12-5

8. What is the key concept for this section? What are the two main approaches used by governments to influence food production? How have governments used subsidies to influence food production and what have been some of their effects? Describe the system used by Jennifer Burney to help people grow crops in parts of sub-Saharan Africa. What are two other ways in which organizations are improving food security? Explain three of the benefits of buying locally grown food. How can urban farming help to increase food security?

Section 12-6

9. What is the key concept for this section? What is **soil conservation**? Describe six ways to reduce topsoil erosion. What is **hydroponics**, and how can use of it help to deal with soil degradation? Distinguish among **organic fertilizer**, **synthetic inorganic fertilizer**, **animal manure**, **green manure**, and **compost**. How does crop rotation help to restore topsoil fertility? What are some ways to prevent and some ways to clean up soil salinization? How can we reduce desertification? Describe three ways to make aquaculture more sustainable. Describe two ways to make meat production and consumption more sustainable. Summarize three important aspects of making a shift to more sustainable food production. Describe the role of organic farming in making this shift. List the advantages of relying more on organic polyculture and perennial crops. What five strategies could help

farmers and consumers to shift to more sustainable agriculture? What are three important ways in which individual consumers can help to promote more sustainable food production?

10. What are the three big ideas of this chapter? Explain how making the transition to more sustainable

food production such as that promoted by the Growing Power farm (**Core Case Study**) will involve applying the six **principles of sustainability**.

Note: Key terms are in bold type.

Critical Thinking

1. Suppose you got a job with Growing Power, Inc. (**Core Case Study**) and were given the assignment to turn an abandoned suburban shopping center and its large parking lot into an organic farm. Write up a plan for how you would accomplish this.

2. Do you think that the advantages of organic agriculture outweigh its disadvantages? Explain. Do you eat or grow organic foods? If so, explain your reasoning for making this choice. If not, explain your reasoning for some of the food choices you do make.

3. Food producers can now produce more than enough food to feed everyone on the planet a healthy diet. Given this fact, why do you think that nearly a billion people are chronically undernourished or malnourished? Assume you are in charge of solving this problem, and write a plan for how you will accomplish it.

4. Explain why you support or oppose greatly increasing the use of **(a)** genetically modified food production and **(b)** organic perennial polyculture.

5. What might happen to industrialized food production if oil prices rise sharply in the next two decades,

as many analysts predict they will? How might this affect your life? How will it affect the lives of any children or grandchildren you might have? List two ways in which you would deal with these changes.

6. You are the head of a major agricultural agency in the area where you live. Weigh the advantages and disadvantages of using synthetic pesticides and explain why you would support or not support the increased use of such pesticides as a way to help farmers raise their yields. What are the alternatives?

7. If the mosquito population in the area where you live were proven to be carrying malaria or some dangerous virus, would you want to spray DDT in your yard, inside your home, or all through the local area to reduce this risk? Explain. What are the alternatives?

8. According to physicist Albert Einstein, "Nothing will benefit human health and increase the chances of survival of life on Earth as much as the evolution to a vegetarian diet." Are you willing to eat less meat or no meat? Explain.

Doing Environmental Science

For a week, weigh the food that is purchased in your home and the food that is thrown out. Record and compare these numbers from day to day. Develop a plan for

cutting your household food waste in half. Consider making a similar study for your school cafeteria and reporting the results and your recommendations to school officials.

Global Environment Watch Exercise

In the *Soil Erosion* portal, look for information on causes of soil erosion and how it affects soil fertility. Write a report on your findings. If you were to overhear a group of farmers complaining about how much money they must spend

on fertilizers, what suggestions would you give them for saving money? Include your answer to this question, along with your reasoning, in your report.

Ecological Footprint Analysis

The following table gives the world's fish harvest and population data.

Years	Fish Catch (million metric tons)	Aquaculture (million metric tons)	Total (million metric tons)	World Population (in billions)	Per Capita Fish Consumption (kilograms/person)
1990	84.8	13.1	97.9	5.27	
1991	83.7	13.7	97.4	5.36	
1992	85.2	15.4	100.6	5.44	
1993	86.6	17.8	104.4	5.52	
1994	92.1	20.8	112.9	5.60	
1995	92.4	24.4	116.8	5.68	
1996	93.8	26.6	120.4	5.76	
1997	94.3	28.6	122.9	5.84	
1998	87.6	30.5	118.1	5.92	
1999	93.7	33.4	127.1	6.00	
2000	95.5	35.5	131.0	6.07	
2001	92.8	37.8	130.6	6.15	
2002	93.0	40.0	133.0	6.22	
2003	90.2	42.3	132.5	6.31	
2004	94.6	45.9	140.5	6.39	
2005	94.2	48.5	142.7	6.46	
2006	92.0	51.7	143.7	6.54	
2007	90.1	52.1	142.2	6.61	
2008	89.7	52.5	142.3	6.69	

Compiled by the authors using data from UN Food and Agriculture Organization and Earth Policy Institute.

1. Use the world fish harvest and population data in the table to calculate the per capita fish consumption from 1990 to 2008 in kilograms per person. (*Hints:* 1 million metric tons equals 1 billion kilograms; the human population data are measured in billions; and per capita consumption can be calculated directly by dividing the total amount consumed by a population figure for any year.)

2. Did per capita fish consumption generally increase or decrease between 1990 and 2008?

3. In what years did per capita fish consumption decrease?

1. Industrialized meat production has harmful environmental consequences. These include all of the following EXCEPT
   a. the use of large amounts of energy.
   b. contributing to the eutrophication of the Gulf of Mexico.
   c. contributing to global warming.
   d. polluting the air, water, and soil with manure.
   e. natural control of pests.

2. Which of the following techniques helps to limit soil erosion?
   a. Cutting down all the trees in a specific area
   b. Growing only wheat, rice, and corn, which provide approximately half the calories people ingest
   c. The creation of steep steps on a hillside, creating a series of flat areas on which to plant crops
   d. The presence of large numbers of animals grazing too long in a grassy area, which hardens the soil they stand on
   e. The use of off-road vehicles because they flatten and harden the soil beneath them

3. Which best describes genetically modified crops (GMOs)?
   a. Genes are spliced from one plant species into another plant species.
   b. Planting a variety of crops in the same area such as native grass, grains, and sunflowers.
   c. Used mostly in tropical developing countries, crops such as bananas, coffee, and sugar cane are grown on large plantations and sold to developed countries.
   d. Fishes are grown in underwater cages, protected from pest species.
   e. Plants are not grown in soil, but instead their roots are in nutrient-filled water so they are protected from pests that would otherwise damage them.

4. One way that farmers can help to reduce soil erosion is by using soil conservation methods. Which of the following is an example of a soil conservation method?
   a. One or two crops are planted together in rows between trees or shrubs, which provide shade as well as fruit or nuts.
   b. Cutting down trees to allow sunlight and rain to reach the soil.
   c. Native plants are plowed up to make room for crops, and after the crops are harvested, land is plowed and left bare until the next growing season.
   d. Frequent flood irrigation of agricultural land, whereby the water flows in ditches through the fields.
   e. Frequent irrigation so water accumulates under the ground and, when the water table rises, supplies constant water to the plant roots.

5. Which of the following would be a disadvantage to using pesticides?
   a. Selective pesticides kill specific organisms, such as acequinocyl, which kills mites.
   b. The financial treadmill occurs, wherein pests become resistant to pesticides so farmers must use higher doses to kill the pests.
   c. By killing the pests that damage agricultural crops, pesticides help keep the price of food down.
   d. Pesticides have a long shelf life, so farmers will not lose money if the pesticides are kept for a long time before being used.
   e. Pesticides save lives by killing the vectors for diseases such as malaria, bubonic plague, and typhus.

6. Which is the best example of persistence?
   a. DDT, because it remained harmful in the environment for years
   b. First-generation pesticides, natural chemicals found in plants such as nicotine sulfate in tobacco leaves
   c. The use of essential oils taken from common species such as clove, mint, rosemary, and thyme
   d. Protecting the enemies of pests by providing habitats for them, such as creating areas where spiders are safe during harvesting so they will persist in eating the pests
   e. The use of genetic engineering to create pest-resistant crops so successive generations of those crops will also be pest resistant

7. The following practices are used to grow organic crops EXCEPT
   a. providing habitat for insects or other small invertebrates that feed on pests.
   b. keeping organic matter and nutrients in the soil as much as possible.
   c. planting crops before or after insect pests are there to cause damage.
   d. bringing in natural enemies to the insect pests, such as releasing ladybugs or praying mantis insects.
   e. spraying pesticides on the roots so that the part of the plant that is eaten is not contaminated.

8. A method of using biological control to control pests is to
   a. bring in natural enemies such as predators, parasites, and bacteria.
   b. utilize synthetic herbicides.
   c. allow wind to carry pesticides from one region to another.
   d. use broad-spectrum agents.
   e. use a second-generation pesticide.

9. The goal of integrated pest management is to
   a. wipe out as many of the pest species as possible to save crops.
   b. wipe out as many of the pest species as possible to save human lives.
   c. genetically engineer as many crop species as possible to eliminate the need for pesticides.
   d. encourage farmers to grow organically, eliminating the need for pesticides.
   e. use biological, chemical, and cultivation techniques in a coordinated way to eliminate pest species.

10. One way to successfully incorporate genetic engineering is to
    a. add pesticides to the DNA of the crop plant.
    b. cross a crop plant with a native variety.
    c. bring in hormones to control developmental processes at different stages.
    d. use biological control by importing natural predators.
    e. get on the pesticide treadmill.

11. All of the following are true about aquaculture EXCEPT
    a. China raises 70% of the world's fishes using aquaculture methods.
    b. farmed fishes are safe from contamination by toxins such as PCBs because their cages do not allow them to swim deep enough in ocean water.
    c. algae and other plants are the most common food eaten by fishes raised in aquaculture environments.
    d. also called the blue revolution, this method involves raising fishes in ponds, lakes, rice paddies, and underwater cages in oceans or estuaries.
    e. about 37% of the wild marine fishes caught are used to provide food to farmed fishes.

12. A more sustainable form of meat production is to
    a. graze beef in larger numbers but on local lands so the meat does not have to travel as far to the consumer.
    b. switch to poultry and farm-raised fish because these are more grain-efficient animals.
    c. inject cattle with hormones to grow larger with more muscle mass so fewer head of cattle would have to be raised.
    d. switch to beef rather than pork or poultry because cattle are larger and can supply meat to more people.
    e. use less land by putting more animals in feedlots and injecting them with antibiotics to keep them healthy.

13 Water Resources

Our liquid planet glows like a soft blue sapphire in the hard-edged darkness of space. There is nothing else like it in the solar system. It is because of water.

JOHN TODD

Key Questions

13-1 Will we have enough usable water?

13-2 Is groundwater a sustainable resource?

13-3 Can surface water resources be expanded?

13-4 Can water transfers be used to expand water supplies?

13-5 Is desalination a useful way to expand water supplies?

13-6 How can we use freshwater more sustainably?

13-7 How can we reduce the threat of flooding?

Each day these women carry water to their village in a dry area of India.

Shiv Ji Joshi/National Geographic Creative

Figure 13-1 The *Colorado River basin:* The area drained by this river system is more than one-twelfth of the land area of the lower 48 states. This map shows 6 of the river's 14 dams.

present populations; and in California's Imperial Valley, vast fields of vegetables would eventually be replaced by cactus and mesquite plants.

So much water is withdrawn from this river to grow crops and support cities in this dry, desertlike climate that very little of it reaches the sea. To make matters worse, since 1999, the system has experienced severe **drought**, a prolonged period in which precipitation is at least 70% lower and evaporation is higher than normal. This overuse of the Colorado River illustrates the challenges faced by governments and people living in arid and semi-arid regions with shared river systems, as population growth and economic growth place increasing demands on limited or decreasing supplies of surface water.

To many analysts, emerging shortages of water for drinking and irrigation in several parts of the world represent one of the major environmental challenges that we will face during this century.

The Colorado River, the major river of the arid southwestern United States, flows 2,300 kilometers (1,400 miles) through seven states to the Gulf of California (Figure 13-1). Most of its water comes from snowmelt in the Rocky Mountains. During the past 50 years, this once free-flowing river has been tamed by a gigantic plumbing system consisting of 14 major dams and reservoirs (Figure 13-2), and canals that help control flooding and supply water and electricity to farmers, ranchers, industries, and cities.

This system of dams and reservoirs provides water and electricity from hydroelectric plants at the major dams for roughly 30 million people in seven states—about one of every ten people in the United States. The river's water is used to produce about 15% of the nation's crops and livestock. The system also supplies water to some of the nation's driest and hottest cities. Take away this tamed river, and Las Vegas, Nevada, would be a mostly uninhabited desert area; San Diego and Los Angeles, California, could not support their

Figure 13-2 The Glen Canyon Dam was built to create the Lake Powell reservoir on the Colorado River.

13-1 Will We Have Enough Usable Water?

CONCEPT 13-1A
We are using available freshwater unsustainably by wasting it, polluting it, and underpricing this irreplaceable natural resource.

CONCEPT 13-1B
Freshwater supplies are not evenly distributed, and one of every six people on the planet does not have adequate access to clean water.

Freshwater Is an Irreplaceable Resource That We Are Managing Poorly

We live on a planet that is unique to our solar system because of a precious layer of water—most of it saltwater—covering about 71% of its surface (see Figure 8-2, p. 169, and Figure 25, pp. S50–S51, Supplement 6). Look in the mirror. What you see is about 60% water, most of it inside your cells.

Water is an amazing and irreplaceable chemical with unique properties that help to keep us and other species alive (see Science Focus 3.2, p. 64). You could survive for several weeks without food, but for only a few days without **freshwater**, or water that contains very low levels of dissolved salts.

It takes huge amounts of water to supply us with food and with most of the other things that we use to meet our daily needs and wants. Water also plays a key role in determining and moderating the earth's climates, in removing and diluting some of the pollutants and wastes that we produce, and in sculpting the planet's surface to produce rock formations such as the Grand Canyon (Figure 13-3).

Freshwater is one of the earth's most important forms of natural capital, but despite its importance, it is also one of our most poorly managed resources. We use it inefficiently and pollute it, and it is available at too low a cost to billions of consumers. This encourages still greater waste and pollution of this resource, for which we have no substitute (Concept 13-1A).

Access to freshwater is a *global health issue*. The World Health Organization (WHO) estimates that an average of 9,300 people die each day from waterborne infectious diseases because they do not have access to safe drinking water.

Access to freshwater is also an *economic issue* because water is vital for producing food and energy and for reducing poverty. According to the United Nations, about 780 million people—almost 2.5 times the entire U.S. population—do not have access to clean water where they live. Because almost half of the world's people do not have water piped to their homes, people in many less-developed countries, usually women and children, have to find and carry daily supplies of freshwater from distant sources (see chapter-opening photo).

Water is also a *national and global security issue* because of increasing tensions within and between some nations over access to limited freshwater resources that they share.

Finally, water is an *environmental issue* because excessive withdrawal of freshwater from rivers and aquifers has resulted in falling water tables, decreasing river flows (**Core Case Study**), shrinking lakes, and disappearing wetlands. This, in combination with water pollution in many areas of the world, has degraded water quality, reduced fish populations, hastened the extinction of some aquatic species, and degraded aquatic ecosystem services (see Figure 8-5, p. 170, and Figure 8-14, p. 178).

Most of the Earth's Freshwater Is Not Available to Us

The earth is often called the blue planet because of its abundance of water in all forms: liquid, solid (ice), and gas (water vapor). However, only a tiny fraction—*about 0.024%* of the planet's enormous water supply—is readily available to us as liquid freshwater, stored in accessible

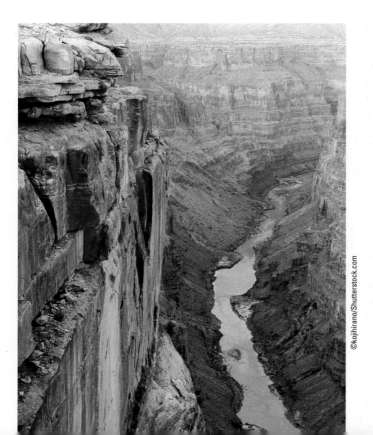

Figure 13-3 Water flowing in the Colorado River (**Core Case Study**) over a period of nearly 2 billion years created the spectacular Grand Canyon in the United States.

underground deposits and in lakes, rivers, and streams. The rest is in the salty oceans (about 96.5% of the earth's volume of liquid water), in frozen polar ice caps and glaciers (1.7%), and in underground aquifers (1.7%) (see Figure 25, pp. S50–S51, Supplement 6).

Fortunately, the world's freshwater supply is continually recycled, purified, and distributed in the earth's *hydrologic cycle* (see Figure 3-15, p. 63). This irreplaceable water recycling and purification system works well, unless we alter it, overload it with pollutants, or withdraw freshwater from underground and surface water supplies faster than it can be replenished.

We can also alter the hydrologic cycle through our influence on projected climate change. A 2012 study by oceanographer Paul Durack and his colleagues suggested that atmospheric warming has sped up the water cycle by about 4%, twice as fast as climate models have been projecting, and that by 2100, this acceleration could make the cycle 16–24% faster. As a result, on average, wet places will get wetter with more frequent flooding, dry places will get drier with more intense drought, and many storms will be more violent.

We have paid little attention to our effects on the water cycle mostly because we have thought of the earth's freshwater as essentially a free and infinite resource—an example of the tragedy of the commons (Chapter 1, p. 12). As a result, we have placed little or no economic value on the irreplaceable ecosystem services that water provides (Concept 13-1A), and this is a violation of the full-cost pricing **principle of sustainability** (see Figure 1-5, p. 9 or back cover).

On a global basis, we have plenty of freshwater, but it is not distributed evenly (Concept 13-1B). Differences in average annual precipitation (Figure 13-4) and economic resources divide the world's countries and people into water *haves* (such as Brazil, Figure 13-5, left) and *have-nots* (such as Saudi Arabia, Figure 13-5, right). For example, Canada, with only 0.5% of the world's population, has 20% of its liquid freshwater, while China, with 19% of the world's people, has only 7% of the supply.

Groundwater and Surface Water Are Critical Resources

Some precipitation infiltrates the ground and percolates downward through spaces in soil, gravel, and rock until an impenetrable layer of rock or clay stops it. The freshwater in these spaces is called **groundwater**—a key component of the earth's natural capital.

The spaces in soil and rock close to the earth's surface hold little moisture. However, below a certain depth, in the **zone of saturation**, these spaces are completely filled with freshwater. The top of this groundwater zone is the **water table**. It falls in dry weather, or when we remove groundwater faster than nature can replenish it, and it rises in wet weather.

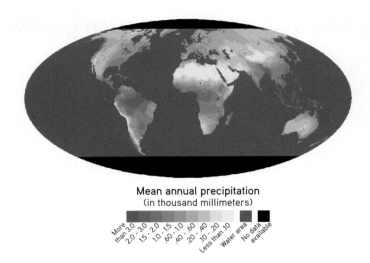

Mean annual precipitation
(in thousand millimeters)

More than 3.0 | 2.0 - 3.0 | 1.5 - 2.0 | 1.0 - 1.5 | .60 - 1.0 | .40 - .60 | .20 - .40 | .10 - .20 | Less than .10 | Water area | No data available

Figure 13-4 Global average annual precipitation.

National Geographic Maps/National Geographic Creative; data from NOAA/NESDIS/NCDC/Satellite Data Service

Deeper down are geological layers called **aquifers**, underground caverns and porous layers of sand, gravel, or rock through which groundwater flows. Mostly because of gravity, groundwater normally moves from points of high elevation and pressure to points of lower elevation and pressure. Some caverns have rivers of groundwater flowing through them. However, the porous layers of sand, gravel, or rock in most aquifers are like large, elongated sponges through which groundwater seeps—typically moving only a meter or so (about 3 feet) per year and rarely more than 0.3 meter (1 foot) per day. Watertight layers of rock or clay below such aquifers keep the freshwater from escaping deeper into the earth.

We use pumps to bring this groundwater to the surface for irrigating crops and supplying households and industries. Most aquifers are replenished naturally by precipitation that percolates downward through exposed soil and rock, a process called *natural recharge*. Others are recharged from the side by *lateral recharge* from nearby lakes, rivers, and streams.

According to the U.S. Geological Survey (USGS), there is 100 times more freshwater underground than there is in all of the world's rivers and lakes. However, most aquifers recharge extremely slowly, and because so much of the earth's landscape has been built on or paved over, freshwater can no longer penetrate the ground to recharge aquifers below many urban areas. In addition, in dry areas of the world, there is little precipitation available to recharge aquifers.

Aquifers that get very little, if any, recharge are called *nonrenewable aquifers*. They are found deep underground and were formed tens of thousands of years ago. Withdrawing freshwater from these aquifers amounts to *mining* a nonrenewable resource—an extreme example of the tragedy of the commons (Chapter 1, p. 12).

Another crucial resource is **surface water**, the freshwater from rain and melted snow that flows across the

Figure 13-5 Some countries have ample freshwater. The Iguassu River, marking part of the border between Brazil and Argentina, flows over the world's largest series of falls (left). Other countries such as Saudi Arabia (right) have very little freshwater.

Left: ©jose marques lopes/Shutterstock.com. Right: David Steele/Shutterstock.com.

earth's land surface and into lakes, wetlands, streams, rivers, estuaries, and the oceans. Precipitation that does not infiltrate the ground or return to the atmosphere by evaporation—about 39% of all precipitation—is called **surface runoff**. The land from which surface runoff drains into a particular stream, lake, wetland, or other body of water is called its **watershed**, or **drainage basin**. For example, the drainage basin for the Colorado River is shown in yellow and green on the map in Figure 13-1 (**Core Case Study**).

CONSIDER THIS. . .

CONNECTIONS Groundwater and Surface Water

There is usually a connection between surface water and groundwater because eventually, much groundwater naturally flows into rivers, lakes, estuaries, and wetlands. Thus, if we disrupt the hydrologic cycle by removing groundwater in a particular location faster than it is replenished, some nearby streams, lakes, and wetlands can dry up. This process degrades aquatic biodiversity and other ecosystem services.

We Are Using Increasing Amounts of the World's Reliable Runoff

According to *hydrologists* (scientists who study water and its movements above, on, and below the earth's surface), two-thirds of the annual surface runoff of freshwater into rivers and streams is lost in seasonal floods and is not available for human use. The remaining one-third is **reliable surface runoff**, which we can generally count on as a source of freshwater from year to year.

During the last century, the human population tripled, global water withdrawals increased sevenfold, and per capita withdrawals quadrupled. As a result, we now withdraw about 34% of the world's reliable runoff. This is a global average. In the arid American Southwest, up to 70% of the reliable runoff is withdrawn for human purposes, mostly for irrigation (**Core Case Study** and Case Study that follows). Some water experts project that because of a combination of population growth, rising rates of water use per person, longer dry periods, and fail-

ure to reduce unnecessary water losses, we may be withdrawing up to 90% of the reliable freshwater runoff by 2025.

Worldwide, we use 70% of the freshwater we withdraw each year from rivers, lakes, and aquifers to irrigate cropland and raise livestock; that figure rises to 90% in arid regions. Industry uses roughly another 20% of the water withdrawn each year, and cities and residences use the remaining 10%. Our **water footprint** is a rough measure of the volume of freshwater that we use directly and indirectly to stay alive and to support our lifestyles. Agriculture accounts for 92% of humanity's water footprint.

According to the American Water Works Association, the average American (see Case Study that follows) each day directly uses about 260 liters (69 gallons) of freshwater—enough water to fill about 1.7 typical bathtubs of water (a bathtub contains about 151 liters or 40 gallons of water). This water is used mostly for flushing toilets (27%), washing clothes (22%), taking showers (17%), and running faucets (16%), or is lost through leaking pipes, faucets, and other fixtures (14%).

We use even larger amounts of freshwater indirectly. This water is called **virtual water**, the freshwater that is not directly consumed but is used to produce food and other products. It makes up a large part of our water footprints, especially in more-developed nations. Producing and delivering a typical hamburger, for example, takes

These items...	take this many bathtubs full of water to produce
Cup of coffee	1
2-liter soda	3
Loaf of bread	4
1 dozen eggs	14
1 pound of cheese	15
Hamburger	16
T-shirt	17
Blue jeans	72
Small car	2,600
Medium-sized house	16,000

Figure 13-6 Producing and delivering a single one of each of the products shown here requires the equivalent of at least one and usually many bathtubs full of freshwater, called *virtual water. Note:* 1 bathtub = 151 liters (40 gallons).

(Compiled by the authors using data from UN Food and Agriculture Organization, UNESCO-IHE Institute for Water Education, World Water Council, and Water Footprint Network.)

about 2,400 liters (630 gallons) of freshwater—most of which is used to grow grain that is fed to cattle. This water would fill about 16 bathtubs.

Figure 13-6 shows one way to measure the amounts of virtual water used for producing and delivering products. These values can vary depending on how much of the supply chain is included, but they give us a rough estimate of the size of our water footprints.

Because of global trade in food and products, the virtual water used to produce and transport the wheat in a loaf of bread or the coffee beans used to make a cup of coffee (Figure 13-6) is often withdrawn as groundwater or surface water in another part of the world. For some water-short countries, it makes sense to save real freshwater by importing virtual water through food imports, instead of producing food domestically.

Large exporters of virtual water—mostly in the form of wheat, corn, soybeans, and other foods—are the European Union, the United States, Brazil, and Australia. Indeed, Brazil's supply of freshwater per person is more than 8 times the U.S. supply per person, 14 times China's supply, and 29 times India's supply. This explains why Brazil is likely to be one of the world's largest exporters of virtual water in the form of crops such as soybeans (see Figure 1-4, p. 8) to countries such as China.

However, prolonged severe droughts in parts of Australia, the United States, and the European Union are stressing the abilities of these countries to meet the growing global demand for their food exports. If climate change threatens freshwater supplies as is projected in some important crop-growing and food-exporting regions, freshwater scarcity will turn into food scarcity as virtual water supplies drop—another example of connections in nature.

CASE STUDY
Freshwater Resources in the United States

The United States has more than enough renewable freshwater to meet its needs. But it is unevenly distributed and much of it is contaminated by agricultural and industrial practices. The eastern states usually have ample precipitation, whereas many western and southwestern states have little (Figure 13-7).

According to a 2010 study by the USGS, the major uses of groundwater and surface freshwater in the United States are the cooling of electric power plants (41% of total water use), irrigation (37%), public water supplies (13%), industry (5%), and raising livestock (4%). In the eastern United States, most water is used for manufacturing and for cooling power plants (with most of the water heated and returned to its source). In many parts of this area, the most serious water problems are flooding, occasional water shortages as a result of pollution, and drought.

In the arid and semiarid regions of the western half of the United States (**Core Case Study**), irrigation counts for as much as 85% of freshwater use. Much of it is lost to evaporation and a great deal of it is used to grow thirsty crops. The major water problem in much of the western United States is a shortage of freshwater runoff caused by low precipitation (Figure 13-7), high evaporation, and recurring prolonged drought.

Water tables in many water-short areas, especially in the arid and semiarid western states, are dropping quickly

Average annual precipitation (centimeters)

Less than 41	81–122
41–81	More than 122

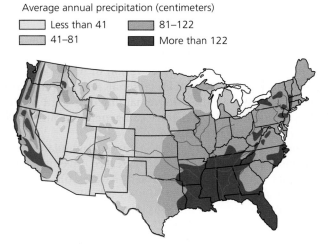

Figure 13-7 This map shows the long-term average annual precipitation and major rivers in the continental United States.

(Compiled by the authors using data from U.S. Water Resources Council and U.S. Geological Survey.)

as farmers and rapidly growing urban areas deplete many aquifers faster than they can be recharged. In 2007, the USGS projected that areas of at least 36 states are likely to face freshwater shortages by 2025 because of a combination of drought, rising temperatures, population growth, urban sprawl, and increased per-person use of water.

The U.S. Department of the Interior has mapped out *water hotspots* in 17 western states (Figure 13-8). In these areas, competition for scarce freshwater to support growing urban areas, irrigation, recreation, and wildlife could trigger intense political and legal conflicts between states and between rural and urban areas within states between now and 2030. Based on 19 global climate models used by the UN Intergovernmental Panel on Climate Change, climate researcher Richard Seager and his colleagues at Columbia University have projected that the southwestern United States and parts of northern Mexico are very likely to have long periods of extreme drought throughout most of the rest of this century.

The Colorado River system (Figure 13-1) will be directly affected by such drought. There are four major problems associated with the use of freshwater from the river (**Core Case Study**). *First,* the Colorado River basin includes some of the driest lands in the United States and Mexico. *Second,* legal pacts signed in 1922 and 1944 between the United States and Mexico allocated more freshwater for human use than the river can supply—even in rare years when there is no drought. In addition, the pacts allocated no water for protecting aquatic and terrestrial wildlife. *Third,* since 1960, the river has rarely flowed all the way to the Gulf of California because of its reduced water flow (due to many dams), increased freshwater withdrawals, and the prolonged drought. And *fourth,* the river receives enormous amounts of pollutants from urban areas, farms, animal feedlots, and industries, as it makes its way toward the sea.

Freshwater Shortages Will Grow

The main factors that cause water scarcity in any particular area are a dry climate, drought, too many people using a freshwater supply more quickly than it can be replenished, and inefficient use of freshwater. Figure 13-9 shows the current degree of *scarcity stress*—a measure based on a comparison of the amount of freshwater available with the amount used by humans—faced by each of the world's major river systems. Like the Colorado River (**Core Case Study**), major rivers including the Nile, Jordan, Yangtze, and Ganges are stressed so highly that their flows regularly dwindle to almost nothing in some locations.

More than 30 countries—most of them in the Middle East and Africa—now face stress from freshwater scarcity. By 2050, some 60 countries, many of them in Asia, with three-fourths of the world's population, are likely to be suffering from such freshwater stress. In 2006, the Chinese government reported that two-thirds of China's 600 major cities faced freshwater shortages.

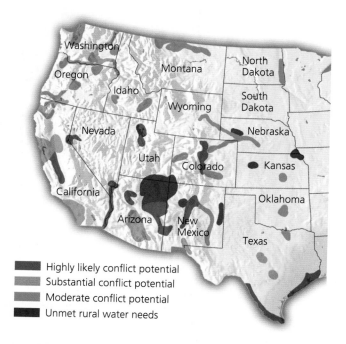

Figure 13-8 This map shows water scarcity hotspots in 17 western states that, by 2025, could face intense conflicts over scarce water needed for urban growth, irrigation, recreation, and wildlife. **Question:** Which, if any, of these areas are found in the Colorado River basin (**Core Case Study**)?

(Compiled by the authors using data from U.S. Department of the Interior and U.S. Geological Survey.)

Currently, about 30% of the earth's land area—a total area roughly 5 times the size of the United States—experiences severe drought. By 2059, as much as 45% of the earth's land surface—about 7 times the area of the United States—could experience extreme drought, mostly as a result of projected climate change caused by a warmer atmosphere, according to a 2007 study by climate researcher David Rind and his colleagues.

In each of 263 of the world's water basins, two or more countries share the available freshwater supplies. However, countries in only 158 of those basins have water-sharing agreements. This explains why conflicts among nations over shared freshwater resources are likely to happen more as populations grow, as demand for water increases, and as supplies shrink in many parts of the world.

In 2010, the United Nations reported that about 780 million people—one of every nine in the world—lacked regular access to enough clean water for drinking, cooking, and washing (Figure 13-10), mostly due to poverty (Concept 1-B). The report also projected that by 2025, at least 3 billion people—a number roughly equal to the current combined populations of China, India, and the United States—will likely lack such access to clean water. This helps to explain why many analysts view the likelihood of expanding water shortages in many parts of the world as one of the most serious environmental, health, and national security problems.

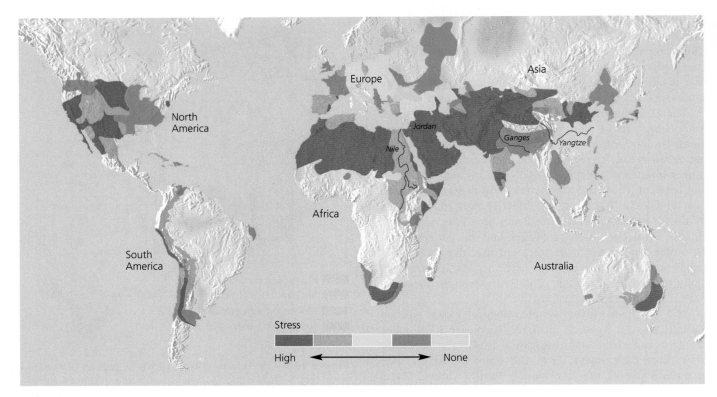

Figure 13-9 **Natural capital degradation:** The world's major river basins differ in their degree of freshwater *scarcity stress* (**Concept 13-1B**). ***Questions:*** If you live in a freshwater-stressed area, what signs of stress have you noticed? In what ways, if any, has it affected your life?

(Compiled by the authors using data from World Commission on Water Use for the 21st Century, UN Food and Agriculture Organization, and World Water Council.)

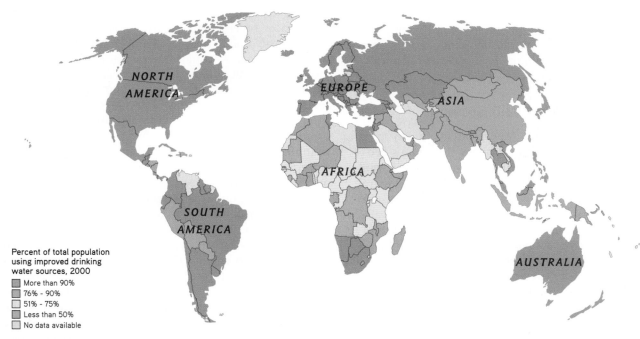

Percent of total population using improved drinking water sources, 2000

- More than 90%
- 76% - 90%
- 51% - 75%
- Less than 50%
- No data available

Figure 13-10 Access to freshwater.

National Geographic Maps/National Geographic Creative; World Health Organization World Health Statistics, 2010

SCIENCE FOCUS 13.1

USING SATELLITES TO MONITOR GROUNDWATER SUPPLIES

Figure 13-A These twin satellites measure slight variations in the earth's gravitational field that scientists use to detect and measure changes in large bodies of groundwater.

Scientists use a variety of methods to measure water supplies. For example, the USGS has drilled more than 7,000 observation wells throughout the United States to find and monitor groundwater supplies. Today some scientists are using satellites to measure changes in bodies of groundwater and surface water.

Since 2002, hydrologist Jay S. Famiglietti and his colleagues have been using two twin satellites (Figure 13-A) to measure small variations in the planet's gravitational pull that tell them about changes in ice cover, snow cover, surface water, soil moisture, and groundwater supplies. The two satellites,

each the size of a small car, travel in the same orbit, one about 217 kilometers (135 miles) behind the other. They constantly beam microwaves at each other and can detect changes of less than the diameter of a human hair in this distance between the satellites. When the leading satellite speeds up, increasing the distance, it means the mass of the earth beneath the satellite has increased, tugging a bit harder on the satellite. Various forms of water content on and under the surface of the earth can cause such an increase.

By comparing thousands of these satellite measurements to data from ground

measurements and computer modeling, Famiglietti and his team have learned what various changes in the data mean. In this way, they have calculated that large aquifers in many areas of the world are being drawn down rapidly. For example, in California's Central Valley, where groundwater is pumped to irrigate crops, Famiglietti estimated that between 2003 and 2010, underlying aquifers were drawn down by nearly 31 cubic kilometers (25 million acre-feet)—more than enough to fill Lake Powell (Figure 13-2), the second-largest reservoir in the United States.

According to the scientists, the satellite data indicate that large aquifers in several other areas, including North Africa, northeastern China, and northern India, are also being overpumped. These are all areas where water shortages are worsening every year.

Critical Thinking

What are some possible problems with relying on satellite data for measuring changes in water supplies?

🔍 CONSIDER THIS. . .

CONNECTIONS Virtual Water, Grain, and Hunger

Some water-short countries are reducing their irrigation water needs by importing grain, thereby freeing up more of their own water supplies for industrial and urban development. The result is a competition for the world's grain, which includes indirect competition for water (virtual water used to grow grain). In this global competition for water and grain, financially strong countries hold the advantage. As a result, grain prices are likely to rise, leading to food shortages and more widespread hunger among the poor.

There Are Several Ways to Increase Freshwater Supplies

What can we do to deal with the projected shortages of freshwater in many parts of the world? An important part of any strategy will be getting accurate information about where the water shortages are and where they are getting worse. Scientists have found a number of ways to obtain this information, including using satellites (Science Focus 13.1).

Several approaches to dealing with water shortages have been tried. They include increasing water supplies by withdrawing groundwater; building dams and reservoirs to store runoff in rivers for release as needed (**Core Case Study**); transporting surface water from one area to another; and converting saltwater to freshwater (desalination). The next four sections of this chapter look at these options. Another important solution is to reduce the unnecessary losses of freshwater, an option we consider in Section 13-6.

13-2 Is Groundwater a Sustainable Resource?

CONCEPT 13-2
Groundwater used to supply cities and grow food is being pumped from aquifers in some areas faster than it is renewed by precipitation.

Groundwater Is Being Withdrawn Faster Than It Is Replenished in Some Areas

Aquifers provide drinking water for nearly half of the world's people. Most aquifers are renewable resources unless the groundwater they contain becomes contaminated or is removed faster than it is replenished. In the United States, aquifers supply almost all of the drinking water in rural areas, one-fifth of that in urban areas, and 37% of the country's irrigation water. Relying more on groundwater has advantages and disadvantages (Figure 13-11).

Test wells and satellite data (Science Focus 13.1) indicate that water tables are falling in many areas of the world because the rate at which most of the world's aquifers are being pumped (mostly to irrigate crops) is greater than the rate of natural recharge from rainfall and snowmelt (Concept 13-2). The world's three largest grain producers—China, the United States, and India—as well as Mexico, Saudi Arabia, Iran, Yemen, Israel, and Pakistan, are overpumping many of their aquifers.

Every day, the world withdraws enough freshwater from aquifers to fill a convoy of large tanker trucks that could stretch 480,000 kilometers (300,000 miles)—well beyond the distance to the moon. Currently, more than 400 million people are being fed by grain produced through this unsustainable use of groundwater, according to the World Bank. This number is expected to grow.

The widespread drilling of inexpensive tube wells by farmers, especially in India and China, has accelerated aquifer overpumping. As water in aquifers is removed faster than it is renewed, water tables fall. Then farmers drill deeper wells, buy larger pumps, and use more energy to pump water to the surface. This process eventually depletes the groundwater in some aquifers or at least removes all the water that can be pumped at an affordable cost.

Saudi Arabia is as water-poor (Figure 13-5, right) as it is oil-rich. Much of its freshwater is pumped from ancient, nonrenewable deep aquifers to irrigate crops such as wheat grown on desert land (Figure 13-12) and to fill large numbers of fountains and swimming pools, which lose a great deal of water through evaporation into the hot, dry desert air. In 2008, Saudi Arabia announced that irrigated wheat production had largely depleted its major deep aquifer and said that it would stop producing wheat by 2016 and import grain (virtual water) to help feed its 30 million people.

In the United States, aquifer depletion is a growing problem, especially in the vast Ogallala aquifer (see Case Study that follows) and in California's arid Central Valley.

CASE STUDY
Overpumping the Ogallala

In the United States, groundwater is being withdrawn from aquifers, on average, 4 times faster than it is replenished, according to the USGS (Concept 13-2). Figure 13-13 shows the areas of greatest aquifer depletion. One of the most serious overdrafts of groundwater is in the lower half of the Ogallala, the world's largest known aquifer, which lies under eight Midwestern states from southern South Dakota to Texas (blowup section of Figure 13-13).

The gigantic Ogallala aquifer supplies about one-third of all the groundwater used in the United States and has

Trade-Offs

Withdrawing Groundwater

Advantages	Disadvantages
Useful for drinking and irrigation	Aquifer depletion from overpumping
Exists almost everywhere	Sinking of land (subsidence) from overpumping
Renewable if not overpumped or contaminated	Some deeper wells are nonrenewable
Cheaper to extract than most surface waters	Pollution of aquifers lasts decades or centuries

© Cengage Learning

Figure 13-11 Withdrawing groundwater from aquifers has advantages and disadvantages. *Questions:* Which two advantages and which two disadvantages do you think are the most important? Why?

Top: ©Ulrich Mueller/Shutterstock.com

Figure 13-12 **Natural capital degradation:** Satellite photos of farmland irrigated by groundwater pumped from a deep aquifer in a vast desert region of Saudi Arabia between 1986 (left) and 2004 (right). Irrigated areas appear as green dots (each representing a circular spray system) and brown dots show areas where wells have gone dry and the land has returned to desert.

Left: ©U.N. Environment Programme and U.S. Geological Survey. Right: ©U.N. Environment Programme and U.S. Geological Survey.

helped to turn the Great Plains into one of world's most productive irrigated agricultural regions (Figure 13-14). The problem is that the Ogallala is essentially a one-time deposit of liquid natural capital with a very slow rate of recharge.

In parts of the southern half of the Ogallala, groundwater is being pumped out 10–40 times faster than the slow natural recharge rate. This has lowered water tables and raised pumping costs, especially in northern Texas. In parts of Texas and Kansas, farmers cannot always pump enough water to meet their irrigation needs.

Government subsidies designed to increase crop production have encouraged farmers to grow water-thirsty crops in dry areas, which has accelerated depletion of the Ogallala. In particular, corn is a very thirsty crop that has been planted widely on fields watered by the Ogallala. Another resulting problem in addition to aquifer depletion is that nitrates from fertilizers and pesticides commonly sprayed on cornfields are now found in high concentrations in some parts of the aquifer.

The Ogallala also helps to support biodiversity. In various places, groundwater from the aquifer flows out of the ground onto land or onto lake bottoms through exit points called *springs*. In some cases, springs feed wetlands, which are vital habitats for many species, especially birds.

When the water tables fall, many of these aquatic oases of biodiversity dry out.

Overpumping of Aquifers Can Have Harmful Effects

Overpumping of aquifers can contribute to limits on food production, rising food prices, and widening gaps between the rich and poor in some areas. As water tables drop, farmers must drill deeper wells, buy larger pumps, and use more electricity to run those pumps. Poor farmers cannot afford to do this and end up losing their land and working for richer farmers, or migrating to cities that are already crowded with poor people struggling to survive.

Withdrawing large amounts of groundwater sometimes causes the sand and rock that is held in place by water pressure in aquifers to collapse. This can cause the land above the aquifer to *subside* or sink, a phenomenon known as *land subsidence*. Extreme, sudden subsidence is sometimes referred to as a *sinkhole*. Once an aquifer becomes compressed by subsidence, recharge is impossible. In addition, land subsidence can damage roadways, water and sewer lines, and building foundations. Since 1925, overpumping of an aquifer to irrigate crops

Animated Figure 13-13 **Natural capital degradation:** This map shows areas of greatest aquifer depletion from groundwater overdraft in the continental United States. The blowup section (right) shows where water levels in the Ogallala aquifer have dropped sharply at its southern end beneath parts of Kansas, Oklahoma, Texas, and New Mexico. **Questions:** Do you depend on any of these aquifers for your drinking water? If so, what is the level of severity of overdraft where you live?

(Compiled by the authors using data from U.S. Water Resources Council and U.S. Geological Survey.)

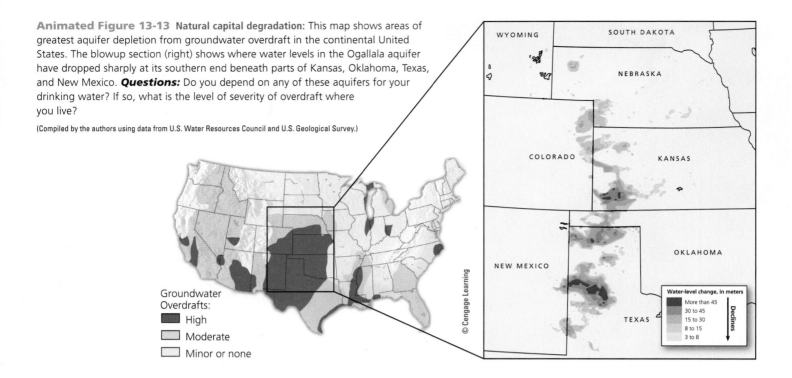

Groundwater Overdrafts:
- ■ High
- ▨ Moderate
- ☐ Minor or none

Water-level change, in meters
- More than 45
- 30 to 45
- 15 to 30
- 8 to 15
- 3 to 8

Declines

in California's San Joaquin Valley has caused half of the valley's land to subside by more than 0.3 meter (1 foot) and, in one area, by more than 8.5 meters (28 feet) (Figure 13-15).

Mexico City, built on an old lakebed, has one of the world's worst subsidence problems because its groundwater is being increasingly overpumped due to rapid population growth and urbanization. Some parts of the city have sunk as much as 10 meters (33 feet). Parts of Beijing, China, are also sinking as its water table has fallen 61 meters (200 feet) since 1965.

Groundwater overdrafts in coastal areas, where many of the world's largest cities and industrial areas are found, can pull saltwater into freshwater aquifers. The resulting contaminated groundwater is undrinkable and unusable for irrigation. This problem is especially serious in coastal areas of the U.S. states of California, Texas, Florida, Georgia, South Carolina, and New Jersey, as well as in coastal areas of Turkey, Thailand, and the Philippines.

If we do not sharply slow the depletion of aquifers, a growing number of the world's people will have to live on rainwater and suffer from food and drinking water shortages. Figure 13-16 lists ways to prevent or slow the problem of aquifer depletion by using this largely renewable resource more sustainably. **GREEN CAREER:** hydrogeologist

Deep Aquifers Might Be Tapped

With global shortages of freshwater looming, scientists are evaluating *deep aquifers* as future sources of freshwater. Preliminary results suggest that some of these aquifers hold enough freshwater to support billions of people for centuries. In addition, the quality of freshwater in these aquifers may be much higher than the quality of the freshwater in most rivers and lakes.

There are four major concerns about tapping these ancient deposits of freshwater. *First,* they are nonrenewable and cannot be replenished on a human time scale.

Figure 13-14 Center-pivot irrigation systems are moved around in circles to irrigate crops with water that is typically pumped from aquifers such as the Ogallala.

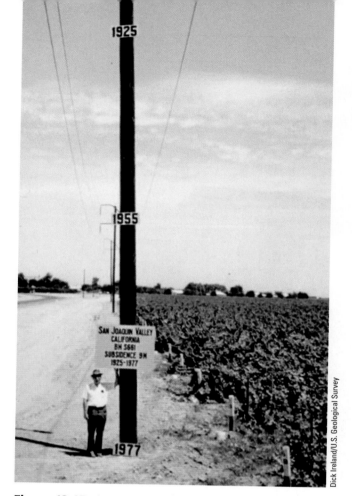

Figure 13-15 This pole shows subsidence from overpumping of an aquifer for irrigation in California's San Joaquin Central Valley between 1925 and 1977. In 1925, this area's land surface was near the top of the pole. Since 1977, this problem has gotten worse.

Solutions

Groundwater Depletion

Prevention	Control
Use water more efficiently	Raise price of water to discourage waste
Subsidize water conservation	Tax water pumped from wells near surface water
Limit number of wells	Build rain gardens in urban areas
Stop growing water-intensive crops in dry areas	Use permeable paving material on streets, sidewalks, and driveways

WATER BILL

Figure 13-16 There are a number of ways to prevent or slow groundwater depletion by using freshwater more sustainably. **Questions:** Which two of these solutions do you think are the most important? Why?

Top: © Anhong | Dreamstime.com. Bottom: Banol2007/Dreamstime.com.

Second, little is known about the geological and ecological impacts of pumping large amounts of freshwater from deep aquifers. *Third*, some deep aquifers flow beneath more than one country and there are no international treaties that govern rights to them. Without such treaties, wars could break out over this resource. *Fourth*, the costs of tapping deep aquifers are unknown and could be high.

13-3 Can Surface Water Resources Be Expanded?

CONCEPT 13-3
Large dam-and-reservoir systems have greatly expanded water supplies in some areas, but have also disrupted ecosystems and displaced people.

Use of Large Dams Provides Benefits and Creates Problems

A **dam** is a structure built across a river to control its flow. Usually, dammed water creates an artificial lake, or **reservoir**, behind the dam (Figure 13-2). The main goals of a dam-and-reservoir system are to capture and store the surface runoff from a river's watershed, and release it as needed to control floods; to generate electricity (hydropower); and to supply freshwater for irrigation and for towns and cities. Reservoirs also provide recreational activities such as swimming, fishing, and boating. Large

dams and reservoirs provide benefits but have drawbacks (Figure 13-17).

The world's 45,000 large dams—those that are 15 meters (49 feet) or higher—capture and store about 14% of the world's surface runoff, provide water for almost half of all irrigated cropland, and supply more than half the electricity used in 65 countries. The United States has more than 70,000 large and small dams, capable of capturing and storing half of the country's entire river flow.

By using dams, we have increased the annual reliable runoff available for our uses by nearly 33%. As a result, the world's reservoirs now hold 3 to 6 times more freshwater than the total amount flowing at any moment in all of the world's natural rivers. On the down side, this engineering approach to river management has displaced 40–80 million people from their homes, flooded an area of mostly productive land totaling roughly the area of the

Provides irrigation water above and below dam

Provides water for drinking

Reservoir useful for recreation and fishing

Can produce cheap electricity (hydropower)

Reduces down-stream flooding of cities and farms

Flooded land destroys forests or cropland and displaces people

Large losses of water through evaporation

Deprives downstream cropland and estuaries of nutrient-rich silt

Risk of failure and devastating downstream flooding

Disrupts migration and spawning of some fish

Power lines

Reservoir

Dam

Intake

Powerhouse

Turbine

© Cengage Learning

Figure 13-17 Trade-offs: Use of large dams and reservoirs has its advantages (green) and disadvantages (orange) (Concept 13-3). *Questions:* Which single advantage and which single disadvantage do you think are the most important? Why?

U.S. state of California, and impaired some of the important ecosystem services that rivers provide (see Figure 8-14, left, p. 178, and Concept 13-3).

A 2007 study by the World Wildlife Fund (WWF) estimated that about one out of five of the world's freshwater fish and plant species are either extinct or endangered, primarily because dams and water withdrawals have sharply

decreased the flows of many rivers such as the Colorado (**Core Case Study**). The study also found that, because of dams, excessive water withdrawals, and in some areas, prolonged severe drought, only 21 of the planet's 177 longest rivers consistently run all the way to their outlets to the sea before running dry (see Case Study that follows).

The reservoirs behind dams eventually fill up with sediments such as mud and silt, typically within 50 years, which makes them useless for storing water or producing electricity. For example, in the Colorado River system (**Core Case Study**), the equivalent of roughly 20,000 dump-truck loads of silt are deposited on the bottoms of the Lake Powell and Lake Mead reservoirs every day. Sometime during this century, these reservoirs will probably be too full of silt to control floods or to store enough water for generating hydroelectric power. About 85% of all U.S. dam-and-reservoir systems will be 50 years old or more by 2025, and some of those aging dams are being removed because their reservoirs have filled with silt.

If climate change occurs as projected during this century, it will intensify shortages of water in many parts of the world. For example, mountain snows that feed the Colorado River system (**Core Case Study**) will melt faster and earlier, making less freshwater available to the river system when it is needed during hot summer months.

Nearly 3 billion people in South America, China, India, and other parts of Asia depend on river flows fed by mountain glaciers, which act like aquatic savings accounts. They store precipitation as ice and snow in wet periods and release it slowly during the dry season for use on farms and in cities. In 2010, according to the World Glacier Monitoring Service, many of these mountain glaciers had been shrinking for 19 consecutive years, mostly due to a warming atmosphere. For a while, their melting will likely increase water supplies. However, if these glaciers eventually disappear, most of those 3 billion people will likely be short of water for all purposes, including food production.

Figure 13-18 The Colorado River delta once contained a rich variety of forests, wetlands, and wildlife. Now it is covered mostly by mud flats and desert.

CASE STUDY

How Dams Can Kill an Estuary

Since 1905, the amount of water flowing to the mouth of the heavily dammed Colorado River (**Core Case Study**) has dropped dramatically. In most years since 1960, the river has dwindled to a small, sluggish stream by the time it reaches the Gulf of California.

This is the subject of an online video by National Geographic Explorer Alexandra Cousteau (see Individuals Matter 8.1, Chapter 8, p. 183), called *Death of a River*. In that film, Cousteau explains that the river once emptied into a vast jungle *delta*, the wetland area at the mouth of a river containing the river's estuary. The Colorado's delta covered an area of more than 800,000 hectares (2 million acres)—the size of the state of Rhode Island. It hosted forests, lagoons, and marshes rich in plant and animal life and supported a thriving coastal fishery for hundreds of years.

Since the damming of the Colorado—within one human lifetime—this rich ecosystem has collapsed and is now covered by mud flats and desert (Figure 13-18). The dams upstream have cut off the water supply that kept the system alive. All but one-tenth of the river's flow was diverted for use in seven U.S. states. Most of the remaining 10% is assigned to farms and to the growing cities of Mexicali and Tijuana in Mexico. The delta and its wildlife are now mostly gone and its coastal fishery that fed many generations of area residents is disappearing.

Historically, about 80% of the water withdrawn from the Colorado has been used to irrigate crops and raise cattle. That is because the government paid for the dams and reservoirs and has supplied many farmers and ranchers with water at low prices. These government subsidies have led to inefficient use of irrigation water for growing crops such as rice, cotton, and alfalfa that need a lot of water. However, much of this water is lost to the river sys-

tem before it can be used. For example, there are huge losses of water in the Lake Mead and Lake Powell reservoirs due to evaporation and seepage of water into porous rock beds under the reservoirs.

To deal with the water supply problems in the dry Colorado River basin, water experts call for the seven states using the river to enact and enforce strict water conservation measures and to slow population growth and urban development. They also call for phasing out state and federal government subsidies for agriculture in this region, shifting water-thirsty crops to less arid areas, and banning or severely restricting the use of surface water and groundwater to keep golf courses and lawns green in the desert areas of the Colorado River basin (Figure 13-1). They suggest that the best way to implement such solutions is to sharply raise the historically low price of the river's freshwater over the next decade—another application of the full-cost pricing **principle of sustainability**. According to Cousteau, if just 1% of the river's flow were restored to the delta area, much of it could be partially restored and its fisheries might begin to rebound.

CONSIDER THIS. . .

THINKING ABOUT The Colorado River

What are three steps you would take to deal with the problems of the Colorado River system?

13-4 Can Water Transfers Be Used to Expand Water Supplies?

CONCEPT 13-4
Transferring water from one place to another has greatly increased water supplies in some areas but has also disrupted ecosystems.

Water Transfers Can Be Inefficient and Environmentally Harmful

In some heavily populated dry areas of the world, governments have tried to solve water shortage problems by transferring water to the dry areas from water-rich areas. For example, northern China contains rapidly growing cities, including Beijing with 17.3 million people. These urban populations have helped to deplete underlying aquifers. According to the Chinese Academy of Sciences, two-thirds of China's 669 major cities have water shortages, more than 40% of its rivers are severely polluted, and about 300 million rural residents—a number almost equal to the size of the U.S. population—do not have access to safe drinking water. To help with this serious problem and to boost food production, the Chinese government is implementing its *South–North Water Diversion Project* to transfer 23 trillion liters (6 trillion gallons of water) per year from the Yangtze River in southern China to the thirsty north.

In other cases, water has been transferred to arid areas primarily to irrigate farm fields. For example, when you have lettuce in a salad in the United States, chances are good that it was grown with the use of heavy irrigation in the arid Central Valley of California. In fact, the *California State Water Project* (see map in Figure 13-19) is one of the world's largest freshwater transfer projects. It uses a maze of giant dams, pumps, and lined canals, or *aqueducts* (photo in Figure 13-19), to transport freshwater from the High Sierra Mountains of northeastern California to heavily populated cities and agricultural regions in water-poor central and southern California.

These water transfers have many benefits. Some 440 million Chinese will be served by the South–North Water Diversion Project. California's Central Valley supplies half of the United States' fruits and vegetables, and sprawling southern California cities including San Diego and Los Angeles use water transferred from the north. These areas would be mostly desert without the water transfers.

However, massive water transfers also involve high economic and social costs, large water losses through evaporation and leaks

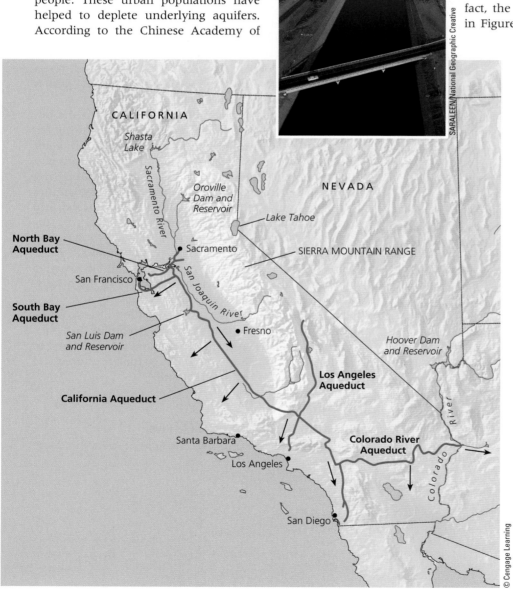

Figure 13-19 The California State Water Project transfers huge volumes of freshwater from one watershed to another. The arrows on the map show the general direction of water flow. The photo shows one of the aqueducts carrying water within the system. **Questions:** What effects might this system have on the areas from which the water is taken?

SARALEEN/National Geographic Creative

© Cengage Learning

in the water-transfer systems, and degraded ecosystems in areas from which the water was taken (**Concept 13-4**). (See the Case Study that follows.) For example, China's massive water transfer will be quite expensive. To make way for the canals that will carry water, more than 350,000 villagers will have to relocate from lands they have farmed for generations to low-grade farmland or to urban areas. Also, scientists worry that removing huge volumes of water from the Yangtze River could severely damage its ecosystem, which is now suffering from its worst drought in 50 years.

In California, sending water south has degraded the Sacramento River, threatening fisheries and reducing the flushing action that helps to cleanse San Francisco Bay of pollutants. The bay has suffered from pollution, and the flow of freshwater to its coastal marshes and other ecosystems has dropped, putting stress on wildlife species that depend on these ecosystems. Water was also diverted from streams that flow into Mono Lake, an important feeding stop for migratory birds on the eastern slope of the High Sierras. This lake experienced an 11-meter (35-foot) drop in its water level until the diversions were stopped. The water level is now rising again, but Mono Lake's entire ecosystem was in jeopardy. Before that time, Owens Lake, another large lake in the area, was completely drained by the water transfer and its ecosystem was essentially destroyed.

One reason why such damaging water transfers have taken place is that governments have subsidized the costs of such transfers for inefficient uses such as the irrigation of lettuce and other crops that need lots of water in desertlike areas. In central California, agriculture consumes three-fourths of the water that is transferred, and much of it is lost through inefficient irrigation systems. Studies have shown that making irrigation just 10% more efficient would provide all the water necessary for domestic and industrial uses in southern California. Yet, inefficient use of the water continues because taxpayer subsidies make the water inexpensive to farmers and other users. Many scientists have argued for replacing these subsidies with subsidies that would encourage more efficient use of water.

In 2012, the U.S. Bureau of Reclamation was evaluating a proposal to build a 970-kilometer-long (600-mile-long) pipeline to transfer water from the Missouri and Mississippi Rivers to reservoirs in Denver, Colorado, to help supplement the water supply in the Colorado River basin (**Core Case Study**). However, the loss of so much water from these rivers, which require high water flows to sustain large vessel navigation, would face strong political opposition, and it would have harmful effects on these river ecosystems.

According to several studies, projected climate change will make matters worse in many areas where water is being removed for transfers. In southern China, climate change could worsen the drought and create a need for an even larger and more expensive transfer of water. In northern California, many people depend on *snowpacks*, bodies of densely packed, slowly melting snow, in the High Sierras for more than 60% of their freshwater during the hot, dry summer months. Projected atmospheric warming could shrink the snowpack by as much as 40% by 2050 and by as much as 90% by the end of this century. This will sharply reduce the amount of freshwater available for transfer to central and southern California.

CASE STUDY

The Aral Sea Disaster: A Glaring Example of Unintended Consequences

The shrinking of the Aral Sea (Figure 13-20) is the result of a large-scale freshwater transfer project in an area of the former Soviet Union with the driest climate in central Asia. Since 1960, enormous amounts of irrigation water have been diverted from the two rivers that supply water to the Aral Sea. The goal was to create one of the world's largest irrigated areas, mostly for raising cotton and rice. The irrigation canal, the world's longest, stretches more than 1,300 kilometers (800 miles)—roughly the distance

1976   **2012**

Figure 13-20 Natural capital degradation: The *Aral Sea*, straddling the borders of Kazakhstan and Uzbekistan, was one of the world's largest saline lakes. These satellite photos show the sea in 1976 (left) and in 2012 (right). *Question:* What do you think should be done to help prevent further shrinkage of the Aral Sea?

Figure 13-21 *Boat cemetery:* These fishing boats, which once plied the Aral Sea, are now stranded in the desert formed by the shrinkage of the sea.

Vladb/Dreamstime.com

between the two U.S. cities of Boston, Massachusetts, and Chicago, Illinois.

This large-scale freshwater diversion project, coupled with drought and high evaporation rates due to the area's hot and dry climate, has caused a regional ecological and economic disaster. Since 1961, the sea's salinity has risen sevenfold and the average level of its water has dropped by an amount roughly equal to the height of a six-story building. The Southern Aral Sea has lost 90% of its volume of water and is a remnant of the original sea lying in a salt-covered desert (Figure 13-20, right). Water withdrawal for agriculture has reduced the two rivers feeding the sea to mere trickles.

About 85% of the area's wetlands have been eliminated and about half the local bird and mammal species have disappeared. A huge area of the former Southern Aral Sea's lake bottom is now a human-made desert covered with glistening white salt. The sea's greatly increased salt concentration—3 times saltier than ocean water—has caused the presumed local extinction of 26 of the area's 32 native fish species. This has devastated the area's fishing industry, which once provided work for more than 60,000 people. Fishing villages and boats once located on the sea's coastline now sit abandoned in a salty desert (Figure 13-21).

Winds pick up the sand and salty dust and blow it onto fields as far as 500 kilometers (310 miles) away. As the salt spreads, it pollutes water and kills wildlife, crops, and other vegetation. Aral Sea dust settling on glaciers in the Himalayas is causing them to melt at a faster than normal rate—a prime example of unexpected connections and unintended consequences.

The shrinkage of the Aral Sea has also altered the area's climate. The once-huge sea acted as a thermal buffer that moderated the heat of summer and the extreme cold of winter. Now there is less rain, summers are hotter and drier, winters are colder, and the growing season is shorter. The combination of such climate change and severe salinization has reduced crop yields by 20–50% on almost one-third of the area's cropland—the opposite of the project's intended consequences.

Since 1999, the United Nations, the World Bank, and the five countries surrounding the lake have worked to improve irrigation efficiency. They have also partially replaced thirsty crops with other crops that require less irrigation water. Because of a dike that blocks the flow of water from the Northern Aral Sea into the southern sea, the level of the northern sea has risen by 2 meters (7 feet), its salinity has dropped, and dissolved oxygen levels are up. It now supports a healthy fishery that produced 5 times as big a catch (by total weight) in 2011 as it had produced in 2005.

However, the formerly much larger southern sea is still shrinking. By 2012, its eastern lobe was essentially gone, and the European Space Agency projects that the rest of the Southern Aral Sea could dry up completely by 2020.

13-5 Is Desalination a Useful Way to Expand Water Supplies?

CONCEPT 13-5
We can convert salty ocean water to freshwater, but the cost is high, and the resulting salty brine must be disposed of without harming aquatic or terrestrial ecosystems.

Removing Salt from Seawater Is Costly and Has Harmful Effects

Desalination is the process of removing dissolved salts from ocean water or from brackish (slightly salty) water in aquifers or lakes. It is another way to increase supplies of freshwater (Concept 13-5).

The two most widely used methods for desalinating water are distillation and reverse osmosis. *Distillation* involves heating saltwater until it evaporates (leaving behind salts in solid form) and condenses as freshwater. *Reverse osmosis* (or *microfiltration*) uses high pressure to force saltwater through a membrane filter with pores small enough to remove the salt and other impurities.

In 2010, according to the International Desalination Association, there were more than 15,000 desalination plants (250 in the United States, half of them in Florida) operating in more than 125 countries, most of them arid nations of the Middle East, North Africa, the Caribbean Sea, and the Mediterranean Sea. Desalination supplies less than 0.3% of the world's demand and only about 0.4% of the U.S. demand for freshwater.

There are three major problems with the widespread use of desalination. *First* is the high cost, because it takes a lot of increasingly expensive energy to remove salt from seawater. A *second* problem is that pumping large volumes of seawater through pipes and using chemicals to sterilize the water and keep down algae growth kills many marine organisms and also requires large inputs of energy (and thus money). A *third* problem is that desalination produces huge quantities of salty wastewater that must go somewhere. Dumping it into nearby coastal ocean waters increases the salinity of those waters, which can threaten food resources and aquatic life, especially if it is dumped near coral reefs, marshes, or mangrove forests. Disposing of it on land could contaminate groundwater and surface water (Concept 13-5).

Currently, desalination is practical only for water-short, wealthy countries and cities that can afford its high cost. However, scientists and engineers are working to develop better and more affordable desalination technologies (Science Focus 13.2).

13-6 How Can We Use Freshwater More Sustainably?

CONCEPT 13-6
We can use freshwater more sustainably by cutting water waste, raising water prices, slowing population growth, and protecting aquifers, forests, and other ecosystems that store and release freshwater.

Reducing Freshwater Losses Can Provide Many Benefits

According to water resource expert Mohamed El-Ashry of the World Resources Institute, about 66% of the freshwater used in the world and about 50% of the freshwater used in the United States is lost through evaporation, leaks, and inefficient use. El-Ashry estimates that it is economically and technically feasible to reduce such losses to 15%, thereby meeting most of the world's freshwater needs for the foreseeable future.

So why do we allow for such large losses of freshwater? According to water resource experts, there are two major reasons. First, the *cost of freshwater to users is low.* Such underpricing is mostly the result of government subsidies that provide irrigation water, or the electricity and diesel fuel used by farmers to pump freshwater from rivers and aquifers, at below-market prices. Because these subsidies keep freshwater prices artificially low, due to a failure to apply the full-cost pricing **principle of sustainability** (see Figure 1-5, p. 9 or back cover), users have little or no financial incentive to invest in water-saving technologies. However, farmers, industries, and others who benefit from government freshwater subsidies argue that the subsidies promote the farming of unproductive land, stimulate local economies, and help to keep the prices of food, manufactured goods, and electricity low.

CONSIDER THIS. . .

THINKING ABOUT Government Freshwater Subsidies
Should governments provide subsidies to farmers and cities to help keep the price of freshwater low? Explain.

Higher prices for freshwater encourage water conservation but make it difficult for low-income farmers and city dwellers to buy enough freshwater to meet their needs. When South Africa raised freshwater prices, it dealt with this problem by establishing *lifeline rates,* which give each household a set amount of free or low-priced freshwater to meet basic needs. When users exceed this amount, they pay higher prices as their freshwater use increases. This is a *user-pays* approach.

THE SEARCH FOR IMPROVED DESALINATION TECHNOLOGY

Reverse osmosis is the favored desalination technology because it requires much less energy than distillation, but it is still energy intensive. In this process, high pressure is applied to seawater in order to squeeze the freshwater out of it. The membrane that filters out the salt must be strong enough to withstand such pressure. Seawater must be pretreated and treated again after desalination to make it pure enough for drinking and for irrigating crops.

Much of the scientific research in this field is aimed at improving the membrane and the pre- and post-treatment processes to make desalination more energy efficient. Scientists are working to develop new, more efficient and affordable membranes that can separate freshwater from saltwater under lower pressure, which would require less energy. According to the World Business Council, since 1995, technological advances have brought the cost of desalination down by 80%, but this is still not affordable enough to make it useful for large-scale irrigation or to meet much of the world's demand for drinking water.

Scientists are considering ways to use solar and wind energy—applying one of the three scientific **principles of sustainability**—in combination with conventional power sources to help bring down the cost of desalinating seawater. In 2012, Saudi Arabia completed the world's largest solar-powered desalination project. It uses concentrated solar energy to power new filtration technology at a plant that will meet the daily water needs of 100,000 people.

Two Australian companies, Energetech and H2AU, have joined forces to build an experimental desalination plant that uses the power generated by ocean waves to drive reverse-osmosis desalination. The plant is located on a barge anchored offshore on the Australian coast. This approach produces no air pollution and conserves energy by using it where it is produced to desalinate water. Thus, costs should be much lower than for other desalination methods.

Some scientists argue for building fleets of such floating desalination plants. These plants would operate out of sight from coastal areas and transfer the water to shore through seabed pipelines or in food-grade shuttle tankers. Because of their distance from shore, the ships could draw water from depths below where most marine organisms are found. The resulting brine could be returned to the ocean and diluted far away from coastal waters, thus helping to solve the brine pollution problem caused by onshore desalination.

Another way to save energy in desalination is to build the plants near coastal electric power plants that use seawater for cooling. The heated water from the power plants could be fed into the desalination process, with some of the waste heat from the power plants used to help power the desalination plants. Scientists estimate that this approach could reduce the cost of today's most efficient desalination process by about 17%.

Despite the still high costs of desalination, analysts expect it to be used more widely in the future as water shortages become worse, especially in arid regions of the world. Given the high costs and other obstacles, however, scientists and engineers still have a lot of work to do before desalination will be a major source of freshwater. **GREEN CAREER:** desalination engineer

Critical Thinking

Do you think that improvements in desalination will justify highly inefficient uses of water, such as maintaining swimming pools, fountains, and golf courses in desert areas? Explain.

The second major cause of unnecessary losses of freshwater is a *lack of government subsidies for improving the efficiency of freshwater use.* Withdrawing some of the subsidies that encourage high rates of freshwater use and replacing them with subsidies for more efficient freshwater use would sharply reduce freshwater losses. Understandably, farmers and industries that receive subsidies for freshwater use have vigorously and successfully opposed efforts to eliminate or reduce them.

We Can Improve Efficiency in Irrigation

Since 1980, the amount of food that can be grown per drop of water has roughly doubled. And since the 1970s, the amount of water used per person in the United States has dropped by about 33%, after rising for decades. Most of these water savings have come from improvements to irrigation efficiency in the United States and other more-developed countries.

However, there is still a long way to go, especially in less-developed countries. Only about 60% of the world's irrigation water reaches crops, which means that most irrigation systems are highly inefficient. The most inefficient irrigation system, commonly used in less-developed countries, is *flood irrigation,* in which water is pumped from a groundwater or surface water source through unlined ditches where it flows by gravity to the crops being watered (Figure 13-22, left). This method delivers far more water than is needed for crop growth and typically, about 45% of this water is lost through evaporation, seepage, and runoff.

Another inefficient system is the traditional spray irrigation system, a widely used tool of industrialized crop pro-

Figure 13-22 Traditional irrigation methods rely on gravity and flowing water (left). Newer systems such as center-pivot, low-pressure sprinkler irrigation (right) and drip irrigation (center) are far more efficient.

© Cengage Learning

Gravity flow
(efficiency 60% and 80% with surge valves)

Water usually comes from an aqueduct system or a nearby river.

Drip irrigation
(efficiency 90–95%)

Above- or belowground pipes or tubes deliver water to individual plant roots.

Center pivot
(efficiency 80% with low-pressure sprinkler and 90–95% with LEPA sprinkler)

Water usually pumped from underground and sprayed from mobile boom with sprinklers.

duction (Figure 13-23). It sprays huge volumes of water onto large fields, and as much as 40% of this water is lost to evaporation, especially in dry and windy areas, according to the U.S. Geological Survey. These systems are commonly used in the Midwestern United States and have helped to draw down the Ogallala aquifer (Figure 13-13).

More efficient irrigation technologies greatly reduce water losses by delivering water more precisely to crops— a *more crop per drop* strategy. For example, the *center-pivot, low-pressure sprinkler* (Figure 13-22, right), which uses pumps to spray water on a crop, allows about 80% of the water to reach crops. *Low-energy, precision application (LEPA) sprinklers,* another form of center-pivot irrigation, put 90–95% of the water where crops need it.

Drip, or *trickle irrigation,* also called *microirrigation* (Figure 13-22, center) is the most efficient way to deliver small amounts of water precisely to crops. It consists of a network of perforated plastic tubing installed at or below the ground level. Small pinholes in the tubing deliver drops of water at a slow and steady rate, close to the roots of individual plants. These systems drastically reduce water waste because 90–95% of the water input reaches the crops.

Drip irrigation is used on just over 1% of the world's irrigated crop fields and on only about 4% of those in the United States, largely because most drip irrigation systems are costly. This percentage rises to 13% in the U.S. state of California, 66% in Israel, and 90% in Cyprus. However, if freshwater were priced closer to the value of the ecosystem services it provides and if government subsidies for inefficient use of freshwater were reduced or eliminated, drip irrigation could more easily be used to irrigate most of the world's crops.

According to the United Nations, reducing the current global withdrawal of water for irrigation by just 10%

Tish1/Shutterstock.com

Figure 13-23 Movable spray irrigation systems, widely used to irrigate crops, are inefficient.

Figure 13-24 Solutions: In areas of Bangladesh and India, where water tables are high, many small-scale farmers use treadle pumps to supply irrigation water to their fields.

Solutions

Reducing Irrigation Water Losses

- Avoid growing thirsty crops in dry areas

- Import water-intensive crops and meat

- Encourage organic farming and polyculture to retain soil moisture

- Monitor soil moisture to add water only when necessary

- Expand use of drip irrigation and other efficient methods

- Irrigate at night to reduce evaporation

- Line canals that bring water to irrigation ditches

- Irrigate with treated wastewater

Figure 13-25 There are a number of ways to reduce freshwater losses in irrigation. **Questions:** Which two of these solutions do you think are the best ones? Why?

would save enough water to grow crops and meet the estimated additional water demands of the earth's cities and industries through 2025.

Poor Farmers Conserve Water Using Low-Tech Methods

Many of the world's poor farmers use small-scale, low-cost traditional irrigation technologies that are far more sustainable than most large-scale irrigation systems. For example, millions of farmers in Bangladesh and other countries where water tables are high use human-powered treadle pumps to bring groundwater up to the earth's surface and into irrigation ditches (Figure 13-24). These wooden devices are inexpensive (about $25) and easy to build from local materials. One such pump developed by the nonprofit International Development Enterprises (IDE) uses 60–70% less water than a conventional gravity-flow system to irrigate the same amount of cropland.

Other farmers in some less-developed countries use buckets, small tanks with holes, or simple plastic tubing systems for drip irrigation. The cost of a small-scale drip irrigation system developed by IDE is one-tenth as much per unit of land area as that of conventional drip systems (see Individuals Matter 12.1, p. 303).

Rainwater harvesting is another simple and inexpensive way to provide water and is used throughout much of the developing world. It involves using pipes from rooftops, and channels dug in the ground, to direct rainwater that would otherwise run off the land. It can be stored in underground or aboveground storage tanks (cisterns), ponds, and plastic barrels for use during dry seasons. This is especially useful in countries such as India, where much of the rain comes in a short monsoon season.

Farmers in the parched agricultural state of Punjab in India are using a newer tool called a *tensiometer,* a simple tool that measures soil moisture. In a 2011 study, Columbia University scientists provided these devices to 500 Punjabi farmers and requested that they irrigate their crops only when the tensiometer showed soil moisture content to be falling. Those who followed these instructions used 30% less water than they had before using the tool with the same results. The experiment is now being expanded to include 5,000 more farmers.

Other strategies used by poor farmers to increase the amount of crop per drop of rainfall include polyculture farming to create more canopy cover and reduce evaporative water losses; planting deep-rooted perennial crop varieties (see Figure 12-C, p. 310); controlling weeds; and mulching fields to retain more moisture.

Figure 13-25 summarizes several ways to reduce freshwater losses in crop irrigation. Since 1950, Israel GOOD NEWS has used many of these techniques to slash irrigation water losses by 84% while irrigating 44% more land. Israel now treats and reuses 30% of its municipal sewage water for crop production and plans to increase this to 80% by 2025. The government also gradually eliminated most freshwater subsidies to raise Israel's price of irrigation water, which is now one of the highest in the world.

We Can Cut Freshwater Losses in Industries and Homes

Producers of chemicals, paper, oil, coal, primary metals, and processed foods consume almost 90% of the freshwater used by industries in the United States. Some of these industries recapture, purify, and recycle water GOOD NEWS to reduce their water use and water treatment costs.

Figure 13-26 This yard in Encinitas, a city in a dry area of southern California (USA), uses a mix of plants that are native to the arid environment and require little watering.

Paul Senyszyn/Escaflowne/iStockphoto.com

For example, more than 95% of the freshwater used to make steel can be recycled. Even so, most industrial processes could be redesigned to use much less freshwater. **GREEN CAREER:** water conservation specialist

Flushing toilets with freshwater (most of it clean enough to drink) is the single largest use of domestic freshwater in the United States and accounts for about one-fourth of home water use. Since 1992, U.S. government standards have required that new toilets use no more than 6.1 liters (1.6 gallons) of freshwater per flush. Even at this rate, just two flushes of such a toilet require more than the daily amount of freshwater available for all uses to many of the world's poor living in arid regions (see chapter-opening photo).

Other water-saving appliances are widely available. Low-flow showerheads can save large amounts of freshwater by cutting the flow of a shower in half. Front-loading clothes washers use 30% less water than top-loading machines use. According to the American Water Works Association, if the typical American household were to stop all water leaks and use these devices, along with low-flow toilets and faucets, it could cut its daily water use by nearly a third.

According to UN studies, 30–60% of the freshwater supplied in nearly all of the world's major cities in less-developed countries is lost, primarily through leakage from water mains, pipes, pumps, and valves. Water experts say that fixing these leaks should be a high priority for water-short countries, because it would cost less than building dams or importing freshwater. Even in advanced industrialized countries such as the United States, these losses to leakage average 10–30%. However,

leakage losses have been reduced to about 3% in Copenhagen, Denmark, and to 5% in Fukuoka, Japan.

CONSIDER THIS. . .

CONNECTIONS Water Leaks and Water Bills
Any water leak wastes freshwater and raises water bills. You can detect a silent toilet water leak by adding a few drops of food coloring to the toilet tank and waiting 5 minutes. If the color shows up in the bowl, you have a leak. Also, a faucet leaking one drop per second can drip about 8,200 liters (3,000 gallons) of water down the drain in a year—enough to fill about 75 bathtubs. This also represents money going down the drain.

Many home owners and businesses in water-short areas are using drip irrigation on their properties and copying nature by replacing green lawns with a mix of native plants that need little or no watering (Figure 13-26). Such water-thrifty landscaping saves money by reducing freshwater use by 30–85% and by sharply reducing labor, fertilizer, and fuel requirements. This example of reconciliation ecology (see Chapter 10, p. 242) also helps to preserve biodiversity by providing habitat for threatened honeybee, butterfly, and songbird species—applying the biodiversity **principle of sustainability** (see Figure 1-2, p. 6 or back cover). And it can help landowners to reduce polluted runoff, air pollution, and yard wastes.

However, in some of the more-developed countries, people living in arid areas maintain green lawns by watering them heavily. Some communities in water-short areas have even passed ordinances that require green lawns and prevent people from planting native vegetation in their yards. But this might change if people who live in such areas ever have to pay much higher water bills.

Water used in homes can also be reused and recycled. About 50–75% of a typical household's *gray water*—used water from bathtubs, showers, sinks, dishwashers, and clothes washers—could be stored in a holding tank and then reused to irrigate lawns and nonedible plants, to flush toilets, and to wash cars. Such efforts mimic the way nature recycles water, and thus they follow the chemical cycling **principle of sustainability**.

The relatively low cost of water in most communities is one of the major causes of excessive water use in homes and industries. Many water utility and irrigation authorities charge an annual flat fee for water use, and some charge less to the largest users of water. About one-fifth of all U.S. public water systems do not use water meters and charge a single low annual rate for almost unlimited use of high-quality freshwater.

However, simply monitoring water use can help people to reduce their use and losses of freshwater. When the U.S. city of Boulder, Colorado, introduced water meters, water use per person dropped by 40%. In some cities in Brazil, people obtain freshwater on a pay-as-you-go basis. They buy *smart cards,* each of which contains a certain number of water credits that entitle their owners to measured amounts of freshwater. Brazilian officials say this approach saves water and typically reduces household water bills by 40%.

Some colleges and universities are taking the lead in finding ways to use freshwater more sustainably. What students on these campuses are learning can be applied to homes and businesses of all types and sizes.

CONSIDER THIS. . .

CAMPUS SUSTAINABILITY
University of California–Santa Barbara

At the University of California–Santa Barbara, a team of students and faculty are leading water conservation efforts. They have calculated that student housing is responsible for more than half of the water used on campus, so one effort involves a competition between dorms to reduce water consumption. The school has installed dozens of low-flow water taps and waterless urinals. It is also experimenting with a closed-loop, chilled-water cooling system that has replaced air conditioning units (which lose water to evaporation) in 14 campus buildings. Another important measure has been the use of recycled gray water for irrigating more than 90% of the campus grounds (see photo).

Figure 13-27 lists various ways to use freshwater more efficiently in industries, homes, and businesses (**Concept 13-6**).

We Can Use Less Water to Remove Wastes

Currently, we use large amounts of freshwater clean enough to drink to flush away industrial, animal, and household wastes. According to the Food and Agriculture Organization of the United Nations (FAO), if cur-

---

Solutions

Reducing Water Losses

- Redesign manufacturing processes to use less water
- Recycle water in industry
- Fix water leaks
- Landscape yards with plants that require little water
- Use drip irrigation on gardens and lawns
- Use water-saving showerheads, faucets, appliances, and toilets (or waterless composting toilets)
- Collect and reuse gray water in and around houses, apartments, and office buildings
- Raise water prices and use meters, especially in dry urban areas

Figure 13-27 There are a number of ways to reduce freshwater losses in industries, homes, and businesses (**Concept 13-6**). ***Questions:*** Which three of these solutions do you think are the best ones? Why?

rent growth trends in population and water use continue, within 40 years, we will need the world's entire reliable flow of river water just to dilute and transport the wastes we produce each year.

We could save much of this freshwater by finding different ways to move the wastes. For example, we could recycle and reuse gray water from homes and businesses and wastewater from sewage treatment plants for purposes such as cleaning equipment and watering lawns and some crops. Israel reuses 70% of its wastewater to irrigate nonfood crops. In Singapore, all sewage water is treated at reclamation plants for reuse by industry. U.S. cities such as Las Vegas, Nevada, and Los Angeles, California, are also beginning to clean up and reuse some of their wastewater. However, only about 7% of the water in the United States is reused, a violation of the chemical cycling **principle of sustainability** that we know how to correct.

Another way to keep freshwater out of the waste stream is to rely more on waterless composting toilets. These devices convert human fecal matter to a small amount of dry and odorless soil-like humus material that can be removed from a composting chamber every year or so and returned to the soil as fertilizer. One of the authors (Miller) used a composting toilet for over a decade with no problems, while living and working deep in the woods in an experimental home and office (see p. 1).

We Can Each Help Out in Using Water More Sustainably

More sustainable water use would include a variety of strategies (Figures 13-16, 13-25, and 13-27) aimed not only at conserving water and using it efficiently, but also at protecting water supplies and the ecosystems that sustain them (Concept 13-6). Such strategies would have to be applied at local and regional levels, as well as national and international levels.

However, in order to be successful, such strategies would also have to be applied at the personal level. Each of us can help to bring about a "blue revolution" in which we could reduce our water footprints by using less freshwater and using water much more efficiently (Figure 13-28).

Finding more sustainable ways to use freshwater is the subject of some major research efforts. One group that is working on this problem is the Global Water Policy Project, founded by the renowned water supply expert, and National Geographic Explorer, Sandra Postel (Individuals Matter 13.1).

Figure 13-28 Individuals matter: You can reduce your use and waste of freshwater. **Questions:** Which of these steps have you taken? Which would you like to take?

What Can You Do?

Water Use and Waste

- Use water-saving toilets, showerheads, and faucets
- Take short showers instead of baths
- Turn off sink faucets while brushing teeth, shaving, or washing
- Wash only full loads of clothes or use the lowest possible water-level setting for smaller loads
- Repair water leaks
- Wash your car from a bucket of soapy water, use gray water, and use the hose for rinsing only
- If you use a commercial car wash, try to find one that recycles its water
- Replace your lawn with native plants that need little if any watering
- Water lawns and gardens only in the early morning or evening and use gray water
- Use drip irrigation and mulch for gardens and flowerbeds

© Cengage Learning

13-7 How Can We Reduce the Threat of Flooding?

CONCEPT 13-7

We can lessen the threat of flooding by protecting more wetlands and natural vegetation in watersheds, and by not building in areas subject to frequent flooding.

Some Areas Get Too Much Water from Flooding

Some areas have too little freshwater, but others sometimes have too much because of natural flooding by streams, caused mostly by heavy rain or rapidly melting snow. A flood happens when freshwater in a stream overflows its normal channel and spills into the adjacent area, called the **floodplain**.

People settle on floodplains to take advantage of their many assets, including fertile soil on flat land suitable for crops, ample freshwater for irrigation, and availability of nearby rivers for transportation and recreation. In efforts to reduce the threat of flooding on floodplains, rivers have been narrowed and straightened (or *channelized*), surrounded by protective dikes and *levees* (long mounds of earth along their banks), and dammed to create reservoirs that store and release water as needed (Figure 13-2). However, in the long run, such measures can lead to greatly increased flood damage when heavy snowmelt or prolonged rains overwhelm them.

Floods provide several benefits. They have created some of the world's most productive farmland by depositing nutrient-rich silt on floodplains. They also help recharge groundwater and refill wetlands that are commonly found on floodplains, thereby supporting biodiversity and aquatic ecosystem services.

At the same time, floods kill thousands of people every year and cost tens of billions of dollars in property damage (see the Case Study that follows). Floods are usually considered natural disasters, but since the 1960s, human activities have contributed to a sharp rise in flood deaths and damages, meaning that such disasters are partly human-made.

One such human activity is the *removal of water-absorbing vegetation*, especially on hillsides (Figure 13-29). Once the trees on a hillside have been cut for timber, fuelwood, livestock grazing, or farming, freshwater from precipitation rushes down the denuded slopes, erodes precious topsoil, and can increase flooding and pollution in local streams. Such deforestation can also make landslides and mudflows more likely. A 3,000-year-old Chinese proverb says, "To protect your rivers, protect your mountains."

The second human activity that increases the severity of flooding is the *draining of wetlands* that naturally absorb floodwaters. These areas are then often covered with pavement and buildings that greatly increase runoff. When Hurricane Katrina struck the Gulf Coast of the United States in August 2005 and contributed to the flooding of the city of New Orleans, Louisiana, and

Sandra Postel: National Geographic Fellow and Freshwater Conservationist

©Mark Thiessen/National Geographic Creative

Sandra Postel is regarded as one of the world's most respected authorities on water issues. In 1994 she founded the Global Water Policy Project, which, through research and education, promotes a deeper knowledge of, and more sustainable use of, the earth's finite freshwater supply. Postel has authored or coauthored four influential books, including *Last Oasis: Facing Water Scarcity* and *Rivers for Life: Managing Water for People and Nature.* She has also written scores of articles for popular and scholarly publications.

In a relentless quest to educate people about water supply issues, Postel has appeared on several television news shows, taken part in numerous environmental documentary films (including the BBC's *Planet Earth*), and addressed the European Parliament. In 2010, she was appointed as Freshwater Fellow of the National Geographic Society, where she serves as the lead water expert for the Society's Freshwater Initiative—a global effort to inspire and empower people to conserve freshwater and to preserve aquatic biodiversity. Postel has also served as a consultant to the Nature Conservancy, the World Bank, and the U.S. National Intelligence Council, and as vice president for research at the Worldwatch Institute.

In 2002, Postel was named one of the "Scientific American 50" for her contributions to science and technology. She has been hailed for her "inspiring, innovative, and practical approach" to promoting the preservation and sustainable use of the earth's freshwater supply.

Background photo: Nikitsin/Dreamstime.com

Forested Hillside

Diverse ecological habitat

Evapotranspiration

Trees reduce soil erosion from heavy rain and wind

Agricultural land

Tree roots stabilize soil

Vegetation releases water slowly and reduces flooding

After Deforestation

Tree plantation

Roads destabilize hillsides

Evapotranspiration decreases

Overgrazing accelerates soil erosion by water and wind

Winds remove fragile topsoil

Agricultural land is flooded and silted up

Gullies and landslides

Heavy rain erodes topsoil

Silt from erosion fills rivers and reservoirs

Rapid runoff causes flooding

© Cengage Learning

Animated Figure 13-29 **Natural capital degradation:** A hillside before and after deforestation. ***Question:*** How might a drought in this area make these effects even worse?

surrounding areas, the damage was intensified because of the degradation or removal of coastal wetlands. These wetlands had historically helped to absorb water and buffer this low-lying land from storm surges.

Another human-related factor that will likely increase flooding is a rise in sea levels, projected to occur during this century. This is one likely result of climate change related to human activities that have led to the warming of the atmosphere. Climate change models project that, by 2075, as many as 150 million people living in the world's largest coastal cities—a number roughly equal to half of the current U.S. population—could be flooded out by rising sea levels.

CASE STUDY

Living Dangerously on Floodplains in Bangladesh

Bangladesh is one of the world's most densely populated countries. In 2012, its 153 million people were packed into an area roughly the size of the U.S. state of Wisconsin (which has a population of less than 6 million). And the country's population is projected to increase to 226 million by 2050. Bangladesh is a very flat country, only slightly above sea level, and it is one of the world's poorest countries.

The people of Bangladesh depend on moderate annual flooding during the summer monsoon season to grow rice and help maintain soil fertility in their country's delta basin region, which is fed by numerous river systems. The annual floods also deposit eroded Himalayan soil on the country's crop fields. Bangladeshis have adapted to mod-

erate flooding. Most of the houses have flat thatch roofs on which families can take refuge with their belongings in case of rising waters. The roofs can be detached from the walls, if necessary, and floated like rafts. After the waters have subsided, the roof can be reattached to the walls of the house. However, great floods can overwhelm such defenses.

In the past, great floods occurred every 50 years or so. But between 1987 and 2007 there were five severe floods, each of which covered a third or more of the country with water. Bangladesh's flooding problems begin in the Himalayan watershed, where rapid population growth, deforestation, overgrazing, and unsustainable farming on steep and easily erodible slopes have led to increased flows of water during the summer monsoon season. Monsoon rains now run more quickly off the denuded Himalayan foothills, carrying vital topsoil with them (Figure 13-29, right).

This increased runoff of topsoil, combined with heavier-than-normal monsoon rains, has led to more severe flooding along Himalayan rivers as well as downstream in Bangladesh's delta areas. In 1998, a disastrous flood covered two-thirds of Bangladesh's land area, in some places for 2 months, drowning at least 2,000 people and leaving 30 million homeless. It also destroyed more than one-fourth of the country's crops, which caused thousands of people to die of starvation. In 2002, another flood left 5 million people homeless and flooded large areas of rice fields. Yet another major flood occurred in 2004.

Living at sea level on Bangladesh's coastal floodplain also means coping with storm surges, cyclones, and tsunamis. In 1970, a tropical cyclone killed as many as 1 million people. Another cyclone in 2003 killed more than a million people and left tens of millions homeless.

Many of the coastal mangrove forests in Bangladesh (and elsewhere; see Figure 8-9, p. 172) have been cleared for fuelwood, farming, and aquaculture ponds created for raising shrimp. The result: more severe flooding because these coastal wetlands had helped to shelter Bangladesh's low-lying coastal areas from storm surges, cyclones, and tsunamis. In areas of Bangladesh still protected by mangrove forests, damages and death tolls from cyclones have been much lower than they were in areas where the forests have been cleared.

Projected rises in sea level and storm intensity during this century, primarily due to projected climate change caused by atmospheric warming, will likely be a major threat to Bangladeshis who live on the flat delta adjacent to the Bay of Bengal. This would create millions of environmental refugees with no place to go in this already densely populated country.

Bangladesh is one of the few less-developed nations that is preparing and implementing plans to adapt to projected rising sea levels. This includes using varieties of rice and other crops that can better tolerate flooding, saltwater, and drought; shifting to new crops such as corn; developing small vegetable gardens in bare patches between houses to help reduce Bangladeshi dependence on rice; building small ponds that will collect monsoon rainwater to use for irrigating vegetable gardens during dry periods; and creating a network of earthen embankments that might protect against high tides and moderate storm surges when cyclones strike.

We Can Reduce Flood Risks

Many scientists argue that we could improve flood control by relying less on engineered devices such as dams and levees and more on nature's systems such as wetlands and forests in watersheds.

One engineering approach is the channelizing of streams, which does reduce upstream flooding. However, it also eliminates the aquatic habitats that lie along a meandering stream by taking the water from those systems and sending it in a faster flow straight down a channel. It also reduces groundwater recharge and often leads to downstream flooding.

Similarly, levees or floodwalls along the banks of a river contain and speed up stream flow and can lead to flooding downstream. They also do not protect against unusually high and powerful floodwaters such as those that occurred in 1993 when two-thirds of the levees along the Mississippi River were damaged or destroyed. Similar flooding occurred along the Mississippi in 2011.

Damming, the most common engineering approach, can reduce the threat of flooding by storing water in a reservoir and releasing it gradually. However, dams also have a number of drawbacks (Figure 13-17).

A more ecologically oriented approach to reducing flooding is to *preserve existing wetlands* and *restore degraded wetlands* that lie in floodplains to take advantage of the natural flood control they provide. We would also be wise to sharply reduce emissions of greenhouse gases that contribute to projected atmospheric warming, which will likely raise sea levels and flood many of the world's coastal areas during this century.

On a personal level, we can *think carefully about where we choose to live.* Many poor people live in flood-prone areas because they have nowhere else to go. Most people, however, can choose not to live in areas especially subject to flooding or to water shortages.

Figure 13-30 summarizes these various ways to reduce flooding risks (Concept 13-7).

Solutions

Reducing Flood Damage

Prevention	Control
Preserve forests in watersheds	Strengthen and deepen streams (channelization)
Preserve and restore wetlands on floodplains	
Tax development on floodplains	Build levees or floodwalls along streams
Increase use of floodplains for sustainable agriculture and forestry	Build dams

© Cengage Learning

Figure 13-30 Methods for reducing the harmful effects of flooding (Concept 13-7). **Questions:** Which two of these solutions do you think are the best ones? Why?

Top: allensima/Shutterstock.com. Bottom: ©Zeljko Radojko/Shutterstock.com.

Big Ideas

- One of the major global environmental problems is the growing shortage of freshwater in many parts of the world.

- We can expand water supplies in water-short areas in a number of ways, but the most important ways are to reduce overall water use and to use water much more efficiently.

- We can use water more sustainably by cutting water losses, raising water prices, and protecting aquifers, forests, and other ecosystems that store and release water.

Atm2003/Shutterstock.com

The **Core Case Study** that opens this chapter discusses the problems and tensions that can occur when a large number of U.S. states share a limited river water resource in a water-short region. Such problems are representative of those faced by many other dry regions of the world, especially areas where the population is growing rapidly and water resources are dwindling for various reasons.

Large dams, river diversions, levees, and other big engineering schemes have helped to provide much of the world with electricity, food from irrigated crops, drinking water, and flood control. However, they have also degraded the aquatic natural capital necessary for long-term economic and ecological sustainability by seriously disrupting rivers, streams, wetlands, aquifers, and other aquatic systems.

The three scientific **principles of sustainability** (see Figure 1-2, p. 6 or back cover) can guide us in using water more sustainably during this century. Scientists hope to use solar energy to desalinate water and expand freshwater supplies. Recycling more water will help us to reduce water losses. Preserving biodiversity by avoiding disruption of aquatic systems and their bordering terrestrial systems is a key factor in maintaining water supplies and water quality.

SUSTAINABILITY

Chapter Review

Core Case Study

1. Summarize the importance of the Colorado River basin in the United States and how human activities are stressing this system (**Core Case Study**). Define **drought** and explain how it has affected the Colorado River system.

Section 11-1

2. What are the two key concepts for this section? Define **freshwater**. Explain why access to water is a health issue, an economic issue, a national and global security issue, and an environmental issue. What percentage of the earth's freshwater is available to us? Describe how water is recycled by the hydrologic cycle and how human activities can interfere with this cycle. Define **groundwater**, **zone of saturation**, **water table**, and **aquifer**, and explain how aquifers are recharged. Define and distinguish between **surface water** and **surface runoff**. What is a **watershed (drainage basin)**?

3. What is **reliable surface runoff**? What percentage of the world's reliable runoff are we using and what percentage are we likely to be using by 2025? How is most of the world's water used? Define and give an example of a **water footprint**. Define **virtual water** and give three examples of it. Describe the availability

and use of freshwater resources in the United States and the water shortages that could occur during this century. What are four major problems resulting from the way people are using water from the Colorado River basin?

4. How many countries face water scarcity today and how many could face water scarcity by 2050? What percentage of the earth's land suffers from severe drought today and how might this change by 2059? How many people in the world lack regular access to clean water today and how high might this number grow by 2025? Why do many analysts view the likelihood of steadily worsening water shortages as one of the world's most serious environmental problems? Explain the connection between water shortages and hunger. How do scientists use satellites to measure changes in water supplies?

Section 11-2

5. What is the key concept for this section? What are the advantages and disadvantages of withdrawing groundwater? Summarize the problem of groundwater depletion in the world and in the United States, especially in the Ogallala aquifer. List two problems that result from the overpumping of aquifers. Describe the potential for using deep aquifers and list the four main concerns about using this resource.

Section 11-3

6. What is the key concept for this section? What is a **dam**? What is a **reservoir**? What are the advantages and disadvantages of using large dams and reservoirs? What has happened to water flows in the Colorado River (**Core Case Study**) since 1960? Explain how the damming of this river has affected its delta. What other problems are likely to further shrink this supply of water? List three possible solutions to the supply problems in the Colorado River basin.

Section 11-4

7. What is the key concept for this section? What is a water transfer? Describe two large water-transfer programs, explain how they came about, and summarize the controversy around each of these programs. Summarize the story of the Aral Sea water-transfer project and its disastrous consequences.

Section 11-5

8. What is the key concept for this section? Define **desalination** and list three major limitations on the widespread use of desalination. What are scientists doing to try to deal with these problems?

Section 11-6

9. What is the key concept for this section? What percentage of available freshwater is lost through inef-ficient use and other causes, in the world and in the United States? What are two major reasons for those losses? Describe three major irrigation methods and list ways to reduce water losses in irrigation. What are three ways in which people in less-developed countries conserve water? List four ways to reduce water waste in industries and homes. How can we use less water to remove wastes? List four ways in which you can reduce your use and waste of water.

Section 11-7

10. What is the key concept for this section? What is a **floodplain** and why do people like to live on floodplains? What are the benefits and harms of flooding? List two human activities that increase the risk of flooding. Describe the flooding risks that many people in Bangladesh face and what they are doing about it. List and compare two engineering approaches to flood control and two ecologically oriented approaches. What are this chapter's *three big ideas*? Explain how the scientific **principles of sustainability** (see Figure 1-2, p. 6 or back cover) can guide us in using water more sustainably during this century.

Note: Key terms are in bold type.

Critical Thinking

1. What do you believe are the three most important priorities for dealing with the water resource problems of the Colorado River basin, as discussed in the **Core Case Study** that opens this chapter? Explain your choices.

2. List three ways in which human activities are affecting the water cycle. How might these changes to the water cycle affect your lifestyle? How might your lifestyle be contributing to these effects?

3. Explain how our current use of the earth's water resources can be viewed as a good example of the tragedy of the commons (see Chapter 1, p. 12).

4. What role does population growth play in water supply problems? Relate this to water supply problems of the Colorado River basin (**Core Case Study**).

5. Explain why you are for or against **(a)** raising the price of water while providing lower lifeline rates for poor consumers, **(b)** withdrawing government subsidies that provide farmers with water at low cost, and **(c)** providing government subsidies to farmers for improving irrigation efficiency.

6. Calculate how many liters (and gallons) of water are lost in 1 month by a toilet or faucet that leaks 2 drops of water per second. (One liter of water equals about 3,500 drops and 1 liter equals 0.265 gallon.) How many bathtubs (each containing about 151 liters or 40 gallons) could be filled with this lost water?

7. List the three most important ways in which you could use water more efficiently. Which, if any, of these measures do you already take?

8. List three ways in which human activities increase the harmful effects of flooding. What is the best way to prevent each of these human impacts? Do you think they should be prevented? Why or why not?

Doing Environmental Science

Investigate water use at your school. Try to determine all specific sources of any water losses, taking careful notes and measurements for each of them, and estimate how much water is lost per hour, per day, and per year from each source. Sum these estimated amounts to arrive at an estimate of total water losses for a year at your school. Develop a water conservation plan for your school and submit it to school officials.

Global Environment Watch Exercise

Search for *Ogallala* and research the overpumping of this aquifer. Plot the decline of this aquifer on a graph, and write a report explaining the causes and possible ways to slow the decline of the Ogallala aquifer.

Ecological Footprint Analysis

In 2005 (latest available data), the population of the U.S. state of Florida consumed 24.5 billion liters (6.5 billion gallons) of freshwater daily. It is projected that in 2025, the daily consumption will increase to 32.1 billion liters (8.5 billion gallons) per day. Between 2005 and 2025, the population of Florida is projected to increase from 17.5 million to 25.9 million.

1. Based on total freshwater use:
   a. Calculate the per capita consumption of water per day in Florida in 2005 and the projected per capita consumption per day for 2025.

   b. Calculate the per capita consumption of water for the year in Florida in 2005 and the projected per capita consumption per year for 2025.

2. In 2005, how did the Florida *average water footprint* (consumption per person per year), based only on water used within the state, compare with the average U.S. water footprint of approximately 249,000 liters (66,000 gallons) per person per year, and with the global average water footprint of 123,770 liters (32,800 gallons) per person per year?

1. Approximately what percentage of the water on the earth is saltwater?
   a. 97%
   b. 93%
   c. 75%
   d. 60%
   e. 45%

2. A drought in the 1930s transformed large parts of the Midwest into a "Dustbowl." This is an example of
   a. what would happen if the giant toad or feral pigs are not controlled.
   b. the results of many inches of rain falling within a few hours.
   c. the conversion of rangeland or cropland into a desertlike area.
   d. evaporation of water from the ground and release of moisture from leaves.
   e. water flowing over the surface when it is unable to infiltrate the ground because of saturation of the soil or a steep slope.

3. One way to conserve water in agriculture is to
   a. plant crops that have adapted to arid climates by sending down deep tap roots and storing water in times of drought.
   b. use a technique similar to the ancient Egyptians', whereby they relied on a river to regularly overflow its banks.
   c. plant crops in places with less sunlight so there are fewer hours during the day for evaporation to occur.
   d. channel gray water from homes and industry to agricultural areas.
   e. use a watering technique that focuses the water in the immediate area of plant roots.

4. The Ogallala Aquifer is a one-time deposit of water and has such a slow recharge rate it is
   a. huge and can accommodate the water needs of the Great Plains for at least several more decades.
   b. not replenishing at the rate it is being used, and one result is land subsidence.
   c. protected by state laws and is therefore underutilized.
   d. suffering from severe salinization as water infiltrates the ground and dissolved minerals are added to the aquifer.
   e. increasing each year with the expectation that the rate of increase will grow as the climate changes and brings more storms to the Great Plains.

5. The Colorado River project is
   a. extremely eutrophic because as water slows down behind the dams, there are more opportunities for nitrates and phosphates to enter the river water.
   b. an underutilized resource that could easily provide water to farmers, ranchers, and cities.
   c. a government-subsidized water utility that provides cities such as Las Vegas, Los Angeles, and San Diego with cheaper water because Lake Powell and Lake Mead are huge bodies of water.
   d. a water diversion project that supplies water to large cities such as Las Vegas, Los Angeles, and San Diego.
   e. a confined aquifer that is being overpumped and polluted because of increasing populations in the cities it serves.

Questions 6–9 refer to the following terms.
   a. Aquifer
   b. Surface water
   c. Confined aquifer
   d. Water table
   e. Transpiration

6. The freshwater from precipitation and snowmelt

7. Evaporation from the leaves of plants into the atmosphere

8. Underground caverns and porous layers of sand, gravel, or bedrock through which groundwater flows

9. Upper surface of the zone of saturation

10. A forested hillside will help eliminate
   a. the need for research such as that done at the Hubbard Brook Experimental Forest in New Hampshire.
   b. the use of pesticides, because the shade of the forest prevents insect pests from entering the area.
   c. droughts, because transpiration ensures that forest areas always have a high moisture level.
   d. flooding, because trees help reduce soil erosion and the flow of water downslope.
   e. unsustainable farming, because crops cannot be grown on forested slopes.

11. One method of desalination is
    a. forcing the water through a filter, which removes salts from the water.
    b. taking more water from the aquifer than is being replaced, so less salts and minerals are entering the water.
    c. saturating the groundwater so the water table rises, and thus decreasing the ratio of salts within the groundwater.
    d. the use of hydroponic agriculture so water and salts are not added to the ground during irrigation.
    e. planting only drought-resistant vegetation, reducing the amount of water needed in home gardens and in agriculture.

Use the graph below to answer questions 12–14.

12. According to the graph, the flow of the Colorado River has
    a. dropped continually since 1910.
    b. gone up and down but continues to trend downward.
    c. stayed steady over the last 10 years.
    d. dropped steadily from 1935 until 1945.
    e. slowly increased in the last 10 years.

13. What can be inferred about the climate from 1960 to 1980?
    a. There was excessive rainfall.
    b. Rainfall led to variation in flow.
    c. There were average rates of precipitation.
    d. There was a drought.
    e. Windstorms deposited large amounts of sediment in the reservoir.

14. Which of the following best explains the general downward trend in Colorado River flow from 1910 until 1970?
    a. More water is being removed from the Colorado River before it reaches any of the dams.
    b. A series of heavy winter snowstorms have changed the water level of the Colorado River.
    c. Residents are using more water-saving devices in their homes.
    d. The dams on the Colorado River divert water to cities and for agriculture.
    e. The underground aquifer is being depleted as more farmers use its water for agricultural irrigation.

14 Nonrenewable Mineral Resources

Civilization exists by geological consent, subject to change without notice.

WILL DURANT

Key Questions

14-1 What are the earth's major geological processes and what are mineral resources?

14-2 How long might supplies of nonrenewable mineral resources last?

14-3 What are the environmental effects of using nonrenewable mineral resources?

14-4 How can we use mineral resources more sustainably?

14-5 What are the earth's major geological hazards?

Climax Molybdenum mine site in the Rocky Mountains near Leadville, Colorado.

Pete Mcbride/National Geographic Creative

Mineral resources extracted from the earth's crust are processed into an amazing variety of products that can make our lives easier and provide us with economic benefits and jobs. Most people are aware of our use of mineral resources such as gold, copper, aluminum, sand, and gravel. But fewer people have heard of *rare-earth metals and oxides* that are crucial to the major technologies that support today's lifestyles and economies.

The 17 rare-earth metals, also called *rare earths,* include scandium, yttrium, and 15 lanthanide chemical elements (see periodic table, Figure 1, p. S12, Supplement 4) such as lanthanum (Figure 14-1). Because of their superior magnetic strength and other unique properties, these elements and their compounds are very important for a number of widely used technologies.

Rare earths are used to make liquid crystal display flat screens for computers and television sets, energy-efficient compact fluorescent and LED lightbulbs, solar cells, fiber-optic cables, cell phones, and

digital cameras. They are also important in the manufacture of batteries and motors for electric and hybrid-electric cars (Figure 14-2), catalytic converters in car exhaust systems, jet engines, and the powerful magnets in wind turbine generators. Rare earths also go into missile guidance systems, smart bombs, aircraft electronics, and satellites.

Without affordable supplies of these metals, industrialized nations could not develop the current versions of cleaner energy technology and other high-tech products that will be major sources of economic growth and profits during this century. Nations also need these metals to maintain their military strength.

In this chapter we look at the nature and supplies of mineral resources, the rock cycle that produces them, the harmful environmental impacts of their use, and how we can use these resources more sustainably. We also look at the earth's natural geological processes and major geological hazards.

©Jurii

Figure 14-1 The highly reactive rare-earth metal lanthanum (La) and its compounds are used in a variety of products, including batteries in electric and hybrid-electric cars (Figure 14-2) and lenses in cameras and video cameras.

Catalytic converter
• Cerium
• Lanthanum

Battery
• Lanthanum
• Cerium

LCD screen
• Europium
• Yttrium
• Cerium

Electric motors and generator
• Dysprosium
• Neodymium
• Praseodymium
• Terbium

Figure 14-2 Manufacturers of all-electric and hybrid-electric cars use a variety of rare-earth metals.

© Cengage Learning

14-1 What Are the Earth's Major Geological Processes and What Are Mineral Resources?

CONCEPT 14-1A
Dynamic processes within the earth and on its surface produce the mineral resources we depend on.

CONCEPT 14-1B
Mineral resources are nonrenewable because they are produced and renewed over millions of years mostly by the earth's rock cycle.

The Earth Is a Dynamic Planet

Geology is the science devoted to the study of dynamic processes taking place on the earth's surface and in its interior. As the primitive earth cooled over eons, its inte-

rior separated into three major concentric zones: the *core,* the *mantle,* and the *crust* (Figure 14-3).

The **core** is the earth's innermost zone. It is extremely hot and has a solid inner part, surrounded by a thick layer of *molten rock,* or hot liquid rock, and semisolid material. Surrounding the core is a thick zone called the **mantle**—a zone made mostly of solid rock that can be soft and pliable at very high temperatures. The outermost part of the mantle is solid rock, and under that part is a zone called the **asthenosphere**—a volume of hot, partly melted rock that flows. Tremendous heat within the core and mantle generates *convection cells* or *currents* that slowly move large volumes of rock and heat in loops within the mantle like gigantic conveyor belts (Figure 14-3). Some of the molten

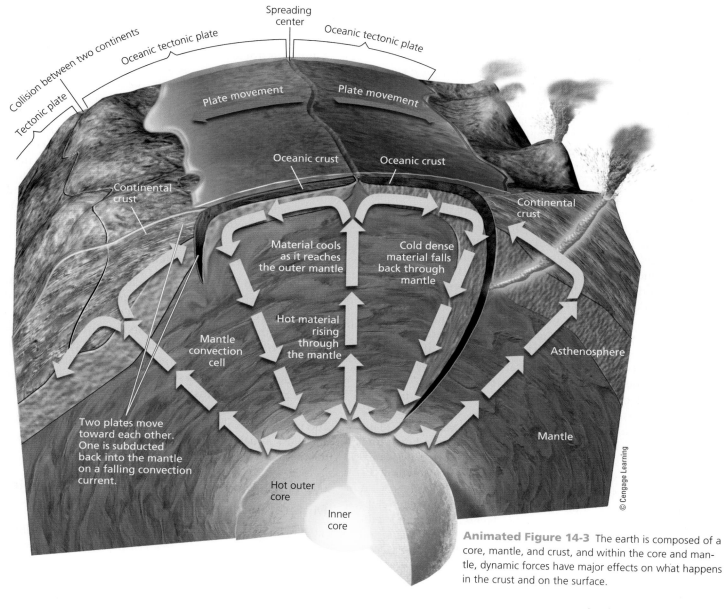

Animated Figure 14-3 The earth is composed of a core, mantle, and crust, and within the core and mantle, dynamic forces have major effects on what happens in the crust and on the surface.

© Cengage Learning

rock in the asthenosphere flows upward into the crust, where it is referred to as *magma*.

The outermost and thinnest zone of solid material is the earth's **crust**. It consists of the *continental crust*, which underlies the continents (including the continental shelves extending into the oceans), and the *oceanic crust*, which underlies the ocean basins and makes up 71% of the earth's crust. The combination of the crust and the rigid, outermost part of the mantle is called the **lithosphere**. It is in this zone that we find the mineral resources on which we depend (Concept 14-1A).

What Are Minerals and Rocks?

The earth's crust beneath our feet consists mostly of minerals and rocks. A **mineral** is a naturally occurring chemical element or inorganic compound that exists as a solid with a regularly repeating internal arrangement of its atoms or ions (a *crystalline solid*). A **mineral resource** is a concentration of one or more minerals in the earth's crust that we can extract and process into raw materials and useful products at an affordable cost. Because minerals take millions of years to form, they are *nonrenewable resources,* and their supplies can be depleted (Concept 14-1B).

A few minerals consist of a single chemical element. They include mercury (see Figure 2-3, left, p. 36), gold (see Figure 2-3, right), and rare-earth metals (**Core Case Study**) such as lanthanum (Figure 14-1). However, most of the more than 2,000 identified mineral resources that we use occur as inorganic compounds formed by various combinations of elements. Examples include salt (sodium chloride, or NaCl; see Figure 2-6, p. 38), quartz (silicon dioxide, or SiO_2), and rare-earth oxides (formed when rare-earth metals combine with oxygen) such as lanthanum oxide (La_2O_3, Figure 14-4).

Rock is a solid combination of one or more minerals found in the earth's crust. Some kinds of rock such as limestone (calcium carbonate, or $CaCO_3$) and quartzite (silicon dioxide, or SiO_2) contain only one mineral. But most rocks consist of two or more minerals. For example, granite is a mixture of mica, feldspar, and quartz crystals. Deposits of rare earths (**Core Case Study**) typically contain a mixture of the metals (Figure 14-1) and their oxides (Figure 14-4) that are difficult and costly to separate from one another and to purify to an acceptable level.

Based on the way it forms, rock is placed in three broad classes: sedimentary, igneous, or metamorphic. **Sedimentary rock** is made of *sediments*—dead plant and animal remains and tiny particles of weathered and eroded rocks. These sediments are transported by water, wind, or gravity to downstream, downwind, downhill, or underwater sites. There they are deposited in layers that accumulate over time. Eventually, the increasing weight and pressure on the underlying layers convert the sedimentary layers into rock. Examples include *sandstone* and *shale* (formed from pressure created by deposited layers made primarily of sand), *dolomite* and *limestone* (formed

Figure 14-4 Oxides of rare-earth metals, such as lanthanum oxide shown here, are often converted to rare-earth metals and used in a variety of products and technologies.

from the compacted shells, skeletons, and other remains of dead aquatic organisms), and *lignite* and *bituminous coal* (derived from compacted plant remains).

Igneous rock forms below or on the earth's surface under intense heat and pressure when magma wells up from the earth's mantle and then cools and hardens. Examples include *granite* (formed underground) and *lava rock* (formed aboveground). Igneous rock forms the bulk of the earth's crust but is usually covered by sedimentary rock.

Metamorphic rock forms when an existing rock is subjected to high temperatures (which may cause it to melt partially), high pressures, chemically active fluids, or a combination of these agents. Examples include *slate* (formed when shale and mudstone are heated) and *marble* (produced when limestone is exposed to heat and pressure).

Earth's Rocks Are Recycled Very Slowly

The interaction of physical and chemical processes that change the earth's rocks from one type to another is called the **rock cycle** (Figure 14-5 and Concept 14-1B). Rocks are recycled over millions of years by three processes—*erosion, melting,* and *metamorphism*—which produce *sedimentary, igneous,* and *metamorphic* rocks, respectively.

In these processes, rocks are broken down, melted, fused together into new forms by heat and pressure, cooled, and sometimes recrystallized within the earth's interior and crust. The rock cycle is the slowest of the earth's cyclic processes and plays a major role in forming concentrated deposits of the nonrenewable mineral resources that we use.

We Depend on a Variety of Nonrenewable Mineral Resources

We know how to find and extract more than 100 different minerals from the earth's crust. They include *metallic minerals* made of single elements, such as aluminum,

Figure 14-5 Natural capital: The rock cycle is the slowest of the earth's cyclical processes.

Left: Dwight Smith/Shutterstock.com. Center: LesPalenik/Shutterstock.com. Right: ©Bragin Alexey/Shutterstock.com.

Erosion

Transportation

Deposition

Weathering

Sedimentary rock
Sandstone, limestone

Igneous rock
Granite, pumice, basalt

© Cengage Learning

Heat, pressure

Heat, pressure, stress

Cooling

Magma (molten rock)

Metamorphic rock
Slate, marble, gneiss, quartzite

Melting

gold, and the rare earths (**Core Case Study**), and those made of compounds such as rare-earth oxides (Figure 14-4). Other mineral resources, including sand and limestone, are *nonmetallic minerals* made of nonmetallic elements and compounds.

An **ore** is rock that contains a large enough concentration of a particular mineral—often a metal—to make it profitable for mining and processing. A **high-grade ore** contains a large concentration of the desired mineral, whereas a **low-grade ore** contains a smaller concentration.

Nonrenewable mineral resources are used for many purposes. Today about 60 of the chemical elements in the periodic table (see Figure 1, p. S12, in Supplement 4) are used for making computer chips. *Aluminum* (Al) is used as a structural material in beverage cans, motor vehicles, aircraft, and buildings. *Steel,* an essential material used in buildings, machinery, and motor vehicles, is a mixture (alloy) of iron (Fe) and other elements that gives it certain physical properties. *Manganese* (Mn), *cobalt* (Co), and *chromium* (Cr) are widely used in steel alloys. *Copper* (Cu), a good conductor of electricity, is used to make electrical and communications wiring and plumbing pipes. *Gold* (Au) is a component of electrical equipment, tooth fillings, jewelry, coins, and some medical implants. Molybdenum (Mo) (see chapter-opening photo) is widely used to harden steel and to make it more resistant to corrosion. And many of today's important new technologies make use of various rare-earth elements (**Core Case Study**).

There are several widely used nonmetallic mineral resources. *Sand,* which is mostly silicon dioxide (SiO_2), is used to make glass, bricks, and concrete for the construction of roads and buildings. *Gravel* is used for roadbeds and to make concrete. Another common nonmetallic mineral is *limestone* (mostly calcium carbonate, or $CaCO_3$), which is crushed to make concrete and cement. Still another is *phosphate salts,* used to make inorganic fertilizers and certain detergents.

14-2 How Long Might Supplies of Nonrenewable Mineral Resources Last?

CONCEPT 14-2A
Nonrenewable mineral resources exist in finite amounts and can become economically depleted when it costs more than it is worth to find, extract, and process the remaining deposits.

CONCEPT 14-2B
There are several ways to extend supplies of mineral resources, but each of them is limited by economic and environmental factors.

Supplies of Nonrenewable Mineral Resources Can Be Economically Depleted

Most published estimates of the supply of a given nonrenewable mineral resource refer to its **reserves**: identified deposits from which we can extract the mineral profitably at current prices. Reserves can be expanded when we find new, profitable deposits or when higher prices or improved mining technologies make it profitable to extract deposits that previously were considered too expensive to remove.

The future supply of any nonrenewable mineral resource depends on the actual or potential supply of the mineral and the rate at which we use it. We have never completely run out of a nonrenewable mineral resource, but a mineral becomes *economically depleted* when it costs more than it is worth to find, extract, transport, and process the remaining deposits (Concept 14-2A). At that point, there are five choices: *recycle or reuse existing supplies, waste less, use less, find a substitute,* or *do without.*

Depletion time is the time it takes to use up a certain proportion—usually 80%—of the reserves of a mineral at a given rate of use. When experts disagree about depletion times, it is often because they are using different assumptions about supplies and rates of use (Figure 14-6).

The shortest depletion-time estimate assumes no recycling or reuse and no increase in reserves (curve A, Figure 14-6). A longer depletion-time estimate assumes that recycling will stretch existing reserves and that better mining technology, higher prices, or new discoveries will increase reserves (curve B). The longest depletion-time estimate (curve C) makes the same assumptions as A and B, but also includes reuse and reduced consumption to further expand reserves. Finding a substitute for a resource leads to a new set of depletion curves for the new resource.

The earth's crust contains fairly abundant deposits of nonrenewable mineral resources such as iron and aluminum. But concentrated deposits of important mineral resources such as manganese, chromium, cobalt, platinum, and some of the rare earths (**Core Case Study**) are relatively scarce. In addition, deposits of many mineral resources are not distributed evenly among countries.

Five nations—the United States, Canada, Russia, South Africa, and Australia—supply most of the nonrenewable mineral resources used by modern societies. South Africa, for example, is the world's largest producer of chromium and platinum.

Since 1900, and especially since 1950, there has been a sharp rise in the total and per capita use of mineral resources in the United States. According to the U.S. Geological Survey, each person in the United States uses an average of 22 metric tons (24 tons) of mineral resources per year. As a result, the United States has economically depleted some of its once-rich deposits of metals such as lead, aluminum, and iron. Currently, the United States imports all of its supplies of 20 key nonrenewable mineral resources and more than 90% of its supplies of 4 other key minerals.

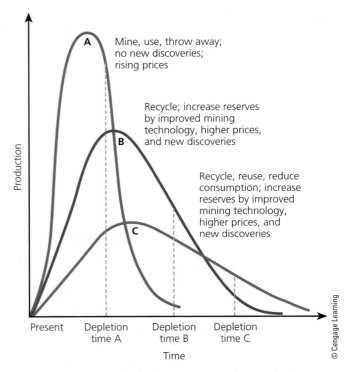

Figure 14-6 Natural capital depletion: Each of these *depletion curves* for a mineral resource (such as aluminum, copper, or a rare earth) is based on a different set of assumptions. Dashed vertical lines represent the times at which 80% depletion occurs.

Most U.S. imports of metallic mineral resources come from reliable and politically stable countries. But there are serious concerns about access to adequate supplies of four *strategic metal resources*—manganese, cobalt, chromium, and platinum—which are essential for the country's economic and military strength. The United States has little or no reserves of these metals.

Global and U.S. Rare-Earth Supplies

Rare-earth elements (**Core Case Study**) are not actually rare, for the most part, but they are hard to find in concentrations high enough to extract and process at an affordable price. There is no global shortage of rare-earth minerals. According to the U.S. Geological Survey, in 2010 China had roughly 50% of the world's known rare-earth reserves, Russia had 15%, and the United States 13%.

Currently, China produces about 97% of the world's rare-earth metals and oxides. This is partly because China does not strictly regulate the environmentally disruptive mining and processing of rare earths. This means that Chinese companies have lower costs than do companies in countries that seek to protect their environments from the disruptions of rare-earth mining and processing. China can thus afford to sell its rare earths at a lower price.

The United States and Japan, which are also heavily dependent on rare earths and their oxides, produce none. Japan has no rare-earth reserves, and U.S. production

Figure 14-7 Improved mining technology has expanded the mining of taconite iron ore in northern Minnesota.

stopped in 2002. A large rare-earth mine in California—the only one in the United States—used to be the world's largest supplier of rare-earth metals. However, it closed down because of the expense of meeting pollution regulations, and because China had driven the prices of rare-earth metals down to a point where the mine was too costly to operate. There are plans to reopen the California mine in the near future, if it can meet government environmental standards at an affordable cost. However, it contains mostly lighter rare earths (with lower atomic numbers), which are more abundant and thus are not as valuable. Meanwhile, the lack of production in the United States makes the country vulnerable to supply interruptions.

Since 2010, China has been reducing its exports of rare-earth metals and their oxides to other countries and has sharply raised its prices on such exports. China's growing use of these resources means that it will use most of the rare earths it produces and may soon need to import some types of rare earths from other countries to meet its industrial needs.

Market Prices Affect Supplies of Nonrenewable Mineral Resources

Geological processes determine the quantity and location of a nonrenewable mineral resource in the earth's crust, but economics determines what part of the known supply is extracted and used. An increase in the price of a scarce mineral resource can often lead to increased supplies and can encourage more efficient use, but there are limits to this effect (**Concept 14-2B**).

According to standard economic theory, in a competitive market system when a resource becomes scarce, its price rises. This can encourage exploration for new deposits, stimulate development of better mining technology, and make it profitable to mine lower-grade ores. It can also encourage a search for substitutes and promote resource conservation. For example, shortly after World War II, high-grade iron ore from rich deposits in northern Minnesota was economically depleted. But by the 1960s, a new process had been developed for mining *taconite* (Figure 14-7), a low-grade and plentiful ore that had been viewed as waste rock in the iron mining process. This improvement in mining technology for iron expanded the reserves in Minnesota and supports a taconite mining industry today.

CONSIDER THIS. . .

CONNECTIONS Metal Prices and Thievery

Resource scarcity can also promote thievery. For example, because of increasing demand, copper prices have risen sharply in recent years. As a result, in several U.S. communities, people have been stealing copper to sell it—stripping abandoned houses of copper pipe and wiring and stealing outdoor central air conditioning units for their copper coils. They have also stolen electrical wiring from beneath city streets and copper piping from farm irrigation systems. In one Oklahoma town, someone cut down several utility poles and stole the copper electrical wiring, causing a blackout.

According to some economists, this price effect may no longer apply very well in most of the more-developed countries. Governments in such countries often use subsidies, tax breaks, and import tariffs to control the supply, demand, and prices of key mineral resources to such an extent that a truly competitive free market does not exist. In the United States, for instance, mining companies get various types of government subsidies, including *depletion allowances*—permission to deduct from their taxable incomes the costs of developing and extracting mineral resources. These allowances amount to 5–22% of their gross income gained from selling the mineral resources.

Mining company representatives insist that they need taxpayer subsidies and low taxes to keep the prices of minerals low for consumers. They also claim that, without the subsidies, their companies might move their operations to other countries where they would not have to pay taxes or comply with strict mining and pollution control regulations.

Can We Expand Reserves by Mining Lower-Grade Ores?

Some analysts contend that we can increase supplies of some minerals by extracting them from lower-grade ores. They point to the development of new earth-moving equipment, improved techniques for removing impurities from ores, and other technological advances in mineral extraction and processing that can make lower-grade ores accessible, sometimes at lower costs. For example, in 1900, the copper ore mined in the United States was typically about 5% copper by weight. Today, it is typically about 0.5%, yet copper costs less (when prices are adjusted for inflation).

However, several factors can limit the mining of lower-grade ores (Concept 14-4B). For one, it requires mining and processing larger volumes of ore, which takes more energy and costs more. Another factor is the dwindling supplies of freshwater needed for the mining and processing of some minerals, especially in arid and semiarid areas. A third limiting factor is the growing environmental impacts of land disruption, along with waste material and pollution produced during mining and processing (see chapter-opening photo). We discuss this further in the next section of this chapter.

One way to improve mining technology and reduce its environmental impact is to use a biological approach, sometimes called *biomining*. Miners use natural or genetically engineered bacteria to remove desired metals from ores through wells bored into the deposits. This leaves the surrounding environment undisturbed and reduces the air and water pollution associated with removing the metal from metal ores. On the down side, biomining is slow. It can take decades to remove the same amount of material that conventional methods can remove within months or years. So far, biomining methods are economically feasible only with low-grade ores for which conventional techniques are too expensive.

Can We Get More Minerals from the Oceans?

Most of the chemical elements and compounds found in seawater occur in such low concentrations that recovering these mineral resources takes more energy and money than they are worth. Currently, only magnesium, bromine, and sodium chloride are abundant enough to be extracted profitably from seawater. On the other hand, in sediments along the shallow continental shelf and adjacent shorelines, there are significant deposits of minerals such as sand, gravel, phosphates, copper, iron, tungsten, silver, titanium, platinum, and diamonds.

Another potential ocean source of some minerals is *hydrothermal ore deposits* that form when superheated, mineral-rich water shoots out of vents in volcanic regions of the ocean floor. As the hot water comes into contact with cold seawater, black particles of various metal sulfides precipitate out and accumulate as chimney-like structures, called *black smokers*, near the hot water vents (Figure 14-8). These deposits are especially rich in minerals such as copper, lead, zinc, silver, gold, and some of the rare-earth metals. A variety of exotic forms of life—including giant clams, six-foot tubeworms, and eyeless shrimp—exist in the dark depths around these black smokers. These life-forms are supported by bacteria that produce food by chemosynthesis from sulfur compounds discharged by the vents.

Because of the rapidly rising prices of many of these metals, there is growing interest in deep-sea mining. In 2011, China began using remote-controlled underwater equipment and a manned deep-sea craft to evaluate the mining of mineral deposits around black smokers in the Indian Ocean near Madagascar as well as in other deep-sea areas. Also that year, Japanese explorers found large deposits of rare-earth minerals (Core Case Study) on the floor of the Pacific Ocean. The UN International Seabed Authority, established to manage seafloor mining in international waters, began issuing mining permits in 2011.

Some analysts say that seafloor mining is less environmentally harmful than mining on land. However, marine biologists are concerned that the sediment stirred up by such mining could harm or kill organisms that feed by filtering seawater. Proponents of the mining say that the

Figure 14-8 Natural capital: *Hydrothermal deposits*, or *black smokers*, are rich in various minerals.

number of potential mining sites is quite small and that many of these organisms can live elsewhere.

Another possible source of metals from the ocean is the potato-size *manganese nodules* that cover large areas of the Pacific Ocean floor and smaller areas of the Atlantic and Indian Ocean floors. They also contain some low concentrations of various rare-earth minerals (**Core Case Study**).

These modules could be sucked up by giant vacuum pipes or scooped up by underwater mining machines.

So far, obtaining minerals from the ocean bottom has been hindered by the high costs involved, the potential threat to marine ecosystems, and arguments over rights to these minerals in deep-ocean areas that belong to no one country.

14-3 What Are the Environmental Effects of Using Nonrenewable Mineral Resources?

CONCEPT 14-3

Extracting minerals from the earth's crust and converting them into useful products can disturb the land, erode soils, produce large amounts of solid waste, and pollute the air, water, and soil.

Mineral Use Creates Environmental Impacts

Every metal product has a *life cycle* that includes mining the mineral, processing it, manufacturing the product, and disposal or recycling of the product (Figure 14-9). This process makes use of large amounts of energy and water, and results in pollution and waste (Concept 14-3).

The environmental impacts of mining a metal ore are determined partly by the ore's percentage of metal content, or *grade*. The more accessible higher-grade ores are usually exploited first. Mining lower-grade ores takes more money, energy, water, and other resources, and leads to more land disruption, mining waste, and pollution.

Removing Mineral Deposits Has Harmful Environmental Effects

Several different mining techniques are used to remove mineral deposits. Shallow mineral deposits are removed by **surface mining**, in which vegetation, soil, and rock overlying a mineral deposit are cleared away. The soil and rock, called **overburden**, are usually deposited in piles of waste material called **spoils** (Figure 14-10). Surface mining is used to extract about 90% of the non-fuel mineral resources and 60% of the coal used in the United States.

The type of surface mining used depends on two factors: the resource being sought and the local topography. In **open-pit mining**, machines are used to dig very large holes and remove metal ores containing copper (Figure 14-11), gold (see Case Study, p. 360), or other metals, or sand, gravel, or stone.

Strip mining is any form of mining involving the extraction of mineral deposits that lie in large horizontal beds close to the earth's surface. In **area strip mining**, used where the terrain is fairly flat, a gigantic earthmover strips away the overburden, and a power shovel—which can be as tall as a 20-story building—removes the mineral deposit or an energy resource such as coal (Figure 14-12). The resulting trench is filled with overburden, and a new cut is made parallel to the previous one. This process is repeated over the entire site.

Contour strip mining (Figure 14-13) is used mostly to mine coal and various mineral resources on hilly or mountainous terrain. Huge power shovels and bulldozers cut a series of terraces into the side of a hill. Then, earth-movers remove the overburden, an excavator or power shovel extracts the coal, and the overburden from each

Mining ⟶ Metal ore ⟶ Separation of ore from waste material ⟶ Smelting ⟶ Melting metal ⟶ Conversion to product ⟶ Discarding of product

Recycling

© Cengage Learning

Figure 14-9 Each metal product that we use has a *life cycle*.

Left: ©kaband/Shutterstock.com. Second to left: ©Andrey N Bannov/Shutterstock.com. Center left: ©Vladimir Melnik/Shutterstock.com. Center: ©mares/Shutterstock.com. Center right: Zhu Difeng/Shutterstock.com. Second to right: ©Michael Shake/Shutterstock.com. Right: Pakhnyushcha/Shutterstock.com.

Figure 14-10 Natural capital degradation: This spoils pile in Zielitz, Germany, is made up of waste material from the mining of potassium salts used to make fertilizers.

new terrace is dumped onto the one below. Unless the land is restored, what is left are a series of spoils banks and a highly erodible bank of soil and rock called a *highwall*.

Another surface mining method is **mountaintop removal**, in which the top of a mountain is removed to expose seams of coal, which are then extracted (Figure 14-14). This method is commonly used in the Appalachian Mountains of the United States.

With this method, after a mountaintop is blown apart, enormous machines plow waste rock and dirt into valleys below the mountaintops. This destroys forests, buries mountain streams, and increases the risk of flooding. Wastewater and toxic sludge, produced when the coal is processed, are often stored behind dams in these valleys, which can overflow or collapse and release toxic substances such as arsenic and mercury.

In the United States, more than 500 mountaintops in West Virginia and other Appalachian states have been removed to extract coal. According to the U.S. Environmental Protection Agency (EPA), the resulting spoils have buried more than 1,100 kilometers (700 miles) of stream—a total roughly equal in length to the distance between the two U.S. cities of New York and Chicago.

The U.S. Department of the Interior estimates that at least 500,000 surface-mined sites dot the U.S. landscape, mostly in the West. Such sites can be cleaned up and restored, but it is costly and is rarely done even when required by law.

Deep deposits of minerals are removed by **subsurface mining**, in which underground mineral resources are removed through tunnels and shafts. This method is

Figure 14-11 Natural capital degradation: Kennecott's Bingham Canyon Mine in the U.S. state of Utah has produced more copper than any mine in history. This gigantic open-pit mine is almost 5 kilometers (3 miles) wide and 1,200 meters (4,000 feet) deep, and it is getting deeper.

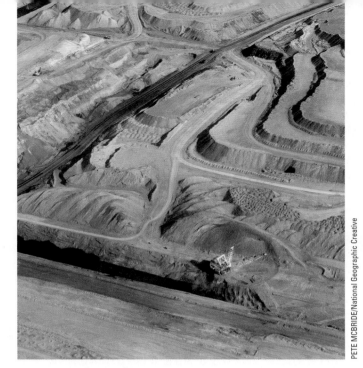

Figure 14-12 Natural capital degradation: This area strip mining for coal is taking place in the U.S. state of Wyoming.

Undisturbed land

Overburden

Highwall

Coal seam

Overburden

Pit

Bench

Coal seam

Spoil banks

Figure 14-13 Natural capital degradation: Contour strip mining is used in hilly or mountainous terrain.

used to remove metal ores and coal that are too deep to be extracted by surface mining. Miners dig a deep, vertical shaft, blast open subsurface tunnels and chambers to reach the deposit, and use machinery to remove the resource and transport it to the surface.

Subsurface mining disturbs less than one-tenth as much land as surface mining disturbs, and it usually produces less waste material. However, it creates hazards such as cave-ins, explosions, and fires. Miners often get lung diseases caused by prolonged inhalation of mineral or coal dust in subsurface mines. Another problem is

subsidence—the collapse of land above some underground mines. It can damage houses, crack sewer lines, break gas mains, and disrupt groundwater systems.

Surface and subsurface mining operations also produce large amounts of solid waste—three-fourths of all U.S. solid waste—and cause major water and air pollution. For example, *acid mine drainage* occurs when rainwater that seeps through an underground mine or a spoils pile from a surface mine carries sulfuric acid (H_2SO_4, pro-

Figure 14-14 Natural capital degradation: Mountaintop removal coal mining near Whitesville, West Virginia.

duced when aerobic bacteria act on minerals in the spoils) to nearby streams (Figure 14-15) and groundwater.

According to the EPA, mining has polluted mountain streams in about 40% of the western watersheds in the United States, and it accounts for 50% of all the country's emissions of toxic chemicals into the atmosphere. In fact, the mining industry produces more of such toxic emissions than any other U.S. industry.

Where environmental regulations and enforcement have not been strict, the mining and processing of rare earths (**Core Case Study**) has had a very harmful environmental impact. For example, in China, the extraction and processing of rare-earth metals and oxides has stripped much land of its vegetation and topsoil while polluting the air, acidifying streams, and leaving toxic and radioactive waste piles.

CASE STUDY
The Real Cost of Gold

Many newlyweds would be surprised to know that mining enough gold to make their wedding rings produced roughly enough mining waste to equal the total weight of more than three mid-size cars. This waste is usually left piled near the mine site and can pollute the air and nearby surface water.

In 2012, the world's top five gold-producing countries were, in order, China, Australia, the United States, South Africa, and Russia. In Australia and North America, mining companies dig up massive amounts of rock containing only small concentrations of gold. At about 90% of the world's gold mines, the mineral is extracted with the use of a solution of highly toxic cyanide salts sprayed onto piles of crushed rock. The solution reacts with the gold and then drains off the rocks, pulling some gold with it, into settling ponds (Figure 14-16). After the solution is recirculated a number of times, the gold is removed from the ponds.

Until sunlight breaks down the cyanide, it is extremely toxic to birds and mammals drawn to these ponds in search of water. The ponds can also leak or overflow, which poses threats to underground drinking water supplies and to fish and other forms of life in nearby lakes and streams. Special liners in the settling ponds can help prevent leaks, but some have failed. According to the EPA, all such liners will eventually leak.

After extracting the gold from a mine, some mining companies have declared bankruptcy. This has allowed them to walk away from cleaning up their mining operations, leaving behind large amounts of cyanide-laden water in leaky holding ponds.

Since 1980, millions of poverty-stricken miners have streamed into tropical forests in search of gold and this influx has increased greatly since 2001 when the price of gold began rising sharply. As a result, illegal deforestation to make way for mines has also increased rapidly in parts of the Amazon Basin. These miners use toxic mercury

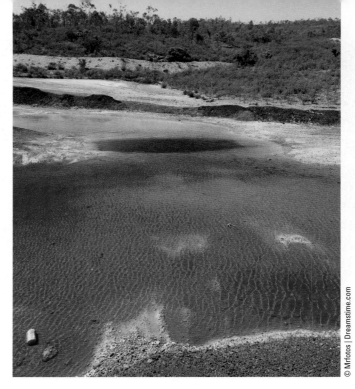

Figure 14-15 Acid mine drainage from an abandoned open-pit coal mine in Portugal.

illegally to separate gold from its ore, and they heat the mixture of gold and mercury to vaporize the mercury and leave the gold, causing dangerous air and water pollution. Many of these miners and villagers near their mines then inhale toxic mercury, drink polluted water, and eat fish contaminated with mercury.

There are nontoxic methods for extracting gold, but the cyanide process is strongly entrenched in the industry. For gold and other mineral resource use to become less environmentally harmful, consumers will have to start demanding more sustainably produced mineral products and substitutes, and producers will have to find ways to produce them—a subject of the next section of this chapter.

Removing Metals from Ores Has Harmful Environmental Effects

Ore extracted by mining typically has two components: the ore mineral, containing the desired metal, and waste material. Removing the waste material from ores produces **tailings** that are left in piles or put into tailings ponds where they settle out. The photo that opens this chapter shows one of two tailings ponds that contain toxic heavy metals at a molybdenum mine in Colorado, located at the headwaters of two mountain streams that flow into the Colorado River (Figure 13-1, p. 318). Particles of toxic metals in tailings piles can be blown by the wind or washed out by rainfall and can contaminate surface water and groundwater. Tailings ponds can also leak and contaminate surface water and groundwater.

Figure 14-16 This gold mine in the Black Hills of the U.S. state of South Dakota produces cyanide leach piles and settling ponds.

ants, including sulfur dioxide and suspended toxic particles, which damage vegetation and acidify soils in the surrounding area. Smelters also cause water pollution and produce liquid and solid hazardous wastes that require safe disposal. A 2012 study by Blacksmith Institute and Green Cross Switzerland found that lead smelting is the world's second most toxic industry after the recycling of lead–acid batteries. Using chemicals to extract metals from their ores can also create numerous problems, as we saw in the case of using cyanide to remove gold (Case Study, p. 360).

After the waste material is removed, heat or chemical solvents are used to extract the metals from mineral ores. Heating ores to release metals is called **smelting** (Figure 14-9). Without effective pollution control equipment, smelters emit enormous quantities of air pollut-

14-4 How Can We Use Mineral Resources More Sustainably?

> **CONCEPT 14-4**
> We can try to find substitutes for scarce resources, reduce resource waste, and recycle and reuse minerals.

We Can Find Substitutes for Some Scarce Mineral Resources

Some analysts believe that even if supplies of key minerals become too expensive or too scarce due to unsustainable use, human ingenuity will find substitutes (Concept 14-4). They point to the current *materials revolution* in which silicon and other materials are replacing some metals for common uses. They also point out the possibilities of finding substitutes for scarce minerals through nanotechnology (Science Focus 14.1), as well as through other emerging technologies.

For example, fiber-optic glass cables that transmit pulses of light are replacing copper and aluminum wires in telephone cables, and nanowires may eventually replace fiber-optic glass cables. High-strength plastics and materials, strengthened by lightweight carbon, hemp, and glass fibers, are beginning to transform the automobile and aerospace industries. These new materials do not need painting (which reduces pollution and costs),

can be molded into any shape, and can increase fuel efficiency by greatly reducing the weights of motor vehicles and airplanes. Such new materials are even being used to build bridges. One such material, called *graphene,* promises to be another new breakthrough material (Science Focus 14.2).

But resource substitution is not a cure-all. For example, platinum is currently unrivaled as a catalyst and is used in industrial processes to speed up chemical reactions, and chromium is an essential ingredient of stainless steel. We can try to find substitutes for such scarce resources, but this may not always be possible.

We Can Use Mineral Resources More Sustainably

Figure 14-17 lists several ways to use mineral resources more sustainably (Concept 14-4). One strategy is to focus on recycling and reuse of nonrenewable mineral resources, especially valuable or scarce metals such as gold, iron, copper, aluminum, and platinum. Recycling, an application of the chemical cycling **principle of sustainability** (see Figure 1-2, p. 6 or back cover), has a much lower environmental impact than that of mining and processing metals from ores. For example,

THE NANOTECHNOLOGY REVOLUTION

Nanotechnology, or *tiny tech,* uses science and engineering to manipulate and create materials out of atoms and molecules at the ultra-small scale of less than 100 nanometers. The diameter of the period at the end of this sentence is about 1 million nanometers.

At the nanometer level, conventional materials have unconventional and unexpected properties. For example, scientists have learned to link carbon atoms together to form carbon nanotubes that are 60 times stronger than high-grade steel. An incredibly thin and nearly invisible thread of this material is strong enough to suspend a pickup truck. Using carbon nanotubes to build cars would make them stronger and safer and would improve gas mileage by making them up to 80% lighter.

Currently, nanomaterials are used in more than 1,300 consumer products and the number is growing rapidly. Such products include stain-resistant and wrinkle-free clothes, odor-eating socks, self-cleaning glass surfaces and exterior surfaces on buildings, self-cleaning sinks and toilets, sunscreens, waterproof coatings for cell phones and other electrical devices, some cosmetics, some processed foods, and food containers that release nanosilver ions to kill bacteria, molds, and fungi. A new one-atom-thick nanotextile material that can conduct electricity could be incorporated into clothing, and this could allow you to charge your cellphone or start your laptop by plugging it into your jeans or your tee shirt.

Nanotechnologists envision technological innovations such as a supercomputer smaller than a grain of rice, thin and flexible solar-cell films that could be attached to or painted onto almost any surface, motors smaller than a human cell, biocomposite materials that would make our bones and tendons super strong, nanovessels filled with medicines that could be delivered to cells anywhere in the body, and nanomolecules specifically designed to seek out and kill cancer cells. Computer-controlled tabletop nanofactories could produce items from computerized blueprints within minutes.

Nanotechnology has many potential environmental benefits. Designing and building existing and new products from the molecular level up would greatly reduce the need to mine many materials. It would also require less material and energy and reduce waste production. We may be able to use nanoparticles to remove industrial pollutants in contaminated air, soil, and groundwater. Nanofilters might someday be used to desalinate seawater at an affordable cost and to purify water by removing chemicals, bacteria, and even viruses. **GREEN CAREER:** environmental nanotechnology

So what's the catch? The main problem is the serious concerns about the possible harmful health effects of nanotechnology on humans. The tiny size and large combined surface area of the huge numbers of these nanoparticles involved in any particular application make them more chemically reactive and potentially more toxic to humans and other animals than are conventional materials composed of much larger particles. Laboratory studies involving mice and other test animals reveal that nanoparticles can

- be inhaled deeply into the lungs and absorbed into the bloodstream and can penetrate cell membranes, including those in the brain;
- move across the placenta from a mother to her fetus and can move from the nasal passage to the brain;
- result in lung damage similar to that caused by mesothelioma, a deadly cancer resulting from the inhalation of asbestos particles;
- and cause changes in the rate and rhythm of a rodent's heartbeat, similar to such changes caused by heart disease.

We know far too little about these and other and risks at a time when the use of untested, unregulated, and unlabeled nanoparticles is growing rapidly. In 2009, an expert panel of the U.S. National Academy of Sciences said that the federal government was not doing enough to evaluate the potential health and environmental risks of using engineered nanomaterials. For example, the U.S. Food and Drug Administration does not even maintain a list of the food products and cosmetics that contain nanomaterials.

By contrast, the European Union (EU) takes a precautionary approach to the use of nanomaterials and untested chemicals. While U.S. regulators generally assume that nanoparticles and chemicals are innocent until shown to be harmful, EU laws require that manufacturers demonstrate the safety of their products before they can enter the marketplace.

Many analysts say we need to take three steps before unleashing nanotechnology more broadly. *First,* greatly increase research on the potential harmful health effects of nanoparticles. *Second,* develop guidelines and regulations for controlling its growing applications until we know more about the potentially harmful effects of this new technology. *Third,* require labeling of all products containing nanoparticles.

Critical Thinking

Do you think the potential benefits of nanotechnology products outweigh their potentially harmful effects? Explain.

GRAPHENE: A REVOLUTIONARY MATERIAL

Graphene is made from graphite—the material in the center of a pencil. Both are made of carbon. However, ultrathin graphene consists of a single layer of carbon atoms packed into a two-dimensional hexagonal lattice somewhat like chicken wire and that can be applied as a transparent film to surfaces (Figure 14-A).

Graphene is one of the world's thinnest and strongest materials. A sheet of graphene is 150,000 times thinner than a human hair and 200 times stronger than structural steel. A piece of this amazing material stretched over a coffee mug could support the weight of a car. It is also a good conductor of electricity and conducts heat better than any known material. In 2010, the two scientists who had discovered how to make graphene 6 years earlier won the Nobel Prize in Physics for their discovery.

The use of graphene could revolutionize the electric car industry by leading to the production of batteries that can be recharged 10 times faster and hold 10 times more power than current car batteries. Graphene composites can also be used to make stronger and lighter plastics, light-

weight aircraft and motor vehicles, flexible computer tablets, and TV screens as thin as a magazine. Within 5 years, it might also be used to make flexible, more efficient, less costly solar cells that can be attached to almost anything. Graphene could also replace the silicon transistors that are used as chips in computers and other electronic devices, and this could make almost any electronic device run much faster and use less power. Indeed, some scientists contend that use of graphene will change the world more than any technological development since the invention of the silicon chip.

Graphene is made from very high purity and expensive graphite. According to the U.S. Geological Survey, China controls about 73% of the world's high-purity graphite production. As with rare-earth metals (**Core Case Study**), the United States mines very little natural graphite and imports most of its graphite from China, which may restrict U.S. product exports as the use of graphene grows.

Geologists are looking for deposits of graphite in the United States. However, in 2011, a team of Rice University chemists,

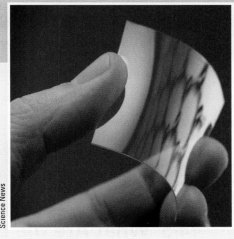

Figure 14-A Graphene, which consists of a single layer of carbon atoms linked together in a hexagonal lattice, is a revolutionary new material.

led by James M. Tour, found ways to make large sheets of high-quality graphene from inexpensive materials found in garbage and from dog feces. If such a process becomes economically feasible, concern over supplies of graphite could vanish, along with any harmful environmental effects of the mining and processing of graphite.

Critical Thinking

Would you invest money in a company that mines and purifies graphite and converts it to graphene? Explain.

---

recycling aluminum beverage cans and scrap aluminum produces 95% less air pollution and 97% less water pollution, and uses 95% less energy, than mining and processing aluminum ore. Cleaning up and reusing items instead of recycling them has an even lower environmental impact.

Researchers are working hard to try to ensure adequate supplies of rare earths on a short-term basis, and on a long-term basis, to find alternatives to these materials (**Core Case Study**). One way to increase supplies is to extract and recycle rare-earth metals from the massive amounts of electronic wastes that are being produced.

Companies that build batteries for electric cars are beginning to switch from making nickel–metal-hydride batteries, which require the rare-earth metal lanthanum, to manufacturing lighter-weight lithium-ion batteries, which researchers are now trying to improve (Individu-

als Matter 14.1). Lithium-ion batteries are also used in cell phones, iPods, iPads, and laptop computers. The problem is that some countries, including the United States, do not have adequate supplies of lithium.

🔍 CONSIDER THIS. . .

CONNECTIONS Lithium and U.S. Energy Dependence

Lithium (Li), the world's lightest metal, is a vital component of lithium-ion batteries, which are being used in a variety of devices on a rapidly growing basis. The South American countries of Bolivia, Chile, and Argentina have about 80% of the global reserves of lithium. Bolivia alone has about 50% of these reserves, while the United States holds only about 3%. Japan, China, South Korea, and the United Arab Emirates are buying up access to global lithium reserves to ensure their ability to sell lithium or batteries to the rest of the world. Within a few decades, the United States may be heavily dependent on expensive imports of lithium and lithium batteries.

Scientists are also searching for substitutes for rare-earth metals that could be used to make increasingly important powerful magnets. In Japan and the United States, researchers are developing a variety of devices that use no rare-earth minerals, are light and compact, and can deliver more power with greater efficiency at a reduced cost. These include electric motors for hybrid-electric vehicles and electromagnets for use in wind turbines.

Figure 14-17 We can use nonrenewable mineral resources more sustainably (Concept 14-4). **Questions:** Which two of these solutions do you think are the most important? Why?

14-5 What Are the Earth's Major Geological Hazards?

CONCEPT 14-5
Dynamic processes move matter within the earth and on its surface and can cause volcanic eruptions, earthquakes, tsunamis, erosion, and landslides.

The Earth Beneath Your Feet Is Moving

We tend to think of the earth's crust, mantle, and core as fairly static. However, according to geologists, the flows of energy and heated material within the earth's convection cells (Figure 14-3) are so powerful that they have caused the lithosphere to break up into a dozen or so huge rigid plates, called **tectonic plates**, which move extremely slowly atop the asthenosphere (Figure 14-18).

These gigantic plates are somewhat like the world's largest and slowest-moving surfboards on which we ride without noticing their movement. Their typical speed is about the rate at which fingernails grow. Throughout the earth's history, continents have split apart and joined as tectonic plates drifted atop the earth's asthenosphere (Figure 4-9, p. 86).

Much of the geological activity at the earth's surface takes place at the boundaries between tectonic plates as they separate, collide, or grind along against each other (Figure 14-19). The tremendous forces produced at these plate boundaries can cause mountains to form, earthquakes to shake parts of the crust, and volcanoes to erupt.

Volcanoes Release Molten Rock from the Earth's Interior

An active **volcano** occurs where magma rising in a plume through the lithosphere reaches the earth's surface through a central vent or a long crack, called a *fissure* (Figure 14-20). Magma that reaches the earth's surface is called *lava* and often builds into a cone.

Many volcanoes form along the boundaries of the earth's tectonic plates when one plate slides under or moves away from another plate. Volcanoes can *erupt,* releasing large chunks of lava rock, glowing hot ash, liquid lava, and gases (including water vapor, carbon dioxide, and sulfur dioxide) into the environment (Concept 14-5). Eruptions can be extremely destructive, causing loss of life and obliterating ecosystems and human communities.

While volcanic eruptions can be destructive, they do provide some benefits. They can result in the formation of majestic mountains and lakes such as Crater Lake (see Figure 8-15, p. 179), and the weathering of lava contributes to fertile soils. Hundreds of volcanoes have erupted on the ocean floor, building cones that have reached the ocean's surface, eventually to form islands that have become favorable for human settlement.

We can reduce the loss of human life and some of the property damage caused by volcanic eruptions in several ways. We use historical records and geological measurements to identify high-risk areas, so that people can avoid living in those areas. We also use monitoring devices that warn us when volcanoes are likely to erupt, and in some areas prone to volcanic activity, evacuation plans have been developed.

Earthquakes Are Geological Rock-and-Roll Events

Forces inside the earth's mantle put tremendous stress on rock within the crust. Such stresses can be great enough to cause sudden breakage and shifting of the rock, producing a *fault,* or fracture in the earth's crust (Figure 14-19). When a fault forms, or when there is abrupt movement on an existing fault, energy that has accumulated over time is released in the form of vibrations, called *seismic waves,* that move in all directions through the surrounding rock—an event called an **earthquake** (Figure 14-21 and Concept 14-5). Most earthquakes occur at the boundaries of tectonic plates.

Seismic waves move upward and outward from the earthquake's *focus* (Figure 14-21, left) like ripples in a pool

Yu-Guo Guo: Designer of Nanotechnology Batteries and National Geographic Emerging Explorer

©Jinsong Hu

Yu-Guo Guo is a professor of chemistry and a nanotechnology researcher at the Chinese Academy of Sciences in Beijing. He has invented nanomaterials that can be used to make lithium-ion battery packs smaller, more powerful, and less costly, which makes them more useful for powering electric cars and electric bicycles. This is an important scientific advance, because the battery pack is the most important and expensive part of any electric vehicle.

Guo's innovative use of nanomaterials has greatly increased the power of lithium-ion batteries by enabling electric current to flow more efficiently through what he calls "3-D conducting nanonetworks." With this promising technology, lithium-ion battery packs in electric vehicles can be fully charged in a few minutes. They also have twice the energy storage capacity of today's batteries, and thus will extend the range of electric vehicles by enabling them to run longer.

Guo predicts that within 5 years, the electric-vehicle market will be well established and, in some cities, will account for 10% of all cars. He is also interested in developing nanomaterials for use in solar cells and fuel cells that could be used to generate electricity and to power vehicles.

Background photo: Pi-Lens/Shutterstock.com

---

of water. Scientists measure the severity of an earthquake by the *magnitude* of its seismic waves. The magnitude is a measure of ground motion (shaking) caused by the earthquake, as indicated by the *amplitude,* or size of the seismic waves when they reach a recording instrument, called a *seismograph.*

Scientists use the *Richter scale,* on which each unit has an amplitude 10 times greater than the next smaller unit. Seismologists rate earthquakes as *insignificant* (less than 4.0 on the Richter scale), *minor* (4.0–4.9), *damaging* (5.0–5.9), *destructive* (6.0–6.9), *major* (7.0–7.9), and *great* (over 8.0). The largest recorded earthquake occurred in Chile on May 22, 1960, and measured 9.5 on the Richter scale. Each year, scientists record the magnitudes of more than 1 million earthquakes, most of which are too small to feel.

The primary effects of earthquakes include shaking and sometimes a permanent vertical or horizontal displacement of a part of the crust. These effects can have serious consequences for people and for buildings, bridges, freeway overpasses, dams, and pipelines. A major earthquake is a very large rock-and-roll geological event.

One way to reduce the loss of life and property damage from earthquakes is to examine historical records and make geological measurements to locate active fault zones. We can then map high-risk areas and establish building codes that regulate the placement and design of buildings in such areas. Then people can evaluate the

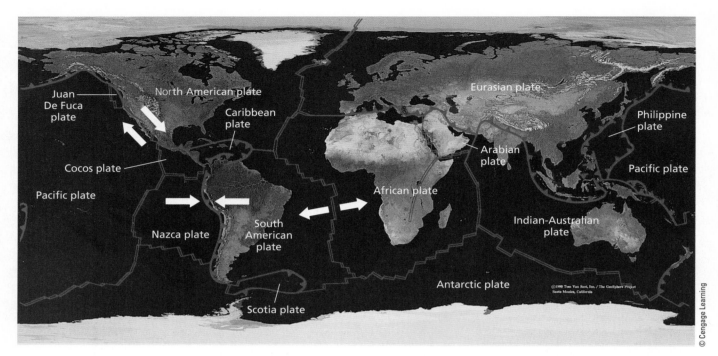

Animated Figure 14-18 The earth's crust has been fractured into several major tectonic plates. White arrows indicate examples of where plates are colliding, separating, or grinding along against each other in opposite directions. **Question:** Which plate are you riding on?

Figure 14-19 The San Andreas Fault, created by the North American Plate and the Pacific Plate sliding very slowly past each other, runs almost the full length of California (see map). It is responsible for earthquakes of various magnitudes, which have caused rifts on the land surface in some areas (photo).

Kevin Schafer/Age Fotostock

risk and factor it into their decisions about where to live. Also, engineers know how to make homes, large buildings, bridges, and freeways more earthquake resistant, although this is costly. See Figure 29, p. S55, in Supplement 6, for a map comparing earthquake risks in various areas of the United States, and Figure 30, p. S55, in Supplement 6 for a map of such areas throughout the world.

Earthquakes on the Ocean Floor Can Cause Huge Waves Called Tsunamis

A **tsunami** is a series of large waves generated when part of the ocean floor suddenly rises or drops (Figure 14-22). Most large tsunamis are caused when certain types of faults in the ocean floor move up or down as a result of a large underwater earthquake. Other causes are landslides

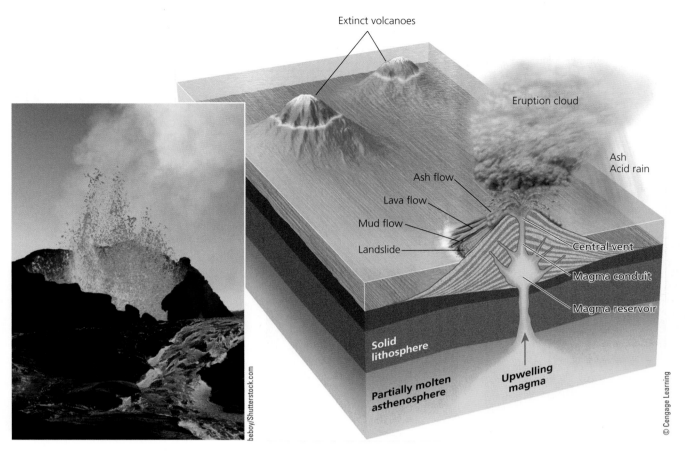

Figure 14-20 Sometimes, the internal pressure in a volcano is high enough to cause lava, ash, and gases to be ejected into the atmosphere (photo inset) or to flow over land, causing considerable damage.

Figure 14-21 An *earthquake* (left) is one of nature's most powerful events. The photo shows damage from a 2010 earthquake in Port-au-Prince, Haiti.

Earthquake in seafloor swiftly pushes water upwards, and starts a series of waves

Waves move rapidly in deep ocean reaching speeds of up to 890 kilometers per hour.

As the waves near land they slow to about 45 kilometers per hour but are squeezed upwards and increased in height.

Waves head inland causing damage in their path.

Undersea thrust fault

Upward wave

Earthquake

© Cengage Learning

Bangladesh
India
Myanmar
Thailand
Sri Lanka
Malaysia
Sumatra
Indonesia

December 26, 2004, tsunami

Figure 14-22 This diagram illustrates how a tsunami forms. The map shows the area affected by a large tsunami in December 2004—one of the largest ever recorded.

generated by earthquakes and volcanic eruptions (**Concept 14-5**).

Tsunamis are often called *tidal waves,* although they have nothing to do with tides. They can travel far across the ocean at the speed of a jet plane. In deep water the waves are very far apart—sometimes hundreds of kilometers—and their crests are not very high. As a tsu-

nami approaches a coast with its shallower waters, it slows down, its wave crests squeeze closer together, and their heights grow rapidly. It can hit a coast as a series of towering walls of water that can level buildings.

The largest recorded loss of life from a tsunami occurred in December 2004 when a great underwater earthquake in the Indian Ocean with a magnitude of 9.15

Figure 14-23 The tsunami of December 2004 killed 168,000 people in Indonesia, as well as tens of thousands more in other countries bordering the Indian Ocean. These photos show the Banda Aceh Shore near Gleebruk in Indonesia on June 23, 2004, before the tsunami (left), and on December 28, 2004, after it was stuck by the tsunami (right) (**Concept 14-5**).

Left: DigitalGlobe. Right: DigtialGlobe.

caused a tsunami with towering waves that killed around 279,900 people and devastated many coastal areas of Indonesia (see map in Figure 14-22 and Figure 14-23), Thailand, Sri Lanka, South India, and even eastern Africa. It also displaced about 1.8 million people (1.3 million of them in India and Indonesia), and destroyed or damaged about 470,000 buildings and houses. There were no recording devices in place to provide an early warning of this tsunami.

In 2011, a large tsunami caused by a powerful earthquake off the coast of Japan generated 3-story-high waves that killed almost 19,000 people, displaced more than 300,000 people, and destroyed or damaged 125,000 buildings. It also heavily damaged three nuclear reactors, which then released dangerous radioactivity into the surrounding environment.

In some areas, scientists have built networks of ocean buoys and pressure recorders on the ocean floor to collect data that can be relayed to tsunami emergency warning centers. However, these networks are far from complete.

Big Ideas

■ Dynamic forces that move matter within the earth and on its surface recycle the earth's rocks, form deposits of mineral resources, and cause volcanic eruptions, earthquakes, and tsunamis.

■ The available supply of a mineral resource depends on how much of it is in the earth's crust, how fast we use it, the mining technology used to obtain it, its market prices, and the harmful environmental effects of removing and using it.

■ We can use mineral resources more sustainably by trying to find substitutes for scarce resources, reducing resource waste, and reusing and recycling nonrenewable minerals.

TYING IT ALL TOGETHER   Rare-Earth Metals and Sustainability

M. Dykstra/Shutterstock.com

In the **Core Case Study** that opened this chapter, we learned about the importance of various rare-earth metals that are used in a variety of modern technologies, including electric motor vehicles and the energy-efficient compact fluorescent lightbulbs (photo at left) that are replacing energy-inefficient incandescent bulbs.

In this chapter, we looked at technological developments that could help us to expand supplies of mineral resources and to use them more sustainably. For example, if we develop it safely, we could use nanotechnology (Science Focus 14.1) to make new materials that could replace scarce mineral resources and greatly reduce the environmental impacts of mining and processing such resources. We looked at how biomining makes use of microbes to extract mineral resources without disturbing the land or polluting air and water as much as conventional mining operations do. Another emerging technology uses graphene to replace conventional transistors and to produce more efficient and affordable solar cells to generate electricity—an application of the solar energy **principle of sustainability** (see Figure 1-2, p. 6 or back cover).

We can also use mineral resources more sustainably by reusing and recycling them, and by reducing unnecessary resource use and waste—applying the chemical cycling **principle of sustainability**. In addition, industries can mimic nature by using a diversity of ways to reduce the harmful environmental impacts of mining and processing mineral resources, thus applying the biodiversity **principle of sustainability**.

Chapter Review

Core Case Study

1. Explain the importance of the rare-earth metals.

Section 14-1

2. What are the two key concepts for this section? Define **geology**, **core**, **mantle**, **asthenosphere**, **crust**, and **lithosphere**. Define **mineral**, **mineral resource**, and **rock**. Define and distinguish among **sedimentary rock**, **igneous rock**, and **metamorphic rock** and give an example of each. Define and describe the nature and importance of the **rock cycle**. Define **ore** and distinguish between a **high-grade ore** and a **low-grade ore**. List five important nonrenewable mineral resources and their uses.

Section 14-2

3. What are the two key concepts for this section? What are the **reserves** of a mineral resource and how can they be expanded? What two factors determine the future supply of a nonrenewable mineral resource? Explain how the supply of a nonrenewable mineral resource can be economically depleted and list the five choices we have when this occurs. What is **depletion time** and what factors affect it?

4. What five nations supply most of the world's nonrenewable mineral resources? How dependent is the United States on other countries for important nonrenewable mineral resources? Explain the concern over U.S. access to rare-earth mineral resources. Describe the conventional view of the relationship between the supply of a mineral resource and its market price. Explain why some economists believe this relationship no longer applies. Summarize the pros and cons of providing government subsidies and tax breaks for mining companies.

5. Summarize the opportunities and limitations of expanding mineral supplies by mining lower-grade ores. What are the advantages and disadvantages of biomining? Describe the opportunities and possible problems that could result from deep-sea mining.

Section 14-3

6. What is the key concept for this section? Summarize the life cycle of a metal product.

7. What is **surface mining**? Define **overburden** and **spoils**. Define **open-pit mining** and **strip mining**, and distinguish among **area strip mining**, **contour strip mining**, and **mountaintop removal mining**. Describe three harmful environmental effects of surface mining. What is **subsurface mining**? Describe the harmful effects of gold mining. Describe the environmental effects of mining and processing rare earths. Define **tailings** and explain why they can be hazardous. What is **smelting** and what are its major harmful environmental effects?

Section 14-4

8. What is the key concept for this section? Describe the opportunities and limitations of finding substitutes for key mineral resources. What is **nanotechnology** and what are some of its potential environmental and other benefits? What are some problems that could arise from the widespread use of nanotechnology? Describe the potential of using graphene as a new resource. Explain the benefits of recycling and reusing valuable metals. List five ways to use nonrenewable mineral resources more sustainably. What are two examples of research into substitutes for rare-earth metals? What is the relationship between lithium and U.S. energy independence?

Section 14-5

9. What is the key concept for this section? What are **tectonic plates**, and what typically happens when they collide, move apart, or grind against one another? Define **volcano** and describe the nature and major effects of a volcanic eruption. Define **earthquake** and describe its nature and major effects. What is a **tsunami** and what are its major effects?

10. What are the three big ideas of this chapter? Explain how we can apply the three scientific **principles of sustainability** to obtain and use rare-earth metals and other nonrenewable mineral resources in more sustainable ways.

Note: Key terms are in bold type.

Critical Thinking

1. Give three reasons why rare-earth metals (**Core Case Study**) are important to your lifestyle.

2. Would you favor giving the owners of the California rare-earth metals (**Core Case Study**) mine significant government tax breaks and subsidies to put the mine back into production? Explain. Would you favor reducing the environmental regulations for the mining and processing of these metals? Explain.

3. You are an igneous rock. Describe what you experience as you move through the rock cycle. Repeat this exercise, assuming you are a sedimentary rock and again assuming you are a metamorphic rock.

4. What are three ways in which the rock cycle benefits your lifestyle?

5. Use the second law of thermodynamics (see Chapter 2, p. 43) to analyze the scientific and economic feasibility of each of the following processes:
   a. Extracting certain minerals from seawater
   b. Mining increasingly lower-grade deposits of minerals
   c. Continuing to mine, use, and recycle minerals at increasing rates

6. Suppose you were told that mining deep-ocean mineral resources would mean severely degrading ocean-bottom habitats and life-forms such as giant tube worms and giant clams. Do you think that such information should prevent or put an end to ocean-bottom mining? Explain.

7. List three ways in which a nanotechnology revolution could benefit you and three ways in which it could harm you.

8. What are three ways to reduce the harmful environmental impacts from the mining and processing of nonrenewable mineral resources? What are three aspects of your lifestyle that contribute to these harmful impacts?

Doing Environmental Science

Do research to determine which nonrenewable mineral resources go into the manufacture of each of the following items and how much of each of these resources are required to make each item: **(a)** a cell phone, **(b)** a wide-screen TV, and **(c)** a large pickup truck.

Global Environment Watch Exercise

Use the Global Environment Watch site to find and read an article that deals with U.S. rare-earth metal supplies (**Core Case Study**). Is there scientific information cited in the article to support the author's point of view? Give specific examples. Do you think there are any types of supporting scientific data not mentioned in the article that would strengthen the author's point of view? For example, would you add statistical data to support a point, or data in a graph indicating possible cause-and-effect relationships? Be specific and give reasons for your suggestions.

Data Analysis

Rare-earth metals are widely used in a variety of important products (**Core Case Study**). In 2010, China produced about 97% of the world's rare earths and the United States produced none. According to the U.S. Geological Survey, China has 50% of the world's reserves of rare-earth metals and the United States has 13%.

1. In 2010, China had 55 million metric tons of rare-earth metals in its reserves and produced 130,000 metric tons of these metals. At this rate of production how long will China's rare-earth reserves last?

2. In 2010, the global demand for rare-earth metals was about 133,600 metric tons. If China produced all of the world's rare-earth metals, how long would their reserves last?

3. Global demand for rare earths is projected to rise to at least 185,000 metric tons by 2015. If China produced all of the world's rare-earth metals how long would its reserves last?

1. The North American and Eurasian plates are moving apart from one another and creating the Mid-Atlantic ridge. This is an example of what type of plate boundary?
   a. Convergent, such as the plate movement that causes earthquakes in Japan and the Philippines
   b. Divergent, such as between the African and Arabian plates
   c. Transform, such as the plate boundary in California that caused the 1906 San Francisco earthquake
   d. Subduction, such as along the West Coast of South America
   e. Convection, which causes molten rock beneath the surface to move in opposite directions

2. Tectonic plates move slowly across the asthenosphere on
   a. ocean currents because the less dense material rises to the surface.
   b. ocean ridges because they are diverging plate boundaries.
   c. plate boundaries because there is tectonic movement along all plate boundaries.
   d. convection currents as a result of differences in density.
   e. groundwater because it is always moving beneath the surface.

Questions 3–5 refer to the terms below.
   a. Oil shale
   b. Igneous
   c. Metamorphic
   d. Sandstone
   e. Leachate

3. Forms when pre-existing rock is subjected to high temperatures, pressure, or chemicals.

4. Forms when existing rock is weathered and deposited as small grains.

5. Forms when magma cools and hardens.

6. Strip mining is a useful and economic way to extract mineral resources. However, strip mining has all of the following harmful environmental effects EXCEPT
   a. regrowth of vegetation where strip mining has occurred is very fast.
   b. toxic substances are left behind.
   c. biodiversity is destroyed in the area.
   d. streams around the area are polluted.
   e. air pollution around the area occurs.

Use the graph below to answer question 7.

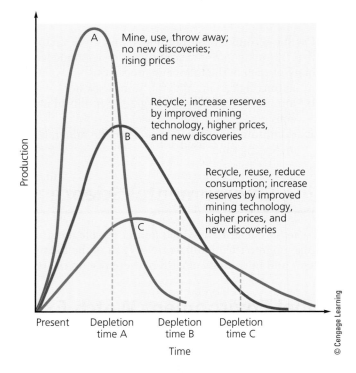

7. According to the graph above,
   a. recycling a resource does not significantly lower depletion time.
   b. mining, using, and throwing away a resource is the best way to lower depletion time.
   c. recycling a resource causes it to be depleted quickest.
   d. 80% depletion occurs slower as recycling, reuse, and reduction are implemented.
   e. as we recycle, reuse, and reduce our consumption rates, we slow down depletion time of nonrenewable resources.

8. Finding substitutes for mineral resources will
   a. shift mining to lower-grade ores.
   b. have no environmental impacts.
   c. help to keep supplies of resources sustainable.
   d. cause thermal water pollution.
   e. cause eutrophication.

9. Gold is obtained from areas containing low concentrations through the process of
   a. subsurface mining.
   b. mining of near-shore sediments.
   c. strip mining and mountaintop removal.
   d. cyanide leeching of crushed rock.
   e. smelting of high-grade ore.

10. When a mineral resource becomes scarce, its
    a. price increases, which can encourage use of substitutes and resource conservation.
    b. demand goes down because fewer people are interested in buying a mineral that is becoming rare.
    c. price stays the same so the remaining minerals will be quickly purchased and the mining companies can begin to mine something else.
    d. virgin material increases and, after a short time, a new supply of the mineral will become available.
    e. environmental impact decreases because mining decreases when a mineral becomes scarce.

11. One big environmental problem with smelting is
    a. it is very costly and results in increased taxes paid by local residents.
    b. subsidence after ore mineral is removed from the land.
    c. air pollution because smelting releases sulfur dioxide and particulates.
    d. shaft mining in which underground tunnels are built to reach minerals that could not be reached any other way.
    e. a result of the Surface Mining Control and Reclamation Act of 1977 because much damage is done to the land during restoration.

12. When a mineral resource is too deep to be surface mined, which technique is used?
    a. Mining spoils, because there is still mineral ore in the soil and rock after initial removal of ore
    b. Subsurface mining, in which miners use elevators to get to and from the mineral ores
    c. Placer mining, because it is safer and easier to remove the minerals from spring water that flows up from deep underground
    d. Subsidence, because it is easier and safer to reach the mineral ore once the land has sunk down to the level of the desired resources
    e. Open-pit mining, because miners keep digging until they reach the depth of the desired mineral resource

15 Nonrenewable Energy

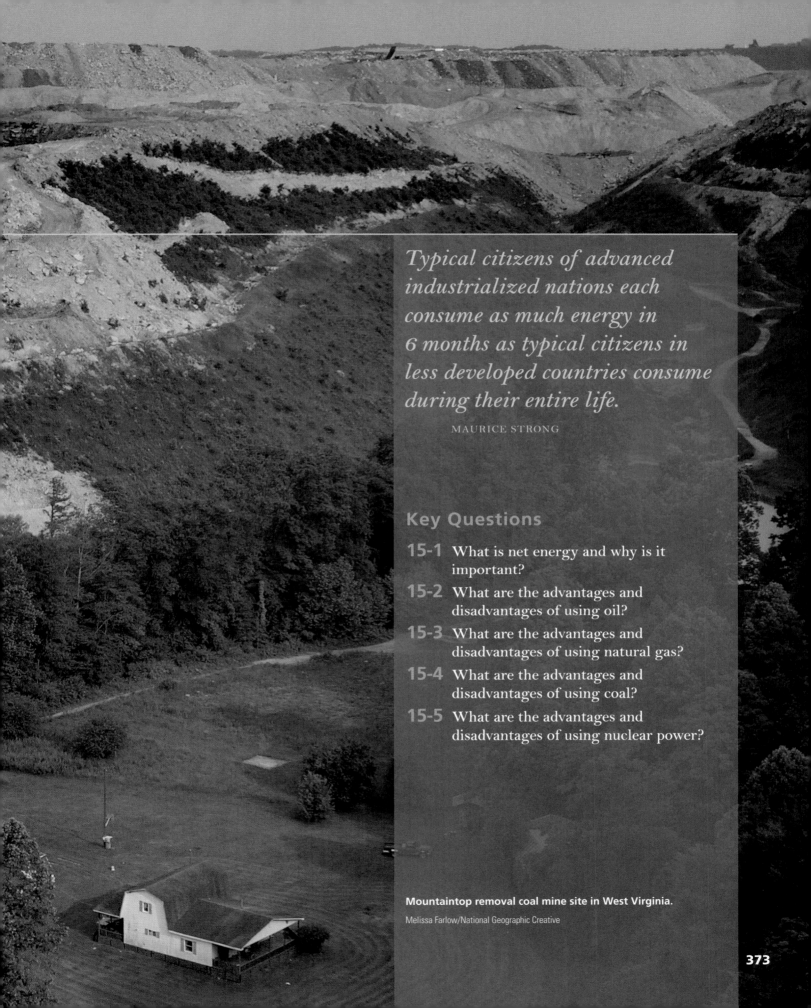

Typical citizens of advanced industrialized nations each consume as much energy in 6 months as typical citizens in less developed countries consume during their entire life.

MAURICE STRONG

Key Questions

15-1 What is net energy and why is it important?

15-2 What are the advantages and disadvantages of using oil?

15-3 What are the advantages and disadvantages of using natural gas?

15-4 What are the advantages and disadvantages of using coal?

15-5 What are the advantages and disadvantages of using nuclear power?

Mountaintop removal coal mine site in West Virginia.

Melissa Farlow/National Geographic Creative

Is the United States Entering a New Oil and Natural Gas Era?

Oil and natural gas are the two most widely used energy resources in the United States (Figure 15-1). There are signs that U.S. oil and natural gas production could increase sharply.

Between 1985 and 2008, oil production in the United States fell while consumption kept rising, and oil imports rose to make up the difference between consumption and production. However, since 2008, U.S. oil production has increased somewhat, largely because high oil prices and improved drilling and extraction technology have made it profitable to extract oil that is dispersed and tightly held in dense formations of shale rock.

According to some oil economists and the International Energy Agency, if oil production increases as projected and oil prices remain at $50 a barrel or higher, the United States could become the world's largest oil producer, probably sometime before 2020. Such a boom in domestic oil production would create large numbers of jobs, stimulate the U.S. economy, and reduce the country's expensive dependence on imported oil.

Since 1999, natural gas drilling and production have also increased dramatically, especially the extraction of natural gas held tightly within shale rock using the same drilling and extraction technology that is used to pull tightly held oil from shale rock. By 2011, this growing supply had led to sharply lower U.S. natural gas prices and made the United States the world's leading natural gas producer. If U.S. natural gas production from shale rock continues to grow as projected, and if natural gas prices do not rise significantly, natural gas could displace environmentally harmful coal as the country's largest source of electricity within three to four decades.

However, there are two major problems with this scenario. One is that the large-scale removal of natural gas and oil held tightly in shale rock requires huge amounts of water and also produces heavily polluted wastewater. This, along with leaks from gas and oil well piping systems, could contaminate shallow aquifers that feed many drinking water wells, as well as deep aquifers, unless the entire drilling, extraction, and wastewater treatment process is strictly monitored and regulated to protect drinking water.

A second problem is that by burning more carbon-containing oil, natural gas, and coal, we will continue to release growing quantities of carbon dioxide (CO_2) and methane (CH_4) into the atmosphere faster than they can be removed by the carbon cycle (see Figure 3-17, p. 66). Computer models project that rising atmospheric levels of these greenhouse gases will play a key role in changing the world's climate in potentially very harmful ways during this century.

In this chapter, we discuss the advantages and disadvantages of using nonrenewable fossil fuels (such as oil, natural gas, and coal) and nuclear power. In the next chapter, we look at the advantages and disadvantages of improving energy efficiency and using a variety of renewable energy resources.

Figure 15-1 Sources of energy used in the United States in 2011. Oil, the most widely used resource, is removed from deposits underground and from deep under the ocean floor in some coastal areas (see photo).

(Compiled by the authors using data from U.S. Energy Information Administration, British Petroleum, and International Energy Agency.)

15-1 What Is Net Energy and Why Is It Important?

CONCEPT 15-1

Energy resources vary greatly in their *net energy yields*— the amount of energy available from a resource minus the amount of energy needed to make it available.

Net Energy Is the Only Energy That Really Counts

It takes energy to produce energy. For example, before oil becomes useful to us, it must be found, pumped up from beneath the ground or ocean floor (see photo in Figure 15-1), transferred to a refinery, converted to gasoline and other fuels and a variety of other widely used chemicals, and delivered to consumers. Each of these steps uses high-quality energy, mostly obtained by burning fossil fuels, especially gasoline and diesel fuel produced from oil. Because of the second law of thermodynamics (see Chapter 2, p. 43), which we cannot violate, some of the high-quality energy used in each step is automatically wasted and degraded to lower-quality energy, mostly heat that ends up in the environment.

The usable amount of high-quality energy available from an energy resource is its **net energy yield.** It is the total amount of high-quality energy available from an energy resource minus the high-quality energy needed to make the energy available (Concept 15-1). It is also related to the *energy return on investment (EROI)*— the energy obtained per unit of energy used to obtain it. Suppose that it takes about 9 units of high-quality energy to produce 10 units of high-quality energy from an energy resource. Then the net energy yield is only 1 unit of energy.

Net energy is like the net profit earned by a business after it deducts its expenses. If a business has $1 million in sales and $900,000 in expenses, its net profit is $100,000.

Figure 15-2 shows generalized net energy yields for energy resources and systems that generate electricity, heat homes and buildings, produce high-temperature heat for industrial processes, and provide transportation. It is based on several sources of scientific data and classifies estimated net energy yields as high, medium, low, or negative (negative being a net energy loss).

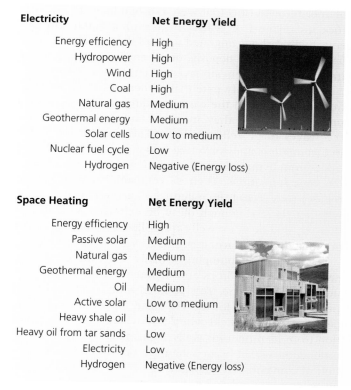

Electricity	Net Energy Yield
Energy efficiency	High
Hydropower	High
Wind	High
Coal	High
Natural gas	Medium
Geothermal energy	Medium
Solar cells	Low to medium
Nuclear fuel cycle	Low
Hydrogen	Negative (Energy loss)

High-Temperature Industrial Heat	Net Energy Yield
Energy efficiency (cogeneration)	High
Coal	High
Natural gas	Medium
Oil	Medium
Heavy shale oil	Low
Heavy oil from tar sands	Low
Direct solar (concentrated)	Low
Hydrogen	Negative (Energy loss)

Space Heating	Net Energy Yield
Energy efficiency	High
Passive solar	Medium
Natural gas	Medium
Geothermal energy	Medium
Oil	Medium
Active solar	Low to medium
Heavy shale oil	Low
Heavy oil from tar sands	Low
Electricity	Low
Hydrogen	Negative (Energy loss)

Transportation	Net Energy Yield
Energy efficiency	High
Gasoline	High
Natural gas	Medium
Ethanol (from sugarcane)	Medium
Diesel	Medium
Gasoline from heavy shale oil	Low
Gasoline from heavy tar sand oil	Low
Ethanol (from corn)	Low
Biodiesel (from soy)	Low
Hydrogen	Negative (Energy loss)

Figure 15-2 Generalized *net energy* yields for various energy systems (Concept 15-1). **Question:** Based only on these data, which two resources in each category should we be using?

(Compiled by the authors using data from the U.S. Department of Energy; U.S. Department of Agriculture; Colorado Energy Research Institute, *Net Energy Analysis,* 1976; Howard T. Odum and Elisabeth C. Odum, *Energy Basis for Man and Nature,* 3rd ed., New York: McGraw-Hill, 1981, and Charles A.S. Hall and Kent A. Klitgaard, *Energy and the Wealth of Nations,* New York: Springer, 2012.)

Top left: Yegor Korzh/Shutterstock.com. Bottom left: Donald Aitken/National Renewable Energy Laboratory. Top right: Serdar Tibet/Shutterstock.com. Bottom right: Michel Stevelmans/Shutterstock.com.

Some Energy Resources Need Help to Compete in the Marketplace

The following general rule can help us to evaluate the long-term economic usefulness of an energy resource based on its net energy yield: *An energy resource with a low or negative net energy yield can have a hard time competing in the marketplace with other energy alternatives that have medium to high net energy yields unless it receives financial support from the government (taxpayers) or other outside sources.* Such financial support is generally referred to as a *subsidy,* and providing it is called *subsidizing.*

For example, electricity produced by nuclear power has a low net energy yield because large amounts of high-quality energy are needed for each step in the *nuclear power fuel cycle:* to extract and process uranium ore, convert it into nuclear fuel, build and operate nuclear power plants, safely store the resulting highly radioactive wastes for thousands of years, and dismantle each plant after its useful life (typically 40–60 years) and safely store its high-level radioactive parts for thousands of years. The resulting low net energy yield for the whole nuclear fuel cycle is one reason why governments throughout the world heavily subsidize nuclear power to make it available to consumers at an affordable price. Such subsidies help to hide the true cost of the nuclear power fuel cycle and thus violate the full-cost pricing **principle of sustainability** (see Figure 1-5, p. 9 or back cover).

15-2 What Are the Advantages and Disadvantages of Using Oil?

CONCEPT 15-2A
Conventional crude oil is abundant and has a medium net energy yield, but using it causes air and water pollution and releases greenhouse gases to the atmosphere.

CONCEPT 15-2B
Unconventional heavy oil from oil shale rock and tar sands exists in potentially large supplies but has a low net energy yield and a higher environmental impact than conventional oil has.

We Depend Heavily on Oil

Oil is the world's most widely used energy resource (Figure 15-3). We use oil to heat our homes, grow most of our food, transport people and goods, make other energy resources available for use, and manufacture most of the things we use every day, from plastics to cosmetics to asphalt on roads.

Crude oil, or **petroleum**, is a black, gooey liquid consisting mostly of a mix of different combustible hydrocarbons along with small amounts of sulfur, oxygen, and nitrogen impurities. It is also known as conventional or light crude oil. It was formed from the decayed remains of ancient organisms that were crushed beneath layers of rock for millions of years. The resulting liquid and gaseous hydrocarbons migrated upward through porous rock layers to collect as deposits of oil and natural gas, often trapped together beneath layers of impermeable rock.

Scientists identify potential oil deposits by using large machines to pound the earth, sending shock waves deep underground, and they measure how long it takes for the waves to be reflected back. This information is fed into computers and converted into *3-D seismic maps* of the underground that show the locations and sizes of various rock formations. Then oil companies drill holes and remove rock cores from potential oil deposit areas to learn whether there is enough oil to be extracted profitably. If there is, one or more wells are drilled and the light oil is pumped to the surface.

After years of pumping, usually a decade or so, the pressure in a well drops and its rate of crude oil production starts to decline. This point in time is referred to as **peak production** for the well. The same thing can happen to a large oil field when the overall rate of production from its numerous wells begins to drop. Global peak production would occur when the rate of global production of conventional oil begins to decline faster than new oil fields are found and put into production. There is disagreement over whether we have reached global peak production of conventional crude oil and when we might reach it if we have not.

Crude oil from a well cannot be used as it is. It is transported to a refinery by pipeline, truck, rail, or ship (oil tanker) where it is heated to separate it into various

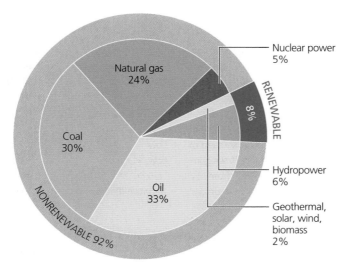

Figure 15-3 Global energy use in 2011.

(Compiled by the authors using data from British Petroleum, U.S. Energy Information Administration, and International Energy Agency.)

Lowest Boiling Point

Gases

Gasoline

Aviation fuel

Heating oil

Diesel oil

Naphtha

Naphtha Solvent

Heated crude oil

Grease and wax

Grease Wax

Furnace

Asphalt

© Cengage Learning

Highest Boiling Point

Figure 15-4 When crude oil is refined, many of its components are removed at various levels of a distillation column, depending on their boiling points. The most volatile components with the lowest boiling points are removed at the top of the column, which can be as tall as a nine-story building. The photo shows an oil refinery in Texas.

age fotostock/Superstock

fuels and other components with different boiling points (Figure 15-4) in a complex process called **refining**. This process, like all other steps in the cycle of oil production and use, requires an input of high-quality energy and decreases the net energy yield of oil. About 2% of the products of refining, called **petrochemicals**, are used as raw materials to make industrial organic chemicals, cleaning fluids, pesticides, plastics, synthetic fibers, paints, medicines, cosmetics, ice cream, and many other products.

Are We Running Out of Conventional Oil?

We use an astonishing amount of oil. Laid end to end, the roughly 32 billion barrels of crude oil used worldwide in 2011 would stretch to about 28 million kilometers (18 million miles)—far enough to reach to the moon and back about 37 times. (One barrel of oil contains 159 liters or 42 gallons of oil.)

According to the 2012 BP Statistical Review of World Energy, in 2011, the world's three largest producers of conventional light oil, in order, were Saudi Arabia (13.2% of world production), Russia (13%), and the United States (8%, see the Case Study that follows). The International Energy Agency projects that by 2017, the United States is likely to be the world's largest oil producer. In 2011, the world's three largest oil consumers were the United States

(using 21% of all oil produced), China (11%), and Japan (5%). The Earth Policy Institute projects that, by 2035, China will be using 4 times more oil than the United States.

How much oil is there? No one knows, although geologists have provided us with estimates of amounts existing in identified deposits. However, not all such deposits can be exploited at a profit, and oil that cannot be extracted profitably is not considered to be available. Availability is determined mostly by five factors that can change over time: **(1)** the demand for the oil, **(2)** the technology used to make it available, **(3)** the rate at which we can remove the oil, **(4)** the cost of making it available, and **(5)** its market price.

Available deposits are called **proven oil reserves**—deposits from which the oil can be extracted profitably at current prices with current technology. Proven oil reserves are not fixed. For example, recently improved oil extraction technology (Science Focus 15.1) and higher oil prices have made it profitable to extract light oil that is tightly held in layers of shale rock, and this has increased proven oil reserves.

The world is not about to run out of conventional light oil in the near future, but the easily extracted cheap oil that supports our economies and lifestyles may be running low. Most of the world's oil comes from huge oil fields that were discovered decades ago. Production from many of these fields has begun to decline and new fields are getting harder to find and more expensive to develop.

We can produce more conventional light oil from far offshore in deep ocean seabed deposits and from areas near the Arctic Circle. We can also rely more on unconventional heavy oil—a type of crude oil that does not flow as easily as light oil—from depleted oil wells and other sources. But use of these sources of oil results in lower net energy yields, higher production costs, and higher environmental impacts. Having to rely more on such sources leaves us with three major options: **(1)** learn to live with much higher oil prices and thus higher prices on many other items; **(2)** extend supplies by using oil much more

SCIENCE FOCUS 15.1

REMOVING TIGHTLY HELD OIL AND NATURAL GAS BY DRILLING SIDEWAYS AND FRACKING

Geologists have known for decades about vast deposits of oil and natural gas that are widely dispersed and tightly held in dense layers of shale rock formations found in many areas of the United States, including North Dakota, Texas, and Pennsylvania (see map in Figure 33, p. S58, in Supplement 6).

Until recently, it cost too much to extract such oil and natural gas from shale rock. This situation has changed because of high oil prices along with the use of two newer extraction technologies (Figure 15-A). One is **horizontal drilling**, a method of drilling first vertically to a certain point, then bending the flexible well bore and drilling horizontally. This method is used to gain greater access to oil and gas deposits located within layers of shale or other rock deposits. Usually, wells are drilled vertically for 1.6–2.4 kilometers (1–1.5 miles) or more and then horizontally for up to 1.6 kilometers (1 mile). Two or three horizontally drilled wells can often produce as much oil as 20 vertical wells, which reduces the area of land damaged by drilling operations.

The second technology, called **hydraulic fracturing** or **fracking**, is then used to free the tightly held oil and natural gas. After perforated tubes with explosive charges create fissures in the rock, high-pressure pumps shoot a mixture of water, sand, and chemicals into the well. When the pressure builds enough to fracture the rock, the mixture of water, sand, and chemicals flows into the cracks and creates more cracks and weak spots. The sand allows the cracks to remain open so that the oil or natural gas can flow out of the well to the surface.

Over a period of time this process is repeated 7–10 times per well. Much of the water that is contaminated with fracking chemicals and various other chemicals released from the rock (some of them hazardous) flows back to the surface. This contaminated water can be cleaned up and recycled. Other options are to store it in lined holding ponds or to inject it into deep underground hazardous waste disposal wells. Currently, about 80% of this contaminated water is injected into disposal wells. The rest is either recycled or stored in lined ponds. Using fracking to extract oil or natural gas costs about 5–10 times more than a conventional oil or natural gas well, and the supply is depleted about twice as fast.

The growing use of these two extraction technologies will be a key to the projected new era of oil and natural gas production in the United States (**Core Case Study**), provided the market prices of oil and natural gas remain high enough to make it profitable. However, there are

Figure 15-A Horizontal drilling and hydraulic fracturing, or fracking, are being used to release large amounts of oil and natural gas that are tightly held in shale rock formations.

some potentially serious environmental problems related to widespread use of these technologies. We discuss these in Section 15-3.

Critical Thinking

Why do you think horizontal drilling allows better access to tightly held oil and natural gas deposits than does drilling vertically into such deposits?

efficiently, for example, by sharply improving vehicle fuel efficiencies; and **(3)** use other energy resources.

The 12 countries that make up the Organization of Petroleum Exporting Countries (OPEC) have about 72% of the world's proven crude oil reserves and thus are likely to control most of the world's oil supplies for many years to come. Today, OPEC's members are Algeria, Angola, Ecuador, Iran, Iraq, Kuwait, Libya, Nigeria, Qatar, Saudi Arabia, United Arab Emirates, and Venezuela. Other countries, including Canada and Russia, also have large oil reserves. According to BP, in 2011, Venezuela had the largest portion (18%) of the world's proven light oil reserves, followed by

Figure 15-5 The amount of conventional light crude oil that might be found in the Arctic National Wildlife Refuge, if developed and extracted over 50 years, is a tiny fraction of the projected U.S. demand for oil.

(Compiled by the authors using data from U.S. Department of Energy and U.S. Geological Survey.)

Projected U.S. oil consumption

Arctic refuge oil output over 50 years

U.S. Fish and Wildlife Service

Saudi Arabia (16%), Canada (11%), Iran (9%), Iraq (9%), Kuwait (6%), the United Arab Emirates (6%), and Russia (5%). The world's three largest users of light oil—the United States, China, and Japan—have, in order, only about 2%, 1%, and 0.003% of the world's proven crude oil reserves.

Based on data from the U.S. Department of Energy and the U.S. Geological Survey, if global consumption of conventional light oil continues to grow at about 2% per year, then:

- Saudi Arabia, with the world's second largest crude oil reserves, could supply the world's demand for oil for about 7 years.
- Estimated unproven crude oil reserves under Alaska's Arctic National Wildlife Refuge (ANWR) (see Figure 31, p. S56, in Supplement 6) would meet the world's demand for 1–5 months and U.S. demand for 7–24 months (Figure 15-5).
- The Arctic Circle holds enough technically recoverable crude oil to meet the global demand for about 3 years at high production costs.

Bottom line: to keep using conventional light oil at the projected rate of increase, we must expand global proven crude oil reserves by an amount equal to Saudi Arabia's current reserves every 7 years. Most oil geologists say this is highly unlikely.

Use of Conventional Oil Has Environmental Costs

The extraction, processing, and burning of conventional crude oil has severe environmental impacts, including land disruption, greenhouse gas emissions, and other forms of air pollution, water pollution, and loss of biodiversity.

A critical and growing problem is that burning oil or any carbon-containing fossil fuel releases the greenhouse gas CO_2 into the atmosphere. According to most of the world's top climate scientists, this has been warming the atmosphere and will contribute to projected climate change during this century. Currently, burning oil, mostly as gasoline and diesel fuel for transportation, accounts for 43% of global CO_2 emissions, which have been increasing rapidly (see Figure 14, p. S70, in Supplement 7).

Another problem is that, as easily accessible deposits are becoming depleted, oil producers are turning to oil that is buried deep underground in sensitive areas and under the ocean floor in certain coastal areas. As was revealed in the catastrophic oil spill in the Gulf of Mexico in 2010, going to these harder-to-reach deposits greatly increases the risk of severe environmental degradation.

Figure 15-6 lists the advantages and disadvantages of using conventional oil as an energy resource.

CASE STUDY

Oil Production and Consumption in the United States

The United States gets about 87% of its commercial energy from fossil fuels, with 37% coming from oil (Figure 15-1). Currently, oil production in the United States, especially from shale rock, is increasing rapidly and could make the country the world's largest oil producer by 2020. However, oil production from shale rock is likely to decline over the next two decades as the richest deposits are depleted. The long-term problem for the United States is that it uses about 21% of the oil produced globally but produces only 9% of the world's oil

Figure 15-6 Using conventional light oil as an energy resource has advantages and disadvantages. **Questions:** Which single advantage and which single disadvantage do you think are the most important? Why? Do the advantages of relying on conventional light oil outweigh its disadvantages? Explain.

Photo: Richard Goldberg/Shutterstock.com

and has only about 2% of the world's proven conventional oil reserves. Also, much of its supply lies under environmentally sensitive land and coastal areas that are expensive to develop.

Since 1982, oil consumption in the United States has greatly exceeded domestic production. This helps to explain why, in 2012, the United States imported 39% of its crude oil (compared to 24% in 1970 and 60% in 2005). The decrease in dependence on oil imports between 2005 and 2012 resulted from a combination of a weak economy and reduced oil consumption due to improvements in fuel efficiency. Also, there was a slight increase in U.S. oil production between 2008 and 2012, mostly because of increased production of oil from shale rock deposits.

In 2012, the five largest suppliers of oil imported into the United States were, in order, Canada, Saudi Arabia, Venezuela, Mexico, and Iraq. By importing oil, the United States transfers a massive amount of its wealth to these oil-producing countries.

Can the United States significantly reduce its dependence on oil imports by producing its own oil faster than its current oil supply is being depleted? Some say "yes" and project that domestic oil production will increase dramatically over the next few decades—especially from oil found in shale rock. They argue that this will lead to a new era of oil production in the United States (**Core Case Study**). The U.S. Energy Information Agency (EIA) estimates that increased production of oil from shale rock could continue to reduce U.S. dependence on imported oil.

However, we do not know how much of the oil held tightly in shale rock deposits can be extracted profitably and at an acceptable environmental cost. For example,

horizontal drilling and fracking produce massive amounts of contaminated wastewater. And there is no guarantee that the well pipes and casing in any fracking operation will not leak the toxic chemicals used in and produced by fracking into underground drinking water supplies.

In addition, experience indicates that production of oil from shale rock beds drops off about twice as fast as it does in most conventional oil fields. Thus, long-term, profitable oil production from these resources may be overestimated. And producing oil from shale rock using current technology will be profitable only as long as the price of oil is at least $50 a barrel. In addition, this oil would be developed with high production costs, lower net energy yields, and potentially high environmental impacts.

Heavy Oil from Oil Shale Rock

A potential supply of heavy oil is *shale oil*. It is produced by mining, crushing, and heating oil shale rock (Figure 15-7, left) to extract a mixture of hydrocarbons called *kerogen* that can be distilled to produce shale oil (Figure 15-7, right). Before the thick shale oil is sent by pipeline to a refinery, it must be heated to increase its flow rate and processed to remove sulfur, nitrogen, and other impurities, which decreases its net energy yield. (These kerogen-containing deposits found in oil shale rock differ from the oil dispersed and tightly held in shale rock, discussed in the **Core Case Study**.)

About 72% of the world's estimated oil shale rock reserves are buried deep in rock formations located primarily under government-owned land in the U.S. states of Colorado, Wyoming, and Utah in an area known as the Green River formation (see map in Figure 31, p. S56, in Supplement 6). The U.S. Bureau of Land Management estimates that these deposits contain an amount of potentially recoverable heavy oil equal to almost 4 times the amount in Saudi Arabia's proven reserves of conventional oil. Estimated potential global supplies of unconventional heavy shale oil are about 240 times larger than estimated global supplies of conventional crude oil.

The problem is that it takes considerable energy, money, and water to extract kerogen from shale rock and convert it to shale oil. Thus, its net energy yield is low. Also, the process pollutes large amounts of water and releases 27–52% more CO_2 into the atmosphere per unit of energy produced than does producing conventional crude oil (**Concept 15-2B**). In 2011, the U.S. Bureau of Land Management stated that unless oil prices rise sharply: "There are no economically viable ways yet known to extract and process oil shale for commercial purposes."

Heavy Oil from Tar Sands

A growing source of heavy oil is **tar sands**, or **oil sands**, which are a mixture of clay, sand, water, and a combustible organic material called *bitumen*—a thick, sticky, tarlike heavy oil (Figure 15-8) with a high sulfur content.

Figure 15-7 Heavy shale oil (right) can be extracted from oil shale rock (left).

Figure 15-8 This gooey tar, or bitumen, with the consistency of peanut butter, was extracted from tar sands in Alberta, Canada, to be converted to heavy synthetic oil.

Northeastern Alberta in Canada has three-fourths of the world's tar sands resource in sandy soil under a vast area of remote boreal forest (see Figure 7-13, bottom photo, p. 156). If we include its conventional light oil and heavy oil from tar sands, Canada has the world's third largest proven oil reserves.

The big drawback to using tar sands is that developing this resource has major harmful impacts on the land (Figure 15-9), air, water, wildlife, and climate, compared to developing conventional light oil and tightly held oil from shale rock (**Concept 15-2B**). Before the mining takes place, the overlying boreal forest is clear-cut, wetlands are drained, and rivers and streams are sometimes diverted. Next the overburden of sandy soil, rocks, peat, and clay is stripped away to expose the tar sands deposits. Then five-story-high electric power shovels dig up the sand and load it into three-story-high trucks, which carry it to an upgrading plant. There, the tar sands are mixed with hot water and steam to extract the bitumen. Next, the bitumen is heated by natural gas in huge cookers and converted into a low-sulfur, synthetic, heavy crude oil that has to be processed further to allow it to flow through pipelines to a refinery.

According to a 2009 study by CERA, an energy consulting group, the process of extracting, processing, and refining bitumen from tar sands into heavy oil releases 3 to 5 times more greenhouse gases per barrel of oil produced than does extracting and producing conventional light oil. Part of this process is the removal of the boreal forest and peat deposits lying above the tar sands deposits. Left undisturbed, these forests and peatlands help to reduce the threat of projected climate change by storing great amounts of carbon. Much of this carbon is released to the atmosphere as CO_2 when the forests and peatlands are removed, which adds to the threat of global climate disruption.

Although tar sand sites can be planted with vegetation after the strip mining, such restoration is expensive and rare and cannot match the capacity of ancient peatlands and boreal forests for absorbing carbon and helping to offset projected climate dis-

Figure 15-9 Tar sands mining operation in Alberta, Canada.

ruption. Also, some studies have estimated that the conversion of tar sands to heavy oil produces only 5–20% more greenhouse gases than does conventional oil production. However, these studies did not take into account the entire production process, including the loss of carbon-absorbing peatlands and boreal forest.

In addition, this process uses huge amounts of water and creates lake-size tailings ponds containing toxic sludge and wastewater. Many migrating birds die trying to get water and food from these ponds. Also, the dikes of compacted sand surrounding the tailings ponds have the potential to leak and release large volumes of toxic sludge onto nearby land and into streams and rivers.

Finally, it takes a great deal of energy to produce oil from tar sands, which involves burning natural gas to provide heat for the bitumen cookers and using diesel fuel to run the massive vehicles and machinery. This too adds to the pollution of air, water, and land and to the emissions of greenhouse gases.

Figure 15-10 lists the major advantages and disadvantages of using heavy oil from tar sands and from oil shale rock as an energy resource.

Trade-Offs

Heavy Oils from Oil Shale and Tar Sand

Advantages	Disadvantages
Large potential supplies	Low net energy yield
Easily transported within and between countries	Releases CO_2 and other air pollutants when produced and burned
Efficient distribution system in place	Severe land disruption and high water use

© Cengage Learning

Figure 15-10 Using heavy oil from tar sands and from oil shale rock as an energy resource has advantages and disadvantages (Concept 15-2). *Questions:* Which single advantage and which single disadvantage do you think are the most important? Why? Do the advantages of relying on heavy oil from these sources outweigh the disadvantages? Explain.

Photo: Christopher Kolaczan/Shutterstock.com

15-3 What Are the Advantages and Disadvantages of Using Natural Gas?

CONCEPT 15-3
Conventional natural gas is more plentiful than oil, has a medium net energy yield and a fairly low production cost, and is a clean-burning fuel, but producing it has created environmental problems.

Natural Gas Is a Useful, Clean-Burning, but Not Problem-Free Fossil Fuel

Natural gas is a mixture of gases of which 50–90% is methane (CH_4). It also contains smaller amounts of heavier gaseous hydrocarbons such as propane (C_3H_8) and butane (C_4H_{10}), and small amounts of highly toxic hydrogen sulfide (H_2S). This versatile fuel has a medium net energy yield (Figure 15-2) and is widely used for cooking, heating space and water, and industrial purposes, including production of most of the world's nitrogen fertilizer. It can also be used as a fuel for cars and trucks and for natural gas turbines used to produce electricity in power plants.

This versatility helps to explain why natural gas provides about 28% of the energy consumed in the United States. It burns cleaner than oil and much cleaner than coal, and when burned completely, it emits about 30% less CO_2 than oil and about 50% less than coal.

Conventional natural gas is often found in deposits lying above deposits of conventional oil. It also exists in tightly held deposits in shale rock and can be extracted through horizontal drilling and fracking (see Science Focus

15.1). See Figure 33, p. S58, in Supplement 6 for a map of major U.S. natural gas shale rock deposits. In the United States and many other countries, natural gas is distributed to users by a large network of underground pipelines.

When a natural gas deposit is tapped, propane and butane gases can be liquefied under high pressure and removed as **liquefied petroleum gas (LPG)**. LPG is stored in pressurized tanks for use mostly in rural areas not served by natural gas pipelines. Natural gas can also be transported across oceans, by converting it to **liquefied natural gas (LNG)** at a high pressure and at a very low temperature. This highly flammable liquid is transported in refrigerated tanker ships. At its destination port, it is heated and converted back to the gaseous state and then distributed by pipeline. LNG has a low net energy yield, because more than a third of its energy content is used to liquefy it, process it, deliver it to users by ship, and convert it back to natural gas.

In 2011, the International Energy Agency estimated that recoverable conventional supplies of natural gas could meet the current global demand for about 120 years, and that potentially recoverable unconventional supplies of natural gas could sustain current global production for 250 years. That year, according to BP's 2012 Annual Review of World Energy, Russia had about 21% of the world's proven conventional natural gas reserves, followed by Iran (16%), Qatar (12%), and Turkmenistan (12%). China and India, with their rapidly growing economies, have only 1.5% and 0.6%, respectively, of the

GOOD NEWS

world's proven natural gas reserves and the United States has only 4%. Japan has no significant natural gas reserves and depends on imports of expensive LNG.

In 2011, the world's three largest producers of natural gas were the United States (with 20% of global total production), Russia (19%), and Canada (5%). In that year, the United States used about 22% of the world's production. (See Figure 5, p. S66, in Supplement 7 for a graph of U.S. natural gas consumption between 1980 and 2012, with projections to 2040.) U.S. natural gas production has been increasing rapidly, mostly because of development of the technology used to extract tightly held natural gas from shale rock (Figure 15-A). This source accounted for about 60% of U.S. natural gas production in 2011 (up from 2% in 2000). Thus, the United States does not have to rely on natural gas imports.

The demand for natural gas in the United States is projected to more than double between 2010 and 2050. If much of this demand is met by increased production of natural gas from shale rock, the United States could continue meeting its needs for natural gas from domestic resources. If natural gas prices remain affordable, such a trend would reduce the use of coal-burning power plants and make new nuclear power plants even more uneconomical than they are now. This could also slow the shift to greater use of renewable solar and wind energy resources.

However, U.S. natural gas producers would like to export natural gas as LPG to countries where natural gas prices are much higher than in the United States. Chemical industries and utilities that use natural gas to provide heat and produce electricity oppose this because it could decrease the supply and raise domestic natural gas prices.

There are some potential problems that could temper this rosy outlook for natural gas. In 2011, the U.S. Geological Survey cut its nationwide estimate of recoverable shale gas by 50% and pointed out that natural gas production from shale rock tends to peak and drop off much faster than does production from conventional natural gas wells. One question is whether the rate of increased production of natural gas from shale rock can exceed the rate of decline in conventional natural gas production from aging fields. Another major potential drawback is the environmental problems related to greatly increasing U.S. production of natural gas from shale rock, as discussed in the Case Study that follows.

Figure 15-11 lists the advantages and disadvantages of using conventional natural gas as an energy resource.

CASE STUDY

Natural Gas Production and Fracking in the United States: Environmental Problems and Solutions

The production of natural gas from shale rock deposits involves drilling wells; using huge amounts of water, sand, and chemicals to frack the gas; bringing up the nat-

Trade-Offs

Conventional Natural Gas

Advantages

Ample supplies

Versatile fuel

Medium net energy yield

Emits less CO_2 and other air pollutants than other fossil fuels when burned

Disadvantages

Low net energy yield for LNG

Production and delivery may emit more CO_2 and CH_4 per unit of energy produced than coal

Fracking uses and pollutes large volumes of water

Potential groundwater pollution from fracking

© Cengage Learning

Figure 15-11 Using conventional natural gas as an energy resource has advantages and disadvantages. **Questions:** Which single advantage and which single disadvantage do you think are the most important? Why? Do you think that the advantages of using conventional natural gas outweigh the disadvantages? Explain.

Photo: Werner Muenzker/Shutterstock.com

ural gas along with the resulting toxic wastewater; dealing with this wastewater; and transporting the natural gas to users through underground pipelines. At several points in this production process, natural gas can leak into the atmosphere and into underground sources of drinking water for nearby homes and communities. Despite the industry's attempts to solve this problem, natural gas has leaked from loose pipe fittings and faulty cement seals in natural gas well bore holes, as well as from pipelines and cutoff valves used to deliver natural gas.

Drinking water contaminated by natural gas can catch fire (Figure 15-12), and some home owners have had to install expensive systems to remove the natural gas to prevent explosions. According to a 2012 study by the National Academy of Sciences, people living within 900 meters (3,000 feet) of a natural gas well are likely to have up to 17 times more methane in their groundwater than those who live farther away.

Within a decade or two, there may be at least 100,000 more natural gas wells using fracking technology, according to the U.S. Energy Information Administration. Without increased monitoring and regulation of the entire natural gas production process, including fracking, the greatly increased production of natural gas (and oil) from shale rock could have several harmful environmental effects:

- Fracking requires enormous volumes of water. In water-short areas this could help to deplete aquifers, degrade aquatic habitats, and diminish the availability of water for other purposes.

Figure 15-12 Natural gas fizzing from this faucet in a Pennsylvania home can be lit like a natural gas stove burner. This began happening after an energy company drilled a fracking well in the area, but the company denies responsibility. The home owners have to keep their windows open year-round to keep the lethal and explosive gas from building up in the house.

MARK THIESSEN/National Geographic Creative

- Fracking fluids can contain several potentially hazardous chemicals that are used to reduce friction, inhibit corrosion, and stop bacterial growth. Each fracked well produces millions of gallons of wastewater that are brought to the surface along with the released natural gas. This slurry contains a mix of naturally occurring salts, toxic heavy metals, and radioactive materials leached from the rock. It also contains chemicals used in the fracking process, which natural gas companies do not have to reveal to the public. After fracking, the slurry is stored in various ways, some more secure than others, and there are several points in this complex process at which some or all of this toxic slurry could be released to contaminate nearby groundwater, surface waters, or land.

- Because fracking requires a special type of sand, some areas of the country, especially western Wisconsin, are seeing a boom in the mining of this sand. This new form of strip mining has expanded rapidly with little regulation and is destroying large areas of wildlife habitat while creating air and noise pollution in the mining areas.

- According to a 2012 study by the National Academy of Sciences and another study by the U.S. Geological Survey, in recent years, one of the major causes of hundreds of small earthquakes in 13 states has been the shifting of bedrock resulting from the high-pressure injection of large amounts of wastewater from fracking and other industrial activities into deep underground storage wells. The pressure involved is typically 9 times higher than the pressure required to crush a submarine on its deepest dive. Such earthquakes could release hazardous wastewater into aquifers and cause breaks in the steel lining and cement seals of oil and gas well pipes. In 2012, the U.S. state of Ohio shut down a deep well used for the disposal of fracking wastewater after several small earthquakes occurred in the vicinity of the well.

- In 2011, ecologist Robert Howarth and geoscientist David Hughes, in separate studies, both estimated that emissions of climate-changing CH_4 and CO_2 from the entire process for supplying and burning natural gas from shale rock are higher than those from supplying and burning conventional natural gas and coal. If these preliminary estimates are verified, the positive environmental image of natural gas—based mainly on the fact that it is relatively clean-burning and not on its entire cycle of production and use—will be tarnished.

Producers maintain that fracking is necessary for exploiting natural gas from shale deposits at an affordable cost. They point out that increased natural gas production from fracking has lowered U.S. natural gas prices and benefitted the 55% of U.S. consumers who burn natural gas. In addition, the natural gas fracking boom has created thousands of jobs and boosted local economies in some areas. And the resulting shift from coal to natural gas for producing electricity has reduced U.S. air pollution.

Producers also argue that no groundwater contamination directly due to fracking has ever been recorded, mostly because the fracking takes place far below drinking water aquifers. However, critics report that the EPA has found at least one example of drinking water contamination that resulted from fracking. They also contend that natural gas producers have squelched numerous reports of drinking water contamination from fracking by offering financial settlements to people who make such claims with the stipulation that they cannot reveal any information about the alleged contamination. Much of the contamination may have come from leaks and faulty cement seals in well pipes, which indicates inadequate inspection and regulation of the entire natural gas production process by states and the federal government.

Currently, people who rely on aquifers and streams for their drinking water in areas affected by the boom in shale gas production have little protection against pollution of their water supplies resulting from natural gas production. This is because, under political pressure from natural gas suppliers, the 2005 Energy Policy Act excluded the fracking process from certain regulations under the federal Safe Drinking Water Act. Other loopholes have also exempted natural gas production from parts of several other federal environmental laws, including the Clean Water Act, the Clean Air Act, and the National Environmental Policy Act.

In addition, people who live near fracking operations must put up with around-the-clock noise and air pollution from drilling equipment, diesel engines, trucks hauling sand, and explosions set off each time a well is fracked. Without stricter regulation and monitoring, the drilling of another 100,000 natural gas wells during the next 10–20 years will increase the risk of harmful environmental effects from the production of natural gas, which could cause a public backlash against this technology.

According to a 2010 MIT study, the harmful environmental impacts of producing natural gas from shale rock are "manageable but challenging." Some energy analysts, along with the U.S. Energy Information Agency, have suggested several ways to reduce the environmental threats arising from shale gas production (Figure 15-13). Making such changes would help us in implementing the full-cost pricing **principle of sustainability** (see Figure 1-5, p. 9 or back cover).

Unconventional Natural Gas

There are two major sources of unconventional natural gas that are both difficult and costly to exploit without high environmental impacts. One source is *coal bed methane gas* found in coal beds near the earth's surface across parts of the United States and Canada (see the map in Figure 32, p. S57, in Supplement 6). The environmental impacts of using this resource would include scarring of land, depletion of some water sources, and possible pollution of aquifers. So far it has not been economical to exploit this resource.

The other source of unconventional natural gas is *methane hydrate*—methane trapped in icy, cage-like structures

Solutions

- Step up research on the environmental impact of natural gas production

- Greatly increase monitoring and legal regulation of natural gas production, including regular inspections of the metal casings and concrete seals in well pipes

- Develop federal regulations on disposal, storage, treatment, and reuse of fracking wastewater

- Require complete disclosure of all chemicals used in fracking

- Require use of the least harmful chemicals available in fracking fluids

- Require testing of aquifers and drinking water wells for any chemical contamination from fracking operations before drilling begins and as long as gas extraction continues

- Overturn all exemptions for oil and natural gas production from any and all federal pollution regulations

Figure 15-13 Solutions: Reforms such as these, recommended by several energy analysts, could reduce the harmful impact of shale gas production. **Questions:** Which three of these steps do you think are the most important ones to take? Explain.

of water molecules buried under arctic permafrost in tundra areas of North America, northern Europe, and Siberia. Methane hydrate is also found lying on the ocean floor in several areas of the world. So far, it costs too much to get natural gas from methane hydrates. Also, scientists warn that the projected large-scale release of methane (a potent greenhouse gas) to the atmosphere during removal and processing of this resource would likely speed up atmospheric warming and the resulting projected climate disruption.

15-4 What Are the Advantages and Disadvantages of Using Coal?

CONCEPT 15-4A
Conventional coal is plentiful and has a high net energy yield at low costs, but using it results in a very high environmental impact.

CONCEPT 15-4B
We can produce gaseous and liquid fuels from coal, but they have lower net energy yields and using them would result in higher environmental impacts than those of conventional coal.

Coal Is a Plentiful but Dirty Fuel

Coal is a solid fossil fuel formed from the remains of land plants that were buried 300–400 million years ago and exposed to intense heat and pressure over millions of years (Figure 15-14).

Coal is burned in power plants (Figure 15-15) to generate about 45% of the world's electricity, according to the International Energy Agency. This includes 93% of the electricity used in South Africa and 73% of that used in China. In the United States, mostly because of cheaper natural gas, the percentage of all electricity used that is produced by burning coal dropped from 53% in 1997 to 37% in 2012 and could drop to 30% by 2020.

Coal is also burned in industrial plants to make steel, cement, and other products. In order, the world's five largest users of coal are China, the United States, India, Russia, and Japan. In 2010, China burned 3 times more coal than the United States burned.

Coal is an abundant fossil fuel. Five countries have three-fourths of the world's proven coal reserves. They are the United States with 28% of global coal reserves (see

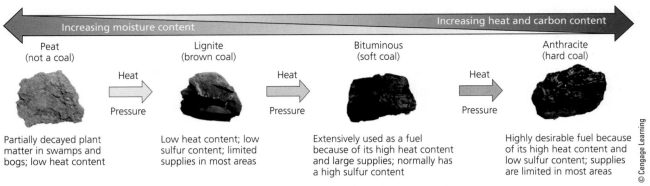

Figure 15-14 Over millions of years, several different types of coal have formed. Peat is a soil material made of moist, partially decomposed organic matter, similar to coal; it is not classified as a coal, although it is used as a fuel. These different major types of coal vary in the amounts of heat, carbon dioxide, and sulfur dioxide released per unit of mass when they are burned.

Figure 31, p. S56, in Supplement 6 for a map of major U.S. coal deposits), followed by Russia (with 18%), China (13%), Australia (9%), and India (7%). The U.S. Geological Survey estimates that identified U.S. coal reserves could last about 250 years at the current consumption rate and that identified and potential global supplies of coal could last for 200–1,100 years, depending on how rapidly they are used.

The problem is that coal is by far the dirtiest of all fossil fuels. Even when costly air-pollution-control technologies are used, burning coal pollutes the air and creates a toxic ash that is difficult to deal with. And the processes of making coal available severely degrade land and pollute water and air (see chapter-opening photo and Figures 14-12 through 14-14, p. 359).

Coal is mostly carbon but contains small amounts of sulfur, which is converted to the air pollutant sulfur dioxide (SO_2) when the coal burns. Burning coal also releases large amounts of black carbon particulates, or *soot* (Figure 15-16), and much smaller, fine particles of air pollutants such as mercury. The fine particles can get past our bodies' natural defenses that help to keep our lungs clean. According to a 2010 study by the Clean Air Task Force, fine-particle pollution in the United States, mostly from the older U.S. coal-burning power plants without the latest air-pollution-control technology, prematurely kills at least 13,000 people a year—an average of nearly 36 people every day. According to a World Bank report, burning coal in China, where air pollution control is far from adequate, causes at least 650,000 deaths a year.

Coal-burning power and industrial plants are among the largest emitters of the greenhouse gas CO_2 (Figure 15-17). China leads the world in such emissions, followed by the United States. Another problem with burning coal is that it emits trace amounts of radioactive materials as well as toxic and indestructible mercury into the atmosphere.

Finally, burning coal and removing some of the pollutants it releases from smokestack emissions produce a highly toxic ash. In the United States, about 57% of the ash is buried in landfills or in active or abandoned mines or is made into a wet slurry that is stored in holding ponds (Figure 15-18). The ash stored underground can slowly leach into groundwater, and the wet slurry can break through a pond's earthen walls, as it did at a coal ash storage pond near Knoxville, Tennessee in 2008. Such a spill can severely pollute nearby surface waters, groundwater, and land.

Coal ash is a major problem in China. According to a 2010 Greenpeace study, coal ash dumped into open landfills (Figure 15-19) is China's largest category of solid industrial waste. From these landfills, toxic chemicals are easily dispersed into the environment by wind and rain.

The Clean Coal Campaign

For decades, economically and politically powerful U.S. coal mining companies, coal-hauling railroad companies, and coal-burning power companies and industries have fought to preserve their profits by opposing measures such as stricter air pollution standards for coal-burning plants and classification of coal ash as a hazardous waste. For more than 30 years, these companies have also led the fight against efforts to classify climate-changing CO_2 as a pollutant that could be regulated by the EPA. Such regulation would likely raise their cost of doing business and make coal less competitive with cheaper sources of electricity such as natural gas and wind.

Since 2008, U.S. coal and electric utility industries have mounted a highly effective, well-financed publicity campaign built around the notion of clean coal. We can burn coal more cleanly by adding costly air-pollution-control devices to power plants. But critics argue that there could never be such a thing as clean coal. Even with stricter air pollution controls, burning coal will always involve some emissions of health-damaging air pollutants and climate-changing CO_2. It will always create indestructible and hazardous coal ash, which will actually increase with better air pollution controls, because such controls

Figure 15-15 This power plant burns pulverized coal to boil water and produce steam that spins a turbine to produce electricity. The largest coal-burning power plant in the United States, located in Indiana, burns three 100-car trainloads of coal per day. **Question:** Does the electricity that you use come from a coal-burning power plant?

involve the creation of more coal ash. Also, mining coal will always involve disrupting land—in many cases, vast areas of land—and polluting water and air.

Coal companies and utilities can get away with talking about clean coal, and can continue to produce electricity cheaply by burning coal, primarily because the harmful environmental and health costs of producing and using coal are not included in the market prices of coal and coal-fired electricity. This violates the full-cost pricing **principle of sustainability** (see Figure 1-5, p. 9 or back cover). According to a 2010 study by Harvard Medical School's Center for Health and the Global Environment and a similar 2009 study by the U.S. National Academy of Sciences, including all such costs would double or triple the price of electricity from coal-fired power plants.

Figure 15-20 lists the advantages and disadvantages of using coal as an energy resource (Concept 15-4A). An important and difficult question for humanity is whether

Figure 15-16 These smokestacks on a coal-burning industrial plant emit large amounts of air pollution because the plant has inadequate air pollution controls.

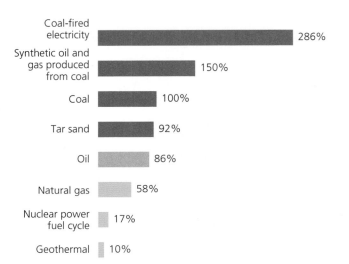

Figure 15-17 CO_2 emissions, expressed as percentages of emissions released by burning coal directly, vary with different energy resources. **Question:** Which of these produces more CO_2 emissions per kilogram: burning coal to heat a house, or heating with electricity generated by coal?

(Compiled by the authors using data from U.S. Department of Energy.)

we should begin shifting from use of abundant coal to using less environmentally harmful energy resources. Part of this shift would be to enact and enforce much stricter regulations on air pollution from coal-burning plants and on the handling of coal ash. In countries such as the United States, Russia, China, and India that have large reserves of coal, this would be a difficult economic and political challenge.

We Can Convert Coal into Gaseous and Liquid Fuels

We can convert solid coal into **synthetic natural gas (SNG)** by a process called *coal gasification,* which removes sulfur and most other impurities from coal. We can also convert it into liquid fuels such as methanol and synthetic gasoline through a process called *coal liquefaction.* These fuels, called *synfuels,* are often referred to as cleaner versions of coal.

However, compared to burning coal directly, producing synfuels requires the mining of 50% more coal. Producing and burning synfuels could also add 50% more carbon dioxide to the atmosphere (Figure 15-17). As a result, synfuels have a lower net energy yield and cost more to produce per unit of energy than does coal production. Also, it takes large amounts of water to produce synfuels. Therefore, greatly increasing the use of these synfuels would worsen two of the world's major environmental problems: projected climate disruption caused mostly by CO_2 emissions and increasing water shortages in many parts of the world (see Figure 13-9, p. 324, and Figure 13-10, p. 324).

Figure 15-21 lists the advantages and disadvantages of using liquid and gaseous synfuels produced from coal (**Concept 15-4B**).

Figure 15-18 Coal sludge impoundment in West Virginia above an elementary school.

Jim West/Alamy

Figure 15-19 Coal ash from a power plant in China is dumped into an open landfill where it can be blown by wind and washed by precipitation onto surrounding lands.

GREG GIRARD/National Geographic Creative

Trade-Offs

Coal

Advantages	Disadvantages
Ample supplies in many countries	Severe land disturbance and water pollution
Medium to high net energy yield	Fine particle and toxic mercury emissions threaten human health
Low cost when environmental costs are not included	Emits large amounts of CO_2 and other air pollutants when produced and burned

© Cengage Learning

Figure 15-20 Using coal as an energy resource has advantages and disadvantages. **Questions:** Which single advantage and which single disadvantage do you think are the most important? Why? Do you think that the advantages of using coal as an energy resource outweigh its disadvantages? Explain.

Photo: El Greco/Shutterstock.com

Trade-Offs

Synthetic Fuels

Advantages	Disadvantages
Large potential supply in many countries	Low to medium net energy yield
Vehicle fuel	Requires mining 50% more coal with increased land disturbance, water pollution, and water use
Lower air pollution than coal	Higher CO_2 emissions than coal

© Cengage Learning

Figure 15-21 The use of synthetic natural gas (SNG) and liquid synfuels produced from coal as energy resources has advantages and disadvantages (Concept 15-2). **Questions:** Which single advantage and which single disadvantage do you think are the most important? Why? Do you think that the advantages of using synfuels produced from coal as an energy source outweigh the disadvantages? Explain.

Photo: ©mironov/Shutterstock.com

15-5 What Are the Advantages and Disadvantages of Using Nuclear Power?

CONCEPT 15-5
Nuclear power has a low environmental impact and a very low accident risk, but its use has been limited by a low net energy yield, high costs, fear of accidents, long-lived radioactive wastes, and its role in spreading nuclear weapons technology.

How Does a Nuclear Fission Reactor Work?

To evaluate the advantages and disadvantages of nuclear power, we must know how a nuclear power plant and its accompanying nuclear fuel cycle work. A nuclear power plant is a highly complex and costly system designed to perform a relatively simple task: to boil water and produce steam that spins a turbine and generates electricity.

Figure 15-22 This water-cooled nuclear power plant, with a pressurized water reactor, produces intense heat that is used to convert water to steam, which spins a turbine that generates electricity. **Question:** How does this plant differ from the coal-burning plant in Figure 15-15?

Robert Llewellyn/SuperStock

© Cengage Learning

What makes a nuclear power plant complex and costly is the use of a controlled nuclear fission reaction (see Figure 2-9, center, p. 40) to provide the heat. The fission reaction takes place in a *reactor*. The most common reactors, called light-water reactors (LWRs; see Figure 15-22), produce 85% of the world's nuclear-generated electricity (100% in the United States).

The fuel for a reactor is made from uranium ore mined from the earth's crust. After it is mined, the ore must be enriched to increase the concentration of its fissionable uranium-235 by 1–5%. The enriched uranium-235 is processed into small pellets of uranium dioxide. Each pellet, about the size of an eraser on a pencil, contains the energy equivalent of about a ton of coal. Large numbers of the pellets are packed into closed pipes, called *fuel rods*,

which are then grouped together in *fuel assemblies*, to be placed in the core of a reactor.

Control rods are moved in and out of the reactor core to absorb neutrons generated in the fission reaction, thereby regulating the rate of fission and the amount of power produced. A *coolant*, usually water, circulates through the reactor's core to remove heat to keep the fuel rods and other reactor components from melting and releasing massive amounts of radioactivity into the environment. An LWR includes an emergency core cooling system as a backup to help prevent such meltdowns. A nuclear reactor cannot explode as an atomic bomb does and cause massive damage. The danger is from smaller explosions that can release radioactivity into the environment or cause a core meltdown.

Some nuclear plants withdraw the large quantities of cooling water they need from a nearby source such as a river or lake and return the heated water to that source. Other nuclear plants transfer the waste heat from the intensely hot water to the atmosphere by using one or more gigantic cooling towers, such as those at the Three Mile Island nuclear power plant near Harrisburg, Pennsylvania (USA), shown in the inset photo of Figure 15-22. There, a serious accident in 1979 caused a partial meltdown of one of the plant's reactors, but no lives were lost.

A *containment shell* with thick, steel-reinforced concrete walls surrounds the reactor core. It is designed to help keep radioactive materials from escaping into the environment, in case there is an internal explosion or a melting of the reactor's core. It also helps protect the core against some external threats such as tornadoes and plane crashes. These essential safety features help to explain why a new nuclear power plant costs as much as $10 billion and why that cost continues to rise.

What Is the Nuclear Fuel Cycle?

Building and running a nuclear power plant is only one part of the **nuclear fuel cycle** (Figure 15-23), which also includes the mining of uranium, processing and enriching the uranium to make fuel, using it in a reactor, safely storing the resulting highly radioactive wastes for thousands of years until their radioactivity falls to safe levels, and retiring the highly radioactive worn-out plant by taking it apart and storing its high- and moderate-level radioactive parts safely for thousands of years.

The final step in the cycle occurs when, after 20–60 years, a reactor comes to the end of its useful life, mostly because of corrosion and radiation damage to its metal parts, and it must be *decommissioned,* or retired. It cannot simply be shut down and abandoned, because its structure contains large quantities of high- and intermediate-level radioactive materials that must be kept out of the environment for thousands of years.

As long as a reactor is operating safely, the power plant itself has a fairly low environmental impact and a very low risk of an accident. However, considering the entire nuclear fuel cycle, the potential environmental impact increases. High-level radioactive wastes must be stored safely for thousands of years, and several points in the cycle are vulnerable to terrorist attack. Also, the uranium-enrichment and other technologies used in the cycle can be used to produce nuclear weapons–grade uranium (Concept 15-5).

Each step in the nuclear fuel cycle adds to the cost of nuclear power and reduces its net energy yield (Concept 15-1). If we add the enormous amount of energy needed to dismantle a plant at the end of its life and transport and safely store its highly radioactive materials, some scientists estimate that using nuclear power will eventually have a negative net energy yield, requiring more energy than it will ever produce.

Proponents of nuclear power tend to focus on the low CO_2 emissions and multiple safety features of the reactors. But in evaluating the safety, economic feasibility, net energy yield, and overall environmental impact of nuclear power, energy experts and economists caution us to look at the entire nuclear fuel cycle, not just the power plant itself. Figure 15-24 lists the major advantages and disadvantages of producing electricity by using the nuclear power fuel cycle (Concept 15-3).

Let's look more closely at some of the challenges involved in using nuclear power.

Storing Radioactive Spent-Fuel Rods Presents Risks

The high-grade uranium fuel in a nuclear reactor lasts for 3–4 years, after which it becomes *spent,* or useless, and must be replaced. On a regular basis, reactors are shut down for refueling, which usually involves replacing about a third of the reactor's fuel rods that contain the spent fuel.

The amount of nuclear waste from nuclear reactors is not huge, but the spent-fuel rods are so intensely hot and highly radioactive that they cannot be simply thrown away. Researchers have found that 10 years after being removed from a reactor, a single spent-fuel rod assembly can still emit enough radiation to kill a person standing 1 meter (39 inches) away in less than 3 minutes.

Thus, after spent-fuel rod assemblies are removed from reactors, they are stored in *water-filled pools* (Figure 15-25, left). After several years of cooling, they can be transferred to *dry casks* made of heat-resistant metal alloys and concrete and filled with inert helium gas (Figure 15-25, right). No one knows how long these casks can be used before they break down. They are licensed for 20 years and could last for 100 or more years—still a tiny fraction of the thousands of years that the waste must be safely stored.

A 2005 study by the U.S. National Academy of Sciences warned that the intensely radioactive waste storage pools and dry casks at 68 nuclear power plants in 31 U.S. states are especially vulnerable to sabotage or terrorist attack because they lie outside of the heavily protected reactor containment buildings. At each of the country's nuclear power plants, a government team of mock terrorists runs a test "attack" about every 3 years to test their security. Government records reveal that, between 2005 and 2010, eight of the roughly 100 attempts to breach security at U.S. nuclear plants were successful.

A 2002 study by the Institute for Resource and Security Studies and the Federation of American Scientists pointed out that in the United States, many millions of people live near aboveground spent-fuel storage sites. For some time, critics have been calling for the construction of much more secure structures to protect spent-fuel storage pools and dry casks and for moving more of the wastes from pools to casks. They charge that this has not been done because it would add billions of dollars to the already high cost of electricity produced by the nuclear power fuel cycle.

Figure 15-23 Using nuclear power to produce electricity involves a sequence of steps and technologies that together are called the *nuclear fuel cycle*. **Question:** Do you think the market price of nuclear-generated electricity should include all the costs of the nuclear fuel cycle, in keeping with the full-cost pricing **principle of sustainability**? (See Figure 1-5, p. 9 or back cover.) Explain.

Trade-Offs

Conventional Nuclear Fuel Cycle

Advantages	Disadvantages
Low environmental impact (without accidents)	Low net energy yield
	High overall cost
Emits 1/6 as much CO$_2$ as coal	Produces long-lived, harmful radioactive wastes
Low risk of accidents in modern plants	Promotes spread of nuclear weapons

Figure 15-24 Using the nuclear power fuel cycle (Figure 15-23) to produce electricity has advantages and disadvantages. **Questions:** Which single advantage and which single disadvantage do you think are the most important? Why? Do you think that the advantages of using nuclear power outweigh its disadvantages? Explain.

Photo: ©Kletr/Shutterstock.com

CONSIDER THIS. . .

THINKING ABOUT Nuclear Waste Security

Do you favor measures to provide better protection for spent-fuel rods, even if they would raise the cost of electricity? Explain.

Dealing with Radioactive Nuclear Wastes Is a Difficult Scientific and Political Problem

The nuclear waste problem begins with spent-fuel rods. They can be processed to remove radioactive plutonium, which can then be used as nuclear fuel, thus closing the nuclear fuel cycle (Figure 15-23). This reprocessing reduces the storage time for the remaining wastes from up to 240,000 years (longer than the current version of the human species has been around) to about 10,000 years.

However, reprocessing is very costly, and the resulting plutonium could also be used by terrorists or nations to make nuclear weapons, as India did in 1974. This is mainly why the United States, after spending billions of dollars, abandoned this fuel recycling approach in 1977.

Figure 15-25 After 3 or 4 years in a reactor, spent-fuel rods are removed and stored in a deep pool of water contained in a steel-lined concrete basin (left) for cooling. After about 5 years of cooling, the fuel rods can be stored upright on concrete pads (right) in sealed dry-storage casks made of heat-resistant metal alloys and thick concrete. **Questions:** Would you be willing to live within a block or two of these casks or have them transported through the area where you live in the event that they were transferred to a long-term storage site? Explain. What are the alternatives?

Photos: U.S. Department of Energy/Nuclear Regulatory Commission

Also, a 2007 study by the nonprofit Institute for Energy and Environmental Research found that nuclear reprocessing increases the volume of nuclear waste sixfold and costs more than using mined uranium, further adding to the high cost of the nuclear fuel cycle. Currently, France, Russia, Japan, India, the United Kingdom, and China reprocess some of their nuclear fuel.

Some analysts have suggested that we could shoot our intensely radioactive wastes into space or into the sun. But the costs of such an effort would be extremely high and a launch accident—such as the 1986 explosion of the Space Shuttle Challenger—could disperse high-level radioactive wastes over large areas of the earth's surface.

Most scientists and engineers agree in principle that deep burial in an underground repository is the safest and cheapest way to store high-level radioactive wastes for thousands of years. Such repositories are in use on a limited basis in the U.S. state of New Mexico, for long-term storage of nuclear waste from the U.S. nuclear weapons program, and in Finland. However, some scientists contend that it is not possible to demonstrate that this or any method will work for thousands of years.

Between 1987 and 2009 the U.S. Department of Energy spent $12 billion on research and testing of a repository for long-term underground storage of high-level radioactive wastes from commercial nuclear reactors on federal land in the Yucca Mountain desert region northwest of Las Vegas, Nevada. In 2010, this project was abandoned for scientific and political reasons and because it was too small to store even the existing radioactive wastes. A government panel is looking for alternative solutions and sites, including two sites for temporary storage of dry casks. Meanwhile these deadly wastes are building up.

Some are calling for reviving the Yucca Mountain site process and for finding another underground storage site to handle future wastes. This presents a political problem, because most states do not want to host a nuclear waste repository, and most people do not want to have highly radioactive wastes transported through their communities on a regular basis from the country's various nuclear reactors (see the map in Figure 34, p. S59 in Supplement 6) to a central nuclear waste repository.

Another radioactive waste problem arises when a nuclear power plant reaches the end of its useful life after about 40 to 60 years and must be closed. Around the world, 285 of the 432 commercial nuclear reactors now operating will need to be decommissioned by 2025.

Eventually all nuclear plants will have to be dismantled and their high-level radioactive materials will have to be stored safely. Scientists have proposed three ways to do this. For any particular plant, one strategy is to store the highly radioactive parts in a permanent, secure repository. A second approach is to install a physical barrier around the plant and set up full-time security for 30–100 years, until the plant can be dismantled after its radioactivity has reached safer levels. These levels would still be high enough to require long-term safe storage of leftover materials.

A third option is to enclose the entire plant in a concrete and steel–reinforced tomb, called a containment structure. This is what was done with a reactor at Chernobyl, Ukraine, that exploded and nearly melted down in 1986, due to a combination of poor reactor design and human operator error. The explosion and the radiation released over a large area killed a number of people and contaminated a vast area of land with long-lasting radioactive fallout in what is viewed as the world's worst nuclear power plant accident. However, within a few years, the containment structure began to crumble, due to the corrosive nature of the radiation inside the damaged reactor, and to leak radioactive wastes. The structure is being rebuilt at great cost and is unlikely to last even several hundred years.

Regardless of the method chosen, the high costs of retiring nuclear plants add to the enormous costs of the nuclear power fuel cycle and reduce its already low net energy yield. Even if all the nuclear power plants in the world were shut down tomorrow, we would still have to find a way to protect ourselves from their high-level radioactive components for thousands of years.

Can Nuclear Power Lessen Dependence on Imported Oil and Help Reduce Projected Climate Change?

Some proponents of nuclear power in the United States claim it will help the country to reduce its dependence on imported crude oil. Other analysts argue that because oil-burning power plants provide only about 1% of the electricity produced in the United States and similarly small amounts in most other countries that use nuclear power, replacing these plants with very costly nuclear power plants would not save much oil.

Nuclear power advocates also contend that increased use of nuclear power will greatly reduce the CO_2 emissions that contribute to projected climate change. Critics argue that the nuclear power industry has mounted a misleading but effective public relations campaign to convince the public that nuclear power does not emit CO_2 and other greenhouse gases.

As scientists point out, this argument is only partially correct. While nuclear plants are operating, they do not emit CO_2. However, during the 10 years that it typically takes to build a plant, especially in the manufacturing of many tons of construction cement, large amounts of CO_2 are emitted. Every other step in the nuclear power fuel cycle (Figure 15-23) also involves CO_2 emissions. Such emissions are much lower than those from coal-burning power plants (Figure 15-17), but they still contribute to atmospheric warming and projected climate disruption.

In 2009, Michael Mariotte, Executive Director of the Nuclear Information and Resource Service, estimated that in order for nuclear power to play an effective role in slowing projected climate disruption over the next 50 years, the world would need to build some 2,000 nuclear reactors—an average of one about every 2 weeks. If power plants could be brought online at such a pace, the amount of high-level radioactive wastes generated would grow dramatically. This would require that new and very costly nuclear waste repositories be built at a similar fast pace.

Experts Disagree about the Future of Nuclear Power

In the 1950s, researchers predicted that by the year 2000, at least 1,800 nuclear power plants would supply 21% of the world's commercial energy (25% of that in the United States) and most of the world's electricity. After almost 60 years of development, a huge financial investment, and enormous government subsidies, some 432 commercial nuclear reactors in 31 countries produced only 5% of the world's commercial energy and 15% of its electricity. In the United States, 104 licensed commercial nuclear power reactors generate about 8% of the country's overall energy and 19% of its electricity.

In 2013, 60 new nuclear reactors were under construction in 13 countries, far from the number needed just to replace the reactors that will have to be decommissioned in coming years. Another 156 reactors are planned, but even if they are completed after a decade or two, they will not replace the 285 aging reactors that must be retired around the world. This helps to explain why nuclear power is now the world's slowest-growing form of commercial energy (see Figure 8, p. S67, in Supplement 7).

The future of nuclear power is a subject of debate. Critics argue that the most serious problem with the nuclear power fuel cycle is that it is uneconomical. They contend that the nuclear power industry could not exist without high levels of financial support from governments and taxpayers, because of the extraordinarily high cost of ensuring safety and the low net energy yield of the nuclear power fuel cycle.

For example, the U.S. government has provided huge research and development subsidies, tax breaks, and loan guarantees to the industry (with taxpayers accepting the risk of any debt defaults) for more than 50 years. It also assumes most of the financial burden of finding ways to store radioactive wastes. In addition, the government provides accident insurance guarantees, because insurance companies have refused to fully insure any nuclear reactor from the consequences of a catastrophic accident. Nuclear power is the only energy resource that receives this government subsidy.

According to the nonpartisan Congressional Research Service, since 1948, the U.S. government has spent more than $95 billion (in 2011 dollars) on nuclear energy research and development (R & D)—more than 4 times the amount spent on R & D for solar, wind, geothermal, biomass, biofuels, and hydropower combined. Critics contend that, without these large and little-known subsidies and tax breaks, the nuclear industry would not exist in the United States. Some question the need for continuing such taxpayer support.

CONSIDER THIS. . .

THINKING ABOUT Government Subsidies for Nuclear Power

Do you think the benefits of nuclear power justify high government (taxpayer) subsidies and tax breaks for the nuclear industry? Explain.

Another obstacle to the growth of nuclear power has been public concerns about the safety of nuclear reactors. Because of the multiple built-in safety features, the risk of exposure to radioactivity from nuclear power plants in the United States and most other more-developed countries is extremely low. However, several explosions and partial or complete meltdowns have occurred (see the Case Study that follows).

CONSIDER THIS. . .

CONNECTIONS Nuclear Power Plants and the Spread of Nuclear Weapons

In the international marketplace, the United States and 14 other countries have been selling commercial and experimental nuclear reactors and uranium fuel-enrichment and purification technology for decades. Much of this information and equipment can be used to produce nuclear weapons. Energy expert John Holdren pointed out that 60 countries that have nuclear weapons or the knowledge to develop them (not including the United States, Great Britain, and the former Soviet Union) have gained most of such information by using civilian nuclear power technology. Some critics see this as the single most important reason for not building more nuclear power plants anywhere in the world.

Proponents of nuclear power argue that governments should continue funding research, development, and pilot-plant testing of potentially safer and less costly new types of reactors. The nuclear industry claims that hundreds of new *advanced light-water reactors (ALWRs)* could be built in just a few years. ALWRs have built-in safety features designed to make meltdowns and releases of radioactive emissions almost impossible. Another proposal is to mix beryllium and uranium to create fuel rods for conventional reactors that would help to prevent meltdowns.

In 2012, the U.S. Nuclear Regulatory Commission approved construction of two new nuclear reactors at an existing power plant in Georgia. Each will have a passive cooling system that can slowly cool the core and prevent a meltdown without the use of pumps when the system has to be shut down. These reactors are being built with the help of loan guarantees from the federal government. Also, the state of Georgia gave permission to the electric utilities to pre-charge their customers for the costs of building these reactors even if they are never completed. Also in 2012, the Department of Energy provided research and development funds for the development and evaluation of small modular reactors that would be built at a factory, hauled to a plant site by train, truck, or barge, placed underground, and wired up.

Some scientists call for replacing today's uranium-based reactors with new ones to be fueled by thorium. They argue that such reactors would be much less costly and safer because they cannot melt down. Also, the nuclear waste they produce cannot be used to make nuclear weapons. To help reduce its dependence on coal, China, which gets less than 2% of its electricity from nuclear power, is building 26 nuclear power plants, some of them fueled with thorium. China is also building a repository for high-level nuclear waste in the country's arid west.

To be environmentally and economically acceptable, some analysts believe that any new-generation nuclear technology should meet the five criteria listed in Figure 15-26. So far, no existing or proposed reactors even come close to doing so. However, even with considerable government financial support and loan guarantees, most U.S. utility companies and money lenders are unlikely to take on the financial risk of building new nuclear plants of any design, as long as electricity can be produced more cheaply with the use of natural gas (**Core Case Study**) and wind power.

CASE STUDY

The 2011 Nuclear Power Plant Accident in Japan

A major accident occurred on March 11, 2011, at the Fukushima Daiichi Nuclear Power Plant on the northeast coast of Japan. The accident was triggered by a major offshore earthquake that caused a severe tsunami (see Figure 14-22, p. 368). A huge wave of seawater washed over the nuclear plant's protective seawalls and knocked out the circuits and backup diesel generators of the emergency core cooling systems for three of the reactors. Then, explosions (presumably from the buildup of hydrogen gas) blew the roofs off three of the reactor buildings (Figure 15-27) and released radioactivity into the atmosphere and nearby coastal waters. Evidence indicates that the cores of these three reactors suffered full meltdowns.

After some initial confusion and conflicting statements about the severity of the accident, the Japanese government evacuated all residents within a 20-kilometer (12-mile) radius of the plant. Two weeks later, as the severity of the accident became more apparent, people within 30 kilometers (19 miles) were urged to evacuate. More than 110,000 people left their homes, and some areas that now still contain high radiation levels will likely remain unsafe to occupy for up to 20 years.

Preliminary studies indicate that four key human-related factors contributed to this accident: **(1)** failure of the utility company to develop worst-case scenarios that

---

Solutions

- Reactors must be built so that a runaway chain reaction is impossible.

- The reactor fuel and methods of fuel enrichment and fuel reprocessing must be such that they cannot be used to make nuclear weapons.

- Spent fuel and dismantled structures must be easy to dispose of without burdening future generations with harmful radioactive waste.

- Taking its entire fuel cycle into account, nuclear power must generate a net energy yield high enough so that it does not need government subsidies, tax breaks, or loan guarantees to compete in the open marketplace.

- Its entire fuel cycle must generate fewer greenhouse gas emissions than other energy alternatives.

© Cengage Learning

Figure 15-26 Some critics of nuclear power say that any new generation of nuclear power plants should meet all of these five criteria. *Question:* Do you agree or disagree with these critics? Explain.

would have helped speed up their reaction to the crisis, (2) the fact that the plant's protective seawalls were not built high enough to withstand huge tsunami waves in this well-known earthquake zone, (3) design flaws that exposed the emergency core cooling system controls and backup generators to flooding and that failed to protect the spent-fuel rod storage pools from the damages they suffered, and (4) a too-cozy relationship between nuclear plant owners and the government's nuclear regulatory officials.

This accident and the flawed response to it on the parts of plant officials and the government greatly damaged the public confidence of Japanese citizens in the safety of nuclear power. It was a serious accident that contaminated a large area with low to moderate levels of radioactivity. Over the long term, this radiation is projected to kill from 100 to 1,500 people by causing cancers, especially thyroid cancer.

This accident could lead to reduced reliance on nuclear power in Japan. It has prompted Germany, Switzerland, and Belgium to announce plans for phasing out nuclear power. Nuclear power proponents see this as an overreaction, arguing that the annual death toll resulting from the burning of coal is much greater than that of nuclear power accidents.

Is Nuclear Fusion the Answer?

Other proponents of nuclear power hope to develop **nuclear fusion**—a nuclear change at the atomic level in which the nuclei of two isotopes of a light element such as hydrogen are forced together at extremely high temperatures until they fuse to form a heavier nucleus, releasing energy in the process (see Figure 2-9, bottom, p. 40). Some scientists hope that controlled nuclear fusion will provide an almost limitless source of energy.

With nuclear fusion, there would be no risk of a meltdown or of a release of large amounts of radioactive materials, and little risk of the additional spread of nuclear weapons. Fusion power might also be used to destroy toxic wastes and to supply electricity for desalinating water and for decomposing water to produce hydrogen fuel as a very clean-burning energy source.

However, in the United States, after more than 50 years of research and a $25 billion investment, controlled nuclear fusion is still in the laboratory stage. None of the approaches tested so far has produced more energy than they use. In 2006, the United States, China, Russia, Japan, South Korea, India, and the European Union agreed to spend at least $12.8 billion in a joint effort to build a large-scale experimental nuclear fusion reactor by 2026 to determine if it can produce a net energy yield. By 2012, the estimated cost of this project had doubled and it was behind schedule.

If everything goes well, the experimental fusion reactor is supposed to produce enough electricity to run the air conditioners in a small city for a few minutes. Some critics view it as a very costly pie-in-the-sky project that

Figure 15-27 An explosion in one of the reactors in the Fukushima Daiichi nuclear power plant severely damaged the reactor.

diverts money from more promising energy alternatives like wind and solar energy. Unless there is an unexpected scientific breakthrough, some skeptics will continue to quip that "nuclear fusion is the power of the future and always will be."

Big Ideas

- A key factor to consider in evaluating the long-term usefulness of any energy resource is its net energy yield.

- Conventional oil, natural gas, and coal are plentiful and have moderate to high net energy yields, but use of these fossil fuels, especially coal, has a high environmental impact.

- The nuclear power fuel cycle has a low environmental impact and a very low accident risk, but high costs, a low net energy yield, long-lived radioactive wastes, and its role in spreading nuclear weapons technology have limited its use.

A New U.S. Oil and Natural Gas Era and Sustainability

Pepik/Shutterstock.com

We began this chapter with a look at the possibility of a new era of oil and natural gas production in the United States. We also learned a guiding scientific principle underlying all energy use—that the long-term usefulness of any energy resource depends on its *net energy yield.* Conventional oil, natural gas, and coal have medium to high net energy yields that will decrease as we use up their easily accessible supplies. Using these energy resources and others derived from them involves high environmental impacts, although these impacts vary. For example, using coal has a very high environmental impact (see photo at left and chapter-opening photo), compared to use of oil and natural gas. Implementing regulations to reduce these harmful environmental impacts will further reduce the net energy yields of these fuels.

We also looked at how we use nuclear power to produce electricity. As we evaluate the net energy yield and environmental impacts of nuclear power, it is important to consider the whole cradle-to-grave nuclear power fuel cycle—from mining the uranium fuel to dismantling and storing the worn-out reactor parts for thousands of years. Considering this whole fuel cycle, nuclear power's environmental impacts are fairly high, and its net energy yield is so low that it is economically unsustainable and must be propped up by various subsidies.

We cannot recycle energy because of the second law of thermodynamics (see Chapter 2, p. 43). However, by reusing and recycling more of the materials we use in our daily lives and in industry and transportation, we could cut our need for energy, thereby raising net energy yields—an application of the chemical cycling **principle of sustainability** (see Figure 1-2, p. 6 or back cover). Also, by using a diversity of energy resources, just as nature relies on the biodiversity **principle of sustainability**, we can further reduce the environmental impacts of our use of energy. Applying the full-cost pricing **principle of sustainability** (see Figure 1-5, p. 9 or back cover) to all energy resources would give us a more realistic understanding of the true economic and environmental costs of using nonrenewable fossils fuels and nuclear power, as well as a variety of renewable energy alternatives that we examine in the next chapter.

Chapter Review

Core Case Study

1. Summarize the potential for greatly increased production of oil and natural gas in the United States and describe two major problems to be overcome.

Section 15-1

2. What is the key concept for this section? What is **net energy yield** and why is it important in evaluating energy resources? Explain why some energy resources need help in the form of subsidies to compete in the marketplace, and give an example.

Section 15-2

3. What are the two key concepts for this section? What is **crude oil (petroleum)**, and how are oil deposits detected and removed? What percentages of the commercial energy used in the world and in the United States are provided by conventional crude oil? What is the **peak production** for an oil well and for the world oil deposits? What is **refining**? What are **petrochemicals** and why are such chemicals important? What countries are the world's three largest producers of oil and what countries are the three largest consumers?

4. What are **proven oil reserves** and what five factors determine such reserves? Define **horizontal drilling** and **hydraulic fracturing** or **fracking** and explain how these two technologies are being used to extract tightly held oil and natural gas from shale rock. What three countries have the largest percentages of the world's proven oil reserves? What percentages are found in the United States and China? Based on data from the U.S. Department of Energy and the U.S. Geological Survey, what are three conclusions that have been drawn concerning consumption of conven-

tional light oil? What are the major environmental costs of using oil? What are the major advantages and disadvantages of using conventional light oil as an energy resource? Describe U.S. dependence on oil and on imported oil. How can U.S. oil production increase dramatically in coming years, according to some analysts? What are two factors that could limit such an increase in domestic oil production?

5. Explain how we can get heavy oil from oil shale rock and from **tar sands (oil sands)**. What are the major advantages and disadvantages of using heavy oils produced from tar sands and from oil shale rock?

Section 5-3

6. What is the key concept for this section? Define **natural gas**, **liquefied petroleum gas (LPG)**, and **liquefied natural gas (LNG)**. What countries have the three largest portions of the world's proven natural gas reserves? What percentages of the world's reserves are held by the United States and China? What are the major advantages and disadvantages of using conventional natural gas as an energy resource? Why has natural gas production risen sharply in the United States and what two factors could hinder this rise? Describe five major problems resulting from increased use of fracking to produce natural gas in the United States and six ways to deal with these problems. What are two other sources of unconventional natural gas and what major problems are related to the use of these resources?

Section 15-4

7. What are the two key concepts for this section? What is **coal** and how is it formed? How does a coal-burning power plant work? What percentage of the electricity used in the world and in the United States comes from burning coal in power plants? What three countries have the largest proven reserves of coal? What are three major problems resulting from the use of coal? Explain why there is no such thing as clean coal. What are the major advantages and disadvantages of using coal as an energy resource? What are the major advantages and disadvantages of using **synthetic natural gas (SNG)** produced from coal?

Section 15-5

8. What is the key concept for this section? How does a nuclear fission reactor work and what are its major safety features? Describe the **nuclear fuel cycle**. What are three ways to decommission a nuclear power plant at the end of its useful life? What are the major advantages and disadvantages of relying on the nuclear fuel cycle as a way to produce electricity? How do nuclear plant operators store highly radioactive spent-fuel rods? Why is dealing with the highly radioactive wastes produced by the nuclear fuel cycle such a difficult problem? Explain why nuclear power is not likely to reduce U.S. dependence on imported oil and why it is not likely to reduce projected climate disruption.

9. Compare the projections for growth of the nuclear industry in the 1950s with its actual role now in generating electricity. What is the role of government subsidies in nuclear power? List and explain three major factors that have kept nuclear power from growing. What is the connection between commercial nuclear power plants and the spread of nuclear weapons? Describe the 2011 nuclear power plant accident in Japan. Define **nuclear fusion**, and summarize the story of attempts to develop it as an energy resource.

10. What are this chapter's *three big ideas*? Explain how the chemical cycling, biodiversity, and full-cost pricing **principles of sustainability** (see Figure 1-2, p. 6, and Figure 1-5, p. 9, or back cover) could be applied to our future energy resource choices.

Note: Key terms are in bold type.

Critical Thinking

1. How might greatly increased production of domestic oil and natural gas in the United States over the next two decades (**Core Case Study**) affect the country's future use of coal, nuclear power, and energy from the sun and wind? How might such an increase affect your life during the next 20 years?

2. Should governments give a high priority to net energy yields when deciding what energy resources to support? What are other factors that should be considered? Explain your thinking.

3. To continue using conventional oil at the current rate, we must discover and add to global oil reserves the equivalent of one new Saudi Arabian reserve every 7 years. Do you think this is possible? If not, what

effects might the failure to find such supplies have on your life and on the lives of any child and grandchild that you might have?

4. List three steps you could take to reduce your dependence on oil and gasoline. Which of these things do you already do or plan to do?

5. Some people see imports of Canadian oil produced from tar sands as a way to reduce U.S. dependence on oil imports from potentially unstable Middle Eastern countries and from Venezuela. Others call for developing unconventional oil from shale rock (Science Focus 15.1) as a domestic resource in order to reduce U.S. oil imports. Still others oppose both of these options because they involve environmentally

harmful processes. What is your view on this issue? Explain.

6. Explain why you agree or disagree with the following proposals made by various energy analysts as ways to solve U.S. energy problems: **(a)** find and develop more domestic supplies of conventional crude oil; **(b)** place a heavy federal tax on gasoline and imported oil to help reduce the consumption of crude oil resources and to encourage use of other alternatives; **(c)** increase dependence on coal; **(d)** phase out the use of coal by 2050; **(e)** increase dependence on nuclear power; **(f)** phase out all nuclear power plants by 2040.

7. Explain why you agree or disagree with each of the following proposals made by the U.S. nuclear power industry: **(a)** provide a new round of large government subsidies and loan guarantees to build a large number of better-designed nuclear fission power plants, in order to reduce dependence on imported oil and slow projected climate change; **(b)** prevent the public from participating in hearings on the licensing of new nuclear power plants and on safety issues at the nation's nuclear reactors; **(c)** allow electric utility companies to begin charging consumers for some of the costs of proposed new nuclear power plants before they are built; **(d)** greatly increase federal subsidies for developing nuclear fusion.

8. In 2011, the top five U.S. oil companies generated $135 billion in profits while receiving $4 billion in government subsidies and $24 billion in tax cuts. In 2011, the U.S. Senate failed to pass legislation to eliminate these subsidies and tax breaks. Do you believe that U.S. oil companies should continue to receive these subsidies and tax breaks? Explain.

Doing Environmental Science

Do a study of energy use at your school. Try to determine how the electricity used by the school is generated. If it comes from more than one source (such as coal-fired, gas-fired, or nuclear power plants, and wind farms), determine or estimate the percentages of the total for the various sources. Similarly, determine how buildings are heated, how water is heated, and how vehicles are powered. Write a report summarizing your findings.

Global Environment Watch Exercise

Search *fracking* and use the results to find information on the latest developments in public opposition to, and regulation of, fracking in the United States. Write a report on your findings.

Data Analysis

Use the graph below, comparing U.S. oil consumption, production, and imports, to answer the questions that follow.

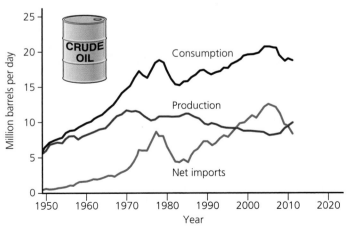

(Compiled by the authors using data from U.S. Energy Information Administration, *Monthly Energy Review*, April 2012, and *Annual Energy Review*, October 2011.)

1. By what percentage did U.S. oil consumption increase between 1982 and 2005?

2. By what percentage did U.S. oil consumption decrease between 2005 and 2011?

3. By what percentage did U.S. oil production decrease between 1985 and 2004?

4. By what percentage did U.S. oil production increase between 2004 and 2011?

5. By what percentage did U.S. net oil imports increase between 1982 and 2005?

6. By what percentage did U.S. net oil imports decrease between 2005 and 2011?

What general conclusions can you draw from these data? Compare notes and discuss your conclusions with your classmates.

The graphs below show total (upper left) and per capita (upper right) energy consumption in the United States and energy consumption by fuel (lower) in the United States. Use the graphs below to answer questions 1–4.

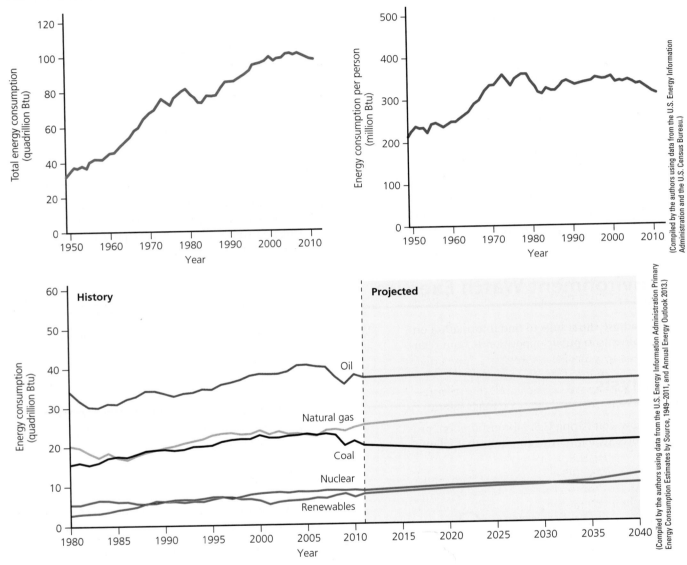

1. According to the graph, in 2000 what was the approximate total energy consumption in BTUs for the U.S.?
   a. 40 quadrillion
   b. 95 quadrillion
   c. 60 quadrillion
   d. 35 quadrillion
   e. 350 million

2. According to the graph, the projected energy consumption for nuclear power is
   a. growing exponentially.
   b. decreasing rapidly.
   c. remaining fairly flat.
   d. an example of logistic growth.
   e. growing rapidly each year.

---

3. Energy use per person in the United States is
   a. growing exponentially.
   b. decreasing logarithmically.
   c. remaining fairly flat.
   d. an example of logistic growth.
   e. growing in a linear trend.

4. The increase in per capita energy consumption from 1950–2010 has risen
   a. 10%.   d. 60%.
   b. 20%.   e. 80%.
   c. 40%.

5. The world's largest oil reserves are found in
   a. Saudi Arabia.   d. Venezuela.
   b. Mexico.   e. the United States.
   c. China.

6. When global demand for oil exceeds the rate at which it is produced
   a. price decreases and demand goes up.
   b. price and demand stay the same.
   c. flow rate is stopped.
   d. flow rate to consumers goes down and price goes up.
   e. flow rate to consumers goes up and price goes down.

7. Advocates of drilling for oil in the Alaska National Wildlife Refuge (ANWR) believe that by using the oil found here we would decrease our dependence on imported oil. However, opponents feel that
   a. tundra ecosystems recover quickly and can handle the stress.
   b. there is relatively little oil so it is not worth degrading this fragile ecosystem.
   c. since there is 20 years of oil found there we should do more research into this resource.
   d. as demand goes up we will not need this resource.
   e. a switch to renewable energy will cause America to not need this resource.

8. The conventional way that oil is extracted has all of the following disadvantages EXCEPT
   a. OPEC controls much of the world's crude oil and is therefore able to control the price of oil.
   b. the possibility of oil tanker leaks and spills such as what occurred with the Exxon Valdez.
   c. the loss or degradation of habitat if a leak develops in the pipelines used to transport oil.
   d. depending on consumption, proven and unproven oil reserves could last more than 100 years.
   e. the possibility of accidents on offshore platforms, which could pollute water from oil spills and pollute air from fires.

9. The reason three-fourths of the world's oil sand is found in Alberta, Canada, is because
   a. Bogota, Colombia, was built on Andes Mountain rock, not on sand.
   b. Austin, Texas, has already removed all of its oil sand from the ground.
   c. Mexico City, Mexico, is on a former sandy lake bed and water is beneath the city, not oil.
   d. there is the right combination of clay, sand, oil, and bitumen.
   e. the Middle Eastern countries have already removed three-fourths of their oil sand.

10. Which of the following is a benefit of using natural gas?
    a. It is four times as energy efficient as coal.
    b. It is cleaner-burning than coal.
    c. Coal power plants are cheaper to build and maintain than natural gas.
    d. Getting natural gas from Canada is inexpensive.
    e. CO_2 is not released when burning natural gas.

11. If America's per capita energy consumption was approximately 350 million BTUs, then how much does the entire population use?
    a. 105×10^5 d. 105×10^8
    b. 105×10^6 e. 105×10^9
    c. 105×10^7

12. The three largest users of coal are
    a. China, the United States, and India.
    b. China, the United States, and Mexico.
    c. the United States and Mexico.
    d. Russia, China, and India.
    e. Saudi Arabia, Japan, and China.

13. Coal has a huge environmental impact because of its
    a. threat to human health due to air pollutants released when coal burns.
    b. release of carbon, which contributes to global climate change.
    c. effect on global climate.
    d. high cost, mostly to cover the transportation of coal to electric power plants.
    e. use of new technology that makes coal safer and cleaner to remove from the earth and to transport to homes and industry.

14. Nuclear power was predicted to supply 21% of the world's commercial energy. This has not been met because of
    a. thermal pollution, which becomes a problem when nuclear power plants break down or must be dismantled.
    b. approximately 75% of the available high-quality energy being given off as waste heat.
    c. the small supply of enriched uranium-235.
    d. the lack of long-term storage facilities for high-level radioactive wastes such as spent fuel rods.
    e. the high risk of accidents such as Fukushima in Japan in 2011, Chernobyl in the U.S.S.R. in 1986, and Three Mile Island in the United States in 1979.

16 Energy Efficiency and Renewable Energy

Just as the 19th century belonged to coal and the 20th century to oil, the 21st century will belong to the sun, the wind, and energy from within the earth.

LESTER R. BROWN

Key Questions

16-1 Why is energy efficiency an important energy resource?

16-2 What are the advantages and disadvantages of using solar energy?

16-3 What are the advantages and disadvantages of using hydropower?

16-4 What are the advantages and disadvantages of using wind power?

16-5 What are the advantages and disadvantages of using biomass as an energy source?

16-6 What are the advantages and disadvantages of using geothermal energy?

16-7 What are the advantages and disadvantages of using hydrogen as an energy source?

16-8 How can we make the transition to a more sustainable energy future?

Wind farm in Oregon (USA).

Jim Oleachea/National Geographic Creative

The Astounding Potential for Wind Power in the United States

Between the earth's equator and its polar regions, the sun's rays strike the earth at different angles. This results in more solar energy at the equator and less at the poles. Together with the earth's rotation, these differences create flows of air called *wind* (see Figure 7-3, p. 146). Because wind power is an indirect form of solar energy, relying more on it is a way to apply the solar energy **principle of sustainability** (see Figure 1-2, p. 6 or back cover).

We can capture this energy with groups of wind turbines called *wind farms* that convert it into electrical energy (Figure 16-1, left) that can be fed into electrical grids. Wind farms can be located on land (see chapter-opening photo) or at sea (Figure 16-1, right).

Unlike nonrenewable fossil fuels, wind power is widely distributed and inexhaustible. The cost of producing electricity from wind is much lower than that of producing it by using the nuclear fuel cycle (Figure 15-23, p. 392) and roughly equal to the cost of producing electricity by burning coal. When we include the harmful environmental and health costs, producing electricity from wind is cheaper than producing it from coal. Using wind power also greatly reduces our emissions of climate-changing carbon dioxide (CO_2) into the atmosphere.

Because today's wind turbine towers can be as tall as a 30-story building and typically have blades as long as 60 meters (200 feet), they can tap into the stronger and more reliable winds found at higher altitudes on land and at sea. The U.S. Department of Energy (DOE) estimates that wind farms at favorable sites in three states—North Dakota, Kansas, and Texas—could more than meet the electricity needs of the lower 48 states. In addition, the National Renewable Energy Laboratory (NREL) estimates that offshore winds—off the Atlantic and Pacific coasts, and off the shores of the Great Lakes—could generate 4 times the electricity currently used in the lower 48 states.

Texas, the nation's leading oil-producing state, is also the nation's leading producer of electricity from wind power, followed by California. These two states produce more electricity from wind than all but five of the world's countries. To some analysts, wind power has more advantages and fewer serious disadvantages than all other energy resources except for energy efficiency, another topic covered in this chapter.

Wind power proponents call for a crash program to develop land-based and offshore wind farms in the United States. Over time, this would help reduce dependence on coal, which, when burned, produces air pollutants that kill at least 24,000 Americans a year and adds huge quantities of CO_2 to the atmosphere. Greatly expanding the U.S. wind turbine industry would also create much-needed new jobs and boost the American economy.

In this chapter, we explore and compare the benefits and drawbacks of improving energy efficiency, relying on wind power, and using a variety of other renewable energy resources.

Wind turbine

Figure 16-1 Wind turbines convert the kinetic energy in wind to electricity, another form of kinetic energy (moving electrons). Wind power is an indirect form of solar energy.

Photo: ssuaphotos/Shutterstock.com

16-1 Why Is Energy Efficiency an Important Energy Resource?

CONCEPT 16-1A
Improvements in energy efficiency could save at least a third of the energy used in the world and up to 43% of the energy used in the United States.

CONCEPT 16-1B
We have a variety of technologies for sharply increasing the energy efficiency of industrial operations, motor vehicles, appliances, and buildings.

We Use Energy Inefficiently

Energy efficiency is a measure of how much useful work we can get from each unit of energy we use. Improving energy efficiency means using less energy to provide the same amount of work in the forms of light, heat, transportation, and other benefits. This amounts to a largely untapped source of energy that is abundant, clean, cheap, and readily available.

You may be surprised to learn that roughly 84% of all commercial energy used in the United States does not produce useful work (Figure 16-2). About 41% of this energy unavoidably ends up as low-quality waste heat in the environment because of the degradation of energy quality imposed by the second law of thermodynamics (see Chapter 2, p. 43). The other 43% is wasted unnecessarily, mostly due to the inefficiency of lightbulbs, industrial motors, motor vehicles, and power plants (**Concept 16-1A**).

Another reason for our inefficient and thus wasteful use of energy is that many people live and work in poorly insulated, badly designed buildings that require excessive heating during cold weather and excessive cooling during hot weather. Also, many Americans live in ever-expanding suburban areas around large cities where they must depend on their cars for getting around. Roughly three of every four Americans commute to work, mostly in energy-inefficient vehicles, and only 5% rely on more energy-efficient mass transit. In addition, many people leave lights, computers, and other electrical devices on for long periods when they are not using these devices.

Energy inefficiency costs the United States an average of about $570,000 per minute, according to energy analyst Amory Lovins (see his online Guest Essay on this topic). Much of this cost is due to heavy reliance on four energy-inefficient devices:

- Huge *data centers* (electronic clouds), filled with racks of electronic servers (Figure 16-3) that process information flowing on the Internet, use only about 10% of the electrical energy they pull off of the grid. The other 90% ends up as low-quality heat that flows into the environment. Most of these centers run 24 hours a day at their maximum capacities regardless of the demand. They also require large amounts of energy for cooling and for providing hundreds to thousands of backup batteries.

- The *internal combustion engine* that propels most motor vehicles wastes about 80% of the high-quality energy in its fuel. In other words, only about 20% of the money that people spend on gasoline gets put to good use.

- A *nuclear power plant* (see Figure 15-22, p. 390), which produces electricity, wastes about 75% of the high-quality energy in its nuclear fuel and probably closer to 82% when we include the additional energy used in the nuclear fuel cycle (see Figure 15-23, p. 392).

- A *coal-fired power plant* (see Figure 15-15, p. 387) wastes about 65% of the energy that is released by burning coal to produce electricity, and probably 75–80% if we include the energy used to dig up the coal and transport it to the plant, and to transport and store the ash by-product.

By improving energy efficiency, we can gain numerous economic and environmental benefits (Figure 16-4). To most energy analysts, *it is the quickest, cleanest, and usually the cheapest way to provide more energy, reduce pollution and environmental degradation, and slow projected climate change.*

GOOD NEWS

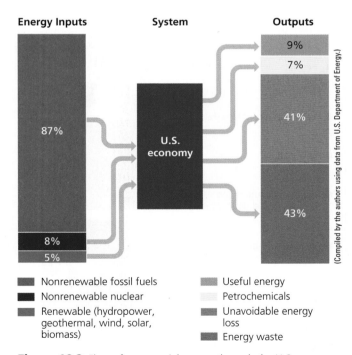

Energy Inputs | **System** | **Outputs**

87%

8%

5%

U.S. economy

9%

7%

41%

43%

(Compiled by the authors using data from U.S. Department of Energy.)

- ■ Nonrenewable fossil fuels
- ■ Nonrenewable nuclear
- ■ Renewable (hydropower, geothermal, wind, solar, biomass)
- ■ Useful energy
- □ Petrochemicals
- ■ Unavoidable energy loss
- ■ Energy waste

Figure 16-2 Flow of commercial energy through the U.S. economy. Only 16% of the country's high-quality energy ends up performing useful tasks. ***Question:*** What are two examples of unnecessary energy waste?

Figure 16-3 Electronic data processing centers that support the web-based global economy typically waste about 90% of the large amount of energy they use.

poetic_disorder/Photos.com

Solutions

Improving Energy Efficiency

- Prolongs fossil fuel supplies

- Reduces oil imports and improves energy security

- Very high net energy yield

- Low cost

- Reduces pollution and environmental degradation

- Buys time to phase in renewable energy

- Creates local jobs

© Cengage Learning

Figure 16-4 Improving energy efficiency has several benefits. **Questions:** Which two of these benefits do you think are the most important? Why?

Top: ©Dmitriy Raykin/Shutterstock.com. Center: ©V. J. Matthew/Shutterstock.com. Bottom: ©andrea lehmkuhl/Shutterstock.com.

We Can Improve Energy Efficiency in Industries and Utilities

Industry accounts for about 30% of the world's energy consumption and 33% of U.S. energy consumption, mostly for producing metals, chemicals, petrochemicals, cement, and paper. Here are several ways in which industries can improve their energy efficiency and save money (**Concept 16-1B**).

One way is to use **cogeneration**, which involves using a *combined heat and power (CHP)* system. For example, the steam used for generating electricity in a CHP system can be captured and used again to heat the power plant or other nearby buildings, rather than released into the environment as waste heat. The energy efficiency of these systems is 75–90% (compared to 30–40% for coal-fired, natural gas, and nuclear power plants). Denmark leads the world by getting 52% of its electricity from CHP systems, compared to just 8% in the United States.

GOOD NEWS

Industries could also use more *energy-efficient electric motors.* Typical inefficient motors use one-fourth of the electricity produced in the United States and 65% of the electricity used in U.S. industry. Most of these motors run only at full speed with their output throttled to match the task—somewhat like keeping one foot on the gas pedal of a car and the other on the brake pedal to control its speed. They can be replaced with more energy-efficient variable-speed motors, which run at the minimum rate needed for each job.

Recycling materials such as steel and other metals is a third way for industry to save energy and money. For example, producing steel from recycled scrap iron uses 75% less high-quality energy than does producing steel from virgin iron ore and emits 40% less CO_2. A fourth way is to use more *energy-efficient lighting* (see Figure 2-14, p. 44).

Yet another way to save energy would be to redesign the software and cooling systems in government and private electronic data processing systems (Figure 16-3). This could be done with available technology. And the servers could be powered down when they are not in use. We could also improve the efficiency of transmitting electricity to industries and communities (see the following Case Study).

A growing number of major corporations are saving money by improving energy efficiency. For example, the CEO of Dow Chemical Company, which operates 165 manufacturing plants in 37 countries, estimates that between 1996 and 2009, energy efficiency improvements cost Dow about $1.1 billion, but resulted in savings of about $9.4 billion.

Figure 16-5 The outdated high-voltage grid system in the United States could someday be replaced with an ultra-high-voltage smart grid system.

CASE STUDY

Building a Smarter and More Energy-Efficient Electrical Grid

Grid systems of high-voltage transmission lines carry electricity from power plants, wind farms (**Core Case Study**), and other electricity producers to users. In the United States, many energy experts place top priority on converting and expanding the outdated regional electrical grid system (Figure 16-5) into what they call a regional and national *smart grid*. This would be an energy-efficient, digitally controlled, ultra-high-voltage (UHV) system with superefficient transmission lines.

A smart grid could adapt to a disaster that causes a power loss in one area by quickly bringing in electricity from other parts of the country. It would also make it easier for power companies and consumers to buy electricity produced from wind, solar, and other renewable forms of energy in different parts of the country. According to the U.S. Department of Energy (DOE), building such a grid would cost the United States up to $800 billion over the next 20 years but would save the U.S. economy up to $2 trillion during that period.

A smart grid system would use smart meters to monitor the amount of electricity used and the patterns of use for each customer. This information would help consumers to reduce their power consumption and power bills. Utilities have learned that when a monthly electric bill lists the average consumption in the neighborhood, this nudges many home owners to significantly reduce their electricity use and waste.

We Can Improve Energy Efficiency and Save Money in Transportation

There is a lot of room for improving energy efficiency in transportation, which accounts for about 28% of the energy consumption and two-thirds of the oil consumption in the United States. In 2011, the U.S. government agreed to raise its fuel economy standards for new cars and light trucks from 12.6 kilometers per liter (kpl) (29.6 miles per gallon or mpg) to 23.3 kpl (54.5 mpg) by 2025. Energy experts such as Joseph Romm call for the government to require all new cars and light trucks sold in the United States to get more than 43 kpl (100 mpg) by 2040.

One reason why many consumers buy large and energy-inefficient motor vehicles is that they do not realize that gasoline costs them much more than the price they pay at the pump. The *hidden costs* not included in the price include government subsidies and tax breaks for oil companies, car manufacturers, and road builders; costs of pollution control and cleanup; time wasted idling in traffic jams; and higher medical bills and health insurance premiums resulting from illnesses caused by air and water pollution.

Consumers pay these hidden costs, but not at the gas pump. According to a 2005 study by the International Center for Technology Assessment, the hidden costs of gasoline for U.S. consumers were about $3.18 per liter ($12 per gallon). Thus, when gas costs $4 a gallon, U.S. consumers are really paying about $16 a gallon.

One way to include more of these hidden costs in the market price is through gasoline taxes, which are widely used in Europe but are politically unpopular in the United States. Some analysts call for increasing U.S. gasoline taxes and reducing payroll and income taxes to balance such increases, thereby relieving consumers of any additional financial burden.

Another way for governments to encourage higher energy efficiency in transportation is to give consumers significant tax breaks or other economic incentives to encourage them to buy more fuel-efficient vehicles. Energy expert Amory Lovins has proposed a *fee-bate* program in which buyers of fuel-inefficient vehicles would pay a high fee, and the resulting revenues would be given to buyers of fuel-efficient vehicles as rebates. For example, the fee on a gas-guzzling vehicle averaging about

Conventional hybrid

Fuel tank

Battery

Internal
combustion
engine

Transmission   Electric motor

Plug-in hybrid

Fuel tank

Battery

Internal
combustion
engine

Transmission   Electric motor

© Cengage Learning

Gyuszko/Dreamstime.com

Figure 16-6 Solutions: A *conventional gasoline–electric hybrid* vehicle (left) is powered mostly by a small internal combustion engine with an assist from a strong battery. A *plug-in hybrid electric* vehicle (right) typically has a smaller internal combustion engine with a larger and more powerful battery that can be plugged into a 110-volt or 220-volt outlet and recharged (see photo). An all-electric vehicle (not shown) runs completely on a rechargeable battery. **Question:** Would you buy one of these vehicles? Explain.

5 kpl (12 mpg) or less might be $10,000. The government would then give that amount as a rebate to a buyer of a fuel-efficient vehicle that averages 17 kpl (40 mpg) or more. Within a short time, such a program—endorsed by the U.S. National Academy of Sciences—would greatly increase sales of gas-sipping vehicles. It would cost the government (taxpayers) very little because the fees are collected from one set of buyers and given to another set of buyers.

Other ways to save energy and money in transportation include building or expanding mass transit systems within cities, constructing high-speed rail lines between cities (as is done in Japan, much of Europe, and China), and carrying more freight by rail instead of in heavy trucks. Another approach is to encourage bicycle use by building bike lanes along highways and on city streets. The University of New England in Maine offers free, high-quality bikes to new students who agree to leave their cars at home. Also, many companies are saving money by using Internet conferencing as an alternative to flying their employees to meetings.

More Energy-Efficient Vehicles Are on the Way

There is growing interest in developing modern, superefficient, ultralight, and ultrastrong cars that could get up to 130 kpl (300 mpg) using existing technology—a concept that Amory Lovins developed in 1991.

One of these vehicles is the energy-efficient, gasoline–electric *hybrid car* (Figure 16-6, left). These cars have a small, traditional gasoline-powered engine and a battery-powered electric motor used to provide the energy needed for acceleration and hill climbing. The most efficient current models of these cars get a combined city/highway

mileage of up to 21 kpl (50 mpg) and emit about 65% less CO_2 per kilometer driven than do comparable conventional cars.

A newer option is the *plug-in hybrid electric vehicle* (Figure 16-6, right). Such vehicles can travel 48–97 kilometers (30–60 miles) on electricity alone. Then the small gasoline motor kicks in, recharges the battery, and extends the driving range to 600 kilometers (370 miles) or more. The battery can be plugged into a conventional outlet and fully charged in 6 to 8 hours. Another option is an all-electric vehicle that runs on a battery only. The problem for the average consumer is that the prices on hybrid, plug-in hybrid, and all-electric cars are high because of the high cost of their batteries.

According to a DOE study, replacing most of the current U.S. vehicle fleet with plug-in hybrid vehicles over two decades would cut U.S. oil consumption by 70–90%, eliminate the need for costly oil imports, save consumers money, and reduce CO_2 emissions by 27%. If the batteries in these cars were recharged mostly by electricity generated by renewable resources such as wind (**Core Case Study**), solar cells, or hydroelectric power, U.S. emissions of CO_2 would drop by 80–90%. This would help to slow projected climate change and save thousands of lives by reducing air pollution from motor vehicles and from coal-burning power plants.

GOOD NEWS

Fully electric cars are promoted as not contributing to climate change because, while operating, they don't emit CO_2. This is correct but misleading, because the process of manufacturing these vehicles does add CO_2 to the atmosphere, requiring about twice as much energy as it takes to build a similar-sized gasoline powered car. Also, when their batteries are recharged with electricity produced by coal-burning, natural gas-burning, or nuclear-power plants, more CO_2 is emitted. Such emis-

THE SEARCH FOR BETTER BATTERIES

The major obstacle standing in the way of wider use of plug-in hybrid-electric and all-electric vehicles is the lack of an affordable, small, lightweight, and easily rechargeable car battery that can store enough energy to power a vehicle over long distances.

One promising type of battery is the *lithium-ion battery,* commonly used in laptop computers and cell phones. These batteries are light (because lithium is the lightest solid chemical element) and can pack a lot of energy into a small space. By hooking many of them together, we can use them to power hybrid, plug-in hybrid, and all-electric motor vehicles.

Much of the current research on batteries involves nanotechnology (see Science Focus 14.1, p. 362). For example, researchers at the Massachusetts Institute of Technology (MIT) developed a new type of lithium-ion battery using such technology. It is less expensive and can be charged 40 times faster than the nickel–metal hydride batteries that are used to power many of today's hybrid vehicles.

GOOD NEWS

In 2010, MIT researcher Michael Strano and his colleagues devel-oped a *thermal wave battery* that produces electricity when a pulse of heat travels through carbon nanotubes. The tubes are coated with a nanosized film of reactive fuel that burns when ignited by a spark or a laser. This creates a fast-moving thermal wave that races along the length of each carbon nanotube 10,000 times faster than the heat produced by the normal burning of the fuel. Prototypes of such batteries put out about 100 times more energy than do lithium-ion batteries of the same weight, and they can be stored indefinitely without leaking any of their charge.

MIT researcher Angela Belcher is working on another new type of battery. She has genetically engineered a virus that can coat itself with electricity-conducting materials to form a miniscule nanowire. Belcher is trying to find ways to link these tiny wires up to form the components of a battery far more compact and powerful than any yet developed. Such viral batteries would essentially grow themselves without producing the toxic wastes often associated with battery production.

Scientists have also developed *ultracapacitors,* which are small mechanical batteries that quickly store and release large amounts of electrical energy, thus providing the power needed for quick acceleration. They can be recharged in minutes and can hold a charge much longer than conventional chemical batteries, which recharge more slowly and have to be replaced more frequently than ultracapacitors. Graphene (see Science Focus 14.2, p. 363) can be used as a component of ultracapacitors. However, ultracapacitors cannot store nearly as much energy as comparable chemical batteries and so far are more expensive per unit of energy produced.

If any one or a combination of these new battery technologies can be mass-produced at an affordable cost, plug-in hybrid and all-electric vehicles could take over the car and truck market. **GREEN CAREER:** battery engineer

Critical Thinking

How might your life change if one or more of the new battery technologies discussed above become a reality?

---

sions drop sharply when the batteries are recharged with electricity produced by wind turbines, solar cells, or hydropower.

The key to greatly increasing the use of hybrid, plug-in hybrid, and all-electric motor vehicles is to ramp up research and development of improved, more affordable batteries (Science Focus 16.1). Another important factor will be to build a network of recharging stations in many convenient locations within and between communities. **GREEN CAREERS:** plug-in hybrid and all-electric car technology

A future stage in the development of superefficient cars might be an electric vehicle that uses a *fuel cell*—a device that uses hydrogen gas (H_2) as a fuel to produce electricity. Fuel cells are at least twice as efficient as internal combustion engines, have no moving parts, and require little maintenance. Their H_2 fuel is usually produced with the use of electricity, and if such electricity could be generated by wind and solar power, the widespread use of fuel cells would drastically reduce our emissions of CO_2 and other pollutants that cause air pollution and climate change—two of the world's most serious environmental, economic, and health problems. However, such cars are unlikely to be widely available until 2020 or later and will probably be very expensive, in part because H_2 has a negative net energy yield, as discussed later in this chapter. **GREEN CAREER:** fuel-cell technology

We can also improve a vehicle's fuel efficiency by reducing its weight. One way to do this is to make car bodies out of *ultralight* and *ultrastrong* composite materials such as fiberglass, carbon fiber, hemp fiber, and graphene (see Science Focus 14.2, p. 363). In addition to being lighter and stronger than some metals, these materials also would not require painting, a process that is costly and environmentally hazardous. The current cost of making such car bodies is high, but mass production would likely help to bring these costs down.

Figure 16-7 This green roof is one of hundreds in New York City. One of the windows shows a reflection of the Empire State Building, which was retrofitted in 2011. Its energy costs were cut by 38% after the renovation.

DIANE COOK & LEN JENSHEL/National Geographic Creative

We Can Design Buildings That Save Energy and Money

Globally, the construction and operation of buildings, most of which have life spans of 50–100 years, accounts for about one-third of the world's resource consumption, 25–40% of its use of energy, 30–40% of all CO_2 emissions, and 30–40% of the solid wastes produced.

We know how to make a transition to using more sustainable buildings over the next few decades. We can apply the solar energy and chemical cycling **principles of sustainability** (see Figure 1-2, p. 6 or back cover) by using energy more efficiently and using more renewable energy resources, reusing and recycling materials, reducing pollution and waste, and in the process, creating jobs and saving money.

For example, if a building is oriented to face the sun, it can get more of its heat from solar energy, which can cut heating costs by up to 20%. This is a simple application of the solar energy **principle of sustainability** that people have been using for centuries. Heating costs can be cut by as much as 75% when the building is well-insulated and airtight.

Green architecture, based on energy-efficient, resource-efficient, and money-saving designs, employs several old and new technologies. Along with natural lighting, direct solar heating, insulated windows, and energy-efficient appliances and lighting, it makes use of solar hot water heaters, electricity from solar cells, windows that darken automatically to deflect heat from the sun, thin sheets of aerogel insulation, and recycling of wastewater. **GREEN CAREERS:** sustainable environmental design and architecture

Some green buildings also have *living roofs,* or *green roofs,* covered with specially designed soil and vegetation (Figure 16-7). Such a roof can reduce the costs of cooling and heating a building by absorbing heat from the summer sun and helping to insulate the structure and retain heat in the winter. A living roof also absorbs precipitation, which would normally become part of a city's storm water runoff, adding to the pollution of its waterways.

Several European countries have used green roofs for more than 30 years. The cities of Stuttgart, Germany, Copenhagen, Denmark, and Toronto, Canada, now require green roofs on all new flat-roofed buildings. Green roofs are also sprouting in many American cities, including New York City (Figure 16-7) and Chicago, Illinois, which has 600 green roofs, more than any other U.S. city. White or light-colored roofs are also used to help reduce cooling costs by reflecting incoming solar radiation, especially in hotter climates—another strategy for working with nature that people have used for centuries.

Figure 16-8 Solutions: This energy-efficient, Victorian-style straw-bale house was designed and built by Alison Gannett in Crested Butte, Colorado. The left photo was taken during construction; the right photo shows the completed house.

Photos: ©Alison Gannett

Superinsulation is very important in energy-efficient design. A house can be so heavily insulated and airtight that heat from direct sunlight, appliances, and human bodies can warm it with little or no need for a backup heating system, even in extremely cold climates. Superinsulated houses in Sweden use 90% less energy for heating and cooling than do typical American homes of the same size. One example of superinsulation is straw-bale construction, in which a house's walls are built of straw bales that are covered on the inside and outside with adobe (Figure 16-8). Such walls can have insulating values of 2 to 6 times those of conventional walls. (See the online Guest Essay on this topic by Nancy Wicks.)

Green building certification standards have now been adopted in 21 countries, thanks to the efforts of the World Green Building Council. Between 1999 and 2012, the U.S. Green Building Council's Leadership in Energy and Environmental Design (LEED) program has awarded silver, gold, and platinum standard certificates to more than 14,200 U.S. buildings that meet certain efficiency and environmental standards. In 2009, the Sustainable Building Coalition proposed that we go further and evaluate the sustainability of buildings over their lifetimes according to measurements of resource use, energy use, water use, CO_2 emissions, outputs of solid and hazardous wastes, thermal comfort, and indoor air quality.

CONSIDER THIS. . .

THINKING ABOUT Efficient Building Design

What are three ways in which the building in which you live or work could have been designed to be more energy efficient?

We Can Save Money and Energy in Existing Buildings

There are many ways to save energy and money in existing buildings. The Department of Energy estimates that U.S. home owners can reduce their energy bills by as much as 25% through simple measures such as adding insulation, sealing air leaks, and using energy-efficient appliances. A good first step toward getting such savings is to have an expert make an *energy audit* of a house or other building to suggest ways to improve energy efficiency. Some utility companies provide energy audits at little or no cost. Such an audit might result in some or all of the following recommendations:

GOOD NEWS

- *Insulate the building and plug leaks.* About one-third of the heated air in typical U.S. homes and other buildings escapes through holes, cracks, and closed, single-pane windows (Figure 16-9). During hot weather, these windows and cracks let heat in, which increases the need for air conditioning. Adding insulation and plugging air leaks are two of the quickest and least expensive ways to improve energy efficiency and save money.

- *Use energy-efficient windows.* Energy-efficient windows can pay for themselves within several years, cut expensive heat losses from a house or other building by two-thirds, lower cooling costs in the summer, and reduce heating system CO_2 emissions. Also, *smart windows* that automatically adjust the tint of glass to reflect or absorb light are now available.

- *Stop other heating and cooling losses.* Leaky heating and cooling ducts in attics and unheated basements allow 20–30% of a home's heating and cooling energy to escape, and those in basements can draw unwanted moisture into homes. Careful sealing of duct joints can reduce this loss and save money. Also, use of white or light-colored roofs and living roofs can cut electricity use for air conditioning and reduce CO_2 emissions. Engineers are working on roofs that can change colors—being white and reflecting solar energy in hot weather and turning dark in order to absorb solar energy during cold weather.

- *Heat houses and other buildings more efficiently.* In order, the most energy-efficient ways to heat indoor space or to conserve indoor heat are: superinsulation (which would include plugging leaks); a geo-

Figure 16-9 A *thermogram*, or infrared photo, of a house in Great Britain before (left) and after it was well insulated (right). Many homes in the United States and other countries are so full of leaks that their heat loss in cold weather and heat gain in hot weather are equivalent to what would be lost through a large, window-sized hole in a wall of the house. **Question:** How do you think the place where you live would compare to the house on the left in terms of heat loss and the resulting waste of money on high heating and cooling bills?

thermal heat pump that transfers heat stored from underground into a home; passive solar heating; a high-efficiency, conventional heat pump (in warm climates only); and a high-efficiency natural gas furnace.

- *Heat water more efficiently.* One approach is to use a roof-mounted solar hot water heater. Another option is a *tankless instant water heater* fired by natural gas or LPG. (Using electricity for heating water is not efficient.) These devices, widely used in many parts of Europe and the United States, heat water instantly, providing hot water only when and for as long as it is needed.

- *Use energy-efficient appliances.* A refrigerator with its freezer on the bottom uses about half as much energy as one with the freezer on top or on the side, because on such older models, the dense cold air flows out whenever the door is opened. Microwave ovens use 25–50% less electricity than electric stoves do for cooking and 20% less than convection ovens use. Clothes dryers with moisture sensors cut energy use by 15%. Front-loading clothes washers use 55% less energy and 30% less water than top-loading models use and cut operating costs in half. LED televisions use about 70% less energy than plasma sets of the same size use.

- *Stop using the standby mode.* According to the DOE, keeping TVs, satellite and cable boxes, computers, and other electronic devices on standby when they are not being used consumes about 10% of the electricity used by a typical household. Consumers can reduce their energy use and their monthly power bills by plugging their standby electronic devices into a smart power strip that cuts off power to a device when it detects that the device has been turned off.

- *Use energy-efficient computers.* According to the EPA, if all computers sold in the United States met its Energy Star requirements, consumers would save $1.8 billion a year in energy costs and reduce greenhouse gas emissions by an amount equal to that of taking about 2 million cars off the road.

- *Use energy-efficient lighting.* As energy-wasting incandescent bulbs are being phased out, home owners, colleges and universities, and businesses are switching to more energy-efficient and longer-lasting compact fluorescent light (CFL) bulbs and LED bulbs (see Figure 2-14, p. 44). Halogen lightbulbs are also available and are less expensive than most LED bulbs, but they are only slightly more efficient than incandescent bulbs and run so hot that they can start a fire when they come into contact with flammable materials. As prices come down, highly efficient LED bulbs that use 85% less energy and last about 17 times as long as incandescent bulbs and 4 times as long as CFLS are expected to take over more of the lighting market. In 2011, the Earth Policy Institute estimated that by shifting to energy-efficient fluorescent lights in homes, office buildings, and factories and using LEDs for all traffic lights, we could save enough electricity to close more than 700 of the world's 2,800 coal-burning power plants. **GOOD NEWS**

- *Use automatic motion sensors to turn lights on and off as needed.* In 2011, the Earth Policy Institute estimated that by shifting from incandescent bulbs to LEDs and installing automatic motion sensors and dimmers, we could cut the energy used for lighting in buildings by more than 90%.

Attic
- Hang reflective foil near roof to reflect heat.
- Use house fan.
- Be sure attic insulation is at least 30 centimeters (12 inches).

Bathroom
- Install water-saving toilets, faucets, and shower heads.
- Repair water leaks promptly.

Kitchen
- Use microwave rather than stove or oven as much as possible.
- Run only full loads in dishwasher and use low- or no-heat drying.
- Clean refrigerator coils regularly.

Outside
Plant deciduous trees to block summer sun and let in winter sunlight.

Other rooms
- Use compact fluorescent lightbulbs or LEDs and avoid using incandescent bulbs.
- Turn off lights, computers, TV, and other electronic devices when they are not in use.
- Use high-efficiency windows; use insulating window covers and close them at night and on sunny, hot days.
- Set thermostat as low as you can in winter and as high as you can in summer.
- Weather-strip and caulk doors, windows, light fixtures, and wall sockets.
- Keep heating and cooling vents free of obstructions.
- Keep fireplace damper closed when not in use.
- Use fans instead of, or along with, air conditioning.

Basement or utility room
- Use front-loading clothes washer. If possible run only full loads with warm or cold water.
- If possible, hang clothes on racks for drying.
- Run only full loads in clothes dryer and use lower heat setting.
- Set water heater at 140°F if dishwasher is used and 120°F or lower if no dishwasher is used.
- Use water heater thermal blanket.
- Insulate exposed hot water pipes.
- Regularly clean or replace furnace filters.

© Cengage Learning

Figure 16-10 **Individuals matter:** You can save energy where you live. ***Question:*** Which of these things do you do?

Figure 16-10 summarizes ways in which you can cut your energy use in the place where you live. By saving energy, you can reduce your ecological footprint and save more of your hard-earned money.

CONSIDER THIS...

CAMPUS SUSTAINABILITY Competing to Save Energy at UNC

In 2010, the Morrison Residence Hall at the University of North Carolina at Chapel Hill won the first prize in the EPA's first Energy Star National Building Competition. Students living there reduced their energy use by 36% in 1 year and saved more than $250,000 in energy costs by keeping track of their energy consumption on a touch-screen computer in the dorm's lobby. Competitions among the dorm floors encouraged students to turn off lights and computers and take shorter showers. The university helped out by improving the building's lighting, ventilation, heating, and cooling systems.

Why Are We Still Wasting So Much Energy and Money?

Considering its impressive array of benefits (Figure 16-4), why is there so little emphasis on improving energy efficiency? One reason is that in the United States and other countries, energy resources such as fossil fuels and nuclear power are artificially cheap, primarily because of the government subsidies and tax breaks they receive and because their market prices do not include the harmful environmental and health costs of their production and use. This is a violation of the full-cost pricing **principle of sustainability** (see Figure 1-5, p. 9 or back cover).

Another reason is that there are too few government tax breaks, rebates, low-interest and long-term loans, and

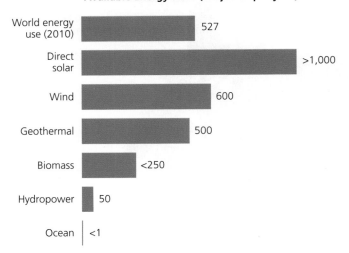

Available Energy Flow (exajoules per year)

- World energy use (2010): 527
- Direct solar: >1,000
- Wind: 600
- Geothermal: 500
- Biomass: <250
- Hydropower: 50
- Ocean: <1

Figure 16-11 Estimated energy potentially available from various renewable energy resources, 2011. (A joule is a basic unit of energy and an exajoule is 1×10^{18} joules.)

(Compiled by the authors using data from Energy Information Agency and the Worldwatch Institute, and T.B. Johansson.)

other economic incentives for consumers and businesses that invest in improving energy efficiency. A third reason is that some governments have not put a high priority on educating the public about the environmental and economic advantages of cutting energy waste.

We Can Use Renewable Energy to Provide Heat and Electricity

The lesson from one of nature's three scientific **principles of sustainability** (see Figure 1-2, p. 6 or back cover) is to *rely mostly on solar energy.* We can get renewable solar energy directly from the sun or indirectly from wind (**Core Case Study**), flowing water, and biomass, none of which would exist without direct solar energy. Another form of renewable energy is geothermal energy from the earth's interior. Figure 16-11 shows the energy potentially available from some major renewable energy resources.

Studies show that with increased and consistent government backing in the form of research and development funds and subsidies and tax breaks, renewable energy could provide 20% of the world's electricity by 2025 and 50% by 2050. In 2012, 13 countries got more than 30% of their electricity from renew-

able energy. Costa Rica gets more than 95% of its electric power from renewable hydroelectric, wind, and geothermal energy. Germany is phasing out nuclear power and gets 22% of its electricity from wind, solar energy, hydropower, and biomass. Germany has plans to get 80% of its electricity from renewable resources by 2050. Canada gets 63% of its electricity from renewable energy, most of it from hydropower. In 2012, the National Renewable Energy Laboratory projected that, with a crash program, the United States could get 50% of its electricity from renewable energy sources by 2050.

There is growing evidence that a gradual shift from an economy powered primarily by climate-disrupting nonrenewable fossil fuels to one powered by a mix of climate-stabilizing renewable energy sources is underway and will gain momentum during this century. This would result in more decentralized and energy-efficient national economies that are less vulnerable to energy supply cutoffs and natural disasters. Also, it would greatly reduce air and water pollution, slow projected climate disruption, create large numbers of jobs, and save consumers money.

If renewable energy is so great, why does it provide only 8% of the world's energy (see Figure 15-3, p. 376) and 5% of the energy used in the United States (see Figure 15-1, p. 374)? There are three major reasons. *First,* since 1950, government tax breaks, subsidies, and funding for research and development of renewable energy resources have been much lower than those for fossil fuels (especially oil) and nuclear power, although subsidies and tax breaks for renewables have increased in recent years.

Second, although subsidies and tax breaks for fossil fuels and nuclear power have essentially been guaranteed for many decades, those for renewable energy in the United States have to be renewed by Congress every few years. The resulting financial uncertainty makes it risky for companies to invest in renewable energy.

Third, the prices we pay for nonrenewable fossil fuels and nuclear power do not include the harmful environmental and human health costs of producing and using them. This helps to shield them from free-market competition with renewable sources of energy.

Energy analysts such as Amory Lovins (Individuals Matter 16.1) say that if these economic handicaps—*inaccurate pricing* and *unbalanced and intermittent subsidies*—were eliminated, many forms of renewable energy would be cheaper than fossil fuels and nuclear energy, and would quickly take over the energy marketplace.

Throughout the rest of this chapter, we evaluate these renewable energy options.

Amory B. Lovins: Sustainable Energy Leader

©Rocky Mountain Institute

For over three decades, world-renowned energy expert Amory B. Lovins has been working to demonstrate that we can greatly improve energy efficiency and rely on a variety of renewable energy resources in order to reduce pollution, help preserve biodiversity, and slow projected climate change. He has estimated that in doing so, the world could save at least $1 trillion a year.

In 1976, Lovins published a paper in *Foreign Affairs* titled "Energy Strategy: The Road Not Taken," in which he made a distinction between two possible energy futures. One is a *hard energy path*, built on dependence on fossil fuels and nuclear power. The other is a *soft energy path*, based on using energy more efficiently and depending on a variety of renewable energy sources such as solar, wind, and geothermal energy and biofuels. At the time, critics viewed Lovins as an unrealistic dreamer. But the ground-breaking ideas in his soft-path energy strategy now form the core of many national and corporate energy strategies.

Lovins is co-founder, chairman, and chief scientist at the Rocky Mountain Institute (RMI) in Snowmass, Colorado. RMI is a nonprofit think tank where more than 90 scientists, designers, and engineers conduct cutting-edge research and consultation on energy resources, energy efficiency, and energy policy.

Lovins and the RMI staff are regularly consulted by hundreds of utility companies and major corporations and 60 of the world's governments. *Newsweek* magazine has described Lovins as "one of the Western world's most influential energy thinkers." He has published more than 450 papers, written 31 books, taught at nine major universities, and won essentially all of the world's major environmental awards. In the mid-1990s, RMI, under Lovins' direction, developed cutting-edge designs for hybrid and plug-in hybrid motor vehicles (Figure 16-6).

In 1984, Lovins applied his research and knowledge in designing and building his home and office in Colorado, where winters are extremely cold. This superinsulated, solar-heated, solar-powered, and partially earth-sheltered building incorporated energy-efficiency ideas that were 20 years ahead of the times. Since it was built, solar energy has provided 99% of its heat and hot water and 90% of its household electricity. The building's heating bill is less than $50 a year. These efficiencies were accomplished through the use of energy-efficient lights, appliances, computers, and other electrical devices, and solar cells that generate electricity, along with the building's passive solar design.

Background photo: Chris Gunn/National Renewable Energy Laboratory

16-2 What Are the Advantages and Disadvantages of Using Solar Energy?

CONCEPT 16-2
Passive and active solar heating systems can heat water and buildings effectively, and the costs of using direct sunlight to produce high-temperature heat and electricity are coming down.

We Can Heat Buildings and Water with Solar Energy

Homes and buildings can get all or most of their heat through a **passive solar heating system** (Figures 16-12, top, and 16-13 and Individuals Matter 16.1). Such a system absorbs and stores heat from the sun directly within a well-insulated structure. Water tanks and walls and floors made of concrete, adobe, brick, or stone can store much of the collected solar energy as heat and release it slowly throughout the day and night. One new form of heat storage technology involves a wallboard that contains microscopic capsules of paraffin wax. In a passively heated building, a small backup heating system such as a vented natural gas or propane heater can be used, but it is not usually necessary in a well-designed building (see the online Guest Essay on this topic by Nancy Wicks).

Use of passive solar energy is not new. For thousands of years, people have intuitively followed this principle of sustainability, by orienting their dwellings to take advantage of the sun's heat and light. They have also built thick walls of adobe and stone that collect and store heat, and in hot climates, people have used light colors on their roofs and walls to reflect incoming solar energy and keep their

PASSIVE

ACTIVE

© Cengage Learning

Figure 16-12 Solutions: A home can be heated with a passive or an active solar system.

Figure 16-13 This passive solar home (right) in Golden, Colorado, collects and stores incoming solar energy to provide much of its heat in a climate with cold winters. Notice the solar hot water heating panels in the yard. Some passive solar houses have sunrooms (see inset photo) to help collect incoming solar energy.

Photos: Alan Ford/National Renewable Energy Laboratory

houses cool. Now we are rediscovering this ancient solar wisdom. According to the U.S. Department of Energy and the Passive House Institute, a typical passive U.S. house uses about one-fifth as much energy per year as the average U.S. house uses.

An **active solar heating system** (Figure 16-12, bottom) captures energy from the sun by pumping a heat-absorbing fluid (such as water or an antifreeze solution) through special collectors, usually mounted on a roof or on special racks to face the sun. Some of the collected heat can be used directly. The rest can be stored in a large insulated container filled with gravel, water, clay, or a heat-absorbing chemical and used as needed.

Rooftop active solar collectors are also used to heat water in many homes and apartment buildings. With systems that cost as little as $200, about one in ten houses and apartment buildings in China currently use the sun to provide hot water. By 2030, half of all households in China may get their hot water in this way—an excellent example of applying the solar energy **principle of sustainability** (see Figure 1-2, p. 6 or back cover). Once the fairly low initial cost is paid, the hot water is essentially free. In Spain and Israel, all new buildings must have rooftop systems for heating water and interior space. According to the U.N. Development Programme, solar water heaters could be used to provide half of the world's hot water.

Figure 16-14 lists the major advantages and disadvantages of using passive or active solar heating systems for heating buildings. They can be used to heat new homes in areas with adequate sunlight (see the maps in Figure 35, p. S60, and Figure 36, p. S61, in Supplement 6). But solar energy cannot be used to heat existing homes and other buildings that are not oriented to receive sunlight or that are blocked from sunlight by other buildings or trees.

We Can Cool Buildings Naturally

Direct solar energy actually works against us when we want to keep a building cool, but we can use indirect solar energy (mainly wind) to help cool buildings. For example, we can open windows to take advantage of breezes and use fans to keep the air moving. When there is no breeze, superinsulation and high-efficiency windows help to keep hot air outside. Here are some other ways to keep cool:

- Block the high summer sun with shade trees, broad overhanging eaves, window awnings, or shades (Figure 16-12, top).
- In warm climates, use a light-colored roof to reflect as much as 90% of the sun's heat (compared to only 10–15% for a dark-colored roof), or use a green roof (Figure 16-7).
- Use geothermal heat pumps for cooling (and for heating in winter).

Figure 16-14 Heating a house with passive or active solar energy system has advantages and disadvantages (Concept 16-2). **Questions:** Which single advantage and which single disadvantage do you think are the most important? Why? Do you think that the advantages of using these technologies outweigh their disadvantages?

We Can Concentrate Sunlight to Produce High-Temperature Heat and Electricity

One of the problems with direct solar energy is that it is dispersed. **Solar thermal systems**, also known as *concentrated solar power (CSP)*, use different methods to collect and concentrate solar energy in order to boil water and produce steam for generating electricity (Figure 16-15). These systems are used mostly in desert areas with ample sunlight (see the map in Figure 35, p. S60, and Figure 36, p. S61, in Supplement 6).

One such system uses troughs of curved collectors that concentrate solar energy and use it to heat synthetic oil in a pipe that runs through the center of each trough (Figure 16-15, left). This concentrated heat—as high as 400°C (750°F)—is used to boil water and produce steam that powers a turbine that drives a generator to produce electricity. Another system (Figure 16-15, right) uses an array of computer-controlled mirrors to track the sun and focus its energy on a central power tower to provide enough heat to boil water that is used to produce electricity. The heat produced by either of these systems can also be used to melt a certain kind of salt stored in a large insulated container. The heat stored in this *molten salt system* can then be released as needed to produce electricity at night or on cloudy days.

Scientists estimate that we could meet all of the world's electricity needs by using solar thermal systems in combination with high-voltage smart electrical grids (see Case Study, p. 405). Together, these systems would

Figure 16-15 *Solar thermal power:* This solar power plant (left) in a California desert uses curved (parabolic) solar collectors to concentrate solar energy to provide enough heat to boil water and produce steam for generating electricity. In another type of system (right), an array of mirrors tracks the sun and focuses reflected sunlight on a central receiver to boil the water for producing electricity.

Left: ©National Renewable Energy Laboratory; Right: Sandia National Laboratories/National Renewable Energy Laboratory

Trade-Offs

Solar Thermal Systems

Advantages	Disadvantages
High potential for growth	Low net energy yield and high costs
No direct emissions of CO_2 and other air pollutants	Needs backup or storage system on cloudy days
Lower costs with natural gas turbine backup	Requires high water use
Source of new jobs	Can disrupt desert ecosystems

© Cengage Learning

Figure 16-16 Using solar energy to generate high-temperature heat and electricity has advantages and disadvantages (**Concept 16-2**). **Questions:** Which single advantage and which single disadvantage do you think are the most important? Why? Do you think that the advantages of using these technologies outweigh their disadvantages?

Top: Sandia National Laboratories/National Renewable Energy Laboratory. Bottom: ©National Renewable Energy Laboratory.

cover less than 1% of the world's deserts—an area roughly the size of Austria or Wisconsin—according to these estimates.

GOOD NEWS

One problem with solar thermal systems is that their net energy yield is low, which means that they need large government subsidies or tax breaks in order to compete in the marketplace with alternatives that have higher net energy yields. Researchers are working on ways to raise the net energy yield for such systems to make them more affordable.

Water shortages in some areas may also limit the production of electricity from solar thermal power plants. These systems require large volumes of cooling water for condensing the steam back to water and for washing the surfaces of the mirrors and parabolic troughs. And they are all built in sunny, arid desert areas where water is scarce.

Figure 16-16 summarizes the major advantages and disadvantages of concentrating solar energy to produce high-temperature heat and electricity.

We can use concentrated solar energy on a smaller scale, as well. In some sunny areas, people use inexpensive *solar cookers* to focus and concentrate sunlight for boiling and sterilizing water (Figure 16-17, left) and for cooking food (Figure 16-17, right). In 2009, inventor Jon Boehner received a $75,000 prize for develop-

Figure 16-17 Solutions: This woman in Tibet is using a solar cooker to boil water (left). The simple solar oven (right) is being used to cook food.

Left: Patstock/Age Fotostock

M. Cornelius/Shutterstock.com

ing a $6 solar cooker made from a cardboard box. GOOD NEWS Solar cookers can replace wood and charcoal fires, which help to reduce indoor air pollution (a major killer for many of the world's poor people) and deforestation by decreasing the need for firewood and charcoal made from firewood. Also, researchers at Rice University have modified a solar stove so that it can be used as an autoclave to sterilize surgical instruments in areas of less-developed countries not connected to an electrical grid.

We Can Use Solar Cells to Produce Electricity

The dream of using solar energy to produce electricity goes way back. In 1931, Thomas Edison (inventor of the electric lightbulb) told Henry Ford, "I'd put my money on the sun and solar energy." This dream is now a reality.

We can convert solar energy directly into electrical energy using **photovoltaic (PV) cells**, commonly called **solar cells**. Most solar cells are thin wafers of purified silicon (Si) or polycrystalline silicon with trace amounts of metals that allow them to conduct electricity. A typical solar cell has a thickness ranging from less than that of a human hair to that of a sheet of paper. When sunlight strikes these transparent cells, they produce electricity (a flow of electrons), and many cells wired together in a panel can produce electrical power. Such systems can be connected to existing electrical grid systems or to batteries that store the electrical energy until it is needed.

We can mount arrays of solar cells on rooftops and incorporate them into almost any type of roofing material, including tiles of different colors and shapes. The material used to make them can be made into paper-thin, rigid or flexible sheets that can be printed like newspapers and attached to a variety of surfaces such as walls, windows, and clothing. Solar cells have no moving parts and they operate safely and quietly with no emissions of pollutants or greenhouse gases. In 2012, Stanford University scientists developed a "peel and stick" thin solar cell that can be attached to a variety of surfaces. A growing number of businesses and some colleges are covering their parking lots with solar-cell arrays (Figure 16-18). Producing electricity from a network of solar cells on rooftops connected by a smart grid is like connecting millions of laptop computers to the Internet.

GREEN CAREER: solar-cell technology

Nearly 1.6 billion people, or one of every four in the world, live in less-developed countries in rural villages that are not connected to an electrical grid, but many of them are now getting electrical service with the use of solar cells (Figure 16-19). As these small, off-grid GOOD NEWS systems reach more rural villages, they will help hundreds of millions of people to lift themselves out of poverty. Eventually, new thin-film solar cells are expected to drastically lower the cost of providing electricity to these areas, eliminating the need for expensive electrical grids of any size.

Large solar-cell power plants are in operation in Portugal, southern Spain, Germany, South Korea, the southwestern United States, and China (Figure 16-20). Energy from some of these plants is stored for use at night or on cloudy days by using the excess electricity they generate to run pumps that compress air into underground caverns. This stored pressure is released as needed to drive turbines and produce electricity.

Another way to store energy is to use electricity from solar-cell power plants and wind farms to recharge the batteries of plug-in hybrid electric cars (Figure 16-6, right)

©Moreno Soppelsa/Shutterstock.com

and all-electric cars. Such electricity could also be used to decompose water and produce hydrogen gas, which can be used as a fuel, as we discuss later in this chapter. Thus, we are applying the solar energy **principle of sustainability** by indirectly using the sun's energy to power motor vehicles.

Solar cells emit no greenhouse gases, although they are not carbon-free, because fossil fuels are used to produce and transport the panels. However, these emissions are small compared to those resulting from the use of fossil fuels and the nuclear power fuel cycle. Conventional solar cells also contain toxic materials that must be recovered when the cells wear out after 20–25 years of use, or when they are replaced with new systems.

One problem with solar cells is that they are not highly energy efficient and thus have a low to medium net energy yield, depending on the design and the production process. Scientists are rapidly improving the efficiency of solar cells, which will increase their net energy yield. Figure 16-21 lists the major advantages and disadvantages of using solar cells (**Concept 16-2**).

The main problem with using solar cells to produce electricity has been their high cost, although prices have been dropping sharply. Since 1980 the price of a watt of electricity from solar cells has dropped by a factor of 20. As a result, solar cells have become the world's fastest-growing way to produce electricity (see the graph in Figure 9, p. S67, in Supplement 7). Indeed, such use of solar cells grew by 74% in 2011. This is because of their advantages (Figure 16-21, left) and because of increased government subsidies and tax breaks for solar developers. Production is likely to grow much more, because within the next decade or two, new thin-film nanotechnology and graphene (Science Focus 14.2, p. 363) solar cells are projected to become inexpensive enough to compete with fossil fuels.

In addition, some businesses and home owners are installing solar-cell systems and spreading the costs out over decades by including them in their mortgages. Others are leasing solar-cell systems from companies that install and maintain them. Some communities and neighborhoods are installing shared solar-cell arrays, known as "community solar" or "solar gardens." They can provide electricity for individuals who rent, live in condos, or have their roofs shaded by buildings or trees. And it is often

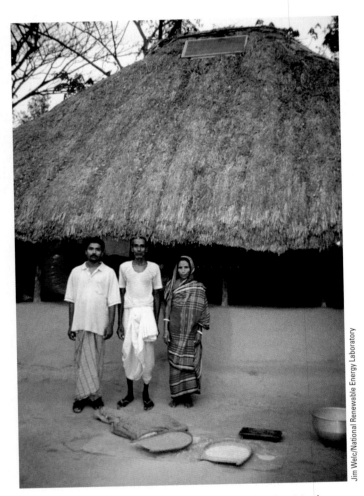

Jim Welc/National Renewable Energy Laboratory

Figure 16-19 Solutions: A solar-cell panel provides electricity for lighting this hut in rural West Bengal, India. *Question:* Do you think your government should provide aid to poor countries for obtaining solar-cell systems? Explain.

Figure 16-20 This solar-cell power plant is located near the city of Shizuishan, China.

cheaper to use solar cells to provide electricity for individual homes in the villages of less-developed countries (Figure 16-19) than to build a central power plant and grid.

Jim Lyons, chief engineer for General Electric, projects that solar cells will be the world's number-one source of electricity by the end of this century. If that happens, it will represent a huge global application of the solar energy **principle of sustainability. GREEN CAREER:** solar-cell technology

Figure 16-21 Using solar cells to produce electricity has advantages and disadvantages (Concept 16-2). *Questions:* Which single advantage and which single disadvantage do you think are the most important? Why? Do you think that the advantages of using this technology outweigh its disadvantages? Why?

Top: ©Martin D. Vonka/Shutterstock.com. Bottom: ©pedrosala/Shutterstock.com.

Trade-Offs		
Solar Cells		
Advantages		**Disadvantages**
Medium net energy yield		Need access to sun
		Some designs have low net energy yield
Little or no direct emissions of CO_2 and other air pollutants		Need electricity storage system or backup
Easy to install, move around, and expand as needed		Costs high for older systems but dropping rapidly
Competitive cost for newer cells		Solar-cell power plants could disrupt desert ecosystems

© Cengage Learning

16-3 What Are the Advantages and Disadvantages of Using Hydropower?

CONCEPT 16-3

We can use water flowing over dams, tidal flows, and ocean waves to generate electricity, but environmental concerns and limited availability of suitable sites may limit the use of these energy resources.

We Can Produce Electricity from Falling and Flowing Water

Hydropower is any technology that uses the kinetic energy of flowing and falling water to produce electricity. It is an indirect form of solar energy because it depends on heat from the sun evaporating water, which is deposited as rain or snow at higher elevations where it can flow to lower elevations in rivers as part of the earth's solar-powered water cycle (see Figure 3-15, p. 63).

The most common approach to harnessing hydropower is to build a high dam across a large river to create a reservoir. Some of the water stored in the reservoir is allowed to flow through large pipes at controlled rates to spin turbines that produce electricity (see Figure 13-17, p. 330).

Hydropower, the leading renewable energy source, produces about 16% of the world's electricity in 150 countries. In order, the world's top four producers of hydropower are China, Canada, Brazil, and the United States. In 2010, hydropower supplied about 6% of the electricity used in the United States (but about 50% of that used on the West Coast). See Figure 12, p. S69, in Supplement 7 for a graph of the growth of hydropower between 1965 and 2011.

According to the United Nations, only about 13% of the world's potential for hydropower has been developed. Much of the untapped potential is in China, India, South America, Central Africa, and parts of the former Soviet Union. China has plans to more than double its hydropower output during the next decade and is also building or funding more than 200 dams around the world.

However, some analysts expect that use of large-scale hydropower plants will fall slowly over the next several decades as many existing reservoirs fill with silt and become useless faster than new systems are built. Also, there is growing concern over emissions of methane, a potent greenhouse gas, from the decomposition of submerged vegetation in hydropower plant reservoirs, especially in warm climates. In fact, scientists at Brazil's National Institute for Space Research estimate that the world's largest dams altogether are the single largest human-caused source of methane.

In addition, if atmospheric temperatures continue to rise and change the world's climate as projected, many mountain glaciers will melt. Several of these melting gla-ciers are the primary sources of water for some hydropower plants, so the electrical outputs of these plants are likely to drop. Figure 16-22 lists the major advantages and disadvantages of using large-scale hydropower plants to produce electricity (Concept 16-3).

The use of *microhydropower generators* may become an increasingly important way to produce electricity. These are floating turbines, each about the size of an overnight suitcase. They can be placed in any stream or river without altering its course to provide electricity at a very low cost with very low environmental impact.

We Can Use Tides and Waves to Produce Electricity

We can also produce electricity from flowing water by tapping into the energy from *ocean tides* and *waves*. In some coastal bays and estuaries, water levels can rise or fall by 6 meters (20 feet) or more between daily high and low tides. Dams can be built across the mouths of such bays and estuaries to capture the energy in these flows for hydropower, but sites with large tidal flows are rare. Only two large tidal energy dams are currently operating, one at La Rance on the northern coast of France, and the other in Nova Scotia's Bay of Fundy.

Several turbines, resembling underwater wind turbines, have also been installed to tap the tidal flow of the

Trade-Offs

Large-Scale Hydropower

Advantages	Disadvantages
High net energy yield	Large land disturbance and displacement of people
Large untapped potential	High CH_4 emissions from rapid biomass decay in shallow tropical reservoirs
Low-cost electricity	
Low emissions of CO_2 and other air pollutants in temperate areas	Disrupts downstream aquatic ecosystems

© Cengage Learning

Figure 16-22 Using large dams and reservoirs to produce electricity has advantages and disadvantages (Concept 16-3). ***Questions:*** Which single advantage and which single disadvantage do you think are the most important? Why? Do you think that the advantages of this technology outweigh its disadvantages? Why?

Photo: ©Andrew Zarivny/Shutterstock.com

East River near New York City. They swivel to face the incoming and outgoing tides, and they have produced electricity efficiently. Such a system powers a town in Norway. However, these systems are limited to the small number of rivers that have adequate tidal flows.

For decades, scientists and engineers have also been trying to produce electricity by tapping wave energy along seacoasts where there are almost continuous waves. Large, snakelike chains of floating steel tubes have been installed off the coast of Portugal. The up-and-down motion of these chains driven by wave action generates electricity. Scientists estimate that learning how to tap into the world's wave power at an affordable cost could provide more than twice the amount of electricity that the world uses.

Production of electricity from tidal and wave systems is limited because of a lack of suitable sites, citizen opposition at some sites, high costs, and equipment damage from saltwater corrosion and storms. Proponents hope that improved technology can greatly increase the production of electricity from tides and waves sometime during this century.

16-4 What Are the Advantages and Disadvantages of Using Wind Power?

CONCEPT 16-4

When we include the environmental costs of using energy resources in their market prices, wind power is the least expensive and least polluting way to produce electricity.

Using Wind to Produce Electricity Is an Important Step toward Sustainability

Because today's wind turbines are so tall (**Core Case Study**) and have very long blades (Figure 16-23), they can extract more energy from the wind more efficiently than smaller turbines nearer the ground can. According to the DOE's National Renewable Energy Laboratory (NREL), wind farms with these giant turbines can generate more energy with fewer turbines and a lower land and ecological footprint than those with the smaller turbines, and this will help to lower the price of wind energy.

So far, most of the world's rapidly growing number of wind farms have been built on land in parts of Europe, China, and the United States. However, the frontier for wind energy is offshore wind farms (Figure 16-1, right), which have been built off the coasts of ten European countries as well as China and Japan. South Korea is planning a number of offshore wind farms.

Since 1990, wind power has been the world's second fastest-growing source of energy after solar cells (see the graph in Figure 10, p. S68, in Supplement 7), with more than 80 countries now harnessing energy from the wind. In order, the countries with the largest installed wind-power capacity in 2012 were China, the United States (**Core Case Study**), Germany, Spain, and India. However, many of China's wind farms are in remote areas and produce no electricity because some of the electrical transmission lines for China's proposed national smart grid (Case Study, p. 405) have yet to be built. Wind power produced about 3.5% of the world's electricity in 2011 and by 2050, could produce about 31% of the world's electricity, according to a 2013 analysis by the market research firm iSUPPLi and the Global Wind Energy Council.

Denmark, the world's most energy-efficient country, got 30% of its electricity from wind in 2012 and is aiming for 50% by 2020. Around the world, more than 400,000 people are employed in the production, installation, and maintenance of wind turbines (Figure 16-24). These job numbers are likely to rise rapidly in coming years.

In 2009, a Harvard University study led by Xi Lu estimated that wind power has the potential to produce 40 times the world's current use of electricity. Even though offshore wind farms are more costly to install, analysts expect to see increasing use of them because wind speeds over water are often stronger and steadier than those over land. Also, locating them offshore lessens the need for negotiations among multiple land owners over the locations of the turbines.

A 2009 study published in the *Proceedings of the U.S. National Academy of Sciences* estimated that the United States has enough wind potential to meet an estimated 16 to 22 times its current electricity needs (**Core Case Study**). Texas leads the nation in wind energy production, followed by California, Iowa, and South Dakota, with Iowa and South Dakota each getting about 24% of their electricity from wind power in 2012. Texas wind farms now supply 8% of the state's electricity and, by 2025, could supply about 90% of the state's current residential electricity needs.

According to a 2009 study by the U.S. Department of the Interior, with expanded and sustained subsidies, wind farms off of the Atlantic and Gulf coasts could generate enough electricity to more than replace all of the country's coal-fired power plants. Many Atlantic and Gulf coast states are making plans to tap into this vast source of energy and create jobs and income in the process. See Figure 37, p. S62, in Supplement 6 for a map of the potential supply of land-based and offshore wind energy in the United States.

Unlike fossil fuels, wind is abundant, widely distributed, and inexhaustible. A wind farm can be built within 9 to 12 months, can be expanded as needed, and has a

Figure 16-23 Tractor trailer hauling two wind turbine blades from Beijing to a wind farm in the Inner Mongolia region of China.

small land footprint. Although wind farms can cover large areas of land, the turbines occupy only about 1% of the land, which can be used for other purposes such as growing crops or raising cattle. Wind farms do not need water for cooling, unlike coal, natural gas, and nuclear power plants. Also, home owners and neighborhoods can use small, quiet wind turbines that operate at low speeds to produce electricity.

In addition, wind power has a high net energy yield. The DOE and the Worldwatch Institute estimate that, if we were to apply the full-cost pricing **principle of sustainability** by including the harmful environmental and health costs of various energy resources in comparative cost estimates, wind energy would be the cheapest way to produce electricity (**Concept 16-4**). In making the transition from climate-disrupting fossil fuels—especially coal—to a mix of renewable energy resources, wind has a wide lead. By 2012, wind turbines in more than 80 countries generated about 3.5 times more electricity than solar cells and about 22 times more electricity than geothermal power plants.

Like any energy source, wind power has some drawbacks. For example, areas with the greatest wind power potential are often sparsely populated and located far from cities. Thus, to take advantage of the potential for electricity from wind energy, the United States and other countries will have to invest in upgrading and expanding their outdated electrical grids (Figure 16-5), converting them to smart grid systems (Case Study, p. 405). However, even if we continue to rely on coal-fired and nuclear power plants, such an upgrade and expansion will still have to take place. Many of the new lines could be run along state-owned interstate highway corridors to avoid legal conflicts and further degradation of natural areas.

Another problem is that winds can die down and thus require a backup source of power, such as natural gas, for generating electricity. However, analysts calculate that a large number of wind farms in different areas connected to a smart electrical grid could take up the slack when winds die down in any one area and thus could make wind power a very stable source of electricity.

Figure 16-24 Maintenance workers get a long-distance view from atop a wind turbine, somewhere in North America.

Figure 16-25 Using wind to produce electricity has advantages and disadvantages (Concept 16-4). **Questions:** Which single advantage and which single disadvantage do you think are the most important? Why? Do you think the advantages outweigh the disadvantages? Why?

Top: ©TebNad/Shutterstock.com. Bottom: Yegor Korzh/Shutterstock.com.

Trade-Offs

Wind Power

Advantages	Disadvantages
High net energy yield	Needs backup or storage system when winds die down
Widely available	
Low electricity cost	Visual pollution for some people
Little or no direct emissions of CO_2 and other air pollutants	Low-level noise bothers some people
Easy to build and expand	Can kill birds if not properly designed and located

CONSIDER THIS. . .

CONNECTIONS Bird and Bat Deaths and Wind Turbines

Wildlife ecologists and bird experts have estimated that collisions with wind turbines kill as many as 440,000 birds each year in the United States, although more recent estimates put the figure at 70,000 to 100,000. Compare this to much larger numbers reported by Defenders of Wildlife and in 2013 by the U.S. Fish and Wildlife Service: domestic and feral cats kill 1.4 billion to 3.7 billion birds a year; hunters, more than 100 million; cars and trucks, about 80 million; pesticide poisoning, 67 million; and collisions with tall structures and windows, at least 10 million every year. Most of the wind turbines involved in bird and bat deaths were built with the use of outdated designs, and some were built in bird migration corridors. Wind power developers now avoid such corridors, as well as areas with large bat colonies, when building wind farms. Newer turbine designs reduce bird deaths considerably by using slower blade rotation speeds and by not providing places for birds to perch or nest.

In windy parts of the U.S. Midwest and in Canada, many farmers and ranchers welcome wind farms and some have become wind developers themselves. For each wind turbine located on a farmer's or rancher's land, the land owner typically receives $3,000 to $10,000 a year in royalties. And the land can still be used for growing crops or grazing cattle. An acre of land in northern Iowa planted in corn can produce about $1,000 worth of ethanol car fuel. The same site used for a single wind turbine can produce $300,000 worth of electricity per year, which is why many land owners are investing in wind farms.

Figure 16-25 lists the major advantages and disadvantages of using wind to produce electricity. According to energy analysts, wind power has more benefits and fewer serious drawbacks than any other energy resource, except for improvements in energy efficiency (Figure 16-4). With sufficient and consistent government incentives, wind power could supply more than 10% of the world's electricity and 20% of the electricity used in the United States by 2030. According to the DOE, this would reduce U.S. greenhouse gas emissions by an amount equivalent to that of taking 140 million motor vehicles off the road. It could also create hundreds of thousands of jobs. **GREEN CAREER:** wind-energy engineering

16-5 What Are the Advantages and Disadvantages of Using Biomass as an Energy Source?

CONCEPT 16-5A
Solid biomass is a renewable resource for much of the world's population, but burning it faster than it is replenished produces a net gain in atmospheric greenhouse gases.

CONCEPT 16-5B
We can use liquid biofuels derived from biomass to lessen our dependence on oil-based fuels, but creating biofuel plantations can degrade soil and biodiversity, increase greenhouse gas emissions, and lead to higher food prices.

We Can Produce Energy by Burning Solid Biomass

Biomass consists of plant materials (such as wood and agricultural waste) that we can burn directly as a solid fuel or convert into gaseous or liquid biofuels. Biomass is another indirect form of solar energy because it consists of combustible organic (carbon-containing) compounds in plant matter produced mainly by photosynthesis.

Solid biomass is burned mostly for heating and cooking, but also for industrial processes and for generating electricity. Wood, wood wastes, charcoal made from wood, and other forms of biomass used for heating and cooking supply 10% of the world's energy, 35% of the energy used in less-developed countries, and 95% of the energy used in the poorest countries. In agricultural areas, *crop residues* (such as sugarcane and cotton stalks, rice husks, straw, and coconut shells) and *animal manure* are collected and burned.

Wood is a renewable resource only if it is not harvested faster than it is replenished. The problem is that about 2.7 billion people in 77 less-developed countries face a *fuelwood crisis* and are often forced to meet their fuel needs by cutting forests faster than they can grow to replace the trees that are removed. One way to deal with this problem is to plant fast-growing trees, shrubs, or perennial grasses in *biomass plantations*. But repeated cycles of growing and harvesting these plantations can deplete the soil of key nutrients. Also, the clearing of forests and grasslands for such plantations destroys or degrades biodiversity. And some plantation tree species such as European poplar and American mesquite are invasive species that can spread from plantations and take over nearby areas.

Another problem with depending on solid biomass as a fuel is that clearing forests to provide the fuel reduces the amount of vegetation that would otherwise capture CO_2, and burning biomass produces CO_2. If the rate of use of biomass does not exceed the rate at which it is replenished by new plant growth, there is no net increase in CO_2 emissions. However, monitoring and managing this balance, globally or throughout any one country, is very difficult.

Burning solid biomass to produce electricity has a moderate net energy yield, but burning it in a cogeneration system almost doubles its net energy yield. Denmark gets almost half of its electricity by burning wood and agricultural wastes for cogeneration.

Figure 16-26 lists the general advantages and disadvantages of burning solid biomass as a fuel (Concept 16-5A).

We Can Convert Plants and Plant Wastes to Liquid Biofuels

Liquid biofuels such as *ethanol* (ethyl alcohol produced from plants and plant wastes) and *biodiesel* (produced from vegetable oils) are being increasingly used to fuel motor vehicles. The biggest producers of liquid biofuels are, in order, the United States (producing mostly ethanol from corn), Brazil (producing mostly ethanol from sugarcane residues), the European Union (producing mostly biodiesel from vegetable oils), and China (producing mostly ethanol from non-grain plant sources to avoid diverting grains from its food supply).

Biofuels have three major advantages over gasoline and diesel fuel produced from oil. *First*, biofuel crops can be grown throughout much of the world, and thus they can help countries to reduce their dependence on imported oil. *Second*, if these crops are not used faster than they are

Trade-Offs

Solid Biomass

Advantages	Disadvantages
Widely available in some areas	Contributes to deforestation
Moderate costs	Clear-cutting can cause soil erosion, water pollution, and loss of wildlife habitat
Medium net energy yield	
No net CO_2 increase if harvested, burned, and replanted sustainably	Can open ecosystems to invasive species
Plantations can help restore degraded lands	Increases CO_2 emissions if harvested and burned unsustainably

Figure 16-26 Burning solid biomass as a fuel has advantages and disadvantages (Concept 16-5A). ***Questions:*** Which single advantage and which single disadvantage do you think are the most important? Why? Do you think the advantages outweigh the disadvantages? Why?

Top: ©Fir4ik/Shutterstock.com. Bottom: Eppic/Dreamstime.com.

replenished by new plant growth, there is no net increase in CO_2 emissions, unless existing grasslands or forests are cleared to plant biofuel crops. *Third*, biofuels are easy to store and transport through existing fuel networks and can be used in motor vehicles at little or no additional cost.

However, in a 2007 UN report on bioenergy, and in another study by R. Zahn and his colleagues, scientists warned that large-scale biofuel-crop farming could degrade biodiversity by leading to more clearing of natural forests and grasslands; increase soil degradation, erosion, and nutrient leaching; push small farmers off their land; and raise food prices if farmers can make more money by growing corn and other crops to fuel cars rather than to feed livestock and people (Concept 16-5B).

CONSIDER THIS. . .

CONNECTIONS Biofuels and Climate Change

In 2007, Nobel Prize–winning chemist Paul Crutzen warned that intensive farming of biofuel crops could speed up atmospheric warming and projected climate change by producing more greenhouse gases than would be produced by burning fossil fuels instead of biofuels. This would happen if nitrogen fertilizers were used to grow corn and other biofuel crops. Such fertilizers, when applied to the soil, release large amounts of the potent greenhouse gas nitrous oxide. A 2008 study by Finn Danielsen and a team of other scientists concluded that keeping tropical rain forests intact is a better way to slow projected climate change than burning and clearing such forests and replacing them with biofuel plantations.

Another problem with biofuel production is that by growing corn and soybeans in dry climates that require irrigation, farmers reduce water supplies in these arid regions. In fact, the two most water-intensive ways to produce a unit of energy are by irrigating soybean crops to produce biodiesel fuel and by irrigating corn to produce ethanol.

The challenge is to grow crops for food and biofuels by using more sustainable agriculture (see Figure 12-33, p. 309) with less irrigation, land degradation, air and water pollution, greenhouse gas emissions, and degradation of biodiversity. Also, any system used to produce a biofuel should have at least a moderate net energy yield so that it can compete in the energy marketplace without large government subsidies.

CASE STUDY

Is Biodiesel the Answer?

If a truck or bus whizzes by and leaves a scent of fast food, it is probably running on *biodiesel*. This biofuel, designed to run in diesel engines, is produced from vegetable oil extracted from soybeans, rapeseeds, sunflowers, oil palms, and jatropha shrubs. It can also be made from used vegetable oils from restaurants. European Union countries (primarily Germany, France, and Italy) produce about 95% of the world's biodiesel, mostly from rapeseeds and sunflower seeds, and these countries hope to get 20% of their diesel fuel from this source within a decade. In Europe, more than half of all cars have diesel engines, primarily because they are as much as 40% more efficient than gasoline engines.

Aided by government subsidies, biodiesel production is growing rapidly in the United States. However, soybean and canola oil crops (rapeseeds or field mustard) grown for biodiesel production require large areas of land. In addition, using industrialized agriculture to produce these crops results in topsoil losses and fertilizer runoff. Biodiesel production also requires energy (mostly from crude oil and natural gas), which reduces its net energy yield and increases emissions of the greenhouse gases nitrous oxide (N_2O) and CO_2.

Brazil, Malaysia, and Indonesia produce biodiesel from palm oil, extracted from large plantations of African oil palm (see Figure 12-5, p. 282), and export much of it to Europe. The net energy yield for biodiesel from oil palm is 5 times that from rapeseeds used in Europe and about 8 to 9 times higher than the yield from soybeans used to produce biodiesel in the United States. However, increased burning and clearing of tropical forests and other wooded lands to establish oil palm plantations in these countries poses a serious threat to their biodiversity and releases CO_2 into the atmosphere. It also reduces CO_2 uptake by eliminating rain forests that store large amounts of carbon and replacing them with crops that store much less carbon. Also, African oil palm is an invasive plant that has taken over adjacent farms and forest areas in parts of Brazil, further degrading their biodiversity.

CASE STUDY

Is Ethanol the Answer?

Ethanol can be made from plants such as sugarcane, corn, and switchgrass, and from agricultural, forestry, and municipal wastes. This process involves converting plant starches or other plant materials into simple sugars, which are processed to produce ethanol.

Brazil, by far the world's largest sugarcane producer, is the world's second largest ethanol producer after the United States. Brazil makes its ethanol from *bagasse,* a residue produced when sugarcane is crushed. This sugarcane ethanol has a medium net energy yield that is about 8 times higher than that of ethanol produced from corn. About 45% of Brazil's motor vehicles run on ethanol or ethanol–gasoline mixtures produced from sugarcane grown on only 1% of the country's arable land.

Within a decade, Brazil could expand its sugarcane production, eliminate oil imports, and greatly increase ethanol exports to other countries. To do this, Brazil plans to clear large areas of its rapidly disappearing Cerrado, a wooded savanna region, to make way for sugarcane plantations. This will further stress biodiversity in this biodiversity hotspot (see Figure 10-28, p. 238). Other harmful effects of producing ethanol from sugarcane include CO_2 emissions from the burning of oil and gasoline during sugarcane production, heavy soil erosion and resulting water pollution after sugarcane plantations are harvested, and stresses on water supplies. Producing 1 liter (0.27 gallons) of ethanol from sugarcane requires the equivalent of about 174 bathtubs full of water.

Figure 16-27 **Natural capital:** The cellulose in this rapidly growing switchgrass can be converted into ethanol, but further research is needed to develop affordable production methods.

Trade-Offs

Liquid Biofuels

Advantages	Disadvantages
Reduced CO_2 emissions for some crops	Fuel crops can compete with food crops for land and raise food prices
Medium net energy yield for biodiesel from oil palms	Fuel crops can be invasive species
Medium net energy yield for ethanol from sugarcane	Low net energy yield for corn ethanol and for biodiesel from soybeans
	Higher CO_2 emissions from corn ethanol

© Cengage Learning

Figure 16-28 There are advantages and disadvantages to using liquid biofuels. (**Concept 16-5B**). ***Questions:*** Which single advantage and which single disadvantage do you think are the most important? Why?

In the United States, most ethanol is made from corn (see Figure 11, p. S68, in Supplement 7). However, studies indicate that corn-based ethanol has a low net energy yield because fossil fuels are used heavily for growing the corn and converting it to ethanol. This is one reason why it has received big government subsidies. It also helps to explain why scientists calculate that using this ethanol results in a net increase in greenhouse gas emissions. Also, raising corn requires a great deal of water and thus adds to stressed water supplies in many areas.

According to a 2007 study by environmental economist Stephen Polasky, processing all of the corn grown in the United States into ethanol each year would meet only about 30 days' worth of the country's current demand for gasoline. This would leave no corn for other uses and would cause sharp increases in the prices of corn-based foods, including cereals, tortillas, and meats, with a devastating impact on the poor in several countries. Replacing all gasoline with corn-based ethanol in the United States would provide no major environmental benefits, and it would likely lead to higher emissions of climate-changing CO_2 and higher food prices for consumers. Also, energy expert Vaclav Smil has calculated that in order to replace all of the gasoline used in the United States with ethanol, about 6 times the country's total area of farmable land would have to be planted in corn.

An alternative to corn-based ethanol is *cellulosic ethanol,* which is produced from the inedible cellulose that makes up most of the biomass of plants—material such as leaves, stalks, and wood chips. A plant that could be used for cellulosic ethanol production is *switchgrass* (Figure 16-27), a tall perennial grass native to North American prairies that grows faster than corn. It is disease resistant and drought tolerant, and can be grown without the use of nitrogen fertilizers on land unfit for other crops. According to a 2008 study led by agronomist Ken Vogel, switchgrass-based ethanol has a medium net energy yield compared to a low net energy yield from corn-based ethanol.

One major problem is that affordable chemical processes for converting cellulosic material to ethanol are still being developed. Also, according to cellulosic-ethanol expert Robert Ranier, growing enough switchgrass to meet half of the U.S. demand for gasoline would require about 7 times the land area currently used for all U.S. corn production. In addition, we need more research to learn more about CO_2 emissions, invasive species effects, and net energy yields related to switchgrass and other potential cellulosic ethanol crops such as arundo and miscanthus plants.

Scientists are also evaluating the use of algae and bacteria to produce biofuels (Science Focus 16.2). Figure 16-28 compares the advantages and disadvantages of using biodiesel and ethanol liquid biofuels.

MAKING GASOLINE AND DIESEL FUEL FROM ALGAE AND BACTERIA

Scientists are looking for ways to produce biofuels that are almost identical to gasoline and biodiesel from various types of existing or genetically engineered oil-rich algae that can grow rapidly at any time of the year in various aquatic environments.

As they grow, the algae remove CO_2 from the atmosphere and convert it to oil, proteins, and other useful products. They also require much less land, water, and other resources than biofuel plantations do and would not affect food prices by competing for cropland. In addition, the algae could be grown in wastewater from sewage treatment plants where they would help to clean up the wastewater while producing biofuel. Another possibility would

be to transfer CO_2 produced by coal-burning power plants into nearby algae ponds or bioreactors for use in making biofuel.

There are four major challenges in converting such technological dreams into reality: (1) cutting the very high cost of producing oil by such methods; (2) learning whether it is economically better to grow the algae in open ponds or in enclosed bioreactors; (3) finding an economically affordable way to get the oil out of the algae (for example, by drying and crushing the algae or removing it with a solvent); and (4) progressing from small-scale pilot projects to large-scale production systems.

A different approach is to make gasoline or diesel fuel from rapidly multiplying

bacteria by using techniques developed in the new field of *synthetic biology*. Producing gasoline and diesel fuels from algae and bacteria could be done almost anywhere. The resulting fuels could be distributed by the world's current gasoline and diesel fuel distribution system. Thus, it is not surprising that some of the world's major oil companies are investing heavily in research on producing oil-like compounds from algae and bacteria, hoping to lock in numerous patents for such processes.

Critical Thinking

What might be some environmental risks of producing biofuels from algae or bacteria?

16-6 What Are the Advantages and Disadvantages of Using Geothermal Energy?

CONCEPT 16-6

Geothermal energy has great potential for supplying many areas with heat and electricity, and has a generally low environmental impact, but the sites where it can be produced economically are limited.

We Can Get Energy by Tapping the Earth's Internal Heat

Geothermal energy is heat stored in soil, underground rocks, and fluids in the earth's mantle. We can tap into this stored energy to heat and cool buildings and to produce electricity. Scientists estimate that using just 1% of the heat stored in the uppermost 5 kilometers (8 miles) of the earth's crust would provide 250 times more energy than that stored in all the earth's proven crude oil and natural gas reserves combined.

One way to capture geothermal energy is by using a *geothermal heat pump* system (Figure 16-29). It can heat and cool a house by exploiting the temperature difference, almost anywhere in the world, between the earth's surface and underground at a depth of 3–6 meters (10–20 feet), where the temperature typically is 10–20°C (50–60°F) year-round. In winter, a closed loop of buried pipes circulates a fluid, which extracts heat from the ground and carries it to a heat pump, which transfers

heat to a home's heat distribution system. In summer, this system works in reverse, removing heat from a home's interior and storing it in the ground. The EPA estimates that such a system can heat or cool a 190-square-meter (2,000-square-foot) house for as little as $1 a day.

According to the EPA, a well-designed geothermal heat pump system is the most energy-efficient, reliable, environmentally clean, and cost-effective way to heat or cool a space. Once installed, it produces no air pollutants and emits no CO_2. Installation costs can be high but are generally recouped within 3–5 years. Thereafter, such systems save money for their owners. Initial costs can be added to a home mortgage to spread the financial burden over two or more decades.

We can also tap into deeper, more concentrated *hydrothermal reservoirs* of geothermal energy (Figure 16-30). This is done by drilling wells into the reservoirs to extract their dry steam (with a low water content), wet steam (with a high water content), or hot water, which are then used to heat homes and buildings, provide hot water, grow vegetables in greenhouses, raise fish in aquaculture ponds, and spin turbines to produce electricity.

The United States is the world's largest producer of geothermal electricity from hydrothermal reservoirs. Most of it is produced in California, Nevada, Utah, and Hawaii (see Figure 39, p. S63, in Supplement 6 for a map of the

Figure 16-29 **Natural capital:** A geothermal heat pump system can heat or cool a house almost anywhere.

2. Heat from underground spins a turbine to power a generator and produce electricity

3. Steam from turbine condenses to water and is pumped back down to geothermal reservoir

1. Hot water or steam is pumped under pressure to the surface from underground

Figure 16-30 Power plants can produce electricity from heat extracted from underground geothermal reservoirs. The photo shows a geothermal power plant in Iceland that produces electricity and heats a nearby spa called the Blue Lagoon.

Andrés Ruzo—Geothermal Energy Sleuth and National Geographic Young Explorer

©Sofia Ruzo

Andrés Ruzo is a geophysicist with a driving passion to learn about geothermal energy and to show how this renewable and clean energy source can help us to solve some of the world's energy problems. As a boy, he spent his summers on the family farm in Nicaragua. Because the farm rests on top of the Casita Volcano, he was able to experience firsthand the power of the earth's internal heat.

As an undergraduate student at Southern Methodist University (SMU) in Dallas, Texas (USA), because of his boyhood experience, he took a course in volcanology. The course awakened his passion for geology along with a desire to learn more about the earth's heat as a source of energy. This led him to pursue a PhD in geophysics at SMU's Geothermal Laboratory.

This National Geographic Young Explorer and his wife and field assistant, Sofia, spent most of 2012 gathering data to develop the first detailed map of the geothermal energy resources of northern Peru. Their field work involved lowering temperature measuring equipment down into abandoned water wells and oil wells in the desert area of northern Peru where daytime temperatures can exceed 54°C (130°F). These data will provide him with a geothermal map showing how heat flows through the upper crust of the earth and describing its potential to be tapped as a source of energy.

Ruzo believes that geothermal energy is a hidden sleeping giant that can be tapped as an important renewable source of heat and electricity. He says that his goal in life is "to be a force of positive change in the world."

Background photo: N.Minton/Shutterstock.com

best geothermal sites in the continental United States). It meets the electricity needs of about 6 million Americans—an amount roughly equal to the combined populations of Los Angeles, California, and Houston, Texas—and supplies almost 6% of California's electricity. By 2012, there were 130 new geothermal power plants under construction or in development in the United States, enough to triple the country's electrical output from geothermal energy.

Iceland gets almost all of its electricity from hydroelectric and geothermal power plants (Figure 16-30, photo).

The Philippines gets 18% of its electricity from geothermal plants—enough electricity for 19 million people. China has a large potential for geothermal power, which could help to reduce its dependence on coal-fired power plants (see Figure 38, p. S63, in Supplement 6 for a map of the world's best geothermal sites). In Peru, a National Geographic Young Explorer is carrying out research to develop that country's geothermal resources (Individuals Matter 16.2).

Another potential source of geothermal energy is *hot, dry rock* found 5 or more kilometers (3 or more miles) underground almost everywhere. It involves drilling a well into the rock and injecting water into the well. Some of this water can absorb enough heat to produce steam that is brought to the surface and used to spin turbines to generate electricity. According to the U.S. Geological Survey, tapping just 2% of this source of geothermal energy in the United States could produce more than 2,000 times as much electricity as is used every year in the country. The limiting factor is cost, which could be brought down by more research and improved technology. There is also concern that tapping this deep source of geothermal energy could increase the frequency of earthquakes in the drilled areas. **GREEN CAREER:** geothermal engineer

Figure 16-31 lists the major advantages and disadvantages of using geothermal energy (Concept 16-6). The biggest problem limiting the widespread use of geothermal energy is the lack of sites with high enough concentrations of heat needed to make this energy resource affordable for widespread use.

Trade-Offs		
Geothermal Energy		
Advantages		**Disadvantages**
Medium net energy yield and high efficiency at accessible sites		High cost except at concentrated and accessible sources
Lower CO_2 emissions than fossil fuels		Scarcity of suitable sites
Low cost at favorable sites		Noise and some CO_2 emissions

© Cengage Learning

Figure 16-31 Using geothermal energy for space heating and for producing electricity or high-temperature heat for industrial processes has advantages and disadvantages. (Concept 16-6). ***Questions:*** Which single advantage and which single disadvantage do you think are the most important? Why? Do you think the advantages of this energy resource outweigh its disadvantages? Why?

Photo: ©N.Minton/Shutterstock.com

16-7 What Are the Advantages and Disadvantages of Using Hydrogen as an Energy Source?

CONCEPT 16-7
Hydrogen is a clean energy source as long as it is not produced with the use of fossil fuels, but it has a negative net energy yield.

Will Hydrogen Save Us?

Hydrogen (H) is the simplest and most abundant chemical element in the universe. The sun produces its energy, which sustains life on the earth, through the nuclear fusion of hydrogen atoms (see Figure 2-9, bottom, p. 40).

Some scientists say that the fuel of the future is hydrogen gas (H_2). Most of their research has been focused on using fuel cells (Figure 16-32) that combine H_2 and oxygen gas (O_2) to produce electricity and water vapor ($2 H_2 + O_2 \rightarrow 2 H_2O$ + energy), which is emitted into the atmosphere.

Widespread use of hydrogen as a fuel would eliminate most of the outdoor air pollution that comes from motor vehicles and coal-burning power plants. It would also greatly reduce the threat of projected climate change, because in using it, we emit no CO_2—as long as the H_2 is not produced with the use of fossil fuels or nuclear power. Hydrogen also provides more energy per gram than does any other fuel, making it a lightweight fuel ideal for aviation.

So what's the catch? There are three challenges in turning the vision of hydrogen as a fuel into reality. *First,* there is hardly any hydrogen gas (H_2) in the earth's atmosphere, so it must be produced from hydrogen, which is chemically locked up in water and in organic compounds

© Cengage Learning

Figure 16-32 A *fuel cell* takes in hydrogen gas and separates the hydrogen atoms' electrons from their protons. The electrons flow through wires to provide electricity, while the protons pass through a membrane and combine with oxygen gas to form water vapor. Note that this process is the reverse of *electrolysis*, the process of passing electricity through water to produce hydrogen fuel.

SCIENCE
FOCUS 16.3

THE QUEST TO MAKE HYDROGEN WORKABLE

Scientists are proposing various ideas for producing and storing hydrogen gas (H_2) as fuel. For example, naturally occurring bacteria and algae can produce H_2 by bio-degrading almost any organic material in a microbiological fuel cell.

The most likely production methods will involve using electricity generated by solar-cell power plants, wind farms, and geothermal energy. In 2008, MIT scientists Daniel Nocera and Matthew Kanan developed a catalyst made from inexpensive cobalt and phosphate salts that can be used to split water into hydrogen and oxygen using a relatively small amount of electricity. This breakthrough could make it affordable to use electricity produced by wind turbines or solar cells to produce H_2 and thus to store the energy produced by the wind and sun in the H_2. Researchers are also working on using green algae to produce H_2 from water (Figure 16-A).

Once produced, H_2 can be stored in a pressurized tank as liquid hydrogen. It can also be stored in solid metal hydride com-

pounds and in sodium borohydride, which releases H_2 when heated. Scientists are also evaluating ways to store H_2 by coating the surfaces of activated charcoal or carbon nanofibers with it. When heated, these coated surfaces release the H_2. Another possibility is to store H_2 inside nano-sized glass microspheres that can easily be filled and refilled. More research is needed to transform these possibilities into realities.

Metal hydrides, sodium borohydride, charcoal powders, ammonia borane, carbon nanotubes, and glass microspheres containing H_2 will not explode or burn if a vehicle's fuel tank is ruptured in an accident. Thus, H_2 stored in such ways is a much safer fuel than gasoline, diesel fuel, natural gas, and concentrated ethanol. Also, the use of ultralight car bodies made of composites with an energy-efficient aerodynamic design would improve vehicle fuel efficiency so that large hydrogen fuel tanks would not be needed.

In 2007, engineering professor Jerry Woodall invented a new way to produce

Figure 16-A Scientists are evaluating the use of green algae that can use sunlight to produce hydrogen gas from water.

hydrogen on demand, without producing toxic fumes, by exposing pellets of an aluminum–gallium alloy to water. If this process is perfected and proves economically feasible, H_2 could be generated as needed inside a tank about the same size as an average car's gasoline tank, and thus the fuel would not have to be transported or stored.

Critical Thinking

Do you think that governments should subsidize research and development of these and other technologies in order to help make H_2 a workable fuel? Explain.

such as methane and gasoline. We can produce H_2 by heating water or passing electricity through it; by stripping it from the methane (CH_4) found in natural gas and from gasoline molecules; and through a chemical reaction involving coal, oxygen, and steam. The problem is that it takes energy and money to produce H_2 using these methods. In other words, H_2 must be produced by using other forms of energy, and therefore has a *negative net energy yield*. The amount of energy it takes to make this fuel will always be more than the amount we can get by burning it.

Second, fuel cells are the best way to use H_2 to produce electricity, but current versions of fuel cells are expensive. However, progress in the development of nanotechnology (see Science Focus 14.1, p. 362) could lead to less expensive and more energy-efficient fuel cells.

Third, whether or not a hydrogen-based energy system produces less outdoor air pollution and CO_2 than a fossil fuel system depends on how the H_2 is produced. We could use electricity from coal-burning and conventional nuclear power plants to decompose water into H_2 and O_2. But this approach does not avoid the harmful environmental effects associated with using coal and the nuclear

fuel cycle. Also, making H_2 from coal or stripping it from methane or gasoline adds much more CO_2 to the atmosphere per unit of heat generated than does burning these carbon-containing fuels directly.

If renewable energy sources such as wind farms and solar-cell power plants were used to make H_2, these CO_2 emissions would be sharply reduced. Making H_2 by decomposing the methane in natural gas could serve as a transition to producing H_2 by using renewable energy, as long as there were a sharp reduction of the methane leaks currently resulting from natural gas production (see Science Focus 15.1, p. 378).

Hydrogen's negative net energy yield is a serious limitation and means that this fuel will have to be subsidized in order for it to compete in the open marketplace with fuels that have medium-to-high net energy yields. However, widespread use of H_2 would greatly reduce air pollution and projected climate change. Some analysts contend that the potential for reducing the huge economic costs and environmental harm related to air pollution and projected climate change could make it worthwhile to subsidize the use of hydrogen (Science Focus 16.3).

GOOD NEWS

In the 1990s, Amory Lovins (Individuals Matter 16.1) and his colleagues at the Rocky Mountain Institute designed a very light, safe, extremely efficient hydrogen-powered car. It is the basis of most prototype hydrogen fuel-cell cars now being tested by major automobile companies. Some analysts project that fuel-cell cars, running on affordable H_2 produced from natural gas, could be in widespread use by 2030 to 2050.

Larger, stationary fuel cells could provide electricity and heat for commercial and industrial users. A 45-story office building in New York City gets much of its heat from two large fuel-cell stacks. Japan has built a large fuel cell that produces enough electricity to run a small town.

A Canadian company is developing a unit about the size of a dishwasher that will allow consumers to use electricity to produce their own H_2 from tap water. The unit could be installed in a garage and used to fuel a hydrogen-powered vehicle overnight, when electricity rates are sometimes lower. In sunny areas, people could install rooftop panels of solar cells to produce and store H_2 for their cars. In Japan, some home owners get their electricity and hot water from refrigerator-sized fuel-cell units, which produce H_2 from the methane in natural gas. Figure 16-33 lists the major advantages and disadvantages of using hydrogen as an energy resource (Concept 16-7). **GREEN CAREER:** hydrogen energy

Trade-Offs

Hydrogen

Advantages	Disadvantages
Can be produced from plentiful water at some sites	Negative net energy yield
No CO_2 emissions if produced with use of renewables	CO_2 emissions if produced from carbon-containing compounds
Good substitute for oil	High costs create need for subsidies
High efficiency in fuel cells	Needs H_2 storage and distribution system

© Cengage Learning

Figure 16-33 Using hydrogen as a fuel for vehicles and for providing heat and electricity has advantages and disadvantages. (Concept 16-7). **Questions:** Which single advantage and which single disadvantage do you think are the most important? Why? Do you think that the advantages of hydrogen fuel outweigh its disadvantages? Why?

Photo: LovelaceMedia/Shutterstock.com

16-8 How Can We Make the Transition to a More Sustainable Energy Future?

CONCEPT 16-8
We can make the transition to a more sustainable energy future by greatly improving energy efficiency, using a mix of renewable energy resources, and including the environmental and health costs of energy resources in their market prices.

Choosing Energy Paths

We need to develop energy policies with the future in mind, because experience shows that it usually takes at least 50 years and huge financial investments to phase in new energy alternatives. Creating energy policy involves trying to answer the following questions for *each* energy alternative:

- How much of the energy resource is likely to be available in the near future (the next 25 years) and in the long term (the next 50 years)?
- What is the estimated net energy yield (p. 375) for the resource?
- What are the estimated costs for developing, phasing in, and using the resource?
- What kinds of government research and development subsidies and tax breaks will be needed to help develop the resource?

- How will dependence on the resource affect national and global economic and military security?
- How vulnerable is the resource to terrorism?
- How will extracting, transporting, and using the resource likely affect the environment, the earth's climate, and human health?
- Does use of the resource produce hazardous, toxic, or radioactive substances that we must safely store for very long periods of time?

In considering possible energy futures, scientists and energy experts who have evaluated energy alternatives have come to three general conclusions. First, *during this century, there will likely be a gradual shift from large, centralized macropower systems to smaller, decentralized micropower systems* (Figure 16-34) such as wind turbines, household solar-cell panels, rooftop solar water heaters, and small natural gas turbines. There will also be a shift from gasoline-powered motor vehicles to hybrid and plug-in electric cars. This may be followed by a shift to fuel cells for cars and to stationary fuel cells for houses and commercial buildings.

A shift to a more decentralized energy system would improve national and economic security, because countries would rely on diverse, dispersed, domestic, renewable energy resources instead of on a smaller number of large power plants that are vulnerable to storm dam-

Bioenergy power plants

Wind farms

Small solar-cell power plants

Fuel cells

Rooftop solar-cell arrays

Solar-cell rooftop systems

Smart electrical and distribution system

Commercial

Small wind turbines

Residential

Industrial

Microturbines

© Cengage Learning

Figure 16-34 Solutions: During the next few decades, we will probably shift from dependence on a *centralized electric power system,* based on a few hundred large coal-burning and nuclear power plants, to a *decentralized power system,* in which electricity is produced by a large number of dispersed, small-scale, local power-generating systems that depend primarily on a variety of locally available, low-carbon renewable energy resources.

age and sabotage. Similarly, states and communities within countries could become more energy-independent and secure by making this shift.

The second general conclusion of experts about the future of energy use is that *a combination of improved energy efficiency and carefully regulated use of natural gas will be the best way to make the transition to using mostly renewable energy resources during this century* (Concept 16-8). By using a variety of locally available renewable energy resources, we could apply the biodiversity **principle of sustainability**.

The third general conclusion is that *because fossil fuels are still abundant and artificially cheap, we will continue to use them in large quantities* (Figure 16-35). This presents two major challenges. One is to find ways to reduce the harmful environmental impacts of widespread use of fossil fuels, with special emphasis on reducing outdoor air pollution and greenhouse gas emissions. The second is to find ways to include more of the harmful environmental and health costs of using fossil fuels in their market prices and thus to implement the full-cost pricing **principle of sustainability**.

China (which uses 20% of the world's energy) and the United States (which uses 19% of the world's energy) are

the key players in making the shift away from fossil fuels and thus in slowing the rate of projected climate change during this century.

Figure 16-36 summarizes these and other strategies for making the transition to a more sustainable energy future over the next 50 years (Concept 16-8). Making such a transition presents exciting challenges as well as big rewards, including major cost savings, stimulating and fulfilling careers for young people, and improvements to the environment that could benefit many generations to come (see Chapter 1, Core Case Study, p. 4).

Economics, Politics, and Education Can Help Us Shift to More Sustainable Energy Resources

To most analysts, economics, politics, and consumer education hold the keys to making a shift to a mix of improved energy efficiency and more sustainable energy resources. Governments can use three strategies to help stimulate or reduce the short-term and long-term use of a particular energy resource.

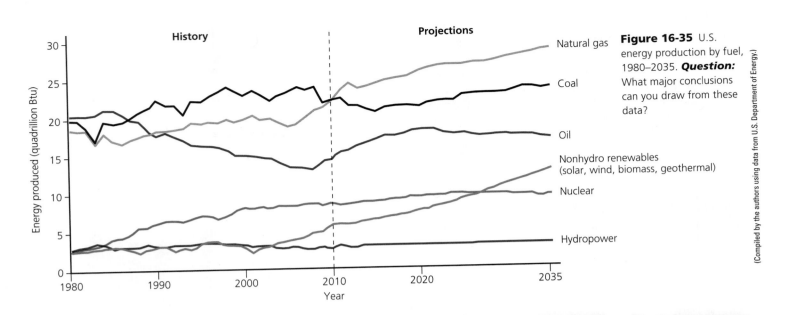

Figure 16-35 U.S. energy production by fuel, 1980–2035. **Question:** What major conclusions can you draw from these data? (Compiled by the authors using data from U.S. Department of Energy.)

History

Projections

Natural gas

Coal

Oil

Nonhydro renewables (solar, wind, biomass, geothermal)

Nuclear

Hydropower

Solutions

Making the Transition to a More Sustainable Energy Future

Improve Energy Efficiency

Increase fuel-efficiency standards for vehicles, buildings, and appliances

Provide large tax credits or fee-bates for buying efficient cars, houses, and appliances

Reward utilities for reducing demand for electricity

Greatly increase energy efficiency research and development

More Renewable Energy

Greatly increase use of renewable energy

Provide large subsidies and tax credits for use of renewable energy

Greatly increase renewable energy research and development

Reduce Pollution and Health Risk

Phase out coal subsidies and tax breaks

Levy taxes on coal and oil use

Phase out nuclear power subsidies, tax breaks, and loan guarantees

© Cengage Learning

Figure 16-36 Energy analysts have made a number of suggestions for helping us make the transition to a more sustainable energy future (**Concept 16-8**). **Questions:** Which five of these solutions do you think are the best ones? Why?

Top: ©andrea lehmkuhl/Shutterstock.com. Bottom: ©pedrosala/Shutterstock.com.

First, they can *keep the prices of selected energy resources artificially low to encourage use of those resources.* They do this by providing research and development subsidies, tax breaks, and loan guarantees to encourage the development of those resources, and by enacting regulations that favor them. For decades, this approach has been employed to stimulate the development and use of fossil fuels and nuclear power in the United States, as well as in most other more-developed countries. This has created an uneven economic playing field that has *encouraged* energy waste and increased use of these nonrenewable energy resources, while it has *discouraged* improvements in energy efficiency and the development of a variety of renewable energy resources.

Many energy analysts argue that one of the most important steps that governments can take to level the economic playing field is to phase out the $250 billion to $300 billion in annual subsidies and tax breaks now provided worldwide for fossil fuels and nuclear energy—both of which are mature industries that could be left to stand on their own economically. Critics say that solar and wind energy receive too many government subsidies compared to fossil fuels and nuclear power. However, in 2010, fossil fuels and nuclear energy together received $6.7 billion in direct government subsidies compared to $6.1 billion for solar and wind combined, according to the Department of Energy. Energy analysts also call for greatly increasing subsidies and tax breaks for developing and phasing in affordable renewable energy and energy-efficiency technologies.

However, making such a shift in energy subsidies is difficult because of the immense political and financial power of the fossil fuel and nuclear power industries. It is not surprising that they vigorously oppose the loss of their subsidies and tax breaks. Generally, they have also opposed significant increases in subsidies and tax breaks for energy efficiency and for competing renewable energy sources.

A *second* major strategy that governments can use is to *allow the prices of selected energy resources to rise to their natural market levels by removing government subsidies, thereby discouraging the use of these resources.* They can do this by eliminating existing tax breaks and other subsidies that favor the use of the targeted resource. They can also tax and regulate the use of such resources as a way of helping to control and pay for the harmful environmental costs of using them. Such measures can increase government revenues, encourage improvements in energy efficiency, and reduce dependence on environmentally harmful energy resources. And they would represent an application of the full-cost pricing **principle of sustainability**.

Third, governments can *emphasize consumer education.* Even if governments offer generous financial incentives for energy efficiency and renewable energy use, people will not make such investments if they are uninformed—or misinformed—about the availability, advantages, disadvantages, and hidden environmental costs of various energy resources.

An excellent example of what a government can do to bring about a more sustainable energy mix is the case of Germany. It is the world's most solar-powered nation, with half of the world's installed capacity. Why does cloudy Germany have more solar water heaters and solar-cell panels than sunny France and Spain have? One reason is that the German government made the public aware of the environmental benefits of these technologies. The other is that the government provided consumers with substantial economic incentives for using the technologies.

We have the creativity, wealth, and most of the technologies we need to make the transition to a more GOOD NEWS sustainable energy future within your lifetime. Making this transition depends primarily on *education, economics,* and *politics*—on how well individuals understand environmental and energy problems and their possible solutions, and on how they vote and thus influence their elected officials. People can also vote with their pocketbooks by refusing to buy energy-inefficient and environmentally harmful products and services, and by letting company executives know about their choices. In 2012, environmental engineer Mark Z. Jacobson wrote: "You could power America with renewables from a technical and economic standpoint. The biggest obstacles are social and political."

Figure 16-37 lists some ways in which you can contribute to making the transition to a more environmentally and economically sustainable energy future.

What Can You Do?

Shifting to More Sustainable Energy Use

- Walk, bike, or use mass transit or a car pool to get to work or school

- Drive only vehicles that get at least 17 kpl (40 mpg)

- Have an energy audit done in the place where you live

- Superinsulate the place where you live and plug all air leaks

- Use passive solar heating

- For cooling, open windows and use fans

- Use a programmable thermostat and energy-efficient heating and cooling systems, lights, and appliances

- Turn down your water heater's thermostat to 43–49°C (110–120°F) and insulate hot water pipes

- Turn off lights, TVs, computers, and other electronics when they are not in use

- Wash laundry in cold water and air dry it on racks

© Cengage Learning

Figure 16-37 Individuals matter: You can make a shift in your own life toward using energy more sustainably. **Questions:** Which three of these measures do you think are the most important ones to take? Why? Which of these steps have you already taken and which do you plan to take?

Big Ideas

- We should evaluate energy resources on the basis of their potential supplies, their net energy yields, and the environmental and health impacts of using them.

- By using a mix of renewable energy sources—especially solar, wind, flowing water, sustainable biofuels, and geothermal energy—we could drastically reduce pollution, greenhouse gas emissions, and biodiversity losses.

- Making the transition to a more sustainable energy future will require sharply increasing energy efficiency, using a mix of environmentally friendly renewable energy resources, and including the harmful environmental and health costs of energy resources in their market prices.

In the **Core Case Study** that opens this chapter, we considered the immense potential of wind—an indirect form of solar energy—as an energy resource, particularly in the United States. Relying mostly on nonrenewable fossil fuels and nuclear power violates the three scientific **principles of sustainability** (see Figure 1-2, p. 6 or back cover). By relying more on a diversity of direct and indirect forms of renewable solar energy coupled with improving energy efficiency to reduce unnecessary waste of energy and money, we could implement these three principles of sustainability.

Making this shift to a more sustainable energy future involves

- relying much more on direct and indirect forms of solar energy for our electricity, heating and cooling, and other needs;
- recycling and reusing more materials and thus reducing wasteful and excessive consumption of energy and matter; and
- mimicking nature's reliance on biodiversity by using a diverse mix of locally and regionally available renewable energy resources.

By making this shift, we would also be applying the three social science **principles of sustainability** (see Figure 1-5, p. 9 or back cover). By including the harmful environmental and health costs of fossil fuels and nuclear energy in their market prices, we would be applying the full-cost pricing principle. This could only be accomplished with the aid of compromise and trade-offs in the political arena by applying the win-win solutions principle. And many analysts argue that such solutions would greatly benefit life on earth immediately and in the long run—a classic application of the principle of responsibility to future generations.

Chapter Review

Core Case Study

1. Describe the potential for using renewable energy from the wind to produce most of the electricity used in the United States.

Section 16-1

2. What are the two key concepts for this section? What is **energy efficiency**? Explain why we can think of energy efficiency as an energy resource. What percentage of the energy used in the United States is unnecessarily wasted? List five widely used energy-inefficient technologies. What are the major advantages of reducing energy waste? List four ways to save energy and money in industry. What is **cogeneration**? How can electric utility companies help people reduce their energy waste? What is a smart grid and why is it important? What are the hidden costs of using gasoline?

3. What are four ways to save energy and money in transportation? Summarize the development of more energy-efficient vehicles. Explain the importance of developing better batteries and list some advances in this area. What are fuel cells and what are their advantages? What are some technologies applied by green architecture? Explain benefits of living roofs. List four ways to improve energy efficiency in new buildings and eight ways to improve energy efficiency in existing buildings. List six ways in which you can save energy where you live. Give three reasons why we waste so much energy. List five advantages of relying more on a variety of renewable energy sources and list three factors that are holding back such a transition.

Section 16-2

4. What is the key concept for this section? Distinguish between a **passive solar heating system** and an **active solar heating system** and discuss the major advantages and disadvantages of using such systems for heating buildings. What are three ways to cool houses naturally? What are **solar thermal systems**, how are they used, and what are the major advantages and disadvantages of using them? What is a **solar cell** (**photovoltaic** or **PV cell**) and what are the major advantages and disadvantages of using such devices to produce electricity?

Section 16-3

5. What is the key concept for this section? Define **hydropower** and summarize the potential for expanding it. What are the major advantages and disadvantages of using hydropower? What is the potential for using tides and waves to produce electricity?

Section 16-4

6. What is the key concept for this section? What are the advantages of using taller wind turbines? Summarize the global potential for wind power. What are the major advantages and disadvantages of using wind to produce electricity?

Section 16-5

7. What are the two key concepts for this section? What is **biomass** and what are the major advantages and disadvantages of using wood to provide heat and electricity? What are the major advantages and disadvantages of using biodiesel and ethanol to power

motor vehicles? Explain how algae and bacteria can be used to produce fuels nearly identical to gasoline and diesel fuel.

Section 16-6

8. What is the key concept for this section? What is **geothermal energy** and what are three sources of such energy? What are the major advantages and disadvantages of using geothermal energy as a source of heat and to produce electricity?

Section 16-7

9. What is the key concept for this section? What are the major advantages and disadvantages of using hydrogen as a fuel to use in producing electricity and powering vehicles?

Section 16-8

10. What is the key concept for this section? List eight questions that policy makers should ask about each source of energy. List three general conclusions that energy experts have come to in considering possible energy futures. List five major strategies suggested by such experts for making the transition to a more sustainable energy future. Explain three strategies that governments can use to encourage or discourage the use of an energy resource. What are this chapter's *three big ideas*? Explain how we would be applying the six **principles of sustainability** by improving energy efficiency and shifting to renewable energy resources.

Note: Key terms are in bold type.

Critical Thinking

1. Suppose that a wind power developer has proposed building a wind farm near where you live (**Core Case Study**). Would you be in favor of the project or opposed to it? Write a letter to your local newspaper or a blog for a website explaining your position and your reasoning. As part of your research, determine how the electricity you use now is generated and where the nearest power plant is located, and include this information in your arguments.

2. List five ways in which you unnecessarily waste energy during a typical day, and explain how these actions violate the three scientific **principles of sustainability**.

3. Congratulations! You have won $500,000 to build a more sustainable house of your choice. With the goal of maximizing energy efficiency, what type of house would you build? How large would it be?

Where would you locate it? What types of materials would you use? What types of materials would you *not* use? How would you heat and cool the house? How would you heat water? What types of lighting, stove, refrigerator, washer, and dryer would you use? Which, if any, of these appliances could you do without?

4. Suppose that a homebuilder installs electric baseboard heat and claims, "It is the cheapest and cleanest way to go." Apply your understanding of the second law of thermodynamics (see Chapter 2, p. 43) and net energy yield (see Figure 15-2, p. 375) to evaluate this claim. Write a response to the homebuilder summarizing your findings.

5. Should buyers of energy-efficient motor vehicles receive large rebates funded by fees levied on gas guzzlers? Explain.

6. Do you think that the estimated hidden costs of gasoline should be included in its price at the pump? Explain. Would you favor much higher gasoline taxes if payroll taxes and income taxes were reduced to balance gasoline tax increases, with no net additional cost to consumers? Explain.

7. Explain why you agree or disagree with the following proposals made by various energy analysts:
   a. Government subsidies for all energy alternatives should be eliminated so that all energy resources can compete on a level playing field in the marketplace.
   b. All government tax breaks and other subsidies for conventional fossil fuels (oil, natural gas, and coal), synthetic natural gas and oil, and nuclear power (fission and fusion) should be eliminated. They should be replaced with subsidies and tax breaks for improving energy efficiency and developing renewable energy alternatives.
   c. Development of renewable energy alternatives should be left to private enterprise and should receive little or no help from the federal government, but the nuclear power and fossil fuels industries should continue to receive large federal government subsidies.

8. What percentages of the U.S. Department of Energy research and development budget would you devote to fossil fuels, nuclear power, renewable energy, and improving energy efficiency? How would you distribute your funds among the various renewable energy options? Explain your thinking.

Doing Environmental Science

Do a survey of energy use at your school, based on the following questions: How is the electricity generated? How are most of the buildings heated? How is water heated? How are most of the vehicles powered? How is the computer network powered? How could energy efficiency be improved, if at all, in each of these areas? How could your school make use of solar, wind, biomass, and other forms of renewable energy if it does not already do so? Write up a proposal for using energy more efficiently and sustainably at your school and submit it to school officials.

Global Environment Watch Exercise

In the *Renewable Energy* portal, search for information on the renewable energy policy in Germany (you may also use the World Map). What are some of the challenges the German government is facing, and how are they being addressed? What challenges would arise if a U.S. government mandate required a similar transition to renewable energies? Write a report answering these questions and discussing the effects of the "German Experiment" on the United States as it looks to revise its energy policy and address climate change.

Data Analysis

Study the table below and then answer the questions that follow it by filling in the blank columns in the table.

Combined City/Highway Fuel Efficiency for 2012 Cars (mpg)				
Model	Miles per Gallon (mpg)	Kilometers per Liter (kpl)	Annual Liters (Gallons) of Gasoline	Annual CO_2 Emissions
Toyota Prius Hybrid	50			
Ford Focus AM-6	31			
Chevrolet Cruze A-S6	30			
Honda Accord A-S	27			
Ford F150 A-S6 Pickup	19			
Jeep Grand Cherokee A-6 SUV	16			
Lamborghini Aventador Coupe	13			

(Compiled by the authors using data from the U.S. Environmental Protection Agency.)

1. Using Supplement 1 (Measurement Units, p. S1), convert the miles per gallon figures in the table to kilometers per liter (kpl).

2. How many liters (and how many gallons) of gasoline would be used annually by each car if it were driven 19,300 kilometers (12,000 miles) per year?

3. How many kilograms (and how many pounds) of carbon dioxide would be released into the atmosphere each year by each car, based on the fuel consumption calculated in question 2, assuming that the combustion of gasoline releases 2.3 kilograms of CO_2 per liter (19 pounds per gallon)?

Questions 1–4 refer to the following approximate energy efficiencies.

    a. 1%
    b. 5%
    c. 8%
    d. 20%
    e. 33%

1. Coal-fired power plant including the energy used to dig up the coal

2. Nuclear power plant including the nuclear fuel cycle

3. Incandescent lightbulb

4. Internal combustion engine used to power most motor vehicles

5. The Corporate Average Fuel Economy Standards (CAFÉ) have
    a. been raised to 35 mpg.
    b. had extreme success in the United States.
    c. increased steadily since 1973.
    d. decreased to about 21 mpg since 1985.
    e. been raised due to hybrid technology.

6. All of the following are ways to help save energy EXCEPT
    a. improve energy efficiency such as in Japan, Germany, and France.
    b. the use of cogeneration in a trash-to-electricity plant.
    c. use heat-efficient windows on the south side of buildings.
    d. use LED bulbs in traffic lights and streetlights.
    e. build houses with walls of superinsulating straw bales.

7. An example of using an active solar heating system would be to
    a. plant a deciduous tree outside a window to keep the sun out.
    b. install a photovoltaic system on the roof.
    c. use super windows.
    d. use heavy blinds on the windows.
    e. use vents to allow hot air to escape in the summer.

Questions 8–11 refer to the following disadvantages of alternative energy sources.

    a. Interferes with migratory birds
    b. Air pollution
    c. Scarcity of suitable sites
    d. Negative net energy
    e. May raise food prices

8. Geothermal

9. Hydrogen fuel cells

10. Ethanol fuel

11. Wind energy

Questions 12 and 13 refer to the diagram below.

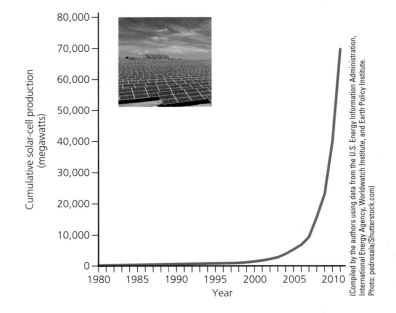

(Compiled by the authors using data from the U.S. Energy Information Administration, International Energy Agency, Worldwatch Institute, and Earth Policy Institute. Photo: pedrosala/Shutterstock.com)

12. How many years did it take to triple the production of solar cells from 1995?
    a. 3
    b. 6
    c. 9
    d. 12
    e. 15

13. What information can be inferred from the graph above?
    a. More people are using solar energy now than in 1980.
    b. Solar energy utilization has surpassed that of wind energy.
    c. By 2010 there were 14,000 megawatts of solar-cell production.
    d. The amount of available solar energy is currently increasing exponentially.
    e. There has been a linear increase in solar-cell production since 1995.

17 Environmental Hazards and Human Health

The dose makes the poison.

PARACELSUS, 1540

Key Questions

17-1 What major health hazards do we face?

17-2 What types of biological hazards do we face?

17-3 What types of chemical hazards do we face?

17-4 How can we evaluate chemical hazards?

17-5 How do we perceive risks and how can we avoid the worst of them?

Many coal-burning factories and power plants release toxic mercury and other air pollutants into the atmosphere.

Dudarev Mikhail/Shutterstock.com

Mercury (Hg) and its compounds are all toxic. Research indicates that long-term exposure to high levels of mercury can permanently damage the human nervous system, brain, kidneys, and lungs. Fairly low levels of mercury can also harm fetuses and cause birth defects.

This toxic metal is released into the air from rocks, soil, and volcanoes and by vaporization from the oceans. Such natural sources account for about one-third of the mercury reaching the atmosphere each year. According to the EPA, the remaining two-thirds come from human activities—primarily from the smokestacks of coal-burning power and industrial plants (chapter-opening photo), cement kilns, and solid-waste incinerators. When it rains or snows these emissions are washed out of the atmosphere onto the soil and into bodies of water.

Because mercury is an element (see Figure 2-3, left, p. 36), it cannot be broken down or degraded. Therefore, this indestructible global pollutant can accumulate in soil, in water, and in the tissues of humans and other animals. In the atmosphere, some elemental mercury is converted to more toxic inorganic and organic mercury compounds that can be deposited in lakes and other aquatic environments and on land. Under certain conditions in aquatic systems, bacteria can convert inorganic mercury compounds to highly toxic *methylmercury* (CH_3Hg^+), which can be biologically magnified in food chains and webs (see Figure 9-13, p. 202). As a result, in such aquatic systems, high levels of methylmercury are often found in the tissues of large fishes that feed at high trophic levels.

Humans are exposed to mercury in two major ways. *First,* we can eat fish and shellfish that are contaminated with methylmercury (Figure 17-1). This accounts for 75% of all human exposures to toxic forms of mercury. *Second,* we inhale vaporized elemental mercury or particles of inorganic mercury salts such as mercury sulfide (HgS) and mercuric chloride ($HgCl_2$)—pollutants in the air coming mostly from coal-burning power plants and solid-waste incinerators.

The greatest risk from exposure to low levels of methylmercury is brain damage in fetuses and young children. Studies estimate that 30,000–60,000 of the children born each year in the United States are likely to have reduced IQs and possible nervous system damage because of such exposure. Also, methylmercury has been shown to harm the hearts, kidneys, and immune systems of some adults.

This problem raises several important questions: How do scientists determine the potential harm from exposure to mercury and other chemicals? How serious is the risk of harm from a particular chemical compared to other risks? What should individuals and government health officials do with any evidence of harm?

In this chapter, we will look at how scientists try to answer these questions about our exposure to chemicals. We also examine health threats from disease-causing bacteria, viruses, and protozoa, and from other hazards that kill millions of people every year.

Figure 17-1 Gloved hands hold a fish contaminated with mercury (left). Fish are contaminated with mercury in many lakes, including this one in Wisconsin (right).

Left: Peter Casolino/Alamy. Right: Drake Fleege/Alamy.

17-1 What Major Health Hazards Do We Face?

CONCEPT 17-1
We face health hazards from biological, chemical, physical, and cultural factors, and from the lifestyle choices we make.

Risks Are Usually Expressed as Probabilities

A **risk** is the *probability* of suffering harm from a hazard that can cause injury, disease, death, economic loss, or damage. It is usually expressed as a mathematical statement about how likely it is that harm will be suffered from a hazard. Scientists often state probability in terms such as, "The lifetime probability of developing lung cancer from smoking one pack of cigarettes per day is 1 in 250." This means that 1 of every 250 people who smoke a pack of cigarettes every day will likely develop lung cancer over a typical lifetime (usually considered to be 70 years). Probability can also be expressed as a percentage, as in a 30% chance of rain.

Risk assessment is the process of using statistical methods to estimate how much harm a particular hazard can cause to human health or to the environment. It helps us to estimate the probability of a risk, compare it with the probability of other risks, and establish priorities for avoiding or managing risks. **Risk management** involves deciding whether and how to reduce a particular risk to a certain level and at what cost. Figure 17-2 summarizes how risks are assessed and managed.

We Face Many Types of Hazards

All of us take risks every day. Examples include choosing to drive or ride in a car through heavy traffic; talking on a phone or texting while driving; eating foods with a high cholesterol or fat content that can contribute to potentially fatal heart attacks; drinking too much alcohol; smoking or being in an enclosed space with a smoker; lying out in the sun or going to a tanning parlor, which increases the risk of getting skin cancer; and living in a tornado-, hurricane-, or flood-prone area.

None of us can live a risk-free life, but we can reduce our exposure to risks. The key questions are, *How serious are the risks we face?* and *Do the benefits of certain activities outweigh the risks?*

Figure 17-2 Risk assessment and risk management are used to estimate the seriousness of various risks and to help reduce such risks. **Question:** What is an example of how you have applied this process in your daily living?

We can suffer harm from five major types of hazards (Concept 17-1):

- *Biological hazards* from more than 1,400 **pathogens**—organisms that can cause disease in another organism—including bacteria, viruses, parasites, protozoa, and fungi.
- *Chemical hazards* from harmful chemicals in our air, water, soil, food, and human-made products (**Core Case Study**).
- *Natural hazards* such as fire, earthquakes, volcanic eruptions, floods, and storms.
- *Cultural hazards* such as unsafe working conditions, unsafe highways, criminal assault, and poverty.
- *Lifestyle choices* such as smoking, making poor food choices, drinking too much alcohol, and having unsafe sex.

 CONSIDER THIS. . .

THINKING ABOUT Health Hazards

Think of a hazard from each of these categories that you may have faced recently. Which one was the most threatening?

17-2 What Types of Biological Hazards Do We Face?

CONCEPT 17-2
The most serious biological hazards we face are infectious diseases such as flu, AIDS, tuberculosis, diarrheal diseases, and malaria.

Some Diseases Can Spread from One Person to Another

An **infectious disease** is a disease caused by a pathogen such as a bacterium, virus, or parasite invading the body and multiplying in its cells and tissues. **Bacteria** are singe-cell organisms that are found everywhere and

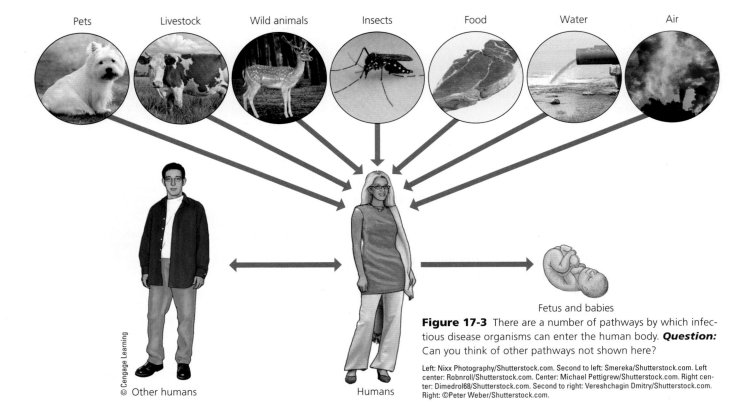

Pets Livestock Wild animals Insects Food Water Air

© Other humans Humans

Fetus and babies

Figure 17-3 There are a number of pathways by which infectious disease organisms can enter the human body. **Question:** Can you think of other pathways not shown here?

that can multiply very rapidly on their own. Most bacteria are harmless or beneficial but some cause diseases such as tuberculosis and strep throat. **Viruses** are smaller than bacteria and work by invading a cell and taking over its genetic machinery to copy themselves and then spreading throughout a body. They cause diseases such as flu and acquired immune deficiency syndrome (AIDS). **Parasites**, organisms that live on or inside other organisms and feed on them, can also cause serious infectious diseases, including malaria.

A **transmissible disease** (also called a *contagious* or *communicable disease*) is an infectious disease that can be transmitted from one person to another. Some are bacterial diseases such as tuberculosis and gonorrhea. Others are viral diseases such as the common cold, flu, and AIDS. A **nontransmissible disease** is caused by something other than a living organism and does not spread from one person to another. Such diseases include cardiovascular (heart and blood vessel) diseases, most cancers, asthma, and diabetes.

In 1900, infectious disease was the leading cause of death in the world. Since then, and especially since 1950, the incidences of infectious diseases and the death rates from them have been greatly reduced. This has been achieved mostly by a combination of better health care, the use of antibiotics to treat diseases caused by bacteria, and the development of vaccines to prevent the spread of some viral diseases. As a result, average life expectancy has increased in most countries.

GOOD NEWS

Despite the declining risk of harm from infectious diseases, they remain as serious health threats, especially in less-developed countries. Figure 17-3 shows major pathways by which infectious disease organisms can enter the human body. Such diseases can then be spread through air, water, food, and body fluids such as feces, urine, blood, and droplets sprayed by sneezing and coughing.

A large-scale outbreak of an infectious disease in an area or a country is called an *epidemic*. A global epidemic such as tuberculosis or AIDS is called a *pandemic*. Figure 17-4 shows the annual death tolls from the world's seven deadliest infectious diseases (**Concept 17-2**).

One reason why infectious disease is still a serious threat is that many disease-carrying bacteria have developed genetic immunity to widely used antibiotics (Science Focus 17.1). Also, many disease-transmitting species of insects such as mosquitoes have become immune to widely used pesticides such as DDT that once helped to control their populations.

Another factor that will likely keep infectious diseases high on the list of environmental health threats is projected climate change due to atmospheric warming. Many scientists warn that warmer temperatures will likely allow some diseases to spread more easily from tropical to temperate areas. A 2011 study showed that warmer ocean waters, also resulting from atmospheric warming, are already causing rapid growth of the *Vibrio* genus of bacteria that can cause a toxic infection of the blood, gastroen-

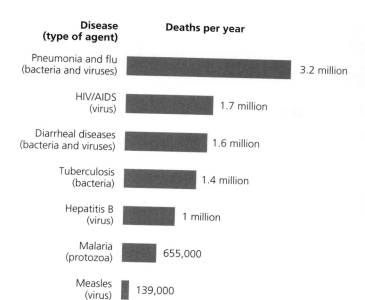

Disease (type of agent)	Deaths per year
Pneumonia and flu (bacteria and viruses)	3.2 million
HIV/AIDS (virus)	1.7 million
Diarrheal diseases (bacteria and viruses)	1.6 million
Tuberculosis (bacteria)	1.4 million
Hepatitis B (virus)	1 million
Malaria (protozoa)	655,000
Measles (virus)	139,000

Figure 17-4 *Global outlook:* The World Health Organization has estimated that the world's seven deadliest infectious diseases killed about 10 million people per year—most of them poor people in less-developed countries (**Concept 17-2**). This averages about 27,400 mostly preventable deaths every day. **Question:** How many people, on average, die prematurely from these diseases every hour?

(Compiled by the authors using data from the World Health Organization, 2010 and 2012 and the U.S. Centers for Disease Control and Prevention.)

Figure 17-5 The colorized red areas in this chest X-ray show where TB bacteria have destroyed lung tissue.

© Michael Abbey/Visuals Unlimited/Corbis

teritis, and cholera. Some *Vibrio* species are also becoming resistant to antibiotics.

CASE STUDY

The Growing Global Threat from Tuberculosis

Since 1990, one of the world's most underreported stories has been the rapid spread of tuberculosis (TB), an extremely contagious bacterial infection of the lungs. According to the World Health Organization (WHO), about one-third of the world's people are infected with the TB bacterium. However, most of them have *latent*, or inactive, TB and only 10% of them will eventually become sick with active TB.

Many TB-infected people do not appear to be sick, and most of them do not know they are infected. Left untreated, each person with active TB typically infects a number of other people. Without treatment, about half of the people with active TB die from bacterial destruction of their lung tissue (Figure 17-5).

According to the WHO, TB strikes about 8.8 million people per year and kills 1.4 million—about 84% of them in less-developed countries (Figure 17-6). On average, someone dies of TB every 23 seconds.

Several factors account for the recent spread of TB. One is that there are too few TB screening and control programs, especially in less-developed countries, where 95% of the new cases occur (Figure 17-6). A second prob-

lem is that most strains of the TB bacterium have developed genetic resistance to the majority of the effective antibiotics (Science Focus 17.1). Also, population growth, urbanization, and air travel have greatly increased person-to-person contacts, and TB is spreading faster in areas where large numbers of poor people crowd together. A person with active TB might infect several people during a single bus or plane ride.

In addition, AIDS greatly weakens its victims' immune systems, making AIDS patients highly susceptible to unchecked growth of the TB bacteria. As a result, people with AIDS are 30 to 50 times more likely to develop active TB.

Slowing the spread of the disease requires early identification and treatment of people with active TB, especially those with a chronic cough, which is the primary way in which the disease is spread from person to person. Treatment with a combination of four inexpensive drugs can cure 90% of individuals with active TB. To be effective, the drugs must be taken every day for 6–9 months. Because the symptoms disappear after a few weeks, many patients, thinking they are cured, stop taking the drugs, allowing the disease to recur in drug-resistant forms and to spread to other people.

In recent years, a deadly and apparently incurable form of tuberculosis, known as *multidrug-resistant TB*, has been spreading. According to the WHO, there are about 440,000 new cases of this form of TB every year, and 150,000 deaths resulting from it. Because this disease cannot be treated effectively with antibiotics, victims must be permanently isolated from the rest of society. Victims also pose a threat to health workers. The WHO has estimated that between 2011 and 2015, more than 2 million people will contract multidrug-resistant TB.

Most people who have active TB live in the world's poorest countries and many of them cannot afford treatments. Thus, there is little financial incentive for large drug companies to invest a great deal of money on developing drugs to treat the disease. However, efforts to

GENETIC RESISTANCE TO ANTIBIOTICS IS INCREASING

Antibiotics are chemicals that can kill bacteria. No such chemicals have been found to kill viruses. We risk falling behind in our efforts to prevent infectious bacterial diseases because of the astounding reproductive rate of bacteria, some of which can grow from a population of 1 to well over 16 million in 24 hours. This allows bacteria to quickly become genetically resistant to an increasing number of antibiotics through natural selection (Figure 17-A).

Other factors can promote such genetic resistance, also called *antibiotic resistance*. One is the spread of bacteria around the globe by human travel and international trade. Airplanes, for example, are incubators for disease-causing viruses and bacteria that are picked up in one country, circulated for hours within a closed environment, and then carried to other parts of the world. Another factor in antibiotic resistance is the overuse of pesticides, which increases populations of pesticide-resistant insects and other carriers of bacterial diseases. In addition, some drug-resistant bacteria can quickly transfer their resistance to nonresistant bacteria by exchanging genetic material.

Yet another factor is the overuse of antibiotics for colds, flu, and sore throats, most of which are viral diseases that cannot be treated with antibiotics. In many countries, antibiotics are available without a prescription, which promotes unnecessary use.

Bacterial resistance to some antibiotics has increased because these drugs are widely used to control disease and to promote growth in dairy and beef cattle, poultry, hogs, and other livestock. In 2011 the U.S. Food and Drug Administration (FDA) estimated that about 70% of all antibiotics used in the United States and 50% of those used worldwide are added to the feed of healthy livestock. Each year in the United States livestock animals are fed about 13 million kilograms (29 million pounds) of antibiotics and people use 3 million kilograms (7 million pounds). In addition, the growing use of antibacterial hand soaps and other antibacterial cleansers is probably promoting antibiotic resistance in bacteria, which is the opposite of the intended effect of these products.

As a result of these factors acting together, every major disease-causing bacterium has now developed strains that resist at least 1 of the roughly 200 antibiotics used to treat bacterial infections. In 2011, infectious disease specialist Brad Spellberg and the Infectious Diseases Society of America estimated that roughly 99,000 people in the United States are dying annually from antibiotic-resistant infections. The U.S. Centers for Disease Control and Prevention (CDC) now considers antibiotic resistance to be one of its top concerns.

For example, a bacterium known as *methicillin-resistant staphylococcus aureus,* commonly known as MRSA (pronounced "mersa"), has become resistant to most common antibiotics. This type of staph infection first appears on the skin as a red, swollen, sometimes painful pimple or boil that will not heal. MRSA can cause a vicious type of pneumonia, flesh-eating wounds, and a quick death if it gets into the bloodstream.

Figure 17-6

Estimated rates of occurrence of TB throughout the world in 2009.

(Compiled by the authors using data from the World Health Organization and the U.S. Centers for Disease Control and Prevention.)

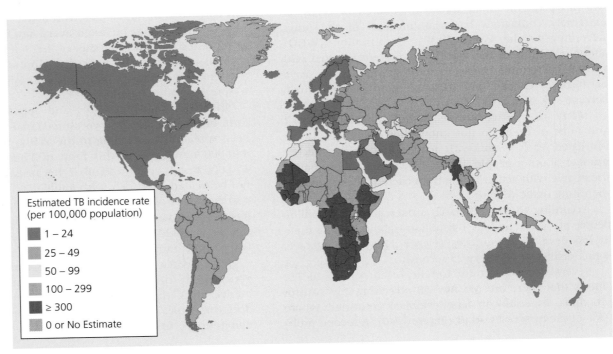

Estimated TB incidence rate (per 100,000 population)

- 1 – 24
- 25 – 49
- 50 – 99
- 100 – 299
- ≥ 300
- 0 or No Estimate

a. A group of bacteria, including genetically resistant ones, are exposed to an antibiotic

b. Most of the normal bacteria die

c. The genetically resistant bacteria start multiplying

d. Eventually the resistant strain replaces the strain affected by the antibiotic

Normal bacterium   Resistant bacterium

© Cengage Learning

Figure 17-A *Genetic resistance:* **(a)** A population of bacteria is exposed to an antibiotic, which **(b)** kills all individuals except those possessing a trait that makes them resistant to the drug. **(c)** The resistant bacteria multiply and eventually **(d)** replace all or most of the nonresistant bacteria.

MRSA can be found in hospitals, although the CDC reported in 2011 that MRSA infections that began in hospitals had declined by 28% between 2005 and 2008 and were continuing to fall. However, such infections still kill more than 18,000 people each year in the United States, according to the CDC. MRSA can also spread in nursing homes, dialysis centers, schools, meeting rooms, gyms, and college dormitories. It can be spread through skin contact, unsanitary use of tattoo needles, and contact with poorly laundered clothing and shared items such as towels, bed linens, athletic equipment, and razors.

In 2013, U.S. health officials issued an alert about the appearance of a class of superbugs, called carbapenem-resistant Enterobacteriaceae, or CRE, spreading in some hospitals and nursing homes among patients with weakened immune systems. CRE are resistant to all major antibiotics and can share their resistant genes with other more common bacteria that cause diarrhea, pneumonia, and urinary tract infections. CRE infections are quite deadly, with fatality rates as high as 50%. If CRE is allowed to spread, common and widespread infections could become untreatable with antibiotics, and this could lead to higher numbers of severe health problems and deaths due to such infections.

Critical Thinking

What are three steps that we could take to slow the rate at which disease-causing bacteria are developing resistance to antibiotics?

---

develop more effective antibiotics and vaccines are being undertaken by governments and private groups. Also, researchers are developing new and easier ways to detect TB in its victims and to monitor victims for loss of liver function as a possible side effect of taking anti-TB drugs (Individuals Matter 17.1).

 CONSIDER THIS. . .

THINKING ABOUT Dealing with Tuberculosis

If you were a public health official, what would you do to try to slow the spread of tuberculosis?

Viral Diseases and Parasites Kill Large Numbers of People

Viruses evolve quickly, are not affected by antibiotics, and can kill large numbers of people. The biggest killer is the *influenza* or *flu* virus (Concept 17-2), which is transmitted by the body fluids or airborne emissions of an infected person. Easily transmitted and especially potent flu viruses could spread around the world in a pandemic that could kill millions of people in only a few months.

The second biggest viral killer is the *human immunodeficiency virus* (HIV) (see the following Case Study). On a global scale, according to the WHO, HIV infects about 2.5 million people every year, and the complications resulting from AIDS kill about 1.7 million people annually.

The third largest viral killer is the *hepatitis B virus* (HBV), which damages the liver and kills about 1 million people each year, according to the WHO. Like HIV, it is transmitted by unsafe sex, sharing of needles by drug users, infected mothers who pass the virus to their offspring before or during birth, and exposure to infected blood.

One virus that has become prevalent fairly recently is the *West Nile virus,* which is transmitted to humans by the bite of a common mosquito that gets infected when

Hayat Sindi: Science Entrepreneur and National Geographic Emerging Explorer

©HAYAT SINDI

Growing up in a home of humble means in Saudi Arabia, Hayat Sindi was determined to get an education, become a scientist, and do something for humanity. Learning English on her own, she became the first Saudi woman to be accepted at Cambridge University where she entered the field of biotechnology. She eventually earned her PhD, taught in the Cambridge's international medical program, and in 2011 was a visiting scholar at Harvard University. In 2012, she was named a National Geographic Emerging Explorer.

As a scientist, Sindi has focused on finding and implementing simple, inexpensive ways to monitor the health of people living in remote, less-developed regions of the world. She knew that many people in these areas are taking strong medicines to combat tuberculosis, hepatitis, or AIDS. She also knew that these medicines, if not closely monitored, can cause liver damage or can kill a patient, and that in many areas, expensive health monitoring is not an option. After becoming aware of an important new diagnostic tool developed by a team of scientists at Harvard, she worked with those scientists to co-found a nonprofit marketing company called Diagnostics for All to bring low-cost health monitoring to remote, poor areas of the world.

The Harvard team's diagnostic tool is an inexpensive, low-tech device that can be used to detect certain illnesses and medical problems, such as declining liver function. It is a piece of paper the size of a postage stamp, with tiny channels and wells etched into it. Technicians load the etched paper with certain chemicals and then put a drop of a patient's blood, urine, or saliva onto this paper where the chemicals react with the fluid to change its color. The colors can be read with the aid of a color scale that indicates different sorts of infections and conditions such as declining liver function. The test takes less than a minute, can be conducted by a technician with minimal training, and requires no electricity, clean water, or special equipment. After the paper testing tool is used, it can be burned on the spot to prevent spreading of infectious agents.

Sindi has also launched a nonprofit foundation known as the Institute for Imagination and Ingenuity to help young people in poor areas of the world to do what she did. It will give funding and guidance to those who wish to attend a university and then return to their home countries to use their talents in helping others.

Sindi is a passionate advocate for science education and scientific careers for women, especially in the Middle East. She also has plans to build a world-class biotechnology center of excellence in Saudi Arabia with the support of Harvard University and MIT.

Background photo: © imagebroker/Alamy

it feeds on birds that carry the virus. Since 1992 when this virus emerged in the United States, it has spread from coast to coast. According to the CDC, between 1999 and 2012, the West Nile virus caused severe illnesses in nearly 26,000 people and killed more than 1,100 people. Such illnesses include viral encephalitis, an acute infection of the brain and spinal cord, and viral meningitis, an infection in the membranes that cover the brain and spinal cord.

In 2012, West Nile virus outbreaks occurred in Europe, Africa, and the United States. The U.S. outbreak was the worst ever recorded. Scientists have not determined the causes for these outbreaks, but some suspect that a long stretch of warm, dry weather in 2012 was a factor. The *Culex* mosquito that carries the virus reproduces and matures faster in such weather.

Scientists believe that more than half of all infectious diseases throughout history were originally transmitted to humans from wild or domesticated animals. The West Nile virus, HIV/AIDS, and a flu strain from birds, called *avian flu*, all fall in this category. The development of such diseases has spurred the growth of the relatively new field of *ecological medicine* (Science Focus 17.2).

You can greatly reduce your chances of getting infectious diseases by practicing good old-fashioned hygiene. Wash your hands thoroughly and frequently with plain soap, avoid sharing personal items such as razors or towels, keep all cuts and scrapes covered with bandages until healed, and stay away from people who have flu or other viral diseases. Some health officials warn that using antibacterial soaps, liquids, and sprays in order to avoid infectious disease may be doing more harm than good, because they can contribute to genetic resistance in infectious bacteria (Science Focus 17.1).

Yet another growing health hazard is infectious diseases caused by parasites, especially malaria (see the second Case Study that follows).

CASE STUDY

The Global HIV/AIDS Epidemic

The global spread of *acquired immune deficiency syndrome (AIDS)*, caused by infection with the *human immunodeficiency virus (HIV)*, is a major global health threat. This virus cripples the immune system and leaves the body vulnerable to infections such as tuberculosis (TB) and rare forms of cancer such as *Kaposi's sarcoma*. HIV is transmitted from one person to another by unsafe sex, sharing of needles by drug users, infected mothers who pass the virus on to their children before or during birth, and exposure to infected blood. A person infected with HIV can live a normal life, but if the infection develops into AIDS, death is likely.

Since HIV was identified in 1981, this viral infection has spread around the globe. According to the WHO, in 2011, a total of about 34.2 million people worldwide (about 1.1 million in the United States) were living with HIV. In recent years, there have been about 2.5 million new cases of AIDS per year (56,000 in the United States)—half of them in people aged 15 to 24. This is down from an estimated 3.2 million new cases in 1997.

Treatment that includes combinations of expensive antiviral drugs can slow the progress of HIV and AIDS, but they are expensive. With such drugs, a person with AIDS, on average, can expect to live about 24 years after being infected at a cost of about $25,200 a year. Such drugs cost too much to be used extensively in less-developed countries where AIDS infections are widespread. Patients taking the drugs should also be monitored for loss of liver function.

For people who do not take such antiviral drugs, it takes an average of 10 to 11 years for an HIV infection to progress to AIDS. This long incubation period means that infected people often spread the virus for several years without knowing they are infected. About one of every five people infected with HIV is not aware of the infection. The CDC estimates that about half of all people who are infected in the United States each year get HIV from someone who is unaware of his or her own infection.

Between 1981 and 2012, more than 30 million people (617,000 in the United States) died of AIDS-related diseases, according to the CDC. Each year, AIDS kills about 1.7 million people (about 18,000 in the United States). AIDS has reduced the life expectancy of the 750 million people living in sub-Saharan Africa from 62 to 47 years, on average, and to 40 years in the seven countries most severely affected by AIDS. The premature deaths of teachers, health-care workers, and other young, productive adults affect the population age structure in African countries such as Botswana (see Figure 6-15, p. 134). These deaths also lead to diminished education and health care, decreased food production and economic development, and disintegrating families.

The fight against AIDS has been vigorous. Worldwide funding for AIDS programs grew tenfold between 2001 and 2009, according to the UN. Scientists are working to find ways to treat and monitor the people most in need—those living in remote regions of less-developed countries (Individuals Matter 17.1). Between 2004 and 2009, AIDS-related deaths fell 19% and new HIV infections dropped by 19% between 2001 and 2010.

CASE STUDY

Malaria—The Spread of a Deadly Parasite

Almost half of the world's people—most of them living in poor African countries—are at risk of getting malaria (Figure 17-7). So is anyone traveling to malaria-prone areas, because there is no vaccine that can prevent this disease.

Malaria is caused by a parasite that is spread by the bites of certain mosquito species. It infects and destroys

ECOLOGICAL MEDICINE: TRACKING HOW HUMANS CAN GET INFECTIOUS DISEASES FROM OTHER ANIMALS

Figure 17-B This deer tick can carry the Lyme disease bacterium from a deer or mouse to a human.

There are a number of infectious diseases that move from one animal species to another and from wild and domesticated animal species to humans, and these infections are on the rise. In a 2008 study Peter Daszak and other scientists found that about 60% of the 325 diseases emerging between 1940 and 2004 were transmitted from animals to humans, and the majority of those came from wild animals. Examples of such diseases and their origins include:

- Avian flu—moved from wild birds to chickens to humans
- Lyme disease—moved from wild deer and mice through ticks to humans
- HIV—moved from chimpanzees to humans (see Case Study, p. 449)
- Plague—moved from rats to rat fleas to humans
- HBV and dengue fever—moved from apes to humans
- African sleeping sicknesses—moved from wild and domestic grazing animals to humans

Ecological medicine is a new interdisciplinary field devoted to tracking down these unwanted connections between animals and humans. Scientists in this field have identified several human practices that encourage the spread of diseases among animals and people. One is the clearing or fragmenting of forests to make way for settlements, farms, and expanding cities. For example, in the United States, the push of suburban development into forests has increased the chances of many suburbanites becoming infected with Lyme disease. The bacterium that causes this disease lives in the bodies of deer and white-footed mice and is passed between these two animals and to humans, mostly by certain types of ticks (Figure 17-B), but also by mosquitoes, spiders, fleas, and mites. Left untreated, Lyme disease can cause debilitating arthritis, heart disease, and nervous disorders.

Expanding suburbs into increasingly fragmented woodland areas has also led to greatly reduced populations of foxes and wildcats, which had kept down populations of the white-footed mouse that carries the Lyme bacterium. The result: white-footed mouse and tick populations have exploded and more people are becoming infected with Lyme disease.

Cutting down tropical forests has displaced fruit bats, causing them to visit both orchards and the fruit trees that grow in people's yards. Bats can be infected with viruses that they can then transfer to humans. In 2007, for example, villagers in the Congo shot thousands of fruit bats. Then, 360 people who came in contact with the blood-soaked bat corpses became infected with the deadly Ebola virus and 168 of them died.

The practice of hunting wild game for food has also increased the spread of infectious diseases. In parts of Africa and Asia, local people who kill monkeys and other animals for bush meat (see Figure 9-16, p. 206) come in regular contact with primate blood and can be exposed to a simian (ape or monkey) strain of HIV, which causes AIDS.

Another important factor in the spread of these diseases is the legal and illegal international trade in wild species. According to the U.S. Fish and Wildlife Service, over 200 million wild animals—from kangaroos to iguanas to tropical fish—are legally imported into the United States each year with little quarantining or screening for disease. Most are imported for commercial sale in the pet trade. In addition, countless live animals, as well as bush meat and animal parts, are smuggled illegally across international borders every year.

Industrialized meat production is another major factor in the spread of food-borne infectious diseases to humans. For example, a deadly form of *E. coli* bacteria sometimes spreads from livestock to humans when people eat meat contaminated by animal manure. Salmonella bacteria found on animal hides and in poorly processed, contaminated meat also can cause food-borne disease.

A number of scientists are also looking at the connections between projected climate change due to atmospheric warming and the spread of infectious diseases such as malaria, meningitis, dengue fever, West Nile virus, and cholera. **GREEN CAREER:** ecological medicine

Critical Thinking

Do you think that air travelers should be more carefully monitored in the face of threats such as the Ebola and bird flu viruses? Explain.

red blood cells, causing intense fever, chills, drenching sweats, severe abdominal pain, vomiting, headaches, and increased susceptibility to other diseases. In 2010, according to the WHO, it killed 655,000 people, or about 75 people every hour. However, some experts believe this total could be as high as 1.2 million, because public health records are incomplete in many areas. More than 80% of all malaria victims live in the area of Africa south of the Sahara Desert, and most of them are children younger than age 5 (**Concept 17-2**). Many of the children who survive suffer brain damage or impaired learning ability.

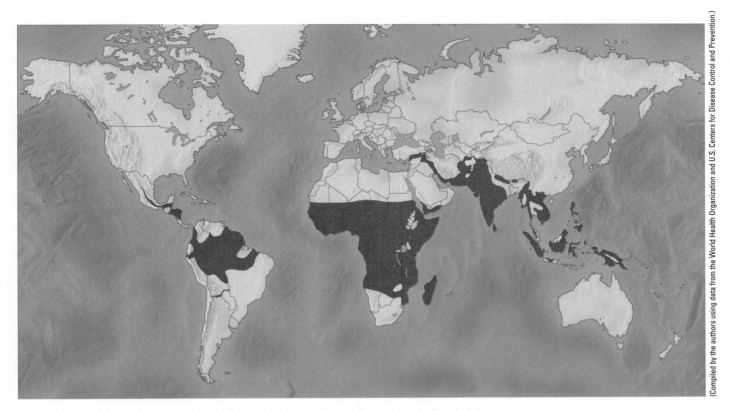

Figure 17-7 *Global outlook:* About 47% of the world's population live in areas in which malaria is prevalent.

Four species of protozoan parasites in the genus *Plasmodium* cause malaria. The cycle of infection begins when an uninfected female of any of about 60 *Anopheles* mosquito species bites a person (usually at night) who is infected with the *Plasmodium* parasite, ingests blood that contains the parasite, and later bites an uninfected person. The parasite then moves out of the mosquito and into the human's bloodstream and liver, where they multiply. Malaria can also be transmitted by contaminated blood transfusions and by drug users sharing needles.

Over the course of human history, malarial protozoa probably have killed more people than all the wars ever fought. During the 1950s and 1960s, the spread of malaria was sharply slowed when swamplands and marshes where mosquitoes breed were drained or sprayed with insecticides, and drugs were used to kill the parasites in victims' bloodstreams. Since 1970, however, malaria has come roaring back. Most species of the *Anopheles* mosquito have become genetically resistant to most insecticides. Worse, the *Plasmodium* parasites have become genetically resistant to common antimalarial drugs as well (Science Focus 17.1).

During this century, if the climate changes as projected by scientists, malaria is likely to spread across a wider area of the globe. As the average atmospheric temperature increases, populations of malaria-carrying mosquitoes will likely spread from current tropical areas to warmer temperate areas of the earth.

CONSIDER THIS. . .

CONNECTIONS Deforestation and Malaria

The clearing and development of tropical forests has led to the spread of malaria among workers and the settlers who follow them. In a 2010 study, Sarah Olson and her colleagues at the University of Wisconsin found that a 5% loss of tree cover in one part of Brazil's Amazon forest led to a 50% increase in malaria in that study area. Olson hypothesized that deforestation creates partially sunlit pools of water that make ideal breeding ponds for malaria-carrying mosquitoes.

Researchers are working to develop new antimalarial drugs and vaccines, as well as biological controls for *Anopheles* mosquitoes. In 2011, for the first time, a vaccine was partially effective in protecting children aged 5–17 months. Three doses of the vaccine for each person studied cut the malaria risk for this group by about half, according to a *New England Journal of Medicine* report. Researchers are hoping for more such positive results and to have the vaccine ready for widespread use in the near future.

Another approach is to provide poor people in malarial regions with free or inexpensive insecticide-treated bed nets (Figure 17-8) and window screens. Also, zinc and vitamin A supplements could be given to children to boost their resistance to malaria. And we can greatly reduce the incidence of malaria at a low cost by spraying

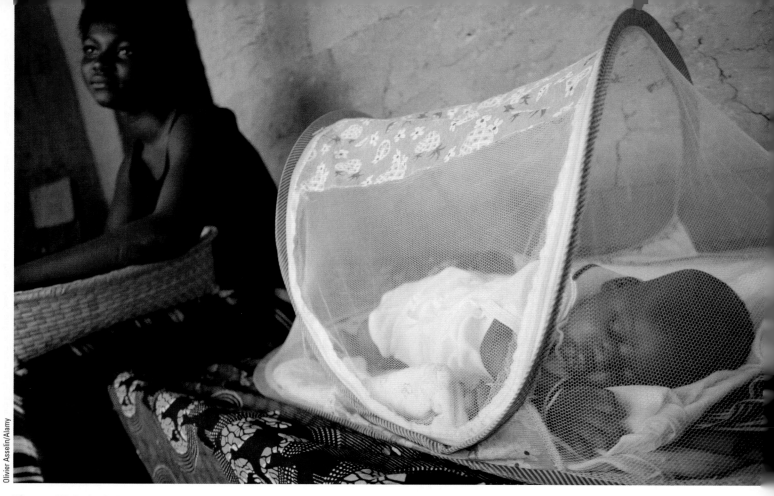

Figure 17-8 This baby in Senegal, Africa, is sleeping under an insecticide-treated mosquito net to reduce the risk of being bitten by malaria-carrying mosquitoes.

the insides of homes with low concentrations of the pesticide DDT twice a year. While DDT is being phased out in most countries, the WHO supports limited use of this sort for malaria control.

Some public health officials believe the world needs to make a more coordinated and aggressive effort to control malaria. This will become more urgent as malaria parasites develop greater genetic resistance to treatments, and as the range of *Anopheles* mosquitoes expands due to projected climate change.

We Can Reduce the Incidence of Infectious Diseases

According to the WHO, the percentage of all deaths worldwide resulting from infectious diseases dropped from 35% to 15% between 1970 and 2008. Also, between 1980 and 2010, the percentage of children in the world who were immunized to prevent various diseases grew as follows: tetanus, diphtheria, and pertussis, 20% to 85%; measles, 16% to 85%; and polio, 22% to 86%. Between 1990 and 2010, the estimated annual number of children younger than age 5 who died

GOOD NEWS

from infectious diseases dropped from nearly 12 million to 4.4 million.

Figure 17-9 lists measures promoted by health scientists and public health officials to help prevent or reduce the incidence of infectious diseases—especially in less-developed countries. An important breakthrough has been the development of simple *oral rehydration therapy* to help prevent death from dehydration for victims of severe diarrhea, which causes about one-fourth of all deaths of children younger than age 5 (Concept 17-2). This therapy involves administering a simple solution of boiled water, salt, and sugar or rice at a cost of only a few cents per person. It was the major factor in reducing the annual number of deaths from diarrhea from 4.6 million in 1980 to 1.6 million in 2008.

The WHO has estimated that implementing the solutions listed in Figure 17-9 could save the lives of as many as 4 million children younger than age 5 each year. Improving the health of people in these countries would help to stabilize their societies and would give them a boost in making the demographic transition (see Figure 6-16, p. 135) toward more sustainable economies. **GREEN CAREER:** infectious disease prevention

CONSIDER THIS. . .

CONNECTIONS Drinking Water, Latrines, and Infectious Diseases

More than a third of the world's people—2.6 billion—do not have sanitary bathroom facilities, and more than 1 billion get their water for drinking, washing, and cooking from sources polluted by animal and human feces. A key to reducing sickness and premature death from infectious disease is to focus on providing these people with simple latrines and access to safe drinking water. The estimated total cost of such improvements is about equal to the total amount that people in more-developed countries spend each year on bottled water.

Figure 17-9 There are a number of ways to prevent or reduce the incidence of infectious diseases, especially in less-developed countries. *Questions:* Which three of these approaches do you think are the most important? Why?

Top: ©Omer N Raja/Shutterstock.com. Bottom: Rob Byron/Shutterstock.com.

Solutions

Infectious Diseases

- Increase research on tropical diseases and vaccines
- Reduce poverty and malnutrition
- Improve drinking water quality
- Reduce unnecessary use of antibiotics
- Sharply reduce use of antibiotics on livestock
- Immunize children against major viral diseases
- Provide oral rehydration for diarrhea victims
- Conduct global campaign to reduce HIV/AIDS

© Cengage Learning

17-3 What Types of Chemical Hazards Do We Face?

CONCEPT 17-3
There is growing concern about chemicals in the environment that can cause cancers and birth defects, and disrupt the human immune, nervous, and endocrine systems.

Some Chemicals Can Cause Cancers, Mutations, and Birth Defects

A **toxic chemical** is an element or compound that can cause temporary or permanent harm or death to humans and animals. The U.S. Environmental Protection Agency (EPA) has listed arsenic, lead, mercury (**Core Case Study**), vinyl chloride (used to make PVC plastics), and polychlorinated biphenyls (PCBs) as the top five toxic substances in terms of human and environmental health.

There are three major types of potentially toxic agents. **Carcinogens** are chemicals, some types of radiation, and certain viruses that can cause or promote *cancer*—a disease in which malignant cells multiply uncontrollably and create tumors that can damage the body and often lead to premature death. Examples of carcinogens are arsenic, benzene, formaldehyde, gamma radiation, PCBs, radon, ultraviolet (UV) radiation, certain chemicals in tobacco smoke, and vinyl chloride.

Typically, 10–40 years may elapse between the initial exposure to a carcinogen and the appearance of detectable cancer symptoms. Partly because of this time lag, many healthy teenagers and young adults have trouble believing that their smoking, drinking, eating, and other habits today could lead to some form of cancer before they reach age 50.

The second major type of toxic agent, **mutagens**, includes chemicals or forms of radiation that cause or increase the frequency of *mutations*, or changes, in the DNA

molecules found in cells. Most mutations cause no harm, but some can lead to cancers and other disorders. For example, nitrous acid (HNO_2), formed by the digestion of nitrite (NO_2^-) preservatives in foods, can cause mutations linked to increases in stomach cancer in people who consume large amounts of processed foods and wine containing such preservatives. Harmful mutations occurring in reproductive cells can be passed on to offspring and to future generations.

Teratogens, a third type of toxic agent, are chemicals that harm or cause birth defects in a fetus or embryo. Ethyl alcohol is a teratogen. Drinking during pregnancy can lead to offspring with low birth weight and a number of physical, developmental, behavioral, and mental problems. Other teratogens are angel dust, benzene, formaldehyde, lead, mercury (**Core Case Study**), PCBs (see the following Case Study), phthalates, thalidomide, and vinyl chloride. Between 2001 and 2006, birth defects in Chinese infants soared by nearly 40%. Officials link this to the country's growing pollution, especially from coal-burning power plants and industries.

CASE STUDY

PCBs Are Everywhere—A Legacy from the Past

PCBs are a class of more than 200 chlorine-containing organic compounds that are very stable and nonflammable. They exist as oily liquids or solids but, under certain conditions, they can enter the air as a vapor. Between 1929 and 1977, PCBs were widely used as lubricants, hydraulic fluids, and insulators in electrical transformers and capacitors. They also became ingredients in a variety of products including paints, fire retardants in fabrics, preservatives, adhesives, and pesticides.

Figure 17-10 PCBs and other persistent toxic chemicals can move through the living and nonliving environment by a number of pathways.

exposure to some chemicals in the environment can disrupt the body's immune, nervous, and endocrine systems (**Concept 17-3**).

The *immune system* consists of specialized cells and tissues that protect the body against disease and harmful substances by forming *antibodies*, or specialized proteins that render invading agents harmless. Some chemicals such as arsenic, methylmercury (**Core Case Study**), and dioxins can weaken the human immune system and leave the body vulnerable to attacks by allergens and infectious bacteria, viruses, and protozoa.

Some natural and synthetic chemicals in the environment, called *neurotoxins*, can harm the human *nervous system* (brain, spinal cord, and peripheral nerves). Effects can include behavioral changes, learning disabilities and delays, attention deficit disorder, paralysis, and death. Examples of neurotoxins are PCBs, arsenic, lead, and certain pesticides.

Methylmercury (**Core Case Study**) is an especially dangerous neurotoxin because it is so persistent in the environment and it can be biologically magnified in food chains and food webs (see Figure 9-13, p. 202). According to the Natural Resources Defense Council, predatory fish such as large tuna, marlin, orange roughy, swordfish, mackerel, Chilean sea bass, grouper, and sharks can have mercury concentrations in their bodies that are 10,000 times higher than the levels in the water around them. A 2009 EPA study found that almost half of the fish tested in 500 lakes and reservoirs across the United States had levels of mercury that exceeded safe levels (Figure 17-1). Similarly, a 2009 study by the U.S. Geological Survey of nearly 300 streams across the United States found mercury-contaminated fish in all of the streams surveyed, with one-fourth of the fish exceeding the safe levels determined by the EPA.

The symptoms of mercury poisoning in adults include loss of fine motor skills, poor balance and coordination, muscle weakness, tremors, memory problems, insomnia, hearing loss, loss of hair, and loss of peripheral vision. A 2006 study led by Leonardo Trasande found that 316,000–647,000 babies are born each year in the United States with mercury levels high enough to cause measurable brain damage (Figure 17-11). Two other large-scale studies found that children born with mercury in their bodies had long-lasting problems with concentration, language, memory, and coordination. A 2012 study of 600 mothers and their children indicated that children, especially males,

The U.S. Congress banned the domestic production of PCBs in 1977 after research showed that they could cause liver cancer and other cancers in test animals. Studies also showed that pregnant women exposed to PCBs gave birth to underweight babies who eventually suffered permanent neurological damage, sharply lower-than-average IQs, and long-term growth problems.

Production of PCBs has also been banned in most other countries, but the potential health threats from these chemicals will be with us for a long time. For decades, PCBs entered the air, water, and soil as they were manufactured, used, and disposed of, as well as through accidental spills and leaks. Because PCBs break down very slowly in the environment, they can travel long distances in the air before landing far away from where they were released. Also, because they are fat-soluble, PCBs can also be biologically magnified in food chains and food webs (see Figure 9-13, p. 202).

As a result, PCBs are now found almost everywhere—in the air, soil, lakes, rivers, fish, birds, most human bodies, and even the bodies of polar bears in the Arctic. PCBs are also present in the milk of some nursing mothers, although most scientists say the health benefits of mother's milk outweigh the risk of infants' exposure to *trace amounts,* or very small but traceable levels, of PCBs in breast milk.

According to the EPA, about 70% of all the PCBs made in the United States are still in the environment. Figure 17-10 shows pathways by which persistent toxic chemicals such as PCBs can move through the living and nonliving environment.

Some Chemicals May Affect Our Immune and Nervous Systems

Since the 1970s, research on wildlife and laboratory animals, along with some studies of humans, have yielded a growing body of evidence that suggests that long-term

Figure 17-11 Mercury poisoning has eroded pockets of tissue in this human brain.

JAMES L. STANFIELD/National Geographic Creative

who are exposed to high levels of mercury in the womb had a 40–70% higher chance of suffering from attention deficit hyperactivity disorder (ADHD). Another 2012 study of 300 Intuit children in Québec, Canada, produced similar results. The EPA estimates that about 1 of every 12 women of childbearing age in the United States has enough mercury in her blood to harm a developing fetus.

CONSIDER THIS. . .

CONNECTIONS Climate Change and Mercury Pollution

Mercury particles (**Core Case Study**) from active volcanoes and from coal-burning power plant emissions are transported through the atmosphere to arctic regions where they can get trapped in arctic ice. Scientists are concerned that as the atmosphere warms and as this arctic ice melts, more of these mercury particles will flow into the ocean food webs, and evidence supports this hypothesis. In some arctic ringed seals and beluga whales, mercury levels have risen fourfold since the early 1980s.

As with all forms of pollution, prevention is the best policy. Figure 17-12 lists ways to prevent or reduce human inputs of mercury (**Core Case Study**) into the environment.

Some Chemicals Affect the Human Endocrine System

The *endocrine system* is a complex network of glands that release tiny amounts of *hormones* into the bloodstreams of humans and other vertebrate animals. Very low levels of these chemical messengers (often measured in parts per billion or parts per trillion) regulate the bodily systems that control sexual reproduction, growth, development, learning ability, and behavior. Each type of hormone has a unique molecular shape that allows it to attach to certain parts of cells called *receptors*, and to transmit its chemical message (Figure 17-13, left). In this "lock-and-key" relationship, the receptor is the lock and the hormone is the key.

Molecules of certain pesticides and other synthetic chemicals such as bisphenol A (BPA; Science Focus 17.3) have shapes similar to those of natural hormones. This allows them to attach to molecules of natural hormones and to disrupt the endocrine systems in humans and in some other animals (Concept 17-3). These molecules

are called *hormonally active agents* (HAAs) or *endocrine-disrupting chemicals* (EDCs).

Examples of HAAs include aluminum, Atrazine™ and several other widely used herbicides, DDT, PCBs, mercury (**Core Case Study**), phthalates, and BPA. Some hormone imposters, or *hormone mimics*, such as BPA are chemically similar to estrogens (female sex hormones) and can disrupt the endocrine system by attaching to estrogen receptor

Solutions

Mercury Pollution

Prevention	Control
Phase out waste incineration	Sharply reduce mercury emissions from coal-burning plants and incinerators
Remove mercury from coal before it is burned	Label all products containing mercury
Switch from coal to natural gas and renewable energy resources	Collect and recycle batteries and other products containing mercury

© Cengage Learning

Figure 17-12 There are a number of ways to prevent or control inputs of mercury (**Core Case Study**) into the environment from human sources—mostly coal-burning power plants and incinerators. **Questions:** Which four of these solutions do you think are the most important? Why?

Top: Mark Smith/Shutterstock.com. Bottom: ©tuulijumala/Shutterstock.com.

THE CONTROVERSY OVER BPA

The estrogen mimic *bisphenol A (BPA)* serves as a hardening agent in certain plastics (especially shatter-proof polycarbonate) that are used in a variety of products, including some baby bottles, sipping cups, and pacifiers, as well as certain reusable water bottles, sports drink and juice bottles, microwave dishes, and food storage containers. BPA is also used to make some dental sealants, as well as the plastic resins that line nearly all food and soft drink cans (Figure 17-C) and cans holding baby formulas and foods. This type of liner allows containers to withstand extreme temperatures, keeps canned food from interacting with the metal in the cans, prevents rust in the cans, and helps to preserve the canned food.

A 2007 study by the Centers for Disease Control and Prevention (CDC) indicated that 93% of Americans older than age 6 had trace levels of BPA in their urine. While these levels were well below the acceptable level set by the EPA, the EPA level was established in the late 1980s,

long before we knew much about the potential effects of BPA on human health. The CDC study also found that children and adolescents had generally higher urinary BPA levels than adults had.

Research indicates that the BPA in plastics can leach into water or food when the plastic is heated to high temperatures, microwaved, or exposed to acidic liquids. A 2009 Harvard University Medical School study found that there was a 66% increase in BPA levels in the urine of participants who drank from polycarbonate bottles regularly for just 1 week. According to lead researcher Karin Michels, the study confirmed that certain containers can release BPA into the liquids stored inside them, even when they are not heated.

Three studies released in 2010 indicated that BPA was commonly found on credit card and other types of receipts printed by thermal imaging printers. The skin on one's fingers, especially when wet, can absorb this BPA. Gloves do not prevent such absorption because the BPA

Figure 17-C A variety of products, including resin liners in cans used to store many foods, contain BPA.

molecules can move through most glove fabrics.

By 2012, more than 90 published studies by independent laboratories had found a number of significant adverse effects on test animals from exposure to very low levels of BPA. These effects include brain damage, early puberty, decreased sperm quality, certain cancers, heart disease, obesity, liver damage, impaired immune function, type 2 diabetes, hyperactivity, impaired learning, impotence in males, and obesity in unborn test animals exposed to BPA.

On the other hand, 12 studies funded by the chemical industry found no evi-

Normal Hormone Process

Hormone Mimic

Hormone Blocker

Figure 17-13 Each type of hormone has a unique molecular shape that allows it to attach to specially shaped receptors on the surface of, or the inside of, a cell and to transmit its chemical message (left). Molecules of certain pesticides and other synthetic chemicals, called *hormonally active agents* (HAAs, center and right), have shapes similar to those of natural hormones, allowing them to attach to the hormone molecules and disrupt the endocrine systems of humans and other animals.

dence or only weak evidence of adverse effects from low-level exposure to BPA in test animals. In 2011, the German Society of Toxicology concluded that BPA in polycarbonate plastic containers and in the linings of food and beverage cans "represents no noteworthy risk to the health of the human population," arguing that not enough studies showing adverse effects had been reproduced to confirm the results.

However, a 2011 study led by researchers at the Harvard School of Public Health related exposure to BPA in the womb to behavior and emotional problems in young girls, including hyperactivity, aggressive behavior, anxiety, and depression. In a 2012 study, researchers from Washington State University and the University of California, Davis, exposed rhesus monkeys to levels of BPA similar to those commonly encountered by humans. The test monkeys suffered serious reproductive abnormalities with increased risks of miscarriage and birth defects. These studies also reinforced concerns about BPA being a possible contributor to breast cancer in women.

In the face of these conflicting and possibly confusing research findings, consumers now have more choices, since most makers of baby bottles, sipping cups, and sports water bottles offer BPA-free alternatives. Also, many consumers are avoiding plastic containers with a #7 recycling code (which indicates that BPA can be present). People are also using powdered infant formula instead of liquid formula from metal cans, choosing glass bottles and food containers instead of those made of plastic or lined with plastic resins, and using glass, ceramic, or stainless steel coffee mugs rather than plastic cups.

However, many manufacturers have replaced BPA with bisphenol S (BPS). Studies indicate that BPS has effects similar to those of BPA, and BPS is now showing up in human urine at levels similar to those of BPA.

It would be difficult to replace the plastic resins containing BPA or BPS that line food containers (Figure 17-C) and that help to prevent food contamination and spoilage. More than 90% of the 130 billion food cans produced annually in the United States contain such resins, according to the

North American Metal Packaging Alliance. There are substitutes for these liners, but they are more expensive, and the potential health effects of some chemicals they contain also need to be evaluated. Meanwhile, companies including Campbell Soup and H.J. Heinz are phasing out the use of BPA in can liners.

In 2010, both Canada and the European Union voted to ban the sale of plastic baby bottles that contain BPA. Six U.S. states have banned the use of BPA in children's products, as has the city of Chicago, Illinois. In 2008, the U.S. Food and Drug Administration (FDA) had concluded that BPA in food and drink containers was not a health hazard. However, in 2012, the FDA, while not reversing its general position, banned the use of BPA in baby bottles and sipping cups.

Critical Thinking

Should plastics that contain BPA or BPS be banned from use in all children's products? Explain. Should such plastics be banned from use in the liners of canned food containers? Explain. What are the alternatives?

---

molecules. Others, called *hormone blockers,* disrupt the endocrine system by preventing natural hormones such as androgens (male sex hormones) from attaching to their receptors.

Estrogen mimics and hormone blockers can have a number of effects on sexual development and reproduction. Numerous studies involving wild animals, laboratory animals, and humans suggest that the males of species that are exposed to hormonal disruption are generally becoming more feminized. A likely culprit is long-term exposure to low levels of synthetic chemicals that we have added to the environment. There is also growing concern about another group of HAAs—pollutants that can act as *thyroid disrupters* and cause growth, weight, brain, and behavioral disorders.

Scientists are also increasingly concerned about the possible harmful effects of exposure to low levels of certain HAAs called *phthalates* (pronounced THALL-eights). These chemicals are used to soften polyvinyl chloride (PVC) plastic found in a variety of products, and they are used as solvents in many consumer products. Phthalates

are found in many detergents, perfumes, cosmetics, baby powders, body lotions for adults and babies, sunscreens, hair sprays, deodorants, soaps, nail polishes, and shampoos for adults and babies. They are also found in PVC products such as soft vinyl toys (Figure 17-14), teething rings, blood storage bags, intravenous drip bags, medical tubing, and shower curtains.

Exposure of laboratory animals to high doses of various phthalates has caused birth defects and liver cancer, kidney and liver damage, premature breast development, immune system suppression, and abnormal sexual development in these animals. Studies at the Mount Sinai School of Medicine have linked exposure of human babies to phthalates with early puberty in girls and eventual fertility problems in men. A 2012 study led by Susan L. Teitelbaum showed an association between childhood obesity and exposure to phthalates found in some personal care products. The European Union and at least 14 other countries have banned several phthalates. But scientists, government regulators, and manufacturers in the

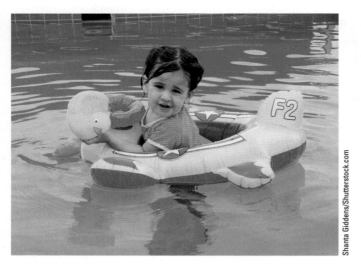

Figure 17-14 The surfaces of many soft plastic products can contain phthalates.

What Can You Do?

Exposure to Hormone Disrupters

- Eat certified organic produce and meats
- Avoid processed, prepackaged, and canned foods
- Use glass and ceramic cookware
- Store food and drinks in glass containers
- Use only natural cleaning and personal care products
- Use natural fabric shower curtains, not vinyl
- Avoid artificial air fresheners, fabric softeners, and dryer sheets
- Use only glass baby bottles and BPA-free, phthalate-free sipping cups, pacifiers, and toys

Figure 17-15 Individuals matter: You can reduce your exposure, and that of any children you care for, to hormone disrupters. *Questions:* Which three of these steps do you think are the most important ones to take? Why?

United States are divided on its risks to human health and reproductive systems.

Some scientists hypothesize that rising levels of hormone disrupters in our bodies may be related to more frequent health problems, including sharp drops in male sperm counts and sperm motility; rising rates of testicular cancer and genital birth defects in men; and rising rates of breast cancer in women. A 2013 report by the U.N. Environment Programme and the WHO linked HAAs to breast, ovarian, testicular, prostate, and thyroid cancers. Some scientists argue that there are not enough scientific study results or statistical evidence to strongly link these medical problems with HAA levels in humans. They call

for more research before HAAs are banned or severely restricted, which would cause huge economic losses for the companies that make them.

Concerns about BPA and phthalates highlight the difficulty in assessing possible harmful health effects from exposure to very low levels of various chemicals. Resolving these uncertainties will take decades of research. Some scientists believe that as a precaution during this period of research, we should sharply reduce our use of potentially harmful hormone disrupters, especially in products frequently used by pregnant women, infants, young children, and teenagers (Figure 17-15).

17-4 How Can We Evaluate Chemical Hazards?

CONCEPT 17-4A

Scientists use live laboratory animals, case reports of poisonings, and epidemiological studies to estimate the toxicity of chemicals, but these methods have limitations.

CONCEPT 17-4B

Many health scientists call for much greater emphasis on pollution prevention to reduce our exposure to potentially harmful chemicals.

Many Factors Determine the Harmful Health Effects of Chemicals

We are exposed to small amounts of potentially harmful chemicals every day in the air we breathe, the water we drink, and the food we eat. **Toxicology** is the study of the harmful effects of chemicals on humans and other organisms. In effect, it is a study of poisonous substances.

Toxicity is a measure of the harmfulness of a substance—its ability to cause injury, illness, or death to a living organism. A basic principle of toxicology is that *any synthetic or natural chemical can be harmful if ingested in a large enough quantity*. But the critical question is this: *At what level of exposure to a particular toxic chemical will the chemical cause harm?* This is a difficult question to answer because of the many variables involved in estimating the effects of human exposure to chemicals (Figure 17-16). A key factor is the **dose**, the amount of a harmful chemical that a person has ingested, inhaled, or absorbed through the skin at any one time. This is the meaning of the chapter-opening quote by the German scientist Paracelsus: *The dose makes the poison.*

The effects of a particular chemical can also depend on the age of the person exposed to it. For example, toxic chemicals usually have a greater effect on fetuses, infants, and children than on adults (see the following Case Study). Toxicity also depends on *genetic makeup,*

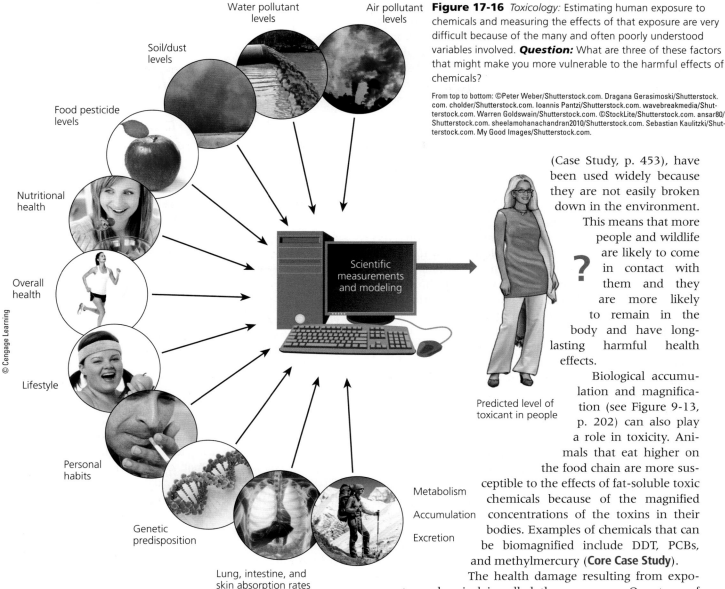

Water pollutant levels

Air pollutant levels

Soil/dust levels

Food pesticide levels

Nutritional health

Overall health

Lifestyle

Personal habits

Genetic predisposition

Lung, intestine, and skin absorption rates

Scientific measurements and modeling

Predicted level of toxicant in people

Metabolism

Accumulation

Excretion

© Cengage Learning

Figure 17-16 *Toxicology:* Estimating human exposure to chemicals and measuring the effects of that exposure are very difficult because of the many and often poorly understood variables involved. **Question:** What are three of these factors that might make you more vulnerable to the harmful effects of chemicals?

From top to bottom: ©Peter Weber/Shutterstock.com. Dragana Gerasimoski/Shutterstock.com. cholder/Shutterstock.com. Ioannis Pantzi/Shutterstock.com. wavebreakmedia/Shutterstock.com. Warren Goldswain/Shutterstock.com. ©StockLite/Shutterstock.com. ansar80/Shutterstock.com. sheelamohanachandran2010/Shutterstock.com. Sebastian Kaulitzki/Shutterstock.com. My Good Images/Shutterstock.com.

which determines an individual's sensitivity to a particular toxin. Some people are sensitive to a number of toxins— a condition known as *multiple chemical sensitivity* (MCS). Another factor is how well the body's detoxification systems, including the liver, lungs, and kidneys, are working.

Several other variables can affect the level of harm caused by a chemical. One is its *solubility.* Water-soluble toxins (which are often inorganic compounds) can move throughout the environment and get into water supplies, as well as the aqueous solutions that surround the cells in our bodies. Oil- or fat-soluble toxins (which are usually organic compounds) can penetrate the membranes surrounding cells, because these membranes allow similar oil-soluble chemicals to pass through them. Thus, oil- or fat-soluble toxins can accumulate in body tissues and cells.

Another factor is a substance's *persistence,* or resistance to breakdown. Many chemicals, including DDT and PCBs

(Case Study, p. 453), have been used widely because they are not easily broken down in the environment. This means that more people and wildlife are likely to come in contact with them and they are more likely to remain in the body and have long-lasting harmful health effects.

Biological accumulation and magnification (see Figure 9-13, p. 202) can also play a role in toxicity. Animals that eat higher on the food chain are more susceptible to the effects of fat-soluble toxic chemicals because of the magnified concentrations of the toxins in their bodies. Examples of chemicals that can be biomagnified include DDT, PCBs, and methylmercury (**Core Case Study**).

The health damage resulting from exposure to a chemical is called the **response**. One type of response, an *acute effect,* is an immediate or rapid harmful reaction ranging from dizziness to death. By contrast, a *chronic effect* is a permanent or long-lasting consequence (kidney or liver damage, for example) of exposure to a single dose or to repeated lower doses of a harmful substance.

Some people have the mistaken idea that all natural chemicals are safe and all synthetic chemicals are harmful. In fact, many synthetic chemicals, including many of the medicines we take, are quite safe if used as intended, while many natural chemicals such as lead and mercury (**Core Case Study**) are deadly.

CASE STUDY

Protecting Children from Toxic Chemicals

In 2005, the Environmental Working Group analyzed umbilical cord blood from ten randomly selected newborns in U.S. hospitals. Of the 287 chemicals detected in that study, 180 have been shown to cause cancers in humans or animals, 217 have damaged the nervous systems of test animals, and 208 have caused birth defects or abnormal development in test animals. Scientists do not know what harm, if any, might be caused by the very low concentrations of these chemicals found in infants' blood.

However, more recent science has caused some experts to suggest that exposure to chemical pollutants in the womb may be related to increasing rates of autism, childhood asthma, and learning disorders. In 2009, researchers for the first time found a connection between the exposure of pregnant women to air pollutants and lower IQ scores in their children as they grew. A team of researchers led by Frederica Perera of Columbia University reported that children exposed to high levels of air pollution before birth scored 4–5 points lower, on average, in IQ tests than did children with less exposure.

Infants and young children are more susceptible to the effects of toxic substances than are adults, for three major reasons. *First,* they generally breathe more air, drink more water, and eat more food per unit of body weight than do adults. *Second,* they are exposed to toxins in dust and soil when they put their fingers, toys, and other objects in their mouths. *Third,* children usually have less well-developed immune systems and body detoxification processes than adults have. Fetuses are also highly vulnerable to trace amounts of toxic chemicals such as methylmercury (**Core Case Study**) that they can receive from their mothers.

In 2003, the EPA proposed that in determining any risk, regulators should assume that children have a 10-times higher risk factor than adults have. Some health scientists suggest that to be on the safe side, we should assume that this risk for children is 100 times the risk for adults.

 CONSIDER THIS...

THINKING ABOUT Toxic Chemical Levels for Children

Do you think environmental regulations should require that the allowed levels of exposure to toxic chemicals for children be 100 times lower than those for adults? Explain your reasoning.

Scientists Use Live Laboratory Animals and Non-Animal Tests to Estimate Toxicity

The most widely used method for determining toxicity is to expose a population of live laboratory animals to measured doses of a specific substance under controlled conditions. Laboratory-bred mice and rats are widely used because, as mammals, their systems function, to some degree, similarly to human systems. Also, they are small and can reproduce rapidly under controlled laboratory conditions.

Animal tests typically take 2–5 years, involve hundreds to thousands of test animals, and can cost as much as $2 million per substance tested. Such tests can be painful to the test animals and can kill or harm them. Animal welfare groups want to limit or ban the use of test animals and, at the very least, to ensure that they are treated in the most humane manner possible.

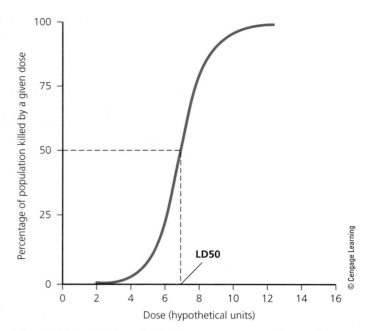

Figure 17-17 This hypothetical *dose-response curve* illustrates how scientists can estimate the LD50, the dosage of a specific chemical that kills 50% of the animals in a test group. Toxicologists use this method to compare the toxicities of different chemicals.

Scientists estimate the toxicity of a chemical by determining the effects of various doses of the chemical on test organisms and plotting the results in a **dose-response curve** (Figure 17-17). One approach is to determine the *lethal dose*—the dose that will kill an animal. A chemical's *median lethal dose (LD50)* is the dose that can kill 50% of the animals (usually rats and mice) in a test population, usually expressed in milligrams of chemical per kilogram of body weight (mg/kg).

Chemicals vary widely in their toxicity (Table 17-1). Some poisons can cause serious harm or death after a single exposure at very low dosages. Others cause such harm only at dosages so huge that it is nearly impossible to get enough into the body to cause injury or death. Most chemicals fall between these two extremes.

There are two general types of dose-response curves. With the *nonthreshold dose-response model* (Figure 17-18, left), any dosage of a toxic chemical causes harm that increases with the dosage. With the *threshold dose-response model* (Figure 17-18, right), a certain level of exposure to the chemical must be reached before any detectable harmful effects occur, presumably because the body can repair the damage caused by low dosages of some substances.

Establishing which of these two models applies at low dosages is extremely difficult and controversial. To be on the safe side, scientists often choose the nonthreshold dose-response model. Fairly high dosages are used to reduce the number of test animals needed, obtain results quickly, and lower costs. Otherwise, chemical manufacturers would need to run tests on millions of laboratory animals for many years, which means they could not afford to test most chemicals.

Table 17-1 Toxicity Ratings and Average Lethal Doses for Humans

Toxicity Rating	LD50 (milligrams per kilogram of body weight)*	Average Lethal Dose†	Examples
Supertoxic	Less than 5	Less than 7 drops	Nerve gases, botulism toxin, mushroom toxin, dioxin (TCDD)
Extremely toxic	5–50	7 drops to 1 teaspoon	Potassium cyanide, heroin, atropine, parathion, nicotine
Very toxic	50–500	1 teaspoon to 1 ounce	Mercury salts, morphine, codeine
Moderately toxic	500–5,000	1 ounce to 1 pint	Lead salts, DDT, sodium hydroxide, sodium fluoride, sulfuric acid, caffeine, carbon tetrachloride
Slightly toxic	5,000–15,000	1 pint to 1 quart	Ethyl alcohol, some cleansers and soaps
Essentially nontoxic	15,000 or greater	More than 1 quart	Water, glycerin, table sugar

*Dosage that kills 50% of individuals exposed.
†Amounts of substances in liquid form at room temperature that are lethal when given to a 70-kilogram (150-pound) human.
© Cengage Learning

For the same reasons, scientists usually use mathematical models to *extrapolate,* or estimate, the effects of low-dose exposures based on the measured results of high-dose exposures. Then they extrapolate these results from test organisms to humans as a way of estimating LD50 values for acute toxicity (Table 17-1). Some scientists challenge the validity of extrapolating data from test animals to humans, because human physiology and metabolism often differ from those of the test animals. Other scientists say that such tests and models work fairly well (especially for revealing cancer risks) when the correct experimental animal is chosen or when a chemical is toxic to several different test-animal species.

More humane methods for toxicity testing are available and are being used more often to replace testing on live animals. They include making computer simulations and using tissue cultures of cells and bacteria, chicken egg membranes, and individual animal cells, instead of whole, live animals. High-speed robot testing devices can now screen the biological activity of more than 1 million compounds a day to help determine their possible toxic effects. GOOD NEWS

Since 2008, U.S. government labs have been making much greater use of such non-animal testing methods because they are faster and less expensive than animal tests, and because of growing public concern over animal testing. For example, government toxicity testing labs can run between 10 and 100 tests a year using live rodents such as rats and mice but can run more than 10,000 tests a day using specialized cells and computerized testing measurements.

The problems with estimating toxicities by using laboratory experiments get even more complicated (**Concept 17-4A**). In real life, each of us is exposed to a variety of chemicals, some of which can interact in ways that decrease or enhance their short- and long-term individual effects. Toxicologists already have great difficulty in estimating the toxicity of a single substance. Adding the problem of evaluating *mixtures of potentially toxic substances,* isolating the culprits, and determining how they can interact with one another is overwhelming from a scientific and economic standpoint. For example, just studying the interactions among 3 of the 500 most widely used industrial chemicals would take 20.7 million experiments—a physical and financial impossibility.

🔍 CONSIDER THIS. . .

THINKING ABOUT Animal Testing

Should laboratory-bred mice, rats, and other animals be used to determine toxicity and other effects of chemicals? Explain.

Figure 17-18 Scientists use two types of *dose-response curves* to help them estimate the toxicity of various chemicals. The linear and nonlinear curves in the graph on the left apply if even the smallest dosage of a chemical has a harmful effect that increases with the dosage. The graph on the right applies if a harmful effect occurs only when the dosage exceeds a certain *threshold level.*

© Cengage Learning

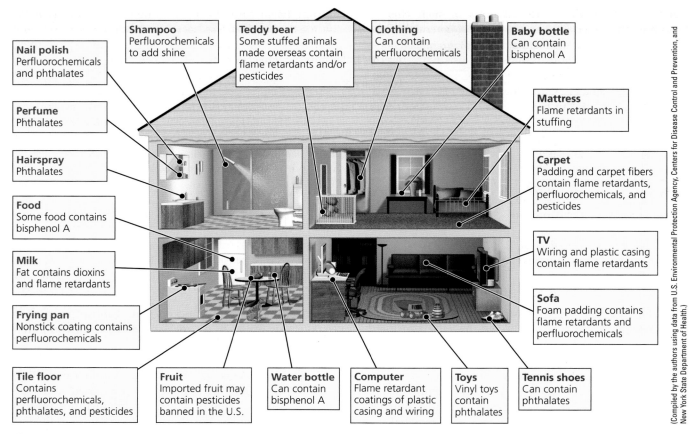

Figure 17-19 A number of potentially harmful chemicals are found in many homes. ***Questions:*** Does the fact that we do not know much about the long-term harmful effects of these chemicals make you more likely or less likely to minimize your exposure to them? Why?

Nail polish
Perfluorochemicals and phthalates

Perfume
Phthalates

Hairspray
Phthalates

Food
Some food contains bisphenol A

Milk
Fat contains dioxins and flame retardants

Frying pan
Nonstick coating contains perfluorochemicals

Shampoo
Perfluorochemicals to add shine

Teddy bear
Some stuffed animals made overseas contain flame retardants and/or pesticides

Clothing
Can contain perfluorochemicals

Baby bottle
Can contain bisphenol A

Mattress
Flame retardants in stuffing

Carpet
Padding and carpet fibers contain flame retardants, perfluorochemicals, and pesticides

TV
Wiring and plastic casing contain flame retardants

Sofa
Foam padding contains flame retardants and perfluorochemicals

Tile floor
Contains perfluorochemicals, phthalates, and pesticides

Fruit
Imported fruit may contain pesticides banned in the U.S.

Water bottle
Can contain bisphenol A

Computer
Flame retardant coatings of plastic casing and wiring

Toys
Vinyl toys contain phthalates

Tennis shoes
Can contain phthalates

(Compiled by the authors using data from U.S. Environmental Protection Agency, Centers for Disease Control and Prevention, and New York State Department of Health.)

There Are Other Ways to Estimate the Harmful Effects of Chemicals

Scientists use several other methods to get information about the harmful effects of chemicals on human health. For example, *case reports,* usually made by physicians, provide information about people who have suffered from adverse health effects or died after exposure to a chemical. Such information often involves accidental or deliberate poisonings, drug overdoses, homicides, or suicide attempts.

Most case reports are not reliable sources for estimating toxicity because the actual dosage and the exposed person's health status are often unknown. But such reports can provide clues about environmental hazards and suggest the need for laboratory investigations.

Another source of information is *epidemiological studies,* which compare the health of people exposed to a particular chemical (the *experimental group*) with the health of a similar group of people not exposed to the agent (the *control group*). The goal is to determine whether the statistical association between exposure to a toxic chemical and a health problem is strong, moderate, weak, or undetectable.

Four factors can limit the usefulness of epidemiological studies. *First,* in many cases, too few people have been exposed to high enough levels of a toxic agent to detect statistically significant differences. *Second,* the studies usually take a long time. *Third,* closely linking an observed effect with exposure to a particular chemical is difficult because people are exposed to many different toxic agents throughout their lives and can vary in their sensitivity to such chemicals. *Fourth,* we cannot use epidemiological studies to evaluate hazards from new technologies or chemicals to which people have not yet been exposed.

Are Trace Levels of Toxic Chemicals Harmful?

Almost everyone who lives in a more-developed country is now exposed to potentially harmful chemicals (Figure 17-19) that have built up to trace levels in their blood and in other parts of their bodies. A 2011 CDC study showed that the blood of an average American contains traces of 212 different chemicals, including potentially harmful chemicals such as arsenic and BPA.

Should we be concerned about trace amounts of various synthetic chemicals in our air, water, food, and bodies? In most cases, we simply do not know because there are too few data and because of the difficulty in determining the effects of exposures to low levels of these chemicals (**Concept 17-4A**).

Some scientists view trace amounts of such chemicals with alarm, especially because of their potential long-term effects on the human immune, nervous, and endocrine systems. In 2012, a team of 12 scientists who studied hormone-disrupting chemicals for 3 years reported that exposure to very small amounts of such chemicals at crucial stages of development can have impacts on individuals that can extend to their offspring. Such impacts can affect reproductive and sexual development, neurological systems, and immune system function, and can lead to diabetes and obesity. The scientists noted that these effects can occur at different stages of life from infancy through old age.

Others scientists view the risks from trace levels of chemicals as minor. They point out that average life expectancy has been increasing in most countries, especially more-developed countries, for decades. Some scientists contend that the concentrations of such chemicals are so low that they are harmless. All agree that there is a need for much more research on the effects of trace levels on human health.

Why Do We Know So Little about the Harmful Effects of Chemicals?

As we have seen, all methods for estimating toxicity levels and risks have serious limitations (Concept 17-4A), but they are all we have. To take this uncertainty into account and to minimize harm, scientists and regulators typically set allowed levels of exposure to toxic substances at 1/100th or even 1/1,000th of the estimated harmful levels.

According to risk assessment expert Joseph V. Rodricks, "Toxicologists know a great deal about a few chemicals, a little about many, and next to nothing about most." The U.S. National Academy of Sciences estimates that only 10% of the more than 80,000 registered synthetic chemicals in commercial use have been thoroughly screened for toxicity. Only 2% have been adequately tested to determine whether they are carcinogens, mutagens, or teratogens. Hardly any of the chemicals in commercial use have been screened for possible damage to the human nervous, endocrine, and immune systems. Because of insufficient data and the high costs of regulation, federal and state governments do not supervise the use of nearly 99.5% of the commercially available chemicals in the United States, and the problem is much worse in some less-developed countries.

How Far Should We Go in Using Pollution Prevention and the Precautionary Principle?

We know little about the potentially toxic chemicals around us and inside of us, and estimating their effects is very difficult, time-consuming, and expensive. So where does this leave us?

Some scientists and health officials, especially those in European Union countries, are pushing for much greater emphasis on *pollution prevention*. They say we should not release chemicals that we know or suspect can cause significant harm into the environment. This means looking for harmless or less harmful substitutes for toxic and hazardous chemicals. Another option is to recycle them within production processes to keep them from reaching the environment, as companies such as DuPont and 3M have been doing (see the following Case Study).

Pollution prevention is a strategy for implementing the *precautionary principle* (see Chapter 9, p. 211). According to this principle, which can be applied to many other problems, when there is substantial preliminary evidence that an activity, technology, or chemical substance can harm humans or the environment, we should take precautionary measures to prevent or reduce such harm, rather than waiting for more conclusive scientific evidence (Concept 17-4B).

There is controversy over how far we should go in using pollution prevention based on the precautionary principle. With this approach, those proposing to introduce a new chemical or technology would bear the burden of establishing its safety. This requires two major changes in the way we evaluate risks. *First,* we would assume that new chemicals and technologies could be harmful until scientific studies could show otherwise. *Second,* we would remove existing chemicals and technologies that appear to have a strong chance of causing significant harm from the market until we could establish their safety. For example, after decades of research revealed the harmful effects of lead, especially on children, lead-based paints and leaded gasoline were phased out in most developed countries.

Some movement is being made in the direction of the precautionary principle, especially in the European Union. In 2000, negotiators agreed to a global treaty that would ban or phase out the use of 12 of the most notorious *persistent organic pollutants (POPs)*, also called the *dirty dozen*. These highly toxic chemicals have been shown to produce numerous harmful effects, including cancers, birth defects, compromised immune systems, and a 50% decline in sperm counts and sperm quality in men in a number of countries. The list includes DDT and eight other pesticides, PCBs, and dioxins. In 2009, nine more POPs were added, some of which are widely used in pesticides and in flame-retardants added to clothing, furniture, and other consumer goods. New chemicals will be added to the list when the harm they could potentially cause is seen as outweighing their usefulness. The POPs treaty went into effect in 2004 but has not been ratified and implemented by the United States.

In 2007, the European Union enacted regulations known as REACH (for *Registration, Evaluation, Authorization, and restriction of CHemicals*). It required the registration of 30,000 untested, unregulated, and potentially harmful chemicals. In the REACH process, the most hazardous substances are not approved for use if safer alternatives exist. In addition, when there is no alternative, producers must present a research plan aimed at finding one.

REACH puts more of the burden on industry to show that chemicals are safe. Conventional regulation has put the burden on governments to show that they are dangerous. For example, the U.S. chemical regulation structure was enacted in 1976. At U.S. Congressional hearings in 2009, experts testified that the current system makes it virtually impossible for the government to limit or ban the use of toxic chemicals. The hearings found that under this system, the EPA had required testing for only 200 of the more than 80,000 chemicals registered for use in the United States—less than one of every 400—and had issued regulations to control fewer than 12 of those chemicals.

Manufacturers and businesses contend that widespread application of the much more precautionary REACH approach would make it too expensive and almost impossible to introduce any new chemical or technology. Proponents of increased reliance on using pollution prevention agree that we can go too far, but argue that we have an ethical responsibility to reduce known or potentially serious risks to human health, to the earth's life-support system, and to future generations, in keeping with one of the social science **principles of sustainability** (see Figure 1-5, p. 9 or back cover).

Proponents also point out that using the precautionary principle focuses the efforts and creativity of scientists, engineers, and businesses on finding solutions to pollution problems based on prevention rather than on cleanup. It also reduces health risks for employees and society, frees businesses from having to deal with pollution regulations, and reduces the threat of lawsuits from injured parties. In some cases, applying this principle helps companies to increase their profits from sales of safer products and innovative technologies, and it improves the public image of businesses operating in this manner.

For almost 40 years, most laws and technologies for dealing with pollution have focused on cleaning up or diluting pollution after it has been produced. Experience shows that such pollution control often is only a temporary solution that can be overwhelmed by a growing population consuming more goods and services and producing more pollution. Environmental and health scientists say we can do better than this by putting much greater emphasis on pollution prevention (**Concept 17-4B**).

In 2011, after a 35-year delay promoted by politically powerful coal companies and utilities that burn coal to produce electricity, the U.S. government took a step in this direction by issuing a rule to control emissions of mercury (**Core Case Study**) and other toxic chemicals from older coal-burning plants in 27 states. Unless it is blocked by the courts or Congress, the EPA estimates that each year, these new air pollution standards could prevent as many as 11,000 premature deaths, 4,700 heart attacks, and 130,000 childhood asthma attacks.

The coal industry and many utility companies have mounted a campaign to have Congress overturn or seriously weaken these new standards, contending that they will cost jobs and increase electricity costs. However, supporters of the new standards argue that they will create jobs and lead to lower electricity bills by encouraging a shift away from older, dirtier coal-burning power plants to power plants that run on cleaner and cheaper natural gas or on renewable energy sources.

CASE STUDY
Pollution Prevention Pays

The 3M Company, based in the U.S. state of Minnesota, makes 60,000 different products in 100 manufacturing plants around the world. In 1975, 3M began a Pollution Prevention Pays (3P) program. It redesigned its equipment and processes, used fewer hazardous raw materials, identified its toxic chemical outputs, and began to recycle and sell many of those outputs as raw materials to other companies. It also put a new focus on making more products that consumers could use without polluting the environment.

Between 1975 and 2008, the 3M 3P program prevented more than 1.4 million metric tons (1.5 million tons) of pollutants from reaching the environment—an amount weighing about the same as 100 empty jumbo jet airliners. The company immediately saved more than $1.2 billion in waste disposal and material costs, and continued to save money every year in lower operating costs for pollution control facilities and reduced raw material and fuel costs. This is an excellent example of how pollution prevention pays.

This program has been successful largely because any person at any level in the company can participate by creating a 3P project. Employees are rewarded if their projects eliminate or reduce a pollutant; reduce the amount of energy, materials, or other resources required in production; or save money through reduced pollution control costs, lower operating costs, or increased sales of new or existing products. Employees at 3M have now completed more than 6,300 3P projects.

Since 1990, a growing number of companies have adopted similar pollution and waste prevention programs that have led to cleaner production. (See the online Guest Essay by Peter Montague on cleaner production.) They are learning that, in addition to saving money by preventing pollution and waste production, they have a much easier job of complying with pollution laws and regulations. Also, they are not subject to lawsuits based on exposure to harmful chemicals, they provide a safer environment for their workers (which can reduce their employee health insurance costs), and they improve their public image.

CONSIDER THIS...

CAMPUS SUSTAINABILITY
Pollution Prevention at Pomona

At southern California's Pomona College, students and administrators have opted to practice pollution prevention in several ways. Campus buildings are now cleaned with only 100% natu-

©Courtesy of Pomona College

rally derived cleaning products and cleaning machines that use ionized water instead of chemicals and solvents, which helps to keep these potentially toxic chemicals out of the environment. The campus also uses integrated pest management (p. 300), maximizing biological methods of pest and weed control and minimizing the use of pesticides and the resulting pollution. And Pomona College works hard to collect all toxic e-waste (see photo on previous page) and uses a qualified sustainable recycling program in order to keep toxic e-waste pollutants out of landfills.

17-5 How Do We Perceive Risks and How Can We Avoid the Worst of Them?

CONCEPT 17-5

We can reduce the major risks we face by becoming informed, thinking critically about risks, and making careful choices.

The Greatest Health Risks Come from Poverty, Gender, and Lifestyle Choices

Risk analysis involves identifying hazards and evaluating their associated risks (*risk assessment;* Figure 17-3, left), ranking risks *(comparative risk analysis),* determining options and making decisions about reducing or eliminating risks (*risk management;* Figure 17-3, right), and informing decision makers and the public about risks *(risk communication).*

Statistical probabilities based on past experience, animal testing, and other assessments are used to estimate risks from older technologies and chemicals. To evaluate new technologies and products, risk evaluators use more uncertain statistical probabilities, based on models rather than on actual experience and testing.

The greatest risks many people face today are rarely dramatic enough to make the daily news. In terms of the number of deaths per year (Figure 17-20), *the greatest risk by far is poverty.* The high death toll ultimately resulting from poverty is caused by malnutrition, increased susceptibility to normally nonfatal infectious diseases, and often-fatal infectious diseases transmitted by unsafe drinking water (Figure 17-21).

In the United States and in most of the more-developed countries, the four greatest risks in terms of shortened life spans are living in poverty, being born male, smoking, and being obese. Some of the greatest risks of premature death are illnesses that result primarily from lifestyle choices that people make (Figure 17-22) (Concept 17-1). For example, overeating can lead to obesity and diabetes and smoking can cause cancer. Also, in the United States, more deaths are due to prescription drug overdoses than to illegal drug overdoses. While the leading cause of death worldwide was once infectious diseases, it is now nontransmissible cardiovascular disease, and the percentage of people dying from cancers is increasing in all countries. To some extent, these causes are related to lifestyle choices.

About two-thirds of all Americans are classified as either overweight or obese. In 2009, the American Cancer Society reported that an 18-year study of 900,000 cancer-free U.S. adults indicated that more than 90,000 cancer deaths could be prevented each year if Americans maintained healthy body weights.

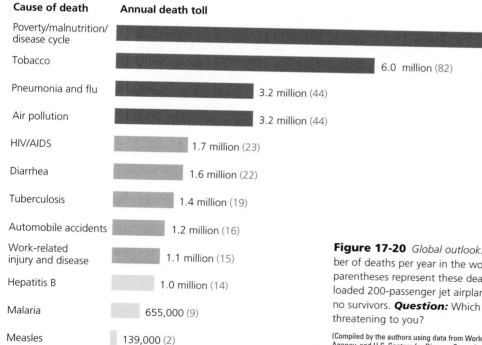

Cause of death   **Annual death toll**

Cause of death	Annual death toll
Poverty/malnutrition/disease cycle	11 million (150)
Tobacco	6.0 million (82)
Pneumonia and flu	3.2 million (44)
Air pollution	3.2 million (44)
HIV/AIDS	1.7 million (23)
Diarrhea	1.6 million (22)
Tuberculosis	1.4 million (19)
Automobile accidents	1.2 million (16)
Work-related injury and disease	1.1 million (15)
Hepatitis B	1.0 million (14)
Malaria	655,000 (9)
Measles	139,000 (2)

Figure 17-20 *Global outlook:* Scientists have estimated the number of deaths per year in the world from various causes. Numbers in parentheses represent these death tolls in terms of numbers of fully loaded 200-passenger jet airplanes crashing *every day of the year* with no survivors. **Question:** Which three of these causes are the most threatening to you?

(Compiled by the authors using data from World Health Organization, Environmental Protection Agency, and U.S. Centers for Disease Control and Prevention.)

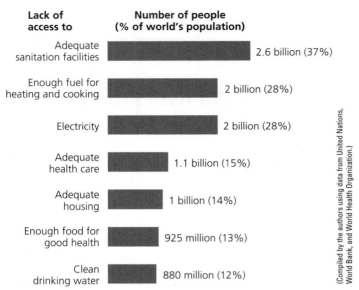

(Compiled by the authors using data from United Nations, World Bank, and World Health Organization.)

Figure 17-21 These are some of the harmful effects that result from living in poverty. **Questions:** Which two of these effects would be the most harmful for you? Why?

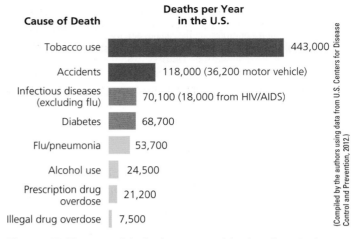

(Compiled by the authors using data from U.S. Centers for Disease Control and Prevention, 2012.)

Figure 17-22 Some of the leading causes of death in the United States are preventable to the extent that they result from lifestyle choices.

The best ways to reduce one's risk of premature death and serious health problems are to avoid smoking and exposure to smoke (see the following Case Study), lose excess weight, reduce consumption of foods containing cholesterol and saturated fats, eat a variety of fruits and vegetables, exercise regularly, drink little or no alcohol (no more than two drinks in a single day), avoid excess sunlight (which ages skin and can cause skin cancer), and practice safe sex (**Concept 17-5**). A 2005 study by Majid Ezzati with participation by 100 scientists around the world estimated that if everyone were to follow these guidelines, one-third of the 7.6 million annual deaths from cancer could be prevented. The number of lives this would save each year equals roughly the entire population of the U.S. city of Chicago, Illinois.

CASE STUDY
Death from Smoking

What is roughly the diameter of a 30-caliber bullet, can be bought almost anywhere, is highly addictive, and kills an average of about 14,800 people every day, or about one every 6 seconds? It is a cigarette. *Cigarette smoking is the world's most preventable major cause of suffering and premature death among adults.*

In 2007, the WHO estimated that tobacco use helped to kill 100 million people during the 20th century and could kill 1 billion people during this century unless individuals and governments act now to dramatically reduce smoking. In 2011, the WHO estimated that each year, tobacco contributes to the premature deaths of about 6 million people (about half from more-developed and half from less-developed countries) from 25 illnesses, including heart disease, stroke, lung and other cancers, and bronchitis.

In 2010, U.S. research scientist Rachel A. Whitmer and her colleagues reported on a study that had tracked the health of 21,123 individuals for 30 years. They found that people between the ages of 50 and 60 who smoked 1–2 packs of cigarettes daily had a 44% higher chance of getting Alzheimer's disease or vascular dementia (which reduces blood flow to the brain and can erode memory) by age 72. Another disease related to smoking is *emphysema*, which results in irreversible lung damage and chronic shortness of breath (Figure 17-23).

According to the WHO, lifelong smokers reduce their life spans by an average of 15 years. By 2030, the annual death toll from smoking-related diseases is projected to reach more than 8 million—an average of 21,900 preventable deaths per day. About 80% of these deaths are expected to occur in less-developed countries, especially China, where 30% of the world's smokers live. The annual death toll from smoking in China is about 1.2 million, an average of about 137 people every hour.

In 2012, the CDC estimated that smoking kills about 443,000 Americans per year—an average of 1,214 deaths per day, or nearly one every minute (Figure 17-23). This death toll accounts for one of every five deaths in the United States every year and is roughly equivalent to six fully loaded 200-passenger jet planes crashing *every day of the year* with no survivors. Smoking also causes about 8.6 million illnesses every year in the United States.

The overwhelming scientific consensus is that the nicotine inhaled in tobacco smoke is highly addictive. Only 7% of the people who try to quit smoking succeed on their first try. Also, a British government study showed that adolescents who smoke more than one cigarette have an 85% chance of becoming smokers.

Studies also show that breathing secondhand smoke poses health hazards for children and adults. Children who grow up living with smokers are more likely to develop allergies and asthma. Among adults, nonsmoking spouses of smokers have a 30% higher risk of both heart attack and lung cancer than spouses of nonsmokers have. In 2010, the CDC and the California Environmental Protection Agency estimated that each year, secondhand smoke causes an estimated 3,400 lung cancer deaths and 46,000 deaths from heart disease in the United States. A 2010 study by British researchers found that, globally, exposure to secondhand smoke contributes to about 600,000 deaths per year.

A study published in 2004 by Richard Doll and Richard Peto found that cigarette smokers die, on average, 10 years earlier than nonsmokers, but that kicking the habit—even at 50 years of age—can cut such a risk in half. If people quit smoking by the age of 30, they can avoid nearly all the risk of dying prematurely. However, the longer one smokes, the harder it is to quit.

In the United States, the percentage of adults who smoke dropped from roughly 40% in the 1960s to 21% in 2009 (24% of males and 18% of females). Such declines can be attributed to media coverage about the harmful health effects of smoking, sharp increases in cigarette taxes in many states, mandatory health warnings on cigarette packs, and the banning of smoking in workplaces, bars, restaurants, and public buildings.

In 2009, the government gave the U.S. Food and Drug Administration the power to classify and regulate tobacco as a drug. This was an important step forward, according to some analysts. These experts also call for eliminating all federal subsidies and tax breaks to tobacco farmers and tobacco companies. Ireland, Norway, Scotland, India, France, and Great Britain have all enacted bans on smoking in all indoor workplaces, bars, and restaurants. By 2013, 28 U.S. states had also enacted such bans.

Despite some encouraging trends, tobacco companies produced more than 5 trillion cigarettes in 2010—an average of about 700 for every person on the planet. In China about 350 million people—a number larger than the entire U.S. population—are hooked on tobacco. One major problem is that the price of a pack of cigarettes does not always reflect their harmful health effects. In 2006, CDC researchers calculated that the cost of treating smoking-related illnesses in the United States and the cost of lost worker productivity resulting from such illnesses amounted to about $10.50 a pack. Some states have raised taxes on cigarettes to include some of these costs—an application of the full-cost pricing **principle of sustainability** (see Figure 1-5, p. 9 or back cover).

Estimating Risks from Technologies Is Not Easy

The more complex a technological system, and the more people needed to design and run it, the more difficult it is to estimate the risks of using the system. The overall *reliability*—the probability (expressed as a percentage) that a person, device, or complex technological system will complete a task without failing—is the product of two factors:

System reliability (%) = Technology reliability (%) × Human reliability (%)

Figure 17-23 There is a startling difference between normal human lungs (left) and the lungs of a person who died of emphysema (right). The major causes of emphysema are prolonged smoking and exposure to air pollutants.

With careful design, quality control, maintenance, and monitoring, a highly complex system such as a nuclear power plant or a deep-sea oil-drilling rig can achieve a high degree of technological reliability. But human reliability usually is much lower than technological reliability and is almost impossible to predict: *To err is human.*

Suppose the technological reliability of a nuclear power plant is 95% (0.95) and human reliability is 75% (0.75). Then the overall system reliability is 71% (0.95 × 0.75 = 71%). Even if we could make the technology 100% reliable (1.0), the overall system reliability would still be only 75% (1.0 × 0.75 = 75%).

One way to make a system more foolproof, or failsafe, is to move more of the potentially fallible elements from the human side to the technological side. However, chance events such as a lightning strike can knock out an automatic control system, and no machine or computer program can completely replace human judgment. Also, the parts in any automated control system (such as the blowout protectors on the BP oil well that ruptured in the Gulf of Mexico in 2010) are manufactured, assembled, tested, certified, inspected, and maintained by fallible human beings. In addition, computer software programs used to monitor and control complex systems can be flawed because of human design error or can be deliberately sabotaged to cause their malfunction.

Most People Do a Poor Job of Evaluating Risks

Most of us are not good at assessing the relative risks from the hazards that surround us. Many people deny or shrug off the high-risk chances of death (or injury) from voluntary activities they enjoy such as *motorcycling* (death by motorcycle crash being 30 times more likely than death by car crash), *smoking* (1 in 250 by age 70 for a pack-a-day smoker), *hang gliding* (1 in 1,250), and *driving* (1 in 3,300 without a seatbelt and 1 in 6,070 with a seatbelt). Indeed, the most dangerous thing that many people do each day is to drive or ride in a car.

Yet some of these same people may be terrified about their chances of being killed by *the flu* (a 1 in 130,000 chance), *a nuclear power plant accident* (1 in 200,000), *West Nile virus* (1 in 1 million), *lightning* (1 in 3 million), *a commercial airplane crash* (1 in 9 million), *snakebite* (1 in 36 million), or *shark attack* (1 in 281 million). Worldwide each year, sharks kill an average of 6 people, while elephants kill at least 200 people.

Five factors can cause people to see a technology or a product as being more or less risky than experts judge it to be. First is *fear.* Research going back three decades shows that fear causes people to overestimate risks and to worry more about unusual risks than they do about common, everyday risks. Studies show that people tend to overestimate numbers of deaths caused by tornadoes, floods, fires, homicides, cancer, and terrorist attacks, and to underestimate numbers of deaths from flu, diabetes, asthma, heart attack, stroke, and automobile accidents.

The second factor in our estimation of risk is the *degree of control* we have in a given situation. Most of us have a greater fear of things over which we do not have personal control. For example, some individuals feel safer driving their own car for long distances than traveling the same distance on a plane. But look at the numbers. The risk of dying in a car accident in the United States while using a seatbelt is 1 in 6,070, whereas the risk of dying in a U.S. commercial airliner crash is about 1 in 9 million.

The third factor is *whether a risk is catastrophic,* not chronic. We usually are more frightened by news of catastrophic accidents such as a plane crash than we are of a cause of death such as smoking, which has a much higher death toll spread out over time.

Fourth, some people suffer from *optimism bias,* the belief that risks that apply to other people do not apply to them. While people get upset when they see others driving erratically while talking on a cell phone or text messaging, they may believe that talking on the cell phone or texting does not impair their own driving ability.

A fifth factor is that many of the risky things we do are highly pleasurable and give *instant gratification,* while the potential harm from such activities comes later. Examples are smoking cigarettes, eating lots of ice cream, and getting a tan.

Certain Principles Can Help Us Evaluate and Reduce Risk

Here are four guidelines for evaluating and reducing risk (Concept 17-5):

- *Compare risks.* Is there a risk of getting cancer by eating a charcoal-broiled steak once or twice a week for a life-

time? Yes, because almost any chemical can harm you if the dose is large enough. The question is whether this danger is great enough for you to worry about. In evaluating a risk, the key question is not "Is it safe?" but rather *"How risky is it compared to other risks?"*

- *Determine how much risk you are willing to accept.* For most people, a 1 in 100,000 chance of dying or suffering serious harm from exposure to an environmental hazard is a threshold for changing their behavior. However, in establishing standards and reducing risk, the EPA generally assumes that a 1 in 1 million chance of dying from an environmental hazard is acceptable. People involuntarily exposed to such risks believe that this standard is too high.

- *Evaluate the actual risk involved.* The news media usually exaggerate the daily risks we face in order to capture our interest and attract more readers, listeners, or viewers. As a result, most people who are exposed to a daily diet of such exaggerated reports believe that the world is much more risk-filled than it really is.

- *Concentrate on evaluating and carefully making important lifestyle choices.* This will help you to have a greater chance of living a longer, healthier, and less fearful life. When you worry about a risk, the most important question to ask is, "Do I have any control over this?" There is no point worrying about risks over which you have no control. But you do have control over major ways to reduce risks from heart attack, stroke, and certain forms of cancer, because you can decide whether to smoke, what to eat, and how much alcohol to drink. Other factors under your control are whether you practice safe sex, how much exercise you get, how safely you drive, and how often you expose yourself to ultraviolet rays from the sun or from a tanning bed.

Big Ideas

- We face significant hazards from infectious diseases such as flu, AIDS, tuberculosis, diarrheal diseases, and malaria, and from exposure to chemicals that can cause cancers and birth defects, and to those that can disrupt the human immune, nervous, and endocrine systems.

- Because of the difficulty of evaluating the harm caused by exposure to chemicals, many health scientists call for much greater emphasis on pollution prevention.

- By becoming informed, thinking critically about risks, and making careful choices, we can reduce the major risks we face.

Robert and Jean Pollock/Science Source

In the **Core Case Study** that opens this chapter, we saw that mercury (Hg) and its compounds that occur regularly in the environment can permanently damage the human nervous system, kidneys, and lungs and harm fetuses and cause birth defects. In this chapter, we also learned of many other chemical hazards, as well as biological, physical, cultural, and lifestyle hazards, in the environment. In addition, we saw how difficult it is to evaluate the nature and severity of threats from these various hazards.

One of the important facts discussed in this chapter is that on a global basis, the greatest threat to human health is the poverty–malnutrition–disease cycle, followed by the threats from smoking, pneumonia and flu, air pollution, and HIV/AIDS. There are some threats that we can do little to avoid, but we can reduce other threats, partly by applying the three scientific **principles of sustainability** (see Figure 1-2, p. 6 or back cover). For example, we can greatly reduce our exposure to mercury and other pollutants by shifting from use of nonrenewable fossil fuels (especially coal) to using a diversity of renewable energy resources, including solar energy. We can reduce our exposure to harmful chemicals used in the manufacturing of various goods by cutting resource use and waste and by reusing and recycling material resources. We can also mimic biodiversity by using diverse strategies for solving environmental and health problems, and especially for reducing poverty and controlling population growth. In so doing, we also help to preserve the earth's biodiversity.

Is this idealistic? Sure. But if creative and caring people had not acted throughout human history to improve the world by pursuing goals that others said were impossible or too idealistic, we would have accomplished very little. Each of us can make a difference.

Chapter Review

Core Case Study

1. Describe the toxic effects of mercury and its compounds and explain how humans are exposed to these toxins (**Core Case Study**).

Section 17-1

2. What is the key concept for this section? Define and distinguish among **risk**, **risk assessment**, and **risk management**. Give an example of how scientists state probabilities. Give an example of a risk from each of the following: biological hazards, chemical hazards, natural hazards, cultural hazards, and lifestyle choices. What is a **pathogen**?

Section 17-2

3. What is the key concept for this section? Define **infectious disease**, distinguish among **bacteria**, **viruses**, and **parasites**, and give examples of diseases that each can cause. Define and distinguish between **transmissible disease** and **nontransmis-** **sible disease**, and give an example of each. List four ways in which infectious organisms can enter the body. Distinguish between an epidemic and a pandemic of an infectious disease. In terms of death rates, what are the world's four most serious infectious diseases? List five factors that have contributed to genetic resistance in bacteria to commonly used antibiotics. What is MRSA and why is it so threatening?

4. Describe the global threat from tuberculosis and list three factors that have helped it to spread. What is the biggest viral killer and how does it spread? Describe and compare the threats from the hepatitis B and West Nile viruses. What is the best way to reduce one's chances of getting an infectious disease? What is the focus of ecological medicine and what are some of its findings regarding the spread of diseases? Summarize the story of health threats from the global HIV/AIDS pandemic.

5. What is malaria and how does it spread? How much of the human population is subject to this threat?

Explain how deforestation can promote the spread of malaria. List six major ways to reduce the global threat from infectious diseases.

Section 17-3

6. What is the key concept for this section? What is a **toxic chemical**? Define and distinguish among **carcinogens**, **mutagens**, and **teratogens**, and give an example of each. What are PCBs and why are they a threat? Describe the human immune, nervous, and endocrine systems and give an example of a chemical that can threaten each of these systems. What is a neurotoxin and why is methylmercury (**Core Case Study**) an especially dangerous one? What are six ways to prevent or control environmental inputs of mercury? What are hormonally active agents (HAAs), what risks do they pose, and how can we reduce those risks? Summarize concerns about exposure to bisphenol A (BPA) and the controversy over what to do about exposure to this chemical. Summarize concerns over exposure to phthalates. List six ways to reduce your exposure to hormone disrupters.

Section 17-4

7. What are the two key concepts for this section? Define **toxicology**, **toxicity**, **dose**, and **response**. What are three factors that affect the level of harm caused by a chemical? Give three reasons why children are especially vulnerable to harm from toxic chemicals. Explain how the toxicity of a substance can be estimated by testing laboratory animals, and explain the limitations

of this approach. What is a **dose-response curve**? Explain how toxicities are estimated through use of case reports and epidemiological studies, and discuss the limitations of these approaches.

8. Summarize the controversy over the effects of trace levels of chemicals. Why do we know so little about the harmful effects of chemicals? Explain why the use of the precautionary principle and pollution prevention to deal with health threats from chemicals is controversial. Describe some efforts to apply this principle on national and international levels. How did pollution prevention pay for the 3M Company?

Section 17-5

9. What is the key concept for this section? What is **risk analysis**? In terms of premature deaths, what are the three greatest threats that people face? What are six ways in which poverty can threaten one's health? Describe the health threats from smoking and how we can reduce these threats. How can we reduce the threats resulting from the use of various technologies? What are five factors that can cause people to misjudge risks? List four guidelines for evaluating and reducing risk.

10. What are this chapter's *three big ideas*? Explain how we can lessen the threats of harm from mercury in the environment by applying the three scientific **principles of sustainability**.

Note: Key terms are in bold type.

Critical Thinking

1. Assume that you are a national official with the power to set policy for controlling environmental mercury pollution (**Core Case Study**). Describe your policy in terms of goals and steps to take and explain how it will work to accomplish the goals. List three or more possible problems that could result from implementing your policy.

2. What are three actions you would take to reduce the global threats to human health and life from each of the following: **(a)** tuberculosis, **(b)** HIV/AIDS, and **(c)** malaria?

3. Evaluate the following statements:
   a. We should not worry much about exposure to toxic chemicals because almost any chemical, at a large enough dosage, can cause some harm.
   b. We should not worry much about exposure to toxic chemicals because, through genetic adaptation, we can develop immunities to such chemicals.
   c. We should not worry much about exposure to toxic chemicals because we can use genetic engineering to reduce our susceptibility to their effects.

   d. We should not worry about exposure to bisphenol A (BPA) because it has not been absolutely proven scientifically that BPA has killed anyone.

4. Should we ban the use of hormone mimics such as bisphenol A in products used by children younger than age 5? Should we ban them for use in all products? Explain.

5. Workers in a number of industries are exposed to higher levels of various toxic substances than the general public are. Should we reduce workplace levels allowed for such chemicals? What economic effects might this have?

6. Explain why you agree or disagree with the proposals for reducing the death toll and other harmful effects of smoking listed in the Case Study on p. 466. Do you believe there should be a ban on smoking indoors in all public places? Explain.

7. What are the three major risks you face from each of the following: **(a)** your lifestyle, **(b)** where you live, and **(c)** what you do for a living? Which of these

risks are voluntary and which are involuntary? List three steps you could take to reduce each of these risks. Which of these steps do you already take or plan to take?

8. In deciding what to do about risks from chemicals in the area where you live, would you support legisla-tion based in the REACH approach, assuming that chemicals are potentially harmful until shown other-wise, and requiring the use of pollution prevention based on the precautionary principle? Explain.

Doing Environmental Science

Pick a potentially harmful chemical and use the library or Internet to learn about **(a)** what it is used for and how widely it is used, **(b)** its potential harm, **(c)** the scientific evidence for such claims, and **(d)** proposed solutions for dealing with this threat. Pick a study area, such as your dorm or apartment building, your block, or your city. In this area, try to determine the level of presence of the chemical you are studying. You could do this by finding four or five examples of items or locations containing the chemical and then estimating the total amount based on your sample. Write a report summarizing your findings.

Global Environment Watch Exercise

Search for *mercury pollution* and research the latest devel-opments in studies of the harmful health effects of mer-cury (**Core Case Study**). Find an example of an effort to prevent or control mercury pollution and write a short report summarizing your findings. Try to find reports of two studies that reach different conclusions about how mercury should be regulated. Summarize the arguments for these conclusions on both sides. Based on what you have found, do you believe that mercury pollution should be regulated more strictly in the state or country where you live? Explain your reasoning.

Data Analysis

The graph below shows the effects of AIDS on life expec-tancy at birth in Botswana, 1950–2000, and projects these effects to 2050. Study this graph and answer the questions that follow.

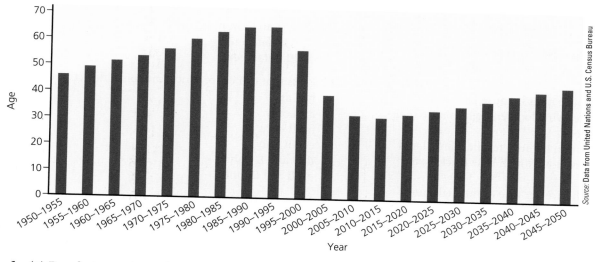

Source: Data from United Nations and U.S. Census Bureau

1. **(a)** By what percentage did life expectancy in Botswana increase between 1950 and 1995? **(b)** By what percentage was life expectancy in Botswana projected to decrease between 1995 and 2015?

2. **(a)** By what percentage is life expectancy in Botswana projected to increase between 2015 and 2050? **(b)** By what percentage was life expectancy in Botswana pro-jected to decrease between 1995 and 2050?

Question 1 refers to the projection of the Botswana population in 2020 below.

(Data from the United Nations.)

With AIDS   Without AIDS

1. What percent of the people aged 20–29 years will be lost if they die from AIDS?
   a. 0.36
   b. 0.64
   c. 36
   d. 64
   e. 76

2. Which of the following is NOT one of the social ramifications of HIV infection and resulting death of young adults?
   a. Decreasing orphanage rates because parents infected with HIV are living long enough to raise their children to adulthood
   b. Diminishing education because teachers are ill and dying, and there are not enough new teachers to replace them
   c. Diminishing health care because of the high number of health-care workers who either die or are too ill to do their jobs
   d. Decreasing food production because of the illness and deaths of adults who would otherwise have planted and harvested crops
   e. Disintegrating families because of the number of parents who have either died or are ill and unable to care for their children

3. Which of the following is a part of risk assessment?
   a. Comparative risk analysis of other risks
   b. Determining the financial cost of reducing the risk
   c. Assessing the probability of risk
   d. Developing a strategy for reducing risk
   e. Evaluating the quantity of risk reduction needed

4. Which of the below diseases is incorrectly paired with its type of organism?
   a. Virus–HIV
   b. Bacteria–Severe Acute Respiratory Syndrome (SARS)
   c. Fungi–Ringworm
   d. Protozoan–Malaria
   e. Prion–Creutzfeldt-Jakob Disease (Mad Cow)

5. A United Nations risk management training program reported that the risk of accidental death from playing field sports is 1 in 25,000. This means that
   a. one out of every 25,000 people that play field sports will die from an accidental death.
   b. people who play field sports are at equal risk of accidental death.
   c. one out of every 25,000 people will likely die from accidental death while playing field sports.
   d. there is a very low possibility that one who plays field sports will die from an accidental death.
   e. accidental deaths from playing field sports are uncommon.

Questions 6 and 7 refer to the three types of chemical hazards below.

    I. Carcinogens
    II. Mutagens
    III. Teratogens

6. PCBs that can cause cancer and birth defects are classified as
   a. I only.
   b. II only.
   c. III only.
   d. I and II only.
   e. II and III only.

7. Ethyl alcohol can cause birth defects and is classified as
   a. I only.
   b. II only.
   c. III only.
   d. I and II only.
   e. II and III only.

8. Which of the following is NOT a factor in the spread of infectious diseases to humans from other animals?
   a. Clearing of forests, which exposes humans to animals such as deer and mice and the ticks on them that carry bacteria
   b. Expanding suburbs and fragmentation of woodlands, which destroy predators' habitats, allowing their prey such as mice and ticks to spread Lyme disease
   c. Limits put on the international trade of livestock animals to protect local agricultural economies
   d. Legal and illegal trade in wild species, including kangaroos and tropical fishes, which are often not sufficiently screened or quarantined
   e. Local people using monkeys and other animals for bushmeat, because they might contain the Ebola or bird flu viruses

9. One of the largest threats of infectious disease to humans next to HIV and malaria is the spread of tuberculosis (TB). This is primarily due to
   a. the overuse of antibiotics treating people infected with TB.
   b. increased urbanization coupled with too few screening and control programs.
   c. identifying people with chronic coughs that may be active carriers.
   d. quarantining people who have an incurable form of TB.
   e. utilizing viruses as transmission vectors for gene therapy.

10. One reason that mercury (Hg) is so persistent and bio-accumulates without being broken down is that Hg is a
    a. potent carcinogen that can cause nervous system damage and birth defects.
    b. radioactive element with a long half-life.
    c. carbon-based organic compound naturally occurring in many organisms.
    d. common molecule whose atoms are loosely held together by chemical bonds and therefore easily broken down.
    e. stable element such as gold.

11. A major limitation of epidemiological studies used to determine the harmful effects of chemicals is
    a. people have not been exposed to new technologies or chemicals.
    b. computer simulations are not always accurate for living processes.
    c. threshold dosages are frequently difficult to determine.
    d. toxicity of single substances is often difficult to sort out from multiple sources.
    e. actual dosages from case reports are frequently unknown.

12. The LD50 of a drug or compound
    a. depends on how much of the drug is required for biological activity.
    b. is based on the number of days it takes to kill the test subjects.
    c. utilizes the response times of individual cells in petri dishes to determine lethal doses.
    d. is based on computer modeling of simulated cell responses.
    e. is the dose that can kill 50% of the animals in a test population.

18 Air Pollution

The atmosphere is the key symbol of global interdependence. If we can't solve some of our problems in the face of threats to this global commons, then I can't be very optimistic about the future of the world.

MARGARET MEAD

Key Questions

18-1 What is the nature of the atmosphere?

18-2 What are the major outdoor air pollution problems?

18-3 What is acid deposition and why is it a problem?

18-4 What are the major indoor air pollution problems?

18-5 What are the health effects of air pollution?

18-6 How should we deal with air pollution?

18-7 How have we depleted ozone in the stratosphere and what can we do about it?

Air pollution from pulp mills released above the clouds.

James P. Blair/National Geographic Creative

Air pollution is no longer viewed as primarily a localized urban problem. Annual satellite images and recent studies by the U.N. Environment Programme (UNEP) have found massive, dark brown clouds of pollution—called the *South Asian Brown Clouds*—stretching across much of India, Bangladesh, and the industrial heart of China, as well as parts of the western Pacific Ocean (Figure 18-1, left).

In most years, this 3-kilometer-thick (2-mile-thick) set of clouds covers an area about the size of the continental United States. The clouds contain small particles of dust, smoke, and ash resulting from drought and from the clearing and burning of forests for planting crops. They also contain particles of soot (mostly from diesel engine exhaust and from the burning of biomass such as wood and animal dung); acidic compounds from vehicle exhaust and coal-burning power plant emissions; and particles of toxic metals such as mercury (see Chapter 17, Core Case Study, p. 442) and lead, produced mostly by coal-burning facilities, metal smelters, and waste incinerators. Instead of blue skies, many of the people living under this cloud see brown or gray polluted skies much of the year (Figure 18-1, right).

These enormous, dense pollution clouds can move across the Asian continent within 3 to 4 days, and satellites have tracked the spread of such clouds from northern China across the Pacific Ocean to the West Coast of the United

States. The U.S. Environmental Protection Agency (EPA), using measurements made by atmospheric scientists, has estimated that on certain days, nearly 25% of the particulate matter, 77% of the black carbon (soot), and 33% of the toxic mercury in the skies above Los Angeles, California,

NASA

can be traced to China's coal-fired power plants, smelters, wood and dung fires, diesel trucks, and dust storms blowing out of drought-stricken, deforested areas.

Satellite measurements show that these and other long-lived air pollutants from China, India, the United States, or anywhere else on the planet can circle the entire globe in about 2 weeks. Thus, air pollution connects us all.

The history of air pollution control in Europe and the United States shows that pollution such as that in the South Asian Brown Clouds can be cleared up fairly quickly by various means. Such cleanups have involved enacting and enforcing air pollution laws that set standards for coal-burning industries and utilities. Another factor in some areas has been a gradual shift from burning coal to generate heat and electricity to using cleaner-burning natural gas, wind, and solar cells. China, India, and the United States are beginning to take such steps, but they have a long way to go.

Frederic Cirou/PhotoAlto/Alamy

Figure 18-1 The *South Asian Brown Clouds* stretch over much of South Asia (left) and contribute to air pollution in large cities such as Shanghai, China (right).

18-1 What Is the Nature of the Atmosphere?

CONCEPT 18-1

The two innermost layers of the atmosphere are the *troposphere*, which supports life, and the *stratosphere*, which contains the protective ozone layer.

The Atmosphere Consists of Several Layers

We live under a thin blanket of gases surrounding the earth, called the *atmosphere*. It is divided into several spherical layers defined mostly by temperature differences (Figure 18-2). Our focus in this book is on the atmosphere's two innermost layers: the troposphere and the stratosphere.

The *density,* or the number of gas molecules per unit of air volume, varies throughout the atmosphere because gravity pulls its gas molecules toward the earth's surface. This means that lower layers have more gases (more weight) in them than upper layers do, and are more densely packed with molecules. Thus, the air we breathe at sea level has a higher density than the air we would inhale on top of a high mountain.

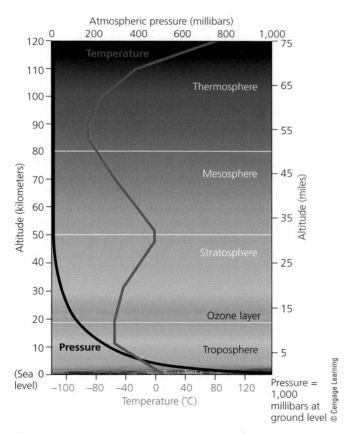

Figure 18-2 **Natural capital:** The earth's atmosphere is a dynamic system that includes four layers. The average temperature of the atmosphere varies with altitude (red line) and with differences in the absorption of incoming solar energy. ***Question:*** Why do you think most of the planet's air is in the troposphere?

Another important atmospheric variable is *atmospheric pressure*—the force, or mass, per unit area of a column of air. This force is caused by the bombardment of a surface such as your skin by the molecules in air. Atmospheric pressure decreases with altitude (see black line in Figure 18-2) because there are fewer gas molecules at higher altitudes. Density and atmospheric pressure are important because they play major roles in the weather.

Air Movements in the Troposphere Play a Key Role in the Earth's Weather and Climate

About 75–80% of the earth's air mass is found in the **troposphere**, the atmospheric layer closest to the earth's surface. This layer extends only about 17 kilometers (11 miles) above sea level at the equator and 6 kilometers (4 miles) above sea level over the poles. If the earth were the size of an apple, this lower layer containing the air we breathe would be no thicker than the apple's skin.

Take a deep breath. About 99% of the volume of air you inhaled consists of two gases: nitrogen (78%) and oxygen (21%). The remainder consists of water vapor (varying from 0.01% at the frigid poles to 4% in the humid tropics, for an average of about 1%), 0.93% argon (Ar), 0.040% carbon dioxide (CO_2), and trace amounts of dust and soot particles, as well as other gases, including methane (CH_4), ozone (O_3), and nitrous oxide (N_2O).

Rising and falling air currents, winds, and concentrations of CO_2 and other greenhouse gases in the troposphere play a major role in the planet's short-term *weather* (see Supplement 5, p. S19) and long-term *climate*. Recall that climate is the average weather in a given area, as measured mostly by temperature and precipitation, for at least three decades.

The Stratosphere Is Our Global Sunscreen

The atmosphere's second layer is the **stratosphere**, which extends from about 17 to about 48 kilometers (from 11 to 30 miles) above the earth's surface (Figure 18-2). Although the stratosphere contains less matter than the troposphere, its composition is similar, with two notable exceptions: its volume of water vapor is about 1/1,000th that of the troposphere and its concentration of ozone (O_3) is much higher.

Much of the atmosphere's small amount of ozone (O_3) is concentrated in a portion of the stratosphere called the **ozone layer**, found roughly 17–26 kilometers (11–16 miles) above sea level. Most stratospheric ozone is produced when some of the oxygen molecules in this layer interact with ultraviolet (UV) radiation emitted by the sun ($3O_2 + UV \rightleftarrows 2O_3$). This "global sunscreen" of ozone in the stratosphere keeps about 95% of the sun's harmful UV radiation (see Figure 2-11, p. 41) from reaching the earth's surface.

The UV filtering effect of ozone in the lower stratosphere allows us and other forms of life to exist on land and helps to protect us from sunburn, skin and eye cancers, cataracts, and damage to our immune systems. It also prevents much of the oxygen in the troposphere from being converted to photochemical ozone, a harmful air pollutant when found near the ground. Thus, the stratospheric ozone layer is a vital part of the earth's natural capital (see Figure 1-3, p. 7) that helps to keep us and other species alive and that supports our economies. Without it much of the earth's life would have to live underground or on the sea bottom to stay alive.

18-2 What Are the Major Outdoor Air Pollution Problems?

CONCEPT 18-2
Pollutants mix in the air to form *industrial smog*, primarily as a result of burning coal, and to form *photochemical smog*, caused by emissions from motor vehicles, industrial facilities, and power plants.

Air Pollution Comes from Natural and Human Sources

Air pollution is the presence of chemicals in the atmosphere in concentrations high enough to harm organisms, ecosystems, or human-made materials, or to alter climate. Note that almost any chemical in the atmosphere can become a pollutant if it occurs in a high enough concentration. The effects of air pollution range from annoying to lethal.

Air pollutants come from natural and human sources. Natural sources include wind-blown dust, pollutants from wildfires and volcanic eruptions, and volatile organic chemicals released by some plants. Most natural air pollutants are spread out over the globe or removed by chemical cycles, precipitation, and gravity. But in areas experiencing volcanic eruptions or forest fires, chemicals emitted by these events can temporarily reach harmful levels.

Most human inputs of outdoor air pollutants occur in industrialized and urban areas where people, cars, and factories are concentrated. These pollutants are generated mostly by the burning of fossil fuels (see Figure 2-13, p. 43) in power plants, industrial facilities, and motor vehicles. Technology is available to reduce such air pollution, but using it is costly and results in higher fuel prices. However, not reducing air pollution is even more costly in that it contributes to premature deaths and serious health problems.

Some Pollutants in the Atmosphere Combine to Form Other Pollutants

Scientists classify outdoor air pollutants into two categories: **Primary pollutants** are chemicals or substances emitted directly into the air from natural processes and human activities (Figure 18-3, center) at concentrations high enough to cause harm. While in the atmosphere, some primary pollutants react with one another and with other natural components of air to form new harmful chemicals, called **secondary pollutants** (Figure 18-3, right).

With their high concentrations of cars and factories, urban areas normally have higher outdoor air pollution levels (Figure 18-1, right) than rural areas have. But prevailing winds can spread long-lived primary and secondary air pollutants from urban and industrial areas to the countryside and to other urban areas. Long-lived pollutants entering the atmosphere in India and China now find their way across the Pacific where they affect the West Coast of North America (**Core Case Study**). Even in arctic regions where very few people live, air pollutants flowing north from Europe, Asia, and North America collect to form *arctic haze*. There is no place on the planet's surface that has not been affected by air pollution.

Over the past 30 years, the quality of outdoor air in most more-developed countries has improved greatly. This has occurred primarily because grassroots pressure from citizens, especially during the 1970s, led governments to pass and enforce air pollution control laws. GOOD NEWS

On the other hand, according to the World Health Organization (WHO), more than 1.1 billion people (one of every six people on the earth) live in urban areas where outdoor air is unhealthy to breathe. Most of them live in densely populated cities in less-developed countries where air pollution control laws do not exist or are poorly enforced. For example, the World Bank estimates that only 1% of China's 688 million urban inhabitants breathe air that meets the European Union's air quality standards.

What Are the Major Outdoor Air Pollutants?

Carbon Oxides

Carbon monoxide (CO) is a colorless, odorless, and highly toxic gas that forms during the incomplete combustion of carbon-containing materials (Table 18-1). Major sources are motor vehicle exhaust, burning of forests and grasslands, smokestacks of fossil fuel–burning power plants and industries, tobacco smoke, and open fires and inefficient stoves used for cooking.

CO can combine with hemoglobin in red blood cells, which reduces the ability of blood to transport oxygen

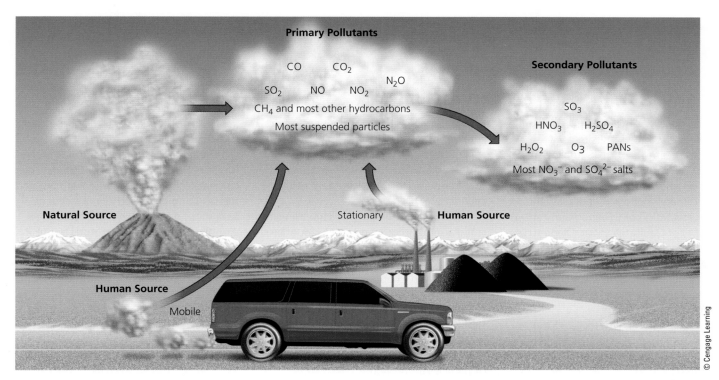

Figure 18-3 Human inputs of air pollutants come from *mobile sources* (such as cars) and *stationary sources* (such as industrial, power, and cement plants). Some *primary air pollutants* react with one another and with other chemicals in the air to form *secondary air pollutants*.

to body cells and tissues. Long-term exposure can trigger heart attacks and aggravate lung diseases such as asthma and emphysema. At high levels, CO can cause headache, nausea, drowsiness, confusion, collapse, coma, and death.

Carbon dioxide (CO_2) is a colorless, odorless gas. About 93% of the CO_2 in the atmosphere is the result of the natural carbon cycle (see Figure 3-17, p. 66). The rest comes from human activities, primarily the burning of fossil fuels and the clearing of CO_2-absorbing forests and grasslands. Until recently, it has not been classified in many countries as an air pollutant. But there is considerable and growing scientific evidence that increasing levels of CO_2 (see Figure 14, p. S70, in Supplement 7) caused by human activities

are contributing to atmospheric warming and projected climate change, because CO_2 is being added to the atmosphere faster than it is removed by the natural carbon cycle. On the other hand, as studies of the South Asian Brown Clouds have shown (**Core Case Study**), some pollutants have a cooling effect (Science Focus 18.1). We examine atmospheric warming and climate change in Chapter 19.

Nitrogen Oxides and Nitric Acid

Nitric oxide (NO) is a colorless gas that forms when nitrogen and oxygen gas in the air react under high-combustion temperatures in automobile engines and coal-burning power and industrial plants (Table 18-1). Lightning and certain bacteria in soil and water also produce NO as part of the nitrogen cycle (see Figure 3-18, p. 67).

In the air, NO reacts with oxygen to form *nitrogen dioxide* (NO_2), a reddish-brown gas. Collectively, NO and NO_2 are called *nitrogen oxides* (NO_x). Some of the NO_2 reacts with water vapor in the air to form *nitric acid* (HNO_3) and nitrate salts (NO_3^-), components of harmful *acid deposition,* which we discuss later in this chapter. Both NO and NO_2 play a role in the formation of *photochemical smog*—a mixture of chemicals formed under the influence of sunlight in cities with heavy traffic (discussed further below). *Nitrous oxide* (N_2O), a greenhouse gas, is emitted from fertilizers and animal wastes, and is produced by the burning of fossil fuels.

At high enough levels, nitrogen oxides can irritate the eyes, nose, and throat, aggravate lung ailments such

Table 18-1 Chemical Reactions that Form Major Air Pollutants	
Pollutant	**Chemical Reaction**
Carbon monoxide (CO)	$2C + O_2 \rightarrow 2CO$
Carbon dioxide (CO_2)	$C + O_2 \rightarrow CO_2$
Nitric oxide (NO)	$N_2 + O_2 \rightarrow 2NO$
Nitrogen dioxide (NO_2)	$2NO + O_2 \rightarrow 2NO_2$
Sulfur dioxide (SO_2)	$S + O_2 \rightarrow SO_2$

© Cengage Learning

THE SOUTH ASIAN BROWN CLOUDS, MELTING GLACIERS, AND CLIMATE CHANGE

In 2008, the U.N. Environment Programme (UNEP) reported the results of a 7-year study of the effects of the South Asian Brown Clouds (**Core Case Study**) carried out by an international team of scientists led by climate scientist V. Ramanathan of Scripps Institution of Oceanography at the University of California in San Diego (USA). According to these scientists, the South Asian Brown Clouds have been a factor in the gradual melting of the Himalayan glaciers, which are the source of water for most of Asia's major rivers.

The researchers hypothesized that soot and some of the other particles in the brown clouds absorb sunlight and heat the air above the glaciers. Black soot falling on the white surface of the gla-

ciers also decreases their ability to reflect sunlight back into space. The glaciers then absorb more solar energy, which adds to the warming of the air above them and also increases the rate of glacial melting.

The very slow melting of some Himalayan glaciers and of many other mountain glaciers is taken as a sign of atmospheric warming and climate change (which we discuss fully in Chapter 19). But the UNEP study also pointed out that the South Asian Brown Clouds, in addition to contributing to such warming, have helped to mask the full impact of global atmospheric warming. This occurs because certain types of particles in the clouds reflect some incoming sunlight back into space, which

has helped to cool part of the earth's surface beneath them.

The net effect is that during this century, the South Asian Brown Clouds have helped to slow atmospheric warming and the resulting projected climate change. At the same time, this severe case of air pollution has had numerous harmful effects on water and food supplies in Asia and is directly linked to the deaths of at least 380,000 people a year in China and India.

Critical Thinking

Do you think the benefit of the lessened atmospheric warming due to the South Asian Brown Clouds outweighs their harmful effects? Explain.

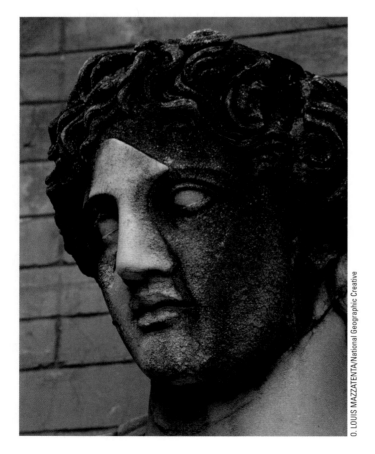

O. LOUIS MAZZATENTA/National Geographic Creative

as asthma and bronchitis, suppress plant growth, and reduce visibility when they are converted to nitric acid and nitrate salts.

Sulfur Dioxide and Sulfuric Acid

Sulfur dioxide (SO_2) is a colorless gas with an irritating odor. About one-third of the SO_2 in the atmosphere comes from natural sources as part of the sulfur cycle (see Figure 3-20, p. 69). The other two-thirds (and as much as 90% in some urban areas) come from human sources, mostly combustion of sulfur-containing coal in power and industrial plants (Table 18-1), oil refining, and smelting of sulfide ores.

In the atmosphere, SO_2 can be converted to *aerosols*, which consist of microscopic suspended droplets of *sulfuric acid* (H_2SO_4) and suspended particles of sulfate (SO_4^{2-}) salts that return to the earth as a component of acid deposition. Sulfur dioxide, sulfuric acid droplets, and sulfate particles reduce visibility and aggravate breathing problems. They can also damage crops, trees, soils, and aquatic life in lakes, and they corrode metals and damage paint, paper, leather, and stone on buildings and statues (Figure 18-4). They are a major component of the South Asian Brown Clouds (**Core Case Study**).

Figure 18-4 Acid deposition and other forms of air pollution have damaged this statue in Rome, Italy. The nose and part of the forehead have been restored.

Figure 18-5 Severe air pollution from an iron and steel factory in Czechoslovakia.

JAMES P. BLAIR/National Geographic Creative

Particulates

Suspended particulate matter (SPM) consists of a variety of solid particles and liquid droplets that are small and light enough to remain suspended in the air for long periods. The EPA classifies particles as fine, or PM-10 (with diameters less than 10 micrometers), and ultrafine, or PM-2.5 (with diameters less than 2.5 micrometers). About 62% of the SPM in outdoor air comes from natural sources such as dust, wildfires, and sea salt. The remaining 38% comes from human sources, including coal-burning power and industrial plants (Figure 18-5), motor vehicles, and road construction. Such particulates are a major component of the South Asian Brown Clouds (**Core Case Study**).

 CONSIDER THIS. . .

CONNECTIONS South Asian Brown Clouds, Food Production, and Solar Power

Soot and sulfate particles in the South Asian Brown Clouds (**Core Case Study**, Figure 18-1) decrease the amount of sunlight that penetrates the atmosphere. This dimming can lead to lower food production by hindering photosynthesis and by cooling land and ocean surfaces, thereby reducing evaporation and rainfall. The UNEP projects that unless the pollutants in these clouds are reduced, they could increase crop losses in South Asia by up to 40%. The dimming of sunlight also lowers the output of centralized solar energy power plants. A 2009 study by Kaicun Wang found that global dimming, mostly from air pollution, is occurring everywhere except in most of Europe, where reductions in air pollution have led to brighter skies.

According to the American Lung Association, more than 2,000 studies published since 1990 link SPM with adverse health effects. These particles can irritate the nose and throat, damage the lungs, aggravate asthma and bronchitis, and shorten life spans. Toxic particulates such as lead (see the following Case Study), cadmium, and polychlorinated biphenyls (PCBs) can cause genetic mutations, reproductive problems, and cancer. Particulates also reduce visibility, corrode metals, and discolor clothing and paints.

Ozone

Ozone (O_3), a colorless and highly reactive gas, is a major ingredient of photochemical smog. It can cause coughing and breathing problems, aggravate lung and heart diseases, reduce resistance to colds and pneumonia, and irritate the eyes, nose, and throat. It also damages plants, rubber in tires, fabrics, and paints.

Ozone in the troposphere near ground level is often referred to as "bad" ozone, while we view ozone in the stratosphere as "good" ozone because it protects us from harmful UV radiation. However, both are the same chemical. Significant evidence indicates that some human activities are *decreasing* the amount of beneficial ozone in the stratosphere and *increasing* the amount of harmful ozone in the troposphere near ground level—especially in some urban areas. We examine the issue of decreased stratospheric ozone in the final section of this chapter.

Volatile Organic Compounds (VOCs)

Organic compounds that exist as gases in the atmosphere or as gases that evaporate from sources on earth into the atmosphere are called *volatile organic compounds* (VOCs). Examples are hydrocarbons emitted by the leaves of many plants, and *methane* (CH_4), a greenhouse gas that is about 25 times more effective per molecule than CO_2 is at warming the atmosphere. About a third of global methane emissions come from natural sources, mostly plants, wetlands, and termites. The rest come from human sources, primarily rice paddies, landfills, natural gas wells and pipelines, landfills, and cows (mostly from their belching).

Other VOCs are liquids that can evaporate into the atmosphere. Examples are benzene and other liquids used as industrial solvents, dry-cleaning fluids, and various components of gasoline, plastics, and other products.

CASE STUDY

Lead Is a Highly Toxic Pollutant

Lead (Pb) is a soft gray metal used to make various products including lead–acid batteries and bullets, and it was once a common ingredient of gasoline and paints. It is also a pollutant that is found in air, water, soil, plants, and animals. Because it is a chemical element, lead does not break down in the environment. This indestructible and potent neurotoxin can harm the nervous system, especially in young children. Each year, 12,000–16,000 American children under age 9 are treated for acute lead poisoning, and about 200 die. About 30% of the survivors suffer from palsy, partial paralysis, blindness, and mental retardation.

Children under age 6 and unborn fetuses, even with low blood levels of lead, are especially vulnerable to nervous system impairment, lowered IQ (by 2 to 5 points), shortened attention span, hyperactivity, hearing damage, and various behavior disorders. A 1993 study by the U.S. National Academy of Sciences along with numerous other more recent studies indicate that there is no safe level of lead in children's blood. As a result, health scientists call for sharply reducing the currently allowed levels for lead in the air and water.

Between 1976 and 2006, the percentage of U.S. children of ages 1 to 5 with blood lead levels above the safety standard dropped from 85% to about 1%, which prevented at least 9 million childhood lead poisonings. The primary reason for this drop was that leaded gasoline was banned by federal law in 1976 and completely phased out by 1986. The law also greatly reduced the allowable levels of lead in paints in 1970 (even though illegal use of leaded paints continued until about 1978). This is an excellent example of the effectiveness of pollution prevention.

Despite this success, in 2013, the U.S. Centers for Disease Control and Prevention (CDC) estimated that about 535,000 U.S. children under age 6 still have unsafe blood levels of lead caused by exposure from a number of sources. The major source is peeling lead-based paint and lead-contaminated dust in some 4 million older homes. Children can inhale or accidentally ingest paint particles from these sources. Another source is soils contaminated with lead emitted by motor vehicles before leaded gasoline was banned. Lead can also leach from water pipes and faucets containing lead parts or lead solder.

Other sources are older coal-burning power plants that have not been required to meet the emission standards of new plants, lead smelters, and waste incinerators. And so far, the U.S. ban on leaded gasoline does not apply to aviation fuel, and several million children living or going to school in areas around the nation's 22,000 airports are exposed to unsafe levels of lead from this source.

In 2012, the CDC used the latest scientific data to come up with new guidelines for identifying children aged 5 and under who have potentially dangerous blood lead levels. These guidelines raised the estimated number of young children at risk from lead poisoning from 77,000 to about 535,000 at a time when there are pressures on Congress to sharply cut funding for lead poisoning prevention.

CONSIDER THIS...

CONNECTIONS Lead and Urban Gardening

Health officials and scientists have urged people who grow some of their own food in urban gardens to have their garden soils tested for lead. For decades, lead particles fell from the air into urban soils, primarily from the exhaust fumes of vehicles burning leaded gasoline. Soil found to have lead in it can be treated or removed from urban gardens and replaced with uncontaminated soil.

Health scientists have proposed a number of ways to help protect children from lead poisoning, as listed in Figure 18-6. Although the threat from lead has been greatly reduced in the United States, this is not the case in many less-developed countries. About 80% of the gasoline sold in the world today is unleaded, but about 100 countries still use leaded gasoline and most countries have not banned lead-based paint.

The WHO estimates that 130 million to 200 million children around the world are at risk of lead poisoning, and 15 million to 18 million children in less-developed countries have permanent brain damage because of lead poisoning. The key sources of lead pollutants in these countries are leaded gasoline and poorly regulated lead smelters and waste incinerators.

Children and adults in China and several African countries are also exposed to dangerous levels of lead when they work in recycling centers extracting lead and other valuable metals from electronic waste (e-waste)—discarded computers, TV sets, cellphones, and other electronic devices (Figure 18-7). Much of this e-waste is shipped to such unsafe recycling centers from the United States and other more-developed countries.

Solutions

Lead Poisoning

Prevention	Control
Replace lead pipes and plumbing fixtures containing lead solder	Remove leaded paint and lead dust from older houses and apartments
Phase out leaded gasoline worldwide	Sharply reduce lead emissions from incinerators
Phase out waste incineration	
Ban use of lead solder	Remove lead from TV sets and computer monitors before incineration or land disposal
Ban use of lead in computer and TV monitors	
Ban lead glazing for ceramicware used to serve food	Test blood for lead by age 1
Wash fresh fruits and vegetables	Test for lead in existing ceramicware used to serve food

© Cengage Learning

Figure 18-6 There are several ways to help protect children from lead poisoning. *Questions:* Which two of these solutions do you think are the best ones? Why?

Top: Ssuaphotos/Shutterstock.com. Center: Mark Smith/Shutterstock.com. Bottom: ©Dmitry Kalinovsky/Shutterstock.com.

PETER ESSICK/National Geographic Creative

Figure 18-7 This worker in Taizhou City in China's Zhejiang Province is heating discarded electronic circuit boards to recover tin and lead.

Environmental and health scientists call for global bans on leaded gasoline and lead-based paints, and for much stricter controls on emissions from coal-burning power plants, smelters, and waste incinerators. China recently phased out leaded gasoline in less than 3 years, a process that took several decades in the United States. Scientists also call for improved health standards for workers in e-waste recycling shops, and they urge manufacturers of electronic goods to find substitutes for lead and other toxic materials that are used in electronics manufacturing. Another priority for many public health scientists is to continually improve the monitoring of outdoor air for the presence of dangerous pollutants (Science Focus 18.2).

GOOD NEWS

Burning Coal Produces Industrial Smog

Sixty years ago, cities such as London, England, and the U.S. cities of Chicago, Illinois, and Pittsburgh, Pennsylvania, burned large amounts of coal in power plants and factories and for heating homes and often for cooking food. People in such cities, especially during winter, were exposed to **industrial smog** consisting mostly of an unhealthy mix of sulfur dioxide, suspended droplets of sulfuric acid, and a variety of suspended solid particles in outside air. Those who burned coal inside their homes were exposed to dangerous levels of indoor air pollutants.

The chemistry of industrial smog is fairly simple. When burned, most of the carbon in coal and oil is converted to carbon monoxide (CO) and carbon dioxide (CO_2). Unburned carbon in coal also ends up in the atmosphere as suspended particulate matter (soot).

When coal and oil are burned (Figure 18-8), the sulfur compounds they contain react with oxygen to produce sulfur dioxide (O_2) gas, some of which is converted to tiny suspended droplets of sulfuric acid (H_2SO_4). Some of these droplets react with ammonia (NH_3) in the atmosphere to form solid particles of ammonium sulfate or ($NH_4)_2SO_4$.

The suspended particles of such salts and soot give the resulting smog a gray color (Figures 18-1, right and 18-5), which is why it is sometimes called *gray-air smog* (Concept 18-2).

Today, urban industrial smog is rarely a problem in most of the more-developed countries where coal and heavy oil are burned only in large power and industrial plants with reasonably good pollution control. However, many of these facilities have tall smokestacks that send the pollutants high into air (see chapter-opening photo) where prevailing winds carry them downwind to rural areas, and this presents problems that we deal with later in this chapter.

Industrial smog remains a problem in industrialized urban areas of China (Figure 18-1), India, Ukraine, and some other eastern European countries (Figure 18-5), where large quantities of coal are still burned in houses, power plants, and factories with inadequate pollution controls. Such coal burning is a major contributor to the South Asian Brown Clouds (**Core Case Study**).

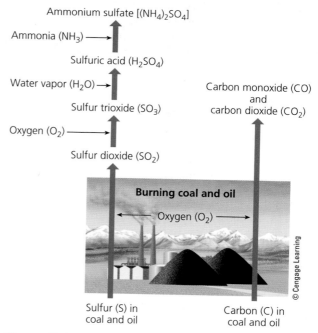

Figure 18-8 A greatly simplified model of how pollutants are formed when coal and oil are burned. The result is industrial smog.

DETECTING AIR POLLUTANTS

We can detect the presence of pollutants in the air with the use of chemical instruments and satellites armed with various sensors. The scientists who discovered the components and effects of the South Asian Brown Clouds (**Core Case Study**) used small unmanned aircraft carrying miniaturized instruments to measure chemical concentrations, temperatures, and other variables within the clouds.

Aerodyne Research in the U.S. city of Boston, Massachusetts, has developed a mobile laboratory that uses sophisticated instruments to make instantaneous measurements of primary and secondary air pollutants from motor vehicles, factories, and other sources. This laboratory can also monitor changes in concentrations of the pollutants throughout a day and under different weather conditions, and it can measure the effectiveness of various air pollution control devices used in cars, trucks, and buses.

Scientists are also using nanotechnology (see Science Focus 14.1, p. 362) to try to develop inexpensive nanodetectors for various air pollutants. Another way to detect air pollutants is through biological indicators, including lichens (Figure 18-A).

A lichen consists of a fungus and an alga living together, usually in a mutually beneficial (mutualistic) relationship. These hardy pioneer species are good biological indicators of air pollution because they continually absorb air as a source of nourishment. A highly polluted area around an industrial plant might have only gray-green crusty lichens or none at all. An area with moderate air pollution might support only orange crusty lichens. In contrast, areas with fairly clean air can support larger varieties of lichens.

Some lichen species are sensitive to specific air-polluting chemicals. Old man's beard (*Usnea trichodea*, Figure 18-A, left) and yellow *Evernia* lichens, for example, can sicken or die in the presence of excessive sulfur dioxide, even if the pollutant originates far away. For example, scientists discovered sulfur dioxide pollution on Isle Royale, Michigan (USA) in Lake Superior, an island where no car or smokestack has ever intruded. They used *Evernia* lichens to point the finger northward to coal-burning facilities in and around the Canadian city of Thunder Bay, Ontario. Damaged leaves on some plants can also be a sign of air pollution.

Critical Thinking

How do you think the science and technology of air pollution detection should be financed? Who should pay for it?

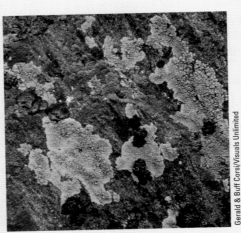

Milton Rand/Tom Stack & Associates

Gerald & Buff Corsi/Visuals Unlimited

Figure 18-A Natural capital: This old man's beard (*Usnea trichodea*) lichen (left) is growing on a branch of a larch tree, and these red and yellow crustose lichens (right) are growing on slate rock.

Because of its heavy reliance on coal, China has some of the world's highest levels of industrial smog and 16 of the world's 20 most polluted cities (Figure 18-1, right). On January 12, 2013 an air pollution monitor atop the U.S. Embassy in Beijing recorded an air quality index of 755, which was almost twice the level of 400, considered to be hazardous for everyone. The WHO and researchers at the University of Washington have estimated that in 2010, outdoor air pollution killed 1.2 million people in China—nearly 40% of the global death toll from outdoor air pollution. In India, outdoor air pollution killed an estimated 620,000 people in 2010.

Sunlight Plus Cars Equals Photochemical Smog

A *photochemical reaction* is any chemical reaction activated by light. **Photochemical smog** is a mixture of primary and secondary pollutants formed under the influence of UV radiation from the sun. In greatly simplified terms,

$$VOCs + NO_x + heat + sunlight \rightarrow \begin{array}{l} \text{ground level ozone } (O_3) \\ + \text{ other photochemical oxidants} \\ + \text{ aldehydes} \\ + \text{ other secondary pollutants} \end{array}$$

The formation of photochemical smog (Figure 18-9) begins when exhaust from morning commuter traffic releases large amounts of NO and VOCs into the air over a city. The NO is converted to reddish-brown NO_2, and this explains why photochemical smog is sometimes called *brown-air smog.* When exposed to ultraviolet radiation from the sun, some of the NO_2 reacts in complex ways with VOCs released by certain trees (such as some oak species, sweet gums, and poplars), motor vehicles, and businesses (such as bakeries and dry cleaners).

The resulting photochemical smog is a mixture of ozone, nitric acid, aldehydes, peroxyacyl nitrates (PANs), and other secondary pollutants. Collectively, NO_2, O_3, and PANs in this chemical brew are called *photochemical oxidants* because these damaging chemicals can react with, and oxidize, certain compounds in the atmosphere or inside our lungs. Hotter days lead to higher levels of ozone and other components of smog. As traffic increases on a sunny day, photochemical smog (dominated by ozone) usually builds up to peak levels by late morning, irritating some people's eyes and respiratory tracts.

All modern cities have some photochemical smog, but it is much more common in cities with sunny, warm, and dry climates and a great number of motor vehicles. Examples are Los Angeles, California (Figure 18-10), and Salt Lake City, Utah, in the United States; Sydney, Australia; São Paulo, Brazil; Bangkok, Thailand; Mexico City, Mexico; and Santiago, Chile. According to a 1999 study, if there were 400 million conventional gasoline-powered cars on the road in China by 2050 as has been projected, the resulting photochemical smog could regularly cover the entire western Pacific ocean, extending to the United States.

The already poor air quality in urban areas of many less-developed countries is worsening as their numbers of motor vehicles rise. Many of these vehicles, especially the older ones, have no pollution control devices and burn leaded gasoline.

CONSIDER THIS...

CONNECTIONS Short Driving Trips and Air Pollution

According to several studies, including one done in Australia in 2011, about 60% of the pollution from motor vehicle emissions occurs in the first minutes of operation before pollution control devices are working at top efficiency. Yet 40% of all U.S. car trips take place within 3 kilometers (2 miles) of drivers' homes, and half of the U.S. working population drives 8 kilometers (5 miles) or less to work.

Several Factors Can Decrease or Increase Outdoor Air Pollution

Five natural factors help *reduce* outdoor air pollution. First, *particles heavier than air* settle out as a result of gravitational attraction to the earth. Second, *rain and snow* help cleanse the air of pollutants. Third, *salty sea spray from the oceans* washes out many pollutants from air that flows from land over the oceans. Fourth, *winds* sweep pollutants away and mix them with cleaner air. Fifth, some pollutants are removed by *chemical reactions*. For example, SO_2

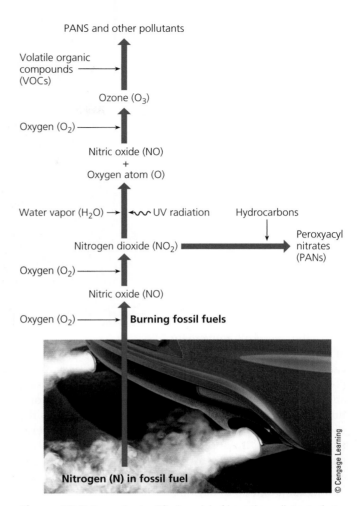

Figure 18-9 A greatly simplified model of how the pollutants that make up photochemical smog are formed.

Photo: © Ssuaphotos/Shutterstock.com

can react with O_2 in the atmosphere to form SO_3, which reacts with water vapor to form droplets of H_2SO_4 that fall out of the atmosphere as acidic precipitation.

Six other factors can *increase* outdoor air pollution. First, *urban buildings* slow wind speed and reduce the dilution and removal of pollutants. Second, *hills and mountains* reduce the flow of air in valleys below them and allow pollutant levels to build up at ground level. Third, *high temperatures* promote the chemical reactions leading to formation of photochemical smog. Fourth, *emissions of volatile organic compounds (VOCs)* from certain trees and plants (including kudzu; see Figure 9-10, p. 200) in heavily wooded urban areas can play a large role in the formation of photochemical smog.

A fifth factor—the so-called *grasshopper effect*—occurs when air pollutants are transported at high altitudes by evaporation and winds from tropical and temperate areas through the atmosphere to the earth's polar areas. This happens mostly during winter. It explains why, for decades, pilots have reported seeing dense layers of reddish-brown haze over the Arctic. It also explains why polar bears,

Figure 18-10 Photochemical smog is a serious problem in Los Angeles, California, although air pollution laws have helped to reduce the average number of severe smog days per year. **Question:** How serious is photochemical smog where you live?

Lee Petter/iStockphoto.com

sharks, and native peoples in remote arctic areas have high levels of various toxic pollutants in their bodies.

The sixth factor has to do with the vertical movement of air. During daylight, the sun warms the air near the earth's surface. Normally, this warm air and most of the pollutants it contains rise to mix with the cooler air above and are dispersed. Under certain atmospheric conditions, however, a layer of warm air can temporarily lie atop the cooler air nearer the ground, and this is called a **temperature inversion**. Because the cooler air is denser than the warmer air above it, the air near the surface does not rise and mix with the air above. If this condition persists, pollutants can build up to harmful and even lethal concentrations in the stagnant layer of cool air near the ground.

Two types of areas are especially susceptible to prolonged temperature inversions. The first is a town or city located in a valley surrounded by mountains where the weather turns cloudy and cold during part of the year (Figure 18-11, left). In such cases, the clouds block much

of the winter sunlight that causes air to heat and rise, and the mountains block winds that could disperse the pollutants. As long as these stagnant conditions persist, pollutants in the valley below will build up to harmful and even lethal concentrations.

The other type of area vulnerable to temperature inversions is a city with many motor vehicles in an area with a sunny climate, mountains on three sides, and an ocean on the fourth side (Figure 18-11, right). Here, the conditions are ideal for the formation of photochemical smog, worsened by frequent thermal inversions. The surrounding mountains prevent the polluted surface air from being blown away by breezes coming off the sea. This describes several cities, including heavily populated Los Angeles, California, which has prolonged temperature inversions (Figure 18-10).

Figure 18-11 A *temperature inversion,* in which a warm air layer sits atop a cooler air layer, can take place in either of the two sets of topography and weather conditions shown here. Normally, the air temperature decreases steadily with increasing altitude within the troposphere, but during an inversion, there is a layer of air that is warmer than the cooler air above and below (see graph).

18-3 What Is Acid Deposition and Why Is It a Problem?

CONCEPT 18-3

Acid deposition is caused mainly by coal-burning power plants and motor vehicle emissions, and in some regions it threatens human health, aquatic life and ecosystems, forests, and human-built structures.

Acid Deposition Is a Serious Regional Air Pollution Problem

Most coal-burning power plants, ore smelters, and other industrial facilities in more-developed countries use tall smokestacks (see chapter-opening photo) to vent the exhausts from burned fuel at a high altitude. The exhausts contain sulfur dioxide, suspended particles, and nitrogen oxides, and the smokestacks send them high into the atmosphere where wind can dilute and disperse them.

Use of such smokestacks reduces *local* air pollution, but it can increase *regional* air pollution downwind. Prevailing winds can transport the primary pollutants (SO_2 and NO_x) as far as 1,000 kilometers (600 miles). During their trip, these compounds form secondary pollutants such as droplets of sulfuric acid (H_2SO_4), nitric acid vapor (HNO_3), and particles of acid-forming sulfate (SO_4^{2-}) and nitrate (NO_3^-) salts (Figure 18-3).

These acidic substances remain in the atmosphere for 2–14 days, depending mostly on prevailing winds, precipitation, and other weather patterns. During this period, they descend to the earth's surface in two forms: *wet deposition*, consisting of acidic rain, snow, fog, and cloud vapor with a pH of less than 5.6 (the acidity level of unpolluted rain; see Figure 4, p. S15, in Supplement 4), and *dry deposition*, consisting of acidic particles. The resulting mixture is called **acid deposition** (Figure 18-12)— sometimes termed *acid rain*. Most dry deposition occurs within 2–3 days of emission, fairly near the industrial sources, whereas most wet deposition takes place within 4–14 days in more distant downwind areas.

Acid deposition has been occurring since the Industrial Revolution began in the mid-1700s. In 1872, British chemist Robert A. Smith coined the term *acid rain* after observing that rain was eating away stone in the walls of buildings in major industrial areas. Acid deposition is the result of human activities that disrupt the natural nitrogen cycle (see Figure 3-18, p. 67) and sulfur cycle (see Figure 3-20, p. 69) by adding excessive amounts of nitrogen oxides and sulfur dioxide to the atmosphere.

Acid deposition is a *regional* air pollution problem (Concept 18-3) in areas that lie downwind from coal-burning facilities and from urban areas with large numbers of cars, as shown in Figure 18-13. In some areas, soils contain *basic* compounds such as calcium carbonate ($CaCO_3$) or limestone that can react with and neutralize, or *buffer*, some inputs of acids. The areas most sensitive to acid deposition are those with thin, already acidic soils that provide no

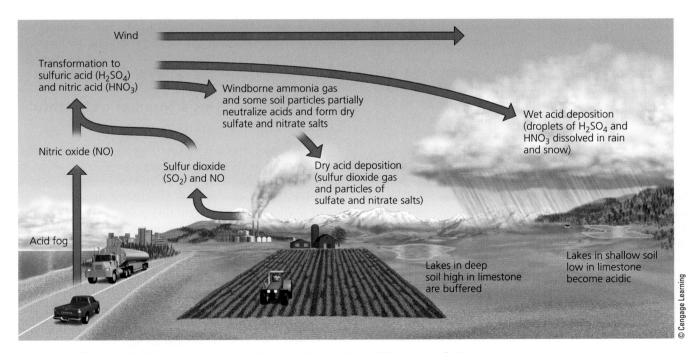

Animated Figure 18-12 Natural capital degradation: *Acid deposition*, which consists of rain, snow, dust, or gas with a pH lower than 5.6, is commonly called acid rain (see Figure 4, p. S15, in Supplement 4). **Question:** What are three ways in which your daily activities contribute to acid deposition?

(Compiled by the authors using data from World Resources Institute and U.S. Environmental Protection Agency.)

Potential problem areas because of sensitive soils

Potential problem areas because of air pollution: emissions leading to acid deposition

Current problem areas (including lakes and rivers)

Figure 18-13 This map shows regions where acid deposition is now a problem and regions with the potential to develop this problem. Such regions have large inputs of air pollution (mostly from power plants, industrial facilities, and ore smelters) or are sensitive areas with naturally acidic soils and bedrock that cannot neutralize (buffer) additional inputs of acidic compounds. **Question:** Do you live in or near an area that is affected by acid deposition or an area that is likely to be affected by acid deposition in the future?

such natural buffering (Figure 18-13, green and most red areas) and those where the buffering capacity of soils has been depleted by decades of acid deposition.

In the United States, older coal-burning power and industrial plants without adequate pollution controls, especially in the Midwest, emit the largest quantities of SO_2 and other pollutants that cause acid deposition. Because of these emissions and those of other urban industries and motor vehicles, as well as the prevailing west-to-east winds, typical precipitation in the eastern United States is at least 10 times more acidic than natural precipitation is. Some mountaintop forests in the eastern United States, as well as east of Los Angeles, California, are bathed in fog and dews that are as acidic as lemon juice—with about 1,000 times the acidity of unpolluted precipitation.

Many acid-producing chemicals generated in one country are exported to other countries by prevailing winds. For example, acidic emissions from the United Kingdom and Germany blow south and east into Switzerland and Austria, and north and east into Norway and other neighboring countries.

The worst acid deposition occurs in Asia, especially in China, which gets 70% of its total energy and 73% of its electricity from burning coal, according to the U.S. Energy Information Administration. According to its government,

China is the world's top emitter of SO_2. The resulting acid precipitation is damaging crops and threatening food security in China, Japan, and North and South Korea. It also contributes to the South Asian Brown Clouds (**Core Case Study**).

Acid Deposition Has a Number of Harmful Effects

Acid deposition damages statues (Figure 18-4) and buildings, contributes to human respiratory diseases, and can leach toxic metals (such as lead and mercury) from soils and rocks into lakes used as sources of drinking water. These toxic metals can accumulate in the tissues of fish eaten by people (especially pregnant women) and other animals. Currently, 45 U.S. states have issued warnings telling people to avoid eating fish caught from waters that are contaminated with toxic mercury (see Chapter 17, Core Case Study, p. 442).

CONSIDER THIS. . .

THINKING ABOUT Acid Deposition and Mercury

Do you live in or near an area where government officials have warned people not to eat fish caught from waters contaminated with mercury? If so, what do you think are the specific sources of the mercury pollution and how could this pollution be cleaned up or prevented?

Animated Figure 18-14 Natural capital degradation: Air pollution is one of several interacting stresses that can damage, weaken, or kill trees and pollute surface and groundwater. The inset photo shows air pollution damage to trees at a high elevation in Mount Mitchell State Park, North Carolina (USA).

Acid deposition also harms aquatic ecosystems. Most fish cannot survive in water with a pH less than 4.5. In addition, as acid precipitation flows through soils, it can release aluminum ions (Al^{3+}) attached to minerals in the soils and carry them into lakes, streams, and wetlands. These ions lead to the suffocation of many kinds of fish by stimulating excessive mucus formation, which clogs their gills.

Because of excess acidity, several thousand lakes in Norway and Sweden, and 1,200 lakes in Ontario, Canada, contain few if any fish. In the United States, several hundred lakes (most in the Northeast) are similarly threatened. In addition, scientists are only just beginning to study the effects of the South Asian Brown Clouds on oceans (**Core Case Study**). Aerosols in the clouds are pulled into thunderstorms that then dump acid rain into the Indian and Pacific Oceans, possibly harming marine life and ecosystems.

Acid deposition (often along with other air pollutants such as ozone) can harm crops, especially when the soil pH is below 5.1. Low pH reduces plant productivity and the ability of soils to buffer or neutralize acidic inputs. An estimated 30% of China's cropland suffers from excess acidity.

This combination of acid deposition and other air pollutants can also affect forests in two ways. One is by leaching essential plant nutrients such as calcium and magnesium from forest soils. The other is by releasing ions of aluminum, lead, cadmium, and mercury, which are toxic to the trees (Figure 18-14), from forest soils. These two

effects rarely kill trees directly, but they can weaken them and leave them vulnerable to stresses such as severe cold, diseases, insect attacks, and drought.

Mountaintop forests are the terrestrial areas hit hardest by acid deposition (see inset photo in Figure 18-14). These areas tend to have thin soils without much buffering capacity and some such areas are bathed almost continuously in highly acidic fog and clouds. As a result, uncontrolled emissions of sulfur dioxide and other pollutants can devastate the vegetation in these sensitive areas.

Most of the world's forests and lakes are not being destroyed or seriously harmed by acid deposition. Rather, this regional problem is harming forests and lakes that lie downwind from coal-burning facilities and from large, motor vehicle–dominated cities without adequate pollution controls (**Concept 18-3**). Also, since 1994, acid deposition as measured by pH has decreased sharply in the United States and especially in the eastern half of the country, partly because of significant reductions in SO_2 and NO_x emissions from coal-fired power and industrial plants under the 1990 amendments to the U.S. Clean Air Act. **GOOD NEWS**

Despite this overall decrease in the United States, soils and surface waters in many areas of the country are still acidic because of the accumulation of acids over decades of acid deposition. Scientists estimate that an additional 80% reduction in SO_2 emissions from coal-burning power and

Prevention	Cleanup
Reduce coal use and burn only low-sulfur coal	Add lime to neutralize acidified lakes
Use natural gas and renewable energy resources in place of coal	Add phosphate fertilizer to neutralize acidified lakes
Remove SO₂ and NOₓ from smokestack gases and remove NOₓ from motor vehicular exhaust	Add lime to neutralize acidified soils
Tax SO₂ emissions	

© Cengage Learning

Figure 18-15 There are several ways to reduce acid deposition and its damage. **Questions:** Which two of these solutions do you think are the best ones? Why?

Top: Brittany Courville/Shutterstock.com. Bottom: Yegor Korzh/Shutterstock.com.

industrial plants in the Midwestern United States will have to occur before northeastern lakes, forests, and streams can recover from past and projected effects of acid deposition.

We Know How to Reduce Acid Deposition

Figure 18-15 summarizes ways to reduce acid deposition. According to most scientists studying the problem, the best solutions are *preventive approaches* that reduce or eliminate emissions of sulfur dioxide, nitrogen oxides, and particulates.

Although we know how to prevent acid deposition (Figure 18-15, left), implementing these solutions is politically difficult. One problem is that the people and ecosystems it affects often are quite far downwind from the sources of the problem. Also, countries with large supplies of coal (such as China, India, Russia, and the United States) have a strong incentive to use it. In addition, owners of coal-burning power plants resist adding the latest pollution control equipment to their facilities, using low-sulfur coal, and removing sulfur from coal before burning

it, arguing that these measures would increase the cost of electricity for consumers.

However, in the United States, the use of affordable and cleaner-burning natural gas (see Chapter 15, Core Case Study, p. 374) and wind (see Chapter 16, Core Case Study, p. 402) for generating electricity is on the rise, and this has reduced the use of coal to some extent. Environmental scientists point out that including the largely hidden, harmful health and environmental costs of burning coal in its market prices would further reduce coal use, spur the use of cleaner ways to generate electricity, and help prevent acid deposition.

As for technological fixes, large amounts of limestone or ground lime are used to neutralize some acidified lakes and surrounding soils—the only cleanup approach now being used. However, this expensive and temporary remedy usually must be repeated annually. Also, it can kill some types of plankton and aquatic plants, and can harm certain wetland plants that need acidic water. Also, it is difficult to know how much lime to put where, whether in the water or along the shore.

According to the U.S. National Emissions Inventory, between 1980 and 2010, air pollution laws in the United States helped to reduce sulfur dioxide emissions from all sources by 69% and nitrogen oxide emissions by 52%. This has helped to reduce the acidity of rainfall in parts of the Northeast, Mid-Atlantic, and Midwest regions. However, there is still room for more reductions of these and other harmful emissions from older coal-burning power and industrial plants. **GOOD NEWS**

China, the world's largest emitter of sulfur dioxide, has one of the world's most serious acid deposition problems. However, between 2006 and 2009, estimated SO₂ emissions in China fell by more than 13%, despite the construction of many new coal-burning power plants. This is encouraging, but China has a long way to go in curtailing acid deposition.

🔍 CONSIDER THIS...

CONNECTIONS Low-Sulfur Coal, Atmospheric Warming, and Toxic Mercury

Some U.S. power plants have lowered SO₂ emissions by switching from high-sulfur to low-sulfur coals. However, this has increased CO₂ emissions that contribute to atmospheric warming and projected climate change. This occurs because low-sulfur coal has a lower heat value, which means that more coal must be burned to generate a given amount of electricity. Low-sulfur coal also has higher levels of toxic mercury and other trace metals, so burning it emits more of these hazardous chemicals into the atmosphere.

18-4 What Are the Major Indoor Air Pollution Problems?

CONCEPT 18-4
The most threatening indoor air pollutants are smoke and soot from the burning of wood and coal in cooking fires (mostly in less-developed countries), cigarette smoke, and chemicals used in building materials and cleaning products.

Indoor Air Pollution Is a Serious Problem

In less-developed countries, the indoor burning of wood, charcoal, dung, crop residues, coal, and other fuels in open fires (Figure 18-16) or in unvented or poorly vented stoves exposes people to dangerous levels of particulate air pollution. Workers, including children, are also exposed to high

Figure 18-16 By burning wood to cook food inside this dwelling in Nepal, this woman is exposing herself and other occupants to dangerous levels of indoor air pollution.

levels of indoor air pollution in countries where there are few if any pollution laws or regulations. According to the WHO and the World Bank, *indoor air pollution is the world's most serious air pollution problem, especially for poor people.*

Indoor air pollution is also a serious problem in developed areas of all countries, mostly because of chemicals used in building materials and products. Figure 18-17 shows some typical sources of indoor air pollution in a modern home.

EPA studies have revealed some alarming facts about indoor air pollution. *First,* levels of 11 common pollutants generally are 2 to 5 times higher inside U.S. homes and commercial buildings than they are outdoors, and in some cases, they are as much as 100 times higher. *Second,* pollution levels inside cars in traffic-clogged urban areas can be up to 18 times higher than outside levels. *Third,* the health risks from exposure to such chemicals are magnified because most people in developed urban areas spend the majority of their time indoors or inside vehicles.

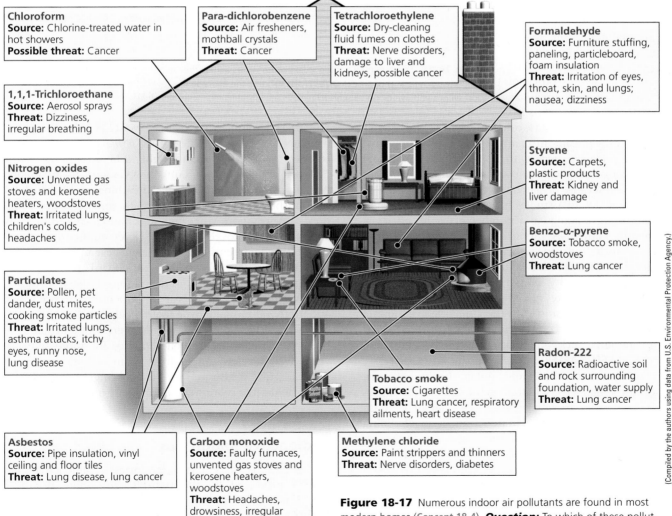

Chloroform
Source: Chlorine-treated water in hot showers
Possible threat: Cancer

1,1,1-Trichloroethane
Source: Aerosol sprays
Threat: Dizziness, irregular breathing

Nitrogen oxides
Source: Unvented gas stoves and kerosene heaters, woodstoves
Threat: Irritated lungs, children's colds, headaches

Particulates
Source: Pollen, pet dander, dust mites, cooking smoke particles
Threat: Irritated lungs, asthma attacks, itchy eyes, runny nose, lung disease

Asbestos
Source: Pipe insulation, vinyl ceiling and floor tiles
Threat: Lung disease, lung cancer

Para-dichlorobenzene
Source: Air fresheners, mothball crystals
Threat: Cancer

Tetrachloroethylene
Source: Dry-cleaning fluid fumes on clothes
Threat: Nerve disorders, damage to liver and kidneys, possible cancer

Formaldehyde
Source: Furniture stuffing, paneling, particleboard, foam insulation
Threat: Irritation of eyes, throat, skin, and lungs; nausea; dizziness

Styrene
Source: Carpets, plastic products
Threat: Kidney and liver damage

Benzo-α-pyrene
Source: Tobacco smoke, woodstoves
Threat: Lung cancer

Radon-222
Source: Radioactive soil and rock surrounding foundation, water supply
Threat: Lung cancer

Tobacco smoke
Source: Cigarettes
Threat: Lung cancer, respiratory ailments, heart disease

Carbon monoxide
Source: Faulty furnaces, unvented gas stoves and kerosene heaters, woodstoves
Threat: Headaches, drowsiness, irregular heartbeat, death

Methylene chloride
Source: Paint strippers and thinners
Threat: Nerve disorders, diabetes

(Compiled by the authors using data from U.S. Environmental Protection Agency.)

Figure 18-17 Numerous indoor air pollutants are found in most modern homes (Concept 18-4). **Question:** To which of these pollutants are you exposed?

Since 1990, the EPA has placed indoor air pollution at the top of its list of 18 sources of cancer risk. The EPA estimates that it causes as many as 6,000 premature cancer deaths per year in the United States. At greatest risk are smokers, children younger than age 5, the elderly, the sick, pregnant women, people with respiratory or heart problems, and factory workers.

Pesticide residues and lead particles brought indoors on shoes can collect in carpets and furnishings. According to the EPA, three of every four U.S. homes use pesticides indoors at least once a year. Many chemicals containing potentially harmful organic solvents are also found in paints and various sprays.

Living organisms and their excrements can also pollute indoor air. Evidence indicates that exposure to allergens such as *dust mites* and *cockroach droppings* found in some homes plays an important role in the growing number of people suffering from asthma in the United States. Another living source of indoor air pollution is toxic *airborne spores of fungal growths* such as *molds* and *mildew* that can cause headaches and allergic reactions, and aggravate asthma and other respiratory diseases. Some evidence suggests that spores from molds and mildew growing underneath houses and on interior walls are the single greatest cause of allergic reactions to indoor air.

Danish and U.S. EPA studies have linked various air pollutants found in buildings to a number of health effects, a phenomenon known as the *sick-building syndrome* (SBS). Such effects include dizziness, headaches, coughing, sneezing, shortness of breath, nausea, burning eyes, sore throats, chronic fatigue, irritability, skin dryness and irritation, respiratory infections, flu-like symptoms, and depression. EPA and Labor Department studies in the United States indicate that almost one in five commercial buildings in the United States is considered "sick" and is exposing employees to these health risks. **GREEN CAREER:** indoor air pollution specialist

According to the EPA and public health officials, the four most dangerous indoor air pollutants in more-developed countries are *tobacco smoke* (see Chapter 17, Case Study, p. 466); *formaldehyde* emitted from many building materials and various household products; *radioactive radon-222 gas*, which can seep into houses from underground rock deposits (see the following Case Study); and *very small (ultrafine) particles* of various substances in emissions from motor vehicles, coal-burning and industrial power plants, wood burning, and forest and grass fires.

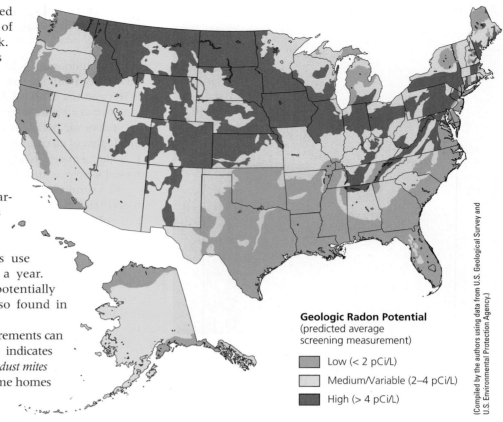

Geologic Radon Potential
(predicted average screening measurement)

- ▨ Low (< 2 pCi/L)
- ▢ Medium/Variable (2–4 pCi/L)
- ▩ High (> 4 pCi/L)

(Compiled by the authors using data from U.S. Geological Survey and U.S. Environmental Protection Agency.)

Figure 18-18 The potential for radon exposure varies across the United States, depending on the types of underlying soils and bedrock. **Question:** What is the average risk level of exposure to radioactive radon where you live or go to school?

The chemical that causes most people in more-developed countries difficulty is *formaldehyde* (CH_2O), a colorless, extremely irritating chemical. According to the EPA and the American Lung Association, 20–40 million Americans suffer from chronic breathing problems, dizziness, rashes, headaches, sore throats, sinus and eye irritation, skin irritation, wheezing, and nausea caused by daily exposure to low levels of formaldehyde emitted from common household items and materials. These include building materials (such as plywood, particleboard, paneling, and high-gloss wood used on floors and for cabinets), furniture, drapes, upholstery, adhesives in carpeting and wallpaper, and urethane-formaldehyde foam insulation. Partly because of this, the EPA estimates that 1 of every 5,000 people who live in manufactured (mobile) homes for more than 10 years will develop cancer from formaldehyde exposure. Formaldehyde is also present in some fingernail hardeners and wrinkle-free coatings on permanent-press clothing.

CASE STUDY

Radioactive Radon Gas

Radon-222 is a colorless, odorless, radioactive gas that is produced by the natural radioactive decay of uranium-238, small amounts of which are contained in most rocks and soils. But this isotope is much more con-

Outlet vents for furnace, dryer, and woodstove

Open window

Cracks in wall

Openings around pipes

Slab joints

Wood stove

Cracks in floor

Sump pump

Clothes dryer

Furnace

Slab

Radon-222 gas

Uranium-238

Soil

(Compiled by the authors using data from U.S. Environmental Protection Agency.)

Figure 18-19 There are a number of ways that radon-222 gas can enter homes and other buildings. *Question:* Has anyone tested the indoor air where you live for radon-222?

Scott Camazine/Alamy

Figure 18-20 CT scan from a 54-year-old man with lung cancer.

centrated in underground deposits of minerals such as uranium, phosphate, shale, and granite (which is why some people have had granite countertops removed from their homes). Figure 18-18 compares the potential geological risk of exposure to radioactive radon across the United States in terms of concentrations of radioactive radon in picocuries per liter (pCi/L).

When radon gas from such deposits seeps upward through the soil and is released outdoors, it disperses quickly in the air and decays to harmless levels. However, in buildings above such deposits, radon gas can enter through cracks in a foundation's slab and walls, as well as through well water, openings around sump pumps and drains, and hollow concrete blocks (Figure 18-19). Once inside, it can build up to high levels, especially in unventilated lower levels of homes and buildings.

Radon-222 gas quickly decays into solid particles of other radioactive elements such as polonium-210, which, if inhaled, expose lung tissue to large amounts of ionizing radiation from alpha particles (see Figure 2-9, top, p. 40). This exposure can damage lung tissue and lead to lung cancer (Figure 18-20) over the course of a 70-year lifetime. In 2012, lung cancer killed about 160,000 Americans. Your chances of getting lung cancer, the leading cancer killer in both men and women in the United States, from radon depend mostly on how much radon is in your home, how much time you spend in your home, and whether you are a smoker or have ever smoked. About 90% of radon-related lung cancers occur among current or former smokers.

According to the EPA, radioactive radon is the second-leading cause of lung cancer after smoking, and each year, radon-induced lung cancer kills an estimated 21,000

people in the United States. Since the late 1980s, about half a million Americans have died from radon-induced lung cancer. Despite this serious risk, less than 20% of U.S. households have followed the EPA's recommendation to conduct radon tests, which can be done with do-it-yourself testing kits that cost $20–$30. Many schools and day-care centers also have not tested for radon, and only a few states have laws that require radon testing for schools.

According to the EPA, prevention of radon contamination typically adds $350–$500 to the cost of a new home, and correcting a radon problem in an existing house runs $800–$2,500, with an average cost of $1,200. Remedies include sealing cracks in the foundation's slab and walls, increasing ventilation by cracking a window or installing vents in the basement, and using a fan to create cross ventilation.

CONSIDER THIS. . .

THINKING ABOUT Preventing Indoor Air Pollution

What are some steps you could take to prevent indoor air pollution, especially regarding the four most dangerous indoor air pollutants listed above?

CONSIDER THIS. . .

CAMPUS SUSTAINABILITY
Clean Air to Breathe at Coyote Village

One of the goals set by the University of South Dakota when its Coyote Village Residence Hall was built was to provide clean indoor air for the students who would live there. To that end, the dorm is built of environmentally friendly building materials and contains a state-of-the-art ventilation system with sensors to help detect

University of South Dakota, Vermillion

and control indoor air contaminants. The university also installed highly efficient heating and cooling systems, partly because for maintaining indoor air quality, it helps to minimize the operation of such systems. The building also includes a well-insulated building envelope and digital thermostats designed to optimize energy efficiency.

18-5 What Are the Health Effects of Air Pollution?

CONCEPT 18-5
Air pollution can contribute to asthma, chronic bronchitis, emphysema, lung cancer, heart attack, and stroke.

Your Body's Natural Defenses against Air Pollution Can Be Overwhelmed

Your respiratory system (Figure 18-21) helps to protect you from air pollution in various ways. Hairs in your nose filter out large particles. Sticky mucus in the lining of your upper respiratory tract captures smaller (but not the smallest) particles and dissolves some gaseous pollutants. Sneezing and coughing expel contaminated air and mucus when pollutants irritate your respiratory system.

In addition, hundreds of thousands of tiny, mucus-coated, hair-like structures, called *cilia*, line your upper respiratory tract. They continually move back and forth and transport mucus and the pollutants it traps to your throat where they are swallowed or expelled.

Prolonged or acute exposure to air pollutants, including tobacco smoke, can overload or break down these natural defenses. Fine and ultrafine particulates can get lodged deep in the lungs and contribute to lung cancer, asthma, heart attack, and stroke. According to the National Institutes of Health, almost 7% of the U.S. population suffers from asthma and an average of about 14 asthma sufferers die each day from asthma attacks. A French study found that asthma attacks increased by about 30% on smoggy days. Years of smoking or breathing polluted air can lead to other lung ailments such as chronic bronchitis and emphysema, which leads to acute shortness of breath and usually to death (see Figure 17-23, p. 467).

Air Pollution Is a Big Killer

According to a 2013 study by the WHO and researchers at the Health Effects Institute of Boston, Massachusetts (USA), at least 3.2 million people worldwide die prematurely each year—an average of nearly 365 deaths every hour—from the effects of outdoor air pollution. About two-thirds of these deaths occur in Asia, including about 1.2 million deaths per year in China and about 620,000 in India. According to a 2009 study by a team of Princeton University researchers led by Junfeng Liu, pollution in the South Asian Brown Clouds (**Core Case Study**) contributes to at least 380,000 premature deaths every year, mostly in China and India.

The EPA estimates that in the United States, the annual number of deaths related to indoor and outdoor air pollution ranges from 150,000 to 350,000 people—a death toll equivalent to that of two to five fully loaded, 200-passenger airliners crashing every day of the year with no survivors. Millions more suffer from asthma attacks and other respiratory disorders. According to a number of recent scientific findings, very small particulates, mostly from older coal-burning power plants, are responsible for an estimated 50,000 deaths a year in the United States

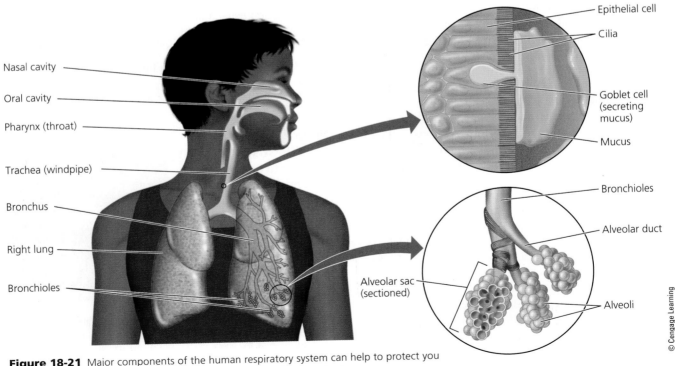

Figure 18-21 Major components of the human respiratory system can help to protect you from air pollution, but these defenses can be overwhelmed or breached.

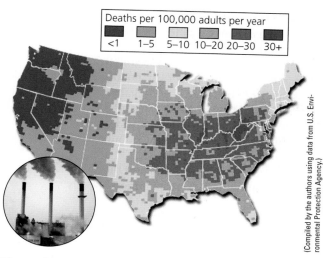

(Compiled by the authors using data from U.S. Environmental Protection Agency.)

Figure 18-22 Distribution of premature deaths from air pollution in the United States, mostly from very small, fine, and ultrafine particles added to the atmosphere by coal-burning power plants. **Questions:** Why do the highest death rates occur in the eastern half of the United States? If you live in the United States, what is the risk at your home or where you go to school?

StonePhotos/Photos.com

Figure 18-23 Cargo ships such as this one operating in Lake Erie near Cleveland, Ohio, are major sources of air pollution and are poorly regulated.

(Figure 18-22). A 2009 study by University of Ottawa scientists examined health data for 500,000 people living in soot-laden areas in 116 U.S. cities. The researchers found that inhalation of soot particles raised the chances of deadly heart attacks by 24% for this group of people.

In 2012, the WHO concluded that prolonged exposure to diesel fumes can cause lung cancer. According to recent EPA studies, each year, more than 125,000 Americans (96% of them living in urban areas) get cancer from breathing soot-laden diesel fumes emitted by buses and trucks. Other sources of these fumes include tractors, bulldozers and other construction equipment, trains, and large ships. According to the EPA, a large diesel truck emits as much particulate matter as 150 cars, and particulate emissions from a diesel train engine equal those from 1,500 cars.

According to a 2009 study by Daniel Lack and other NOAA scientists, the world's 100,000 or more diesel-powered cargo ships emit almost half as much particulate pollution as do the world's 760 million cars (Figure 18-23). This is largely because many cargo ships burn low-grade oil called *bunker fuel* in which the concentration of polluting sulfur is 30 times higher than that of diesel fuel sold at the pumps of U.S. gas stations.

An international team of 31 atmospheric scientists spent 4 years studying the effects of fine-particle soot emissions produced mostly by diesel engines, primitive cook stoves, kerosene lamps, the burning of coal, and forest fires. In 2013, they reported that these black carbon particles contribute almost 3 times as much to the atmospheric warming as had been estimated in earlier studies.

Together the ports of Los Angeles and nearby Long Beach, California, form the largest port in the western hemisphere. Between 2005 and 2011, these ports reduced their emissions of particulate matter from the burning of diesel fuel by 73%. This was done by using cleaner-burning cranes, other machinery, and trucks, and implementing rules requiring the use of cleaner, low-sulfur fuel.

GOOD NEWS

18-6 How Should We Deal with Air Pollution?

CONCEPT 18-6

Legal, economic, and technological tools can help us to clean up air pollution, but the best solution is to prevent it.

Laws and Regulations Can Reduce Outdoor Air Pollution

The United States provides an excellent example of how a regulatory approach can reduce air pollution (**Concept 18-6**). The U.S. Congress passed the Clean Air Acts in 1970, 1977, and 1990. With these laws, the federal government established air pollution regulations for key pollutants that are enforced by states and major cities.

Congress directed the EPA to establish air quality standards for six major outdoor pollutants—carbon monoxide (CO), nitrogen dioxide (NO_2), sulfur dioxide (SO_2), suspended particulate matter (SPM, smaller than PM-10), ozone (O_3), and lead (Pb). One limit, called a *primary standard,* is set to protect human health. Another limit, called a *secondary standard,* is intended to prevent environmental and property damage. Each standard specifies the maximum allowable level for a pollutant, averaged over a specific period.

The EPA has also established national emission standards for more than 188 *hazardous air pollutants (HAPs)* that may cause serious health and ecological effects. Most of these chemicals are chlorinated hydrocarbons, volatile organic compounds, or compounds of toxic metals. An important public source of information about hazardous air pollutants is the annual *Toxic Release Inventory* (TRI). The TRI law (passed in 1990 as part of the Pollution Prevention Act) requires more than 20,000 refineries, power plants, mines, chemical manufacturers, and factories to report their releases and waste management methods for 667 toxic chemicals. The TRI, which is available on the Internet, lists this information by community, allowing individuals and local groups to evaluate potential threats to their health. Since the first TRI report was released in 1988, reported emissions of toxic chemicals have dropped sharply.

According to a 2012 EPA report, the combined emissions of the six major pollutants decreased by about 63% between 1980 and 2010, even with significant increases during the same period in gross domestic product (up 128%), vehicle miles traveled (up 94%), energy consumption (up 26%), and population (up 37%). Emissions during this period dropped by 97% for lead (Pb), 83% for PM-10, 71% for CO, 69% for SO_2, 52% for NO_2, and 14% for ground-level ozone (O_3). In addition, VOC levels dropped by 63%.

However, the EPA has warned that projected atmospheric warming during this century is likely to increase the rates of the chemical reactions that produce photochemical smog. If this happens, there will be a significant increase in ground-level concentrations of ozone in many urban areas of the world, including the eastern half of the United States.

In 2010, the EPA announced a new rule designed to tackle regional air pollution problems, including acid deposition. The rule requires power plants in 28 states, most of them in the eastern United States where emissions are the highest, to reduce their emissions of SO_2 and nitrogen oxides (NO_x). However, in 2012, a federal appeals court struck down the EPA rule saying that the agency had exceeded its authority under the Clean Air Act.

In 2012, the government proposed a new rule that would require coal-burning power plants built after 2013 to keep their CO_2 emissions below a specified level. The new rule does not impose such standards on existing coal-burning power plants, which are responsible for nearly 40% of CO_2 emissions in the United States. To meet these standards, operators of new coal-burning power plants will have to install expensive carbon-capture technology, which could make the construction of new coal-fired plants less likely. On the other hand, new natural gas power plants could meet the new standards without the need for expensive pollution controls. The coal industry hopes to have these new standards thrown out by the courts.

In 2013, the EPA proposed stricter motor vehicle emission standards that would reduce emissions of VOCs and nitrogen oxides by 80% and particulate emissions by 70%. The EPA estimates that each year, these new standards would cut the death toll from outdoor air pollution by 2,000 and reduce the number of cases of respiratory ailments in children by 23,000. They would also lead to estimated savings of $7 in health-care costs for every $1 spent to implement the new standards. The resulting increase in the cost of a gallon of gasoline would be 1 cent. Oil companies oppose the new standards, saying they would cost too much and would hinder economic growth.

The reduction of outdoor air pollution in the United States since 1970 has been a remarkable success story, primarily because of two factors. *First*, U.S. citizens insisted that laws be passed and enforced to improve air quality. *Second*, the country was affluent enough to afford such controls and improvements.

GOOD NEWS

Environmental scientists applaud the success, but they call for strengthening U.S. air pollution laws by:

- Putting much greater emphasis on preventing air pollution. The power of prevention is clear (**Concept 18-6**). In the United States, the air pollutant with the largest drop in its atmospheric level was lead, which was banned as a component of gasoline and other products (see Case Study, p. 479).
- Sharply reducing emissions from older coal-burning power and industrial plants, cement plants, oil refineries, and waste incinerators. Approximately 20,000 older coal-burning power plants, industrial facilities, and oil refineries in the United States have not been required to meet the air pollution standards for new facilities under the Clean Air Acts.
- Continuing to improve fuel efficiency standards for motor vehicles.
- Regulating more strictly the emissions from motorcycles and two-cycle gasoline engines used in devices such as chainsaws, lawnmowers (Figure 18-24), generators, scooters, and snowmobiles. The EPA estimates that using a typical riding gas-powered lawn mower for 1 hour creates as much air pollution as driving 34 cars for an hour.
- Setting much stricter air pollution regulations for airports and oceangoing ships (Figure 18-23) in U.S. waters.
- Sharply reducing indoor air pollution.

Figure 18-24 The typical two-cycle engine that powers many lawnmowers emits many times more air pollutants per hour than a typical car with up-to-date pollution controls emits.

Executives of companies that would be affected by implementing stronger air pollution regulations claim that correcting these problems would cost too much and would hinder economic growth. Proponents of stronger regulations contend that history has shown that most industry cost estimates for implementing U.S. air pollution control standards have been much higher than the costs actually proved to be. In addition, implementing such standards has helped some companies and created jobs by stimulating these companies to develop new pollution control technologies.

CONSIDER THIS. . .

CONNECTIONS Scooters and Smog

Many of the motorized scooters so commonly found on most college campuses, especially those with two-cycle engines, produce more nitrogen oxides and hydrocarbons—pollutants that contribute to photochemical smog—per unit of distance driven, than the average car produces. Older scooters and poorly maintained scooters emit many times more of these pollutants than cars emit. Thus, even though they are more fuel-efficient than most cars, as a group, scooters are major contributors to urban air pollution.

We Can Use the Marketplace to Reduce Outdoor Air Pollution

One approach to reducing pollutant emissions has been to allow producers of air pollutants to buy and sell government air pollution allotments in the marketplace. For example, with the goal of reducing SO_2 emissions, the Clean Air Act of 1990 authorized an *emissions trading,* or *cap-and-trade program,* which enables the 110 most polluting coal-fired power plants in 21 states to buy and sell SO_2 pollution rights.

Under this system, each power plant is annually given a number of pollution credits, which allow it to emit a certain amount of SO_2. A utility that emits less than its allotted amount has a surplus of pollution credits. That utility can use its credits to offset SO_2 emissions at another of its plants, keep them for future plant expansions, or sell them to other utilities or to private citizens or groups. Between 1990 and 2010, this emissions trading program helped to reduce SO_2 emissions from power plants in the United States by 69%, at a cost of less than one-tenth of the cost projected by the industry. GOOD NEWS

Proponents of this approach say it is cheaper and more efficient than government regulation of air pollution. Critics of this approach contend that it allows utilities with older, dirtier power plants to buy their way out of their environmental responsibilities and to continue to pollute. In addition, without strict government oversight, this approach makes cheating possible, because cap-and-trade is based largely on self-reporting of emissions.

The ultimate success of any emissions trading approach depends on two factors: how low the initial cap is set and how often it is lowered in order to promote continuing innovation in air pollution prevention and control. Without these two elements, emissions trading programs mostly shift pollution problems from one area to another without achieving any overall improvement in air quality.

An emissions trading program is also being used to control NO_x emissions. However, environmental and health scientists strongly oppose the use of a cap-and-trade program for controlling emissions of toxic mercury from coal-burning power plants and industries (see Chapter 17, Core Case Study, p. 442). They warn that operators of coal-burning plants who choose to buy permits instead of sharply reducing their mercury emissions will create toxic hotspots with unacceptably high levels of mercury.

The jury is still out for emissions trading programs in general. In 2002, the EPA reported results from the country's oldest and largest emissions trading program in southern California, in effect since 1993. According to the EPA, this cap-and-trade program fell far short of projected emissions reductions. The same study also found accounting inaccuracies. This highlights the need for more careful monitoring of all cap-and-trade programs.

There Are Many Ways to Reduce Outdoor Air Pollution

Figure 18-25 summarizes ways to reduce emissions of sulfur oxides, nitrogen oxides, and particulate matter from stationary sources such as coal-burning power plants and industrial facilities.

Figure 18-26 lists ways to prevent and reduce emissions from motor vehicles, the primary factor in the formation of photochemical smog. However, the already poor air quality in urban areas of many less-developed countries is worsening as the number of motor vehicles

Solutions

Stationary Source Air Pollution

Prevention	Reduction or Dispersal
Burn low-sulfur coal or remove sulfur from coal	Disperse emissions (which increase downwind pollution) using tall smokestacks
Convert coal to a liquid or gaseous fuel	Remove pollutants from smokestack gases
Switch from coal to natural gas and renewables	Tax each unit of pollution produced

© Cengage Learning

Figure 18-25 There are several ways to prevent, reduce, or disperse emissions of sulfur oxides, nitrogen oxides, and particulate matter from stationary sources, especially coal-burning power plants and industrial facilities (**Concept 18-6**). ***Questions:*** Which two of these solutions do you think are the best ones? Why?

Top: Brittany Courville/Shutterstock.com. Bottom: Yegor Korzh/Shutterstock.com.

Solutions

Motor Vehicle Air Pollution

Prevention	Reduction
Walk or bike or use mass transit	Require emission control devices
Improve fuel efficiency	Inspect car exhaust systems twice a year
Get older, polluting cars off the road	Set strict emission standards

Figure 18-26 There are a number of ways to prevent or reduce emissions from motor vehicles (Concept 18-6). *Questions:* Which two of these solutions do you think are the best ones? Why?

Top: egd/Shutterstock.com. Bottom: Tyler Olson/Shutterstock.com.

Solutions

Indoor Air Pollution

Prevention	Reduction and Dilution
Ban indoor smoking	Use adjustable fresh air vents for work spaces
Set stricter formaldehyde emissions standards for carpet, furniture, and building materials	Circulate air more frequently
Prevent radon infiltration	Circulate a building's air through rooftop greenhouses
Use naturally based cleaning agents, paints, and other products	Use solar cookers and efficient, vented wood-burning stoves

Figure 18-27 There are several ways to prevent or reduce indoor air pollution (Concept 18-6). *Questions:* Which two of these solutions do you think are the best ones? Why?

Tribalium/Shutterstock.com

in these countries rises because of lack of air-pollution-control laws or weak enforcement of existing laws.

However, because of the Clean Air Acts, a new car today in the United States emits 75% less pollution than did a pre-1970 car. Over the next 10–20 years, technology will bring more gains through improved engine and emis-

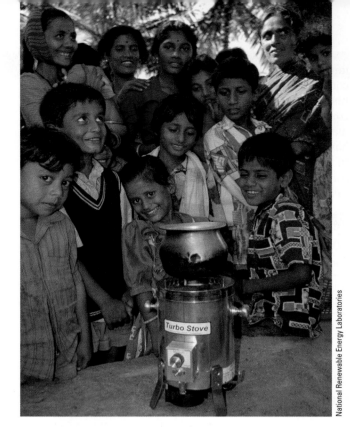

Figure 18-28 The energy-efficient Turbo Stove™ can greatly reduce indoor air pollution, thereby helping to cut the number of premature deaths in less-developed countries such as India.

sion systems, hybrid-electric vehicles (see Figure 16-6, left, p. 406), plug-in hybrid vehicles (see Figure 16-6, right, p. 406), and all-electric vehicles. **GOOD NEWS**

Reducing Indoor Air Pollution Should Be a Priority

Little effort has been devoted to reducing indoor air pollution even though it poses a much greater threat to human health than does outdoor air pollution. Air pollution experts suggest several ways to prevent or reduce indoor air pollution, as shown in Figure 18-27.

In less-developed countries, indoor air pollution from open fires and leaky, inefficient stoves that burn wood, charcoal, or coal could be reduced. More people could use inexpensive clay or metal stoves (Figure 18-28) that burn fuels (including straw and other crop wastes) more efficiently and vent their exhausts to the outside, or they could use stoves that use solar energy to cook food (see Figure 16-17, p. 417). This would also reduce deforestation by cutting the demand for fuelwood and charcoal. **GOOD NEWS**

Figure 18-29 lists some ways in which you can reduce your exposure to indoor air pollution.

We Can Emphasize Pollution Prevention

Environmental and health scientists are putting much more emphasis on *preventing* air pollution. With this approach, the question is not *"What can we do about the air*

What Can You Do?

Indoor Air Pollution

- Test for radon and formaldehyde inside your home and take corrective measures as needed

- Do not buy furniture and other products containing formaldehyde

- Test your home or workplace for asbestos fiber levels and check for any crumbling asbestos materials

- If you smoke, do it outside or in a closed room vented to the outside

- Make sure that wood-burning stoves, fireplaces, and kerosene- and gas-burning heaters are properly installed, vented, and maintained

- Install carbon monoxide detectors in all sleeping areas

- Use fans to circulate indoor air

- Grow house plants, the more, the better

- Do not store gasoline, solvents, or other volatile hazardous chemicals inside a home or attached garage

- Remove your shoes before entering your house to reduce inputs of dust, lead, and pesticides

Figure 18-29 Individuals matter: You can reduce your exposure to indoor air pollution. **Questions:** Which three of these actions do you think are the most important ones to take? Why?

pollutants we produce?" but rather *"How can we avoid producing these pollutants in the first place?"*

Like the shift to *controlling outdoor air pollution* between 1970 and 2012, this new shift to *preventing outdoor and indoor air pollution* will not take place unless individual citizens and groups put political pressure on elected officials to enact the appropriate regulations. In addition, citizens can, through their purchases, put economic pressure on companies to get them to manufacture and sell only products and services that do not add to air pollution problems.

18-7 How Have We Depleted Ozone in the Stratosphere and What Can We Do about It?

CONCEPT 18-7A

Our widespread use of certain chemicals has reduced ozone levels in the stratosphere and allowed more harmful ultraviolet radiation to reach the earth's surface.

CONCEPT 18-7B

To reverse ozone depletion, we must stop producing ozone-depleting chemicals and adhere to the international treaties that ban such chemicals.

Our Use of Certain Chemicals Threatens the Ozone Layer

A layer of ozone in the lower stratosphere keeps about 95% of the sun's harmful ultraviolet (UV-A and UV-B) radiation from reaching the earth's surface and harming us and a number of other species (Figure 18-2). However, measurements taken over several decades by researchers show a considerable seasonal depletion (thinning) of ozone concentrations in the stratosphere above Antarctica and the Arctic over the past 40 years. Similar measurements reveal a lower overall ozone thinning everywhere except over the tropics.

In 1984, researchers analyzing satellite data unexpectedly discovered that each year, 40–50% of the ozone in the upper stratosphere over Antarctica (100% in some places) was disappearing during September and October. This annual observed loss of ozone has been called an *ozone hole*. A more accurate term is *ozone thinning*. Colorized satellite data show a sharp drop in total ozone above Antarctica every September that got more pronounced every year from 1980 to 2000 (Figure 18-30, left and center). This trend had generally leveled off by 2011 (Figure 18-30, right).

When the seasonal thinning ends each year, huge masses of ozone-depleted air above Antarctica flow northward, and these masses linger for a few weeks over parts of Australia, New Zealand, South America, and South Africa. This has raised biologically damaging UV-B levels in these areas by 3–10%, and in some years, by as much as 20%. Based on scientific measurements and on mathematical and chemical models, the overwhelming consensus of researchers in this field is that ozone depletion in the stratosphere poses a serious threat to humans, other animals, and some primary producers (mostly plants) that use sunlight to support the earth's food webs (Concept 18-7A).

In 1988, scientists discovered that similar but usually less severe ozone thinning occurs over the Arctic from February to June, resulting in a typical ozone loss of 11–38%

| 1980 | 2000 | 2011 |

Photos: NASA

Figure 18-30 **Natural capital degradation:** These colorized satellite images show ozone thinning over Antarctica during September of 1980 (left), 2000 (center), and 2011 (right). Ozone depletion of 50% or more occurred in the dark blue and purple areas.

(compared to a typical 50% loss above Antarctica). When this body of air above the Arctic breaks up each year, large masses of ozone-depleted air flow south to linger over parts of Europe, North America, and Asia. However, models indicate that the Arctic is unlikely to develop the large-scale ozone thinning found over the Antarctic.

The origin of this serious environmental threat began with the discovery of the first chlorofluorocarbon (CFC) in 1930. Chemists soon developed similar compounds to create a family of highly useful CFCs, known by their trade name Freon. These chemically unreactive, odorless, nonflammable, nontoxic, and noncorrosive compounds seemed to be dream chemicals. Inexpensive to make, they became popular as coolants in air conditioners and refrigerators, propellants in aerosol spray cans, cleaners for electronic parts such as computer chips, fumigants for granaries and ships' cargo holds, and gases used to make foam insulation and packaging.

It turned out that CFCs were too good to be true. Starting in 1974 with the work of chemists Sherwood Rowland and Mario Molina (Individuals Matter 18.1), scientists demonstrated that CFCs are persistent chemicals that destroy protective ozone in the stratosphere. Satellite data and other measurements and models indicate that 75–85% of the observed ozone losses in the stratosphere since 1976 resulted from people releasing CFCs and other ozone-depleting chemicals into the atmosphere, beginning in the 1950s. These long-lived chemicals entered the troposphere and eventually reached the stratosphere where they began destroying ozone faster than it was being formed by the interaction of stratospheric oxygen and incoming solar radiation.

CFCs are not the only ozone-depleting chemicals. Others are *halons* and *hydrobromoflurocarbons* (HBFCs) (used in fire extinguishers), *methyl bromide* (a widely used fumi-

gant), *hydrogen chloride* (emitted into the stratosphere by the launches of certain space vehicles), and cleaning solvents such as *carbon tetrachloride, methyl chloroform, n-propyl bromide*, and *hexachlorobutadiene*.

Why Should We Worry about Ozone Depletion?

Why is ozone depletion something that should concern us? Figure 18-31 lists some of the demonstrated effects of decreased ozone levels in the stratosphere. One effect is that more biologically damaging UV-A and UV-B radiation reaches the earth's surface (Concept 18-7A) and is a likely contributor to rising numbers of eye cataracts, damaging sunburns, and skin cancers. Figure 18-32 lists ways in which you can protect yourself from harmful UV radiation.

Another serious threat from ozone depletion that some scientists have pointed out is that the resulting increase in UV radiation reaching the planet's surface could impair or destroy phytoplankton, especially in Antarctic waters (see Figure 3-12, p. 60). These tiny marine plants play a key role in removing CO_2 from the atmosphere. They also form the base of ocean food webs. By destroying them, we would be eliminating the vital ecosystem services they provide.

We Can Reverse Stratospheric Ozone Depletion

According to researchers in this field, we should immediately stop producing all ozone-depleting chemicals (Concept 18-7B). However, models indicate that even with immediate and sustained action, it will take about 60 years for the earth's ozone layer to recover the levels of

Sherwood Rowland and Mario Molina—A Scientific Story of Expertise, Courage, and Persistence

University of California

DONNA COVE, MIT/UNIVERSITY OF CALIFORNIA, SAN DIEGO

In 1974, calculations by the late Sherwood Rowland (upper photo) and Mario Molina (lower photo), chemists at the University of California–Irvine, indicated that chlorofluoro-carbons (CFCs) were lowering the average concentration of ozone in the stratosphere. These scientists decided that they had an ethical obligation to go public with the results of their research. They shocked both the scientific community and the $28-billion-per-year CFC industry by calling for an immediate ban of CFCs in spray cans, for which substitutes were available.

The research of these two scientists led them to four major conclusions. *First,* once CFCs are injected into the atmosphere, these persistent chemicals remain there.

Second, over 11–20 years, these compounds rise into the stratosphere through convection, random drift, and the turbulent mixing of air in the lower atmosphere.

Third, once they reach the stratosphere, the CFC molecules break down under the influence of high-energy UV radiation. This releases highly reactive chlorine atoms (Cl), as well as atoms of fluorine (F) and bromine (Br), all of which accelerate the breakdown of ozone (O_3) into O_2 and O in a cyclic chain of chemical reactions. As a consequence, ozone is destroyed faster than it forms in some parts of the stratosphere.

Fourth, each CFC molecule can last in the stratosphere for 65–385 years, depending on its type. During that time, each chlorine atom released during the breakdown of CFC can convert hundreds of O_3 molecules to O_2.

The CFC industry (led by DuPont) was a powerful, well-funded adversary with a lot of profits and jobs at stake. It attacked Rowland's and Molina's calculations and conclusions, but the two researchers held their ground, expanded their research, and explained their results to other scientists, elected officials, and the media. After 14 years of delaying tactics, DuPont officials acknowledged in 1988 that CFCs were depleting the ozone layer, and they agreed to stop producing them and to sell higher-priced alternatives that their chemists had developed.

In 1995, Rowland and Molina received the Nobel Prize in chemistry for their work on CFCs.

Background photo: NASA

Natural Capital

Effects of Ozone Depletion

Human Health and Structures

- Worse sunburns

- More eye cataracts and skin cancers

- Immune system suppression

Food and Forests

- Reduced yields for some crops

- Reduced seafood supplies due to smaller phytoplankton populations

- Decreased forest productivity for UV-sensitive tree species

Wildlife

- More eye cataracts in some species

- Shrinking populations of aquatic species sensitive to UV radiation

- Disruption of aquatic food webs due to shrinking phytoplankton populations

Air Pollution and Climate Change

- Increased acid deposition

- Increased photochemical smog

- Degradation of outdoor painted surfaces, plastics, and building materials

- While in troposphere, CFCs act as greenhouse gases

© Cengage Learning

Figure 18-31 Decreased levels of ozone in the stratosphere can have a number of harmful effects. (**Concept 18-7A**). **Questions:** Which three of these effects do you think are the most threatening? Why?

---

ozone it had in 1980, and it could take about 100 years for it to recover to pre-1950 levels.

In 1987, representatives of 36 nations met in Montreal, Canada, and developed the *Montreal Protocol*. This treaty's goal was to cut emissions of CFCs (but not other ozone-depleting chemicals) by about 35% between 1989 and 2000. After hearing more bad news about seasonal ozone thinning above Antarctica in 1989, representatives of 93 countries had more meetings and in 1992 adopted the *Copenhagen Amendment*, which accelerated the phase-out of key ozone-depleting chemicals.

What Can You Do?

Reducing Exposure to UV Radiation

- Stay out of the sun, especially between 10 A.M. and 3 P.M.

- Do not use tanning parlors or sunlamps

- When in the sun, wear clothing and sunglasses that protect against UV-A and UV-B radiation

- Be aware that overcast skies do not protect you

- Do not expose yourself to the sun if you are taking antibiotics or birth control pills

- When in the sun, use a sunscreen with a protection factor of at least 15

© Cengage Learning

Figure 18-32 Individuals matter: You can reduce your exposure to harmful UV radiation. **Question:** Which of these precautions do you already take?

These agreements set an important precedent because nations and companies worked together, using a *prevention approach* to try to solve a serious environmental problem. This approach worked for three reasons. *First*, there was convincing and dramatic scientific evidence of a serious problem. *Second*, CFCs were produced by a small number of international companies and this meant there was less corporate resistance to finding a solution. *Third*, the certainty that CFC sales would decline over a period of years because of government bans unleashed the economic and creative resources of the private sector to find even more profitable substitute chemicals.

Substitutes are available for most uses of CFCs, and others are being developed. However, the most widely used substitutes are hydrofluorocarbons (HFCs), which act as greenhouse gases. An HFC molecule can be up to 10,000 times more potent in warming the atmosphere than a molecule of CO_2 is. Currently, HFCs account for only 2% of all greenhouse gas emissions, but the Intergovernmental Panel on Climate Change (IPCC) has warned that global use of HFCs is growing rapidly. In 2009, the U.S. government asked the UN to enact mandatory reductions in HFC emissions through the Montreal Protocol.

In addition, there is a growing consensus among scientists that the Montreal Protocol should also be used to regulate the greenhouse gas nitrous oxide (N_2O), which is released from fertilizers and livestock manure. It remains in the troposphere for about 100 years and then migrates to the stratosphere where it can destroy ozone.

Progress in reducing ozone depletion could be set back by projected climate change. Scientists from Johns Hopkins University reported in 2009 on their discovery that warming of the troposphere makes the stratosphere cooler, which slows down its rate of ozone repair. Thus, as the atmosphere warms and the global climate changes, ozone levels may take much longer to return to where they were before ozone thinning began, or they may never recover completely. In 2012, Harvard University chemist James G. Anderson and his colleagues suggested that updrafts from large and more intense thunderstorms resulting from projected climate change could inject water vapor into the stratosphere (which is normally drier than a desert) and set off chemical reactions that could accelerate ozone depletion.

The landmark international agreements on dealing with stratospheric ozone depletion, now signed by all 196 of the world's countries, are important examples of global cooperation in response to a serious environmental problem. If all nations continue to follow these agreements while acting to slow the rate of climate change (discussed in Chapter 19), ozone levels could return to 1980 levels by 2068 (18 years later than originally projected) and to 1950 levels by 2108 (**Concept 18-7B**).

Big Ideas

- Outdoor air pollution, in the form of industrial smog, photochemical smog, and acid deposition, and indoor air pollution are serious global problems.

- Each year, at least 2.4 million people die prematurely from the effects of air pollution; indoor air pollution, primarily in less-developed countries, causes about two-thirds of these deaths.

- We need to give top priority status to the prevention of outdoor and indoor air pollution throughout the world and the reduction of stratospheric ozone depletion.

TYING IT ALL TOGETHER

The South Asian Brown Clouds and Sustainability

El Greco/Shutterstock.com

The South Asian Brown Clouds (**Core Case Study**) are a striking example of how bad air pollution can get. They result from large and dense populations of people relying mostly on wood, animal dung, and fossil fuels for their energy. It is an example of what can happen when people violate all three scientific **principles of sustainability** (see Figure 1-2, p. 6 or back cover) on a massive scale.

However, we can use these three principles to help reduce air pollution. *First,* we can reduce inputs of conventional air pollutants into the atmosphere by relying more on direct and indirect forms of solar energy than on fossils fuels. *Second,* we can greatly reduce the use and waste of matter and energy resources by recycling and reusing much more of what we use. *Third,* we can mimic nature's biodiversity by using a diversity of nonpolluting or low-polluting renewable energy resources.

A key strategy for solving air pollution problems is to focus on preventing or sharply reducing outdoor and indoor air pollution at global, regional, national, local, and individual levels. In addition, we must continue to strengthen international efforts to sharply reduce our use of chemicals that can deplete life-sustaining ozone in the stratosphere. Each of us has an important role to play in protecting the atmosphere that sustains life and supports our economies.

Chapter Review

Core Case Study

1. Describe the nature, origins, and harmful effects of the massive South Asian Brown Clouds (**Core Case Study**).

Section 18-1

2. What is the key concept for this section? Define density, as it relates to the atmosphere, and atmospheric pressure and explain why both are two important atmospheric variables. Define **troposphere**, **stratosphere**, and **ozone layer**. What are the major differences between the troposphere and stratosphere?

Section 18-2

3. What is the key concept for this section? What is **air pollution**? Distinguish between **primary pollutants** and **secondary pollutants** and give an example of each. List the major outdoor air pollutants and their harmful effects. What is the role of the South Asian Brown Clouds (**Core Case Study**) in atmospheric warming, according to some scientists? Explain the connections between South Asian Brown Clouds, food production, and solar power. Describe the effects of lead as a pollutant and how we can reduce our exposure to this harmful chemical. Give examples of a chemical method and a biological method for detecting air pollutants.

4. Distinguish between **industrial smog** and **photochemical smog** in terms of their chemical composition and formation. List and briefly describe five natural factors that help to reduce outdoor air pollution and six natural factors that help to worsen it. What is a **temperature inversion** and how can it affect outdoor air pollution levels?

Section 18-3

5. What is the key concept for this section? What is **acid deposition** and how does it form? What are its major environmental impacts on lakes, forests, human-built structures, and human health? List three major ways to reduce acid deposition. Explain the connections among low-sulfur coal, atmospheric warming, and toxic mercury.

Section 18-4

6. What is the key concept for this section? What is the major indoor air pollutant in many less-developed countries? What are the top three indoor air pollutants in the United States? Give three reasons why they present a serious threat to human health. Explain why radon-222 is an indoor air pollution threat, how and where it occurs, and what can be done about it.

Section 18-5

7. What is the key concept for this section? Describe the human body's defenses against air pollution, how they can be overwhelmed, and the illnesses that can result. Approximately how many people die prematurely from air pollution each year in the world and in the United States? What percentage of these deaths are caused by indoor air pollution? Describe the health threat from diesel fumes.

Section 18-6

8. What is the key concept for this section? Summarize the U.S. air pollution laws and how they have worked to reduce pollution. Explain how these laws can be strengthened, according to some scientists.

9. List the advantages and disadvantages of using an emissions trading program to control pollution. Summarize the major ways to reduce emissions from power plants and motor vehicles. What are four ways to reduce indoor air pollution? Why is preventing air pollution more important than controlling it?

Section 18-7

10. What are the two key concepts for this section? How have human activities depleted ozone in the stratosphere? List five harmful effects of such depletion. Explain how Sherwood Rowland and Mario Molina alerted the world to this threat. What has the world done to reduce the threat of ozone depletion in the stratosphere? How might projected climate change undermine this progress? What are the *three big ideas* for this chapter? Discuss the relationship between the South Asian Brown Clouds (**Core Case Study**) and the ways in which people have violated the three scientific **principles of sustainability**. Explain how we can apply these principles to the problems of air pollution.

Note: Key terms are in bold type.

Critical Thinking

1. The South Asian Brown Clouds (**Core Case Study**) form in southern and eastern Asia, and some people argue that they are the Asians' problem to solve. Others argue that their causes and effects are global in nature and must be dealt with through a global effort. What is your position on this? Explain.

2. China relies on coal for two-thirds of its commercial energy usage, partly because the country has abundant supplies of this resource. Yet China's coal burning has caused innumerable and growing problems for the country and for its neighboring nations. Also, because of the South Asian Brown Clouds (**Core Case Study**), the Pacific Ocean and the West Coast of North America are experiencing air pollution resulting from China's use of coal. Do you think China is justified in developing this resource to the maximum, as other countries— including the United States—have done with their coal resources? Explain. What are China's alternatives?

3. Considering your use of motor vehicles, now and in the future, what are three ways in which you could reduce your contribution to photochemical smog?

4. Should tall smokestacks (see chapter-opening photo) be banned to help prevent downwind air pollution and acid deposition? Explain.

5. Explain how sulfur impurities in coal can lead to increased acidity in rainwater and to the subsequent depletion of soil nutrients. Write an argument for or against requiring the use of low-sulfur coal in all coal-burning facilities.

6. If you live in the United States, list three important ways in which your life would be different if citizen-led actions during the 1970s and 80s had not led to the Clean Air Acts of 1970, 1977, and 1990, despite strong political opposition by the affected industries. List three important ways in which your life in the future might be different if such actions do not lead now to the strengthening of the U.S. Clean Air Act. If you do not live in the United States, research the air pollution laws in your country and explain if and how they could be strengthened.

7. List three ways in which you could apply **Concept 18-6** to making your lifestyle more environmentally sustainable.

8. Congratulations! You are in charge of the world. Explain your strategy for dealing with each of the following problems: **(a)** indoor air pollution, **(b)** outdoor air pollution, **(c)** acid deposition, and **(d)** ozone depletion.

Doing Environmental Science

Find out whether or not the buildings at your school have been tested for radon. If so, what were the results? What has been done about any areas with unacceptable levels of radon? If this testing has not been done, talk with school officials about having it done. You could also test for radon in your room or apartment or in the main living area of the house or building where you live. (Radon testing kits are available at affordable prices in most hardware stores, drug stores, and home centers.)

Global Environment Watch Exercise

Go to the *Ozone Depletion* portal and use the information to determine the latest research on the effectiveness of international efforts to reduce the use of CFCs. Research the use of substitutes such as HFC-22 and report on any potential pitfalls related to using such substitutes. What, if anything, is being done to address these problems?

Data Analysis

The U.S. Clean Air Act limits sulfur emissions from large coal-fired boilers to 0.54 kilograms (1.2 pounds) of sulfur per million Btus (British thermal units) of heat generated. (1 metric ton = 1,000 kilograms = 2,200 pounds = 1.1 ton; 1 kilogram = 2.20 pounds.)

1. Given that coal used by power plants has a heating value of 27.5 million Btus per metric ton (25 million Btus per ton), determine the number of kilograms (and pounds) of coal needed to produce 1 million Btus of heat.

2. About 10,000 Btus of heat input are required for an electric utility to produce 1 kilowatt-hour (kwh) of electrical energy. How many metric tons (and how many tons) of coal must be supplied each hour to provide the input heat requirements for a 1,000-megawatt (1-million-kilowatt) power plant?

3. Assuming that this power plant uses coal with 1.00% sulfur and operates at full capacity 24 hours per day, how many metric tons (and how many tons) of sulfur will be released into the atmosphere each year?

Questions 1–4 refer to the diagram of the earth's atmosphere below.

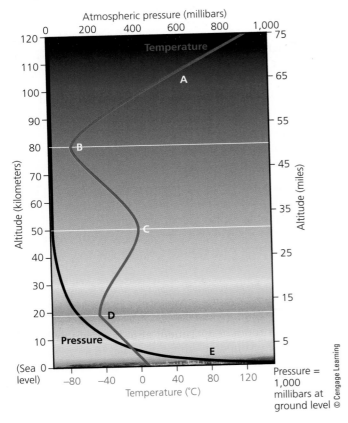

1. The ozone layer is found at this location.

2. The atmosphere is at its warmest point at this location.

3. The section of the atmosphere responsible for our daily weather.

4. The section of the atmosphere with 75–80% of the mass of the atmosphere.

5. A large cloud of dust and pollution that originates over Southeast Asia is
   a. now a permanent fixture in the atmosphere.
   b. often a part of pollution over California.
   c. being ignored by the governments of India and China.
   d. cooling the lower atmosphere by reflecting sunlight.
   e. intermittent and only found during the summertime.

6. The source of the ozone in the ozone layer is
   a. oxygen molecules gas interacting with UV light to produce ozone.
   b. ozone molecules formed at the earth's surface drifting upwards.
   c. ozone formed by plants during photosynthesis moving upwards.
   d. oxygen molecules forming the more stable compound of ozone through catalysis.
   e. carbon dioxide molecules breaking down in the upper atmosphere when struck by UV light.

Questions 7 and 8 refer to the air pollutants below.
   a. SO_2
   b. O_3
   c. CO
   d. hydrocarbons
   e. suspended particles

7. A secondary pollutant and component of photochemical smog.

8. Can be solid or liquid in form and can often aggravate asthma or bronchitis.

9. Which of the following is NOT an area in which U.S. air pollution laws are deficient and could be addressed to improve air pollution in the United States?
   a. Airports are exempt from most air pollution regulations.
   b. The United States relies on pollution cleanup rather than prevention.
   c. Ultrafine particles are not regulated.
   d. The release of hazardous air pollutants (HAPs) has increased over the last century.
   e. Little to no regulation of air pollution from ocean-going ships in our ports.

10. All of the following methods are designed to prevent indoor air pollution EXCEPT
   a. preventing pesticide residues and lead particles accumulating on carpets and furniture by taking off shoes upon entering a house.
   b. banning smoking indoors.
   c. heating the house with wood stoves instead of conventional oil burners.
   d. having the house checked for radon-222.
   e. avoiding the use of mothballs, air fresheners, or aerosol sprays.

11. What do all photochemical reactions have in common?
    a. Release photons of light called fluorescence as byproducts
    b. Absorb light and release heat in the process
    c. They are activated by light
    d. Utilize red light to boost electrons to higher orbitals
    e. Couple visible light with heat to drive reactions

12. The air pollutant lead causes which of the following?
    a. Damage to the nervous system, especially in children
    b. Respiratory problems and heart disease, especially in urban areas
    c. Respiratory damage, and when found indoors is mostly the result of cigarette smoke and burning wood
    d. Interference with the ability of blood to transport oxygen, resulting in possible heart attacks
    e. Damage to aquatic systems, forests, and some crops, as well as statues and buildings

13. Aquatic ecosystems are harmed from air pollution primarily through
    a. deposition of particulates that block the sunlight and prevent photosynthesis.
    b. radon gas dissolving in the water, which can cause explosions.
    c. mercury leaches into lakes from gold mines and bonds with water molecules.
    d. photochemical reactions of water and sunlight that occur when ozone is released into the atmosphere.
    e. deposition of acid rain, which releases aluminum ions, causing many kinds of fishes to die of clogged gills.

19 Climate Disruption

Civilization has evolved during a period of remarkable climate stability, but this era is drawing to a close. We are entering a new era, a period of rapid and often-unpredictable climate change.

LESTER R. BROWN

Key Questions

19-1 How is the earth's climate changing?

19-2 Why is the earth's climate changing?

19-3 What are the possible effects of a warmer atmosphere?

19-4 What can we do to slow projected climate disruption?

19-5 How can we adapt to climate change?

Areas that could be flooded by the end of this century (shown in red) by a 1-meter (3-foot) rise in sea level due to projected climate change.

NASA

Greenland is a largely ice-covered island in the Arctic Circle. It is the world's largest island, nearly the size of Mexico, and has a population of just under 58,000 people.

Greenland's ice, which formed during the last ice age, lies in glaciers that are as deep as 3.2 kilometers (2 miles) and cover about 80% of this mountainous island. Scientists are studying this ice closely because it appears to be melting at an accelerating rate during Greenland's summers (Figure 19-1). Some of this ice is replaced by snow during winter months, but the annual net loss of Greenland's ice has increased during recent years.

A number of climate scientists hypothesize that a key factor in this melting is *atmospheric warming*—the gradual rise of the average temperature of the atmosphere near the earth's surface. This warming is projected to increase significantly during this century. It could lead to rapid *climate disruption,* which is likely to have harmful effects on most of the earth's ecosystems, people, and economies.

Greenland's glaciers contain about 10% of the world's freshwater—enough water to raise the global sea level by as much as 7 meters (23 feet) if all of it were to melt and drain into the sea. It is highly unlikely that this will happen, but even a moderate loss of this ice over one or several centuries would raise sea levels considerably (see chapter-opening photo).

A glacier is a huge, moving ice mass that scrapes along extremely slowly, but scientists have demonstrated that it can pick up speed when meltwater on its surface, flowing down through its crevices, lubricates its bottom layer, which sits on bedrock. Some scientists hypothesize this is what is happening to some of northern Greenland's glaciers. In 2009, scientists found that some of Greenland's glaciers are also getting thinner at their ocean outlets. Such thinning can allow a glacier to float free of the ground at the shore, which has the effect of taking the brakes off and speeding the glacier's flow toward the sea. In some areas, the rate of Greenland's ice moving toward the sea has doubled since 1982, according to the U.S. National Aeronautics and Space Administration (NASA).

The Intergovernmental Panel on Climate Change (IPCC) is an international body of more than 2,500 scientists that was created in 1988 to study climate change. After reviewing tens of thousands of research studies, the IPCC and most of the world's major scientific bodies, including the National Academy of Sciences and the British Royal Society, have come to three major conclusions: **(1)** climate change is real and is happening now, **(2)** it is caused primarily by human activities such as the burning of fossil fuels and the clearing of forests, and **(3)** it is projected to accelerate during this century.

In its 2007 report, the IPCC made several projections. One of them was that the average global sea level is likely to rise by 18–59 centimeters (7–23 inches) by the end of this century. Since then, however, several scientists studying the melting of Greenland's ice have concluded that the rise in the average global sea level by 2100 will likely be closer to 100 centimeters (39 inches) and perhaps twice that. They estimate that Greenland's ice loss will make up half to two-thirds of that rise.

This change in projections about the rising average global sea level and Greenland's contribution to it is one example of how scientists are revising some IPCC projections upward due to the acceleration of rising temperatures at the earth's poles. It is an excellent example of how science advances its knowledge by taking into account new data and hypotheses.

Most climate scientists and researchers working in related disciplines consider the likely prospect of serious climate disruption during this century to be the single most serious environmental problem that humanity faces. Throughout this chapter, we examine the findings of climate scientists along with their projections related to atmospheric warming and climate disruption and their likely causes and effects. We also look at some possible ways to deal with this serious environmental, economic, and political challenge.

1982

2012

Figure 19-1 The total area of Greenland's glacial ice that melted by July 8, 2012 (red area in right image) was much greater than the amount that melted during the summer of 1982 (left).

(Compiled by the authors using data from Konrad Steffen and Russell Huff, University of Colorado, Boulder.)

19-1 How Is the Earth's Climate Changing?

CONCEPT 19-1
Considerable scientific evidence indicates that the earth's atmosphere is warming at a rapid rate that is likely to lead to significant climate disruption during this century.

Weather and Climate Are Not the Same

In thinking about climate change, it is very important to distinguish between weather and climate. *Weather* consists of short-term changes in atmospheric variables such as the temperature, precipitation, wind, and barometric pressure in a given area over a period of hours or days (see Supplement 5, p. S19).

By contrast, *climate* is determined by the *average* weather conditions of the earth or of a particular area, especially temperature and precipitation, over periods of at least three decades to thousands of years. According to the World Meteorological Organization, the minimum period considered for purposes of measuring climate is three decades. Scientists have used such long-term measurements to divide the earth into various climate zones (see Figure 7-2, p. 145).

During any period of 30 or more years, in a given area of the planet, there will often be hotter years and cooler years, and wetter years and drier years, as weather often fluctuates widely from day to day and from year to year. However, such short-term changes in weather do not necessarily tell us that the earth's climate or the climate of a particular area is warming or cooling. Climate scientists look at data on the normally fluctuating weather conditions for a particular area of the earth to see if there has been a general rise or fall in any measurements such as average temperature or precipitation over a period of at least 30 years. This is the only way they that they can determine whether and how the climate of an area has changed.

Climate Change Is Not New but Has Accelerated

Climate change is neither new nor unusual. Over the past 3.5 billion years, the planet's climate has been altered many times by several factors, including volcanic emissions, changes in solar input, the shifting of tectonic plates (Figure 4-9, p. 86), and impacts by meteors and asteroids.

The earth's climate is also affected by global air circulation patterns (see Figure 7-3, p. 146), large areas of ice (**Core Case Study**) that reflect incoming solar energy and help cool the atmosphere, varying concentrations of the different gases that make up the atmosphere, and occasional slight changes in ocean currents (see Figure 7-5, p. 147). We look at these factors in this chapter. Average temperature and precipitation are the two main factors affecting global, regional, and local climates, but temperature is the key variable that climate specialists watch.

Over the past 900,000 years, the atmosphere has experienced prolonged periods of global cooling and global warming (Figure 19-2). These alternating cycles of freezing and thawing are known as *glacial* and *interglacial* (between ice ages) *periods.*

For roughly 10,000 years, we have had the good fortune to live in an interglacial period characterized by a fairly stable climate based mostly on a generally steady global average surface temperature. This allowed the human population to grow as agriculture developed, and later as cities grew. For the past 1,000 years, the average temperature of the atmosphere near the earth's surface has remained fairly stable but began to rise during the 19th and 20th centuries when a growing population began clearing large areas of forest and burning more fossil fuels. Most of the recent increase in the global average atmospheric temperature has taken place since 1975 (Figure 19-3).

Figure 19-2 Over the past 900,000 years, the average global temperature of the atmosphere near the earth's surface has fluctuated widely. This graph is based on a body of scientific evidence that contains gaps, but the data clearly indicate general trends. ***Question:*** What are two conclusions you can draw from this graph?

(Compiled by the authors using data from Goddard Institute for Space Studies, Intergovernmental Panel on Climate Change, National Academy of Sciences, National Aeronautics and Space Administration, National Center for Atmospheric Research, and National Oceanic and Atmospheric Administration.)

Figure 19-3 Atmospheric warming has trended upward since 1975 with a slight slowing of this trend between 2002 and 2012. ***Question:*** What are two conclusions you can draw from this graph?

(Compiled by the authors using data from Goddard Institute for Space Studies, Intergovernmental Panel on Climate Change, National Academy of Sciences, National Aeronautics and Space Administration, National Center for Atmospheric Research, and National Oceanic and Atmospheric Administration.)

Figure 19-4 *Ice cores* are extracted by drilling deep holes into ancient glaciers at various sites such as this one near the South Pole in Antarctica. Analysis of ice cores yields information about the past composition of the lower atmosphere, temperature trends such as those shown in Figure 19-2, solar activity, snowfall, and forest fire frequency.

Scientists estimate past temperature changes such as those depicted in Figure 19-2 by analyzing different types of evidence. These include radioisotopes in rocks and fossils; plankton and radioisotopes in ocean sediments; tiny bubbles, layers of soot, and other materials trapped in different layers of ancient air found in ice cores from glaciers (Figure 19-4); pollen from the bottoms of lakes and bogs; tree rings; and temperature measurements taken regularly since 1861. These temperature measurements now include data from more than 40,000 measuring stations around the world. Such measurements have some limitations, but they show general changes in atmospheric temperature, which in turn can affect the earth's climate.

Figure 19-5 Melting of Himalayan glaciers between 1921 (top) and 2009 (bottom).

The scientific evidence gathered in recent years indicates that atmospheric warming has been accelerating since about 1975 (Figure 19-3). In 1988, a number of nations, working through the World Meteorological Organization and the United Nations Environment Programme, established the Intergovernmental Panel on Climate Change (IPCC) to document past climate changes and project future changes. The IPCC network includes more than 2,500 climate scientists and scientists in disciplines related to climate studies from more than 130 countries.

In 2007, the IPCC issued a report based on more than 29,000 sets of data. This report, along with a set of reports issued in 2010 by the U.S. National Academy of Sciences, reports from the British Royal Society and the U.S. National Atmospheric and Oceanic Administration (NOAA), a 2011 analysis of global temperature measurements since 1750 by physicist Richard Muller and his colleagues, a 2012 report from NASA, and a 2012 report from the World Bank include the following findings:

- Between 1906 and 2012, the average global surface temperature has risen by about 0. 8C° (1.4F°), and over land, the average rise has been 1.4C° (2.5F°). (Land warms more than ocean waters do because heat can stay concentrated near the land surface while waves mix ocean water and transfer some of the heat to deeper water.) Most of this increase has taken place since 1975 (Figure 19-3).
- The first decade in this century (2000–2009) was the warmest decade since 1881, and 2012 was the warmest year on record. Of the 12 hottest years on record, 11 have occurred since 2000, and the twelfth was 1998.
- In some parts of the world, glaciers are melting at increasing rates (Figure 19-5 and **Core Case Study**).
- Floating arctic sea ice is shrinking. In 2012, it had shrunk to the smallest area ever recorded in 34

years of satellite measurements, according to NASA scientists.

- During the 20th century, the world's average sea level rose by 19 centimeters (7.6 inches)—an average of 1.9 cm (0.76 inches) per decade—mostly because of the expansion of ocean water as its temperature increased and to a lesser extent by runoff from melting land-based ice. By comparison, sea levels rose about 2 centimeters (three-quarters of an inch) in the 18th century and 6 centimeters (2 inches) in the 19th century.

Based on such evidence, the IPCC and other scientific bodies such as the U.S. National Academy of Sciences and the British Royal Society have concluded that the earth's lower atmosphere has warmed, especially since 1975, and that this warming is beginning to change the earth's climate.

19-2 Why Is the Earth's Climate Changing?

CONCEPT 19-2

Scientific evidence strongly indicates that the earth's atmosphere has been warming at a rapid rate since 1975 and that human activities, especially the burning of fossil fuels and deforestation, have played a major role in this warming.

The Natural Greenhouse Effect Plays a Key Role in Climate

The earth's climate is strongly influenced by changes in the amount of solar radiation reaching the earth. But a natural process called the *greenhouse effect* (see Figure 3-3, p. 54) also plays a key role in determining the earth's climate. It occurs when some of the solar energy absorbed by the earth radiates into the atmosphere as infrared radiation (heat) at various wavelengths. About 1% of the earth's lower atmosphere is composed of *greenhouse gases*, primarily water vapor (H_2O), carbon dioxide (CO_2), methane (CH_4), and nitrous oxide (N_2O). Heat radiated into the atmosphere by the earth causes molecules of these gases to vibrate and release infrared radiation with longer wavelengths into the lower atmosphere. As this radiation interacts with molecules in the air, it increases their kinetic energy and warms the lower atmosphere and the earth's surface.

There are several greenhouse gases, but we focus here on three major gases—CO_2, CH_4, and N_2O—that play varying roles in atmospheric warming because of their varying lifetimes in the atmosphere and varying warming potentials (Figure 19-6). Other greenhouse gases are hydrofluorocarbons (HFCs), chlorofluorocarbons (CFCs), and hydrochlorofluorocarbons (HCFCs), which are used as refrigerator coolants, cleaners, and fire retardants. Ozone (O_3) also has a warming effect.

Swedish chemist Svante Arrhenius first recognized the greenhouse effect in 1896. Since then, numerous laboratory experiments and measurements of temperatures at different altitudes have confirmed this effect—now one of the most widely accepted theories in the atmospheric sciences. Life on the earth and the world's economies are totally dependent on the greenhouse effect—one of the planet's most important forms of natural capital. Without it, the planet would be a frigid, uninhabitable place.

The atmospheric concentration of CO_2, as part of the carbon cycle (see Figure 3-17, p. 66), plays an important role in determining the average temperature of the atmosphere. Measurements of CO_2 in bubbles at various depths in ancient glacial ice indicate that changes in the levels of these gases in the lower atmosphere have correlated fairly closely with changes in the global average temperature near the earth's surface during the past 400,000 years (Figure 19-7). Scientists have noted a similar correlation between atmospheric temperatures and methane (CH_4) emissions.

Water vapor (H_2O), another greenhouse gas, accounts for about 66% of the earth's greenhouse effect. However, it stays in the atmosphere for only about 1 to 3 weeks,

Atmospheric lifetime

Atmospheric Warming Potential
(over 100 years, as multiples of CO_2 warming potential)

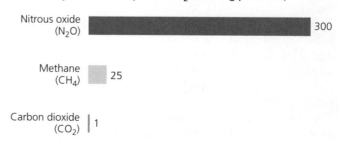

Figure 19-6 Atmospheric lifetimes and warming potentials for three major greenhouse gases.

(Compiled by the authors using data from National Oceanic and Atmospheric Administration and U.N. Framework Convention on Climate Change.)

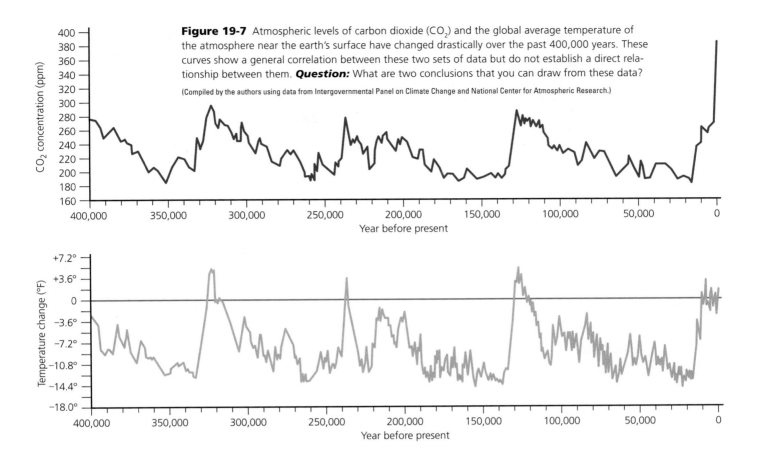

Figure 19-7 Atmospheric levels of carbon dioxide (CO_2) and the global average temperature of the atmosphere near the earth's surface have changed drastically over the past 400,000 years. These curves show a general correlation between these two sets of data but do not establish a direct relationship between them. **Question:** What are two conclusions that you can draw from these data?

(Compiled by the authors using data from Intergovernmental Panel on Climate Change and National Center for Atmospheric Research.)

on average, compared to years for the other gases (Figure 19-6). As the air warms, any given volume of air can hold more water vapor. Thus, the warming effect of other greenhouse gases can lead to a higher level of water vapor, which in turn amplifies the greenhouse effect. For this reason, scientists consider CO_2 to be the main engine of the atmospheric warming related to greenhouse gases, while water vapor plays a secondary role in response to CO_2 levels.

Human Activities Play a Key Role in Atmospheric Warming

Since the beginning of the Industrial Revolution in the mid-1700s, human activities—mainly the burning of fossil fuels, deforestation, and agriculture, in that order—have led to significant increases in the levels of greenhouse gases in the lower atmosphere, especially since 1960 (see Figure 14, p. S70, in Supplement 7).

Analyses of ice cores from Antarctica done in 2007 indicate that current levels of CO_2 in the atmosphere are higher than at any time during the past 800,000 years. Based on these studies, long-term changes in levels of CO_2 correlate fairly closely with changes in the global average temperature of the atmosphere near the earth's surface (Figures 19-7 and 19-8) (**Concept 19-2**). However, temperature increases generally lag behind increases in CO_2.

The burning of fossil fuels tops the list of human activities that emit CO_2, with coal being by far the highest emit-

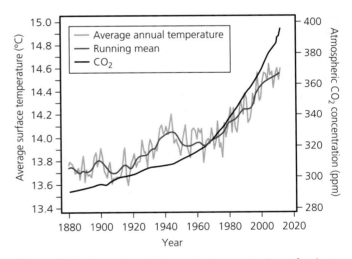

Figure 19-8 Comparison of atmospheric concentrations of carbon dioxide (CO_2) and the atmosphere's average temperature for the period 1880–2011. Temperature data show the fluctuating seasonal average temperatures (orange curve) as well as the *mean*, or average of these data (red curve). While the data show an apparent relationship between the mean temperature and CO_2 concentrations, they do not establish that these two variables are related.

(Compiled by the authors using data from NASA Goddard Institute of Space Studies, National Oceanic and Atmospheric Administration Earth System Research Laboratory, Earth Policy Institute, and Worldwatch Institute.)

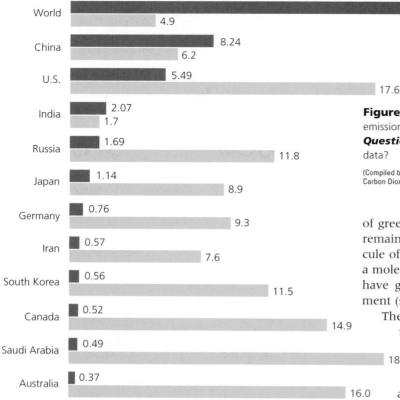

World	33.51	
	4.9	
China	8.24	
	6.2	
U.S.	5.49	
	17.6	
India	2.07	
	1.7	
Russia	1.69	
	11.8	
Japan	1.14	
	8.9	
Germany	0.76	
	9.3	
Iran	0.57	
	7.6	
South Korea	0.56	
	11.5	
Canada	0.52	
	14.9	
Saudi Arabia	0.49	
	18.2	
Australia	0.37	
	16.0	

■ Total (million metric tons)
■ Per capita (metric tons)

Figure 19-9 Total CO_2 emissions (in million metric tons) and CO_2 emissions per person (in metric tons) for selected countries in 2010. **Question:** What are two conclusions that you can draw from these data?

(Compiled by the authors using data from International Energy Agency and U.S. Department of Energy Carbon Dioxide Information Analysis Center.)

ter of all fossil fuels (see Figure 15-17, p. 388). The average atmospheric concentration of CO_2 rose from a level of 280 parts per million (ppm) at the start of the Industrial Revolution to 400 ppm in 2013, according to data from the NOAA (see Figure 14, p. S70, in Supplement 7). This amounts to a 43% increase in the amount of CO_2 in the atmosphere, and this is a long-term increase, because CO_2 typically remains in the atmosphere for about 100 years (Figure 19-6).

Ice core analysis reveals that about 70% of methane (CH_4) emissions during the last 275 years are the result of human activities, including livestock production, rice production, natural gas production, leaky coal mines, landfills, and the flooding of land behind large dams. In that time, methane levels in the atmosphere have tripled. Methane remains in the atmosphere for about 12 years and each molecule of methane has 25 times the warming potential of a molecule of carbon dioxide (Figure 19-6).

Methane emissions are expected to rise in the future if the atmosphere warms as projected and melts some of the permafrost in the arctic tundra and in the Arctic Ocean seabed, which will release a large volume of CH_4 into the atmosphere. Also, a 2012 study led by NASA's Eric Kort found surprising levels of methane being released through cracks in floating arctic ice. This methane is thought to be generated by marine bacteria, and emissions from this source will likely increase as arctic seawater warms and as more cracks appear in the melting sea ice.

Nitrous oxide (N_2O) levels have risen about 20% during the last 275 years, mostly as a result of the increased use of nitrogen fertilizers. This gas accounts for about 9%

of greenhouse gas emissions from human activities, but it remains in the atmosphere for 114 years and each molecule of N_2O has nearly 300 times the warming potential of a molecule of CO_2. Fertilizers and other sources of nitrogen have greatly increased nitrogen inputs into the environment (see the graph in Figure 16, p. S70, in Supplement 7).

The IPCC scientists, in their 2007 report, concluded that the earth's lower atmosphere has warmed, especially since 1975, due mostly to increased levels of CO_2 (Figure 19-8) and other greenhouse gases. They also concluded that human activities, especially the burning of fossil fuels and deforestation, played an important role in this warming (**Concept 19-2**).

In 2010, China and the United States together accounted for about 41% of the world's annual greenhouse gas emissions (Figure 19-9). China has been the leading emitter of CO_2 since 2007. However, China's per capita CO_2 emissions are low because most members of its huge population are too poor to use large amounts of fossil fuels and other resources. The United States emits nearly 3 times as much CO_2 per person as China emits and about 10 times as much CO_2 per person as India emits. Figure 19-9 shows total and per capita CO_2 emissions for the world's highest-emitting countries.

In 2012, according to a survey by researchers at Stanford University, nearly 98% of climate scientists who were actively publishing their research agreed that the burning of fossil fuels has played a key role in atmospheric warming since 1975 (Figure 19-8). In 2013, the third National Climate Assessment written by 240 scientists, business leaders, and other climate experts concluded that "climate change, once considered an issue for a distant future, has moved firmly into the present."

Climate models (Science Focus 19.1) project that rising CO_2 emissions and atmospheric temperatures could bring about significant changes in the earth's climate that are likely to cause major ecological and economic disruption in the latter half of this century. IPCC scientists have identified several natural and human-influenced factors that might *amplify* or *dampen* the projected changes in the average atmospheric temperature. We now examine some of those factors.

What Role Does the Sun Play?

The energy output of the sun plays the key role in the earth's temperature, and this output has varied over millions of years. However, in 2007, climate researchers Claus Froehlich, Mike Lockwood, and Ben Santer all concluded in separate studies that most of the rise in global average atmospheric temperatures since 1975 (Figure 19-3) could not be the result of increased solar output. Instead, they determined that the energy output of the sun has dropped slightly during the past several decades. In 2011, a detailed data analysis by physicist Richard Muller and his colleagues confirmed this conclusion.

Froehlich noted that, according to satellite and weather balloon measurements since 1975, the troposphere has warmed while the stratosphere has cooled. This is the opposite of what a hotter sun would do. According to Santer, "If you increase output from the sun, you increase the amount of energy that arrives at the top of the Earth's atmosphere and get heating throughout the atmosphere." Instead, the data show that the atmosphere is now heating from the bottom up, which indicates that inputs at the earth's surface (most likely from human activities) play the more important role in atmospheric warming.

What Role Do Oceans Play?

The world's oceans absorb CO_2 from the atmosphere as part of the carbon cycle (see Figure 3-17, p. 66) and thus help to moderate the earth's average surface temperature and its climate. It is estimated that the oceans remove roughly a third of the CO_2 pumped into the lower atmosphere by human activities. The oceans also absorb heat from the lower atmosphere. Then, partly driven by this heat, ocean currents slowly transfer some of the CO_2 to the deep ocean (see Figure 7-5, p. 147), where it is buried in carbon compounds in bottom sediments for several hundred million years.

The ability of the oceans to absorb CO_2 decreases as water temperatures rise. As the oceans warm up, some of their dissolved CO_2 is released into the lower atmosphere, like CO_2 bubbling out of a warm carbonated soft drink. According to scientific measurements from a 2007 study, the upper portion of the oceans warmed by an average of 0.32–0.67°C (0.6–1.2°F) during the last century—an astounding increase considering the huge volume of water involved—most likely due to increasing atmospheric temperatures. A 2012 study by oceanographers Dean R. Roemmich and John Gibson found that the upper ocean warmed by about 0.2°C (0.1°F) between 1990 and 2010—a sharp increase for such a massive body of water in a short time.

As the oceans have absorbed more CO_2 and become warmer, they have also become more acidic, and this could result in some potentially harmful effects on marine ecosystems (see Science Focus 11.2, p. 252). We discuss this process and its effects in the next section.

There Is Uncertainty about the Effects of Cloud Cover on Atmospheric Warming

The role played by clouds in the warming of the atmosphere is uncertain. Warmer temperatures increase evaporation of surface water, which increases the relative humidity of the atmosphere in various parts of the world and creates more clouds that can either cool or warm the atmosphere.

An increase in thick and continuous *cumulus clouds* at low altitudes (Figure 19-10, left) could decrease surface warming by reflecting more sunlight back into space. But an increase in thin, wispy *cirrus clouds* at high altitudes (Figure 19-10, right) could cause more warming of the lower atmosphere by preventing some heat from escaping into space. Climate modelers have not been able to show which of these effects is more important and are working hard to understand more about the role of clouds in their climate models. One thing that current climate models do agree on is that any cooling effect that clouds might have will not be enough to offset the long-term heating effects of human-generated greenhouse gas emissions.

In 2010, Axel Lauer and other researchers developed a new model for evaluating the effects of cloud cover on atmospheric warming. According to their model, by the end of this century, higher CO_2 levels and a warmer atmosphere will likely shrink the size of low cumulus clouds. This will allow more solar radiation to reach the lower atmosphere, which will likely accelerate atmospheric warming toward the higher estimated temperatures shown in Figure 19-C (p. 515).

Tatiana Grozetskaya/Shutterstock.com

Figure 19-10 Cumulus clouds (left) are thick, relatively low-lying clouds that tend to decrease surface warming by reflecting some incoming solar radiation back into space. Cirrus clouds (right) are thin and float at high altitudes; they tend to warm the earth's surface by preventing some heat from flowing into space.

Cheryl Casey/Shutterstock.com

USING MODELS TO PROJECT FUTURE CHANGES IN ATMOSPHERIC TEMPERATURES

To project the effects of increasing levels of greenhouse gases on average global temperatures, scientists have developed complex *mathematical models,* which simulate interactions among incoming sunlight, clouds, landmasses, oceans, ocean currents, concentrations of greenhouse gases and air pollutants, and other factors within the earth's complex climate system. They run these continually improving models on supercomputers and compare the results to known past climate changes, from which they project future changes in the earth's average temperature. Figure 19-A gives a greatly simplified summary of some of the key interactions in the global climate system.

Such models provide *projections* of what is *likely* or *very likely* to happen to the average temperature of the lower atmosphere. How well these projections match what actually happens in the real world depends on the validity of the assumptions and variables built into the models and on the accuracy of the data used. Recall that no scientific research can give us absolute proof or certainty, but instead provides us with varying levels of certainty. However, when most experts in a particular scientific field generally agree that the level of certainty involved in a set of measurements or a model is at least 90%, they are very likely to be correct.

In 1990, 1995, 2001, and 2007, the IPCC published reports on how global temperatures have changed in the past (Figure 19-2) and made projections on how they are likely to change during this century and how such changes are likely to affect the earth's climate. According to the 2007 report, based on analysis of past climate data and the use of 19 climate models, it is very likely that human activities, especially the burning of fossil fuels, have played an important role in the observed atmospheric warming since 1975 (Figure 19-3 and Concept 19-2). The researchers based this conclusion on the fact that, after thousands of times running the models, the only way they can get the model results to match actual measurements is by including the human activities factor (Figure 19-B).

Similar to the 2007 IPCC report, more recent runs of several of the more than two dozen climate models now in use suggest that it is very likely that the earth's mean surface temperature will increase by 2–4.5°C (3.6–8.1°F) between now and 2100 (Figure 19-C) unless the world halts deforesta-

Figure 19-A A simplified model of some major processes that interact to determine the average temperature and greenhouse gas content of the lower atmosphere and thus the earth's climate. Red arrows show processes that warm the atmosphere and blue arrows show those that cool it. **Question:** Why do you think a decrease in snow and ice cover is adding to the warming of the atmosphere?

© Cengage Learning

🔍 **CONSIDER THIS. . .**

CONNECTIONS Air Travel and Atmospheric Warming

Infrared satellite images indicate that wispy condensation trails (contrails) left behind by jet planes expand and turn into cirrus clouds, which tend to heat the atmosphere. A 2010 study led by Ulrike Burkhardt and Bernd Kärcher found that contrails are a significant contributor to atmospheric warming, more so than the total amount of CO_2 that is released by jet airplanes, at least in the short term.

Figure 19-B Comparison of actual climate data with modeled projections for the period between 1860 and 2000 using natural factors only (left) and a combination of natural and human factors (right). Scientists have found that actual data match projections far more closely when human factors are included in the models.

(Compiled by the authors using data from Intergovernmental Panel on Climate Change, 2001 and 2007.)

tion, makes drastic cuts in greenhouse gas emissions from fossil fuel–burning power plants, factories, and motor vehicles, and switches to improved energy efficiency and low-carbon energy resources such as wind and solar energy (see Figure 16-36, p. 434).

In any scientific modeling, there is some degree of uncertainty, as indicated by the fairly wide range of projected temperature changes in Figure 19-C. Climate experts are working to improve these models and to narrow the projected range of temperatures. Also, current climate models are better at projecting likely general changes at the global level than at making projections for particular areas, which can have local climates that vary significantly from the average global climate. In other words, as the atmosphere warms globally, some areas will be hotter and some will be cooler. However, current climate models project that a global average temperature increase of 2–3°C (3.6–5.4°F) would be enough to overcome local cooling and lead to a warmer climate almost everywhere.

Critical Thinking

If the projected temperature increases in Figure 19-C take place, what are three major ways in which this will affect your lifestyle and that of any children or grandchildren you might have?

Figure 19-C Estimated changes (yellow area) and measured changes (black curve) in the average temperature of the atmosphere near the earth's surface between 1860 and 2011, and the projected range of temperature increase during the rest of this century.

(Compiled by the authors using data from U.S. National Academy of Sciences, National Center for Atmospheric Research, Intergovernmental Panel on Climate Change, and Hadley Center for Climate Prediction and Research.)

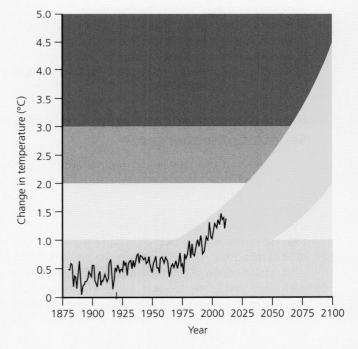

Outdoor Air Pollution Can Temporarily Slow Atmospheric Warming

There is considerable evidence that human activities have played a role in atmospheric warming, but there is also evidence that aerosol air pollution from human activities has slowed the rate of atmospheric warming. Aerosols (suspended microscopic droplets and solid particles) of various air pollutants are released or formed in the troposphere by volcanic eruptions and human activities (see

Figure 18-3, p. 477). They can hinder or enhance both the greenhouse warming effect and cloud formation, depending on factors such as their size and reflectivity.

Most aerosols, such as light-colored sulfate particles produced by fossil fuel combustion, tend to reflect incoming sunlight and cool the lower atmosphere as they seem to have done in the case of the South Asian Brown Clouds (see Chapter 18, Core Case Study, p. 474). A 2012 study led by Harvard University's Eric Leibensperger estimated that atmospheric temperatures in the eastern United States had dropped slightly between 1930 and 1990—due largely to pollution from coal-fired power and industrial plants—while worldwide average temperatures had increased slightly.

While sulfate particles reflect sunlight, black carbon particles, or *soot*, have the opposite effect. A 2008 study by atmospheric scientist Veerabhadran Ramanathan and his colleagues found that soot emitted into the air by coal-burning power and industrial plants, diesel exhaust, open cooking fires, and burning forests has a warming effect that is 4 times greater than was estimated earlier. These findings were supported by another study reported in 2009 by meteorologist Gunnar Myhre.

Climate scientists do not expect aerosols and soot pollutants to counteract or enhance projected climate change very much in the next 50 years for two reasons. *First,* aerosols and soot fall back to the earth or are washed out of the lower atmosphere within weeks or months, whereas CO_2 typically remains in the lower atmosphere for about 100 years (Figure 19-6). *Second,* aerosol and soot inputs into the lower atmosphere are being reduced—especially in more-developed countries—because of their harmful impacts on plants and human health.

According to the IPCC, the decline in atmospheric sulfate concentrations in some countries since 1970 has played a role in the warming of the atmosphere, especially since 1990, and this effect will grow as sulfate concentrations drop further due to improved air pollution regulations. We could possibly slow projected climate disruption by increasing outdoor sulfate air pollution. However, because such pollution kills more than 800,000 people a year (about 50,000 people in the United States), this would not be a desirable or ethically acceptable way to try to deal with projected climate disruption.

19-3 What Are the Possible Effects of a Warmer Atmosphere?

CONCEPT 19-3

The projected rapid change in the atmosphere's temperature could have severe and long-lasting consequences, including increased drought and flooding, rising sea levels, and shifts in the locations of croplands and wildlife habitats.

Rapid Atmospheric Warming Could Have Serious Consequences

Most historical changes in the temperature of the lower atmosphere took place over thousands of years (Figure 19-2). What makes the current problem urgent is that we face *a rapid projected increase in the average temperature of the lower atmosphere during this century* (Figure 19-C, p. 515). This, in turn, is very likely to rapidly change the fairly mild climate that we have had for the past 10,000 years.

Climate models indicate that, in the worst-case scenario, we could face floods in low-lying coastal cities, forests being consumed in vast wildfires, grasslands turning into dust bowls, rivers drying up, ecosystems collapsing, the extinction of up to half of the world's species, more intense and longer-lasting heat waves, more destructive storms and flooding, and the rapid spread of some infectious tropical diseases. If the models are correct, we will have to deal with many of these disruptive effects within this century—an incredibly short time span in terms of the earth's overall climate history.

Climate models also show that atmospheric warming and its harmful effects will be unevenly distributed,

according to a 2012 World Bank report. Temperatures in the tropics could reach extremes that people there have not seen in over 10,000 years. The sea-level rise also would be 15–20% larger than the global average in tropical areas, and droughts could also be much worse in such areas. In our interconnected global economy, such effects would ripple strongly to other, less-affected areas of the world.

In anticipation of such possibilities, scientists have identified a number of *tipping elements,* or components of the climate system that could pass **climate change tipping points**—those thresholds beyond which natural systems could change for hundreds to thousands of years, with possibly catastrophic effects. One such tipping element is atmospheric CO_2 levels. A number of scientific studies and major climate models indicate that we need to prevent CO_2 levels from exceeding 450 ppm—an estimated tipping point beyond which we might experience large-scale climate changes that would last for hundreds to thousands of years. We have already reached 400 ppm, and without strong efforts to improve energy efficiency and replace fossil fuels with low-carbon energy resources, we could exceed 450 ppm within a couple of decades.

Another tipping element is global average atmospheric temperature. Several findings related to atmospheric warming are leading some scientists to suggest that we can no longer avoid a global temperature rise of more than 2C° (3.6F°), the threshold beyond which many say climate change will be very dangerous. Several of the effects of climate change that we will now examine could lead to these and other tipping points.

Average minimum ice cover, 1979–2010

Minimum ice cover, September 16, 2012

NASA/Goddard Scientific Visualization Studio

Figure 19-11 *The big melt:* Rising average atmospheric and ocean temperatures have caused more and more arctic sea ice to melt during the summer months. The yellow line added to this satellite image (left) shows the average summer minimum area of ice during the period 1979–2010, in contrast to the white, ice-covered summer minimum in 2012. The graphs (right) show that the sea ice melt has been accelerating. **Question:** Which is declining faster, the summer sea ice area, or the summer sea ice volume?

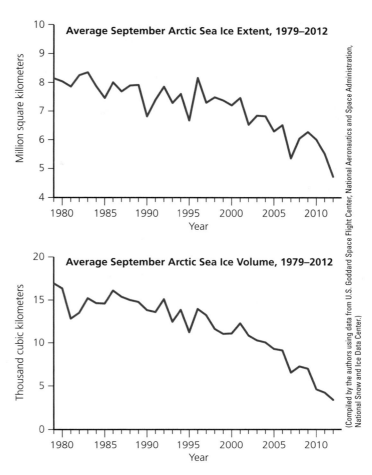

(Compiled by the authors using data from U.S. Goddard Space Flight Center, National Aeronautics and Space Administration, National Snow and Ice Data Center.)

More Ice and Snow Are Likely to Melt

Each summer, some of the floating ice in the Arctic Sea melts and then refreezes during winter. But in recent years, rising average atmospheric and ocean temperatures have caused more and more ice to melt during the summer months. Satellite data show a 49% drop in the average area covered by summer arctic sea ice between 1979 and 2012 (Figure 19-11, left). In 2012, scientists measured the largest loss of arctic sea ice ever recorded in 34 years of satellite measurements—an area of sea ice larger than the continental United States.

According to National Snow and Ice Data scientist Julienne Stroeve, this ice area declined at an average of 6.5% per decade between 1979 and 2001 and since then has been falling twice as fast, on average (Figure 19-11, right, upper graph). Similarly, the volume of arctic sea ice has declined even more, due to dramatic thinning of the ice. In 2012, the total volume of arctic sea ice was less than one-fourth of what it was in 1979 (Figure 19-11, right, lower graph).

The melting of this ice is due to both warmer air above the ice and warmer water below the ice. Scientists are not sure what causes periodic rises and falls in the tem-

perature of arctic seawater, occurring over 20- to 30-year cycles. These cycles are poorly understood and their impact on sea ice is largely unknown.

However, according to the 2007 IPCC report, arctic air temperatures have risen almost twice as fast as average temperatures in the rest of the world during the past 50 years. Historically, light-colored ice and snow in these regions has helped to cool the earth by reflecting incoming solar energy. However, the melting of ice and snow has exposed much darker land and sea areas, which reflect significantly less sunlight and absorb more solar energy. In addition, soot generated by North American, European, and Asian industries is darkening arctic ice and lessening its ability to reflect sunlight.

Recall from Chapter 2 our discussion of *positive feedback loops*—cycles of change that cause systems to change further in a certain direction (see Figure 2-16, p. 45). An important example of such a loop is the accelerating melting of arctic ice, which causes further warming, which, in turn, speeds the melting. As such a feedback loop kicks in, the rate of change and the resulting harmful effects can get out of control.

One of the climate-change tipping points that concern scientists is the complete melting of summer arctic sea ice. If the current trend continues, this summer ice may be gone by 2040, according to researchers Muyin Wang and

James Overland, and perhaps earlier according to others, including NASA climate scientist Jay Zwally and ocean physics professor Peter Wadhams. This would open these waters to shipping and would allow access to oil and mineral deposits in the arctic region. However, it could also lead to dramatic and long-lasting changes in weather and climate that could affect the whole planet.

For example, according to a 2012 study led by atmospheric scientists Jennifer A. Francis and Stephen J. Vavrus, changes in arctic seawater temperatures have already slowed the *jet stream*—a high, rapidly moving flow of air that affects global weather patterns (see Figure 2, p. S20, in Supplement 5). As a result, weather conditions can remain in certain areas for longer periods of time. Rainy weather can hold its position long enough to cause flooding, and dry weather can turn to drought. Another result, according to a 2010 report by NOAA scientist James Overland, could be colder and snowier winters in Europe, eastern North America, and eastern Asia because of a slower jet stream sinking southward and bringing cold air into those areas.

Another effect of this warming cycle is faster melting of polar land-based ice including that in Greenland (**Core Case Study** and Figure 19-12). This melting adds freshwater to the northern seas, as could an increase in rainfall, which is another projected effect of atmospheric warming. Some scientists hypothesize that at some point, this additional water could slow or stop ocean currents that move heat around the globe, which could ramp up the warming in Greenland. Some scientists are dedicating much of their time to studying Greenland's melting ice (Individuals Matter 19.1).

Another great storehouse of ice is the earth's mountain glaciers, which expand with added snowfall in winter (called the *snowpack*) and slowly melt during summer months. During the past 25 years, many of the world's mountain glaciers have been slowly shrinking (Figure 19-5) wherever summer melting exceeds the winter snowpack due to a warming atmosphere. For example, Glacier National Park in the U.S. state of Montana once had 150 glaciers, but by the end of 2010, only 25 remained, according to park officials.

Mountain glaciers play a vital role in the water cycle (see Figure 3-15, p. 63) by storing water as ice during cold seasons and releasing it slowly to streams during warmer seasons. A prime example of these high-elevation reservoirs is the glaciers of the Himalayan Mountains in Asia (Figure 19-5). They are a major source of water for large rivers such as the Ganges, which provides water for more than 400 million people in India and Bangladesh. They also feed China's Yangtze and Yellow Rivers, whose basins are home to more than 500 million people. In 2009, Swiss scientists reported that some Himalayan glaciers, as well as some glaciers in the European Alps, shrank farther and faster in the last decade than they had in the previous 150 years.

CONSIDER THIS. . .

CONNECTIONS Tibetan Plateau Glaciers and World Food Prices

Environmental analyst Lester Brown points out that as ice on the Tibetan Plateau in China melts, irrigation water will grow short, and China will need to import much more grain to feed its population. As China becomes more active in the international grain market, its 1.3 billion people, whose average incomes are on the rise, will be competing with U.S. consumers for grains, and this could drive up food prices in the United States and elsewhere.

About 80% of the mountain glaciers in South America's Andes range are also slowly shrinking. By 2010, Peru had lost 22% of its glaciers and about half of Bolivia's glacial ice had melted, according to the World Glacier Monitoring Service. If this continues, 53 million people in Bolivia, Peru, and Ecuador who rely on meltwater from the glaciers for irrigation and hydropower could face severe water, power, and food shortages at some point during the next 50 to 200 years. In the United States, people living in the Columbia, Sacramento, and Colorado River basins could face similar threats as the winter snowpack that feeds these rivers is projected to shrink by 70% by 2050, according to climate models.

Permafrost Is Likely to Melt: Another Dangerous Scenario

Permafrost occurs in soils found beneath 24% of the exposed land in the northern hemisphere. The amount of carbon locked up in permafrost soils is 50–60 times the

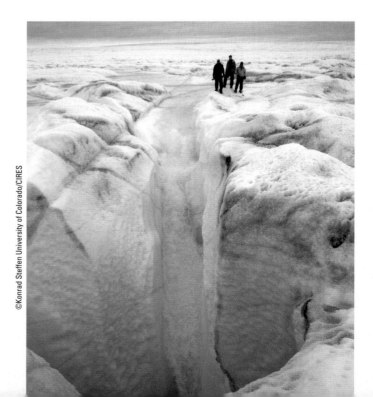

Figure 19-12 Accelerating melting of Greenland's ice has created flows such as this one, called *moulins*, which run from the surface to the base of the glaciers, leading to more dramatic melting and movement of the glacier toward the sea.

©Konrad Steffen University of Colorado/CIRES

Konrad Steffen: Studying Melting Ice in Greenland

Photo by Konrad Steffen, University of Colorado/CIRES

In 2012, Konrad Steffen, formerly the Director of the Cooperative Institute for Research in Environmental Sciences (CIRES) at the University of Colorado, became Director of the Swiss Federal Institute for Forest, Snow and Landscape Research. He is one of the world's most respected experts on Greenland's melting ice (**Core Case Study**).

Since 1990, he and his colleagues have conducted extensive research on the Greenland ice sheet. This led to the building of the Greenland Climate Network—22 weather stations that use solar-powered monitoring equipment to automatically record the amounts of precipitation in Greenland and the melting and changes in the surface height of Greenland ice. Scientists correlate this data with satellite observations in order to estimate ice coverage in Greenland (Figure 19-1).

Steffen personally adapted and installed much of the equipment in these monitoring stations that provide scientists around the world with information on hourly conditions on Greenland's ice sheet and how they are changing. His research revealed that the melting of Greenland's ice accelerates when meltwater flows downward to bedrock (Figure 19-12). There it acts as a lubricant that accelerates the slide of ice into the ocean, an effect of atmospheric warming that is occurring more rapidly than was projected by the IPCC.

Steffen and his colleagues have also studied ice cores to look at the climate history of Greenland for the past 800 years. Their research has revealed that between 1979 and 2010, the area of ice that melted expanded by 30% because of rising summer air and water temperatures. This research also indicated that about 50% of Greenland's ice loss is due to melting, and the other 50% is the result of icebergs breaking off into the ocean. Using statistical modeling, Steffen has projected that the rise in sea level due to rising water temperatures and to the melting of polar ice might be 1 to 2 meters (3.3 to 6.6 feet) by the end of this century.

Background photo: ©Sofiaworld/Shutterstock.com

Current

2090–2100

Boreal Forest

RUSSIA   ARCTIC TUNDRA

Boreal Forest

RUSSIA

Figure 19-13 Projected thaw of permafrost in the arctic tundra of eastern Russia between 2004 and 2100 as a result of atmospheric warming. Tall shrubs and trees are likely to take over a growing area of the much warmer arctic tundra by the end of this century.

(Compiled by the authors using data from Intergovernmental Panel on Climate Change and 2004 Arctic Climate Impact Assessment.)

amount emitted as CO_2 each year from the burning of fossil fuels.

A 2012 report by the United Nations Environment Programme (UNEP) estimated that human-caused climate change is expected to thaw out significant amounts of permafrost. According to the 2004 Arctic Climate Impact Assessment, 10–20% of the Arctic's current permafrost might thaw during this century over a vast area of arctic tundra (Figure 19-13, left). If this happens, a great deal of stored organic material will rot and release large amounts of CH_4 and CO_2 into the atmosphere. This would accelerate projected warming, which would in turn melt more permafrost in yet another positive feedback loop.

Some scientists are concerned about another methane source—a layer of permafrost on the Arctic Sea floor and also on many arctic lake bottoms. In 2009, Natalia Shakhova and other researchers found areas of the seafloor to be leaking bubbles of methane. In 2012, aquatic ecologist Katey Walter Anthony (a National Geographic Emerging Explorer) found methane bubbling up from thousands of seeps on some arctic lake bottoms. In warmer parts of the world, yet another possible source of increasing methane emissions is tropical wetlands. If rainfall increases as projected in tropical areas, these wetlands will expand and produce more plants, which will eventually decay and release methane through anaerobic decomposition (see p. 57). Some scientists refer to such sources as "methane time bombs."

Sea Levels Are Rising

In 2007, the IPCC estimated that sea levels would rise by about 0.5 meters (20 inches) by the end of this century, primarily due to the expansion of seawater as it warms and to the melting of land-based ice as a result of atmospheric warming (**Core Case Study**). However, in 2011, the U.S. National Academy of Sciences (NAS) reported that, based on more recent studies, sea levels could rise by as

much as 1.5 meters (5 feet) by 2100—about 3 times the increase estimated in the 2007 IPCC report. The NAS panel also warned that sea levels would probably keep rising for centuries after 2100.

The rise in sea level is projected to be more dramatic if the vast land-based glaciers in Greenland (**Core Case Study**) and western Antarctica, as well as many of the world's mountaintop glaciers, continue to melt at their current, or higher, rates as the atmosphere warms. Taking this into account, Konrad Steffen (Individuals Matter 19.1) projects a 1- to 2-meter (3.3- to 6.6-foot) rise in the average sea level by the end of this century. A 2012 analysis by Richard Alley and 46 other glaciologists concluded that ice losses from Greenland and West Antarctica due to a warming atmosphere and warming ocean have been accelerating.

According to the IPCC and the NAS study, a 1-meter (3.3-foot) rise in sea level during this century (excluding the additional effects of storm surges) could cause the following serious effects:

- Degradation or destruction of at least one-third of the world's coastal estuaries, wetlands, coral reefs, and deltas where much of the world's rice is grown.
- Disruption of many of the world's coastal fisheries.
- Flooding and erosion of low-lying barrier islands and gently sloping coastlines, especially in U.S. coastal states such as Texas, Louisiana, North and South Carolina, and Florida (Figure 19-14).
- Saltwater contamination of freshwater coastal aquifers resulting in degraded supplies of groundwater.
- Submersion of low-lying islands in the Indian Ocean (Figure 19-15), the Pacific Ocean, and the Caribbean Sea, which are home to 1 of every 20 of the world's people.

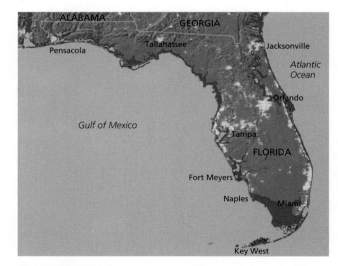

Figure 19-14 If the average sea level rises by 1 meter (3.3 feet), the areas shown here in red in the U.S. state of Florida will be flooded.

(Compiled by the authors using data from Jonathan Overpeck, Jeremy Weiss, and the U.S. Geological Survey.)

Figure 19-15 For a low-lying island nation like the Maldives in the Indian Ocean, even a small rise in sea level could spell disaster for most of its 295,000 people. About 80% of the 1,192 small islands making up this country lie less than 1 meter (3.2 feet) above sea level. Rising sea levels and higher storm surges during this century could flood most of these islands and their coral reefs.

- Flooding of some of the world's largest coastal cities, including Kolkata (India), Mumbai (India), Dhaka (Bangladesh), Guangzhou (China), Ho Chi Minh City (Vietnam), Shanghai (China), Bangkok (Thailand), Rangoon (Myanmar), Haiphong (Vietnam), and the U.S. cities of Miami, Florida, and New York City (see red areas in chapter-opening photo and Figure 19-14), and displacement of at least 100 million people.

The sea-level rise will not be uniform around the world, according to climate models, because of factors such as ocean currents and winds. This is already apparent, according to a 2012 study by the U.S. Geological Survey, which reported that the sea level is rising as much as 4 times faster than the global average along parts of the U.S. Atlantic coast.

The Oceans Are Becoming More Acidic

According to a 2005 report by the United Kingdom's Royal Society and a 2010 report by the U.S. National Academy of Sciences, rising levels of CO_2 in the ocean have increased the acidity of its surface waters by 30% since about 1800, and ocean acidity could reach dangerous levels before 2050. This is because much of the CO_2 absorbed by the ocean reacts with water to produce carbonic acid (H_2CO_3)—the same weak acid found in carbonated drinks—in a process called **ocean acidification** (see Science Focus 11.2, p. 252).

Scientists warn that this higher acidity threatens corals, snails, oysters, and other organisms with shells and body structures composed of calcium carbonate, because it hinders their ability to build and repair such structures. Oceanographer Carl Safina notes that acidification will slow the growth and repair of damaged coral reefs and could begin to dissolve some reefs by the end of the century.

Another problem is that increased acidity has been shown to decrease populations of phytoplankton that are the primary producer species of ocean food webs and that also remove CO_2 from the atmosphere. Phytoplankton are at the base of the ocean food webs, which include the human seafood supply, and could be severely degraded and might possibly even collapse if these producer organisms decline dramatically.

Severe Drought Is Likely to Become More Common

Recall that *drought* occurs when evaporation due to higher temperatures greatly exceeds precipitation for a prolonged period. A 2005 study by National Center for Atmospheric Research scientist Aiguo Dai and his colleagues found that severe and prolonged drought was affecting at least 30% of the earth's land (excluding Antarctica)—an area the size of Asia (Figure 19-16).

Some scientists hypothesize that higher temperatures due to atmospheric warming are intensifying the water cycle and adding more water vapor to the atmosphere. This would increase the warming effect (because water vapor is a greenhouse gas) and lead to more evaporation followed by greater warming in another example of a positive feedback loop. One effect of this intensified water cycle, according to the scientists, is that in some areas droughts will last much longer and will be more widespread. As a result, according to a 2007 study by climate researchers at NASA's Goddard Institute for Space Studies, by 2059, up to 45% of the world's land area could be experiencing extreme drought.

As droughts become worse and more widespread, the growth of trees and other plants in some areas has declined, which has reduced the removal of CO_2 from the atmosphere. Soils in some areas have dried out, making surface temperatures hotter. As some forests and grasslands are also drying out, wildfires are becoming more

Figure 19-16 These people were trying to get water from a well that was running dry in the drought-prone state of Gujarat in western India in 2010.

©REUTERS/Amit Dave

frequent in some areas, and this adds CO_2 to the atmosphere. Some streams and other surface waters are drying up. In some areas, water tables are falling, partly because farmers are irrigating more to make up for drier conditions. Climate scientists project that these combined effects from the projected worsening of prolonged droughts are likely to speed up atmospheric warming and lead to more prolonged droughts—one more example of a positive feedback loop.

Other Forms of Extreme Weather Will Be More Likely

Climate models show that along with increasing drought, a more intensified water cycle with more water vapor in the atmosphere will also bring more flooding to some areas due to heavy snowfall or rainfall (Figure 19-17).

In 2011, the IPCC issued a special report authored by Maarten van Aalst and other climate scientists on the increased likelihood of extreme weather events due to atmospheric warming. It suggested that heat waves that now occur once in a lifetime in areas of Latin America, Africa, and Asia, will become more frequent and hotter than any recorded so far. Another study projected that heavy rainstorms now occurring every 20 years in areas of North America will become 3 times more frequent by the end of the century and that floods in Southeast Asia will likely happen 4 times as often as they do now. And although there may be less snow in some areas, there could be more blizzards. Data from the U.S. Global Change Research Program indicate that the annual number of days of extreme heat in parts of the United States will rise dramatically by the end of the century.

There is controversy among scientists over whether projected atmospheric warming will increase the frequency and intensity of tropical storms and hurricanes. In 2008, climatologists Mark Saunders and Adam Lea analyzed data collected since 1950 and found that for every increase of about 0.8°C (1°F) in the water temperature of the Atlantic Ocean, the overall number of hurricanes and tropical storms increased by about a third, and the number of intense hurricanes increased by 45%. Other studies have supported these results.

However, some researchers blame the ups and downs in the intensity of tropical Atlantic hurricanes on natural climate cycles that can change ocean circulation patterns. It is important to note that we cannot conclusively tie any specific hurricane or severe winter storm, which are weather events, to atmospheric warming and climate change. However, in 2010, a World Meteorological Organization panel of experts on hurricanes and climate change from both sides of this scientific debate reached the following consensus view: projected atmospheric warming is likely to lead to fewer but stronger hurricanes that could cause more damage.

Climate Disruption Is a Threat to Biodiversity

According to the IPCC, projected climate disruption is likely to upset ecosystems and take a toll on biodiversity in areas of every continent. For example, the Amazon rain forest could be devastated, even by moderate warming. According to a 2009 British study led by Chris Jones, up to 85% of this forest could be lost if atmospheric warming increases as projected. The scientists noted that this would involve extensive losses in biodiversity and other ecosystem services as the rain forest dries out and becomes more prone to burning and the resulting conversion to tropical savanna (see Chapter 10, p. 226)—the result of exceeding an ecological tipping point.

According to the 1997 IPCC study, approximately 30% of the land-based plant and animal species assessed so far could disappear if the average global temperature change reaches 1.5–2.5°C (2.7–4.5°F). This percentage could grow to 70% if the temperature change exceeds 3.5°C (6.3°F) (Concept 19-5). The hardest hit will be plant and animal species in colder climates such as the polar bear in the Arctic (see Case Study, p. 203) and penguins in Antarctica; species that live at higher elevations; species with limited ranges such as some amphibians (see Chapter 4, Core Case Study, p. 78); and those with limited toler-

Palash Khan/Age Fotostock

Figure 19-17 An elderly woman seeks a bag of rice in Dhaka, Bangladesh, after a 2007 flood that killed at least 198 people and drove 19 million people from their homes.

ance for temperature change. The primary cause of such extinctions would be loss of habitat.

Several species, including the comma butterfly of Great Britain, the golden-winged warbler in North America, and the mouse-like pica of Yosemite National park, are known to have migrated northward or upward within mountain ranges in recent decades to avoid the warmer conditions of climate change. A 2012 study by University of Washington scientists found that as the climate warms, about 9% of the western hemisphere's mammals (up to 40% in some areas) likely will not be able to move fast enough to find safe habitats that are cool enough for their survival. The researchers pointed out that while animals have moved to new ranges in the distant past to avoid climate change effects, today's landscape includes vast areas of farm fields, four-lane highways, and urban areas that will prevent some species from reaching new ranges.

One ecosystem that is already feeling the effects of climate change is the Southern Ocean near the Antarctic Peninsula. There, oceanographer Martin Montes-Hugo has been studying the phytoplankton that form the base of the Antarctic food web (see Figure 3-12, p. 60). His team compared phytoplankton production during two periods, 1978–1986 and 1998–2006, and found that it had declined nearly 90%. They said the main reason for this drop was the net shrinkage of sea ice during this period, which has resulted in less freshwater on which the phytoplankton depend. The researchers see this as part of the reason for the drop in numbers of Adélie penguins (top consumers in this food web; Figure 19-18) of more than 80% since 1975. Other vulnerable ecosystems are coral reefs (see Chapter 8, Core Case Study, p. 168), coastal wetlands, high-elevation mountaintops, and alpine and arctic tundra.

A warmer climate could also greatly increase populations of insects and fungi that damage trees. In the Canadian province of British Columbia, for example, warmer winters have led to surges in mountain pine beetle populations that have infected huge areas of lodgepole pine forests, which are now dying (Figure 19-19). Since 2000, pine beetles have also damaged about 60% of the lodgepole pines in the U.S. state of Colorado, which has been experiencing warmer winters that are more favorable to the beetles. Large numbers of dry and dead trees can increase the risk of severe forest fires that would further threaten biodiversity.

Because of such damage to trees in combination with drier conditions, forest fires might become more frequent and intense in some areas such as the southeastern and western United States. This would severely degrade some forest ecosystems, add more CO_2 to the atmosphere, reduce total CO_2 uptake by trees and plants, and help accelerate projected climate disruption through still another positive feedback loop.

Food Production Could Face an Overall Decline

Farmers will face dramatic changes due to shifting climates and a faster hydrologic cycle, if the earth warms as projected (Figure 19-C). In a warmer world, agricultural productivity will drop in some areas. This decrease is likely to be worse in areas where farmers depend on monocultures of plants bred to thrive at certain temperatures and with certain levels of rainfall.

Some scientists say that climate change is already affecting food production in certain areas. In Russia, the 2010 heat wave and drought destroyed one-third of the country's wheat crop. In the United States, the 2012 corn harvest was down 13% from 2011, even though the total area planted in corn was the largest since the 1930s, according to the U.S. Department of Agriculture. Some of these losses were related to intense heat that scientists in two 2012 studies argued was highly unlikely without the human influence on climate change.

On the other hand, crop productivity will likely increase in some areas in the short run. According to the IPCC, crop productivity is projected to increase slightly at middle to high latitudes in areas such as midwestern Canada, Russia, and Ukraine, with moderate warming, but decrease if warming goes too far. However, this effect may be limited because soils generally are more lacking in plant nutrients in these northern regions.

Figure 19-18 Declining populations of Adélie penguins in Antarctica have been linked to declines in floating sea ice and phytoplankton because of atmospheric warming.

Keith Szafranski/iStockphoto.com

Climate change models predict a decline in agricultural productivity in tropical and subtropical regions, especially in Southeast Asia and Central America. Also, flooding of river deltas due to rising sea levels could reduce crop production, partly because aquifers that supply irrigation water will be infiltrated by saltwater. Fish production in coastal aquaculture ponds could also be reduced.

Areas especially vulnerable to these effects will be the populous nations of Bangladesh, Egypt, and Vietnam, and parts of the African coast, according to a 2012 World Bank report. Food production could also drop in farm regions that are dependent on rivers fed by glaciers that are slowly disappearing, and in any arid and semiarid areas where droughts become more prolonged. Scientists warn that by 2050, some 200–600 million of the world's poorest and most vulnerable people could face starvation and malnutrition due to these projected effects of climate disruption.

Climate Disruption Threatens Human Health and Economies

According to IPCC and other reports, more frequent and prolonged heat waves in some areas will raise the numbers of deaths and illnesses, especially among older people, people in poor health, and the urban poor who cannot afford air conditioning. Fewer people will die from cold weather, but a 2007 study led by Mercedes Medina-Ramon of the Harvard University School of Public Health suggests that the projected rise in the number of heat-related deaths will exceed the projected drop in the number of cold-related deaths.

A 2012 World Bank report notes that climate change will likely hinder the fight against childhood mortality and malnutrition in many areas. While economic growth has helped areas of sub-Saharan Africa and South Asia in this regard, even the warming that now seems unavoidable is likely to reverse such gains, and warming by up to 4°C (7.2°F) could make these problems far worse than they are today.

A warmer, CO_2-rich world will likely favor rapidly multiplying insects, microbes, toxic molds, and fungi that make us sick, as well as plants that produce pollens that cause allergies and asthma attacks. Longer and more intense pollen seasons will mean more itchy eyes, runny noses, and asthma attacks. Insect pests and weeds will likely multiply, spread, and reduce crop yields.

Microbes that cause infectious tropical diseases such as dengue fever and yellow fever are likely to expand their ranges and numbers if mosquitoes that carry them spread to warmer temperate and higher-elevation areas, as they have begun to do (Figure 19-20). Latin America and other regions are seeing a rise in the incidence of dengue fever. The *Economist* reported in 2007 that Mexico's number of cases had grown fourfold since 2001.

Higher atmospheric temperatures and higher levels of water vapor in urban areas will also increase some forms of air pollution. The greatest effect will be to speed up the rate of the chemical reactions that produce ozone and other harmful chemicals in photochemical smog in urban areas (see Figure 18-9, p. 483). This in turn will likely cause more pollution-related deaths from heart ailments and respiratory problems.

Climate disruption is also likely to take a toll on human economies. Economist Nicholas Stern has estimated that the cost of projected climate change, if current trends continue, would be up to 20% per year of the world's economic activity as measured by gross domestic output. Other researchers say the cost could be higher, because most models used to calculate such costs have not accounted for ocean acidification, loss of glaciers, sea-level rise, or biodiversity losses.

For these reasons, projected climate disruption will very likely force many millions of people to move, in search of better conditions. Environmental scientist Norman Myers estimates that projected climate change during this century could produce at least 150 million, and perhaps 250 million, environmental refugees who will be forced to migrate because of greater hunger, flooding, and drought. (See the online Guest Essay on this topic by Myers.)

Considering the speed with which these and other effects of climate change are occurring, some scientists and analysts now argue that the IPCC and other climate science groups have been too conservative in their warnings of climate disruption (Science Focus 19.2).

Figure 19-19 With warmer winters, exploding populations of mountain pine beetles have munched their way through large areas of lodgepole pine forest in the Canadian province of British Columbia. In this photo, the orange-colored trees are those that are dead or dying—killed by these beetles.

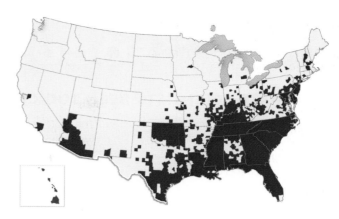

Figure 19-20 Areas in blue show counties in 28 U.S. states where one or both species of mosquitoes that transmit dengue fever had been found as of 2005. This supports the claim by some scientists that projected climate disruption will expand the range of mosquitoes that carry dengue fever, a debilitating and potentially deadly disease.

(Compiled by the authors using data from Natural Resources Defense Council, 2009.)

19-4 What Can We Do to Slow Projected Climate Disruption?

CONCEPT 19-4

We can reduce greenhouse gas emissions and the threat of climate disruption while saving money and improving human health if we cut energy waste and rely more on cleaner renewable energy resources.

Dealing with Climate Disruption Is Difficult

Many scientists argue that our most urgent priority is *to do all we can to avoid any and all estimated climate change tipping points.* Once we reach such a point, there is no going back. The problem is that we do not know how close we are to such tipping points, although there are no indications that we will cross any of them in the immediate future. However, it is estimated that if greenhouse gas emissions continue to rise, we could cross some of these thresholds within a few decades or by the end of this century. Figure 19-21 lists the tipping points that scientists are most concerned about.

Partly because of these possibilities, many analysts believe that addressing climate change is one of the most urgent scientific, political, economic, and ethical issues that humanity faces. But the following characteristics of this complex problem make it difficult to tackle:

- *The problem is global.* Dealing with this threat will require unprecedented and prolonged international cooperation.
- *The problem is a long-term political issue.* Voters and elected officials generally respond better to short-term problems than to long-term threats. Most of the people who could suffer the most serious harm from projected climate disruption during the latter half of this century have not been born.

- *The projected harmful and beneficial impacts of climate disruption are not spread evenly.* For example, higher-latitude nations such as Canada, Russia, and New Zealand could have higher crop yields, fewer deaths in winter, and lower heating bills. Other nations such as Bangladesh could see more flooding and higher death tolls.

- Atmospheric carbon level of 450 ppm

- Melting of all arctic summer sea ice

- Collapse and melting of the Greenland ice sheet

- Severe ocean acidification, collapse of phytoplankton populations, and a sharp drop in the ability of the oceans to absorb CO_2

- Massive release of methane from thawing arctic permafrost and from the arctic seafloor

- Collapse and melting of most of the western Antarctic ice sheet

- Severe shrinkage or collapse of Amazon rain forest

Figure 19-21 Climate scientists have come up with this list of possible climate change tipping points.

SCIENCE FOCUS 19.2

HAVE CLIMATE SCIENTISTS BEEN TOO CONSERVATIVE?

Some people who are skeptical about climate change have accused the IPCC scientists of being alarmists who are over-stating the possible effects of atmospheric warming. However, a growing number of scientists are suggesting the opposite—that the IPCC has been too conservative in its projections.

One scientist who was once quite skep-tical about climate change has become con-vinced not only that it is occurring but also that the IPCC projections are too conserva-tive. Physicist Richard A. Muller founded the Berkeley Earth Surface Temperature project to monitor atmospheric temperatures. In 2011, he and his colleagues found that the average temperature over the earth's land had risen higher and faster than the IPCC had projected. Using 5 times the amount of temperature data available to the IPCC in 2007 along with newer climate models and statistical analysis tools, the team also concluded it is likely that "essentially all of this increase results from the human emis-sion of greenhouse gases." The IPCC had concluded only that most of this warming was due to human activities.

Similarly, according to models used by scientists at the Massachusetts Institute of Technology and other institutions, the trend in atmospheric temperatures will likely take us to an average increase of 3.5 to 7.4C° (6.3 to 13.3F°) by 2100, whereas the IPCC in 2007 projected a rise of 2.4 to 6.4C° (4.3 to 11.5F°) (Figure 19-B). Other IPCC projections, compared with updated esti-mates produced by newer research and model runs, include the following:

- The IPCC projected that the Arctic would not be ice-free in summer until 2070; current projections estimate the ice-free summer could occur by around 2030.
- In 2007, the IPCC projected that in a worst-case scenario, humans would be emitting 30 billion tons of carbon per year by 2100. In fact, we have already exceeded that total, emitting about 31.6 billion tons in 2012, according to the International Energy Agency.
- The losses of ice in Greenland (**Core Case Study**) and Antarctica are at lev-els that the IPCC projected would be occurring about 100 years from now.
- Permafrost is melting faster in areas of Alaska and Siberia than models used by IPCC scientists had projected.
- IPCC projections on the rise in sea level, based mostly on expansion of the oceans' waters, estimated a rise of up to 0.6 meter (1.9 feet) by 2100. Current models that include land-based ice melt project that the sea-level rise will range between 0.7 and 1.9 meters (2.4 to 6.2 feet) by 2100.
- A 2011 study led by University of York researcher Chris Thomas found that plants and animals were migrating to higher elevations twice as fast as projected and that others were mov-ing north 3 times as fast. University of Exeter scientists found in 2011 that extinction rates were more than twice as high as computer models had projected.

No model is perfect (Science Focus 19.1), and there are a number of reasons why the IPCC underestimated some effects of climate change. Generally, they have to do with unforeseen developments such as faster-than-expected melting of Green-land's ice and the sharp rise in emissions from coal-burning facilities, especially in China. Also, some scientists criticize the IPCC for not including a few important fac-tors in their modeling, including the melt-ing of arctic permafrost and the dynamics of melting glaciers.

Other scientists defend the IPCC, say-ing that science must necessarily be a painstaking process governed not by public opinion about dramatic events but by sci-entific principles. In other words, science is and must be conservative by its nature, according to these defenders.

Some critics argue that because of this fundamentally conservative approach to climate science, the consequences of climate change could be more severe because the public might not gain an important sense of urgency from the IPCC reports. Important decisions could be delayed until it is too late to succeed in slowing climate disruption.

Critical Thinking

Do you think that the need for scientists to be conservative in reporting their findings outweighs any need to urgently warn the public about possible threats indicated by their research? Explain.

- *Many proposed solutions, such as sharply reducing or phas-ing out the use of fossil fuels, are controversial* because they could disrupt economies and lifestyles and threaten the profits of economically and politically powerful oil, coal, and natural gas companies.
- *Some social scientists contend that humans are not "hard-wired" to respond to long-term threats.* Scientists and edu-cators are frustrated by how difficult it is to convince the public that projected climate disruption could pose

such a great threat to many people living on earth today, and even more so to their children, grandchil-dren, and great grandchildren.

For several years, many environmental analysts have been calling for urgent action at the national and interna-tional levels to curb greenhouse gas emissions, primarily from the top down, by regulating and taxing such emis-sions. Now a growing number of analysts are contending that this approach, by itself, will not work because, *first,* psychological research indicates that using fear and guilt and promoting sacrifice to change behavior rarely works. *Second,* people are primarily interested in the short-term, not long-term, benefits of changing their behavior. *Third,* for elected officials, spending their efforts on long-term problems is most often not in their best short-term political interests in getting reelected.

What Are Our Options?

There are two basic approaches to dealing with the projected harmful effects of global climate disruption. One, called *mitigation,* is to act to slow it and to avoid climate change tipping points. The other approach, called *adaptation,* is to recognize that some climate change is unavoidable and to try to reduce some of its harmful effects. (We discuss this option in Section 19-5.)

Regardless of the approach we take, many scientists and business leaders see the climate change problem as an urgent challenge, but also as one that is filled with economic opportunity. In 2009, a group of investors who collectively manage more than $13 trillion of the world's financial assets warned that climate disruption poses a threat to the global economy, but they projected that the challenge of making the transition to a low-carbon world presents an enormous global investment opportunity. For example, in the United States, electric cars are being made in Michigan, wind generators are being produced in Illinois, Oregon, and Iowa, and Texas produces more electricity from wind power than all but five of the world's countries.

We now look more specifically at mitigation options and at the opportunities they present.

We Could Control and Prevent Greenhouse Gas Emissions

One way to measure CO_2 emissions is by their global total weight. According to the International Energy Agency, the human population in 2012 emitted about 31.6 billion metric tons, or *gigatons,* of carbon into the atmosphere. Climate scientists have generally agreed that in order to avoid some of the severely harmful effects of climate change, we need to limit the global average temperature increase to 2°C (3.6°F) over the preindustrial global average temperature. To do that, we would have to limit our total carbon emissions to 565 gigatons between now and 2050, according to the latest climate models. However, at the current rate of emissions (which is on the rise), we will easily surpass that total by around 2030.

The problem is that the world's nations and energy companies all together hold reserves of fossil fuels that represent 2,795 gigatons—nearly 5 times what scientists say we can safely emit—according to the Carbon Tracker Initiative, a team of financial and environmental analysts. Currently, the survival of the politically powerful fossil fuel companies and most of the world's national economies

depend on using all or most of these fossil fuel reserves. This is one of the major forces pushing climate change.

Between 2000 and 2008, CO_2 emissions in the United States leveled off, and between 2008 and 2012, they declined. This was due to the combination of an economic slowdown, improved fuel efficiency for new cars, and the partial shift from burning coal to using natural gas to produce the country's electricity. At the same time, CO_2 emissions in China were rising rapidly and by 2010 were twice as high as those in the United States. Analysts argue that unless China, by far the world's largest CO_2 emitter since 2007, can sharply cut its emissions, primarily by shifting from coal to a mixture of natural gas and low-carbon renewable energy sources, the world will likely experience a sharp rise in the global atmospheric temperature during this century.

Figure 19-22 lists a number of ways to slow the rate and degree of atmospheric warming and the resulting climate disruption caused by human activities (**Concept 19-4**). Among these solutions are four major *prevention* strategies that, by 2050, could reduce human greenhouse gas emissions by 57–83%, according to a 2010 study by the U.S. National Academy of Sciences:

1. Improve energy efficiency to reduce fossil fuel use, especially the use of coal.
2. Shift from carbon-based fossil fuels to low-carbon renewable energy resources based on local and regional availability.
3. Stop cutting down tropical forests and plant trees to help remove more CO_2 from the atmosphere.
4. Shift to more sustainable and climate-friendly agriculture (Figure 12-33, p. 309).

CONSIDER THIS. . .

THINKING ABOUT Climate Disruption Prevention Strategies

How are each of the four strategies listed above related to the three scientific **principles of sustainability** (see Figure 1-2, p. 6 or back cover)?

The solutions listed on the right side of Figure 19-22 are *output,* or *cleanup,* strategies, which focus on dealing with CO_2 after it has been produced. Many analysts argue that, rather than spending huge amounts of time and resources on such strategies, we should focus on reducing and preventing CO_2 emissions, as soon as possible.

In addition, some scientists urge us to ramp up efforts to reduce emissions of other greenhouse gases. For example, methane (CH_4) is emitted from landfills, rice paddies, some coal mines, older natural-gas pipelines, and the overall production of natural gas. In 2010, atmospheric scientists Veerabhadran Ramanathan and David Victor estimated that, just by collecting CH_4 from landfills, minimizing the rotting of rice in paddies, and replacing old leaky natural-gas pipelines, we could cut global CH_4 emissions by 40%.

Another climate-changing pollutant is soot generated by inefficient coal-burning power plants, diesel engines, brick-making kilns, and cook stoves, among many other sources. Soot is accumulating on Himalayan Mountain glaciers and arctic ice fields and contributing to the melting of this ice and to atmospheric warming. Ramanathan and Victor report that in China and India, programs to make power plants more efficient, to filter soot from diesel engines, to cut emissions from kilns, and to replace leaky stoves with efficient ones (see Figure 18-28, p. 496) could cut soot levels by almost two-thirds.

Some scientists argue that we could quickly slow atmospheric warming by focusing on reducing black carbon (a component of soot), methane, and hydrofluorocarbons (HFCs), partly because these chemicals are short-lived in the atmosphere, compared to CO_2 emissions, and because we have the technologies to accomplish such cuts fairly quickly. According to a 2011 study by the UNEP, major cuts in these greenhouse gas emissions could potentially reduce the rate of atmospheric warming in half by 2050, while also preventing up to 4 million deaths from air pollution and billions of dollars in crop losses every year.

In earlier chapters, we looked at several ways to prevent or reduce the release of greenhouse gases to the atmosphere in order to slow projected climate disruption. These have involved energy efficiency and the use of renewable energy (Chapter 16), sharply reducing tropical deforestation (Chapter 10), and shifting to more sustainable agriculture (Chapter 12).

We Could Try to Clean Up Greenhouse Gas Emissions

Analysts have proposed several output solutions—strategies for removing some of the CO_2 from the atmosphere or from smokestacks and storing (sequestering) it in other parts of the environment (Figure 19-22, right).

One way to increase the uptake of CO_2 is by quickly implementing a massive, global tree-planting program, especially on degraded land in the tropics. A similar approach is to restore wetlands that have been drained for farming. Wetlands are very efficient at taking up CO_2.

A third cleanup approach is to plant large areas of degraded land with fast-growing perennial plants such as switchgrass (see Figure 16-27, p. 426), which remove CO_2 from the air and store it in the soil. To be sustainable, however, this approach could not involve clearing existing forests.

Along these lines, a fourth carbon sequestration measure that is getting more attention is the use of *biochar*. Scientists have rediscovered this fertilizer that Amazon Basin natives used for growing crops several centuries ago. Producing biochar involves burning biomass such as chicken waste or wood in a low-oxygen environment to make a charcoal-like material—a process that produces no smoke and no odor. The resulting carbon-rich biochar makes an excellent organic fertilizer that helps to keep carbon in the soil for as long as 1,000 years. While making biochar does release CO_2, burying it stores considerably more CO_2 in the soil. Farmers could plant fast-growing trees on

degraded land and use them to make biochar. They could then make money selling some of it as fertilizer.

A fifth cleanup approach is to help the natural uptake and storage of carbon by preserving and restoring natural forests. In 2007, biologist Renton Righelato and climate scientist Dominick Spracklen concluded that the amount of carbon that we could sequester by protecting and restoring forests is greater than the amount of carbon in CO_2 emissions that we would avoid by using biofuels such as biodiesel and ethanol (see Chapter 16 Case Studies, p. 425).

Another approach that has received a lot of attention is to remove CO_2 from the smokestacks of coal-burning power and industrial plants and to store it somewhere. This approach is generally referred to as **carbon capture and storage**, or **CCS** (Science Focus 19.3).

 CONSIDER THIS...

CONNECTIONS Sea Creatures, Carbon Dioxide, and Cement

In 2010, earth science professor Brent Constantz developed a process for removing CO_2 from the smokestack emissions of a power plant and converting it to cement by spraying it with mineral-rich seawater. He developed this CCS method by mimicking sea creatures that convert CO_2 to calcium carbonate to form their skeletons and shells. He envisions converting CO_2 emissions into stone and locking them away essentially forever in the concrete foundations of our cities.

Solutions

Slowing Climate Disruption

Prevention		Cleanup
Cut fossil fuel use (especially coal)		Sequester CO_2 by planting trees and preserving forests and wetlands
Shift from coal to natural gas		
Repair leaky natural gas pipelines and facilities		Sequester carbon in soil using biochar
Improve energy efficiency		Sequester CO_2 deep underground (with no leaks allowed)
Shift to renewable energy resources		
Reduce deforestation		Sequester CO_2 in the deep ocean (with no leaks allowed)
Use more sustainable agriculture and forestry		
Put a price on greenhouse gas emissions		Remove CO_2 from smokestack and vehicle emissions

© Cengage Learning

Figure 19-22 Some ways to slow atmospheric warming and the resulting projected climate disruption during this century (**Concept 19-3**). **Questions:** Which five of these solutions do you think are the best ones? Why?

Top: Mark Smith/Shutterstock.com. Center: Yegor Korzh/Shutterstock.com. Bottom: ©pedrosala/Shutterstock.com.

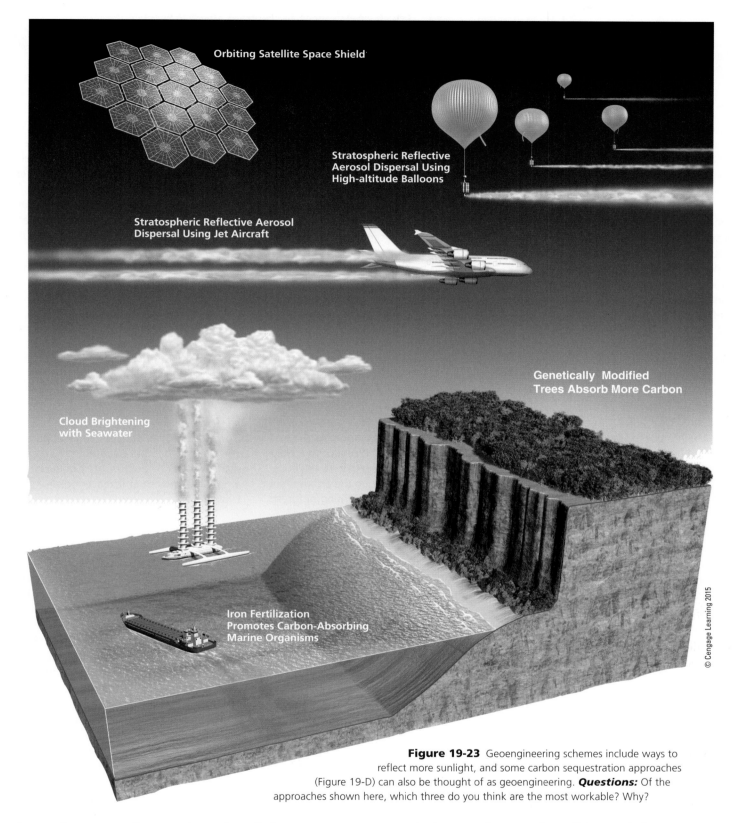

Figure 19-23 Geoengineering schemes include ways to reflect more sunlight, and some carbon sequestration approaches (Figure 19-D) can also be thought of as geoengineering. **Questions:** Of the approaches shown here, which three do you think are the most workable? Why?

Labels in figure:
- Orbiting Satellite Space Shield
- Stratospheric Reflective Aerosol Dispersal Using High-altitude Balloons
- Stratospheric Reflective Aerosol Dispersal Using Jet Aircraft
- Genetically Modified Trees Absorb More Carbon
- Cloud Brightening with Seawater
- Iron Fertilization Promotes Carbon-Absorbing Marine Organisms

© Cengage Learning 2015

Some Promote Geoengineering Schemes to Ward Off Climate Change

Many scientists and engineers want to use strategies that fall under the umbrella of *geoengineering*, or trying to manipulate certain natural conditions to help counter an enhanced greenhouse effect (Figure 19-23). In recent years, some of these scientists and engineers have become discouraged by the slow response of governments to what they see as the global emergency of projected climate dis-

IS CAPTURING AND STORING CO_2 THE ANSWER?

Figure 19-D shows several techniques for removing some of the CO_2 from the atmosphere and from smokestacks, and storing (sequestering) it in other parts of the environment.

Some researchers are studying ways to remove CO_2 from the exhaust coming out of smokestacks at coal-burning power and industrial plants. They hope to pump it deep underground into abandoned coal beds and oil and gas fields or to liquefy it and inject it into sediments under the seafloor (Figure 19-D). Another idea is to turn the collected CO_2 into charcoal or into carbonate rock. These schemes are generally referred to as *carbon capture and storage (CCS)*.

One problem with this approach is that it would take many years. The International Energy Agency estimates that we would need more than 200 CCS power plants by 2030 in order to keep atmospheric warming below 3°C, and currently, no commercial-scale CCS plant has been built.

Also, it is becoming clear to scientists that stored CO_2 would have to remain sealed from the atmosphere forever. Any large-scale leaks caused by earthquakes or other shocks, as well as any number of smaller continuous leaks from CO_2 storage sites around the world, could dramatically increase atmospheric warming and climate disruption in a very short time. Some scientists doubt that we can develop the technology to guarantee that all such stored CO_2 would remain safely sequestered without some of it leaking out.

There are other serious problems. Removing all or most of the CO_2 from smokestack emissions and transporting and storing it takes a lot of energy. This would greatly reduce the net energy yield and significantly raise the cost of burning coal to supply electricity. Also, these measures would do nothing to reduce the massive amounts of CO_2 and other greenhouse gases emitted by using fossil fuels in transportation and food production. Nor would they do anything to curb land-use practices that add CO_2 to the atmosphere, such as the cutting and burning of forests.

Most worrisome of all to some scientists, CCS could promote more use of coal, the world's most environmentally harmful fuel.

Environmental scientists including Peter Montague argue that CCS is a costly and extremely risky *output* solution to a serious problem that can be dealt with by using a variety of cheaper, quicker, and safer *input*, or *prevention*, approaches (Figure 19-22, left). They argue for producing the electricity we need by using low-carbon energy resources such as energy efficiency, solar cells, and wind power. To these scientists, when we face a problem such as CO_2 coming out of a smokestack or exhaust pipe, the most important question to ask is not *What do we do with it?* but, *How do we avoid producing the CO_2 in the first place?*

Critical Thinking

Do you think that a government-supported crash program to build coal-burning power plants with CCS components would be a good investment of taxpayer dollars? Explain.

---

ruption with its potentially serious harmful effects. They suggest that we look at the implications and costs of using large-scale geoengineering schemes as a last resort, if humanity fails to deal with the world's projected climate change emergency soon enough.

For example, some scientists have studied the 1991 eruption of Mount Pinatubo in the Philippines and the fact that the average global atmospheric temperature cooled for about 15 months after the eruption. The volcano pumped massive amounts of SO_2 into the atmosphere where chemical reactions converted it to sulfate particles, which were thought to have reflected some incoming sunlight into space, thus cooling the troposphere. These scientists have suggested using balloons, large jet planes, or giant cannons to inject sulfate particles into the stratosphere to get the same effect (Figure 19-23). Other scientists have called for placing a series of giant mirrors in orbit above the earth to reflect incoming sunlight for the same purpose.

Some scientists reject the idea of launching sulfates into the stratosphere as being too risky because of our limited knowledge about possible unknown effects. For example, if the sulfates reflected too much sunlight, they could reduce evaporation enough to alter the water cycle and worsen the already dangerous droughts in Asia and Africa. Also, a 2008 study by atmospheric scientist Simone Tilmes indicated that chlorine released by reactions involved in this scheme could speed up the thinning of the earth's vital ozone layer (see Chapter 18, p. 497).

Even if it worked, this approach would allow CO_2 levels in the lower atmosphere to continue rising and adding to climate disruption and ocean acidification. Some scientists would deal with the latter problem by building a global network of thousands of chemical plants that would remove acid from seawater to reduce its acidity. But this would be quite costly and also could have unpredictable and possibly harmful ecological effects on ocean ecosystems.

🔍 CONSIDER THIS...

THINKING ABOUT Tinkering with the Stratosphere

Do you think the projected benefits of injecting large quantities of sulfate particles into the stratosphere to help cool the troposphere would outweigh any projected negative results? Explain.

Figure 19-D Solutions: Several proposed methods for removing carbon dioxide from the atmosphere or from smokestacks and storing it in soil, plants, and deep underground reservoirs, as well as in sediments beneath the ocean floor. These proposed solutions are based on an *output approach*. **Questions:** Which two of these solutions do you think would work best? Which two would be the least effective? Why?

Another geoengineering proposal involves seeding the oceans with iron to promote the growth of more marine algae and other phytoplankton, which absorb huge amounts of CO_2 from the atmosphere as they grow (Figure 19-23). When they die, they sink to the seafloor, carrying the carbon with them. The proposal is controversial because the long-term effects of adding large amounts of iron to ocean ecosystems are unknown.

In a 2009 study involving an Atlantic Ocean test site seeded with iron, it was found that shrimp-like animals called copepods had quickly devoured the large swarm of algae produced by the seeding. If this were to happen regularly, more carbon would move up the food chain instead of sinking to the ocean floor, and eventually, it would be released by large organisms as CO_2, thus adding to the CO_2 problem. And according to a 2013 study by marine scientist Daniel Harrison, the amount of CO_2 removed from the atmosphere by this method is largely countered by the amount added to the atmosphere when fossil fuels are burned to produce the iron salts and to power the ships used in the seeding process. Thus, the net effect of this process is unknown.

Scientist James Lovelock has proposed another geoengineering idea. It would involve anchoring large vertical pipes in the oceans as part of a system that would allow wave motion to pump nutrient-rich water up from the deep-ocean layer to fertilize algae on the ocean surface. Lovelock contends that the resulting algal blooms would remove CO_2 from the atmosphere and emit dimethyl sulfide (see Figure 3-20, p. 69), which would contribute to the formation of low clouds that would reflect sunlight.

According to skeptical scientists, one major problem with most of these technological fixes and with some carbon capture and storage schemes (Science Focus 19.3) is that they require huge investments of energy and materials, and there is no guarantee that they will work. Geoengineering schemes all depend on complex machinery running constantly, flawlessly, and essentially forever, primarily to pump something from one place to another in the environment. If we rely on these systems and continue emitting greenhouse gases, and if the systems then fail, atmospheric temperatures will likely soar at a rapid rate and essentially ensure severe climate disruption.

Another problem with this approach is the political and ethical aspect. One country or one company could undertake a geoengineering project that could be harmful to whole regions or to the entire planet. Critics ask how any decision to employ a geoengineering project should be made and by whom. Who will take responsibility for controlling it, and who will pay for any damages? Such questions have yet to be answered, and critics such as climate scientist Ken Caldeira argue that they should be explored before any major projects get started.

Critics of large-scale geoengineering schemes argue for slowing climate change by using prevention approaches, such as improving energy efficiency, replacing fossil fuels with already available renewable energy resources, and drastically reducing tropical deforestation. They say this would be a better investment of limited resources than would gambling on large-scale, costly changes to the global environment that could have unknown and potentially long-lasting harmful effects.

Greenhouse Gases Can Be Regulated as Pollutants

One approach to promoting the preventive solutions listed in Figure 19-22 is to strictly regulate carbon dioxide and methane as air pollutants. In Chapter 1, we defined a pollutant as a chemical or other agent such as noise or heat in the environment that proves harmful to the health, survival, or activities of humans or other organisms.

Carbon dioxide is now classified as an air pollutant, but you might wonder how CO_2 can possibly be classified as such, because it is a natural and important part of the carbon cycle (see Figure 3-17, p. 66). Humans and most animal life-forms constantly emit CO_2 into the atmosphere, while plants remove it and use it for photosynthesis. However, one characteristic of any chemical that determines whether or not it should be classified as a pollutant is its *concentration* in the atmosphere, soil, water, or plant or animal tissues. For example, if we were to fill up a room with CO_2, everyone there would die very quickly from lack of oxygen. When we add CO_2 to the atmosphere faster than the carbon cycle can remove it, we help to warm the atmosphere and contribute to projected climate disruption. At some point in that process (at some atmospheric concentration), CO_2 becomes an air pollutant.

This is why, in 2009, the U.S. Environmental Protection Agency found that CO_2, CH_4, N_2O, and three other greenhouse gases are a danger to public health and welfare and should be listed as pollutants under the Clean Air Act. However, politically and economically powerful coal, oil, natural gas, and utility companies that make their money by producing and burning carbon-containing fossil fuels have successfully opposed regulating carbon dioxide for more than three decades and hope to overturn or greatly weaken any government efforts to regulate emissions of carbon dioxide and other greenhouse gases.

We Can Put a Price on Carbon Emissions

Many economists argue that it is important to put a price on carbon emissions in order to include some of the harmful environmental and health costs of using fossil fuels in fuel prices, in keeping with the full-cost pricing **principle of sustainability** (see Figure 1-5, p. 9 or back cover). One way to do this is to phase in *carbon taxes* or *fees* on each unit of CO_2 or CH_4 emitted by fossil fuel use, or to levy *energy taxes* or *fees* on each unit of fossil fuel that is burned (Figure 19-24). Lowering taxes on income, wages, and profits to offset such taxes could help to make such a strategy more politically acceptable, and it might also stimulate the economy. In other words, *tax pollution, not payrolls and profits.*

In British Columbia, Canada, this approach has had good results. The province levied a tax of $30 per metric ton of CO_2 emitted, and then lowered corporate and personal income tax rates, which are now among the lowest of such rates among the world's wealthier nations. In the 4 years after the carbon tax was created, British Columbia's greenhouse gas emissions dropped by 4.5% while its population grew, and gasoline sales dropped by 2% while rising by 5% overall in Canada.

Some European countries are gradually phasing in such a tax shift. For example, since 2008, Ireland has taxed carbon dioxide emissions and used the revenue to help reduce the country's staggering debt and by 2012 had seen a 15% drop in its CO_2 emissions. However, economists warn that the biggest global players in climate change—the United States, Europe, and China—would need to tax carbon emissions at basically the same rates in order for any carbon tax to be effective. A 2012 study by the Congressional Budget Office estimated that between 2012 and 2022, a modest carbon tax could generate about $1.25 trillion in U.S. tax revenues.

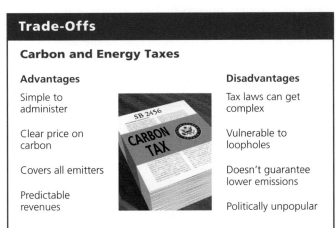

Trade-Offs

Carbon and Energy Taxes

Advantages	Disadvantages
Simple to administer	Tax laws can get complex
Clear price on carbon	Vulnerable to loopholes
Covers all emitters	Doesn't guarantee lower emissions
Predictable revenues	Politically unpopular

© Cengage Learning

Figure 19-24 Using carbon and energy taxes or fees to help reduce greenhouse gas emissions has advantages and disadvantages. ***Question:*** Which two advantages and which two disadvantages do you think are the most important and why?

Trade-Offs

Cap-and-Trade Policies

Advantages	Disadvantages
Clear legal limit on emissions	Revenues not predictable
Rewards cuts in emissions	Vulnerable to cheating
Record of success	Rich polluters can keep polluting
Low expense for consumers	Puts variable price on carbon

© Cengage Learning

Figure 19-25 Using a cap-and-trade policy to help reduce greenhouse gas emissions has advantages and disadvantages. **Question:** Which two advantages and which two disadvantages do you think are the most important and why?

Another way to include some costs of carbon emissions in market prices is to place a cap on total human-generated CO_2 and CH_4 emissions in a country or region, issue permits to emit these pollutants, and then let polluters trade their permits in the marketplace. This *cap-and-trade* or *pollution reduction approach* (Figure 19-25) has a political advantage over taxes, but it is difficult to manage because there are so many different emitters of greenhouse gases. In addition, this approach works only if the original caps are set low enough to encourage a serious reduction in greenhouse gas emissions and are lowered on a regular basis to promote further cuts. A related and more politically acceptable approach is to use a *cap-and-rebate* scheme in which all money made by any trading of permits would go back to the public in the form of cash payments.

Environmental economists argue that, regardless of whether governments use taxes or a cap-and-trade or cap-and-rebate policy, the most critical goal is to put a price on carbon emissions to get all emitters to come as close as possible to paying the total estimated harmful environmental and health costs of their carbon emissions. Economist Gary Yohe and others contend that the resulting higher costs for fossil fuels will spur innovation in finding ways to reduce carbon emissions, improve energy efficiency, and phase in a mix of low-carbon renewable energy alternatives.

Government Subsidies Can Be Used to Address Climate Disruption

Another way in which governments can address climate disruption is to level the economic playing field by greatly increasing *government subsidies* (direct payments and other forms of financial support) to businesses and individuals to encourage their use of energy-efficiency technologies, low-carbon renewable energy sources, and more sustain-

able food production methods. This would also include phasing out or sharply reducing subsidies and tax breaks that encourage the use of fossil fuels and nuclear power, unsustainable agriculture, and the clearing of forests. In other words, during the next 2 to 3 decades, we could shift from environmentally degrading to environmentally sustaining subsidies and tax breaks.

One important form of government support is funding of research and development, a great deal of which has gone to the fossil fuel and nuclear energy industries in decades past. Governments could shift this research and development focus to clean energy industries. With such help, these industries could develop innovations that would lower the costs of energy resources such as wind, solar power, and geothermal energy so that they could compete more favorably in the marketplace with fossil fuels, which have been heavily subsidized for decades.

Governments Can Cooperate Internationally

Governments have entered into international climate negotiations. In December 1997, delegates from 161 nations met in Kyoto, Japan, to negotiate a treaty to slow atmospheric warming and its projected climate disruption. The first phase of the resulting *Kyoto Protocol* went into effect in February 2005 with 187 of the world's 194 countries (not including the United States) ratifying the agreement by late 2009.

This agreement required the 36 participating more-developed countries to cut their emissions of CO_2, CH_4, and N_2O to certain levels by 2012, when the treaty was to expire. Less-developed countries were excluded from this requirement, because such reductions would curb their economic growth. In 2005, participating countries began negotiating a second phase of the treaty, which was supposed to go into effect after 2012. That year, at the annual UN climate conference, held in Doha, Qatar, participating nations agreed to keep the Kyoto Protocol in effect until at least 2015 and to commit to more intensive but unspecified actions to reduce greenhouse gases, which were to be explored in 2013.

Some analysts see current and proposed international climate agreements as weak and slow responses to an urgent global problem, pointing out that atmospheric carbon dioxide levels have increased 35% faster than was expected under the Kyoto agreement. Many have also concluded that a top-down international agreement will be too slow and ineffective. Some call for the nations that produce most of the world's greenhouse gas emissions—especially the United States, China, India, Brazil, and the European Union nations (Figure 19-9)—to get together and work out an agreement to sharply lower their emissions.

Another way in which governments can cooperate is through *technology transfer*. Governments of more-developed countries could help to fund the transfer of the latest green technologies to less-developed countries so that they could bypass older, energy-inefficient and polluting technologies.

Some analysts argue that helping poorer countries to deal with the harmful effects of projected climate disruption makes sense, because many of these countries will suffer the most from such effects, which have been caused mostly by more-developed countries. They also point out that even if all countries do not suffer the worst effects of climate disruption, the whole world will eventually suffer very harmful effects due to rampant widespread poverty and homelessness, mass migrations of environmental refugees, and sharply decreased international security. Environmental analyst Lester Brown has suggested that increasing the current tax on each international currency transaction by a quarter of a penny could finance effective technology transfer.

Another important form of international cooperation involves efforts to protect large forests, especially tropical forests, which are important players in the carbon cycle. The effects of deforestation account for 12–17% of global greenhouse gas emissions, according to the Environmental Policy Institute. One such effort is the UN's Reducing Emissions from Deforestation and Forest Degradation (REDD) program, which helps to fund efforts by farmers and others who depend on forests and who seek to use them more sustainably. Since it was founded in 2008, the program has helped 44 countries in Africa, Asia, and Latin America to reduce deforestation. Brazil has slowed deforestation by 40% since 2008 with the help of the REDD program, and is projected to reach its reduction goal of 80% by 2020.

GOOD NEWS

Some Countries, States, and Localities Are Leading the Way

Some nations are leading others in facing the challenges of projected climate disruption. Costa Rica aims to be the first country to become *carbon neutral* by cutting its net carbon emissions to zero by 2030. The country generates 78% of its electricity with renewable hydroelectric power and another 18% from renewable wind and geothermal energy.

Some analysts are urging rapidly developing countries such as China and India to shift toward more sustainable, low-carbon economic development by leapfrogging over the traditional forms of economic development that have led to the threat of climate disruption. If China and India continue their extensive coal burning and other high-carbon activities, the world will not be able to avoid the projected harmful effects of climate disruption.

While China emits more CO_2 into the atmosphere than any other country, it also has one of the world's most intensive energy efficiency programs, according to ClimateWorks, a global energy consulting firm. Chinese automobile fuel-efficiency standards are higher than U.S. standards, but enforcement of the standards is still weak. China's government is working with the country's top 1,000 industries

Figure 19-26 Portland, Oregon, has reduced car use by developing an efficient and widely used light-rail mass transit system. For more than 40 years, Portland has been consistently near the top of U.S. green city lists.

to implement tough energy efficiency goals. China is also rapidly becoming the world's leader in developing and selling solar cells, solar water heaters, wind turbines, advanced batteries, and plug-in hybrid electric cars.

By 2010, at least 30 U.S. states had set goals for reducing greenhouse gas emissions. California plans to get 33% of its electricity from low-carbon renewable energy sources by 2030. Since 1990, local governments in more than 650 cities around the world (including more than 450 U.S. cities) have established programs to reduce their greenhouse gas emissions.

The first major U.S. city to do this was Portland, Oregon. Between 1993 and 2012, the city cut its per capita greenhouse gas emissions below 1990 levels by 26%—more than any other U.S. city. Portland promotes energy-efficient buildings and the use of electricity from wind and solar sources. It has also built a network of bicycle trails and has greatly expanded its mass transit system (Figure 19-26). In 2009, the city implemented a Climate Action Plan with the goal of cutting its greenhouse gas emissions to 40% below 1990 levels by 2030 and to 80% below those levels by 2050. Far from hurting Portland's economy, these strategies have produced an economic boom and have saved the city $2 million a year on its energy bills.

Some Companies Are Reducing Their Carbon Footprints

Leaders of some of the most prominent U.S. companies, including Alcoa, DuPont, Ford Motor Company, General Electric, Johnson & Johnson, PepsiCo, and Shell Oil, have joined with leading environmental organizations to form the U.S. Climate Action Partnership. They have called on the government to enact strong national climate change legislation, realizing that significantly reducing their own greenhouse gas emissions will save them money.

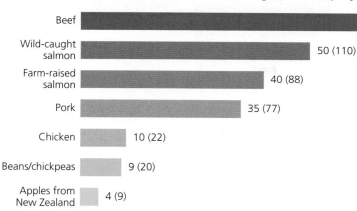

Estimated kilograms (pounds) of CO_2 equivalents per year per person eating one serving per week

Beef	120 (264)
Wild-caught salmon	50 (110)
Farm-raised salmon	40 (88)
Pork	35 (77)
Chicken	10 (22)
Beans/chickpeas	9 (20)
Apples from New Zealand	4 (9)

Figure 19-27 Comparison of diet items in terms of greenhouse gas emissions per year by a person eating one serving per week, in CO_2 equivalents. (Compiled by the authors using data from the Swedish Environmental Protection Agency.)

For example, between 1990 and 2006, DuPont slashed its energy usage and cut its greenhouse gas emissions by 72% and saved $3 billion, while the company grew its business by almost a third.

Leaders of these and many other major companies see an enormous profit opportunity in developing or using energy-efficient and cleaner technologies. For example, the production of building materials is energy-intensive and produces 10–12% of all U.S. CO_2 emissions. Companies such as the brick maker Calstar Products have found ways to produce building materials with much lower carbon outputs. Calstar recycles fly ash from coal-burning facilities and it has reduced the carbon footprint of brick-making by 80%. Many companies now recognize that "there is gold in going green."

Colleges and Universities Are Going Green

Many colleges and universities are also taking action. The Tempe campus of Arizona State University (ASU) boasts the largest collection of solar panels of any American university. ASU was the first university in the country to establish an entire academic program with its School of Sustainability.

The College of the Atlantic in Bar Harbor, Maine (USA) has been carbon neutral since 2007. It gets all of its electricity from renewable hydropower, and many of its buildings are heated with the use of renewable wood pellets. Some students there built a wind turbine that powers a nearby organic farm, which offers organic produce to the campus, to local schools, and to food banks.

EARTH University in Costa Rica has a mission to promote sustainable development in tropical countries. Its sustainable agriculture degree program has attracted students from more than 20 different nations.

The number of examples such as these is large and growing. Here are just a few more: Students and fac-

ulty at Oberlin College in the U.S. state of Ohio created a Web-based system in some of the school's dorms that monitors use of energy and water, giving students real-time feedback that can help them to reduce their waste of these resources. The University of Wisconsin–River Falls hosts the St. Croix Institute for Sustainable Community Development—a program for demonstrating sustainable development principles that has served many U.S. companies and communities. In 2012, some students at Swarthmore College asked their administrators to sell the college endowment's holdings in large fossil fuel companies, and students on dozens of other campuses have followed their lead by making similar requests. (See the Campus Sustainability features that appear in several other chapters of this book.)

Every Individual Choice Makes a Difference

Each of us will play a part in the projected acceleration of atmospheric warming and climate disruption during this century. Whenever we use energy generated by fossil fuels, for example, we add a certain amount of CO_2 to the atmosphere. Each use of energy adds to an individual's *carbon footprint,* the amount of CO_2 generated directly and indirectly by any one person or group. You can calculate your own carbon footprint by using a footprint calculator, several of which are available online.

One important aspect of your carbon footprint is your diet. Foods vary greatly in the greenhouse gases that result from their production and delivery (Figure 19-27). For example, processed foods require much more energy to produce than do whole foods such as fresh fruit and vegetables. Meat production, especially as it is done on factory farms, involves far higher greenhouse gas emissions than production of grains and vegetables. By choosing foods carefully, you can reduce your carbon footprint.

Another major factor in the average carbon footprint is nonfood purchases of common goods such as clothing and electronic devices. The Union of Concerned Scientists has estimated that such purchases represent 26% of the average American's carbon footprint, noting that this is an area where there is a great opportunity for cutting carbon emissions. There are many sources of information on the Internet and elsewhere comparing the carbon emissions involved with the production of various goods and services.

Some websites suggest ways for you to reduce, or *offset,* some of your carbon dioxide emissions by helping to fund or participate in projects such as forest restoration and renewable energy development. However, buyers need to do their homework, because some carbon-offset programs are not real and do not actually produce emissions cuts. Critics of such carbon-offset schemes argue that they can encourage people to continue producing greenhouse gases instead of making changes to reduce their carbon footprints by, for example, driving less, using mass transit, or consuming less electricity.

Figure 19-28 lists some ways in which you can cut your CO_2 emissions. One person taking each of these steps makes a small contribution to reducing global greenhouse gas emissions, but when millions of people take such steps, the cumulative result is quite large. In a 2009 study, human ecologist Thomas Dietz estimated that if each American took steps such as those listed in Figure 19-28, the cumulative impact of such individual actions could cut U.S. emissions of CO_2 by more than 7% within a decade.

Figure 19-28 Individuals matter: You can reduce your annual emissions of CO_2. **Question:** Which of these steps, if any, do you take now or plan to take in the future?

What Can You Do?

Reducing CO₂ Emissions

- Calculate your carbon footprint (there are several helpful websites)
- Drive a fuel-efficient car, walk, bike, carpool, and use mass transit
- Reduce garbage by reducing consumption, recycling, and reusing more items
- Use energy-efficient appliances and compact fluorescent or LED lightbulbs
- Wash clothes in warm or cold water and hang them up to dry
- Close window curtains to keep heat in or out
- Use a low-flow showerhead
- Eat less meat or no meat
- Heavily insulate your house and seal all air leaks
- Use energy-efficient windows
- Set your hot-water heater to 49°C (120°F)
- Plant trees
- Buy from businesses working to reduce their emissions

© Cengage Learning

19-5 How Can We Adapt to Climate Change?

CONCEPT 19-5
While we can prepare for some climate change that is now inevitable, we could realize important economic, ecological, and health benefits by drastically reducing greenhouse gas emissions with the goal of slowing projected climate disruption.

We Can Prepare for Climate Disruption

According to global climate models, the world will have to make a 50–85% cut in emissions of greenhouse gases by 2050 to stabilize concentrations of these gases in the atmosphere. This would help to prevent the planet from heating up by more than 2C° (3.6F°) and to head off rapid changes in the world's climate and the projected harmful effects that would result.

However, because of the difficulty of making such large reductions, many analysts believe that while we work to slash emissions, we should also begin to prepare for the likely harmful effects of projected climate disruption. Figure 19-29 shows some ways to do so.

For example, relief organizations such as the International Red Cross are turning their attention to projects such as expanding mangrove forests as buffers against storm surges, building shelters on high ground, and planting trees on slopes to help prevent landslides. In the face of projected higher levels of precipitation and rising sea levels, seawall design and construction will likely be a major growth industry. And low-lying countries such as Bangladesh are trying to figure out what to do with millions of environmental refugees who will be flooded out by rising sea levels and more intense storms. Some cities plan to establish cooling centers to shelter residents during increasingly intense heat waves.

Several cities have developed adaptation plans. In London, England, a flood-control barrier is being strengthened at a cost of about $1 billion (Figure 19-30). Other cities, including the U.S. cities of New York City and Seattle, Washington, and several states have also developed adaptation plans. California is beefing up its forest firefighting capabilities and is proposing desalination plants to help relieve projected water shortages, which will worsen as some of its mountain glaciers melt. In the aftermath of flooding caused by Hurricane Sandy in 2012, New York City is considering new floodwalls and flood gates for its subway system. In anticipation of rising sea levels, Boston has elevated one of its sewage treatment plants.

Some people fear that emphasizing these adaptation approaches will distract us from the more urgent need to reduce greenhouse gas emissions. However, to some analysts, projected climate change is already such a serious threat that we have no alternative but to implement both prevention and adaptation strategies, and we have no time to lose in doing this on an emergency basis.

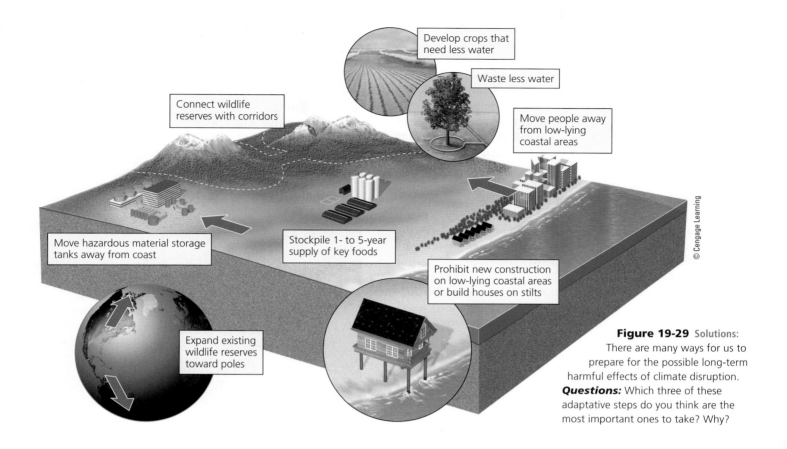

Figure 19-29 Solutions: There are many ways for us to prepare for the possible long-term harmful effects of climate disruption. **Questions:** Which three of these adaptative steps do you think are the most important ones to take? Why?

Labels in figure:
- Develop crops that need less water
- Waste less water
- Connect wildlife reserves with corridors
- Move people away from low-lying coastal areas
- Move hazardous material storage tanks away from coast
- Stockpile 1- to 5-year supply of key foods
- Prohibit new construction on low-lying coastal areas or build houses on stilts
- Expand existing wildlife reserves toward poles

A No-Regrets Strategy

The threat of climate disruption pushes us to look for preventive solutions such as those listed in Figure 19-22. But suppose we had the ability to look into the future and, in doing so, we found that the climate models were wrong and that climate disruption was not a serious threat after all. Should we then return to the present and abandon the search for preventive solutions?

A number of climate and environmental scientists, economists, and business leaders say *no* to this question, and they call for us to begin implementing such changes now as a *no-regrets strategy*. They argue that actions such as those listed in Figure 19-22 should be implemented in any case because they will lead to very important environmental, health, and economic benefits.

For example, improving energy efficiency has numerous economic and environmental advantages (see Figure 16-4, p. 404). Individuals, companies, and local governments can save large amounts of money and improve their economic security by cutting their energy bills. In addition, the use of fossil fuels, especially coal, causes a

Figure 19-30 This high-tech, movable flood-control barrier—the second largest in the world—was built across the Thames River to prevent flooding from very high tides and storm surges in London, England.

great deal of air pollution, which harms or kills millions of people every year. Thus, a sharp decrease in the use of these fuels will improve the overall health of the population, save lives, and save money now spent on health care. In addition, cutting coal use will greatly reduce the land disruption resulting from surface mining.

Another benefit of the no-regrets strategy is that, by relying more on a mix of renewable, domestic energy resources, many countries will cut their costly dependence on imports of fossil fuels. The money that they now spend for fuel imports will then be available for investing in health care, education, and jobs, which will further bolster their economic security. In addition, sharply decreasing or halting the destruction of tropical forests would help to preserve the increasingly threatened biodiversity that makes up an important part of the earth's vital natural capital (see Figure 1-3, p. 7).

Thus, leaving out any concerns about climate disruption, there are plenty of reasons for implementing a no-regrets strategy for sharply reducing our carbon outputs on an urgent basis.

Big Ideas

- Considerable scientific evidence indicates that the earth's atmosphere is warming, mostly because of human activities, and that this is likely to lead to significant climate disruption during this century that could have severe and long-lasting harmful consequences.

- Reducing the projected harmful effects of rapid climate disruption during this century requires emergency action to increase energy efficiency, sharply reduce greenhouse gas emissions, and rely more on renewable energy resources.

- While we can prepare for some climate change that is now inevitable, we could realize important economic, ecological, and health benefits by drastically reducing greenhouse gas emissions with the goal of slowing climate change.

TYING IT ALL TOGETHER Melting Ice in Greenland and Sustainability

Jan Martin Will/Shutterstock.com

In this chapter, we have seen that human activities such as the burning of fossil fuels and the widespread clearing and burning of forests for planting crops have contributed to higher levels of greenhouse gases in the atmosphere. This has contributed to increased atmospheric warming, which in turn is projected to result in climate disruption during this century.

The effects of climate disruption include rapid, increased melting of land-based ice (**Core Case Study**) and sea ice, worsening drought, rising sea levels, ocean acidification, declining biodiversity, and severe threats to human health and economies. Many of these effects could further accelerate climate disruption in worsening spirals of change known as positive feedback loops.

We can apply the six **principles of sustainability** (see Figure 1-2, p. 6, Figure 1-5, p. 9, or back cover) to help reduce the harmful effects of projected climate disruption. We can reduce inputs of pollutants and greenhouse gases into the atmosphere by relying more on direct and indirect forms of solar energy than on fossil fuels; recycling and reusing matter resources much more widely than we do now; and mimicking biodiversity by using a variety of low-carbon renewable energy resources in place of fossil fuels and nuclear power. We can advance toward these goals by including the harmful environmental and health costs of fossil fuel use in market prices; seeking win-win solutions that will benefit both the economy and the environment; and giving high priority to passing along to future generations an environment in which they too can thrive.

Chapter Review

Core Case Study

1. Summarize the story of Greenland's melting glaciers (**Core Case Study**) and the possible effects of this process. Explain how it fits into the IPCC projections about climate change.

Section 19-1

2. What is the key concept for this section? Explain why weather and climate are not the same. Summarize the trends in atmospheric warming and cooling over the past 900,000 years and during the last century. How do scientists get information about past temperatures and climates? List eight pieces of evidence that indicate that the atmosphere has been warming since 1975.

Section 19-2

3. What is the key concept for this section? What is the greenhouse effect and why is it so important to life on the earth? What are greenhouse gases and how do they affect the atmosphere? What are three key greenhouse gases and how do they differ in atmospheric lifetimes and warming potentials? What is the role of water vapor in the atmosphere and how does it relate to other greenhouse gases in the warming process?

4. How have human activities affected atmospheric greenhouse gas levels during the last 275 years and especially in the last 30 years? List the major human activities that add CO_2, CH_4, and N_2O to the atmosphere. How do scientists use models to make projections about future temperature changes? What do a number of climate models project about temperature changes during this century? Explain how each of the following might contribute to projected atmospheric warming and resulting global climate disruption: **(a)** a hotter sun, **(b)** the oceans, **(c)** cloud cover, and **(d)** air pollution.

Section 19-3

5. What is the key concept for this section? What makes the current atmospheric warming problem urgent? Define **climate change tipping points** and give three examples. Briefly describe the possible effects of a warmer atmosphere on arctic sea ice and land-based ice, permafrost, and sea levels. Define **ocean acidification** and explain how and why it is occurring. Explain how atmospheric warming, global drought, extreme weather, and an intensified water cycle are related. What are the possible effects of climate disruption on biodiversity, food production, and human health and economies? Pick three of these effects and describe how each can become part of a positive

feedback loop leading to increased climate disruption. Summarize the science community's discussion of whether or not the IPCC scientists were too conservative in their projections.

Section 19-4

6. What is the key concept for this section? What are five factors that make it difficult to deal with the problem of projected climate disruption? What are two basic approaches to dealing with the projected harmful effects of global climate disruption? List four major prevention strategies for reducing greenhouse gas emissions. Why do some scientists argue for focusing on reducing greenhouse gases other than carbon dioxide? List seven strategies for trying to clean up carbon emissions. What is **carbon capture and storage (CCS)**? Describe three problems associated with capturing and storing carbon dioxide emissions. What are geoengineering schemes (give two examples) and what are two major problems with most of them?

7. Summarize the argument for regulating greenhouse gases as pollutants. List two ways to put a price on carbon emissions. What are the advantages and disadvantages of using taxes on carbon emissions or energy use to help reduce greenhouse gas emissions? What is cap and trade and what are the advantages and disadvantages of using it to help reduce greenhouse gas emissions? What is the Kyoto Protocol and how has it progressed? What are three ways in which governments can cooperate internationally to deal with projected climate change?

8. Give two examples of what some countries are doing on their own to deal with climate change. Give two examples of what some companies are doing and two examples of what some colleges and universities have done to reduce their carbon footprints. List five ways in which you can reduce your carbon footprint.

Section 19-5

9. What is the key concept for this section? List five ways in which we can prepare for the possible long-term harmful effects of climate disruption. Describe the no-regrets strategy for dealing with the problems associated with energy waste and fossil fuel use, regardless of climate change.

10. What are this chapter's *three big ideas*? Explain how we can apply the six **principles of sustainability** (see Figure 1-2, p. 6, Figure 1-5, p. 9, or back cover) to the problem of projected climate disruption.

Note: Key terms are in bold type.

Critical Thinking

1. If you had convincing evidence that at least half of Greenland's glaciers (**Core Case Study**) were sure to melt during this century, would you argue for taking serious actions now to slow projected climate disruption? If so, summarize the arguments you would use. If not, explain why you would be opposed to such actions.

2. A top U.S. economic adviser to a president once gave a speech in Williamsburg, Virginia (USA), to representatives of governments from a number of countries. He told his audience not to worry about atmospheric warming because the average global temperature increases predicted by scientists were much less than the temperature increase he had experienced that day in traveling from Washington, DC, to nearby Williamsburg. What was the flaw in his reasoning? Write an argument that you could use to counter his claim.

3. China relies on coal for two-thirds of its commercial energy usage, partly because the country has abundant supplies of this resource. As a result, China's use of coal is playing a large role in the warming of the global atmosphere. Do you think China is justified in expanding its use of this resource as other countries, including the United States, have done with their coal resources? Explain. What are China's alternatives?

4. Explain why you would support or oppose each of the strategies listed in Figure 19-22 for slowing projected climate disruption caused by atmospheric warming.

5. Suppose someone tells you that carbon dioxide (CO_2) should not be classified as an air pollutant because it is a natural chemical that is part of the carbon cycle (see Figure 3-17, p. 66) and a chemical that we add to the atmosphere every time we exhale. Explain the faulty reasoning in this statement about CO_2.

6. One way to slow the rate of CO_2 emissions is to reduce the clearing of forests—especially in less-developed tropical countries where intense deforestation is taking place. Should the United States and other more-developed countries pay poorer countries to stop cutting their forests? Explain.

7. Some scientists have suggested that, in order to help cool the warming atmosphere, we could annually inject huge quantities of sulfate particles into the stratosphere. This might have the effect of reflecting some incoming sunlight back into space. Explain why you would support or oppose this geoengineering scheme.

8. What are three consumption patterns or other aspects of your lifestyle that directly add greenhouse gases to the atmosphere? Which, if any, of these habits would you be willing to give up in order to help slow projected climate disruption? Write up an argument that you could use to convince others to do the same. Compare your answers to those of your classmates.

Doing Environmental Science

Gather data on the average annual temperature and average annual precipitation for the area where you live over the last 30 years. Try to find data for as many of these years as possible, and plot these data to determine whether the average temperature and precipitation during this period has increased, decreased, or stayed about the same. Write a report summarizing your search for data, your results, and your conclusions.

Global Environment Watch Exercise

Within the GREENR database, use the World Map feature. Under "Browse" select *Climate Change* and click on the pin for Greenland. Find the latest information on the melting of glaciers in Greenland (**Core Case Study**). What is the rate of melting of the glaciers and how might this affect sea levels during this century? Briefly summarize the evidence used by scientists to support their statements about melting ice in Greenland. Summarize any information you find about ongoing and future studies of Greenland's ice.

Ecological Footprint Analysis

According to the International Energy Agency, the average American adds 19.6 metric tons (21.6 tons) of CO_2 per year to the atmosphere, compared with a world average of 4.23 metric tons (4.65 tons). The table on the next page is designed to help you understand the sources of your personal inputs of CO_2 into the atmosphere. You will be making calculations to fill in the blanks in this table.

Some typical numbers are provided in the "Typical Quantity per Year" column of the table. However, your calculations will be more accurate if you can use information based on your own personal lifestyle, which you can enter in the blank "Personal Quantity per Year" column. For example, you could add up your monthly utility bills for a year and divide the total by the number of persons in your household to get a rough estimate of your own utility use.

	Units per Year	Typical Quantity per Year	Personal Quantity per Year	Multiplier	Emissions per Year (lb. CO_2)
Residential Utilities					
Electricity	kwh	4,500		1.5	
Heating oil	gallons	37		22	
Natural gas	hundreds of cubic feet (ccf)	400		12	
Propane	gallons	8		13	
Coal	tons	—		4,200	
Transportation					
Automobiles	gallons	600		19	
Air travel	miles	2,000		0.6	
Bus, urban	miles	12		0.07	
Bus, intercity	miles	0		0.2	
Rail or subway	miles	28		0.6	
Taxi or limousine	miles	2		1	
Other motor fuel	gallons	9		22	
Household Waste					
Trash	pounds	780		0.75	
Recycled items	pounds	337		−2	
				Total (pounds)	
				Total (tons)	
				Total (kilograms)	
				Total (metric tons)	

Source: Thomas B. Cobb at Ohio's Bowling Green State University developed this CO_2 calculator.

1. Calculate your carbon footprint. To calculate your emissions, first complete the blank "Personal Quantity per Year" column as described above. Wherever you cannot provide personal data, use that listed in the "Typical Quantity per Year" column. Enter each number in the "Personal Quantity per Year" column to represent your annual consumption (using the units specified in the "Units per Year" column). Now multiply your annual consumption for each activity by the associated number in the "Multiplier" column to obtain an estimate of the pounds of CO_2 resulting from that activity, which you will enter in the "Emissions per Year" column. Finally, add the numbers in that column to find your carbon footprint, and express the final CO_2 result in both pounds and tons (1 ton = 2,000 pounds) and in kilograms and metric tons (1 kilogram = 2.2 pounds; 1 metric ton = 1.1 tons).

2. Compare your emissions with those of your classmates and with the per capita U.S. average of 19.6 metric tons (21.6 tons) of CO_2 per person per year. Actually, your answer should be considerably less—roughly about half the per capita value—because this computation only accounts for direct emissions. For instance, CO_2 resulting from driving a car is included, but the CO_2 emitted in manufacturing and disposing of the car is not. You can find more complete carbon footprint calculators at several sites on the Web.

3. Consider and list actions you could take to reduce your carbon footprint by 20%.

1. In 1991, Mount Pinatubo exploded, throwing tons of pollutants and ash into the atmosphere and giving climatologists and scientists a real-world laboratory to study. What was one result for climatologists of the Mount Pinatubo eruption?
   a. Climatologists were unable to make any predictions about the effect of the eruption on global climate.
   b. Climatologists were able to assess their predictions of the eruption on global temperatures.
   c. Climatologists predicted no global changes from the eruption.
   d. Climatologists were not making predictions when Pinatubo erupted.
   e. Climatologists had to radically revise their predictions based on the eruption.

Questions 2–4 are based on the diagram below.

AVERAGE TEMPERATURE (over past 130 years)

AVERAGE TEMPERATURE (over past 900,000 years)

(Compiled by the authors using data from the Goddard Institute for Space Studies, Intergovernmental Panel on Climate Change, National Academy of Sciences, National Aeronautics and Space Administration, National Center for Atmospheric Research, and National Oceanic and Atmospheric Administration.)

2. Which of the statements below about the earth's current warming trend is true based on the diagrams?
   a. There has been a uniform warming of the earth since 1900.
   b. The temperature of the earth has been stable and unchanging over the previous centuries.
   c. Greenhouse gases have been slowly accumulating in the atmosphere, promoting global warming.
   d. The earth has experienced many warming and cooling trends over the previous centuries.
   e. Humans have been influencing global temperatures for the last few centuries.

3. Calculate the average annual change in temperature from 1900 to 2000.
   a. 0.006
   b. 0.06
   c. 0.6
   d. 6.
   e. 60.

4. Which of the below was not utilized to generate the data shown in the graphs?
   a. Bubbles of ancient air found in ice cores
   b. Pollen samples taken from the bottoms of lakes
   c. Dust samples from the ozone layer
   d. Historical records
   e. Pollen and minerals in different layers of bat dung

5. Which of the below has a cooling effect on the earth's atmosphere?
   a. Particulate matter from burning grasslands and forests
   b. Aerosols such as sulfate particles
   c. Carbon particles from diesel exhaust
   d. Carbon dioxide from burning fossil fuels
   e. Condensation trails left by jet planes

6. Which of the following is an unavoidable effect of the current warming trend of the earth?
   a. Flooding of low-lying cities and island nations
   b. Melting of the permafrost and tundra, releasing methane gas
   c. Premature extinction of 20–30% of plant and animal species
   d. Bleaching of coral reefs and spread of deserts
   e. Collapse of the Amazon rain forest

7. Which of the below is a positive feedback mechanism that could be accelerating the warming of the earth's atmosphere?
   a. Absorption of CO_2 by the oceans
   b. Increase in severe weather in certain areas
   c. Severe drought, causing the earth to be browner and absorb more radiation
   d. Spraying of sulfate particles in the atmosphere, reflecting light
   e. Changing of ocean currents, carrying heat to different places

8. A 2007 Intergovernmental Panel on Climate Change (IPCC) report stated that the world's oceans are very likely to rise 18–59 cm during this century. The majority of this rise will occur as a result of
   a. increased rainfall in areas that normally have droughts, as a result of El Niño.
   b. newly opened shipping lanes that cause more melting of the arctic ice sheet.
   c. more icebergs calving and flowing to lower latitudes to melt.
   d. additional water reaching the oceans as permafrost melts.
   e. thermal expansion of ocean water.

9. Which of the below is NOT a way of reducing the release of greenhouse gases into the atmosphere?
   a. Utilizing cleaner burning natural gas with less sulfur
   b. Increasing the production of energy with nuclear power
   c. Use conservation tillage on all cropland
   d. Increase efficiency of coal-fired plants
   e. Shift to renewable energy sources

10. Which of the following is NOT an effect of ozone depletion in the stratosphere?
   a. Increased sunburns, resulting in a possible increase in skin cancers
   b. An increase in cataracts in both humans and some species of wildlife
   c. Disrupted populations of phytoplankton, which could reduce populations of certain marine organisms
   d. Increased asthma and bronchitis problems caused by inhalation of propellants in aerosol cans
   e. Increased atmospheric warming as oceans become unable to remove as much carbon dioxide from the air

11. Which of the below diagrams is an appropriate representation of the ozone molecule?

(A)   (B)   (C)   (D)   (E)

12. Which of the below is NOT true of CFC molecules as they rise through the atmosphere?
   a. CFCs remain in the atmosphere for extended periods of time.
   b. CFCs rise through the atmosphere via convection currents.
   c. CFCs are broken down by UV radiation, releasing chlorine atoms.
   d. CFC atoms can last in the atmosphere hundreds of years.
   e. CFC atoms are no longer a danger to ozone after UV light breaks them down.

13. Many climate change analysts believe that because of the difficulty in making large reductions in greenhouse gases, we need to prepare for the harmful effects of large-scale climate warming and change. Which of the following is a method of preparing for long-term harmful effects of climate change?
   a. Replace incandescent lightbulbs with compact fluorescents.
   b. Plant trees to shade houses and absorb CO_2 from the atmosphere.
   c. Buy products from companies that are actively reducing their carbon footprints.
   d. Walk, bike, carpool, or utilize mass transit.
   e. Connect wildlife reserves with corridors and move people inland.

20 Water Pollution

It is a hard truth to swallow, but nature does not care if we live or die. We cannot survive without the oceans, for example, but they can do just fine without us.

ROGER ROSENBLATT

Key Questions

20-1 What are the causes and effects of water pollution?

20-2 What are the major water pollution problems in streams and lakes?

20-3 What are the major pollution problems affecting groundwater?

20-4 What are the major water pollution problems affecting oceans?

20-5 How can we deal with water pollution?

Brown pelican severely oiled by the 2010 BP *Deepwater Horizon* oil well rupture in the Gulf of Mexico.

Joel Sartore/National Geographic Creative

The Mississippi River basin (Figure 20-1, top) lies within 31 states and contains almost two-thirds of the continental U.S. land area. With more than half of all U.S. croplands, it is one of the world's most productive agricultural regions. However, water draining into the Mississippi River and its tributaries from farms, cities, factories, and sewage treatment plants in this huge basin contains sediments and other pollutants that end up in the Gulf of Mexico (Figure 20-1, bottom)—a major supplier of the country's fish and shellfish.

Each spring and summer, this huge input of plant nutrients, mostly nitrates from crop fertilizers, enters the north-ern Gulf of Mexico and overfertilizes the coastal waters of the U.S. states of Mississippi, Louisiana, and Texas. One result is an explosion of populations of phytoplankton (mostly algae) that eventually die and fall to the seafloor. Hordes of oxygen-consuming bacteria go to work decomposing the phytoplankton remains and, in the process, deplete the dissolved oxygen in the Gulf's bottom layer of water.

The huge volume of *oxygen-depleted water* resulting from this seasonal event is called a *dead zone* because it contains little animal marine life. Its low oxygen levels (called *hypoxia*) drive away faster-swimming marine organisms and suffocate bottom-dwelling fish, crabs, oysters, and shrimp that cannot move to less polluted areas. Large amounts of sediment, mostly from soil eroded from the Mississippi River basin, can also kill bottom-dwelling forms of aquatic life. The dead zone appears each spring and grows until fall when storms churn the water and redistribute dissolved oxygen to the Gulf bottom.

The size of the Gulf of Mexico's annual dead zone depends primarily on the amount of water flowing into the Mississippi River each year. In years with ample rainfall and snowmelt, such as 2003, it has covered an area as large as the state of Massachusetts—27,300 square kilometers (10,600 square miles). In 2012, a severe drought year, it covered a much smaller area of 6,400 square kilometers (2,500 square miles)—about the size of the state of Delaware.

The annual Gulf of Mexico dead zone is one of about 400 oxygen-depleted zones scattered around the world. The largest of these zones lies in the Baltic Sea to the south of Sweden in Northern Europe. Others have appeared in coastal waters of Australia, New Zealand, Japan, China, and South America.

Oxygen-depleted zones represent a disruption of the nitrogen cycle (Figure 3-18, p. 67) caused primarily by human activities. This is because huge quantities of nitrogen from nitrate fertilizers are added to ecosystems such as the Mississippi River and Gulf of Mexico much faster than the nitrogen cycle can remove them. As a result, by producing crops for food and ethanol fuel in the vast Mississippi basin, we end up disrupting coastal aquatic life and seafood production in the Gulf of Mexico.

This overfertilization of coastal waters is one of the world's many forms of water pollution. Some good news is that we know how to reduce the sizes of dead zones and how to reduce other forms of water pollution that we explore in this chapter.

Figure 20-1 Water containing sediments, dissolved nitrate fertilizers, and other pollutants drains from the Mississippi River basin (top) into the Mississippi River and from there into the northern Gulf of Mexico (bottom). The boxed area at the bottom of the map shows the dead zone with the red area having the lowest oxygen level.

20-1 What Are the Causes and Effects of Water Pollution?

CONCEPT 20-1A
Water pollution causes illness and death in humans and other species, and disrupts ecosystems.

CONCEPT 20-1B
The chief sources of water pollution are agricultural activities, industrial facilities, and mining, but the growth of both the human population and our rate of resource use makes it increasingly worse.

Water Pollution Comes from Point and Nonpoint Sources

Water pollution is any change in water quality that can harm living organisms or make the water unfit for human uses such as drinking, irrigation, and recreation. It can come from a single (point) source, or from a larger, more dispersed (nonpoint) source. **Point sources** discharge pollutants into bodies of surface or underground water at specific locations through drain pipes (Figure 20-2), ditches, or sewer lines. Examples include factories, sewage treatment plants (which remove some, but not all, pollutants), underground mines (Figure 14-15, p. 360), and oil tankers.

Because point sources are located at specific places, they are fairly easy to identify, monitor, and regulate. Most of the world's more-developed countries have laws that help control point-source discharges of harmful chemicals and disease organisms into aquatic systems. However, in most of the less-developed countries, there is little control of such discharges.

Nonpoint sources are broad and diffuse areas where rainfall or snowmelt washes pollutants off the land into bodies of surface water. Examples include runoff of eroded soil and chemicals such as fertilizers and pesticides from cropland, livestock feedlots, logged forests (Figure 10-12, p. 224), urban streets, parking lots, lawns, and golf courses. We have made little progress in controlling water pollution from nonpoint sources because of the difficulty and expense of identifying and controlling discharges from so many diffuse sources. According to the U.S. Environmental Protection Agency (EPA), nonpoint source pollution is the main reason why 40% of all U.S. rivers, lakes, and estuaries are still not clean enough for uses such as fishing and swimming, despite the enactment of major water pollution control laws more than 35 years ago.

Agricultural activities are by far the leading cause of water pollution. Sediment eroded from agricultural lands (Figure 20-3) is the most common pollutant. Other major agricultural pollutants include fertilizers and pesticides, bacteria from livestock and food-processing wastes, and excess salts from the soils of irrigated cropland. *Industrial facilities*, which emit a variety of harmful inorganic and organic chemicals, are a second major source of water pollution.

 CONSIDER THIS. . .

CONNECTIONS Cleaning Up the Air and Polluting the Water
Stricter air pollution control laws (see Chapter 18, p. 493) have forced coal-burning power plants in more-developed countries to remove many of the harmful gases and particles from their smokestack emissions. But this results in hazardous ash, which is typically placed in slurry ponds (see Figure 15-18, p. 388) that can rupture and cause water pollution. Thus, the problem of coal ash illustrates one way in which air and water pollution are connected.

Mining is the third biggest source of water pollution (Figure 20-4). Surface mining disturbs the land (see Figure 14-11, p. 358; Figure 14-12, p. 359; and 14-14, p. 359), creating major erosion of sediments and runoff of toxic chemicals into surface waters.

Another form of water pollution is caused by the widespread use of human-made materials such as plastics that are used to make millions of products. Much of the plastic that is improperly discarded eventually ends up in waterways and in the oceans (Figure 20-5). Such discarded plastic products can harm various forms of wildlife (see Figure 11-4, p. 253).

CONSIDER THIS. . .

CONNECTIONS Atmospheric Warming and Water Pollution
Projected climate change from atmospheric warming will likely contribute to water pollution in some areas of the globe. In a warmer world, some regions

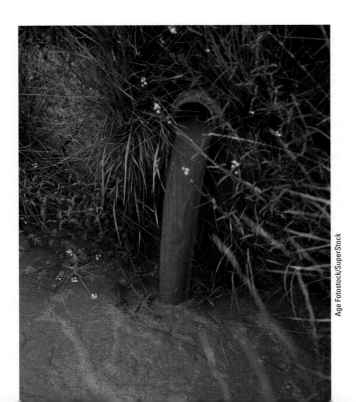

Age Fotostock/SuperStock

Figure 20-2 This is a point source of water pollution in Gargas, France.

Figure 20-3 Nonpoint sediment from farmland runoff flows into streams and sometimes changes their courses or dams them up. As measured by weight, it is the largest source of water pollution. **Question:** What do you think the owner of this farm could have done to prevent such sediment pollution?

Figure 20-4 Water pollution flows from the wastes of the abandoned Climax Mine, a molybdenum mining site in Colorado that is one of the major U.S. waste sites yet to be cleaned up.

will get more precipitation and other areas will get less. More intense downpours will flush more harmful chemicals, plant nutrients, and disease-causing microorganisms into some waterways, while prolonged drought will reduce river flows that dilute wastes in other areas.

Major Water Pollutants Have Harmful Effects

Table 20-1 lists the major types of water pollutants along with examples of each and their harmful effects and sources (Concept 20-1A).

One of the major water pollution problems that we face is exposure to infectious disease organisms (pathogens) primarily through contaminated drinking water. Scientists have identified more than 500 types of disease-causing bacteria, viruses, and parasites that can be transferred into water from the wastes of humans and animals. They can get into drinking water and cause painful, debilitating, and often life-threatening diseases, including typhoid fever, cholera, hepatitis B, giardiasis, and cryptosporidium.

Figure 20-5 Plastics and other forms of waste pollute this mountain lake, as well as many other bodies of water around the world, and can release harmful chemicals into the water.

Stephane Bidouze/Shutterstock.com

The World Health Organization (WHO) estimates that about 1 billion people—one of every seven in the world—do not have access to clean drinking water. The WHO estimates that each year, more than 1.6 million people die from largely preventable waterborne infectious diseases that they get by drinking contaminated water or by not having enough clean water for adequate hygiene. This adds up to an average of nearly 4,400 premature deaths a day, 90% of them among children younger than age 5. This death toll could be reduced with more widespread use of well-developed methods for detecting pollutants (Science Focus 20.1).

Table 20-1 Major Water Pollutants and Their Sources

Type/*Effects*	Examples	Major Sources
Infectious agents (pathogens) *Cause diseases*	Bacteria, viruses, protozoa, parasites	Human and animal wastes
Oxygen-demanding wastes *Deplete dissolved oxygen needed by aquatic species*	Biodegradable animal wastes and plant debris	Sewage, animal feedlots, food-processing facilities, paper mills
Plant nutrients *Cause excessive growth of algae and other species*	Nitrates (NO_3^-) and phosphates (PO_4^{3-})	Sewage, animal wastes, inorganic fertilizers
Organic chemicals *Add toxins to aquatic systems*	Oil, gasoline, plastics, pesticides, cleaning solvents	Industry, farms, households
Inorganic chemicals *Add toxins to aquatic systems*	Acids, bases, salts, metal compounds	Industry, households, surface runoff, mining sites
Sediments *Disrupt photosynthesis, food webs, other processes*	Soil, silt	Land erosion
Heavy metals *Cause cancer, disrupt immune and endocrine systems*	Lead, mercury, arsenic	Unlined landfills, household chemicals, mining refuse, industrial discharges
Thermal *Make some species vulnerable to disease*	Heat	Electric power and industrial plants

© Cengage Learning

SCIENCE FOCUS 20.1

TESTING WATER FOR POLLUTANTS

Scientists use a variety of methods to measure water quality. For example, they test samples of water for the *presence of various infectious agents* such as certain strains of the coliform bacteria *Escherichia coli,* or *E. coli* (Figure 20-A), which live in the colons and intestines of humans and other animals and thus are present in their fecal wastes. Although most strains of coliform bacteria do not cause disease, their presence indicates that water has been exposed to human or animal wastes that are likely to contain disease-causing agents.

To be considered safe for drinking, a 100-milliliter (about 1/2-cup) sample of water should contain no colonies of coliform bacteria. To be considered safe for swimming, such a water sample should contain no more than 200 colonies of coliform bacteria. By contrast, a similar sample of raw sewage may contain several million coliform bacterial colonies.

Another indicator of water quality is its *level of dissolved oxygen (DO).* Excessive inputs of oxygen-demanding wastes can deplete DO levels in water. Figure 20-B shows the relationship between dissolved oxygen content and water quality. In the dead zone formed in the Gulf of Mexico (**Core Case Study**), dissolved oxygen levels near the seafloor typically drop below 3 ppm.

Scientists can use *chemical analysis* to determine the presence and concentrations of specific organic chemicals in polluted water. They can also monitor water pollution by using living organisms as *indicator species.* For example, they remove aquatic plants such as cattails from areas contaminated with fuels, solvents, and other organic chemicals, and analyze them to determine the exact pollutants contained in their tissues. Scientists also determine water quality by analyzing bottom-dwelling species such as mussels, which feed by filtering water through their bodies.

Genetic engineers are working to develop bacteria and yeasts (single-celled fungi) that glow in the presence of specific pollutants such as toxic heavy metals in the ocean, toxins in the air, and carcinogens in food.

Scientists measure the amount of sediment in polluted water by evaporating the water in a sample and weighing the resulting sediment. They also use instruments called colorimeters, which measure specific wavelengths of light shined through a water sample to determine the concentrations of pollutants in the water. Another tool is the turbidimeter, used to measure the *turbidity,* or cloudiness, of water samples containing sediment.

The technology for testing waters is evolving. For example, in 2010, researchers in Spain used five robot fish, each one about as long as a seal, to detect potentially hazardous pollutants in a river.

Figure 20-A Colonies of coliform bacteria growing on a petri dish.

Critical Thinking

Runoff of fertilizer into a lake from farmland, lawns, and sewage treatment plants can overload the water with nitrogen and phosphorus—plant nutrients that can cause algae population explosions. How could this lower the dissolved oxygen level of the water and lead to fish kills?

Water Quality	DO (ppm) at 20°C
Good	8–9
Slightly polluted	6.7–8
Moderately polluted	4.5–6.7
Heavily polluted	4–4.5
Gravely polluted	Below 4

Figure 20-B Scientists measure dissolved oxygen (DO) content in parts per million (ppm) at 20°C (68°F) as an indicator of water quality. Only a few fish species can survive in water with less than 3 ppm of dissolved oxygen at this temperature. **Question:** Would you expect the dissolved oxygen content of polluted water to increase or decrease if the water were heated? Explain.

20-2 What Are the Major Water Pollution Problems in Streams and Lakes?

CONCEPT 20-2A
Streams and rivers around the world are extensively polluted, but they can cleanse themselves of many pollutants if we do not overload them or reduce their flows.

CONCEPT 20-2B
Adding excessive nutrients to lakes from human activities can disrupt their ecosystems, and prevention of such pollution is more effective and less costly than cleaning it up.

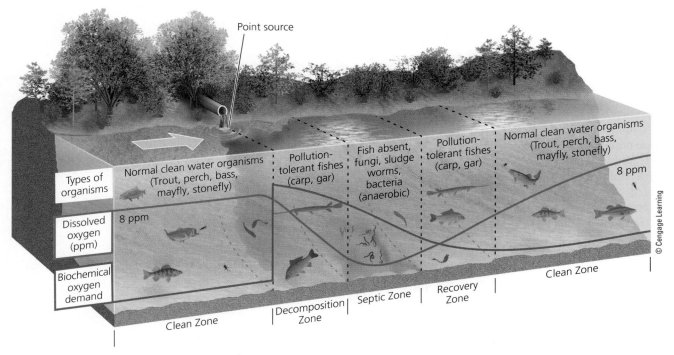

Point source

| Types of organisms | Normal clean water organisms (Trout, perch, bass, mayfly, stonefly) | Pollution-tolerant fishes (carp, gar) | Fish absent, fungi, sludge worms, bacteria (anaerobic) | Pollution-tolerant fishes (carp, gar) | Normal clean water organisms (Trout, perch, bass, mayfly, stonefly) |

Dissolved oxygen (ppm)

8 ppm   8 ppm

Biochemical oxygen demand

Clean Zone | Decomposition Zone | Septic Zone | Recovery Zone | Clean Zone

© Cengage Learning

Animated Figure 20-6 **Natural capital:** A stream can dilute and decay degradable, oxygen-demanding wastes, and it can also dilute heated water. This figure shows the oxygen sag curve (blue) and the curve of oxygen demand (red). Streams recover from oxygen-demanding wastes and from the injection of heated water if they are given enough time and are not overloaded (**Concept 20-2A**). **Question:** What would be the effect of putting another discharge pipe emitting biodegradable waste to the right of the one in this picture?

Streams Can Cleanse Themselves, If We Do Not Overload Them

Flowing rivers and streams can recover rapidly from moderate levels of degradable, oxygen-demanding wastes through a combination of dilution and bacterial biodegradation of such wastes. But this natural recovery process does not work when streams become overloaded with such pollutants or when drought, damming, or water diversion reduce their flows (**Concept 20-2A**). Also, while this process can remove biodegradable wastes, it does not eliminate slowly degradable and nondegradable pollutants. In a flowing stream, the breakdown of biodegradable wastes by bacteria depletes dissolved oxygen and creates an *oxygen sag curve* (Figure 20-6).

Stream Pollution in More-Developed Countries

Laws enacted in the 1970s to control water pollution have greatly increased the number and quality of wastewater treatment plants in the United States and in most other more-developed countries. Such laws also require industries to reduce or eliminate their point-source discharges of harmful chemicals into surface waters. This is an impressive accomplishment given the

GOOD NEWS

country's growth in population, economic activity, and resource consumption since the passage of these laws.

One success story is the cleanup of the U.S. state of Ohio's Cuyahoga River. It was so polluted that it caught fire several times and, in 1969, was photographed while burning as it flowed through the city of Cleveland toward Lake Erie. The highly publicized image of this combustible river prompted elected officials to enact laws to limit the discharge of industrial wastes into the river and into local sewage systems and to provide funds for upgrading sewage treatment facilities. Today, the river is cleaner, is no longer flammable, and is widely used by boaters and anglers. This accomplishment illustrates the power of bottom-up pressure by citizens, who prodded elected officials to change a severely polluted river into an economically and ecologically valuable public resource.

Another spectacular cleanup occurred in Great Britain. In the 1950s, the Thames River was little more than a flowing, smelly sewer. Now, after 50 years of effort and large inputs of money from British taxpayers and private industry, the Thames has made a remarkable recovery. Commercial fishing is thriving and the number of fish species in the river has increased 20-fold since 1960. In addition, many species of waterfowl and wading birds have returned to their former feeding grounds along the banks of the Thames.

Figure 20-7 Natural capital degradation: This portion of China's Yangtze River near the Three Gorges dam is heavily polluted with chemicals, sediment, and trash.

© CHINA DAILY/Reuters/Corbis

Fish kills and drinking water contamination still occur occasionally in some of the rivers and lakes of more-developed countries such as the United States. Some of these problems are caused by the accidental or deliberate release of toxic inorganic and organic chemicals by industries and mining operations (see Figure 14-15, p. 360, and Figure 20-4). Another cause is malfunctioning sewage treatment plants. A third cause is nonpoint runoff of pesticides and excess plant nutrients from cropland and animal feedlots.

Stream Pollution in Less-Developed Countries

In most less-developed countries, stream pollution from discharges of untreated sewage and industrial wastes is a serious and growing problem. According to the Global Water Policy Project, most cities in less-developed countries discharge 80–90% of their untreated sewage directly into rivers, streams, and lakes whose waters are often used also for drinking, bathing, and washing clothes.

According to the World Commission on Water for the 21st Century, half of the world's 500 major rivers are heavily polluted, and most of these polluted waterways run through less-developed countries. A majority of these countries cannot afford to build waste treatment plants and do not have, or do not enforce, laws for controlling water pollution.

In Latin America and Africa, most streams passing through urban or industrial areas suffer from severe pollution. Industrial wastes and sewage pollute more than two-thirds of India's water resources and 54 of the 78 rivers and streams monitored in China. In some parts of China, river water is too toxic to touch, much less drink. Liver and stomach cancer, linked in some cases to water pollution, are among the leading causes of death in the Chinese countryside where many industries have been relocated.

In 2010, Chinese officials reported that huge islands of garbage are threatening to jam the flow of water over the country's massive Three Gorges Dam on the Yangtze River (Figure 20-7). As it decays, this garbage will pollute the water.

Too Little Mixing and Low Flow Rates Make Lakes and Reservoirs Vulnerable to Water Pollution

Lakes and reservoirs are generally less effective at diluting pollutants than streams are, for two reasons. *First,* lakes and reservoirs often contain stratified layers (see Figure 8-16, p. 179) that undergo little vertical mixing. *Second,* they have low flow rates or no flow at all. The flushing and changing of water in lakes and large artificial reservoirs can take from 1 to 100 years, compared with several days to several weeks for streams.

As a result, lakes and reservoirs are more vulnerable than streams are to contamination by runoff or discharge of plant nutrients, oil, pesticides, and nondegradable toxic substances, such as lead, mercury, and arsenic. These contaminants can kill bottom-dwelling organisms and fish, as well as birds that feed on contaminated aquatic organisms. Many toxic chemicals and acids also enter lakes and reservoirs from the atmosphere (see Figure 18-12, p. 485).

Due to the stratified layers and reduced flows in lakes and reservoirs, the concentrations of some harmful chemicals are biologically magnified as they pass through food webs in these waters. Examples include DDT (see Figure 9-13, p. 202), PCBs (see Chapter 17, Case Study, p. 453), some radioactive isotopes, and some mercury compounds.

Cultural Eutrophication Is Too Much of a Good Thing

Eutrophication is the name given to the natural nutrient enrichment of a body of water such as a lake, coastal areas at the mouth of a river (**Core Case Study**), or a slow-

Figure 20-8 In 2010, severe cultural eutrophication affected Chaohu Lake near Hefei City in China's Anhui Province.

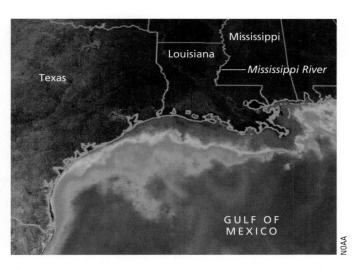

Figure 20-9 The colorized area in this satellite image of the northern Gulf of Mexico represents the seasonal dead zone of 2012 with the red area having the lowest oxygen level.

moving stream. It is caused mostly by runoff of plant nutrients such as nitrates and phosphates from land bordering such bodies of water.

An *oligotrophic lake* is low in nutrients and its water is clear (see Figure 8-15, p. 179). Over time, some lakes become more eutrophic (see Figure 8-17, p. 180) as nutrients are added from natural and human sources in the surrounding watersheds. Near urban or agricultural areas, human activities can greatly accelerate the input of plant nutrients to a lake—a process called **cultural eutrophication**. Such inputs involve mostly nitrate- and phosphate-containing effluents from various sources, including farmland, feedlots, urban streets and parking lots, chemically fertilized suburban yards, mining sites, and municipal sewage treatment plants. Some nitrogen also reaches lakes by deposition from the atmosphere (Figure 18-12, p. 485).

During hot weather or drought, this nutrient overload can produce dense growths, or "blooms," of organisms such as algae and cyanobacteria in slow-moving surface waters (see Figure 8-17, p. 180). When the algae die, they are decomposed by swelling populations of aerobic bacteria, which deplete the dissolved oxygen in the surface layer of water near the shore, as well as in the bottom layer of a lake or coastal area. This can kill fish, shellfish, and other aerobic aquatic animals that cannot move to safer waters. If excess nutrients continue to flow into a lake, anaerobic bacteria take over and produce gaseous products such as smelly, highly toxic hydrogen sulfide and methane.

According to the EPA, about one-third of the 100,000 medium to large lakes and 85% of the large lakes near major U.S. population centers have some degree of cultural eutrophication. The International Water Association estimates that more than half of the lakes in China, along with some of its coastal zones, suffer from cultural eutrophication (Figure 20-8).

There are several ways to *prevent* or *reduce* cultural eutrophication. We can use advanced (but expensive) waste treatment processes to remove nitrates and phosphates from wastewater before it enters a body of water. We can also use a preventive approach by banning or limiting the use of phosphates in household detergents and other cleaning agents, and by employing soil conservation (see Chapter 12, p. 304) and other controls to reduce nutrient runoff (**Concept 20-2B**).

There are several ways to *clean up* waters suffering from cultural eutrophication. They include mechanically removing excess weeds, controlling undesirable plant growth with herbicides and algaecides, and pumping air into lakes and reservoirs to prevent oxygen depletion, all of which are expensive and energy-intensive methods.

As usual, pollution prevention is more effective and quite often cheaper in the long run than cleanup. Most lakes and other surface waters can recover from cultural eutrophication, if excessive inputs of plant nutrients are stopped. In addition to cultural eutrophication, pollution causes other problems that plague many lakes (Case Study that follows).

Revisiting the Gulf of Mexico: An Extreme Case of Cultural Eutrophication

Since the 1950s, the level of nitrates discharged from the Mississippi River into the northern Gulf of Mexico has nearly tripled (**Core Case Study**). This has disrupted the nitrogen cycle and animal life in the area's affected coastal waters. Figure 20-9 provides a closer look at the severe oxygen depletion (shown in red) in 2012 in the dead zone of the northern Gulf of Mexico, where dissolved oxygen levels were typically less than 2 parts per million (Figure 20-B)—low enough to kill bottom-dwelling shellfish.

Excess nutrients fueling an explosive growth of algae (eutrophication) is a natural process that often occurs in some areas of the ocean where upwellings bring nutrients from the ocean bottom to surface waters (see Figure 7-2, p. 145). However, excessive inputs of plant nutrients, especially nitrates, from human activities (cultural eutrophication) is the primary cause of the annual dead zone in the northern Gulf of Mexico and in many other hypoxic zones around the world, including parts of Chesapeake Bay (see Figure 8-13, p. 177) and the San Francisco Bay.

In more detail, here is how the Gulf's seasonal dead zone forms. During the spring and summer, nitrate-laden freshwater flowing into the Gulf forms an oxygen-rich layer on top of the Gulf's cooler and more dense saltwater. Because there are few storms at this time of year, this sun-heated upper layer of water remains fairly calm and does not mix with the bottom layer of low-oxygen water. The combination of sunlight and large inputs of nitrate plant nutrients from fertilizer and sewage into the freshwater layer leads to massive blooms of phytoplankton, mostly blue-green algae.

When these algae die, they sink into the saltier water below where they are decomposed by oxygen-consuming bacteria, which use up nearly all of the dissolved oxygen, leaving less than 2 parts per million in the deeper water. Mobile species can survive this lack of oxygen by migrating to oxygen-rich waters, but certain species of fish, shellfish, and other organisms are not able to escape, and they die off.

Thus, the oxygen-depleted bottom layer of water becomes a dead zone for certain species. This disrupts the Gulf's food web, because die-offs of such species lead to the deaths of seabird and marine mammal species that depend on the dying fish and shellfish for their survival. The dead zone breaks up, beginning in early fall when cooler weather, storms, and hurricanes mix the top and lower layers of water and distribute dissolved oxygen throughout the layers.

In addition to greatly increased nitrate levels from human inputs, other human factors have contributed to the formation of the dead zone. In efforts to control flooding along the upper Mississippi River, engineers have dredged and straightened parts of the river and raised its banks with levees in many places. This has had the effect of speeding the river's flow of nutrients and sediment pollution into the Gulf. Also, most of the river basin's original freshwater wetlands, which acted as natural filters that helped to remove excess nutrients and sediments from flood water, have been drained for farming and urban development.

The seasonal formation of dead zones in the northern Gulf of Mexico, mostly resulting from human activities, is a reminder that in nature, everything is connected. Plant nutrients flowing into the Mississippi from a farm in Iowa or a sewage treatment plant in Wisconsin help to cause fish kills (Figure 20-10) a thousand miles away on the gulf coast of Texas. Researchers fear that if the size of the Gulf's annual dead zone is not sharply reduced, its long-term effects could permanently alter the ecological makeup of these coastal waters.

Figure 20-10 On August 13, 2012, hundreds of thousands of dead shad fish washed ashore on Jamaica Beach near Galveston, Texas. The likely cause of this die-off was low oxygen levels in the Gulf of Mexico.

CASE STUDY

Pollution in the Great Lakes

The five interconnected Great Lakes of North America (Figure 20-11) contain about 95% of the fresh surface water in the United States and one-fifth of the world's fresh surface water. At least 38 million people in the United States and Canada obtain their drinking water from these lakes.

Despite their enormous size, these lakes are vulnerable to pollution from point and nonpoint sources. One reason is that each year, less than 1% of the water entering these lakes flows east into the St. Lawrence River and then out to the Atlantic Ocean, meaning that pollutants can take as long as 100 years to be flushed out to sea.

By the 1960s, many areas of the Great Lakes were suffering from severe cultural eutrophication, huge fish kills, and contamination from bacteria and a variety of toxic industrial wastes. The impact on Lake Erie was particularly intense because it is the shallowest of the Great Lakes and has the highest concentrations of people and industrial activity along its shores.

In 1972, the United States and Canada signed the Great Lakes Water Quality Agreement, which is considered a model of international cooperation. They agreed to spend more than $20 billion to maintain and restore the chemical, physical, and biological integrity of the Great Lakes basin ecosystem. This program has helped to cut the number and sizes of algal blooms, raised dissolved oxygen levels, boosted sport and commercial fishing catches in Lake Erie, and allowed most swimming beaches to reopen—making this one of the nations' greatest environmental success stories. These improvements resulted mainly from the use of new or upgraded

GOOD NEWS

Figure 20-11 The five Great Lakes of North America make up the world's largest freshwater system. Dozens of major cities in the United States and Canada are located on their shores, and water pollution in the lakes is a growing problem.

sewage treatment plants, better treatment of industrial wastes, and bans on the use of detergents, household cleaners, and water conditioners that contain phosphates. Most of these measures were instituted largely as a result of bottom-up citizen pressure.

Despite this important progress, many problems remain. Increasing nonpoint runoff of pesticides and fertilizers resulting from urban sprawl, fueled by population growth, now surpasses industrial pollution as the greatest threat to the lakes. Bottom sediments in 26 toxic hotspots remain heavily polluted. In addition, *biological pollution* in the form of growing populations of zebra mussels (Figure 11-20, p. 268) and more than 180 other invasive species threaten some native aquatic species and cause at least $200 million a year in damages (see Chapter 11, Case Study, p. 268).

Air quality over the Great Lakes has generally improved, but about half of the toxic compounds entering the lakes still come from atmospheric deposition of pesticides, mercury from coal-burning plants, and other toxic chemicals from as far away as Mexico and Russia.

A survey done by Wisconsin biologists found that one of every four fish taken from the Great Lakes was unsafe for human consumption.

Current efforts are aimed at improving the water quality of the lakes and other aquatic systems within the Great Lakes basin. To that end, the major goals are to clean up contaminated river bottoms, prevent erosion and runoff into lakes and streams, protect and restore wetlands, clean up toxic hotspots, and reduce the number of invasive species entering the lakes.

Some environmental and health scientists call for taking a prevention approach and banning the use of toxic chlorine compounds, such as bleach used in the pulp and paper industry, which is prominent around the Great Lakes. They would also ban new waste incinerators, which can release toxic chemicals into the atmosphere, and they would stop the discharge into the lakes of 70 toxic chemicals that threaten human health and wildlife. So far, officials in the industries involved have successfully opposed such bans.

20-3 What Are the Major Pollution Problems Affecting Groundwater?

CONCEPT 20-3A

Chemicals used in agriculture, industry, transportation, and homes can spill and leak into groundwater and make it undrinkable.

CONCEPT 20-3B

There are both simple ways and complex ways to purify groundwater used as a source of drinking water, but protecting it through pollution prevention is the least expensive and most effective strategy.

Groundwater Cannot Cleanse Itself Very Well

Aquifers provide drinking water for about half of the U.S. population and 95% of Americans who live in rural areas. According to many scientists, groundwater pollution is a serious threat. Common pollutants such as fertilizers, pesticides, gasoline, and organic solvents can seep into groundwater from numerous sources (Figure 20-12). People who dump or spill gasoline, oil, and paint thinners and other organic solvents onto the ground also help to contaminate groundwater (Concept 20-3A).

Scientists are now concerned about a new and growing potential threat to groundwater—the drilling of thousands of new natural gas wells in parts of the United States involving a process called *hydraulic fracturing*, or *fracking* (see Chapter 15, Case Study, p. 383). Groundwater contamination could result from leaky well pipes and pipe fittings and from contaminated wastewater brought to the surface during fracking operations. Without strict monitoring and enforcement of pollution regulations (which at this point do not exist), fracking could become a serious groundwater pollution threat.

Once a pollutant from a leaking underground storage tank or other source contaminates groundwater, it fills the aquifer's porous layers of sand, gravel, or bedrock like water saturates a sponge. This makes removal of the contaminant difficult and costly. The slowly flowing groundwater disperses the pollutant in a widening *plume* of contaminated water. If the plume reaches a well used to extract groundwater, the toxic pollutants can get into drinking water and into water that is used to irrigate crops.

When groundwater becomes contaminated, it cannot cleanse itself of degradable wastes as quickly as flowing surface water can. Groundwater flows so slowly—usually less than 0.3 meter (1 foot) per day—that contaminants

Figure 20-12 Natural capital degradation: Principal sources of groundwater contamination in the United States (Concept 20-3A). Another source in coastal areas is saltwater intrusion from excessive groundwater withdrawal. (Figure is not drawn to scale.) **Question:** What are three sources shown in this figure that might be affecting groundwater in your area?

© Cengage Learning

are not diluted and dispersed effectively. In addition, groundwater usually has much lower concentrations of dissolved oxygen (which helps decompose many contaminants) and smaller populations of decomposing bacteria. The usually cold temperatures of groundwater also slow down chemical reactions that decompose wastes.

Groundwater Pollution Is a Serious Hidden Threat in Some Areas

On a global scale, we do not know much about groundwater pollution because few countries go to the great expense of locating, tracking, and testing aquifers. But the results of scientific studies in scattered parts of the world are alarming.

Groundwater provides about 70% of China's drinking water. According to the Chinese Ministry of Land and Resources, in 2010, about 90% of China's shallow groundwater was polluted with chemicals such as toxic heavy metals, organic solvents, nitrates, petrochemicals, and pesticides. About 37% of this groundwater is so polluted that it cannot be treated for use as drinking water. Every year, according to the WHO and the World Bank, water pollution causes an estimated 190 million Chinese to become ill and kills about 60,000.

In the United States, an EPA survey of 26,000 industrial waste ponds and lagoons found that one-third of them had no liners to prevent toxic liquid wastes from seeping into aquifers. One-third of these sites are within 1.6 kilometers (1 mile) of a drinking water well. In addition, almost two-thirds of America's liquid hazardous wastes are injected into the ground in disposal wells (Figure 20-12), some of which leak their contents into aquifers that are used as sources of drinking water.

By 2008, the EPA had cleaned up about 357,000 of the more than 479,000 underground tanks in the United States that were leaking gasoline, diesel fuel, home heating oil, or toxic solvents into groundwater. During this century, scientists expect many of the millions of such tanks, which have been installed around the world, to become corroded and leaky, possibly contaminating groundwater and becoming a major global health problem. Determining the extent of a leak from a single underground tank can cost $25,000–$250,000, and cleanup costs range from $10,000 to more than $250,000. If the chemical reaches an aquifer, effective cleanup is often not possible or is too costly.

Groundwater that is used as a source of drinking water can also be contaminated with *nitrate ions* (NO_3^-), especially in agricultural areas where nitrates in fertilizer can leach into groundwater. Nitrite ions (NO_2^-) in the stomach, colon, and bladder can convert some of the nitrate ions in drinking water to organic compounds that have been shown in tests to cause cancer in more than 40 animal species. The conversion of nitrates in tap water to nitrites in infants under 6 months old can cause a potentially fatal condition known as "blue baby syndrome," in which blood lacks the ability to carry sufficient oxygen to body cells.

It can take decades to thousands of years for contaminated groundwater to cleanse itself of *slowly degradable wastes* (such as DDT). On a human time scale, *nondegradable wastes* (such as toxic lead and arsenic, see the following Case Study) remain in the water permanently. Although there are ways to clean up contaminated groundwater (Figure 20-13, right), such methods are very expensive. Cleaning up a single contaminated aquifer can cost anywhere from $10 million to $10 billion. Thus, preventing groundwater contamination (Figure 20-13, left) is the only effective and affordable way to deal with this serious water pollution problem.

CASE STUDY

Arsenic in Drinking Water

Arsenic contaminates drinking water when a well is drilled into an aquifer where the soil and rock are naturally rich in arsenic, or when human activities such as mining and ore processing release arsenic into drinking water supplies. Some rivers used for drinking water also are contaminated naturally, having originated in springs that have high levels of arsenic.

The World Health Organization (WHO) estimates that more than 140 million people in 70 countries are drinking water with arsenic concentrations of 5–100 times the accepted safe level of 10 parts per billion (ppb). Some scientists from the WHO and other organizations argue that even the 10 ppb standard is not safe.

Figure 20-13 There are ways to prevent and ways to clean up contaminated groundwater, but prevention is the only effective approach (Concept 20-3B). **Question:** Which two of these preventive solutions do you think are the most important? Why?

Arsenic levels are especially high in Bangladesh, China, India's state of West Bengal, and parts of northern Chile. The WHO estimates that long-term exposure to nondegradable arsenic in drinking water is likely to cause hundreds of thousands of premature deaths from cancer of the skin, bladder, and lungs.

There is also concern over arsenic levels in drinking water in parts of the United States. According to the EPA, some 13 million people in several thousand communities, mostly in the western half of the country, are exposed to arsenic levels of 3–10 ppb in their drinking water.

In 2006, researchers from Rice University in Houston, Texas (USA), reported that by suspending nanoparticles of rust in arsenic-contaminated water and then drawing them out with handheld magnets, they had removed enough arsenic from the water to make it safe to drink. Use of this technique could greatly reduce the threat of arsenic in drinking water for many families at a cost of a few cents a day.

There Are Many Ways to Purify Drinking Water

Most of the more-developed countries have laws establishing drinking water standards. But most of the less-developed countries do not have such laws or, if they do have them, they do not enforce them. There are both simple ways and complex ways to purify drinking water obtained from surface and groundwater sources (Concept 20-3B).

In more-developed countries, wherever people depend on surface water sources, water is usually stored in a reservoir for several days. This improves its clarity and taste by increasing its dissolved oxygen content and allowing suspended matter to settle. The water is then pumped to a purification plant and treated to meet government drinking water standards. In areas with very pure groundwater or surface water sources, little treatment is necessary. Several major U.S. cities, including New York City, Boston, Seattle, and Portland, Oregon, have avoided building expensive water treatment facilities by investing in protection of the forests and wetlands in the watersheds that provide their water supplies.

CONSIDER THIS. . .

THINKING ABOUT Protecting the Sources of Drinking Water

Where does the community in which you live get its drinking water? If the source is surface water, could the community save money and help to preserve biodiversity by finding ways to protect the watershed for this source?

We have the technology to convert sewer water into pure drinking water, which could help to reduce depletion of some water supplies. However, recycling wastewater is expensive and it faces opposition from citizens and from some health officials who are unaware of the advances in this technology, much of it developed to recycle wastewater on the International Space Station. In a world where we will face deepening shortages of drinking water, wastewater purification is likely to become a major growth business and is much cheaper than using desalination to turn ocean water into drinking water (p. 335). Wastewater is being recycled to provide freshwater in Israel, Singapore, Orange County and San Diego in California, and El Paso, Texas, among a growing number of places.

In many areas of both more-developed and less-developed countries, people have turned to bottled water, hoping that it is a safe alternative to local drinking water sources (see the following Case Study). However, there are several simple ways to purify drinking water that are far less costly than using bottled water. For example, in tropical countries that lack centralized water treatment systems, the WHO urges people to purify drinking water by exposing a clear plastic bottle filled with contaminated water to intense sunlight. The sun's heat and ultraviolet (UV) rays can kill infectious microbes in as little as 3 hours. Painting one side of the bottle black improves heat absorption in this simple solar disinfection method, which applies the solar energy **principle of sustainability**.

Where this measure has been used, the incidence of dangerous childhood diarrhea has decreased by 30–40%. In 2012, researchers found they could speed up the disinfection process by adding lime juice to the bottles of water. In 2011, scientists in Brazil found that adding dried bits of banana peels is an inexpensive way to remove toxic metals such as mercury, arsenic, and lead from drinking water polluted by mining and industrial wastes.

In 2007, Danish inventor Torben Vestergaard Frandsen developed the LifeStraw™, an inexpensive, portable water filter that eliminates many viruses and parasites from water that is drawn through it (Figure 20-14). This filter has been particularly useful in Africa, where aid agencies are distributing it.

Another product that could be helpful for the 1 billion people who do not have access to safe drinking water is PUR, a powder containing chlorine and iron sulfate. A user empties a packet of PUR into a large container of dirty water, then stirs the mixture and lets it sit for about 20 minutes. After the water is poured through a piece of clean cloth to filter out the particles, it is ready to drink.

CASE STUDY

Is Bottled Water a Good Option?

Bottled water can be a useful (but expensive) option in countries and areas where people do not have access to safe and clean drinking water. However, despite some problems, experts say the United States has some of the world's cleanest drinking water. Municipal water systems in the United States are required to test their water regularly for a number of pollutants and to make the results available to citizens.

Yet about half of all Americans worry about getting sick from tap water contaminants, and many drink high-

Vestergaard Frandsen

Figure 20-14 The *LifeStraw™* is a personal water purification device that gives many poor people access to safe drinking water. Here, four young men in Uganda demonstrate its use. **Question:** Do you think the development of such devices should make prevention of water pollution less of a priority? Explain.

priced bottled water or install expensive water purification systems. Americans are the world's largest consumers of bottled water, followed by Mexico, China, and Brazil. In 2011, Americans spent about $15 billion on bottled water—enough to meet the annual drinking water needs of the roughly 1 billion people in the world who routinely lack access to safe, clean drinking water.

Studies by the Natural Resources Defense Council (NRDC) reveal that in the United States, a bottle of water costs between 240 and 10,000 times as much as the same volume of tap water. In 2008, water expert Peter Gleick estimated that more than 40% of the expensive bottled water that Americans drink is really bottled tap water. And a 4-year study by the NRDC concluded that most bottled water is of good quality. However, they found traces of bacteria and synthetic organic chemicals in 23 of the 123 brands tested. Bottled water is less regulated than tap water and the EPA contamination standards that apply to public water supplies do not apply to bottled water.

About a third of all bottled water suppliers do not disclose anything about the treatment or purity of the water they sell. However, the International Bottled Water Association and the nonprofit National Sanitation Foundation International certify bottled water companies that meet their standards. Cautious buyers use only bottled water that is certified by such organizations.

Use of bottled water also causes environmental problems. In the United States, according to the Container Recycling Institute, more than 67 million plastic water bottles are discarded every day. Each year, this amounts to enough bottles to, if lined up end-to-end, wrap around the planet 149 times. Most water bottles are made of recyclable polyethylene terephthalate (PET) plastic, but in the United States, only about 29% of these bottles get recycled. Many of the rest end up in landfills, where they can remain for hundreds of years, or in incinerators, which release some of their harmful chemicals into the atmo-

sphere. Also, many millions of discarded bottles get scattered on the land and wind up in rivers, lakes (Figure 20-5), and oceans.

It takes huge amounts of energy to manufacture bottled water and to transport it across countries and around the world, as well as to refrigerate much of it in stores. Toxic gases and liquids are released during the manufacture of plastic water bottles, and greenhouse gases and other air pollutants are emitted by the fossil fuels burned to make them and to deliver bottled water to suppliers. In addition, withdrawing water for bottling is helping to deplete some aquifers. There is also concern about health risks from chemicals such as bisphenol A (BPA, see Science Focus 17.3, p. 456) that can leach into the water from the plastic in some water bottles, especially if they are exposed to the hot sun.

Because of these harmful environmental impacts and the high cost of bottled water, there is a growing *back-to-the-tap* movement. From San Francisco to New York to Paris, city governments, restaurants, schools, religious groups, and many consumers are refusing to buy bottled water.

Also, people are increasingly using refillable glass and metal bottles, filling them with tap water and using simple filters to improve the taste and color of water where necessary. In Germany and other countries, most bottled water is sold in returnable and reusable glass bottles. Some health officials suggest that before drinking expensive bottled water or buying costly home water purifiers, consumers have their water tested by local health departments or private labs (but not by companies trying to sell water purification equipment).

Using Laws to Protect Drinking Water Quality

About 54 countries, most of them in North America and Europe, have legal standards for safe drinking water. For example, the U.S. Safe Drinking Water Act of 1974 requires the EPA to establish national drinking water standards, called *maximum contaminant levels*, for any pollutants that could have adverse effects on human health. Currently, this act strictly limits the levels of 91 potential contaminants in U.S. tap water. However, in most of the less-developed countries, such laws do not exist or are not enforced.

Health scientists call for strengthening the U.S. Safe Drinking Water Act in several ways. Here are three of their recommendations:

- Combine many of the drinking water treatment systems that serve fewer than 3,300 people with nearby larger systems to make it easier for these smaller systems to meet federal standards.
- Strengthen and enforce requirements concerning public notification of violations of drinking water standards.
- Ban the use of any toxic lead in new plumbing pipes, faucets, and fixtures. Current law allows for fixtures with up to 10% lead content to be sold as lead-free.

According to the Natural Resources Defense Council (NRDC), making these three improvements would cost U.S. taxpayers an average of about $30 a year per household.

However, certain industries are pressuring elected officials to weaken the Safe Drinking Water Act, because complying with it increases their costs. One proposal is to eliminate national testing of drinking water and requirements for public notification about violations of drinking water standards. Another proposal would allow states to give providers of drinking water a permanent right to violate the standard for a given contaminant if they claim they cannot afford to meet the standard. Also, some critics call for greatly reducing the EPA's already-low budget for enforcing the Safe Drinking Water Act.

20-4 What Are the Major Water Pollution Problems Affecting Oceans?

CONCEPT 20-4A
Most ocean pollution originates on land and includes oil and other toxic chemicals, as well as solid waste, which threaten fish and wildlife and disrupt marine ecosystems.

CONCEPT 20-4B
The key to protecting the oceans is to reduce the flow of pollution from land and air and from streams emptying into ocean waters.

Ocean Pollution Is a Growing and Poorly Understood Problem

Coastal areas—especially wetlands, estuaries, coral reefs, and mangrove swamps—bear the brunt of our enormous inputs of pollutants and wastes into the ocean (Figure 20-15). Roughly 40% of the world's people (53% in the United States) live on or near coastlines, which helps to explain why 80% of marine pollution originates on land (**Concept 20-4A**), and coastal populations are projected to double by 2050.

According to a study by the U.N. Environment Programme (UNEP), 80–90% of the municipal sewage from coastal areas of less-developed countries is dumped into oceans without treatment. This often overwhelms the ability of the coastal waters to degrade these biodegradable wastes. Many areas of China's coastline, for example, are so choked with algae growing on the nutrients provided by sewage that some scientists have concluded that large areas of China's coastal waters can no longer sustain marine ecosystems.

In deeper waters, the oceans can dilute, disperse, and degrade large amounts of raw sewage and other types of degradable pollutants. Some scientists suggest that it is safer to dump sewage sludge, toxic mining wastes, and most other harmful wastes into the deep ocean than to bury them on land or burn them in incinerators. Other scientists disagree, pointing out that we know less about

the deep ocean than we do about the moon. They add that dumping harmful wastes into the ocean would delay urgently needed pollution prevention measures and promote further degradation of this vital part of the earth's life-support system.

Recent studies of some U.S. coastal waters have found vast colonies of viruses thriving in raw sewage and in effluents from sewage treatment plants (which do not remove viruses) and leaking septic tanks. According to one study, on average, one-fourth of the people using coastal beaches in the United States develop ear infections, sore throats, eye irritations, respiratory disease, or gastrointestinal disease from swimming in seawater containing infectious viruses and bacteria.

🔍 **CONSIDER THIS. . .**

CONNECTIONS Cruise Ships and Water Pollution
A cruise liner can carry as many as 6,300 passengers and 2,400 crewmembers, and it can generate as much waste (toxic chemicals, garbage, sewage, and waste oil) as a small city. Many cruise ships dump these wastes at sea. In U.S. waters, such dumping is illegal, but some ships continue dumping secretively, usually at night. Some environmentally aware vacationers are refusing to go on cruise ships that do not have sophisticated systems for dealing with the wastes they produce.

Runoff of sewage and agricultural wastes into coastal waters can carry large quantities of nitrate (NO_3^-) and phosphate (PO_4^{3-}) plant nutrients, which can cause explosive growths of harmful algae and lead to dead zones (**Core Case Study**) during summer months. These *harmful algal blooms* are called red, brown, or green toxic tides (Figure 20-16). They can release waterborne and airborne toxins that poison seafood, kill fish and some fish-eating birds, and discourage tourism in the affected coastal areas. Each year, harmful algal blooms lead to the poisoning of about 60,000 Americans who eat shellfish contaminated by the algae.

Harmful algal blooms occur annually in about 400 oxygen-depleted zones around the world, mostly in temperate coastal waters and in large bodies of water with

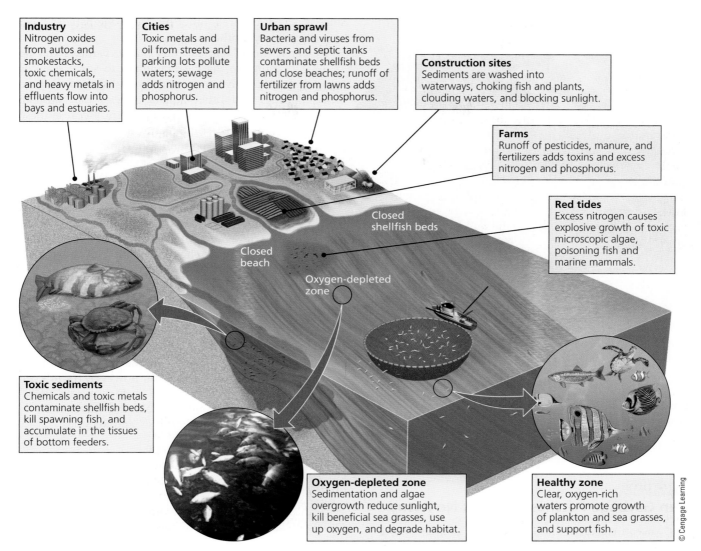

Industry
Nitrogen oxides from autos and smokestacks, toxic chemicals, and heavy metals in effluents flow into bays and estuaries.

Cities
Toxic metals and oil from streets and parking lots pollute waters; sewage adds nitrogen and phosphorus.

Urban sprawl
Bacteria and viruses from sewers and septic tanks contaminate shellfish beds and close beaches; runoff of fertilizer from lawns adds nitrogen and phosphorus.

Construction sites
Sediments are washed into waterways, choking fish and plants, clouding waters, and blocking sunlight.

Farms
Runoff of pesticides, manure, and fertilizers adds toxins and excess nitrogen and phosphorus.

Red tides
Excess nitrogen causes explosive growth of toxic microscopic algae, poisoning fish and marine mammals.

Closed shellfish beds

Closed beach

Oxygen-depleted zone

Toxic sediments
Chemicals and toxic metals contaminate shellfish beds, kill spawning fish, and accumulate in the tissues of bottom feeders.

Oxygen-depleted zone
Sedimentation and algae overgrowth reduce sunlight, kill beneficial sea grasses, use up oxygen, and degrade habitat.

Healthy zone
Clear, oxygen-rich waters promote growth of plankton and sea grasses, and support fish.

© Cengage Learning

Figure 20-15 Natural capital degradation: Residential areas, factories, and farms all contribute to the pollution of coastal waters. **Question:** What do you think are the three worst pollution problems shown here? For each one, how does it affect two or more of the ecosystem and economic services listed in Figure 8-5 (p. 170)?

restricted outflows, such as the Baltic and Black seas. The largest of these zones in U.S. coastal waters forms each year in the northern Gulf of Mexico (**Core Case Study**). A 2008 study by Luan Weixin, of China's Dalain Maritime University, found that water pollutants such as nitrates and phosphates seriously contaminated about half of China's shallow coastal waters. Warmer ocean water temperatures, thought by most scientists to be due to climate change, are extending the size and duration of dead zones in the world's oceans.

© Purestock/Alamy

Figure 20-16 A red tide: This harmful algal bloom in Washington state's Puget sound contains organisms that give the water a red tint and can be toxic to fish, wildlife, and people. **Question:** What are two ways in which this sort of pollution could be prevented?

© Cengage Learning

Figure 20-17 The North Pacific Garbage Patch is actually two vast, slowly swirling masses of small plastic particles floating just under the water. Four other huge garbage patches have been discovered in the world's other major oceans.

Ocean Garbage Patches: There Is No Away

In 1997, ocean researcher and sea captain Charles Moore accidentally discovered two gigantic, slowly rotating masses of small pieces of plastic and other solid wastes in the middle of the North Pacific Ocean near the Hawaiian Islands. These wastes, floating on or just beneath the ocean's surface and known as the North Pacific Garbage Patch, are trapped there by a vortex where rotating ocean currents called *gyres* meet (Figure 20-17).

Roughly 80% of this trash comes from the land—washed or blown off beaches, pouring out of storm drains, and floating down streams and rivers that empty into the sea from the west coast of North America and the east coast of Asia (Concept 20-4A). Most of the rest is dumped into the ocean from cargo and cruise ships.

The North Pacific Garbage Patch is estimated by some scientists to occupy an area at least the size of Texas. However, such estimates are difficult to verify because this plastic-laden soup consists mostly of small rice-grain- to pencil-eraser-sized particles of plastic suspended slightly underwater, and thus it is difficult to see and measure.

Figure 20-18 It is likely that this plastic debris found in the stomach of this albatross caused it to starve to death.

Since it was discovered, other huge garbage patches have been discovered in other oceans.

Research shows that the tiny plastic particles can be harmful to marine mammals, to some seabirds such as albatrosses (Figure 20-18), and to some fishes and other aquatic species that mistake them for food and swallow them. Because these animals cannot digest the plastic, it can cause them to die from starvation, poisoning, or choking.

Fish that feed on plankton ingest these tiny plastic particles, which can contain PCBs, DDT, hormone-mimicking BPA, and other harmful chemicals. These long-lived toxins can build up to high concentrations in food chains and webs (see Figure 9-13, p. 202), and they can end up in fish sandwiches and seafood dinners. Thus, toxic chemicals from a discarded plastic grocery bag, water bottle, or chips bag could end up in our stomachs. Everything is connected.

Who is to blame for these patches? Each of us risks making a contribution every time we use and discard a plastic item. We may think we have simply thrown it away, but there is no away.

What can be done? There is no practical or affordable way to clean up this mess. We can, however, try to keep the problem from growing worse by focusing on pollution prevention, which means practicing the four Rs of resource use: Refuse, Reduce, Reuse, and Recycle.

Ocean Pollution from Oil

Crude petroleum (oil as it comes out of the ground) and *refined petroleum* (fuel oil, diesel, gasoline, and other processed petroleum products; see Figure 15-4, p. 377) reach

©Frans Lanting/National Geographic Creative

U.S. Coast Guard

Figure 20-19 The *Deepwater Horizon* drilling platform burned and sank in the Gulf of Mexico after exploding on April 20, 2010.

fairly rapid currents. But in cold and calm waters, full recovery can take decades. In 2009, some 20 years after the *Exxon Valdez* spill, researchers found patches of oil remaining on some stretches of the shoreline of the U.S. state of Alaska's Prince William Sound. In addition, recovery from exposure to refined oil, especially in estuaries and salt marshes, can take 10–20 years or longer. Oil slicks that wash onto beaches can have a serious economic impact on coastal residents, who lose income normally gained from fishing and tourism.

the ocean from a number of sources and become highly disruptive pollutants (Concept 20-4A).

The most visible sources are tanker accidents, such as the *Exxon Valdez* oil spill in the U.S. state of Alaska in 1989, and blowouts at offshore oil drilling rigs, such as that of the BP *Deepwater Horizon* rig in the Gulf of Mexico in 2010 (see the Case Study that follows). However, studies show that the largest source of ocean oil pollution is urban and industrial runoff from land. Much of this waste oil comes from leaks in pipelines, refineries, and other oil-handling and storage facilities. An estimated one-third to one-half of all ocean oil pollution comes from oil and oil products that are intentionally dumped or accidentally spilled or leaked onto the land or into sewers by home owners and industries.

Volatile organic hydrocarbons in oil kill many aquatic organisms immediately upon contact, especially if these animals are in their vulnerable larval forms. Other chemicals in oil form tar-like globs that float on the surface and coat the feathers of seabirds (see chapter-opening photo) and the fur of marine mammals. This oil coating destroys their natural heat insulation and buoyancy, causing many of them to drown or die of exposure from loss of body heat.

Heavy oil components that sink to the ocean floor or wash into estuaries and coastal wetlands can smother bottom-dwelling organisms such as crabs, oysters, mussels, and clams, or make them unfit for human consumption. Some oil spills have killed coral reefs.

Research shows that, within about 3 years, populations of many forms of marine life can recover from exposure to large amounts of crude oil in warm waters with

Some oil spills that are not too large can be partially cleaned up by mechanical means including floating booms, skimmer boats, and absorbent devices such as large pillows filled with feathers or hair. But scientists estimate that current cleanup methods can recover no more than 15% of the oil from a major spill.

Thus, *preventing* oil pollution is the most effective and, in the long run, the least costly approach (Concept 20-4B). One of the best ways to prevent tanker spills is to use oil tankers with double hulls. Stricter safety standards and inspections could help to reduce oil well blowouts at sea. Most importantly, businesses, institutions, and citizens living in coastal areas must take care to prevent leaks and spillage of even the smallest amounts of oil and oil products such as paint thinners and gasoline.

 CONSIDER THIS. . .

THINKING ABOUT Ocean Oil Pollution

What are three ways in which you might be contributing to ocean oil pollution? How could you reduce your contribution to this environmental problem?

CASE STUDY

The BP *Deepwater Horizon* Oil-Rig Spill

On April 20, 2010, the world learned a harsh lesson about the possible environmental impacts of deep-sea oil drilling when the BP Company's *Deepwater Horizon* offshore oil-drilling rig exploded (Figure 20-19). The accident occurred in the Gulf of Mexico, 64 kilometers (40 miles) off the Louisiana coast, after the wellhead on the ocean bottom, almost 1.6 kilometers (1 mile) below the surface, ruptured and released natural gas. The resulting fire and explosion killed 11 of the rig's crewmates and injured 17 more. After burning and belching oil smoke into the air for 36 hours, the rig sank.

During the 3 months following this accident, the ruptured wellhead on the ocean floor released about 4.9 million barrels (206 million gallons) of crude oil before it was capped. It was the largest accidental oil spill that ever occurred in U.S. waters. After the rupture, engineers worked frantically, trying to cap the gushing well from almost a mile above the wellhead, using robot-controlled submarines. It took them several attempts before they succeeded in stopping the flow of oil.

The oil contaminated some ecologically vital coastal marshes, mangrove forests, sea-grass beds, fish nurseries, and deep coral reefs along Louisiana's coast and along the barrier island coastlines of Mississippi, Alabama, and parts of Florida. Scientists suspect that deep-ocean aquatic organisms were affected, but the true extent of the ecological damage will not be known for years. According to estimates from the U.S. Coast Guard, the National Oceanic and Atmospheric Administration (NOAA) and BP, the oil spill killed at least 6,100 seabirds and oiled another 2,000 (see chapter-opening photo) and killed more than 600 sea turtles and oiled another 460.

The spill also disrupted the livelihoods of people who depend on the Gulf Coast's fisheries, and it caused large economic losses for the area's tourism businesses (Figure 20-20). By 2012, BP had spent or pledged to spend roughly $40 billion for cleanup costs, damages, restoration costs, and fines for felony misconduct and neglect. In 2013, the company was in U.S. federal court facing the possibility of being held accountable for several billion dollars more in damages and additional penalties.

Several studies, including that of a Presidential commission, have concluded that the main causes of the accident were failure of equipment that could have detected the leak earlier, a faulty blowout preventer, failure of several safety valves, and a number of poor decisions made by workers and managers. This event also revealed flaws in federal oversight of offshore drilling, which many argue has included too cozy a relationship between the oil industry and its Department of Interior regulators.

Since the accident, the U.S. government has developed new standards for each step in the offshore drilling process and agreed to give a higher priority to environmental

Figure 20-20 This cleanup crew in protective suits was combing the beach in Gulf Shores, Alabama after the 2010 BP oil spill, looking for deposits of toxic crude oil.

Tyrone Turner/National Geographic Creative

concerns in the leasing process. However, the Presidential oil spill commission called for close oversight by Congress to ensure that such reforms are put in place and carefully monitored.

The 2012 report of the Presidential oil spill commission faulted the U.S. Congress for not providing funding to help prevent another catastrophe and to implement a major program of ecosystem restoration of the Gulf of Mexico's wetlands and barrier islands. The report also included a strong warning about the risks of drilling in arctic waters, where harsh conditions could greatly increase the risk of oil spills that would likely cause much longer-lasting damage than that experienced in the Gulf of Mexico.

The widespread global publicity following the BP accident helped to educate the public about the dangers of accidents that can release large quantities of oil into the oceans. However, there has been little attention focused on the numerous smaller oil spills that have been occurring more-or-less continually for almost 50 years in Nigeria's oil-producing Niger Delta, where there is little or no government oversight of oil company operations. Each year, an amount of oil roughly equal to that of the Exxon Valdez disaster leaks and spills from rusted and aging pipes and other facilities in the delta region. Decades of such leaks have caused widespread ecological destruction and threats to human health. Some scientists argue that these spills altogether are more damaging than the large, well-known spills.

20-5 How Can We Deal with Water Pollution?

CONCEPT 20-5
Reducing water pollution requires that we prevent it, work with nature to treat sewage, and use natural resources far more efficiently.

Reducing Ocean Water Pollution

Most ocean pollution occurs in coastal waters and comes from human activities on land (**Concept 20-4A**). Figure 20-21 lists ways to prevent pollution of coastal waters and ways to reduce it.

Solutions

Coastal Water Pollution

Prevention		Cleanup
Separate sewage and storm water lines	Ban dumping of wastes and sewage by ships in coastal waters	Improve oil-spill cleanup capabilities
Require secondary treatment of coastal sewage	Strictly regulate coastal development, oil drilling, and oil shipping	Use nanoparticles on sewage and oil spills to dissolve the oil or sewage (still under development)
Use wetlands and other natural methods to treat sewage	Require double hulls for oil tankers	

Figure 20-21 Methods for preventing excessive pollution of coastal waters and methods for cleaning it up (**Concept 20-4B**). **Question:** Which two of these solutions do you think are the best ones? Why?

Top: Rob Byron/Shutterstock.com. Bottom: Nightman1965/Shutterstock.com.

Experience shows that we can sharply reduce the formation of dead zones. For example, a vast dead zone in the Black Sea largely disappeared in the 1990s when the breakup of the Soviet Union led to a sharp increase in fertilizer prices. Reducing nutrient inputs has also shrunk the sizes of dead zones in New York's Hudson River and in California's San Francisco Bay. Also, countries along Europe's Rhine River have reduced nitrate levels in the North Sea's dead zone by about 35%, primarily by reducing nitrate inputs from sewage treatment and industrial plants.

In addition to reducing nitrate inputs, it would be helpful to capture nutrients after they leave farm fields. This would involve two steps. The first would be to protect the

The key to protecting the oceans is to reduce the flow of pollution from land and air and from streams emptying into these waters (**Concept 20-4B**). Thus, ocean pollution control must be linked with land-use and air pollution control policies, which in turn are linked to energy policies (see Figure 16-36, p. 434) and climate policies (Figure 19-22, p. 528).

Reducing Surface Water Pollution from Nonpoint Sources

There are a number of ways to reduce nonpoint sources of water pollution, most of which come from agricultural practices. Examples include the following:

- Reducing soil erosion and fertilizer runoff by keeping cropland covered with vegetation and using conservation tillage (see Chapter 12, p. 305) and other soil conservation methods (see Figure 12-28, p. 305).
- Using fertilizers that release plant nutrients slowly.
- Using no fertilizers on steeply sloped land.
- Relying more on organic farming (see Chapter 12, Core Case Study, p. 278) and more sustainable food production (see Figure 12-33, p. 309) to reduce the use and runoff of plant nutrients and pesticides.
- Planting buffer zones of vegetation between cultivated fields (Figure 20-22) and between animal waste storage sites and nearby surface waters.
- Setting discharge standards for nitrate chemicals from sewage treatment and industrial plants.

The annual formation of the dead zone in the Gulf of Mexico (**Core Case Study** and Figure 20-C) will be difficult to prevent because of the importance of the Mississippi River basin for growing crops. However, nutrient inputs can be reduced with the widespread use of the fertilizer management practices listed above.

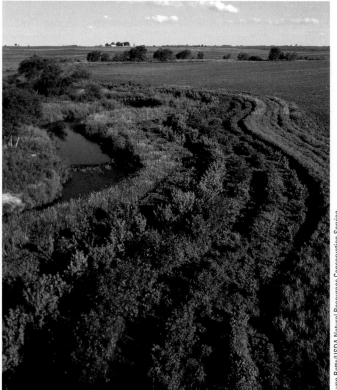

Figure 20-22 This buffer zone of vegetation on an Iowa farm helps to reduce runoff of fertilizer and pesticides into waterways from nearby crop fields. However, such zones are rare.

remaining inland and coastal wetlands, vegetation in *riparian zones* (zones along river banks; see Figure 10-23, right, p. 232), and floodplains, all of which act as natural filters to pull nutrients from waters that are running off the land and spilling out of streams. However, many of these natural features have been replaced by cropland. For example, in Ohio, Indiana, and Iowa, landowners have drained more than 80% of these states' natural wetlands, mostly to plant crops. Also, many wetlands and floodplains have been cut off from their sources of freshwater by flood-control levees along the rivers that feed them.

The second step would involve restoring key wetlands that have been destroyed or degraded and, where feasible, restoring and reconnecting rivers to their natural flood-plains. In addition to reducing water pollution, this would restore natural habitats for a variety of species in keeping with the biodiversity **principle of sustainability** (see Figure 1-2, p. 6 or back cover).

Point-source water pollution is considerably easier to control than non-point-source pollution, and there have been a number of success stories around the world in this area. One of the best examples is that of U.S. efforts to deal with this problem (see the following Case Study).

CASE STUDY

The U.S. Experience with Reducing Point-Source Pollution

Efforts to control pollution of surface waters in the United States are based on the Federal Water Pollution Control Act of 1972 (renamed the Clean Water Act when it was amended in 1977) and the 1987 Water Quality Act. The Clean Water Act sets standards for allowed levels of 100 key water pollutants and requires polluters to get permits that limit the amounts of these various pollutants that they can discharge into aquatic systems.

The EPA has also been experimenting with a *discharge trading policy*, which uses market forces to reduce water pollution as has been done with sulfur dioxide for air pollution control (see Chapter 18, p. 495). Under this program, a permit holder can pollute at higher levels than allowed in its permit by buying credits from permit holders who are polluting below their allowed levels. Environmental scientists warn that the effectiveness of such a system depends on how low the cap on total pollution levels in any given area is set and on how regularly the cap is lowered. They also warn that discharge trading could allow water pollutants to build up to dangerous levels in areas where credits are bought.

According to the EPA, the Clean Water Act of 1972 led to numerous improvements in U.S. water quality, including the following:

- 95% of all Americans are served by public drinking water systems that must meet federal health standards.

- 60% of all tested U.S. streams, lakes, and estuaries can be used safely for fishing and swimming, compared to 33% in 1972.
- 75% of the U.S. population is served by sewage treatment plants.
- Annual losses of U.S. wetlands that naturally absorb and purify water have been reduced by 80% since 1992.

These are impressive achievements, given the increases in the U.S. population and its per capita consumption of water and other resources since 1972. However, according to the EPA and recent studies by the *New York Times,* there is still considerable room for improvement. Those reports noted the following:

- About 40% of the nation's surveyed streams, lakes, and estuaries are still too polluted for swimming or fishing.
- Runoff of animal wastes from hog, poultry, and cattle feedlots and meat processing facilities pollutes seven of every ten U.S. rivers.
- Tens of thousands of gasoline storage tanks in 43 states are leaking.
- One in five U.S. water treatment systems violated the Safe Drinking Water Act between 2003 and 2008, releasing sewage and chemicals such as arsenic and radioactive uranium.
- There are currently no federal drinking water standards for hexavalent chromium, which studies indicate is a potent carcinogen. In 2011, residents of Midland, Texas, sued Dow Chemical for allegedly putting dangerous levels of this chemical into their drinking water.
- About 45% of the country's largest water polluters have declared that the Clean Water Act no longer applies to waters that they are polluting. This results from a U.S. Supreme Court decision that created uncertainty over which waterways are protected by the law. About 117 million Americans (32% of the population) get some or all of their drinking water from sources that are being polluted in this way.

Some environmental scientists call for strengthening the Clean Water Act. Suggested improvements include shifting the focus of the law to water pollution prevention instead of focusing mostly on end-of-pipe removal of specific pollutants; greatly increased monitoring for violations of the law and much larger mandatory fines for violators; and regulating irrigation water quality (for which there is no federal regulation). Another suggestion is to expand the rights of citizens to bring lawsuits to ensure that water pollution laws are enforced. Still another suggestion is to rewrite the Clean Water Act to clarify that it covers all waterways (as Congress originally intended). This would eliminate confusion about which waterways are covered and stop major polluters from using this confusion to keep polluting in many areas.

Manhole cover (for cleanout)

Gas

Scum

Wastewater

Sludge

Septic tank

Distribution box

Drain field (gravel or crushed stone)

Vent pipe

Perforated pipe

© Cengage Learning

Figure 20-23 Solutions: *Septic tank systems* are often used for disposal of domestic sewage and wastewater in rural and suburban areas.

Many businesses and legislators oppose these proposals, contending that the Clean Water Act's regulations are already too restrictive and costly. Some state and local officials argue that in many communities, it is unnecessary and too expensive to test all the water for pollutants as required by federal law.

Some members of Congress, under pressure from regulated industries, go further and suggest seriously weakening or repealing the Clean Water Act. They contend that such regulations hinder economic growth and prevent job growth. In 2012, William K. Reilly, who headed the EPA from 1989 to 1993 and served as a co-chairman of the Presidential commission on offshore drilling, said: "If we buy into the misguided notion that reducing protection of our waters will somehow ignite the economy, we will shortchange our, health, environment, and economy."

Sewage Treatment Reduces Water Pollution

In rural and suburban areas of more-developed countries, sewage from each house usually is discharged into a **septic tank** with a large drainage field (Figure 20-23). In such a system, household sewage and wastewater is pumped into a settling tank, where grease and oil rise to the top and solids fall to the bottom and are decomposed by bacteria. The partially treated wastewater that results is discharged in a large drainage field through small holes in perforated pipes embedded in porous gravel or crushed stone just below the soil's surface. As these wastes drain from the pipes and percolate downward, the soil filters out some potential pollutants and soil bacteria decompose biodegradable materials.

About one-fourth of all homes in the United States are served by septic tanks. If these systems are not installed correctly and pumped out regularly, they can cause sewage to back up into homes or to pollute nearby groundwater and surface water. Chlorine bleaches, drain cleaners, and antibacterial soaps should not be used in these systems, because they can kill the bacteria that decompose

the wastes. Kitchen sink garbage disposals should not be used either, because they can overload septic systems.

In urban areas in the United States and other more-developed countries, most waterborne wastes from homes, businesses, and storm runoff flow through a network of sewer pipes to *wastewater* or *sewage treatment plants*. Raw sewage reaching a treatment plant typically undergoes one or two levels of wastewater treatment. The first is **primary sewage treatment**—a *physical* process that uses screens and a grit tank to remove large floating objects and to allow solids such as sand and rock to settle out. Then the waste stream flows into a primary settling tank where suspended solids settle out as sludge (Figure 20-24, left).

By itself, primary treatment removes about 60% of the suspended solids and 30–40% of the oxygen-demanding organic wastes from sewage. It removes no phosphates, nitrates, salts, radioisotopes, trace amounts of medications, pesticides, or pathogens such as viruses.

The second level is **secondary sewage treatment**—a *biological* process in which aerobic bacteria remove as much as 90% of dissolved and biodegradable, oxygen-demanding organic wastes (Figure 20-24, right). A combination of primary and secondary treatment removes 95–97% of the suspended solids and oxygen-demanding organic wastes, 70% of most toxic metal compounds and nonpersistent synthetic organic chemicals, 70% of the phosphorus, and 50% of the nitrogen. However, this process removes only a tiny fraction of the persistent and potentially toxic organic substances found in some pesticides and in discarded medicines that people put into sewage systems, and it does not kill pathogens.

A third level of cleanup, *advanced* or *tertiary sewage treatment*, uses a series of specialized chemical and physical processes to remove specific pollutants left in the water after primary and secondary treatment. In its most common form, advanced sewage treatment uses special filters to remove phosphates and nitrates from wastewater before it is discharged into surface waters. This third

Primary

Secondary

Bar screen | Grit chamber | Settling tank | Aeration tank | Settling tank | Chlorine disinfection tank

Raw sewage from sewers

Sludge

Activated sludge

Air pump

Sludge digester

To river, lake, or ocean

(kills bacteria)

Sludge drying bed

Disposed of in landfill or ocean or applied to cropland, pasture, or rangeland

© Cengage Learning

Figure 20-24 Solutions: Primary and secondary sewage treatment systems help to reduce water pollution. *Question:* What do you think should be done with the sludge produced by sewage treatment plants?

stage would help a great deal in reducing nutrient overload from nitrates and phosphates, but because of its high costs, it is not widely used.

Before discharge, water from sewage treatment plants usually undergoes *bleaching,* to remove water coloration, and *disinfection,* to kill disease-carrying bacteria and some (but not all) viruses. The usual method for accomplishing this is *chlorination.* But chlorine can react with organic materials in water to form small amounts of chlorinated hydrocarbons. Some of these chemicals cause cancers in test animals, can increase the risk of miscarriages, and can damage the human nervous, immune, and endocrine systems. The use of other disinfectants, such as ozone and ultraviolet light, is increasing, but they cost more and their effects do not last as long as those of chlorination.

Federal law in the United States requires primary and secondary treatment in all municipal sewage treatment plants, but exemptions from secondary treatment are possible for towns and cities that can show that the cost of installing such systems poses an excessive financial burden. In addition, the EPA says that at least two-thirds of the country's sewage treatment plants have sometimes violated water pollution regulations. At least 500 U.S. cities have failed to meet federal standards for sewage treatment plants, and 34 East Coast cities simply screen out large floating objects from their sewage before discharging it into the coastal waters of the Atlantic Ocean.

Some cities have two separate networks of pipes, one for carrying storm-water runoff from streets and parking lots, and the other for carrying sewage. But about 1,200 U.S. cities have combined these two systems into one set of pipes to cut costs. Heavy rains or a high volume of sewage from too many users hooked up to such com-

bined systems can cause them to overflow and discharge untreated sewage (called *effluents*) directly into surface waters such as the Great Lakes (see Case Study, p. 552).

According to the EPA, at least 40,000 of these overflows occur each year in the United States, and about 7.1 million people get sick every year from swimming in waters contaminated by overflows that contain sewage. The best way to deal with this overflow problem is to prevent it by using separate systems to carry storm water and sewage (**Concept 20-5**).

We Can Improve Conventional Sewage Treatment

Some scientists call for redesigning the conventional sewage treatment system shown in Figure 20-24. The idea is to prevent toxic and hazardous chemicals from reaching sewage treatment plants and thus from getting into sludge and water discharged from such plants.

They suggest several ways to implement this pollution prevention approach. One is to require industries and businesses to remove toxic and hazardous wastes from water sent to municipal sewage treatment plants. Another is to encourage industries to reduce or eliminate their use and waste of toxic chemicals.

A major problem is that building and maintaining sewage treatment plants costs too much for governments in less-developed countries where untreated or poorly treated wastewater causes environmental and human health problems. Some individuals are trying to address this problem by showing that wastewater and sewage sludge can be thought of as valuable resources to be recycled (Individuals Matter 20.1).

Ashley Murray: Wastewater Engineer, Entrepreneur, and National Geographic Emerging Explorer

Ashley Murray grew up in a family of small business owners and earned MS and PhD degrees in engineering from the University of California. She has worked in several countries, including China, India, Senegal, Kenya, Uganda, and Ghana, where she now lives. She has used her background in business, science, and engineering to found and run Waste Enterprisers, a company devoted to revolutionizing the way we deal with wastewater and fecal sludge, especially in less-developed countries.

Her approach is to promote cleaning up and reusing wastewater by viewing the energy it contains as a financially valuable resource that can be harnessed and sold. Her business relies on using human waste as a raw material and converting it into a renewable solid fuel that can be used by industries such as cement plants. Revenue from the sale of fuel more than covers the processing costs, essentially turning sanitation into the byproduct of manufacturing fuel.

After 18 months of research to validate the concept and conduct bench-scale experiments, Waste Enterprisers has built a demonstration plant in Accra, Ghana, where they have the capacity to produce 2 tons of fuel per day. This is produced from approximately 100 cubic meters of raw human waste from the city's pit latrines and septic tanks (see background photo). Waste Enterprisers are conducting burning trials with local industries and plan to build a commercial scale plant in Kenya next year.

Waste Enterprisers' fuel can be used as a replacement for coal in kilns and boilers. When cement industries burn the fuel they can even incorporate the residual ash into their cement product, making the process a zero-waste, closed loop solution to sanitation.

Murray hopes to build human waste-to-fuel plants across sub-Saharan Africa. She says, "the real goal is improving basic sanitation, health, and environmental conditions for some of the world's poorest populations."

Background photo: ©Ashley Murray

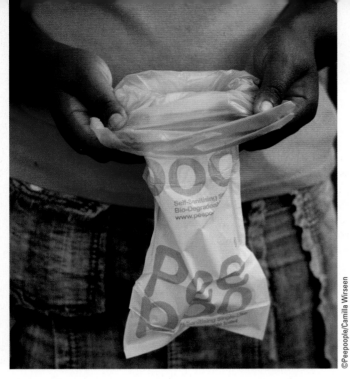

Figure 20-25 The Peepoo is a single-use biodegradable plastic bag that can be used as a toilet in the urban slums of less-developed countries. Each bag costs 3¢ and the company buys back the used bags for 1¢ and turns the waste into fertilizer.

Another suggestion is to require or encourage more households, apartment buildings, and offices to eliminate sewage outputs by switching to waterless, odorless *composting toilet systems,* to be installed, maintained, and managed by professionals. These systems, pioneered several decades ago in Sweden, convert nutrient-rich human fecal matter into a soil-like humus that can be used as a fertilizer supplement.

By using such systems, we can return the plant nutrients in human waste to the soil, thus mimicking the chemical cycling **principle of sustainability**. This would reduce the need for energy-intensive, water-polluting commercial fertilizers. Composting toilets also save large amounts of water, reduce water bills, and decrease the amount of energy used to pump and purify water along with the resulting greenhouse gas emissions. For example, in the United States the collection, distribution, and treatment of drinking water and wastewater (Figure 20-24) releases as much CO_2 each year as 10 million cars, and this could be reduced if more people used waterless composting toilets. The downside of these systems is the initial cost for home owners, which is much higher than that of conventional toilets.

One of the authors of this book (Miller) used a waterless compositing toilet for over a decade with no problems in his office-home facility. This more environmentally sustainable replacement for the conventional toilet is now being used in parts of more than a dozen countries, including China, India, Mexico, Syria, and South Africa.

In 2009, a Swedish architect and professor developed a biodegradable single-use plastic bag, called a Peepoo (Fig-

Solutions

Water Pollution

- Prevent groundwater contamination
- Reduce nonpoint runoff
- Work with nature to treat sewage and reuse treated wastewater
- Find substitutes for toxic pollutants
- Practice the four Rs of resource use (refuse, reduce, reuse, recycle)
- Reduce air pollution
- Reduce poverty
- Slow population growth

© Cengage Learning 2015

Figure 20-26 Ways to prevent or reduce water pollution (Concept 20-5). *Question:* Which two of these solutions do you think are the best ones? Why?

ure 20-25), that can be used as a toilet in urban slums and in other areas where 40% of the world's people do not have access to toilets. After it is used, the bag is knotted and buried. A thin layer of urea in this bag kills the disease-producing pathogens in the feces and helps break down the waste into plant nutrients that are then recycled—a simple and inexpensive, low-tech application of the chemical cycling **principle of sustainability**.

Many communities are using unconventional, but highly effective, *wetland-based sewage treatment systems,* which work with nature (Science Focus 20.2). Such systems would be ideal in some less-developed countries where water is short and disease is rampant.

🔍 **CONSIDER THIS. . .**

CAMPUS SUSTAINABILITY
Oberlin College

Oberlin College in Ohio is consistently listed as one of the greenest colleges in the United States. A group of Oberlin students worked with faculty members and architects to design a more sustainable environmental studies building powered by solar panels, which produce 30% more electricity than the building uses. A living machine (Science Focus 20.2) in the building's greenhouse purifies all of its wastewater. The building collects rainwater that is used to irrigate the campus's grasses, gardens, and meadow, which contain a diversity of plant and animal species. The college also gets half of its electricity from renewable energy resources and has a car-sharing program.

There Are Sustainable Ways to Reduce or Prevent Water Pollution

It is encouraging that since 1970, most of the world's more-developed countries have enacted laws and regulations that have significantly reduced point-source water

SCIENCE FOCUS 20.2

TREATING SEWAGE BY WORKING WITH NATURE

Some communities and individuals are seeking better ways to purify sewage by working with nature (Concept 20-5). Biologist John Todd has developed an ecological approach to treating sewage, which he calls *living machines* (Figure 20-C).

This purification process begins when sewage flows into a passive solar greenhouse or outdoor site containing rows of large open tanks populated by an increasingly complex series of organisms. In the first set of tanks, algae and microorganisms decompose organic wastes, with sunlight speeding up the process. Water hyacinths, cattails, bulrushes, and other aquatic plants growing in the tanks take up the resulting nutrients.

After flowing though several of these natural purification tanks, the water passes through an artificial marsh made of sand, gravel, and bulrushes, which filters out algae and remaining organic waste. Some of the plants also absorb, or *sequester,* toxic metals such as lead and mercury and secrete natural antibiotic compounds that kill pathogens.

Next, the water flows into aquarium tanks, where snails and zooplankton consume microorganisms and are in turn consumed by crayfish, tilapia, and other fish that can be eaten or sold as bait. After 10 days, the clear water flows into a second artificial marsh for final filtering and cleansing. The water can be made pure enough to drink by treating it with ultraviolet light or by passing the water through an ozone generator, usually immersed out of sight in an attractive pond or wetland habitat. Operating costs are about the same as those of a conventional sewage treatment plant. These systems are widely used on a small scale. However, they have been difficult to maintain on a scale large enough to handle the typical variety of chemicals in the sewage wastes from more-developed urban areas.

More than 800 cities and towns around the world (150 in the United States) use natural or artificially created wetlands to treat sewage as a lower-cost alternative to expensive waste treatment plants. For example, in Arcata, California—a coastal town of about 17,000 people—scientists and workers created some 65 hectares (160 acres) of wetlands between the town and the adjacent Humboldt Bay. The marshes and ponds, developed on land that was once a dump, act as a natural waste treatment plant. The project cost was less than half the estimated price of a conventional treatment plant.

This system returns purified water to Humboldt Bay, and the sludge that is removed is processed for use as fertilizer. The marshes and ponds also serve as an Audubon Society bird sanctuary, which provides habitats for thousands of seabirds, otters, and other marine animals.

Figure 20-C Solutions: Ecological wastewater purification system called a *living machine* at the Solar Sewage Treatment Plant in the U.S. city of Providence, Rhode Island. Biologist John Todd is demonstrating this ecological process that he invented for purifying wastewater by using the sun and a series of tanks containing living organisms.

The town even celebrates its natural sewage treatment system with an annual "Flush with Pride" festival.

This approach and the living machine system developed by John Todd apply all three scientific **principles of sustainability**: using solar energy, employing natural processes to remove and recycle nutrients and other chemicals, and relying on a diversity of organisms and natural processes.

Critical Thinking

Can you think of any disadvantages of using such a nature-based system instead of a conventional sewage treatment plant? Do you think any such disadvantages outweigh the advantages? Why or why not?

---

pollution. These improvements were largely the result of *bottom-up* political pressure on elected officials from individuals and groups.

On the other hand, little has been done to reduce water pollution in most of the less-developed countries. However, China plans to provide all of its cities with small sewage treatment plants that will make wastewater clean enough to be recycled back into the urban water supply systems. If China is successful in this ambitious plan, it could become a world leader in developing ways to tackle both water pollution and water scarcity through systems that recycle water.

Many environmental and health scientists argue that the next step is to increase efforts to reduce and prevent water pollution in both more- and less-developed countries. They would begin by asking the question: *How can we avoid producing water pollutants in the first place?* (Concept 20-5). Figure 20-26 lists ways to achieve this goal over the next several decades.

This shift to pollution prevention will not take place unless citizens put political pressure on elected officials and also take actions to reduce their own daily contributions to water pollution. Figure 20-27 lists some actions you can take to help reduce water pollution.

Figure 20-27 Individuals matter: You can help reduce water pollution. **Question:** Which three of these steps do you think are the most important ones to take? Why?

Big Ideas

■ There are a number of ways to purify drinking water, but the most effective and least costly strategy is pollution prevention.

■ The key to protecting the oceans is to reduce the flow of pollution from land and air, and from streams emptying into ocean waters.

■ Reducing water pollution requires that we prevent it, work with nature in treating sewage, and use natural resources far more efficiently.

TYING IT ALL TOGETHER Dead Zones and Sustainability

Bull's-Eye Arts/Shutterstock.com

The **Core Case Study** that opens this chapter explains how certain farming practices can violate the chemical cycling **principle of sustainability** (see Figure 1-2, p. 6 or back cover) by disrupting the nitrogen cycle. This occurs every year when nitrate plant nutrients in fertilizers flow in runoff from farmland into streams in the Mississippi River basin. Much of this flow goes into the Mississippi River and ends up overfertilizing an area of the Gulf of Mexico, depleting the dissolved oxygen and reducing seafood production in these coastal waters. This process also violates the biodiversity **principle of sustainability** by disrupting ecological interactions among species in river and coastal ecosystems.

The three scientific **principles of sustainability** can guide us in reducing and preventing water pollution. We can use solar energy to purify much more of the water we use, and this will help us to reduce water waste. We can also use natural nutrient cycles (chemical cycling) to treat our wastes in wetland-based sewage treatment systems (Science Focus 20.2). In addition, we can view human waste as an important and potentially profitable source of nutrients. Finally, by preventing the pollution and degradation of aquatic systems and their bordering terrestrial systems, we will help to preserve biodiversity—a key component of the life-support system that we depend on for maintaining water supplies and water quality.

Chapter Review

Core Case Study

1. Describe the nature and causes of the annual dead (oxygen-depleted) zone in the Gulf of Mexico.

Section 20-1

2. What are the two key concepts for this section? What is **water pollution**? Distinguish between **point sources** and **nonpoint sources** of water pollution and give two examples of each. Explain how certain efforts to clean up the air can contribute to water pollution. Explain how atmospheric warming will likely increase water pollution.

3. List seven major types of water pollutants and three diseases that can be transmitted to humans through polluted water. Describe some of the chemical and biological methods that scientists use to measure water quality.

Section 20-2

4. What are the two key concepts for this section? Explain how streams can cleanse themselves and how this cleansing process can be overwhelmed. Describe the varying states of stream pollution in more- and less-developed countries. Give two reasons why lakes cannot cleanse themselves as readily as streams can. Distinguish between **eutrophication** and **cultural eutrophication**. List ways to prevent or reduce cultural eutrophication. Explain how seasonal dead zones form in the Gulf of Mexico. Describe the pollution of the Great Lakes and the progress being made in reducing this pollution.

Section 20-3

5. What are the two key concepts for this section? Explain why groundwater cannot cleanse itself very well. What are the major sources of groundwater contamination in the United States? List three harmful effects of groundwater pollution. Describe the threat from arsenic in groundwater. List three ways to prevent groundwater contamination and three ways to clean up groundwater contamination. Explain why prevention is the most important approach.

6. Explain how drinking water is purified typically in more-developed countries. List three ways to provide safe drinking water in poor countries. Describe the environmental problems caused by the widespread use of bottled water. Summarize the U.S. laws for protecting drinking water quality. List three ways to strengthen the U.S. Safe Drinking Water Act.

Section 20-4

7. What are the two key concepts for this section? How are coastal waters and deeper ocean waters most often polluted? What causes algal blooms and what are their harmful effects? Describe an ocean garbage patch and explain how it can harm marine life and how the growth of such patches could be prevented. How serious is oil pollution of the oceans? What are its effects and what can be done to reduce such pollution? Describe the 2010 BP *Deepwater Horizon* oil blowout and its causes and effects. How might such accidents be prevented in the future?

Section 20-5

8. What is the key concept for this section? List four ways to reduce water pollution from nonpoint sources. Describe the U.S. successes and failures with using laws and regulations to reduce point-source water pollution. What is a **septic tank** and how does it work? Define **primary sewage treatment** and **secondary sewage treatment** and explain how they are used to treat wastewater.

9. What are three ways to improve conventional sewage treatment? Summarize Ashley Murray's plans for improving sanitation in less-developed countries. What is a waterless composting toilet system? Explain how we can use wetlands to treat sewage. Describe John Todd's use of living machines to treat sewage.

10. List six ways to prevent or reduce water pollution. List five things you can do to help reduce water pollution. What are this chapter's *three big ideas*? Explain how we can use the three scientific **principles of sustainability** (see Figure 1-2, p. 6 or back cover) to prevent and reduce water pollution.

Note: Key terms are in bold type.

Critical Thinking

1. How might you be contributing directly or indirectly to the annual dead zone that forms in the Gulf of Mexico (**Core Case Study**)? What are three things you could do to reduce your impact?

2. Suppose a large number of dead fish are found floating in a lake. How would you determine whether they died from cultural eutrophication or from exposure to toxic chemicals?

3. Assume that you are a regulator charged with drawing up plans for controlling water pollution, and briefly describe your strategy for controlling water pollution from each of the following sources: **(a)** a pipe from a factory discharging effluent into a stream, **(b)** a parking lot at a shopping mall bordered by a stream, and **(c)** a farmer's field on a slope next to a stream.

4. How might you be contributing directly or indirectly to groundwater pollution? What are three things you could do to reduce your impact?

5. When you flush your toilet, where does the wastewater go? Trace the actual flow of this water in your community from your toilet through sewers to a wastewater treatment plant (or to a septic system) and from there to the environment. Try to visit a local sewage treatment plant to see what it does with wastewater. Compare the processes it uses with those shown in Figure 20-24. What happens to the sludge

produced by this plant? What improvements, if any, would you suggest for this plant?

6. In your community, **(a)** what is the source of drinking water? **(b)** How is drinking water treated? **(c)** What problems related to drinking water, if any, have arisen in your community? **(d)** What actions, if any, has your local government taken to solve them?

7. List three ways in which you could apply Concept 20-5 in order to make your lifestyle more environmentally sustainable.

8. Congratulations! You are in charge of the world. What are three actions you would take to **(a)** sharply reduce point-source water pollution in more-developed countries, **(b)** sharply reduce nonpoint-source water pollution throughout the world, **(c)** sharply reduce groundwater pollution throughout the world, and **(d)** provide safe drinking water for the poor and for other people in less-developed countries?

Doing Environmental Science

Do some research using the library and/or the Internet and evaluate the relative effectiveness of three different home water purification devices. Try to determine the costs of these systems, including purchase, installation, and maintenance costs, and record these costs for each system. Determine the type or types of water pollutants that each device is supposed to remove and the effectiveness of each such device. You might also consider looking up reviews of each product and finding some users of these products to interview. Write a report summarizing your findings.

Global Environment Watch Exercise

Visit the *Gulf of Mexico Oil Spill 2010* topic portal and research the ecological effects of the oil well blowout and spill. Write a report that answers the following questions:

1. Is oil from the spill in the Gulf still considered to be a problem?

2. How has it affected the coastal wetlands along the gulf?

3. How has it affected the gulf's bottom-dwelling species?

4. How has it affected bird life in the gulf?

5. How has it affected the fishing industry there?

Data Analysis

In 2006, scientists assessed the overall condition of the estuaries on the western coasts of the U.S. states of Oregon and Washington. To do so, they took measurements of various characteristics of the water, including dissolved oxygen (DO), in selected locations within the estuaries. The concentration of DO for each site was measured in terms of milligrams (mg) of oxygen per liter (L) of water sampled. The scientists used the following DO concentration ranges and quality categories to rate their water samples: water with greater than 5 mg/L of DO was considered *good* for supporting aquatic life; water with 2 to 5 mg/L of DO was rated as *fair*; and water with less than 2 mg/L of DO was rated as *poor*.

The following graph shows measurements taken in bottom water at 242 locations. Each triangle mark represents one or more measurements. The x-axis on this graph represents DO concentrations in mg/L. The y-axis represents percentages of the total area of estuaries studied (estuarine area).

Concentrations of dissolved oxygen in bottom waters of estuaries in Washington and Oregon.

To read this graph, pick one of the triangles and observe the values on the x- and y-axes. For example, note that the circled triangle lines up approximately with the 5-mg/L mark on the x-axis and with a value of about 34% on the y-axis. This means that waters at this particular measurement station (or stations), along with about 34% of the total area being studied, are estimated to have a 5% or lower DO concentration.

Use this information, along with the graph, to answer the following questions:

1. Half of the estuarine area has waters falling below a certain DO concentration level, and the other half has levels above that level. What is that level, in mg/L?

2. Give your estimate of the highest DO concentration measured and your estimate of the lowest concentration.

3. Approximately what percentage of the estuarine area studied is considered to have poor DO levels? About what percentage has fair DO levels, and about what percentage has good DO levels?

1. Below are parts of a river associated with a point source pollutant. Which of the below is incorrectly paired?
   a. Clean zone–high levels of dissolved oxygen
   b. Point of input–rapid increase of biological oxygen demand
   c. Decomposition zone–only pollution tolerant fishes found
   d. Septic zone–most fish absent, low levels of dissolved oxygen
   e. Recovery zone–normal levels of dissolved oxygen

2. Frequently, lakes near areas with high human population densities will experience large growths of algae or other unwanted bacteria and phytoplankton, often due to a lack of secondary waste treatment. This is the result of
   a. power companies dumping heated water from cooling towers.
   b. development of lakefront shorelines with beaches.
   c. sediment runoff from construction sites.
   d. nonpoint source pollution runoff from sidewalks and parking lots.
   e. large nutrient influxes from sewage treatment plants.

3. Which of the below water pollutants is **incorrectly** paired with its source?
   a. Oxygen demanding wastes–sewage
   b. Organic chemicals–industry and farms
   c. Sediments–unlined landfills
   d. Heavy metals–mining refuse
   e. Thermal–electric power plants

4. Which of the below is an example of a point source pollutant?
   a. Ballast water discharge from a ship offshore
   b. Runoff from parking lots
   c. Organic waste from cow feed lots
   d. Sediments form land erosion
   e. Nutrients from golf courses and lawns

5. How is cholera transmitted?
   a. A person inhales bacteria expelled into the air by someone coughing
   b. A mosquito transmits the parasite when it bites a human
   c. Ingesting water contaminated by human waste material from a person who already has the disease
   d. Coming in contact with skin pustules present on a person who already has the disease
   e. Ingesting water contaminated by leaking gasoline tanks

6. Which process best explains eutrophication?
   a. The downward motion of water through soil, dissolving minerals along the way
   b. The increase of a chemical such as DDT in the bodies of animals at higher trophic levels
   c. A process of nature that helps support life, such as rivers purifying the water
   d. The natural aging process of a shallow lake or slow-moving stream as nutrients enrich the water
   e. The collection of sediment behind a dam caused by the slowing of the river water

Questions 7 and 8 refer to the diagram of the Mississippi River drainage basin below.

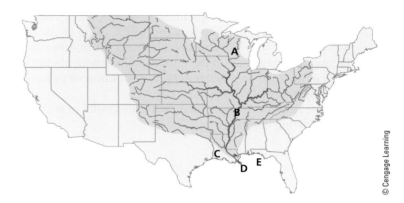

© Cengage Learning

7. Which letter best shows the location of depleted oxygen caused by the Mississippi River?

8. The depleted oxygen zone caused by the Mississippi River is caused primarily by
   a. sediment runoff from construction sites.
   b. wakes from boats traveling up and down the Mississippi River.
   c. lack of current flow in the main stem of the river.
   d. nitrate runoff from agricultural fields and sewage treatment plants.
   e. freshwater of the river dissolving in the Gulf of Mexico and organic chemicals precipitating out.

9. Which law sets standards for the allowed levels of key water pollutants and requires polluters to get permits limiting how much of various pollutants they can discharge into aquatic systems?
   a. National Environmental Policy Act (NEPA) 1969
   b. Water Pollution Control Act of 1972
   c. Clean Water Act 1977
   d. Coastal Zone Management Act 1980
   e. Water Quality Act 1987

Questions 10–12 are based on the steps of waste-water treatment below.

   a. Primary treatment
   b. Secondary treatment
   c. Disinfection
   d. Advanced treatment
   e. Discharge

10. Water flows slowly through grit chambers, allowing sand and small particles to settle out.

11. Biological processes are used to remove oxygen-demanding organic wastes before chlorine is added to kill bacteria.

12. Because this step removes nitrates and phosphates with special filters, there is less nutrient overload where the water is discharged.

21 Solid and Hazardous Waste

Follow nature's example, realize waste's potential.

Key Questions

21-1 What are solid waste and hazardous waste, and why are they problems?

21-2 How should we deal with solid waste?

21-3 Why are refusing, reducing, reusing, and recycling so important?

21-4 What are the advantages and disadvantages of burning or burying solid waste?

21-5 How should we deal with hazardous waste?

21-6 How can we make the transition to a more sustainable low-waste society?

Recycling computer monitors in Grand Prairie, Texas.

Peter Essick/National Geographic Creative

Electronic waste, or *e-waste,* consists of discarded television sets, cell phones, computers, and other electronic devices (see chapter-opening photo). It is the fastest-growing solid waste problem in the United States and in the world. Each year, Americans discard several hundred million cell phones, personal computers, television sets, iPods, Blackberries, fax machines, printers, and other electronic products.

Only about 14% of all U.S. e-waste is recycled, according to the Consumer Electronics Association. Most of the rest of it ends up in landfills and incinerators, even though many of the components in electronic devices contain valuable materials that could be recycled or reused. These wasted resources include high-quality plastics and valuable metals such as aluminum, copper, platinum, silver, gold, and rare-earth metals that are used to build most of the world's electronic devices and military weapons systems (see Chapter 14, Core Case Study, p. 350). E-waste is also a source of toxic and hazardous chemicals that can contaminate air, surface water, groundwater, and soil and cause serious and even life-threatening health problems for e-waste workers.

Much of the e-waste in the United States that is not buried or incinerated is shipped to China (Figure 21-1), India, and other Asian and African countries where labor is cheap and environmental regulations are weak. Workers there—many of them children—dismantle, burn, and treat e-waste products with acids to recover valuable metals and reusable parts. As they do this, they are exposed to toxic metals such as lead and mercury and other harmful chemicals. The remaining scrap is dumped into waterways and fields or burned in open fires, exposing many people to highly toxic chemicals called dioxins.

Transfer of such hazardous waste from more-developed to less-developed countries was banned by the International Basel Convention, a treaty now signed by 179 countries to reduce and control the movement of hazardous waste across international borders. Despite this ban, much of the world's e-waste is not officially classified as hazardous waste or is illegally smuggled out of some countries. The United States can export its e-waste legally because it is the only industrialized nation that has not ratified the Basel Convention.

The European Union (EU) has led the way in dealing with e-waste, taking a *cradle-to-grave* approach, which requires manufacturers to take back electronic products at the end of their useful lives and repair, remanufacture, or recycle them. In the EU, e-waste is banned from landfills and incinerators. Japan is also taking this approach. To cover the costs of these programs, consumers pay a recycling tax on electronic products, an example of implementing the full-cost pricing **principle of sustainability** (see Figure 1-5, p. 9 or back cover). However, because of the high costs involved in complying with these laws, much of this waste that is supposed to be recycled domestically is shipped illegally to other countries.

Electronic waste is just one of many types of solid and hazardous waste that we discuss in this chapter.

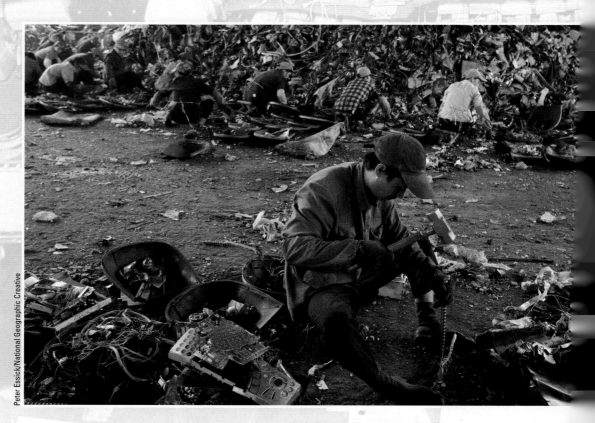

Figure 21-1 These workers in Taizhou City in China's Zhejiang Province are recovering valuable materials from scrap computers shipped from the United States. **Question:** Have you disposed of an electronic device lately? If so, how did you dispose of it?

Peter Essick/National Geographic Creative

21-1 What Are Solid Waste and Hazardous Waste, and Why Are They Problems?

CONCEPT 21-1A
Solid waste contributes to pollution and includes valuable resources that could be reused or recycled.

CONCEPT 21-1B
Hazardous waste contributes to pollution, as well as to natural capital degradation, health problems, and premature deaths.

We Throw Away Huge Amounts of Useful Things

In the natural world, wherever humans are not dominant, there is essentially no waste because the wastes of one organism become nutrients or raw materials for others (see Figure 3-10, p. 59). This natural cycling of nutrients is the basis for one of the three scientific **principles of sustainability** (see Figure 1-2, p. 6 or back cover).

Modern humans violate this principle in a big way. We produce huge amounts of waste materials that go unused and that pollute the environment. For example, the manufacturing of a single desktop computer requires 700 or more different materials obtained from mines, oil wells, and chemical factories all over the world, along with large amounts of energy. For every 0.5 kilogram (1 pound) of electronics it contains, approximately 3,600 kilograms (8,000 pounds) of waste were created—an amount roughly equal to the weight of a large pickup truck.

Because of the law of conservation of matter (see Chapter 2, p. 40) and the nature of human lifestyles, we will always produce some waste. However, studies and experience indicate that we could reduce this waste of potential resources and the resulting environmental harm it causes by up to 80%.

One major category of waste is **solid waste**—any unwanted or discarded material we produce that is not a liquid or a gas. Solid waste can be divided into two types. One type is **industrial solid waste** produced by mines (see Figure 14-10, p. 358), farms, and industries that supply people with goods and services. The other is **municipal solid waste (MSW)**, often called *garbage* or *trash*, which consists of the combined solid waste produced by homes and workplaces other than factories. Examples include paper and cardboard, food wastes, cans, bottles, yard wastes, furniture, plastics, metals, glass, wood, and e-waste (**Core Case Study**). Some resource experts believe that we should change the name of the trash we produce from MSW of MWR—mostly wasted resources (**Concept 21-1A**).

Much of this waste ends up in rivers (see Figure 20-7, p. 550), lakes (Figure 20-5, p. 547), the ocean (Figure 20-17, p. 560), and natural landscapes (Figure 21-2). Discarded plastic items are a threat to many terrestrial animal species, as well as millions of seabirds (see Figure 20-18, p. 560), marine mammals (see Figure 11-4, p. 253), and sea turtles, which can mistake a floating plastic sandwich bag for a jellyfish or get caught in plastic fishing nets (see Figure 11-1, p. 248). About 80% of the plastics in the ocean are blown or washed in from beaches, rivers, storm drains, and other sources, and the rest get dumped into the ocean from ocean-going garbage barges, ships, and fishing boats.

In more-developed countries, most MSW is buried in landfills or burned in incinerators. In many less-developed countries, much of it ends up in open dumps, where poor people eke out a living finding items they can use or sell (Figure 21-3). The United States is the world's largest producer of MSW (see the following Case Study).

Rechitan Sorin/Shutterstock.com

Figure 21-2 Various types of solid waste have been dumped in this isolated mountain area.

William Albert Allard/National Geographic Creative

Figure 21-3 This child and woman survive by searching this open trash dump in India for useful items.

Hazardous Waste Is a Serious and Growing Problem

Another major category of waste is **hazardous**, or **toxic**, **waste**—any discarded material or substance that threatens human health or the environment because it is poisonous, dangerously chemically reactive, corrosive, or flammable. Examples include industrial solvents, hospital medical waste, car batteries (containing lead and acids), household pesticide products, dry-cell batteries (containing mercury and cadmium), and ash and sludge from incinerators and coal-burning power and industrial plants (see Figure 15-18, p. 388, and Figure 15-19, p. 389). Improper handling of these wastes can lead to pollution of air and water, degradation of ecosystems, and severe health threats (Concept 21-1B).

Figure 21-4 lists some of the harmful chemicals found in many homes. The two largest classes of hazardous wastes are *organic compounds* (such as various solvents, pesticides, PCBs, and dioxins) and nondegradable *toxic heavy metals* (such as lead, mercury, and arsenic).

Another form of extremely hazardous waste is the highly radioactive waste produced by nuclear power plants and nuclear weapons facilities (see Chapter 15, p. 391). Such wastes must be stored safely for 10,000 to 240,000 years, depending on what radioactive isotopes are present in them. After 60 years of research, scientists and governments have not found a scientifically and politically acceptable way to safely isolate these dangerous wastes for such long periods of time.

According to the U.N. Environment Programme (UNEP), more-developed countries produce 80–90% of the world's hazardous wastes, and the United States is the largest producer. However, as China continues to industrialize rapidly, largely without adequate pollution controls, it may soon take over the number-one spot. In order, the top three U.S. producers of hazardous waste are the military, the chemical industry, and the mining industry.

CASE STUDY

Solid Waste in the United States

The United States leads the world in total solid waste production and in solid waste per person. With only 4.6% of the world's people, the United States produces about 25% of the world's solid waste. According to the U.S. Environmental Protection Agency (EPA), about 98.5% of all solid waste produced in the United States is industrial

---

What Harmful Chemical Are in Your Home?

Cleaning
Disinfectants
Drain, toilet, and window cleaners
Spot removers
Septic tank cleaners

Paint Products
Paints, stains, varnishes, and lacquers
Paint thinners, solvents, and strippers
Wood preservatives
Artist paints and inks

General
Dry-cell batteries (mercury and cadmium)
Glues and cements

Gardening
Pesticides
Weed killers
Ant and rodent killers
Flea powders

Automotive
Gasoline
Used motor oil
Antifreeze
Battery acid
Brake and transmission fluid

© Cengage Learning 2015

Figure 21-4 Harmful chemicals are found in many homes. The U.S. Congress has exempted the disposal of many of these household chemicals and other items from government regulation. **Question:** Which of these chemicals could you find in the place where you live?

Top: ©tuulijumala/Shutterstock.com. Center: Katrina Outland/Shutterstock.com. Bottom: Karramba Production/Shutterstock.com.

Jose Azel/Aurora Photos

Figure 21-5 A year's worth of solid waste produced by family of four in San Diego, California. Recyclable items are on the left and the rest of their MSW is on the right.

In 2010, the largest categories of this MSW were paper and cardboard (29% of total), food waste (14%), yard waste (13%), plastics (12%), and metals (9%). According to the U.S. Energy Information Administration, packaging makes up 32% of all the country's MSW—the single greatest category of MSW in the United States. Every year, the United States generates enough MSW to fill a bumper-to-bumper convoy of garbage trucks long enough to encircle the globe almost 6 times, or to reach more than halfway to the moon. About 67% of it is dumped in landfills and about 9% is incinerated, but much of it ends up as litter.

Consider some of the solid wastes that consumers throw away in the high-waste economy of the United States:

- Enough tires each year to encircle the planet almost 3 times.
- An amount of disposable diapers each year that, if linked end to end, would reach to the moon and back 7 times.
- Enough carpet each year to cover the U.S. state of Delaware.
- About 2.8 million nonreturnable plastic bottles *every hour*. If laid end-to-end, the number of these bottles produced each year would reach from the earth to the moon and back about 6 times.
- About 274 million plastic shopping bags per day, an average of about 3,200 every second.
- Enough office paper each year to build a wall 3.5 meters (11 feet) high across the country from New York City to San Francisco, California.
- Enough plastic wrap each year to shrink-wrap Texas.
- 25 billion nonrecyclable plastic foam cups a year—enough, if lined up end to end, to circle the earth at its equator 436 times.

Most of these wastes break down very slowly, if at all. Lead, mercury, glass, plastic foam, and most plastic bottles essentially take forever to break down; an aluminum can takes 500 years; plastic bags, 400 to 1,000 years; and a plastic six-pack holder, 100 years (Science Focus 21.1).

solid waste from mining (76%), agriculture (13%), and industry (9.5%). The remaining 1.5% of U.S. solid waste is MSW.

While the EPA reports regularly on the amount and composition of U.S. MSW, Columbia University researchers and *BioCycle* magazine have gathered actual nationwide MSW data and found that it differs significantly from the EPA solid waste data, which is based on theoretical matter-flow models that are being reevaluated. According to this data-based analysis published in 2010, Americans produce much more waste and recycle and compost less waste than the EPA has estimated.

The Columbia and *BioCycle* researchers estimated that the total amount of MSW produced by the United States in 2008 (the latest year for which they had data) was about 354 million metric tons (389 million tons). This means that in 2008 each American threw away an average of about 3.2 kilograms (7 pounds) of trash every day (about a third more than what the EPA estimated), or about 1,168 kilograms (2,570 pounds) in a year (Figure 21-5). The researchers also found that about 24% of these wastes were recycled (which is lower than the EPA estimate of about 34%). These estimates do not include the much larger numbers for industrial solid waste—about 98.5% of all solid waste—generated to support the American economy and lifestyles.

GARBOLOGY AND TRACKING TRASH

How do we know about the composition of trash in landfills? Much of that information comes from research by *garbologists* such as William Rathje, an anthropologist who pioneered the field of garbology in the 1970s at the University of Arizona. These scientists work in the fashion of archaeologists, training their students to sort, weigh, and itemize people's trash, and to bore holes in garbage dumps and analyze what they find.

Many people think of landfills as huge compost piles where biodegradable wastes are decomposed within a few months. But garbologists looking at the contents of landfills found 50-year-old newspapers that were still readable and hot dogs and

pork chops buried for decades that had not decayed. In landfills (as opposed to open dumps), trash can resist decomposition for perhaps centuries because it is tightly packed and protected from sunlight, water, and air, and from the bacteria that could digest and decompose much of these wastes in keeping with the chemical cycling **principle of sustainability**.

In 2009, a team of researchers, led by Carlo Riatti, at the Massachusetts Institute of Technology (MIT) conducted a project called "Trash Track." The project's goals were to find out where urban trash goes and to help New York City increase its recycling rate from the current 30% to 100% by 2030.

The researchers attached wireless transmitters about the size of a matchbook to several thousand different items of trash produced by volunteer participants in New York City, Seattle, Washington, and London, England. They are now using these battery-operated electronic tags to track the trash items and show their real-time movements whether they go to recycling plants, landfills, or incinerators. For example, in Seattle, Washington, 75% of the tracked waste items reached recycling plants.

Critical Thinking

Should landfills be exposed to more air and water to hasten decomposition of their wastes? Explain.

21-2 How Should We Deal with Solid Waste?

CONCEPT 21-2
A sustainable approach to solid waste is first to reduce it, then to reuse or recycle it, and finally to safely dispose of what is left.

We Can Burn, Bury, or Recycle Solid Waste or Produce Less of It

We can deal with the solid wastes we create in two ways. One method is **waste management**, in which we attempt to control wastes in ways that reduce their environmental harm without seriously trying to reduce the amount of waste produced. It typically involves mixing wastes together and then transferring them from one part of the environment to another, usually by burying them, burning them, or shipping them to another location. The second approach is **waste reduction**, in which we produce much less waste and pollution, and the wastes we do produce are considered to be potential resources that we can reuse, recycle, or compost (**Concept 21-2**).

There is no single solution to the solid waste problem. Most analysts call for using **integrated waste management**—a variety of coordinated strategies for both waste disposal and waste reduction (Figure 21-6).

Environmental scientists and the EPA call for much greater emphasis on waste prevention and reduction and

pollution prevention than on waste disposal. Figure 21-7 compares the science-based waste-management goals of the EPA and National Academy of Sciences with actual waste-management trends based on analysis of waste data by Columbia University and *BioCycle* researchers.

Some scientists and economists estimate that we could eliminate up to 80% of the solid waste we produce by rigorously applying the strategies shown in Figure 21-7, left. Let us look more closely at these options in the order of priorities suggested by scientists.

We Can Cut Solid Wastes by Refusing, Reducing, Reusing, and Recycling

Waste reduction (**Concept 21-2**) is based on the four Rs of resource use:

- **Refuse**: Don't use it.
- **Reduce**: Use less.
- **Reuse**: Use it over and over.
- **Recycle**: Convert it to useful items and buy products made from recycled materials.

From an environmental standpoint, the first three Rs are preferred because they are *input,* or *waste prevention,* approaches that tackle the problem of waste production at the front end—before it occurs. Recycling is important

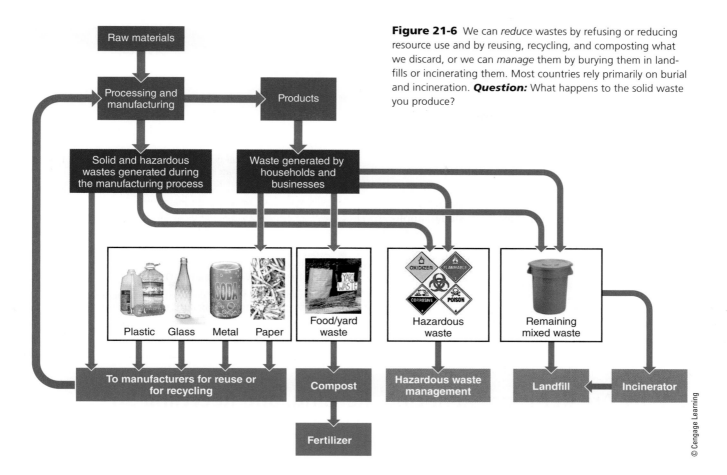

Figure 21-6 We can *reduce* wastes by refusing or reducing resource use and by reusing, recycling, and composting what we discard, or we can *manage* them by burying them in landfills or incinerating them. Most countries rely primarily on burial and incineration. **Question:** What happens to the solid waste you produce?

but it deals with wastes after they have been produced. An important form of recycling is **composting**, which involves using bacteria to decompose yard trimmings, vegetable food scraps, and other biodegradable organic wastes into materials than can increase soil fertility.

By refusing, reducing, reusing, recycling, and composting, we save matter and energy resources, reduce pollution (including greenhouse gas emissions), help protect biodiversity, and save money. Figure 21-8 lists some ways in which you can use the four Rs of waste reduction to reduce your output of solid waste. GOOD NEWS

Here are six strategies that industries and communities have used to reduce resource use, waste, and pollution.

First, *change industrial processes to eliminate or reduce the use of harmful chemicals*. Since 1975, the 3M company has taken this approach and in the process saved $1.2 billion (see Chapter 17, Case Study, p. 464).

Second, *redesign manufacturing processes and products to use less material and energy*. For example, the weight of a typical car has been reduced by about one-fourth since the 1960s through the use of lighter steel and lightweight plastics and composite materials.

Third, *develop products that are easy to repair, reuse, remanufacture, compost, or recycle*. For example, Xerox photocopiers that are leased by businesses are made of reusable or recyclable parts that allow for easy remanufacturing and

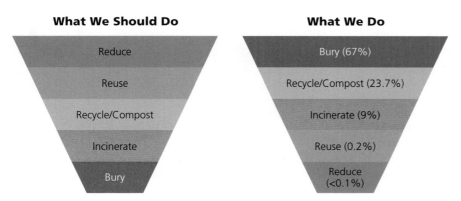

Figure 21-7 Priorities recommended by the U.S. National Academy of Sciences for dealing with municipal solid waste (left) compared with actual waste-handling practices in the United States based on data (right). **Question:** Why do you think most countries do not follow most of the scientific-based priorities listed on the left?

(Compiled by the authors using data from U.S. Environmental Protection Agency, U.S. National Academy of Sciences, Columbia University, and *BioCycle*.)

are projected to save the company $1 billion in manufacturing costs.

Fourth, *eliminate or reduce unnecessary packaging.* Use the following hierarchy for product packaging: no packaging, reusable packaging, and recyclable packaging. The 37 European Union countries require the recycling of 55–80% of all packaging waste.

Fifth, *use fee-per-bag waste collection systems* that charge consumers for the amount of waste they throw away but provide free pickup of recyclable and reusable items.

Sixth, *establish cradle-to-grave responsibility laws* that require companies to take back various consumer products such as electronic equipment (**Core Case Study**), appliances, and motor vehicles, as Japan and many European countries do.

Figure 21-8 Individuals matter: You can save resources by reducing your output of solid waste and pollution. ***Questions:*** Which three of these steps do you think are the most important ones to take? Why? Which of these things do you already do?

21-3 Why Are Refusing, Reducing, Reusing, and Recycling So Important?

CONCEPT 21-3

By refusing and reducing resource use and by reusing and recycling what we use, we decrease our consumption of matter and energy resources, reduce pollution and natural capital degradation, and save money.

There Are Alternatives to the Throwaway Economy

Today's more-developed economies are based largely on getting people to buy more and more stuff whether they need it or not. Such buying helps to support businesses that sell us these materials, and it supports jobs. However, there is growing evidence that in the long run, economies based on steadily increasing consumption of resources are not environmentally or economically sustainable.

Refusing to buy some things, especially those with a significant environmental impact, is one way to live more sustainably. It begins by asking important questions such as: "Do I really need this stuff—this piece of clothing or this electronic gadget? Do I really need to replace my mobile phone, computer, or other electronic device just because a newer model has come out?"

Reducing is the next level of living more sustainably. It means buying less, and involves asking questions such as: "How many pairs of shoes or pants or how many blouses or skirts do I need? How many electronic devices do I need?"

Reusing is the third level of more sustainable living. It involves cleaning and using items such as containers,

dishes, utensils, and razor blades over and over, and thus increasing the typical life span of such products. This form of waste reduction decreases the use of matter and energy resources, cuts waste and pollution (including greenhouse gases), creates local jobs, and saves money (**Concept 21-3**).

The problem is that in today's industrialized societies, we have increasingly substituted throwaway items for reusable ones, which has resulted in more pollution and growing masses of solid waste. For example, if the 1 billion throwaway paper coffee cups (Figure 21-9, left) used by one famous chain of donut shops each year were lined

Levent Konuk/Shutterstock.com

ffolas/Shutterstock.com

Figure 21-9 *Reuse:* Instead of using throwaway paper (left) or plastic cups, many people take their own refillable cups (right) to workplaces, coffee shops, restaurants, and school cafeterias.

Figure 21-10 *Reuse.* A growing number of people are using reusable cloth bags instead of throwaway plastic and paper bags.

Brenda Carson/Shutterstock.com

What Can You Do?

Reuse

- Buy beverages in refillable glass containers
- Use reusable lunch containers
- Store refrigerated food in reusable containers
- Use rechargeable batteries and recycle them when their useful life is over
- When eating out, bring your own reusable container for leftovers
- Carry groceries and other items in a reusable basket or cloth bag
- Buy used furniture, cars, and other items, whenever possible

Figure 21-11 Individuals matter: There are many ways to reuse the items we purchase. *Question:* Which of these suggestions have you tried and how did they work for you?

up end-to-end, they would encircle the earth twice. Many coffee shops are reducing such waste and cutting their costs by discounting their hot drink prices for customers who bring their own refillable mugs (Figure 21-9, right).

In general, reuse is on the rise, although there is great potential for more of it. Denmark, Finland, and the Canadian province of Prince Edward Island have banned all beverage containers that cannot be reused. In Finland, 95% of the soft drink, beer, wine, and spirits containers are refillable. The latest rechargeable batteries come fully charged, can hold a charge for up to 2 years when they are not used, and can be recharged in about 15 minutes. We could greatly reduce toxic waste by replacing all conventional batteries with such rechargeables.

Instead of using throwaway paper or plastic bags to carry home groceries and other items they purchase, many people have switched to reusable cloth bags (Figure 21-10). Both paper and plastic bags are environmentally harmful, and the question of which is more damaging has no clear-cut answer. In many countries, the landscape is littered with plastic bags, as are many areas of the ocean. They can take 400 to 1,000 years to break down. Less than 1% of the estimated 102 billion plastic bags used each year in the United States are recycled. In addition to being an eyesore, they often block drains and sewage systems and can kill wildlife and livestock that try to eat them or become ensnared in them.

To encourage people to carry reusable bags, the governments of Ireland, Taiwan, and the Netherlands tax plastic shopping bags. In Ireland, a tax of about 25¢ per bag helped to cut plastic bag litter by 90% as people switched to reusable bags. Kenya, Uganda, Rwanda, South Africa, Bangladesh, Bhutan, parts of India, Taiwan, China, Australia, France, and Italy also have banned the use of all or most types of plastic shopping bags. By 2012,

the U.S. cities of San Francisco and Los Angeles, California; Austin, Texas; and Portland, Oregon, along with the state of Hawaii, had instituted complete bans or phase-outs of plastic bags. At least 20 other U.S. communities had followed suit.

There are many other ways to reuse various items, some of which are listed in Figure 21-11. 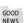 GOOD NEWS

There Is Great Potential for Recycling

Households and workplaces produce five major types of materials that can be recycled into new, useful products: paper products, glass, aluminum, steel, and some plastics. We can reprocess such materials in two ways. In **primary**, or **closed-loop**, **recycling**, materials such as aluminum cans are recycled into new products of the same type. In **secondary recycling**, waste materials are converted into different products.

Scientists also distinguish between two types of recyclable wastes: *preconsumer* or *internal waste* generated in a manufacturing process, and *postconsumer* or *external waste* generated by consumers' use of products. Preconsumer waste makes up more than three-fourths of the total.

Recycling involves three steps: collecting materials for recycling, converting the recycled materials to new products, and the selling and buying of products containing recycled material. Recycling does not work unless all three of these steps are taken consistently.

In 2010, the United States recycled or composted less than a quarter of its MSW. According to the EPA, this included the recycling of about 96% of all lead–acid batteries, 72% of all newspapers, directories, and newspaper inserts, 67% of all steel cans, and 50% of all aluminum cans discarded, according to EPA estimates. Recycling rates for some MSW categories have increased, partly because the number of curbside-pickup recycling pro-

Figure 21-12 When the compost in this backyard composter drum is ready to be used, the device can be wheeled to a garden and emptied into the soil.

Some cities in Canada and in many European Union countries collect and compost 85% or more of their biodegradable wastes in centralized community facilities. In the United States, about 3,000 municipal composting programs recycle about 60% of the yard wastes in the country's MSW (Figure 21-13), and in 2012, at least 25 states had some version of a ban on yard wastes in sanitary landfills. Compost from such programs can be used as organic soil fertilizer, topsoil, or landfill cover, and as a source of revenues for the communities.

To be successful, a large-scale composting program must be located carefully and odors must be controlled, especially near residential areas. Sometimes, composting takes place in huge indoor buildings. In the Canadian city of Edmonton, Alberta, an indoor facility the size of eight football fields composts 50% of the city's organic solid waste. Composting programs must also exclude toxic materials that can contaminate the compost and make it unsafe for fertilizing crops and lawns.

grams has risen to about 9,000, now serving about half of the U.S. population. Experts say that with education and proper incentives, the United States could recycle and compost 80% of its MSW, in keeping with the chemical cycling **principle of sustainability**.

Composting is another form of recycling that mimics nature's recycling of nutrients. It involves using bacteria to decompose yard trimmings, vegetable food scraps, and other biodegradable organic wastes. The resulting organic material can be added to soil to supply plant nutrients, slow soil erosion, retain water, and improve crop yields. Home owners can compost such wastes in simple backyard containers, in composting piles that must be turned over occasionally, or in small composting drums (Figure 21-12) that are rotated to mix the wastes and speed up the decomposition process.

We Can Mix or Separate Household Solid Wastes for Recycling

One way to recycle is to send mixed household and business wastes to centralized *materials-recovery facilities* (MRFs, see Figure 21-6). There, machines or workers separate the mixed waste to recover valuable materials for sale to manufacturers as raw materials. The remaining paper, plastics, and other combustible wastes are recycled or burned to produce steam or electricity, which is used to run the recovery plant or is sold to nearby industries or homes.

Such plants are expensive to build, operate, and maintain. If not operated properly, they can emit CO_2 and toxic air pollutants, and they produce a toxic ash that must be disposed of safely, usually in landfills. Because MRFs

Figure 21-13 Large-scale municipal composting site.

require a steady diet of garbage to make them financially successful, their owners have a vested interest in increasing the flow of matter and energy resources. Thus, use of MRFs encourages people to produce more trash—the reverse of what many scientists urge us to do (Figure 21-7, left).

To many experts, it makes more environmental and economic sense for households and businesses to separate their trash into recyclable categories such as glass, paper, metals, certain types of plastics, and compostable materials. This *source separation* approach produces much less air and water pollution and costs less to implement than do MRFs. It also saves more energy, provides more jobs per unit of material, and yields cleaner and usually more valuable recyclables.

To promote separation of wastes for recycling, about 7,000 communities in the United States use a *pay-as-you-throw* or *fee-per-bag* waste collection system. They charge households and businesses for the amount of garbage that is picked up, but do not charge them for picking up materials separated for recycling or reuse. When the U.S. city of Ft. Worth, Texas, instituted such a program, the proportion of households recycling their trash went from 21% to 85%. In addition, the city went from losing $600,000 in its recycling program to making $1 million a year because of increased sales of recycled materials to industries.

Recycling Paper

About 55% of the world's industrial tree harvest is used to make paper. However, according to the U.S. Department of Agriculture, we could make tree-free paper from straw and other agricultural residues and from the fibers of rapidly growing plants such as kenaf (see Figure 10-17, p. 228) and hemp.

The pulp and paper industry is the world's fifth largest energy consumer and uses more water to produce a metric ton of its product than any other industry. In both Canada and the United States, it is the third-largest industrial energy user and polluter, and paper is the dominant material in the MSW of both countries.

Paper (especially newspaper and cardboard) is easy to recycle. Recycling newspaper involves removing its ink, glue, and coating and then reconverting the paper to pulp, which is pressed into new paper. Making recycled paper uses 64% less energy and produces 35% less water pollution and 74% less air pollution than does making paper from wood pulp, and, of course, no trees are cut down.

In 2009, the Natural Resources Defense Council (NRDC) mounted a campaign to pressure paper companies to stop cutting down trees in some of North America's old-growth forests in order to make toilet paper. The NRDC estimates that nearly 425,000 trees would be saved each year if every U.S. household used just one 500-sheet roll of toilet paper made from 100% recycled paper in place of a roll made from virgin timber.

Recycling Plastics

Plastics consist of various types of large polymers, or *resins*—organic molecules made by chemically linking organic chemicals produced mostly from oil and natural gas. About 46 different types of plastics are used in consumer products, and some products contain several kinds of plastic.

Currently, only about 7% by weight of all plastic wastes in the United States (and 13% of plastic containers and packaging) is recycled. These percentages are low because there are many different types of plastic resins, which are difficult to separate from products that contain them. However, progress is being made in the recycling of plastics (Individuals Matter 21.1) and in the development of more degradable bioplastics (Science Focus 21.2).

Recycling Has Advantages and Disadvantages

Figure 21-14 lists the advantages and disadvantages of recycling (Concept 21-3). Whether recycling makes economic sense depends on how we look at its economic and environmental benefits and costs.

Critics of recycling programs argue that recycling is costly and adds to the taxpayer burden in communities where recycling is funded through taxation. They say that recycling may make economic sense for valuable and easy-to-recycle materials in MSW such as paper and paperboard, steel, and aluminum, but probably not for cheap or plentiful resources such as glass made from silica.

Proponents of recycling point to studies showing that the net economic, health, and environmental benefits of recycling (Figure 21-14, left) far outweigh the costs. For example, the EPA estimates that recycling and composting in the United States in 2010 reduced emissions of climate-changing carbon dioxide by an amount roughly equal to

Trade-Offs

Recycling

Advantages	Disadvantages
Reduces energy and mineral use and air and water pollution	Can cost more than burying in areas with ample landfill space
Reduces greenhouse gas emissions	Reduces profits for landfill and incinerator owners
Reduces solid waste	Inconvenient for some

© Cengage Learning 2015

Figure 21-14 Recycling solid waste has advantages and disadvantages (Concept 21-3). **Questions:** Which single advantage and which single disadvantage do you think are the most important? Why?

Photo: Jacqui Martin/Shutterstock.com

Mike Biddle's Contribution to Plastics Recycling

In 1994, Mike Biddle, a former PhD engineer with Dow Chemical, and his business partner Trip Allen founded MBA Polymers, Inc. Their goal was to develop a commercial process for recycling high-value plastics from complex streams of manufactured goods such as computers, electronics, appliances, and automobiles. They succeeded by designing a 21-step automated process that separates plastics from nonplastic items in mixed waste streams, and then separates plastics from each other by type and grade. The process then converts them to pellets that can be used to make new products.

The pellets are cheaper than virgin plastics because the company's process uses 90% less energy than that needed to make a new plastic and because the raw material is discarded junk that is cheap or free. In addition, greenhouse gas emissions from this process are much lower than those from the process of making virgin plastics. And recycling plastic wastes reduces the need to incinerate them or bury them in landfills.

The company is considered a world leader in plastics recycling. It operates a large, state-of-the-art research and recycling plant in Richmond, California (USA), and recently opened the world's two most advanced plastics recycling plants in China and Austria. MBA Polymers has won many awards, including the Thomas Alva Edison Award for Innovation, and it was selected by *Inc.* magazine as one of "America's Most Innovative Companies."

Maybe you can become an environmental entrepreneur by using your brainpower to develop an environmentally beneficial and financially profitable process or business.

Background photo: justasc/Shutterstock.com

that emitted by 36 million passenger vehicles. According to the Steel Recycling Institute, recycling a metric ton of steel eliminates the need to mine 2.3 metric tons (2.5 tons) of iron ore and 1.3 metric tons (1.4 tons) of coal. Proponents also argue that the U.S. recycling industry employs about 1.1 million people—more than the number of people working in the U.S. auto industry—and that its annual revenues are much larger than those of the waste-management industry.

Cities that make money by recycling and that have higher recycling rates tend to use a *single-pickup system* for both recyclable and nonrecyclable materials, instead of a more expensive dual-pickup system. Successful systems also tend to use a pay-as-you-throw approach, and they have citizens and businesses sort their trash more carefully. San Francisco, California (USA) uses such a system, and in 2012, the city recycled, composted, or reused 78% of its MSW.

GOOD NEWS

SCIENCE FOCUS 21.2

BIOPLASTICS

Most of today's plastics are made from organic polymers produced from petroleum-based chemicals. This may change as scientists shift to developing plastics made from biologically based chemicals.

Henry Ford, who developed the first Ford motorcar, supported research on the development of a bioplastic made from soybeans and another made from hemp. A 1914 photograph shows him using an ax to strike the body of a car made from soy bioplastic to demonstrate its strength and resistance to denting.

However, as oil became widely available, petrochemical plastics took over the market. Now, with projected climate change and other environmental problems associated with the use of oil, chemists are stepping up efforts to make biodegradable and more environmentally sustainable plas-

tics. These *bioplastics* can be made from corn, soy, sugarcane, switchgrass (see Figure 16-27, p. 426), chicken feathers, and some components of garbage.

Compared to conventional oil-based plastics, properly designed bioplastics are lighter, stronger, and cheaper, and the process of making them requires less energy and produces less pollution per unit of weight. Instead of being sent to landfills, some packaging made from bioplastics (Figure 21-A) can be composted to produce a soil conditioner, in keeping with the chemical cycling **principle of sustainability**.

Some bioplastics are more ecofriendly than others. For example, some are made from corn raised by industrial agricultural methods, which require great amounts of energy, water, and petrochemical fertil-

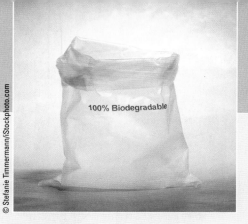

Figure 21-A Biodegradable plastic bag.

izers and thus have a very large ecological footprint. In evaluating and choosing bioplastics, scientists urge consumers to learn how they were made and whether or not they are truly recyclable or compostable.

Critical Thinking

What might be some disadvantages of bioplastics? Do you think they outweigh the advantages?

21-4 What Are the Advantages and Disadvantages of Burning or Burying Solid Waste?

CONCEPT 21-4
Technologies for burning and burying solid wastes are well developed, but burning contributes to air and water pollution and greenhouse gas emissions, and buried wastes eventually contribute to the pollution and degradation of land and water resources.

Burning Solid Waste Has Advantages and Disadvantages

Globally, MSW is burned in more than 600 large *waste-to-energy incinerators*, which use the heat they generate to boil water and make steam for heating water or interior spaces, or for producing electricity (Figure 21-15). They are intended to burn trash that cannot be recycled or reused. There are 115 of these incinerators in the United States, according to Columbia University and *BioCycle* researchers.

The United States incinerates about 9% of its MSW. One reason for this fairly low percentage is that incineration has a bad reputation stemming from past use of highly polluting and poorly regulated incinerators. Also,

incineration competes with an abundance of low-cost landfills in many parts of the country.

By contrast, Denmark incinerates 54% of its MSW in waste-to-energy incinerators with newer designs. The country also recycles or composts 42% of its MSW (compared to 24% in the United States) and buries only 4% of it in landfills (compared to 67% in the United States). Denmark's incinerators use dozens of filters to remove pollutants such as toxic mercury and dioxins. They run so cleanly that they release just a fraction of the toxic dioxins released by typical home fireplaces and backyard barbecues, and they exceed European air pollution standards by a factor of 10. A 2009 EPA study concluded that landfills emit more air pollutants than modern waste-to-energy incinerators. On the other hand, the toxic chemicals that are filtered out still have to be stored somewhere.

Denmark and most other European nations burn their trash in small, less costly, community-based incinerators instead of hauling the trash long distances by truck to large incinerators, as is done in the United States. Germany incinerates 34% of its MSW and recycles and composts most of the rest.

Electricity

Smokestack

Turbine

Generator

Steam

Crane

Furnace

Boiler

Wet scrubber

Electrostatic precipitator

Waste pit

Water added

Conveyor

Bottom ash

Dirty water

Fly ash

To waste treatment plant

© Cengage Learning

Ash for treatment, disposal in landfill, or use as landfill cover

Figure 21-15 Solutions: A waste-to-energy incinerator with pollution controls burns mixed solid wastes and recovers some of the energy to produce steam to use for heating or producing electricity. **Questions:** Would you invest in such a project? Why or why not?

Trade-Offs

Waste-to-Energy Incineration

Advantages	Disadvantages
Reduces trash volume	Expensive to build
Produces energy	Produces a hazardous waste
Concentrates hazardous substances into ash for burial	Emits some CO_2 and other air pollutants
Sale of energy reduces cost	Encourages waste production

© Cengage Learning 2015

Figure 21-16 Incinerating solid waste has advantages and disadvantages (**Concept 21-4**). These trade-offs also apply to the incineration of hazardous waste. **Questions:** Which single advantage and which single disadvantage do you think are the most important? Why?

Top: Ulrich Mueller/Shutterstock.com. Bottom: Dmitry Kalinovsky/Shutterstock.com.

In the United States, incinerators are used in some areas, including the city of Fort Myers, Florida, which has closed its landfills, recycles or composts half of its trash, and burns the rest to provide electricity for 36,000 homes. However, many U.S. citizens, local governments, and environmental scientists oppose any use of waste incinerators because it undermines efforts to increase reuse and recycling. Once an incinerator is built, it must be fed large volumes of trash in order to be profitable for its owners. Figure 21-16 lists the advantages and disadvantages of using incinerators to burn solid waste.

Burying Solid Waste Has Advantages and Disadvantages

About 67%, by weight, of the MSW in the United States is buried in sanitary landfills, compared to 80% in Canada, 15% in Japan, and 4% in Denmark. There are two types of landfills. In newer landfills, called **sanitary landfills** (Figure 21-17), solid wastes are spread out in thin layers, compacted, and covered daily with a fresh layer of clay or plastic foam. This process helps keep the material dry and reduces leakage of contaminated water (leachate)

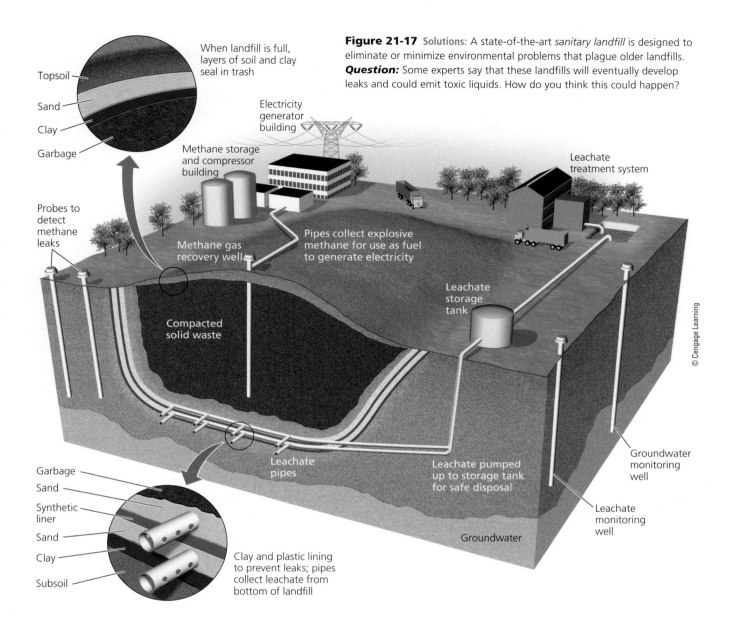

Figure 21-17 Solutions: A state-of-the-art *sanitary landfill* is designed to eliminate or minimize environmental problems that plague older landfills. **Question:** Some experts say that these landfills will eventually develop leaks and could emit toxic liquids. How do you think this could happen?

Labels in figure:

When landfill is full, layers of soil and clay seal in trash

Topsoil
Sand
Clay
Garbage

Electricity generator building

Methane storage and compressor building

Probes to detect methane leaks

Methane gas recovery well

Pipes collect explosive methane for use as fuel to generate electricity

Leachate treatment system

Compacted solid waste

Leachate storage tank

Garbage
Sand
Synthetic liner
Sand
Clay
Subsoil

Leachate pipes

Clay and plastic lining to prevent leaks; pipes collect leachate from bottom of landfill

Leachate pumped up to storage tank for safe disposal

Groundwater

Groundwater monitoring well

Leachate monitoring well

© Cengage Learning

---

from the landfill. This covering also lessens the risk of fire, decreases odor, and reduces accessibility to vermin.

The bottoms and sides of these landfills have strong double liners and containment systems that collect the liquids leaching from them. Some landfills also have systems for collecting and burning methane, the potent greenhouse gas that is produced when the wastes decompose in the absence of oxygen. Figure 21-18 lists the advantages and disadvantages of using sanitary landfills to dispose of solid waste.

Figure 21-18 Using sanitary landfills to dispose of solid waste has advantages and disadvantages (**Concept 21-4**). **Questions:** Which single advantage and which single disadvantage do you think are the most important? Why?

Photo: © redbrickstock.com/Alamy

Trade-Offs

Sanitary Landfills

Advantages	Disadvantages
Low operating costs	Noise, traffic, and dust
Can handle large amounts of waste	Releases greenhouse gases (methane and CO_2) unless they are collected
Filled land can be used for other purposes	Output approach that encourages waste production
No shortage of landfill space in many areas	Eventually leaks and can contaminate groundwater

© Cengage Learning

Figure 21-19 Some open dumps like this one near the city of Matamoros, Mexico, are routinely burned, releasing pollutants into the atmosphere.

Joel Sartore/National Geographic Creative

The second type of landfill is an **open dump**, essentially a field or large pit where garbage is deposited and sometimes burned (Figure 21-19). Open dumps are rare in more-developed countries, but are widely used near major cities in many less-developed countries (Figure 21-3). China disposes of about 85% of its rapidly growing mountains of solid waste mostly in rural open dumps or in poorly designed and poorly regulated landfills that do not have most of the features of sanitary land-

fills. Also, some solid wastes end up in rivers, lakes, and oceans.

Some analysts say we should think of landfills and open dumps as future mines for metals, plastics, paper, and other useful materials that we throw away. However, critics of this idea point out that it takes significant amounts of energy and is expensive to dig up and separate the materials in landfills and dumps. They argue that we should not mix different types of waste that we may want to separate later.

21-5 How Should We Deal with Hazardous Waste?

CONCEPT 21-5

A more sustainable approach to hazardous waste is first to produce less of it, then to reuse or recycle it, then to convert it to less-hazardous materials, and finally to safely store what is left.

We Can Use Integrated Management of Hazardous Waste

Figure 21-20 shows an integrated management approach suggested by the U.S. National Academy of Sciences that establishes three priority levels for dealing with hazardous waste: produce less; convert as much of it as possible to less-hazardous substances; and put the rest in long-term, safe storage (Concept 21-5). Denmark follows these priorities, but most countries do not.

As with solid waste, the top priority should be pollution prevention and waste reduction. With this approach,

industries try to find substitutes for toxic or hazardous materials, reuse or recycle the hazardous materials within industrial processes, or use them as raw materials for making other products. (See the online Guest Essays on this subject by Lois Gibbs and Peter Montague.)

At least 33% of industrial hazardous wastes produced in the European Union are exchanged through clearinghouses where they are sold as raw materials for use by other industries. The producers of these wastes do not have to pay for their disposal and recipients get low-cost raw materials. About 10% of the hazardous waste in the United States is exchanged through such clearinghouses, a figure that could be raised significantly.

GOOD NEWS

We can also recycle e-waste (**Core Case Study**). However, most e-waste recycling efforts create further hazards, especially for workers in some less-developed countries (see the following Case Study).

Produce Less Hazardous Waste	Convert to Less Hazardous or Nonhazardous Substances	Put in Perpetual Storage
■ Change industrial processes to reduce or eliminate hazardous waste production ■ Recycle and reuse hazardous waste	■ Natural decomposition ■ Incineration ■ Thermal treatment ■ Chemical, physical, and biological treatment ■ Dilution in air or water	■ Landfill ■ Underground injection wells ■ Surface impoundments ■ Underground salt formations

© Cengage Learning

Figure 21-20 *Integrated hazardous waste management:* The U.S. National Academy of Sciences has suggested these priorities for dealing with hazardous waste (**Concept 21-5**). **Question:** Why do you think most countries do not follow these priorities?

Recycling E-waste

In some countries, workers in e-waste recycling operations—many of them children—are often exposed to toxic chemicals as they dismantle the electronic trash to extract its valuable metals or other parts that can be sold for reuse or recycling (**Core Case Study** and Figure 21-21).

According to the United Nations, more than 70% of the world's e-waste is shipped to China. A center for such waste is the small port city of Guiyu, where the air reeks of burning plastic and acid fumes. There, more than 5,500 small-scale e-waste businesses employ over 30,000 people (including some children) who work at very low wages in dangerous conditions to extract valuable metals like gold and copper and various rare-earth metals from millions of discarded computers, television sets, and cell phones.

These workers usually wear no masks or gloves, often work in rooms with no ventilation, and are usually exposed to a cocktail of toxic chemicals. They carry out dangerous activities such as smashing TV picture tubes with large hammers to recover certain components—a method that releases large amounts of toxic lead dust into the air. They also burn computer wires to expose copper, melt circuit boards in metal pots over coal fires to extract lead and other metals, and douse the boards with strong acid to extract gold.

After the valuable metals are removed, leftover parts are burned or dumped into rivers or onto the land. Atmospheric levels of deadly dioxin in Guiyu are up to 86 times higher than World Health Organization safety standards, and an estimated 82% of the area's children younger than age 6 suffer from lead poisoning.

The United States produces roughly 50% of the world's e-waste (**Core Case Study**) and recycles only about 14% of it, according to the Consumer Electronics Association (CEA). However, e-waste recycling is on the rise. In 2012, at least 20 states had banned the disposal of computers and TV sets in landfills and incinerators, and these measures set the stage for an emerging, highly profitable *e-cycling* industry. According to a CEA report, electronics recycling in the United States increased by 53% between

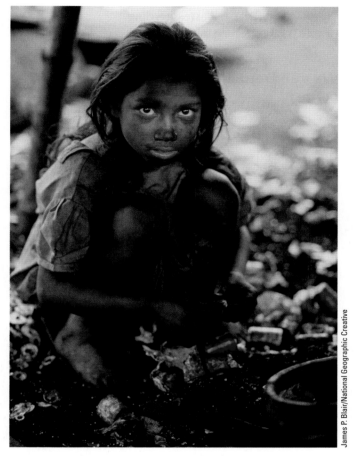

James P. Blair/National Geographic Creative

Figure 21-21 This young girl in Dhaka, Bangladesh, is recycling batteries by hammering them apart to extract tin and lead. The workers at this shop are mostly women and children.

2010 and 2011 and the number of e-waste drop-off sites rose from 5,000 to 7,500. Also, 13 states as well as New York City make manufacturers responsible for recycling most electronic devices.

Some U.S. electronics manufacturers have free recycling programs. A few will arrange to pick up discarded electronic products or will pay shipping costs to have them returned. Some have called for a standardized U.S. federal law that makes manufacturers responsible for taking back all electronic devices they produce and recycling them domestically with the process paid for by a recycling fee.

Recycling probably will not keep up with the explosive growth of e-waste, and there is money to be made from illegally sending such materials to other countries. According to Jim Puckett, coordinator of the Basel Action Network, an organization working to stem the flow of toxic waste, the only real long-term solution is a *prevention* approach through which electrical and electronic products are designed without the use of toxic materials and in such a way that they can be easily repaired, remanufactured, and recycled.

 CONSIDER THIS. . .

THINKING ABOUT E-Waste Recycling

Would you support a recycling fee on all electronic devices? Why or why not?

We Can Detoxify Hazardous Wastes

The first step in dealing with hazardous wastes is to collect them. In Denmark, all hazardous and toxic waste from industries and households is delivered to any of 21 transfer stations throughout the country. From there it is taken to a large processing facility, where three-fourths of the waste is detoxified by physical, chemical, and biological methods. The rest is buried in a carefully designed and monitored landfill.

Physical methods for detoxifying hazardous wastes include using charcoal or resins to filter out harmful solids, distilling liquid wastes to separate out harmful chemicals, and precipitating (allowing natural processes to separate) such chemicals from solution. Especially deadly wastes can be encapsulated in glass, cement, or ceramics and then put in secure storage sites.

Chemical methods are used to convert hazardous chemicals to harmless or less harmful chemicals through chemical reactions. Currently, chemists are testing the use of *cyclodextrin*—a type of sugar made from cornstarch—to remove toxic materials such as solvents and pesticides from contaminated soil and groundwater. To clean up a site contaminated by hazardous chemicals, a solution of cyclodextrin is applied. After this molecular sponge-like material moves through the soil or groundwater, picking up various toxic chemicals, it is pumped out of the ground, stripped of its contaminants, and reused.

Another approach is the use of *nanomagnets*, magnetic nanoparticles coated with certain compounds that can remove various pollutants from water. Magnetic fields are used to remove the pollutant-coated nanomagnets. The pollutants can then be separated out and disposed of or recycled, and the magnetic nanoparticles can be reused.

Some scientists and engineers consider *biological methods* for treatment of hazardous waste to be the wave of the future. One such approach is *bioremediation,* in which bacteria and enzymes help to destroy toxic or hazardous substances, or convert them to harmless compounds. For example, microorganisms are used at some sites to break down hazardous chemicals, such as PCBs, pesticides, and oil, leaving behind harmless substances such as water and water-soluble chloride salts. Bioremediation takes a little longer to work than most physical and chemical methods, but it costs much less. (See the online Guest Essay by John Pichtel on this topic.)

Another approach is *phytoremediation*, which involves using natural or genetically engineered plants to absorb, filter, and remove contaminants from polluted soil and water (Figure 21-22). Various plants have been identified as "pollution sponges" that can help to clean up soil and water contaminated with chemicals such as pesticides, organic solvents, and radioactive or toxic metals. Phytoremediation is still being evaluated and is slow (it can take several growing seasons) compared to other alternatives. However, it can be used to help clean up toxic hotspots.

We can incinerate hazardous wastes to break them down and convert them to harmless or less harmful chemicals such as carbon dioxide and water. This has the same combination of advantages and disadvantages as does burning solid wastes (Figure 21-16). However, incinerating hazardous waste without effective and expensive air pollution controls can release air pollutants such as highly toxic dioxins, and it produces an extremely toxic ash that must be safely and permanently stored in a landfill or vault especially designed for this hazardous end product.

We can also detoxify hazardous wastes by using *plasma gasification,* a technology that uses arcs of electrical energy to produce very high temperatures in order to vaporize trash in the absence of oxygen. This process reduces the volume of a given amount of waste by 99%, produces a synthetic gaseous fuel, and encapsulates toxic metals and other materials in glassy lumps of rock. Figure 21-23 lists the major advantages and disadvantages of using this process.

So why are we not making widespread use of the plasma arc torch to detoxify hazardous wastes? The main reason is its high cost. Plasma arc companies are working to bring the cost down.

We Can Store Some Forms of Hazardous Waste

Ideally, we should use burial on land or long-term storage of hazardous and toxic wastes in secure vaults only as the third and last resort after the first two priorities have been exhausted (Figure 21-20 and **Concept 21-5**). Currently, however, burial on land is the most widely used method in the United States and in most countries, largely because it is the least expensive of all methods.

The most common form of burial is *deep-well disposal,* in which liquid hazardous wastes are pumped under pressure through a pipe into dry, porous rock formations far beneath aquifers that are tapped for drinking and irrigation water (see Figure 20-12, p. 554). Theoretically, these

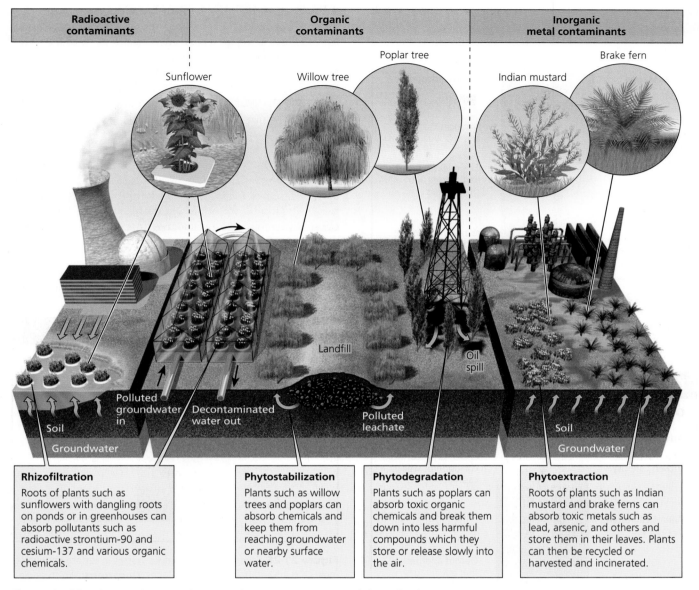

| Radioactive contaminants | Organic contaminants | Inorganic metal contaminants |

Rhizofiltration

Roots of plants such as sunflowers with dangling roots on ponds or in greenhouses can absorb pollutants such as radioactive strontium-90 and cesium-137 and various organic chemicals.

Phytostabilization

Plants such as willow trees and poplars can absorb chemicals and keep them from reaching groundwater or nearby surface water.

Phytodegradation

Plants such as poplars can absorb toxic organic chemicals and break them down into less harmful compounds which they store or release slowly into the air.

Phytoextraction

Roots of plants such as Indian mustard and brake ferns can absorb toxic metals such as lead, arsenic, and others and store them in their leaves. Plants can then be recycled or harvested and incinerated.

Figure 21-22 Solutions: Phytoremediation involves using various types of plants that function as pollution sponges to clean up contaminants such as radioactive substances (left), organic compounds (center), and toxic metals (right) from soil and water.

(Compiled by the authors using data from American Society of Plant Physiologists, U.S. Environmental Protection Agency, and Edenspace.)

liquids soak into the porous rock material and are isolated from overlying groundwater by essentially impermeable layers of clay and rock. The cost is low and the wastes can often be retrieved if problems develop.

However, this fairly inexpensive, out-of-sight and out-of-mind approach presents some problems. There are a limited number of such sites and limited space within them. Sometimes the wastes can leak into groundwater from the well shaft or migrate into groundwater in unexpected ways. Also, this is an output approach that encourages the production of hazardous wastes. In the United States, almost two-thirds of liquid hazardous wastes are injected into deep disposal wells, and this amount will

increase sharply as the country relies more on fracking to produce natural gas and oil trapped in shale rock (see Figure 15-A, p. 378). Many scientists believe that current regulations for deep-well disposal in the United States are inadequate and should be improved. Figure 21-24 lists the advantages and disadvantages of deep-well disposal of liquid hazardous wastes.

Surface impoundments are lined ponds, pits, or lagoons in which liquid hazardous wastes are stored (Figure 21-25). Sometimes they include liners to help contain the waste. As the water evaporates, the waste settles and becomes more concentrated. But using no liner, using leaking single liners, and failing to use double liners can

Trade-Offs

Plasma Arc

Advantages	Disadvantages
Produces a mixture of CO and H$_2$ that can be used as a fuel	High cost
	Produces CO$_2$ and CO
Mobile. Easy to move to different sites	Can release particulates and chlorine gas
	Can vaporize and release toxic metals and radioactive elements
Produces no toxic ash	

Figure 21-23 Using *plasma gasification* to detoxify hazardous wastes has advantages and disadvantages. **Questions:** Which single advantage and which single disadvantage do you think are the most important? Why?

Trade-Offs

Deep-Well Disposal

Advantages	Disadvantages
Safe if sites are chosen carefully	Leaks can occur from corrosion of well casing
Wastes can often be retrieved	Emits CO$_2$ and other air pollutants
Low cost	Output approach that encourages waste production

Figure 21-24 Injecting liquid hazardous wastes into deep underground wells has advantages and disadvantages. **Questions:** Which single advantage and which single disadvantage do you think are the most important? Why?

Figure 21-25 A surface impoundment for storing liquid hazardous wastes located in Niagara Falls, New York (USA). Such sites can pollute the air and nearby groundwater and surface water.

Trade-Offs

Surface Impoundments

Advantages	Disadvantages
Low cost	Water pollution from leaking liners and overflows
Wastes can often be retrieved	Air pollution from volatile organic compounds
Can store wastes indefinitely with secure double liners	Output approach that encourages waste production

Figure 21-26 Storing liquid hazardous wastes in surface impoundments has advantages and disadvantages. **Questions:** Which single advantage and which single disadvantage do you think are the most important? Why?

allow such wastes to percolate into the groundwater, and because these impoundments are not covered, volatile harmful chemicals can evaporate into the air. Also, powerful storms can cause them to overflow. Figure 21-26 lists the advantages and disadvantages of this method.

EPA studies have found that 70% of the storage ponds in the United States have no liners and could threaten groundwater supplies. According to the EPA, eventually all impoundment liners are likely to leak and could contaminate groundwater.

There are some highly toxic materials (such as mercury; see Chapter 17, Core Case Study, p. 442) that we cannot destroy, detoxify, or safely bury. The best way to deal with such materials is to prevent or reduce their use and store them in sealed containers. These containers could be placed above ground in specially designed storage buildings or underground in salt mines or bedrock caverns where they can be inspected on a regular basis and retrieved if necessary.

Sometimes both liquid and solid hazardous wastes are put into drums or other containers and buried in carefully designed and monitored *secure hazardous waste landfills* (Figure 21-27). This is the least-used method because of the expense involved.

Figure 21-28 lists some ways in which you can reduce your output of hazardous waste—the first step in dealing with it.

Bulk waste | Gas vent | Topsoil | Earth | Plastic cover | Sand | Impervious clay cap | Clay cap

Impervious clay

Water table

Earth

Groundwater

Leak detection system

Double leachate collection system | Plastic double liner | Reactive wastes in drums | Groundwater monitoring well

Figure 21-27 Solutions: Hazardous wastes can be isolated and stored in a secure hazardous waste landfill.

CASE STUDY

Hazardous Waste Regulation in the United States

About 5% of all hazardous waste produced in the United States is regulated under the Resource Conservation and Recovery Act (RCRA, pronounced "RICK-ra"), passed by the U.S. Congress in 1976 and amended in 1984. Under this act, the EPA sets standards for the management of several types of hazardous waste and issues permits to companies that allow them to produce and dispose of a certain amount of those wastes by approved methods. Permit holders must use a *cradle-to-grave* system to keep track of waste they transfer from a point of generation (cradle) to an approved off-site disposal facility (grave), and they must submit proof of this disposal to the EPA.

RCRA is a good start, but about 95% of the hazardous and toxic wastes, including e-waste, produced in the United States is not regulated. In most other countries, especially less-developed countries, the amount of regulated waste is even smaller.

CONSIDER THIS. . .

THINKING ABOUT Hazardous Waste

Why is it that 95% of the hazardous waste, including the growing amounts of e-waste (**Core Case Study**) produced in the United States, is not regulated? Do you favor regulating such wastes? What do you think would be the economic consequences of doing so?

What Can You Do?

Hazardous Waste

- Avoid using pesticides and other hazardous chemicals, or use them in the smallest amounts possible

- Use less harmful substances instead of commercial household cleaners. For example, use vinegar to polish metals, clean surfaces, and remove stains and mildew, and baking soda to clean utensils and to deodorize and remove stains.

- Do not dump pesticides, paints, solvents, oil, antifreeze, or other hazardous chemicals down the toilet, down the drain, into the ground, into the garbage, or down storm drains. Free hazardous waste disposal services are available in many cities.

- Do not throw old fluorescent lightbulbs (which contain mercury) into regular trash. Many communities and home product retailers offer free recycling of these bulbs.

Figure 21-28 Individuals matter: You can reduce your output of hazardous wastes (Concept 21-5). **Questions:** Which two of these measures do you think are the most important ones to take? Why?

In 1976, the U.S. Congress passed the Toxic Substances Control Act, which was intended to regulate and ensure the safety of the thousands of chemicals used in the manufacture of many products and contained in many products. Under this law, companies must notify the EPA before introducing a new chemical into the marketplace, but they are not required to provide any data about its safety. In other words, any new chemical is viewed as safe unless the EPA, with a very limited budget, can show that it is harmful.

Since 1976, the EPA has used this act to ban only 5 of the roughly 80,000 chemicals in use. Environmental and health scientists call for Congress to reform this almost-40-year-old law by requiring manufacturers to provide data showing that a chemical or product containing a certain chemical is safe before it can be sold in the marketplace.

In 1980, the U.S. Congress passed the *Comprehensive Environmental Response, Compensation, and Liability Act,* commonly known as the *CERCLA* or *Superfund* program, supervised by the EPA. Its goals are to identify sites, commonly called Superfund sites, where hazardous wastes have contaminated the environment (Figure 21-29) and to clean them up on a priority basis. The worst sites—those that represent an immediate and severe threat to human health—are put on a *National Priorities List* and scheduled for cleanup using EPA-approved methods.

By 2013, there were 1,320 sites on the Superfund list, along with 54 proposed new sites, and 365 sites had been cleaned up and removed from the list. The Waste Management Research Institute estimates that at least 10,000 sites should be on the priority list and that cleanup of these sites would cost about $1.7 trillion, not including legal fees. This is a glaring example of the economic and environmental value of emphasizing waste reduc-

Figure 21-29 Leaking barrels of toxic waste found at a Superfund site that has since been cleaned up.

©Phototake, Inc.

23,800 large manufacturing facilities to report their annual releases into the environment of any of nearly 650 toxic chemicals. If you live in the United States, you can find out what toxic chemicals are being stored and released in your neighborhood by going to the EPA's *Toxic Release Inventory* website.

The Superfund Act, designed to make polluters pay for cleaning up abandoned hazardous waste sites, greatly reduced the number of illegal dumpsites around the country. It also forced waste producers who were fearful of liability claims to reduce their production of such waste and to recycle or reuse much more of it. However, under pressure from polluters, the U.S. Congress refused to renew the tax on oil and chemical companies that had financed the Superfund legislation after it expired in 1995. The Superfund is now broke, and taxpayers, not polluters, are footing the bill for cleanups when the responsible parties cannot be found. As a result, the pace of cleanup has slowed.

The U.S. Congress and several state legislatures have also passed laws that encourage the cleanup of *brownfields*—abandoned industrial and commercial sites such as factories, junkyards, older landfills, and gas stations. GOOD NEWS In most cases, they are contaminated with hazardous wastes. Brownfields can be cleaned up and reborn as parks, nature reserves, athletic fields, ecoindustrial parks (see the Case Study at the end of this chapter), and residential neighborhoods.

tion and pollution prevention over the *end-of-pipe* cleanup approach that most countries rely on.

In 1984, Congress amended the Superfund Act to give citizens the right to know what toxic chemicals are being stored or released in their communities. This required

21-6 How Can We Make the Transition to a More Sustainable Low-Waste Society?

CONCEPT 21-6

Shifting to a low-waste society requires individuals and businesses to reduce resource use and to reuse and recycle wastes at local, national, and global levels.

Grassroots Action Has Led to Better Solid and Hazardous Waste Management

In the United States, individuals have organized grassroots (bottom-up) citizen movements to prevent the construction of hundreds of incinerators, landfills, treatment plants for hazardous and radioactive wastes, and polluting chemical plants in or near their communities. Health risks from incinerators and landfills, when averaged over the entire country, are quite low, but the risks for people living near such facilities are much higher.

Manufacturers and waste industry officials point out that something must be done with the toxic and hazardous wastes created in the production of certain goods and services. They contend that even if local citizens adopt a "not in my back yard" (NIMBY) approach, the waste will always end up in someone's back yard.

Many citizens do not accept this argument. To them, the best way to deal with most toxic and hazardous waste is to produce much less of it, as suggested by the U.S. National Academy of Sciences (Figure 21-20). For such materials, they believe that the goal should be "not in anyone's back yard" (NIABY) or "not on planet Earth" (NOPE), which calls for drastically reducing production of such wastes by emphasizing pollution prevention and using the *precautionary principle* (see Chapter 9, p. 211).

Providing Environmental Justice for Everyone Is an Important Goal

Environmental justice is an ideal whereby every person is entitled to protection from environmental hazards regardless of race, gender, age, national origin, income, social class, or any political factor. (See the online Guest Essay on this subject by Robert Bullard.)

Studies have shown that a lopsided share of polluting factories, hazardous waste dumps, incinerators, and landfills in the United States are located in communities populated mostly by African Americans, Asian Americans, Latinos, and Native Americans. Studies have also shown that, in general, toxic waste sites in white communities have been cleaned up faster and more completely than similar sites in African American and Latino communities have.

Such environmental discrimination in the United States and in other parts of the world has led to a growing grassroots approach to this problem that is known as the *environmental justice movement*. Supporters of this group have pressured governments, businesses, and environmental organizations to become aware of environmental injustice and to act to prevent it. They have made some progress toward their goals, but have a long way to go.

🔍 CONSIDER THIS. . .

THINKING ABOUT Environmental Injustice

Have you or anyone in your family ever been a victim of environmental injustice? If so, what happened? What would you do to help prevent such environmental injustice?

We Can Encourage Reuse and Recycling

Three factors hinder reuse and recycling. *First,* the market prices of almost all products do not include the harmful environmental and health costs associated with producing, using, and discarding them.

Second, the economic playing field is uneven, because in most countries, resource-extracting industries receive more government tax breaks and subsidies than reuse and recycling industries get.

Third, the demand and thus the price paid for recycled materials fluctuates, mostly because buying goods made with recycled materials is not a priority for most governments, businesses, and individuals.

How can we encourage reuse and recycling? Proponents say that leveling the economic playing field is the best way to start. Governments can *increase* subsidies and tax breaks for reusing and recycling materials, and *decrease* subsidies and tax breaks for making items from virgin resources.

One way to include some of the harmful environmental costs of products in their prices, while encouraging recycling, is to attach a small deposit fee to the price of recyclable items, as is done in many European countries, several Canadian provinces, and ten U.S. states that have *bottle bills.* Such laws place a deposit fee of 5 or 10 cents on each beverage container, and consumers can recover that fee by returning their empty containers to the store. In 2012, these ten states recycled at least 70% of their bottles and cans, compared to 28% in states with no bottle bills. This approach is in keeping with the full-cost pricing **principle of sustainability**.

Another strategy is to greatly increase use of the fee-per-bag waste collection system. When the U.S. city of Fort Worth, Texas, instituted such a program, the proportion of households recycling their trash went from 21% to 85%. The city went from losing $600,000 in its recycling program to making $1 million a year because of increased sales of recycled materials to industries.

Governments can also pass laws requiring companies to take back and recycle or reuse packaging and electronic waste discarded by consumers (**Core Case Study**), as is done in Japan and some European Union countries. Another important strategy is to encourage or require government purchases of recycled products to help increase demand for and lower prices of these products. Also, citizens can pressure governments to require product labeling that lists the recycled content of products, as well as the types and amounts of any hazardous materials they contain. This would help consumers to make more informed choices about the environmental consequences of buying certain products. It would also help to expand the market for recycled materials by spurring demand for them.

Reuse, Recycling, and Composting Present Economic Opportunities

A growing number of people are saving money through reuse, regularly going to yard sales, flea markets, secondhand stores, and online sites such as eBay and craigslist. The Freecycle Network links people who want to give away household belongings to people who want or need them. Its nearly 7 million members reuse an average of 640 metric tons (700 tons) of items every day, which is about the amount of solid waste that arrives at a mid-size landfill every day.

For many, recycling has become a business opportunity. In particular, *upcycling,* or recycling materials into products of a higher value (Figure 21-30), is a growing field. For example, a British company called Worn Again is converting discarded textiles such as old hot-air balloons and worn-out seat covers into windbreaker jackets and other products. And Professor Na Lu in North Carolina has found a way to upcycle plastic bottles to make a building material that could outperform composite lumber and wood lumber. Entrepreneurs see upcycling as an area of great opportunity.

Figure 21-30 *Upcycling:* This handbag was made from an old airline seat.

Rebecca Hale/National Geographic Creative

Companies are also realizing that they can save money by minimizing packaging or by using *dual-use packaging*. For example, Hewlett-Packard has packaged some of its laptop computers in padded laptop bags that can be used indefinitely. In doing so, the company reduced packaging waste for those computers by 97%. As an added benefit, HP made these bags out of recycled materials.

International Treaties Have Reduced Hazardous Waste

Environmental justice also applies at the international level. For decades, some more-developed countries had been shipping hazardous wastes to less-developed countries. However, since 1992, an international treaty known as the Basel Convention has been in effect. It bans participating countries from shipping hazardous waste (including e-waste; see **Core Case Study**) to or through other countries without their permission. In 1995, the treaty was amended to outlaw all transfers of hazardous wastes from industrial countries to less-developed countries. By 2012, this agreement had been ratified (formally approved and implemented) by 179 countries. The United States, Afghanistan, and Haiti have signed but have not ratified the convention.

This ban will help, but it will not wipe out the very profitable illegal shipping of hazardous wastes. Hazardous waste smugglers evade the laws by using an array of tactics, including bribes, false permits, and mislabeling of hazardous wastes as recyclable materials.

In 2000, delegates from 122 countries completed a global treaty known as the Stockholm Convention on Persistent Organic Pollutants (POPs). It regulates the use of 12 widely used persistent organic pollutants that can accumulate in the fatty tissues of humans and other animals that occupy high trophic levels in food webs. At such levels, these hazardous chemicals can reach levels hundreds of thousands of times higher than their levels in the general environment (see Figure 9-13, p. 202). Because they persist in the environment, POPs can also be transported long distances by wind and water.

The original list of 12 hazardous chemicals, called the *dirty dozen*, includes DDT and 8 other chlorine-containing persistent pesticides, PCBs, dioxins, and furans. Using blood tests and statistical sampling, medical researchers at New York City's Mount Sinai School of Medicine found that it is likely that nearly every person on earth has detectable levels of POPs in their bodies. The long-term health effects of this involuntary global chemical experiment are largely unknown.

By 2012, 178 countries had ratified a strengthened version of the POPs treaty that seeks to ban or phase out the use of these hazardous chemicals and to detoxify or isolate existing stockpiles. It allows 25 countries to continue using DDT to combat malaria until safer alternatives are found. The United States has not yet ratified this treaty. The list of regulated POPs is expected to grow.

In 2000, the Swedish Parliament enacted a law that, by 2020, will ban all potentially hazardous chemicals that are persistent in the environment and can accumulate in living tissue. This law also requires industries to perform risk assessments on the chemicals they use and to show that these chemicals are safe to use, as opposed to requiring the government to show that they are dangerous. In other words, chemicals are assumed to be guilty until proven innocent—the reverse of the current policy in the United States and most other countries. There is strong opposition to this approach in the United States, especially from most of the industries that produce and use potentially hazardous chemicals.

We Can Make the Transition to Low-Waste Societies

According to physicist Albert Einstein, "A clever person solves a problem; a wise person avoids it." Many people are taking these words seriously. The governments of Norway, Austria, and the Netherlands have committed to reducing their resource waste by 75%. Many school cafeterias, restaurants, national parks, and corporations are participating in a rapidly growing "zero waste" movement to reduce, reuse, and recycle, and some have lowered their waste outputs by up to 80%, with the ultimate goal of eliminating their waste outputs.

Many environmental scientists argue that we can prevent pollution, reduce waste, and make a transition to a low-waste society by understanding and following these key principles:

1. Everything is connected.
2. There is no *away*, as in *to throw away*, for the wastes we produce.
3. Producers and polluters should pay for the wastes they produce.
4. We can mimic nature by reusing, recycling, composting, or exchanging most of the municipal solid wastes we produce (see the following Case Study).

CASE STUDY

Industrial Ecosystems: Copying Nature

An important goal for a more sustainable society is to make its industrial manufacturing processes cleaner and more sustainable by redesigning them to mimic how nature deals with wastes—an approach called *biomimicry*. In nature, according to the chemical cycling **principle of sustainability**, the waste outputs of one organism become the nutrient inputs of another organism, so that all of the earth's nutrients are endlessly recycled. This explains why there is essentially no waste in undisturbed ecosystems.

One way for industries to mimic nature is to reuse or recycle most of the minerals and chemicals they use,

Figure 21-31 Solutions: This *industrial ecosystem* in Kalundborg, Denmark, reduces waste production by mimicking a natural ecosystem's food web. The wastes of one business become the raw materials for another, thus mimicking the way that nature recycles chemicals. ***Question:*** Is there an industrial ecosystem near where you live or go to school? If not, think about where and how such a system could be set up.

instead of burying or burning them or shipping them somewhere. Another method that industries can use to mimic nature is to interact with each other through *resource exchange webs* in which the wastes of one manufacturer become the raw materials for another—similar to food webs in natural ecosystems (see Figure 3-12, p. 60).

This is happening in Kalundborg, Denmark, where an electric power plant and nearby industries, farms, and homes are collaborating to save money and to reduce their outputs of waste and pollution within what GOOD NEWS is called an *ecoindustrial park,* or *industrial ecosystem.* They exchange waste outputs and convert them into resources, as shown in Figure 21-31. This cuts pollution and waste and reduces the flow of nonrenewable mineral and energy resources through the local economy.

Today, more than 40 ecoindustrial parks (18 of them in the United States) operate in various places around the world, and more are being built or planned—some of them on brownfield sites. A number of people who

work in the rapidly growing field of industrial ecology are focusing on developing a global network of industrial ecosystems over the next few decades, and this could lead to an important *ecoindustrial revolution.* **GREEN CAREER:** industrial ecology

These and other industrial forms of biomimicry provide many economic benefits for businesses. By encouraging recycling and waste-reduction prevention, they reduce the costs of managing solid wastes, controlling pollution, and complying with pollution regulations. They also reduce a company's chances of being sued because of damages to people and the environment caused by their actions. In addition, companies improve the health and safety of workers by reducing their exposure to toxic and hazardous materials, thereby reducing company health insurance costs. Biomimicry also encourages companies to come up with new, environmentally beneficial, and less resource-intensive chemicals, processes, and products that they can sell worldwide.

Biomimicry involves two major steps. The first is to observe certain changes in nature and to study how natural systems have responded to such changing conditions over many millions of years. The second step is to try to copy or adapt these responses within human systems in order to help us deal with various environmental challenges. In the case of solid and hazardous wastes, the food web serves as a natural model for responding to the growing problem of these wastes. This is in keeping with the three scientific **principles of sustainability** (see Figure 1-2, p. 6 or back cover) that nature has used for billions of years.

Big Ideas

- The order of priorities for dealing with solid waste should be to produce less of it, reuse and recycle as much of it as possible, and safely burn or bury what is left.

- The order of priorities for dealing with hazardous waste should be to produce less of it, reuse or recycle it, convert it to less-hazardous material, and safely store what is left.

- We need to view solid wastes as wasted resources, and hazardous wastes as materials that we should not be producing in the first place.

TYING IT ALL TOGETHER   E-Waste and Sustainability

Bakalusha/Shutterstock.com

One of the problems of maintaining a high-waste society is the growing mass of e-waste (**Core Case Study**) and other types of solid and hazardous waste discussed in this chapter. The challenge is to make the transition from an unsustainable high-waste, throwaway economy to a more sustainable low-waste, reducing–reusing–recycling economy as soon as possible.

Such a transition will require applying the six **principles of sustainability** (see Figure 1-2, p. 6, Figure 1-5, p. 9, or back cover). We can reduce our outputs of solid and hazardous waste by relying much less on fossil fuels and nuclear power (which produces long-lived, hazardous radioactive wastes) while relying much more on renewable energy from the sun, wind, and flowing water. We can mimic nature's chemical cycling processes by reusing and recycling materials as much as possible. Integrated waste management, which uses a diversity of approaches and emphasizes waste reduction and pollution prevention, is a way to mimic nature's use of biodiversity.

By including more of the harmful environmental and health costs of the consumer economy in market prices, we would be applying the full-cost pricing **principle of sustainability** while encouraging people to refuse, reduce, reuse, and recycle. In doing so, we would benefit the environment, create new jobs and businesses capitalizing on the four Rs, and gain health and environmental benefits for ourselves, thus finding win-win solutions. This could also lead to lower levels of resource use per person, and thus a lower demand for materials that eventually become solid and hazardous wastes. All these measures together would help us to pass along to future generations a world that is at least as livable as the one we have enjoyed.

Chapter Review

Core Case Study

1. Explain how and why electronic waste (e-waste) has become a growing solid waste problem.

Section 21-1

2. What are the two key concepts for this section? Distinguish among **solid waste, industrial solid waste, municipal solid waste (MSW)**, and **hazardous (toxic) waste**, and give an example of each. Summarize the types and sources of solid waste generated in the United States and explain what happens to it. What is garbology?

Section 21-2

3. What is the key concept for this section? Distinguish among **waste management, waste reduction**, and **integrated waste management**. Summarize the priorities that prominent scientists believe we should use for dealing with solid waste and compare them to actual practices in the United States. Distinguish among **refusing, reducing, reusing**, and **recycling** in dealing with the solid wastes we produce. Why are the first three Rs preferred from an environmental standpoint? What is **composting**? List six ways in which industries and communities can reduce resource use, waste, and pollution.

Section 21-3

4. What is the key concept for this section? Explain why refusing, reducing, reusing, and recycling are so important and give examples of each. List five ways to reuse various items.

5. Distinguish between **primary (closed-loop) recycling** and **secondary recycling**. What are three important steps that must occur for any recycling program to work? What are some benefits of composting? What is the benefit of having households and businesses separate their trash into recyclable categories? What is the fee-per-bag approach? Explain how paper and some plastics are being recycled. What are bioplastics? What are the major advantages and disadvantages of recycling?

Section 21-4

6. What is the key concept for this section? What are the major advantages and disadvantages of using incinerators to burn solid and hazardous waste?

Distinguish between **sanitary landfills** and **open dumps**. What are the major advantages and disadvantages of burying solid waste in sanitary landfills?

Section 21-5

7. What is the key concept for this section? What are the priorities that scientists believe we should use in dealing with hazardous waste? Summarize the problems involved in sending e-wastes to less-developed countries for recycling. Describe three ways to detoxify hazardous wastes. What is bioremediation? What is phytoremediation? What are the major advantages and disadvantages of incinerating hazardous wastes? What are the major advantages and disadvantages of using plasma gasification to detoxify hazardous wastes?

8. What are the major advantages and disadvantages of disposing of liquid hazardous wastes (a) in deep underground wells and (b) in surface impoundments? What is a secure hazardous waste landfill? List four ways to reduce your output of hazardous waste. Summarize how laws in the United States regulate hazardous wastes. What is a brownfield?

Section 21-6

9. What is the key concept for this section? How has grassroots action improved solid and hazardous waste management in the United States? What is **environmental justice** and how well has it been applied in locating and cleaning up hazardous waste sites in the United States? What are three factors that discourage recycling? What are three ways to encourage recycling and reuse? Give three examples of how people are saving or making money through reuse, recycling, and composting. Describe regulation of hazardous wastes at the global level through the Basel Convention and the treaty to control persistent organic pollutants (POPs). What is biomimicry? What is an industrial ecosystem?

10. What are this chapter's three big ideas? Explain how we could deal with the growing problems of e-waste and other wastes (**Core Case Study**) by applying the six **principles of sustainability**.

Note: Key terms are in bold type.

Critical Thinking

1. Do you think that manufacturers of computers, television sets, and other forms of e-waste (**Core Case Study**) should be required to take their products back at the end of their useful lives for repair, remanufacture, or recycling in a manner that is environmentally responsible and that does not threaten the health of recycling workers? Explain. Would you be willing to pay more for these products to cover the costs of such a take-back program? If so, what percentage more per purchase would you be willing to pay for electronic products?

2. Find three items that you regularly use once and then throw away. Are there other reusable items that you could use in place of these disposable items? For each item, calculate and compare the cost of using the disposable option for a year versus the cost of using the reusable alternative. Write a brief report summarizing your findings.

3. Do you think that you could consume less by refusing to buy some of the things you regularly buy? If so, what are three of those things? Do you think that this is something you ought to do? Explain your reasoning.

4. A company called Changing World Technologies has built a pilot plant to test a process it has developed for converting a mixture of discarded computers, old tires, turkey bones and feathers, and other wastes into oil by mimicking and speeding up natural processes for converting biomass into oil. Explain why this recycling process, if it turns out to be technologically and economically feasible, could lead to increased waste production.

5. Would you oppose having **(a)** a hazardous waste landfill, **(b)** a waste treatment plant, **(c)** a deep-injection well, or **(d)** an incinerator in your community? For each of these facilities, explain your answer. If you oppose having such disposal facilities in your community, how do you believe the hazardous waste generated in your community should be managed?

6. How does your school dispose of its solid and hazardous wastes? Does it have a recycling program? How well does it work? Does your school encourage reuse? If so, how? Does it have a hazardous waste collection system? If so, describe it. List three ways in which you would improve your school's waste-reduction and management systems.

7. List three ways in which you could apply Concept 21-6 to making your lifestyle more environmentally sustainable.

8. Congratulations! You are in charge of the world. List the three most important components of your strategy for dealing with **(a)** solid waste and **(b)** hazardous waste.

Doing Environmental Science

Collect the trash (excluding food waste) that you generate in a typical week. Measure its total weight and volume. Sort it into major categories such as paper, plastic, metal, and glass. Then weigh each category and calculate its percentage by weight of the total amount of trash that you have measured. What percentage by weight of this waste consists of materials that could be recycled? What percentage consists of materials for which you could have used a reusable substitute, such as a coffee mug instead of a disposable cup? What percentage by weight of the items could you have done without? Compare your answers to these questions with those of your classmates. Together with your classmates, combine all the results and do the same analysis for the entire class. Use these results to estimate the same values for the entire student population at your school. (If you wish, you could include food wastes in your research, but you would need to find a way to store it such that it won't become a nuisance or a health hazard.)

Global Environment Watch Exercise

Within the GREENR database, go to the *E-Waste* topic portal. Research and find statistics on how rapidly the world's production of e-waste (**Core Case Study**) is growing and how rapidly e-waste production is growing in the United States. Write a brief report on what the United States and one other country of your choice are doing to deal with this growing waste problem. Compare the two approaches in terms of how successful they are.

Ecological Footprint Analysis

Researchers estimate that the average daily municipal solid waste production per person in the United States is 3.2 kilograms (4.34 pounds). Use the data in the pie chart below to get an idea of a typical annual MSW ecological footprint for each American by calculating the total weight in kilograms (and pounds) for each category generated during 1 year (1 kilogram = 2.20 pounds). Use the table below to enter your answers.

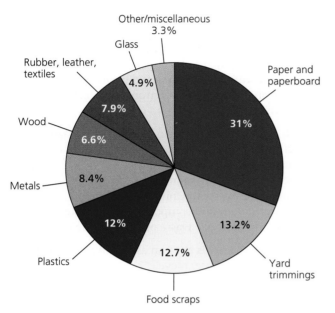

Composition of a typical sample of U.S. municipal solid waste, 2010.

(Compiled by the authors using data from U.S. Environmental Protection Agency.)

Waste Category	Annual MSW Footprint per Person
Paper and paperboard	
Yard trimmings	
Food scraps	
Plastics	
Metals	
Wood	
Rubber, leather, and textiles	
Glass	
Other/miscellaneous	

© Cengage Learning

1. Recycling aluminum (Al) cans saves 95% of the energy needed to produce new aluminum cans from raw materials. How many Al cans can be made out of recycled Al with the same amount of energy that it takes to make one Al can out of new material?
   a. 1
   b. 2
   c. 5
   d. 10
   e. 20

2. E-waste (electronic waste) is often labeled as material to be recycled and sent to undeveloped countries, where workers recover a small amount of valuable metals from the waste and the remainder is then burned or dumped in landfills, releasing toxic pollutants such as dioxins. This is frequently used as a means of bypassing the
   a. Clean Air Act.
   b. Kyoto Protocol.
   c. International Basel Convention.
   d. Resource Conservation and Recovery Act.
   e. Non-Proliferation Treaty.

3. One method to remove contaminants such as cesium-137 from surface soil would be
   a. to use phytoremediation, allowing plants' roots to absorb toxins by rhizofiltration.
   b. to use a plasma arc torch to incinerate the toxins *in situ*.
   c. create a surface impoundment with a liner.
   d. spraying the surface with a genetically engineered bacteria that breaks down cesium.
   e. to burn and replace the soil.

4. One of the primary concerns of the 12 POPs (persistent organic pollutants) such as dioxins or PCBs is that they
   a. are soluble in water, causing large-scale problems.
   b. can be concentrated in the fatty tissues of organisms.
   c. are rarely found but very toxic.
   d. quickly break down into other more harmful pollutants.
   e. are found only in plants and disrupt photosynthesis.

5. The majority of solid waste generated in the United States is in what form?
   a. Solid matter produced by mines, farms, and industries supplying people with goods
   b. Radioactive waste generated by nuclear power plants
   c. Toxic waste such as car batteries, hospital medical waste, and industrial solvents
   d. Solid matter such as food, yard wastes, plastics, and paper discarded by homes and businesses
   e. Electronic waste from electronic devices such as televisions, computers, and cell phones

6. Which of the below is a form of integrated waste reduction?
   a. Burying MSW in a landfill
   b. Incinerating MSW to produce electricity
   c. Shipping E-waste to China for disposal
   d. Separating out and composting yard waste
   e. Dumping nontoxic waste offshore in the oceans

7. Which of the below is NOT an aspect of the popular phrase, "reduce, reuse, recycle"?
   a. Use e-mail or text in place of conventional paper mail
   b. Read newspapers or magazines online
   c. Buy products in bulk form whenever possible
   d. Utilize plastic in place of paper bags
   e. Donate or sell used items

8. One of the major disadvantages of recycling goods is that
   a. landfill operators must charge more money for trash delivery because they are receiving less of it.
   b. approximately 55% of the trees cut down to make paper are saved when paper is recycled.
   c. marine mammals and sea turtles mistake floating plastic for food or get caught in it.
   d. energy is saved and less pollution is produced when we recycle than when raw materials are removed from the earth.
   e. it may cost more than burying in landfills in areas with ample space.

9. Why should you remove plastic caps from bottles when you recycle them?
   a. A small amount of nonrecyclable plastic can contaminate and make useless recycled plastic.
   b. It is impossible to remove plastic caps from bottles before recycling.
   c. Plastic bottle caps cost too much to dispose of after their removal.
   d. The coloring of bottle caps makes them difficult to recycle.
   e. Bottles with caps on cannot be recycled and are thrown away.

10. Many environmental scientists advocate that the best and cheapest way to deal with solid waste is to
    a. truck the waste to landfills and bury it.
    b. reduce the amount produced in the first place.
    c. incinerate the waste to generate electricity.
    d. establish grass roots organizations advocating for fewer landfills.
    e. pass more laws and treaties with tighter controls on emissions of pollutants.

Questions 11–13 refer to the diagram below.

© Cengage Learning

11. What is the primary concern of most people when a trash incinerator is proposed to be built in their neighborhood?
    a. Air pollution from the incinerator
    b. The use of waste ash in products such as road construction or cement blocks
    c. Because incinerators need large amounts of trash to be profitable, people will be encouraged to produce more trash
    d. The initial cost of building an incinerator
    e. What to do with tires that are not incinerated

12. What is the function of the part of the diagram labeled "D"?
    a. Incinerate the trash
    b. House the control rods
    c. Cooling system to prevent contamination
    d. Produce electricity
    e. Receive waste for the plant

13. How is the leftover ash from the incinerator disposed of?
    a. Spread on agricultural fields as fertilizer
    b. Composted to return to the environment
    c. Sent to a Superfund site for disposal
    d. Dumped in the ocean
    e. Dumped in a sanitary landfill

14. What is a brownfield?
    a. Mineral resources that are mined on the leeward sides of mountains
    b. A processing facility where hazardous wastes are taken to be detoxified
    c. Deep underground wells where liquid hazardous wastes are injected and stored
    d. An agricultural area that has turned brown because of a summer drought
    e. An abandoned industrial site that has soil, water, or air pollution

22 Urbanization and Sustainability

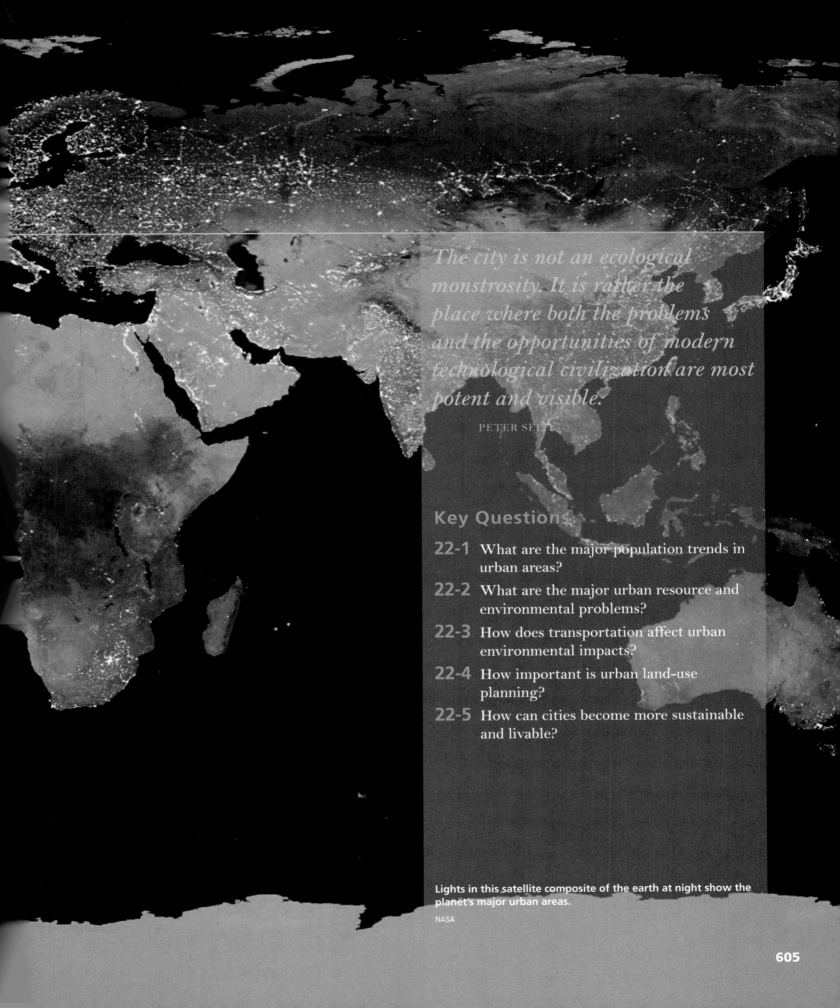

The city is not an ecological monstrosity. It is rather the place where both the problems and the opportunities of modern technological civilization are most potent and visible.

PETER SELF

Key Questions

22-1 What are the major population trends in urban areas?

22-2 What are the major urban resource and environmental problems?

22-3 How does transportation affect urban environmental impacts?

22-4 How important is urban land-use planning?

22-5 How can cities become more sustainable and livable?

Lights in this satellite composite of the earth at night show the planet's major urban areas.

NASA

Portland, Oregon, with almost 600,000 people, lies on the banks of the Willamette River (Figure 22-1). For over four decades Portland has consistently ranked at or near the top in several lists of the most sustainable U.S. cities.

Since the 1970s, Portland has used smart growth strategies and strong land-use policies to control sprawl, reduce dependence on automobiles, and provide green space. In 1978, the city demolished a six-lane highway and replaced it with a waterfront park.

Portland encourages clustered, mixed-use neighborhood development, with stores, light industries, professional offices, high-density housing, and access to mass transit, which allows most people to meet most of their daily needs without a car. The city has excellent light-rail and bus lines, and has further reduced car use by developing an extensive network of bike lanes and walkways. By promoting these alternatives, along with its compact growth policy, Port-

land has reduced the need to drive more than most other major American cities have done, and this has saved its residents more than $1 billion a year in transportation costs. It has also contributed to public health by cutting air pollution and encouraging higher levels of physical activity.

Portland implemented a recycling system in 1987 and, by 2012, was recycling or composting 67% of its municipal solid waste, achieving one of the highest recycling rates in the country. In 1993, Portland became the first U.S. city to develop a plan to reduce its greenhouse gas emissions. By 2012, it had reduced its per capita greenhouse gas emissions to 26% below 1990 levels—more than any other major U.S. city had done.

In 2007, the city adopted a goal of cutting its use of fossil fuels in half by 2030. And in 2009, Portland implemented a Climate Action Plan with the goal of cutting its greenhouse gas emissions to 40% below 1990 levels by 2030 and to 80% below those levels by 2050.

Portland has more than 20 farmers markets and 35 community gardens, which provide fresh, locally produced food for residents and chefs. The city also has many vegetarian-friendly restaurants.

So why should we care about Portland or any other urban area? One reason is that more than half of the world's people live in urban areas and by 2050, two of every three people are likely to be urban dwellers—most of them in rapidly growing cities in less-developed countries. Another reason is that, because urban areas use most of the world's resources and produce most of the world's pollution and wastes, urban areas have huge environmental impacts that extend far beyond their boundaries. The environmental quality of the future depends largely on whether we can make urban areas more sustainable and livable during the next few decades. Portland and a growing number of other cities are leading the way.

Jit Lim/cutcaster

Figure 22-1 Portland, Oregon, is one of the most environmentally friendly and sustainable cities in the United States.

22-1 What Are the Major Population Trends in Urban Areas?

CONCEPT 22-1
Urbanization continues to increase steadily, and the numbers and sizes of urban areas are growing rapidly, especially in less-developed countries.

More Than Half of the World's People Live in Urban Areas

People live in *urban areas,* or cities (Figure 22-2, top), suburbs that spread out around city centers (Figure 22-2, center), and rural areas or villages (Figure 22-2, bottom). **Urbanization** is the creation and growth of urban and suburban areas. It is measured as the percentage of the people in a country or in the world living in such areas. **Urban growth** is the *rate* of increase of urban populations.

We live on an increasingly urbanized planet. About 52% of the world's people live in urban areas, and every day there are about 200,000 more urban dwellers. Urban areas grow in two ways—by *natural increase* (more births than deaths) and by *immigration,* mostly from rural areas. Rural people are *pulled* to urban areas in search of jobs, food, housing, educational opportunities, better health care, and entertainment. Some are also *pushed* from rural to urban areas by factors such as poverty, lack of land for growing food, famine, war, and religious, racial, and political conflicts.

Three major trends in urban population dynamics are important for understanding the problems and challenges of urban growth:

1. *The percentage of the global population living in urban areas has increased sharply, and this trend is projected to continue.* Between 1850 and 2012, the percentage of the world's people living in urban areas increased from 2% to 52% (see chapter-opening photo) and is likely to reach 67% by 2050. Between 2012 and 2050, the world's urban population is projected to grow from 3.6 billion to 6.3 billion. The great majority of these 2.7 billion new urban dwellers will live in less-developed countries.

2. *The numbers and sizes of urban areas are mushrooming.* In 2012, about 48% of the world's people lived in nearly 850 urban areas with 500,000 or more people (see chapter-opening photo). Also, there were 26 *megacities*—cities with 10 million or more people—19 of them in less-developed countries (Figure 22-3). Nine of these urban areas are *hypercities* with more

Figure 22-2 About 52% of the world's people live in urban areas, or cities, such as Shanghai, China (top), and their surrounding suburban areas such as this one in Phoenix, Arizona (center). The other 48% live in rural areas—in villages such as this one in the southern African country of Malawi (bottom), in small towns, or in the countryside.

Figure 22-3 *Global outlook:* Megacities, or major urban areas with 10 million or more people, in 2012. **Question:** In order, what were the world's five most populous urban areas in 2012?

(Compiled by the authors using data from National Geophysics Data Center, Demographia, National Oceanic and Atmospheric Administration, and United Nations Population Division.)

than 20 million people. The largest hypercity is Tokyo, Japan, with 37.1 million—more than the entire population of Canada. Some of the world's megacities and hypercities are merging into vast urban *megaregions*, each with more than 100 million people. The largest megaregion is the Hong Kong–Shenzhen–Guangzhou region in China with about 120 million people.

3. *Poverty is becoming increasingly urbanized, mostly in less-developed countries.* The United Nations estimates that at least 1 billion people (almost 3 times the current U.S. population) live in the slums and shantytowns that lie within or around most of the major cities of less-developed countries. This number may triple by 2050.

🔍 **CONSIDER THIS. . .**

THINKING ABOUT Urban Trends

If you could reverse one of the three urban trends discussed here, which one would it be? Explain.

If you visit a poor, overcrowded area of a large city in a less-developed country, your senses may be overwhelmed by a vibrant but chaotic crush of people, vehicles of all types, traffic jams (Figure 22-4), noise, and odors, including raw sewage and smoke from burning trash, as well as from wood and coal cooking fires. Many people sleep on the streets or live in crowded, unsanitary, rickety, and unsafe slums and shantytowns with little or no access to safe drinking water or modern sanitation facilities.

CASE STUDY

Urbanization in the United States

Between 1800 and 2012, the percentage of the U.S. population living in urban areas rose from 5% to 82%. This population shift has occurred in three phases.

First, *people migrated from rural areas to large central cities.* In 2012, about 71% of Americans lived in urban areas with at least 50,000 people, and about 54% lived in 51 urban areas with 1 million or more residents (Figure 22-5).

Second, *many people migrated from large central cities to smaller cities and suburbs.* Currently, about half of urban Americans live in the suburbs (Figure 22-2, center photo), nearly a third in central cities, and the rest mostly in rural housing developments beyond suburbs.

Third, *many people migrated from the North and East to the South and West.* Between 1980 and 2009, about 80% of the U.S. population growth occurred in the South and West, especially in states such as Nevada, Colorado, and Florida. This migration slowed in 2010 and 2011.

Since 1920, many of the worst urban environmental problems in the United States have been reduced significantly (see Figure 6-7, p. 127). Most people have GOOD NEWS better working and housing conditions, while air and water quality have improved. Better sanitation, clean public water supplies, and expanded medical care have slashed death rates and incidences of sickness from infectious diseases. Also, the concentration of most of the population in urban areas has helped to protect the country's biodiversity by reducing the destruction and degradation of wildlife habitat.

Figure 22-4 A typical traffic jam of people and motor vehicles in Kolkata, India, a megacity of 14.4 million people.

Figure 22-5 Urban areas in the United States with more than 1 million people (shaded areas) where about 54% of Americans live. **Questions:** Why do you think many of the largest urban areas are located near water? What effect might projected climate change have on these urban areas? (See Chapter 19 opening photo, p. 504.)

(Compiled by the authors using data from National Geophysical Data Center/National Oceanic and Atmospheric Administration, U.S. Census Bureau.)

1973

2009

Animated Figure 22-6 These satellite images show the growth of *urban sprawl* in and around the U.S. city of Las Vegas, Nevada, between 1973 and 2009. **Question:** What might be a limiting factor on population growth in Las Vegas?

However, a number of U.S. central cities—especially older ones—have deteriorating services and aging *infrastructures* (streets, bridges, dams, power lines, schools, water supply pipes, and sewers). Funds for repairing and upgrading urban infrastructure have declined in many urban areas as the flight of people and businesses to the suburbs and beyond has decreased central city revenues from property taxes, although this trend, too, has been

reversed in some cities, including Portland, Oregon (**Core Case Study**).

Urban Sprawl Gobbles Up the Countryside

In the United States and some other countries, **urban sprawl**—the growth of low-density development on the edges of cities and towns—is eliminating surrounding agricultural and wild lands (Figure 22-6). It results in a dispersed jumble of housing developments (Figure 22-2, center), shopping malls, parking lots, and office complexes that are loosely connected by multilane highways and freeways.

Urban sprawl is largely the product of ample affordable land, automobiles, federal and state funding of highways, and inadequate urban planning. Many people prefer living in suburbs and *exurbs*—housing developments scattered over vast areas that lie beyond suburbs and have no socioeconomic centers. Compared to central cities, these areas provide lower-density living and access to larger lot sizes and single-family homes. Often these areas also have newer public schools and lower crime rates.

On the other hand, urban sprawl has caused or contributed to a number of environmental problems. Because of nonexistent or inadequate mass transportation in most such areas, sprawl forces people to drive everywhere, emitting climate-changing greenhouse gases and other forms of air pollution in the process. Partly for this reason, sprawl has caused people to use energy less efficiently while increasing traffic congestion and destroying prime cropland (Figure 22-7), forests, and wetlands. Urban sprawl also has led to the economic deaths of many

Figure 22-7 This suburban development in Ventura County, California, was once prime cropland and is likely to expand and take over more of such land.

Figure 22-8 Some of the undesirable impacts of urban sprawl, or car-dependent development. **Question:** Which five of these effects do you think are the most harmful?

Left: Condor 36/Shutterstock.com. Left center: spirit of america/Shutterstock.com. Right center: ssuaphotos/ Shutterstock.com. Right: ronfromyork, 2009/Used under license from Shutterstock.com.

Natural Capital Degradation

Urban Sprawl

Land and Biodiversity

Loss of cropland

Loss and fragmentation of forests, grasslands, wetlands, and wildlife habitat

Water

Increased use and pollution of surface water and groundwater

Increased runoff and flooding

Energy, Air, and Climate

Increased energy use and waste

Increased emissions of carbon dioxide and other air pollutants

Economic Effects

Decline of downtown business districts

More unemployment in central cities

© Cengage Learning 2015

central cities, as people and businesses have moved out of these areas. Figure 22-8 summarizes these and other undesirable consequences of urban sprawl.

 CONSIDER THIS...

THINKING ABOUT Urban Sprawl

Do you think the advantages of urban sprawl outweigh its disadvantages? Explain.

22-2 What Are the Major Urban Resource and Environmental Problems?

CONCEPT 22-2
Most cities are unsustainable because of high levels of resource use, waste, pollution, and poverty.

Urbanization Has Advantages

Urbanization has many benefits. From an *economic standpoint,* cities are centers of economic development, innovation, education, technological advances, social and cultural diversity, and jobs. Urban residents in many parts of the world tend to live longer than do rural residents and to have lower infant mortality and fertility rates. They also have better access to medical care, family planning, education, and social services than do their rural counterparts.

GOOD NEWS

Urban areas also have some environmental advantages. Recycling is more economically feasible because of the high concentrations of recyclable materials in urban areas. Concentrating people in cities can help to preserve biodiversity by reducing the stress on wildlife habitats. Multistory apartment and office buildings in central cities require less energy per person to heat and cool them than do single-family homes and smaller office buildings in suburbs. Also, central-city dwellers tend to drive less and rely more on mass transportation, walking, and bicycling. This helps to explain why New York City and many other large U.S. cities have some of the lowest per capita carbon dioxide emissions of all cities.

Urbanization Has Disadvantages

Intense population pressure and high population densities make most of the world's cities more environmentally unsustainable every year (**Concept 22-2**). Even in more sustainable cities such as Portland, Oregon (**Core Case Study**), it is a challenge to maintain a high level of sustainability in the face of a growing population and higher rates of resource use per person. Here we take a closer look at some of the significant problems faced by most of the world's cities.

Most Urban Areas Have Huge Ecological Footprints

Urban populations occupy only about 3% of the earth's land area, but they consume about 75% of its resources and produce about 75% of the world's pollution and wastes. Because of this high input of food, water, and other resources, and the resulting high waste output (Figure 22-9), most of the world's cities have huge ecological footprints that extend far beyond their boundaries, and they are not self-sustaining systems (**Concept 22-2**).

For example, according to an analysis by Mathis Wackernagel and William Rees, developers of the ecological footprint concept (see pp. 12–13), London, England, requires an area 58 times as large as the city to supply its residents with resources. These researchers estimate that if all of the world's people used resources at the same rate as Londoners do, it would take at least three more planet Earths to meet their needs.

Inputs
- Energy
- Food
- Water
- Raw materials
- Manufactured goods
- Money
- Information

Outputs
- Solid wastes
- Waste heat
- Air pollutants
- Water pollutants
- Greenhouse gases
- Manufactured goods
- Noise
- Wealth
- Ideas

© Cengage Learning

Figure 22-9 Natural capital degradation: The typical city depends on nonurban areas for huge inputs of matter and energy resources, while it generates and concentrates large outputs of pollution, waste matter, and heat. **Question:** How would you apply the three scientific **principles of sustainability** (see Figure 1-2, p. 6 or back cover) to lessen some of these impacts?

CONSIDER THIS...

CONNECTIONS Urban Living and Biodiversity Awareness

Recent studies reveal that urban dwellers tend to live most or all of their lives in an artificial environment that isolates them from forests, grasslands, streams, and other natural areas. They are thus separated from the biodiversity that provides most of the ecosystem services that sustain their lives and economies. As a result, many urban residents tend to be unaware of the importance of protecting not only the earth's increasingly threatened biodiversity but also its other forms of natural capital that support their lives and the cities in which they live.

Most Cities Lack Vegetation

In urban areas, most trees, shrubs, grasses, and other plants are cleared to make way for buildings, roads, parking lots, and housing developments. Thus, most cities do not benefit from vegetation that would absorb air pollutants, give off oxygen, dampen urban noise, and provide shade, wildlife habitat, and aesthetic pleasure.

Many Cities Have Water Problems

As cities grow and their water demands increase, expensive reservoirs and canals must be built and deeper wells must be drilled. This can deprive rural and wild areas of surface water and deplete groundwater supplies.

Flooding also tends to be greater in cities that are built on floodplains near rivers or along low-lying coastlines subject to natural flooding. In addition, covering land with buildings, asphalt, and concrete causes precipitation to run off quickly and overload storm drains. Urban development has often destroyed or degraded large areas of wetlands that have served as natural sponges to help absorb excess storm water. Many of the world's largest coastal cities (Figure 22-3) will face a new flooding threat at some time in this century as sea levels rise because of projected climate change due to a warmer atmosphere (see Chapter 19 opening photo, p. 504). Projected climate change is also expected to melt some mountaintop glaciers, and cities that depend on this ice for their water supplies will face severe water shortages.

Cities Tend to Concentrate Pollution and Health Problems

Because of their high population densities and high rates of resource consumption, cities produce most of the world's air pollution, water pollution, and solid and hazardous wastes. Pollutant levels are generally higher because pollution is produced in a smaller area and cannot be dispersed and diluted as readily as pollution produced in rural areas can.

The concentration of motor vehicles and industrial facilities in urban centers, with just over two-thirds of the world's emissions of CO_2 from human-related sources, causes disruption of local and regional portions of the carbon cycle (see Figure 3-17, p. 66). This concentration of urban air pollution also disrupts the nitrogen cycle (see Figure 3-18, p. 67) because of large quantities of nitrogen oxide emissions, which play a key role in the formation of photochemical smog.

Also, urban nitric acid and nitrate emissions are major components of acid deposition in urban areas and beyond (see Figure 18-12, p. 485, and Figure 18-13, p. 486). Nitrogen nutrients in urban runoff and discharges from urban sewage treatment plants can also disrupt the nitrogen cycle in nearby lakes and other bodies of water, and cause excessive eutrophication (see Figure 20-1, p. 544, and Figure 20-8, p. 551).

In addition, high population densities in urban areas can promote the spread of infectious diseases, especially if drinking water and sewage systems are inadequate or not available.

Cities Have Excessive Noise

Most urban dwellers are subjected to **noise pollution**: any unwanted, disturbing, or harmful sound that damages, impairs, or interferes with hearing, causes stress, hampers concentration and work efficiency, or causes accidents. Noise levels are measured in decibel-A (dbA) sound pressure units that vary with different human activities (Figure 22-10).

Sound pressure becomes damaging at about 85 dbA and painful at around 120 dbA. At 180 dbA, sound can

Figure 22-10 *Noise levels* (in decibel-A [dbA] sound pressure units) of some common sounds. **Question:** How often are your ears subjected to noise levels of 85 or more dbA?

kill. Prolonged exposure to sound levels above 85 dbA can cause permanent hearing damage. Just one-and-a-half minutes of exposure to 110 decibels or more can cause such damage. You are being exposed to a sound level high enough to cause permanent hearing damage if a noise requires you to raise your voice to be heard, if a noise causes your ears to ring, or if a noise makes nearby speech seem muffled. Prolonged exposure to lower noise levels and occasional loud sounds may not damage your hearing, but these sound levels can be very stressful.

Cities Affect Local Climates and Cause Light Pollution

On average, cities tend to be warmer, rainier, foggier, and cloudier than suburbs and nearby rural areas. In cities, the enormous amount of heat generated by cars, factories, furnaces, lights, air conditioners, and heat-absorbing dark roofs and streets creates an *urban heat island* that is surrounded by cooler suburban and rural areas. In areas with hot climates, this can accelerate the formation of photochemical smog (see Figure 18-9, p. 483, and Figure 18-10, p. 484) by speeding up the chemical reactions involved in its formation. As cities grow and merge, their heat islands merge, which can reduce the natural dilution and cleansing of polluted air that takes place in open areas.

The urban heat island effect can also greatly increase dependence on air conditioning. This in turn leads to higher energy consumption, greenhouse gas emissions, and other forms of air pollution in a positive feedback loop.

The artificial light created by cities (see chapter-opening photo and Figure 22-2, top) affects some plant and animal species. For example, some endangered sea turtles lay their eggs on beaches at night and require darkness to do so. In addition, each year, large numbers of migrating birds, lured off course by the lights of high-rise buildings, fatally collide with these structures. Excessive light also makes it difficult for astronomers and urban dwellers to study and enjoy the night sky.

🔍 **CONSIDER THIS. . .**

THINKING ABOUT Disadvantages of Urbanization

Which two of these disadvantages of urbanization do you think are the most serious? Explain.

Life Is a Desperate Struggle for the Urban Poor in Less-Developed Countries

Poverty is a way of life for many urban dwellers in less-developed countries. According to a 2006 UN study, the number of urban dwellers living in poverty—now about 1 billion—could reach 1.4 billion by 2020. Some of these people live in *slums*—areas dominated by dilapidated tenements, or rooming houses where the rooms are often small and numerous and where several people might live in a single room.

Other poor people live in *shantytowns* on the outskirts of cities. They build shacks from corrugated metal, plastic sheets, scrap wood, cardboard, and other scavenged building materials, or they live in rusted shipping containers and junked cars. Some shantytowns are illegal subdivisions where landowners rent land to the poor without city approval. Others are illegal *squatter settlements* where people take over unoccupied land without the owners' permission, simply because it is their only option for survival. Consider Mumbai, India, with a population of 16.9 million. Between 1970 and 2010, the proportion of its residents living in slums and shantytowns grew from 17% to about 62% (Figure 22-11).

Poor people living in shantytowns and squatter settlements or on the streets (see Figure 6-20, p. 137) usually lack clean water supplies, sewers, electricity, and roads, and are subject to severe air and water pollution and hazardous wastes from nearby factories. Many of these settlements are in locations especially prone to landslides, flooding, or earthquakes. Some city governments regularly bulldoze squatter shacks and send police to drive illegal settlers out. The people usually move back in within a few days or weeks, or develop another shantytown elsewhere.

Figure 22-11 Slum in Mumbai, India.

Some governments have addressed these problems. The governments of Brazil and Peru legally recognize existing slums *(favelas)* and grant legal titles to the land. They base this on evidence that poor people usually improve their living conditions once they know they have a permanent place to live. They can then become productive working citizens who contribute to tax revenues that, in turn, help to pay for the government programs that assist the poor. In some cases, impoverished people with titles to land have developed their own schools, day-care centers, and other such structures for social improvement.

CASE STUDY

Mexico City

With 19.5 million people, Mexico City is one of the world's megacities (Figure 22-3) and will soon become a hypercity with more than 20 million people. More than one-third of its residents live in slums called *barrios* or in squatter settlements that lack running water and electricity. As a result, the city suffers from serious problems.

For example, at least 3 million people in the barrios have no sewage facilities, so human waste from these slums is deposited in gutters, vacant lots, and open ditches every day, attracting armies of rats and swarms of flies. When the winds pick up dried excrement, a *fecal snow* blankets parts of the city. This bacteria-laden fallout leads to widespread salmonella and hepatitis infections, especially among children.

Mexico City has serious air pollution problems (Figure 22-12) because of a combination of factors: too many

cars, polluting factories, a warm, sunny climate and thus more smog, and topographical bad luck. The city sits in a high-elevation, bowl-shaped valley surrounded on three sides by mountains—conditions that trap air pollutants at ground level (see Figure 18-11, left, p. 484).

The city also has one of the world's highest water consumption rates and suffers from chronic water shortages. Large-scale water withdrawals from the city's aquifer, which supplies about two-thirds of its water, have

Figure 22-12 Photochemical smog in Mexico City, Mexico, is caused by cars and other motor vehicles that generate pollutants and is made worse by the city's location within a bowl-shaped valley that traps emissions, causing them to accumulate to dangerous levels.

caused parts of the city to subside by 9 meters (30 feet) during the last century. Some areas are now subsiding as much as 30 centimeters (1 foot) a year. The city's growing population increasingly relies on pumping water from as far away as 150 kilometers (93 miles), which requires large amounts of energy, and then pumping it another 1,000 meters (3,300 feet) uphill to reach the city.

In 1992 the United Nations named Mexico City "the most polluted city on the planet." Since then Mexico City has made progress in reducing the severity of some of its air pollution problems. The percentage of days each year in which air pollution standards are violated has fallen from 50% to 20% and ozone and other air pollutants are now at about the same levels as those found in Los Angeles, California.

This is partly due to the fact that the city government has moved refineries and factories out of the city, banned cars in its central zone, and required air pollution controls on all cars made after 1991. It has also phased out the use of leaded gasoline, expanded public transportation, and replaced some old buses, taxis, and trucks with vehicles that produce fewer emissions. In 2013, the Institute for Transportation and Development awarded Mexico City its Sustainable Transportation Award in recognition of the city's expansion of its bus rapid transit system, rebuilding of its public parks and plazas, reduction of crime rates, and expansion of its bike lane network and bike-sharing program that had been launched in 2010.

In addition, Mexico City has instituted a program to reduce water use and waste and implemented a water pricing system designed to promote water conservation. It also bought land for use as green space and planted more than 25 million trees to help absorb pollutants. It is working to build state-of-the-art waste processing centers that, within a few years, will process as much as 85% of the city's solid waste through recycling and composting, while burning some of it for energy.

Mexico City still has a long way to go as its human population grows along with its number of cars. However, this story shows what can be done to improve environmental quality once a community decides to act.

GOOD NEWS

22-3 How Does Transportation Affect Urban Environmental Impacts?

CONCEPT 22-3

In some countries, many people live in widely dispersed urban areas and depend mostly on motor vehicles for their transportation, which greatly expands their ecological footprints.

Cities Can Grow Outward or Upward

If a city cannot spread outward, it must grow vertically—upward and downward (below ground)—so that it occupies a small land area with a high population density. Most people living in *compact cities* such as Hong Kong, China, and Tokyo, Japan, get around by walking, biking, or using mass transit such as rail and bus systems. Some high-rise apartment buildings in these Asian cities contain everything from grocery stores to fitness centers that reduce the need for their residents to travel for food, entertainment, and other services.

In other parts of the world, a combination of plentiful land, relatively cheap gasoline, and networks of highways has produced *dispersed cities* whose residents depend on motor vehicles for most travel (Concept 22-3). Such car-centered cities are found in the United States, Canada, Australia, and other countries where ample land often is available for these cities to expand outward. The resulting urban sprawl (Figure 22-6) can have a number of undesirable effects (Figure 22-8).

The United States is a prime example of a car-centered nation. With 4.4% of the world's people, the United States has about 25% of the world's more than 1 billion motor vehicles, according to the U.S. Department of Transportation. In its dispersed urban areas, U.S. passenger vehicles are used for 86% of all transportation and 76% of residents drive alone to work every day. Largely because of urban sprawl, Americans altogether drive about the same distance each year as the total distance driven by all other drivers in the world combined, and in the process, use about 43% of the world's gasoline.

Use of Motor Vehicles Has Advantages and Disadvantages

Motor vehicles provide mobility and offer a convenient and comfortable way to get from one place to another. For many people, they are symbols of power, sex appeal, social status, and success. Also, much of the world's economy is built on producing motor vehicles and supplying fuel, roads, services, and repairs for them.

Despite their important benefits, motor vehicles have many harmful effects on people and the environment. Globally, automobile accidents kill about 1.3 million people a year—an average of nearly 3,600 deaths per day—and injure another 50 million people. They also kill about 50 million wild animals and family pets every year.

In the United States, motor vehicle accidents killed 36,200 people in 2012 and injured another 2.2 million, at least 300,000 of them severely. *Car accidents have killed more Americans than have all the wars in the country's history.*

Motor vehicles are the world's largest source of outdoor air pollution, which causes 30,000–60,000 premature deaths per year in the United States, according to the Environmental Protection Agency. They are also

Figure 22-13 Cloverleafs like this tangled network of thruways in the U.S. city of Los Angeles, California, are found in most of the world's increasingly car-dependent cities.

the fastest-growing source of climate-changing CO_2 emissions.

Motor vehicles have helped to create urban sprawl and the car commuter culture. At least a third of the world's urban land and half of that in the United States is devoted to roads, parking lots, gasoline stations, and other automobile-related uses. This prompted urban expert Lewis Mumford to suggest that the U.S. national flower should be the concrete cloverleaf (Figure 22-13).

Another problem is congestion (Figure 22-14). If current trends continue, U.S. motorists will spend an average of 2 years of their lives in traffic jams, as streets and freeways will more often resemble parking lots. Traffic congestion in some cities in less-developed countries is much worse. Building more roads may not be the answer. Various commentators have noted that historically, the number of cars has always expanded to fill new roadways.

Reducing Automobile Use Is Not Easy, but It Can Be Done

Some environmental scientists and economists suggest that we can reduce the harmful effects of automobile use by making drivers pay directly for most of the environmental and health costs caused by their automobile use—a *user-pays* approach.

One way to phase in such *full-cost pricing*, in keeping with one of the social science **principles of sustainability** (see Figure 1-5, p. 9 or back cover), would be to charge a tax or fee on gasoline to cover the estimated harmful costs of driving. According to a study by the International Center for Technology Assessment, such a tax would amount to about $3.18 per liter ($12 per gallon) of gasoline in the United States (see Chapter 16, p. 405). Gradually phasing in such a tax, as has been done in many European nations, would spur the use of

Figure 22-14 Traffic jam in Los Angeles, California.

more energy-efficient motor vehicles and mass transit, lessen dependence on imported oil, and thus strengthen economic and national security. It would also reduce pollution and environmental degradation and help to slow projected climate change.

Proponents of higher gasoline taxes urge governments to do two major things. *First,* fund programs to educate people about the hidden costs they are paying for gasoline. *Second,* use gasoline tax revenues to help finance mass-transit systems, bike lanes, and sidewalks as alternatives to cars, and to reduce taxes on income, wages, and wealth to offset the increased taxes on gasoline. Such a *tax shift* would help to make higher gasoline taxes more politically and economically acceptable.

Taxing gasoline heavily would be difficult in the United States, for three reasons. *First,* it faces strong opposition from two groups. One group is made up of those people who feel they are already overtaxed, many of whom are largely unaware of the hidden costs they are paying for gasoline. The other group is made up of the powerful transportation-related industries such as carmakers, oil and tire companies, road builders, and many real estate developers. *Second,* the dispersed nature of most U.S. urban areas makes people dependent on cars, and thus higher taxes would be an economic burden for them. *Third,* fast, efficient, reliable, and affordable mass-transit options, bike lanes, and sidewalks are not widely available in the United States. These factors make it politically difficult to raise gasoline taxes. But U.S. taxpayers might accept sharp increases in gasoline taxes if a tax shift were employed, as mentioned above.

Another way to reduce automobile use and urban congestion is to raise parking fees and charge tolls on roads, tunnels, and bridges leading into cities—especially during peak traffic times. Densely populated Singapore is rarely congested because it auctions the rights to buy a car, and its cars carry electronic sensors that automatically charge the drivers a fee every time they enter the city. Several European cities have also imposed stiff fees for motor vehicle use in their central cities. Shanghai, China,

discourages car use by charging more than $9,000 for a license plate. Copenhagen, Denmark, has encouraged the use of bicycles and mass transit by banning on-street parking of cars and establishing a safe network of bike lanes. Paris, France, has removed 200,000 parking spaces.

In Germany, Austria, Italy, Switzerland, and the Netherlands, more than 300 cities have *car-sharing* networks that provide short-term rental of cars. Portland, Oregon (**Core Case Study**), was the first U.S. city to develop a car-sharing system. These systems can reduce car ownership and also help to reduce congestion and air pollution. Network members reserve a car in advance or contact the network and are directed to the closest car. They are billed monthly for the time they use a car and the distance they travel. In Berlin, Germany, car sharing has cut car ownership by 75%. According to the Worldwatch Institute, car sharing in Europe has reduced the average driver's CO_2 emissions by 40–50%. In the United States, car-sharing networks have sprouted on some college campuses and in several cities, and some large car-rental companies have begun renting cars by the hour.

Some Cities Promote Alternatives to Cars

There are several alternatives to motor vehicles, each with its own advantages and disadvantages. Figure 22-15 shows the transportation hierarchy in more-sustainable cities such as Portland, Oregon (**Core Case Study**).

One widely used alternative is the *bicycle* (Figure 22-16). There is also a growing use of bicycles with lightweight electric motors. Bicycling accounts for about a third of all urban trips in the Netherlands and in Copenhagen, Denmark. By contrast, bicycling accounts for less than 1% of urban trips in the United States, but this percentage is higher in bike-friendly cities such as Minneapolis, Minnesota, and Portland, Oregon, which has the nation's highest percentage (8%) of bicycle commuters. Some 25% of Americans polled on this issue say they would bike to work or school if safe bike lanes and secure bike storage were available.

Transportation Priorities

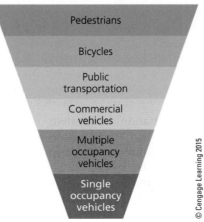

Pedestrians

Bicycles

Public transportation

Commercial vehicles

Multiple occupancy vehicles

Single occupancy vehicles

© Cengage Learning 2015

Figure 22-15 Transportation priorities in more-sustainable cities.

Trade-Offs

Bicycles

Advantages	Disadvantages
Are quiet and nonpolluting	Provide little protection in an accident
Take few resources to manufacture	Provide no protection from bad weather
Burn no fossil fuels	Are impractical for long trips
Require little parking space	Bike lanes and secure bike storage not yet widespread

© Cengage Learning 2015

Figure 22-16 Bicycle use has advantages and disadvantages. **Question:** Which single advantage and which single disadvantage do you think are the most important?

Photo: Tyler Olson/Shutterstock.com

Trade-Offs

Buses

Advantages	Disadvantages
Reduce car use and air pollution	Can lose money because they require affordable fares
Can be rerouted as needed	Can get caught in traffic and add to noise and pollution
Cheaper than heavy-rail system	Commit riders to transportation schedules

© Cengage Learning 2015

Figure 22-17 Bus rapid-transit systems and conventional bus systems in urban areas have advantages and disadvantages. **Question:** Which single advantage and which single disadvantage do you think are the most important?

Photo: Isaak/Shutterstock.com

Trade-Offs

Mass Transit Rail

Advantages	Disadvantages
Uses less energy and produces less air pollution than cars do	Expensive to build and maintain
Uses less land than roads and parking lots use	Cost-effective only in densely populated areas
Causes fewer injuries and deaths than cars	Commits riders to transportation schedules

© Cengage Learning 2015

Figure 22-18 Mass-transit rail systems in urban areas have advantages and disadvantages. **Question:** Which single advantage and which single disadvantage do you think are the most important?

Photo: Steve Rosset/Shutterstock.com

More than 500 cities in 49 countries, including Portland, have bike-sharing systems that allow individuals to rent bikes as needed from widely distributed stations. The combined global fleet of these bicycles numbers more than 500,000 bikes (358,000 of them in China).

Portland, with more than 510 kilometers (317 miles) of bikeways, has a goal of increasing its percentage of bike commuters to 25% by 2030 (**Core Case Study**). This biking culture has created many local jobs related to bicycle manufacturing, sales, and service. In addition, a Portland-based delivery company uses electric tricycles instead of trucks and can transport parcels weighing as much as 270 kilograms (600 pounds).

Another alternative to cars is *buses,* which are the most widely used form of mass transit in urban areas worldwide (Figure 22-17). In some cities, bus rapid-transit (BRT) systems are making bus use more convenient by including fast express routes, allowing riders to pay at machines located at each bus stop so that they can board a bus more quickly, and having three or four doors for quicker boarding. Newer systems are also using diesel-electric hybrid motors to cut pollution and noise from buses. BRT systems are used in a growing number of cities, including Mexico City (Case Study, p. 614); Curitiba, Brazil (Case Study, p. 625); Bogotá, Columbia; Seoul, South Korea; Istanbul, Turkey; and 11 cities in China, including Beijing. Each year, Portland's bus system (**Core Case Study**) carries more than 66 million riders.

Another urban transportation alternative is *mass-transit rail systems* (Figure 22-18). They include *heavy-rail* systems (subways, elevated railways, and metro trains) and *light-rail* systems (streetcars, trolley cars, and tramways). Portland (**Core Case Study**) has a widely used light-

rail system (Figure 22-19). It also has an aerial tramway that transports large numbers of people from its downtown waterfront development to a hilltop medical center and university campus.

A *rapid-rail system between urban areas* is another option for reducing the use of cars (Figure 22-20). Western Europe, Japan, and China have high-speed bullet trains that travel between cities at up to 306 kilometers (190 miles) per hour. For decades, many analysts in the United States have talked about building high-speed rail systems between cities, but so far, none have been built.

© Ken Hawkins/Alamy

Figure 22-19
Widespread bicycle use and a light-rail system, in operation since 1986, have helped to reduce car use in Portland, Oregon (**Core Case Study**).

Trade-Offs

Rapid Rail

Advantages

Much more energy efficient per rider than cars and planes are

Produces less air pollution than cars and planes

Can reduce need for air travel, cars, roads, and parking areas

Disadvantages

Costly to run and maintain

Causes noise and vibration for nearby residents

Adds some risk of collision at car crossings

© Cengage Learning 2015

Figure 22-20 Rapid-rail systems between urban areas have advantages and disadvantages. **Question:** Which single advantage and which single disadvantage do you think are the most important?

Photo: Alfonso d'Agostino/Shutterstock.com

22-4 How Important Is Urban Land-Use Planning?

CONCEPT 22-4
Urban land-use planning can help to reduce uncontrolled sprawl and slow the resulting degradation of air, water, land, biodiversity, and other natural resources.

Conventional Land-Use Planning

Most urban and some rural areas use various forms of **land-use planning** to determine the best present and future uses of various parcels of land. Once a land-use plan is developed and adopted, governments can control the uses of certain parcels of land by legal and economic methods. The most widely used approach is **zoning**, in which parcels of land are designated for certain uses such as residential, commercial, or mixed use. Zoning can be used to control growth and to protect areas from certain types of development. For example, Portland, Oregon (**Core Case Study**), and other cities have used zoning to encourage high-density development along major mass-transit corridors to reduce automobile use and air pollution.

Despite its usefulness, zoning has several drawbacks. One problem is that some developers can influence or modify zoning decisions in ways that threaten or destroy wetlands, prime cropland, forested areas, and open space. Another problem is that zoning often favors high-priced housing, factories, hotels, and other businesses over protecting environmentally sensitive areas and providing low-cost housing. This is largely because most local governments depend on property taxes for their revenue.

In addition, overly strict zoning can discourage innovative approaches to solving urban problems. For example, the pattern in the United States and in some other countries has been to prohibit businesses in residential areas, which causes separate business and residential developments and encourages suburban sprawl. Some urban planners have returned to *mixed-use zoning* to help reduce this problem. In the 1970s, Portland, Oregon (**Core Case Study**), managed to cut driving and gasoline consumption and to promote walking and bicycling by using mixed zoning to encourage the establishment of neighborhood groceries and other small stores in residential areas.

Smart Growth Can Work

Smart growth is one way to encourage more environmentally sustainable development that requires less dependence on cars, controls and directs sprawl, and reduces wasteful resource use (see the Case Study that follows). It recognizes that urban growth will occur, but at the same time, it uses zoning laws and other tools to channel that growth into areas where it can cause less harm.

Smart growth can discourage sprawl, reduce traffic, protect ecologically sensitive and important lands and waterways, and develop neighborhoods that are more enjoyable places in which to live. Figure 22-21 lists popular smart growth tools that cities are using to prevent and control urban growth and sprawl. Portland, Oregon (**Core Case Study**), has used many of these tools.

China has taken the strongest stand of any country against urban sprawl. The government has designated 80% of the country's arable land as *fundamental land*. Building on such land requires approval from local and provincial governments and from the State Council, the central government's chief administrative body.

Many European countries have been successful in discouraging urban sprawl in favor of compact cities. They have controlled development at the national level and imposed high gasoline taxes to discourage car use and to encourage people to live closer to workplaces and shops. High taxes on heating fuel have encouraged some people to live in apartments or smaller houses. These governments have used most of the resulting gasoline and heating fuel tax revenues to develop efficient train systems and other mass-transit options within and between cities.

Solutions

Smart Growth Tools

Limits and Regulations

Limit building permits

Draw urban growth boundaries

Create greenbelts around cities

Zoning

Promote mixed use of housing and small businesses

Concentrate development along mass transportation routes

Planning

Ecological land-use planning

Environmental impact analysis

Integrated regional planning

Protection

Preserve open space

Buy new open space

Prohibit certain types of development

Taxes

Tax land, not buildings

Tax land on value of actual use instead of on highest value as developed land

Tax Breaks

For owners agreeing not to allow certain types of development

For cleaning up and developing abandoned urban sites

Revitalization and New Growth

Revitalize existing towns and cities

Build well-planned new towns and villages within cities

© Cengage Learning 2015

Figure 22-21 We can use these *smart growth tools* to prevent and control urban growth and sprawl. ***Questions:*** Which five of these tools do you think would be the best methods for preventing or controlling urban sprawl? Which, if any, of these tools are used in your community?

Top: Tungphoto/Shutterstock.com. Bottom: © Richard Schmidt-Zuper/iStockphoto.com.

CONSIDER THIS. . .

THINKING ABOUT Smart Growth

Do you think that the approach taken by China for controlling urban growth is reasonable? If so, explain. If not, what do you think would be a more reasonable approach?

Preserving and Using Open Space

One way to preserve open space outside a city is to draw a boundary around the city and to prohibit urban development outside that boundary. This *urban growth boundary* approach is used in the U.S. states of Oregon, Washington, and Tennessee. In 1979, Portland, Oregon (**Core Case Study**),

Figure 22-22 With almost 344 hectares (850 acres) that include woodlands, lawns, and small lakes and ponds, New York City's Central Park is a dramatic example of a large open space in the center of a major urban area.

Age Fotostock/SuperStock

established a strict urban growth boundary, and it worked. Although Portland's population increased by 38% between 1980 and 2011, its urban area expanded by only 2%.

Another approach is to surround a large city with a *greenbelt*—an open area reserved for recreation, sustainable forestry, or other nondestructive uses. In many cases, satellite towns have been built outside these greenbelts. The best of these are self-contained, not sprawling, and are linked to the central city by a public transport system that does minimal damage to the greenbelt. Many cities in western Europe and the Canadian cities of Toronto and Vancouver have used this approach. Greenbelts can provide vital ecosystem services such as absorption of CO_2

and other air pollutants, which can make urban air more breathable and help to cut a city's contribution to climate change.

A more traditional way to preserve large blocks of open space is to create municipal parks. Examples of large urban parks in the United States are Central Park in New York City (Figure 22-22); Golden Gate Park in San Francisco, California; and Grant Park in Chicago, Illinois. Portland, Oregon (**Core Case Study**), has 288 parks along with trails and natural areas distributed throughout the city that give its citizens easy access to nature and recreational opportunities. Currently, urban forests occupy about 26% of Portland, and the city has plans to increase this to at least 33%.

22-5 How Can Cities Become More Sustainable and Livable?

CONCEPT 22-5

An *eco-city* allows people to choose walking, biking, or mass transit for most transportation needs; to recycle or reuse most of their wastes; to grow much of their food; and to protect biodiversity by preserving surrounding land.

New Urbanism Is Growing

Since World War II, the typical approach to suburban housing development in the United States has been to bulldoze a tract of woods or farmland and build rows of houses on standard-size lots (Figure 22-23, center). Many of these developments and their streets, with names like

Oak Lane, Cedar Drive, Pheasant Run, and Fox Valley, are named after the trees and wildlife that were removed to make room for them.

In recent years, builders have increasingly used a pattern known as *cluster development* (Figure 22-23, bottom), in which houses, town houses, condominiums, and two-to six-story apartments are built on parts of the tract. The rest, typically 30–50% of the area, is left as open space for wildlife preserves, parks, and walking and biking paths. When this is done properly, residents can enjoy recreational space set within aesthetically pleasing surroundings. They might also see lower heating and cooling costs because some walls are shared in multiple-family dwell-

Undeveloped land
Creek
Marsh

Typical housing development

Cluster housing development
Cluster
Creek
Cluster
Pond

© Cengage Learning

Figure 22-23 These models compare a *conventional housing development* (center) with a *cluster housing development* (bottom). **Question:** What are the comparative effects of each type of development on the undeveloped land (top)?

ings. Developers, too, can cut their costs for site preparation, roads, utilities, and other forms of infrastructure.

Some communities are going further and using the goals of *new urbanism,* which is a modern form of what could be called *old villageism,* to develop entire villages and promote mixed-use neighborhoods near and within existing cities. These goals include:

- *Walkable and bike-friendly neighborhoods,* with nearby stores, recreational activities, and access to mass transit.
- *Mixed-use and diversity,* which provides a blend of pedestrian-friendly shops, offices, apartments, and homes to attract people of different ages, classes, cultures, and races.
- *Quality urban design,* emphasizing green space, beauty, aesthetics, and architectural diversity.
- *Environmental sustainability* based on development with minimal environmental impact.
- *Smart transportation,* with well-designed train and bus systems connecting neighborhoods and nearby towns and cities to one another.

Examples of new urban villages in the United States are Mayfaire, in Wilmington, North Carolina; Mizner Place in Boca Raton, Florida; Middleton Hills near Madison, Wisconsin; Yardley, Pennsylvania; Kentlands in Gaithersburg, Maryland; Valencia, California (near Los Angeles); and Stapleton, Colorado (built on an old airport site in Denver).

The Eco-City Concept: Cities for People, Not Cars

According to most environmentalists and urban planners, the primary problem with large cities is our failure to make them more sustainable and enjoyable places to live through good ecological design. (See the online Guest Essay by David Orr on this topic.) New urbanism is a step in the right direction, but many scientists call for going further by employing *sustainable community development*—an evolving science based on the idea that sustainable urban development, in order to be effective, must be based at the local level (Science Focus 22.1).

An important result of these trends is the *eco-city,* or *green city,* model for new urban development and for renovation of existing cities. It is being applied in cities around the world, including Portland, Oregon (**Core Case Study**). Such cities are people-oriented, not car-oriented, and their residents are able to walk, bike, or use low-polluting mass transit for most of their travel.

Residents of today's eco-cities seek to apply all of the scientific **principles of sustainability** (see Figure 1-2, p. 6 or back cover). Much of their energy comes from solar cells on the rooftops and walls of buildings (see Figure 16-12, p. 414), solar hot-water heaters on rooftops, and micro-wind turbines, as well as from geothermal heating and cooling systems (see Figure 16-29, p. 428). They work to ensure that the buildings, vehicles, and appliances they use meet high energy-efficiency standards. They typically reuse, recycle, and compost 60–80% of their municipal solid waste.

Eco-city residents apply the biodiversity principle by preserving or planting trees and plants that are adapted to their local climate and soils. They work to clean up and restore all abandoned lots and industrial sites and to preserve forests, grasslands, wetlands, and farms on the city's outskirts. They often make use of *green roofs* (Figure 22-24) to reduce energy use and help cool city buildings. And in the ideal eco-city, parks are plentiful and easily available to everyone.

Finally, much of the food that eco-city dwellers eat comes from urban or regional organic farms, solar greenhouses, community gardens, and small gardens on rooftops, in yards, and in window boxes. In the future, large-scale rooftop greenhouses and vertical high-rise urban farms (Science Focus 22.2) might contribute to urban food supplies.

SUSTAINABLE COMMUNITY DEVELOPMENT—A KEY TO GLOBAL SUSTAINABILITY

Sustainable community development (SCD) is a multidisciplinary approach to economic development that incorporates scientific methods and principles and focuses on a triple bottom line—people, planet, and profit—to work toward sustainability at the community level. Advocates of this approach argue that we probably cannot achieve sustainability on national and world regional levels unless we can accomplish it at the local level.

One prominent SCD researcher is environmental scientist Kelly D. Cain, Director of the St. Croix Institute for Sustainable Community Development at the University of Wisconsin–River Falls. Cain notes that the major objectives of SCD for any community are to achieve economic and environmental health and social equity. He contends that in order to meet such objectives, a sustainable community must have broad-based citizen participation, agree on a vision based on the best available science, and establish clear and measurable goals for benchmarking and tracking performance. Such goals for a community can include

- growing as much of its own food as possible;
- generating all or most of its own carbon-neutral energy;
- restoring the carrying capacity of local ecosystems that are damaged in the building and maintenance of the community;

- recycling pollutants and wastes as resources within industrial ecosystems (see Figure 21-31, p. 599) or as biodegradable wastes, and returning extracted resources such as water and minerals to the ecosystem; and
- doing all of the above in order to become carbon-negative as quickly as possible.

Becoming *carbon-negative*—absorbing more carbon than the community generates in order to shrink its carbon footprint—is one of the most important goals, according to Cain. "Becoming carbon-neutral is no longer a good enough goal," he notes. He points out that if we all simply become carbon-neutral, it is likely that we will still see severely damaging climate change because of the huge amount of carbon already in the atmosphere in the form of CO_2 that will remain there for 80 to 120 years, and these CO_2 levels are rising (see Figure 19-8, p. 511).

Research by Cain and others indicates that communities can become carbon-negative by greatly improving energy efficiency and by relying on low- and no-carbon energy resources such as wind, solar power, and geothermal energy. To become carbon-negative, it also helps if a community can protect and nurture enough native soils and vegetation in its grasslands, forests, gardens, and other green spaces to absorb and sequester more carbon than it generates on most days.

Cain and his team of researchers evaluate the progress of communities in becoming more sustainable by using various indicators, including energy bills, water-use rates, recycling rates, composting rates, local food production, green space and other common areas, and percentages of electric and hybrid cars among a community's total number of vehicles. They also measure community outputs such as volumes of municipal solid waste and carbon emissions, and they track locally invested carbon offsets.

An important SCD concept is the *triple bottom line:* a three-part way to measure the success of SCD efforts. With this approach, the goal for businesses and other organizations is to achieve positive results for people, the planet, and profits. However, Cain argues that in order to be truly sustainable, we will need a *quadruple bottom line* approach in which we add a fourth factor—the community factor—which leads to a strengthened community based on the core values of ingenuity, creativity, innovation, entrepreneurship, and responsibility for self, family, neighborhood, and community.

Critical Thinking

List three steps that your community or school could take, and three steps that you could take, to become carbon-negative.

The eco-city is not a futuristic dream, but a reality in many cities that are striving to become more environmentally sustainable and livable. In addition to Portland, Oregon (**Core Case Study**), other examples include Curitiba, Brazil (see the following Case Study); Bogotá, Colombia; Waitakere City, New Zealand; Stockholm, Sweden; Helsinki, Finland; Copenhagen, Denmark;

GOOD NEWS

Figure 22-24 This green roof on Chicago's City Hall is one of many green roofs in this city. While Chicago is not considered an eco-city, it is working hard to apply the eco-city model in many ways.

Diane Cook & Len Jenshel/National Geographic Creative

URBAN INDOOR FARMING

Scientists are studying various ways to grow more food indoors in urban areas. One method that city dwellers could implement quickly is the use of rooftop greenhouses to provide them with most of their fruits and vegetables. Researchers have found that such greenhouses use as little as 10% of the water and 5% of the area occupied by conventional farms to produce similar yields.

In 2007, Sun Works, a company in New York City that designs energy-efficient urban greenhouses, built a demonstration greenhouse on a barge floating in the Hudson River. It used solar power and recycled water to grow vegetables and fruits (see Figure 22-A) and is now operated by the nonprofit Groundwork Hudson Valley group as an educational center for thousands of annual visitors. In 2008, Sun Works installed a greenhouse on top of a school in the city. It serves as a science teaching area and supplies produce for the school cafeteria. Ted Caplow, executive director of Sun Works, projects that the rooftops of New York City could provide roughly twice the greenhouse area needed to supply the entire city with fruits and vegetables.

Some scientists envision *vertical farms* in multistory buildings (Figure 12-27, p. 304, and Figure 22-B). These scientists estimate that such a farm could provide vegetables, fruits, chickens, eggs, and fish (grown in aquaculture tanks) for up to 50,000 people. Hydroponic crops (see Science Focus 12.2, p. 306) would be grown on the upper floors. Chickens (and the eggs they produce) and fish that feed on plant wastes would be grown on lower floors.

Crops on these farms could be fed nitrogen and other plant nutrients extracted from the animal wastes and perhaps from city sewage treatment plants. Electricity for heat and lighting could be provided by geothermal, solar, or wind energy, energy from composted plant and animal wastes, and fuel cells powered by hydrogen produced from such forms of renewable energy. Thus, an urban vertical farm would mimic nature by applying all three scientific **principles of sustainability**.

Critical Thinking

In terms of environmental effects, what are three advantages and three disadvantages of widespread use of urban farming? Do you believe that the advantages outweigh the disadvantages? Explain.

Figure 22-B Vertical farms in cities may provide food for many urban dwellers in the future.

Figure 22-A The Science Barge located in Yonkers, New York, is a prototype for a sustainable urban farm that raises vegetables hydroponically with no use of pesticides, no runoff of polluted water, and no net emissions of carbon dioxide.

Melbourne, Australia; Vancouver, Canada; Leicester, England; and in the United States, Davis, California; Olympia, Washington; and Chattanooga, Tennessee. According to a 2012 survey by London's University of Westminster, China is building more eco-cities than any country, followed by the United States.

The Eco-City Concept in Curitiba, Brazil

An example of an eco-city is Curitiba ("koor-i-TEE-ba"), a city of 3.2 million people, known as the "ecological capital" of Brazil. In 1969, planners in this city decided to focus on an inexpensive and efficient mass-transit system rather than on the car.

Curitiba's superb bus rapid-transit system efficiently moves large numbers of passengers, including 72% of the city's commuters. Each of the system's five major "spokes," connecting the city center with outlying districts (map in Figure 22-25), has two express lanes used only by buses. Double- and triple-length bus sections are coupled together as needed to carry up to 300 passengers. Boarding is speeded up by the use of extra-wide bus doors and boarding platforms under glass tubes where passengers can pay before getting on the bus (photo in Figure 22-25). Only high-rise apartment buildings are allowed near major bus routes, and each building must devote its bottom two floors to stores—a practice that reduces the need for residents to travel.

Cars are banned from 49 blocks in the center of the downtown area, which has a network of pedestrian walkways connected to bus stations, parks, and bicycle paths running throughout most of the city. Consequently, Curitiba uses less energy per person and has lower emissions of greenhouse gases and other air pollutants and less traffic congestion than do most comparable cities.

The city removed most buildings from flood-prone areas along its six rivers and replaced them with a series of interconnected parks. Volunteers have planted more than 1.5 million trees throughout the city, none of which can be cut down without a permit, which also requires that two trees must be planted for each one that is cut down.

Curitiba recycles roughly 70% of its paper and 60% of its metal, glass, and plastic. Recovered materials are sold mostly to the city's 500 or more major industries, which must meet strict pollution standards.

Curitiba's poor residents receive free medical and dental care, child care, and job training, and 40 feeding centers are available for street children. People who live in areas not served by garbage trucks can collect garbage and exchange filled garbage bags for surplus food, bus tokens, and school supplies. The city uses its retired buses as roving classrooms to train its poor in the basic skills needed for jobs. Other retired buses have become health clinics, soup kitchens, and day-care centers that are free for low-income parents.

About 95% of Curitiba's citizens can read and write and 83% of its adults have at least a high school education. All school children study ecology. Polls show that

Figure 22-25 Solutions: Curitiba, Brazil's, bus rapid-transit system has greatly reduced car use.

99% of the city's inhabitants would not want to live anywhere else.

Curitiba does face challenges, as do all cities, mostly due to a fivefold increase in its population since 1965. Its once-clear streams are often overloaded with pollutants. The bus system is nearing capacity, and car ownership is on the rise. The city is considering building a light-rail system to relieve some of the pressure.

This internationally acclaimed model of urban planning and sustainability is the brainchild of architect and former college professor Jaime Lerner, who has served as the city's mayor 3 times since 1969. In the face of new challenges, Lerner and other leaders in Curitiba argue that education is still a key to making cities more sustainable, and they want Curitiba to continue serving as an educational example for that purpose.

The Eco-Village Movement Is Growing

Another innovative approach is the growing global *eco-village movement*, in which people come together to design and live in more sustainable villages in rural and suburban areas, and in neighborhoods or "eco-hoods" within cities.

Eco-villagers use diverse methods to live more sustainably and to decrease their ecological footprints. Strategies

Figure 22-26 The Eco-Village plan includes a community center (left end of map), 18 LEED platinum, energy-efficient homes, and a field of photovoltaic solar collectors (right end of map) that will provide most if not all of the village's electrical needs. Some of the homes share a wall, which helps to make them energy efficient. **Question:** Can you think of any disadvantages to living in such a community? Do you think they would outweigh the advantages? Explain.

© Frisbee Architects

include generating electricity from solar cells, small wind turbines, and small hydropower systems; collecting rainwater; and using passive solar design, energy-efficient houses, green roofs, rooftop solar collectors to provide hot-water, waterless composting toilets, and organic farming plots. In Findhorn, Scotland, residents have reduced their ecological footprint per person by about 40% by using some of these methods.

The St. Croix Institute for Sustainable Community Development (Science Focus 22.1) has teamed with Habitat for Humanity to design and build an Eco-Village (Figure 22-26) on a vacant parcel of land (called a *grey field*) in River Falls, Wisconsin. It will include 18 highly energy-efficient (and possibly carbon-negative) homes with roof-mounted solar hot-water and photovoltaic collectors, designed to be affordable for low-income families. These home designs have received the LEED platinum certification (see p. 409). Construction on the village began in 2012 and is scheduled to be completed by 2014.

The River Falls Eco-Village will embody the goals of sustainable community development (Science Focus 22.1). Home owners will share community vegetable gardens, to be fed by a rainwater collection system, and a fleet of electric cars. Edible landscaping will include fruit trees and perennial shrubs available as a local, inexpensive source of nutrients to all residents. Construction wastes will be recycled at the highest possible rate (with a goal to exceed 90%). The village's community center will include a solar electric car battery charging system. Pedestrian paths will connect the village to nearby neighborhoods and parks to promote walking and biking. And driveways and paths will be designed and built to allow precipitation to flow into the soil beneath them.

The village will also serve social equity goals. In partial payment for their new homes, low-income residents will be able to contribute hours of labor in maintaining the community's homes, gardens, and landscaping. And the community center will be open to people in surrounding neighborhoods to link village residents with those of the larger community.

Another example of this trend is the Los Angeles Eco-Village, started in 1993 to show how people in the middle of a sprawling and unsustainable city such as Los Angeles, California, can establish small pockets of sustainability. It consists of two small apartment buildings with about 55 residents and a common courtyard. Working together, its members use solar panels to heat much of their hot water. They compost their organic wastes and use the compost in a large courtyard garden that provides vegetables and fruits, and serves as a community commons area for relaxation and social interaction. The eco-village is within walking distance of subway and bus stops, and members who live without a car pay a lower rent. The group has also established a bicycle repair shop.

By 2011, there were more than 380 eco-villages. While half or more of them are located in Europe and North America, they are scattered on every continent in the world except for Antarctica.

Big Ideas

- Urbanization is increasing steadily and the numbers and sizes of urban areas are growing rapidly, especially in less-developed countries.

- Most urban areas are unsustainable with their large and growing ecological footprints and high levels of poverty.

- Urban areas can be made more sustainable and livable just as some cities and villages already are.

Stephen Rees/Cutcaster

Urban areas, where more than half of the world's people live, have large environmental impacts. Most urban areas are unsustainable and are rapidly growing centers of poverty, especially in less-developed countries. Thus, the global environmental quality of the future depends on making urban areas more sustainable and livable as cities such as Portland, Oregon (**Core Case Study**), and Curitiba, Brazil, are doing.

In order to become more sustainable during the next few decades, urban areas will have to apply the three scientific **principles of sustainability** (see Figure 1-2, p. 6 or back cover). This will involve greatly improving energy efficiency and relying much more on solar, wind, and geothermal energy for electricity, heating, and cooling. It will also require that most solid wastes be reused, recycled, or composted. And cities will need to preserve their parks and wooded areas, set aside more parks, and plant more trees and vegetation. All of these steps will further the goal of becoming carbon-negative.

Also, in eco-cities of the future, people will move about primarily by walking, biking, and using low-polluting mass transit. Much of the food for eco-city residents will be produced through sustainable agriculture on urban and regional farms. Some of these goals are already being met in many cities around the world.

Making this transition toward sustainability is also in keeping with the social science **principles of sustainability** (see Figure 1-5, p. 9 or back cover). Full-cost pricing requires that the harmful environmental costs of urbanization be included in market prices of urban goods and services. For example, some cities are making it more expensive to drive a car within their borders. Such changes are good for the economy, in the long run, as well as for the environment, and are thus win-win solutions. And these solutions will be an application of the ethical principle that calls for leaving the world in a sustainable condition for future generations.

Chapter Review

Core Case Study

1. Explain how Portland, Oregon, has attempted to become a more sustainable city (**Core Case Study**).

Section 22-1

2. What is the key concept for this section? Distinguish between **urbanization** and **urban growth**. What percentage of the world's people live in urban areas? List two ways in which urban areas grow. What are some of the reasons why people move to cities?

3. List three major trends in global urban growth. Describe the three phases of urban growth in the United States. What is **urban sprawl**? List five factors that have promoted urban sprawl in the United States. List five undesirable effects of urban sprawl.

Section 22-2

4. What is the key concept for this section? What are the major advantages and disadvantages of urbanization? Explain why most urban areas are unsustainable systems and how these factors contribute to their unsustainability: lack of vegetation, water supply problems and flooding, pollution, health problems, the heat island effect, and light pollution. What is **noise pollution** and why is it an urban problem? Describe the major aspects of poverty in urban areas. Summarize Mexico City's major urban and environmental problems and what government officials are doing about them.

Section 22-3

5. What is the key concept for this section? Distinguish between compact and dispersed cities, and give an

example of each. What are the major advantages and disadvantages of using motor vehicles? List four ways to reduce dependence on motor vehicles. List the major advantages and disadvantages of relying more on **(a)** bicycles, **(b)** bus rapid-transit systems, **(c)** mass-transit rail systems within urban areas, and **(d)** rapid-rail systems between urban areas.

Section 22-4

6. What is the key concept for this section? What is **land-use planning**? What is **zoning** and what are its limitations? Define **smart growth** and list five tools that are used to implement it. What are three ways to preserve open spaces around a city?

Section 22-5

7. What is the key concept for this section? What is cluster development? What are the five key goals of new urbanism? What is sustainable community development and what are six indicators that scientists

study to assess a community's level of sustainability? Explain why becoming carbon-negative is an important goal for communities.

8. Describe the eco-city model and how it applies the **principles of sustainability**. Describe the potential for urban indoor farming. Give five examples of how Curitiba, Brazil, has attempted to become a more sustainable and livable eco-city.

9. What is the eco-village movement? How do eco-villages apply each of the three scientific **principles of sustainability**?

10. What are this chapter's three big ideas? Explain how Portland, Oregon, and other cities are applying the six **principles of sustainability** (see Figure 1-2, p. 6, Figure 1-5, p. 9, or back cover) to become more sustainable urban areas.

Note: Key terms are in bold type.

Critical Thinking

1. Portland, Oregon (**Core Case Study**), has made significant progress in becoming a more environmentally sustainable and desirable place to live. If you live in an urban area, what steps, if any, has your community taken toward becoming more environmentally sustainable? Given the story of Portland, what steps could your community take that it has not taken to follow Portland's example?

2. Do you think that urban sprawl is a problem and something that should be controlled? Develop an argument to support your answer. Compare your argument with those of your classmates.

3. Write a brief essay that includes at least three reasons why you **(a)** enjoy living in a large city, **(b)** would like to live in a large city, or **(c)** do not wish to live in a large city. Be specific in your reasoning. Compare your essay with those of your classmates.

4. One issue debated at a UN conference was the question of whether housing is a universal *right* (a position supported by most less-developed countries) or just a *need* (supported by the United States and several other more-developed countries). What is your position on this issue? Defend your choice.

5. If you own a car or hope to own one, what conditions, if any, would encourage you to rely less on your car and to travel to school or work by bicycle, on foot, by mass transit, or by carpool?

6. Do you believe the United States or the country in which you live should develop a comprehensive and integrated mass-transit system over the next 20 years, including an efficient rapid-rail network for travel within and between its major cities? How would you pay for such a system?

7. Consider the characteristics of an eco-city listed on p. 622. How close to this eco-city model is the city in which you live or the city nearest to where you live? Pick what you think are the five most important characteristics of an eco-city and, for each of these characteristics, describe a way in which your city could attain it.

8. Congratulations! You are in charge of the world. List the three most important components of your strategy for dealing with urban growth and sustainability in **(a)** more-developed countries and **(b)** less-developed countries.

Doing Environmental Science

The campus where you go to school is something like an urban community. Choose five eco-city characteristics (p. 622) and apply them to your campus. For each of the five characteristics:

1. Create a scale of 1 to 10 in order to rate the campus on how well it does in having that characteristic. (For example, how well does it do in giving students options for getting around, other than by using a car?

A rating of 1 could be *not at all,* while a rating of 10 could be *excellent.*)

2. Do some research and rate your campus for each characteristic.

3. Write an explanation of your research process and why you chose each rating.

4. Write a proposed plan for how the campus could improve its rating.

Global Environment Watch Exercise

Using the GREENR database, search global population by 2050. (You may need to utilize other related portals to obtain the depth of information necessary to complete your research.) Find three different projections for the global urban population in 2050 and explain how the projections were made. To do this, try to find out the assumptions behind each of the projections with regard to urbanization trends and other factors. Based on your reading, choose the projection that you believe to be the closest to reality and explain why you chose this projection.

Ecological Footprint Analysis

Under normal driving conditions an internal combustion engine in a car produces approximately 200 grams (0.2 kilograms, or 0.44 pounds) of CO_2 per kilometer driven. Assume that a city you are studying has 7 million cars and that each car is driven an average of 20,000 kilometers (12,400 miles) per year.

1. Calculate the carbon footprint per car, that is, the number of metric tons (and tons) of CO_2 produced by a typical car in 1 year. Calculate the total carbon footprint for all cars in the city, that is, the number of metric tons (and tons) of CO_2 produced by all the cars in 1 year. (*Note:* 1 metric ton = 1,000 kilograms = 1.1 tons.)

2. If 20% of the city's residents were to walk, bike, or take mass transit instead of driving their cars in the city, how many metric tons (and tons) of the annual emissions of CO_2 would be eliminated?

3. By what percentage, and by what average distance driven, would use of cars in the city have to be cut in order to reduce by half the annual carbon footprint calculated in question 1?

Questions 1–3 refer to the labeled locations on the map below.

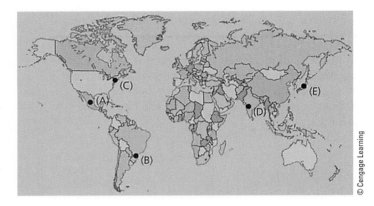

© Cengage Learning

1. This city, internationally acclaimed for its green design and eco-city concept for its 3.2 million residents, is found in which location?

2. The only hypercity in the world, boasting the largest population in the world with over 20 million residents, is found in which location?

3. This city of 18 million has frequent air inversions because of the mountains that surround it and its residents experience some of the worst air pollution in the world. In which location is it found?

4. Light-rail systems are effective means to reduce reliance on cars as mass transit systems. Why did the United States move away from light-rail systems toward an increasing reliance on the automobile in the mid-1900s?
   a. Light-rail systems became too expensive to run.
   b. The automobile was cheaper and out-competed the light-rail system.
   c. A holding company purchased the light-rail systems and dismantled them.
   d. Labor shortages made it too expensive to produce railroad cars.
   e. Gasoline became cheaper than the diesel used to run the light-rail engines.

5. Which of the following is an input and not an output of large cities?
   a. Educated people, because many college graduates leave the cities once they get their degrees
   b. Greenhouse gases, because they remain in the air above the cities
   c. Waste heat from industry, because it contributes to the higher temperature of cities compared with nearby rural areas
   d. Because cities require large amounts of food and water, they must be imported
   e. The stress of being exposed to noise pollution from a construction site

6. Which of the below is a major advantage of urbanization?
   a. Cities serve as centers of economic development and jobs.
   b. Less land is directly affected by the concentration of people in cities.
   c. Pollution is concentrated in cities.
   d. Cities are generally warmer and rainier than surrounding countrysides.
   e. Cities have larger ecological footprints.

7. Which of the following alternatives to automobiles has the advantages of less noise, produces no pollution, and takes few resources to produce?
   a. Rapid rail, because it reduces car usage
   b. Bicycles, despite offering no protection from bad weather and little protection in case of an accident
   c. Mass transit, because, especially in urban areas, the tracks are underground
   d. Buses, especially in urban areas, have the benefit of bus lanes that other vehicles cannot use
   e. Walking, because, although no noise is produced, it adds carbon dioxide to the air

Questions 8–10 are based on the terms below.

    a. Urban sprawl
    b. Natural capital
    c. Carbon footprint
    d. Eco-city
    e. Cluster development

8. Can easily be reduced by individuals on a daily basis.

9. High-density housing units are concentrated in a small portion of a parcel with the remainder being used for shared open space.

10. Protect undeveloped land, encourage farmers markets, and use efficient energy.

11. Which of the factors below is an advantage of the automobile?
    a. Fastest-growing source of greenhouse gases
    b. Helps to create urban sprawl
    c. Takes few resources to build
    d. Provides mobility and ease of transportation
    e. Causes few injuries and deaths per year

12. Which of the below is NOT an aspect of urban sprawl?
    a. Increased runoff and flooding
    b. Fragmentation of wetland habitats
    c. Loss of tax base in the central city
    d. Increased greenbelts
    e. De-emphasis of downtown business districts

13. Which of the following common sounds after an 8-hour exposure does NOT cause permanent ear damage?
    a. Vacuum cleaner
    b. Lawn mower
    c. Boom cars
    d. Ear phones at loud level
    e. Rock music

23

Economics, Environment, and Sustainability

When it is asked how much it will cost to protect the environment, one more question should be asked: How much will it cost our civilization if we do not?

Key Questions

23-1 How are economic systems related to the biosphere?

23-2 How can we estimate the values of natural capital, pollution control, and resource use?

23-3 How can we use economic tools to deal with environmental problems?

23-4 How can reducing poverty help us to deal with environmental problems?

23-5 How can we make the transition to more environmentally sustainable economies?

Rapeseed farm and wind farm in Plön, Schleswig-Holstein, Germany.

© Caro/Alamy

Germany, one of the world's most industrialized nations with intensive energy needs, is undergoing a renewable energy revolution. The country aims to phase out its dependence on fossil fuels and nuclear power and, by 2050, to get 80% of its electricity from renewable energy resources.

In 2012, Germany generated 20% of its electricity using wind farms on land (see chapter-opening photo) and at sea (Figure 23-1, left), as well as solar energy (Figure 23-1, right), and other renewable sources. Since 2000, this shift to renewable energy has created more than 380,000 jobs and a multibillion-dollar industry that includes renewable energy production and sales of renewable energy technology around the world.

This transition was spurred by government legislation that allows homeowners, businesses, and communities that produce electricity from solar cells, wind, and other renewable energy systems to sell the electricity they produce to Germany's four major power companies at a fixed rate that guarantees that their investments will at least break even. With this *feed-in*

tariff system, the government ensures that these renewable energy producers will not lose money. In fact, they often make a profit, and no taxpayer money is involved, because the utilities pass their extra costs on to consumers. And even with these added costs, as of 2012, German consumers' energy bills had increased by less than 5%.

The German government has also promoted the building of wind farms on land and offshore along the North Sea and Baltic Sea coasts (Figure 23-1, left). It projects an increase in the number of offshore wind turbines from 27 to 10,000 between 2012 and 2030. Offshore wind farms can be more productive than land-based systems because offshore winds tend to be more steady and stronger. There are also plans to lay more than 3,700 kilometers (2,300 miles) of high-voltage electrical cables throughout parts of the country and under the North Sea as part of a new state-of-the-art electrical grid. Such a grid would be far more efficient than conventional grids, and would help to make renewable energy sources more dependable (see Chapter 16, Case Study, p. 405).

In 2011, solar energy production in Germany, much of it through rooftop solar collectors, rose by 60%—enough to power more than 5 million homes—while the nation's total carbon emissions fell. Wind energy production rose by 30%, and even in a sagging economy, solar and wind energy production continued to grow in 2012.

Germany's shift to renewable energy has had some challenges that we discuss in this chapter. Even critics of the feed-in tariff agree that it has done its job in helping to establish a vibrant renewable energy industry in Germany. And Germany's example shows that economic improvements and improvements in environmental quality can go hand in hand. Some economists argue that shifting to cleaner energy resources, cleaner industrial production, and more sustainable agriculture would create more environmentally sustainable economies that would benefit more people. They say that the best way to reduce poverty, create new jobs, and provide profits is to invest in efforts to sustain the natural capital that supports all life and all economies.

Figure 23-1 This wind farm (left) is located off Germany's coast, and this rural village (right) in Germany's Rhineland-Palatinate is typical for German towns in having several rooftop solar panels.

23-1 How Are Economic Systems Related to the Biosphere?

CONCEPT 23-1
Ecological economists and most sustainability experts regard human economic systems as subsystems of the biosphere.

Economic Systems Vary but All Depend on Natural Capital

Economics is a social science that deals with the production, distribution, and consumption of goods and services to satisfy people's needs and wants. An **economic system** is a social institution through which goods and services are produced, distributed, and consumed to satisfy people's needs and wants, ideally in the most efficient way possible.

In a truly *free-market* economic system (Figure 23-2), all economic decisions are governed solely by the competitive interactions of *supply* (the amount of a good or service that is available), *demand* (the amount of a good or service that people want), and *price* (the market cost of a good or service) with little or no government control or interference in these interactions.

In Figure 23-2, *supply* is represented by the blue line (commonly called a *curve*) showing how much a producer of any good or service is willing to supply (measured on the horizontal *quantity* axis) for different prices (measured on the vertical *price* axis). *Demand* is represented by the red curve showing how much consumers will pay for different quantities of the good or service. The point at which the curves intersect is called the *market price equilibrium point,* where the supplier's price matches what buyers are willing to pay for some quantity and a sale is made.

Changes in supply and demand can shift one or both curves back and forth, and thus change the equilibrium price. For example, when supply is increased (shifting the blue curve to the right) and demand remains the same, the equilibrium price will go down. Similarly, when demand is increased (shifting the red curve to the right) and supply remains the same, the equilibrium price will rise.

A truly free-market economy rarely exists in today's capitalist market systems because factors other than supply and demand influence prices and sales. The primary goal of any business is to make as large a profit as possible for its owners or stockholders. To do so, most businesses try to take business away from their competitors and to exert as much control as possible over the prices of the goods and services they provide.

For example, many companies push for government support such as **subsidies**, or payments intended to help a business grow and thrive, along with tax breaks, trade barriers, and regulations that will give their products an advantage in the market over their competitors' products. When companies or industries receive large subsidies, it can create an uneven playing field.

Also, some companies withhold information from consumers about the costs and dangers that their products may pose to human health or to the environment, unless the government requires them to provide such information. Thus, buyers often do not get complete information about the harmful environmental impacts of the goods and services they buy.

Most economic systems use three types of capital, or resources, to produce goods and services (Figure 23-3). **Natural capital** (see Figure 1-3, p. 7) includes resources and services produced by the earth's natural processes, which support all economies and all life. **Human capital**, or **human resources**, includes people's physical and mental talents that provide labor, organizational and management skills, and innovation. **Manufactured capital**, or **manufactured resources**, are tools and materials such as machinery, materials, and factories made from natural resources with the help of human resources.

Figure 23-2 Supply and demand curves for a saleable product in a free market economic system. If all factors except supply, demand, and price are held fixed, *market equilibrium* occurs at the point where the supply and demand curves intersect. **Question:** How would a decrease in the available supply of oil shift the market equilibrium point on this diagram?

Governments Intervene to Help Correct Market Failures

Markets usually work well in guiding the efficient production and distribution of *private goods*. But experience shows that they cannot be relied upon to provide adequate levels of *public services*, such as national security and environmental protection. Economists generally refer to

Natural Capital   **Manufactured Capital**   **Human Capital**   **Goods and Services**

Figure 23-3 In an economic system, we use three types of resources to produce goods and services.

Center: ©Elena Elisseeva/Shutterstock.com. Right center: ©Michael Shake/Shutterstock.com. Right: Yuri Arcurs/Shutterstock.com.

such deficiencies as *market failures*. An important example of a market failure is the inability of markets to prevent the degradation of *open-access resources,* such as clean air, the open ocean, and the earth's overall life-support system. Such resources are not bought and sold in the marketplace, because they are owned by no one and available for use by everyone at little or no charge. Yet these natural resources and ecosystem services are an important part of the natural capital that supports all economies.

Governments intervene in market systems to provide various public services, such as national security and police and fire protection, and to help correct other market failures. An important application of this is the intervention by governments to protect open-access resources. For example, governments have passed laws to control air and water pollution.

One reason why markets often fail in the area of environmental protection is that they usually do not assign monetary value to the benefits provided by the earth's natural capital or to the harmful effects of various human activities on the environment and on human health. For example, the benefits of leaving an old-growth forest undisturbed (ecosystem services such as water purification and soil erosion reduction) usually are not weighed against the monetary value of cutting the timber in the forest. Thus, many old-growth forests have been clear-cut for their timber, while their non-timber natural capital value, which can be much higher than the value of their timber (Figure 10-A, p. 221), is lost. Governments can use economic tools such as regulations, taxes, subsidies, and tax breaks to correct this market failure and include at least some of the value of natural capital in the market prices of goods and services.

Economists Disagree over the Importance of Natural Capital and Whether Economic Growth Is Sustainable

Economic growth is an increase in the capacity of a nation, state, city, or company to provide goods and services to people. (See Figure 3, p. S28, in Supplement 6 for a comparison of economic growth in the world's countries as measured by per capita income.)

Today, a typical advanced industrialized country depends on a **high-throughput economy**, which attempts to boost economic growth by increasing the flow of matter and energy resources through the economic system to produce more goods and services (Figure 23-4). Such an economy produces valuable goods and services, but it also converts some high-quality matter and energy resources into waste, pollution, and low-quality heat, which tend to flow into planetary *sinks* (air, water, soil, and organisms), where some pollutants and wastes can accumulate to harmful levels. This process seriously degrades various forms of natural capital.

Economic development is any set of efforts focused on creating economies that can serve to improve human well-being—meeting basic human needs for items such as food, shelter, physical and economic security, and good health. The conventional form of economic development has relied mostly on economic growth, with the goal of establishing high-throughput economies.

However, many analysts are now calling for a much greater emphasis on **environmentally sustainable economic development**. Its goal is to use political and economic systems to encourage environmentally beneficial and more sustainable forms of economic improvement, and to discourage environmentally harmful and unsustainable forms of economic growth. This is in keeping with the win-win **principle of sustainability** (see Figure 1-5, p. 9), which calls for economic

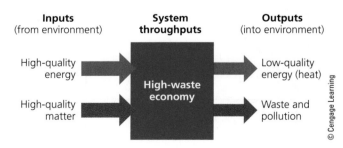

Inputs (from environment)   **System throughputs**   **Outputs** (into environment)

High-quality energy → **High-waste economy** → Low-quality energy (heat)

High-quality matter → **High-waste economy** → Waste and pollution

Animated Figure 23-4 The *high-throughput economies* of most of the world's more-developed countries rely on continually increasing the flow of energy and matter resources to promote economic growth. ***Question:*** What are three ways in which you regularly add to this throughput of matter and energy through your daily activities?

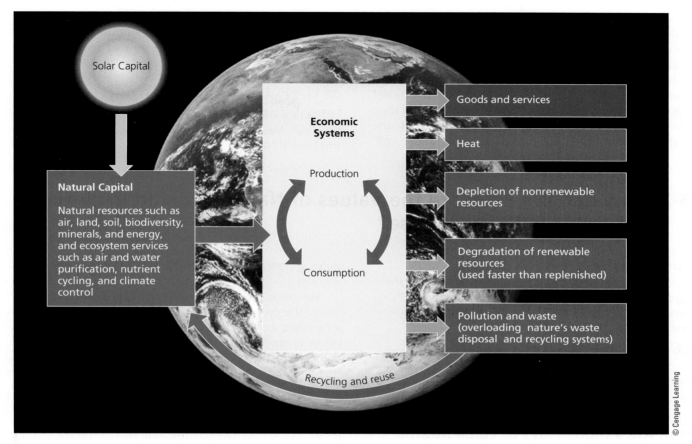

Animated Figure 23-5 *Ecological economists* see all human economies as subsystems of the biosphere that depend on natural resources and services provided by the sun and earth. **Question:** Do you agree or disagree with this model? Explain.

solutions that benefit the environment as well as human societies (see Science Focus 22.1, p. 623). We discuss several of these possible solutions later in this chapter.

Neoclassical economists, such as the late Alfred Marshall (1842–1924) and the late Milton Friedman (1912–2006), view the earth's natural capital as a subset, or part, of a human economic system. They assume that the potential for economic growth is essentially unlimited and is necessary for providing profits for businesses and jobs for workers. They also consider natural capital to be important but not indispensable because they believe we can find substitutes for essentially any resource.

Ecological economists such as Herman Daly (see his online Guest Essay on this topic) and Robert Costanza disagree with this model. They point out that there are no substitutes for many vital natural resources, such as air, water, fertile soil, biodiversity, and nature's free ecosystem services such as climate control, air and water purification, pest control, pollination, and nutrient cycling. In contrast to neoclassical economists, they view human economic systems as subsystems of the biosphere that depend heavily on the earth's irreplaceable natural resources and ecosystem services (Concept 23-1) (Figure 23-5).

Ecological economists also contend that conventional economic growth eventually will become unsustainable because it can deplete or degrade various irreplaceable forms of natural capital, on which all human economic systems depend. Also, they warn that it could exceed the capacity of the environment to handle the pollutants and wastes we produce. In other words, our ecological footprints are expanding beyond the carrying capacity of the earth (see Figure 1-13, p. 14, and Figure 4, p. S29, Supplement 6).

The world's population may level off at around 8 billion sometime this century, which would reduce the overall environmental impact of population (see Figure 1-14, p. 15). However, the environmental impact per person is likely to increase sharply as more people become middle-class consumers and consumption per person rises. According to a 2013 United Nations study, the global middle class is likely to expand to 3.2 billion people by 2020 and to 4.9 billion people by 2030, with 66% of these consumers living in Asia, especially China and India, and in Brazil.

In general, the models of ecological economists are built on the assumption that supplies of many resources are limited and that there are no substitutes for most types of natural capital. Therefore, they argue, we should encourage environmentally beneficial and sustainable forms of economic development, and discourage environmentally harmful and unsustainable forms of economic growth that will deplete or degrade our natural capital.

Taking the middle ground in this debate between neoclassical economists and ecological economists are *environmental economists*. They generally agree with the model proposed by ecological economists (Figure 23-5), and they argue that some forms of economic growth are not sustainable and should be discouraged. However, unlike ecological economists, they would accomplish this by fine-tuning existing economic systems and tools, rather than by replacing some of them with new or redesigned systems or tools.

 CONSIDER THIS. . .

THINKING ABOUT Economic Growth

Do you think that the economy of the country where you live is sustainable or unsustainable? Explain.

23-2 How Can We Estimate the Values of Natural Capital, Pollution Control, and Resource Use?

CONCEPT 23-2A

Economists have developed several ways to estimate the present and future values of a resource or ecosystem service, and optimum levels of pollution control and resource use.

CONCEPT 23-2B

Comparing the likely costs and benefits of an environmental action is useful, but it involves many uncertainties.

There Are Various Ways to Value Natural Capital

Environmental and ecological economists have developed various tools for estimating the values of the earth's natural capital. One approach involves estimating the monetary worth of the earth's natural ecosystem services (Figure 23-6). For example, using this approach, ecological economist Robert Costanza and his colleagues (see Science Focus 10.1, p. 221) have estimated the value of pollination to be $1.2 trillion per year. They estimated that the ecosystem services provided by the earth's forests are worth at least $4.7 trillion a year. This is hundreds of times greater than the estimated value of their economic services.

The basic problem is that the estimated economic values of the ecosystem services provided by forests, oceans, and other ecosystems are not included in the market prices of timber, fish, and other goods that we get from them. Ecological economists point out that until this underpricing is corrected, we will continue our unsustainable use of forests, oceans, the atmosphere, and many of nature's other irreplaceable forms of natural capital.

However, according to neoclassical economists, a product or service has no economic value until it is sold in the marketplace, and thus because they are not sold in the marketplace, ecosystem services have no economic value. Ecological economists view this as a misleading circular argument designed to exclude serious evaluation of ecosystem services in the marketplace.

Ecological and environmental economists have developed ways to estimate *nonuse values* of natural resources and ecosystem services that are not represented in market transactions (Concept 23-2A). One such value is an *existence value*—a monetary value placed on a resource such as an old-growth forest or endangered species just because it exists, even though we may never see it or use it. Another is *aesthetic value*—a monetary value placed on a forest, species, or a part of nature because of its beauty. A third type, called a *bequest* or *option value,* is based on the willingness of people to pay to protect some forms of natural capital for use by future generations.

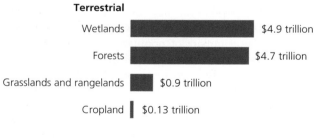

Terrestrial

Wetlands	$4.9 trillion
Forests	$4.7 trillion
Grasslands and rangelands	$0.9 trillion
Cropland	$0.13 trillion

Aquatic

Coastal waters	$12.6 trillion
Open ocean	$8.4 trillion
Lakes and rivers	$1.7 trillion

Figure 23-6 Natural capital: Estimated monetary value of ecosystem services provided each year by the world's major terrestrial and aquatic ecosystems.

(Compiled by the authors using data from Robert Costanza et al., "The Value of the World's Ecosystem Services and Natural Capital," *Nature,* 1997.)

Estimating the Future Value of a Resource Is Controversial

One tool used by economists, businesses, and investors to determine the value of a resource is the **discount rate**, which is an estimate of a resource's future economic value compared to its present value (Concept 23-2A). It is based on the idea that today's value of a resource may be higher than its value in the future. Thus, its future value should be discounted. The size of the discount rate (usually given as a percentage) is a key factor affecting how a resource such as a forest or fishery is used or managed.

At a zero discount rate, for example, the timber from a stand of redwood trees (Figure 23-7) worth $1 million today will still be worth $1 million 50 years from now. However, the U.S. Office of Management and Budget, the World Bank, and most businesses typically use a 10% annual discount rate to estimate the future value of a resource. At this rate, within 45 years, the timber in the stand of redwood trees mentioned above will be worth less than $10,000. Using this discount rate, it makes sense from an economic standpoint for the owner of this resource to cut these trees down as quickly as possible and invest the money in something else.

However, this economic analysis does not take into account the immense economic value of the ecosystem services provided by forests. Such services include the absorption of precipitation and gradual release of water and other nutrients, natural flood control, water and air purification, prevention of soil erosion, removal and storage of atmospheric carbon, and protection of biodiversity within a variety of forest habitats (see Figure 10-4, left, p. 220).

This one-sided, incomplete economic evaluation loads the dice against sustaining these important ecosystem services. If these economic values were included, it would make more sense now and in the future to preserve large areas of redwoods for the ecosystem services they provide and to find substitutes for redwood products. However, while these ecosystem services are vital for the earth as a whole and for future generations, they do not provide the current owner of the redwoods with any monetary return.

Setting discount rates can be difficult and controversial. Proponents cite several reasons for using high (5–10%) discount rates. One argument is that inflation can reduce the value of future earnings on a resource. Another is that innovation or changes in consumer preferences can make a product or resource obsolete. For example, the plastic composites made to look like redwood may reduce the future use and market value of this timber. In addition, owners of resources such as forests argue that without a high discount rate, they can make more money by investing their capital in some other venture.

Critics point out that high discount rates encourage rapid exploitation of resources for immediate payoffs, thus making long-term sustainable use of most renewable natural resources virtually impossible. These critics believe that a 0% or even a negative discount rate should be used

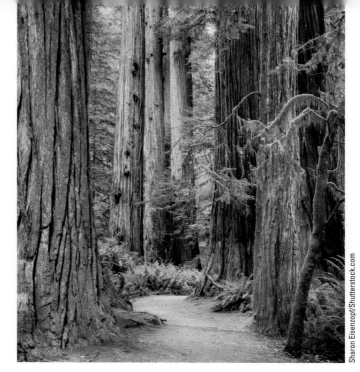

Figure 23-7 Economists have tried several methods for estimating the economic value of ecosystem services, recreation opportunities, and beauty in ecosystems such as this patch of redwood forest. *Question:* What discount rate, if any, would you assign to this stand of trees?

to protect unique, scarce, and irreplaceable resources such as old-growth forests. They also point out that moderate discount rates of 1–3% would make it profitable to use nonrenewable and renewable resources more slowly and in more sustainable ways.

 CONSIDER THIS. . .

THINKING ABOUT Discount Rates

If you owned a forested area, would you want the discount rate for resources such as trees from the forest to be high, moderate, or zero? Explain.

We Can Estimate Optimum Levels of Pollution Control and Resource Use

An important concept in environmental economics is that of *optimum levels* for pollution control and resource use (Concept 23-2A). In the early days of a new coal mining operation, for example, the cost of extracting coal is typically low enough to make it easy for developers to recover their investments by selling their product. However, the cost of removal goes up with each additional unit of coal taken—what economists call its *marginal cost*. After most of the more readily accessible coal has been removed from a mine, taking what is left can become too costly, unless some factor such as scarcity raises the value of the coal remaining in the mine.

Figure 23-8 shows this in terms of supply, demand, and equilibrium. The point at which removing more coal is not worth the marginal cost is where the demand curve

High

Cost

Marginal benefit of resource use

Marginal cost of resource production

Optimum level of resource use

Low

0   25   50   75   100

Coal removed (%)

© Cengage Learning

Figure 23-8 *Optimum resource use:* The cost of extracting coal (blue line) from a particular mine rises with each additional unit removed. Mining a certain amount of coal is profitable, but at some point, the marginal cost of further removal exceeds the monetary benefits (red line). **Question:** How would the location of the optimum level of resource use shift if the price of coal doubled?

Photo: © nito/Shutterstock.com

crosses the supply curve, theoretically the *optimum level of resource use.*

You might think that the best solution for pollution is total cleanup. In fact, there are optimum levels for various kinds of pollution. This is because the cost of pollution control goes up for each additional unit of a pollutant removed from the environment. The main reason for this is that, as concentrations of a pollutant from the air, water, or soil get lower, it takes larger amounts of energy to remove the pollutant. At some point, the cost of removing more pollutants (the marginal cost) is greater than the harmful costs of the pollution to society. That point is the equilibrium point, or the *optimum level* for pollution cleanup.

Cost–Benefit Analysis Is a Useful but Crude Tool

Another widely used tool for making economic decisions about how to control pollution and manage resources is **cost–benefit analysis**. In this process, analysts compare estimated costs and benefits of actions such as implementing a pollution control regulation, building a dam on a river, and preserving an area of forest. Economists also use cost–benefit analysis in trying to estimate the optimum level of pollution cleanup or resource use (Figure 23-8).

Making a cost–benefit analysis involves determining who or what might be affected by a particular regulation or project, projecting potential outcomes, evaluating alternative options, and establishing who benefits and who is harmed. Then an attempt is made to assign monetary values (costs and benefits) to each of the factors and components involved.

Direct costs involving land, labor, materials, and pollution-control technologies are often fairly easy to estimate. However, estimates of indirect costs, such as a project's effects on air and water, are not considered in the marketplace and are therefore controversial and difficult to make. We can put estimated price tags on human life, good health, clean air and water, and natural capital such as an endangered species, a forest, or a wetland. However, such monetary value estimates vary widely depending on the assumptions, value judgments, and the discount factors used by the estimators.

Because of these drawbacks, a cost–benefit analysis can lead to a wide range of benefits and costs with a lot of room for error, and this is a source of controversy. For example, one cost–benefit analysis sponsored by a U.S. industry estimated that compliance with a regulation written to protect American workers from vinyl chloride would cost $65 billion to $90 billion. In the end, complying with the regulation cost the industry less than $1 billion. A study by the Economic Policy Institute of Washington, D.C. found that the estimated costs projected by industries for complying with proposed U.S. environmental regulations are almost always more (and often much more) than the actual costs of complying with the regulations. In such cases, inflated cost–benefit analyses are used as an economic and political ploy by industries to prevent, weaken, or delay compliance with pollution standards.

If conducted fairly and accurately, cost–benefit analysis can be a helpful tool for making economic decisions, but it always includes uncertainties (Concept 23-2B). Environmental economists advocate using the following guidelines in order to minimize possible abuses and errors in any cost–benefit analysis involving some part of the environment:

- Clearly state all assumptions used.
- Include estimates of the ecosystem services provided by the ecosystems involved.
- Estimate short- and long-term benefits and costs for all affected population groups.
- Compare the costs and benefits of alternative courses of action.

23-3 How Can We Use Economic Tools to Deal with Environmental Problems?

CONCEPT 23-3
We can use resources more sustainably by including the harmful environmental and health costs of producing goods and services in their market prices *(full-cost pricing)*; by subsidizing environmentally beneficial goods and services; and by taxing pollution and waste instead of wages and profits.

We Can Apply the Principle of Full-Cost Pricing

The *market price*, or *direct price*, that we pay for a product or service usually does not include all of the *indirect*, or *external, costs* of harm to the environment and human health associated with its production and use. For this reason, such costs are often called *hidden costs*.

For example, if we buy a car, the price we pay includes the *direct*, or *internal, costs* of raw materials, labor, shipping, and a markup for dealer profit. In using the car, we pay additional direct costs for gasoline, maintenance, repairs, and insurance. However, in order to extract and process raw materials to make a car, manufacturers use energy and mineral resources, produce solid and hazardous wastes, disturb land, pollute the air and water, and release greenhouse gases into the atmosphere. These are the hidden external costs that can have short- or long-term harmful effects on us, on other people, on future generations, and on the earth's life-support systems.

Because these harmful external costs are not included in the market price of a car, most people do not connect them with car ownership. Still, the car buyer and other people in a society pay these hidden costs sooner or later, in the forms of poorer health, higher expenses for health care and insurance, higher taxes for pollution control, traffic congestion, and the damage done to ecosystems during the building of highways and parking lots. Similarly, when a farmer practices high-input industrialized farming, or a power plant pollutes the air or the water in a river, those costs are also not added to your food prices or your electric bill.

Many economists and environmental experts believe that these harmful external costs should be included in the market prices of goods—a practice called **full-cost pricing** (Concept 23-3A and one of the **principles of sustainability**; see Figure 1-5, p. 9 or back cover). They cite this failure to include the harmful environmental costs in the market prices of goods and services as one of the major causes of the environmental problems we face (see Figure 1-15, p. 16). They also argue that full-cost pricing is one of the characteristics of a truly free-market economy.

Full-cost pricing includes the estimated costs related to the harmful environmental and health effects of producing and using the goods and services that we buy. For example, if the harmful environmental and health costs of mining and burning coal to produce electricity (see Figure 15-16, p. 387, and Figure 23-9) were included in the market prices of coal and coal-fired electricity, coal would be too expensive to use and would be replaced by less environmentally harmful resources such as solar and wind power (Figure 23-1, **Core Case Study**).

Figure 23-9 Most of the harmful environmental and health effects of strip mining coal and burning it to produce electricity are not included in the cost of electricity.

According to environmental and ecological economists, full-cost pricing would reduce resource waste, pollution, and environmental degradation and improve human health by encouraging producers to invent more resource-efficient and less-polluting methods of production. It would also enable consumers to make more informed decisions about the goods and services they buy. Jobs and profits would be lost in environmentally harmful businesses as consumers would more often choose green products, but jobs and profits would be created in environmentally beneficial businesses.

Such shifts in job markets and profits and changes in the structure of the marketplace are normal developments in creative market-based capitalism. This is why we replaced whale oil lamps with lightbulbs in the 1800s and are now gradually replacing inefficient incandescent lightbulbs with newer, more efficient bulbs. If a shift to full-cost pricing were phased in over a decade or two, some environmentally harmful businesses would have time to transform themselves into profitable, environmentally beneficial businesses. Consumers would also have time to adjust their buying habits to favor more environmentally beneficial products and services.

Full-cost pricing seems to make a lot of sense, so why is it not used more widely? *First,* most producers of environmentally harmful products and services would have to charge more for them, and some would go out of business. Naturally, these producers oppose such pricing. *Second,* it is difficult to estimate many environmental and health costs, and to know how they might change in the future. However, ecological and environmental economists argue that making the best possible estimates is far better than continuing with the current misleading and eventually unsustainable system, which excludes such costs. *Third,* many environmentally harmful businesses have used their political and economic power to obtain environmentally harmful government subsidies that help them to avoid true free-market competition and to retain their economic power.

Subsidies Can Be Environmentally Harmful or Beneficial

Subsidies that lead to environmental damage are called **perverse subsidies**—a term coined by environmental scientist Norman Myers. (See Myer's online Guest Essay on this topic.) For example, some fishing fleets have received government subsidies to help them boost their catches, thereby bolstering seafood supplies. However, some fisheries, including Newfoundland's Atlantic cod fishery, were then overfished and they collapsed because the subsidies enabled too many fishing fleets to take too many fish (see Figure 11-7, p. 255).

Myers, who has done an extensive study on perverse subsidies, points to six sectors of the economy—agriculture, fossil fuels, highway transportation, water supplies, forestry, and fisheries—as the areas where perverse subsidies and tax breaks have been extensively awarded and have caused severe environmental damage. Other examples include depletion subsidies and tax breaks for extracting minerals and fossil fuels, cutting timber on public lands, and irrigating with low-cost water. These subsidies and tax breaks distort the economic playing field and create a huge economic incentive for unsustainable resource waste, depletion, and environmental degradation.

Myers estimates that these subsidies to these sectors total at least $2 trillion per year, globally, which amounts to $3.8 million a minute. This amount is larger than all but a few of the national economies in the world and twice as large as all military spending. It dwarfs the estimated $20 billion per year that is spent by governments worldwide on protecting natural areas. Myers also estimates that perverse subsidies cost the average American taxpayer $2,000 per year.

Many scientists call for phasing out perverse subsidies and tax breaks. However, the economically and politically powerful interests receiving them spend a lot of time and money *lobbying,* or trying to influence governments to continue and even to increase their subsidies. For example, the fossil fuel and nuclear power industries in the United States are mature and highly profitable industries that get billions of dollars in government subsidies and tax breaks every year. Such industries also lobby against subsidies and tax breaks for their more environmentally beneficial competitors such as solar and wind energy.

Some countries have reduced perverse subsidies. Japan, France, and Belgium have phased out all coal subsidies, and Germany plans to do so by 2018. China has cut coal subsidies by about 73% and has imposed a tax on high-sulfur coals. **GOOD NEWS**

Subsidies can also be used for environmentally beneficial purposes. They could be awarded to companies that are involved in pollution prevention, sustainable forestry and agriculture, conservation of water supplies, energy efficiency improvements, renewable energy use, and measures to slow projected climate change (Concept 23-3). Germany is leading the way (**Core Case Study**) by phasing out nuclear power and using its feed-in tariff system to spur a shift to renewable solar and wind energy. Making such subsidy shifts on a broad basis, globally, would encourage businesses to make the transition from environmentally harmful to more environmentally beneficial practices.

Environmental Economic Indicators Could Help Us Reduce Our Environmental Impact

Economic growth is usually measured by the percentage change in a country's **gross domestic product (GDP)**: the annual market value of all goods and services produced by all firms and organizations, foreign and domestic, operating within a country. Changes in a country's economic growth per person are measured by **per capita GDP**: the GDP divided by a country's total population at midyear.

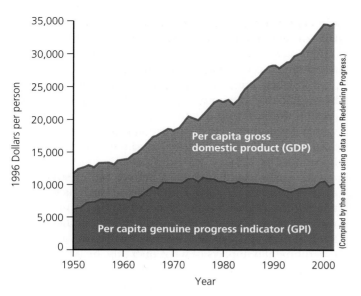

Figure 23-10 *Monitoring environmental progress:* The per capita gross domestic product (GDP) compared with the per capita genuine progress indicator (GPI) in the United States between 1950 and 2004. ***Questions:*** Would you favor making widespread use of this or similar green economic indicators? Why do you think this hasn't been done?

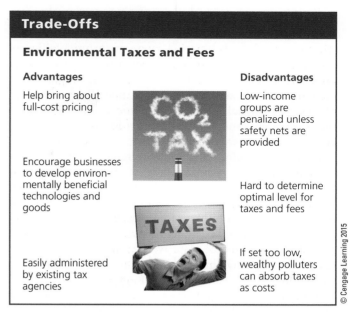

Figure 23-11 Using green taxes to help reduce pollution and resource waste has advantages and disadvantages. ***Questions:*** Which single advantage and which single disadvantage do you think are the most important? Why?

Top: chuong/Shutterstock.com. Bottom: EduardSV/Shutterstock.com.

GDP and per capita GDP indicators provide a standardized and useful method for measuring and comparing the economic outputs of nations. The GDP is deliberately designed to measure such outputs without distinguishing between goods and services that are environmentally or socially beneficial and those that are harmful.

Environmental and ecological economists and environmental scientists call for the development and widespread use of new indicators—called *environmental indicators*—to help monitor environmental quality and human well-being. One such indicator is the **genuine progress indicator (GPI)**—the GDP plus the estimated value of beneficial transactions that meet basic needs, but in which no money changes hands, minus the estimated harmful environmental, health, and social costs of all transactions. The *per capita GPI* would be the GPI for a country divided by that country's population at midyear. Redefining Progress, a nonprofit organization that develops economic and policy tools to help promote environmental sustainability, introduced the GPI in 1995. (This group also developed the concept of ecological footprints; see Figure 1-11, p. 13.)

Examples of beneficial transactions included in the GPI include unpaid volunteer work, health care and child care provided by family members, and housework. Harmful costs that are subtracted to arrive at the GPI include costs of pollution, resource depletion and degradation, and crime. The GPI is calculated as follows:

$$
\begin{array}{cccc}
\text{Genuine} & & \text{Benefits not} & \text{Harmful} \\
\text{progress} & = \text{GDP} + & \text{included in market} - & \text{environmental} \\
\text{indicator} & & \text{transactions} & \text{and social costs}
\end{array}
$$

Figure 23-10 compares the per capita GDP and GPI for the United States between 1950 and 2004 (the latest data available). While the per capita GDP rose sharply over this period, the per capita GPI stayed flat, or in some cases even declined slightly. This shows that even if a nation's economy is growing, its people are not necessarily better off.

The GPI and other environmental indicators under development are far from perfect, which is true of the GDP as well. However, without such environmental indicators, we will not know much about what is happening to people, to the environment, or to the planet's natural capital, nor will we have a way to evaluate which policies are likely to work best in improving environmental quality and life satisfaction.

Taxing Pollution and Wastes Instead of Wages and Profits

Another way to discourage pollution and resource waste is to tax them. This would involve using *green taxes,* or *ecotaxes,* to help include many of the harmful environmental and health costs of production and consumption in market prices (**Concept 23-3**). Taxes could be levied on a per-unit basis on the amount of pollution and hazardous waste produced by a farm, business, or industry, and on the use of fossil fuels, nitrogen fertilizer, timber, minerals, water, and other resources they use. Figure 23-11 lists some of the advantages and disadvantages of using green taxes.

To many analysts, the tax systems in most countries are backward. They *discourage* what we want more of—

jobs, income, and profit-driven innovation—and *encourage* what we want less of—pollution, resource waste, and environmental degradation. A more environmentally sustainable economic and political system would lower taxes on labor, income, and wealth, and raise taxes on environmentally harmful activities that produce pollution, wastes, and environmental degradation. Some 2,500 economists, including eight Nobel Prize winners in economics, have endorsed this *tax shifting* concept.

Proponents point out three requirements for successful implementation of green taxes. *First,* they would have to be phased in over 10–20 years to allow businesses to plan for the future. *Second,* income, payroll, or other taxes would have to be reduced by the same amount as the green tax so that there would be no net increase in taxes. *Third,* the poor and lower-middle class would need a safety net to reduce the regressive nature of any new taxes on essentials such as fuel, water, electricity, and food.

In Europe and the United States, polls indicate that once such tax shifting is explained to voters, 70% of them support the idea. Germany's green tax on fossil fuels, introduced in 1999, has reduced pollution and greenhouse gas emissions, helped to create up to 250,000 new jobs, lowered taxes on wages, and greatly increased use of renewable energy resources (**Core Case Study**). Sweden, Denmark, Spain, and the Netherlands have raised taxes on several environmentally harmful activities while cutting taxes on income, wages, or both.

To help reduce carbon dioxide emissions, since 1997, Costa Rica has imposed a 3.5% tax on the market values of any fossil fuels that are burned in the country. The tax revenues go into a national forest fund set up for paying indigenous communities to help protect the forests around them, thereby helping to reverse deforestation. The fund is also intended to help them work their way out of poverty. Costa Rica has also taxed water use to reduce water waste and pollution, and the tax revenues are used to pay villagers living upstream to reduce their inputs of water pollutants.

The U.S. Congress has not enacted green taxes, mostly because politically powerful industries, including the automobile, fossil fuel, mining, and chemical industries, claim that such taxes will reduce their competitiveness and harm the economy by forcing them to raise the prices of their goods and services. In addition, most voters have been conditioned to oppose any new taxes and have not been educated about the economic and environmental benefits of a tax shifting system designed to prevent a net increase in taxes.

We Could Label Environmentally Beneficial Goods and Services

Product eco-labeling and *certification* can encourage companies to develop green products and services and can help consumers to select more environmentally beneficial products and services. Eco-labeling programs have been developed in Europe, Japan, Canada, and the United States. The U.S. *Green Seal* labeling program has certified more than 335 products and services as environmentally friendly based on life-cycle analysis (see Figure 14-9, p. 357).

Eco-labels are also being used to identify fish caught by sustainable methods (certified by the Marine Stewardship Council) and to certify timber produced and harvested by sustainable methods (evaluated by organizations such as the Forest Stewardship Council; see Science Focus 10.2, p. 228).

Eco-labels could include a simple rating scale such as 0–10 that could be applied to factors such as environmental damage, climate impact, carbon footprint, air and water pollution, and energy, water, and pesticide use. Such eco-labeling would inform consumers about the environmental impacts of what they buy, helping them to vote with their wallets.

Providing such easily understandable ratings on the sustainability of goods and services could help to expose and reduce **greenwashing**: a deceptive practice that some businesses use to spin environmentally harmful products and services as green, clean, or environmentally beneficial. For example, in 2008, the U.S. coal industry spent about $45 million on a successful public relations campaign to imbed the words "clean coal" in the minds of Americans, even though certain harmful aspects of mining and using coal will always make it by far the dirtiest fossil fuel (see p. 385 and Figure 23-9). EnviroMedia Marketing and the University of Oregon have developed a "greenwashing index" with ratings from "authentic" to "bogus" to help consumers evaluate greenwashing advertisements.

Environmental Regulations Can Discourage or Encourage Innovation

Environmental regulation is a form of government intervention in the marketplace that is widely used to help control or prevent pollution and to reduce resource waste and environmental degradation. It involves enacting and enforcing laws that set pollution standards, regulate harmful activities such as the release of toxic chemicals into the environment, and protect certain irreplaceable or slowly replenished resources such as public forests and wilderness areas (Figure 23-12) from unsustainable use.

So far, most environmental regulation in the United States and many other more-developed countries has involved passing laws that are enforced through a *command-and-control* approach. Critics say that this strategy can unnecessarily increase costs and discourage innovation, because many of these regulations concentrate on enforcing cleanup instead of prevention. Some regulations also set compliance deadlines that are too short to allow companies to find innovative solutions to pollution and excessive resource use. They may also require the use of specific technologies where less costly but equally effective alternatives might be available or are in the process of being developed.

Figure 23-12 Environmental regulations have helped us to preserve irreplaceable resources such as this mountainous National Wilderness Area near Aspen, Colorado.

Charles Kogod/National Geographic Creative

A different approach favored by many economists, as well as environmental and business leaders, is to use *incentive-based environmental regulations*. Rather than to require all companies to follow the same fixed procedures or use the same technologies, this approach uses the economic forces of the marketplace to encourage businesses to be innovative in reducing pollution and resource waste. The experiences of several European nations show that *innovation-friendly environmental regulation* sets goals, frees industries to meet them in any way that works, and allows enough time for innovation. This can motivate companies to develop green products and industrial processes that create jobs. It can also increase company profits and make the companies more competitive in national and international markets.

For years, many companies generally fought against environmental regulation. But in recent years, a growing number of companies have realized the economic and competitive advantages of making environmental improvements. They now recognize that the value of their shareholders' investments depends in part on having a good environmental record. They also see that improvements to energy efficiency and resource conservation can save money and boost profits. And they realize that whenever they avoid producing a waste or pollutant, they can reduce government-mandated paperwork and avoid lawsuits by parties that might have suffered damage, had the pollutant been produced.

Using the Marketplace to Reduce Pollution and Resource Waste

In one incentive-based regulation system, the government decides on acceptable levels of total pollution or resource use, sets limits, or *caps,* to maintain these levels, and gives or sells companies a certain number of *tradable pollution* or *resource-use permits* governed by the caps.

With this *cap-and-trade* approach, a permit holder that does not use its entire allocation can save credits for future expansion, use them in other parts of its operation, or sell them to other companies. The United States has used this *cap-and-trade* approach to reduce the emissions of sulfur dioxide (see Chapter 18, p. 495) and other air pollutants. Tradable rights can also be established among countries to help preserve biodiversity and to reduce emissions of greenhouse gases and other regional and global pollutants.

Figure 23-13 lists the advantages and disadvantages of using tradable pollution and resource-use permits. The effectiveness of such programs depends on how high or low the initial cap is set, and on the rate at which the cap is reduced to encourage further innovation.

Reducing Pollution and Resource Waste by Selling Services Instead of Goods

One approach to working toward more sustainable economies is to shift from the current *material-flow economy* (Figure 23-4) to a *service-flow economy*. Instead of buying many goods outright, customers lease or rent the *services* that such goods provide. In a service-flow economy, a manufacturer makes more money if its product uses the minimum amount of materials, lasts as long as possible, is energy efficient, produces as little pollution (including greenhouse gases) and wastes as possible in its production and use, and is easy to maintain, repair, remanufacture, reuse, or recycle.

Such an economic shift is under way in some businesses. Since 1992, Xerox has been leasing most of its copy machines as part of its mission to provide *document services* instead of selling photocopiers. When a customer's service contract expires, Xerox takes the machine back for reuse or remanufacture. It has a goal of sending no material to landfills or incinerators. To save money, Xerox designs machines to have few parts, be energy efficient, and emit as little noise, heat, ozone, and chemical waste as possible. Canon, a Japanese manufacturer of optical imaging products, and Fiat, the Italian carmaker, are taking similar measures.

In Europe, the furnace maker Carrier has begun shifting from selling heating and air conditioning equipment to providing indoor heating and cooling services. It makes higher profits by leasing and installing energy-efficient heating and air conditioning equipment that lasts as long as possible and is easily rebuilt or recycled. Carrier also makes money through helping clients to save energy by adding insulation, eliminating heat losses, and boosting energy efficiency in offices and homes.

CONSIDER THIS. . .

THINKING ABOUT Selling Services

Do you favor a shift in some cases from selling goods to selling the services that the goods provide? List three services that you would be willing to lease.

Trade-Offs

Tradable Environmental Permits

Advantages	Disadvantages
Flexible and easy to administer	Wealthy polluters and resource users can buy their way out
Encourage pollution prevention and waste reduction	Caps can be too high and not regularly reduced to promote progress
Permit prices to be determined by market transactions	Self-monitoring of emissions can allow cheating

Figure 23-13 Using tradable pollution and resource-use permits to reduce pollution and resource waste has advantages and disadvantages. **Question:** Which two advantages and which two disadvantages do you think are the most important? Why?

Top: M. Shcherbyna/Shutterstock.com. Bottom: Lasse Kristensen/Shutterstock.com.

23-4 How Can Reducing Poverty Help Us to Deal with Environmental Problems?

CONCEPT 23-4

Reducing poverty can help us to reduce population growth, resource use, and environmental degradation.

We Can Reduce Poverty

Poverty is defined as the condition under which people cannot meet their basic economic needs. According to the World Bank, poverty is the way of life for nearly half of the world's people who have to live on an income equivalent to less than $2.25 per day. One-fifth of the world's people live in extreme poverty, struggling to survive on an income equivalent to less than $1.25 a day (Figure 23-14).

Poverty has numerous harmful health and environmental effects and has been identified as one of the four major causes of the environmental problems we face (see Figure 1-15, p. 16). Reducing poverty can benefit individuals, economies, and the environment, while empowering many poor women and helping to slow population growth (Concept 23-4).

Some nations have climbed out of abject poverty within a short period of time. For example, South Korea and Singapore made the journey from being poor, less-developed nations to becoming world-class industrial powers in only two decades. They did this by focusing on education, hard work, and discipline, all of which attracted investment capital. China is on a similar path; between 1990 and 2012, the estimated number of people living in extreme poverty in China dropped from 685 million to 213 million, according to the World Bank.

Analysts point out that in order to reduce poverty, the governments of most of the less-developed countries will have to make policy changes. As part of this, these analysts recommend a strong emphasis on education to develop a trained workforce that can participate in an increasingly globalized economy.

To assist in this process, governments, businesses, international lending agencies, and wealthy individuals could also undertake the following measures:

- Mount a massive global effort to combat malnutrition and the infectious diseases that kill millions of people prematurely (see Figure 17-20, p. 465, and Individuals Matter 23.1).
- Provide universal primary school education for the world's nearly 800 million illiterate adults (a number that is 2.5 times the size of the U.S. population). Illiteracy can foster terrorism and strife within countries by contributing to the creation of large numbers of unemployed individuals who have little hope of improving their lives or those of their children.
- Provide assistance to help less-developed countries in stabilizing their population growth as soon as possible,

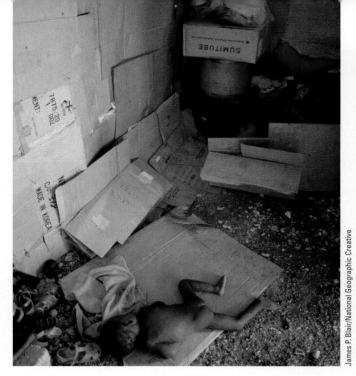

Figure 23-14 This 3-year-old girl was sleeping in her family's shack in a slum in Port-Au-Prince, Haiti.

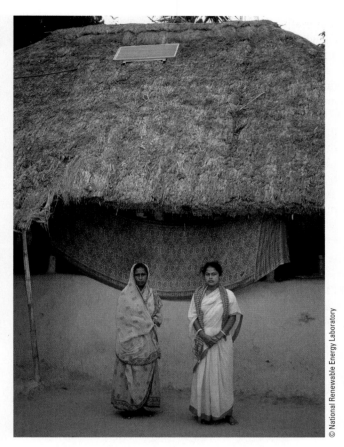

Figure 23-15 A microloan helped these women in a rural village in India to buy a small solar-cell panel (installed on the roof behind them) that provides electricity to help them make a living, thus applying the solar energy **principle of sustainability**.

mostly by investing in family planning, reducing poverty, and elevating the social and economic status of women.

- Focus on sharply reducing the total and per capita ecological footprints of their own countries, as well as those of rapidly growing less-developed countries such as China and India (see Figure 1-11, p. 13).

- Make large investments in small-scale infrastructure such as solar-cell power facilities in rural villages, as well as sustainable agriculture projects that would enable less-developed nations to work towards more energy-efficient and sustainable economies.

- Encourage lending agencies to make small loans to poor people who want to increase their income (see the following Case Study).

In 2012, the United Nations reported that, for the first time since poverty trends were first monitored, the number of people living in extreme poverty fell in all regions of the less-developed world. This meant that the world was on track to cut in half its numbers of people living in poverty by 2015.

GOOD NEWS

CASE STUDY
Microlending

Most of the world's poor people want to work and earn enough to climb out of poverty. But few of them have credit records or assets that they could use for collateral to secure loans. With loans, they could buy what they would need to start farming or to start small businesses.

For almost three decades, an innovation called *microlending*, or *microfinance*, has helped a number of people living in poverty to deal with this problem. For example, in 1983, economist Muhammad Yunus started the Grameen (Village) Bank in Bangladesh, a country with a high poverty rate and a rapidly growing population. Unlike commercial banks, the Grameen Bank is essentially owned and run by borrowers and by the Bangladeshi government. Its board of directors is made up largely of the people who use the bank. Since it was founded, the bank has provided a total of $7.4 billion in *microloans* of $50 to $500 at very low interest rates to 7.6 million impoverished people in Bangladesh who do not qualify for loans at traditional banks.

About 97% of these loans have been used by women, mostly to start small businesses, to plant crops, to buy small irrigation pumps, to buy cows and chickens for producing and selling milk and eggs, or to buy bicycles for transportation. Grameen Bank microloans are also being used to develop day-care centers, health-care clinics, reforestation projects, drinking water supply projects, literacy programs, and small-scale solar- and wind-power systems in rural villages (Figure 23-15).

To promote loan repayment, the bank puts borrowers into groups of five. If a group member fails to make a weekly payment, other members must pay it. The average repayment rate on its microloans has been 95% or higher—much

Sasha Kramer: Ecologist and National Geographic Emerging Explorer

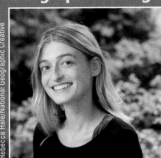

©Rebecca Hale/National Geographic Creative

In the ravaged island nation of Haiti, the soil is so depleted that farmers are greatly challenged to grow anywhere near enough food to feed the population. Most farmers are so poor that they cannot afford the large amounts of synthetic fertilizers they would need to restore the soil. As a result of this and other economic factors, Haiti imports more than 60% of its food and 90% of the rice that serves as a staple for its people. This means that, while its GDP is extremely low, its cost of living is high—comparable to that of the United States.

Working in Haiti, ecologist Sasha Kramer made a connection between this problem and another issue that plagues the people of Haiti—the widespread lack of basic sanitary services. When she arrived in 2004, only 16% of rural Haitians and only 35% of city dwellers had access to toilets. Those without such access disposed of their wastes in rivers, in the ocean, and in abandoned lots. Kramer saw this as a major contributor to the spread of infectious diseases and an offense to the dignity of Haitians, and she decided to do something about it.

In 2006, Kramer and engineer Sarah Brownell found a way to attack these two major problems with one strategy. They founded Sustainable Organic Integrated Livelihoods (SOIL), a nonprofit company dedicated to transforming Haiti's human wastes into a valuable soil conditioner. They knew that fertilizer made from human wastes through devices such as the dry composting toilet could provide nutrients essential to rebuilding Haiti's soil.

The SOIL team designed and built composting toilets and distributed them to areas of Haiti where they were needed. SOIL workers now regularly collect the toilet receptacles when they are full and take them to a central site where a lengthy composting process kills pathogens in the waste and produces a rich fertilizer. Thus human feces that once polluted Haiti's coastal waters and spread disease are now enriching and helping to rebuild its soil, while farmers are using a locally produced, natural, and affordable product to help rebuild Haiti's domestic food supply base.

Kramer believes one of the greatest benefits of this process is that Haitians, who have tremendous national pride and a desire for self-sufficiency, have regained a measure of economic independence. With the help of Kramer and others, Haitians have reduced the threat of infectious diseases while making their local economy more sustainable.

Background photo: © Arindam Banerjee/iStockphoto.com

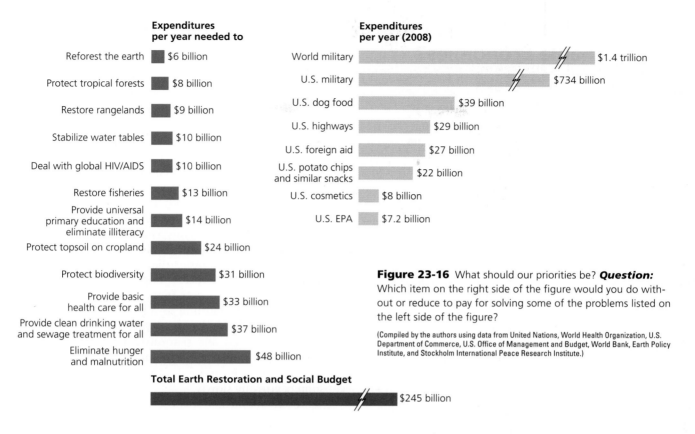

Expenditures per year needed to

Reforest the earth	$6 billion
Protect tropical forests	$8 billion
Restore rangelands	$9 billion
Stabilize water tables	$10 billion
Deal with global HIV/AIDS	$10 billion
Restore fisheries	$13 billion
Provide universal primary education and eliminate illiteracy	$14 billion
Protect topsoil on cropland	$24 billion
Protect biodiversity	$31 billion
Provide basic health care for all	$33 billion
Provide clean drinking water and sewage treatment for all	$37 billion
Eliminate hunger and malnutrition	$48 billion

Total Earth Restoration and Social Budget — $245 billion

Expenditures per year (2008)

World military	$1.4 trillion
U.S. military	$734 billion
U.S. dog food	$39 billion
U.S. highways	$29 billion
U.S. foreign aid	$27 billion
U.S. potato chips and similar snacks	$22 billion
U.S. cosmetics	$8 billion
U.S. EPA	$7.2 billion

Figure 23-16 What should our priorities be? **Question:** Which item on the right side of the figure would you do without or reduce to pay for solving some of the problems listed on the left side of the figure?

(Compiled by the authors using data from United Nations, World Health Organization, U.S. Department of Commerce, U.S. Office of Management and Budget, World Bank, Earth Policy Institute, and Stockholm International Peace Research Institute.)

greater than the average repayment rate for loans by conventional banks—and the bank has made a consistent profit. Typically, about half of Grameen's borrowers move above the poverty line within 5 years of receiving their loans.

Between 1975 and 2005, the Grameen Bank's innovative approach helped to reduce the poverty rate in Bangladesh from 74% to 40%, primarily because of the hard work of the people receiving the microloans. In addition, birth rates are lower among most of the borrowers, a majority of whom are women, and as a result, they gain more freedom and control over their lives.

In 2006, Yunus and his colleagues at the bank jointly won the Nobel Peace Prize for their pioneering use of microcredit. Banks based on the Grameen microcredit model have spread to 58 countries, including the United States. By 2009, this approach had helped at least 133 million people worldwide to work their way out of poverty—85% of them in Asia.

With microlending, Yunus and his colleagues developed a new business model that combined the power of market capitalism with the desire for a more humane and sustainable world. One of the bank's goals was to help protect borrowers from loan sharks who were charging very high interest rates and bankrupting many people. Unfortunately, some loan sharks and commercial companies have moved into the microfinance sector and turned it to their advantage, which has given microlending a bad name in some areas. However, Yunus and his supporters still argue that microlending, when done properly, can help people to escape poverty and improve their lives.

Working Toward the Millennium Development Goals

In 2000, the world's nations set goals—called *Millennium Development Goals*—for sharply reducing hunger and poverty, improving health care, achieving universal primary education, empowering women, and moving toward environmental sustainability by 2015. More-developed countries pledged to donate 0.7%—or $7 of every $1,000—of their annual national income to less-developed countries to help them in achieving these goals. But so far, only five countries—Denmark, Luxembourg, Sweden, Norway, and the Netherlands—have donated what they had promised.

In fact, the average amount donated in most years has been 0.25% of national income. The United States—the world's richest country—gives only 0.16% of its national income to help poor countries, and Japan, another wealthy country, gives only 0.18%, compared with the 0.9% given by Sweden. For any country, deciding whether or not to commit 0.7% of annual national income toward the Millennium Development Goals is an ethical issue that requires individuals and nations to evaluate their priorities (Figure 23-16).

🔍 **CONSIDER THIS. . .**

THINKING ABOUT The Millennium Development Goals

Do you think the country where you live should devote at least 0.7% of its annual national income toward achieving the Millennium Development goals? Explain.

23-5 How Can We Make the Transition to More Environmentally Sustainable Economies?

CONCEPT 23-5

We can use the principles of sustainability, as well as various economic and environmental strategies, to develop more environmentally sustainable economies.

We Are Living Unsustainably

What might happen if growing populations continue to use and waste more energy and matter resources at an increasing rate in today's high-throughput economies? The law of conservation of matter (see Chapter 2, p. 40) and the two laws of thermodynamics (see Chapter 2, p. 43) tell us that eventually, this resource consumption and waste will exceed the capacity of the environment to sufficiently renew those resources, to dilute and degrade waste matter, and to absorb waste heat. Indeed, according to some estimates our ecological footprints are already using the renewable resources of 1.5 planet Earths and probably will be using that of two planet Earths by 2030 (see Figure 1-13, p. 14).

In other words, we are living unsustainably and depleting the earth's natural capital. No one knows how long we can continue on this path, but environmental alarm bells are going off.

A temporary solution to this problem is to convert our linear throughput economy (Figure 23-4) into a circular **matter recycling and reuse economy**. Such an economy mimics nature by recycling and reusing most matter outputs instead of dumping them into the environment. This involves applying the chemical cycling **principle of sustainability**.

Low-Throughput Economies Are More Sustainable

The three scientific laws governing matter and energy changes (see Chapter 2, pp. 40 and 43) and the six **principles of sustainability** (see Figure

1-2, p. 6, Figure 1-5, p. 9, or back cover) suggest that the best long-term solution to our environmental and resource problems is to shift from a high-throughput (high-waste) economy based on ever-increasing matter and energy flow (Figure 23-4) to a more sustainable **low-throughput (low-waste) economy**—an economic system based on energy flow and matter recycling that works with nature to reduce excessive throughput and the unnecessary waste of matter and energy (Figure 23-17).

A low-throughput economy works by **(1)** reusing and recycling most nonrenewable matter resources; **(2)** using renewable resources no faster than natural processes can replenish them; **(3)** reducing resource waste by using matter and energy resources more efficiently; **(4)** reducing environmentally harmful forms of consumption; and **(5)** promoting pollution prevention and waste reduction. Some experts would add that such an economy works best in the long run wherever population growth can be slowed in order to keep the number of matter and energy consumers growing slowly.

Some environmental scientists suggest that an important step in shifting to a low-throughput economy will be to *relocalize some economic engines and the benefits they provide*. For example, Kelly Cain and his colleagues (see Science Focus 22.1, p. 623) have created a computer model for estimating the amount of money and other resources that leave any community that imports most of its food, usually through large retailers. Cain argues that such a community can save large amounts of money and shrink its ecological footprint by relocalizing much of its food and energy production—learning how to produce much more of its own food and energy.

One highly successful example of relocalizing an economy (and of Germany's shift to renewable energy; see **Core Case Study**) is the small windswept rural village of Feldheim, south of Berlin, with a population of about 150. There a young energy entrepreneur, interested in relocalizing energy production, invested in a small number of wind turbines. The village followed his lead and built its own power grid,

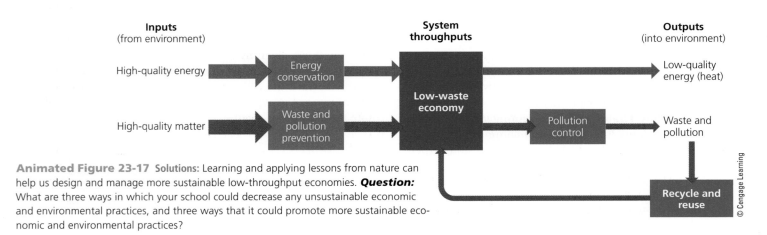

Animated Figure 23-17 Solutions: Learning and applying lessons from nature can help us design and manage more sustainable low-throughput economies. ***Question:*** What are three ways in which your school could decrease any unsustainable economic and environmental practices, and three ways that it could promote more sustainable economic and environmental practices?

Figure 23-18 Solutions: Some of the components of more environmentally sustainable economic development favored by ecological and environmental economists. **Question:** What are three new types of jobs that could be generated by such an economy?

along with a biogas plant that produces natural gas from corn cobs, pig manure, and other farm wastes. Today the village produces all of its own heat and electricity and has a zero-carbon footprint and full employment. It makes a profit by selling the excess energy it produces to major power companies for use in Germany's electrical grid system.

We Can Shift to More Sustainable Economies

Figure 23-18 shows some of the components of societies that have more sustainable economic systems. A common goal of such systems is to put more emphasis on conserving and sustaining the air, water, soil, biodiversity, and other natural resources and ecosystem services that in turn sustain all life and all economies.

In making the shift to more sustainable economies, some industries and businesses will disappear or remake themselves, and new ones will appear—a normal process in a dynamic and creative capitalist economy. *Eco-*

logical succession occurs when changes in environmental conditions enable certain species to move into an area and replace other species that are no longer favored by the changing environmental conditions (see Figure 5-12, p. 110). By analogy, *economic succession* in a dynamic capitalist economy occurs as new and more innovative businesses replace older ones that can no longer thrive under changing economic conditions.

The drive to improve environmental quality and to work toward environmental sustainability has created major growth industries along with new-found profits and large numbers of new *green jobs* (Figure 23-19). Examples of such jobs include those devoted to protecting natural capital, expanding organic agriculture, making homes and other buildings more energy efficient, modernizing the electrical grid system, and developing low-carbon renewable energy resources. According to Ethan Pollack of the Economic Policy Institute, by 2010, there were more than 3.1 million green jobs in the United States and that number has since been increasing faster than the overall economy.

Ray Anderson

Interface

Ray Anderson (1934–2011), founder of Interface, a company based in Atlanta, Georgia (USA), was one of the world's most respected and effective leaders in the movement to make businesses more sustainable. The company is the world's largest commercial manufacturer of carpet tiles, with 25 factories in 7 countries, and customers in 110 countries.

Anderson, who was once called "the greenest CEO in America," had changed the way he viewed the world—and his business—after reading Paul Hawken's book *The Ecology of Commerce*. In 1994, he announced plans to develop the nation's first totally sustainable green corporation. He then implemented hundreds of projects with goals of producing zero waste, greatly reducing energy use, reducing fossil fuel use, relying increasingly on solar energy, and copying nature. Anderson also wanted to influence other business leaders to adopt similar goals by providing a successful example.

Observing that there is no material waste in nature, Anderson began the transition of his business with a focus on sharply cutting waste in his company. Within 16 years of setting this and other environmental goals, Interface had decreased water usage by 74%, cut its net greenhouse gas emissions by 32%, reduced solid waste by 63%, reduced fossil fuel use by 60%, and lowered energy use by 44%. The company now gets 31% of its total energy from renewable resources. These efforts have saved the company more than $433 million.

Anderson also sent his carpet design team into the forest and told them to learn how nature would design floor covering. What the team discovered is nature's biodiversity **principle of sustainability**. They observed that there were no regularly repeating patterns on the forest floor, that instead, there was disorder and diversity. With this in mind, the design team created a line of carpet tiles, no two of which had the same design. Within 18 months after it was introduced, this new product line was the company's top-selling design. Buyers like it because it seems to bring nature indoors. Similar product lines based on the biodiversity principle now make up 52% of the company's sales.

Interface has also applied nature's chemical cycling **principle of sustainability**. Its carpet tiles are made from recycled fibers, and because the random design tiles can be taken out of a box and laid down randomly, there is no need for large orders of any one style, which helps reduce waste. Also, the carpet tile factory runs partly on solar energy, thus applying another of nature's sustainability principles. In addition, the company invented a new carpet recycling process that does not emit carbon dioxide to the atmosphere.

Anderson was one of a growing number of business leaders committed to finding more economically and ecologically sustainable, yet profitable, ways to do business. Under his leadership, annual sales increased to $1 billion, making Interface the world's largest seller of carpet tiles, and profits tripled. Anderson also created a new consulting group as part of Interface to help other businesses start on the path toward being more sustainable. He believed that businesses can and should practice environmental stewardship and social responsibility in order to benefit their profit margins, as well as people and the planet.

Since his death, Anderson's work has been carried on as vigorously as before at Interface. In 2012, the company announced plans to form a partnership with the Zoological Society of London to create Net-Works, a trial project for recycling discarded fishing nets—one of the major solid waste problems now plaguing the oceans—into new carpet tiles. The company will pay qualified communities that collect the discarded nets on the condition that the funds will be used for economic development programs in the communities.

Environmentally Sustainable Businesses and Careers

Aquaculture	Environmental law
Biodiversity protection	Environmental nanotechnology
Biofuels	Fuel-cell technology
Climate change research	Geographic information systems (GIS)
Conservation biology	Geothermal geologist
	Hydrogen energy
Ecotourism management	Hydrologist
Energy-efficient product design	Marine science
	Pollution prevention
Environmental chemistry	Reuse and recycling
Environmental design and architecture	Selling services in place of products
	Solar-cell technology
Environmental economics	Sustainable agriculture
	Sustainable forestry
Environmental education	Urban gardening
	Urban planning
Environmental engineering	Waste reduction
	Watershed hydrologist
Environmental entrepreneur	Water conservation
Environmental health	Wind energy

© Cengage Learning

Figure 23-19 *Green careers:* Some key environmental businesses and careers are expected to flourish during this century, while environmentally harmful, or *sunset,* businesses are expected to decline. See the website for this book for more information on various environmental careers. **Question:** How could some of these careers help you to apply the three **principles of sustainability**?

Top: Goodluz/Shutterstock.com. Second from top: Goodluz/Shutterstock.com. Second from bottom: Dusit/Shutterstock.com. Bottom: Corepics VOF/Shutterstock.com.

Older industries such as the fossil fuels industry have claimed repeatedly that a switch to a more sustainable economy will lead to massive job losses. However, in 2009, a study by University of California–Berkeley scientists led by Max Wei reviewed 15 studies on job creation in the energy sector. They found that the production and use of renewable energy sources created more jobs per unit of energy generated than did the production and use of fossil fuels (Figure 23-20).

Making the shift to more sustainable economies will require governments and industries to greatly increase their spending on research and development—especially in the areas of energy efficiency and renewable energy—as Germany has done in recent years (**Core Case Study**). The shift toward sustainability will also require a new type of business leader who understands why such a shift is important ecologically and economically (Individuals Matter 23.2).

Lessons from Nature Will Help Us in Making the Transition

In this chapter, we have considered how certain **principles of sustainability** can guide us in shifting to more sustainable economic systems. This has revealed a sharp contrast between the hypotheses of neoclassical economists and those of ecological economists. We close this chapter with the words of the highly regarded environmental scientist Donella Meadows (1941–2001). In 1996, she contrasted the views of neoclassical and ecological economists as follows:

- The first commandment of economics is: Grow. Grow forever. . . . The first commandment of the earth is: Enough. Just so much and no more. . . .
- Economics says: Compete. . . . The earth says: Compete, yes, but keep your competition in bounds. Don't annihilate. Take only what you need. Leave your competitor enough to live. Wherever possible, don't compete, cooperate. . . . You're not in a war, you're in a community. . . .

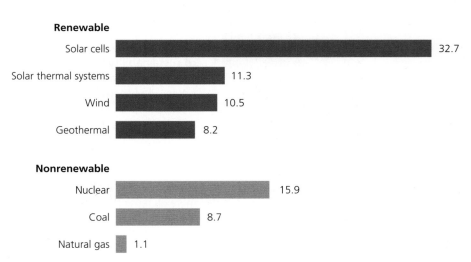

Figure 23-20 Comparison of job-creation potential for various energy resources. The number following each bar represents the number of jobs created per unit of energy produced.

(Compiled by the authors using data from Max Wei, Shana Patadia, and Daniel Kammen, *Putting Renewables and Energy Efficiency to Work,* 2009.)

- Economics says: Use it up fast. Don't bother with repair; the sooner something wears out, the sooner you'll buy another. This makes the gross national product go round. Throw things out when you get tired of them. . . . Get the oil out of the ground and burn it now. . . . The earth says: What's the hurry? . . . When any part wears out, don't discard it, turn it into food for something else. . . .

- Economics discounts the future. . . . a resource ten years from now is worth only half of what it's worth now. Take it now. Turn it into dollars. The earth says: Nonsense. . . . give to the future. . . . Never take more in your generation than you give back to the next.

- The economic rule is: Do whatever makes sense in monetary terms. The earth says: Money measures nothing more than the relative power of some humans over other humans, and that power is puny compared with the power of the climate, the oceans, the uncounted multitudes of one-celled organisms that created the atmosphere, that recycle the waste, and that have lasted for 3 billion years. The fact that the economy, which has lasted for maybe 200 years, puts zero values on these things means only that the economy knows nothing about value—or about lasting.

Big Ideas

- Making a transition to more sustainable economies will require finding ways to estimate and include the harmful environmental and health costs of producing goods and services in their market prices.

- Making this economic transition will also mean phasing out environmentally harmful subsidies and tax breaks, and replacing them with environmentally beneficial subsidies and tax breaks.

- Another way to further this transition would be to tax pollution and wastes instead of wages and profits and to use most of the revenues from these taxes to promote environmental sustainability and reduce poverty.

TYING IT ALL TOGETHER — Germany's Transition and Sustainability

Richard Schmidt-Zuper/iStockphoto

The case study that opens this chapter is about how Germany has used economic tools to spur a shift from using fossil fuels and nuclear energy to relying increasingly on renewable energy (see photo at left). It shows how a country can use its economic policy tools to affect the energy market. As energy use and environmental quality in a country or region are closely intertwined, this story also shows how economics can be used directly to affect a country's environmental impact—how economics can play a major role in determining the size of a country's ecological footprint.

This story and others in this chapter show how several of the **principles of sustainability** (see Figure 1-2, p. 6, Figure 1-5, p. 9, or back cover) can be applied to help us shift to more sustainable economies in the near future. The full-cost pricing principle will play a major role in such a shift, because if consumers have to pay the harmful environmental and health costs of the goods and services they use, they will be inclined to choose those that have lower costs and thus lower impacts on the environment and human health. Germany is finding that renewable energy resources—the sun, wind, biogas, and other resources—have lower environmental and health costs, and the country is thus applying the solar energy principle. And several companies are developing products and services based more on reuse and recycling, in accordance with the chemical cycling principle.

Think about the other **principles of sustainability** and see if you can find ways in which Germany and other subjects of case studies in this chapter are applying those principles as they take part in an historic effort to shift to a more sustainable economy.

Chapter Review

Core Case Study

1. Explain how Germany's feed-in tariff has spurred a shift to renewable energy use in that country (**Core Case Study**). What are some of the elements of Germany's changing energy economy?

Section 23-1

2. What is the key concept for this section? Define **economics** and **economic system**. Describe the interactions among supply, demand, and market prices in a market economic system. Explain why a truly free-market economy rarely exists in today's capitalist market systems. What are **subsidies**? Distinguish among **natural capital**, **human capital (human resources)**, and **manufactured capital (manufactured resources)**.

3. What is a market failure? Give an example of how a government can correct a market failure. Define **economic growth**. What is a **high-throughput economy**? Distinguish between **economic development** and **environmentally sustainable economic development**. Compare how neoclassical economists and ecological and environmental economists view economic systems. What is the general assumption underlying ecological economists' models?

Section 23-2

4. What are the two key concepts for this section? Describe three ways in which economists estimate the economic values of natural goods and services. Why are such values not included in the market prices of goods and services? Define **discount rate** and explain the controversy over how to assign such rates. Explain how economists can estimate the optimal levels for pollution control and resource use. Define **cost–benefit analysis** and list its advantages and limitations.

Section 23-3

5. What is the key concept for this section? Distinguish between direct (internal) and indirect (external) costs of goods and services and give an example of each that is related to a specific product. What is **full-cost pricing** and what are some benefits of using it to determine the market values of goods and services? Give three reasons why it is not widely used.

6. What are **perverse subsidies**? How can subsidies be used for environmentally beneficial purposes? Give three examples of perverse subsidies and three examples of environmentally beneficial government subsidies and tax breaks. Define **gross domestic product (GDP)** and **per capita GDP**. What is the **genuine progress indicator (GPI)** and how does it differ from the GDP economic indicator?

7. What are the major advantages and disadvantages of using green taxes? Why do many analysts consider the tax systems in most countries to be backward? What are three requirements for implementing green taxes successfully? What are the benefits of providing consumers with eco-labels on the goods and services they buy? Define **greenwashing** and give an example of it.

8. Distinguish between command-and-control and incentive-based government regulations, and describe the advantages of the second approach. What is the cap-and-trade approach to implementing environmental regulations, and what are the major advantages and disadvantages of this approach? What are some environmental benefits of selling services instead of goods? Give two examples of this approach.

Section 23-4

9. What is the key concept for this section? Define **poverty**. List six ways in which governments, businesses, international lending agencies, and wealthy individuals could help to reduce poverty. Describe how Sasha Kramer is trying to improve basic sanitation services for the poor in Haiti. What is microlending, and how has it helped people? What are millennium development goals and what role should more-developed countries play in helping the world achieve these goals?

Section 23-5

10. What is the key concept for this section? What is a **matter recycling and reuse economy**? What is a **low-throughput (low-waste) economy** and how does it work? What does it mean to relocalize economic benefits? Give an example of how this has benefited people. List ten components of more environmentally sustainable economic development. Name six new businesses and careers that would be important in more sustainable economies. Summarize the story of Ray Anderson and his work to build a more environmentally sustainable carpet business. What are this chapter's *three big ideas*? Explain how we can apply the **principles of sustainability** in making a shift toward a more sustainable economic system, as Germany has done (**Core Case Study**).

Note: Key terms are in bold type.

Critical Thinking

1. Explain how Germany's transition to broader use of renewable energy systems (**Core Case Study**) shows that the economy and the environment are linked. How is the country's economic development policy changing its environmental quality? What are some ways in which Germany's example could be applied to improve the environment and the economy where you live?

2. Is it a good idea to maximize economic growth by producing and consuming more and more economic goods and services? Explain. What are some alternatives?

3. According to one definition, *environmentally sustainable economic development* involves meeting the needs of the present human generation without compromising the ability of future generations to meet their needs. What do you believe are the needs referred to in this definition? Compare this definition with the characteristics of a low-throughput economy depicted in Figure 23-17.

4. Is environmental regulation bad for the economy? Explain. Assume you are a government official and devise an incentive-based regulation for an industry of your choice. (It could be a coal mine, a power plant, a fishing fleet, a chemical plant, or any other business that has a large effect on the environment.) Explain how your regulatory plan will benefit both the industry and the environment.

5. Suppose that over the next 20 years, the environmental and health costs of goods and services are gradually added to market prices until their market prices more closely reflect their total costs. What harmful effects and what beneficial effects might such a full-cost pricing process have on your lifestyle and on the lives of any children, grandchildren, and great grandchildren you might eventually have?

6. Do you believe that reducing poverty should be a major environmental goal? Explain. List three ways in which reducing poverty could benefit you and any children, grandchildren, and great grandchildren you might eventually have. Why do you think the world has not focused more intense efforts on reducing poverty?

7. Are you for or against shifting to a *service-flow economy* based on buying services instead of things, and on leasing instead of buying various products? Explain. If you are for such a shift, what do you think is the best strategy for making it happen? If you are opposed, what are your main objections to the idea?

8. Congratulations! You are in charge of the world. Write up a five- to ten-point strategy for shifting the world to more environmentally sustainable economic systems over the next 50 years.

Doing Environmental Science

Go online or to a library and find a tool for estimating the full cost (including harmful environmental and health costs) of common products. Choose five products that you regularly buy and use this tool to estimate the full cost of each. Record these data in a table, along with the price you paid for each product. (Estimate this price if you don't remember what you paid.) Now do some market research and try to find alternatives to these products that have lower full costs. Record these data in your table. Do some calculations to learn **(a)** the differences between the prices you paid for your common products and their full costs; **(b)** for each product, the difference between its price and the price of the alternative substitute you found; and **(c)** for each product, the difference between its full cost and the full cost of the alternative substitute you found.

Finally, for each product pair, use your data to answer these questions:

1. Without knowledge of the full costs, would the price differences be large enough to keep you from buying the alternative product and sticking with your commonly used product? Explain.

2. Comparing their full costs, does this change your mind about whether the price difference is high enough to keep you from switching products? Explain.

3. How high would the full cost of your commonly used product have to be to get you to pay the higher price for the alternative?

Global Environment Watch Exercise

Search for *global subsidies* and find information on government subsidies for the oil, coal, and nuclear industries and for renewable energy industries. Write a summary of trends in these subsidies, either globally or in a country of your choice. Find an article that indicates a positive view of these trends on the part of the author and one that indicates a negative view. Compare the arguments in these articles and decide which argument you believe to be the stronger.

Data Analysis

Use this graph to compare the per capita gross domestic product (GDP) with the per capita genuine progress indicator (GPI) in the United States between 1950 and 2004 (the latest year for which this data is available).

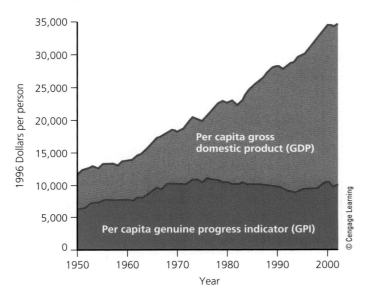

1. What was the monetary difference between per capita GDP and per capita GPI in **(a)** 1950 and **(b)** 2004?

2. **(a)** By what percentage did the per capita GDP increase between 1950 and 1975, and between 1975 and 2004? **(b)** By what percentage did the per capita GPI increase between 1950 and 1975, and decrease between 1975 and 2004?

3. What conclusion can you draw from the answers to questions 1 and 2?

Questions 1–4 refer to the following economic concepts.

- a. Economic system
- b. Natural capital
- c. Human capital
- d. Manufactured capital
- e. Market equilibrium

1. The condition that exists when one price and one quantity prevail in the market.

2. Consists of land, forests, pastures, mineral deposits, and animal life.

3. The organized way that society attains all that it needs or wants.

4. The tools, factories, equipment, and machinery used in the production process.

Use the supply and demand graph of the oil market to answer questions 5–7.

5. Which of the following would not necessarily lead to an increase in the quantity of oil bought and sold in the market?
   - a. Increased demand for oil
   - b. Increased supply of oil
   - c. Simultaneous increases in the demand for and supply of oil
   - d. A simultaneous increase in demand and decrease in supply
   - e. An effective price ceiling on oil

6. What condition would exist in the oil market if the actual price was greater than the equilibrium price?
   - a. Shortage
   - b. Surplus
   - c. Scarcity
   - d. Market equilibrium
   - e. Human capital

7. The graph assumes all of the following about buyers and sellers EXCEPT
   - a. both have equal access to information.
   - b. buyers and sellers act independently.
   - c. buyers and sellers compete with one another.
   - d. government exerts no influence over prices.
   - e. companies lobby politicians for favorable treatment.

8. Which is not a criticism of the neoclassical model?
   - a. Markets efficiently allocate private goods and services.
   - b. Neoclassical economics does not place a value on unused natural resources.
   - c. Markets inefficiently allocate public goods.
   - d. The costs of pollution and waste are not considered in the neoclassical model.
   - e. Neoclassical economics ignores aesthetic value.

9. Which is NOT a disadvantage to tradable environmental permits?
   - a. Flexibility
   - b. Ease of administration
   - c. Permit prices are determined in the market
   - d. Confronts ethical issue of how much pollution/waste is acceptable
   - e. Can reduce pollution through direct government regulation

10. Which of the following is NOT included in both market prices and full-cost prices?
    - a. Costs of production, because many raw materials must be transported long distances
    - b. Labor costs, because employees must be paid a fair salary and benefits
    - c. Cost of air and water pollution that result from the industrial process
    - d. The cost of taxes paid by the manufacturer
    - e. The cost of shipping the goods to markets where they are sold

Use the graph below to answer the following questions.

11. What is the relationship between marginal cost and marginal benefit?
    a. Inversely proportional
    b. Directly proportional
    c. Platonic
    d. Square root
    e. Intermittent

12. At what cost of mining coal does the marginal cost exceed the marginal benefit?
    a. 40%
    b. 55%
    c. 70%
    d. 90%
    e. 100%

You see things, and you say "Why?" But I dream things that never were; and I say "Why not?"

GEORGE BERNARD SHAW

Key Questions

24-1 What is the role of government in making the transition to more sustainable societies?

24-2 How is environmental policy made?

24-3 What is the role of environmental law in dealing with environmental problems?

24-4 What are the major roles of environmental groups?

24-5 How can we improve global environmental security?

24-6 How can we implement more sustainable and just environmental policies?

Marchers in New York City in 2011 protesting against natural gas fracking.

© Richard Levine/Alamy

657

Since the mid-1980s, there has been a boom in environmental awareness on college campuses and in public and private schools around the world. In the United States, hundreds of colleges and universities have now taken the lead in a quest to become more sustainable.

For example, at Northland College in Ashland, Wisconsin, students helped to design a green living and learning center (Figure 24-1) that houses 150 students and features a wind turbine, solar panels, furniture made of recycled materials, and waterless (composting) toilets. Northland students voted to impose a *green fee* of $40 per semester on themselves to help finance the college's sustainability programs.

Berea College in Kentucky boasts an innovative environmental science curriculum including a Sustainable Appalachian Communities course. The school also has a 50-unit eco-village, an experimental residence complex that uses passive solar heating, solar panels, and filtered rainwater.

Green Mountain College in Poultney, Vermont, gets its power from renewable biomass and biogas and plans to become one of the country's first carbon-neutral schools by 2020. At the University of California, Santa Cruz, students reused, recycled, or composted more than 70% of their waste in 2012, with a goal of reaching 100% by 2020. And in Ashville, North Carolina, Warren Wilson College gets more than a third of its food from regional farms, including its own large on-campus organic garden.

There are hundreds of other examples of this greening process on college campuses, some of which have been high-lighted in *Campus Sustainability* pieces throughout this book.

In addition to making campuses greener, some call for more extensive efforts to provide students with a basic education in environmental sustainability. To that end, in 2008, Arizona State University opened the country's first School of Sustainability, which employs more than 60 faculty members from more than 40 disciplines. At the Georgia Institute of Technology, one goal is for every student to take at least one of the more than 100 available courses having a sustainability emphasis. And at the University of Wisconsin–Madison, the Nelson Institute for Environmental Studies seeks to integrate sustainability content throughout disciplines and programs on campus, as well as to serve communities outside of the university.

In this chapter, we examine how politics and environmental science interact. And we see how citizens can affect environmental policies, just as many college students have done, at local, regional, national, and global levels.

Figure 24-1 The Environmental Living and Learning Center is an eco-friendly residence hall and meeting space at Northland College in Ashland, Wisconsin. Northland students had a major role in the design of the building.

Courtesy of Northland College

24-1 What Is the Role of Government in Making the Transition to More Sustainable Societies?

CONCEPT 24-1
Through its policies, a government can help to protect environmental and public interests, and to encourage more environmentally sustainable economic development.

Government Can Serve Environmental and Other Public Interests

Business and industry thrive on change and innovations that lead to new technologies, products, and opportunities for profits. This process, often referred to as *free enterprise*, can lead to jobs and higher living standards for many people, but it can also create harmful impacts on other people and on the environment.

Government, on the other hand, especially democratic government working in the interest of the public, can act as a brake on business enterprises that might result in harm to people or to the environment. Achieving the right balance between free enterprise and government regulation is not easy. Too much government intervention can strangle enterprise and innovation. Too little can lead to environmental degradation and social injustices, and even to a weakening of the government by business interests and global trade policies.

Analysts point out that in today's global economy, some multinational corporations, which often have budgets larger than the budgets of many countries, have greatly increased their economic and political power over national, state, and local governments, and ordinary citizens. However, businesses can also serve environmental and public interests. Green businesses create products and services that help to sustain or improve environmental quality while improving people's lives. They make up one of the world's fastest growing business sectors and are increasingly a source of new jobs.

Many argue that government is the best mechanism for dealing with some of the broader economic and political issues we face, some of which we have discussed in this book. These include:

- *Full-cost pricing* (see Chapter 23, p. 639): Governments can provide subsidies and levy taxes that have the effect of adding harmful environmental and health costs to the market prices of some goods and services and thus implement a social science **principle of sustainability** (see Figure 1-5, p. 9 or back cover).
- *Market failures* (see Chapter 23, p. 633): Governments can use taxes and subsidies to level the playing field wherever the marketplace is not operating freely due to unfair advantages held by some players.

- *The tragedy of the commons:* Government plays a key role in preserving common or open-access renewable resources (see Chapter 1, p. 12) such as clean air and groundwater, the ozone layer in the stratosphere, and our life-support system. This is also the case in the cap-and-trade market approach to solving a problem such as air pollution (see Chapter 18, p. 495), because government oversight is necessary to administer such a program.

The roles played by a government are determined by its **policies**—the set of laws and regulations it enacts and enforces, and the programs it funds (Concept 24-1). **Politics** is the process by which individuals and groups try to influence or control the policies and actions of governments at local, state, national, and international levels. One important application of this process is the development of **environmental policy**—environmental laws, regulations, and programs that are designed, implemented, funded, and enforced by one or more government agencies.

According to social scientists, the development of public policy in democracies often goes through a *policy life cycle* consisting of four stages (Figure 24-2):

- *Problem recognition.* A problem is identified by members of the public or by a policy maker.
- *Policy formulation.* A cause or causes of the problem are identified and a solution such as a law or program to help deal with the problem is proposed and developed.
- *Policy implementation.* A law is passed or a regulation written to put the solution into effect.
- *Policy adjustment.* The new program is monitored, evaluated, and adjusted as necessary.

Another term for this process is *adaptive management*. Ideally, policies are revised and fine-tuned until they succeed in serving all or most of the affected parties in a reasonably balanced way.

Democracy Does Not Always Allow for Quick Solutions

Democracy is government by the people through elected officials and representatives (Figure 24-3). In a *constitutional democracy*, a constitution (a document recording the rights of citizens and outlining the government's responsibilities and limitations) provides the basis of government authority and, in most cases, limits government power by mandating free elections (Figure 24-3) and guaranteeing the right of free speech.

Political institutions in most constitutional democracies are designed to allow gradual change that helps

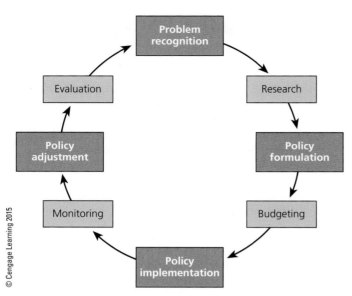

Figure 24-2 The *policy life cycle* has been defined in several ways but generally includes these four phases (listed in the orange boxes).

Figure 24-3 Open election of government officials by citizens is a fundamental right in a constitutional democracy.

ensure economic and political stability. In the United States, for example, rapid and destabilizing change is curbed by a system of checks and balances that distributes power among three branches of government—*legislative, executive,* and *judicial*—and among federal, state, and local governments.

In passing laws, developing budgets, and formulating regulations, elected and appointed government officials must deal with pressure from many competing *special-interest groups.* Each of these groups advocates passing laws, providing subsidies or tax breaks, or establishing regulations favorable to its cause, while attempting to weaken or repeal laws, subsidies, tax breaks, and regulations unfavorable to its position. Some special-interest groups such as corporations are *profit-making organizations.* Others are *nongovernmental organizations (NGOs),* most of which are nonprofit, such as labor unions and environmental organizations.

The design for stability and gradual change in democracies is highly desirable. But several features of democratic governments hinder their ability to deal with environmental problems. For example, problems such as climate change and biodiversity loss are complex and difficult to understand. Such problems also have long-lasting effects, are interrelated, and require integrated, long-term solutions that emphasize prevention. However, because local and national elections are held as often as every 2 years in most democracies, most politicians spend much of their time seeking reelection and tend to focus on short-term, isolated issues rather than on long-term, complex, and time-consuming problems.

One of the greatest challenges for those who want to change an environmental policy is to educate political leaders and the public about the need for long-range thinking and policy formulation. Another problem is that many

political leaders, with hundreds of issues to deal with, have too little understanding of how the earth's natural systems work and how those systems support all life, economies, and societies. Again, there is an urgent need to educate political leaders and voters about these vital matters.

Certain Principles Can Guide Us in Making Environmental Policy

Analysts suggest that when evaluating existing or proposed environmental policies, legislators and individuals should be guided by several principles that can help them to minimize environmental harm:

- *The humility principle:* Our understanding of nature and of how our actions affect nature is quite limited.
- *The reversibility principle:* It is best to avoid making decisions that cannot be reversed later if they turn out to be harmful. For example, two essentially irreversible actions affecting the environment are the production of indestructible hazardous and toxic waste in coal-burning power plants, which we must try to store safely, essentially forever; and production of deadly radioactive wastes throughout the nuclear power fuel cycle, which must be stored safely for 10,000–240,000 years (see Chapter 15, p. 392).
- *The net energy principle:* It is best to avoid the widespread use of energy resources and technologies with low net energy yields (see Figure 15-2, p. 375), which cannot compete in the open marketplace without government subsidies. Examples of energy alternatives with fairly low or negative net energy yields include nuclear power (considering the whole fuel cycle), tar sands, and shale oil, as discussed in Chapter 15, and hydrogen fuel and ethanol made from corn, as discussed in Chapter 16.
- *The precautionary principle:* When substantial evidence indicates that an activity threatens human health or the

environment, take precautionary measures to prevent or reduce such harm, even if some of the cause-and-effect relationships are not well established, scientifically.

- *The prevention principle:* Whenever possible, make decisions that help to prevent a problem from occurring or becoming worse.
- *The polluter-pays principle:* Develop regulations and use economic tools such as green taxes to ensure that polluters bear the costs of dealing with the pollutants and wastes they produce. This is an important way to include some of the harmful environmental and health effects of goods and services in their market prices in accordance with the full-cost pricing **principle of sustainability**.
- *The environmental justice principle:* In the implementation of environmental policy, no group of people

should bear an unfair share of the burden created by pollution, environmental degradation, or the execution of environmental laws. (See the online Guest Essay on this subject by Robert D. Bullard.)

Implementing such principles is not easy and will require that policy makers throughout the world, especially in more-developed countries, become more environmentally literate, based on the latest scientific information about environmental problems and possible solutions to them.

🔍 **CONSIDER THIS. . .**

THINKING ABOUT Environmental Political Principles
Which three of the seven principles listed here do you think are the most important? Why? Which ones do you think could influence legislators in your city, state, or country?

24-2 How Is Environmental Policy Made?

CONCEPT 24-2A
Policy making involves enacting laws, funding programs, writing rules, and enforcing those rules with government oversight—a complex process that is affected at each stage by political processes.

CONCEPT 24-2B
Individuals can work together to become part of political processes that influence how environmental policies are made and whether or not they succeed (*Individuals matter*).

How Democratic Government Works: The U.S. Model

The U.S. federal government consists of three separate but interconnected branches: legislative, executive, and judicial. The *legislative branch,* called the Congress, consists of the House of Representatives and the Senate, which jointly have two main duties. One is to approve and oversee government policy by passing laws that establish government agencies or instruct existing agencies to take on new tasks or programs. The other is to oversee the functioning and funding of agencies in the executive branch concerned with carrying out government policies.

The *executive branch* consists of the president and a staff who oversee the agencies authorized by Congress to carry out government policies. The president proposes annual budgets, legislation, and appointees for major executive positions, which must be approved by Congress. The president also tries to persuade Congress and the public to support executive policy proposals. Citizens use the ballot box (Figure 24-3) to elect the president and vice president in the executive branch and members of Congress in the legislative branch.

The *judicial branch* consists of the Supreme Court and lower federal courts. These courts, along with state and local courts, enforce and interpret different laws passed by legislative bodies. They are to see that the laws preserve the rights and responsibilities of government and citizens as established by the U.S. Constitution. Decisions made by the various courts make up a body of law known as *case law.* Previous court rulings are used as legal guidelines, or *precedents,* to help make new legal decisions and rulings. The president appoints judges at the federal level with the advice and consent of the Senate, and judges at state and local levels of government are variously appointed by executives or elected by voters.

The major function of the federal government in the United States (and in other democratic countries) is to develop and implement policies for dealing with various issues. The important components of policy are the *laws* passed by the legislative branch, *regulations* instituted by the agencies of the executive branch to put laws and programs into effect, and *funding* approved by Congress and the president to finance the executive agencies' programs and to implement and enforce the laws and regulations (Concept 24-2A).

Converting a bill introduced in the U.S. Congress into a law is a complex process (Figure 24-4). An important factor in this process is **lobbying**, in which individuals or groups contact legislators in person, or hire *lobbyists* (representatives) to do so, in order to persuade legislators to vote or act in their favor. The opportunity to lobby elected representatives is an important right for everyone in a democracy. However, some critics of the American system believe lobbyists of large corporations and other organizations have grown too powerful and that their influence overshadows the input that legislators get from ordinary citizens.

Figure 24-4 Greatly simplified overview of how individuals, corporations, and environmental organizations interact with each other and with the legislative, executive, and judicial branches of the U.S. government.

Top row: Left: ©Mesut Dogan/Shutterstock.com. Center: ©Orhan Cam/Shutterstock.com. Right: ©Andrey Burkmakin/Shutterstock.com. Second row: Left: ©James Steidl/Shutterstock.com. Center: © Cameron Witman/Shutterstock.com. Right: public domain. Third row: Left: Marish/Shutterstock.com. Right: ©jl661227/Shutterstock.com. Fourth row: Tyler Olsen/Shutterstock.com.

CONSIDER THIS. . .

CONNECTIONS Lobbying and Perverse Subsidies

Environmental scientist Norman Myers contends that lobbying helps to create subsidies that lead to environmental harms (*perverse subsidies;* see Chapter 23, p. 640). He estimates that roughly 15,000 registered corporate lobbyists spend $3.6 billion per year—an average total of nearly $10 million per day—on efforts to influence the 538 members of the U.S. Congress. That amounts to an average of more than $18,000 per day per member.

Corporations are also the source of billions of dollars used to finance election campaigns. In 2010, the

U.S. Supreme Court ruled that corporations can spend as much money as they want on ads for or against specific candidates running for election.

Most environmental bills are evaluated by as many as ten committees in the U.S. House of Representatives and the Senate. Effective proposals often are weakened by this fragmentation and by lobbying from groups opposing these laws. Nonetheless, since the 1970s, a number of important environmental laws have been passed in the United States, as we discuss in Section 24-3 of this chapter.

Developing Environmental Policy Is a Controversial Process

In the United States, passing a law is not enough to make policy (Concept 24-2A). The next step involves trying to get Congress to appropriate enough funds to implement and enforce each law. Indeed, developing and adopting a budget to finance government agencies, the programs for which they are responsible, and the enforcement of laws and regulations within those programs is the most important and controversial activity of the executive and legislative branches.

Once Congress has passed a law and funded a program, the appropriate government department or agency must draw up regulations for implementing it. A group affected by the program and its regulations may take the agency to court for failing to implement and enforce the regulations effectively or for enforcing them too rigidly.

Businesses facing environmental regulations often put political pressure on regulatory agencies and executives to appoint people from the regulated industries or groups to high positions within the agencies. In other words, the regulated try to take over the regulatory agencies and become the regulators—described by some as "putting foxes in charge of the henhouse."

In addition, people in regulatory agencies work closely with officials in the industries they are regulating, often developing friendships with them. Some industries and other regulated groups offer high-paying jobs to regulatory agency employees in an attempt to influence their regulatory decisions.

Environmental science should play a major role in the formulation of environmental policy, according to many analysts. However, politics usually plays a bigger role, and the scientific and political processes are quite different (Science Focus 24.1).

Individuals Can Influence Environmental Policy

A major theme of this book is that *individuals matter*. History shows that significant change usually comes from the *bottom up* when individuals join with each other to bring about change. Without previous bottom-up (grassroots)

political action by millions of individual citizens and organized citizen groups, the air that many people breathe today and the water they drink would be much more polluted, and much more of the earth's biodiversity would have disappeared.

With the growth of the Internet and digital technology, individuals have become more empowered. For example, in a highly unusual chain of events in 2007, Chinese citizens using mobile phone text messaging organized to oppose construction of a chemical plant that would threaten the safety of 1.5 million people in a port city. By building opposition from the ground up—circulating nearly a million phone messages—they persuaded the Chinese government to freeze the construction project and to consider less hazardous alternatives.

Figure 24-5 lists ways in which you can influence and change government policies in constitutional democracies. *At a fundamental level, all politics is local.* What we do to improve environmental quality in our own neighborhoods, schools, and work places can have national and global implications, much like the ripples spreading outward from a pebble dropped in a pond. And when people work together, starting at the local level, they can influence environmental policy at all levels.

Environmental Leaders Can Make a Big Difference

Each of us can provide environmental leadership in several different ways. First, we can *lead by example*, using our own lifestyles and values to show others that change is possible and can be beneficial. For example, we can use

---

What Can You Do?

Influencing Environmental Policy

- Become informed on issues
- Make your views known at public hearings
- Make your views known to elected representatives and understand their positions on environmental issues
- Contribute money and time to candidates who support your views
- Vote
- Run for office
- Form or join nongovernment organizations (NGOs) seeking change
- Support reform of election campaign financing that reduces undue influence by corporations and wealthy individuals

© Cengage Learning

Figure 24-5 Individuals matter: Some ways in which you can influence environmental policy (Concept 24-2B). *Questions:* Which three of these actions do you think are the most important? Which ones, if any, do you take?

SCIENCE AND POLITICS—PRINCIPLES AND PROCEDURES

The rules of inquiry and debate in science are quite different from those of politics. Science is based on a set of principles designed to make scientific investigations completely open to critical review and testing. Four such principles are:

1. *Any scientific claim must be based on hard evidence and subject to peer review.* This helps to prevent scientists from lying about procedures or falsifying evidence.

2. *Scientists can never establish absolute proof about anything.* Instead, they seek to establish a high degree of certainty about the results of their research.

3. *Scientists vigorously debate the validity of scientific research.* Ideally, such debates focus on the scientific evidence and results, not on personalities involved.

4. *Science advances through the open sharing and peer review of research methods, results, and conclusions.* There are two exceptions to this: first, some scientists who own or work for companies need to protect their

research until legal patents can be obtained. Second, government scientists whose work involves national security often keep their research secret.

In politics, on the other hand, there are no such established and respected principles. In order to win elections and gain influence, politicians use unwritten rules that change frequently. While many politicians would like to base their decisions and actions on facts, others suggest that what matters more than facts is how the public perceives what they do and say. This makes the political process far less open to review and criticism than the scientific process.

Without such openness, the political process often involves tactics that most scientists would reject. For example, some politicians pick and choose facts to support a claim that is not supported by the whole of a body of evidence. They then repeat such a claim over and over until it becomes part of the news media cycle. If this misuse of evidence is not exposed, as it usually is in science, these unsupported claims can become accepted as truth.

Another political tactic that often goes unchallenged is to change a debate about facts to a discussion focused on personal attacks. Such a tactic is meant to make one's opponents look weak, and it helps a politician to avoid serious discussion of issues. In scientific debate, most participants do not tolerate such a shift away from a fact-based discussion.

People have learned how to spread disinformation quickly in this media age of almost instant global news coverage, text messaging, social networking, and Internet blogs and videos. While the Internet allows almost anyone to quickly spread disinformation, it also allows almost anyone to check the validity of much information and to detect and publicize lies and distortions. Learning how to detect and evaluate disinformation is one of the most important purposes of education. This is no easy task, and it requires an open mind and critical thinking.

Critical Thinking

What are two examples of widely accepted results of scientific research that have been politicized to the point where they are largely doubted or ignored by the public?

---

fewer disposable products, eat foods that have been more sustainably produced (see Figure 12-33, p. 309), and walk, take a bus, or ride a bike to work or school (Figure 24-6).

Second, we can *work within existing economic and political systems to bring about environmental improvement* by campaigning and voting for informed, ecologically literate candidates and by communicating with elected officials. We can also send a message to companies that we feel are harming the environment through their production processes or products by *voting with our wallets*—not buying their products or services—and letting them know why. Another way to work within the system is to choose one of the many rapidly growing green careers highlighted throughout this book and described in Figure 23-19 (p. 651) and on this book's companion website.

Third, we can *run for some sort of local office.* Look in the mirror. Maybe you are one who can make a difference as an officeholder.

Figure 24-6 Bicycling to school or work is one way to lead by example. In addition to reducing pollution, it saves you money and provides exercise.

Bryan Hoybook/Shutterstock.com

Fourth, we can *propose and work for better solutions to environmental problems.* Leadership is much more than just taking a stand for or against something. It also involves coming up with solutions to problems and persuading people to work together to achieve them. If we care enough, each of us can make a difference, as environmental leaders Wangari Maathai (see Figure 10-20, p. 230), Muhammad Yunus (see Chapter 23, Case Study, p. 645), and Denis Hayes (Individuals Matter 24.1) have done.

Some environmental leaders are motivated by two important findings: *First,* research by social scientists indicates that social change requires active support by only 5–10% of the population, which often is enough to lead to a political tipping point. *Second,* experience has shown that reaching such a critical mass can bring about social change much faster than most people think.

24-3 What Is the Role of Environmental Law in Dealing with Environmental Problems?

CONCEPT 24-3
We can use environmental laws and regulations to help control pollution, set safety standards, encourage resource conservation, and protect species and ecosystems.

Environmental Law Forms the Basis for Environmental Policy

Environmental law is a body of laws and treaties that broadly define what is acceptable environmental behavior for individuals, groups, businesses, and nations. This body of laws and treaties has evolved through legislative and judicial processes at various levels of government that have usually included attempts to balance competing private, social, and commercial interests. This section of the chapter deals primarily with the U.S. legal system as a model that reveals the advantages and disadvantages of using a legal and regulatory approach to dealing with environmental problems.

One way in which environmental law has evolved is through court cases involving lawsuits, most of which are **civil suits** brought to settle disputes or damages between one party and another. For example, a home owner may bring a nuisance suit against a nearby factory because of the noise it generates. In such a suit, the **plaintiff**, the party bringing the charge (in this case, the home owner), seeks to collect damages from the **defendant**, the party being charged (in this case, the factory), for injuries to health or for economic loss.

The plaintiff may also seek an *injunction,* by which the court hearing the case would order the defendant to stop whatever action is causing the nuisance. Short of closing the factory, often the court tries to find a reasonable or balanced solution to the problem. For example, it may order the factory to reduce the bothersome noise to certain levels or to eliminate it at night.

A *class action suit* is a civil suit filed by a group, often a public interest, consumer, or environmental group, on behalf of a larger number of citizens, all of whom claim to have experienced similar damages from a product or an action, but who need not be listed and represented individually.

Another concept used in environmental law cases is *negligence,* in which a party causes damage by deliberately acting in an unlawful or unreasonable manner. For example, a company may be found negligent if it fails to handle hazardous waste in a way that it knows is required by a *statutory law* (a law, or *statute,* passed by a legislature). A court may also find a company negligent if it fails to do something a reasonable person would do, such as testing waste for certain harmful chemicals before dumping it into a sewer, landfill, or river (Figure 24-7). Generally, negligence is hard to prove.

Environmental Lawsuits Are Difficult to Win

Several factors can limit the effectiveness of environmental lawsuits. *First,* plaintiffs bringing the suit must establish that they have the legal right, or *legal standing,* to do so in a particular court. To have such a right, plaintiffs must show that they have suffered health or financial losses from some alleged environmental harm. *Second,* bringing any lawsuit costs too much for most individuals.

Third, public interest law firms cannot recover their attorneys' fees unless Congress has specifically authorized that they be compensated within the laws that they seek to have enforced. By contrast, corporations can reduce their taxes by deducting their legal expenses—in effect getting a government (taxpayer) subsidy to pay for part of their legal fees. In other words, the legal playing field is uneven and puts individuals and groups that are filing environmental lawsuits at a disadvantage.

Fourth, to stop a nuisance or to collect damages from a nuisance or an act of negligence, plaintiffs must establish that they have been harmed in some significant way and that the defendant caused the harm. Doing this can be difficult and costly. Suppose a company (the defendant) is alleged to have caused cancer in certain individuals (the plaintiffs) by polluting a river (Figure 24-7). If hundreds

Denis Hayes—A Practical Environmental Visionary

As a college student, Denis Hayes took some time to explore the world on foot. He

©Mark Thiessen/National Geographic Creative

had studied ecology and political science in college, and one day he grabbed his backpack, filled it with books, and started traveling to see "what was actually going on" in the countries he had studied.

One night while hitchhiking in Africa, he rested on a hillside and started to put together what he had learned and what he had seen. He thought about how the principles of ecology apply to everything from amoeba to orangutans. Sometime shortly after that, Hayes decided to spend his life figuring out how human societies could benefit from organizing themselves around ecological principles.

At age 25, Hayes was enrolled in the Kennedy School of Government at Harvard University in Cambridge, Massachusetts (USA), at the same time that Wisconsin Senator Gaylord Nelson was organizing environmental teach-ins on college campuses. Hayes approached Nelson about organizing such an event at Harvard, and later the senator asked Hayes to organize an event for the whole country.

What started out as a plan for a number of teach-ins became the first Earth Day—April 22, 1970—thought of by many analysts as the beginning of the modern environmental movement. That day involved teach-ins and much more—thousands of public demonstrations focused on pollution, toxic waste, nuclear power, coal mining, lead contamination, and other urgent environmental issues. More than 20 million people took part. Later, Hayes worked on building the Earth Day Network, which now includes more than 180 nations. As a result, each year, Earth Day is now celebrated globally.

Since the first Earth Day, Hayes has worked tirelessly for the environment in a variety of positions. He is currently president and CEO of the Bullitt Foundation of Seattle, Washington (USA). The foundation seeks to deal with key environmental problems in the U.S. Pacific Northwest by focusing on urban ecological issues and on restoring and protecting ecosystem services in the surrounding environment.

The Bullitt Foundation also seeks to provide a model for more sustainable urban building design. In 2013, Hayes began construction on the foundation's headquarters, which will be one of the world's most sustainable commercial buildings when it is completed. It will qualify as one of the world's 16 certified "living buildings," having been built of locally sourced building materials, including only lumber certified by the Forest Stewardship council as sustainably produced.

The building will get all of its electricity from solar cells (similar to those shown in background photo) and will transmit its unused electricity into the regional grid (making it a zero-net-energy building), and it will get 80% of its daytime interior light from natural daylight. It will also include a system for storing rainwater and purifying it for use by building occupants. Its other major features will include windows and shutters that open or close automatically as needed to keep the interior temperature steady, a small green roof, a system for composting human waste into fertilizer, and bicycle garages and showers. The building is also located near a public transit station.

Hayes has received numerous awards, including the Jefferson Medal for Outstanding Public Service and the highest awards given by the Sierra Club, the Humane Society of the United States, and the Global Environment Facility of the World Bank. In 2008, the Audubon Society listed him as one of the 100 Environmental Heroes of the Twentieth Century. As a model environmental leader, he has dedicated himself to helping people understand and carefully consider ecological principles in deciding what policies they support and in taking political action to support sustainability.

Background photo: Spire Solar Chicago/National Renewable Energy Laboratory

Figure 24-7 This body of water was polluted by a copper mining operation. Such pollution can form the basis for an environmental lawsuit.

Given the influence of corporate wealth and lobbying power, it is unlikely that these reforms will be implemented without strong bottom-up (grassroots) political pressure from concerned citizens.

of other industries and cities dump waste into that river, establishing that one specific company is the culprit is very difficult and requires expensive investigation, scientific research, and expert testimony.

Fifth, most states have *statutes of limitations,* laws that limit how long a plaintiff can take to sue after a particular event occurs. These statutes often make it essentially impossible for victims of cancer, which may take 10–20 years to develop, to file or win a negligence suit.

Sixth, courts can take years to reach a decision. During that time a defendant may continue the allegedly damaging action unless the court issues a temporary injunction against it until the case is decided.

Yet another problem is that corporations and developers sometimes file *strategic lawsuits against public participation (SLAPPs)* targeting citizens who publicly criticize a business for some activity, such as polluting or filling in a wetland. Judges throw out about 90% of the SLAPPs that go to court. But individuals and groups hit with SLAPPs must hire lawyers, and typically spend 1–3 years defending themselves. Most SLAPPs are not meant to be won, but are intended to intimidate and discourage individuals and activist groups.

Analysts have suggested three major reforms to help level the legal playing field for citizens suffering environmental damage. *First,* pressure Congress to pass a law allowing juries and judges to award citizens their attorney fees, to be paid by the defendants, in successful lawsuits.

Second, establish rules and procedures for identifying frivolous SLAPP suits so that cases without factual or legal merit can be dismissed quickly.

Third, raise the fines for violators of environmental laws and punish more violators with jail sentences. Polls indicate that 80% or more of Americans consider damaging the environment to be a serious crime.

U.S. Environmental Laws

Concerned citizens have persuaded Congress to enact a number of important federal environmental and resource protection laws (Figure 24-8). One type of such legislation *sets standards for pollution levels* (as in the Clean Air Acts, Chapter 18, p. 493). A second type *screens new substances for safety and sets standards* (as in the Safe Drinking Water Act, Chapter 20, p. 557). A third type of legislation *encourages resource conservation* (the Resource Conservation and Recovery Act, Chapter 21, p. 595). A fourth type *sets aside or protects certain species, resources, and ecosystems* (the Endangered Species Act, Chapter 9, p. 208, and the Wilderness Act, Chapter 10, p. 236) (**Concept 24-3**).

A fifth type of legislation *requires evaluation of the environmental impact of an activity proposed by a federal agency,* as in the National Environmental Policy Act, or NEPA, passed in 1970. Under NEPA, an *environmental impact statement (EIS)* must be developed for every major federal project likely to have an effect on environmental quality. The EIS must explain why the proposed project is needed and identify its beneficial and harmful environmental impacts. For example, an environmental impact study typically considers a project's likely effects on wildlife habitat, soils, water quality, air quality, stream flows, and other factors. The EIS must also suggest ways to lessen any harmful impacts, and it must present an evaluation of alternatives to the project. The EIS documents must be published and are open to public comment.

NEPA does not prohibit environmentally harmful government projects. But more than one-third of the country's land is under federal management, and NEPA requires the managing agencies to take environmental consequences into account in making decisions. It also exposes proposed projects and their possible harmful

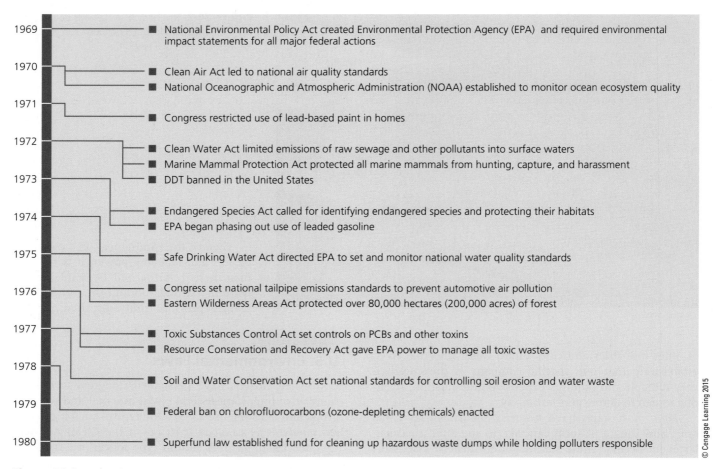

1969 — National Environmental Policy Act created Environmental Protection Agency (EPA) and required environmental impact statements for all major federal actions

1970 — Clean Air Act led to national air quality standards
— National Oceanographic and Atmospheric Administration (NOAA) established to monitor ocean ecosystem quality

1971 — Congress restricted use of lead-based paint in homes

1972 — Clean Water Act limited emissions of raw sewage and other pollutants into surface waters
— Marine Mammal Protection Act protected all marine mammals from hunting, capture, and harassment
1973 — DDT banned in the United States

1974 — Endangered Species Act called for identifying endangered species and protecting their habitats
— EPA began phasing out use of leaded gasoline

1975 — Safe Drinking Water Act directed EPA to set and monitor national water quality standards

1976 — Congress set national tailpipe emissions standards to prevent automotive air pollution
— Eastern Wilderness Areas Act protected over 80,000 hectares (200,000 acres) of forest

1977 — Toxic Substances Control Act set controls on PCBs and other toxins
— Resource Conservation and Recovery Act gave EPA power to manage all toxic wastes

1978 — Soil and Water Conservation Act set national standards for controlling soil erosion and water waste

1979 — Federal ban on chlorofluorocarbons (ozone-depleting chemicals) enacted

1980 — Superfund law established fund for cleaning up hazardous waste dumps while holding polluters responsible

© Cengage Learning 2015

Figure 24-8 Some of the major environmental laws and their amended versions enacted in the United States since 1969. No major new environmental laws have been passed since the 1970s, although some existing laws have been amended.

effects to public scrutiny. Opponents have targeted NEPA as a law to weaken or repeal.

CONSIDER THIS...

THINKING ABOUT Environmental Impact Statements

Are you in favor of requiring government agencies to develop environmental impact statements such as those required by NEPA? Explain.

Environmental laws in the United States have been highly effective, especially in controlling pollution. In 2012, economist Gernot Wagner estimated that the benefits of the U.S. Clean Air Act, including healthier work environments, fewer cases of sickness and death, and higher worker productivity, exceeded the costs of complying with the law by 30 to 1.

In 2012, NEPA had been in effect for 40 years. In that time, the Environmental Protection Agency (EPA), created by NEPA to enforce federal environmental laws, had helped to sharply reduce atmospheric levels of sulfur dioxide and nitrogen oxides, ban the use of leaded gasoline, ban the widespread use of toxic DDT, get secondhand

tobacco smoke classified as a carcinogen, and regulate the use of many other toxic chemicals. Instead of hindering economic growth, as its critics feared, NEPA sparked a domestic environmental protection industry that now employs more than 1.5 million people. It has also spurred profitable industry innovations, including the catalytic converter, a pollution control device for motor vehicles, and chemicals that serve as substitutes for ozone-depleting chemicals.

However, since 1980 a well-organized and well-funded movement has mounted a strong campaign to weaken or repeal existing U.S. environmental laws and regulations. Three major groups are strongly opposed to various environmental laws and regulations: some corporate leaders and other powerful people who see them as threats to their profits, wealth, and power; citizens who see them as threats to their private property rights and jobs; and state and local government officials who resent having to implement federal laws and regulations with little or no federal funding (unfunded mandates), or who disagree with certain federal regulations.

One problem working against effective regulations is that the focus of environmental issues has shifted from easy-to-see dirty smokestacks and filthy rivers to more complex, long-term, and less visible environmental problems such as climate change, biodiversity loss, and groundwater pollution. Explaining such complex issues to the public and mobilizing support for often controversial, long-range solutions to such problems is difficult. (See the online Guest Essay on environmental reporting by Andrew C. Revkin.)

Another problem is that some environmentalists have primarily brought bad news about the state of the environment to the general public. History shows that bearers of bad news are not well received, and opponents of the environmental movement have used this to undermine environmental concerns. History also shows that people are moved to bring about change mostly by an inspiring, positive, and hopeful vision of what the world could be like (see Chapter 1, Core Case Study, p. 4).

Since 2000, efforts to weaken environmental laws and regulations have escalated. Nevertheless, independent polls show that more than 80% of the U.S. public strongly support environmental laws and regulations and do not want them weakened. However, polls also show that less than 10% of the U.S. public (and in hard economic times only about 2–3%) consider the environment to be one of the nation's most pressing problems. As a result, environmental concerns often do not get transferred to the ballot box.

In order to make a transition to a more environmentally sustainable society, U.S. citizens (and citizens in other democratic countries) will have to elect ecologically literate and environmentally concerned leaders. A rapidly growing number of citizens are insisting that elected leaders work across party lines to end the political deadlock that has virtually immobilized the U.S. Congress since 1980, with respect to environmental issues and other key concerns. To prevent any further weakening of U.S. environmental laws and regulations, concerned citizens will need to exert political pressure on several fronts—the type of pressure that led to passage of important environmental laws in the first place.

24-4 What Are the Major Roles of Environmental Groups?

CONCEPT 24-4

Grassroots groups are growing and combining their efforts with those of large environmental organizations in a global sustainability movement.

Citizen Environmental Groups Play Important Roles

The spearheads of the global conservation, environmental, and environmental justice movements are the tens of thousands of nonprofit nongovernmental organizations (NGOs) working at the international, national, state, and local levels. The growing influence of these organizations is one of the most important changes influencing environmental decisions and policies (Concept 24-4).

NGOs range from grassroots groups with just a few members to organizations like the World Wildlife Fund (WWF), a 5-million-member global conservation organization, which operates in 100 countries. Other international groups with large memberships include Greenpeace, the Nature Conservancy, and Conservation International.

Using e-mail, social networks, text messages, and the Internet, some environmental NGOs have organized themselves into an array of influential international networks. Examples include the Pesticide Action, Climate Action, International Rivers, and Women's Environment and Development Networks. They collaborate across national borders, monitor the environmental activities of governments, corporations, and international agencies such as the World Bank and the World Trade Organization (WTO), and attend international conferences to try to influence negotiations and agreements (Individuals Matter 24.2). They also help to expose corruption and violations of national and international environmental agreements, such as the Convention on International Trade in Endangered Species (CITES), which prohibits international trade of endangered species (see Chapter 9, p. 208).

In recent years, *global public policy networks* have formed in response to rapidly changing conditions in a globalizing world. Each of these innovative groups focuses on a particular environmental problem by bringing together governments, the private sector, international organizations, and NGOs. Since the 1990s, more than 50 such networks have emerged. Examples include the *International Forum on Forests*, which develops proposals for sustainable forest management; the *Global Water Partnership*, which works toward integrated water resources management; the *Renewable Energy Policy Network*, which develops policies to spur development of renewable energy; and the *Earth Day Network*.

In the United States, more than 8 million citizens belong to more than 30,000 NGOs that deal with environmental issues. They range from small grassroots groups to large heavily funded *mainline* groups, the latter usually staffed by expert lawyers, scientists, economists, lobbyists, and fund-raisers. The largest of these groups are the World Wildlife Fund, the Sierra Club, the National Wild-

Anjali Appadurai: A College Student Who Electrified a United Nations Conference

Courtesy of Greta Rybus and TedXDirigo

In 2011, Anjali Appadurai was a third-year student at the College of the Atlantic in Maine (USA). That year, she was chosen by her fellow students to represent a wide-ranging international youth group network at the United Nations Climate Change Conference in Durban, South Africa. There she delivered a speech that not only electrified the conference attendees but also went viral on YouTube.

Having moved with her family at age 6 from southern India to Vancouver, Canada, Appadurai wasn't just representing her entire generation. She also felt a special responsibility to the residents of many less-developed countries where the most brutal effects of climate change are now being felt. In her address, referring to the need for major cuts to greenhouse gas emissions, Appadurai said:

> I speak for more than half of the world's population. We are the silent majority. You've given us a seat in this hall, but our interests are not on the table. . . . You've been negotiating all my life. The International Energy Agency tells us we have five years until the window to avoid irreversible climate change closes. You're saying 'Give us ten.' . . . In the long run this will be seen as the defining moment of an era in which narrow self-interest prevailed over science, reason, and common compassion. . . . Respect the future of your descendants. Mandela said, 'It always seems impossible, until it's done.' . . . So distinguished delegates and governments around the world. . . get it done!

After her speech, she took the bold step of engaging her audience in an unplanned call-and-response session, in which she got many of them to yell "Get it done!" loudly and repeatedly. After her speech, the conference administrators put restrictions on Appadurai to prevent her from continuing what they considered to be disruptive behavior.

A year later, Appadurai was again chosen to travel to the annual UN climate change conference, held in Doha, Qatar, in 2012. Recalling the events of 2011, the conference administrators first denied her entry to the conference. However, after a large number of her supporters created a publicity storm on Twitter and Facebook, the administrators were forced to admit her to the conference where she again represented youth from all over the world. While the conference was ineffective in bringing about significant immediate action to curb climate change, Appadurai, who graduated from the College of the Atlantic in 2013, made it known that she comes from a group—and from a whole generation—that will not wait patiently for change.

Background photo: Jan Martin Will/Shutterstock.com

Figure 24-9 These Greenpeace protesters used an inflatable motorized boat to try to hinder the hunting of whales by a Japanese whaling fleet. For several decades Greenpeace has engaged in such environmental actions and in environmental education activities.

life Federation, the Audubon Society, Greenpeace (Figure 24-9), Friends of the Earth, and the Natural Resources Defense Council (see the Case Study that follows).

The largest groups have become powerful and important forces within the U.S. political system. They have helped to persuade Congress to pass and strengthen environmental laws (Figure 24-8), and they fight attempts to weaken or repeal these laws.

Some industries and environmental groups are working together to find solutions to environmental problems. For example, the Environmental Defense Fund worked with McDonald's to redesign its packaging system to eliminate its plastic hamburger containers. Greenpeace worked with a German manufacturer to build a refrigerator that does not use the potent greenhouse gases called HFCs as coolants. Now, there are more than 300 million of these GreenFreeze refrigerators in homes around the world.

Some environmental groups have shifted resources from demonstrating and litigating to publicizing research on innovative solutions to environmental problems. For example, to promote the use of chlorine-free paper, Greenpeace Germany printed a magazine using such paper and encouraged readers to demand that magazine publishers switch to chlorine-free paper. Shortly thereafter, several major magazines made that shift.

Some environmental groups and networks are already beginning to broaden the focus of their efforts. A good example of an extremely active group is the Natural Resources Defense Council.

CASE STUDY
The Natural Resources Defense Council

One of the stated purposes of the Natural Resources Defense Council (NRDC) is "to establish sustainability and good stewardship of the Earth as central ethical imperatives of human society. . . . We work to foster the fundamental right of all people to have a voice in decisions that affect their environment. . . . Ultimately, NRDC strives to help create a new way of life for humankind, one that can be sustained indefinitely without fouling or depleting the resources that support all life on Earth."

To those ends, NRDC goes to court to stop environmentally harmful practices. It also informs and organizes millions of environmental activists, through its website, magazines, and newsletters, to take actions to protect the environment—globally, regionally, and locally. For example, its BioGems network, accessible through the NRDC website, regularly informs subscribers about environmental threats all over the world, and helps people to take action by donating money, signing petitions, and writing letters to corporate and government officials and newspaper editors.

In 2005, with NRDC's help, U.S. citizens organized massive opposition to a proposed government policy that would have allowed sewer operators to routinely dump virtually untreated sewage into the nation's lakes, rivers, and streams. Because of well-informed, very vocal opposition to this proposal, the U.S. House of Representatives voted overwhelmingly to block the EPA from finalizing this so-called "blending" proposal.

In another case, in 2001, NRDC helped forge an agreement among Canadian timber companies, environmentalists, native peoples, and the provincial government of British Columbia (Canada) to protect a vast area of the Great Bear Rainforest from destructive logging. This followed years of pressure from NRDC activists on logging companies, their U.S. corporate customers, and provincial officials to protect the habitats of eagles, grizzly bears, wild salmon, and the rare spirit bear, a subspecies of the American black bear with a creamy white fur coat (Figure 24-10).

Grassroots Environmental Groups Bring about Change from the Bottom Up

The base of the environmental movement in the United States and throughout the world consists of thousands of grassroots citizens' groups organized to improve environmental quality, often at the local level. According to political analyst Konrad von Moltke, "There isn't a government in the world that would have done anything for the environment if it weren't for the citizen groups." Taken together, a loosely connected worldwide network of grassroots NGOs working for bottom-up political, social, economic, and environmental change can be viewed as an emerging citizen-based *global sustainability movement* (**Concept 24-4**).

Since this movement got its start in the 1970s, many grassroots groups have worked with individuals and communities to oppose harmful projects such as landfills, waste incinerators, and nuclear waste dumps, as well as to fight against the clear-cutting of forests and pollution from factories and power plants. They have also taken action against environmental injustice (Figure 24-11). (See the

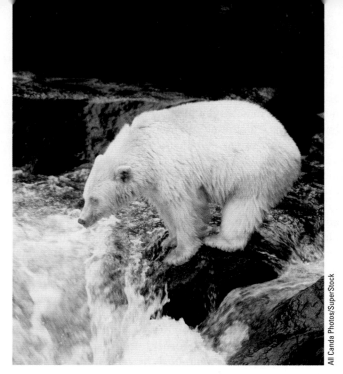

Figure 24-10 The Natural Resources Defense Council (NRDC) has worked to protect the habitat of the rare spirit bear in the Canadian province of British Columbia. The bear's coastal rain forest habitat is threatened by logging. This bear is waiting to catch a salmon.

Figure 24-11 These residents of a neighborhood in Detroit, Michigan (USA), wanted a nearby hospital to shut down its medical waste incinerator, which was polluting the air in their community.

online Guest Essay on this topic by Robert D. Bullard.) And they have worked to make many communities more sustainable (see the Case Study that follows).

Grassroots groups have organized *conservation land trusts* wherein property owners agree to protect their land from development or other harmful environmental activities, often in return for tax breaks on the land's value. These groups have also spurred other similar efforts to save wetlands, forests, farmland, and ranchland from development, while helping to restore clear-cut forests, degraded grasslands, and wetlands and rivers that have been degraded by pollution.

The Internet, social networking, and text messaging have become important tools for grassroots groups. With these tools, they can expand their membership, raise funds, and quickly plan and execute actions such as demonstrations and rallies. The Internet also helps these groups to be more efficient and to become interconnected within networks, which can make them even more effective.

Most grassroots environmental groups use nonviolent and nondestructive tactics such as protest marches, sitting in trees to help prevent the clear-cutting of old-growth forests, and other devices (Figure 24-9) for generating publicity to help educate and encourage the public to oppose various environmentally harmful activities. Such tactics often work because they produce bad publicity for practices and businesses that threaten or degrade the environment. For example, after 2 years of pressure and

protests, Home Depot agreed to sell only wood products made from certified sustainably grown timber. Within a few months, Lowes and eight other major building supply chains in the United States developed similar policies.

Much more controversial are militant environmental groups that use violent means, such as breaking into labs to free animals used in drug testing and destroying property such as bulldozers and SUVs. Most environmentalists strongly oppose such tactics.

CASE STUDY

The Environmental Transformation of Chattanooga, Tennessee

Local officials, business leaders, and citizens have worked together to transform Chattanooga, Tennessee, from a highly polluted city to one of the most sustainable and livable cities in the United States (Figure 24-12).

During the 1960s, U.S. government officials rated Chattanooga as one of the dirtiest cities in the United States. Its air was so polluted by smoke from its industries that people sometimes had to turn on their vehicle headlights in the middle of the day. The Tennessee River, flowing through the city's industrial center, bubbled with toxic waste. People and industries fled the downtown area and left a wasteland of abandoned and polluting factories, boarded-up buildings, high unemployment, and crime.

In 1984, the city decided to get serious about improving its environmental quality. Civic leaders started a *Vision 2000* process with a 20-week series of community meetings in which more than 1,700 citizens from all walks of life gathered to build a consensus about what the city could be at the turn of the century. Citizens identified the

Figure 24-12 Since 1984, citizens have worked together to make the city of Chattanooga, Tennessee, one of the best and most sustainable places to live in the United States.

©Chattanooga Area Convention and Visitors Bureau

city's main problems, set goals, and brainstormed thousands of ideas for solutions.

By 1995, Chattanooga had met most of its original goals. The city had encouraged zero-emission industries to locate there and replaced its diesel buses with a fleet of quiet, zero-emission electric buses, made by a new local firm. The city also launched an innovative recycling program after environmentally concerned citizens blocked construction of a new garbage incinerator that would have emitted harmful air pollutants. These efforts paid off. Since 1989, the levels of the seven major air pollutants in Chattanooga have been lower than the levels required by federal standards.

Another project involved renovating much of the city's low-income housing and building new low-income rental units. Chattanooga also built the nation's largest freshwater aquarium, which became the centerpiece for downtown renewal. The city developed a riverfront park along both banks of the Tennessee River, which runs through downtown. The park typically draws more than 1 million visitors per year. As property values and living conditions have improved, people and businesses have moved back downtown.

In 1993, the community began the second stage of this process in *Revision 2000*. Goals included transforming an abandoned and blighted area in South Chattanooga into a mixed community of residences, retail stores, and zero-emission industries where employees can live near their workplaces. By 2009, most of these goals were met. Between 2009 and 2012, in the face of a general economic downturn, Chattanooga had one of the nation's strongest local economies, with a lower-than-average unemployment rate and rising property values. And this city of 170,000 has focused on and succeeded in attracting young people from other areas.

Chattanooga's environmental success story is a shining example of people working together to produce a more livable and economically and environmentally sustainable city. It is an example of the win-win **principle of sustainability** (see Figure 1-5, p. 9 or back cover).

Student Environmental Groups and Researchers Are Taking Leadership Roles

Campus environmental groups have been leading the way on many campuses as they seek to make their schools more sustainable (**Core Case Study**). Most of these groups work with members of their school's faculty and administration to bring about environmental improvements in their schools and local communities.

For example, students at Middlebury College in Vermont worked with the school staff in switching the college from burning oil to heat its buildings to burning wood chips in a state-of-the-art boiler that is part of a cogeneration system. The system heats about 100 college buildings and spins a turbine to generate about 20% of the electricity used by the college, helping the school to reduce its carbon dioxide emissions by 40%. The college has also planted an experimental patch of fast-growing willow trees as a source of some of the wood chips for the system.

Many student groups make *environmental audits* of their campuses or schools. They gather data on practices affecting the environment and use them to propose changes that will make their campuses or schools more environmentally sustainable while usually saving money in the process. Such audits have focused on implementing or improving recycling programs, convincing university food services to buy more food from local organic farms, shifting from fossil fuels to renewable energy, retrofitting buildings to make them more energy efficient, and implementing concepts of environmental sustainability throughout the curriculum.

🔍 CONSIDER THIS. . .

THINKING ABOUT Environmental Groups
What environmental groups exist at your school? Do you belong to such a group? Why or why not? (See the **Core Case Study** at the beginning of this chapter.)

24-5 How Can We Improve Global Environmental Security?

CONCEPT 24-5
Environmental security is necessary for economic security and is at least as important as national security.

Why Is Global Environmental Security Important?

Countries are legitimately concerned with *national security* and *economic security*. However, ecologists and many economists point out that all economies are supported by the earth's natural capital (see Figure 1-3, p. 7, and Figure 23-5, p. 635). Thus, environmental security, economic security, and national security are interrelated (**Concept 24-5**).

According to environmental scientist Norman Myers,

If a nation's environmental foundations are degraded or depleted, its economy may well decline, its social fabric deteriorate, and its political structure become destabilized as growing numbers of people seek to sustain themselves from declining resource stocks. Thus, national security is no longer about fighting forces and weaponry alone. It relates increasingly to watersheds, croplands, forests, genetic resources, climate, and other factors that, taken together, are as crucial to a nation's security as are military factors. (See Myers' online Guest Essay on this subject.)

For example, Haiti has suffered from a severe loss of environmental and economic security as a result of rapid population growth combined with deforestation, severe soil erosion, rampant poverty, and political insecurity. In a desperate struggle for survival, its people have stripped away most of the country's trees and other vegetation for use as firewood (see Figure 10-18, p. 229 and Figure 24-13). Because of this, along with several damaging hurricanes, the percentage of Haiti's land that was forested dropped from more than 60% in 1923 to about 2% in 2006. This major loss of vegetation led to severe soil erosion. Taken together, these factors along with a severe earthquake in 2010 have greatly reduced food production and led to greater poverty, malnutrition, and social unrest.

Research by Thomas Homer-Dixon, director of Canada's Trudeau Centre for Peace and Conflict Studies, has revealed a strong correlation between growing scarcities of resources, such as cropland, water, and forests, and the spread of civil unrest and violence that can lead to failing states. These countries have dysfunctional governments that can no longer provide security and basic services such as education and health care. They tend to suffer from a breakdown of law and order and general deterioration and often end up in civil wars as groups compete for power, and these wars can spread to nearby countries. Many failing states also become training grounds for terrorists (Afghanistan and Iraq), weapons traders (Nigeria and Somalia), and drug producers (Afghanistan and Myanmar). Together, they also generate millions of refugees who are displaced from their homes and land, often while fleeing for their lives.

Myers and other analysts call for all countries to make environmental security a major focus of diplomacy and government policy at all levels. In 2012, environmental policy expert Lester Brown summarized it this way:

The time when military forces were the prime threat to security has faded into the past. The threats now are climate volatility, spreading water shortages, continuing population growth, spreading hunger, and failing states. . . . We can most effectively achieve our security goals by helping to expand food production, by filling the family planning gap, by building wind farms and solar power plants, and by building schools and clinics.

We Can Develop Stronger International Environmental Policies

A number of international environmental organizations help shape and set global environmental policy. Perhaps the most influential is the United Nations, which houses a large family of organizations including the U.N. Environment Programme (UNEP), the World Health Organization (WHO), the U.N. Development Programme (UNDP), and the Food and Agriculture Organization (FAO).

Other organizations that make or influence environmental decisions are the World Bank, the Global Environment Facility (GEF), and the World Conservation Union (also known as the IUCN). Despite their often limited funding, these and other organizations have played important roles in

• expanding understanding of environmental issues,
• gathering and evaluating environmental data,
• developing and monitoring international environmental treaties,
• providing grants and loans for sustainable economic development and reduction of poverty, and

ROBIN MOORE/National Geographic Creative

Figure 24-13 Hillsides stripped of vegetation near Haiti's capital city of Port-au-Prince.

- helping more than 100 nations to develop environmental laws and institutions.

In 1992, governments of more than 178 nations and hundreds of NGOs met at the U.N. Conference on Environment and Development (UNCED) in Rio de Janeiro, Brazil. The major policy outcome of this conference was Agenda 21, a global agenda for sustainable development in the 21st century, with goals for addressing the world's social, economic, and environmental problems. The conference also established the Commission on Sustainable Development to monitor progress toward the Agenda 21 goals.

Recently, certain groups in the United States have been arguing that Agenda 21 is a secret plan to force people to give up their single-family homes, move to cities, and surrender their cars. They say it is internationally controlled and threatens the American way of life. Defenders of Agenda 21 have debunked these claims, pointing out that trends toward alternatives to the car, sustainable food production, more sustainable urban design, and protection of forests, wetlands, and farmlands through smart growth are local movements springing from the grassroots and are certainly not imposed by any global authority. They note that such local planning processes allow people to shape their own communities, instead of allowing outside interests such as multinational corporations to do so.

Despite the good intentions of Agenda 21, little progress has been made toward its 20-year-old goals. In 2012, the UNEP evaluated progress and found that, of the 90 most crucial goals, only four had been approached and none had been achieved. The removal of lead from gasoline, the

phasing out of ozone-depleting chemicals, improvements to drinking water supplies in poor countries, and research on pollution of the oceans were the four areas where significant progress had been made. In other areas, such as carbon dioxide emissions, extinction threats, overfishing, ocean dead zones, and harm to coral reefs, the world has actually slipped farther away from the Agenda 21 goals.

In 2012, on the 20th anniversary of UNCED, the UN hosted Rio+20, another Earth Summit conference to revisit the issues addressed in 1992. It was the largest conference ever, with 50,000 attendees, and as it kicked off, there were up to 50,000 non-attendees demonstrating in the streets of Rio de Janeiro (Figure 24-14). After three days, the conferees produced a nonbinding document that they referred to as a roadmap for sustainable development. The representatives of more than 190 nations, including the United States, ratified it.

However, some analysts and organizations criticized the agreement harshly for its failings. The document contained no enforceable commitment on climate change. It did not address any proposal to end fossil fuel subsidies and it failed to promote any sort of shift to renewable energy sources. Critics also argued that, because of excessive influence by corporate sponsors, the conference was unable to make real progress toward shifting the world onto a more sustainable path. Other critics, while expressing disappointment, argued that the conference outcome could serve as a foundation for more meaningful changes.

Figure 24-15 summarizes some of the successes and failures from long-term international efforts to deal with global environmental problems.

Figure 24-14 These native Brazilians, part of the Landless People's Movement, demonstrated outside the Rio+20 Conference on Sustainable Development in 2012. They argued that economic development that pushes native people off the land they have lived on sustainably for centuries is an example of environmental injustice.

© Sue Cunningham Photographic/Alamy

Trade-Offs

Global Efforts to Solve Environmental Problems

Successes	Failures
Over 500 international environmental treaties and agreements	Most international environmental treaties lack criteria for evaluating their effectiveness
1992 Copenhagen Ozone Protocol has helped reduce ozone-depleting chemicals	1992 Rio Earth Summit led to nonbinding agreements, inadequate funding, and limited improvements
1992 Rio Earth Summit adopted principles for handling global environmental problems	2012 Rio+20 Earth Summit failed to deal with climate change, energy policy, and biodiversity loss
2012 Rio+20 Earth Summit included small-scale policy improvements	Climate change conferences have all failed to deal with projected climate change

© Cengage Learning 2015

Figure 24-15 There have been successes and failures in international efforts to deal with global environmental problems. ***Questions:*** In weighing these successes and failures, do you believe that international conferences are valuable and should be continued? If not, what are some alternatives?

Photo: NASA

The primary focus of the international community on environmental problems has been the development of various international environmental laws and nonbinding policy declarations called *conventions*. And there are more than 500 international environmental treaties and agreements—known as *multilateral environmental agreements* (MEAs). Environmental issues can also be used to resolve disputes between nations (Individuals Matter 24.3).

To date, the Montreal Protocol and the Copenhagen Amendment for protecting the ozone layer (see Chapter 18, p. 497) are the most successful examples of such agreements (see images in Figure 24-15). The MEA process faces a number of challenges. MEAs typically take years to develop and require full consensus to implement. There is often a lack of funding and it becomes difficult to monitor and enforce these agreements, and they sometimes conflict with one another.

Corporations Can Play a Key Role in Promoting Environmental Sustainability

In our increasingly globalized economy, it has become clear that governments and corporations must work together to achieve goals for increased environmental sustainability. Governments can set environmental standards and goals through legislation and regulations, and corporations generally have highly efficient ways of accomplishing such goals. Making a transition to more sustainable societies and economies will require huge amounts of investment capital and research and development funding. Most of this money will likely have to come from profitable corporations, especially considering the recurring budgetary pressures faced by most governments. Thus, corporations could play a vital role in achieving a more sustainable future.

The good news is that some thoughtful business and political leaders are realizing that "business as usual" is no longer a viable option (see Individuals Matter 23.2, p. 650). A growing number of corporate chief executive officers (CEOs) and investors are aware that there is considerable money to be made from developing and selling green products and services during this century. This switch to new product lines is guided by the concept of *eco-efficiency,* which is about finding ways to create more economic value with less harmful health and environmental impacts. Improving eco-efficiency can also save businesses money and help them to meet their financial responsibilities to stockholders and investors.

In 2012, at the Rio+20 Earth Summit, the Nature Conservancy announced a program to help companies include the monetary value of natural capital in their global goals and strategies. This will include estimating the value of nature's ecosystem services not only to the company but also to the communities in which it works. For example, companies can learn how planting trees can improve their properties and how maintaining coastal wetlands can reduce the costs of hurricane damage. Companies such as Dow Chemical have found that such investments can improve their bottom lines considerably.

GOOD NEWS

Saleem Ali: Environmental Scientist, Mediator, and National Geographic Emerging Explorer

©Courtesy of Denis Hayes/The Bullitt Foundation

Saleem Ali believes that the environment can serve as a powerful tool for reaching peaceful agreements among conflicting groups or nations. As an environmental scientist, he works as a professional mediator, helping governments, corporations, and native groups to resolve their differences.

"When people discuss the high politics of war and peace, the environment can often bring all sides together around the shared goals of conserving resources," says Ali. He points to the Antarctic Treaty, signed by the United States and Soviet Union at the height of the Cold War. Both sides saw the value of Antarctica and agreed to make it a protected scientific common ground.

Ali serves as an advisor to the United Nations on environmental conflicts between nations and promotes the use of *Transboundary Conservation Areas,* or *peace parks*—environmentally protected areas along borders between nations. By agreeing to protect these areas, warring nations can come closer to peaceful agreements on a broader scale, as happened between Ecuador and Peru. Ali would like to see similar progress between India and his native Pakistan.

Ali also promotes the inclusion of corporate interests in solutions to environmental problems, arguing that unless they are part of the solution, the problems likely cannot be solved. He points to Denmark's eco-industrial park (see Figure 21-31, p. 599), which makes use of a sort of industrial food web, recycling industrial wastes as resources among several businesses in the park. This cooperative enterprise was spurred by Denmark's pollution laws that allowed companies to find their own innovative solutions, rather than imposing top-down solutions on them.

Ali works to ensure that the agreements he brokers will not harm future generations, in keeping with one of the social science **principles of sustainability** (see Figure 1-5, p. 9 or back cover). He says, "I remain optimistic about human resilience . . . our ability to adapt and confront new circumstances."

Background photo: Pressmaster/Shutterstock.com

24-6 How Can We Implement More Sustainable and Just Environmental Policies?

CONCEPT 24-6

Making the transition to more sustainable societies will require that nations and groups within nations cooperate and make the political commitment to achieve this transition.

We Can Shift to More Environmentally Sustainable Societies

Scientists and other experts have suggested guidelines that we can follow as we work toward making our societies more environmentally sustainable. *First,* work on *preventing or minimizing* environmental problems instead of letting them build up to crisis levels. *Second,* use well-designed and carefully monitored *marketplace solutions* (see Chapter 23, p. 643) to help prevent or reduce the harmful impact of most environmental problems. *Third,* cooperate and innovate to find *win-win solutions* or *trade-offs* to environmental problems and injustices, in keeping with one of the social science **principles of sustainability**. *Fourth,* be *honest and objective.* People on both sides of thorny environmental issues should take a vow not to exaggerate or distort their positions in attempts to play win-lose or winner-take-all games. Some environmental scientists have specialized in helping people and organizations to apply these principles (Individuals Matter 24.3).

Making the transition to a more sustainable world will also require governments and their citizens to rethink their priorities. Figure 23-16 (p. 647) shows that it would take about $245 billion a year for the world to meet basic social and health goals and to provide environmental security for all. This amounts to about one-eighth of the total amount that countries spend each year on environmentally harmful subsidies.

The world has the knowledge, technologies, and financial resources to make the shift to more equitable and environmentally sustainable global and national policies. Thus, making this shift is primarily an economic, political, and ethical decision (see Chapter 25 for a discussion of environmental ethics). It involves shifting to more sustainable forestry (see Figure 10-16, p. 227), agriculture (see Figure 12-33, p. 309), water resource use (see Figure 13-27, p. 340), energy resource use (see Figure 16-36, p. 434), and economies (see Figure 23-18, p. 649), while slowing projected climate change (see Figure 19-22, p. 528) and educating the public and elected officials about the urgent need to make this shift over the next several decades.

Some say that the call for making this shift is idealistic and unrealistic. Others say that it is unrealistic and dangerous to keep assuming that our present course is sustainable, and they warn that we have precious little time to change.

Big Ideas

- An important outcome of the political process is environmental policy—the body of laws, regulations, and programs that are designed, implemented, funded, and enforced by one or more government agencies.

- All politics is local, and individuals can work with each other to become part of political processes that influence environmental policies *(Individuals matter).*

- Environmental security is necessary for economic security and is at least as important as national security; making the transition to more environmentally sustainable societies will require that nations cooperate just as they do for national security purposes.

Greening College Campuses and Sustainability

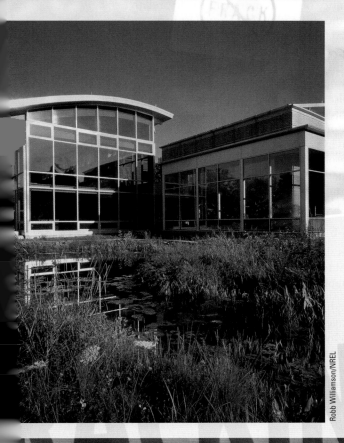

Robb Williamson/NREL

College students around the world have shown that it is possible to create sustainable environmental policies, at least in the communities in and around many college campuses (**Core Case Study**). The world has the abilities and resources to implement policies that would help to eradicate poverty and malnutrition, eliminate illiteracy, sharply reduce infectious diseases, stabilize human populations, and protect the earth's natural capital. We can do this by applying the scientific **principles of sustainability** (see Figure 1-2, p. 6 or back cover)—relying much more on solar energy and other renewable energy sources, reusing and recycling much more of what we produce, and respecting, restoring, and protecting as much as possible of the biodiversity that supports our lives and economies.

National and international policy makers could also be guided by the three social science **principles of sustainability** (see Figure 1-5, p. 9 or back cover). In the political arena, they will have to try harder to find win-win solutions that benefit the largest numbers of people while also benefiting the environment. Such solutions will likely have to include internalizing the harmful environmental and health costs of producing and using goods and services (full-cost pricing). And if they are truly interested in long-term sustainability, these decision makers, as well as all the rest of us, must make each decision with future generations in mind—seeking to leave the world in at least as good a condition as what we now enjoy.

Positive change toward sustainability can occur much more rapidly than we think. Any or all of us can choose to take part in the change by becoming politically aware, informed, and active with regard to issues that affect our environmental and political futures.

Chapter Review

Core Case Study

1. Give four examples of how colleges and universities are playing a leading role in shifting to more environmentally sustainable operations and policies (**Core Case Study**).

Section 24-1

2. What is the key concept for this section? List three broad issues that government is the best equipped to handle and for each one, explain why this is so. What are government's **policies**? Define **politics** and **environmental policy**. What are the four stages of a policy life cycle? Why is the process usually cyclical and what is another name for it? What is a **democracy**? What are two types of special interest groups? Describe two features of democratic governments that hinder their ability to deal with environmental problems. List seven principles that decision makers can use in making environmental policy.

Section 24-2

3. What are the two key concepts for this section? What are the three branches of government in the United States and what major role does each play? Describe three important components of policy. What is **lobbying**? Why are some analysts concerned about the growing power of some lobbyists? Explain why developing environmental policy is a difficult and controversial process. What are three ways in which scientific and political processes differ? Why are these differences important for policy making?

4. What are four ways in which individuals can help to develop or change environmental policy in a democracy? What does it mean to say that *all politics is local*? What are four ways to provide environmental leadership?

Section 24-3

5. What is the key concept for this section? What is **environmental law**? What is a **civil suit**? Distinguish between the **plaintiff** and the **defendant** in a lawsuit. List and explain six factors that can limit the effectiveness of environmental lawsuits. What is a SLAPP?

6. List five general types of U.S. environmental laws and give an example of each. Briefly explain what NEPA is, what it required, and how effective it has been. What is an environmental impact statement? Explain how and why U.S. environmental laws have been under attack since 1980. Explain two problems that work against effective environmental regulations.

Section 24-4

7. What is the key concept for this section? Describe the roles of grassroots and mainstream environmental organizations and give an example of each type of organization. What is a global policy network? Give an example. Describe the role and effectiveness of the Natural Resources Defense Council (NRDC) in the formation of U.S. environmental policy. Give two examples of grassroots environmental groups and of how they can bring about change. How does the story of Chattanooga's transformation illustrate the potential effectiveness of citizen groups in affecting urban environmental policy? Give two examples of how students have played successful roles in affecting environmental policy on their campuses (**Core Case Study**).

Section 24-5

8. What is the key concept for this section? Explain the importance of environmental security, relative to economic and national security. List three examples of international environmental organizations and five ways in which they have played environmental policy making roles. Summarize the stories of the UNCED and Rio+20. What is Agenda 21? List two examples of successes and two examples of failures from international efforts to deal with global environmental problems. What are conventions and multilateral environmental agreements (MEAs) and what are three problems that have arisen with MEAs? Describe roles that corporations can play in helping to achieve environmental sustainability, and give an example of such an effort.

Section 24-6

9. What is the key concept for this section? What are four guidelines for shifting to more environmentally sustainable societies?

10. Explain how many college students have managed to change the environmental policies of their institutions by applying the three social science **principles of sustainability** (**Core Case Study**). How can these principles be applied to guiding national and international environmental policy making processes?

Note: Key terms are in bold type.

Critical Thinking

1. Consider the various actions taken by college students and their institutions as described in the **Core Case Study** that opens this chapter, as well as in other parts of this chapter. Which one or more of these actions would be appropriate and effective on your campus? Explain. Pick one of these actions and write a brief plan for implementing it where you go to school.

2. Pick an environmental problem that affects the area where you live and decide where in the policy life cycle (Figure 24-2) the problem could best be placed. Apply the cycle to this problem and describe how the problem has progressed (or will likely progress) through each stage. If your problem has not progressed to the policy adjustment stage, explain how you think the policy dealing with the problem could be adjusted, if at all.

3. Explain why you agree or disagree with each of the seven principles listed on pp. 660–661, which are recommended by some analysts for use in making environmental policy decisions. Which three of these principles do you think are the most important? Why?

4. What are two ways in which the scientific process described in Chapter 2 (see Figure 2-2, p. 31) parallels the policy life cycle (Figure 24-2)? What are two ways in which they differ?

5. Do you think that corporations and government bodies are ever justified in filing SLAPP lawsuits? Give three reasons for your answer. Do you think that potential defendants of SLAPP suits should be protected in any way from such suits? Explain.

6. Government agencies can help to keep an economy growing or to boost certain types of economic development by, for example, building or expanding a major highway through an undeveloped area. Proponents of such development have argued that requiring environmental impact statements for these projects interferes with efforts to help the economy. Do you agree? Is this a problem? Why or why not?

7. Congratulations! You are in charge of the country where you live. List the five most important components of your environmental policy.

8. List three ways in which you could apply the material in Section 24-2 to try to have an effect on an environmental policy making process.

Doing Environmental Science

Polls have identified five categories of citizens in terms of their concern over environmental quality: **(a)** those involved in a wide range of environmental activities, **(b)** those who do not want to get involved but are willing to pay more for a cleaner environment, **(c)** those who are not involved because they disagree with many environmental laws and regulations, **(d)** those who are concerned but do not believe individual action will make much difference, and **(e)** those who strongly oppose the environmental movement. To which group do you belong? Compile your answer and those of your classmates and determine what percentage of the total number of people in your class represents each category. As a class, conduct a similar poll of your entire school and compile the results.

Global Environment Watch Exercise

Within the GREENR database, go to the *Ozone Depletion* portal. Research ozone depletion and find the latest research on the effectiveness of international efforts to reduce the use of CFCs. Write a summary of how negotiations among nations have progressed since they began.

What were the most difficult stumbling blocks for the negotiators? How did the negotiators overcome these obstacles, if they were able to do so? Include the answers to these questions in your summary report.

Data Analysis

Choose an environmental issue that you have studied in this course, such as climate change, population growth, or biodiversity loss. Conduct a poll of students, faculty, staff, and local residents in your community by asking them the questions that follow, relating to your particular environmental issue. Poll as many people as you can in order to get a large sample. Create categories. For example, note whether each respondent is male or female. By creating such categories, you are placing each person into a *respondent pool.* You can add other questions about age, political leaning, and other factors to refine your pools.

Poll Questions

Question 1   On a scale of 1 to 10, how knowledgeable are you about environmental issue X?

Question 2   On a scale of 1 to 10, how aware are you of ways in which you, as an individual, impact policy making related to environmental issue X?

Question 3   On a scale of 1 to 10, how important is it for you to learn more about environmental issue X?

Question 4   On a scale of 1 to 10, how sure are you that an individual can have a positive influence on policy making related to environmental issue X?

Question 5   On a scale of 1 to 10, how sure are you that the government is providing the appropriate level of leadership with regard to environmental issue X?

1. Collect your data and analyze your findings to measure any differences among the respondent pools.

2. List any major conclusions you would draw from the data.

3. Publicize your findings on your school's website or in the local newspaper.

1. Environmental policy is best defined as
   a. the study of the connections between human systems and natural systems.
   b. what humans consider right or wrong regarding how they treat the environment.
   c. laws, rules, and regulations that pertain to environmental problems.
   d. the study of costs and benefits of environmental policies and regulations.
   e. a social movement to protect the environment.

2. Subsidizing corn farmers in order to produce ethanol violates which guiding principle of environmental policy?
   a. Humility principle
   b. Precautionary principle
   c. Environmental justice principle
   d. Human rights principle
   e. Net energy principle

3. Which of the following is NOT a reason that environmental lawsuits are difficult to win?
   a. It is difficult to establish legal standing.
   b. It is expensive.
   c. The judiciary is indifferent.
   d. Plaintiffs must show they have been harmed.
   e. The process can take years.

4. Which of the following is a method of solving environmental disputes that allows for a larger number of citizens, all of whom claim to have experienced similar damages from a product or an action, but who need not be listed and represented individually?
   a. Civil suit
   b. Arbitration
   c. Nuisance suit
   d. Class action suit
   e. Injunction

5. What are groups that focus on a particular environmental problem by bringing together governments, the private sector, international organizations, and NGOs?
   a. Iron triangles
   b. Global public policy networks
   c. Political action committees
   d. Political parties
   e. Think tanks

6. Most states failing due to growing scarcities of resources occur in
   I. Europe.
   II. Africa.
   III. Asia.
   a. I only
   b. II only
   c. III only
   d. I and II
   e. II and III

7. Environmental interest groups use all of the following tactics to influence environmental policy EXCEPT
   a. lobbying, in which legislators are contacted in person.
   b. getting money or property by using threats.
   c. filing suit so the issue may be decided in court.
   d. providing documents stating facts that an attorney will use in court.
   e. taking part in supporting or opposing a person or issue.

Questions 8–12 refer to the terms below.
   a. Policy life
   b. Recognition
   c. Formulation
   d. Implementation
   e. Control

8. Putting solutions into effect.

9. The process through which laws and regulations are developed and then evaluated.

10. Identifying problems.

11. Monitoring progress and making adaptations.

12. Developing a solution to the problem.

25 Environmental Worldviews, Ethics, and Sustainability

The sustainability revolution is nothing less than a rethinking and remaking of our role in the natural world.

DAVID W. ORR

Key Questions

25-1 What are some major environmental worldviews?

25-2 What is the role of education in living more sustainably?

25-3 How can we live more sustainably?

A view within Yellowstone National Park in the U.S. state of Wyoming.

In 1991, eight scientists (four men and four women) were sealed inside Biosphere 2, a $200 million glass and steel enclosure designed to be a self-sustaining life-support system (Figure 25-1) that would increase our understanding of Biosphere 1: the *earth's* life-support system.

This sealed system of interconnected domes was built in the desert near Tucson, Arizona. It contained artificial ecosystems including a tropical rain forest, savanna, and desert, as well as lakes, streams, freshwater and saltwater wetlands, and a mini-ocean with a coral reef.

Biosphere 2 was designed to mimic the earth's natural chemical recycling systems. Water evaporated from its ocean and other aquatic systems and then condensed to provide rainfall over the tropical rain forest. The precipitation trickled through soil filters into the marshes and back into the ocean before beginning the cycle again.

The facility was stocked with more than 4,000 species of plants and animals, including small primates, chickens, cats,

and insects, selected to help maintain life-support functions. Human and animal excrement and other wastes were treated and recycled as fertilizer to help support plant growth. Sunlight and external natural gas–powered generators provided energy. The Biospherians were to be isolated for 2 years and raise their own food, using intensive organic agriculture. They were to breathe air that was purified by plants and to drink water cleansed by natural chemical cycling processes.

From the beginning, many unexpected problems cropped up and the life-support system began to unravel. The level of oxygen in the air declined with soil organisms converting it to carbon dioxide. Additional oxygen had to be pumped in from the outside to keep the Biospherians from suffocating.

Tropical birds died after the first freeze. An ant species got into the enclosure, proliferated, and killed off most of the system's original insect species. In total, 19 of the Biosphere's 25 small animal spe-

cies (76%) became extinct. Before the 2-year period was over, all plant-pollinating insects went extinct, thereby dooming to extinction most of the plant species.

Despite many problems, the facility's waste and wastewater were recycled. With much hard work, the Biospherians were also able to produce 80% of their food supply, despite rampant weed growths, spurred by higher CO_2 levels, that crowded out food crops. However, the scientists suffered from persistent hunger and weight loss.

In the end, an expenditure of $200 million failed to maintain a life-support system for eight people for 2 years. Ecologists Joel E. Cohen and David Tilman, who evaluated the project, concluded, "No one yet knows how to engineer systems that provide humans with life-supporting services that natural ecosystems provide for free." Biosphere 2 is an example of how we can view the earth's life-support system in different ways, based on our worldviews and ethical frameworks, which is the topic of this chapter.

Figure 25-1 Biosphere 2, constructed near Tucson, Arizona, was designed to be a self-sustaining life-support system.

PR News Foto/Huran Valley Travel

25-1 What Are Some Major Environmental Worldviews?

CONCEPT 25-1

Major environmental worldviews differ on which is more important—human needs and wants, or the overall health of ecosystems and the biosphere.

There Are a Variety of Environmental Worldviews

People disagree on how serious different environmental problems are, as well as on what we should do about them. These conflicts arise mostly out of differing **environmental worldviews**—ways of thinking about how the world works and beliefs that people hold about their roles in the natural world. Another factor is the widespread lack of understanding of how the earth's life-support system works, keeps us alive, and supports our economies.

An environmental worldview is determined partly by a person's **environmental ethics**—what one believes about what is right and what is wrong in our behavior toward the environment. According to environmental ethicist Robert Cahn:

The main ingredients of an environmental ethic are caring about the planet and all of its inhabitants, allowing unselfishness to control the immediate self-interest that harms others, and living each day so as to leave the lightest possible footprints on the planet.

People with widely differing environmental worldviews can take the same data, be logically consistent in their analysis of those data, and arrive at quite different conclusions, because they start with different assumptions and values. Figure 25-2 summarizes the four major beliefs of each of three major environmental worldviews. Some environmental worldviews are human-centered (anthropocentric) worldviews, focusing primarily on the needs and wants of people; others are life- or earth-centered (biocentric) worldviews, focusing on individual species, ecosystems, and the entire biosphere.

Most People Have Human-Centered Environmental Worldviews

It is not surprising that most environmental worldviews are human centered. One such worldview held by many people is the **planetary management worldview**. Figure 25-2 (left) summarizes its major assumptions.

According to this view, humans are the planet's most important and dominant species, and we can and should manage the earth mostly for our own benefit. The values of other species and parts of nature are based primarily on how useful they are to us. According to this view of nature, human well-being depends on the degree of control that we have over natural processes. It also holds that, as the world's most important and intelligent species, we

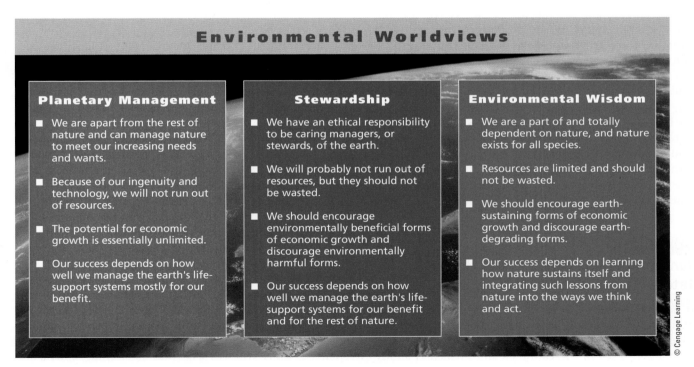

Environmental Worldviews

Planetary Management

- We are apart from the rest of nature and can manage nature to meet our increasing needs and wants.

- Because of our ingenuity and technology, we will not run out of resources.

- The potential for economic growth is essentially unlimited.

- Our success depends on how well we manage the earth's life-support systems mostly for our benefit.

Stewardship

- We have an ethical responsibility to be caring managers, or stewards, of the earth.

- We will probably not run out of resources, but they should not be wasted.

- We should encourage environmentally beneficial forms of economic growth and discourage environmentally harmful forms.

- Our success depends on how well we manage the earth's life-support systems for our benefit and for the rest of nature.

Environmental Wisdom

- We are a part of and totally dependent on nature, and nature exists for all species.

- Resources are limited and should not be wasted.

- We should encourage earth-sustaining forms of economic growth and discourage earth-degrading forms.

- Our success depends on learning how nature sustains itself and integrating such lessons from nature into the ways we think and act.

© Cengage Learning

Figure 25-2 Comparison of three major environmental worldviews (**Concept 25-1**). **Questions:** Which of these descriptions most closely fits your worldview? Which of them most closely fits the worldviews of your parents?

can redesign the planet and its life-support systems to support our ever-growing population and economies.

Here are three variations of the planetary management environmental worldview:

- *The no-problem school:* We can solve any environmental, population, or resource problem with more economic growth and development, better management, and better technology.
- *The free-market school:* The best way to manage the planet for human benefit is through a free-market global economy with minimal government interference and regulation. All public property resources should be converted to private property resources, and the global marketplace, governed only by free-market competition, should decide essentially everything.
- *The spaceship-earth school:* The earth is like a spaceship: a complex machine that we can understand, dominate, change, and manage, in order to provide a good life for everyone without overloading natural systems. This view developed after people saw photographs taken from outer space showing the earth as a finite planet, or an island in space (Figure 25-3).

Another human-centered environmental worldview is the **stewardship worldview**. It assumes that we have an ethical responsibility to be caring and responsible managers, or stewards, of the earth. Figure 25-2 (center) summarizes the major beliefs of this worldview. This worldview forms the basis for how public resources are managed in many parts of the world, including the United States (see the following Case Study).

According to the stewardship view, as we use the earth's natural capital, we are borrowing from the earth and from future generations. We have an ethical responsibility to pay this debt by leaving the earth in at least as good a condition as what we now enjoy. When thinking about our responsibility toward future generations, some analysts suggest that we consider the wisdom expressed in a law of the 18th-century Iroquois Six Nations Confederacy of Native Americans: In our every deliberation, we must consider the impact of our decisions on the next seven generations. In other words, we must learn to be good and wise ancestors, in keeping with one of the social science **principles of sustainability**.

CASE STUDY

Managing Public Lands in the United States—Stewardship in Action

No nation has set aside as much of its land for public use, resource extraction, enjoyment, and wildlife habitat as has the United States. The federal government manages roughly 35% of the country's land, which is jointly owned by all U.S. citizens. About three-fourths of this federal public land is in Alaska and another fifth is in the western states (Figure 25-4).

Figure 25-3 The blue marble in space that we call Earth is our only home.

Sailorr/Shutterstock.com

Some federal public lands are used for many different purposes. For example, the *National Forest System* consists of 155 national forests (see Figure 10-14, p. 225) and 22 national grasslands. These lands, managed by the U.S. Forest Service (USFS), are used for logging, mining, livestock grazing, farming, oil and gas extraction, recreation, and conservation of watershed, soil, and wildlife resources.

The Bureau of Land Management (BLM) manages large areas of land—40% of all land managed by the federal government and 13% of the total U.S. land surface—mostly in the western states and Alaska. These lands are used primarily for mining, oil and gas extraction, logging, and livestock grazing.

The U.S. Fish and Wildlife Service (USFWS) manages 560 *national wildlife refuges*. Most refuges protect habitats and breeding areas for waterfowl (see Figure 9-20, p. 210) and big game to provide a harvestable supply of these species for hunters. Permitted activities in most refuges include hunting, trapping, fishing, oil and gas development, mining, logging, grazing, some military activities, and farming.

The uses of some other public lands are more restricted. The *National Park System,* managed by the National Park Service (NPS), includes 59 major parks (see chapter-opening photo) and 342 national recreation areas, monuments, memorials, battlefields, historic sites, parkways, trails, rivers, seashores, and lakeshores. Only camping, hiking, sport fishing, and boating can take place in the national parks, whereas sport hunting, mining, and oil and gas drilling are allowed in national recreation areas.

The most restricted public lands are 756 roadless areas that make up the *National Wilderness Preservation System*

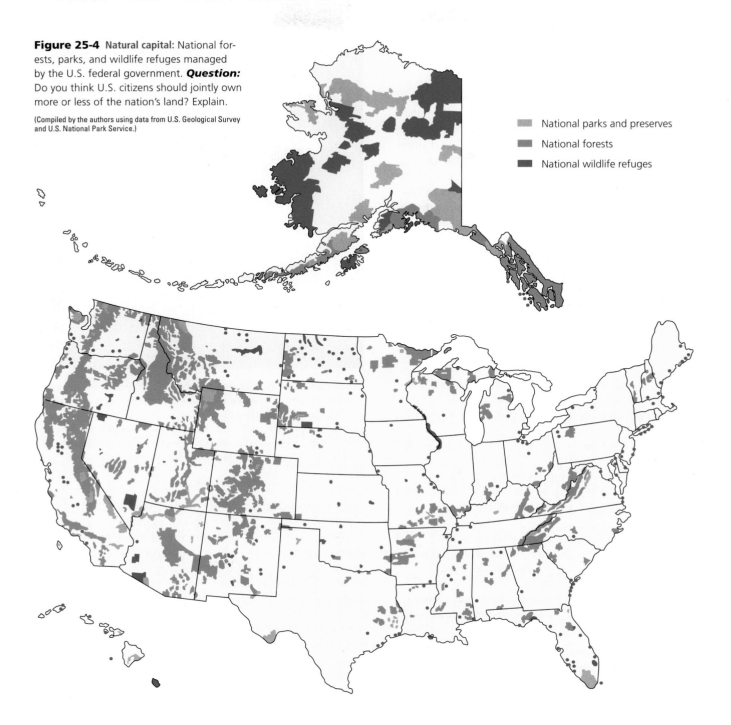

Figure 25-4 Natural capital: National forests, parks, and wildlife refuges managed by the U.S. federal government. ***Question:*** Do you think U.S. citizens should jointly own more or less of the nation's land? Explain.

(Compiled by the authors using data from U.S. Geological Survey and U.S. National Park Service.)

National parks and preserves

National forests

National wildlife refuges

(see Figure 10-27, p. 237). These areas lie within the other public lands and are managed by the agencies in charge of those lands. Most of these areas are open only for recreational activities such as hiking, sport fishing, camping, and non-motorized boating.

Many federal lands contain valuable oil, natural gas, coal, geothermal, timber, and mineral resources. Since the 1800s, there has been intense controversy over how to use and manage the resources on these public lands. This conflict provides an interesting contrast between stewardship and planetary management worldviews.

Most conservation biologists, environmental economists, and many free-market economists believe that four principles should govern the use of public lands:

1. They should be used primarily for protecting biodiversity, wildlife habitats, and ecosystems.
2. No one should receive government subsidies or tax breaks for using or extracting resources on public lands.
3. The American people deserve fair compensation for the use of their property.

4. All users or extractors of resources on public lands should be fully responsible for any environmental damage they cause.

There is strong and effective opposition to these ideas. Developers, resource extractors, many economists, and many citizens tend to view public lands in terms of their usefulness in providing mineral, timber, and other resources, and increasing short-term economic growth. They have succeeded in blocking implementation of the four principles listed above. For example, analyses of budgets and spending reveal that in recent years, the government has given an average of $1 billion a year—more than $2.7 million a day—in subsidies and tax breaks to privately owned interests that use U.S. public lands for activities such as mining, fossil fuel extraction, logging, and grazing.

Some developers and resource extractors have sought to go further in opening up more federal lands (owned jointly by all citizens) for development. Here are five of the proposals that such interests have made to the U.S. Congress in recent years:

1. Sell public lands or their resources to corporations or individuals, usually at proposed prices that are less than market value, or turn over their management to state and local governments.
2. Slash federal funding for the administration of regulations related to public lands.
3. Cut diverse old-growth forests in the national forests for timber and for making biofuels, and replace them with tree plantations to be harvested for the same purposes.
4. Open national parks, national wildlife refuges, and wilderness areas to oil and natural gas drilling, mining, off-road vehicles, and commercial development.
5. Eliminate or take regulatory control away from the National Park Service and launch a 20-year construction program in the parks to build new concessions and theme parks that would be run by private firms.

Although this case study has focused on the debate over the use of public lands in the United States, the same issues apply to the use of government-owned or publicly owned lands in other countries.

 CONSIDER THIS. . .

THINKING ABOUT U.S. Public Lands

Explain why you agree or disagree with the five proposals of developers for changing the use of U.S. public lands, listed above. Are these proposals in keeping with, or in conflict with, your environmental worldview?

Can We Manage the Earth?

Some people believe that any human-centered worldview will eventually fail because it wrongly assumes we now have or can gain enough knowledge to become effective managers or stewards of the earth. As biologist and

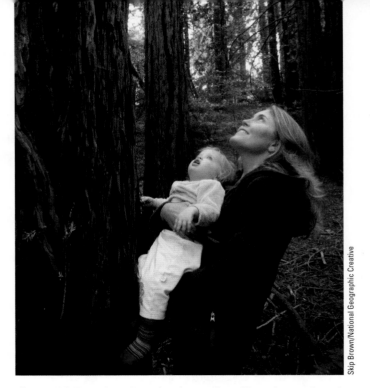

Figure 25-5 We have limited understanding of how these redwood trees in California's Yosemite National Park and the plants and animals in the surrounding forest ecosystem survive, interact, and change in response to different environmental conditions. **Questions:** How does this lack of knowledge relate to the planetary management worldview? Does this mean that we should never cut such trees? Explain.

environmental philosopher René Dubos (1901–1982) observed, "The belief that we can manage the earth and improve on nature is probably the ultimate expression of human conceit." According to environmental leader Gus Speth, "This view of the world—that nature belongs to us rather than we to nature—is powerful and pervasive—and it has led to much mischief."

According to some critics of human-centered worldviews, the unregulated global free-market approach will not work because it is based on continually increasing the use of the earth's natural capital, and it focuses on short-term economic benefits with little regard for the degradation and depletion of natural capital and the resulting long-term harmful environmental, health, and social consequences.

The image of the earth as an island (Figure 25-3) or spaceship has played an important role in raising global environmental awareness. But critics argue that thinking of the earth as a spaceship that we can manage is an oversimplified and misleading way to view an incredibly complex and ever-changing planet.

Critics of human-centered worldviews point out that we do not even know how many plant and animal species live on the earth, much less what their roles are, how they interact with one another and with their nonliving environment, and how they support our lives and economies. We still have much to learn about what goes on in a handful of soil, a patch of forest (Figure 25-5), the bottom of the ocean, and most other parts of the planet.

At the same time, our ecological footprints are spreading (see Figure 1-12, p. 13, and Figure 8, p. S33, Supplement 6). According to the 2005 Millennium Ecosystem Assessment, human activities have degraded or overused about 60% of the earth's ecosystem services (see Figure 1-3, orange boxes, p. 7). As biologist David Ehrenfeld puts it, "In no important instance have we been able to demonstrate comprehensive successful management of the world, nor do we understand it well enough to manage it even in theory." This conclusion is supported by the failure of the Biosphere 2 project (**Core Case Study**).

Some analysts argue that in taking over most of the earth's land and water, we are changing its climate, acidifying the global ocean, and bringing about a sixth mass extinction. They say that in this new Anthropocene era (see Science Focus 3.3, p. 72) we now have no choice but to become much better managers of the earth, if only to help preserve our own species and cultures by not exceeding certain planetary boundaries (see Figure 3-B, p. 73). This would involve learning how to work with nature instead of trying to conquer it primarily for our own use.

Some Environmental Worldviews Are Life-Centered and Others Are Earth-Centered

Critics of human-centered environmental worldviews argue that they should be expanded to recognize that all forms of life have value as participating members of the biosphere, regardless of their potential or actual use to humans. However, most people disagree over how far we should extend our ethical concerns for various forms of life (Figure 25-6).

Eventually all species become extinct. However, most people with a life-centered worldview believe we have an ethical responsibility to avoid hastening the extinction of species through our activities, for two reasons. First, each species is a unique storehouse of genetic information that should be respected and protected simply because it exists. Second, every species has potential for providing economic benefits directly or indirectly through its participation in providing ecosystem services.

Some people think we should go beyond focusing mostly on species. They believe we have an ethical responsibility to take a wider view and work to prevent degradation of the earth's ecosystems, its biodiversity of species and ecosystem services, and the biosphere (Figure 25-6). This earth-centered environmental worldview (Figure 25-3) is devoted to avoiding degradation of the life-support systems for all forms of life now and in the future.

People with earth-centered worldviews believe that humans are not in charge of the world and that human economies and other systems are subsystems of the earth's life-support systems (see Figure 23-5, p. 635). They understand that the earth's natural capital (see Fig-

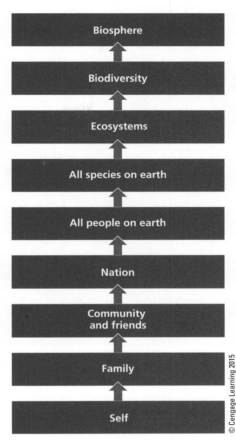

Figure 25-6 Levels of ethical concern: People disagree about how far we should extend our ethical concerns on this scale. **Question:** How far up this scale would you extend your ethical concerns?

© Cengage Learning 2015

ure 1-3, p. 7) keeps us and other species alive and supports our economies. They also understand that preventing the depletion and degradation of this natural capital is a key to promoting environmental and economic sustainability. Regardless of one's worldview, by protecting earth's life-support system, we act in our own self-interest (Figure 25-7), and we can do so by applying the **principles of sustainability** (see Figure 1-2, p. 6, Figure 1-5, p. 9, or back cover) in our daily activities.

One earth-centered worldview is called the **environmental wisdom worldview** (Figure 25-2, right). According to this view, we are within and part of—not apart from—the community of life and the ecological processes that sustain all life. Chief Seattle (1786–1866), leader of the Suquamish and Duwamish Native American tribes in what is now the U.S. state of Washington, summarized this ethical belief: "The earth does not belong to us. We belong to the earth." This worldview recognizes that the sustainability of our species, civilizations, and economies depends on the sustainability of the biosphere, of which we are just one part. Thus, promoting global sustainability helps each of us to safeguard our own individual health and safety, as well as our future as a species. In many respects, the environmental wisdom worldview is the opposite of the planetary management worldview (Figure 25-2, left).

Figure 25-7 The earth flag is a symbol of commitment to promoting environmental and economic sustainability by working with the earth at the individual, local, national, and international levels. ***Question:*** Explain why you agree or disagree with the earth-centered worldview.

©Earth Flag Company

The environmental wisdom worldview holds that the earth does not need us managing it in order to survive, whereas we depend on the earth for our survival. The earth has existed for billions of years and does not need saving. Rather, according to this view, what we need to save is our own species and cultures—which have been around for less than an eye-blink in the 3.5-billion-year history of life on the earth—along with the numerous other species that may become extinct because of our activities. (See the online Guest Essay on this topic by sustainability expert Lester W. Milbrath.)

 CONSIDER THIS. . .

THINKING ABOUT Environmental Worldviews and Biosphere 2

Which environmental worldview best explains the failure of Biosphere 2 (**Core Case Study**)?

25-2 What Is the Role of Education in Living More Sustainably?

CONCEPT 25-2
The first step to living more sustainably is to become environmentally literate.

We Can Become More Environmentally Literate

There is widespread evidence and agreement that we are a species in the process of degrading our own life-support system and that during this century, this behavior will very likely threaten human civilization and the existence of up to half of the world's species. Part of the problem stems from ignorance about how the earth works, how our actions affect its life-sustaining systems, and how we can change our behavior toward the earth and thus toward ourselves. Correcting this begins by understanding three important ideas that form the foundation of environmental literacy:

1. Natural capital matters because it supports the earth's life and our economies.

2. Our ecological footprints are immense and are expanding rapidly and degrading natural capital; in fact, they already exceed the earth's estimated ecological capacity (see Figure 1-13, p. 14).

3. In such degradation, we should not exceed planetary boundaries or ecological tipping points (see Figure 3-B, p. 73, and Figure 19-21, p. 525). Once we cross such a point, neither money nor technology can save us from the resulting consequences that could last for thousands of years.

According to the environmental wisdom worldview, learning how to live more sustainably requires a foundation of environmental education aimed at producing environmentally literate citizens. Acquiring environmental literacy involves being able to answer certain key questions and having a basic understanding of certain key topics, as summarized in Figure 25-8.

Science writer Janine M. Benyus, environmental entrepreneur Paul Hawken, and various scientists have

Questions to answer

- How does life on earth sustain itself?

- How am I connected to the earth and other living things?

- Where do the things I consume come from and where do they go after I use them?

- What is environmental wisdom?

- What is my environmental worldview?

- What is my environmental responsibility as a human being?

Components

- Basic concepts: sustainability, natural capital, exponential growth, carrying capacity

- Principles of sustainability

- Environmental history

- The two laws of thermodynamics and the law of conservation of matter

- Basic principles of ecology: food webs, nutrient cycling, biodiversity, ecological succession

- Population dynamics

- Sustainable agriculture and forestry

- Soil conservation

- Sustainable water use

- Nonrenewable mineral resources

- Nonrenewable and renewable energy resources

- Climate disruption and ozone depletion

- Pollution prevention and waste reduction

- Environmentally sustainable economic and political systems

- Environmental worldviews and ethics

© Cengage Learning

Figure 25-8 Achieving environmental literacy involves being able to answer certain questions and having an understanding of certain key topics (Concept 25-2). **Question:** After taking this course, do you feel that you can answer the questions asked here and have a basic understanding of each of the key topics listed in this figure?

been pioneering a science called *biomimicry*. They urge scientists, engineers, business executives, and entrepreneurs to study the earth's living systems to find out what works and what lasts and how we might copy such earth wisdom. For example, gecko lizards have pads on their feet that allow them to cling to rocks, glass, and other surfaces. Researchers are using this information to develop a new type of adhesive tape. Ray Anderson (see Individuals Matter 23.2, p. 650) spurred the creation of a best-selling carpet tile that mimics the diverse nature of the floor of a tropical rain forest. Scientists and planners are exploring

how to use biomimicry to design eco-villages and more sustainable communities (see Science Focus 22.1, p. 623).

In a broad application of biomimicry, we as authors have studied how nature works and how life has sustained itself, and we arrived at three scientific **principles of sustainability** (see Figure 1-2, p. 6 or back cover), which we have used throughout this book to help explain environmental problems and possible solutions to those problems. For example, in mimicking nature's use of solar energy, scientists and engineers have created solar cells that can be used to provide electricity to people living in remote locations (Figure 25-9).

We Can Learn from the Earth

Formal environmental education is important, but is it enough? Many analysts say no. They call for us to appreciate not only the economic value of nature, but also its ecological, aesthetic, and spiritual values. To these analysts, the problem is not just a lack of environmental literacy but also, for many people, a lack of intimate contact with nature and a limited understanding of how nature works and sustains us. This can reduce our ability to act more responsibly toward the earth and thus toward ourselves.

A growing chorus of analysts suggest that we have much to learn from nature. They argue that we can kindle a sense of awe, wonder, mystery, excitement, and humility by standing under the stars, enjoying a forest, or taking in the majesty and power of the sea. We might pick up a handful of topsoil and try to sense the teeming microscopic life within it that helps to produce most of the food we eat. We might look at a tree (Figure 25-5), a mountain (Figure 25-10), a rock, or a bee, or listen to the sound of a bird and try to sense how each of them is connected to us and we to them, through the earth's life-sustaining processes.

Such direct experiences with nature reveal parts of the complex web of life that cannot be bought, re-created through technology (**Core Case Study**), or reproduced with genetic engineering. Understanding and directly experiencing the precious and free gifts we receive from nature can help us to muster an ethical commitment to live more sustainably on this earth and thus to preserve our own species and cultures.

Best-selling author Richard Louv warns of widespread "nature-deficit disorder" among children who play only or mostly indoors and at best view the natural world digitally—something new in the history of humankind. This disorder can also affect adults who live in cities and seldom or never see natural settings. For the first time in human history, more than half of the world's people now live in rapidly growing urban areas.

In his 2011 book, *The Nature Principle*, Louv argues that a reconnection to the natural world is fundamental to human well-being, happiness, and longevity, as well as to a more sustainable world. He also believes that

Figure 25-9 Solar cells provide this Mongolian family with electricity in their *yurt*, or hut. The yurt and solar cells can be moved easily to support their nomadic way of life.

© Qilai Shen/In Pictures/Corbis

electronic social networking can help individuals, nature experts, families, groups, and communities to connect with each other and engage in outdoor activities such as hiking, bird-watching, restoring damaged natural areas, and practicing reconciliation ecology (see Chapter 10, p. 242). Louv argues that schools and colleges could stimulate teachers in all disciplines to integrate direct experiences with nature into their courses.

CONSIDER THIS. . .

CONNECTIONS Disconnecting from Technology and Reconnecting with Nature

Many of us who venture into the natural world want to carry our GPS units, cell phones, iPods, and other devices that keep us in touch with the world we have temporarily left behind. However, if our goal is to connect with and experience the natural world, then these devices that distract us from that world can defeat our purpose.

Earth-focused philosophers say that to be rooted, each of us needs to find a sense of place—a stream, a mountain, a patch of forest, a yard, a neighborhood lot—any piece of the earth that we know, experience emotionally, and love. According to biologist Stephen Jay Gould, "We will not fight to save what we do not love." When we become part of a place, it becomes a part of us. Then we are driven to defend it from harm and to help heal its wounds (Figure 25-11).

This might lead us to recognize that the healing of the earth and the healing of the human spirit are one and the same. We might discover and tap into what conservationist Aldo Leopold (Individuals Matter 25.1) calls "the green fire that burns in our hearts" and use this as a force for respecting and working with the earth and with one another.

Figure 25-10 An important way to learn about and appreciate nature is to experience its beauty, power, and complexity firsthand. This involves understanding that we are part of—and not apart from or in charge of—nature.

Pichugin Dmitry/Shutterstock.com

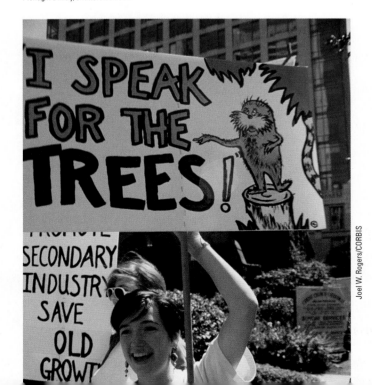

Joel W. Rogers/CORBIS

Figure 25-11 This woman and others in Vancouver, Canada, are protesting the clear-cutting of old-growth forests for timber in the Canadian province of British Columbia.

individuals matter 25.1

Aldo Leopold

Robert McCabe/Courtesy of the University of Wisconsin-Madison Archives

According to the renowned American forester, ecologist, and writer Aldo Leopold (1887–1948), the role of the human species should be to protect nature, not to conquer it. His book, *A Sand County Almanac* (published after his death), is considered to be an environmental classic that helped to inspire the modern environmental and conservation movements.

In 1933, Leopold became a professor at the University of Wisconsin, and in 1935, he helped to found the U.S. Wilderness Society. Through his writings and teachings, he became one of the foremost leaders of the conservation and environmental movements during the 20th century. His energy and foresight helped to lay the critical groundwork for the field of environmental ethics.

Leopold's weekends of planting trees, hiking, and directly observing nature on his Wisconsin farm provided much of the material for *A Sand County Almanac.* Since then, more than 2 million copies of this environmental classic have been sold.

The following quotations from his writings reflect Leopold's land ethic, and they form the basis for many of the beliefs and principles of the modern stewardship and environmental wisdom worldviews:

All ethics so far evolved rest upon a single premise: that the individual is a member of a community of interdependent parts.

To keep every cog and wheel is the first precaution of intelligent tinkering.

That land is a community is the basic concept of ecology, but that land is to be loved and respected is an extension of ethics.

The land ethic changes the role of Homo sapiens from conqueror of the land-community to plain member and citizen of it.

We abuse land because we regard it as a commodity belonging to us. When we see land as a community to which we belong, we may begin to use it with love and respect.

A thing is right when it tends to preserve the integrity, stability, and beauty of the biotic community. It is wrong when it tends otherwise.

Background photo: Aleksander Bolbot/Shutterstock.com

25-3 How Can We Live More Sustainably?

CONCEPT 25-3
We can live more sustainably by becoming environmentally literate, learning from nature, living more simply and lightly on the earth, and becoming active environmental citizens.

We Can Live More Simply and Lightly on the Earth

On a time scale of hundreds of thousands to millions of years, the earth is very resilient and can survive the environmental wounds we are inflicting. However, on the typical human time horizon of a decade or a lifetime (70 years), there is considerable scientific evidence that we are living unsustainably. Scientists warn that if we continue on this path, within your lifetime, it is very likely that we will be living on a changed planet with a warmer and sometimes harsher climate, less dependable supplies of water, more acidic oceans, extensive soil degradation, a sixth mass extinction, degradation of key ecosystem services, and widespread ecological and economic disruption.

With that in mind, environmental ethicists have developed a number of ethical guidelines for developing more sustainable and compassionate societies by converting environmental concerns, literacy, and wisdom into environmentally responsible actions. Here are five of those guidelines:

1. Apply principles such as our six **principles of sustainability** (see Figure 1-2, p. 6, Figure 1-5, p. 9, or back cover) to mimic the ways in which nature sustains itself and to consider the effects that any activity will have on all people and other life-forms, present and future.
2. Protect the earth's natural capital and repair ecological damage caused by human activities.
3. Use matter and energy resources as efficiently as possible.
4. Protect biodiversity.
5. Leave the earth in as good a condition as we found it, or better.

Analysts urge people who have a habit of consuming excessively to learn how to live more simply and sustainably. Seeking happiness through the pursuit of material things is considered folly by almost every major religion and philosophy. Yet, today's advertising messages persistently encourage people to buy more and more things to fill a growing list of wants as a way to achieve happiness (Figure 25-12). As American humorist and writer Mark Twain (1835–1910) observed: "Civilization is the limitless multiplication of unnecessary necessities." American comedian George Carlin (1937–2008) put it another way: "A house is just a pile of stuff with a cover on it. It is a place to keep your stuff while you go out and get more stuff."

According to research by psychologists, what a growing number of people really want, deep down, is more community, not more stuff. They want greater and more fulfilling interactions with family, friends, and neighbors, and a greater opportunity to express their creativity and to have more fun.

Some affluent people in more-developed countries are adopting a lifestyle of voluntary simplicity, in which they seek to learn how to live with much fewer possessions than people in their position typically have. They are also looking for products and services that have a smaller environmental impact (Concept 25-3) and would prefer to become part of a low-throughput economy (see Chapter 23, p. 648). Many of these people view voluntary simplicity not as a sacrifice but as a way to have a more fulfilling and satisfying life. Instead of working longer to pay for bigger vehicles and houses, they are spending more time with their loved ones, friends, and neighbors.

Practicing voluntary simplicity is a way to apply the Indian philosopher and leader Mahatma Gandhi's *principle of enoughness:* "The earth provides enough to satisfy every

Figure 25-12 Shopping has become a major pastime for billions of people. This shopping center in Denver, Colorado, is just one of many thousands throughout the world.

Richard Nowitz/National Geographic Creative

- Avoid buying something just because a friend has bought it

- Go on an ad diet by not watching or reading advertisements

- Avoid shopping for recreation and buying on impulse

- Stop using credit and buy only with cash to avoid overspending

- Borrow and share things like books, tools, and other consumer goods

Figure 25-13 Five ways to withdraw from an addiction to buying more and more stuff.

person's need but not every person's greed. . . . When we take more than we need, we are simply taking from each other, borrowing from the future, or destroying the environment and other species." Most of the world's major religions have similar teachings.

Living more lightly starts with asking the question: How much is enough? Similarly, one can ask: What do I really need? These are not easy questions to answer, because people in affluent societies are conditioned to want more and more material possessions and to view them as needs instead of wants. As a result, many people are addicted to buying more and more stuff as a way to find meaning in their lives, and they often run up large personal debts to feed their habits. Figure 25-13 lists five steps that some psychologists have advised people to take to help them withdraw from this addiction.

CONSIDER THIS. . .

THINKING ABOUT **Your Basic Needs**

Make a list of your basic needs. Is your list of needs compatible with your environmental worldview?

Throughout this text, you have encountered lists of steps we can take to live more lightly by reducing the size

and impact of our ecological footprints on the earth. It would be difficult for most of us to do all or even most of these things. So which ones are the most important? The human activities that have the greatest harmful impacts on the environment are food production, transportation, home energy use, and overall resource use. Based on this fact, Figure 25-14 lists eight key ways in which some people are choosing to live more simply and to reduce their ecological footprints.

CONSIDER THIS. . .

THINKING ABOUT **Living More Lightly**

Which three of the eight steps in Figure 25-14 do you think are the most important? Which three of these things do you already do? Which of them are you thinking about doing? How do your answers to these questions relate to **(a)** the **principles of sustainability**, and **(b)** to your environmental worldview?

Living more sustainably is not easy, and we will not make this transition by relying primarily on technological fixes such as recycling, changing to energy-efficient light-bulbs, and driving energy-efficient cars. These are, of course, important things to do, and they can help us to shrink our ecological footprints and to feel less guilty about our harmful impacts on our life-support system. But these efforts still do not solve the problems of excessive consumption and unnecessary waste of matter and energy resources.

Many of the more effective, far-reaching, and sustainable solutions to these problems will involve changing the way we think about, and act in, the world. This involves each of us examining our behavior and perhaps adjusting our environmental worldview. Changing all or part of our worldview often involves altering our behavior as well—changing habits that no longer serve our purposes that are defined by our worldviews.

Some analysts have suggested that the environmental movement has focused too much on laying blame, which

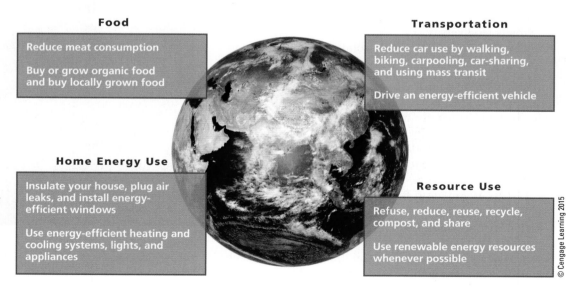

Figure 25-14 *Living more lightly:* Eight ways to shrink our ecological footprints (Concept 25-3). ***Questions:*** Which of these things do you already do? Which, if any, do you hope to do?

has then led people to feel guilty, fearful, and apathetic. They suggest that we move beyond these immobilizing feelings by recognizing and avoiding the following two common mental traps that lead to denial, indifference, and inaction:

- *Gloom-and-doom pessimism (it is hopeless)*
- *Blind technological optimism (science and technofixes will save us)*

Avoiding these traps helps us to hold on to, and be inspired by, empowering feelings of realistic hope, rather than to be immobilized by feelings of despair and fear.

CONSIDER THIS. . .

THINKING ABOUT Mental Traps

Have you fallen into either of these traps? If so, are you aware that you have, and how do you think you could free yourself from either of them?

Here is what business entrepreneur and environmental writer Paul Hawken told the 2009 graduating class at the University of Portland:

> *When asked if I am pessimistic or optimistic about the future, my answer is always the same: If you look at the science about what is happening on the earth and aren't pessimistic, you don't understand the data. But if you meet the people who are working to restore this earth and the lives of the poor, and you aren't optimistic, you haven't got a pulse. . . . You join a multitude of caring people. . . . This is your century. Take it and run as if your life depends on it.*

It is important to recognize that there is no single correct or best solution to each of the environmental problems we face. Indeed, one of nature's three scientific **principles of sustainability** holds that preserving diversity—in this case, being flexible

and adaptable in trying a variety of cultural and technological solutions to our problems—is the best way to adapt to the earth's largely unpredictable, ever-changing environmental conditions.

Finally, we should have fun and take time to enjoy life. Laugh every day and enjoy and celebrate nature, beauty, connectedness, friendship, sharing, caring, and love. This can empower us to become good earth citizens who practice good earthkeeping (see the following Case Study). As Mahatma Gandhi reminded us, "Power based on love is a thousand times more effective and permanent than power derived from fear."

CASE STUDY

Rebuilding Greensburg, Kansas: From Ruin to Sustainability

In 2007 a massive tornado destroyed or severely damaged about 95% of the buildings in the small town of Greensburg, Kansas (Figure 25-15, left). Instead of seeing the town die, school superintendent Darin Headrick and other town officials proposed rebuilding the town and its K–12 school (Figure 25-15, right) as a more sustainable community.

Town officials, residents, local businesses, students, and government agencies, including the U. S. Department of Energy and the National Renewable Energy Laboratory, worked together in a series of town meetings to develop a master plan for a model green community. Residents then formed a nonprofit organization, Greensburg GreenTown, to find funding and to guide the rebuilding process.

By 2012, the town had built more LEED Platinum–certified (see p. 409) new buildings per capita than any other U.S. city, each using half or less of the energy that

Figure 25-15 After a tornado devastated most of Greensburg, Kansas (left), the town rebuilt its houses and buildings with a focus on energy efficiency, reliance on wind power, and other elements of sustainability. One result was the new LEED Platinum–certified K–12 school (right).

Left: Federal Emergency Mgmt. Agency/National Renewable Energy Laboratory

Lester R. Brown: Champion of Sustainability

Lester R. Brown is president of the Earth Policy Institute, which he founded in 2001. The purpose of this nonprofit, interdisciplinary research organization is to provide a plan for a more sustainable future and a roadmap showing how we could get there. In 1974, he founded the Worldwatch Institute, which publishes annual reports on environmental issues and has helped to educate citizens and national leaders about the serious environmental problems we face.

Brown is an interdisciplinary thinker and one of the pioneers of the global sustainability movement. For decades, he has been researching and describing the complex environmental issues we face and proposing concrete strategies for dealing with them. The *Washington Post* has called him as "one of the world's most influential thinkers," and in 2011, *Foreign Policy* named him one of the Top Global Thinkers.

Brown's Plan B for shifting to a more environmentally and economically sustainable future has four main goals: **(1)** stabilize population growth, **(2)** stabilize climate change, **(3)** eradicate poverty, and **(4)** restore the earth's natural support systems. Many of his writings are based on, or form the background for, this general set of strategies.

Brown has written or coauthored more than 50 books, which have been translated into more than 40 languages. His most recent books include *Plan B 3.0: Mobilizing to Save Civilization* (2008); *World on the Edge: How to Prevent Environmental and Economic Collapse* (2011); *Full Planet, Empty Plates: The New Geopolitics of Food Scarcity* (2012); and *Breaking New Ground: A Personal History* (2013). He has also received numerous prizes and awards, including 25 honorary degrees, a MacArthur genius Fellowship, the United Nations Environment Prize, and Japan's Blue Planet Prize. In 2012, he was inducted into the Earth Hall of Fame in Kyoto, Japan. He also holds three honorary professorships in China.

Despite the serious environmental challenges we face, Brown sees reasons for hope. They include his understanding that social change can sometimes occur very quickly. He is also encouraged by improvements in fuel efficiency, the emerging shift from using coal to using solar and wind energy to produce electricity, and a growing public understanding of our need to live more sustainably.

Background photo: Petr Kopka/Shutterstock.com

Figure 25-16 Solutions: Some of the cultural shifts in emphasis that scientists say will be necessary to bring about a sustainability revolution. **Questions:** Which of these shifts do you think are most important? Why?

Current Emphasis	Sustainability Emphasis
Energy and Climate	
Fossil fuels	Direct and indirect solar energy
Energy waste	Energy efficiency
Climate disruption	Climate stabilization
Matter	
High resource use and waste	Less resource use
Consume and throw away	Reduce, reuse, and recycle
Waste disposal and pollution control	Waste prevention and pollution prevention
Life	
Deplete and degrade natural capital	Protect natural capital
Reduce biodiversity	Protect biodiversity
Population growth	Population stabilization

© Cengage Learning

a comparable conventional building uses. Its new K–12 school (Figure 25-15, right)—the town's largest building—uses 72% less energy than a comparable conventional building uses and gets 23% of its energy from renewable sources. It achieves this through a combination of passive solar design with natural heating and lighting, superinsulation, energy-efficient lighting, a wind turbine, and a geothermal heating and cooling system. The school was built partly of recycled building materials, and it includes large storage tanks that collect rainwater for irrigating the school's grass and plants.

Architects, students, and other visitors from around the world come to Greensburg to see this state-of-the-art, energy-efficient school building and the town's other energy-efficient homes and buildings, as well as its wind farm and other applications of renewable energy. The town's goal is to use energy efficiency and renewable energy to supply all of its energy by 2030. Greensburg is a shining example of how people can work together by applying the **principles of sustainability** (see Figure 1-2, p. 6, Figure 1-5, p. 9, or back cover) to overcome adversity by finding more environmentally sustainable ways to live for current and future generations.

We Can Bring about a Sustainability Revolution in Your Lifetime: A Vision for Sustainability

In the first Core Case Study in this book (see Chapter 1, p. 4), we sketched out one possible future for our planet. It is a hopeful and promising vision, and we believe it is attainable, probably within your lifetime.

The Industrial Revolution was a remarkable global transformation that has taken place over the past 275 years. Now in this century, environmental leaders say it is time for another sort of global transformation—a *sustainability revolution* that could lead to the kind of world we envisioned in the Chapter 1 Core Case Study. Figure 25-16 lists some of the major cultural shifts in emphasis that we will need to make in order to bring about such a sustainability revolution. Proponents of such a shift call for us to **(1)** increase energy efficiency, **(2)** shift to renewable energy resources, **(3)** stabilize climate change by

sharply reducing our greenhouse gas emissions, **(4)** stop destroying forests, **(5)** produce food more sustainably, **(6)** reuse or recycle at least 80% of the solid wastes we produce, and **(7)** reconnect and work with the biosphere that sustains all life and all human economies.

One of the leaders in the movement to develop and promote detailed plans for making the shift to more sustainable ways of living is Lester R. Brown (Individuals Matter 25.2). He is a prolific researcher and writer whom we have quoted several times throughout this textbook.

Change can occur in surprising ways. Exponential growth starts off slowly, but at some point, it increases at a very rapid rate. We can use the astonishing power of exponential growth, guided by the **principles of sustainability**, to promote more sustainable environmental, social, economic, and technological changes in a very short time (Figure 25-17).

We know what needs to be done and we can change the way we treat the earth and thus ourselves. According to social science research, in order for a major social change to occur, only 5–10% of the people in the world, or in a country or locality, must be convinced that change must take place and then act to bring about such change. History also shows that we can bring about change faster than we might think, once we have the courage to leave behind ideas and practices that no longer work and to nurture new trends toward positive change such as the rapidly growing seedlings of sustainability listed in Figure 25-17.

While some skeptics say the idea of a sustainability revolution is idealistic and unrealistic, entrepreneur Paul Hawken, in a 2009 graduation address, observed that "the most unrealistic person in the world is the cynic, not the dreamer." And according to the late Steve Jobs, cofounder of Apple Inc., "The people who are crazy enough to think

Figure 25-17 The concerns, trends, tools, and technologies listed under this curve of exponential growth could all be parts of a major shift toward sustainability. These agents of change are growing slowly, but at some point, some or all of them could take off, round the J-curve of exponential growth, and bring about a sustainability revolution within your lifetime. *Questions:* Which two items in each of these four categories do you believe are the most important to promote? What other items would you add to this list?

More Sustainable Living

Change →

Unsustainable Living

Time →

© Cengage Learning 2015

Environmental Concerns

Protecting natural capital

Sustaining biodiversity

Repairing ecological damage

Addressing climate change

Social Trends

Reducing waste

Using less

Living more simply

Reusing and recycling

Growth of eco-cities and eco-neighborhoods

Environmental justice

Environmental literacy

Economic Tools

Full-cost pricing

Micro-lending

Green subsidies

Green taxes

Cap and trade

Net energy analysis

Technologies

Pollution prevention

Organic farming

Drip irrigation

Solar desalinization

Energy efficiency

Solar energy

Wind energy

Geothermal energy

Environmental nanotechnology

Ecoindustrial parks

they can change the world are the ones who do." If these and other individuals had not had the courage to forge ahead with ideas that others called idealistic and unrealistic, very few of the human achievements that we now celebrate would have come to pass.

Big Ideas

- Our environmental worldviews play a key role in how we treat the earth that sustains us and thus in how we treat ourselves.

- We need to become more environmentally literate about how the earth works, how we are affecting its life-support systems that keep us and other species alive, and what we can do to live more sustainably.

- Living more sustainably means learning from nature, living more lightly, and becoming active environmental citizens who leave small environmental footprints on the earth.

© Ragnarock | Dreamstime.com

In Biosphere 2 (**Core Case Study**), scientists and engineers tried to create a microcosm of the earth that would help us understand how to live more sustainably. What they learned was that nature is so complex that predicting and controlling what will happen in the environment is essentially impossible.

As we explore different paths toward sustainability, we must first understand that our lives and societies depend on natural capital and that one of the biggest threats to our way of life is our active role in natural capital degradation. For example, the failure of the Biosphere 2 experiment taught the world about the vital importance of natural capital and the need to put economic values on the ecosystem services that nature provides for us every day.

One way to live more sustainably is to apply the scientific and social science **principles of sustainability** (see Figure 1-2, p. 6, Figure 1-5, p. 9, or back cover). For the energy to keep our lives and economies going, we will need to depend largely on a mix of direct and indirect forms of renewable solar energy. We will need to reuse and recycle as much of our material resources as we can, in keeping with nature's chemical cycling principle, and drastically cut our waste of matter and energy resources by using them more efficiently. We will benefit by preserving as much as we possibly can of the natural resources and ecosystem services that make up the earth's natural biodiversity. The solutions that work will also likely mimic nature's biodiversity by consisting of a variety of localized solutions.

Such solutions will include paying more of the harmful environmental and health costs of our goods and services, thereby reducing the damage we do to the biosphere by making such damage more expensive and in some cases unaffordable. These solutions would also be win-win solutions, satisfying the largest number of individuals while minimizing environmental harms. And these solutions would be beneficial not only to us now, but also to future generations.

All of this requires our understanding that individuals matter. Virtually all of the environmental progress we have made during the last few decades occurred because individuals banded together to insist that we can do better. This journey begins in our own communities and with our own lifestyles, because in the final analysis, all sustainability is local and personal. This is an exciting time to be alive as we learn how to live more lightly on our planetary home.

Chapter Review

Core Case Study

1. Describe the Biosphere 2 project (**Core Case Study**) and summarize the major lessons learned from this project.

Section 25-1

2. What is the key concept for this section? What is an **environmental worldview**? What are **environmental ethics**? What is the **planetary management worldview**? List three variations of this worldview. What is the **stewardship worldview**? What are four major types of public lands in the United States? Summarize the controversy over how we should manage these lands.

3. Summarize the debate over whether we can effectively manage the earth. Summarize the various beliefs about how far we should extend our concern for various forms of life (Figure 25-6). What is the **environmental wisdom worldview**? Explain the thinking within this worldview related to saving the earth.

Section 25-2

4. What is the key concept for this section? List three ideas that form the foundation of environmental literacy. What are five questions that an environmentally literate person should be able to answer? How do some scientists and other thinkers urge us to apply

biomimicry? Give two examples of how this has been done.

5. Explain how we can learn from direct experiences with nature. What is a sense of place and why is it important? Explain why Aldo Leopold is highly regarded and give some examples of the beliefs that made up his land ethic.

Section 25-3

6. What is the key concept for this section? List six ethical guidelines for developing more sustainable and compassionate societies. Describe the lifestyle of voluntary simplicity now being adopted by some affluent people. What is Gandhi's principle of enoughness? List five steps that some psychologists have advised people to take to help them withdraw from an addiction to buying.

7. List eight ways in which people can choose to live more lightly on the earth. List two mental traps that

can lead to denial, indifference, and inaction concerning environmental problems. Describe how the town of Greensburg, Kansas, went from being nearly ruined to getting on a path toward sustainability.

8. List six major shifts that scientists say will be necessary to bring about a sustainability revolution. Describe Lester R. Brown's contributions to helping us make the transition to a more economically and environmentally sustainable world. Explain how exponential growth could apply to our shifting onto a path toward more sustainable societies.

9. What are this chapter's three big ideas?

10. Explain the connections between Biosphere 2, the transition to more environmentally sustainable societies, and the scientific and social science **principles of sustainability** (see Figure 1-2, p. 6, Figure 1-5, p. 9, or back cover).

Note: Key terms are in bold type.

Critical Thinking

1. Some analysts argue that the problems with Biosphere 2 (**Core Case Study**) resulted mostly from inadequate design, and that a better team of scientists and engineers could make it work. Explain why you agree or disagree with this view.

2. Do you believe that we have an ethical responsibility to leave the earth's natural systems in as good, or better, a condition as they are now? Explain. List three aspects of your lifestyle that hinder the implementation of this ideal and three aspects that promote this ideal.

3. This chapter summarized several different environmental worldviews. Go through these worldviews and find the beliefs you agree with and then describe your own environmental worldview. Which of your beliefs, if any, were added or modified as a result of taking this course? Compare your answer with those of your classmates.

4. Explain why you agree or disagree with the following statements: **(a)** everyone has the right to have as many children as they want; **(b)** all people have a right to use as many resources as they want; **(c)** individuals should have the right to do whatever they want with land they own, regardless of whether such actions harm the environment, their neighbors, or the local community; **(d)** other species exist to be used by humans; **(e)** all forms of life have a right to exist; **(f)** we have ethical obligations to maintain a livable world for future generations of humans and other species. Are your answers consistent with the beliefs that make up your environmental worldview, which you described in question 3?

5. Explain why you agree or disagree with **(a)** each of the four principles that biologists and some economists have suggested for using public lands in the United States (p. 687), and **(b)** each of the five suggestions made by developers and resource extractors for managing and using these public lands (p. 688).

6. The American theologian, Thomas Berry (1914–2009), called the industrial–consumer society, built on the human-centered, planetary management environmental worldview, the "supreme pathology of all history." He said, "We can break the mountains apart; we can drain the rivers and flood the valleys. We can turn the most luxuriant forests into throwaway paper products. We can tear apart the great grass cover of the western plains, and pour toxic chemicals into the soil and pesticides onto the fields, until the soil is dead and blows away in the wind. We can pollute the air with acids, the rivers with sewage, the seas with oil. We can invent computers capable of processing 10 million calculations per second. And why? To increase the volume and speed with which we move natural resources through the consumer economy to the junk pile or the waste heap. If, in these activities, the topography of the planet is damaged, if the environment is made inhospitable for a multitude of living species, then so be it. We are, supposedly, creating a technological wonderworld. But our supposed progress is bringing us to a wasteworld instead of a wonderworld." Explain why you agree or disagree with this assessment. If you disagree, answer at least five of Berry's charges with your own arguments as to why you think he is wrong. If you agree, cite evidence as to why.

7. Some analysts believe that trying to gain environmental wisdom by becoming familiar with some part of the natural world and forming an emotional bond with its life-forms and processes is unscientific, mystical nonsense based on a romanticized view of nature. They believe that having a better scientific understanding of how the earth works and inventing or improving technologies to solve environmental problems are the best ways to achieve sustainability. Do you agree or disagree? Explain.

8. Revisit the Core Case study in Chapter 1 (p. 4) titled "A Vision of a More Sustainable World in 2065." Now that you are near the end of this textbook and course, do you feel that we have a reasonable chance of making such a transition? Explain. Is your view more hopeful or less hopeful than it was when you began this course? Compare your answers with those of your classmates.

Doing Environmental Science

Increase your environmental knowledge and awareness by tracing the water you drink from precipitation to tap; finding out what type of soil is beneath your feet; naming five plants and five birds that live in the natural environment around you; finding out what species in your area are threatened with extinction; learning where your garbage goes; and learning where the wastes you flush down the toilet go. Write a report summarizing your findings. Of your findings, which two were the most surprising to you and why? Compare your answer to this question with those of your classmates.

Global Environment Watch Exercise

Within the GREENR database, use the World Map feature, and under "Browse," select *Sustainability*. Click on the pins for the United States, Japan, China, India, and one other country of your choice and research what each of them is doing to try to become more sustainable. Which of these sustainability programs are working well? Which ones are not working well? Write a report comparing these programs.

Ecological Footprint Analysis

Working with classmates, conduct an ecological footprint analysis of your campus. Work with a partner, or in small groups, to research and investigate an aspect of your school such as recycling or composting; water use; food service practices; energy use; building management and energy conservation; transportation for both on- and off-campus trips; or grounds maintenance. Depending on your school and its location, you may want to add more areas to the investigation. You can also decide to study the campus as a whole, or to break it down into smaller research areas, such as dorms, administrative buildings, classroom buildings, grounds, and other areas.

1. After deciding on your group's research area, conduct your analysis. As part of your analysis, develop a list of questions that will help to determine the ecological impact related to your chosen topic. For example, with regard to water use, you might ask how much water is used, what is the estimated amount that is wasted through leaking pipes and faucets, and what is the average monthly water bill for the school, among other questions. Use such questions as a basis for your research.

2. Analyze your results and share them with the class to determine what can be done to shrink the ecological footprint of your school within the area you have chosen.

3. Arrange a meeting with school officials to share your action plan with them.

1. What one believes about what is right or wrong in our behavior toward the environment is
   a. environmental law.
   b. environmental economics.
   c. resource management.
   d. environmental ethics.
   e. consumerism.

2. We will probably not run out of resources, but they should not be wasted best fits in which environmental worldview?
   a. Environmental wisdom
   b. Planetary management
   c. Stewardship
   d. Free market
   e. Spaceship Earth

3. Which of the following is not a guideline for sustainable living?
   a. Protect biodiversity
   b. Do not waste matter and energy resources
   c. Avoid climate changing activities
   d. Repair ecological damage that we have caused
   e. Help reduce the earth's capacity for self-repair

4. The idea that science will solve *all* our problems best describes which mental trap?
   a. Gloom-and-doom pessimism
   b. Blind technological optimism
   c. Paralysis by analysis
   d. Faith and simple, easy answers
   e. Environmental nihilism

5. All of the following are ways to attain a more sustainable world EXCEPT
   a. purchase goods that use less packaging material and recycle that whenever possible.
   b. use incandescent rather than fluorescent lightbulbs.
   c. use energy-efficient appliances and temperature-control systems.
   d. use mass transit, bicycle, or walk whenever possible.
   e. buy food grown within 200 miles of where they are purchased.

Questions 6–9 refer to the environmental ideas below.
   a. A Sand County Almanac
   b. Formation of national parks
   c. Spaceship Earth
   d. Calculation of ecological footprints
   e. Biosphere 2

6. An experience that was considered to be a failure by some but taught us many lessons about the interconnectedness of nature and the difficulty of replicating nature.

7. The view of the earth as an island and a complex machine that we need to understand and manage.

8. One of the early environmental science writings detailing the beliefs and principles of modern stewardship.

9. A sealed system of interconnected domes containing multiple ecosystems.

10. Which of the below is NOT an economic tool that could be used to bring about a sustainable environmental revolution?
    a. Environmental justice
    b. Full-cost pricing
    c. Micro-lending
    d. Green taxes
    e. Green subsidies

11. Religion can play an important role in shaping environmental viewpoints. If 10% of the world's five largest religions mobilized to the cause of building more sustainable societies, how many people would that be?
    a. Approximately 480 million
    b. Approximately 48 million
    c. Approximately 500 million
    d. Approximately 4.8 billion
    e. Approximately 50 million

12. Which of the environmental worldviews below is **incorrectly** paired with its focus?
    a. Self-centered–Planetary management
    b. Anthropocentric–Planetary management
    c. Biocentric–Stewardship
    d. Ecosystem-centered–Environmental wisdom
    e. Biosphere–Stewardship

Length

Metric

1 kilometer (km) = 1,000 meters (m)
1 meter (m) = 100 centimeters (cm)
1 meter (m) = 1,000 millimeters (mm)
1 centimeter (cm) = 0.01 meter (m)
1 millimeter (mm) = 0.001 meter (m)

English

1 foot (ft) = 12 inches (in)
1 yard (yd) = 3 feet (ft)
1 mile (mi) = 5,280 feet (ft)
1 nautical mile = 1.15 miles

Metric–English

1 kilometer (km) = 0.621 mile (mi)
1 meter (m) = 39.4 inches (in) or 3.28 feet (ft)
1 inch (in) = 2.54 centimeters (cm)
1 foot (ft) = 0.305 meter (m)
1 yard (yd) = 0.914 meter (m)
1 nautical mile = 1.85 kilometers (km)

Area

Metric

1 square kilometer (km^2) = 1,000,000 square meters (m^2)
1 square meter (m^2) = 1,000,000 square millimeters (mm^2)
1 square meter (m^2) = 10,000 square centimeters (cm^2)
1 hectare (ha) = 10,000 square meters (m^2)
1 hectare (ha) = 0.01 square kilometer (km^2)

English

1 square foot (ft^2) = 144 square inches (in^2)
1 square yard (yd^2) = 9 square feet (ft^2)
1 square mile (mi^2) = 27,880,000 square feet (ft^2)
1 acre (ac) = 43,560 square feet (ft^2)

Metric–English

1 hectare (ha) = 2.471 acres (ac)
1 square kilometer (km^2) = 0.386 square mile (mi^2)
1 square meter (m^2) = 1.196 square yards (yd^2)
1 square meter (m^2) = 10.76 square feet (ft^2)
1 square centimeter (cm^2) = 0.155 square inch (in^2)

Volume

Metric

1 cubic kilometer (km^3) = 1,000,000,000 cubic meters (m^3)
1 cubic meter (m^3) = 1,000,000 cubic centimeters (cm^3)
1 cubic meter (m^3) = 1,000 liters (L)
1 liter (L) = 1,000 milliliters (mL) = 1,000 cubic centimeters (cm^3)
1 cubic meter (m^3) = 1,000 liters (L)
1 milliliter (mL) = 0.001 liter (L)
1 milliliter (mL) = 1 cubic centimeter (cm^3)

English

1 gallon (gal) = 4 quarts (qt)
1 quart (qt) = 2 pints (pt)

Metric–English

1 liter (L) = 0.265 gallon (gal)
1 liter (L) = 1.06 quarts (qt)
1 liter (L) = 0.0353 cubic foot (ft^3)
1 cubic meter (m^3) = 35.3 cubic feet (ft^3)
1 cubic meter (m^3) = 1.30 cubic yards (yd^3)
1 cubic kilometer (km^3) = 0.24 cubic mile (mi^3)
1 barrel (bbl) = 159 liters (L)
1 barrel (bbl) = 42 U.S. gallons (gal)

Mass

Metric

1 kilogram (kg) = 1,000 grams (g)
1 gram (g) = 1,000 milligrams (mg)
1 gram (g) = 1,000,000 micrograms (μg)
1 milligram (mg) = 0.001 gram (g)
1 microgram (μg) = 0.000001 gram (g)
1 metric ton (mt) = 1,000 kilograms (kg)

English

1 ton (t) = 2,000 pounds (lb)
1 pound (lb) = 16 ounces (oz)

Metric–English

1 metric ton (mt) = 2,200 pounds (lb) = 1.1 tons (t)
1 kilogram (kg) = 2.20 pounds (lb)
1 pound (lb) = 454 grams (g)
1 gram (g) = 0.035 ounce (oz)

Energy and Power

Metric

1 kilojoule (kJ) = 1,000 joules (J)
1 kilocalorie (kcal) = 1,000 calories (cal)
1 calorie (cal) = 4.184 joules (J)

Metric–English

1 kilojoule (kJ) = 0.949 British thermal unit (Btu)
1 kilojoule (kJ) = 0.000278 kilowatt-hour (kW-h)
1 kilocalorie (kcal) = 3.97 British thermal units (Btu)
1 kilocalorie (kcal) = 0.00116 kilowatt-hour (kW-h)
1 kilowatt-hour (kW-h) = 860 kilocalories (kcal)
1 kilowatt-hour (kW-h) = 3,400 British thermal units (Btu)
1 quad (Q) = 1,050,000,000,000,000 kilojoules (kJ)
1 quad (Q) = 293,000,000,000 kilowatt-hours (kW-h)

Temperature Conversions

Fahrenheit (°F) to Celsius (°C): °C = (°F − 32.0) ÷ 1.80
Celsius (°C) to Fahrenheit (°F): °F = (°C × 1.80) + 32.0

Graphs and Maps Are Important Visual Tools

A **graph** is a tool for conveying information that we can summarize numerically by illustrating that information in a visual format. This information, called *data,* is collected in experiments, surveys, and other information-gathering activities. Graphing can be a powerful tool for summarizing and conveying complex information.

In this textbook, we use three major types of graphs: *line graphs, bar graphs,* and *pie graphs.* Here, you will explore each of these types of graphs and learn how to read them.

An important visual tool used to summarize data that vary over small or large areas is a **map**. We discuss some aspects of reading maps relating to environmental science at the end of this supplement.

Line Graphs

Line graphs usually represent data that fall in some sort of sequence such as a series of measurements over time or distance. In most such cases, units of time or distance lie on the horizontal *x-axis.* The possible measurements of some quantity or variable such as temperature that changes over time or distance usually lie on the vertical *y-axis.*

In Figure 1, the x-axis shows the years between 1950 and 2009, and the y-axis displays the possible values for the annual amounts of oil consumed worldwide during that time in millions of tons, ranging from 0 to 4,000 million (or 4 billion) tons. Usually, the y-axis appears on the left end of the x-axis, although y-axes can appear on the right end, in the middle, or on both ends of the x-axis.

The curving line on a line graph represents the measurements taken at certain time or distance intervals. In Figure 1, the curve represents changes in oil consumption between 1950 and 2009. To find the oil consumption for any year, find that year on the x-axis (a point called the *abscissa*) and run a vertical line from the axis to the curve. At the point where your line intersects the curve, run a horizontal line to the y-axis. The value at that point on the y-axis, called the *ordinate,* is the amount you are seeking. You can go through the same process in reverse to find a year in which oil consumption was at a certain point.

Questions

1. What was the total amount of oil consumed in the world in 1990? In about what year between 1950 and 2000 did oil consumption first start declining?
2. About how much oil was consumed in 2009? Roughly how many times more oil was consumed in 2009 than in 1970? How many times more oil was consumed in 2009 than in 1950?

Line graphs have several important uses. One of the most common applications is to compare two or more variables. Figure 2 compares two variables: monthly temperature and precipitation (rain and snowfall) during a typical year in a temperate deciduous forest. However, in this case the variables are measured on two different scales, so there are two y-axes. The y-axis on the left end of the graph shows a Centigrade temperature scale, and the y-axis on the right shows the range of precipitation measurements in millimeters. The x-axis displays the first letters of each of the 12 months' names.

Questions

1. In which month does most precipitation fall? Which is the driest month of the year? Which is the hottest month?

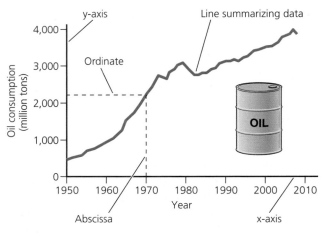

Figure 1 World oil consumption, 1950–2009.

(Compiled by the authors using data from U.S. Energy Information Administration, International Energy Agency, and United Nations.)

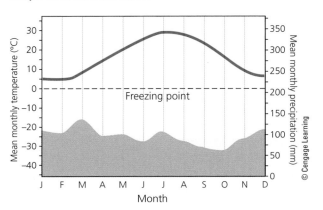

Figure 2 Climate graph showing typical variations in annual temperature (red) and precipitation (blue) in a temperate deciduous forest.

2. If the temperature curve were almost flat, running throughout the year at roughly its highest point of about 30°C, how do you think this forest would change from what it is now (see Figure 7-13, center, p. 156)? If the annual precipitation suddenly dropped and remained under 25 centimeters all year, what do you think would eventually happen to this forest?

It is also important to consider what aspect of a set of data is being displayed on a graph. The creator of a graph can take two different aspects of one data set and create two very different-looking graphs that would give two different interpretations of the same phenomenon. For example, when talking about any type of growth we must be careful to distinguish the question of whether something is growing from the question of how fast it is growing. While a quantity can keep growing continuously, its rate of growth can go up and down.

One of many important examples of growth used in this book is human population growth. For example, the graph in Figure 1-16 (p. 17) gives you the impression that human population growth has, for the most part, been continuous and uninterrupted. However, consider Figure 3, which plots the rate of growth of the human population since 1950. Note that all of the numbers on the y-axis, even the smallest ones, represent growth. The lower end of the scale represents slower growth and the higher end faster growth. Thus, while one graph tracks population growth in terms of numbers of people, the other tracks the rate of growth.

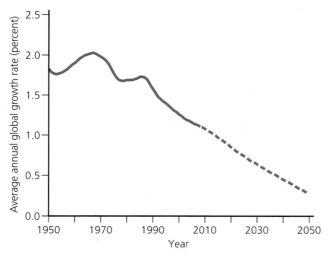

Figure 3 Annual growth rate in world population, 1950–2010, with projections to 2050.

(Compiled by the authors using data from UN Population Division, U.S. Census Bureau, and Population Reference Bureau.)

Questions

1. If this graph were presented to you as a picture of human population growth, what would be your first impression?
2. Do you think that reaching a growth rate of 0.5% would relieve those who are concerned about over-population? Why or why not?

Bar Graphs

The *bar graph* is used to compare measurements for one or more variables across categories. Unlike the line graph, a bar graph typically does not involve a sequence of measurements over time or distance. The measurements compared on a bar graph usually represent data collected at some point in time or during a well-defined period. For instance, we can compare the *net primary productivity (NPP)*, a measure of chemical energy produced by plants in an ecosystem, for different ecosystems, as represented in Figure 4.

In most bar graphs, the categories to be compared are laid out on the x-axis, and the range of measurements for the variable under consideration lies along the y-axis. In our example in Figure 4, the categories (ecosystems) are on the y-axis, and the variable range (NPP) lies on the x-axis. In either case, reading the graph is straightforward. Simply run a line perpendicular to the bar you are reading from the top of that bar (or the right or left end, if it lies horizontally) to the variable value axis. In Figure 4, you can see that the NPP for continental shelf, for example, is close to 1,600 kcal/m²/yr.

Questions

1. What are the two terrestrial ecosystems that are closest in NPP value of all pairs of such ecosystems? About how many times greater is the NPP in a tropical rain forest than the NPP in a savanna?
2. What is the most productive of aquatic ecosystems shown here? What is the least productive?

An important application of the bar graph used in this book is the *age-structure diagram* (see Figure 6-11, p. 131), which describes a population by showing the numbers of males and females in certain age groups (see Chapter 6, pp. 131–132).

Pie Graphs

Like bar graphs, *pie graphs,* or *pie charts,* illustrate numerical values for two or more categories. But in addition to that, they can also show each category's proportion of the total of all measurements. The categories are usually ordered on the graph from largest to smallest, for ease of comparison, although this is not always the case. Also, as with bar graphs, pie graphs are generally snapshots of a

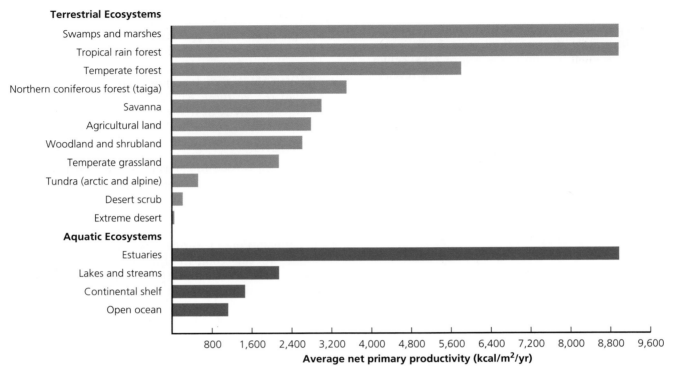

Figure 4 Estimated annual average net primary productivity (NPP) in major life zones and ecosystems, expressed as kilocalories of energy produced per square meter per year (kcal/m²/yr).

(Compiled by the authors using data from R. H. Whittaker, *Communities and Ecosystems,* 2nd ed., New York: Macmillan, 1975.)

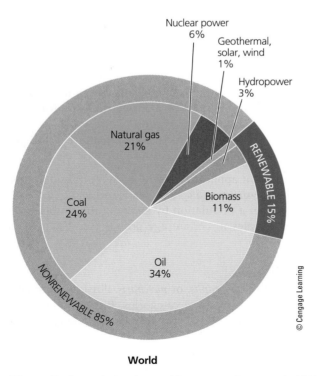

World

Figure 5 Pie graph showing world energy use by source in 2010.

data set at a point in time or during a defined time period. Unlike line graphs, one pie graph cannot show changes over time.

For example, Figure 5 shows how much each major energy source contributed to the world's total amount of energy used in 2010. This graph includes the numerical data used to construct it: the percentages of the total taken up by each part of the pie. But we can use pie graphs without including the numerical data and we can roughly estimate such percentages. The pie graph thereby provides a generalized picture of the composition of a data set.

Questions

1. Can you tell from this graph whether the use of renewable energy sources is growing or shrinking? Explain.
2. About how many times bigger is oil use than biomass use?

Reading Maps

We can use maps for considerably more than showing where places are relative to one another. For example, in environmental science, maps can be very helpful in comparing how people in different areas are affected by environmental problems such as air pollution and acid deposition (a form of air pollution). Figure 6 is a map of the United States showing the relative numbers of premature deaths due to air pollution in the various regions of the country.

Questions

1. Which part of the country generally has the lowest level of premature deaths due to air pollution?
2. Which part of the country has the highest level? What is the level in the area where you live or go to school?

In some of the data analysis exercises that appear at the ends of the chapters in this book, you will have opportunities to apply much of the information from this supplement.

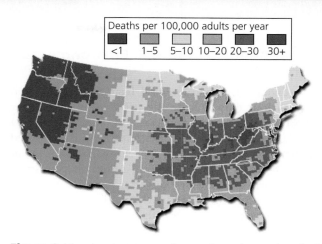

Figure 6 Map showing comparative numbers of premature deaths from air pollution in the United States.

(Compiled by the authors using data from U.S. Environmental Protection Agency.)

The Four Major Eras of U.S. Environmental History

We can divide the environmental history of the United States into four eras. During the *tribal era,* people (now called Native Americans or members of the First Nations) belonging to several hundred different tribes distinguished by language and culture occupied North America for at least 13,000 years before European settlers began arriving in the early 1600s. Some of the tribes were made up of nomadic hunter–gatherers, while others settled into communities centered on agriculture. Some of the more populous Native American societies were more sophisticated than their counterparts in Europe in terms of agricultural practices and other cultural aspects. Yet these North American tribes generally had more sustainable, low-impact ways of life because of their relatively limited numbers and modest rates of resource use per person.

Next was the *frontier era* (1607–1890) when European colonists began settling North America. Faced with a continent offering seemingly inexhaustible resources, the early colonists developed a *frontier environmental worldview.* That is, they saw a wilderness to be conquered and managed for human use.

Next came the *early conservation era* (1832–1870), which overlapped the end of the frontier era. During this period, some people became alarmed at the scope of resource depletion and degradation in the United States. They argued that part of the unspoiled wilderness on public lands should be protected as a legacy for future generations. Most of these warnings and ideas were not taken seriously.

This period was followed by an era—lasting from 1870 to the present—featuring an increased role by the federal government and private citizens in resource conservation, public health, and environmental protection.

The Frontier Era (1607–1890)

During the frontier era, European settlers spread across the land by clearing forests for cropland and settlements. In the process, they displaced the Native Americans who, for the most part, had lived on the land sustainably for thousands of years.

The U.S. government accelerated this settling of the continent and the use of its resources by transferring vast areas of public land to private interests. To encourage settlement, between 1850 and 1890, more than half of the country's public land was given away or sold cheaply by the government to railroad, timber, and mining companies, land developers, states, schools, universities, and homesteaders. This era came to an end when the government declared the frontier officially closed in 1890.

Early Conservationists (1832–1870)

Between 1832 and 1870, some citizens became alarmed at the scope of resource depletion and degradation in the United States. They urged the government to preserve part of the unspoiled wilderness on public lands owned jointly by all people (but managed by the government), and to protect it as a legacy for future generations.

Two of these early conservationists were Henry David Thoreau (1817–1862) and George Perkins Marsh (1801–1882). Thoreau (Figure 1) was alarmed at the loss of numerous wild species from his native eastern Massachusetts. To gain a better understanding of nature, he built a cabin in the woods on Walden Pond near Concord, Massachusetts, lived there alone for 2 years, and wrote *Life in the Woods,* an environmental classic.*

In 1864, Marsh, a scientist and member of Congress from Vermont, published *Man and Nature,* an extensive study of how human activities were altering the environment, which helped legislators and citizens see the need for resource conservation. Marsh questioned the idea that the country's resources were inexhaustible. He also used scientific studies and case studies to show how the rise and fall of past civilizations were linked to the use and

---

*I (Miller) can identify with Thoreau. I spent 10 years living in the deep woods studying and thinking about how nature works and writing early editions of the book you are reading. See the whole story of this adventure on p. 1.

© Cengage Learning

Figure 1 Henry David Thoreau (1817–1862) was an American writer and naturalist who kept journals about his excursions into wild areas in parts of the northeastern United States and Canada, and at Walden Pond in Concord, Massachusetts. He sought self-sufficiency, a simple lifestyle, and a harmonious coexistence with nature.

misuse of their soils, water supplies, and other resources. Some of his resource conservation principles are still used today.

What Happened between 1870 and 1930?

Between 1870 and 1930, a number of events increased the role of the federal government and private citizens in resource conservation and public health. The *Forest Reserve Act of 1891* was a turning point in establishing the responsibility of the federal government for protecting public lands from resource exploitation.

In 1892, nature preservationist and activist John Muir (1838–1914) (Figure 2) founded the Sierra Club. He had been deeply troubled by the rapid deforestation that he had witnessed in large areas of North America. By 1870, the vast forests of the Northeast had been cut down, and between then and 1920, loggers essentially clear-cut the great pine forests of the upper Midwest—an area the size of Europe.

Muir became the leader of the *preservationist movement,* which called for protecting large areas of wilderness on public lands from human exploitation, except for low-impact recreational activities such as hiking and camping. However, this idea was not enacted into law until 1964.

Muir also proposed and lobbied for creation of a national park system on public lands. In addition, he founded the Sierra Club, which is to this day a political force working on behalf of the environment.

Primarily because of political opposition, effective protection of forests and wildlife on federal lands did not begin until Theodore Roosevelt (1858–1919) (Figure 3), an ardent conservationist, became president. His term of office, 1901–1909, has been called the country's *Golden Age of Conservation.*

While in office, Roosevelt persuaded Congress to give the president power to designate public land as federal wildlife refuges. He was motivated partly by the fates of prominent wildlife species such as the passenger pigeon (extinct by 1900) and the North American bison, or American buffalo, which was nearly wiped out. In 1903, Roosevelt established the first federal refuge at Pelican Island (see Figure 9-20, p. 210) off the east coast of Florida for preservation of the endangered brown pelican, and he added 35 more wildlife reserves by 1904. He also more than tripled the size of the national forest reserves.

In 1905, Congress created the U.S. Forest Service to manage and protect the forest reserves. Roosevelt appointed Gifford Pinchot (1865–1946) as its first chief. Pinchot

© Cengage Learning

Figure 2 John Muir (1838–1914) was a Scottish geologist, explorer, and naturalist. He spent 6 years studying, writing journals, and making sketches in the wilderness of California's Yosemite Valley and then went on to explore wilderness areas in Utah, Nevada, the Northwest, and Alaska. He was largely responsible for establishing Yosemite National Park in 1890. He also founded the Sierra Club and spent 22 years lobbying actively for conservation laws.

© Cengage Learning

Figure 3 Theodore (Teddy) Roosevelt (1858–1919) was a writer, explorer, naturalist, avid birdwatcher, and twenty-sixth president of the United States. He was the first national political figure to bring conservation issues to the attention of the American public. According to many historians, Theodore Roosevelt contributed more than any other U.S. president to natural resource conservation in the United States.

pioneered scientific management of forest resources on public lands. In 1906, Congress passed the *Antiquities Act*, which allows the president to protect areas of scientific or historical interest on federal lands as national monuments. Roosevelt used this act to protect the Grand Canyon and other wilderness areas that would later become national parks.

Congress became upset with Roosevelt in 1907, because by then, he had added vast tracts to the national forest reserves. Congress passed a law banning further executive withdrawals of public forests. However, on the day before the bill became law, Roosevelt defiantly reserved another large block of land. Most environmental historians view Roosevelt as the country's best environmental president.

Early in the 20th century, the U.S. conservation movement split into two factions over how public lands should be used. The *wise-use,* or *conservationist,* school, led by Roosevelt and Pinchot, believed all public lands should be managed wisely and scientifically to provide needed resources. The *preservationist* school, led by Muir, wanted wilderness areas on public lands to be left untouched. This controversy over use of public lands continues today.

In 1916, Congress passed the *National Park Service Act,* which declared that parks are to be maintained in a manner that leaves them unimpaired for future generations. The act also established the National Park Service (within the Department of the Interior) to manage the park system. Under its first head, Stephen T. Mather (1867–1930), the dominant park policy was to encourage tourist visits by allowing private concessionaires to operate facilities within the parks.

After World War I, the country entered a new era of economic growth and expansion. To stimulate economic growth during the administrations of Presidents Harding, Coolidge, and Hoover, the federal government promoted the increased sales, at low prices, of timber, energy, mineral, and other resources found on public lands.

President Herbert Hoover went even further and proposed that the federal government return all remaining federal lands to the states or sell them to private interests for economic development. But the Great Depression (1929–1941) made owning such lands unattractive to state governments and private investors. The Depression was bad news for the country. But some say that without it, we might have little if any of the public lands that now make up about one-third of the total land area of the United States (see Figure 25-4, p. 687).

What Happened between 1930 and 1960?

Along with a second wave of national resource conservation, improvements in public health also began in the early 1930s as President Franklin D. Roosevelt (1882–1945) strove to bring the country out of the Great Depression. He persuaded Congress to enact federal programs to provide jobs and to help restore the country's degraded environment.

During this period, the government purchased large tracts of land from cash-poor landowners, and established the *Civilian Conservation Corps* (CCC) in 1933. According to the U.S. Forest Service, it eventually put more than 3 million unemployed people to work planting trees and developing and maintaining parks and recreation areas. During its 10 years of existence, the CCC planted an estimated 3 billion trees. It also restored silted waterways and built levees and dams for flood control.

The government also built and operated many large dams in the Tennessee River Valley, as well as in the arid western states, including Hoover Dam on the Colorado River (see Figure 13-1, p. 318). The goals were to provide jobs, flood control, cheap irrigation water, and cheap electricity for industry.

In 1935, Congress passed the Soil Conservation Act. During the Great Depression, erosion problems had ruined many farms in the Great Plains states and created a large area of degraded land known as the *Dust Bowl.* To correct these devastating erosion problems, the new law established the *Soil Erosion Service* as part of the Department of Agriculture. Its name was later changed to the *Soil Conservation Service*, and now it is called the *Natural Resources Conservation Service*. Many environmental historians praise Franklin D. Roosevelt for his efforts to get the country out of a major economic depression, partly by helping to restore environmentally degraded areas.

Also, in 1935, Aldo Leopold (1887–1948) (Figure 4), wildlife manager, professor, writer, and conservationist, helped to found the U.S. Wilderness Society. Largely through his writings, especially his 1949 essay "The Land Ethic" and his 1949 book *A Sand County Almanac,* he became one of the foremost leaders of the *conservation* and *environmental movements.* His energy and foresight helped to lay the critical groundwork for the field of environmental ethics. Leopold contended that the role of the human species should be to protect nature, not to conquer it.

Federal resource conservation policy changed little during the 1940s and 1950s, mostly because of preoccupation with World War II (1941–1945) and economic recovery after the war.

Between 1930 and 1960, improvements in public health included establishment of public health boards and agencies at the municipal, state, and federal levels; increased public education about health issues; introduction of vaccination programs; and a sharp reduction in the incidence of waterborne infectious diseases, mostly because of improved sanitation and garbage collection.

Figure 4 Aldo Leopold (1887–1948) worked in game management early in his career, but then grew interested in the emerging scientific field of ecology. He became a leading conservationist and his book, *A Sand County Almanac* (published after his death), is considered an environmental classic that helped to inspire the modern environmental and conservation movements.

What Happened during the 1960s?

A number of milestones in American environmental history occurred during the 1960s. In 1962, biologist Rachel Carson (1907–1964) wrote *Silent Spring*, which documented the pollution of air, water, and wildlife from the use of pesticides such as DDT (see Chapter 12, pp. 296–300). This influential book helped to broaden the concept of resource conservation to include preservation of the *quality* of the planet's air, water, soil, and wildlife.

Many environmental historians mark Carson's wake-up call as the beginning of the modern *environmental movement* in the United States. It consists of citizens organized to demand that political leaders enact laws and develop policies to curtail pollution, clean up polluted environments, and protect unspoiled areas from environmental degradation.

In 1964, Congress passed the *Wilderness Act*, inspired by the vision of John Muir more than 80 years earlier. It authorized the government to protect undeveloped tracts of public land as part of the National Wilderness System. Any land in this system is to be used only for nondestructive forms of recreation such as hiking and camping, and it remains in the system unless and until Congress decides that it is needed for the national good.

Between 1965 and 1970, the emerging science of *ecology* received widespread media attention. At the same time, the popular writings of biologists such as Leopold, Carson, Paul Ehrlich, Barry Commoner, and Garrett Hardin awakened Americans to the interlocking relationships among population growth, resource use, and pollution.

During that period, a number of events increased public awareness of pollution, habitat loss, and other forms of environmental degradation. For example, many people learned about these problems when well-known wildlife species such as the American bald eagle, the grizzly bear, the whooping crane, and the peregrine falcon became endangered. Also, the stretch of the Cuyahoga River running near Cleveland, Ohio, was so polluted with oil and other flammable pollutants that it caught fire several times. Another major event was a devastating oil spill off the California coast in 1969.

In 1968, when U.S. astronauts aboard Apollo 8 became the first humans to fly to the moon, they photographed the entire earth for the first time from lunar orbit. This allowed people to see the earth as a tiny blue and white planet in the black void of space, and it led to the development of the *spaceship–earth environmental worldview*. According to this view, we live on a planetary spaceship that we should not harm because it is the only home we have.

What Happened during the 1970s? The Environmental Decade

During the 1970s, media attention, public concern about environmental problems, scientific research, and action to address environmental concerns grew rapidly. This period is sometimes called the *environmental decade*, or the *first decade of the environment*.

During the 1970s, several major U.S. environmental laws were passed. The first and possibly most important of these laws, signed by President Richard M. Nixon on January 1, 1970, was the *National Environmental Policy Act* (NEPA). It established the *environmental impact statement* (EIS) process. An EIS is now required for any action taken by the federal government that could significantly affect the environment. The EIS process forces the government and businesses that receive government contracts to carefully evaluate environmental impacts and avoid those with the potential for significant environmental damage. Many activities, including road building, dam construction, and logging on federal lands, are affected. NEPA set in motion a legislative process that produced several major environmental laws (see Figure 24-8, p. 668).

The first annual *Earth Day* was held on April 20, 1970. This event was proposed by Senator Gaylord Nelson (1916–2005) and organized by then Harvard graduate student Denis Hayes (see Individuals Matter 24.1, p. 666). Some 20 million people in more than

2,000 U.S. communities took to the streets to heighten the nation's environmental awareness and to demand improvements in environmental quality. There were also many environmental efforts that took place on a smaller scale that were greatly instructive for the environmental movement.

The *Environmental Protection Agency* (EPA) was established in 1970. In addition, the *Endangered Species Act of 1973* greatly strengthened the role of the federal government in protecting endangered species and their habitats. In addition, Congress created the Department of Energy in 1977, charging this new agency with developing a long-range energy strategy to help reduce the country's heavy dependence on imported oil.

During the 1970s, the area of land in the National Wilderness System tripled and the area in the National Park System doubled (primarily because vast tracts of land in the state of Alaska were added to these systems). In 1978, the *Federal Land Policy and Management Act* gave the *Bureau of Land Management* (BLM) its first real authority to manage the public land under its control, 85% of which is in 12 western states. This law angered a number of western interests whose use of these public lands was restricted for the first time.

In response, a coalition of ranchers, miners, loggers, developers, farmers, some elected officials, and other citizens in the affected states launched a political campaign known as the *sagebrush rebellion.* It had two major goals. *First,* sharply reduce government regulation of the use of public lands. *Second,* remove most public lands in the western United States from federal ownership and management and turn them over to the states. After that, the plan was to persuade state legislatures to sell or lease the resource-rich lands at low prices to ranching, mining, timber, land development, and other private interests. This represented a return to President Herbert Hoover's plan to get rid of all public land, which had been thwarted by the Great Depression. This political movement continues to exist.

What Happened during the 1980s? Environmental Backlash

In 1980, Congress created the *Superfund* as part of the *Comprehensive Environmental Response, Compensation, and Liability Act* (CERCLA; see Chapter 21, p. 595). Its goal was to clean up abandoned hazardous waste sites, including the Love Canal housing development in Niagara Falls, New York, which had to be abandoned when hazardous wastes from the site of a former chemical company began leaking into school grounds, yards, and basements.

During the decade of the 1980s, farmers, ranchers, and leaders of the oil, coal, automobile, mining, and timber industries strongly opposed many of the environmental laws and regulations developed in the 1960s and 1970s. They organized and funded multiple efforts to defeat environmental laws and regulations. In 1988, these industries backed a new coalition called the *wise-use movement.* Its major goals were to weaken or repeal most of the country's environmental laws and regulations, and destroy the effectiveness of the environmental movement in the United States. Backers of this movement argued that environmental laws had gone too far and were hindering economic growth.

Congress, influenced by this growing backlash, allowed for much more private energy and mineral development and timber cutting on public lands. Congress also lowered automobile gas mileage standards and relaxed federal air and water quality pollution standards.

In 1980, the United States had led the world in research and development of wind and solar energy technologies. Between 1981 and 1983, however, Congress slashed by 90% government subsidies for renewable energy research and for energy efficiency research, and eliminated tax incentives for the residential solar energy and energy conservation programs enacted in the late 1970s. As a result, the United States lost its lead in developing and selling the wind turbines and solar cells to Denmark, Germany, Japan, and China. These businesses are rapidly becoming two of the biggest and most profitable industries in the world.

At the same time, the 1980s saw rising public interest in some environmental issues, largely due to the influence of some prominent environmental scientists. For example, oceanographer Jacques Cousteau continued to tell the story of his four decades of undersea explorations and his concerns for the ocean environment through dozens of TV programs, books, and films. In Africa, zoologist Dian Fossey studied mountain gorillas and raised public awareness about these endangered animals, their shrinking habitats, and the threat of poachers. Her murder in 1985, possibly at the hands of poachers, made her story even more compelling.

What Happened from 1990 to 2013?

Between 1990 and 2013, opposition to environmental laws and regulations gained strength. This occurred because of continuing political and economic support from corporate backers, who not only argued that environmental laws were hindering economic growth, but also helped elect many members of Congress who were generally unsympathetic to environmental concerns. Since 1990, leaders and supporters of the environmental movement have had to spend much of their time and funds fighting efforts to discredit the movement and efforts to weaken or eliminate most environmental laws passed during the 1960s and 1970s.

During the 1990s, many small and mostly local grassroots environmental organizations sprang up to help deal with environmental threats in their local communities. Interest in environmental issues increased on many college and university campuses, resulting in the expansion of environmental studies courses and programs at these institutions. In addition, there was growing awareness of critical and complex environmental issues, such as sustainability, population growth, biodiversity protection, and threats from atmospheric warming and projected climate change. Whereas 20 million Americans took part in the first Earth Day in 1970, an estimated 200 million people from more than 140 countries participated in Earth Day 1990.

During the first decade of the 21st century, some environmental progress was made as a number of business leaders began to argue for a gradual shift from using fossil fuels to using a mix of renewable energy resources, including solar energy, wind power, and geothermal energy. Many of these leaders also put a high priority on saving their companies money by improving their energy efficiency and reducing their unnecessary energy waste.

In addition, during this decade, many analysts as well as business and government leaders have called for developing an effective U.S. policy for slowing projected climate change caused mostly by the burning of fossil fuels. However, the politically and economically powerful fossil fuel industries have been able to block efforts to develop a comprehensive energy policy shift toward renewables, as well as any major effort to address the projected threat of climate disruption.

Chemists Use the Periodic Table to Classify Elements on the Basis of Their Chemical Properties

Matter consists of elements and compounds (see Chapter 2, pp. 35–38). Chemists have developed a way to classify the elements according to their chemical behavior in what is called the *Periodic Table of Elements* (Figure 1). Each horizontal row in the table is called a *period*. Each vertical column lists elements with similar chemical properties and is called a *group*.

The Periodic Table in Figure 1 shows how the elements can be classified as *metals, nonmetals,* and *metalloids*. Exam-

ples of metals are sodium (Na), calcium (Ca), aluminum (Al), iron (Fe), lead (Pb), silver (Ag), and mercury (Hg).

Atoms of metals tend to lose one or more of their electrons to form positively charged ions such as Na^+, Ca^{2+}, and Al^{3+}. For example, an atom of the metallic element sodium (Na, atomic number 11) with 11 positively charged protons and 11 negatively charged electrons can lose one of its electrons. It then becomes a sodium ion with a positive charge of 1 (Na^+) because it now has 11 positive charges (protons) but only 10 negative charges (electrons).

Examples of *nonmetals* are hydrogen (H), carbon (C), nitrogen (N), oxygen (O), phosphorus (P), sulfur (S),

Figure 1 *Periodic Table of Elements.* Elements in the same vertical column, called a *group,* have similar chemical properties.

© Cengage Learning

chlorine (Cl), and fluorine (F). Atoms of some nonmetals such as chlorine, oxygen, and sulfur tend to gain one or more electrons lost by metallic atoms to form negatively charged ions such as O^{2-}, S^{2-}, and Cl^-. For example, an atom of the nonmetallic element chlorine (Cl, atomic number 17) can gain an electron and become a chlorine ion. The ion has a negative charge of 1 (Cl^-) because it has 17 positively charged protons and 18 negatively charged electrons. Atoms of nonmetals can also combine with one another to form molecules in which they share one or more pairs of their electrons. Hydrogen, a nonmetal, is placed by itself above the center of the table because it does not fit very well into any of the groups.

The elements arranged in a diagonal staircase pattern between the metals and nonmetals have a mixture of metallic and nonmetallic properties and are called *metalloids*. Examples are germanium (Ge) and arsenic (As).

Figure 1 also identifies the elements required as *nutrients* (marked by small black squares) for all or some forms of life, and elements that are moderately or highly toxic (marked by small red squares) to all or most forms of life. Note that some elements such as copper (Cu) serve as nutrients, but can also be toxic at high enough doses. Six nonmetallic elements—carbon (C), oxygen (O), hydrogen (H), nitrogen (N), sulfur (S), and phosphorus (P)—make up about 99% of the atoms of all living things.

🔍 CONSIDER THIS. . .

THINKING ABOUT The Periodic Table

Use the Periodic Table to identify by name and symbol two elements that should have chemical properties similar to those of **(a)** Ca, **(b)** potassium, **(c)** S, **(d)** lead.

Ionic and Covalent Bonds Hold Compounds Together

Sodium chloride (NaCl) consists of a three-dimensional network of oppositely charged *ions* (Na^+ and Cl^-) held together by the forces of attraction between opposite charges (see Figure 2-6, p. 38). The strong forces of attraction between such oppositely charged ions are called *ionic bonds*. They are formed when an electron is transferred from a metallic atom such as sodium (Na) to a nonmetallic element such as chlorine (Cl). Because ionic compounds consist of ions formed from atoms of metallic elements (positive ions) and nonmetallic elements (negative ions), they can be described as *metal–nonmetal compounds*.

Sodium chloride and many other ionic compounds tend to dissolve in water and break apart into their individual ions (Figure 2).

$$NaCl \longrightarrow Na^+ + Cl^-$$
sodium chloride (in water)   sodium ion   chloride ion

Figure 2   This model illustrates how a salt dissolves in water.

Water, a *covalent compound*, consists of molecules made up of uncharged atoms of hydrogen (H) and oxygen (O). Each water molecule consists of two hydrogen atoms chemically bonded to an oxygen atom, yielding H_2O molecules. The bonds between the atoms in such molecules are called *covalent bonds* and form when the atoms in the molecule share one or more pairs of their electrons. Because they are formed from atoms of nonmetallic elements (see Figure 1), covalent compounds can be described as *nonmetal–nonmetal compounds*. Figure 3 shows the chemical formulas and shapes of the molecules that are the building blocks for several common *covalent compounds*.

What Makes Solutions Acidic? Hydrogen Ions and pH

The *concentration*, or number of hydrogen ions (H^+) in a specified volume of a solution (typically a liter), is a measure of its acidity. Pure water (not tap water or rainwater) has an equal number of hydrogen (H^+) and hydroxide (OH^-) ions. It is called a **neutral solution**. An **acidic solution** has more hydrogen ions than hydroxide ions per liter. A **basic solution** has more hydroxide ions than hydrogen ions per liter.

Scientists use **pH** as a measure of the acidity of a solution based on its concentration of hydrogen ions (H^+). By definition, a neutral solution has a pH of 7, an acidic solution has a pH of less than 7, and a basic solution has a pH greater than 7.

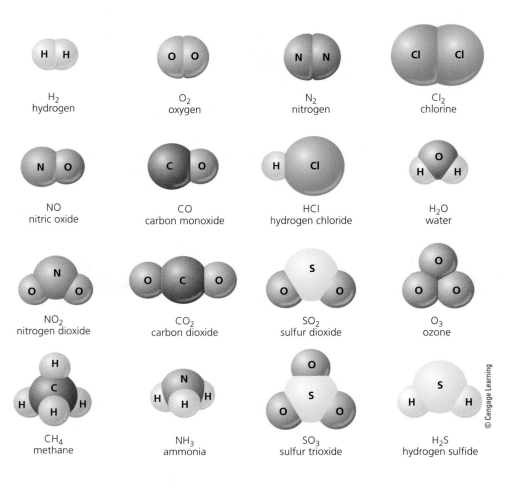

Figure 3 Chemical formulas and shapes for some *covalent compounds* formed when atoms of one or more nonmetallic elements combine with one another. The bonds between the atoms in such molecules are called *covalent bonds*.

H₂ hydrogen	H_2 hydrogen

H₂
hydrogen

O₂
oxygen

N₂
nitrogen

Cl₂
chlorine

NO
nitric oxide

CO
carbon monoxide

HCl
hydrogen chloride

H₂O
water

NO₂
nitrogen dioxide

CO₂
carbon dioxide

SO₂
sulfur dioxide

O₃
ozone

CH₄
methane

NH₃
ammonia

SO₃
sulfur trioxide

H₂S
hydrogen sulfide

© Cengage Learning

Each single unit change in pH represents a tenfold increase or decrease in the concentration of hydrogen ions per liter. For example, an acidic solution with a pH of 3 is 10 times more acidic than a solution with a pH of 4. Figure 4 shows the approximate pH and hydrogen ion concentration per liter of solutions for various common substances.

🔍 **CONSIDER THIS. . .**

THINKING ABOUT pH

A solution has a pH of 2. How many times more acidic is this solution than one with a pH of 6?

The measurement of acidity is important in the study of environmental science, because environmental changes involving acidity can have serious environmental impacts. For example, when coal and oil are burned, they give off acidic compounds that can return to the earth as *acid deposition* (see Figure 18-12, p. 485, and Figure 18-13, p. 486), which has become a major environmental problem.

There Are Weak Forces of Attraction between Some Molecules

Ionic and covalent bonds form between the ions or atoms *within* a compound. There are also weaker forces of attraction *between* the molecules of covalent compounds (such as water) resulting from an unequal sharing of electrons by two atoms.

For example, an oxygen atom has a much greater attraction for electrons than does a hydrogen atom. Thus, the electrons shared between the oxygen atom and its two hydrogen atoms in a water molecule are pulled closer to the oxygen atom, but not actually transferred to the oxygen atom. As a result, the oxygen atom in a water molecule has a slightly negative partial charge and its two hydrogen atoms have a slightly positive partial charge.

The slightly positive hydrogen atoms in one water molecule are then attracted to the slightly negative oxygen atoms in another water molecule. These forces of attraction *between* water molecules are called *hydrogen bonds* (see Figure 3-A, p. 64). They account for many of water's unique properties (see Chapter 3, Science Focus 3.2, p. 64). Hydrogen bonds also form between other covalent molecules or between portions of such molecules containing hydrogen and nonmetallic atoms with a strong ability to attract electrons.

Four Types of Large Organic Compounds Are the Molecular Building Blocks of Life

Larger and more complex organic compounds, called *polymers,* consist of a number of basic structural or molecular units *(monomers)* linked by chemical bonds, somewhat like rail cars linked in a freight train. Four types of macromolecules—complex carbohydrates, proteins, nucleic acids, and lipids—are the molecular building blocks of life.

Complex carbohydrates consist of two or more monomers of *simple sugars* (such as glucose, Figure 5) linked together. One example is the starches that plants use to store energy and also to provide energy for animals that feed on plants. Another is cellulose, the earth's most abundant organic compound, which is found in the cell walls of bark, leaves, stems, and roots.

Proteins are large polymer molecules formed by linking together long chains of monomers called *amino acids* (Figure 6). Living organisms use about 20 different amino acid molecules to build a variety of proteins, which play different roles. Some help to store energy. Some are components of the *immune system* that protects the body against diseases and harmful substances by forming antibodies that make invading agents harmless. Others are *hormones* that are used as chemical messengers in the bloodstreams of animals to turn various bodily functions on or off. In animals, proteins are also components of hair, skin, muscle, and tendons. In addition, some proteins act as *enzymes* that catalyze or speed up certain chemical reactions.

Nucleic acids are large polymer molecules made by linking hundreds to thousands of four types of monomers called *nucleotides*. Two nucleic acids—DNA (**d**eoxyribo**n**ucleic **a**cid) and RNA (**r**ibo**n**ucleic **a**cid)—participate in the building of proteins and carry hereditary information used to pass traits from parent to offspring. Each nucleotide consists of a *phosphate group*, a *sugar molecule* containing five carbon atoms (deoxyribose in DNA molecules and ribose in RNA molecules), and one of four different *nucleotide bases* (represented by A, G, C, and T, the first letter in each of their names, or A, G, C, and U in RNA) (Figure 7). The four basic nucleotides used to make various forms of DNA molecules differ in the types of nucleotide bases they contain—adenine (A), guanine (G), cytosine (C), and thymine (T). (Uracil, labeled U, occurs instead of thymine in RNA.) In the cells of living organisms, these nucleotide units combine in different numbers and sequences to form *nucleic acids* such as various types of DNA and RNA (Figure 8, p. S17).

Hydrogen bonds formed between parts of the four nucleotides in DNA hold two DNA strands together like a spiral staircase, forming a double helix (see Figure 8). DNA molecules can unwind and replicate themselves.

The total weight of the DNA needed to reproduce all of the world's people is only about 50 milligrams—the weight of a small match. If the DNA coiled in your body were unwound, it would stretch about 960 million kilometers (600 million miles)—more than 6 times the distance between the sun and the earth.

The different molecules of DNA that make up the millions of species found on the earth are like a vast and diverse

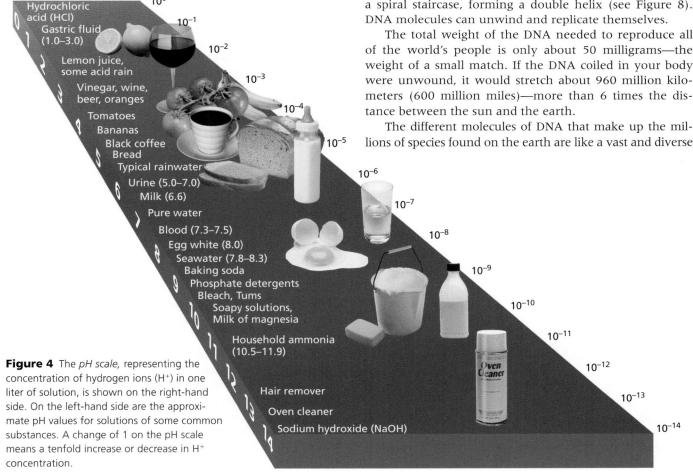

Figure 4 The *pH scale*, representing the concentration of hydrogen ions (H^+) in one liter of solution, is shown on the right-hand side. On the left-hand side are the approximate pH values for solutions of some common substances. A change of 1 on the pH scale means a tenfold increase or decrease in H^+ concentration.

© Cengage Learning

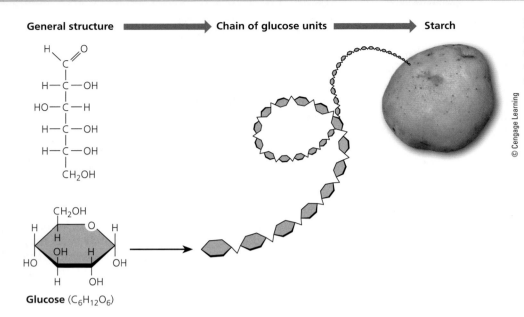

General structure ➡️ **Chain of glucose units** ➡️ **Starch**

Glucose ($C_6H_{12}O_6$)

Figure 5 Straight-chain and ring structural formulas of glucose, a simple sugar that can be used to build long chains of complex carbohydrates such as starch and cellulose.

© Cengage Learning

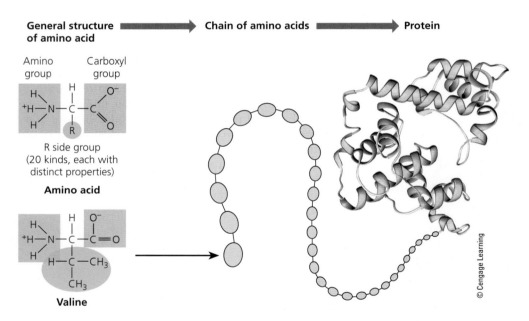

General structure of amino acid ➡️ **Chain of amino acids** ➡️ **Protein**

Amino group   Carboxyl group

R side group (20 kinds, each with distinct properties)

Amino acid

Valine

Figure 6 General structural formula of amino acids and a specific structural formula of one of the 20 different amino acid molecules that can be linked together in chains to form proteins that fold up into more complex shapes.

© Cengage Learning

Deoxyribose in DNA
Ribose in RNA

Phosphate   5-Carbon sugar   Nucleotide base

© Cengage Learning

Figure 7 Generalized structures of the nucleotide molecules linked in various numbers and sequences to form large nucleic acid molecules such as various types of deoxyribonucleic acid (DNA) and ribonucleic acid (RNA). In DNA, the five-carbon sugar in each nucleotide is deoxyribose; in RNA it is ribose.

genetic library. Each species is a unique book in that library. The *genome* of a species is made up of the entire sequence of DNA "letters" or base pairs that combine to "spell out" the chromosomes in typical members of each species. In 2002, scientists were able to map out the genome for the human species by using powerful computers to help them in analyzing the 3.1 billion base sequences in human DNA.

Lipids, a fourth building block of life, are a chemically diverse group of large organic compounds that do not dissolve in water. Examples are *fats and oils* for storing energy (Figure 9), *waxes* for structure, and *steroids* for producing hormones.

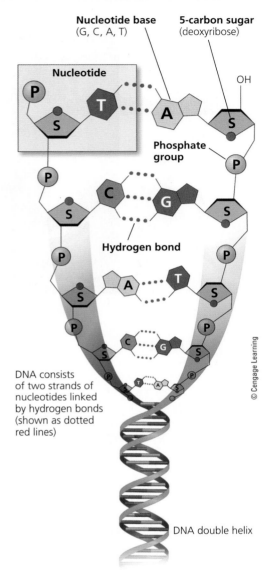

Figure 8 A portion of the double helix of a DNA molecule, which is composed of two spiral (helical) strands of nucleotides. Each nucleotide contains a unit of phosphate (P), deoxyribose (S), and one of four nucleotide bases: adenine (A), guanine (G), cytosine (C), and thymine (T). The two strands are held together by hydrogen bonds formed between various pairs of the nucleotide bases. Guanine (G) bonds with cytosine (C), and adenine (A) with thymine (T).

Figure 10 shows the relative sizes of simple and complex molecules, cells, and multi-celled organisms.

Certain Molecules Store and Release Energy in Cells

Chemical reactions occurring in plant cells during photosynthesis (see Chapter 3, p. 55) release energy that is absorbed by adenosine diphosphate (ADP) molecules

Figure 9 Structural formula of fatty acid, which is one form of lipid (left). Fatty acids are converted into more complex fat molecules (center) that are stored in adipose cells (right).

and stored as chemical energy in adenosine triphosphate (ATP) molecules (Figure 11, left). When cellular processes require energy, ATP molecules release it to form ADP molecules (Figure 11, right).

Chemists Balance Chemical Equations to Keep Track of Atoms

Chemists use a shorthand system, or equation, to represent chemical reactions. These chemical equations are also used as an accounting system to verify that no atoms are created or destroyed in a chemical reaction as required by the law of conservation of matter (see Chapter 2, p.40). As a consequence, each side of a chemical equation must have the same number of atoms or ions of each element involved. Ensuring that this condition is met leads to what chemists call a *balanced chemical equation*. The equation for the burning of carbon ($C + O_2 \rightarrow CO_2$) is balanced because one atom of carbon and two atoms of oxygen are on both sides of the equation.

Consider the following chemical reaction. When electricity passes through water (H_2O), the latter can be broken down into hydrogen (H_2) and oxygen (O_2), as represented by the following equation:

$$H_2O \longrightarrow H_2 + O_2$$

2 H atoms   2 H atoms   2 O atoms
1 O atom

This equation is unbalanced because one atom of oxygen is on the left side of the equation but two oxygen atoms are on the right side. We cannot change the subscripts of any of the formulas to balance this equa-

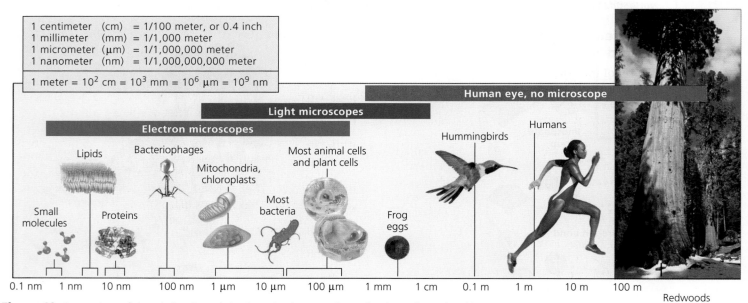

1 centimeter	(cm)	= 1/100 meter, or 0.4 inch
1 millimeter	(mm)	= 1/1,000 meter
1 micrometer	(μm)	= 1/1,000,000 meter
1 nanometer	(nm)	= 1/1,000,000,000 meter

1 meter = 10^2 cm = 10^3 mm = 10^6 μm = 10^9 nm

Human eye, no microscope

Light microscopes

Electron microscopes

Lipids

Bacteriophages

Mitochondria, chloroplasts

Most animal cells and plant cells

Humans

Hummingbirds

Small molecules

Proteins

Most bacteria

Frog eggs

0.1 nm   1 nm   10 nm   100 nm   1 μm   10 μm   100 μm   1 mm   1 cm   0.1 m   1 m   10 m   100 m

Redwoods

Figure 10 Comparison of the relative sizes of simple molecules, complex molecules, cells, and multi-cellular organisms. This scale is exponential, not linear. Each unit of measure is 10 times larger than the unit preceding it.

(Used by permission from STARR/TAGGART, Biology, 11E. © 2006 Cengage Learning.)

ATP synthesis:
Energy is stored in ATP

ADP   Phosphate   Energy

ATP

ATP breakdown:
Energy stored in ATP is released

ATP

ADP   Phosphate   Energy

© Cengage Learning

Figure 11 Models representing energy storage and release in cells.

tion because that would change the arrangements of the atoms, leading to different substances. Instead, we must use different numbers of the molecules involved to balance the equation. For example, we could use two water molecules:

$2 H_2O$ ⟶ H_2 + O_2
4 H atoms   2 H atoms   2 O atoms
2 O atoms

This equation is still unbalanced. Although the numbers of oxygen atoms on both sides of the equation are now equal, the numbers of hydrogen atoms are not. We can correct this problem by having the reaction produce two hydrogen molecules:

$2 H_2O$ ⟶ $2 H_2$ + O_2
4 H atoms   4 H atoms   2 O atoms
2 O atoms

Now the equation is balanced, and the law of conservation of matter has been observed. For every two molecules of water through which we pass electricity, two hydrogen molecules and one oxygen molecule are produced.

CONSIDER THIS...

THINKING ABOUT Chemical Equations

Try to balance the chemical equation for the reaction that combines nitrogen gas (N_2) and hydrogen gas (H_2) to form ammonia gas (NH_3).

Weather Is Affected by Moving Masses of Warm and Cold Air

Weather is the set of short-term atmospheric conditions—typically those occurring over hours or days—for a particular area. Examples of atmospheric conditions include temperature, pressure, moisture content, precipitation, sunshine, cloud cover, and wind direction and speed.

Meteorologists use equipment mounted on weather balloons, aircraft, ships, and satellites, as well as radar and stationary sensors, to obtain data on weather variables. They then feed these data into computer models to draw weather maps. Other computer models project the weather for a period of several days by calculating the probabilities that air masses, winds, and other factors will change in certain ways.

Much of the weather we experience results from interactions between the leading edges of moving masses of warm and cold air driven largely by uneven heating of the earth's surface by the sun (see Figure 7-3, p. 146). Weather changes as one air mass replaces or meets another. The most dramatic changes in weather occur along a **front**, the boundary between two air masses with different temperatures and densities.

A **warm front** is the boundary between an advancing warm air mass and the cooler one it is replacing (Figure 1, left). Because warm air is less dense (weighs less per unit of volume) than cool air, an advancing warm front rises up over a mass of cool air. As the warm front rises, its moisture begins condensing into droplets, forming layers of clouds at different altitudes. Gradually, the clouds thicken, descend to a lower altitude, and often release their moisture as rainfall. A moist warm front can bring days of cloudy skies and drizzle.

A **cold front** (Figure 1, right) is the leading edge of an advancing mass of cold air. Because cold air is denser than warm air, an advancing cold front stays close to the ground and wedges underneath less dense warmer air. An approaching cold front produces rapidly moving, towering clouds called *thunderheads,* with flat, anvil-like tops.

As a cold front passes through, it may cause high surface winds and thunderstorms. After it leaves the area, it usually results in cooler temperatures and a clear sky.

Near the top of the troposphere, hurricane-force winds circle the earth. These powerful winds, called *jet streams,* follow rising and falling paths that have a strong influence on weather patterns (Figure 2).

Weather Is Affected by Changes in Atmospheric Pressure

Changes in atmospheric pressure also affect weather. *Atmospheric pressure* results from molecules of gases (mostly nitrogen and oxygen) in the atmosphere zipping around at very high speeds and hitting and bouncing off everything they encounter.

Atmospheric pressure is greater near the earth's surface because the molecules in the atmosphere are squeezed together under the weight of the air above them. An air mass with high pressure, called a **high**, contains cool, dense air that descends slowly toward the earth's surface and becomes warmer. Because of this warming, condensation of moisture usually does not take place and clouds usually do not form. Fair weather with clear skies follows as long as this high-pressure air mass remains over the area.

In contrast, a low-pressure air mass, called a **low**, produces cloudy and sometimes stormy weather. Because of

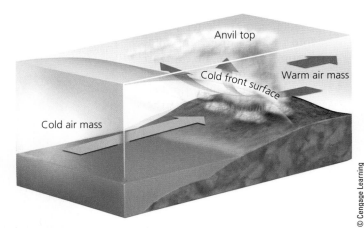

© Cengage Learning

Figure 1 *Weather fronts:* A *warm front* (left) occurs when an advancing mass of warm air meets and rises up over a mass of denser cool air. A *cold front* (right) forms when a moving mass of cold air wedges beneath a mass of less dense warm air.

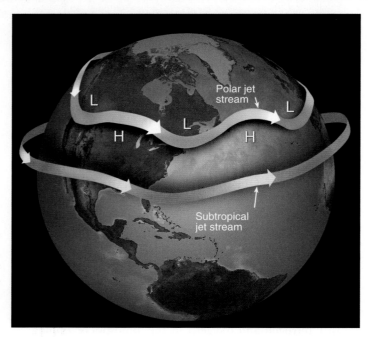

Figure 2 A *jet stream* is a rapidly flowing air current that moves west to east in a wavy pattern. This figure shows a polar jet stream and a subtropical jet stream in winter. In reality, jet streams are discontinuous and their positions vary from day to day.

(Used by permission from AHRENS, Meteorology Today, 8E. © 2007 Cengage Learning.)

its low pressure and low density, the center of a low rises, and its warm air expands and cools. When the temperature drops below a certain level where condensation takes place, called the *dew point,* moisture in the air condenses and forms clouds.

If the droplets in the clouds coalesce into larger drops or snowflakes heavy enough to fall from the sky, precipitation occurs. The condensation of water vapor into water drops usually requires that the air contain suspended tiny particles of material such as dust, smoke, sea salts, or volcanic ash. These so-called *condensation nuclei* provide surfaces on which the droplets of water can form and coalesce.

Figure 3 A shore upwelling occurs when deep, cool, nutrient-rich waters are drawn up to replace surface water moved away from a steep coast by wind flowing along the coast toward the equator.

Every Few Years Major Wind Shifts in the Pacific Ocean Affect Global Weather Patterns

Ocean currents can have a strong effect on the weather. Water moves vertically, as well as horizontally from one area of the ocean to another. An **upwelling**, or upward movement of ocean water, can mix upper levels of seawater with lower levels, bringing cool and nutrient-rich water from the bottom of the ocean to the warmer surface where it supports large populations of phytoplankton, zooplankton, fish, and fish-eating seabirds.

Figure 7-2, p. 145, shows the oceans' major upwelling zones. Upwellings far from shore occur when surface currents move apart and draw water up from deeper layers. Strong upwellings are also found along the steep western coasts of some continents when winds blowing along the coasts push surface water away from the land and draw water up from the ocean bottom (Figure 3).

Every few years, normal shore upwellings in the Pacific Ocean (Figure 4, left) are affected by changes in weather patterns called the *El Niño–Southern Oscillation,* or *ENSO* (Figure 4, right). In an ENSO, often called simply *El Niño,* prevailing winds called tropical trade winds blowing east to west weaken or reverse direction. This allows the warmer waters of the western Pacific to move toward the coast of South America, which suppresses the normal upwellings of cold, nutrient-rich water (Figure 4, right). The decrease in nutrients reduces primary productivity and causes a sharp decline in the populations of some fish species.

When an ENSO lasts 12 months or longer, it can severely disrupt populations of plankton, fish, and sea-

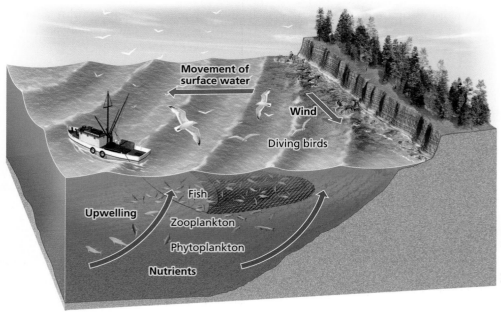

Movement of surface water

Wind

Diving birds

Fish

Zooplankton

Phytoplankton

Upwelling

Nutrients

© Cengage Learning

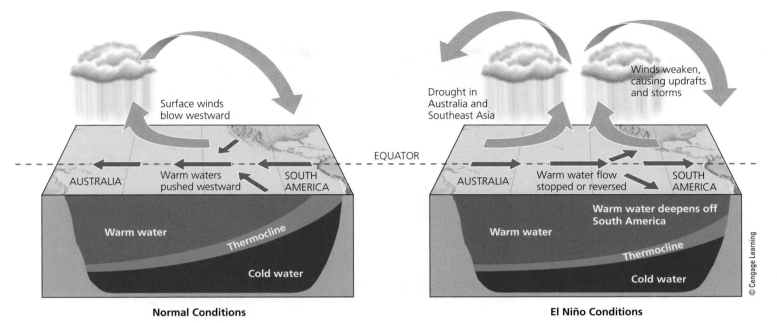

Normal Conditions   El Niño Conditions

Figure 4 Normal trade winds blowing east to west cause shore upwellings of cold, nutrient-rich bottom water in the tropical Pacific Ocean near the coast of Peru (left). A zone of gradual temperature change called the *thermocline* separates the warm and cold water. Every few years, a shift in trade winds known as the *El Niño–Southern Oscillation (ENSO)* disrupts this pattern.

birds in upwelling areas. A strong ENSO can also alter weather conditions over at least two-thirds of the globe (Figure 5)—especially in lands along the Pacific and Indian Oceans. Scientists do not know exactly what causes an ENSO, but they do know how to detect its formation and track its progress.

La Niña, the reverse of El Niño, cools some coastal surface waters and brings back upwellings. Typically, La Niña brings more Atlantic Ocean hurricanes, colder winters in Canada and the northeastern United States, and warmer and drier winters in the southeastern and southwestern United States. It also usually leads to wetter winters in the Pacific Northwest, torrential rains in Southeast Asia, lower wheat yields in Argentina, and more wildfires in Florida.

Tornadoes and Tropical Cyclones Are Violent Weather Extremes

Sometimes we experience *weather extremes*. Two examples are violent storms called *tornadoes* (which form over land) and *tropical cyclones* (which form over warm ocean waters and sometimes pass over coastal areas).

Tornadoes, or *twisters,* are swirling, funnel-shaped clouds that form over land. They can destroy houses

Figure 5 Typical global weather effects of an El Niño–Southern Oscillation. ***Questions:*** How might an ENSO affect the weather where you live or go to school?

(Compiled by the authors using data from United Nations Food and Agriculture Organization.)

Drought

Unusually high rainfall

Unusually warm periods

El Niño

Figure 6 Formation of a *tornado,* or *twister.* Although twisters can form at any time of the year, the most active tornado season in the United States is usually March through August. Meteorologists cannot yet forecast exactly where tornadoes will form at any given time, but research on tornados and advanced computer modeling can help them to identify areas at risk each day for the formation of these deadly storms.

Descending cool air

Severe thunderstorm

Rising warm air

Severe thunderstorms can trigger a number of smaller tornadoes

Tornado forms when cool downdraft and warm updraft of air meet and interact

Rising updraft of air

Warm moist air drawn in

© Cengage Learning

and cause other serious damage in areas where they touch down on the earth's surface. The United States is the world's most tornado-prone country, followed by Argentina and Bangladesh. Tornadoes occur on every continent in the world except for Antarctica.

Tornadoes in the plains of the Midwestern United States usually occur when a large, dry, cold-air front moving southward from Canada runs into a large mass of warm humid air moving northward from the Gulf of Mexico. Most tornadoes occur in the spring and summer when fronts of cold air from the north penetrate deeply into the Great Plains and the Midwest.

As the large warm-air mass moves rapidly over the more dense cold-air mass, it rises swiftly and forms strong vertical convection currents that suck air upward, as shown in Figure 6. Scientists hypothesize that the inter-

action of the cooler air nearer the ground and the rapidly rising warmer air above causes a spinning, vertically rising air mass, or vortex.

Figure 7 shows the areas of greatest risk from tornadoes in the continental United States.

Tropical cyclones are spawned by the formation of low-pressure cells of air over warm tropical seas. Figure 8

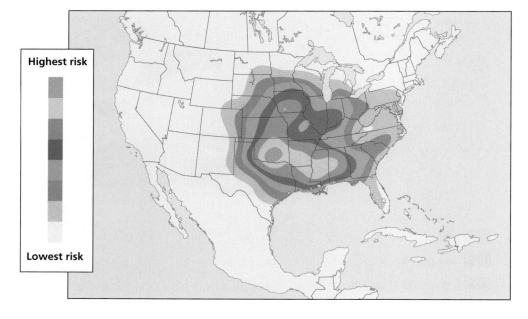

Highest risk

Lowest risk

Figure 7 Comparison of the relative risk of tornados across the continental United States.

(Compiled by the authors using data from NOAA.)

shows the formation and structure of a tropical cyclone. *Hurricanes* are tropical cyclones that form in the Atlantic Ocean; those forming in the Pacific Ocean usually are called *typhoons*. Tropical cyclones take a long time to form and gain strength. As a result, meteorologists can track their paths and wind speeds, and warn people in areas likely to be hit by these violent storms.

For a tropical cyclone to form, the temperature of ocean water has to be at least 27°C (80°F) to a depth of 46 meters (150 feet). A tropical cyclone forms when areas of low pressure over the warm ocean draw in air from surrounding higher-pressure areas. The earth's rotation makes these winds spiral counterclockwise in the northern hemisphere and clockwise in the southern hemisphere (see Figure 7-3, p. 146). Moist air, warmed by the heat of the ocean, rises in a vortex through the center of the storm until it becomes a tropical cyclone (Figure 8).

The intensities of tropical cyclones are rated in different categories, based on their sustained wind speeds: *Category 1,* 119–153 kilometers per hour, or kph (74–95 miles per hour, or mph); *Category 2,* 154–177 kph (96–110 mph); *Category 3,* 178–209 kph (111–130 mph); *Category 4,* 210–249 kph (131–155 mph); and *Category 5,* greater than 249 kph (155 mph). The longer a tropical cyclone stays over warm waters, the stronger it gets. Significant hurricane-force winds can extend 64–161 kilometers (40–100 miles) from the center, or eye, of a tropical cyclone.

Hurricanes and typhoons kill and injure people and damage property and agricultural production. Sometimes, however, the long-term ecological and economic benefits of a tropical cyclone exceed its short-term harmful effects.

For example, in parts of the U.S. state of Texas along the Gulf of Mexico, coastal bays and marshes, because of their unique natural formations and the barrier islands that protect them, normally receive very limited freshwater and saltwater inflows. In August 1999, Hurricane Brett struck this coastal area. According to marine biologists, the storm flushed out excess nutrients from land runoff and swept dead sea grasses and rotting vegetation from the coastal bays and marshes. It also carved out 12 channels through the barrier islands along the coast, allowing huge quantities of fresh seawater to flood the bays and marshes.

This flushing of the bays and marshes reduced brown tides consisting of explosive growths of algae feeding on excess nutrients. It also increased growth of sea grasses, which serve as nurseries for shrimp, crabs, and fish, and provide food for millions of ducks wintering in Texas bays. Production of commercially important species of shellfish and fish also increased.

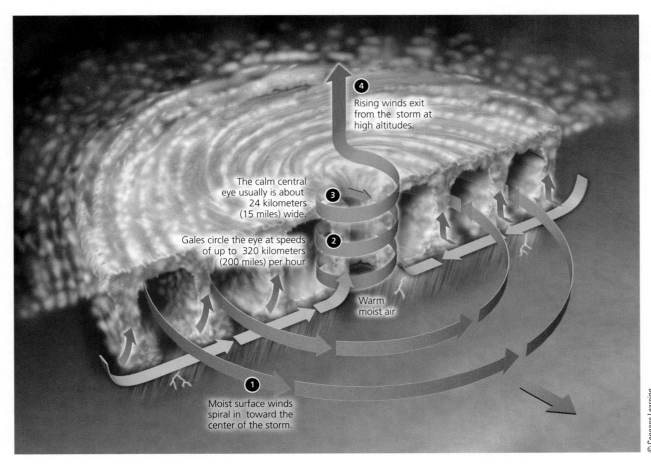

Figure 8 The formation of a *tropical cyclone.* Those forming in the Atlantic Ocean are called *hurricanes;* those forming in the Pacific Ocean are called *typhoons.*

4 Rising winds exit from the storm at high altitudes.

3 The calm central eye usually is about 24 kilometers (15 miles) wide.

2 Gales circle the eye at speeds of up to 320 kilometers (200 miles) per hour

Warm moist air

1 Moist surface winds spiral in toward the center of the storm.

© Cengage Learning

Figure 1 Countries of the world.
Map Analysis
1. Name three countries that border the Arctic Ocean.
2. Which countries surround **(a)** China, **(b)** Mexico, **(c)** Germany, and **(d)** Sudan?

ARCTIC REGION

0   600 km
0   600 mi
Azimuthal Equidistant Projection

Longitude East of Greenwich (London)

WINKEL TRIPEL PROJECTION, CENTRAL MERIDIAN 0°

0   500 1000 1500 2000 2500
KILOMETERS
0   500 1000 1500 2000 2500
STATUTE MILES

ANTARCTIC REGION

0   600 km
0   600 mi
Azimuthal Equidistant Projection
● Research station

Figure 2 Composite satellite view of the earth showing its major terrestrial and aquatic features.

ARCTIC OCEAN

Meridian of Greenwich (London)
Longitude East of Greenwich

British Isles

Iceland

North Sea

Scandinavia

BARENTS SEA

Svalbard

Franz Josef Land

Novaya Zemlya

North Land

EUROPE

A L P S

Black Sea

El'brus
5642

Caspian Sea

MEDITERRANEAN SEA

SAHARA

AFRICA

Gulf of Guinea

Congo Basin

Lake Victoria

Lake Tanganyika

Kilimanjaro
5895

Lake Malawi

Kalahari Desert

Madagascar

Cape of Good Hope

U R A L   M O U N T A I N S

S I B E R I A

ARCTIC CIRCLE

SEA OF OKHOTSK

Kamchatka Peninsula

Lake Baikal

Aral Sea

Tien Shan

GOBI

A S I A

Plateau of Tibet

HIMALAYA

Mt. Everest
8850

INDIA

ARABIAN SEA

ARABIAN PENINSULA

Red Sea

BAY OF BENGAL

NINETYEAST RIDGE

INDIAN

OCEAN

SOUTHWEST INDIAN RIDGE

SOUTHEAST INDIAN RIDGE

NORTH

PACIFIC

OCEAN

TROPIC OF CANCER

PHILIPPINE SEA

SOUTH CHINA SEA

Philippine Islands

Challenger Deep
-10994

INDONESIA

New Guinea

M I C R O N E S I A

M E L A N E S I A

EQUATOR

CORAL SEA

Fiji Islands

New Caledonia

TROPIC OF CAPRICORN

AUSTRALIA

Great Dividing Range

Mt. Kosciuszko
2228

Bass Strait

Tasmania

SOUTH

PACIFIC

OCEAN

TASMAN SEA

North Island

South Island

NEW ZEALAND

ANTARCTIC CIRCLE

Wilkes Land

A N T A R C T I C A

Winkel Tripel Projection, Central Meridian 0°

0 500 1000 1500 2000 2500
KILOMETERS

0 500 1000 1500 2000 2500
STATUTE MILES

Elevations in meters

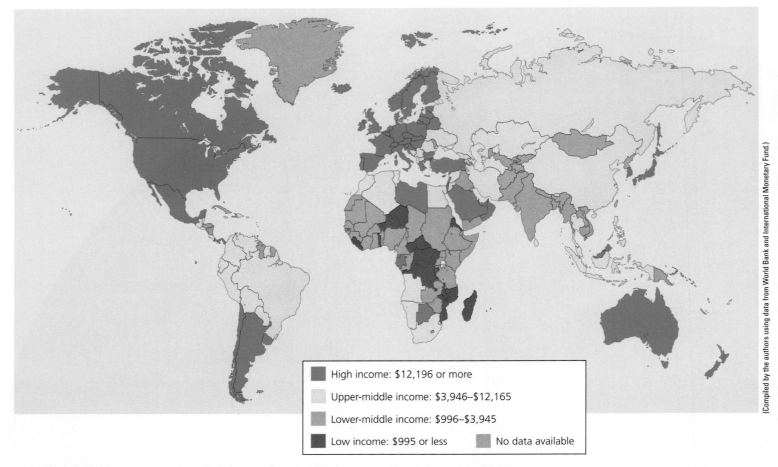

High income: $12,196 or more

Upper-middle income: $3,946–$12,165

Lower-middle income: $996–$3,945

Low income: $995 or less   No data available

Figure 3 High-income, upper-middle-income, lower-middle-income, and low-income countries in terms of gross national income (GNI) purchasing power parity (PPP) per capita (U.S. dollars) in 2010.

Data and Map Analysis

1. In how many countries is the per capita average income $995 or less? Look at Figure 1 and find the names of three of these countries.

2. In how many instances does a lower-middle- or low-income country share a border with a high-income country? Look at Figure 1 and find the names of the countries that reflect three examples of this situation.

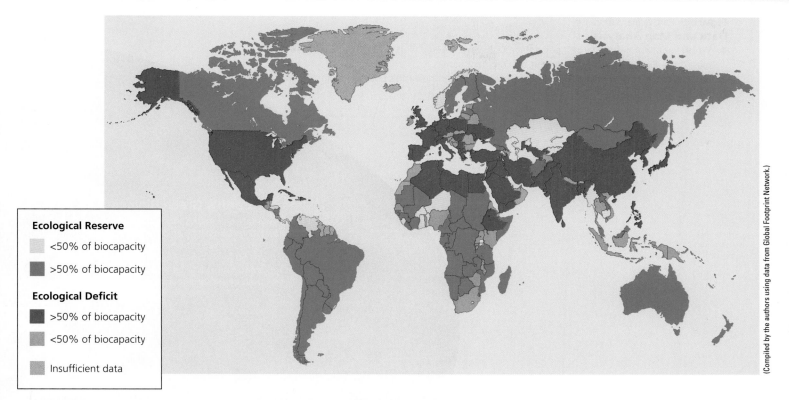

Ecological Reserve

<50% of biocapacity

>50% of biocapacity

Ecological Deficit

>50% of biocapacity

<50% of biocapacity

Insufficient data

Figure 4 *Ecological debtors and creditors:* The ecological footprints of some countries exceed their biocapacity, while other countries still have ecological reserves.

Data and Map Analysis

1. List five countries, including the three largest, in which the ecological deficit is greater than 50% of biocapacity. (See Figure 1 of this supplement for country names.)

2. On which two continents does land with ecological reserves of more than 50% of biocapacity occupy the largest percentage of total land area? (See Figure 2 of this supplement for continent names.)

Figure 5 Earth's plant biomass.

Data and Map Analysis

1. Which continent has the largest percentage of its area covered with minimal biomass? (See Figure 2 of this supplement for continent names.)

2. Is ocean chlorophyll concentration, in general, higher near the earth's poles or near its equator?

National Geographic

Earth's Green Biomass

Earth's vegetative biomass, the foundation of most life on the planet, is measured by chlorophyll-producing plants. Both land and sea process an equal amount of carbon—50 to 60 billion metric tonnes per year. Photoplankton provides the basis of measurement in the oceans; green-leaf mass on land.

KEY TO IMAGES
Ocean: Chlorophyll concentration

>.01   .05   .2   1   2   5   20 50 (a (mg/m^3))

Land cover: Normalized Difference Vegetation Index (NDVI)

Max.   Min.

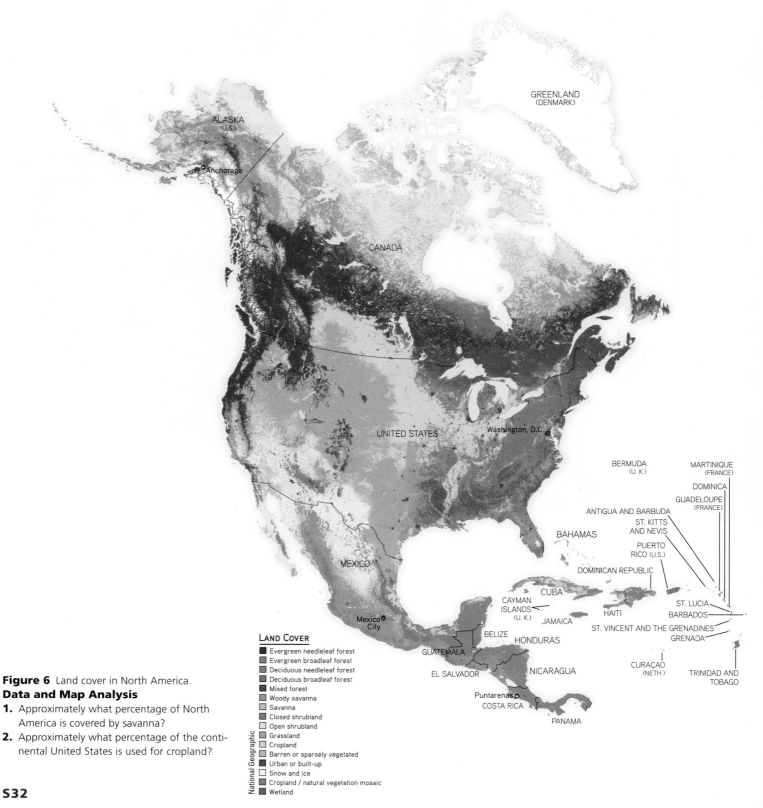

Figure 6 Land cover in North America.
Data and Map Analysis

1. Approximately what percentage of North America is covered by savanna?
2. Approximately what percentage of the continental United States is used for cropland?

LAND COVER

- Evergreen needleleaf forest
- Evergreen broadleaf forest
- Deciduous needleleaf forest
- Deciduous broadleaf forest
- Mixed forest
- Woody savanna
- Savanna
- Closed shrubland
- Open shrubland
- Grassland
- Cropland
- Barren or sparsely vegetated
- Urban or built-up
- Snow and ice
- Cropland / natural vegetation mosaic
- Wetland

National Geographic

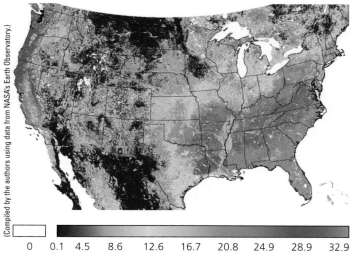

(Compiled by the authors using data from NASA's Earth Observatory.)

Gross primary productivity
(grams of carbon per square meter)

0   0.1   4.5   8.6   12.6   16.7   20.8   24.9   28.9   32.9

Figure 7 **Natural capital:** Gross primary productivity across the continental United States, based on remote satellite data. The differences roughly correlate with variations in moisture and soil types.

Data and Map Analysis

1. Comparing the five northwestern-most states with the five southeastern-most states, which of these regions has the greater variety in levels of gross primary productivity? Which of the regions has the highest levels overall?

2. Compare this map with that of Figure 6. Which biome in the United States is associated with the highest level of gross primary productivity?

Figure 8 **Natural capital degradation:** The human ecological footprint in North America. Colors represent the percentage of each area influenced by human activities.

(Compiled by the authors using data from Wildlife Conservation Society and the Center for International Earth Science Information Network at Columbia University.)

Data and Map Analysis

1. Which general area of the United States has the highest human footprint values?

2. What is the relative value of the human ecological footprint in the area where you live or go to school?

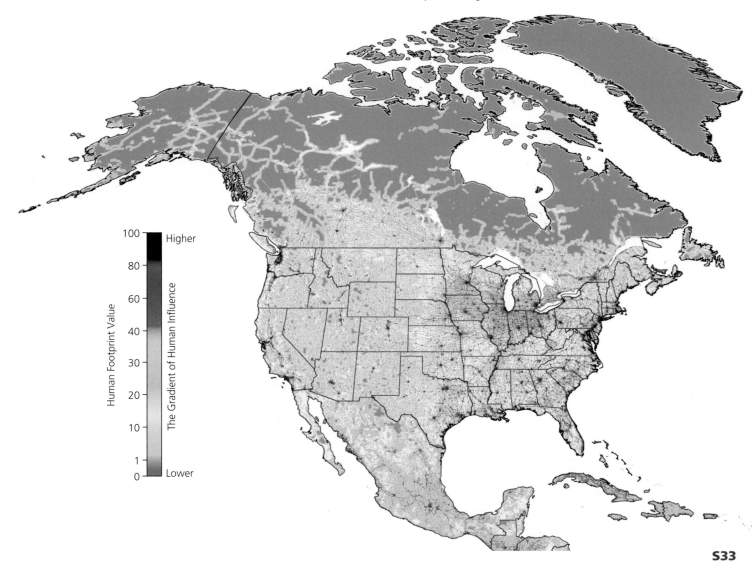

Human Footprint Value

The Gradient of Human Influence

100 — Higher
80
60
40
30
20
10
1
0 — Lower

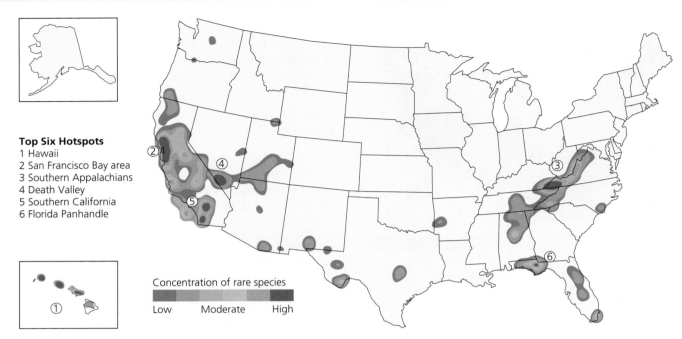

Top Six Hotspots
1 Hawaii
2 San Francisco Bay area
3 Southern Appalachians
4 Death Valley
5 Southern California
6 Florida Panhandle

Concentration of rare species

Low   Moderate   High

Figure 9 **Endangered natural capital:** Major biodiversity hotspots in the United States that need emergency protection. The shaded areas contain the largest concentrations of rare and potentially endangered species.

(Compiled by the authors using data from State Natural Heritage Programs, the Nature Conservancy, and Association for Biodiversity Information.)

Data and Map Analysis

1. If you live in the United States, which of the top six hotspots is closest to where you live or go to school?

2. Which general part of the country has the highest overall concentration of rare species? Which part has the second-highest concentration?

LAND COVER

- ■ Evergreen needleleaf forest
- ■ Evergreen broadleaf forest
- ■ Deciduous needleleaf forest
- ■ Deciduous broadleaf forest
- ■ Mixed forest
- ■ Woody savanna
- □ Savanna
- ■ Closed shrubland
- □ Open shrubland
- ■ Grassland
- □ Cropland
- ■ Barren or sparsely vegetated
- ■ Urban or built-up
- □ Snow and ice
- ■ Cropland / natural vegetation mosaic
- ■ Wetland

Figure 10 Land cover in South America.
Data and Map Analysis

1. Which type of land cover would you say occupies the largest area in South America?
2. Brazil was once almost completely covered by rain forest (evergreen broadleaf), much of which has been converted to savanna as it has been cleared by humans and has not grown back as forest. About what percentage of Brazil would you say is occupied by savanna?

Figure 11 Land cover in Europe.
Data and Map Analysis
1. What are the two most common types of land cover in Europe?
2. What percentage of European land would you estimate to be used for growing crops?

LAND COVER

- Evergreen needleleaf forest
- Evergreen broadleaf forest
- Deciduous needleleaf forest
- Deciduous broadleaf forest
- Mixed forest
- Woody savanna
- Savanna
- Closed shrubland
- Open shrubland
- Grassland
- Cropland
- Barren or sparsely vegetated
- Urban or built-up
- Snow and ice
- Cropland / natural vegetation mosaic
- Wetland

National Geographic

Figure 12 Land cover in Asia.
Data and Map Analysis
1. About what percentage of China's land area would you estimate to be classified as barren or sparsely vegetated?
2. About what percentage of India's land would you say is used for growing crops?

LAND COVER

- Evergreen needleleaf forest
- Evergreen broadleaf forest
- Deciduous needleleaf forest
- Deciduous broadleaf forest
- Mixed forest
- Woody savanna
- Savanna
- Closed shrubland
- Open shrubland
- Grassland
- Cropland
- Barren or sparsely vegetated
- Urban or built-up
- Snow and ice
- Cropland / natural vegetation mosaic
- Wetland

National Geographic

National Geographic

LAND COVER
- ■ Evergreen needleleaf forest
- ■ Evergreen broadleaf forest
- ■ Deciduous needleleaf forest
- ■ Deciduous broadleaf forest
- ■ Mixed forest
- ■ Woody savanna
- □ Savanna
- ■ Closed shrubland
- □ Open shrubland
- ■ Grassland
- □ Cropland
- ■ Barren or sparsely vegetated
- ■ Urban or built-up
- □ Snow and ice
- ■ Cropland / natural vegetation mosaic
- ■ Wetland

Figure 13 Land cover in Africa.

Data and Map Analysis

1. Name six countries in Africa that are more than half covered with barren or sparsely vegetated land.

2. Which African country has the largest area of forest land?

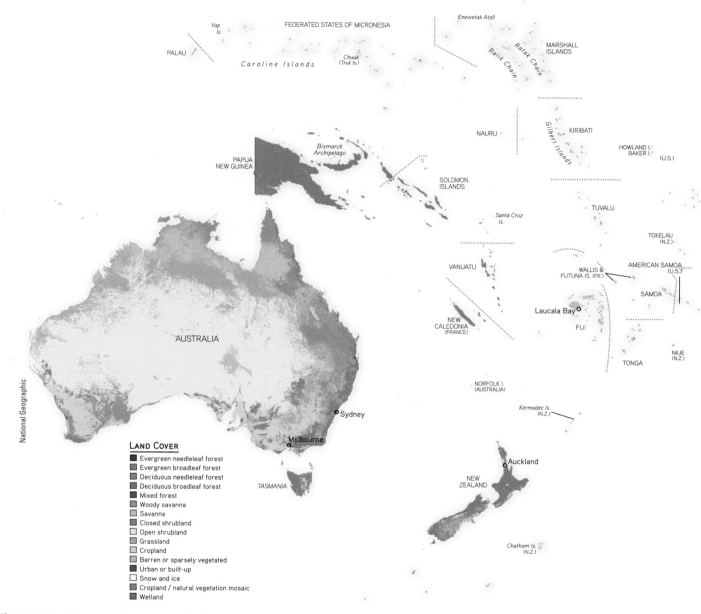

Figure 14 Land cover in Oceania and Australia.

Data and Map Analysis

1. What type of land cover occupies the most land in Australia?

2. What do all urban areas shown in red on this map have in common?

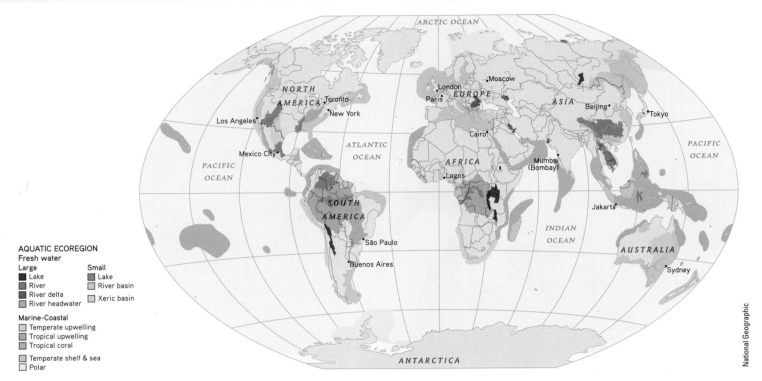

Figure 15 Major types of marine and freshwater habitats.

Data and Map Analysis

1. On which continent is the largest area of river habitat?
2. List the locations of three tropical upwelling areas.

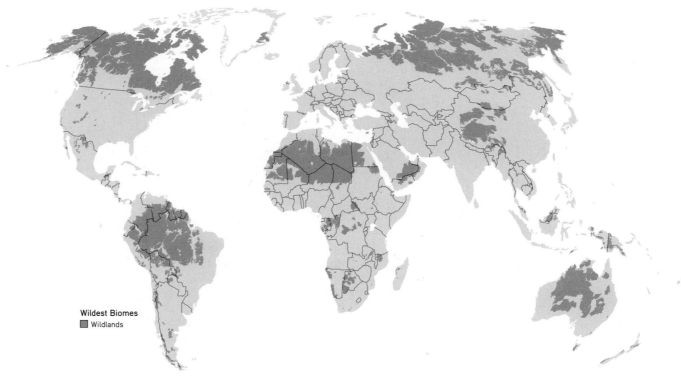

National Geographic

Figure 16 World's remaining wildlands.

Data and Map Analysis

1. Comparing this map with those in Figures 12–14, list the locations of three wild areas that are also classified as barren or sparsely vegetated. (See Figures 1 and 2 of this supplement for country and continent names.)

2. What continent appears to have the largest percentage of wildland?

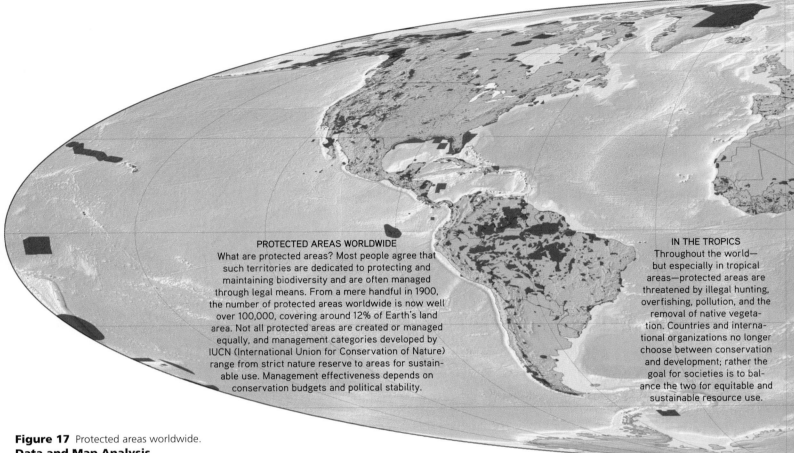

PROTECTED AREAS WORLDWIDE
What are protected areas? Most people agree that such territories are dedicated to protecting and maintaining biodiversity and are often managed through legal means. From a mere handful in 1900, the number of protected areas worldwide is now well over 100,000, covering around 12% of Earth's land area. Not all protected areas are created or managed equally, and management categories developed by IUCN (International Union for Conservation of Nature) range from strict nature reserve to areas for sustainable use. Management effectiveness depends on conservation budgets and political stability.

IN THE TROPICS
Throughout the world—but especially in tropical areas—protected areas are threatened by illegal hunting, overfishing, pollution, and the removal of native vegetation. Countries and international organizations no longer choose between conservation and development; rather the goal for societies is to balance the two for equitable and sustainable resource use.

Figure 17 Protected areas worldwide.

Data and Map Analysis

1. List the locations of three relatively large protected areas. (See Figures 1 and 2 of this supplement for country and continent names.)
2. What continent appears to have the largest percentage of protected land? What continent appears to have the smallest percentage of protected land?

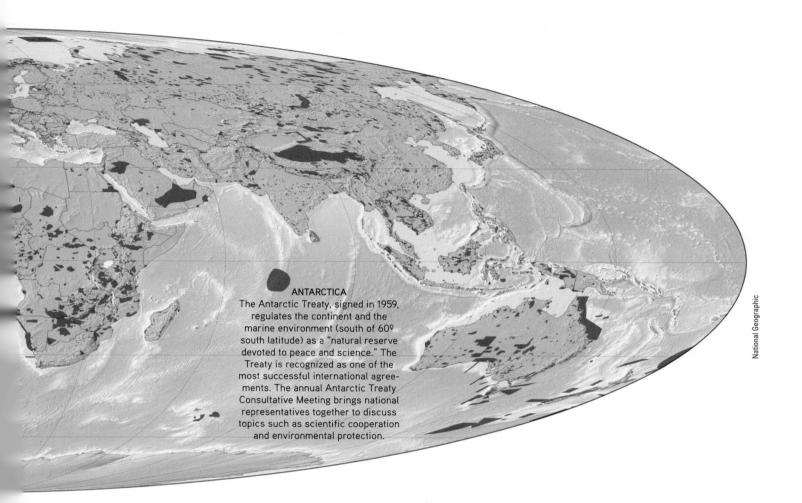

ANTARCTICA

The Antarctic Treaty, signed in 1959, regulates the continent and the marine environment (south of 60º south latitude) as a "natural reserve devoted to peace and science." The Treaty is recognized as one of the most successful international agreements. The annual Antarctic Treaty Consultative Meeting brings national representatives together to discuss topics such as scientific cooperation and environmental protection.

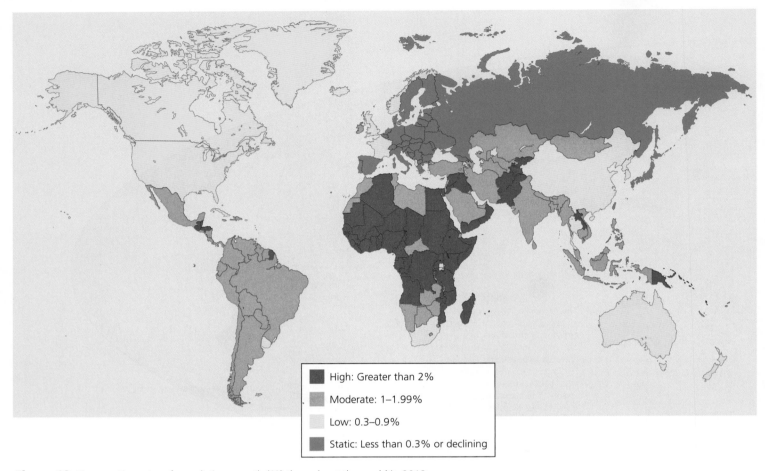

Figure 18 Comparative rates of population growth (%) throughout the world in 2012.

(Compiled by the authors using data from Population Reference Bureau and United Nations Population Division.)

Data and Map Analysis

1. Which continent has the greatest number of countries with high rates of population increase? Which continent has the greatest number of countries with static rates? (See Figure 2 of this supplement for continent names.)

2. For each category on this map, name the two countries that you think are largest in terms of total area (see Figure 1 for country names).

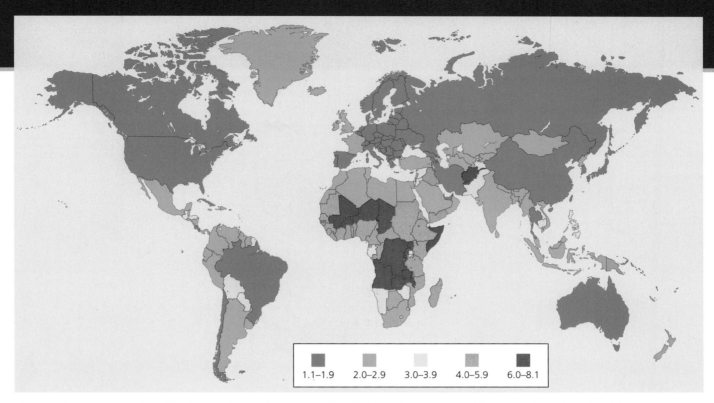

Figure 19 Comparison of total fertility rate (TFR), or average number of children born to the world's women throughout their lifetimes, around the world as measured in 2012.

(Compiled by the authors using data from Population Reference Bureau and United Nations Population Division.)

Data and Map Analysis

1. Which country in the highest TFR category borders two countries in the lowest TFR category? What are those two countries? (See Figure 1 of this supplement for country names.)
2. Do you see any geographic patterns on this map? Explain.

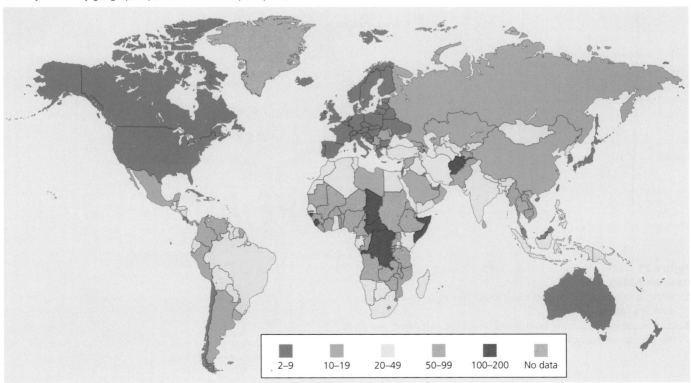

Figure 20 Comparison of global infant mortality rates around the world in 2012.

(Compiled by the authors using data from Population Reference Bureau and United Nations Population Division.)

Data and Map Analysis

1. Do you see a geographic pattern related to infant mortality rates as reflected on this map? Explain.
2. List any similarities that you see in geographic patterns between this map and the one in Figure 19.

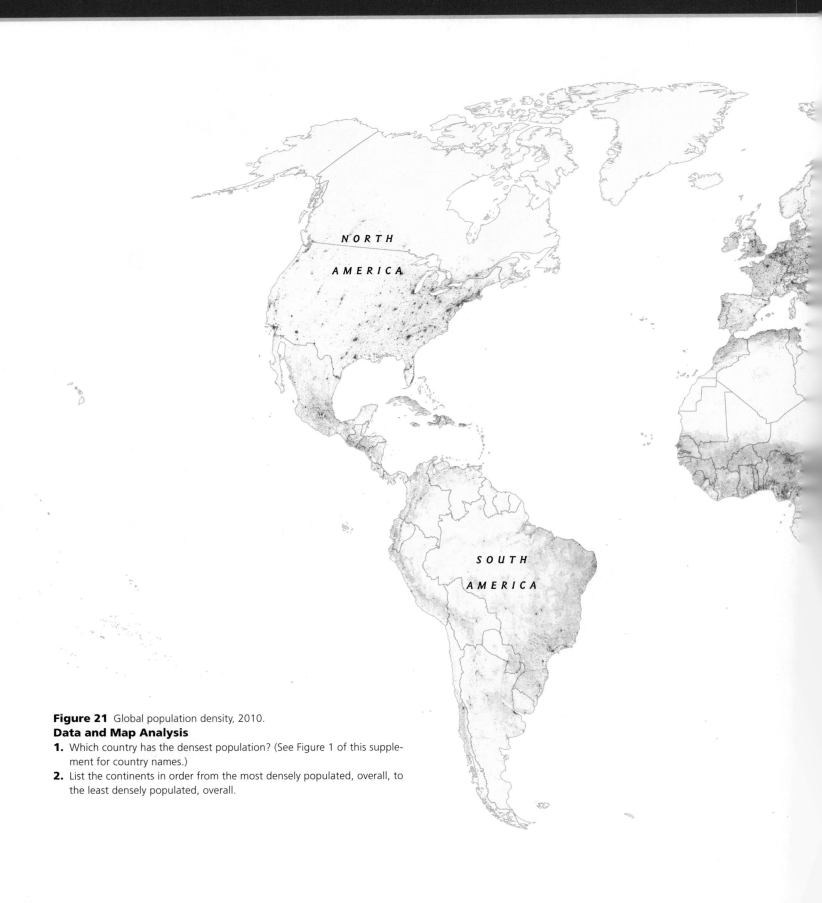

Figure 21 Global population density, 2010.
Data and Map Analysis
1. Which country has the densest population? (See Figure 1 of this supplement for country names.)
2. List the continents in order from the most densely populated, overall, to the least densely populated, overall.

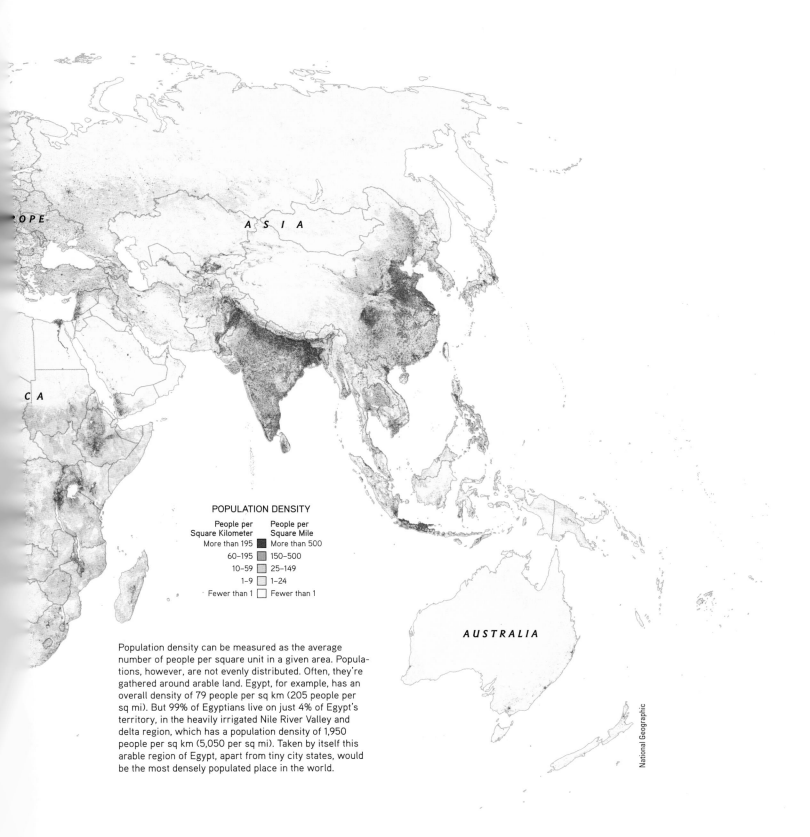

ASIA

AUSTRALIA

POPULATION DENSITY

People per Square Kilometer	People per Square Mile
More than 195	More than 500
60–195	150–500
10–59	25–149
1–9	1–24
Fewer than 1	Fewer than 1

Population density can be measured as the average number of people per square unit in a given area. Populations, however, are not evenly distributed. Often, they're gathered around arable land. Egypt, for example, has an overall density of 79 people per sq km (205 people per sq mi). But 99% of Egyptians live on just 4% of Egypt's territory, in the heavily irrigated Nile River Valley and delta region, which has a population density of 1,950 people per sq km (5,050 per sq mi). Taken by itself this arable region of Egypt, apart from tiny city states, would be the most densely populated place in the world.

National Geographic

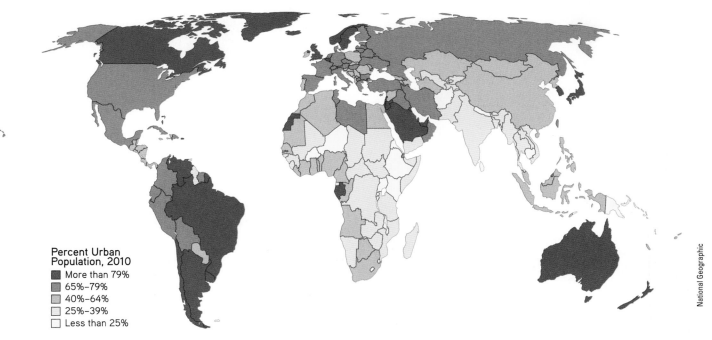

Percent Urban
Population, 2010
- More than 79%
- 65%–79%
- 40%–64%
- 25%–39%
- Less than 25%

National Geographic

Figure 22 Urban population, 2010.
Data and Map Analysis
1. Name six countries whose populations are more than 79% urban. (See Figure 1 of this supplement for country names.)
2. List three pairs of bordering countries in which the population of one country is more than 79% urban and the population of the other is 39% or less urban.

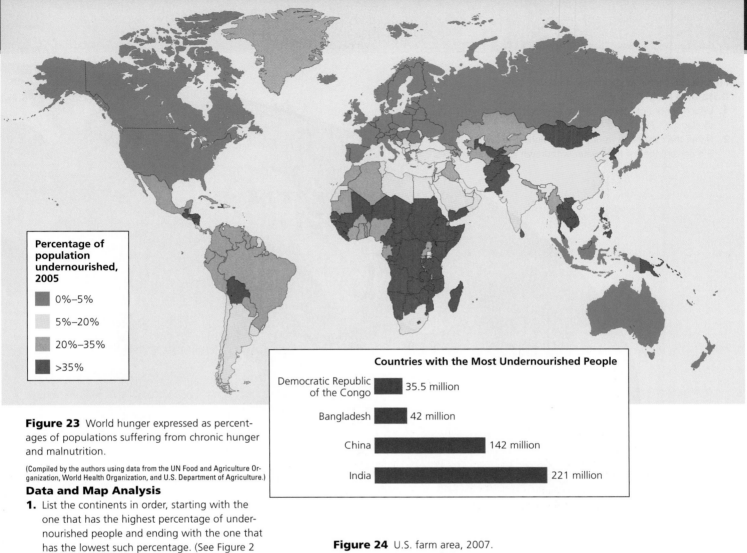

Figure 23 World hunger expressed as percentages of populations suffering from chronic hunger and malnutrition.

(Compiled by the authors using data from the UN Food and Agriculture Organization, World Health Organization, and U.S. Department of Agriculture.)

Percentage of population undernourished, 2005

- 0%–5%
- 5%–20%
- 20%–35%
- >35%

Countries with the Most Undernourished People

Democratic Republic of the Congo	35.5 million
Bangladesh	42 million
China	142 million
India	221 million

Data and Map Analysis

1. List the continents in order, starting with the one that has the highest percentage of undernourished people and ending with the one that has the lowest such percentage. (See Figure 2 of this supplement for continent names.)
2. On which continent is the largest block of countries that suffer the highest levels of undernourishment? List five of these countries.

Figure 24 U.S. farm area, 2007.

Data and Map Analysis

1. Name six states that are primarily more than 70% covered by farmland.
2. Name six states where there is a high variation in farmland cover, ranging from less than 10% to more than 70%.

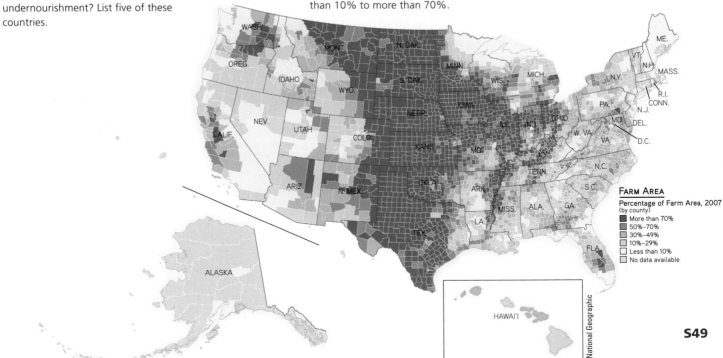

FARM AREA
Percentage of Farm Area, 2007 (by county)

- More than 70%
- 50%–70%
- 30%–49%
- 10%–29%
- Less than 10%
- No data available

National Geographic

Figure 25 Global availability of freshwater.
Data and Map Analysis
1. Which continent has the largest area with the highest water availability?
2. Name the continents in order from highest to lowest, in terms of the percentages of freshwater withdrawn for industrial uses.

Greenland

NORTH AMERICA

ATLANTIC OCEAN

EUROPE

16%
29%
55%

National Geographic

NORTH AMERICA

15%
44%
41%

PACIFIC OCEAN

1982
177 killed in protest over dam

2007
Water conflicts between herders and farmers in Burkina Faso

AFRICA

10% 4%
86%

SOUTH AMERICA

Water Availability
(in millimeters per person per year)
- More than 750
- 251–750
- 26–250
- Less than 26
- — Major river
- — Primary watershed boundary
- ● Selected water dispute

SOUTH AMERICA

19%
13%
68%

Percent of total population using improved drinking water sources, 2000
- More than 90%
- 76% - 90%
- 51% - 75%
- Less than 50%
- No data available

NORTH AMERICA

EUROPE

ASIA

AFRICA

SOUTH AMERICA

AUSTRALIA

ARCTIC OCEAN

EUROPE

ASIA

AFRICA

INDIAN
OCEAN

PACIFIC
OCEAN

AUSTRALIA

ANTARCTICA

ASIA
9%
9%
82%

OCEANIA
17%
10%
73%

2010
Pakistan irrigation
dispute between
tribes kills 116

2007
Injuries at protest over allocation
of water to industry in India

1978 onward
Egypt threatens
Ethiopia over
Nile plans

1999
Villagers killed in
Yemen water clash

2006
Sri Lankan rebels cut
water supplies to villages

2004–2006
250 killed in violence over
water shortage in Ethiopia

2005
90 killed in dispute
over water rights
in Kenya

2007
Australian man murdered in
fight over water restrictions

Freshwater Withdrawal
(as a percentage
of total water utilization)

Agricultural   Domestic

Industrial

**SURFACE AND
ATMOSPHERIC WATER**

Freshwater lakes
67.4%

Biota
0.8%

Rivers
1.6%

Wetlands
8.5%

Soil moisture
12.2%

Atmosphere
9.5%

FRESH WATER

Surface and
atmospheric water
0.4%

Permafrost
0.8%

Glaciers
68.7%

Groundwater
30.1%

TOTAL WATER

Salt water
97.5%

Fresh water
2.5%

WATER WITHDRAWAL
(from groundwater
and surface water)

Agriculture
70%

Domestic
11%

Industrial
19%

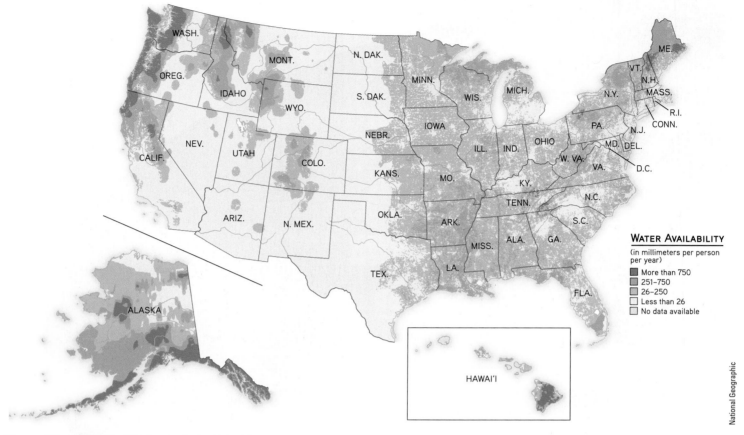

Figure 26 Availability of freshwater in the United States.

Data and Map Analysis

1. Name the three states with the largest areas of land where water availability is more than 750 milliliters per person per year.

2. Name six states where there are no areas of water availability of more than 250 milliliters per person per year.

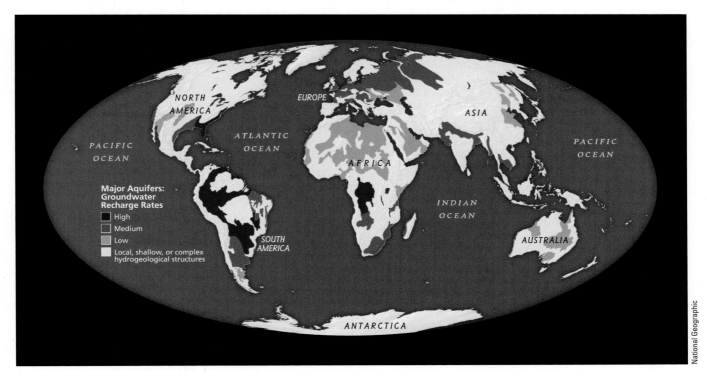

Figure 27 Recharge rates of the world's major aquifers.

Data and Map Analysis

1. Which continent has the largest area of aquifers with high recharge rates?

2. Which continents have no aquifers with high recharge rates?

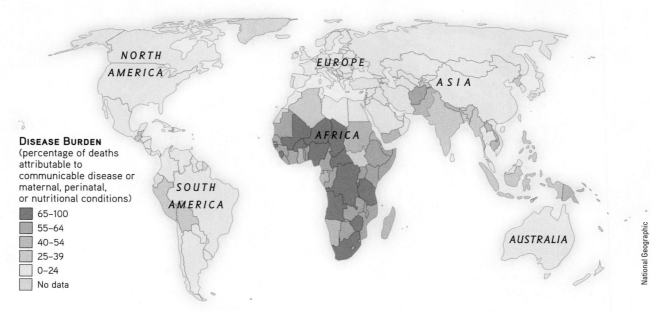

DISEASE BURDEN
(percentage of deaths
attributable to
communicable disease or
maternal, perinatal,
or nutritional conditions)

- 65–100
- 55–64
- 40–54
- 25–39
- 0–24
- No data

National Geographic

Figure 28 Global disease burden.

Data and Map Analysis

1. Name five countries that have the highest rates of communicable diseases. (See Figure 1 to find the names of countries.)

2. Name five countries that have the highest rates of noncommunicable diseases.

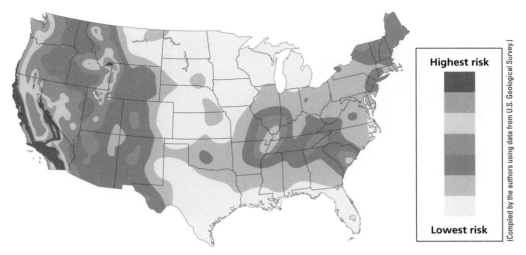

Figure 29 Earthquake (seismic) risk in various areas of the continental United States.

Data and Map Analysis

1. Speaking in general terms (northeast, southeast, central, west coast, etc.) which area has the high-est earthquake risk and which area has the lowest risk?

2. For each of the categories of risk, list the number of states that fall into the category.

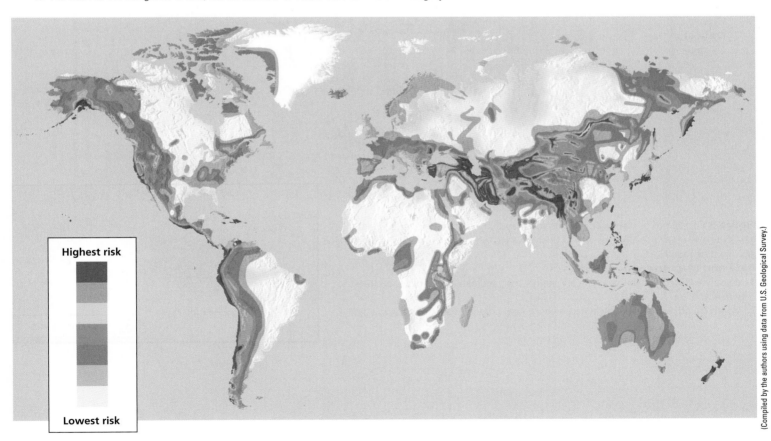

Figure 30 Earthquake (seismic) risk in the world.

Data and Map Analysis

1. How are these areas related to the boundaries of the earth's major tectonic plates as shown in Figure 14-18, p. 366?

2. Which continent has the longest coastal area subject to the highest possible risk? Which continent has the second longest such coastal area? (See Figure 2 of this supplement for continent names.)

Figure 31 Locations of the major, known proven and unproven reserves of oil, natural gas, and coal in North America, as well as offshore areas where more crude oil and natural gas might be found.

Data and Map Analysis

1. If you live in North America, where are the oil, coal, and natural gas deposits closest to where you live?

2. Which country borders on the largest areas of high potential for oil and natural gas?

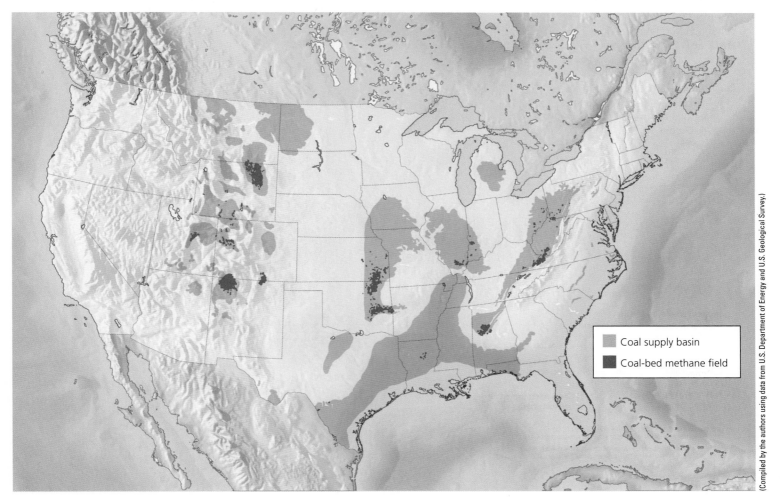

(Compiled by the authors using data from U.S. Department of Energy and U.S. Geological Survey.)

Figure 32 Major coal supply basins and coal-bed methane fields in the lower 48 states of the United States.

Data and Map Analysis

1. If you live in the United States, where are the coal-bed methane deposits closest to where you live?

2. Removing these deposits requires lots of water. Compare the locations of the major deposits of coal-bed methane with water-deficit areas shown in Figure 13-7, p. 322, and Figure 13-8, p. 323.

Figure 33 Major natural gas shale deposits in North America.

Data and Map Analysis

1. What state has the largest area of natural gas shale deposits?
2. Describe the locations of three areas where two states or two countries share a border over a natural gas shale deposit.

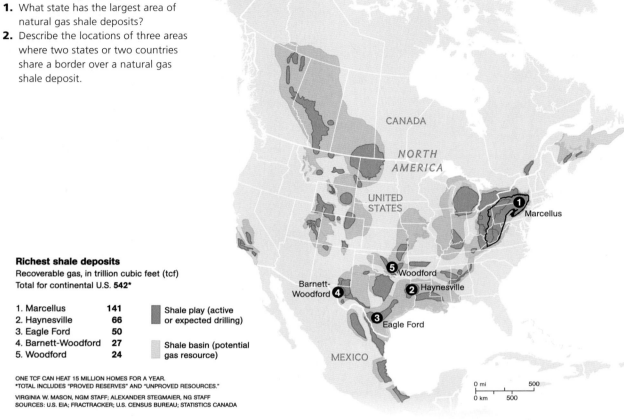

Richest shale deposits

Recoverable gas, in trillion cubic feet (tcf)

Total for continental U.S. **542***

1. Marcellus	**141**
2. Haynesville	**66**
3. Eagle Ford	**50**
4. Barnett-Woodford	**27**
5. Woodford	**24**

Shale play (active or expected drilling)

Shale basin (potential gas resource)

ONE TCF CAN HEAT 15 MILLION HOMES FOR A YEAR.
*TOTAL INCLUDES "PROVED RESERVES" AND "UNPROVED RESOURCES."

VIRGINIA W. MASON, NGM STAFF; ALEXANDER STEGMAIER, NG STAFF
SOURCES: U.S. EIA; FRACTRACKER; U.S. CENSUS BUREAU; STATISTICS CANADA

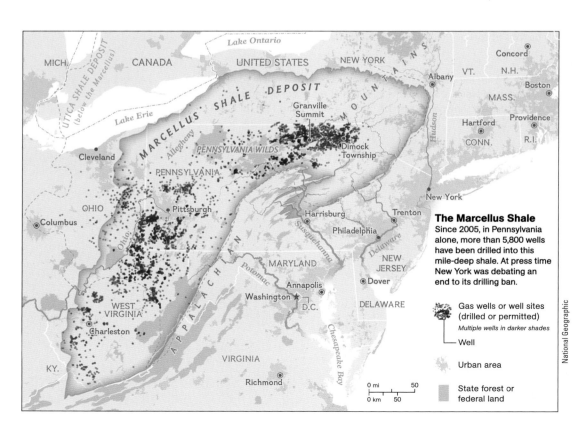

The Marcellus Shale

Since 2005, in Pennsylvania alone, more than 5,800 wells have been drilled into this mile-deep shale. At press time New York was debating an end to its drilling ban.

Gas wells or well sites (drilled or permitted)
Multiple wells in darker shades

Well

Urban area

State forest or federal land

National Geographic

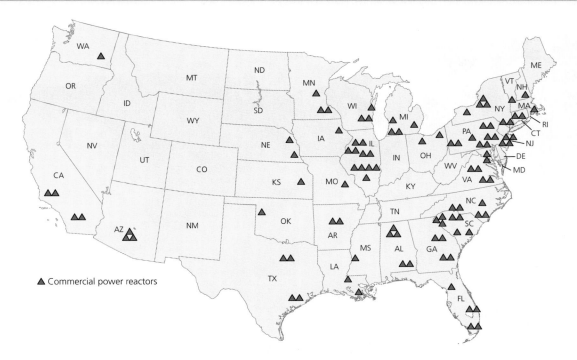

Figure 34 Locations of the 100 commercial nuclear power reactors in the United States.

(Compiled by the authors using data from U.S. Nuclear Regulatory Commission and U.S. Department of Energy.)

Data and Map Analysis

1. If you live in the United States, do you live or go to school within about 97 kilometers (60 miles) of a commercial nuclear power reactor?

2. Which state has the largest number of commercial nuclear power reactors?

▲ Commercial power reactors

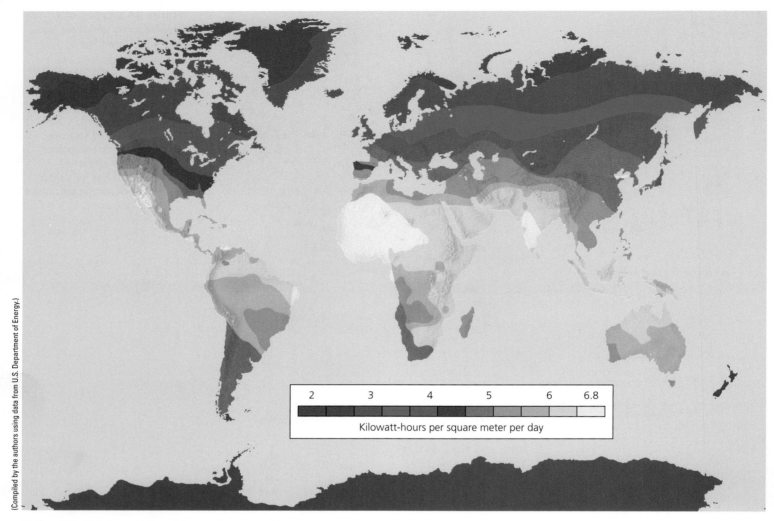

(Compiled by the authors using data from U.S. Department of Energy.)

Figure 35 Global availability of direct solar energy. Areas with more than 3.5 kilowatt-hours per square meter per day (see scale) are good candidates for passive and active solar heating systems and use of solar cells to produce electricity.

Data and Map Analysis

1. What is the potential for making greater use of solar energy to provide heat and produce electricity (with solar cells) where you live or go to school?

2. List the continents in order of overall availability of direct solar energy, from those with the highest to those with the lowest. (See Figure 2 of this supplement for continent names.)

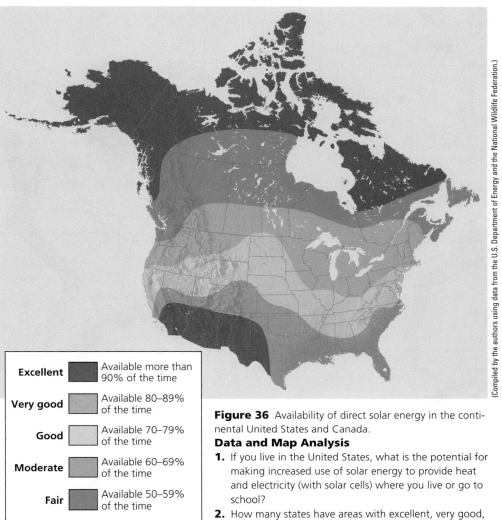

(Compiled by the authors using data from the U.S. Department of Energy and the National Wildlife Federation.)

Excellent		Available more than 90% of the time
Very good		Available 80–89% of the time
Good		Available 70–79% of the time
Moderate		Available 60–69% of the time
Fair		Available 50–59% of the time
Poor		Available less than 50% of the time

Figure 36 Availability of direct solar energy in the continental United States and Canada.

Data and Map Analysis

1. If you live in the United States, what is the potential for making increased use of solar energy to provide heat and electricity (with solar cells) where you live or go to school?

2. How many states have areas with excellent, very good, or good availability of direct solar energy?

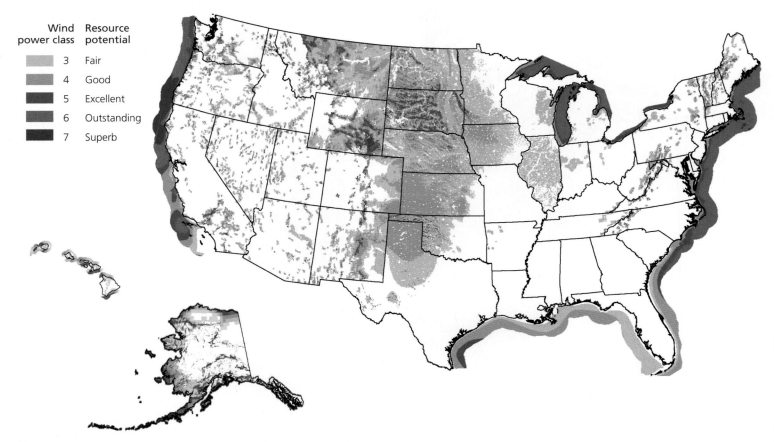

Figure 37 Potential supply of land- and ocean-based wind energy in the United States.

(Compiled by the authors using data from U.S. Geological Survey and U.S. Department of Energy.)

Data and Map Analysis

1. If you live in the United States, what is the general wind energy potential where you live or go to school?

2. How many states have areas with good or excellent potential for wind energy?

Figure 38 Known global reserves of moderate- to high-temperature geothermal energy.

(Compiled by the authors using data from Canadian Geothermal Resources Council, U.S. Geological Survey, and U.S. Department of Energy.)

Data and Map Analysis

1. Between North and South America, which continent appears to have the greater total potential for geothermal energy? (See Figure 2 of this supplement for continent names.)

2. Which country in Asia has the greatest known reserves of geothermal energy? (See Figure 1 of this supplement for country names.)

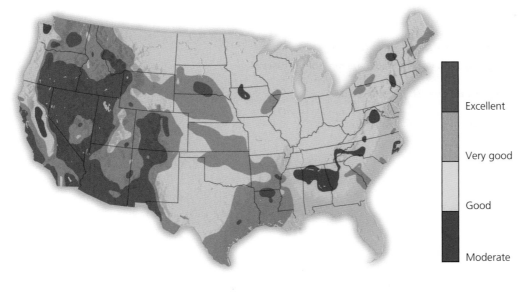

Excellent

Very good

Good

Moderate

Figure 39 Potential geothermal energy resources in the continental United States.

(Compiled by the authors using data from U.S. Department of Energy and U.S. Geological Survey.)

Data and Map Analysis

1. If you live in the United States, what is the potential for using geothermal energy to provide heat or to produce electricity where you live or go to school?

2. How many states have areas with very good or excellent potential for using geothermal energy?

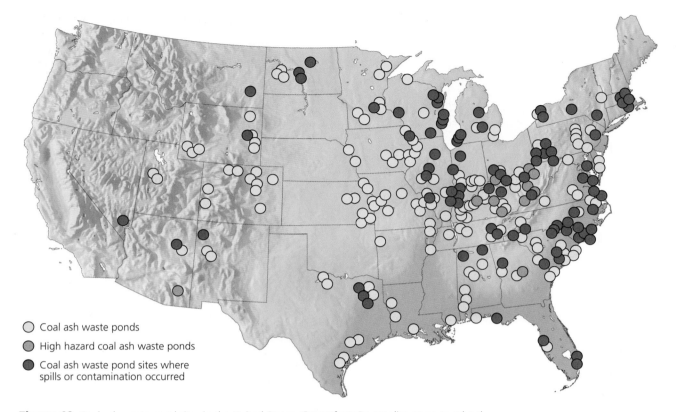

Figure 40 Coal ash waste pond sites in the United States. ***Question:*** Do you live or go to school near any of these ponds?

Key:
- ○ Coal ash waste ponds
- ◐ High hazard coal ash waste ponds
- ● Coal ash waste pond sites where spills or contamination occurred

(Compiled by the authors using data from U.S. Environmental Protection Agency and the Sierra Club.)

Data and Map Analysis

1. In what part of the country (eastern third, central third, or western third) have most of the cases of contamination and spills occurred?

2. In what part of the country are most of the high-hazard coal ash ponds located?

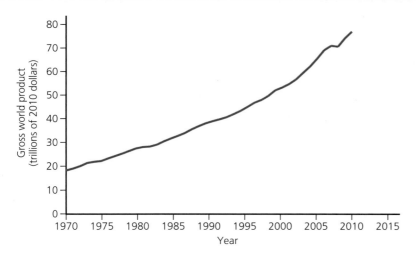

Figure 1 Gross world product (GWP), 1970–2011, in 2010 dollars.

(Compiled by the authors using data from the International Monetary Fund, World Bank, World Economic Fund, UN Population Division, and Earth Policy Institute, using purchasing power parity terms.)

Data and Graph Analysis

1. Roughly how many times bigger than the gross world product of 1985 was the gross world product of 2011?

2. If the current trend continues, about what do you think the gross world product will be in 2015, in trillions of dollars?

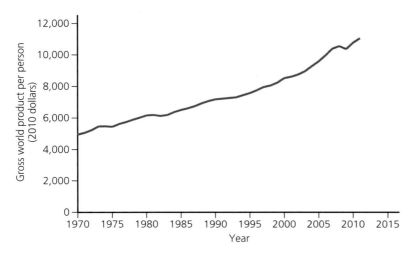

Figure 2 Gross world product (GWP) per person, 1970–2011, in 2010 dollars, using purchasing power parity terms.

(Compiled by the authors using data from the International Monetary Fund, World Bank, World Economic Fund, UN Population Division, and Earth Policy Institute.)

Data and Graph Analysis

1. Roughly how many years did it take for the GWP per person value in 1970 to double?

2. If the current trend continues, about what do you think the GWP per person will be in 2015?

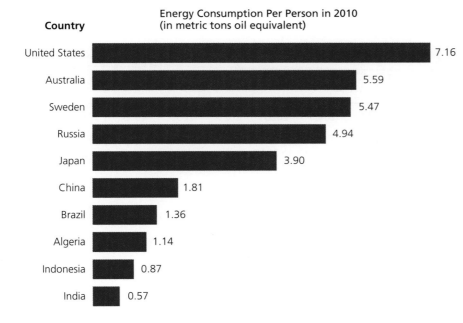

Figure 3 Energy consumption per person in 2010 for selected countries.

(Compiled by the authors using data from the World Bank.)

Data and Graph Analysis

1. On average, how many times more energy does an American consume per year than does a person in **(a)** China, **(b)** India, **(c)** Japan, and **(d)** Brazil?

2. On average, how many times more energy does a Chinese citizen consume per year than does a person in **(a)** India and **(b)** Algeria?

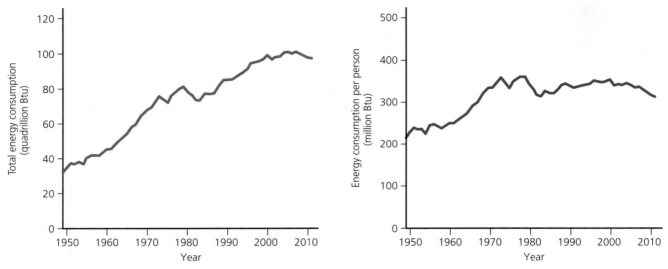

Figure 4 Total (left) and per capita (right) energy consumption in the United States, 1950–2011.

(Compiled by the authors using data from U.S. Energy Information Administration and the U.S. Census Bureau.)

Data and Graph Analysis

1. In what year or years did total U.S. energy consumption reach 80 quadrillion Btus?

2. In what year did energy consumption per person reach its highest level shown on this graph, and about what was that level of consumption?

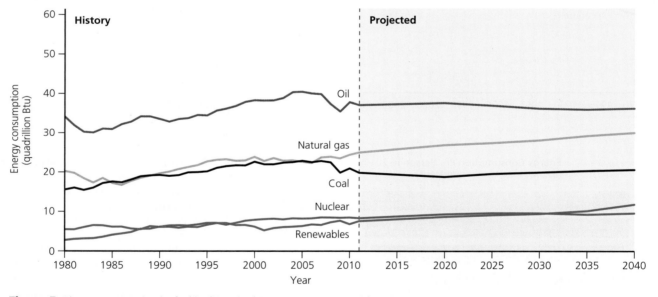

Figure 5 Energy consumption by fuel in the United States, 1980–2011, with projections to 2040.

(Compiled by the authors using data from U.S. Energy Information Administration Primary Energy Consumption Estimates by Source, 1949–2011, and Annual Energy Outlook 2013.)

Data and Graph Analysis

1. Usage of which energy source grew the most between 1990 and 2011? How much did it grow (in quadrillion Btus)?

2. For which two energy sources is usage expected to grow the fastest between 2011 and 2040?

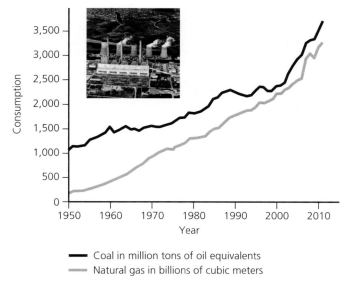

Figure 6 World coal and natural gas consumption for the period 1950–2011.

(Compiled by the authors using data from British Petroleum and International Energy Agency.)

Photo: airphoto.gr/Shutterstock.com

Data and Graph Analysis

1. Which energy source has grown more steadily—coal or natural gas? In what years has coal use grown most sharply?
2. In what year did coal use reach a level twice as high as it was in 1960?

Figure 8 Global electrical generating capacity of nuclear power plants, 1960–2012.

(Compiled by the authors using data from International Energy Agency, Worldwatch Institute, and Earth Policy Institute.)

Photo: SpaceKris/Shutterstock.com

Data and Graph Analysis

1. After 1980, how long did it take to double that year's generating capacity?
2. Considering the decades of the 1970s, 1980s, 1990s, and 2000s, which decade saw the sharpest growth in generating capacity? During which decade did this growth level off?

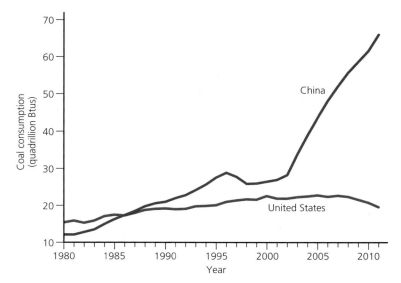

Figure 7 Coal consumption in China and the United States, 1980–2011.

(Compiled by the authors using data from Earth Policy Institute and British Petroleum.)

Data and Graph Analysis

1. By what percentage did coal consumption increase in China between 1980 and 2011?
2. By what percentage did coal consumption increase in the United States between 1980 and 2011?

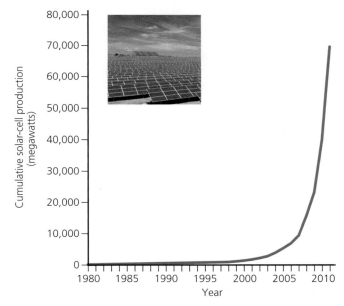

Figure 9 Global cumulative solar (photovoltaic) installations, 1980–2011.

(Compiled by the authors using data from U.S. Energy Information Administration, International Energy Agency, Worldwatch Institute, and Earth Policy Institute.)

Photo: pedrosala/Shutterstock.com

Data and Graph Analysis

1. About how many times more photovoltaic capacity was there in 2011, compared to the capacity in 2000?
2. How long did it take the world to go **(a)** from 0 to 5,000 megawatts in its cumulative installation of photovoltaic capacity and **(b)** from 5,000 to 40,000 megawatts?

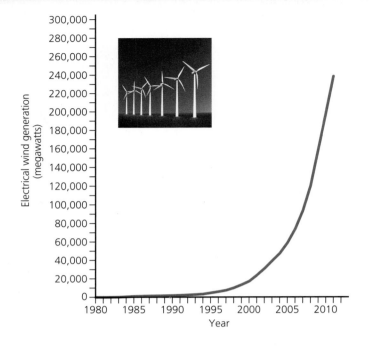

Figure 10 Global installed capacity for generation of electricity by wind energy, 1980–2011.

(Compiled by the authors using data from Global Wind Energy Council, European Wind Energy Association, American Wind Energy Association, Worldwatch Institute, World Wind Energy Association, and Earth Policy Institute.)

Photo: TebNad/Shutterstock.com

Data and Graph Analysis

1. How long did it take for the world to go from 0 to 100,000 megawatts of installed capacity for wind-generated electricity?
2. In 2011, the world's installed capacity for generating electricity by wind power was about how many times more than it was in 1995?

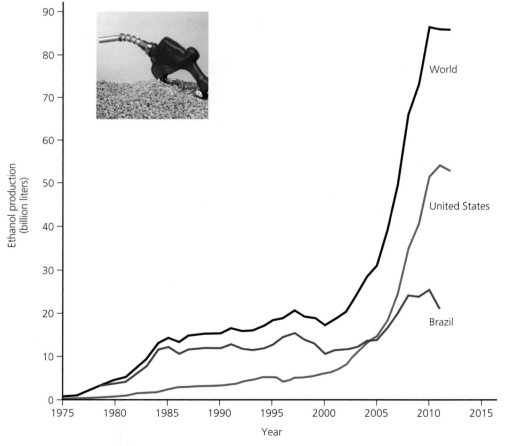

Figure 11 Production of ethanol motor fuel in the world, in Brazil, and in the United States, 1975–2012.

(Compiled by the authors using data from USDA, U.S. Energy Information Administration, Worldwatch Institute, and Earth Policy Institute.)

Photo: Jim Barber/Shutterstock.com

Data and Graph Analysis

1. By roughly what percentage did global ethanol production increase between 2000 and 2010?
2. In what years did ethanol production peak in **(a)** the United States, **(b)** Brazil, and **(c)** the world?

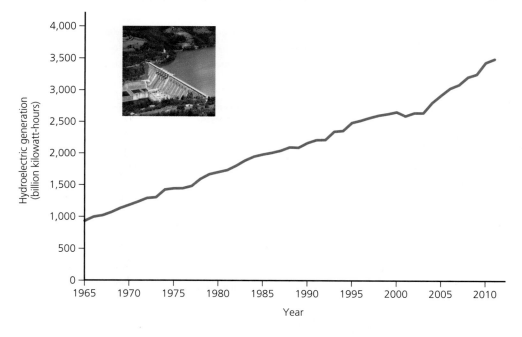

Figure 12 World hydroelectric generation, 1965–2011.

(Compiled by the authors using data from International Energy Agency, British Petroleum, Worldwatch Institute, and Earth Policy Institute.)

Photo: Zeljko Radokjo/Shutterstock.com

Data and Graph Analysis

1. After 1965, how many years did it take for the world to double its hydroelectric capacity?

2. Between what years was growth in hydroelectric capacity the sharpest?

Figure 13 Global cumulative installed geothermal electricity–generating capacity, 1950–2010.

(Compiled by the authors using data from International Energy Agency, Worldwatch Institute, Earth Policy Institute, and Ruggero Bertani.)

Photo: N.Minton/Shutterstock.com

Data and Graph Analysis

1. About how many times more geothermal electricity–generating capacity was available in 2010 as there was in 1965?

2. Between what years was growth in this power source the sharpest?

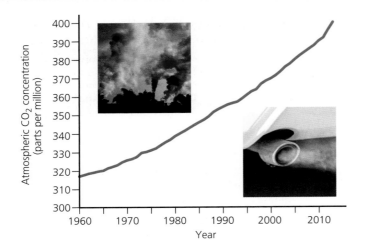

Figure 14 Atmospheric concentration of carbon dioxide (CO_2) measured at a major atmospheric research center in Mauna Loa, Hawaii, 1960–2013.

(Compiled by the authors using data from Scripps Institute of Oceanography, U.S. Energy Information Agency, and Earth Policy Institute.)

Top photo: Peter Weber/Shutterstock.com

Bottom photo: INSAGO/Shutterstock.com

Data and Graph Analysis

1. By how much did atmospheric CO_2 concentrations grow between 1960 and 2011 (in parts per million)?
2. Assuming that atmospheric CO_2 concentrations continue growing as is reflected on this graph, estimate the year in which such concentrations will reach 450 parts per million.

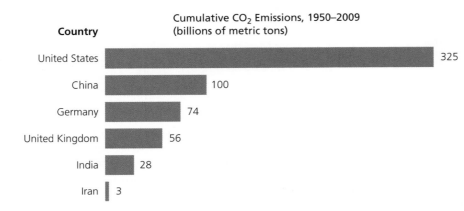

Figure 15 Cumulative carbon dioxide (CO_2) emissions from use of fossil fuels for selected countries, 1950–2009.

(Compiled by the authors using data from International Panel on Climate Change, World Resources Institute, Earth Policy Institute, and British Petroleum.)

Data and Graph Analysis

1. How many times higher are the cumulative CO_2 emissions of the United States than those of **(a)** China and **(b)** India?
2. How many times higher are the cumulative CO_2 emissions of China than those of **(a)** the United Kingdom and **(b)** India?

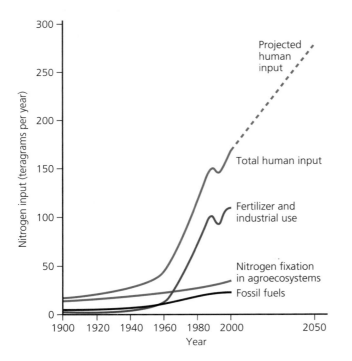

Figure 16 Global trends in the annual inputs of nitrogen into the environment from human activities, with projections to 2050.

(Compiled by the authors using data from Millennium Ecosystem Assessment and International Fertilizer Industry Association.)

Data and Graph Analysis

1. About how many times higher than the total human input of nitrogen to the environment in 1900 was the human input in 2000?
2. If nitrogen inputs continue as projected, about how many times higher will the total inputs be in 2050 than they were in 1980?

1. Corn requires 22 inches of water for a successful crop. A farmer that has a 200-acre farm is in a drought and receives only 2 inches of rain throughout the year.

   a. **Calculate** how many gallons of water the farmer will need to purchase to be able to irrigate his crops. Recall that 60% of the water applied to a field through irrigation is lost. (An acre-foot is a common unit for measuring volumes of water, typically used in irrigation calculations. One acre-foot is the volume of water sufficient to cover an acre of land to a depth of 1 foot, approximately 325,851 U.S. gallons.) Show all work.

   b. **Discuss** *two* effects on a small stream if the farmer were to pump irrigation water from it, decreasing the flow rate of the stream.

   c. **Discuss** *three* ways the farmer could lessen the 60% loss of water while irrigating the farm.

2. Read the following article from the *Belmont Examiner* and answer the questions that follow.

   ## Dump Truck Overturns Spilling Toxic Ash

   Last Tuesday the Belmont Fire Department was dispatched to the westbound lane of State Highway 123 for an accident involving a dump truck on fire.

   Engine 2 responded with a crew of five, followed shortly after by Engine 3 with a crew of four. Engine 2 arrived to find an overturned dump truck with the cab fully involved and the driver self-extricated. Engine 2's crew pulled one of the front bumper lines and started their fire attack. They were followed closely behind by Engine 3's crew, who pulled the other line and helped with the fire suppression.

   The operator of the vehicle suffered minor abrasions and declined further treatment.

   The dump truck was carrying ash from a local incinerator that burns trash to generate electricity. Much of the ash was carried into a nearby stream during the fire suppression. Local authorities and the EPA were notified of a spill of ash with potential toxins into the local watershed.

   a. **Explain** where the truck was most likely headed.

   b. **Identify** *three* potential toxins in the ash and their probable source.

   c. For *two* of the toxins you identified, **discuss** their potential effects on the stream ecology downstream.

3. Most traditional top-loading washing machines use 50 gallons of water per load. Front-loading washing machines use 50% less water, can hold up to 25% more clothes, and spin out more water than top loaders.

   a. If a family of 5 generates 20 loads of laundry each week, **calculate** how much water would be saved if the family were to install a front-loading washing machine.

   b. If the average cost of water in the U.S. is $1.50 per 1,000 gallons, **calculate** the cost saved by installing a front-loading washing machine.

   c. **Identify** *two* other costs the family would also reduce.

   d. Most soaps and detergents are phosphate free. **Explain** why phosphates were banned from detergents.

4. Humans have been altering natural systems for thousands of years. Choose *three* of the below, and for each **discuss** the effects humans have on the natural systems and the potential means of mitigation.

- Slash and burn agriculture
- Mountaintop removal of coal for electricity generation
- Damming of rivers for reservoirs
- Drilling for oil undersea
- Building roads and multilane divided highways

5. The management of common resources is essential to achieve sustainable systems. Choose *three* of the below, and for each **identify** the threats facing the resource and **discuss** the conservation measures necessary to preserve it.

- Grand Banks fishery off Newfoundland, Canada
- Pasture lands for grazing in the western United States
- Fish stocks (e.g., swordfish) in international waters of the Pacific Ocean
- Large underground aquifers (e.g., Ogallala Aquifer)
- Air in large urban cities

6. Below is a food web of a typical North American forest.

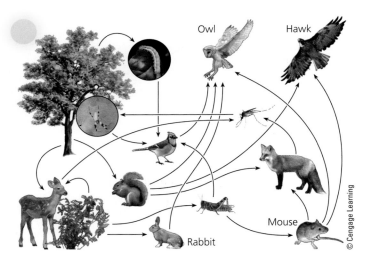

© Cengage Learning

a. Use the information shown in the food web above to **calculate** how large an area of a forest needs to be to support one owl feeding only on rabbits if the owl has a daily caloric requirement of 800 Cal. Express your answer in acres.

- Temperate deciduous forests have a net primary productivity of 1,200 g/m²/yr, and on average the nutrients found in primary producers provide .25 Cal/gram to primary consumers.
- 10,000 m² = 1 hectare = 2.5 acres

b. If hawks have the same daily caloric demand as owls, **calculate** how the area of a hawk would differ if the hawks were feeding only on mice. **Explain** why.

c. A local farmer uses a new pesticide that eliminates all of the grasshoppers in the food web. Discuss the effects on the remaining members of the community.

7. Below are two sets of water quality data for a stream, measured one month apart.

Measurements of Water Quality Parameters					
Bottle	Temp (°C)	pH	D.O. (O_2 – ppm)	Nitrates (NO_3 – ppm)	Turbidity (NTU)
Reading 1	20.	7.1	7.9	0.5	2.2
Reading 2	25.	6.5	6.9	1.0	2.7

a. Using the measurements above and the charts below and on the next page, **calculate** a water quality index (WQI) for the water in the bottle at the *second* reading. Show clearly how you determined your answer.

b. Choose *three* of the water quality parameters and **discuss** why each one is important to measure and how it affects the water quality of the stream.

Determining Q Values		Maximum Solubility of O_2 in Fresh Water	
Ecosystem Parameter	Weighting Factor	°C	DO (mg/L)
T (°C)	0.19	20	9.2
pH	0.19	21	9.0
DO (O_2) % saturation	0.30	22	8.8
		23	8.7
Nitrate (NO_3 – ppm)	0.17	24	8.5
		25	8.4
Turbidity (NTU)	0.14		

Source: Water Quality with Calculators by R. Johnson, K. Redding, and D. Holmquist (Beaverton, Ore.: Vernier Software & Technology, 2000).

Change in Temperature

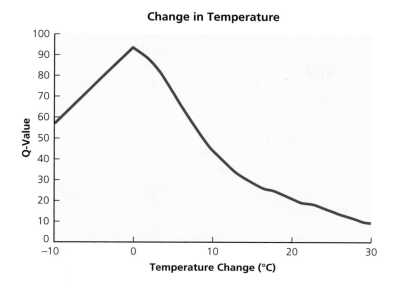

Nitrate Nitrogen (mg/L NO3-N)

NO3-N > 20, Q = 1.0

pH Test

pH < 2.1, Q = 0   pH > 12.0, Q = 0

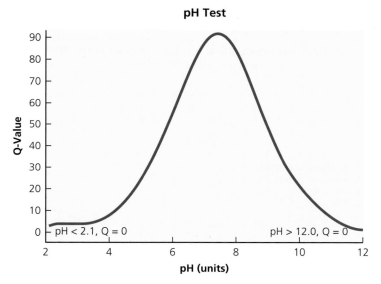

Turbidity

Turbidity > 100, Q = 5

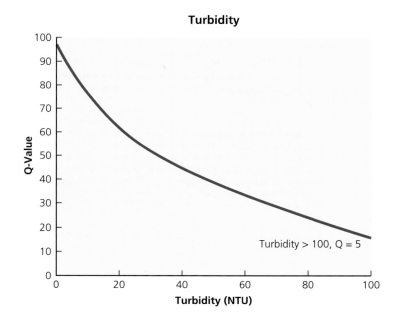

Dissolved Oxygen

DO % Saturation > 140.0, Q = 50.0

Source: Indiana Department of Natural Resources

8. Below is a population sample from two different field ecosystems.

$$D = 1 - \sum \bullet \frac{n_i}{N} \bullet^2$$

Field Ecosystem A (# of individuals)	Species	Field Ecosystem B (# of individuals)
0	A	10
30	B	10
50	C	10
10	D	20
10	E	10
0	F	5
0	G	10
0	H	25

a. **Calculate** the species diversity for the two forest ecosystems above, using the Simpson's Diversity Index formula shown. Show clearly how you determined your answer.

b. Which forest ecosystem is more diverse? **Explain** why.

c. **Discuss** why diversity is important in an ecosystem and **identify** *two* ways humans can help to foster diversity.

9.

Native species of North America are often defined as those species that were present in North America before Europeans arrived in the early 1600s. When Europeans arrived, they found extensive forests dominated by the American Chestnut tree. These chestnut trees were extensively cultivated and planted by the North American natives for food and other wood products. In the early 1900s the chestnut blight (a fungus) virtually exterminated all of the American Chestnut trees from North America.

a. Is the American Chestnut considered a native species despite its intensive cultivation? **Explain** why.

b. Is the chestnut blight considered a native or invasive species? **Explain** why.

c. **Discuss** why invasive species tend to proliferate at very high rates.

d. Many people are trying to grow hybrid chestnut trees that will be immune to the chestnut blight so that forests can be restored to those found by the Europeans in the 1600s. Do you agree or disagree that this should be done? **Justify** your answer.

10. One of the central themes of environmental science is energy flow.

a. **Explain** why it is incorrect to say that producers create energy for an ecosystem.

b. Draw a simple food chain for a marine ecosystem.

c. **Explain** why there are very few large predators in an ecosystem.

d. Some people advocate eating lower on the food chain as a means of solving the world food crisis. **Explain** why this would work.

11. A farmer has a 200-acre farm divided into four separate fields. The farmer recently read some research showing that locally produced natural organic fertilizers are better for the environment than chemical fertilizers, and wishes to determine which produces a better crop yield.

a. **Design** a controlled experiment for the farmer to determine the effectiveness of natural versus chemical fertilizers.

b. **Discuss** why natural fertilizers are better for the environment than chemical fertilizers.

12. Read the following article from the *Belmont Examiner* and answer the questions that follow.

Are Our Recycled Plastics Headed to a Landfill?

Have you ever wondered what happens to your recycled plastics when they are picked up at the curb by the trash truck? It gives everybody a good feeling to recycle, but what if recycling means more trash fees?

Michael Dumper, CEO of Belmont Waste Hauling, recently told citizens of Belmont at a recent Town Council meeting that because of the closure of the closest recycling center, he will now need to haul recycled plastics almost 200 miles to be recycled. Recyclables make up almost 40% of the trash load from Belmont and as a result Belmont Waste Hauling will lose money on hauling recyclables.

The city of Belmont requires all municipal buildings and all newly constructed public buildings to provide recycling containers for use. Belmont Waste Hauling will continue to haul the city's waste and recycling this year, but the company's contract with the city expires next June. At that point the city will need to renegotiate a new contract, maybe with a new company.

Many of the larger trash hauling giants are not willing to deal with the many issues involved with recycling plastics and simply bury it in landfills. It takes a lot of money up front to recycle, and they are not willing to lose any amount of profit on any load of trash.

Michael Dumper maintains that he wants to keep the Belmont trash contract. He hopes that with a small increase in fees he can continue to serve the city of Belmont. Competition forces companies to stop recycling programs unless consumers are willing to pay for them, he says. And yes, finding recycling centers that are still sorting and recycling is a challenge, but Michael is not yet conceding defeat and will continue to face the challenge of finding the plants that are still operating.

   a. **Discuss** the reasons some trash haulers dump recyclable plastics into landfills.

   b. **Explain** why aluminum cans are recycled more often than most other recyclables.

   c. Give *three* reasons why it is important to recycle as much as possible.

   d. **Identify** *three* materials that can be recycled, other than plastic or aluminum, and explain how they are recycled.

13. Indoor air pollution can cause anything from minor discomfort to major illness.

   a. **Discuss** *three* potential sources of indoor air pollution and identify the agent that causes the pollution.

   b. For each pollutant you identified, **explain** how the pollutant affects humans.

   c. Name and describe a governmental agency that works to reduce indoor air pollution for workers.

14. Many scientists believe that ocean acidification will be one of the next major environmental science issues to face humanity.

   a. **Discuss** the primary cause of ocean acidification.

   b. Show with equations how ocean acidification occurs.

   c. **Discuss** *three* effects of ocean acidification on ocean organisms.

15. As oil prices have risen during the past few years, interest in alternative energy sources has risen as well.

   a. **Explain** why, as oil prices rise, this drives a renewed interest in alternative energy sources.

   b. **Identify** *two* alternative energy sources, and for each identify *two* advantages and *one* disadvantage.

   c. **Discuss** *two* methods the state or federal governments could use to make it more attractive for consumers to utilize alternative energy sources.

16. Heavy metals typically have densities above 5 g/cm^3 and are persistent in all parts of the environment. Human activity affects the natural geological and biological redistribution of heavy metals through pollution of the air, water, and soil.

   a. **Identify** *two* heavy metals that are found as pollutants in the environment.

   b. For each of the *two* heavy metals you identified:
   - **Identify** a source from where it enters the environment.
   - Give one toxic effect on humans.

   c. **Explain** why heavy metals are so persistent in all parts of the environment.

   d. Name and describe *one* law that prevents or minimizes the release of heavy metals into the environment.

17. Below is an excerpt from the Chesapeake Bay Program's website. Read the article and answer the questions that follow.

The way we live and use the land greatly influences the health of the Bay and its network of streams and rivers....

The percentage of forested land is slipping as the rate of development increases. In the Bay watershed, approximately 100 acres of forest are lost each day, primarily to development of new roads and buildings to accommodate a growing population....

How has land in the Chesapeake Bay watershed changed over time?

Data published in 2009 by the U.S. Geological Survey (USGS) show how three types of land cover—farmland/open space, urban areas and tree canopy—in the Chesapeake Bay watershed changed between four time periods: 1984, 1992, 2001 and 2006. "Land cover" is measured by interpretation of satellite images that capture the sun's reflection off the earth's surface, while "land use" is the actual way an area of land is used (such as agriculture, forest or residential development).

Farmland/open space decreased by 8,700 acres per year between 1984 and 1992, 2,110 acres per year between 1992 and 2001, and 941 acres per year between 2001 and 2006.

The total amount of urban area in the Bay watershed increased by 14 percent, or 355,146 acres, between 1984 and 2006. The annual rate of increase between 1984 and 1992 was twice the annual rate of increase between 1992 and 2006.

Tree canopy decreased from 62.6 percent of the watershed in 1984 to 61.5 percent in 2006, a loss of 439,080 acres. The highest rate of decline took place between 2001 and 2006, when the watershed lost 37,403 acres of tree canopy per year.

Source: Chesapeake Bay Program

---

a. What percent did the tree canopy decrease from 1984 to 2006? Use this value to **calculate** the approximate total acreage of the Chesapeake Bay watershed.

b. With urban sprawl often comes flash flooding. **Identify** *two* reasons urban sprawl increases flash flooding.

c. **Explain** how urban planning can lessen flash flooding.

d. **Identify** and **describe** one federal law that could help prevent developers from developing a forested piece of land.

18. Economics, society, and various laws, treaties, and protocols are frequently the driving force of environmental science. Choose *three* of the below and for each **identify** what the goal or outcome was and **discuss** the success or failure of the legislation.

- Endangered Species Act
- Superfund Act
- Montreal Protocol
- Kyoto Protocol
- Antarctic Treaty

19. In today's global economy, the right balance between government and free enterprise can be difficult to find.

a. **Discuss** how a new environmental policy/law could be produced, including in your discussion the three branches of the government, individuals, and corporations.

b. **Discuss** what grassroots movements are and how they can be effective at motivating large corporations to change their practices.

c. **Explain** why it is difficult for the U.S. government to enter into treaties with foreign countries to solve global problems.

20. The graphs below depict the price of crude oil from 1970 to 2010 and the top consuming countries of petroleum for 1960–2008.

a. If the five countries continue on the same trend shown for the past 40 years, **calculate** how much oil will be needed in the year 2020.

b. **Calculate** the total expenditure on petroleum for the five countries in the years 1975 and 2008. **Calculate** the percent increase this represents.

c. **Explain** why there was very little interest in identifying and developing the technology for renewable resources during the 1970s.

d. **Explain** why the United States showed a sharp decline in the amount of petroleum consumed from 1980–1985 and **discuss** how this was achieved.

Petroleum Consumption of Top Consuming Countries

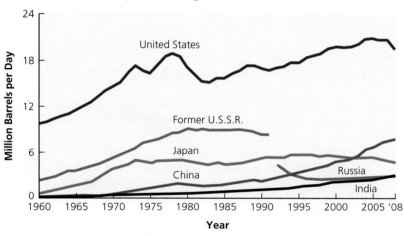

Top Consuming Countries, 1960–2008

Source: U.S. Energy Information Administration

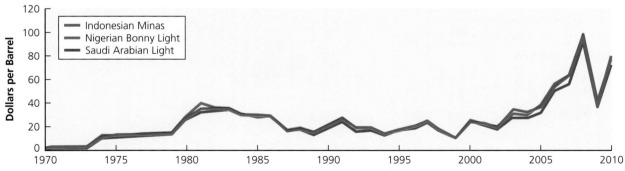

Crude Oil Prices by Selected Types, 1970–2010

Source: U.S. Energy Information Administration

GLOSSARY

abiotic Nonliving. Compare *biotic*.

acid See *acid solution*.

acid deposition The falling of acids and acid-forming compounds from the atmosphere to the earth's surface. Acid deposition is commonly known as *acid rain*, a term that refers to the wet deposition of droplets of acids and acid-forming compounds.

acidity Chemical characteristic that helps determine how a substance dissolved in water (a solution) will interact with and affect its environment; based on the comparative amounts of hydrogen ions (H^+) and hydroxide ions (OH^-) contained in a particular volume of the solution. See *pH*.

acid rain See *acid deposition*.

acid solution Any water solution that has more hydrogen ions (H^+) than hydroxide ions (OH^-); any water solution with a pH less than 7. Compare *basic solution, neutral solution*.

active solar heating system System that uses solar collectors to capture energy from the sun and store it as heat for space heating and water heating. Liquid or air pumped through the collectors transfers the captured heat to a storage system such as an insulated water tank or rock bed. Pumps or fans then distribute the stored heat or hot water throughout a dwelling as needed. Compare *passive solar heating system*.

adaptation Any genetically controlled structural, physiological, or behavioral characteristic that helps an organism survive and reproduce under a given set of environmental conditions. It usually results from a beneficial mutation. See *biological evolution, differential reproduction, mutation, natural selection*.

adaptive trait See *adaptation*.

aerobic respiration Complex process that occurs in the cells of most living organisms, in which nutrient organic molecules such as glucose ($C_6H_{12}O_6$) combine with oxygen (O_2) to produce carbon dioxide (CO_2), water (H_2O), and energy. Compare *photosynthesis*.

affluence wealth that results in high levels of consumption and unnecessary waste of resources, based mostly on the assumption that buying more and more material goods will bring fulfillment and happiness.

age structure Percentage of the population (or number of people of each sex) at each age level in a population.

agrobiodiversity The genetic variety of plant and animal species used on farms to produce food. Compare *biodiversity*.

agroforestry Planting trees and crops together.

air pollution One or more chemicals in high enough concentrations in the air to harm humans, other animals, vegetation, or materials. Excess heat is also considered a form of air pollution. Such chemicals or physical conditions are called air pollutants. See *primary pollutant, secondary pollutant*.

albedo Ability of a surface to reflect light.

alien species See *nonnative species*.

alley cropping Planting of crops in strips with rows of trees or shrubs on each side.

alpha particle Positively charged matter, consisting of two neutrons and two protons, which is emitted as radioactivity from the nuclei of some radioisotopes. See also *beta particle, gamma rays*.

altitude Height above sea level. Compare *latitude*.

anaerobic respiration Form of cellular respiration in which some decomposers get the energy they need through the breakdown of glucose (or other nutrients) in the absence of oxygen. Compare *aerobic respiration*.

animal manure Dung and urine of animals used as a form of organic fertilizer. Compare *green manure*.

annual plant Plant that grows, sets seed, and dies in one growing season. Compare *perennial*.

Anthropocene A new era in which humans have become major agents of change in the functioning of the earth's life-support systems as their ecological footprints have spread over the earth. See *ecological footprint*. Compare *Holopocene*.

anthropocentric Human-centered.

aquaculture Growing and harvesting of fish and shellfish for human use in freshwater ponds, irrigation ditches, and lakes, or in cages or fenced-in areas of coastal lagoons and estuaries or in the open ocean. See *fish farming, fish ranching*.

aquatic Pertaining to water. Compare *terrestrial*.

aquatic life zone Marine and freshwater portions of the biosphere. Examples include freshwater life zones (such as lakes and streams) and ocean or marine life zones (such as estuaries, coastlines, coral reefs, and the open ocean).

aquifer Porous, water-saturated layers of sand, gravel, or bedrock that can yield an economically significant amount of water.

arable land Land that can be cultivated to grow crops.

area strip mining Type of surface mining used where the terrain is flat. An earthmover strips away the overburden, and a power shovel digs a cut to remove the mineral deposit. The trench is then filled with overburden, and a new cut is made parallel to the previous one. The process is repeated over the entire site. Compare *mountaintop removal, open-pit mining, subsurface mining*.

arid Dry. A desert or other area with an arid climate has little precipitation.

artificial selection Process by which humans select one or more desirable genetic traits in the population of a plant or animal species and then use *selective breeding* to produce populations containing many individuals with the desired traits. Compare *genetic engineering, natural selection*.

asthenosphere Zone within the earth's mantle made up of hot, partly melted rock that flows and can be deformed like soft plastic.

atmosphere Whole mass of air surrounding the earth. See *stratosphere, troposphere*. Compare *biosphere, geosphere, hydrosphere*.

atmospheric pressure Force or mass per unit area of air, caused by the bombardment of a surface by the molecules in air.

atom Minute unit made of subatomic particles that is the basic building block of all chemical elements and thus all matter; the smallest unit of an element that can exist and still have the unique characteristics of that element. Compare *ion, molecule*.

atomic number Number of protons in the nucleus of an atom. Compare *mass number*.

atomic theory Idea that all elements are made up of atoms; the most widely accepted scientific theory in chemistry.

autotroph See *producer*.

background extinction rate Normal extinction of various species as a result of changes in local environmental conditions. Compare *mass extinction*.

bacteria Prokaryotic, one-celled organisms. Some transmit diseases. Most act as decomposers and get the nutrients they need by breaking down complex organic compounds in the tissues of living or dead organisms into simpler inorganic nutrient compounds.

basic solution Water solution with more hydroxide ions (OH^-) than hydrogen ions (H^+); water solution with a pH greater than 7. Compare *acid solution, neutral solution*.

benthos Bottom-dwelling organisms. Compare *decomposer, nekton, plankton*.

beta particle Swiftly moving electron emitted by the nucleus of a radioactive isotope. See also *alpha particle, gamma ray*.

bioaccumulation An increase in the concentration of a chemical in specific organs or tissues at a level higher than would normally be expected. Compare *biomagnification*.

biocentric Life-centered. Compare *anthropocentric*.

biodegradable Capable of being broken down by decomposers.

biodegradable pollutant Material that can be broken down into simpler substances (elements and compounds) by bacteria or other decomposers. Paper and most organic wastes such as animal manure are biodegradable but can take decades to biodegrade in modern landfills. Compare *nondegradable pollutant*.

biodiversity Variety of different species (*species diversity*), genetic variability among individuals within each species (*genetic diversity*), variety of ecosystems (*ecological diversity*), and functions such as energy flow and matter cycling needed for the survival of species and biological communities (*functional diversity*).

biodiversity hotspot An area especially rich in plant species that are found nowhere else and are in great danger of extinction. Such areas suffer serious ecological disruption, mostly because of rapid human population growth and the resulting pressure on natural resources.

biofuel Gas (such as methane) or liquid fuel (such as ethyl alcohol or biodiesel) made from plant material (biomass).

biogeochemical cycle Natural processes that recycle nutrients in various chemical forms from the nonliving environment to living organisms and then back to the nonliving environment. Examples include the carbon, oxygen, nitrogen, phosphorus, sulfur, and hydrologic cycles.

biological amplification See *biomagnification*.

biological community See *community*.

biological diversity See *biodiversity*.

biological evolution Change in the genetic makeup of a population of a species in successive generations. If continued long enough, it can lead to the formation of a new species. Note that populations, not individuals, evolve. See also *adaptation, differential reproduction, natural selection, theory of evolution*.

biological extinction Complete disappearance of a species from the earth. It happens when a species cannot adapt and successfully reproduce under new environmental conditions or when a species evolves into one or more new species. Compare *speciation*. See also *endangered species, mass extinction, threatened species*.

biological pest control Control of pest populations by natural predators, parasites, or disease-causing bacteria and viruses (pathogens).

biomagnification Increase in concentration of DDT, PCBs, and other slowly degradable, fat-soluble chemicals in organisms at successively higher trophic levels of a food chain or web. Compare *bioaccumulation*.

biomass Organic matter produced by plants and other photosynthetic producers; total dry weight of all living organisms that can be supported at each trophic level in a food chain or web; dry weight of all organic matter in plants and animals in an ecosystem; plant materials and animal wastes used as fuel.

biome Terrestrial regions inhabited by certain types of life, especially vegetation. Examples include various types of deserts, grasslands, and forests.

biomimicry Process of observing certain changes in nature, studying how natural systems have responded to such changing conditions over many millions of years, and applying what is learned to dealing with some environmental challenge.

biosphere Zone of the earth where life is found. It consists of parts of the atmosphere (the troposphere), hydrosphere (mostly surface water and groundwater), and lithosphere (mostly soil and surface rocks and sediments on the bottoms of oceans and other bodies of water) where life is found. Compare *atmosphere, geosphere, hydrosphere*.

biotic Living organisms. Compare *abiotic*.

biotic pollution The effect of invasive species that can reduce or wipe out populations of many native species and trigger ecological disruptions.

biotic potential Maximum rate at which the population of a given species can increase when there are no limits on its rate of growth. See *environmental resistance*.

birth rate See *crude birth rate*.

bitumen Gooey, black, high-sulfur, heavy oil extracted from tar sand and then upgraded to synthetic fuel oil. See *oil sand*.

broadleaf deciduous plants Plants such as oak and maple trees that survive drought and cold by shedding their leaves and becoming dormant. Compare *broadleaf evergreen plants, coniferous evergreen plants*.

broadleaf evergreen plants Plants that keep most of their broad leaves year-round. An example is the trees found in the canopies of tropical rain forests. Compare *broadleaf deciduous plants, coniferous evergreen plants*.

buffer Substance that can react with hydrogen ions in a solution and thus hold the acidity or pH of a solution fairly constant. See *pH*.

calorie Unit of energy; amount of energy needed to raise the temperature of 1 gram of water by 1 C° (unit on Celsius temperature scale). See also *kilocalorie*.

cancer Group of more than 120 different diseases, one for each type of cell in the human body. Each type of cancer produces a tumor in which cells multiply uncontrollably and invade surrounding tissue.

carbon capture and storage (CCS) Process of removing carbon dioxide gas from coal-burning power and industrial plants and storing it somewhere (usually underground or under the seabed) so that it is not released into the atmosphere, essentially forever.

carbon cycle Cyclic movement of carbon in different chemical forms from the environment to organisms and then back to the environment.

carcinogen Chemicals, ionizing radiation, and viruses that cause or promote the development of cancer. See *cancer*. Compare *mutagen, teratogen*.

carnivore Animal that feeds on other animals. Compare *herbivore, omnivore*.

carrying capacity (*K*) Maximum population of a particular species that a given habitat can support over a given period. Compare *cultural carrying capacity*.

CCS See *carbon capture and storage*.

cell Smallest living unit of an organism. Each cell is encased in an outer membrane or wall and contains genetic material (DNA) and other parts to perform its life function. Organisms such as bacteria consist of only one cell, but most organisms contain many cells.

cell theory The idea that all living things are composed of cells; the most widely accepted scientific theory in biology.

CFCs See *chlorofluorocarbons*.

chain reaction Multiple nuclear fissions, taking place within a certain mass of a fissionable isotope, which release an enormous amount of energy in a short time.

chemical One of the millions of different elements and compounds found naturally and synthesized by humans. See *compound, element*.

chemical change Interaction between chemicals in which the chemical composition of the elements or compounds involved changes. Compare *nuclear change, physical change*.

chemical cycling The continual cycling of chemicals necessary for life through natural processes such as the water cycle and feeding interactions; processes that evolved due to the fact that the earth gets essentially no new inputs of these chemicals.

chemical formula Shorthand way to show the number of atoms (or ions) in the basic structural unit of a compound. Examples include H_2O, $NaCl$, and $C_6H_{12}O_6$.

chemical reaction See *chemical change*.

chemosynthesis Process in which certain organisms (mostly specialized bacteria) extract inorganic compounds from their environment and convert them into organic nutrient compounds without the presence of sunlight. Compare *photosynthesis*.

chlorinated hydrocarbon Organic compound made up of atoms of carbon, hydrogen, and chlorine. Examples include DDT and PCBs.

chlorofluorocarbons (CFCs) Organic compounds made up of atoms of carbon, chlorine, and fluorine. An example is Freon-12 (CCl_2F_2), which is used as a refrigerant in refrigerators and air conditioners and in making plastics such as Styrofoam. Gaseous CFCs can deplete the ozone layer when they slowly rise into the stratosphere and their chlorine atoms react with ozone molecules. Their use is being phased out.

CHP (Combined heat and power) See *cogeneration*.

chromosome A grouping of genes and associated proteins in plant and animal cells that carry certain types of genetic information. See *genes*.

chronic malnutrition Faulty nutrition, caused by a diet that does not supply an individual with enough protein, essential fats, vitamins, minerals, and other nutrients needed for good health. Compare *overnutrition, chronic undernutrition*.

chronic undernutrition Condition suffered by people who cannot grow or buy enough food to meet their basic energy needs. Most chronically undernourished children live in developing countries and are likely to suffer from mental retardation and stunted growth and to die from infectious diseases. Compare *chronic malnutrition, overnutrition*.

civil suit Court case brought to settle disputes or damages between one party and another.

clear-cutting Method of timber harvesting in which all trees in a forested area are removed in a single cutting. Compare *selective cutting, strip cutting*.

climate Physical properties of the troposphere of an area based on analysis of its weather records over a long period (at least 30 years). The two main factors determining an area's climate are its average *temperature*, with its seasonal variations, and the average amount and distribution of *precipitation*. Compare *weather*.

climate tipping point Point at which an environmental problem reaches a threshold level where scientists fear it could cause irreversible climate disruption.

climax community See *mature community*.

closed-loop recycling See *primary recycling*.

coal Solid, combustible mixture of organic compounds with 30–98% carbon by weight, mixed with various amounts of water and small amounts of sulfur and nitrogen compounds. It forms in several stages as the remains of plants are subjected to heat and pressure over millions of years.

coal gasification Conversion of solid coal to synthetic natural gas (SNG).

coal liquefaction Conversion of solid coal to a liquid hydrocarbon fuel such as synthetic gasoline or methanol.

coastal wetland Land along a coastline, extending inland from an estuary that is covered with saltwater all or part of the year. Examples include marshes, bays, lagoons, tidal flats, and mangrove swamps. Compare *inland wetland*.

coastal zone Warm, nutrient-rich, shallow part of the ocean that extends from the high-tide mark on land to the edge of a shelflike extension of continental land masses known as the continental shelf. Compare *open sea*.

coevolution Evolution in which two or more species interact and exert selective pressures on each other that can lead each species to undergo adaptations. See *evolution, natural selection*.

cogeneration Production of two useful forms of energy, such as high-temperature heat or steam and electricity, from the same fuel source.

cold front Leading edge of an advancing mass of cold air. Compare *warm front*.

colony collapse disorder (CCD) Loss through death or disappearance of all or most of the European honeybees in a particular colony due to unknown causes; a phenomenon that has resulted in large losses of European honeybees in the United States and in parts of Europe.

combined heat and power (CHP) production See *cogeneration*.

commensalism An interaction between organisms of different species in which one type of organism benefits and the other type is neither helped nor harmed to any great degree. Compare *mutualism*.

commercial extinction Depletion of the population of a wild species used as a resource to a level at which it is no longer profitable to harvest the species.

commercial forest See *tree plantation*.

commercial inorganic fertilizer Commercially prepared mixture of inorganic plant nutrients such as nitrates, phosphates, and potassium applied to the soil to restore fertility and increase crop yields. Compare *organic fertilizer*.

common law A body of unwritten rules and principles derived from court decisions along with commonly accepted practices, or norms, within a society. Compare *statutory law*.

common-property resource Resource that is owned jointly by a large group of individuals. One example is the roughly one-third of the land in the United States that is owned jointly by all U.S. citizens and held and managed for them by the government. Another example is an area of land that belongs to a whole village and that can be used by anyone for grazing cows or sheep. Compare *open-access renewable resource*. See *tragedy of the commons*.

community Populations of all species living and interacting in an area at a particular time.

competition Two or more individual organisms of a single species (*intraspecific competition*) or two or more individuals of different species (*interspecific competition*) attempting to use the same scarce resources in the same ecosystem.

compost Partially decomposed organic plant and animal matter used as a soil conditioner or fertilizer.

composting See *compost*.

compound Combination of atoms, or oppositely charged ions, of two or more elements held together by attractive forces called chemical bonds. Examples are $NaCl$, CO_2, and $C_6H_{12}O_6$. Compare *element*.

concentration Amount of a chemical in a particular volume or weight of air, water, soil, or other medium.

coniferous evergreen plants Cone-bearing plants (such as spruces, pines, and firs) that keep some of their narrow, pointed leaves (needles) all year. Compare *broadleaf deciduous plants, broad-leaf evergreen plants*.

coniferous trees Cone-bearing trees, mostly evergreens, that have needle-shaped or scalelike leaves. They produce wood known commercially as softwood. Compare *deciduous plants*.

conservation Sensible and careful use of natural resources by humans. People with this view are called *conservationists*.

conservation biology Multidisciplinary science created to deal with the crisis of maintaining the genes, species, communities, and ecosystems that make up earth's biological diversity. Its goals are to investigate human impacts on biodiversity and to develop practical approaches to preserving biodiversity.

conservationist Person concerned with using natural areas and wildlife in ways that sustain them for current and future generations of humans and other forms of life.

conservation-tillage farming Crop cultivation in which the soil is disturbed little (minimum-tillage farming) or not at all (no-till farming) in an effort to reduce soil erosion, lower labor costs, and save energy. Compare *conventional-tillage farming*.

consumer Organism that cannot synthesize the organic nutrients it needs and gets its organic nutrients by feeding on the tissues of producers or of other consumers; generally divided into *primary consumers* (herbivores), *secondary consumers* (carnivores), *tertiary* (*higher-level*) *consumers*, *omnivores*, and *detritivores* (decomposers and detritus feeders). In economics, one who uses economic goods. Compare *producer*.

contour farming Plowing and planting across the changing slope of land, rather than in straight lines, to help retain water and reduce soil erosion.

contour strip mining Form of surface mining used on hilly or mountainous terrain. A power shovel cuts a series of terraces into the side of a hill. An earthmover removes the overburden, and a power shovel extracts the coal. The overburden from each new terrace is dumped onto the one below. Compare *area strip mining, mountaintop removal, open-pit mining, subsurface mining*.

controlled burning Deliberately set, carefully controlled surface fires that reduce flammable litter and decrease the chances of damaging *crown fires*. See *ground fire, surface fire*.

conventional-tillage farming Crop cultivation method in which a planting surface is made by plowing land, breaking up the exposed soil, and then smoothing the surface. Compare *conservation-tillage farming*.

convergent plate boundary Area where the earth's lithospheric plates are pushed together. See *subduction zone*. Compare *divergent plate boundary, transform fault*.

coral reef Formation produced by massive colonies containing billions of tiny coral animals, called polyps, that secrete a stony substance (calcium carbonate) around themselves for protection. When the corals die, their empty outer skeletons form layers and cause the reef to grow. Coral reefs are found in the coastal zones of warm tropical and subtropical oceans.

core Inner zone of the earth. It consists of a solid inner core and a liquid outer core. Compare *crust, mantle*.

corrective feedback loop See *negative feedback loop*.

cost–benefit analysis A comparison of estimated costs and benefits of actions such as implementing a pollution control regulation, building a dam on a river, or preserving an area of forest.

crop rotation Planting a field, or an area of a field, with different crops from year to year to reduce soil nutrient depletion. A plant such as corn, tobacco, or cotton, which removes large amounts of nitrogen from the soil, is planted one year. The next year a legume such as soybeans, which adds nitrogen to the soil, is planted.

crown fire Extremely hot forest fire that burns ground vegetation and treetops. Compare *controlled burning, ground fire, surface fire*.

crude birth rate Annual number of live births per 1,000 people in the population of a geographic area at the midpoint of a given year. Compare *crude death rate*.

crude death rate Annual number of deaths per 1,000 people in the population of a geographic area at the midpoint of a given year. Compare *crude birth rate*.

crude oil Gooey liquid consisting mostly of hydrocarbon compounds and small amounts of compounds containing oxygen, sulfur, and nitrogen. Extracted from underground accumulations, it is sent to oil refineries, where it is converted to heating oil, diesel fuel, gasoline, tar, and other materials.

crust Solid outer zone of the earth. It consists of oceanic crust and continental crust. Compare *core, mantle*.

cultural carrying capacity The limit on population growth that would allow most people in an area or the world to live in reasonable comfort and freedom without impairing the ability of the planet to sustain future generations. Compare *carrying capacity*.

cultural eutrophication Overnourishment of aquatic ecosystems with plant nutrients (mostly nitrates and phosphates) because of human activities such as agriculture, urbanization, and discharges from industrial plants and sewage treatment plants. See *eutrophication*.

culture Whole of a society's knowledge, beliefs, technology, and practices.

dam A structure built across a river to control the river's flow or to create a reservoir. See *reservoir*.

data Factual information collected by scientists.

DDT Dichlorodiphenyltrichloroethane, a chlorinated hydrocarbon that has been widely used as an insecticide but is now banned in some countries.

death rate See *crude death rate*.

debt-for-nature swap Agreement in which a certain amount of foreign debt is canceled in exchange for local currency investments that will improve natural resource management or protect certain areas in the debtor country from environmentally harmful development.

deciduous plants Trees, such as oaks and maples, and other plants that survive during dry or cold seasons by shedding their leaves. Compare *coniferous trees, succulent plants*.

decomposer Organism that digests parts of dead organisms, and cast-off fragments and wastes of living organisms by breaking down the complex organic molecules in those materials into simpler inorganic compounds and then absorbing the soluble nutrients. Producers return most of these chemicals to the soil and water for reuse. Decomposers consist of various bacteria and fungi. Compare *consumer, detritivore, producer*.

defendant The party in a court case being charged with creating a harm. See *plaintiff* and *civil suit*.

deforestation Removal of trees from a forested area.

degree of urbanization Percentage of the population in the world, or in a country, living in urban areas. Compare *urban growth*.

democracy Government by the people through their elected officials and appointed representatives. In a *constitutional democracy*, a constitution provides the basis of government authority and puts restraints on government power through free elections and freely expressed public opinion.

demographic transition Hypothesis that countries, as they become industrialized, have declines in death rates followed by declines in birth rates.

density Mass per unit volume.

depletion time The time it takes to use a certain fraction (usually 80%) of the known or estimated supply of a nonrenewable resource at an assumed rate of use. Finding and extracting the remaining 20% usually costs more than it is worth.

desalination Purification of saltwater or brackish (slightly salty) water by removal of dissolved salts.

desert Biome in which evaporation exceeds precipitation and the average amount of precipitation is less than 25 centimeters (10 inches) per year. Such areas have little vegetation or have widely spaced, mostly low vegetation. Compare *forest, grassland*.

desertification Conversion of rangeland, rain-fed cropland, or irrigated cropland to desertlike land, with a drop in agricultural productivity of 10% or more. It usually is caused by a combination of overgrazing, soil erosion, prolonged drought, and climate change.

detritivore Consumer organism that feeds on detritus, parts of dead organisms, and cast-off fragments and wastes of living organisms. Examples include earthworms, termites, and crabs. Compare *decomposer*.

detritus Parts of dead organisms and cast-off fragments and wastes of living organisms.

detritus feeder See *detritivore*.

deuterium (D; hydrogen-2) Isotope of the element hydrogen, with a nucleus containing one proton and one neutron and a mass number of 2.

developed country See *more-developed country*.

developing country See *less-developed country*.

dieback Sharp reduction in the population of a species when its numbers exceed the carrying capacity of its habitat. See *carrying capacity*.

differential reproduction Phenomenon in which individuals with adaptive genetic traits produce more living offspring than do individuals without such traits. See *natural selection*.

dioxins Family of 75 chlorinated hydrocarbon compounds formed as unwanted by-products in chemical reactions involving chlorine and hydrocarbons, usually at high temperatures.

discount rate An estimate of a resource's future economic value compared to its present value; based on the idea that having something today may be worth more than it will be in the future.

dissolved oxygen (DO) content Amount of oxygen gas (O_2) dissolved in a given volume of water at a particular temperature and pressure, often expressed as a concentration in parts of oxygen per million parts of water.

disturbance An event that disrupts an ecosystem or community. Examples of *natural disturbances* include fires, hurricanes, tornadoes, droughts, and floods. Examples of *human-caused disturbances* include deforestation, overgrazing, and plowing.

divergent plate boundary Area where the earth's lithospheric plates move apart in opposite directions. Compare *convergent plate boundary, transform fault*.

DNA (deoxyribonucleic acid) Large molecules in the cells of living organisms that carry genetic information.

domesticated species Wild species tamed or genetically altered by crossbreeding for use by humans for food (cattle, sheep, and food crops), as pets (dogs and cats), or for enjoyment (animals in zoos and plants in botanical gardens). Compare *wild species*.

dose Amount of a potentially harmful substance an individual ingests, inhales, or absorbs through the skin. Compare *response*. See *dose-response curve, median lethal dose*.

dose-response curve Plot of data showing the effects of various doses of a toxic agent on a group of test organisms. See *dose, median lethal dose, response*.

doubling time Time it takes (usually in years) for the quantity of something growing exponentially to double. It can be calculated by dividing the annual percentage growth rate into 70.

drainage basin See *watershed*.

drift-net fishing Catching fish in huge nets that drift in the water.

drought Condition in which an area does not get enough water because of lower-than-normal precipitation or higher-than-normal temperatures that increase evaporation.

earthquake Shaking of the ground resulting from the fracturing and displacement of subsurface rock, which produces a fault, or from subsequent movement along the fault.

ecological diversity The variety of forests, deserts, grasslands, oceans, streams, lakes, and other biological communities interacting with one another and with their nonliving environment. See *biodiversity*. Compare *functional diversity, genetic diversity, species diversity*.

ecological footprint Amount of biologically productive land and water needed to supply a population with the renewable resources it uses and to absorb or dispose of the wastes from such resource use. It is a measure of the average environmental impact of populations in different countries and areas. See *per capita ecological footprint*.

ecological niche Total way of life or role of a species in an ecosystem. It includes all physical, chemical, and biological conditions that a species needs to live and reproduce in an ecosystem. See *fundamental niche, realized niche*.

ecological restoration Deliberate alteration of a degraded habitat or ecosystem to restore as much of its ecological structure and function as possible.

ecological succession Process in which communities of plant and animal species in a particular area are replaced over time by a series of different and often more complex communities. See *primary ecological succession, secondary ecological succession*.

ecological tipping point Point at which an environmental problem reaches a threshold level, which causes an often irreversible shift in the behavior of a natural system.

ecologist Biological scientist who studies relationships between living organisms and their environment.

ecology Biological science that studies the relationships between living organisms and their environment; study of the structure and functions of nature.

economic depletion Exhaustion of 80% of the estimated supply of a nonrenewable resource. Finding, extracting, and processing the remaining 20% usually costs more than it is worth. May also apply to the depletion of a renewable resource, such as a fish or tree species.

economic development Improvement of human living standards by economic growth. Compare *economic growth, environmentally sustainable economic development*.

economic growth Increase in the capacity to provide people with goods and services; an increase in gross domestic product (GDP). Compare *economic development, environmentally sustainable economic development*. See *gross domestic product*.

economic resources Natural resources, capital goods, and labor used in an economy to produce material goods and services. See *natural resources*.

economics Social science that deals with the production, distribution, and consumption of goods and services to satisfy people's needs and wants.

economic system Method that a group of people uses to choose which goods and services to produce, how to produce them, how much to produce, and how to distribute them to people.

economy System of production, distribution, and consumption of economic goods.

ecosphere See *biosphere*.

ecosystem One or more communities of different species interacting with one another and with the chemical and physical factors making up their nonliving environment.

ecosystem services Natural services or natural capital that support life on the earth and are essential to the quality of human life and the functioning of the world's economies. Examples are the chemical cycles, natural pest control, and natural purification of air and water. See *natural resources*.

electromagnetic radiation Forms of kinetic energy traveling as electromagnetic waves. Examples include radio waves, TV waves, microwaves, infrared radiation, visible light, ultraviolet radiation, X-rays, and gamma rays. Compare *ionizing radiation, nonionizing radiation*.

electron (e) Tiny particle moving around outside the nucleus of an atom. Each electron has one unit of negative charge and almost no mass. Compare *neutron, proton*.

element Chemical, such as hydrogen (H), iron (Fe), sodium (Na), carbon (C), nitrogen (N), or oxygen (O), whose distinctly different atoms serve as the basic building blocks of all matter. Two or more elements combine to form the compounds that make up most of the world's matter. Compare *compound*.

elevation Distance above sea level.

emigration Movement of people out of a specific geographic area. Compare *immigration, migration*.

endangered species Wild species with so few individual survivors that the species could soon become extinct in all or most of its natural range. Compare *threatened species*.

endemic species Species that is found in only one area. Such species are especially vulnerable to extinction.

energy Capacity to do work by performing mechanical, physical, chemical, or electrical tasks or to cause a heat transfer between two objects at different temperatures.

energy conservation Reducing or eliminating the unnecessary waste of energy.

energy efficiency Percentage of the total energy input that does useful work and is not converted into low-quality, generally useless heat in an energy conversion system or process. See *energy quality, net energy*. Compare *material efficiency*.

energy productivity See *energy efficiency*.

energy quality Ability of a form of energy to do useful work. High-temperature heat and the chemical energy in fossil fuels and nuclear fuels are concentrated high-quality energy. Low-quality energy such as low-temperature heat is dispersed or diluted and cannot do much useful work. See *high-quality energy, low-quality energy*.

environment All external conditions, factors, matter, and energy, living and nonliving, that affect any living organism or other specified system.

environmental degradation Depletion or destruction of a potentially renewable resource such as soil, grassland, forest, or wildlife that is used faster than it is naturally replenished. If such use continues, the resource becomes nonrenewable (on a human time scale) or nonexistent (extinct). See also *sustainable yield.*

environmental ethics Human beliefs about what is right or wrong with how we treat the environment.

environmentalism Social movement dedicated to protecting the earth's life-support systems for us and other species.

environmentalist Person who is concerned about the impacts of human activities on the environment.

environmental justice Fair treatment and meaningful involvement of all people, regardless of race, color, sex, national origin, or income, with respect to the development, implementation, and enforcement of environmental laws, regulations, and policies.

environmental law A body of laws and treaties that broadly define what is acceptable environmental behavior for individuals, groups, businesses, and nations.

environmentally sustainable economic development Development that meets the basic needs of the current generations of humans and other species without preventing future generations of humans and other species from meeting their basic needs. It is the economic component of an *environmentally sustainable society.* Compare *economic development, economic growth.*

environmentally sustainable society Society that meets the current and future needs of its people for basic resources in a just and equitable manner without compromising the ability of future generations of humans and other species from meeting their basic needs.

environmental movement Citizens organized to demand that political leaders enact laws and develop policies to curtail pollution, clean up polluted environments, and protect unspoiled areas from environmental degradation.

environmental policy Laws, rules, and regulations related to an environmental problem that are developed, implemented, and enforced by a particular government body or agency.

environmental resistance All of the limiting factors that act together to limit the growth of a population. See *biotic potential, limiting factor.*

environmental revolution Cultural change that includes halting population growth and altering lifestyles, political and economic systems, and the way we treat the environment with the goal of living more sustainably. It requires working with the rest of nature by learning more about how nature sustains itself.

environmental science Interdisciplinary study that uses information and ideas from the physical sciences (such as biology, chemistry, and geology) with those from the social sciences and humanities (such as economics, politics, and ethics) to learn how nature works, how we interact with the environment, and how we can to help deal with environmental problems.

environmental scientist Scientist who uses information from the physical sciences and social sciences to understand how the earth works, learn how humans interact with the earth, and develop solutions to environmental problems. See *environmental science.*

environmental wisdom worldview Worldview holding that humans are part of and totally dependent on nature and that nature exists for all species, not just for us. Our success depends on learning how the earth sustains itself and integrating such environmental wisdom into the ways we think and act. Compare *frontier worldview, planetary management worldview, stewardship worldview.*

environmental worldview Set of assumptions and beliefs about how people think the world works, what they think their role in the world should be, and what they believe is right and wrong environmental behavior (environmental ethics). See *environmental wisdom worldview, frontier worldview, planetary management worldview, stewardship worldview.*

EPA U.S. Environmental Protection Agency; responsible for managing federal efforts to control air and water pollution, radiation and pesticide hazards, environmental research, hazardous waste, and solid waste disposal.

epidemiology Study of the patterns of disease or other harmful effects from exposure to toxins and diseases caused by pathogens within defined groups of people to find out why some people get sick and some do not.

epiphyte Plant that uses its roots to attach itself to branches high in trees, especially in tropical forests.

erosion Process or group of processes by which loose or consolidated earth materials, especially topsoil, are dissolved, loosened, or worn away and removed from one place and deposited in another. See *weathering.*

estuary Partially enclosed coastal area at the mouth of a river where its freshwater, carrying fertile silt and runoff from the land, mixes with salty seawater.

eukaryotic cell Cell that is surrounded by a membrane and has a distinct nucleus. Compare *prokaryotic cell.*

euphotic zone Upper layer of a body of water through which sunlight can penetrate and support photosynthesis.

eutrophication Physical, chemical, and biological changes that take place after a lake, estuary, or slow-flowing stream receives inputs of plant nutrients—mostly nitrates and phosphates—from natural erosion and runoff from the surrounding land basin. See *cultural eutrophication.*

eutrophic lake Lake with a large or excessive supply of plant nutrients, mostly nitrates and phosphates. Compare *mesotrophic lake, oligotrophic lake.*

evaporation Conversion of a liquid into a gas.

evergreen plants Plants that keep some of their leaves or needles throughout the year. Examples include cone-bearing trees (conifers) such as firs, spruces, pines, redwoods, and sequoias. Compare *deciduous plants, succulent plants.*

evolution See *biological evolution.*

exhaustible resource See *nonrenewable resource.*

exotic species See *nonnative species.*

experiment Procedure a scientist uses to study some phenomenon under known conditions. Scientists conduct some experiments in the laboratory and others in nature. The resulting scientific data or facts must be verified or confirmed by repeated observations and measurements, ideally by several different investigators.

exponential growth Growth in which some quantity, such as population size or economic output, increases at a constant rate per unit of time. An example is the growth sequence 2, 4, 8, 16, 32, 64, and so on, which increases by 100% at each interval. When the increase in quantity over time is plotted, this type of growth yields a curve shaped like the letter J. Compare *linear growth.*

external benefit Beneficial social effect of producing and using an economic good that is not included in the market price of the good. Compare *external cost, full cost.*

external cost Harmful environmental, economic, or social effect of producing and using an economic good that is not included in the market price of the good. Compare *external benefit, full cost, internal cost.*

extinction See *biological extinction.*

extinction rate Percentage or number of species that go extinct within a certain period of time such as a year.

family planning Providing information, clinical services, and contraceptives to help people choose the number and spacing of children they want to have.

famine Widespread malnutrition and starvation in a particular area because of a shortage of food, usually caused by drought, war, flood, earthquake, or other catastrophic events that disrupt food production and distribution.

feedback Any process that increases (positive feedback) or decreases (negative feedback) a change to a system.

feedback loop Occurs when an output of matter, energy, or information is fed back into the system as an input and leads to changes in that system. See *positive feedback loop* and *negative feedback loop.*

feedlot Confined outdoor or indoor space used to raise hundreds to thousands of domesticated livestock.

fermentation See *anaerobic respiration.*

fertility rate Number of children born to an average woman in a population during her lifetime. Compare *replacement-level fertility.*

fertilizer Substance that adds inorganic or organic plant nutrients to soil and improves its ability to grow crops, trees, or other vegetation. See *commercial inorganic fertilizer, organic fertilizer.*

first law of thermodynamics Whenever energy is converted from one form to another in a physical or chemical change, no energy is created or destroyed, but energy can be changed from one form to another; you cannot get more energy out of something than you put in; in terms of energy quantity, you cannot get something for nothing. This law does not apply to nuclear changes, in which large amounts of energy can be produced from small amounts of matter. See *second law of thermodynamics.*

fishery Concentration of particular aquatic species suitable for commercial harvesting in a given ocean area or inland body of water.

fish farming See *aquaculture.*

fishprint Area of ocean needed to sustain the consumption of an average person, a nation, or the world. Compare *ecological footprint.*

fissionable isotope Isotope that can split apart when hit by a neutron at the right speed and thus undergo nuclear fission. Examples include uranium-235 and plutonium-239.

floodplain Flat valley floor next to a stream channel. For legal purposes, the term often applies to any low area that has the potential for flooding, including certain coastal areas.

flows See *throughputs.*

food chain Series of organisms in which each eats or decomposes the preceding one. Compare *food web.*

food insecurity Condition under which people live with chronic hunger and malnutrition that threatens their ability to lead healthy and productive lives. Compare *food security.*

food security Condition under which every person in a given area has daily access to enough nutritious food to have an active and healthy life. Compare *food insecurity.*

food web Complex network of many interconnected food chains and feeding relationships. Compare *food chain.*

forest Biome with enough average annual precipitation to support the growth of tree species and smaller forms of vegetation. Compare *desert, grassland.*

fossil fuel Products of partial or complete decomposition of plants and animals; occurs as crude oil, coal, natural gas, or heavy oils as a result of exposure to heat and pressure in the earth's crust over millions of years. See *coal, crude oil, natural gas.*

fossils Skeletons, bones, shells, body parts, leaves, seeds, or impressions of such items that provide recognizable evidence of organisms that lived long ago.

foundation species Species that plays a major role in shaping a community by creating and enhancing a habitat that benefits other species. Compare *indicator species, keystone species, native species, nonnative species.*

fracking Freeing oil or natural gas that is tightly held in rock deposits by using perforated drilling well tubes with explosive charges to create fissures in rock and then using high pressure pumps to shoot a mixture of water, sand, and chemicals into the well to hold the rock fractures open and release the oil or natural gas, which flows back to the surface along with a mixture of water, sand, fracking chemicals, and other chemicals (some of them hazardous) that are released from the rock. See *horizontal drilling.*

free-access resource See *open-access renewable resource.*

Freons See *chlorofluorocarbons.*

freshwater Water that contains very low levels of dissolved salts.

freshwater life zones Aquatic systems where water with a dissolved salt concentration of less than 1% by volume accumulates on or flows through the surfaces of terrestrial biomes. Examples include *standing* (lentic) bodies of freshwater such as lakes, ponds, and inland wetlands and *flowing* (lotic) systems such as streams and rivers. Compare *biome.*

front The boundary between two air masses with different temperatures and densities. See *cold front, warm front.*

frontier science See *tentative science.*

frontier worldview View held by European colonists settling North America in the 1600s that the continent had vast resources and was a wilderness to be conquered by settlers clearing and planting land.

full-cost pricing Finding ways to include the harmful environmental and health costs of producing and using goods in their market prices. See *external cost, internal cost.*

functional diversity Biological and chemical processes or functions such as energy flow and matter cycling needed for the survival of species and biological communities. See *biodiversity, ecological diversity, genetic diversity, species diversity.*

fungicide Chemical that kills fungi.

game species Type of wild animal that people hunt or fish as a food source or for sport or recreation.

gamma ray Form of ionizing electromagnetic radiation with a high energy content emitted by some radioisotopes. It readily penetrates body tissues. See *alpha particle, beta particle.*

GDP See *gross domestic product.*

gene mutation See *mutation.*

gene pool Sum total of all genes found in the individuals of the population of a particular species.

generalist species Species with a broad ecological niche. They can live in many different places, eat a variety of foods, and tolerate a wide range of environmental conditions. Examples include flies, cockroaches, mice, rats, and humans. Compare *specialist species.*

genes Coded units of information about specific traits that are passed from parents to offspring during reproduction. They consist of segments of DNA molecules found in chromosomes.

gene splicing See *genetic engineering.*

genetic adaptation Changes in the genetic makeup of organisms of a species that allow the species to reproduce and gain a competitive advantage under changed environmental conditions. See *differential reproduction, evolution, mutation, natural selection.*

genetically modified organism (GMO) Organism whose genetic makeup has been altered by genetic engineering.

genetic diversity Variability in the genetic makeup among individuals within a single species. See *biodiversity.* Compare *ecological diversity, functional diversity, species diversity.*

genetic engineering Insertion of an alien gene into an organism to give it a beneficial genetic trait. Compare *artificial selection, natural selection.*

genuine progress indicator (GPI) GDP plus the estimated value of beneficial transactions that meet basic needs, but in which no money changes hands, minus the estimated harmful environmental, health, and social costs of all transactions. Compare *gross domestic product.*

geographic isolation Separation of populations of a species into different areas for long periods of time.

geology Study of the earth's dynamic history. Geologists study and analyze rocks and the features and processes of the earth's interior and surface.

geosphere Earth's intensely hot core, thick mantle composed mostly of rock, and thin outer crust that contains most of the earth's rock, soil, and sediment. Compare *atmosphere, biosphere, hydrosphere.*

geothermal energy Heat transferred from the earth's underground concentrations of dry steam (steam with no water droplets), wet steam (a mixture of steam and water droplets), or hot water trapped in fractured or porous rock.

global climate change Broad term referring to long-term changes in any aspects of the earth's climate, especially temperature and precipitation. Compare *weather.*

global warming Warming of the earth's lower atmosphere (troposphere) because of increases in the concentrations of one or more greenhouse gases. It can result in climate change that can last for decades to thousands of years. See *greenhouse effect, greenhouse gases, natural greenhouse effect.*

GMO See *genetically modified organism.*

GPI See *genuine progress indicator.*

GPP See *gross primary productivity.*

grassland Biome found in regions where there is enough annual average precipitation to support the growth of grass and small plants but not enough to support large stands of trees. Compare *desert, forest.*

greenhouse effect Natural effect that releases heat in the atmosphere near the earth's surface. Water vapor, carbon dioxide, ozone, and other gases in the lower atmosphere (troposphere) absorb some of the infrared radiation (heat) radiated by the earth's surface. Their molecules vibrate and transform the absorbed energy into longer-wavelength infrared radiation in the troposphere. If the atmospheric concentrations of these greenhouse gases increase and other natural processes do not remove them, the average temperature of the lower atmosphere will increase. Compare *global warming.*

greenhouse gases Gases in the earth's lower atmosphere (troposphere) that cause the greenhouse effect. Examples include carbon dioxide, chlorofluorocarbons, ozone, methane, water vapor, and nitrous oxide.

green manure Freshly cut or still-growing green vegetation that is plowed into the soil to increase the organic matter and humus available to support crop growth. Compare *animal manure.*

green revolution Popular term for the introduction of scientifically bred or selected varieties of grain (rice, wheat, maize) that, with adequate inputs of fertilizer and water, can greatly increase crop yields.

greenwashing Deceptive practice that some businesses use to spin environmentally harmful products as green, clean, or environmentally beneficial.

gross domestic product (GDP) Annual market value of all goods and services produced by all firms and organizations, foreign and domestic, operating within a country. See *per capita GDP.* Compare *genuine progress indicator (GPI).*

gross primary productivity (GPP) Rate at which an ecosystem's producers capture and store a given amount of chemical energy as biomass in a given length of time. Compare *net primary productivity.*

ground fire Fire that burns decayed leaves or peat deep below the ground's surface. Compare *crown fire, surface fire.*

groundwater Water that sinks into the soil and is stored in slowly flowing and slowly renewed underground reservoirs called *aquifers;* underground water in the zone of saturation, below the water table. Compare *runoff, surface water.*

habitat Place or type of place where an organism or population of organisms lives. Compare *ecological niche.*

habitat fragmentation Breakup of a habitat into smaller pieces, usually as a result of human activities.

hazard Something that can cause injury, disease, economic loss, or environmental damage. See also *risk.*

hazardous chemical Chemical that can cause harm because it is flammable or explosive, can irritate or damage the skin or lungs (such as strong acidic or alkaline substances), or can cause allergic reactions of the immune system (allergens). See also *toxic chemical.*

hazardous waste Any solid, liquid, or containerized gas that can catch fire easily, is corrosive to skin tissue or metals, is unstable and can explode or release toxic fumes, or has harmful concentrations of one or more toxic materials that can leach out. These substances are usually by-products of manufacturing processes. See also *toxic waste.*

heat Total kinetic energy of all randomly moving atoms, ions, or molecules within a given substance, excluding the overall motion of the whole object. Heat always flows spontaneously from a warmer sample of matter to a colder sample of matter. This is one way to state the *second law of thermodynamics.* Compare *temperature.*

herbicide Chemical that kills a plant or inhibits its growth.

herbivore Plant-eating organism. Examples include deer, sheep, grasshoppers, and zooplankton. Compare *carnivore, omnivore.*

heterotroph See *consumer.*

high Air mass with a high pressure. Compare *low.*

high-grade ore Ore containing a large amount of a desired mineral. Compare *low-grade ore.*

high-input agriculture See *industrialized agriculture.*

high-quality energy Energy that is concentrated and has great ability to perform useful work. Examples include high-temperature heat and the energy in electricity, coal, oil, gasoline, sunlight, and nuclei of uranium-235. Compare *low-quality energy.*

high-quality matter Matter that is concentrated and contains a high concentration of a useful resource. Compare *low-quality matter.*

high-throughput economy Economic system in most advanced industrialized countries, in which ever-increasing economic growth is sustained by maximizing the rate at which matter and energy resources are used, with little emphasis on pollution prevention, recycling, reuse, reduction of unnecessary waste, and other forms of resource conservation. Compare *low-throughput economy, matter-recycling economy.*

high-waste economy See *high-throughput economy.*

HIPPCO Acronym used by conservation biologists for the six most important secondary causes of premature extinction: Habitat destruction, degradation, and fragmentation; Invasive (nonnative) species; Population growth (too many people consuming too many resources); Pollution; Climate change; and Overexploitation.

Holocene A geological period of relatively stable climate and other environmental conditions following the last glacial period. It began about 12,000 years ago. Compare *Anthropocene.*

horizontal drilling Method for extracting oil or natural gas from underground deposits by first drilling down and then using a flexible drilling bore to drill horizontally to gain greater access to oil and gas deposits. See *fracking.*

host Plant or animal on which a parasite feeds.

human capital People's physical and mental talents that provide labor, innovation, culture, and organization. Compare *manufactured capital, natural capital.*

human resources See *human capital.*

humus Slightly soluble residue of undigested or partially decomposed organic material in topsoil. This material helps retain water and water-soluble nutrients, which can be taken up by plant roots.

hunger See *chronic undernutrition.*

hunter–gatherers People who get their food by gathering edible wild plants and other materials and by hunting wild animals and catching fish.

hydraulic fracturing See *fracking.*

hydrocarbon Organic compound made of hydrogen and carbon atoms. The simplest hydrocarbon is methane (CH_4), the major component of natural gas.

hydroelectric power plant Structure in which the energy of falling or flowing water spins a turbine generator to produce electricity.

hydrologic cycle Biogeochemical cycle that collects, purifies, and distributes the earth's fixed supply of water from the environment to living organisms and then back to the environment.

hydroponics Form of agriculture in which farmers grow plants by exposing their roots to a nutrient-rich water solution instead of soil.

hydropower Electrical energy produced by falling or flowing water. See *hydroelectric power plant.*

hydrosphere Earth's *liquid water* (oceans, lakes, other bodies of surface water, and underground water), *frozen water* (polar ice caps, floating ice caps, and ice in soil, known as permafrost), and *water vapor* in the atmosphere. See also *hydrologic cycle.* Compare *atmosphere, biosphere, geosphere.*

igneous rock Rock formed when molten rock material (magma) wells up from the earth's interior, cools, and solidifies into rock masses. Compare *metamorphic rock, sedimentary rock.* See *rock cycle.*

immature community Community at an early stage of ecological succession. It usually has a low number of species and ecological niches and cannot capture and use energy and cycle critical nutrients as efficiently as more complex, mature communities. Compare *mature community.*

immigrant species See *nonnative species.*

immigration Migration of people into a country or area to take up permanent residence. Compare *emigration.*

indicator species Species whose decline serves as early warnings that a community or ecosystem is being degraded. Compare *foundation species, keystone species, native species, nonnative species.*

industrialized agriculture Production of large quantities of crops and livestock for domestic and foreign sale; involves use of large inputs of energy from fossil fuels (especially oil and natural gas), water, fertilizer, and pesticides. Compare *subsistence farming.*

industrial smog Type of air pollution consisting mostly of a mixture of sulfur dioxide, suspended droplets of sulfuric acid formed from some of the sulfur dioxide, and suspended solid particles. Compare *photochemical smog.*

industrial solid waste Solid waste produced by mines, factories, refineries, food growers, and businesses that supply people with goods and services. Compare *municipal solid waste.*

inertia See *persistence.*

inexhaustible resource See *perpetual resource.* Compare *nonrenewable resource, renewable resource.*

infant mortality rate Number of babies out of every 1,000 born each year who die before their first birthday.

infectious disease Disease caused when a pathogen such as a bacterium, virus, or parasite invades the body and multiplies in its cells and tissues. Examples are flu, HIV, malaria, tuberculosis, and measles. See *transmissible disease.* Compare *nontransmissible disease.*

infiltration Downward movement of water through soil.

inherent value See *intrinsic value.*

inland wetland Land away from the coast, such as a swamp, marsh, or bog, that is covered all or part of the time with freshwater. Compare *coastal wetland.*

inorganic compounds All compounds not classified as organic compounds. See *organic compounds.*

inorganic fertilizer See *commercial inorganic fertilizer.*

input Matter, energy, or information entering a system. Compare *output, throughput.*

input pollution control See *pollution prevention*.

insecticide Chemical that kills insects.

instrumental value Value of an organism, species, ecosystem, or the earth's biodiversity based on its usefulness to humans. Compare *intrinsic value*.

integrated pest management (IPM) Combined use of biological, chemical, and cultivation methods in proper sequence and timing to keep the size of a pest population below the level that causes economically unacceptable loss of a crop or livestock animal.

integrated waste management Variety of strategies for both waste reduction and waste management designed to deal with the solid wastes we produce.

intercropping Growing two or more different crops at the same time on a plot. For example, a carbohydrate-rich grain that depletes soil nitrogen and a protein-rich legume that adds nitrogen to the soil may be intercropped. Compare *monoculture, polyculture*.

internal cost Direct cost paid by the producer and the buyer of an economic good. Compare *external benefit, external cost, full cost*.

interspecific competition Attempts by members of two or more species to use the same limited resources in an ecosystem. See *competition, intraspecific competition*.

intertidal zone The area of shoreline between low and high tides.

intraspecific competition Attempts by two or more organisms of a single species to use the same limited resources in an ecosystem. See *competition, interspecific competition*.

intrinsic rate of increase (*r*) Rate at which a population could grow if it had unlimited resources. Compare *environmental resistance*.

intrinsic value Value of an organism, species, ecosystem, or the earth's biodiversity based on its existence, regardless of whether it has any usefulness to humans. Compare *instrumental value*.

invasive species See *nonnative species*.

inversion See *temperature inversion*.

invertebrates Animals that have no backbones. Compare *vertebrates*.

ion Atom or group of atoms with one or more positive (+) or negative (−) electrical charges. Examples are Na^+ and Cl^-. Compare *atom, molecule*.

ionizing radiation Fast-moving alpha or beta particles or high-energy radiation (gamma rays) emitted by radioisotopes. They have enough energy to dislodge one or more electrons from the atoms they hit, thereby forming charged ions in tissue that can react with and damage living tissue. Compare *nonionizing radiation*.

IPM See *integrated pest management*.

irrigation Mix of methods used to supply water to crops by artificial means.

isotopes Two or more forms of a chemical element that have the same number of protons but different mass numbers because they have different numbers of neutrons in their nuclei.

J-shaped curve Curve with a shape similar to that of the letter J; can represent prolonged exponential growth. See *exponential growth*.

junk science See *unreliable science*.

kerogen Solid, waxy mixture of hydrocarbons found in oil shale rock. Heating the rock to high temperatures causes the kerogen to vaporize. The vapor is condensed, purified, and then sent to a refinery to produce gasoline, heating oil, and other products. See also *oil shale, shale oil*.

keystone species Species that play roles affecting many other organisms in an ecosystem. Compare *foundation species, indicator species, native species, nonnative species*.

kilocalorie (kcal) Unit of energy equal to 1,000 calories. See *calorie*.

kilowatt (kW) Unit of electrical power equal to 1,000 watts. See *watt*.

kinetic energy Energy that matter has because of its mass and speed, or velocity. Compare *potential energy*.

lake Large natural body of standing freshwater formed when water from precipitation, land runoff, or groundwater flow fills a depression in the earth created by glaciation, earth movement, volcanic activity, or a giant meteorite. See *eutrophic lake, mesotrophic lake, oligotrophic lake*.

land degradation Decrease in the ability of land to support crops, livestock, or wild species in the future as a result of natural or human-induced processes.

landfill See *sanitary landfill*.

land-use planning Planning to determine the best present and future uses of each parcel of land.

latitude Distance from the equator. Compare *altitude*.

law of conservation of energy See *first law of thermodynamics*.

law of conservation of matter In any physical or chemical change, matter is neither created nor destroyed but merely changed from one form to another; in physical and chemical changes, existing atoms are rearranged into different spatial patterns (physical changes) or different combinations (chemical changes).

law of nature See *scientific law*.

law of tolerance Existence, abundance, and distribution of a species in an ecosystem are determined by whether the levels of one or more physical or chemical factors fall within the range tolerated by the species. See *threshold effect*.

LD50 See *median lethal dose*.

LDC See *less-developed country*. Compare more-developed country.

leaching Process in which various chemicals in upper layers of soil are dissolved and carried to lower layers and, in some cases, to groundwater.

less-developed country Country that has low to moderate industrialization and low to moderate per capita GDP. Most are located in Africa, Asia, and Latin America. Compare *more-developed country*.

life-cycle cost Initial cost plus lifetime operating costs of an economic good. Compare *full cost*.

life expectancy Average number of years a newborn infant can be expected to live.

limiting factor Single factor that limits the growth, abundance, or distribution of the population of a species in an ecosystem. See *limiting factor principle*.

limiting factor principle Too much or too little of any abiotic factor can limit or prevent growth of a population of a species in an ecosystem, even if all other factors are at or near the optimal range of tolerance for the species.

linear growth Growth in which a quantity increases by some fixed amount during each unit of time. An example is growth that increases by 2 units in the sequence 2, 4, 6, 8, 10, and so on. Compare *exponential growth*.

liquefied natural gas (LNG) Natural gas converted to liquid form by cooling it to a very low temperature.

liquefied petroleum gas (LPG) Mixture of liquefied propane (C_3H_8) and butane (C_4H_{10}) gas removed from natural gas and used as a fuel.

lithosphere Outer shell of the earth, composed of the crust and the rigid, outermost part of the mantle outside the asthenosphere; material found in the earth's plates. See *crust, geosphere, mantle*.

LNG See *liquefied natural gas*.

lobbying Process in which individuals or groups use public pressure, personal contacts, and political action to persuade legislators to vote or act in their favor.

logistic growth Pattern in which exponential population growth occurs when the population is small, and population growth decreases steadily with time as the population approaches the carrying capacity. See *S-shaped curve*.

low Air mass with a low pressure. Compare *high*.

low-grade ore Ore containing a small amount of a desired mineral. Compare *high-grade ore*.

low-input agriculture See *sustainable agriculture*.

low-quality energy Energy that is dispersed and has little ability to do useful work. An example is low-temperature heat. Compare *high-quality energy*.

low-quality matter Matter that is dilute or dispersed or contains a low concentration of a useful resource. Compare *high-quality matter*.

low-throughput economy Economy based on working with nature by recycling and reusing discarded matter; preventing pollution; conserving matter and energy resources by reducing unnecessary waste and use; and building things that are easy to recycle, reuse, and repair. Compare *high-throughput economy, matter-recycling economy*.

low-waste economy See *low-throughput economy*.

LPG See *liquefied petroleum gas*.

magma Molten rock below the earth's surface.

malnutrition See *chronic malnutrition*.

mangrove swamps Swamps found on the coastlines in warm tropical climates. They are dominated by mangrove trees, any of about 55 species of trees and shrubs that can live partly submerged in the salty environment of coastal swamps.

mantle Zone of the earth's interior between its core and its crust. Compare *core, crust*. See *geosphere, lithosphere*.

manufactured capital See *manufactured resources*.

manufactured inorganic fertilizer See *commercial inorganic fertilizer*.

manufactured resources Manufactured items made from natural resources and used to produce and distribute economic goods and services bought by consumers. They include tools, machinery, equipment, factory buildings, and transportation and distribution facilities. Compare *human resources, natural resources*.

manure See *animal manure, green manure*.

marine life zone See *saltwater life zone*.

mass Amount of material in an object.

mass extinction Catastrophic, widespread, often global event in which major groups of species are wiped out over a short time compared with normal (background) extinctions. Compare *background extinction*.

mass number Sum of the number of neutrons (n) and the number of protons (p) in the nucleus of an atom. It gives the approximate mass of that atom. Compare *atomic number*.

mass transit Buses, trains, trolleys, and other forms of transportation that carry large numbers of people.

material efficiency Total amount of material needed to produce each unit of goods or services. Also called *resource productivity*. Compare *energy efficiency*.

matter Anything that has mass (the amount of material in an object) and takes up space. On the earth, where gravity is present, we weigh an object to determine its mass.

matter quality Measure of how useful a matter resource is, based on its availability and concentration. See *high-quality matter, low-quality matter*.

matter-recycling-and-reuse economy Economy that emphasizes recycling the maximum amount of all resources that can be recycled and reused. The goal is to allow economic growth to continue without depleting matter resources and without producing excessive pollution and environmental degradation. Compare *high-throughput economy, low-throughput economy*.

mature community Fairly stable, self-sustaining community in an advanced stage of ecological succession; usually has a diverse array of species and ecological niches; captures and uses energy and cycles critical chemicals more efficiently than simpler, immature communities. Compare *immature community*.

maximum sustainable yield See *sustainable yield*.

MDC See *more-developed country*.

median lethal dose (LD50) Amount of a toxic material per unit of body weight of test animals that kills half the test population in a certain time.

megacity City with 10 million or more people.

meltdown Melting of the highly radioactive core of a nuclear reactor.

mesotrophic lake Lake with a moderate supply of plant nutrients. Compare *eutrophic lake, oligotrophic lake*.

metabolism Ability of a living cell or organism to capture and transform matter and energy from its environment to supply its needs for survival, growth, and reproduction.

metamorphic rock Rock produced when a preexisting rock is subjected to high temperatures (which may cause it to melt partially), high pressures, chemically active fluids, or a combination of these agents. Compare *igneous rock, sedimentary rock*. See *rock cycle*.

metastasis Spread of malignant (cancerous) cells from a tumor to other parts of the body.

metropolitan area See *urban area*.

microorganisms Organisms such as bacteria that are so small that it takes a microscope to see them.

micropower systems Systems of small-scale decentralized units that generate 1–10,000 kilowatts of electricity. Examples include microturbines, fuel cells, wind turbines, and household solar-cell panels and solar-cell roofs.

migration Movement of people into and out of specific geographic areas. Compare *emigration* and *immigration*.

mineral Any naturally occurring inorganic substance found in the earth's crust as a crystalline solid. See *mineral resource*.

mineral resource Concentration of naturally occurring solid, liquid, or gaseous material in or on the earth's crust in a form and amount such that extracting and converting it into useful materials or items is currently or potentially profitable. Mineral resources are classified as *metallic* (such as iron and tin ores) or *nonmetallic* (such as fossil fuels, sand, and salt).

minimum-tillage farming See *conservation-tillage farming*.

mixture Combination of one or more elements and compounds.

model Approximate representation or simulation of a system being studied.

molecule Combination of two or more atoms of the same chemical element (such as O_2) or different chemical elements (such as H_2O) held together by chemical bonds. Compare *atom, ion*.

monoculture Cultivation of a single crop, usually on a large area of land. Compare *polyculture*.

more-developed country Country that is highly industrialized and has a high per capita GDP. Compare *less-developed country*.

mountaintop removal mining Type of surface mining that uses explosives, massive power shovels, and large machines called draglines to remove the top of a mountain and expose seams of coal underneath a mountain. Compare *area strip mining, contour strip mining*.

MSW See *municipal solid waste*.

multiple use Use of an ecosystem such as a forest for a variety of purposes such as timber harvesting, wildlife habitat, watershed protection, and recreation. Compare *sustainable yield*.

municipal solid waste (MSW) Solid materials discarded by homes and businesses in or near urban areas. See *solid waste*. Compare *industrial solid waste*.

mutagen Chemical or form of radiation that causes inheritable changes (mutations) in the DNA molecules in genes. See *carcinogen, mutation, teratogen*.

mutation Random change in DNA molecules making up genes that can alter anatomy, physiology, or behavior in offspring. See *mutagen*.

mutualism Type of species interaction in which both participating species generally benefit. Compare *commensalism*.

nanotechnology The use of science and engineering to manipulate and create materials out of atoms and molecules at the ultra-small scale of less than 100 nanometers. A nanometer is one-millionth of a meter.

native species Species that normally live and thrive in a particular ecosystem. Compare *foundation species, indicator species, keystone species, nonnative species*.

natural capital Natural resources and natural services that keep us and other species alive and support our economies. See *natural resources, natural services*.

natural capital degradation See *environmental degradation*.

natural gas Underground deposits of gases consisting of 50–90% by weight methane gas (CH_4) and small amounts of heavier gaseous hydrocarbon compounds such as propane (C_3H_8) and butane (C_4H_{10}).

natural greenhouse effect See *greenhouse effect*.

natural income Renewable resources such as plants, animals, and soil provided by natural capital.

natural law See *scientific law*.

natural radioactive decay Nuclear change in which unstable nuclei of atoms spontaneously shoot out particles (usually alpha or beta particles) or energy (gamma rays) at a fixed rate.

natural rate of extinction See *background extinction*.

natural recharge Natural replenishment of an aquifer by precipitation, which percolates downward through soil and rock. See *recharge area*.

natural resources Materials such as air, water, and soil and energy in nature that are essential or useful to humans. See *natural capital*.

natural selection Process by which a particular beneficial gene (or set of genes) is reproduced in succeeding generations more than other genes. The result of natural selection is a population that contains a greater proportion of organisms better adapted to certain environmental conditions. See *adaptation, biological evolution, differential reproduction, mutation*.

natural services Processes of nature, such as purification of air and water and pest control, which support life and human economies. See *natural capital*.

negative feedback loop Feedback loop that causes a system to change in the opposite direction from which is it moving. Compare *positive feedback loop*.

nekton Strongly swimming organisms found in aquatic systems. Compare *benthos, plankton*.

net energy yield Total amount of useful energy available from an energy resource or energy system over its lifetime, minus the amount of energy *used* (the first energy law), *automatically wasted* (the second energy law), and *unnecessarily wasted* in finding, processing, concentrating, and transporting it to users.

net primary productivity (NPP) Rate at which all the plants in an ecosystem produce net useful chemical energy; equal to the difference between the rate at which the plants in an ecosystem produce useful chemical energy (gross primary productivity) and the rate at which they use some of that energy through cellular respiration. Compare *gross primary productivity*.

neurotoxin Chemical that can harm the human *nervous system* (brain, spinal cord, peripheral nerves).

neutral solution Water solution containing an equal number of hydrogen ions (H^+) and hydroxide ions (OH^-); water solution with a pH of 7. Compare *acid solution, basic solution*.

neutron (n) Elementary particle in the nuclei of all atoms (except hydrogen-1). It has a relative mass of 1 and no electric charge. Compare *electron, proton*.

niche See *ecological niche*.

nitric oxide (NO) Colorless gas that forms when nitrogen and oxygen gas in air react at the high-combustion temperatures in automobile engines and coal-burning plants. Lightning and certain bacteria in soil and water also produce NO as part of the *nitrogen cycle*.

nitrogen cycle Cyclic movement of nitrogen in different chemical forms from the environment to organisms and then back to the environment.

nitrogen dioxide (NO_2) Reddish-brown gas formed when nitrogen oxide reacts with oxygen in the air.

nitrogen fixation Conversion of atmospheric nitrogen gas, by lightning, bacteria, and cyanobacteria, into forms useful to plants; it is part of the nitrogen cycle.

nitrogen oxides (NO_x) See *nitric oxide* and *nitrogen dioxide*.

noise pollution Any unwanted, disturbing, or harmful sound that impairs or interferes with hearing, causes stress, hampers concentration and work efficiency, or causes accidents.

nondegradable pollutant Material that is not broken down by natural processes. Examples include the toxic elements lead and mercury. Compare *biodegradable pollutant*.

nonionizing radiation Forms of radiant energy such as radio waves, microwaves, infrared light, and ordinary light that do not have enough energy to cause ionization of atoms in living tissue. Compare *ionizing radiation*.

nonnative species Species that migrate into an ecosystem or are deliberately or accidentally introduced into an ecosystem by humans. Compare *native species*.

nonpoint sources Broad and diffuse areas, rather than points, from which pollutants enter bodies of surface water or air. Examples include runoff of chemicals and sediments from cropland, livestock feedlots, logged forests, urban streets, parking lots, lawns, and golf courses. Compare *point source*.

nonrenewable energy Energy from resources that can be depleted and are not replenished by natural processes within a human time scale. Examples are energy produced by the burning of oil, coal, and natural gas, and nuclear energy released when the nuclei of heavy elements such as uranium are split apart (nuclear fission) or when the nuclei of light atoms such as hydrogen are forced together (nuclear fusion). Compare *renewable energy*.

nonrenewable resource Resource that exists in a fixed amount (stock) in the earth's crust and has the potential for renewal by geological, physical, and chemical processes taking place over hundreds of millions to billions of years. Examples include copper, aluminum, coal, and oil. We classify these resources as exhaustible because we are extracting and using them at a much faster rate than they are formed. Compare *renewable resource*.

nontransmissible disease Disease that is not caused by living organisms and does not spread from one person to another. Examples include most cancers, diabetes, cardiovascular disease, and malnutrition. Compare *transmissible disease*.

no-till farming See *conservation-tillage farming*.

NPP See *net primary productivity*.

nuclear change Process in which nuclei of certain isotopes spontaneously change, or are forced to change, into one or more different isotopes. The three principal types of nuclear change are natural radioactivity, nuclear fission, and nuclear fusion. Compare *chemical change, physical change*.

nuclear energy Energy released when atomic nuclei undergo a nuclear reaction such as the spontaneous emission of radioactivity, nuclear fission, or nuclear fusion.

nuclear fission Nuclear change in which the nuclei of certain isotopes with large mass numbers (such as uranium-235 and plutonium-239) are split apart into lighter nuclei when struck by a neutron. This process releases more neutrons and a large amount of energy. Compare *nuclear fusion*.

nuclear fuel cycle Includes the mining of uranium, processing and enriching the uranium to make nuclear fuel, using it in the reactor, safely storing the resulting highly radioactive wastes for thousands of years until their radioactivity falls to safe levels, and retiring the highly radioactive nuclear plant by taking it apart and storing its high- and moderate-level radioactive material safely for thousands of years.

nuclear fusion Nuclear change in which two nuclei of isotopes of elements with a low mass number (such as hydrogen-2 and hydrogen-3) are forced together at extremely high temperatures until they fuse to form a heavier nucleus (such as helium-4). This process releases a large amount of energy. Compare *nuclear fission*.

nucleus Extremely tiny center of an atom, making up most of the atom's mass. It contains one or more positively charged protons and one or more neutrons with no electrical charge (except for a hydrogen-1 atom, which has one proton and no neutrons in its nucleus).

nutrient Any chemical an organism must take in to live, grow, or reproduce.

nutrient cycle See *biogeochemical cycle*.

nutrient cycling The circulation of chemicals necessary for life, from the environment (mostly from soil and water) through organisms and back to the environment.

ocean acidification Increasing levels of acid in world's oceans due to their absorption of much of the CO_2 emitted into the atmosphere by human activities, especially the burning of carbon-containing fossil fuels. The CO_2 reacts with ocean water to form a weak acid and decreases the levels of carbonate ions (CO_3^{2-}) needed to form coral and the shells and skeletons of organisms such as crabs, oysters, and some phytoplankton.

ocean currents Mass movements of surface water produced by prevailing winds blowing over the oceans.

oil See *crude oil*.

oil reserves See *proven oil reserves*.

oil sand See *tar sand*.

oil shale Fine-grained rock containing various amounts of kerogen, a solid, waxy mixture of hydrocarbon compounds. Heating the rock to high temperatures converts the kerogen into a vapor that can be condensed to form a slow-flowing heavy oil called shale oil. See *kerogen, shale oil*.

old-growth forest Virgin and old, second-growth forests containing trees that are often hundreds—sometimes thousands—of years old. Examples include forests of Douglas fir, western hemlock, giant sequoia, and coastal redwoods in the western United States. Compare *second-growth forest, tree plantation*.

oligotrophic lake Lake with a low supply of plant nutrients. Compare *eutrophic lake, mesotrophic lake*.

omnivore Animal that can use both plants and other animals as food sources. Examples include pigs, rats, cockroaches, and humans. Compare *carnivore, herbivore*.

open-access renewable resource Renewable resource owned by no one and available for use by anyone at little or no charge. Examples include clean air, underground water supplies, the open ocean and its fish, and the ozone layer. Compare *common-property resource*.

open dump Fields or holes in the ground where garbage is deposited and sometimes covered with soil. They are rare in developed countries, but are widely used in many developing countries, especially to handle wastes from megacities. Compare *sanitary landfill*.

open-pit mining Removing minerals such as gravel, sand, and metal ores by digging them out of the earth's surface and leaving an open pit behind. Compare *area strip mining, contour strip mining, mountaintop removal, subsurface mining*.

open sea Part of an ocean that lies beyond the continental shelf. Compare *coastal zone.*

ore Part of a metal-yielding material that can be economically extracted from a mineral; typically containing two parts: the ore mineral, which contains the desired metal, and waste mineral material (gangue). See *high-grade ore, low-grade ore.*

organic agriculture Growing crops with limited or no use of synthetic pesticides and synthetic fertilizers; genetically modified crops, raising livestock without use of synthetic growth regulators and feed additives; and using organic fertilizer (manure, legumes, compost) and natural pest controls (bugs that eat harmful bugs, plants that repel bugs and environmental controls such as crop rotation). See *sustainable agriculture.*

organic compounds Compounds containing carbon atoms combined with each other and with atoms of one or more other elements such as hydrogen, oxygen, nitrogen, sulfur, phosphorus, chlorine, and fluorine. All other compounds are called *inorganic compounds.*

organic farming See *organic agriculture* and *sustainable agriculture.*

organic fertilizer Organic material such as animal manure, green manure, and compost applied to cropland as a source of plant nutrients. Compare *commercial inorganic fertilizer.*

organism Any form of life.

output Matter, energy, or information leaving a system. Compare *input, throughput.*

output pollution control See *pollution cleanup.*

overburden Layer of soil and rock overlying a mineral deposit. Surface mining removes this layer.

overfishing Harvesting so many fish of a species, especially immature individuals, that not enough breeding stock is left to replenish the species and it becomes unprofitable to harvest them.

overgrazing Destruction of vegetation when too many grazing animals feed too long on a specific area of pasture or rangeland and exceed the carrying capacity of a rangeland or pasture area.

overnutrition Diet so high in calories, saturated (animal) fats, salt, sugar, and processed foods, and so low in vegetables and fruits that the consumer runs a high risk of developing diabetes, hypertension, heart disease, and other health hazards. Compare *malnutrition, undernutrition.*

oxygen-demanding wastes Organic materials that are usually biodegraded by aerobic (oxygen-consuming) bacteria if there is enough dissolved oxygen in the water.

ozone (O$_3$) Colorless and highly reactive gas and a major component of photochemical smog. Also found in the ozone layer in the stratosphere. See *photochemical smog.*

ozone depletion Decrease in concentration of ozone (O$_3$) in the stratosphere. See *ozone layer.*

ozone layer Layer of gaseous ozone (O$_3$) in the stratosphere that protects life on earth by filtering out most harmful ultraviolet radiation from the sun.

PANs Peroxyacyl nitrates; group of chemicals found in photochemical smog.

parasite Consumer organism that lives on or in, and feeds on, a living plant or animal, known as the host, over an extended period. The parasite draws nourishment from and gradually weakens its host; it may or may not kill the host. See *parasitism.*

parasitism Interaction between species in which one organism, called the parasite, preys on another organism, called the host, by living on or in the host. See *host, parasite.*

particulates Also known as suspended particulate matter (SPM); variety of solid particles and liquid droplets small and light enough to remain suspended in the air for long periods. About 62% of the SPM in outdoor air comes from natural sources such as dust, wild fires, and sea salt. The remaining 38% comes from human sources such as coal-burning electric power and industrial plants, motor vehicles, plowed fields, road construction, unpaved roads, and tobacco smoke.

parts per billion (ppb) Number of parts of a chemical found in 1 billion parts of a particular gas, liquid, or solid.

parts per million (ppm) Number of parts of a chemical found in 1 million parts of a particular gas, liquid, or solid.

parts per trillion (ppt) Number of parts of a chemical found in 1 trillion parts of a particular gas, liquid, or solid.

passive solar heating system System that, without the use of mechanical devices, captures sunlight directly within a structure and converts it into low-temperature

heat for space heating or for heating water for domestic use. Compare *active solar heating system.*

pasture Managed grassland or enclosed meadow that usually is planted with domesticated grasses or other forage to be grazed by livestock. Compare *feedlot.*

pathogen Living organism that can cause disease in another organism. Examples include bacteria, viruses, and parasites.

PCBs See *polychlorinated biphenyls.*

peak production Point in time when the pressure in an oil well drops and its rate of conventional crude oil production starts declining, usually a decade or so; for a group of wells or for a nation, the point at which all wells on average have passed peak production.

peer review Process of scientists reporting details of the methods and models they used, the results of their experiments, and the reasoning behind their hypotheses for other scientists working in the same field (their peers) to examine and criticize.

per capita ecological footprint Amount of biologically productive land and water needed to supply each person or population with the renewable resources they use and to absorb or dispose of the wastes from such resource use. It measures the average environmental impact of individuals or populations in different countries and areas. Compare *ecological footprint.*

per capita GDP Annual gross domestic product (GDP) of a country divided by its total population at midyear. It gives the average slice of the economic pie per person. Used to be called per capita gross national product (GNP). See *gross domestic product.* Compare *genuine progress indicator (GPI).*

per capita GDP PPP (Purchasing Power Parity) Measure of the amount of goods and services that a country's average citizen could buy in the United States.

percolation Passage of a liquid through the spaces of a porous material such as soil.

perennial Plant that can live for more than 2 years. Compare *annual.*

permafrost Perennially frozen layer of the soil that forms when the water there freezes. It is found in arctic tundra.

perpetual resource Essentially inexhaustible resource on a human time scale because it is renewed continuously. Solar energy is an example. Compare *nonrenewable resource, renewable resource.*

persistence Ability of a living system such as a grassland or forest to survive moderate disturbances. Compare *resilience.*

perverse subsidy Subsidy that leads to environmental damage or damage to human health or well-being. See *subsidy.*

pest Unwanted organism that directly or indirectly interferes with human activities.

pesticide Any chemical designed to kill or inhibit the growth of an organism that people consider undesirable. See *fungicide, herbicide, insecticide.*

petrochemicals Chemicals obtained by refining (distilling) crude oil. They are used as raw materials in manufacturing most industrial chemicals, fertilizers, pesticides, plastics, synthetic fibers, paints, medicines, and many other products.

petroleum See *crude oil.*

pH Numeric value that indicates the relative acidity or alkalinity of a substance on a scale of 0 to 14, with the neutral point at 7. Acid solutions have pH values lower than 7; basic or alkaline solutions have pH values greater than 7.

phosphorus cycle Cyclic movement of phosphorus in different chemical forms from the environment to organisms and then back to the environment.

photochemical smog Complex mixture of air pollutants produced in the lower atmosphere by the reaction of hydrocarbons and nitrogen oxides under the influence of sunlight. Especially harmful components include ozone, peroxyacyl nitrates (PANs), and various aldehydes. Compare *industrial smog.*

photosynthesis Complex process that takes place in cells of green plants. Radiant energy from the sun is used to combine carbon dioxide (CO_2) and water (H_2O) to produce oxygen (O_2), carbohydrates (such as glucose, $C_6H_{12}O_6$), and other nutrient molecules. Compare *aerobic respiration, chemosynthesis.*

photovoltaic (PV) cell Device that converts radiant (solar) energy directly into electrical energy. Also called a solar cell.

physical change Process that alters one or more physical properties of an element or a compound without changing its chemical composition. Examples include changing the size and shape of a sample of matter (crushing ice and cutting aluminum foil) and changing a sample of matter from one physical state to another (boiling and freezing water). Compare *chemical change, nuclear change.*

phytoplankton Small, drifting plants, mostly algae and bacteria, found in aquatic ecosystems. Compare *plankton, zooplankton.*

pioneer community First integrated set of plants, animals, and decomposers found in an area undergoing primary ecological succession. See *immature community, mature community.*

pioneer species First hardy species—often microbes, mosses, and lichens—that begin colonizing a site as the first stage of ecological succession. See *ecological succession, pioneer community.*

plaintiff Party in a court case bringing charges or seeking to collect damages for injuries to health or for economic loss; may also seek an injunction, by which the party being charged would be required to stop whatever action is causing harm. See *defendant* and *civil suit.*

planetary management worldview Worldview holding that humans are separate from nature, that nature exists mainly to meet our needs and increasing wants, and that we can use our ingenuity and technology to manage the earth's life-support systems, mostly for our benefit. It assumes that economic growth is unlimited. Compare *environmental wisdom worldview, stewardship worldview.*

plankton Small plant organisms (phytoplankton) and animal organisms (zooplankton) that float in aquatic ecosystems.

plantation agriculture Growing specialized crops such as bananas, coffee, and cacao in tropical developing countries, primarily for sale to developed countries.

plates See *tectonic plates.*

plate tectonics Theory of geophysical processes that explains the movements of lithospheric plates and the processes that occur at their boundaries. See *lithosphere, tectonic plates.*

point source Single identifiable source that discharges pollutants into the environment. Examples include the smokestack of a power plant or an industrial plant, drainpipe of a meatpacking plant, chimney of a house, or exhaust pipe of an automobile. Compare *nonpoint source.*

poison Chemical that adversely affects the health of a living human or animal by causing injury, illness, or death.

policies programs, and the laws and regulations through which they are enacted, that a government enforces and funds.

politics Process through which individuals and groups try to influence or control government policies and actions that affect the local, state, national, and international communities.

pollutant Particular chemical or form of energy that can adversely affect the health, survival, or activities of humans or other living organisms. See *pollution.*

pollution Undesirable change in the physical, chemical, or biological characteristics of air, water, soil, or food that can adversely affect the health, survival, or activities of humans or other living organisms.

pollution cleanup Device or process that removes or reduces the level of a pollutant after it has been produced or has entered the environment. Examples include automobile emission control devices and sewage treatment plants. Compare *pollution prevention.*

pollution prevention Device, process, or strategy used to prevent a potential pollutant from forming or entering the environment or to sharply reduce the amount entering the environment. Compare *pollution cleanup.*

polychlorinated biphenyls (PCBs) Group of 209 toxic, oily, synthetic chlorinated hydrocarbon compounds that can be biologically amplified in food chains and webs.

polyculture Complex form of intercropping in which a large number of different plants maturing at different times are planted together. See also *intercropping.* Compare *monoculture.*

population Group of individual organisms of the same species living in a particular area.

population change Increase or decrease in the size of a population. It is equal to (Births + Immigration) − (Deaths + Emigration).

population crash Dieback of a population that has used up its supply of resources, exceeding the carrying capacity of its environment. See *carrying capacity.*

population density Number of organisms in a particular population found in a specified area or volume.

population dispersion General pattern in which the members of a population are arranged throughout its habitat.

population distribution Variation of population density over a particular geographic area or volume. For example, a country has a high population density in its urban areas and a much lower population density in its rural areas.

population dynamics Major abiotic and biotic factors that tend to increase or decrease the population size and affect the age and sex composition of a species.

population size Number of individuals making up a population's gene pool.

positive feedback loop Feedback loop that causes a system to change further in the same direction. Compare *negative feedback loop*.

potential energy Energy stored in an object because of its position or the position of its parts. Compare *kinetic energy*.

poverty Inability of people to meet their basic needs for food, clothing, and shelter.

ppb See *parts per billion*.

ppm See *parts per million*.

ppt See *parts per trillion*.

prairie See *grassland*.

precautionary principle When there is significant scientific uncertainty about potentially serious harm from chemicals or technologies, decision makers should act to prevent harm to humans and the environment. See *pollution prevention*.

precipitation Water in the form of rain, sleet, hail, and snow that falls from the atmosphere onto land and bodies of water.

predation Interaction in which an organism of one species (the predator) captures and feeds on some or all parts of an organism of another species (the prey).

predator Organism that captures and feeds on some or all parts of an organism of another species (the prey).

predator–prey relationship Relationship that has evolved between two organisms, in which one organism has become the prey for the other, the latter called the predator. See *predator, prey*.

prey Organism that is killed by an organism of another species (the predator) and serves as its source of food.

primary consumer Organism that feeds on some or all parts of plants (herbivore) or on other producers. Compare *detritivore, omnivore, secondary consumer*.

primary ecological succession Ecological succession in a area without soil or bottom sediments. See *ecological succession*. Compare *secondary ecological succession*.

primary forest See *old-growth forest*.

primary pollutant Chemical that has been added directly to the air by natural events or human activities and occurs in a harmful concentration. Compare *secondary pollutant*.

primary productivity See *gross primary productivity, net primary productivity*.

primary recycling Process in which materials are recycled into new products of the same type—turning used aluminum cans into new aluminum cans, for example.

primary sewage treatment Mechanical sewage treatment in which large solids are filtered out by screens and suspended solids settle out as sludge in a sedimentation tank. Compare *secondary sewage treatment*.

principles of sustainability See *scientific principles of sustainability, social science principles of sustainability*.

probability Mathematical statement about how likely it is that something will happen.

producer Organism that uses solar energy (green plants) or chemical energy (some bacteria) to manufacture the organic compounds it needs as nutrients from simple inorganic compounds obtained from its environment. Compare *consumer, decomposer*.

prokaryotic cell Cell containing no distinct nucleus or organelles. Compare *eukaryotic cell*.

proton (p) Positively charged particle in the nuclei of all atoms. Each proton has a relative mass of 1 and a single positive charge. Compare *electron, neutron*.

proven oil reserves Identified deposits from which conventional crude oil can be extracted profitably at current prices with current technology.

PV cell See *photovoltaic cell*.

pyramid of energy flow Diagram representing the flow of energy through each trophic level in a food chain or food web. With each energy transfer, only a small part (typically 10%) of the usable energy entering one trophic level is transferred to the organisms at the next trophic level.

radiation Fast-moving particles (particulate radiation) or waves of energy (electromagnetic radiation). See *alpha particle, beta particle, gamma ray*.

radioactive decay Change of a radioisotope to a different isotope by the emission of radioactivity.

radioactive isotope See *radioisotope*.

radioactive waste Waste products of nuclear power plants, research, medicine, weapon production, or other processes involving nuclear reactions. See *radioactivity*.

radioactivity Nuclear change in which unstable nuclei of atoms spontaneously shoot out "chunks" of mass, energy, or both at a fixed rate. The three principal types of radioactivity are gamma rays and fast-moving alpha particles and beta particles.

radioisotope Isotope of an atom that spontaneously emits one or more types of radioactivity (alpha particles, beta particles, gamma rays).

rain shadow effect Low precipitation on the leeward side of a mountain when prevailing winds flow up and over a high mountain or range of high mountains, creating semiarid and arid conditions on the leeward side of a high mountain range.

rangeland Land that supplies forage or vegetation (grasses, grasslike plants, and shrubs) for grazing and browsing animals and is not intensively managed. Compare *feedlot, pasture*.

range of tolerance Range of chemical and physical conditions that must be maintained for populations of a particular species to stay alive and grow, develop, and function normally. See *law of tolerance*.

recharge area Any area of land allowing water to percolate down through it and into an aquifer. See *aquifer, natural recharge*.

reconciliation ecology Science of inventing, establishing, and maintaining habitats to conserve species diversity in places where people live, work, or play.

recycle To collect and reprocess a resource so that it can be made into new products; one of the four R's of resource use. An example is collecting aluminum cans, melting them down, and using the aluminum to make new cans or other aluminum products. See *primary recycling, secondary recycling*. Compare *reduce and reuse*.

reduce To consume less of a good or service in order to reduce one's environmental impact and to save money. Compare *recycle, refuse, reuse*.

refining Complex process in which crude oil is heated and vaporized in giant columns and separated, by use of varying boiling points, into various products such as gasoline, heating oil, and asphalt. See *petrochemicals*.

reforestation Renewal of trees and other types of vegetation on land where trees have been removed; can be done naturally by seeds from nearby trees or artificially by planting seeds or seedlings.

refuse To refrain from buying or using a good or service in order to reduce one's ecological impact and to save money. Compare *recycle, reduce, reuse*.

reliable runoff Surface runoff of water that generally can be counted on as a stable source of water from year to year. See *runoff*.

reliable science Concepts and ideas that are widely accepted by experts in a particular field of the natural or social sciences. Compare *tentative science, unreliable science*.

renewable energy Energy that comes from resources that are replenished by natural processes continually or in a relatively short time. Examples are solar energy (sunlight), wind, moving water, heat from the earth's interior (geothermal energy), firewood from trees, tides, and waves. Compare *nonrenewable energy*.

renewable resource Resource that can be replenished rapidly (hours to several decades) through natural processes as long as it is not used up faster than it is replaced. Examples include trees in forests, grasses in grasslands, wild animals, fresh surface water in lakes and streams, most groundwater, fresh air, and fertile soil. If such a resource is used faster than it is replenished, it can be depleted and converted into a nonrenewable resource. Compare *nonrenewable resource* and *perpetual resource*. See also *environmental degradation*.

replacement-level fertility rate Average number of children a couple must bear to replace themselves. The average for a country or the world usually is slightly higher than two children per couple (2.1 in the United States and 2.5 in some developing countries) mostly because some children die before reaching their reproductive years. See also *total fertility rate*.

reproduction Production of offspring by one or more parents.

reproductive isolation Long-term geographic separation of members of a particular sexually reproducing species.

reproductive potential See *biotic potential*.

reserves Resources that have been identified and from which a usable mineral can be extracted profitably at present prices with current mining or extraction technology.

reservoir Artificial lake created when a stream is dammed. See *dam*.

resilience Ability of a living system such as a forest or pond to be restored through secondary ecological succession after a severe disturbance. See *secondary ecological succession*. Compare *persistence*.

resource Anything obtained from the environment to meet human needs and wants. It can also be applied to other species.

resource partitioning Process of dividing up resources in an ecosystem so that species with similar needs (overlapping ecological niches) use the same scarce resources at different times, in different ways, or in different places. See *ecological niche*.

resource productivity See *material efficiency*.

respiration See *aerobic respiration*.

response Amount of health damage caused by exposure to a certain dose of a harmful substance or form of radiation. See *dose, dose-response curve, median lethal dose*.

restoration ecology Research and scientific study devoted to restoring, repairing, and reconstructing damaged ecosystems.

reuse To use a product over and over again in the same form. An example is collecting, washing, and refilling glass beverage bottles. One of the 4 Rs. Compare *recycle, reduce, and refuse*.

riparian zone A thin strip or patch of vegetation that surrounds a stream. These zones are very important habitats and resources for wildlife.

risk Probability that something undesirable will result from deliberate or accidental exposure to a hazard. See *risk analysis, risk assessment, risk management*.

risk analysis Identifying hazards, evaluating the nature and severity of risks associated with the hazards (*risk assessment*), ranking risks (*comparative risk analysis*), using this and other information to determine options and make decisions about reducing or eliminating risks (*risk management*), and communicating information about risks to decision makers and the public (*risk communication*).

risk assessment Process of gathering data and making assumptions to estimate short- and long-term harmful effects on human health or the environment from exposure to hazards associated with the use of a particular product or technology.

risk communication Communicating information about risks to decision makers and the public. See *risk, risk analysis*.

risk management Use of risk assessment and other information to determine options and make decisions about reducing or eliminating risks. See *risk, risk analysis, risk communication*.

rock Any solid material that makes up a large, natural, continuous part of the earth's crust. See *mineral*.

rock cycle Largest and slowest of the earth's cycles, consisting of geologic, physical, and chemical processes that form and modify rocks and soil in the earth's crust over millions of years.

rule of 70 Doubling time (in years) = 70/(percentage growth rate). See *doubling time, exponential growth*.

runoff Freshwater from precipitation and melting ice that flows on the earth's surface into nearby streams, lakes, wetlands, and reservoirs. See *reliable runoff, surface runoff, surface water*. Compare *groundwater*.

salinity Amount of various salts dissolved in a given volume of water.

salinization Accumulation of salts in soil that can eventually make the soil unable to support plant growth.

saltwater intrusion Movement of saltwater or brackish (slightly salty) water into freshwater aquifers in coastal and inland areas as groundwater is withdrawn faster than it is recharged by precipitation.

saltwater life zones Aquatic life zones associated with oceans: oceans and their accompanying bays, estuaries, coastal wetlands, shorelines, coral reefs, and mangrove forests.

sanitary landfill Waste disposal site on land in which waste is spread in thin layers, compacted, and covered with a fresh layer of clay or plastic foam each day. Compare *open dump*.

scavenger Organism that feeds on dead organisms that were killed by other organisms or died naturally. Examples include vultures, flies, and crows. Compare *detritivore*.

science Attempts to discover order in nature and use that knowledge to make predictions about what is likely to happen in nature. See *reliable science, scientific data, scientific hypothesis, scientific law, scientific methods, scientific model, scientific theory, tentative science, unreliable science.*

scientific data Facts obtained by making observations and measurements. Compare *scientific hypothesis, scientific law, scientific methods, scientific model, scientific theory.*

scientific hypothesis An educated guess that attempts to explain a scientific law or certain scientific observations. Compare *scientific data, scientific law, scientific methods, scientific model, scientific theory.*

scientific law Description of what scientists find happening in nature repeatedly in the same way, without known exception. See *first law of thermodynamics, law of conservation of matter, second law of thermodynamics.* Compare *scientific data, scientific hypothesis, scientific methods, scientific model, scientific theory.*

scientific methods The ways scientists gather data and formulate and test scientific hypotheses, models, theories, and laws. See *scientific data, scientific hypothesis, scientific law, scientific model, scientific theory.*

scientific model A simulation of complex processes and systems. Many are mathematical models that are run and tested using computers.

scientific principles of sustainability To live more sustainably we need to rely on solar energy, preserve biodiversity, and recycle the chemicals that we use. These three principles of sustainability are scientific lessons from nature based on observing how life on the earth has survived and thrived for 3.5 billion years. See *biodiversity, chemical cycling, solar energy.* Compare *social science principles of sustainability.*

scientific theory A well-tested and widely accepted scientific hypothesis. Compare *scientific data, scientific hypothesis, scientific law, scientific methods, scientific model.*

secondary consumer Organism that feeds only on primary consumers. Compare *detritivore, omnivore, primary consumer.*

secondary ecological succession Ecological succession in an area in which natural vegetation has been removed or destroyed but the soil or bottom sediment has not been destroyed. See *ecological succession.* Compare *primary ecological succession.*

secondary pollutant Harmful chemical formed in the atmosphere when a primary air pollutant reacts with normal air components or other air pollutants. Compare *primary pollutant.*

secondary recycling A process in which waste materials are converted into different products; for example, used tires can be shredded and turned into rubberized road surfacing. Compare *primary recycling.*

secondary sewage treatment Second step in most waste treatment systems in which aerobic bacteria decompose as much as 90% of degradable, oxygen-demanding organic wastes in wastewater. It usually involves bringing sewage and bacteria together in trickling filters or in the activated sludge process. Compare *primary sewage treatment.*

second-growth forest Stands of trees resulting from secondary ecological succession. Compare *old-growth forest, tree farm.*

second law of thermodynamics Whenever energy is converted from one form to another in a physical or chemical change, we end up with lower-quality or less usable energy than we started with. In any conversion of heat energy to useful work, some of the initial energy input is always degraded to lower-quality, more dispersed, less useful energy—usually low-temperature heat that flows into the environment; you cannot break even in terms of energy quality. See *first law of thermodynamics.*

sedimentary rock Rock that forms from the accumulated products of erosion and in some cases from the compacted shells, skeletons, and other remains of dead organisms. Compare *igneous rock, metamorphic rock.* See *rock cycle.*

selective cutting Cutting of intermediate-aged, mature, or diseased trees in an uneven-aged forest stand, either singly or in small groups. This encourages the growth of younger trees and maintains an uneven-aged stand. Compare *clear-cutting, strip cutting.*

septic tank Underground tank for treating wastewater from a home in rural and suburban areas. Bacteria in the tank decompose organic wastes, and the sludge settles to the bottom of the tank. The effluent flows out of the tank into the ground through a field of drainpipes.

sexual reproduction Reproduction in organisms that produce offspring by combining sex cells or *gametes* (such as ovum and sperm) from both parents. It produces offspring that have combinations of traits from their parents. Compare *asexual reproduction.*

shale oil Slow-flowing, dark brown, heavy oil obtained when kerogen in oil shale is vaporized at high temperatures and then condensed. Shale oil can be refined to yield gasoline, heating oil, and other petroleum products. See *kerogen, oil shale.*

shelterbelt See *windbreak.*

slash-and-burn agriculture Cutting down trees and other vegetation in a patch of forest, leaving the cut vegetation on the ground to dry, and then burning it. The ashes that are left add nutrients to the nutrient-poor soils found in most tropical forest areas. Crops are planted between tree stumps. Plots must be abandoned after a few years (typically 2–5 years) because of loss of soil fertility or invasion of vegetation from the surrounding forest.

sludge Gooey mixture of toxic chemicals, infectious agents, and settled solids removed from wastewater at a sewage treatment plant.

smart growth Form of urban planning that recognizes that urban growth will occur but uses zoning laws and other tools to prevent sprawl, direct growth to certain areas, protect ecologically sensitive and important lands and waterways, and develop urban areas that are more environmentally sustainable and more enjoyable places to live.

smelting Process in which a desired metal is separated from the other elements in an ore mineral.

smog Originally a combination of smoke and fog but now used to describe other mixtures of pollutants in the atmosphere. See *industrial smog, photochemical smog.*

SNG See *synthetic natural gas.*

social capital Result of getting people with different views and values to talk and listen to one another, find common ground based on understanding and trust, and work together to solve environmental and other problems.

social science principles of sustainability To live more sustainably we (1) need to include the harmful health and environmental costs of producing the goods and services in their market prices *(full-cost pricing)*, (2) learn to work together to focus on solutions to environmental problems that will benefit the largest number of people and the environment now and in the future *(win-win solutions)*, and (3) accept our responsibility to future generations to leave the planet's life-support systems in at least as good a shape as what we now enjoy *(responsibility to future generations)*.

soil Complex mixture of inorganic minerals (clay, silt, pebbles, and sand), decaying organic matter, water, air, and living organisms.

soil conservation Methods used to reduce soil erosion, prevent depletion of soil nutrients, and restore nutrients previously lost by erosion, leaching, and excessive crop harvesting.

soil erosion Movement of soil components, especially topsoil, from one place to another, usually by wind, flowing water, or both. This natural process can be greatly accelerated by human activities that remove vegetation from soil. Compare *soil conservation.*

soil horizons Horizontal zones, or layers, that make up a particular mature soil. Each horizon has a distinct texture and composition that vary with different types of soils. See *soil profile.*

soil profile Cross-sectional view of the horizons in a soil. See *soil horizon.*

soil salinization Gradual accumulation of salts in upper soil layers that can stunt crop growth, lower crop yields, and can eventually kill plants and ruin the land.

solar capital Solar energy that warms the planet and supports photosynthesis, the process that plants use to provide food for themselves and for us and other animals. This direct input of solar energy also produces indirect forms of renewable solar energy such as wind and flowing water. Compare *natural capital.*

solar cell See *photovoltaic cell.*

solar collector Device for collecting radiant energy from the sun and converting it into heat. See *active solar heating system, passive solar heating system.*

solar energy Direct radiant energy from the sun and a number of indirect forms of energy produced by the direct input of such radiant energy. Principal indirect forms of solar energy include wind, falling and flowing water (hydropower), and biomass (solar energy converted into chemical energy stored in the chemical bonds of organic compounds in trees and other plants)—none of which would exist without direct solar energy.

solar thermal system System that uses any of various methods to collect and concentrate solar energy in order to boil water and produce steam for generating electricity. Compare *solar cell.*

solid waste Any unwanted or discarded material that is not a liquid or a gas. See *industrial solid waste, municipal solid waste.*

sound science See *reliable science.*

spaceship-earth worldview View of the earth as a spaceship: a machine that we can understand, control, and change at will by using advanced technology. See *planetary management worldview.* Compare *environmental wisdom worldview, stewardship worldview.*

specialist species Species with a narrow ecological niche. They may be able to live in only one type of habitat, tolerate only a narrow range of climatic and other environmental conditions, or use only one type or a few types of food. Compare *generalist species.*

speciation Formation of two species from one species because of divergent natural selection in response to changes in environmental conditions; usually takes thousands of years. Compare *extinction.*

species Group of similar organisms, and for sexually reproducing organisms, they are a set of individuals that can mate and produce fertile offspring. Every organism is a member of a certain species.

species diversity Number of different species (species richness) combined with the relative abundance of individuals within each of those species (species evenness) in a given area. See *biodiversity, species evenness, species richness.* Compare *ecological diversity, genetic diversity.*

species equilibrium model See *theory of island biogeography.*

species evenness Degree to which comparative numbers of individuals of each of the species present in a community are similar. See *species diversity.* Compare *species richness.*

species richness Variety of species, measured by the number of different species contained in a community. See *species diversity.* Compare *species evenness.*

spoils Unwanted rock and other waste materials produced when a material is removed from the earth's surface or subsurface by mining, dredging, quarrying, or excavation.

S-shaped curve Leveling off of an exponential, J-shaped curve when a rapidly growing population reaches or exceeds the carrying capacity of its environment and ceases to grow.

statistics Mathematical tools used to collect, organize, and interpret numerical data.

statutory laws Laws developed and passed by legislative bodies such as federal and state governments. Compare *common law.*

stewardship worldview Worldview holding that we can manage the earth for our benefit but that we have an ethical responsibility to be caring and responsible managers, or *stewards,* of the earth. It calls for encouraging environmentally beneficial forms of economic growth and discouraging environmentally harmful forms. Compare *worldview, environmental wisdom worldview, planetary management worldview.*

stratosphere Second layer of the atmosphere, extending about 17–48 kilometers (11–30 miles) above the earth's surface. It contains small amounts of gaseous ozone (O_3), which filters out about 95% of the incoming harmful ultraviolet radiation emitted by the sun. Compare *troposphere.*

stream Flowing body of surface water. Examples are creeks and rivers.

strip-cropping Planting regular crops and close-growing plants, such as hay or nitrogen-fixing legumes, in alternating rows or bands to help reduce depletion of soil nutrients.

strip cutting Variation of clear-cutting in which a strip of trees is clear-cut along the contour of the land, with the corridor being narrow enough to allow natural regeneration within a few years. After regeneration, another strip is cut above the first, and so on. Compare *clear-cutting, selective cutting.*

strip mining Form of surface mining in which bulldozers, power shovels, or stripping wheels remove large chunks of the earth's surface in strips. See *area strip mining, contour strip mining, surface mining.* Compare *subsurface mining.*

subatomic particles Extremely small particles—electrons, protons, and neutrons—that make up the internal structure of atoms.

subduction zone Area in which the oceanic lithosphere is carried downward (subducted) under an island arc or continent at a convergent plate boundary. A trench ordinarily forms at the boundary between the two converging plates. See *convergent plate boundary.*

subsidence Slow or rapid sinking of part of the earth's crust that is not slope-related.

subsidy Payment intended to help a business grow and thrive; typically provided by a government in the form of a grant or tax break, in order to give the company an advantage in the marketplace over its competitors. See *perverse subsidy.*

subsistence farming See *traditional subsistence agriculture.*

subsurface mining Extraction of a metal ore or fuel resource such as coal from a deep underground deposit. Compare *surface mining.*

succession See *ecological succession, primary ecological succession, secondary ecological succession.*

succulent plants Plants, such as desert cacti, that survive in dry climates by having no leaves, thus reducing the loss of scarce water through *transpiration.* They store water and use sunlight to produce the food they need in the thick, fleshy tissue of their green stems and branches. Compare *deciduous plants, evergreen plants.*

sulfur cycle Cyclic movement of sulfur in various chemical forms from the environment to organisms and then back to the environment.

sulfur dioxide (SO₂) Colorless gas with an irritating odor. About one-third of the SO₂ in the atmosphere comes from natural sources as part of the sulfur cycle. The other two-thirds comes from human sources, mostly combustion of sulfur-containing coal in electric power and industrial plants and from oil refining and smelting of sulfide ores.

superinsulated house House that is heavily insulated and extremely airtight. Typically, active or passive solar collectors are used to heat water, and an air-to-air heat exchanger prevents buildup of excessive moisture and indoor air pollutants.

surface fire Forest fire that burns only undergrowth and leaf litter on the forest floor. Compare *crown fire, ground fire.* See *controlled burning.*

surface mining Removing soil, subsoil, and other strata and then extracting a mineral deposit found fairly close to the earth's surface. See *area strip mining, contour strip mining, mountaintop removal, open-pit mining.* Compare *subsurface mining.*

surface runoff Water flowing off the land into bodies of surface water. See *reliable runoff.*

surface water Precipitation that does not infiltrate the ground or return to the atmosphere by evaporation or transpiration. See *runoff.* Compare *groundwater.*

survivorship curve Graph showing the number of survivors in different age groups for a particular species.

suspended particulate matter See *particulates.*

sustainability Ability of earth's various systems, including human cultural systems and economies, to survive and adapt to changing environmental conditions indefinitely.

sustainability revolution Major cultural change in which people learn how to reduce their ecological footprints and live more sustainably, largely by copying nature and using the six principles of sustainability to guide their lifestyles and economies. See *principles of sustainability.*

sustainable agriculture Method of growing crops and raising livestock based on organic fertilizers, soil conservation, water conservation, biological pest control, and minimal use of nonrenewable fossil-fuel energy.

sustainable development See *environmentally sustainable economic development.*

sustainable living Taking no more potentially renewable resources from the natural world than can be replenished naturally and not overloading the capacity of the environment to cleanse and renew itself by natural processes.

sustainable society Society that manages its economy and population size without doing irreparable environmental harm by overloading the planet's ability to absorb environmental insults, replenish its resources, and sustain human and other forms of life over a specified period, indefinitely. During this period, the society satisfies the needs of its people without depleting natural resources and thereby jeopardizing the prospects of current and future generations of humans and other species.

sustainable yield (sustained yield) Highest rate at which a potentially renewable resource can be used indefinitely without reducing its available supply. See also *environmental degradation.*

synergistic interaction Interaction of two or more factors or processes so that the combined effect is greater than the sum of their separate effects.

synergy See *synergistic interaction.*

synfuels Synthetic gaseous and liquid fuels produced from solid coal or sources other than natural gas or crude oil.

synthetic natural gas (SNG) Gaseous fuel containing mostly methane produced from solid coal.

system Set of components that function and interact in some regular and theoretically predictable manner.

tailings Rock and other waste materials removed as impurities when waste mineral material is separated from the metal in an ore.

tar sand Deposit of a mixture of clay, sand, water, and varying amounts of a tarlike heavy oil known as bitumen. Bitumen can be extracted from tar sand by heating. It is then purified and upgraded to synthetic crude oil. See *bitumen.*

tectonic plates Various-sized areas of the earth's lithosphere that move slowly around with the mantle's flowing asthenosphere. Most earthquakes and volcanoes occur around the boundaries of these plates. See *lithosphere, plate tectonics.*

temperature Measure of the average speed of motion of the atoms, ions, or molecules in a substance or combination of substances at a given moment. Compare *heat.*

temperature inversion Layer of dense, cool air trapped under a layer of less dense, warm air. It prevents upward-flowing air currents from developing. In a prolonged inversion, air pollution in the trapped layer may build up to harmful levels.

tentative science Preliminary scientific data, hypotheses, and models that have not been widely tested and accepted. Compare *reliable science, unreliable science.*

teratogen Chemical, ionizing agent, or virus that causes birth defects. Compare *carcinogen, mutagen.*

terracing Planting crops on a long, steep slope that has been converted into a series of broad, nearly level terraces with short vertical drops from one to another that run along the contour of the land to retain water and reduce soil erosion.

terrestrial Pertaining to land. Compare *aquatic.*

tertiary (higher-level) consumers Animals that feed on animal-eating animals. They feed at high trophic levels in food chains and webs. Examples include hawks, lions, bass, and sharks. Compare *detritivore, primary consumer, secondary consumer.*

theory of evolution Widely accepted scientific idea that all life-forms developed from earlier life-forms. It is the way most biologists explain how life has changed over the past 3.6–3.8 billion years and why it is so diverse today.

theory of island biogeography Widely accepted scientific theory holding that the number of different species (species richness) found on an island is determined by the interactions of two factors: the rate at which new species immigrate to the island and the rate at which species become *extinct,* or cease to exist, on the island. See *species richness.*

thermal energy The energy generated and measured by heat. See *heat.*

thermal inversion See *temperature inversion.*

threatened species Wild species that is still abundant in its natural range but is likely to become endangered because of a decline in numbers. Compare *endangered species.*

threshold effect Harmful or fatal effect of a small change in environmental conditions that exceeds the limit of tolerance of an organism or population of a species. See *law of tolerance.*

throughput Rate of flow of matter, energy, or information through a system. Compare *input, output.*

time delay In a complex system, the period of time between the input of a feedback stimulus and the system's response to it. See *tipping point.*

tipping point Threshold level at which an environmental problem causes a fundamental and irreversible shift in the behavior of a system. See *climate tipping point, ecological tipping point.*

tolerance limits Minimum and maximum limits for physical conditions (such as temperature) and concentrations of chemical substances beyond which no members of a particular species can survive. See *law of tolerance.*

topsoil The uppermost layer of soil as a soil's A-horizon layer. It contains the organic and inorganic nutrients that plants need for their growth and development.

total fertility rate (TFR) Estimate of the average number of children who will be born alive to a woman during her lifetime if she passes through all her childbearing years (ages 15–44) conforming to age-specific fertility rates of a given year. More simply, it is an estimate of the average number of children that women in a given population will have during their childbearing years.

toxic chemical See *poison, carcinogen, hazardous chemical, mutagen, teratogen.*

toxicity Measure of the harmfulness of a substance.

toxicology Study of the adverse effects of chemicals on health.

toxic waste Form of hazardous waste that causes death or serious injury (such as burns, respiratory diseases, cancers, or genetic mutations). See *hazardous waste.*

toxin See *poison.*

traditional intensive agriculture Production of enough food for a farm family's survival and a surplus that can be sold. This type of agriculture uses higher inputs of labor, fertilizer, and water than traditional subsistence agriculture. See *traditional subsistence agriculture.* Compare *industrialized agriculture.*

traditional subsistence agriculture Production of enough crops or livestock for a farm family's survival. Compare *industrialized agriculture, traditional intensive agriculture.*

tragedy of the commons Depletion or degradation of a potentially renewable resource to which people have free and unmanaged access. An example is the depletion of commercially desirable fish species in the open ocean beyond areas controlled by coastal countries. See *common-property resource, open-access renewable resource.*

trait Characteristic passed on from parents to offspring during reproduction in an animal or plant.

transform fault Area where the earth's lithospheric plates move in opposite but parallel directions along a fracture (fault) in the lithosphere. Compare *convergent plate boundary, divergent plate boundary.*

transgenic organisms See *genetically modified organisms.*

transmissible disease Disease that is caused by living organisms (such as bacteria, viruses, and parasitic worms) and can spread from one person to another by air, water, food, or body fluids (or in some cases by insects or other organisms). Compare *nontransmissible disease.*

transpiration Process in which water is absorbed by the root systems of plants, moves up through the plants, passes through their leaves or other parts, and evaporates into the atmosphere as water vapor.

tree farm See *tree plantation.*

tree plantation Site planted with one or only a few tree species in an even-aged stand. When the stand matures it is usually harvested by clear-cutting and then replanted. These farms normally raise rapidly growing tree species for fuelwood, timber, or pulpwood. Compare *old-growth forest, second-growth forest.*

trophic level All organisms that are the same number of energy transfers away from the original source of energy (for example, sunlight) that enters an ecosystem. For example, all producers belong to the first trophic level and all herbivores belong to the second trophic level in a food chain or a food web.

troposphere Innermost layer of the atmosphere. It contains about 75% of the mass of earth's air and extends about 17 kilometers (11 miles) above sea level. Compare *stratosphere.*

true cost See *full cost.*

tsunami Series of large waves generated when part of the ocean floor suddenly rises or drops.

turbidity Cloudiness in a volume of water; a measure of water clarity in lakes, streams, and other bodies of water.

undernutrition See *chronic undernutrition.*

unreliable science Scientific results or hypotheses presented as reliable science without having undergone the rigors of the peer review process. Compare *reliable science, tentative science.*

upwelling Movement of nutrient-rich bottom water to the ocean's surface. It can occur far from shore but usually takes place along certain steep coastal areas where the warm surface layer of ocean water is pushed away from shore and replaced by cold, nutrient-rich bottom water.

urban area Geographic area containing a community with a population of 2,500 or more. The number of people used in this definition may vary, with some countries setting the minimum number of people at 10,000–50,000.

urban growth Rate of growth of an urban population. Compare *degree of urbanization.*

urbanization Creation or growth of urban areas, or cities, and their surrounding developed land. See *degree of urbanization, urban area.*

urban sprawl Growth of low-density development on the edges of cities and towns. See *smart growth.*

virtual water Water that is not directly consumed but is used to produce food and other products.

virus Microorganism that can transmit an infectious disease by invading a cell and taking over its genetic machinery to copy itself and then spread throughout the body. Compare *bacteria.*

volatile organic compounds (VOCs) Organic compounds that exist as gases in the atmosphere and act as pollutants, some of which are hazardous.

volcano Vent or fissure in the earth's surface through which magma, liquid lava, and gases are released into the environment.

warm front Boundary between an advancing warm air mass and the cooler one it is replacing. Because warm air is less dense than cool air, an advancing warm front rises over a mass of cool air. Compare *cold front.*

waste management Managing wastes to reduce their environmental harm without seriously trying to reduce the amount of waste produced. See *integrated waste management.* Compare *waste reduction.*

waste reduction Reducing the amount of waste produced; wastes that are produced are viewed as potential resources that can be reused, recycled, or composted. See *integrated waste management.* Compare *waste management.*

water cycle See *hydrologic cycle.*

water footprint A rough measure of the volume of water that we use directly and indirectly to keep a person or group alive and to support their lifestyles.

waterlogging Saturation of soil with irrigation water or excessive precipitation so that the water table rises close to the surface.

water pollution Any physical or chemical change in surface water or groundwater that can harm living organisms or make water unfit for certain uses.

watershed Land area that delivers water, sediment, and dissolved substances via small streams to a major stream (river).

water table Upper surface of the zone of saturation, in which all available pores in the soil and rock in the earth's crust are filled with water. See *zone of aeration, zone of saturation.*

watt Unit of power, or rate at which electrical work is done. See *kilowatt.*

weather Short-term changes in the temperature, barometric pressure, humidity, precipitation, sunshine, cloud cover, wind direction and speed, and other conditions in the troposphere at a given place and time. Compare *climate.*

weathering Physical and chemical processes in which solid rock exposed at earth's surface is changed to separate solid particles and dissolved material, which can then be moved to another place as sediment. See *erosion.*

wetland Land that is covered all or part of the time with saltwater or freshwater, excluding streams, lakes, and the open ocean. See *coastal wetland, inland wetland.*

wilderness Area where the earth and its ecosystems have not been seriously disturbed by humans and where humans are only temporary visitors.

wildlife All free, undomesticated species. Sometimes the term is used to describe animals only.

wildlife resources Wildlife species that have actual or potential economic value to people.

wild species Species found in the natural environment. Compare *domesticated species.*

windbreak Row of trees or hedges planted to partially block wind flow and reduce soil erosion on cultivated land.

wind farm Cluster of wind turbines in a windy area on land or at sea, built to capture wind energy and convert it into electrical energy.

worldview How people think the world works and what they think their role in the world should be. See *environmental wisdom worldview, planetary management worldview, stewardship worldview.*

zone of aeration Zone in soil that is not saturated with water and that lies above the water table. See *water table, zone of saturation.*

zone of saturation Zone where all available pores in soil and rock in the earth's crust are filled by water. See *water table, zone of aeration.*

zoning Designating parcels of land for particular types of use.

zooplankton Animal plankton; small floating herbivores that feed on plant plankton (phytoplankton). Compare *phytoplankton.*

Wood
  as fuel, 220, 229–230
  harvesting trees, 220–221, 220*f*, 222*f*
  industrial, 220
Woodall, Jerry, 431*b*, 431*f*
World Bank, 130*b*
  on air pollution, 476, 489
  environmental policy development and, 674
  human health and climate change, 524
  on indoor air pollution, 489
  on water pollution in China, 555
World Business Council, 336*b*
World Commission on Water for the 21st Century, 550
World Conservation Union, environmental policy development and, 674
World Glacier Monitoring Service, 331, 518
World Green Building Council, 409
World Health Organization (WHO), 19, 133
  on air pollution, 476, 482, 492
  on arsenic in drinking water, 555–556
  on DDT use for malaria control, 452
  on deadliest infectious diseases globally, 445*f*
  environmental policy development and, 674

on global infectious diseases, 452
on hormonally active agents (HAAs), 458
on indoor air pollution, 489
on lead poisoning in children, 480
on nutritional deficiencies, 280
obesity estimates by, 280
on smoking, 466
on tuberculosis, 445
on waterborne illnesses, 319
on water pollution in China, 555
World maps. *See* Earth, maps of
World Meterological Organization, 507, 509, 522
World Resources Institute (WRI), 223
  coral reefs and, 249
  on topsoil erosion, 289
Worldwatch Institute, 617
World Wildlife Fund (WWF), 17, 669
  on dams and species extinction, 330
  giant panda and, 92–93
  Living Planet Index, 191
  on orangutans, 194
  sea turtles and, 259
  tiger poaching and, 204
Worm, Boris, 79, 176, 252, 255
WRI. *See* World Resources Institute
WWF. *See* World Wildlife Fund

Wyoming
  Grand Teton National Park, 234*f*
  strip mine in, 9*f*
  Yellowstone National Park, 233, 235*b*, 682*f*–683*f*

Y

Yale University, 302
Yangtze River, China, 332–333, 550, 550*f*
Yellowstone National Park, U.S., 233, 235*b*, 682*f*–683*f*
Yield, crop, 281, 283–284, 294, 310
Yohe, Gary, 533
Yosemite National Park, California, 688*f*
Yunus, Muhammad, 645, 647, 665

Z

Zahn, R., 425
Zaradic, Patricia, 236
Zebra mussel, 199*f*, 268–269, 268*f*, 553
Zebras, 153*f*
Zone of saturation, **320**
Zoning, **619**
Zooplankton, 55, 169, 178
Zoos, 210–211
Zooxanthellae, 168*b*
Zwally, Jay, 518